Astrocytes in (Patho)Physiology of the Nervous System

Vladimir Parpura • Philip G. Haydon
Editors

Astrocytes in (Patho)Physiology of the Nervous System

Springer

Editors
Vladimir Parpura
Department of Neurobiology
Center for Glial Biology in Medicine
Civitan International Research Center
Atomic Force Microscopy &
Nanotechnology Laboratories
Evelyn F. McKnight Brain Institute
University of Alabama
Birmingham, AL, USA
vlad@uab.edu

Philip G. Haydon
Department of Neuroscience
Tufts University School of Medicine
136 Harrison Avenue
Boston, MA 02111
philip.haydon@tufts.edu

ISBN: 978-1-4899-7911-7 ISBN: 978-0-387-79492-1 (eBook)
DOI: 10.1007/978-0-387-79492-1

Softcover reprint of the hardcover 1st edition 2009

Printed on acid-free paper

springer.com

To Vedrana, Vuga and Ivan
Vladimir Parpura

To Yolande, Rachel, Daniel and Julia
Philip G. Haydon

Preface

Astrocytes were the original neuroglia that Ramón y Cajal visualized in 1913 using a gold sublimate stain. This stain targeted intermediate filaments that we now know consist mainly of glial fibrillary acidic protein, a protein used today as an astrocytic marker. Cajal described the morphological diversity of these cells with some astrocytes surrounding neurons, while the others are intimately associated with vasculature. We start the book by discussing the heterogeneity of astrocytes using contemporary tools and by calling into question the assumption by classical neuroscience that neurons and glia are derived from distinct pools of progenitor cells. Astrocytes have long been neglected as active participants in intercellular communication and information processing in the central nervous system, in part due to their lack of electrical excitability. The follow up chapters review the "*nuts and bolts*" of astrocytic physiology; astrocytes possess a diverse assortment of ion channels, neurotransmitter receptors, and transport mechanisms that enable the astrocytes to respond to many of the same signals that act on neurons. Since astrocytes can detect chemical transmitters that are released from neurons and can release their own extracellular signals there is an increasing awareness that they play physiological roles in regulating neuronal activity and synaptic transmission. In addition to these physiological roles, it is becoming increasingly recognized that astrocytes play critical roles during pathophysiological states of the nervous system; these states include gliomas, Alexander disease, and epilepsy to mention a few. The goal of this book is to integrate the body of information that has accumulated in recent years revealing the active role of astrocytes in physiological processing in the central nervous system and to use this as a basis for identifying pathological roles for these glial cells in the brain.

Birmingham, AL — Vlad Parpura
Boston, MA — Phil Haydon

Acknowledgment

We would like to thank all the authors for their contributions. This book would not exist without you. This was an exciting journey. We made it.

Contents

Contributors

Cendra Agulhon
Department of Pharmacology, University of North Carolina at Chapel Hill, Chapel Hill, NC, USA

Arturo Alvarez-Buylla
Department of Neurosurgery and Institute for Regeneration Medicine, University of California San Francisco, San Francisco, CA, USA

Alfonso Araque
Instituto Cajal, CSIC, Doctor Arce 37, Madrid, Spain

Issa P. Bagayogo
Department of Neuroscience and Cell Biology, UMDNJ-Robert Wood Medical School, Piscataway, NJ, USA

Devin K. Binder
Department of Neurological Surgery, University of California, Irvine, CA, USA

Michael Brenner
Department of Neurobiology, Evelyn F. McKnight Brain Institute, Center for Glial Biology in Medicine, University of Alabama Birmingham, Birmingham, AL, USA

Roger F. Butterworth
Neuroscience Research Unit, CHUM, University of Montreal, Montreal, QC, Canada

Giorgio Carmignoto
Istituto CNR di Neuroscienze and Dipartimento di Scienze Biomediche Sperimentali, Università di Padova, Viale G. Colombo 3, Padova 35121, Italy

Kristi Casper
Department of Pharmacology, University of North Carolina at Chapel Hill, Chapel Hill, NC, USA

Joachim W. Deitmer
Abteilung für Allgemeine Zoologie, FB Biologie, TU Kaiserslautern, Kaiserslautern, Germany

Cheryl F. Dreyfus
Department of Neuroscience and Cell Biology, UMDNJ-Robert Wood Medical School, Piscataway, NJ, USA

Todd Fiacco
Department of Cell Biology & Neuroscience, University of California, Riverside, CA, USA

R. Douglas Fields
Nervous System Development and Plasticity Section, National Institutes of Health, NICHD, Bethesda, MD, USA

James E. Goldman
Department of Pathology and The Center for Neurobiology and Behavior, Columbia University, New York, NY, USA

Grant R.J. Gordon
Department of Psychiatry and the Brain Research Centre, University of British Columbia, Vancouver, BC, Canada

Michael M. Halassa
Department of Neuroscience, Tufts University School of Medicine, Boston, MA, USA

Philip G. Haydon
Department of Neuroscience, Tufts University School of Medicine, Boston, MA, USA

Rebecca A. Ihrie
Department of Neurosurgery and Institute for Regeneration Medicine, University of California San Francisco, San Francisco, CA, USA

Ying Y. Jean
Department of Neuroscience and Cell Biology, UMDNJ-Robert Wood Medical School, Piscataway, NJ, USA

Harold K. Kimelberg
Neural and Vascular Biology, Ordway Research Institute, Inc., Albany, NY, USA

Paulo Kofuji
Department of Neuroscience, University of Minnesota, Minneapolis, MN, USA

Suzanne Kurtzer-Minton
Department of Pharmacology, University of North Carolina at Chapel Hill, Chapel Hill, NC, USA

Charles P.K. Lai
Department of Cellular and Physiological Sciences, The Faculty of Medicine, The University of British Columbia, Vancouver V6T 1Z3, BC, Canada

James D. Lechleiter
Department of Cellular and Structural Biology, University of Texas Health Science Center at San Antonio, San Antonio, TX, USA

Christian Lohr
Abteilung für Allgemeine Zoologie, FB Biologie, TU Kaiserslautern, Kaiserslautern, Germany

Alejandro Lomniczi
Division of Neuroscience, Oregon National Primate Research Center/Oregon Health & Science University, Beaverton, OR, USA

Brian A. MacVicar
Department of Psychiatry and the Brain Research Centre, University of British Columbia, Vancouver, BC, Canada,

Erik B. Malarkey
Department of Neurobiology, Center for Glial Biology in Medicine, Civitan International Research Center, Atomic Force Microscopy & Nanotechnology Laboratories, and Evelyn F. McKnight Brain Institute, University of Alabama, Birmingham, AL, USA

Magalie Martineau
Centre de Recherche INSERM, U862, Université Victor Segalen Bordeaux 2, Bordeaux, France

Ken D. McCarthy
Department of Pharmacology, University of North Carolina at Chapel Hill, Chapel Hill, NC, USA

Albee Messing
Waisman Center and Department of Comparative Biosciences, University of Wisconsin Madison, Madison, WI, USA

Jean-Pierre Mothet
Centre de Recherche INSERM, U862, Université Victor Segalen Bordeaux 2, Bordeaux, France

Sean J. Mulligan
Department of Physiology, University of Saskatchewan, Saskatoon, SK, Canada

Christian C. Naus
Department of Cellular and Physiological Sciences, The Faculty of Medicine, The University of British Columbia, Vancouver V6T 1Z3, BC, Canada

Eric A. Newman
Department of Neuroscience, University of Minnesota, Minneapolis, MN, USA

Sergio R. Ojeda
Division of Neuroscience, Oregon National Primate Research Center/Oregon Health & Science University, Beaverton, OR, USA

Stéphane H.R. Oliet
Centre de Recherche INSERM, U862, Université Victor Segalen Bordeaux 2, Bordeaux, France

Vladimir Parpura
Department of Neurobiology, Center for Glial Biology in Medicine, Civitan International Research Center, Atomic Force Microscopy & Nanotechnology Laboratories, and Evelyn F. McKnight Brain Institute, University of Alabama, Birmingham, AL, USA

Gertrudis Perea
Instituto Cajal, CSIC, Doctor Arce 37, Madrid, Spain

Roy A. Quinlan
School of Biological and Biomedical Sciences, The University, Durham, UK

Andreas Reichenbach
Paul Flechsig Institute of Brain Research, Leipzig University, Leipzig, Germany

Jeffrey D. Rothstein
Departments of Neurology and Neuroscience, Johns Hopkins University, Baltimore, MD, USA

Eliana Scemes
The Dominick P. Purpura Department of Neuroscience, Albert Einstein College of Medicine, Bronx, NY, USA

Arne Schousboe
Department of Pharmacology and Pharmacotherapy, Faculty of Pharmaceutical Sciences, University of Copenhagen, Copenhagen, Denmark

Karthika Singaravelu
Abteilung für Allgemeine Zoologie, FB Biologie, TU Kaiserslautern, Kaiserslautern, Germany

Ursula Sonnewald
Department of Neurosciences, Norwegian University of Science and Technology, Trondheim, Norway

Harald Sontheimer
Department of Neurobiology & Center for Glial Biology in Medicine, The University of Alabama at Birmingham, Birmingham, AL, USA

David C. Spray
The Dominick P. Purpura Department of Neuroscience, Albert Einstein College of Medicine, Bronx, NY, USA

Christian Steinhäuser
Institute of Cellular Neurosciences, Medical Faculty, University of Bonn, Bonn, Germany

Elizabeth Sweger
Department of Pharmacology, University of North Carolina at Chapel Hill, Chapel Hill, NC, USA

Eva Sykova
Department of Neuroscience and Center for Cell Therapy and Tissue Repair, Charles University, 2nd Medical Faculty; Department of Neuroscience, Institute of Experimental Medicine AS CR, Prague, Czech Republic

Sarah Taves
Department of Pharmacology, University of North Carolina at Chapel Hill, Chapel Hill, NC, USA

Lydia Vargova
Department of Neuroscience and Center for Cell Therapy and Tissue Repair, Charles University, 2nd Medical Faculty; Department of Neuroscience, Institute of Experimental Medicine AS CR, Prague, Czech Republic

Alexei Verkhratsky
Faculty of Life Sciences, The University of Manchester, Manchester, UK

Helle S. Waagepetersen
Department of Pharmacology and Pharmacotherapy, Faculty of Pharmaceutical Sciences, University of Copenhagen, Copenhagen, Denmark

Lora T. Watts
Department of Cellular and Structural Biology, University of Texas Health Science Center at San Antonio, San Antonio, TX, USA

Hartwig Wolburg
Institute of Pathology, University of Tübingen, Tübingen, Germany

Yongjie Yang
Departments of Neurology and Neuroscience, Johns Hopkins University, Baltimore, MD, USA

Micaela Zonta
Istituto CNR di Neuroscienze and Dipartimento di Scienze Biomediche Sperimentali, Università di Padova, Viale G. Colombo 3, Padova 35121, Italy

Chapter 1
Astrocyte Heterogeneity or Homogeneity?

Harold K. Kimelberg

Contents

The history of the morphology and electrophysiology of the *neuroglia*, which was the historical term used for what are now termed *astroglia* or *astrocytes*, is briefly reviewed. The interpretation of these data around 1970 was that astroglia in situ represented a homogeneous electrophysiological phenotype with a major function, based on this, in maintaining a constant extracellular concentration of potassium ions ($[K^+]_o$). It was soon found that astroglia in situ played a major role in the uptake and inactivation of the synaptically released amino acid transmitters glutamate and γ-aminobutyric acid. Subsequent studies in isolated systems, such as primary astrocyte cultures, greatly expanded this view to a more protean cell, with much wider properties in terms of transmitter uptake systems and release, a variety of voltage-dependent ion channels and varying membrane potentials and electrophysiological behaviour, and receptors for a large number of neurotransmitters.

H.K. Kimelberg
Neural and Vascular Biology, Ordway Research Institute, Inc., Albany, New York, USA
hkimelberg@ordwayresearch.org

V. Parpura and P.G. Haydon (eds.), *Astrocytes in (Patho)Physiology of the Nervous System*,
DOI: 10.1007/978-0-387-79492-1_1, © Springer Science+Business Media, LLC 2009

Thus, it seemed quite reasonable that the astroglia would form a functionally as well as morphologically heterogeneous population, with far more varied properties reflective of different roles in different brain regions and sub-regions. However, our recent in situ data has suggested, at least for astrocytes in the *stratum radiatum* of the hippocampus of the adult rat, that the original homogeneous electrophysiological phenotype for mature astrocytes is likely correct, or at minimum provides a characteristic signature for mature astrocytes because of several interpretative problems inherent in applying the whole-cell voltage-clamp technique to these low resistance cells, which are discussed. It remains to be seen whether such cells represent the true mature astrocyte population in other brain regions, but if so then these electrophysiologically defined astroglia can be systematically examined as a function of region for a number of other important characteristics to accurately determine the degree of heterogeneity within this defined cell population in situ.

1.1 The Classification Problem

1.1.1 Morphology and Classification

The discovery of a non-neuronal element, the neuroglia, in the central nervous system (CNS) is generally attributed to Rudolph Virchow around 1850, but in reality should be attributed to anatomists such as Golgi, Ramón y Cajal and others who applied the Golgi potassium dichromate/silver staining method (*reazione nera)* (Golgi, 1985) to brain tissue in the last two decades of the nineteenth century (reviewed in Kettenmann and Ransom (2005), Somjen (1988) and Kimelberg (2004)). This staining revealed a considerable morphological heterogeneity among these neuroglia as illustrated for the mammalian cerebellum in Fig. 1.1 and for the cerebral cortex in Fig. 1.2a. Two decades later other glial classes, the oligodendroglia and microglia, were identified and the astroglia with oligodendroglia were classified as the macroglia (reviewed in Kettenmann and Ransom (2005), Somjen (1988) and Kimelberg (2004)). All the cells illustrated in Figs. 1.1 and 1.2a and the green glial fibrillary acidic protein (GFAP)(+) cells in Fig. 1.2b are now referred to as astroglia based on the fancy that the morphology of the most dominant types, classically referred to as protoplasmic (grey matter) and fibrous (white matter) resembled stars seen in the night sky, whereas before 1920 they were more usually referred to as neuroglia, although the term astroglia had been used sporadically since around 1890 (reviewed in Kettenmann and Ransom (2005)). Why some of these cells with an obviously different morphology, namely the Bergmann glia (or originally the Bergmann fibres plus Golgi epithelial cells) are also included as astroglia is treated in detail in other reviews (Kettenmann and Ransom, 2005; Somjen, 1988; Reichenbach and Wolburg, 2005), but is often used to support morphological heterogeneity. This is logically an unacceptable circular argument, unless there are other criteria that define these cells as astroglia.

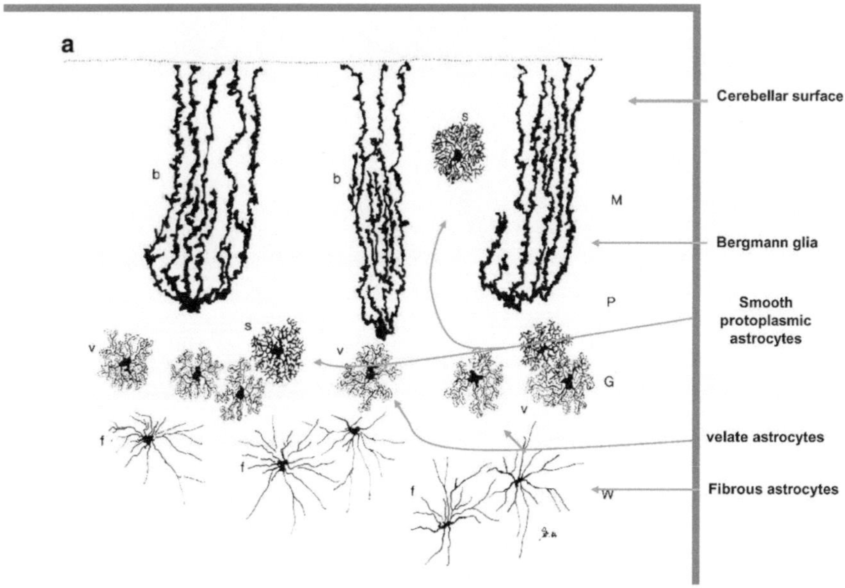

Fig. 1.1 Golgi staining of astroglia in the human cerebellum. M, molecular layer; P, Purkinje-cell layer; G, granule-cell layer; W, white matter. From Ramón y Cajal (1913).

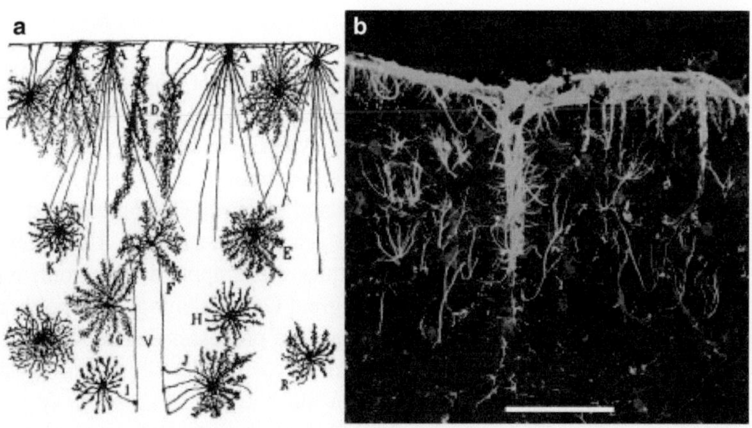

Fig. 1.2 Glial cells in the human and rat brains are morphologically heterogeneous. (a) Astrocytes in the cerebral cortex of a 2-month-old infant stained with the Golgi method. (A–D) are cells in the first cortical lamina and (E–H) are cells in the second and third lamina. (I–J) are cells with end-feet contacting blood vessels. V, blood vessel. From Ramon y Cajal (1913). (b) Staining of cerebral cortex from adult rat. *Top* is surface of cortex showing intense GFAP (green) staining due to the *glia limitans*. *Red* is for NG2, which stains both NG2(+) cells and blood vessels. Scale bar, 100 μm. Unpublished work of G. Schools. (*See Color Plates*)

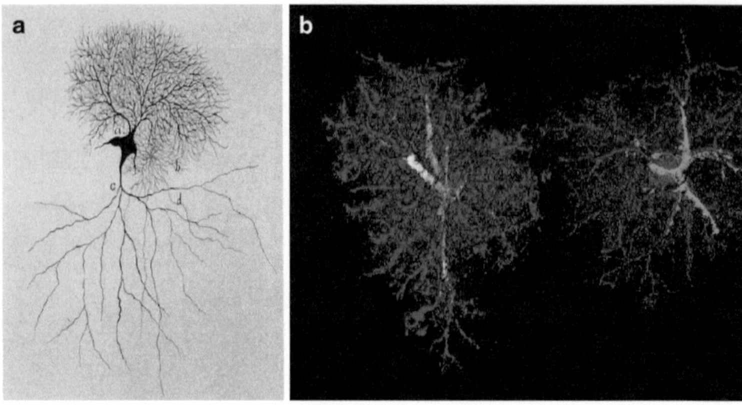

Fig. 1.3 (a) Picture from a chapter on sleep in a book on anesthesia published in 1894 by Carl Ludwig Schleich (1859–1922) in which the author proposed modulation of neuronal currents by swelling and contraction of glia into and out of synapses. A neuron (a), likely a pyramidal cell, in *black* in close contact with a glial cell, presumably an astrocyte, in *red*. See (Dierig, 1994) for further details. (b) Two filled hippocampal astroglial cells from adult rat hippocampus illustrate the domain concept of astroglia. *Yellow* shows GFAP staining, while *blue* and *red* are two dye-filled contiguous astrocytes. From Bushong et al., 2002. (*See Color Plates*)

To break this circular reasoning one needs to know how other more functionally related biochemical and physiological properties define the astroglia, but up to the present there is an insufficient body of systematic work leading to a resolution of this question. There was a gap of around 40 years from 1920 before any physiological studies on glia were done to flesh out the morphological studies, and these were in the amphibian optic nerve where the only penetratable cell bodies were glia (Kuffler et al., 1966). This was due to methodological limitations for studying CNS tissue of other regions of vertebrates as these nervous systems are an extremely intricate mosaic of neurons and glia and their processes. Apart from histology there was little more that could be done on a cellular basis, for the current cell-specific methods of antibody staining, imaging of dye-filled cells and electrophysiological methods for small cells in a complex tissue mosaic were yet to be developed. Parenthetically, when these more advanced imaging techniques were applied they confirmed in greater detail what was apparent from the original Golgi staining; that the radiating processes form an extremely complex framework of processes that ends in finer and finer extensions, as shown in Fig. 1.3b, to compare with older Golgi staining shown in Figs. 1.1, 1.2 and 1.3a. It was also known that the end of these processes surrounded many synapses and surrounded all blood vessels in the mammalian CNS, which early on led to hypotheses of function, such as taking up transmitters (Lugaro, 1907), affecting synaptic activity (see Dierig, (1994)) and bringing nutrients from the blood to neurons (Golgi, 1885); themes echoed today but now with more accurate details and the essential experimental support.

Rather than drawing up detailed competing balance sheets showing how different properties vary between astroglia in different experimental systems I will concern myself with major issues and techniques that bear on this question. I do not consider heterogeneity or homogeneity in regard to developmental changes within the astrocyte population, i.e. I am excluding immature and developing astrocytes, since this is a separate issue. Also cells positive for NG2 (see red cells in Fig. 1.2b), which morphologically resemble astroglia to the extent that they were initially termed smooth protoplasmic astrocytes (Levine and Card, 1987) but are not now considered astrocytes as they do not show a number of defining properties of astrocytes such as excitatory amino acid (EAA) transporters and are not gap-junction-coupled (Nishiyama et al., 2005) (also see below).

1.1.2 Functional Properties and Classification

The intricate process-bearing structure of astroglia cannot per se be taken to indicate functional complexity, in the sense for the CNS of involvement in information processing. Certainly morphology can give clues for basic physiological processes and the speculations of Golgi and Lugaro noted above have been borne out by later more defined hypotheses and experiments (Magistretti et al., 1999; Berl et al., 1961; Rothstein et al., 1996; Danbolt et al., 1992). One clear feature from Figs. 1.1–1.3 is that astrocytes have massive arborizations of finer and finer processes. Thus when one considers the question of heterogeneity or homogeneity it is actually far from clear whether there is a greater heterogeneity within cells in regard to varying properties among the multitudinous processes of single astrocytes compared with the aggregate properties of individual astrocytes, and therefore is it really meaningful to speak of aggregate properties? There could well be spatial segregation between different processes or between the processes and the soma, so that this variation is greater than the differences of aggregate properties between different astrocytes. One of the drawbacks of cell-selective patch-clamp electrophysiology is that it will mainly record the membrane electrophysiological properties of the cell soma as the command voltage and dependent currents likely will not penetrate far and rapidly enough into the processes for the electrophysiological properties of the process tips to be measured. This is still a major technical drawback and we cannot be confident that electrophysiology can see the processes, although recently effective cell–cell current transfer has been reported for mature astrocytes in situ (D'Ascenzo et al., 2007). On the other hand, fluorescent imaging can now discern differences in the Ca^{2+} responses in different processes of Bergmann glia to stimulation of the parallel fibres, and these have been referred to as functional microdomains (Grosche et al., 1999).

1.2 Neurons and Glia

Of course the most fundamental cellular classification in the CNS is into neurons and glia. The true structure of neurons is due to the work of Ramón y Cajal and others in the last decades of the nineteenth century, again using Golgi's stain.

The attribution to their axons of excitability was an extension to these cells (see Katz (1966)) of studies of the injury potentials of nerve tracts by du Bois-Reymond, Bernstein and others from 1850 to 1900, and the elucidation of the ionic basis of the action potentials worked out for the large axons of the giant squid by Hodgkin and Huxley in 1939. This of course has been amply justified by the large body of data acquired since then, and it seems rationally unchallengeable, i.e. beyond reasonable doubt, that the regenerative passage of the polarity change of the axonal membrane potential of neurons (the "action potential") is the fundamental currency of brain information processing. But this "information processing" is so varied, from control of motor function and processing of the activity of our sensory apparatus, through emotions to abstract thinking and "consciousness," that real understanding of how this neuronal electrical activity forms a general substrate for the higher brain functions eludes us (Koch, 2004). Much of cellular neuroscience related to this topic is devoted to how neuronal electrical activity is controlled by the action of transmitters at the circuits' switches, the synapses, to activate or inhibit the switches and thereby control the existence or frequency of trains of action potentials.

1.3 Some Basic Principles of Astroglial Classification

Within the context of the last two sections how can we approach classification and the roles of the astroglia, which bears on the question of homogeneity or heterogeneity? First, we must absolutely distinguish between their roles in the embryological and postnatal development of the nervous system and their roles in the mature nervous system. Then, how can we experimentally explore the functional properties of astroglia, identified morphologically and by markers. Finally, how do we use these properties to reasonably classify these cells so that we will all be talking about the same entities? What number of properties is sufficient to define a cell as astrocytic will be unclear until we study their properties and, because of the empirical nature of the scientific method, will always be a work in progress because, simply put, it depends on observations. This process should be no different from classical classification systems for plants and animals and perhaps we are simply in the early days of our observations. But those classifications are for macroscopic, unmodified characteristics. For cellular classification we always have to select and amplify the characteristics we will use. We generally start with staining or filling for morphology and try to correlate this with selected proteins, which we hope are specific markers and with cellular physiological measurements to arrive at some idea of function. For individual organisms we also have the guiding principle of evolution that all their characteristics have evolved towards survival and procreation. For individual cells this is the larger objective but their specific tasks are to enable the tissue community of which they are members to function optimally so that the organism of which the tissue is a part can survive and procreate.

The issue of the characteristics needed to define astrocytes can be illustrated in the form of a Venn diagram (see Fig. 1.4). The area A is the total of the basic

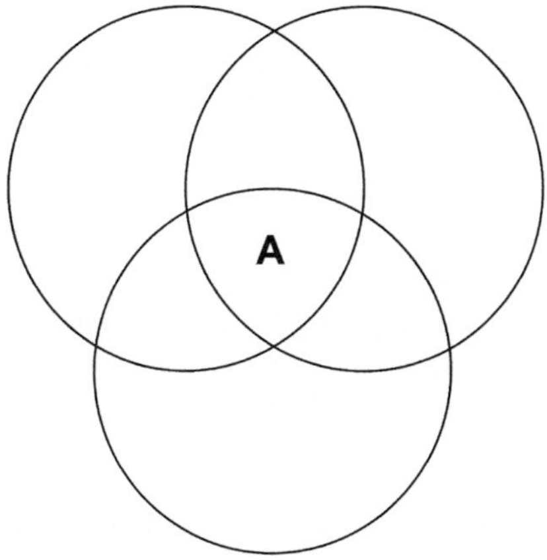

Fig. 1.4 Venn diagram of the astrocyte properties. "A" represents the core astrocyte properties. See text for details.

properties that are needed to define a cell as an astrocyte and will be shared by *all* astrocytes. For convenience I depict three different subclasses of astrocytes but these could be more numerous. The partially overlapping parts depict properties that are shared by some astrocytes and the non-overlapping areas depict properties that are unique to only one subclass of astrocytes. At the start there has to be agreement whether we wish to define the class of astrocytes in this way or as a class of cells that all share the same properties. I propose the former because I think it is more realistic. Thus if all the cells shown in Figs. 1.1 and 1.2 are classified as astroglia they can be divided, rather crudely, into different subclasses based on morphology.

Astrocytes are also often defined on a biochemical basis as expressing an astrocyte-specific protein such as (to date) GFAP, glutamine synthetase (GS), the astrocyte specific EAA transporters GLAST or GLT-1 and the calcium binding protein S100β (also see Reichenbach and Wolburg (2005)). These are very practical because one can use immunocytochemistry to identify the cell more precisely. But even long-used markers such as GFAP are not shared by all cells (Bignami and Dahl, 1974), that would otherwise on the basis of morphology and that their processes abut blood vessels and other distinctive relationships, be characterized as astroglia (Walz, 2000). Also immunohistochemical identification of a cell as being positive is a matter of subjective visual judgment and the sensitivity of the technique, such as whether one uses an amplified or non-amplified antibody-based detection system. Now use of these proteins is being extended to use of DNA constructs artificially incorporated into the cell's genome, which include presumed cell-specific promoters for these proteins linked to a gene expressing a fluorescent protein to define and to mark living cells for further study. Further extension of such genetic engineering includes promoter-specific knockin and knockout of proteins to determine their functions in astrocytes. See Slezak et al. (2007), Djukic et al. (2007) and Chap. 14 for recent references on these topics, which are also beginning to uncover to-be-expected problems in the outcomes of these complex genome-altering procedures.

Likely, morphology nor markers will be sufficient to characterize A in Fig. 1.4. A list was drawn up by the organizer and audience at the 2006 American Society of Neurochemistry in a workshop organized by Dr. Steven Levinson, which I reproduce in Table 1.1, as best I can from my notes with some additions (see also Table 2.1 in Reichenbach and Wolburg (2005)). It also compares immature and mature astrocytes and also mature NG2(+) cells with which astrocytes are still sometimes confused. Obviously, neurons, oligodendroglia and microglia are so different that there would seem to be no useful purpose in including them. Logically we need at least one characteristic and preferably more in region A (Fig. 1.4) to define astrocytes; otherwise if the class of astrocytes is heterogeneous what is it that makes them all members of the astrocyte class? Some might say that it would comprise all

Table 1.1 Some astrocytic and NG2(+) cell characteristics

	Cell type	
Young astrocytes	**Mature astrocytes**	**Mature NG2(+) cells**
	Property	
	A) Electrophysiology	
Varied V_m VDCs Varied input resistances Varied degree of cell–cell coupling	$Vm \approx E_{K^+}$ VDCs not apparent; linear *I–V* plot Very low input resistance Extensive cell–cell coupling	Vm always $< E_{K^+}$ VDCs High input resistances No cell–cell coupling
	B) Markers	
AOs No Glutathione transport No D-serine racemase EAA transporters (e.g., GLAST) GFAP sometimes GS	AOs Glutathione transport D-serine racemase EAA transporters Some strongly GFAP (+) GS	AOs not excessive No Glutathione transport No D-serine racemase No EAA transporters No GFAP No GS
	C) Morphology	
Extensive processes	More extensive processes and arborizations	Extensive, but less-branched processes
Processes contact blood vessels and partition the CNS, e.g., glial limitans, or parts thereof	More extensive processes and arborizations and also contact mature synapses	Processes directly contact nodes of Ranvier
Ionotropic and metabotropic receptors for EAAs and other transmitters	Mainly metabotropic receptors	As for young astrocytes
No direct synaptic inputs	No direct synaptic inputs	Receives direct synaptic inputs

AOs antioxidants, E_{K^+} Equilibrium potential for K^+ *GFAP* glial fibrillary acidic protein, *VDCs* voltage-dependent channels, V_m membrane potential, *EAA* excitatory amino acid, *GS* glutamine synthetase

cells that are not neurons, oligodendroglia, microglia or ependyma. But such a definition by exclusion is not satisfactory and not acceptable to taxonomists. The first problem is to identify the core characteristic or characteristics that represent A. A combination of properties is safer from Bayesian logic, as the author pointed out in a previous publication (Kimelberg, 2004). For example if there are two characteristics instead of one, each with an independent 95% probability of being expressed in astrocytes in a population of cells of which the astrocytes represent 25%, this raises the probability that the dual stained cell is an astrocyte from 86.4% for each of the single characteristics, to 99.2% for both characteristics. The increase in probability becomes greater as astrocytes represent a progressively smaller proportion of the total cell population; if the astrocytes represent 10% of the total cell population the probability that expression denotes an astrocyte is 67.8% with one marker vs. 97.6% for two markers.

A multi-properties definition, as an example three groupings from Table 1.1, groups A, B and C, could be used as a current working definition of a cell as an astrocyte. The morphology would be that some of the processes contact blood vessels and the extensive arborization of processes is contained within a limited volume of tissue; i.e., no projections beyond this volume termed the domain of each cell (Bushong et al., 2002; but see Oberheim et al. (2006) and references therein for extradomain projection of some processes in the human brain). Electrophysiology would be a low membrane resistance and a linear current–voltage (*I*–*V*) relationship (see section 1.4.2.3 for a discussion of what this means for low resistance astrocytes). Markers to date would include GFAP, GS, GLAST and serine racemase. However, these markers will be expanded or modified by the new emerging microarray work on freshly isolated astrocytes and other neural cells, which for example, has unexpectedly shown that message for an aldehyde dehydrogenase 1 family, member L1 (Aldh1L1), is one of the messages most widely expressed in astrocytes (Cahoy et al., 2008).

1.4 Experiments Relevant to Heterogeneity or Non-Heterogeneity

To answer the question "are astrocytes heterogeneous," the first order of business is to determine what properties should be considered common or homogeneous to all astrocytes, which is no means a simple task as just discussed. Then what other properties will we examine to see if they vary enough to conclude that the class is heterogeneous, and can be divided into subclasses. This is the basic issue in the logic of the method of classification, as illustrated in the Venn diagram in Fig. 1.4. As noted in the preceding sections both issues and the defining characteristics are still being debated and therefore ongoing; so here, as an example, I discuss mainly how the electrophysiological properties of astrocytes have been shown to vary. These were the first studies that were not purely morphological and were the first used to attempt to define astroglia on other than morphological grounds.

1.4.1 *Early In Situ Electrophysiology Studies*

These first studies were those of Kuffler and colleagues on glial cell bodies in the amphibian optic nerve (Kuffler et al., 1966; Orkand et al., 1966). This was followed by work using sharp electrode impalements in living mammalian brains as shown in Fig. 1.5 (Picker et al., 1981), and here the cells were post-stained using an injection of horseradish peroxidase from the pipette and visualizing with benzidine, an important step as you can check the morphology of the cell that you recorded. It was therefore hypothesized that the following properties were basic characteristics of all astrocytes. Namely, that all these cells were electrically non-excitable with no capacity for generating action potentials upon injection of positive current, but rather showed a linear relationship between injected current and the change in membrane potential. This was likely due to K^+ channels as a Nernstian relationship between the measured membrane potential and changes in $[K^+]_o$, in both amphibian and mammalian tissues, was found (see Fig. 1.5 for mammals) and the membrane potential measured at zero current was close to the K^+ equilibrium (Nernst) potential. This was a considerable advance. Further blind impalements of glia in the cortex of anesthetized mammals, which were defined by the general criterion of lack of electrical excitability, also responded to the limited range of endogenous $[K^+]$ increases due to neuronal stimulation, by a Nernstian relation (Somjen, 1995).

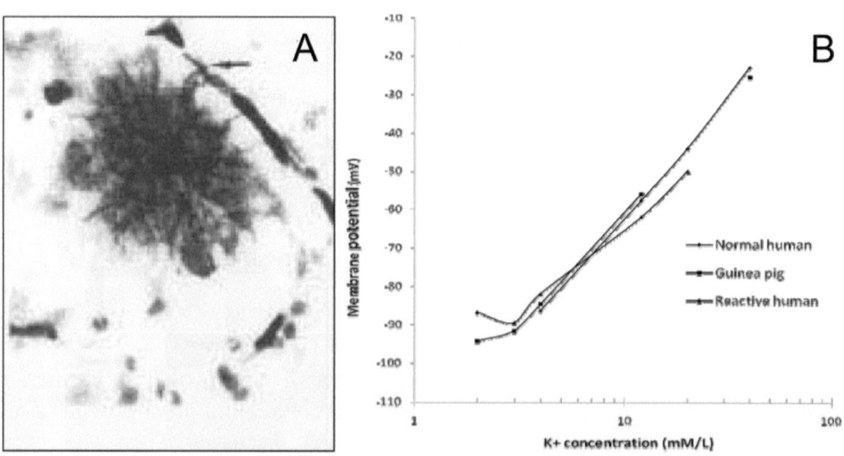

Fig. 1.5 (A) Typical appearance of bushy protoplasmic astrocytes. Cells were visualized after electrophysiological recordings (B) by injecting horseradish peroxidase from the electrode and subsequent histochemistry. Arrow indicates an astrocyte process touching a capillary. (B) A Nernst plot of the changes in membrane potential in millivolts (*y* axis) plotted against the logarithm of imposed K^+ concentrations in the bath solution, in astroglia in normal or epileptic (reactive) human biopsy, and guinea pig cerebrocortical, slices. For this type of plot a slope of 60 shows that the changes in membrane potential can be completely explained by only K^+ carrying the transmembrane currents according to the Nernst equation $V = 60\ \text{mV} \times \log([K^+]_o/[K^+]_i)$. Also when $V = 0$ mV external $[K^+]$ = internal $[K^+]$. The slopes from normal human and guinea pig tissue were ~60 with $[K^+]_i = 120\text{–}130$ mM. Recordings from epileptic tissue showed a smaller slope (~40) indicating some permeability to other ions. From Picker et al. (1981).

Because post-staining was usually not done these cells can only be referred to as "glia." On these bases the K^+ channels were termed *leak channels*, showing no voltage and time-dependent changes, and showing a linear voltage response to injected current. Therefore these neuroglia, which were presumably often astroglia, were homogeneous by electrophysiology but there was always the problem of selection since cells with a membrane potential equal to –50 mV or less were excluded, as there was no independent way of establishing that this was not due to a low electrode seal resistance or even other damage. These were, however, reported to be a minority of the cells sampled, and so damage and/or imperfect seals were a reasonable explanation for such low potential cells. When separate electrodes were used for injecting current and measuring voltage a contribution of the high resistance of the sharp microelectrodes to the linearity of the membrane voltage change was not a factor. I will discuss this issue later in relation to current injection using a single low resistance patch electrode where it is a problem because the membrane voltage change in response to injected current is measured at the top of the electrode and the voltage drop is therefore across both the electrode and membrane resistance (Sontheimer, 1995).

Other defining characteristics soon followed, and since in these microscope-based studies the morphology was checked, they could more safely be referred to as astroglia. Astroglia were found to be a major site of uptake of synaptically released glutamate and its conversion to glutamine based on GS being astrocyte-specific (Martinez-Hernandez et al., 1977). Also that they were extensively linked by gap junctions (Massa and Mugnaini, 1982), which prima facie seemed to fit Kuffler and colleagues' (Orkand et al., 1966) hypothesis of K^+ spatial buffering. However, the short space constant of the cells (=60 to 200 μm) because of their low membrane resistances was always considered to limit the process to only localized increases in extracellular K^+ for most astrocytes, and spatial buffering over limited distances over which K^+ can be transferred of only a few hundred micrometers (Newman, 1995; Gardner-Medwin, 1983). However, regional localization of K^+ channels and the thin planar structure of the retina allowed a form of spatial buffering localized to operate in the more cylindrical and less-branched retinal Muller cells (Newman, 1984). This again is a topic that needs further clarification and may be better resolved when the K^+ channels of astrocytes and their spatial segregations within the cell body and its processes are fully resolved. Other characteristics of the astrocytic processes is that they surround synapses and collections of synapses (glomeruli), and blood vessels and form limiting interfaces, such as the *glial limitans* that is the last region between the CNS and the meninges, as has already been mentioned.

1.4.2 Studies in Different Astroglia Cell Preparations Subsequent to the Early In Situ Studies

1.4.2.1 Cultured Astroglia

Because of the methodological difficulties inherent to work in tissue, the small field of "astrocytology" from the late 1970s, adopted the use of primary cultures prepared from 1- to 2-day-old rodent brains for more detailed electrophysiological, imaging and biochemistry studies. Such cultures grow as monolayers that express GFAP, show glutamate uptake and GS activity, and still have predominantly K^+-based membrane potentials. With studies such as transmitter-receptor effects the omnipresent and vexing problem of indirect effects via neurons was neatly sidestepped. They quickly became the major experimental model for studying astrocyte properties (Kimelberg, 1983, 2001). In contrast to the earlier in situ data these astrocytic primary cultures showed a quite different electrophysiological phenotype, expressing voltage-gated K^+ channels, Ca^{2+} channels, Na^+ channels and Cl^- channels among others (Barres et al. (1990a); Barres (1991b); also see Table 9.1 in Olsen and Sontheimer (2005)). The presence of voltage-gated Na^+ channels in these astrocytes was particularly confusing since they are always found to be electrically non-excitable. The reviews just mentioned can be consulted for the original papers, where full details of the preparations and techniques are given. The clear discrepancy between the newer and the older in situ data was suggested to be due to limitations with the older electrophysiological techniques (Barres, 1991a). That this does not appear to be the case will be argued in the following sections, so that much of the discrepancy is likely due to modified gene expression in the cultures. This is not surprising since a basic biological principle is that gene expression is plastic and of course varies with development and responds to environmental cues via receptors affecting transcription factors. Even the properties of the cultures that were correct in principle, i.e. such as the expression of a number of transmitter uptake systems and ionotropic and metabotropic receptors, were wrong in some of the details (Kimelberg, 2001). *Note added in proof*: a recent microarray study of cultured and isolated astrocytes (Cahoy et al., 2008) has confirmed this principle. A more focused microarray study of gene expression in isolated cells (Lovatt et al., 2007) makes the same point.

1.4.2.2 Astrocytes Freshly Isolated from Brain Slices

It seemed possible that the atypical gene expression problem of primary cultures, yet their amenability to precise experimental control and measurements, could be combined by using acutely isolated astrocytes. This had been done for biochemical studies as early as 1965 using density gradient centrifugation but these preparations were impure and appeared quite damaged and were never examined electrophysiologically to see if they were even viable (see Hamberger et al. (1975) and references therein).

A method that better preserved the cells' integrity was simply to triturate an enzymatically softened tissue, or in some cases mechanically dissociated to avoid enzymatic degradation of exposed surface proteins, and examine the cells individually by electrophysiology and fluorescence indicators for dynamic measurements, or autoradiography for uptake, and then immunocytochemistry for identification. Patch clamp electrophysiological studies on these cells also showed a heterogeneity of electrophysiological phenotypes and the isolated cells never showed the characteristic linear *I–V* curves of astroglia in situ (Steinhauser, 1993; Steinhauser et al., 1994; Verkhratsky and Steinhauser, 2000; Barres et al., 1990b; Zhou and Kimelberg, 2000; Zhou et al., 2000). Figure 1.6a, b shows the two different types found in isolated cells and termed by Zhou and Kimelberg (2000) as outwardly rectifying (a) and variably rectifying (b) astrocytes, respectively. These were also termed glutamate receptor and transporter astrocytes (Glu-R and Glu-T, respectively) on the basis that the former showed α-amino-3-hydroxy-5-methyl-isoxazole propionate (AMPA)-type currents and not transporter currents, and the latter the converse. This was taken as evidence of heterogeneity but, as I will argue in the next section, mature hippocampal astrocytes in situ are always Glu-T, and the Glu-Rs are likely to be NG2(+) cells in mature tissue. Freshly isolated cells were also used to show the presence of metabotropic receptors, especially metabotropic glutamate receptors by measuring changes in intracellular Ca^{2+} concentration using fluorescent probes, and there was some degree of heterogeneity between cells in their responses (Kimelberg et al., 2000). Thus a picture of emerging heterogeneity based mainly on the expression of voltage-dependent currents representing different ion channels and glutamate receptor and transporter currents emerged from the studies on the acutely isolated cells.

1.4.2.3 Recordings from Astrocytes in Brain Slices

The next question was obviously whether even the isolated cells reflected the cells present in situ. No cells with a purely "passive" (i.e. linear *I–V* curves) electrophysiological phenotype (see Fig. 1.6c) were reported in acutely isolated cells, but there had been reports of such cells in astrocytes recorded in freshly cut slices (Steinhauser et al., 1994; Wallraff et al., 2004; D'Ambrosio, 2004). A dye-filled cell in situ from which a recording as shown in Fig. 1.6c would be obtained is shown in Fig. 1.6i, which also shows its dye coupling to other astrocytes. There had been comments that such cells appeared to be more frequent in slices from older animals (Matthias et al., 2003). However, the only systematic development study in slices up to the year 2003 excluded such cells on the basis that they could not be adequately voltage clamped because of their very low input resistances (R_i) of 10–20 MΩ (Bordey and Sontheimer, 1997), and obviously came to different conclusions than if the passive cells, which have Ri values in the range just mentioned, were included. This is the inherent limitation of the scientific method, which only deals with doable observations and was illustrated in a fable by the well-regarded astrophysicist and relativist Sir Arthur Eddington (1882–1944). Namely, that an ichthyologist seeking to classify fishes caught his samples in a net as a scientist would start off doing, and analyzing them concluded that all fish have gills but none were less than 2 inches long (Taylor, 1949)!

It seems reasonable to ask that if the voltage clamp technique has problems for mature, passive, low-resistance astrocytes why use it? In brief, the whole-cell voltage-clamp technique is an excellent way of studying electrophysiological changes in small, high-resistance cells, especially rapidly changing voltage-dependent currents. As is well-known, the techniques involve a relatively large diameter open-tip glass microelectrode, which first requires a high-resistance gigaOhm seal ($\sim 10^9 \Omega$) to be

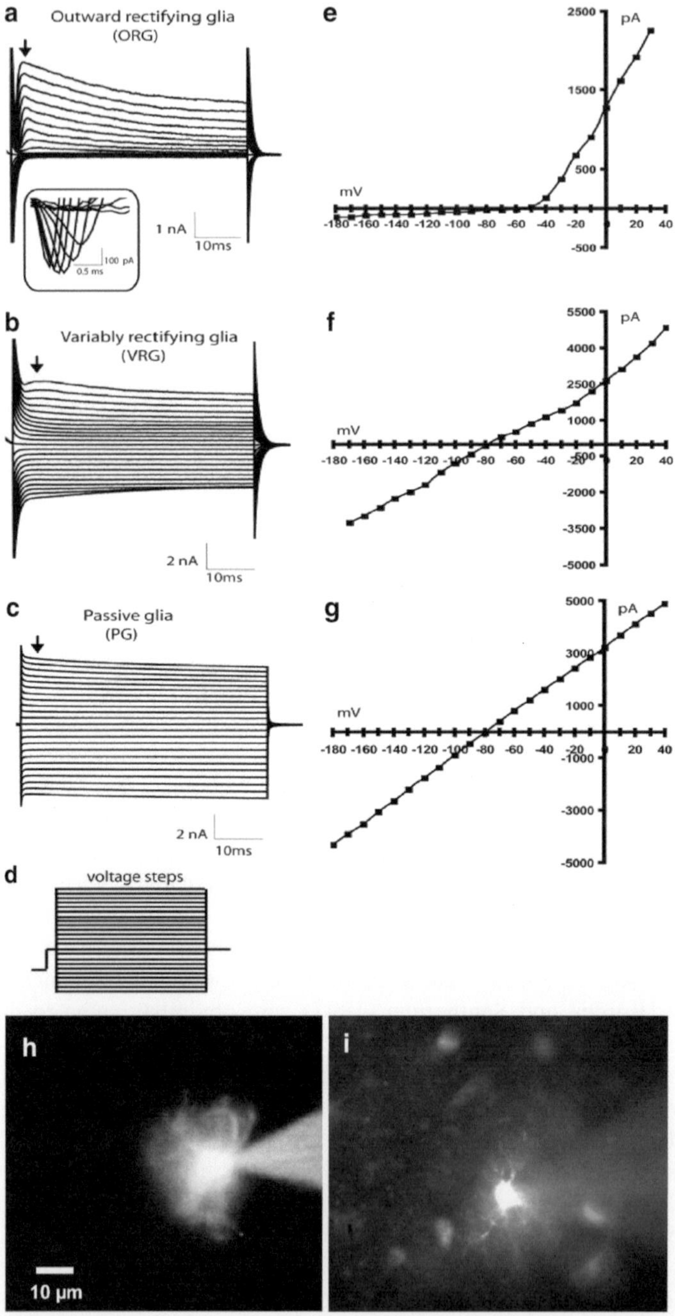

Fig. 1.6 (a–c) show the whole-cell recordings in response to the voltage steps (d, -180 to -40 mV in 10 mV increments) shown in (d) for the three types of electrophysiological phenotype using the nomenclature of Zhou et al. (2000, 2006) (also see text). (e–g) show the resultant *I–V* plots. Only (a) and (b) were found in cells isolated from the hippocampus of 1–35 PN rats (Zhou et al., 2000). (a) also shows Na^+ channels (*inset*).

formed on the intact cell surface. Then suction is applied to break this patch of membrane to access the interior of the cell, with a resultant electrode access resistance (R_a) of ~10 MΩ . This was first applied to neurons and it showed that action potentials, and even synaptic potentials, could be measured in the cell body, which is the only part of the cell big enough to be routinely recorded from (Neher and Sakmann, 1984). The cell's membrane resistance (R_m) had to be at least 100 MΩ for 90% of the changes in potential to be across the cell membrane because it is in series with the R_a of ~10 MΩ and V_c, the clamp (command) potential, is across the total voltage drop (V_t) of $R_a + R_m$ (Sherman-Gold 1993; Sontheimer, 1995). Therefore the voltage drop across R_m is ($V_t - V_a$), where V_a is the voltage drop across R_a and V_t is the total voltage drop. The current pCLAMP 9 program has a membrane test protocol that estimates values for R_a, R_m and membrane capacitance (C_m) based on the value for the total charge (Q_t) delivered to the capacitance, taking into account the offset of the steady-state current, which will rapidly begin to flow across $R_t = R_a + R_m$, where R_t is total resistance. R_a is initially estimated from the time constant of the decay of the capacitance transient, which is small (see Fig. 1.6c) because of the rapidly developing and substantial current flow across the low $R_a + R_m$. Values for R_a and R_m from this analysis have been reported as ~15 and ~5 MΩ, respectively (D'Ascenzo et al., 2007; Djukic et al., 2007), and we find the same values (Zhou et al., in preparation). Since V_c is $V_a + V_m$, V_m is considerably less than V_c; but if the channels are not voltage dependent, then V_m can be calculated from the R_a estimates. Such analysis needed to parse the continuous single electrode whole cell *I–V* data does not preclude some reasonable interpretations. It also adds to the characteristics that are diagnostic of the mature astrocyte and raises the important question of what the extraordinary low R_m is due to. Potentially, these include a large surface area extended to other cells by gap junctions and/or a high density of leak K^+ channels. In terms of the former a critical question is how far does the current and voltage changes spread, i.e., the space clamp problem, and in terms of the latter what are the K^+ channels that could contribute to this extraordinarily low R_m. Continuously open, voltage-independent (leak) potassium channels seems a good bet. Parenthetically, one could also note that freshly isolated GFAP or EAA transporter current positive cells (i.e. characteristics of the passive astrocyte when isolated from older animals) have much higher mean R_t values of several hundred megaOhms (Lalo et al., 2006; Zhou and Kimelberg, 2000). This means either there is a major loss of processes upon isolation, which seems likely and/or loss of the syncytium if the currents indeed travel that far (D'Ascenzo et al., 2007).

Fig. 1.6 (continued) The passive cell that gives a linear *I–V* plot (g) is only observed when astrocytes are recorded in slices (see Fig. 1.7 for how these different phenotypes change with age of the animal from which the slices were obtained). The lines intercepting the voltage coordinate at 0 current, and therefore called the reversal potential (reverses from inward to outward at 0 current) corresponds to the membrane potential under the ion gradients of the experiment, which are designed to duplicate the physiological ion concentrations. (h) shows an isolated astrocyte, dye-filled from the recording pipette while (i) is a cell in a hippocampal slice showing dye spread to other astrocytes. Scale bar, 10 μm (h).

As already noted sharp electrode recordings, where a high seal resistance is obtained by penetrating the cell, were first used for glial cells (the patch-clamp system had not yet been invented) and they were found to be electrically non-excitable. The patch clamp came up with the same thing but could also record voltage-dependent currents (but not in the mature passive astroglia). It is important to understand the problems of this technique applied to low-resistance cells when one is trying to clamp them at a potential and measure the current required to do that, if anything is to make sense. When the resistance is low it might take more time to deliver the current than the time frame in which the channel activates. More likely, when the resistance at the tip of the electrode is also around 10 MΩ, and so the voltage drop is about equal across both the electrode resistance and the cell membrane, and therefore, V_c is around 2-fold greater than the voltage drop across the cell membrane. Third, it may be only the cell body and proximal processes that are clamped, as the current has to pass through 1,000–10,000 processes (Bushong et al., 2002, 2004) that get smaller and smaller, i.e. their cross-sectional resistance gets larger and larger, which will also, of course, limit spatial buffering. In the current-clamp mode (with $I = 0$) one is using the system as a voltage follower as in sharp electrodes (Purves, 1981), and so there is no problem. Further, if the pCLAMP 9 analysis gives reasonably accurate R_a values for these cells then they can be studied with suitable corrections. The discontinuous single electrode voltage-clamp technique should avoid the R_a problem in measuring V_m (Sherman-Gold, 1993), but is not widely used now. Recordings with two electrodes, one for passing current and the other for measuring V_m, are possible but technically difficult given the small size (diameter, ~10 μm) of the astrocyte cell body in situ.

Our group decided to systematically study passive cells by whole-cell voltage clamp, by including all the cell bodies seen with differential interference optics as likely be "glia" from the *stratum radiatum* in hippocampal slices (cell bodies of ~10-μm diameter) from 1- to 105-day-old animals. Their "glial" nature could then be confirmed by their non-excitability in current clamp passing sufficient current to cause activation of voltage-gated Na^+ channels. It turned out that although cells with voltage-dependent currents could be recorded in slices from younger animals some passive cells could also be seen but most significantly, as the age of the animals increased, these represented about 90% of the glial cells in the adult *stratum radiatum.* The original paper (Zhou et al., 2006) can be consulted for the details and the major results are reproduced in Figs. 1.7 and 1.8. There is clearly a development transition around post-natal day 20 (P20), which corresponds to the completion of synaptogenesis in this region, a reasonable criterion for maturity (see Fig. 1.7). This was supplemented by post-recording staining of a separate and smaller group of cells (Fig. 1.8), which showed that we had recorded a shifting population of GLAST + astrocytes and NG2 + glia (cells) that led us to propose the developmental relationships shown in Fig. 1.9.

Thus the mature protoplasmic astroglia does seem to be homogeneous in terms of their electrophysiological characteristic, with the necessary caveat of what is measured in the CA1 region of mature rats. It is interesting that this corresponds to the original model after a 40-year digression into the electrophysiology of primary cultures, acutely isolated cells and astrocytes recorded in slices from immature rats as

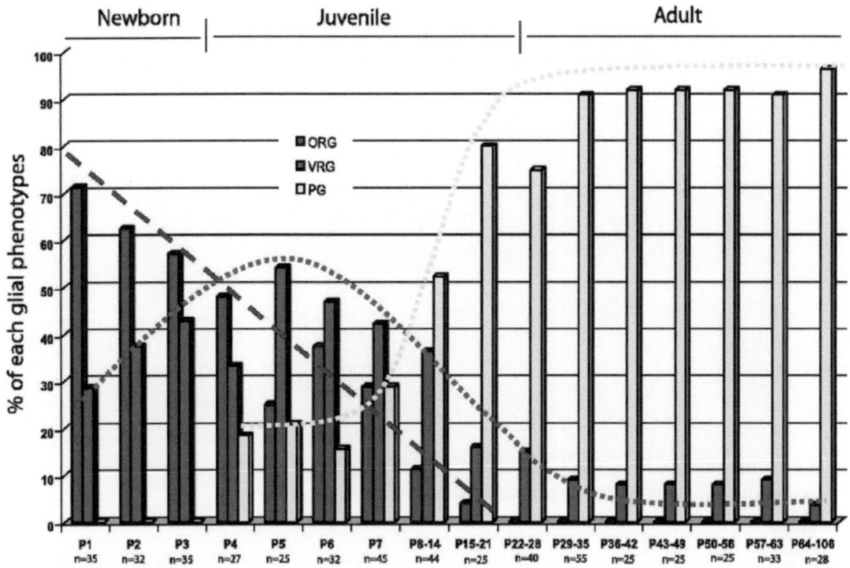

Fig. 1.7 The percentage of glial cells recognized by differential interference contrast optics in a living hippocampal slice, which corresponded to the three electrophysiological phenotypes shown in Fig. 1.6. The numbers under the *x* axis show the post-natal age in days and the number of cells recorded (*n*). On *top* the ages are broadly classified into newborn, juvenile and adult groups. From Zhou et al. (2006). (*See Color Plates*)

models for the mature protoplasmic astrocyte. Of course a huge number of questions remain, and perhaps instead of me laying these down like some litany, which in any case will be my views, the interested reader can think of them for themselves.

However, an obvious one to start the ball rolling is does this emergence of passive astrocytes upon maturity apply to all brain regions? For the purists, and we should all be purists in scientific studies, this will need to be systematically studied in the different regions. In spite of the interpretative problems, the linear *I–V* plots seen by continuous single electrode voltage clamp are a signature of the mature cells, but the V_c is greater (by a factor of at least 2) than the actual V_m, as discussed above. The reversal potential (E_r), at $I = 0$ current, will equal the membrane potential (see Fig. 1.6e–f), but the effect of inhibitors on conductance will need to be corrected for the fact that one is measuring both an affected R_m and an unaffected R_a in series. An interesting aspect of this is that if we can identify the channels we can selectively inhibit different ones to get a "clampable" cell because R_m will increase relative to R_a. To what extent, if at all, the channels in the end-feet will be measured by an electrode located in the cell soma is unknown, and until this technical question is resolved by measuring directly from the processes we will be restricted to describing what is there by immunocytochemistry, for some time.

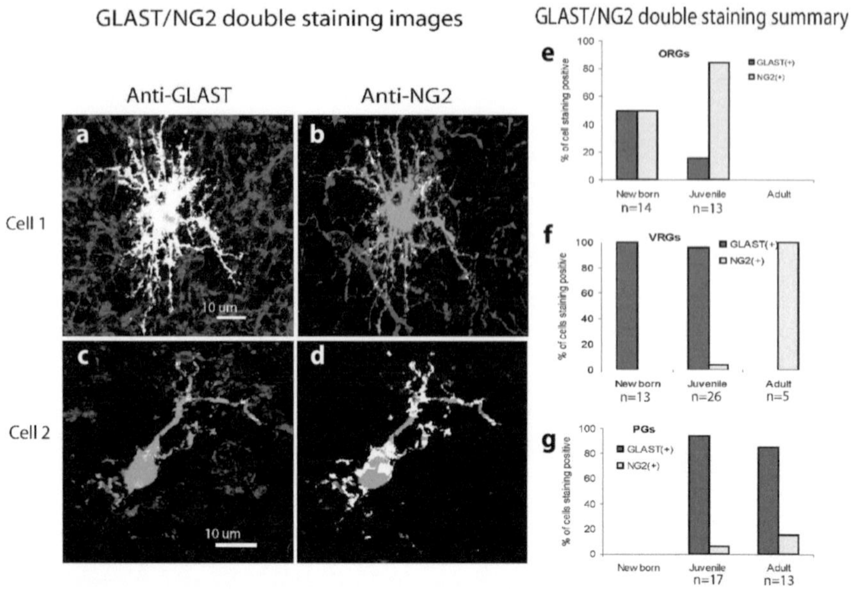

Fig. 1.8 Correlation of electrophysiological phenotype with cell type by immunocytochemical identification of recorded cells. This continues the study shown in Fig. 1.7 by post-recording staining a smaller number of cells (given by *n* for each case) within the three broad age groups identified in Fig. 1.7. The cells were stained for GLAST to identify astrocytes and NG2 to identify this non-astrocytic glial class. (a)–(d) shows examples of staining in a GLAST(+) (cell 1) and an NG2 (+) cell (cell 2). Green is the filling dye and red represents either antibody staining. White represents colour-coded colocalization. As shown in (e), NG2(+) cells shown as *yellow bars* represent outwardly rectifying glial cells (ORGs) equally in the newborn stage and predominantly in the juvenile. *Red* represents GLAST(+) cells. There are no ORGs in the adult. (f) shows that variably rectifying glial cells (VRGs) are all and then predominantly astrocytes, but are only NG2(+) cells in the adult. (g) shows that passive cells are only seen in the juvenile and adult animals and represent mainly astrocytes, although 5–10% are NG2(+) in the adult, but unlike the passive astrocyte these have small Na^+ currents (not shown). See Zhou et al. (2006) for further details. (*See Color Plates*)

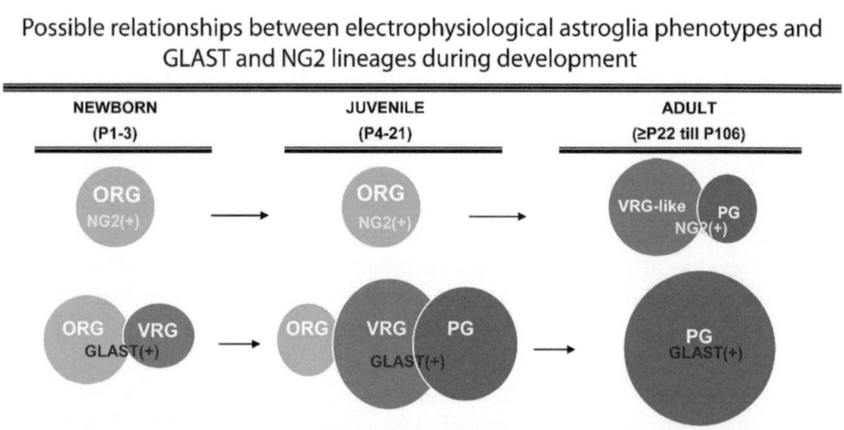

Fig. 1.9 Possible relationships between electrophysiological astroglia phenotypes and GLAST and NG2 lineages during development. From Zhou et al. (2006); based on data in Fig. 1.8. (*See Color Plates*)

A study from our laboratory (Schools et al., 2006) showed that passivity correlated with the extent of cell–cell coupling, and membrane patches pulled from the cell showed a linear *I–V* relation that corresponded more to the parent cell but a minority of around 30% were variably rectifying (Schools et al., 2006). In an excised patch of several hundred megaOhm resistance V_m essentially equals V_c. The issue of different electrophysiological types seems to have been partially solved by the above studies in the hippocampus, where the heterogeneous cells were restricted to earlier development, and mature astrocytes after ~20 days were electrophysiologically passive, but there are at least three characteristics that develop in parallel that could contribute to such behaviour; open K^+ channels, increased syncytium and $R_a > R_m$. Note that the first two also contribute to the third characteristic, and so they are all interrelated. There were also around 20% NG2(+) cells that either showed a variably rectifying or passive electrophysiological phenotype, but with small Na^+ currents that were never seen in passive astrocytes.

The electrophysiologically passive, mature astrocytes can then be individually studied to determine whether they are heterogeneous for transporters and different enzymes and other components that are important for astrocyte functions. To examine these systematically also means defining age, lamina and region from which the cells are obtained. On the basis of such data the field should be reasonably able to answer the question posed in the title of this chapter.

1.5 Envoi

For the question posed in the title we do seem to be at the beginning of a journey rather than even well on our way. We need to correlate the well-defined morphological heterogeneity with other properties. With respect to electrophysiology we seem at least to know where we need to go; see if the linear *I–V* curve correlates with maturity in all astrocytes and what does it represent. For biochemical properties we seem to be restricted to establishing the occurrence by immunocytochemistry as we cannot rely on isolated cells due to likely massive cell process loss plus other still-to-be-defined damage. Genetic engineering techniques linked to specific promoters for astrocytes using so far the GFAP promoter are still in their infancy (e.g., see Pascual et al. (2005), Chap. 14) and will require a large amount of preliminary work with different promoters to understand their specificities (Slezak et al., 2007). A classification could well emerge from these studies of astrocytes defined by a particular promoter-construct activity. Some illustrative and experimentally addressable questions are as follows:

1. Does development of the linear *I–V* characteristics of mature astrocytes differ in different regions and lamina and what precisely does this linear *I–V* curve mean?
2. Does morphological or physiological heterogeneity depend on species?

3. Is there spatial segregation of transporters and channels within the astrocyte, much as in epithelial cells? If these are mainly on the ends of the processes that surround blood vessels and synapses, one cannot study these electrophysiologically, at present, but can do so histologically and dynamically by calcium imaging.

The ultimate aim, of course, is to uncover the functions of astrocytes in the brain; support or an integral part of the information-processing system and how one can tell the difference. The latter has been theoretically ruled out by some scientists interested in information processing, consciousness and other such like big questions on grounds that the astroglial responses lack sufficient "specificity and celerity" (Koch, 2004). Whether these objections are valid underlies a lot of current work on astroglia (rather than "glia" in general, a term neuroscientists really should no longer use; see also Cahoy et al. (2008)) and no doubt increasingly in the future. One message of this chapter is really that we should first adequately define what the term *astroglia* represents.

1.5.1 Experimental Approaches to Heterogeneity of Mature Astrocytes

What other approaches can we use to study astrocyte heterogeneity? If one catalogues all the gene messages significantly expressed by mature astrocytes (*Note added in proof*: the first studies in this area have just been published for sorted, isolated astrocytes from different aged animals (Lovatt et al., 2007; Cahoy et al., 2008), and the major mRNAs do correspond to the major proteins known to be expressed by immunostaining of functional studies.), will this enable us to say whether they are heterogeneous, and give us insight to the outstanding questions for astrocytes in the mammalian brain? However, this would be an example of data-gathering, fishing expeditions and all the other pejorative descriptors applied to what was once considered fundamental to scientific inquiry but is currently non-fashionable; the systematic acquisition of data that precedes hypotheses to explain the phenomena observed, an approach Isaac Newton advocated as the "safest method of philosophizing" (Christianson, 1984). But there are also significant methodological problems to this approach. For example the mRNA microarray approach allows one to assess at one time all the mRNAs expressed by the genome. However this requires an amount of RNA that is about 1,000 times that expressed by a single cell. Thus, we would only get an average and this would not then address the question of cellular astrocyte heterogeneity below the microregional level. Nonetheless, with this type of information we would obtain clues about which proteins to look for, and by using immunocytochemistry we could determine cell-to-cell heterogeneity and just as importantly heterogeneity of location within a single astrocyte, but only at present confidently at the electron microscope level. It is quite likely that improvements in techniques, including the use of linear RNA amplification will, hopefully in the not too distant future, make it possible to perform microarray studies at the single-cell level, and this will solve that problem at the message level.

At present then the question raised of whether mature astrocytes are heterogeneous in the sense of Fig. 1.4, that there is a core constellation of properties "A," which defines astrocytes and then a varying degree of partially overlapping and non-overlapping properties that confers heterogeneity, still needs to be determined. If we ascertain that mature astrocytes in every brain region show linear *I–V* relations, then using this as a signature, together with positivity for unambiguous astrocyte markers such as GLAST, GFAP and others (*Note added in proof*: see Cahoy et al. (2008) for other markers disclosed by global gene expression.), we can see whether there is heterogeneity of protein expression and mRNAs between different brain regions. Finally, as I think is likely and partially supported experimentally, there should be clear spatial heterogeneity such as between the astrocyte membranes that surround blood vessels and/or synapses and other regions, at a minimum. One might assume all such processes equivalent and then test that null hypothesis. This can only be disclosed by very precise microscopic studies the very technique, but of course now far more advanced, with which the cellular nature of the neuroglia was revealed by application of the Golgi staining technique over 100 years ago. In the absence of strong evidence to the contrary I would also propose that another and more basic null hypothesis to be disproved for the fundamental function of astrocytes is that it is limited to homeostasis; to provide a controlled environment that allows the information-processing part of the CNS, the different neuronal circuits made from the heterogeneous neuronal populations, to function optimally. However, other current hypotheses concerning astroglial function that they can influence synaptic activity on the basis that they may show exocytotic release of neurotransmitters, supported by the presumed astrocyte-specific elimination of a vesicle fusion protein resulting in suppression of synaptic transmission and increasing the dynamic range of long-term potentiation (Montana et al., 2006; Volterra and Meldolesi, 2005; Haydon and Carmignoto, 2006; Pascual et al., 2005), can be tested further in the sense that the proteins and such-like needed for this process are present in mature astrocytes (Cahoy et al. (2008) did not find mRNA for many of these). Another hypothesis is that the greater complexity of astroglial structure in different lamina of the human cortex, compared with a much simpler pattern for the rat cortex, but the comparable structure of neurons in the two very different mammals, supports an hypothesis that some of the indisputable greater complexity of function of the human brain compared with the rodent may, in part, reside in the astroglia (Oberheim et al., 2006). Here it would be useful to do cross-species studies.

1.5.2 Domain Concept for Mature Astroglia

The morphological complexity that allows one protoplasmic astrocyte to control a wide expanse of territory, the cellular domain concept first put forward by Bushong et al. (2002, 2004), and the independent functioning of individual processes as shown by calcium imaging (Grosche et al., 1999) can be explained on the basis of cell theory where each part of a cell has to be linked to a cell body containing the cell nucleus. These linkages are the processes emanating from the cell body and target to synapses and blood vessels where they develop systems specialized to sustain, modulate or maintain these targets. When these processes meet other astrocytic processes they form gap junctions, perhaps not so much as a method of communication at all, but as a way of preventing further growth of these processes, defining their boundaries and thus leading to the separate domains (Bushong et al., 2004). On this basis each process ending has to be autonomous and operate independently by simple feedback or feedforward principles. This is further required by there so far being no polarity or clear differences in astrocytic processes, as is known for neurons. More so than other tissues the mammalian brain has limited space, being encased in rigid bone for protection, and the process boundary design prevents the unnecessary multiplication of astrocytic cell bodies beyond what is needed to most parsimoniously perform their functions. Thus we come back (see section 1.1.2) to more heterogeneity within an individual astrocyte than between astrocytes.

Acknowledgements I thank Drs. Gary Schools and Min Zhou for discussions and reading the manuscript and for supplying some of the figures.

References

Barres BA (1991a) Five electrophysiological properties of glial cells. *Ann N Y Acad Sci* 633: 248–254.

Barres BA (1991b) Glial ion channels. *Cur Opin Neurobiol* 1: 354–359.

Barres BA, Chun LLY, Corey DP (1990a) Ion channels in vertebrate glia. *Annual Rev Neurosci* 13: 441–474.

Barres BA, Koroshetz WJ, Chun LLY, Corey DP (1990b) Ion channel expression by white matter glia: The type-1 astrocyte. *Neuron* 5: 527–544.

Berl S, Lajtha A, Waelsch H (1961) Amino acid and protein metabolism – VI cerebral compartments of glutamic acid metabolism. *J Neurochem* 7: 186–197.

Bignami A, Dahl D (1974) Astrocyte-specific protein and neuroglial differentiation. An immunofluorescence study with antibodies to the glial fibrillary acidic protein. *J Comp Neurol* 153: 27–38.

Bordey A, Sontheimer H (1997) Postnatal development of ionic currents in rat hippocampal astrocytes in situ. *J Neurophysiol* 78: 461–477.

Bushong EA, Martone ME, Jones YZ, Ellisman MH (2002) Protoplasmic astrocytes in CA1 stratum radiatum occupy separate anatomical domains. *J Neurosci* 22: 183–192.

Bushong EA, Martone ME, Ellisman MH (2004) Maturation of astrocyte morphology and the establishment of astrocyte domains during postnatal hippocampal development. *Int J Dev Neurosci* 22: 73–86.

Cahoy JD, Emery B, Kausha lA, FooL C, Zamanian JL, Christopherson KS, Xing Y, Lubischer JL, Krieg PA, Krupenko SA, Thompson WJ, Barres BA (2008) A transcriptome database for astrocytes, neurons, and oligodendrocytes: a new resource for understanding brain development and function. *J Neurosci* 28: 264–278.

Christianson GE (1984) In the presence of the creator: Isaac Newton and his Times. 165 pp. New York: The Free Press.

D'Ambrosio R (2004) The role of glial membrane ion channels in seizures and epileptogenesis. *Pharmacol Ther* 103: 95–108.

D'Ascenzo M, Fellin T, Terunuma M, Revilla-Sanchez R, Meaney DF, Auberson YP, Moss SJ, Haydon PG (2007) mGluR5 stimulates gliotransmission in the nucleus accumbens. *Proc Natl Acad Sci U S A* 104: 1995–2000.

Danbolt NC, Storm-Mathisen J, Kanner BI (1992) An [Na^{++} K^{+}]coupled l-glutamate transporter purified from rat brain is located in glial cell processes. *Neurosci* 51: 295–310.

Dierig S (1994) Extending the neuron doctrine: Carl Ludwig Schleich (1859–1922) and his reflections on neuroglia at the inception of the neural-network concept in 1894. *Trends Neurosci* 17: 449–452.

Djukic B, Casper KB, Philpot BD, Chin LS, McCarthy KD (2007) Conditional knock-out of Kir4.1 leads to glial membrane depolarization, inhibition of potassium and glutamate uptake, and enhanced short-term synaptic potentiation. *J Neurosci* 27: 11354–11365.

Gardner-Medwin AR (1983) Analysis of potassium dynamics in mammalian brain tissue. *J Physiol* 335: 393–426.

Golgi C (1885) Sulla fina anatomia degli organi centrali del sisterma nervoso. *Riv Sper Fremiat Med Leg Alienazione Ment* 11: 72–123.

Grosche J, Matyash V, Moller T, Verkhratsky A, Reichenbach A, Kettenmann H (1999) Microdomains for neuron–glia interaction: parallel fiber signaling to Bergmann glial cells. *Nat Neurosci* 2: 139–143.

Hamberger A, Hansson H-A, Sellstrom A (1975) Scanning and transmission electron microscopy on bulk prepared neuronal and glial cells. *Exp Cell Res* 92: 1–10.

Haydon PG, Carmignoto G (2006) Astrocyte control of synaptic transmission and neurovascular coupling. *Physiol Rev* 86: 1009–1031.

Katz B (1966) Nerve, muscle and synapse. New York: McGraw-Hill.

Kettenmann H, Ransom B (2005) The concept of neuroglia: a historical perspective. In: Neuroglia (Kettenmann H, Ransom B, eds), pp. 1–16. New York: Oxford University Press.

Kimelberg HK (1983) Primary astrocyte cultures – a key to astrocyte function. *Cell Mol Neurobiol* 3: 1–16.

Kimelberg HK (2001) Glia–neuronal culture models – do we need to change the paradigms?*Trends Neurosci* 24: 205–206.

Kimelberg HK (2004) The problem of astrocyte identity. *Neurochem Int* 45: 191–202.

Kimelberg HK, Schools GP, Zhou M (2000) Freshly isolated astrocyte (FIA) preparations; a useful single cell system for studying astrocyte properties. *J Neurosci Res* 61: 577–587.

Koch C (2004) The quest for consciousness; a neurobiological approach. pp. 17 and 89 pp. Englewood, CO: Roberts.

Kuffler SW, Nicholls JG, Orkand RK (1966) Physiological properties of glial cells in the central nervous system of amphibia. *J Neurophysiol* 29: 768–787.

Lalo U, Pankratov Y, Kirchhoff F, North RA, Verkhratsky A (2006) NMDA receptors mediate neuron-to-glia signaling in mouse cortical astrocytes. *J Neurosci* 26: 2673–2683.

Levine JM, Card JP (1987) Light and electron microscopic localization of a cell surface antigen (NG2) in the rat cerebellum: association with smooth protoplasmic astrocytes. *J Neurosci* 7: 2711–2720.

Lovatt D, Sonnewald U, Waagepetersen HS, Schousboe A, He W, Lin JH, Han X, Takano T, Wang S, Sim FJ, Goldman SA, Nedergaard M (2007) The transcriptome and metabolic gene signature of protoplasmic astrocytes in the adult murine cortex. *J Neurosci* 27: 12255–12266.

Lugaro E (1907) Sulle Funzioni Della Nevroglia. *Riv D Pat Nerv Ment* 12: 225–233.

Magistretti PJ, Pellerin L, Rothman DL, Shulman RG (1999) Energy on demand. *Science* 283: 496–497.

Martinez-Hernandez A, Bell K, Norenberg MD (1977) Gltutamine synthetase: glial localization in brain. *Science*195: 1356–1358.

Massa PT, Mugnaini E (1982) Cell junctions and intramembrane particles of astrocytes and oligodendrocytes: a freeze-fracture study. *Neuroscience* 7: 523–538.

Matthias K, Kirchhoff F, Seifert G, Huttmann K, Matyash M, Kettenmann H, Steinhauser C (2003) Segregated expression of AMPA-type glutamate receptors and glutamate transporters defines distinct astrocyte populations in the mouse. *J Neurosci* 23: 1750–1758.

Montana V, Malarkey EB, Verderio C, Matteoli M, Parpura V (2006) Vesicular transmitter release from astrocytes. *Glia* 54: 700–715.

Neher E, Sakmann B (1984)Patch clamp techniques for studying ionic channels in excitable membranes. *Ann Rev Physiol* 46: 455–472.

Newman EA (1984)Regional specialization of retinal glial cell membrane. *Nature*309: 155–158.

Newman EA (1995) Glial cell regulation of extracellular potassium. In: Neuroglia (KettenmannH, Ransom B, eds), pp. 717–731. Oxford: Oxford University Press.

Nishiyama A, Yang Z, Butt A (2005) Astrocytes and NG2-glia: what's in a name? *J Anat* 207: 687–693.

Oberheim NA, Wang X, Goldman S, Nedergaard M (2006) Astrocytic complexity distinguishes the human brain. *Trends Neurosci* 29: 547–553.

Olsen ML, Sontheimer H (2005) Voltage-activated ion channels in glial cells. In: Neuroglia (Kettenmann H, Ransom BR, eds), pp. 112–130. Oxford: Oxford University Press.

Orkand RK, Nicholls JG, Kuffler SW (1966) Effect of nerve impulses on the membrane potential of glial cells in the central nervous system of amphibia. *J Neurophysiol* 29: 788–806.

Pascual O, Casper KB, Kubera C, Zhang J, Revilla-Sanchez R, SulJ Y, Takano H, Moss SJ, McCarthy K, Haydon PG (2005) Astrocytic purinergic signaling coordinates synaptic networks. *Science* 310: 113–116.

Picker S, Pieper CF, Goldring S (1981) Glial membrane potentials and their relationship to [K+]o in man and guinea pig. *J Neurosurg* 55: 347–363.

Purves RD (1981) Microelectrode methods for intracellular recording and ionophoresis. London: Academic.

Ramon y Cajal S (1913) Contribucion al conocimento de la neuroglia del cerebro humano. *Trab Lab Invest Biol Univ Madrid* 11: 255–315.

Reichenbach A, Wolburg H (2005)Astrocytes and ependymal glia. In: Neuroglia (Kettenmann H, Ransom B, eds), pp. 19–35. New York: Oxford University Press.

Rothstein JD, Dykes-Hoberg M, Pardo CA, Bristol LA, Jin L, Kuncl RW, Kanai Y, Hediger MA, Wang YF, Schielke JP, Welty DF (1996) Knockout of glutamate transporters reveals a major role for astroglial transport in excitotoxicity and clearance of glutamate. *Neuron*16: 675–686.

Schools GP, Zhou M, Kimelberg HK (2006) Development of gap junctions in hippocampal astrocytes: evidence that whole cell electrophysiological phenotype is an intrinsic property of the individual cell. *J Neurophysiol* 96: 1383–1392.

Sherman-Gold S. (ed.) (1993) The Axon guide for electrophysiology and biophysics laboratory techniques. Foster City, CA: Axon Instruments.

Slezak M, Goritz C, Niemiec A, Frisen J, Chambon P, Metzger D, Pfrieger FW (2007) Transgenic mice for conditional gene manipulation in astroglial cells. *Glia* 55: 1565–1576.

Somjen GG (1988) Nervenkitt: notes on the history of the concept of neuroglia. *Glia* 1: 2–9.

Somjen GG (1995) Electrophysiology of mammalian glial cells in situ. In: Neuroglia (Kettenmann H, Ransom BR, eds), pp. 319–331. Oxford: Oxford University Press.

Sontheimer H (1995) Whole-cell patch-clamp recordings. In: Patch-clamp applications and protocols (Boulton A, Baker GB, Walz W, eds), pp. 37–74. Totowa, New Jersey: Humana.

Steinhauser C (1993) Electrophysiologic characteristics of glial cells. *Hippocampus* 3: 113–124.

Steinhauser C, Kressin K, Kuprijanova E, Weber M, Seifert G (1994)Properties of voltage-activated Na^+ and K^+ currents in mouse hippocampal glial cells in situ and after acute isolation from tissue slices. *Pflugers Arch* 428: 610–620.

Taylor FS (1949) Science: past and present. pp. 348–349 Melbourne: William Heinemann.

Verkhratsky A, Steinhauser C (2000) Ion channels in glial cells. *Brain Res Brain Res Rev* 32: 380–412.

Volterra A, Meldolesi J (2005) Astrocytes, from brain glue to communication elements: the revolution continues. *Nat Rev Neurosci* 6: 626–640.

Wallraff A, Odermatt B, Willecke K, Steinhauser C (2004) Distinct types of astroglial cells in the hippocampus differ in gap junction coupling. *Glia* 48: 36–43.

Walz W (2000) Controversy surrounding the existence of discrete functional classes of astrocytes in adult gray matter. *Glia* 31: 95–103.

Zhou M, Kimelberg HK (2000) Freshly isolated astrocytes from rat hippocampus show two distinct current patterns and different $[K^+]_o$ uptake capabilities. *J Neurophysiol* 84: 2746–2757.

Zhou M, Schools GP, Kimelberg HK (2000) GFAP mRNA positive glia acutely isolated from rat hippocampus predominantly show complex current patterns. *Molec Brain Res* 76: 121–131.

Zhou M, Schools GP, Kimelberg HK (2006) Development of GLAST(+) astrocytes and NG2(+) glia in rat hippocampus CA1: mature astrocytes are electrophysiologically passive. *J Neurophysiol* 95: 134–143.

Abbreviations

C_m	Membrane capacitance
CNS	Central nervous system
EAA	Excitatory amino acid
E_r	Reversal potential
GFAP	Glial fibrillary acidic protein
GS	Glutamine synthetase
I–V	Current–voltage
Q_t	Total charge
R_a	Electrode access resistance
R_m	Membrane resistance
R_t	Total resistance
V_a	Voltage drop across *R*
V_c	Clamp (command) potential
V_t	Total voltage drop
$[K^+]_o$	Extracellular concentration of potassium ions
$[K^+]_i$	Intracellular concentration of potassium ions

Chapter 2
Neural Stem Cells Disguised as Astrocytes

Rebecca A. Ihrie and Arturo Alvarez-Buylla

Contents

2.1 Identification of Neural Stem Cells in the Central Nervous System

2.1.1 Astrocytes and Neurogenesis in the Adult Brain

As a major subclass of glial cells, astrocytes fulfill a diverse array of functional and architectural roles in the brain. These cells were originally classified as "support cells" of the nervous system. A common assumption of classical neuroscience was

A. Alvarez-Buylla
Department of Neurosurgery and Institute for Regeneration Medicine,
University of California San Francisco, San Francisco, CA, USA
abuylla@stemcell.ucsf.edu

V. Parpura and P.G. Haydon (eds.), *Astrocytes in (Patho)Physiology of the Nervous System*,
DOI: 10.1007/978-0-387-79492-1_2,

that neurons and glia are derived from distinct pools of progenitor cells (His, 1889). This idea arose partly due to the sequential developmental patterning of the nervous system; during cortical development, neurons are generated prior to glial cells. The supposed division between neuronal and glial lineages was also employed to explain observations of rare proliferating cells in the adult brain: It was thought that proliferation in the mature brain reflected the generation of new glial cells, and did not correspond to the production of new neurons. However, a number of experiments have now demonstrated ongoing neurogenesis in the adult brain and called into question the idea of separate developmental lineages for neurons and glia (Goldman and Nottebohm, 1983; Galileo et al., 1990; Alvarez-Buylla et al., 2001; Temple, 2001; Gage, 2002).

The disproving of the "no new neuron" dogma suggested that a reservoir of multipotent stem cells might exist within the adult central nervous system (CNS) and support ongoing neurogenesis. The identification and isolation of mammalian neural stem cells was therefore the focus of intense investigation. Tritiated thymidine incorporation studies in multiple animal models demonstrated that proliferation persists in specific regions of the adult brain, suggesting that these regions might contain immature progenitors of neurons or glia (Altman, 1962, 1963; Altman and Gopal, 1965; Altman and Das, 1966). Two germinal regions within the adult mammalian brain have since been shown to contain neural progenitor cells: the subventricular zone (SVZ), along the walls of the lateral ventricles, and the subgranular zone (SGZ) within the dentate gyrus of the hippocampus (Doetsch et al., 1997; Seri et al., 2004). Surprisingly, when the primary progenitors of the new neurons in these regions were identified, they exhibited structural and biological markers typical of differentiated astrocytes (Doetsch et al., 1999b; Seri et al., 2001). Why the stem cells of the CNS closely resemble differentiated glia, and how the proliferation and differentiation of these cells is controlled, continues to be an avenue for many exciting investigations.

2.1.2 Neural Stem Cells and the Architecture of Germinal Regions

2.1.2.1 The Subventricular Zone

The SVZ is located mostly on the lateral walls of the lateral ventricles and is the largest germinal region in the adult brain (Fig. 2.1). The cellular composition of this region has been described through both immunohistochemical studies and electron microscopic analysis (Doetsch et al., 1997, 1999b, 2002; Peretto et al., 1999). The SVZ contains relatively quiescent neural stem cells, known as type B cells, which give rise to actively proliferating type C cells. Type C cells in turn give rise to immature neuroblasts, also called type A cells. These neuroblasts migrate to the olfactory bulb, where they differentiate into interneurons. Remarkably, type B cells, despite their stem-like properties in vitro and in vivo (discussed later), express the intermediate filament component glial fibrillary acidic protein (GFAP), a marker that typically distinguishes mature astrocytes in other brain regions. These cells, when examined via electron microscopy, also have the ultrastructural characteristics typical of astrocytes, including bundles of intermediate filaments, multiple processes intercalating between other cells, and gap junction complexes.

In the SVZ, type A neuroblasts form a network of tangentially oriented chains, many of which join in the anterior and dorsal SVZ to form the rostral migratory stream (RMS) (Lois and Alvarez-Buylla, 1994; Jankovski and Sotelo, 1996; Lois et al., 1996; Peretto et al., 1997). These chains of A cells are ensheathed by the processes of type B cells. Interestingly, some type B cells also contact the ventricle via a process extended between ependymal cells. This process includes a short primary cilium extending into the ventricle (Doetsch et al., 1999a and Mirzadeh et al., 2008). The SVZ also contains blood vessels and a substantial extracellular matrix (Mercier et al., 2002), and is adjacent to the layer of ciliated ependymal cells that line the ventricle. The architecture of this specialized germinal zone allows for extensive cell–cell interaction as well as the propagation of signals from the cerebrospinal fluid in the ventricle, the surrounding extracellular matrix, and local blood vessels. As discussed below, the specialized environment of the SVZ, and signaling within the local niche, is likely to be an important determinant of the proliferative and regenerative potential of the astrocytes that act as neural stem cells in this region.

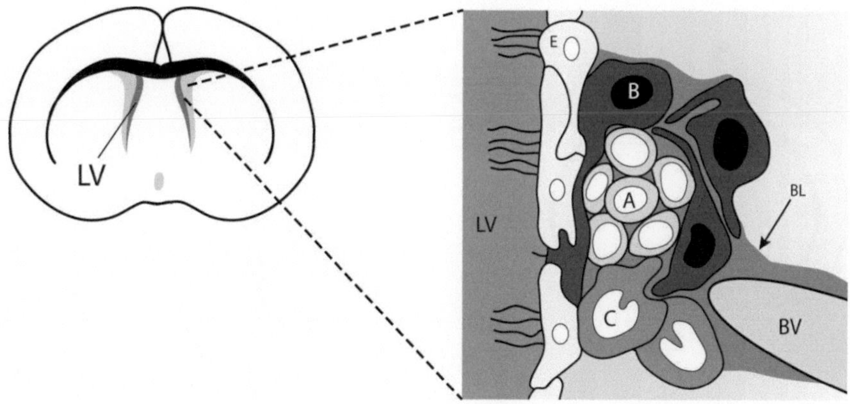

Fig. 2.1 The architecture of the subventricular zone in mouse brain. The SVZ is localized to the walls of the lateral ventricles (LV) in the brain, indicated at *left* and shown in detail at *right*. The SVZ contains type B cells (shown with *dark nuclei*), which are astrocyte-like neural stem cells. Also present are rapidly dividing type C cells, the transit-amplifying progeny of B cells. C cells in turn give rise to neuroblasts, or type A cells, which migrate to the olfactory bulb and give rise to neurons. The cells of the SVZ have extensive contact with the basal lamina (BL, indicated with *arrow*) and are also near blood vessels (BV). SVZ cells also have contact with the ciliated ependymal cells (E) that line the lateral ventricle.

2.1.2.2 The Subgranular Zone

The primary precursors of new neurons in the SGZ within the dentate gyrus have been described using methods similar to those used in the SVZ (Fig. 2.2). It remains controversial whether these cells are *bona fide* neural stem cells, as experiments in vitro have failed to demonstrate multipotentiality and self-renewal capabilities in isolated SGZ cells (Seaberg and van der Kooy, 2002; Bull and Bartlett, 2005). However, similar to their counterparts in the SVZ, the primary precursors of new neurons in the dentate gyrus express GFAP and exhibit ultrastructural characteristics that are typical of astrocytes (Seri et al., 2001). The cytoarchitecture of the SGZ is also reminiscent of the SVZ, with the primary progenitors appearing to interact closely with their progeny (Seri et al., 2004). In the SGZ, radially oriented astrocytes extend a process across the granule cell layer (GCL), as well as tangentially oriented processes at the base of the SGZ. These latter processes appear to act as a "nest" for the immature progeny generated by the division of these astrocytes. Radial astrocytes have also been identified as primary progenitors in additional recent studies (Filippov et al., 2003; Fukuda et al., 2003; Steiner et al., 2006), and are also called type I progenitors. The immature progeny of radial astrocytes are known as type D cells or type II progenitors. These cells form clusters that are closely associated with the processes of radial astrocytes. However, unlike the cells of the SVZ, the progeny of SGZ astrocytes do not migrate a long distance through the brain before maturation. Instead, maturing type D cells migrate a short distance into the GCL to form new granule neurons (Seri et al., 2004). The SGZ also differs from the SVZ with respect to its location within the CNS: The dentate gyrus does not have substantial contact with the ventricles or cerebrospinal fluid.

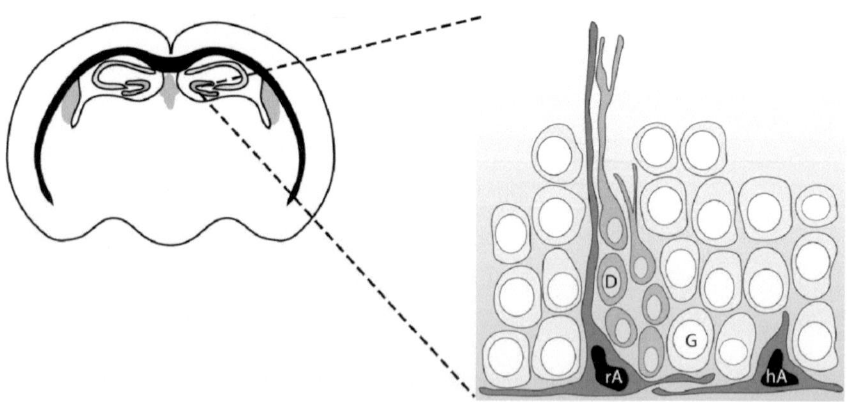

Fig. 2.2 The architecture of the subgranular zone in mouse brain. The SGZ is located within the dentate gyrus of the hippocampus, shown in coronal section at *left* and in detail at *right*. The SGZ contains radial (rA) and horizontal astrocytes (hA). Radial astrocytes (rA) have long radial processes that penetrate the granular layer as well as tangential ones that parallel this layer. These astrocytes give rise to type D immature precursors, which divide and mature into new granule neurons (G). D cells develop apical processes that become the dendrites of the new granule neurons.

Intriguingly, in addition to radial astrocytes, the SGZ also contains astrocytes that lack a radial process, termed horizontal astrocytes (Filippov et al., 2003; Seri et al., 2004). It is unclear whether these astrocytes can also act as primary precursors for new neurons. The protein nestin, a marker of immature neural precursor cells, is expressed in radial astrocytes and not in horizontal astrocytes in the SGZ, while S100β, a calcium-binding protein expressed in some astrocytes, exhibits the opposite pattern (Seri et al., 2004). Direct targeting of the nestin-expressing cells in the SGZ, the radial astrocytes, shows that these cells generate new neurons. Recent evidence indicates that S100β expression is present in oligodendrocyte progenitors (Deloulme et al., 2004; Hachem et al., 2005) and these could perhaps function as precursors for new oligodendrocytes in the hilus. However, in the absence of experimental methods for selectively marking horizontal astrocytes, the potential functional differences between these populations have not yet been investigated.

2.1.3 Experimental Identification of Stem Cells

A series of experiments in vitro and in vivo have determined that the astrocytes present in the SVZ and SGZ function as primary progenitors for the generation of new neurons in the adult brain. The first evidence that some of these progenitors had characteristics of stem cells came from reports that SVZ cells, when cultured with high concentrations of growth factors under nonadherent conditions, could form self-renewing colonies called neurospheres (Reynolds and Weiss, 1992; Morshead et al., 1994; Gage et al., 1995; Weiss et al., 1996). These cells, if subjected to growth factor removal, formed neurons, astrocytes, and oligodendrocytes, suggesting that these self-renewing cells were also multipotent. This ability to behave as a stem cell in vitro was later demonstrated specifically with astrocytes derived from the SVZ in rodents and humans (Doetsch et al., 1999b; Laywell et al., 2000; Sanai et al., 2004). However, formation of neurospheres may not necessarily reflect a role as a primary progenitor in vivo: Multiple neurospheres can also be derived from transit-amplifying (type C) progenitors (discussed in Sect. 2.2.3) and oligodendrocyte precursors (Kondo and Raff, 2000; Doetsch et al., 2002; Nunes et al., 2003). These experiments indicated that both primary (type B cells) and secondary progenitors (type C cells and oligodendrocyte precursors) could function as neural stem cells in vitro when exposed to exogenous growth factors. As indicated earlier, whether similar cells exist in the adult SGZ remains controversial (Gage et al., 1998; Seaberg and van der Kooy, 2002; Bull and Bartlett, 2005).

SVZ and SGZ astrocytes were shown to be primary precursors in vivo through the following experiments: (1) long-term proliferation marker retention; (2) retroviral labeling and fate mapping; (3) survival of antimitotic treatment and regeneration of the germinal layer; and (4) ablation via genetic targeting of astrocytes. Doetsch et al. examined the cells within the SVZ that retain markers of DNA synthesis over long periods of time, indicating the relatively slow cycling time expected of stem cells (Doetsch et al., 1999b). By coupling the administration of tritiated thymidine with electron microscopy analysis, they demonstrated that the only label-retaining cells in the SVZ at long time intervals after administration were GFAP-positive astrocytes. To further demonstrate that these astrocytes can act as neuronal precursors in normal brain, the authors carried out lineage tracing experiments using adult GFAP-Tva mice, which express the receptor for avian leukosis virus under the GFAP promoter. In these mice, the injection of replication-competent avian leukosis virus encoding alkaline phosphatase (RCAS-AP) resulted in specific labeling of GFAP-expressing cells. The subsequent progeny of these cells could then be studied via their expression of alkaline phosphatase. After the administration of this virus, AP-positive SVZ astrocytes were observed at 1 day after infection, suggesting that initial infection was limited to this astrocytic cell population. AP-positive migrating neuroblasts and olfactory bulb interneurons were found at 3.5 and 14 days after infection, respectively, showing that SVZ astrocytes can give rise to new neurons in the adult brain.

Additional experiments utilizing the antimitotic treatment cytosine-β-d-arabinofuranoside (Ara-C) also showed that SVZ astrocytes are capable of regenerating this germinal region after injury. The administration of Ara-C results in the elimination of fast-dividing precursor cells (type C cells) and neuroblasts (type A cells), leaving the ependymal cells and slow-dividing astrocytes (type B cells) (Doetsch et al., 1999a). After Ara-C administration, GFAP-positive astrocytes in the SVZ divide and give rise first to type C cells and subsequently to type A cells, regenerating the germinal zone over a period of 14 days. While SVZ astrocytes and their progeny incorporate bromodeoxyuridine (BrdU), indicating DNA synthesis and division, the ependymal cells that remain after Ara-C administration are not BrdU positive, arguing that these cells do not act as neural stem cells (Doetsch et al., 1999b). Other more recent studies also indicate that ependymal cells do not divide or function as neural stem cells (Chiasson et al., 1999; Capela and Temple, 2002; Spassky et al., 2005).

The characterization of SGZ astrocytes was carried out using methods similar to those described earlier. Seri et al. demonstrated that after antimitotic administration, dividing type D cells are absent from the SGZ, while some type B astrocytes remain (Seri et al., 2001). These cells subsequently divide, giving rise to type D cells, which act as transient secondary precursors for new GCL neurons. In addition, retroviral lineage tracing studies using the GFAP-Tva/RCAS-AP system demonstrated that infected GFAP-positive progenitors give rise to immature precursors, which ultimately become fully differentiated granule neurons in the dentate gyrus. Some SGZ astrocytes also retain proliferation markers at long time intervals after administration, again suggesting that some of these cells normally divide slowly to maintain the stem-cell pool. In the SGZ, radial astrocytes were observed in mitoses; these cells appear to divide asymmetrically with one of the daughter cells (the putative self-renewing progenitor) retaining the radial process (Seri et al., 2004).

The positive identification of SVZ and SGZ astrocytes as neural precursor cells in vivo was also complemented by a genetic loss-of-function analysis. Sofroniew and colleagues generated transgenic mice in which Herpes simplex virus thymidine kinase (HSV-TK) was expressed under the control of the GFAP promoter (Imura et al., 2003; Morshead et al., 2003; Garcia et al., 2004). In these mice, administration of the antiviral agent ganciclovir resulted in specific ablation of dividing GFAP-expressing cells.

Upon daily injections of ganciclovir both the SVZ and SGZ in these mice showed a progressive decrease in the number of BrdU-positive cells and the number of cells staining for polysialic acid-neural cell adhesion molecule (PSA-NCAM), a marker of neuroblasts. After 21 days of ganciclovir administration and a 14-day recovery period, the total number of newly generated mature neurons was reduced to 0% and 1.8% of normal levels in the SVZ and SGZ respectively. These results demonstrate that the GFAP-positive cells in these regions are required for constitutive neurogenesis, again suggesting that these GFAP-expressing cells are the progenitors of new neurons.

The identification of GFAP-expressing cells as the progenitors of new neurons in the adult brain was surprising in part because of the many characteristics these cells share with astrocytes in other regions of the brain. It was initially expected that these multipotent cells might instead resemble immature, undifferentiated cells. However, as previously noted, B cells in both the SVZ and SGZ appear to have many of the classical features of astrocytes, including the expression of GFAP and the ultrastructural details that distinguish these cells. What differences exist between the astrocytes in germinal regions and those in nonneurogenic regions of the brain, how these apparently differentiated cells generate their transit-amplifying progeny, and the other structural roles these astrocytes may fulfill in the germinal zone are discussed in the following section.

2.2 Interactions Within the Stem-Cell Niche

2.2.1 Generation of Intermediate Progenitors

Although neural progenitors resemble developmentally committed astrocytes, they are capable of giving rise to transit-amplifying progeny, which can (in the SVZ) divide rapidly to expand the available pool of neural precursors. How this process occurs is unclear. Do these astrocytes dedifferentiate to divide and produce immature precursors? Alternatively, can these cells divide while retaining their processes within the SVZ or SGZ? The answers to these questions have not yet been fully investigated. It has been suggested that astrocytes that function as neural progenitors are smaller, have fewer processes or have a less electron-dense, lighter appearance under the electron microscope (Alvarez-Buylla and Garcia-Verdugo, 2002; Garcia et al., 2004). Immunostaining of germinal region astrocytes after Ara-C administration, when higher numbers of these cells incorporate proliferation markers, finds that these progenitors have many morphological characteristics (Doetsch et al., 1999a; Seri et al., 2001) previously considered markers of mature astrocytes (Privat, 1977). Electron microscopic analysis in the SGZ by Seri et al. indicates that radial astrocytes retain their long processes, as well as their contacts with blood vessels and neighboring cells, upon division (Seri et al., 2004). The ability to divide while maintaining a specialized morphology is reminiscent of radial glia, the cell type that serves as the stem cell of developing forebrain and gives rise to the stem cells of the adult SVZ (discussed in Sect. 2.3). The pattern of division of adult neural stem cells is also unclear

– unlike studies delineating the pattern of division of embryonic neuroepithelial cells, it is not known whether astrocytes in the SVZ or SGZ undergo symmetric division to expand the stem or transit-amplifying cell pool, asymmetric division to produce a single stem and single transit-amplifying cell, or both.

2.2.2 *Effects of the Niche on Proliferative Potential*

Cell-extrinsic factors within the niche are thought to have significant effects on the proliferation and self-renewal of neural stem cells (Alvarez-Buylla and Lim, 2004). A large number of pathways regulating proliferation and differentiation are thought to be active in the SVZ and/or SGZ and play various roles in the lineage commitment and proliferation of primary progenitors, transit-amplifying cells, or neuroblasts. The precursor cells of the SVZ were first identified by their ability to proliferate in culture when exposed to high concentrations of epidermal growth factor (EGF) and fibroblast growth factor (FGF) (Reynolds and Weiss, 1992; Gage et al., 1995; Craig et al., 1996; Weiss et al., 1996; Gritti et al., 1999). Subsequently, multiple growth factor receptors have been shown to be expressed in different cell types in the SVZ. The receptors for FGF and platelet-derived growth factor (PDGF) are thought to be expressed by stem-cell astrocytes, while the EGF receptor (EGFR) is primarily expressed by transit-amplifying C cells (Doetsch et al., 2002; Zheng et al., 2004; Jackson et al., 2006).

Other pathways with important functions in development have likewise been implicated in control of neural progenitor proliferation. Lim et al. demonstrated that specific bone morphogenetic proteins (BMPs) and their cognate receptors are expressed by cells in the SVZ. Interestingly, ependymal cells that interact closely with SVZ Type B cells, and SVZ astrocytes themselves, produce the BMP antagonist Noggin (Lim et al., 2000; Piccirillo et al., 2006). As BMP signaling appears to inhibit neurogenesis, the presence of Noggin and BMPs in the SVZ provides a potential mechanism for controlling the balance between neuronal production and glial differentiation by neuronal precursors. In the dentate gyrus, signaling via secreted Wnt proteins has also been suggested to control neuroblast proliferation and commitment to a neuronal fate (Lie et al., 2005). Recent studies have also implicated the Notch and Sonic hedgehog (Shh) pathways in the control of SVZ and/or SGZ cell proliferation, but the cell type-specific pattern of expression for components of these pathways has not been fully described (Machold et al., 2003; Ahn and Joyner, 2005; Palma et al., 2005; Androutsellis-Theotokis et al., 2006; Givogri et al., 2006).

2.2.3 *Prolonged EGF Signaling Confers Stem-Like Characteristics on Transit-Amplifying Cells*

Experiments in which EGF was infused into the lateral ventricle had two distinct effects on the cells in the SVZ, and offer some insight into how the properties of neural precursors may be altered in response to elevated mitogenic signaling (Kuhn et al., 1997; Tropepe et al., 1999; Doetsch et al., 2002; Aguirre et al., 2005). First, EGF infusion resulted in a significant increase in the proliferation of type C transit-amplifying cells, the primary EGFR-expressing cell type in this region. This increase in proliferation was accompanied by an arrest in neuroblast production and the invasion of labeled SVZ precursors into adjoining tissue along blood vessels and white matter tracts. Contrary to the assumption that only SVZ stem cells can form neurospheres in culture, dividing EGFR-expressing precursors were further shown to be responsible for the majority of neurosphere production upon culture (Doetsch et al., 2002). These results suggested that, upon prolonged exposure to elevated EGF, the transit-amplifying cells of the SVZ undergo self-renewing divisions rather than neuroblast production and more closely resemble the astrocytic stem cells of this region. In addition to these effects on type C cells, EGF infusion had a second noticeable effect in the SVZ. Although most EGFR-expressing cells correspond to type C cells, a limited number of GFAP-expressing cells also appear to express this receptor. Although GFAP-positive B cells do not exhibit enhanced proliferation upon EGF infusion, in coronal sections a greater number of B cells appear to extend processes to touch the lateral ventricle. More recent data from en-face imaging of the ventricular surface has demonstrated that many B cells normally extend a small apical process to the ventricular surface (Mirzadeh et al., 2008). These new observations suggest that the perceived increase in ventricle-contacting astrocytes may be due to an expansion of the existing apical surface areas of B cells, rather than the formation of new contacts with the ventricle. The ability of a secondary progenitor-like, transit-amplifying type C cell to generate neurospheres and behave like a stem cell in vitro is interesting. It suggests that the transition from type B cell to C cells may not represent irreversible changes in cell-fate determination. This may also occur with other secondary progenitors like those present in white matter or in the oligodendrocyte lineage (Kondo and Raff, 2000; Nunes et al., 2003).

2.2.4 Can All Astrocytes Act as Progenitors?

These studies of the germinal niche environments raise the question of whether the ability of SVZ and SGZ astrocytes to act as primary progenitors in the generation of new neurons is due primarily to the effects of their microenvironment. Do all GFAP-expressing glial cells have latent multipotent potential that can be unlocked given the proper environmental cues? Experimental evidence indicates that this prospect is unlikely. While the architecture of the stem-cell niche clearly has significant effects on stem-cell proliferation and differentiation, niche-derived signals do not appear to be sufficient to confer progenitor-like capabilities on nongerminal astrocytes. Transplantation of parenchymal tissue to the SVZ of adult mice does not cause these astrocytes to become neurogenic (A. Alvarez-Buylla, unpublished). Recent evidence also indicates that specific cell-intrinsic differences between germinal center astrocytes and other astrocytes may exist (Imura et al., 2006). Imura and colleagues demonstrated heterogeneity in marker expression and neurogenic potential between two populations of GFAP-expressing cells: astrocytes derived from the adult SVZ, and astrocytes derived from the cerebral cortex. While SVZ astrocytes were able to form multipotent neurospheres or exhibit neurogenic potential under the appropriate culture conditions, astrocytes derived from the cerebral cortex lacked these capabilities. A similar observation has been made with astrocytes derived from adult human brain (Sanai et al., 2004). A subpopulation of germinal zone GFAP-positive astrocytes expresses the cell-surface marker Lewis antigen (LeX), also known as CD15, a marker previously associated with SVZ neural stem cells (Capela and Temple, 2002). This LeX-positive, GFAP-positive subpopulation appears to be the primary source of multipotent NSCs, as sorted populations of LeX-negative GFAP-expressing astrocytes form few multipotent neurospheres in culture when compared with LeX-positive, GFAP-positive astrocytes (Imura et al., 2006). LeX-expressing astrocytes were not found in the cerebral cortex, suggesting that this marker may serve as a means to identify astrocytes with multipotent neurogenic potential and distinguish further functional or phenotypic differences between germinal zone astrocytes and the larger population of CNS astrocytes. In addition to LeX, other proteins are also emerging as potential markers of germinal zone astrocytes, including brain-lipid-binding protein (BLBP), Nestin, and Sox2 (Filippov et al., 2003; Steiner et al., 2006). However, as with stem-cell niches in other tissues, a combination of markers may be required to define the subpopulation of adult astrocytes that function as neural stem cells.

2.2.5 Functional Heterogeneity Within Germinal Astrocyte Populations

It is possible that further subdivisions within the population of neural stem cells may exist, and that the cell fates of the astrocytes within these germinal regions may be restricted. There is clear heterogeneity in the pool of intermediate precursors within the SVZ. Most (or all) C cells are marked by their expression of the transcription factor Mash1, and go on to produce neuroblasts (Parras et al., 2004). Apparently, a subpopulation of B cells also express Mash1. A smaller subpopulation of B cells and C cells express Olig2, a basic helix-loop-helix transcription factor required for the production of oligodendrocytes and motor neurons (Hack et al., 2005; Menn et al., 2006).

The progenitors of the SVZ normally give rise to multiple types of interneurons that integrate into different layers in the olfactory bulb. By combining lineage tracing techniques with markers for distinct olfactory interneurons, investigators have begun to address the question of how different neuronal subtypes are specified in neuronal progenitors. To date, these studies have focused primarily on Pax6, a transcription factor that is important in the developmental patterning of multiple tissues (Gotz et al., 1998; Kohwi et al., 2005). In adult mice, Pax6 is expressed by a subset of olfactory interneurons, and is also present in the SVZ and RMS. Intriguingly, SVZ cells lacking Pax6 or expressing a dominant-negative Pax6/engrailed chimeric protein are largely unable to form dopaminergic periglomerular cells (PGCs) in the olfactory bulb. Conversely, overexpression of Pax6 in neuronal precursors results in the formation of greater numbers of PGCs. It appears that Pax6 has an instructive role in directing neural precursors toward this neuronal fate, suggesting that other factors may exist that lead to the generation of other neuronal subtypes. The heterogeneous expression of Pax6 in the SVZ and RMS suggests that neuronal fate specification may occur relatively early in the stem cell – transit amplifying cell – neuroblast lineage, and is complemented by work showing that the location of particular SVZ cells within the brain affects the type of neurons they produce (Merkle et al., 2007). How the location and specific genetic makeup of particular neural stem cells may affect the fate of their progeny is an exciting area for future investigation.

In addition to the regulation of cell fate by transcription factors, external signaling may also affect the cell fate of stem cells and their progeny. It has recently been shown that a subpopulation of SVZ B cells expressing PDGF receptor alpha (PDGFRα) can generate neurons and oligodendrocytes (Jackson et al., 2006). Interestingly, these PDGFRα+ cells are capable of hyperproliferation upon the introduction of ectopic PDGF ligand, generating large masses of Olig2-positive intermediate progenitor-like cells. Upon withdrawal of PDGF, these masses regress and an increase in oligodendrocyte production is observed, suggesting that PDGF signaling may affect the balance between neuroblast production and oligodendrocyte precursor cell production in specific SVZ astrocytes.

2.3 Developmental Lineage of Neural Stem Cells

The origins of adult CNS stem cells, and their relationship to earlier progenitor cells in the brain, may offer clues to the unique characteristics that distinguish these germinal astrocytes from other astroglial cells in the brain parenchyma. In addition, the mechanisms controlling the multipotency and proliferation of these cell types may suggest how these processes are disrupted in the development of brain tumors, which exhibit many phenotypic similarities to neural precursor cells.

2.3.1 Stem Cells of the Immature Brain Give Rise to SVZ Astrocytes

The CNS is generated from the neuroepithelium, which begins as a sheet of primary progenitor cells that folds together at its edges to form the neural tube. The center of the neural tube later becomes the ventricular system and spinal canal, and successive divisions of neuroepithelial cells at the ventricular surface thicken and expand the developing brain. Throughout this process, neuroepithelial cells maintain contacts with both the ventral and pial surfaces, resulting in radial stretching of these cells as the brain develops (Haubensak et al., 2004) (Fig. 2.3). At this point

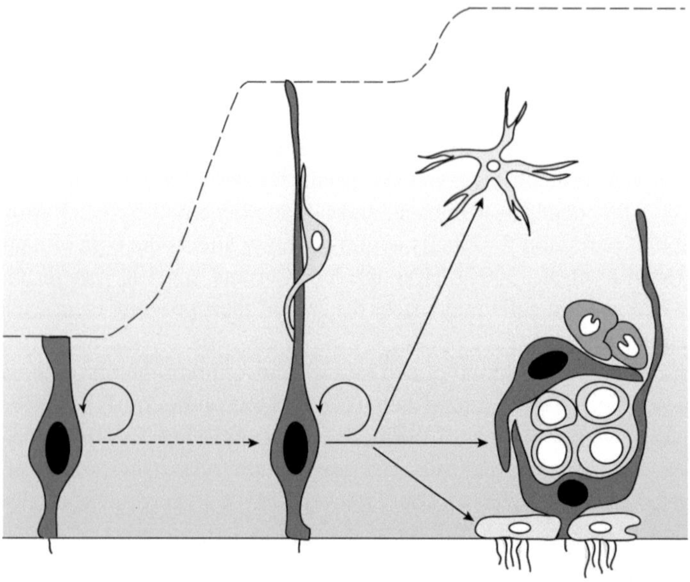

Fig. 2.3 Neural stem cells throughout development. The stem cells of the developing brain change their shape and produce distinct progeny as the brain develops. Neuroepithelial cells (shown at *left*) are the principal progenitors of the early developing brain. These cells are thought to give rise to radial glia (*center*), which begin expressing GFAP during neurogenesis. Radial glia in turn give rise to the germinal zone astrocytes of the mature brain (shown at *right*) in addition to parenchymal astrocytes, oligodendrocytes, and ependymal cells. The astrocytes of the SVZ, whose architecture has been characterized in the greatest detail, are shown here. Both neuroepithelial cells and radial glia maintain contacts with both the ventral surface (*solid line*) and pial surface (*dashed line*) of the brain, and project a single cilium into the developing ventricle. In contrast, SVZ astrocytes, which also often project a single cilium, do not contact the pial surface, although many appear to extend a radial process into adjoining tissue. Instead, as shown in Fig. 2.1, these cells often contact the basal lamina of blood vessels.

in development, the progenitor cells do not express GFAP. At the time of neurogenesis, the neuroepithelial cells are thought to gradually transform into radial glial cells, the principal progenitor of the forebrain. These cells also have a long radial process that contacts the pial surface of the brain, and divide in the ventricular zone much like neuroepithelial cells (Noctor et al., 2001, 2002, 2004; Anthony et al., 2004; Gotz and Huttner, 2005). During the onset of neurogenesis, these cells also begin to express cytoskeletal and cell-surface markers typical of astrocytes, including GFAP (Imura et al., 2003). Radial glia and neuroepithelial cells share many characteristics, including the maintenance of some features of apical–basal polarity and the expression of the intermediate filament protein nestin (Alvarez-Buylla et al., 2001). Because of these shared characteristics and similar patterns of division, it is likely that neuroepithelial cells transform directly into radial glial cells. However, this transformation has not yet been experimentally demonstrated.

Radial glial cells also share many features with germinal zone astrocytes, particularly the astrocytes of the SVZ. Both radial glia and germinal astrocytes occupy the same region of the brain at different developmental times, and some SVZ astrocytes maintain a long radial process similar to that of radial glia. In songbirds and other organisms, a subset of radial glia remain neurogenic during adult life (Alvarez-Buylla et al., 1990; Garcia-Verdugo et al., 2002; Russo et al., 2004; Zupanc, 2006). In mammals, this function appears to be carried out instead by the germinal zone astrocytes, which are derived from radial glia. Experiments using a Cre-lox-based strategy to specifically label neonatal radial glia showed that these cells give rise to multiple cell types, including the astrocytes of the SVZ (Merkle et al., 2004). These results suggest that adult neural stem cells are part of a continuous lineage that begins with neuroepithelial cells, continues through radial glia, and results in germinal zone astrocytes.

The origin of SGZ astrocytes that function as neural progenitors in the adult brain has not been determined experimentally. However, here too a connection to radial glial cells has been hypothesized (Seri et al., 2004). In fact work in the 1970s and 1980s, when the astrocytic nature of primary precursors in the postnatal brain had not been yet recognized, suggested that radial astrocytes in the dentate gyrus are derived from radial glia within the part of the ventricular zone that contributes to the hippocampus (Eckenhoff and Rakic, 1984).

2.3.2 Restriction of Stem-Cell Potential Over Time

Although neuroepithelial cells and radial glia have common characteristics with adult neural stem cells, one fundamental question with significant therapeutic implications is whether adult stem cells retain the capacity to generate earlier-born cell types. Throughout development, the various neural stem-cell types change their morphology and give rise to different types of progeny. At least some of this commitment appears to be specified by an intrinsic developmental program, as progenitors grown in culture proceed along a schedule of development that is synchronous with what has been described in vivo (Shen et al., 2006). This program does appear to be unidirectional, as coculture of older progenitors with younger cells does not allow these progenitors to generate earlier-born cell types. Likewise, older progenitors appear to produce a more limited repertoire of cell types than younger progenitors upon transplantation into the embryonic brain. The mechanisms that contribute to restriction of neural stem-cell potential over time are largely unknown, but likely include both genetic and epigenetic changes as well as potential cell-extrinsic cues. In *Drosophila,* a series of transcription factors that are sequentially expressed during neuroblast lineage development have been identified, illuminating a potential cell-intrinsic mechanism for restricting progenitor fate over time (Pearson and Doe, 2003, 2004). However, in the development of the mammalian cortex, heterochronic transplants of younger progenitors into older tissue indicate that the specification of particular neural progeny is controlled at least in part by environmental, cell-extrinsic factors (reviewed in Pearson and Doe (2004)). How the fate specification of adult germinal zone astrocytes might be reprogrammed for therapeutic use is a question that is open to investigation.

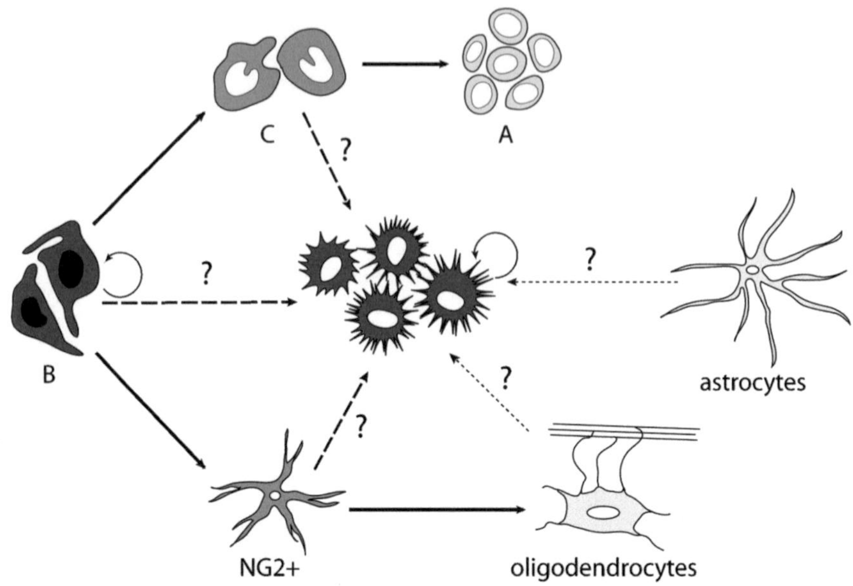

Fig. 2.4 Neural precursors and tumorigenesis. Although the cell-of-origin for adult neural tumors has not been identified, the identification of "tumor stem cells," which share many characteristics with neural precursors, has led to speculation that neural tumors (*center*), such as astrocytomas or oligodendrogliomas, may occur when oncogenic events affect cells in the neural precursor lineage (B,C, and NG2+ precursors). At present, it is unclear whether tumors may arise from mutations affecting primary precursors (such as type B cells in the SVZ), secondary precursors (such as type C or NG2+ cells), or both. It is also possible that these tumors are derived from mature cells in the brain (i.e., astrocytes and oligodendrocytes), which "dedifferentiate" as a result of oncogenic events leading to tumor development, and therefore resemble precursor cells. To date there is little direct evidence favoring either of these two models, and both therefore remain speculative.

2.3.3 Astrocytic Stem Cells and Tumor Stem Cells

Tumors of the CNS share several characteristics with adult neural stem cells and their immediate progeny, transit amplifying cells: expression of similar immunohistochemical markers, high proliferative potential, and the ability to migrate and invade tissue adjacent to the germinal zone (Sanai et al., 2005; Vescovi et al., 2006) (Fig. 2.4). The identification of SVZ and SGZ astrocytes as neural precursors within the adult brain introduced the possibility that, rather than involving dedifferentiation of mature cells, brain tumors might arise when neural stem or precursor cells sustained mutations that resulted in uncontrolled proliferation. Infusion of EGF or PDGF into the lateral ventricle of the adult brain, as noted above, results in elevated proliferation of progenitor cells and the production of highly invasive cells or glioma-like masses (Doetsch et al., 2002; Jackson et al., 2006). In addition, mouse models in which progenitor cells express high levels of PDGF or EGF result in the formation of tumors that resemble human gliomas (Holland et al., 1998; Dai et al., 2001; Assanah et al., 2006). Finally, it has been observed that mice lacking the tumor suppressors p53 and Nf1 in the CNS develop early lesions associated with the SVZ that progress to tumors resembling human malignant astrocytomas (Zhu et al., 2005).

Multiple groups have also reported the isolation of a particular subpopulation of cells from human gliomas. These cells are self-renewing and multipotent in vitro, and have (somewhat controversially) been named tumor stem cells (Singh et al., 2003, 2004; Galli et al., 2004; Yuan et al., 2004). This subpopulation efficiently gives rise to tumors when transplanted into mice, suggesting that these cells may be responsible for tumor growth. Further, traditional therapies such as radiation, which target rapidly dividing cells, appear to spare this relatively quiescent stem-cell population, thereby failing to prevent tumor recurrence (Bao et al., 2006). In contrast, treatment of this tumor stem-cell population with BMPs, which limit the proliferation of normal stem cells, also blocks the ability of these cells to form tumors upon transplantation (Piccirillo et al., 2006). Studies of human ependymomas also identified a subpopulation of radial glia-like cells within these tumors that efficiently form tumors upon transplantation. Expression profiling of these ependymomas suggests that tumors isolated from specific locations retain or reproduce the developmental profile of radial glia originating from different levels of the neuroaxis (Taylor et al., 2005). Recent investigations into the effects of aging on tissue-specific precursor cells, including the astrocyte-like cells of the SVZ, appear to indicate that adult stem cells may maintain a precarious balance between proliferation and tumor development. As the organism ages, neural precursor proliferation appears to be restrained by increased expression of cell-cycle inhibitors. While cell-cycle inhibition may act to prevent hyperproliferation and tumor development as oncogenic mutations accumulate in an aging organism, this inhibition also results in decreased proliferative capacity within the neural progenitor compartment. In aging mice, the cell-cycle inhibitor p16/INK4A is expressed at detectable levels in the SVZ, but this expression is absent in younger mice (Molofsky et al., 2006). Expression of p16/INK4A has not been observed in astrocytes in other regions of the brain, although these studies generally have not focused on samples derived from aging mice. The SVZ precursors in aging mice deficient for p16 appear to have increased proliferative potential when compared with wild-type counterparts. p16/INK4A is frequently lost or mutated in human gliomas, indicating that disruption of the regulation of proliferative potential, when coupled with other oncogenic events, can result in cancer (Sanai et al., 2005). Based on these lines of evidence, it is tempting to speculate that tumor stem cells may arise from neural precursors that sustain oncogenic mutations. However, to date no direct evidence about the glioma cell-of-origin exists. It is unclear whether tumors arise from an aberrant stem cell or whether oncogenic events in more differentiated progeny endow tumor cells with stem-like characteristics. A better understanding of the origins and properties of tumor stem cells will be essential to developing therapies that target these elusive cells.

2.4 Conclusion

Identification of the neural stem cells in the adult brain revealed a surprising fact: These important progenitors have many of the features of mature astrocytes. In fact, these cells were identified as astrocytes before their progenitor function was recognized. The astrocytes of the SVZ and SGZ appear to be strongly influenced by their local microenvironment. Yet, environment alone does not seem to be sufficient to induce nongerminal astrocytes to behave as neural stem cells, and germinal astrocytes also contribute significantly to the niche. Although emerging evidence suggests that functional differences do exist between germinal zone astrocytes and the larger population of CNS astrocytes, it is still unclear how these differences are encoded, and how much heterogeneity exists within the germinal zone population. The identification of neural stem cells as glia is a significant departure from classical concepts in neuroscience that separated the glial and neuronal lineages early in development. The new view of these cells forces us to redraw lineages and place neural stem cells within what could be considered the core neuroepithelial lineage; neuroepithelium-radial glia-germinal astrocytes. Our recently improved understanding of neural stem cells should facilitate molecular studies to further characterize these cells, which escaped identification for many decades with their glial disguise. The description of the glial properties of neural progenitors raises many interesting questions: What makes these cells unique? How do these cells change with time? How is their proliferation and differentiation regulated? Answers to these questions will likely have important implications for understanding brain development, brain cancer, and the treatment of neurodegenerative disease.

Acknowledgments The authors wish to thank the members of the Alvarez-Buylla laboratory for helpful discussions and comments on the manuscript. Work in the Alvarez-Buylla laboratory is supported by grants from the NIH and the Goldhirsh Foundation and the John G. and Frances F. Bowes Developmental and Stem Cell Biology Fund. Rebecca Ihrie is a Damon Runyon Fellow supported by the Damon Runyon Cancer Research Foundation. Arturo Alvarez-Buylla holds the Heather and Melanie Muss Endowed Chair in Neurosurgery.

References

Aguirre A, Rizvi TA, Ratner N, Gallo V (2005) Overexpression of the epidermal growth factor receptor confers migratory properties to nonmigratory postnatal neural progenitors. *J Neurosci* 25:11092–11106.

Ahn S, Joyner AL (2005) In vivo analysis of quiescent adult neural stem cells responding to Sonic hedgehog. *Nature* 437:894–897.

Altman J (1962) Are new neurons formed in the brains of adult mammals? *Science* 135:1127–1128.

Altman J (1963) Autoradiographic investigation of cell proliferation in the brains of rats and cats. *Anat Rec* 145:573–591.

Altman J, Gopal DD (1965) Post-natal origin of microneurones in the rat brain. *Nature* 207:953–956.

Altman J, Das GD (1966) Autoradiographic and histological studies of postnatal neurogenesis. I. A longitudinal investigation of the kinetics, migration and transformation of cells incorporating tritiated thymidine in neonate rats, with special reference to postnatal neurogenesis in some brain regions. *J Comp Neurol* 126:337–390.

Alvarez-Buylla A, Garcia-Verdugo JM (2002) Neurogenesis in adult subventricular zone. *J Neurosci* 22:629–634.

Alvarez-Buylla A, Lim DA (2004) For the long run: maintaining germinal niches in the adult brain. *Neuron* 41:683–686.

Alvarez-Buylla A, Theelen M, Nottebohm F (1990) Proliferation "hot spots" in adult avian ventricular zone reveal radial cell division. *Neuron* 5:101–109.

Alvarez-Buylla A, Garcia-Verdugo JM, Tramontin AD (2001) A unified hypothesis on the lineage of neural stem cells. *Nat Rev Neurosci* 2:287–293.

Androutsellis-Theotokis A, Leker RR, Soldner F, Hoeppner DJ, Ravin R, Poser SW, Rueger MA, Bae SK, Kittappa R, McKay RD (2006) Notch signalling regulates stem cell numbers in vitro and in vivo. *Nature* 442:823–826.

Anthony TE, Klein C, Fishell G, Heintz N (2004) Radial glia serve as neuronal progenitors in all regions of the central nervous system. *Neuron* 41:881–890.

Assanah M, Lochhead R, Ogden A, Bruce J, Goldman J, Canoll P (2006) Glial progenitors in adult white matter are driven to form malignant gliomas by platelet-derived growth factor-expressing retroviruses. *J Neurosci* 26:6781–6790.

Bao S, Wu Q, McLendon RE, Hao Y, Shi Q, Hjelmeland AB, Dewhirst MW, Bigner DD, Rich JN (2006) Glioma stem cells promote radioresistance by preferential activation of the DNA damage response. *Nature* 444:756–760.

Bull ND, Bartlett PF (2005) The adult mouse hippocampal progenitor is neurogenic but not a stem cell. *J Neurosci* 25:10815–10821.

Capela A, Temple S (2002) LeX/ssea-1 is expressed by adult mouse CNS stem cells, identifying them as nonependymal. *Neuron* 35:865–875.

Chiasson BJ, Tropepe V, Morshead CM, Van der Kooy D (1999) Adult mammalian forebrain ependymal and subependymal cells demonstrate proliferative potential, but only subependymal cells have neural stem cell characteristics. *J Neurosci* 19:4462–4471.

Craig CG, Tropepe V, Morshead CM, Reynolds BA, Weiss S, Van der Kooy D (1996) In vivo growth factor expansion of endogenous subependymal neural precursor cell populations in the adult mouse brain. *J Neurosci* 16:2649–2658.

Dai C, Celestino JC, Okada Y, Louis DN, Fuller GN, Holland EC (2001) PDGF autocrine stimulation dedifferentiates cultured astrocytes and induces oligodendrogliomas and oligoastrocytomas from neural progenitors and astrocytes in vivo. *Genes Dev* 15:1913–1925.

Deloulme JC, Raponi E, Gentil BJ, Bertacchi N, Marks A, Labourdette G, Baudier J (2004) Nuclear expression of S100B in oligodendrocyte progenitor cells correlates with differentiation toward the oligodendroglial lineage and modulates oligodendrocytes maturation. *Mol Cell Neurosci* 27:453–465.

Doetsch F, Garcia-Verdugo JM, Alvarez-Buylla A (1997) Cellular composition and three-dimensional organization of the subventricular germinal zone in the adult mammalian brain. *J Neurosci* 17:5046–5061.

Doetsch F, Garcia-Verdugo JM, Alvarez-Buylla A (1999a) Regeneration of a germinal layer in the adult mammalian brain. *Proc Natl Acad Sci USA* 96:11619–11624.

Doetsch F, Caille I, Lim DA, García-Verdugo JM, Alvarez-Buylla A (1999b) Subventricular zone astrocytes are neural stem cells in the adult mammalian brain. *Cell* 97:1–20.

Doetsch F, Petreanu L, Caille I, Garcia-Verdugo JM, Alvarez-Buylla A (2002) EGF converts transit-amplifying neurogenic precursors in the adult brain into multipotent stem cells. *Neuron* 36:1021–1034.

Eckenhoff MF, Rakic P (1984) Radial organization of the hippocampal dentate gyrus: a Golgi, ultrastructural, and immunocytochemical analysis in the developing rhesus monkey. *J Comp Neurol* 223:1–21.

Filippov V, Kronenberg G, Pivneva T, Reuter K, Steiner B, Wang LP, Yamaguchi M, Kettenmann H, Kempermann G (2003) Subpopulation of nestin-expressing progenitor cells in the adult murine hippocampus shows electrophysiological and morphological characteristics of astrocytes. *Mol Cell Neurosci* 23:373–382.

Fukuda S, Kato F, Tozuka Y, Yamaguchi M, Miyamoto Y, Hisatsune T (2003) Two distinct subpopulations of nestin-positive cells in adult mouse dentate gyrus. *J Neurosci* 23:9357–9366.

Gage FH (2002) Neurogenesis in the adult brain. *J Neurosci* 22:612–613.

Gage FH, Ray J, Fisher LJ (1995) Isolation, characterization, and use of stem cells from the CNS. *Annu Rev Neurosci* 18:159–192.

Gage FH, Kempermann G, Palmer T, Peterson DA, Ray J (1998) Multipotent progenitor cells in the adult dentate gyrus. *J Neurobiol* 36:249–266.

Galileo DS, Gray GE, Owens GC, Majors J, Sanes JR (1990) Neurons and glia arise from a common progenitor in chicken optic tectum: Demonstration with two retroviruses and cell type-specific antibodies. *Proc Natl Acad Sci USA* 87:458–462.

Galli R, Binda E, Orfanelli U, Cipelletti B, Gritti A, De Vitis S, Fiocco R, Foroni C, Dimeco F, Vescovi A (2004) Isolation and characterization of tumorigenic, stem-like neural precursors from human glioblastoma. *Cancer Res* 64:7011–7021.

Garcia AD, Doan NB, Imura T, Bush TG, Sofroniew MV (2004) GFAP-expressing progenitors are the principal source of constitutive neurogenesis in adult mouse forebrain. *Nat Neurosci* 7:1233–1241.

Garcia-Verdugo JM, Ferron S, Flames N, Collado L, Desfilis E, Font E (2002) The proliferative ventricular zone in adult vertebrates: a comparative study using reptiles, birds, and mammals. *Brain Res Bull* 57:765–775.

Givogri MI, de Planell M, Galbiati F, Superchi D, Gritti A, Vescovi A, de Vellis J, Bongarzone ER (2006) Notch signaling in astrocytes and neuroblasts of the adult subventricular zone in health and after cortical injury. *Dev Neurosci* 28:81–91.

Goldman SA, Nottebohm F (1983) Neuronal production, migration, and differentiation in a vocal control nucleus of the adult female canary brain. *Proc Natl Acad Sci USA* 80:2390–2394.

Gotz M, Huttner WB (2005) The cell biology of neurogenesis. *Nat Rev Mol Cell Biol* 6:777–788.

Gotz M, Stoykova A, Gruss P (1998) Pax6 controls radial glia differentiation in the cerebral cortex. *Neuron* 21:1031–1044.

Gritti A, Frolichsthal-Schoeller P, Galli R, Parati EA, Cova L, Pagano SF, Bjornson CRR, Vescovi AL (1999) Epidermal and fibroblast growth factors behave as mitogenic regulators for a single multipotent stem cell-like population from the subventricular region of the adult mouse forebrain. *J Neurosci* 19:3287–3297.

Hachem S, Aguirre A, Vives V, Marks A, Gallo V, Legraverend C (2005) Spatial and temporal expression of S100B in cells of oligodendrocyte lineage. *Glia* 51:81–97.

Hack MA, Saghatelyan A, de Chevigny A, Pfeifer A, Ashery-Padan R, Lledo PM, Gotz M (2005) Neuronal fate determinants of adult olfactory bulb neurogenesis. *Nat Neurosci* 8:865–872.

Haubensak W, Attardo A, Denk W, Huttner WB (2004) Neurons arise in the basal neuroepithelium of the early mammalian telencephalon: a major site of neurogenesis. *Proc Natl Acad Sci USA* 101:3196–3201.

His W (1889) Die Neuroblasten und deren Entstehung im embryonalen Mark. *Abh kgl sachs Ges Wissensch math phys Kl* 15:311–372.

Holland EC, Hively WP, DePinho RA, Varmus HE (1998) A constitutively active epidermal growth factor receptor cooperates with disruption of G1 cell-cycle arrest pathways to induce glioma-like lesions in mice. *Genes Dev* 12:3675–3685.

Imura T, Kornblum HI, Sofroniew MV (2003) The predominant neural stem cell isolated from postnatal and adult forebrain but not early embryonic forebrain expresses GFAP. *J Neurosci* 23:2824–2832.

Imura T, Nakano I, Kornblum HI, Sofroniew MV (2006) Phenotypic and functional heterogeneity of GFAP-expressing cells in vitro: differential expression of LeX/CD15 by GFAP-expressing multipotent neural stem cells and non-neurogenic astrocytes. *Glia* 53:277–293.

Jackson EL, Garcia-Verdugo JM, Gil-Perotin S, Roy M, Quinones-Hinojosa A, Vandenberg S, Alvarez-Buylla A (2006) PDGFR -positive B cells are neural stem cells in the adult SVZ that form glioma-like growths in response to increased PDGF signaling. *Neuron* 51:187–199.

Jankovski A, Sotelo C (1996) Subventricular zone–olfactory bulb migratory pathway in the adult mouse: cellular composition and specificity as determined by heterochronic and heterotopic transplantation. *J Comp Neurol* 371:376–396.

Kohwi M, Osumi N, Rubenstein JL, Alvarez-Buylla A (2005) Pax6 is required for making specific subpopulations of granule and periglomerular neurons in the olfactory bulb. *J Neurosci* 25:6997–7003.

Kondo T, Raff M (2000) Oligodendrocyte precursor cells reprogrammed to become multipotential CNS stem cells. *Science* 289:1754–1757.

Kuhn HG, Winkler J, Kempermann G, Thal LJ, Gage FH (1997) Epidermal growth factor and fibroblast growth factor-2 have different effects on neural progenitors in the adult rat brain. *J Neurosci* 17:5820–5829.

Laywell ED, Rakic P, Kukekov VG, Holland EC, Steindler DA (2000) Identification of a multipotent astrocytic stem cell in the immature and adult mouse brain. *Proc Natl Acad Sci USA* 97:13883–13888.

Lie DC, Colamarino SA, Song HJ, Desire L, Mira H, Consiglio A, Lein ES, Jessberger S, Lansford H, Dearie AR, Gage FH (2005) Wnt signalling regulates adult hippocampal neurogenesis. *Nature* 437:1370–1375.

Lim DA, Tramontin AD, Trevejo JM, Herrera DG, García-Verdugo JM, Alvarez-Buylla A (2000) Noggin antagonizes BMP signaling to create a niche for adult neurogenesis. *Neuron* 28:713–726.

Lois C, Alvarez-Buylla A (1994) Long-distance neuronal migration in the adult mammalian brain. *Science* 264:1145–1148.

Lois C, Garcia-Verdugo JM, Alvarez-Buylla A (1996) Chain migration of neuronal precursors. *Science* 271:978–981.

Machold R, Hayashi S, Rutlin M, Muzumdar MD, Nery S, Corbin JG, Gritli-Linde A, Dellovade T, Porter JA, Rubin LL, Dudek H, McMahon AP, Fishell G (2003) Sonic hedgehog is required for progenitor cell maintenance in telencephalic stem cell niches. *Neuron* 39:937–950.

Menn B, Garcia-Verdugo JM, Yaschine C, Gonzalez-Perez O, Rowitch D, Alvarez-Buylla A (2006) Origin of oligodendrocytes in the subventricular zone of the adult brain. *J Neurosci* 26:7907–7918.

Mercier F, Kitasako JT, Hatton GI (2002) Anatomy of the brain neurogenic zones revisited: fractones and the fibroblast/macrophage network. *J Comp Neurol* 451:170–188.

Merkle FT, Tramontin AD, Garcia-Verdugo JM, Alvarez-Buylla A (2004) Radial glia give rise to adult neural stem cells in the subventricular zone. *Proc Natl Acad Sci USA* 101:17528–17532.

Merkle FT, Mirzadeh Z, Alvarez-Buylla A (2007) Mosaic organization of neural stem cells in the adult brain. *Science* 317:381–384.

Mirzadeh Z, Merkle F, Soriano-Navarro M, Garcia-Verdugo JM, Alvarez-Buylla A (2008) Neural stem cells confer unique pinwheel architecture to the ventricular surface in neurogenic regions of the adult brain. *Cell Stem Cell* 3:265–278.

Molofsky AV, Slutsky SG, Joseph NM, He S, Pardal R, Krishnamurthy J, Sharpless NE, Morrison SJ (2006) Increasing p16INK4a expression decreases forebrain progenitors and neurogenesis during ageing. *Nature* 443:448–452.

Morshead CM, Reynolds BA, Craig CG, McBurney MW, Staines WA, Morassutti D, Weiss S, Van der Kooy D (1994) Neural stem cells in the adult mammalian forebrain: a relatively quiescent subpopulation of subependymal cells. *Neuron* 13:1071–1082.

Morshead CM, Garcia AD, Sofroniew MV, van Der Kooy D (2003) The ablation of glial fibrillary acidic protein-positive cells from the adult central nervous system results in the loss of forebrain neural stem cells but not retinal stem cells. *Eur J Neurosci* 18:76–84.

Noctor SC, Flint AC, Weissman TA, Dammerman RS, Kriegstein AR (2001) Neurons derived from radial glial cells establish radial units in neocortex. *Nature* 409:714–720.

Noctor SC, Flint AC, Weissman TA, Wong WS, Clinton BK, Kriegstein AR (2002) Dividing precursor cells of the embryonic cortical ventricular zone have morphological and molecular characteristics of radial glia. *J Neurosci* 22:3161–3173.

Noctor SC, Martinez-Cerdeno V, Ivic L, Kriegstein AR (2004) Cortical neurons arise in symmetric and asymmetric division zones and migrate through specific phases. *Nat Neurosci* 7:136–144.

Nunes MC, Roy NS, Keyoung HM, Goodman RR, McKhann G, II, Jiang L, Kang J, Nedergaard M, Goldman SA (2003) Identification and isolation of multipotential neural progenitor cells from the subcortical white matter of the adult human brain. *Nat Med* 9:439–447.

Palma V, Lim DA, Dahmane N, Sanchez P, Brionne TC, Herzberg CD, Gitton Y, Carleton A, Alvarez-Buylla A, Ruiz i Altaba A (2005) Sonic hedgehog controls stem cell behavior in the postnatal and adult brain. *Development* 132:335–344.

Parras CM, Galli R, Britz O, Soares S, Galichet C, Battiste J, Johnson JE, Nakafuku M, Vescovi A, Guillemot F (2004) Mash1 specifies neurons and oligodendrocytes in the postnatal brain. *Embo J* 23:4495–4505.

Pearson BJ, Doe CQ (2003) Regulation of neuroblast competence in Drosophila. *Nature* 425:624–628.

Pearson BJ, Doe CQ (2004) Specification of temporal identity in the developing nervous system. *Annu Rev Cell Dev Biol* 20:619–647.

Peretto P, Merighi A, Fasolo A, Bonfanti L (1997) Glial tubes in the rostral migratory stream of the adult rat. *Brain Res Bull* 42:9–21.

Peretto P, Merighi A, Fasolo A, Bonfanti L (1999) The subependymal layer in rodents: a site of structural plasticity and cell migration in the adult mammalian brain. *Brain Res Bull* 49:221–243.

Piccirillo SG, Reynolds BA, Zanetti N, Lamorte G, Binda E, Broggi G, Brem H, Olivi A, Dimeco F, Vescovi AL (2006) Bone morphogenetic proteins inhibit the tumorigenic potential of human brain tumour-initiating cells. *Nature* 444:761–765.

Privat A (1977) The ependyma and subependymal layer of the young rat: a new contribution with freeze-facture. *Neuroscience* 2:447–457.

Reynolds B, Weiss S (1992) Generation of neurons and astrocytes from isolated cells of the adult mammalian central nervous system. *Science* 255:1707–1710.

Russo RE, Fernandez A, Reali C, Radmilovich M, Trujillo-Cenoz O (2004) Functional and molecular clues reveal precursor-like cells and immature neurones in the turtle spinal cord. *J Physiol* 560:831–838.

Sanai N, Tramontin AD, Quinones-Hinojosa A, Barbaro NM, Gupta N, Kunwar S, Lawton MT, McDermott MW, Parsa AT, Manuel-Garcia Verdugo J, Berger MS, Alvarez-Buylla A (2004) Unique astrocyte ribbon in adult human brain contains neural stem cells but lacks chain migration. *Nature* 427:740–744.

Sanai N, Alvarez-Buylla A, Berger MS (2005) Neural stem cells and the origin of gliomas. *N Engl J Med* 353:811–822.

Seaberg RM, van der Kooy D (2002) Adult rodent neurogenic regions: the ventricular subependyma contains neural stem cells, but the dentate gyrus contains restricted progenitors. *J Neurosci* 22:1784–1793.

Seri B, Garcia-Verdugo JM, McEwen BS, Alvarez-Buylla A (2001) Astrocytes give rise to new neurons in the adult mammalian hippocampus. *J Neurosci* 21:7153–7160.

Seri B, Garcia-Verdugo JM, Collado-Morente L, McEwen BS, Alvarez-Buylla A (2004) Cell types, lineage, and architecture of the germinal zone in the adult dentate gyrus. *J Comp Neurol* 478:359–378.

Shen Q, Wang Y, Dimos JT, Fasano CA, Phoenix TN, Lemischka IR, Ivanova NB, Stifani S, Morrisey EE, Temple S (2006) The timing of cortical neurogenesis is encoded within lineages of individual progenitor cells. *Nat Neurosci* 9:743–751.

Singh SK, Clarke ID, Terasaki M, Bonn VE, Hawkins C, Squire J, Dirks PB (2003) Identification of a cancer stem cell in human brain tumors. *Cancer Res* 63:5821–5828.

Singh SK, Hawkins C, Clarke ID, Squire JA, Bayani J, Hide T, Henkelman RM, Cusimano MD, Dirks PB (2004) Identification of human brain tumour initiating cells. *Nature* 432:396–401.

Spassky N, Merkle FT, Flames N, Tramontin AD, Garcia-Verdugo JM, Alvarez-Buylla A (2005) Adult ependymal cells are postmitotic and are derived from radial glial cells during embryogenesis. *J Neurosci* 25:10–18.

Steiner B, Klempin F, Wang L, Kott M, Kettenmann H, Kempermann G (2006) Type-2 cells as link between glial and neuronal lineage in adult hippocampal neurogenesis. *Glia* 54:805–814.

Taylor MD, Poppleton H, Fuller C, Su X, Liu Y, Jensen P, Magdaleno S, Dalton J, Calabrese C, Board J, Macdonald T, Rutka J, Guha A, Gajjar A, Curran T, Gilbertson RJ (2005) Radial glia cells are candidate stem cells of ependymoma. *Cancer Cell* 8:323–335.

Temple S (2001) The development of neural stem cells. *Nature* 414:112–117.

Tropepe V, Sibilia M, Ciruna BG, Rossant J, Wagner EF, van der Kooy D (1999) Distinct neural stem cells proliferate in response to EGF and FGF in the developing mouse telencephalon. *Dev Biol* 208:166–188.

Vescovi AL, Galli R, Reynolds BA (2006) Brain tumour stem cells. *Nat Rev Cancer* 6:425–436.

Weiss S, Reynolds BA, Vescovi AL, Morshead C, Craig CG, Van der Kooy D (1996) Is there a neural stem cell in the mammalian forebrain? *Trends Neurosci* 19:387–393.

Yuan X, Curtin J, Xiong Y, Liu G, Waschsmann-Hogiu S, Farkas DL, Black KL, Yu JS (2004) Isolation of cancer stem cells from adult glioblastoma multiforme. *Oncogene* 23:9392–9400.

Zheng W, Nowakowski RS, Vaccarino FM (2004) Fibroblast growth factor 2 is required for maintaining the neural stem cell pool in the mouse brain subventricular zone. *Dev Neurosci* 26:181–196.

Zhu Y, Guignard F, Zhao D, Liu L, Burns DK, Mason RP, Messing A, Parada LF (2005) Early inactivation of p53 tumor suppressor gene cooperating with NF1 loss induces malignant astrocytoma. *Cancer Cell* 8:119–130.

Zupanc GK (2006) Neurogenesis and neuronal regeneration in the adult fish brain. *J Comp Physiol A Neuroethol Sens Neural Behav Physiol* 192:649–670.

Abbreviations

Ara-C	Cytosine-β-d-arabinofuranoside
BMP	Bone morphogenetic protein
BrdU	Bromodeoxyuridine
CNS	Central nervous system
EGF	Epidermal growth factor
EGFR	EGF receptor
FGF	Fibroblast growth factor
GCL	Granule cell layer
GFAP	Glial fibrillary acidic protein
HSV-TK	Herpes simplex virus thymidine kinase
LeX/CD15	Lewis antigen
PGCs	periglomerular cells
PDGF	Platelet-derived growth factor
PDGFRα	PDGF receptor alpha
PSA-NCAM	Polysialic acid-neural cell adhesion molecule
RCAS-AP	Replication-competent avian leukosis virus encoding alkaline phosphatase
RMS	Rostral migratory stream
SGZ	Subgranular zone
SVZ	Subventricular zone

Chapter 3
Neurotransmitter Receptors in Astrocytes

Alexei Verkhratsky

Contents

Astrocytes are the most numerous glial cells. They fulfill a wide variety of vital functions, being in essence the wardens and governors of brain homeostasis. Astrocytes are integrated into a syncytium, being thus able to exchange molecules, and produce long-range signaling in a form of propagating Ca^{2+} waves. Astroglial cells are potentially capable to express virtually all types of neurotransmitter receptors known so far. These receptors can be activated by synaptically released neurotransmitters, by "glio" transmitters or by molecules diffusing in the brain extracellular space (volume transmitters). This chapter provides a concise summary of the properties of the main types of neurotransmitter receptors operative in astroglial cells.

A. Verkhratsky
Faculty of Life Sciences, The University of Manchester, Manchester, UK
alex.verkhratsky@manchester.ac.uk

V. Parpura and P.G. Haydon (eds.), *Astrocytes in (Patho)Physiology of the Nervous System*,
DOI: 10.1007/978-0-387-79492-1_3,

3.1 Introduction

The nervous system is built by two cellular circuits represented by synaptically connected neuronal network and a complex web of glial cells (Retzius, 1890–1916; Golgi, 1903; Ramon y Cajal, 1909; Kettenmann and Ransom, 2005; Volterra and Meldolesi, 2005; Verkhratsky, 2006a). Neurons communicate via rapidly propagating electrical signals, the action potentials, which are generated by their excitable plasmalemma. At the level of synaptic contacts action potentials are converted into chemical signals; this process is accomplished through Ca^{2+}-dependent vesicular release of neurotransmitters from the presynaptic terminal (Katz and Miledi, 1967a, b, 1970). On the postsynaptic level this chemical signal carried by neurotransmitter is once more converted into either electrical excitation or into metabolic cytoplasmic signals, thus realizing information transfer in neuronal networks.

Glial cells, although being unable to generate plasmalemmal action potentials, communicate through intracellular routes, utilizing excitability of endoplasmic reticulum (ER) membrane, which underlies propagating Ca^{2+} waves (Verkhratsky and Kettenmann, 1996; Verkhratsky et al., 1998; Verkhratsky, 2006b; Verkhratsky and Toescu, 2006) or else directly communicating through gap junctions, which integrate glia into three-dimensional web (Dermietzel and Spray, 1993; Dermietzel, 1998). At the same time, glial cells are endowed with a full complement of membrane channels and neurotransmitter receptors (Verkhratsky and Steinhauser, 2000); further, glia are capable of releasing neurotransmitters via several regulated pathways, including exocytosis (Volterra and Meldolesi, 2005). These mechanisms are central for integration within neuronal–glial circuits.

Astrocytes are the most numerous cells in the human brain. The evolution of the central nervous system (CNS) went not only through an increase in the brain volume and numbers of neurons, but also through an incredible advance in the number and complexity of astroglia. The glial to neuron ratio in human cortex is ~1.65 : 1, whereas in rodents the same index is barely reaching 0.3 : 1 (Nedergaard et al., 2003; Sherwood et al., 2006). Similarly, complexity of human protoplasmic astrocytes, each of which enwraps up to 2-million synapses, is immensely higher comparing to rodents, where every astrocyte is contacting ~100,000 synapses (Oberheim et al., 2006). As a result, the astroglial syncytium is controlling and influencing neuronal networks, being responsible for as yet unknown but certainly quite important part of integrative processes in the CNS.

Discovery of glial expression of neurotransmitter receptors with first observations published at the beginning of 1980s (Bowman and Kimelberg, 1984; Kettenmann et al., 1984a, b) was fundamental for the development of the neurobiology of glia. In the present essay I provide a concise overview of the main types of receptors expressed in astrocytes; some features of these receptors are summarized in Table 3.1.

Table 3.1 Neurotransmitter receptors in astroglial cells

Receptor type	Properties/physiological effect	Localization *in situ*	References
Ionotropic receptors			
A. Glutamate receptors			
AMPA receptors	Na^+/K^+ channels $Na^+/K^+/Ca^{2+}$ channels Activation triggers cationic current and cell depolarization	Ubiquitous (grey matter in hippocampus, cortex, cerebellum, white matter), Bergmann glial cells, immature astrocytes	(Steinhäuser and Gallo, 1996; Gallo and Ghiani, 2000; Verkhratsky and Steinhauser, 2000; Seifert and Steinhauser, 2001)
NMDA receptors	$Na^+/K^+/Ca^{2+}$ channels Activation triggers inward Ca^{2+}/Na^+ current, cell depolarization and substantial Ca^{2+} entry	Cortex, spinal cord	(Conti et al., 1996; Ziak et al., 1998; Schipke et al., 2001; Lalo et al., 2006)
B. $GABA_A$ receptors	Cl^- channel Activation triggers Cl^- efflux and cell depolarization	Ubiquitous (hippocampus, cortex, cerebellum, optic nerve, spinal cord, pituitary gland)	(MacVicar et al., 1989; von Blankenfeld and Kettenmann, 1991; Fraser et al., 1994; Muller et al., 1994; Pastor et al., 1995)
C. P2X (ATP) purinoreceptors	$Na^+/K^+/Ca^{2+}$ channels	$P2X_{1-4,6}$ receptor molecules expressed in the hippocampus and nucleus accumbens; functional currents found in cultured astrocytes, in acutely isolated cortical astrocytes	(Walz et al., 1994; Franke et al., 2001; Kukley et al., 2001)
	Activation triggers cationic current, cell depolarization and may cause substantial Ca^{2+} entry	$P2X_7$ receptors are found in retinal Müller cells and in many types of cultured astrocytes	(Ballerini et al., 1996; Sun et al., 1999; Chakfe et al., 2002; Suadicani et al., 2006)
D. Glycine receptors	Cl^- channel Activation triggers Cl^- efflux and cell depolarization	Spinal cord	(Kirchhoff et al., 1996; Oertel et al., 2007)
E. Nicotinic cholinoreceptors	$Na^+/K^+/Ca^{2+}$ channels	Hippocampus, cortex, cerebellum(?)	(Sharma and Vijayaraghavan, 2001; Teaktong et al., 2003, 2004a, 2004b; Yu et al., 2005)

(continued)

Table 3.1 (continued)

Receptor type	Properties/physiological effect	Localization *in situ*	References
Metabotropic receptors			
A. Glutamate receptors (mGluRs)	Group I (mGluRs1, 5) control PLC, IP_3 production and Ca^{2+} release from the ER Group II (mGluRs2, 3) and Group III (mGluRs4, 6, 7) control synthesis of cAMP	Ubiquitous; mGluR3 and mGluR5 are the most abundant	(Kirischuk et al., 1999; Tamaru et al., 2001)
B. $GABA_B$ receptors	Control PLC, IP_3 production and Ca^{2+} release from the ER(?)	Hippocampus	(Kang et al., 1998; Charles et al., 2003)
C. Adenosine receptors A_1, A_2, A_3	A_1 receptors control PLC, IP_3 production and Ca^{2+} release from the ER A_2 receptor increase cAMP	Hippocampus, cortex	(Porter and McCarthy, 1995b; Pilitsis and Kimelberg, 1998)
D. P2Y (ATP) purinoreceptors	Control PLC, IP_3 production and Ca^{2+} release from the ER	Ubiquitous	(Kirischuk et al., 1995b; Verkhratsky and Kettenmann, 1996; Verkhratsky et al., 1998)
E. Adrenergic receptors α_1AR, α_2AR	Control PLC, IP_3 production and Ca^{2+} release from the ER	Hippocampus, Bergmann glial cells	(Shao and McCarthy, 1993; Kirischuk et al., 1996; Kulik et al., 1999)
β_1AR, β_2AR	Control glial-cell proliferation and astrogliosis; $_2$ARs are upregulated in pathology	Cortex, optic nerve	(Sutin and Griffith, 1993; Roy and Sontheimer, 1995; Griffith and Sutin, 1996)
F. Muscarinic cholinoreceptors mChR M_1–M_5	Control PLC, IP_3 production and Ca^{2+} release from the ER	Hippocampus, amygdala	(Catlin et al., 2000; Shelton and McCarthy, 2000; Araque et al., 2002)
G. Oxytocin and vasopressin receptors	Control PLC, IP_3 production and Ca^{2+} release from the ER; may regulate water channel (aquaporin)	Hypothalamus, supraoptic nucleus, other brain regions(?)	(Mittaud et al., 2002; Hatton, 2004)
H. Vasoactive intestinal polypeptide receptors (VIPR1, 2, 3)	Control PLC, IP_3 production and Ca^{2+} release from the ER; control over cAMP production; may regulate energy metabolism, expression of glutamate transporters, induce release of cytokines and promotes proliferation	Supraoptic nucleus; other brain regions(?)	(Olah et al., 1994; Ashur-Fabian et al., 1997; Grimaldi and Cavallaro, 1999)

(continued)

Table 3.1 (continued)

Receptor type	Properties/physiological effect	Localization *in situ*	References
I. Serotonin receptors 5-HT$_{1A}$, 5-HT$_{2A}$, 5-HT$_{5A}$	Increase in cAMP, energy metabolism	Neocortex, corpus callosum, hippocampal fissure and hilus, amygdala and spinal cord	(Carson et al., 1996; Xu and Pandey, 2000; Maxishima et al., 2001)
J. Angiotentsin receptors AT$_1$, AT$_2$	Control PLC, IP$_3$ production and Ca^{2+} release from the ER; inhibition of K^+ channels; modulation of Na^+/K^+ ATPase	White matter (optic nerve, corpus callosum, white mater tracts in cerebellum and subcortical areas)	(Sumners et al., 1994; Gebke et al., 1998; Muscella et al., 2000; Montiel-Herrera et al., 2006)
K. Bradykinin receptors B$_1$, B$_2$	Control PLC, IP$_3$ production and Ca^{2+} release from the ER	Only in cultured astrocytes	(Gimpl et al., 1992)
L. Thyrotropic-releasing hormone receptors, TRH$_1$	?	Spinal cord	(Fernandez-Agullo, 2001)
M. Opiod receptors, μ, δ, κ	Inhibition of DNA synthesis, proliferation and growth, inhibition of cAMP production, regulation of Ca^{2+} channels	Hippocampus	(Eriksson et al., 1993; Hauser et al., 1996)
N. Histamine receptors, H$_1$ H$_2$	Control PLC, IP$_3$ production and Ca^{2+} release from the ER Control synthesis of cAMP	Hippocampus, cerebellum	(Inagaki et al., 1989; Fukui et al., 1991; Kirischuk et al., 1996)
O. Dopamine receptors D$_1$ D$_2$	Control synthesis of cAMP Trigger Ca^{2+} signals	Cortex	(Khan et al., 2001)

3.2 Glutamate Receptors

3.2.1 Ionotropic Receptors

Initial observation of glutamate-mediated activation of glial cells was made in 1984, when electrophysiological recordings showed that externally applied excitatory amino acids (glutamate and aspartate) depolarized astrocytes and oligodendrocytes maintained in cell culture (Bowman and Kimelberg, 1984; Kettenmann et al., 1984a, b). Subsequently glial glutamate receptors were identified in astroglia throughout the brain (Steinhauser and Gallo, 1996; Condorelli et al., 1999; Seifert and Steinhauser, 2001). Ionotropic glutamate receptors (GluRs) are represented by three main families, which have a distinct molecular structure and specific pharmacological properties. These three families are designated as -amino-3-hydroxy-5-methyl-4-isoxazolepropionic acid (AMPA), kainate (KA) and *N*-methyl-d-aspartate (NMDA) receptors, which all belong to a broad family of cationic ligand-operated channels (for review see Wisden and Seeburg (1993), Mayer and Armstrong (2004) and Mayer (2005)).

AMPA receptors were the first to be identified in astroglia, and they represent the dominant ionotropic glutamate receptor present in astrocytes. The functional properties of AMPA receptors are determined by their assembly from four receptor subunits, GluR1–GluR4, encoded by distinct genes (Wisden and Seeburg, 1993; Hollmann and Heinemann, 1994); further diversity is brought by alternative splicing and mRNA editing (Seeburg et al., 1998). AMPA receptors constructed from four main subunits were found in astroglial cells throughout the brain, including hippocampus, cerebellum, neocortex and retina (Gallo and Ghiani, 2000; Verkhratsky and Steinhauser, 2000; Seifert and Steinhauser, 2001). Importantly many types of glial cells do not express the GluR2 subunit, which determines the Ca^{2+} impermeability of the receptor. As a result astroglial AMPA receptors are often Ca^{2+} permeable, with fractional Ca^{2+} currents reaching ~4% (Burnashev, 1998). Activation of these Ca^{2+} permeable receptors triggers thus appreciable Ca^{2+} signals, which were characterized in several types of astrocytes (Enkvist et al., 1989; Glaum et al., 1990; Muller et al., 1992; Jabs et al., 1994; Porter and McCarthy, 1995a).

The second type of ionotropic glutamate receptor, the *kainate receptor*, is constructed from five subunits, the KA1 and KA2 and GluR5–7 (Lerma, 2003). All five subunits were identified in certain types of astroglial cells (e.g. in bovine corpus callosum or in rodent perivascular astrocytes) at either the mRNA or the protein level (Garcia-Barcina and Matute, 1996; Brand-Schieber et al., 2004), although there are no evidence for functional activation of these receptors in astroglia.

Astroglial expression of the third type of ionotropic glutamate receptors, the *NMDA receptors* was denied for a long time. Conceptually the NMDA receptors were believed to be exclusively present in neurons, where they act as a molecular substrate for learning and memory through their established role in controlling synaptic plasticity (Malenka and Nicoll, 1993). This belief had a solid foundation, as indeed NMDA receptors, by virtue of Mg^{2+} block (Mayer et al., 1984; Nowak et al., 1984) being unavailable for activation at negative membrane potentials. This block can be relieved by cell depolarization into the region of ~−40 mV, which makes neuronal NMDA receptors perfect coincidence detectors. Glial cell-membrane potential, however, is characteristically set at about −80 mV; high densities of K^+ channels make substantial depolarization almost impossible. As a consequence it was generally believed that NMDA receptors in astrocytes cannot be operational.

Nonetheless, reports on astroglial NMDA receptor-mediated responses sporadically appeared. Several groups had identified presumed NMDA receptor-mediated activation of cultured radial glial cells, cultured astrocytes (Puro et al., 1996; Lopez et al., 1997; Nishizaki et al., 1999; Kondoh et al., 2001) and in some in situ preparations. For example, applications of exogenous NMDA to brain slices triggered electrical or $[Ca^{2+}]_i$ responses in the cortical (Schipke et al., 2001), the spinal cord (Ziak et al., 1998), in a subpopulation of hippocampal astrocytes (Steinhauser et al., 1994; Porter and McCarthy, 1995a) and in cerebellar Bergmann glial cells (Muller et al., 1993). At the same time expression of NMDA receptor-specific mRNA and NMDA receptor proteins were detected in cortical astrocytes (Conti et al., 1996; Schipke et al., 2001).

Only recently, however, astroglial expression of functional NMDA receptors was confirmed in experiments on cortical astrocytes isolated from genetically modified mice, in which astrocytes expressed green fluorescent protein. Such a model allows unambiguous identification of astrocytes either acutely isolated from or residing in brain slices. Individual astrocytes, obtained by nonenzymatic vibrodissection procedure, were voltage-clamped almost immediately after isolation. In these cells externally applied NMDA activated currents sensitive to glycine and NMDA receptor antagonists MK-801 and d-2-amino-phosphonopentanoic acid (D-AP-5) (Lalo et al., 2006). The same antagonists inhibited a sizable fraction of currents triggered by the application of glutamate. When "green" astrocytes were voltage-clamped in slices, the NMDA-mediated postsynaptic currents activated by electrical stimulation of neuronal afferents were recorded (Lalo et al., 2006). Importantly, NMDA receptors expressed in astrocytes were also able to produce spontaneous ("miniature") excitatory postsynaptic currents in slice preparation, indicating that some of these receptors are clustered in a close proximity to the sites of glutamate release from presynaptic terminals (Lalo et al., 2006; Verkhratsky and Kirchhoff, 2007). Thus, it is without question that cortical astrocytes express functional NMDA receptors. The degree to which this is a uniform property of these glia in different brain regions requires further examination.

Astroglial NMDA receptors were fundamentally different from neuronal ones, as they had a very weak (if any) Mg^{2+} block. Both NMDA-activated currents in isolated cells and synaptically activated NMDA currents in astrocytes in cortical slices were recorded at negative membrane potentials (−80 mV) in the presence of physiological concentrations of Mg^{2+}; furthermore elevation of extracellular Mg^{2+} up to 4–10 mM did not affect these current responses (Lalo et al., 2006). Incidentally, NMDA-induced currents and intracellular Ca^{2+} responses were also recorded from oligodendrocytes (Karadottir et al., 2005; Salter and Fern, 2005; Micu et al., 2006), where they also showed weak Mg^{2+} block. The molecular basis for low Mg^{2+} sensitivity of glial NMDA receptors remains unexplained; it may result, for example, from a specific expression of NR3 NMDA receptor subunits (expression of these subunits was identified in oligodendroglia, but hitherto not in astrocytes).

Physiological and pathological potential of astroglial NMDA receptors is yet to be explored. They can be important for neuronal–glial communications, especially keeping in mind close apposition of astroglial NMDA receptors and sites of glutamate release, and much higher sensitivity of NMDA receptors to glutamate comparing with AMPA receptors (see Verkhratsky and Kirchhoff (2007)), and for astroglial excitotoxicity.

3.2.2 Metabotropic Receptors

Metabotropic glutamate receptors (mGluRs) are classical seven-transmembrane-domain, G-protein-coupled receptors, represented by eight genetically distinct members, mGluR1–mGluR8 (Nakanishi, 1994; Ferraguti and Shigemoto, 2006). These eight receptors are classified into three functionally different groups. Metabotropic glutamate receptors of group I include mGluR1 and mGluR5, which are coupled to phospholipase C (PLC) and synthesis of 1,4,5-inositol-trisphosphate (IP_3) and diacylglycerol (DAG). Metabotropic receptors of group II (mGluRs 2 and 3) and group III (mGluRs 4, 6, 7, 8) control adenylate cyclase. Astroglial cells express mGluR3 and 5 in abundance; although other types of metabotropic receptors are also present in astroglial cells throughout the brain, they are much less characterized. The mGluRs3 and 5 were identified in astrocytes in situ, in astroglial processes (Petralia et al., 1996; Aronica et al., 2000; Tamaru et al., 2001). Activation of mGluR5 triggers cytosolic Ca^{2+} signaling through the stimulation of IP_3-induced Ca^{2+} release from the ER Ca^{2+} store; these Ca^{2+} signals do not require extracellular Ca^{2+} and are sensitive to inhibitors of ER Ca^{2+} accumulation via store-specific Ca^{2+}-adenosine 5′-triphosphate (ATP)ase, thapsigargin or cyclopiazonic acid, and to heparin, which blocks IP_3 receptors residing in the ER membrane (Kirischuk et al., 1999). In fact, in Bergmann glial cells, the mGluR5 represents the main route for Ca^{2+} signal generation following stimulation with glutamate, as Ca^{2+} entry through AMPA receptors is rather limited because of rapid desensitization (Kirischuk et al., 1999).

3.3 GABA Receptors

3.3.1 Ionotropic Receptors

The ionotropic γ-aminobutyric acid (GABA) receptors of $GABA_A$ type are present in many types of astrocytes in culture and in situ (Fraser et al., 1994; Verkhratsky and Steinhauser, 2000). In the latter preparation GABA-mediated currents were identified throughout the brain, which included hippocampus, cerebellum, pituitary gland, optic nerve, retina and spinal cord (Verkhratsky and Steinhauser, 2000). Astroglial $GABA_A$ receptors are of the classical pentameric type, being in essence a GABA-gated Cl^- channels. Biophysical and pharmacological properties of glial $GABA_A$ receptors are similar to neuronal ones; although, in contrast to neurons, benzodiazepine inverse agonists potentiated GABA responses in cultured astrocytes (Backus et al., 1988; Bormann and Kettenmann, 1988). From the point of view of physiological function, however, glial $GABA_A$ receptors are remarkably different from $GABA_A$ receptors in neurons, as GABA-induced activation invariably produces depolarization of astrocytes. This difference results from the peculiar ion distribution across astrocyte membrane, as astroglial cells contain much more Cl^+ than do mature neurons (~35 mM vs. ~3–5 mM; this high Cl^- concentration is maintained by the activity of the $Na^+/K^+/2Cl^-$ cotransporter and the Cl^-/HCO_3^- exchanger in glial membranes (Kettenmann, 1990; Kimelberg, 1990), and therefore the equilibrium potential for Cl^- in astrocytes and oligodendrocytes lies around −40 mV, (the ECl^- in neurons is close to −70 mV). As a consequence, activation of $GABA_A$ receptors in glial cells triggers efflux of Cl^- ions and cell depolarization (MacVicar et al., 1989; von Blankenfeld and Kettenmann, 1991). The functional significance of astroglial $GABA_A$ receptors remains enigmatic. These receptors may be involved in neuronal–glial cross-talk at synaptic level; at least in Bergmann glial cells $GABA_A$ receptors are clustered in membranes enwrapping inhibitory synapses, thus allowing the glial cell to recognize incoming GABA-mediated signaling. In addition $GABA_A$ receptors may be involved in the regulation of glial proliferation and differentiation (Fraser et al., 1994) and in the modulation of other ion channels; for example activation of $GABA_A$ receptors was reported to inhibit K^+ channels, thus facilitating depolarization (Muller et al., 1994; Pastor et al., 1995).

3.3.2 Metabotropic Receptors

There is sporadic evidence indicating astroglial expression of metabotropic $GABA_B$ receptors. All three $GABA_B$ receptor subtypes ($GABA_{B1a}$, $GABA_{B1b}$ and $GABA_{B2}$), for example, were detected in astroglial processes in CA1 area of hippocampus (Charles et al., 2003). Presumed $GABA_B$-mediated Ca^{2+} signals originating from intracellular stores were detected in cultured astrocytes (Nilsson et al., 1993) and in astrocytes in hippocampal slices (Kang et al., 1998).

3.4 Purinoreceptors

3.4.1 Ionotropic Receptors

The ATP, discovered in 1929 by Karl Lohman, Cyrus Hartwell Fiske and Yellagaprada SubbaRow (Fiske and SubbaRow, 1929; Lohmann, 1929), acts as an important extracellular signaling molecule. In the CNS, ATP can be released from synaptic terminals, either on its own or together with other neurotransmitters (Bodin and Burnstock, 2001; North and Verkhratsky, 2006), alternatively ATP can also be released via large pores formed by volume-sensitive Cl^- channels, hemichannels or $P2X_7$ purinoreceptors (Darby et al., 2003; North and Verkhratsky, 2006; Suadicani et al., 2006).

The ATP receptors, generally known as P2 receptors (Burnstock, 1978) are represented by two large families, the ionotropic P2X and metabotropic P2Y receptors (Abbracchio and Burnstock, 1994; North, 2002). The P2X receptors are classical ligand-gated cationic channels, which upon ATP binding undergo rapid conformational change that allows the passage of Na^+, K^+ and Ca^{2+} through the channel pore

(North, 2002). The P2X receptor's subfamily comprises seven subunits ($P2X_1$–$P2X_7$) encoded by distinct genes. These subunits may form homo- or heteromeric receptors, with each functional receptors containing at least three monomers (Barrera et al., 2005; Egan et al., 2006; Roberts et al., 2006). The $P2X_7$ receptor is different from the rest of the subunits as it does not form heteromers, and its activation may result in the appearance of a big pore, which allows passage of relatively large (up to 1 kDa) molecules. Functional P2X receptors display relatively high Ca^{2+} permeability ($P_{Ca^{2+}}/P_{monovalents}$ ~ 2–12:1 – Pankratov et al. (2002) and Egan et al. (2006)).

Purinergic transmission is particularly important for glia as both microglia and macroglia can be stimulated by ATP (Kirischuk et al., 1995a, b; Haas et al., 1996; Cotrina et al., 2000; Fields and Stevens, 2000; Moller et al., 2000). Moreover, ATP acts as a powerful "glio" transmitter. ATP released from astrocytes can signal onto the neighboring cells, thus, assisting propagating Ca^{2+} waves in astroglial syncytium (Guthrie et al., 1999; Cotrina et al., 2000). The ATP released from astrocytes can also affect neurons either via activating neuronal P2 receptors (Zhang et al., 2003) or by providing adenosine, which in turn stimulates neuronal P1 adenosine receptors (Pascual et al., 2005).

Expression and functional importance of astroglial $P2X_{1-6}$ receptors remains very unclear. The ATP-mediated ion currents were detected in cultured astrocytes (Walz et al., 1994), and mRNA specific for $P2X_{1-4}$ and $P2X_6$ receptors were found in the astrocytes from hippocampus and nucleus accumbens (Franke et al., 2001; Kukley et al., 2001). In our recent experiments (Lalo, Pankratov, Kirchhoff and Verkhratsky, unpublished) we succeeded in recording P2X-mediated currents in acutely isolated cortical astrocytes; these currents were sensitive to broad P2X antagonist PPADS and could be mimicked by P2X agonist α,β-methylene-ATP. To the contrary, when we repeated the same experiments on isolated hippocampal astrocytes no ATP-evoked currents were detected; similar results were also obtained by K. Matthias and C. Steinhauser (personal communication).

Astrocytes also express $P2X_7$ receptors, which are implicated in numerous pathological reactions. These $P2X_7$ receptors represent a special class of ionotorpic purinoreceptors as (i) they do not form oligomeres with other P2X subunits; (ii) they are activated by very high (>100 μM) concentrations of ATP and (iii) upon prolonged activation $P2X_7$ receptors form a large pore that is permeable for molecules with molecular weight of 800–1,000 Da (Surprenant et al., 1996; Sperlagh et al., 2006). Expression of $P2X_7$ receptors in neural cells was for a long time somewhat controversial, as it was generally believed that these receptors are confined to immune cells and epithelia (Collo et al., 1997). Recent data, however, demonstrated functional expression of $P2X_7$ receptors in many types of cells from the nervous system, including both peripheral and central neurons as well as astroglial and microglial cells (see Sperlagh et al. (2006) for a comprehensive review). Astrocytic expression of $P2X_7$ receptors was confirmed on many levels from mRNA to the functional proteins (see e.g. Panenka et al., 2001; Fumagalli et al., 2003; Dixon et al., 2004). Activation of $P2X_7$ receptors results in robust cytosolic Ca^{2+} elevation (Ballerini et al., 1996; Sun et al., 1999; Suadicani et al., 2006), triggers interleukin-1β release (Chakfe et al., 2002) and stimulates AKT phopshorylation (Jacques-Silva et al., 2004). Most importantly, however, activated $P2X_7$ receptor may form a pathway for release of glutamate and ATP from astrocytes (Ballerini et al., 1996; Duan et al., 2003), which may have important pathological consequences (e.g., exacerbation of excitotoxicity).

3.4.2 *Metabotropic Receptors*

Metabotropic P2Y receptors are classical seven-transmembrane-domain metabotropic receptors coupled to G proteins. These receptors are represented by at least ten subtypes, out of which $P2Y_1$, $P2Y_2$, $P2Y_6$, $P2Y_{11}$, $P2Y_{12}$, $P2Y_{13}$ and $P2Y_{14}$ are detected in the mammalian brain (Illes and Ribeiro, 2004). In astrocytes the ATP-mediated signaling predominantly occurs through metabotropic P2Y receptors, which control intracellular IP_3-induced Ca^{2+} release and are instrumental for producing propagating Ca^{2+} waves, which serve as a substrate for glial excitability (Kirischuk et al., 1995b; Verkhratsky et al., 1998). Astroglial expression of various P2Y subunits has not been investigated in detail.

3.5 Glycine Receptors

Glycine receptors are the members of superfamily of Cys-loop receptors (other members include nicotinic cholinoreceptors, $GABA_A$ and $GABA_C$ receptors and $5HT_3$ ionotropic receptors (Lester et al., 2004)); they are assembled from five subunits, which create a Cl^--selective channel. Glycine receptors are expressed in the astrocytes from spinal cord, where their activation triggers Cl^- efflux (Kirchhoff et al., 1996), and cell depolarization, very similar to $GABA_A$-mediated responses). Astrocytes from the spinal cord express an unusual $\beta\Delta 7$ subunit, encoded by exon 7 of the *Glrb* gene (Oertel et al., 2007). Expression of this subunit does not affect the channel properties and its functional significance, while physiological role of glial glycine receptors remains unknown.

3.6 Cholinoreceptors

Main types of acetylcholine receptors (ChRs), the nicotinic (nChRs) and muscarinic (mChRs) were detected in astroglial preparations. Functional nChRs, activation of which triggered Ca^{2+} influx and secondary Ca^{2+}-induced Ca^{2+} release, were hitherto found only in cultured astroglia (Sharma and Vijayaraghavan, 2001). The nChRs in cultured astrocytes contained $\alpha 7$ subunit, which underlie their Ca^{2+} permeability (Sharma and Vijayaraghavan, 2001; Oikawa, 2005). Expression of $\alpha 7$ subunit was also detected in brain tissue, in hippocampus and temporal cortex, but not in ventral tegmental area (Jones and Wonnacott, 2004). Astroglial nChRs might play some, yet unidentified, role in the pathogenesis of Alzheimer's disease. Treatment of cultured astrocytes with β-amyloid(1–42) peptide led to an upregulation of α7, α4 and β2 subunits (Xiu et al., 2005). The total number of astrocytes positive for α7 subunit and levels of astroglial α7 subunit expression was significantly higher in the brains of patients, suffering from both family and sporadic Alzheimer's disease (Teaktong et al., 2003, 2004b; Yu et al., 2005). Incidentally, smoking produced an opposite effect, and a significant reduction in α7 immunoreactive astrocytes was detected in the brain tissue of smokers and ex-smokers (Teaktong et al., 2004a).

Metabotropic ChRs were also detected in the astrocytes both in culture and in situ. These receptors are generally coupled with PLC and trigger IP_3-induced Ca^{2+} release from intracellular stores (Catlin et al., 2000; Shelton and McCarthy, 2000). In the slice preparation, mChRs can be activated following presynaptic release of acetylcholine; stimulation of cholinergic terminals in the hippocampus triggered mChR-mediated Ca^{2+} signals in astrocytes (Araque et al., 2002).

3.7 Adrenergic Receptors

Cultured astrocytes and astrocytes in situ express both α- (αARs) and β- (βARs) adrenergic receptors (Lerea and McCarthy, 1989; Porter and McCarthy, 1997; Verkhratsky et al., 1998). The α_1ARs are coupled to PLC and trigger IP_3 formation and subsequent Ca^{2+} release from the ER (Shao and McCarthy, 1993; Kirischuk et al., 1996). These receptors can be stimulated synaptically, e.g., in Bergmann glial cells in cerebellar slices (Kulik et al., 1999). The α_2ARs are present in hippocampal astrocytes, being concentrated on their perisynaptic processes (Milner et al., 1998). The βARs, and especially β_2ARs, are somehow connected with astrogliosis, as the latter upregulates their expression. This upregulation seems to be functionally relevant, as pharmacological inhibition of β_2AR interfered with scar formation (Sutin and Griffith, 1993; Griffith and Sutin, 1996). The β_1ARs are somehow connected with glycogen synthesis and may also provide for cyclic adenosine monophosphate (cAMP)-dependent inhibition of astrocytic K^+ channels (Roy and Sontheimer, 1995).

3.8 Concluding Remarks

Astrocytes are potentially able to express virtually all types of neurotransmitter receptors known so far (Table 3.1). Nonetheless in the brain proper this expression is usually tightly controlled to match transmitters, released in particular brain regions (Verkhratsky and Kettenmann, 1996; Verkhratsky et al., 1998). This allows astrocytes to sense the neuronal activity, thus accomplishing neuronal–glial signaling. Furthermore, astroglial receptors can be activated by "glio" transmitters, being therefore involved in the integrative processes within astroglial syncytium.

References

Abbracchio MP, Burnstock G (1994) Purinoceptors: are there families of P2X and P2Y purinoceptors? *Pharmacol Ther* 64:445–475.

Araque A, Martin ED, Perea G, Arellano JI, Buno W (2002) Synaptically released acetylcholine evokes Ca2+ elevations in astrocytes in hippocampal slices. *J Neurosci* 22:2443–2450.

AronicaE, van VlietEA, MayborodaOA, TroostD, da SilvaFH, Gorter JA (2000) Upregulation of metabotropic glutamate receptor subtype mGluR3 and mGluR5 in reactive astrocytes in a rat model of mesial temporal lobe epilepsy. *Eur J Neurosci* 12:2333–2344.

Ashur-Fabian O, Giladi E, Brenneman DE, Gozes I (1997) Identification of VIP/PACAP receptors on rat astrocytes using antisense oligodeoxynucleotides. *J Mol Neurosci* 9:211–222.

Backus KH, Kettenmann H, Schachner M (1988) Effect of benzodiazepines and pentobarbital on the GABA-induced depolarization in cultured astrocytes. *Glia* 1:132–140.

Ballerini P, Rathbone MP, Di Iorio P, Renzetti A, Giuliani P, D'Alimonte I, Trubiani O, Caciagli F, Ciccarelli R (1996) Rat astroglial P2Z (P2X7) receptors regulate intracellular calcium and purine release. *Neuroreport* 7:2533–2537.

Barrera NP, Ormond SJ, Henderson RM, Murrell-Lagnado RD, Edwardson JM (2005) Atomic force microscopy imaging demonstrates that P2X2 receptors are trimers but that P2X6 receptor subunits do not oligomerize. *J Biol Chem* 280:10759–10765.

Bodin P, Burnstock G (2001) Purinergic signalling: ATP release. *Neurochem Res*26:959–969.

Bormann J, Kettenmann H (1988) Patch-clamp study of -aminobutyric acid receptor Cl– channels in cultured astrocytes. *Proc Natl Acad Sci U S A* 85:9336–9340.

Bowman CL, Kimelberg HK (1984) Excitatory amino acids directly depolarize rat brain astrocytes in primary culture. *Nature* 311:656–659.

Brand-Schieber E, Lowery SL, Werner P (2004) Select ionotropic glutamate AMPA/kainate receptors are expressed at the astrocyte–vessel interface. *Brain Res* 1007:178–182.

Burnashev N (1998) Calcium permeability of ligand-gated channels. *Cell Calcium* 24:325–332.

Burnstock G (1978) A basis for distinguishing two types of purinergic receptor. In: Cell membrane receptors for drugs and hormones: A multidisciplinary approach. (StraubRW,BolisL, eds), pp 107–118. New York: Raven.

Carson MJ, Thomas EA, Danielson PE, Sutcliffe JG (1996) The 5HT5A serotonin receptor is expressed predominantly by astrocytes in which it inhibits cAMP accumulation: a mechanism for neuronal suppression of reactive astrocytes. *Glia* 17:317–326.

Catlin MC, Kavanagh TJ, Costa LG (2000) Muscarinic receptor-induced calcium responses in astroglia. *Cytometry* 41:123–132.

ChakfeY, SeguinR, AntelJP, MorissetteC, MaloD, HendersonD, SeguelaP(2002)ADP and AMP induce interleukin-1β release from microglial cells through activation of ATP-primed P2X7 receptor channels. *J Neurosci* 22:3061–3069.

CharlesKJ, DeucharsJ, DaviesCH, PangalosMN(2003)GABA B receptor subunit expression in glia. *Mol Cell Neurosci* 24:214–223.

Collo G, Neidhart S, Kawashima E, Kosco-Vilbois M, North RA, Buell G (1997) Tissue distribution of the P2X7 receptor. *Neuropharmacology* 36:1277–1283.

Condorelli DF, Conti F, Gallo V, Kirchhoff F, Seifert G, Steinhauser C, Verkhratsky A, Yuan X (1999)Expression and functional analysis of glutamate receptors in glial cells. *Adv Exp Med Biol* 468:49–67.

Conti F, DeBiasi S, Minelli A, Melone M (1996) Expression of NR1 and NR2A/B subunits of the NMDA receptor in cortical astrocytes. *Glia* 17:254–258.

Cotrina ML, Lin JH, Lopez-Garcia JC, Naus CC, Nedergaard M (2000) ATP-mediated glia signaling. *J Neurosci* 20:2835–2844.

Darby M, Kuzmiski JB, Panenka W, Feighan D, MacVicar BA (2003) ATP released from astrocytes during swelling activates chloride channels. *J Neurophysiol* 89:1870–1877.

Dermietzel R (1998) Gap junction wiring: a 'new' principle in cell-to-cell communication in the nervous system?*Brain Res Brain Res* Rev 26:176–183.

Dermietzel R, Spray DC (1993) Gap junctions in the brain: where, what type, how many and why?*Trends Neurosci* 16:186–192.
Dixon SJ, Yu R, Panupinthu N, Wilson JX (2004) Activation of P2 nucleotide receptors stimulates acid efflux from astrocytes. *Glia* 47:367–376.
Duan S, Anderson CM, Keung EC, Chen Y, Swanson RA (2003) P2X7 receptor-mediated release of excitatory amino acids from astrocytes. *J Neurosci* 23:1320–1328.
Egan TM, Samways DS, Li Z (2006) Biophysics of P2X receptors. *Pflugers Arch* 452:501–512.
Enkvist MO, Holopainen I, Akerman KE (1989) Glutamate receptor-linked changes in membrane potential and intracellular Ca2+ in primary rat astrocytes. *Glia* 2:397–402.
Eriksson PS, Nilsson M, Wagberg M, Hansson E, Ronnback L (1993)κ-opioid receptors on astrocytes stimulate L-type Ca2+ channels. *Neuroscience* 54:401–407.
Fernandez-Agullo T (2001) Thyrotropin-releasing hormone and its receptor in glia. *Glia* 33:267–276.
Ferraguti F, Shigemoto R (2006) Metabotropic glutamate receptors. *Cell Tissue Res* 326:483–504.
Fields RD, Stevens B (2000) ATP: an extracellular signaling molecule between neurons and glia. *Trends Neurosci* 23:625–633.
Fiske CH, SubbaRow Y (1929) Phosphorous compounds of muscle and liver. *Science* 70:381–382.
Franke H, Grosche J, Schadlich H, Krugel U, Allgaier C, Illes P (2001) P2X receptor expression on astrocytes in the nucleus accumbens of rats. *Neuroscience* 108:421–429.
Fraser DD, Mudrick-Donnon LA, MacVicar BA (1994) Astrocytic GABA receptors. *Glia* 11:83–93.
Fukui H, Inagaki N, Ito S, Kubo A, Kondoh H, Yamatodani A, Wada H (1991) Histamine H1-receptors on astrocytes in primary cultures: a possible target for histaminergic neurones. *Agents Actions Suppl* 33:161–180.
Fumagalli M, Brambilla R, D'Ambrosi N, Volonte C, Matteoli M, Verderio C, Abbracchio MP (2003)Nucleotide-mediated calcium signaling in rat cortical astrocytes: Role of P2X and P2Y receptors. *Glia* 43:218–203.
Gallo V, Ghiani CA (2000) Glutamate receptors in glia: new cells, new inputs and new functions. *Trends Pharmacol Sci* 21:252–258.
Garcia-Barcina JM, Matute C (1996) Expression of kainate-selective glutamate receptor subunits in glial cells of the adult bovine white matter. *Eur J Neurosci* 8:2379–2387.
Gebke E, Muller AR, Jurzak M, Gerstberger R (1998) Angiotensin II-induced calcium signalling in neurons and astrocytes of rat circumventricular organs. *Neuroscience* 85:509–520.
Gimpl G, Walz W, Ohlemeyer C, Kettenmann H (1992) Bradykinin receptors in cultured astrocytes from neonatal rat brain are linked to physiological responses. *Neurosci Lett* 144:139–142.
Glaum SR, Holzwarth JA, Miller RJ (1990) Glutamate receptors activate Ca2+ mobilization and Ca2+ influx into astrocytes. *Proc Natl Acad Sci U S A* 87:3454–3458.
Golgi C (1903) Opera Omnia. Milano: Hoepli Editore.
Griffith R, Sutin J (1996) Reactive astrocyte formation in vivo is regulated by noradrenergic axons. *J Comp Neurol* 371:362–375.
Grimaldi M, Cavallaro S (1999) Functional and molecular diversity of PACAP/VIP receptors in cortical neurons and type I astrocytes. *Eur J Neurosci* 11:2767–2772.
Guthrie PB, Knappenberger J, Segal M, Bennett MV, Charles AC, Kater SB (1999) ATP released from astrocytes mediates glial calcium waves. *J Neurosci* 19:520–528.
Haas S, Brockhaus J, Verkhratsky A, Kettenmann H (1996) ATP-induced membrane currents in ameboid microglia acutely isolated from mouse brain slices. *Neuroscience* 75:257–261.
Hatton GI (2004) Morphological plasticity of astroglial/neuronal interactions: functional implications In: Glial óneuronal signaling (HattonGI,ParpuraV, eds), pp 365–395. Boston, MA: Kluwer.
Hauser KF, Stiene-Martin A, Mattson MP, Elde RP, Ryan SE, Godleske CC (1996) μ-opioid receptor-induced Ca2+ mobilization and astroglial development: morphine inhibits DNA synthesis and stimulates cellular hypertrophy through a Ca2+-dependent mechanism. *Brain Res* 720:191–203.

Hollmann M, Heinemann S (1994) Cloned glutamate receptors. *Annu Rev Neurosci* 17:31–108.
Illes P, Ribeiro JA (2004) Neuronal P2 receptors of the central nervous system. *Curr Top Med Chem* 4:831–838.
Inagaki N, Fukui H, Taguchi Y, Wang NP, Yamatodani A, Wada H (1989) Characterization of histamine H1-receptors on astrocytes in primary culture: [3H]mepyramine binding studies. *Eur J Pharmacol* 173:43–51.
Jabs R, Kirchhoff F, Kettenmann H, Steinhauser C (1994) Kainate activates Ca2+-permeable glutamate receptors and blocks voltage-gated K+ currents in glial cells of mouse hippocampal slices. *Pflugers Arch* 426:310–319.
Jacques-Silva MC, Rodnight R, Lenz G, Liao Z, Kong Q, Tran M, Kang Y, Gonzalez FA, Weisman GA, Neary JT (2004) P2X7 receptors stimulate AKT phosphorylation in astrocytes. *Br J Pharmacol* 141:1106–1117.
Jones IW, Wonnacott S (2004) Precise localization of α7 nicotinic acetylcholine receptors on glutamatergic axon terminals in the rat ventral tegmental area. *J Neurosci* 24:11244–11252.
Kang J, Jiang L, Goldman SA, Nedergaard M (1998)Astrocyte-mediated potentiation of inhibitory synaptic transmission. *Nat Neurosci* 1:683–692.
Karadottir R, Cavelier P, Bergersen LH, Attwell D (2005) NMDA receptors are expressed in oligodendrocytes and activated in ischaemia. *Nature* 438:1162–1166.
Katz B, Miledi R (1967a) Ionic requirements of synaptic transmitter release. *Nature* 215:651.
Katz B, Miledi R (1967b) The timing of calcium action during neuromuscular transmission. *J Physiol* 189:535–544.
Katz B, Miledi R(1970)Further study of the role of calcium in synaptic transmission. *J Physiol* 207:789–801.
Kettenmann H (1990) Chloride channels and carriers in cultured glial cells. In: Chloride channels and carriers in nerve, muscle, and glial cells (Alvarez-Leefmans FJ,Russel JM, eds), pp 193–208. New York: Plenum.
Kettenmann H, Ransom BR (2005) Neuroglia. Oxford, UK: Oxford University Press.
Kettenmann H, Backus KH, Schachner M (1984a)Aspartate, glutamate and γ-aminobutyric acid depolarize cultured astrocytes. *Neurosci Lett* 52:25–29.
Kettenmann H, Gilbert P, Schachner M (1984b) Depolarization of cultured oligodendrocytes by glutamate and GABA. *Neurosci Lett*47:271–276.
Khan ZU, KoulenP, RubinsteinM, GrandyDK, Goldman-RakicPS(2001)An astroglia-linked dopamine D2-receptor action in prefrontal cortex. *Proc Natl Acad Sci USA* 98:1964–1969.
Kimelberg HK (1990) Chloride transport across glial membranes. In: Chloride channels and carriers in nerve, muscle, and glial cells (Alvarez-Leefmans FJ, Russel JM, eds), pp 159–191. NY: Plenum.
Kirchhoff F, Mulhardt C, Pastor A, Becker CM, Kettenmann H (1996) Expression of glycine receptor subunits in glial cells of the rat spinal cord. *J Neurochem* 66:1383–1390.
Kirischuk S, Scherer J, Kettenmann H, Verkhratsky A (1995a) Activation of P2-purinoreceptors triggered Ca2+ release from InsP3-sensitive internal stores in mammalian oligodendrocytes. *J Physiol Lond* 483:41–57.
Kirischuk S, Moller T, Voitenko N, Kettenmann H, VerkhratskyA(1995b)ATP-induced cytoplasmic calcium mobilization in Bergmann glial cells. *J Neurosci* 15:7861–7871.
Kirischuk S, Tuschick S, Verkhratsky A, Kettenmann H (1996) Calcium signalling in mouse Bergmann glial cells mediated by α_1-adrenoreceptors and H1 histamine receptors. *Eur J Neurosci* 8:1198–1208.
Kirischuk S, Kirchhoff F, Matyash V, Kettenmann H, Verkhratsky A (1999) Glutamate-triggered calcium signalling in mouse Bergmann glial cells in situ: role of inositol-1,4,5-trisphosphate-mediated intracellular calcium release. *Neuroscience* 92:1051–1059.
Kondoh T, Nishizaki T, Aihara H, Tamaki N (2001) NMDA-responsible, APV-insensitive receptor in cultured human astrocytes. *Life Sci* 68:1761–1767.
Kukley M, Barden JA, Steinhauser C, Jabs R (2001)Distribution of P2X receptors on astrocytes in juvenile rat hippocampus. *Glia* 36:11–21.
Kulik A, Haentzsch A, Luckermann M, Reichelt W, Ballanyi K (1999) Neuron–glia signaling via α_1 adrenoceptor-mediated Ca2+ release in Bergmann glial cells in situ. *J Neurosci* 19: 8401–8408.

Lalo U, Pankratov Y, Kirchhoff F, North RA, Verkhratsky A (2006) NMDA receptors mediate neuron-to-glia signaling in mouse cortical astrocytes. *J Neurosci* 26:2673–2683.

Lerea LS, McCarthy KD (1989) Astroglial cells in vitro are heterogeneous with respect to expression of the α_1-adrenergic receptor. *Glia* 2:135–147.

Lerma J (2003)Roles and rules of kainate receptors in synaptic transmission. *Nat Rev Neurosci* 4:481–495.

Lester HA, Dibas MI, Dahan DS, Leite JF, Dougherty DA (2004)Cys-loop receptors: new twists and turns. *Trends Neurosci* 27:329–336.

Lohmann K (1929) Uber die Pyrophosphatfraktion im Muskel. *Naturwissenschaften* 17:624–625.

Lopez T, Lopez-Colome AM, Ortega A (1997) NMDA receptors in cultured radial glia. *FEBS Lett* 405:245–248.

MacVicar BA, Tse FW, Crichton SA, Kettenmann H (1989) GABA-activated Cl– channels in astrocytes of hippocampal slices. *J Neurosci*9:3577–3583.

Malenka RC, Nicoll RA (1993) NMDA-receptor-dependent synaptic plasticity: multiple forms and mechanisms. *Trends Neurosci* 16:521–527.

Maxishima M, Shiga T, Shutoh F, Hamada S, Maeshima T, Okado N (2001) Serotonin 2A receptor-like immunoreactivity is detected in astrocytes but not in oligodendrocytes of rat spinal cord. *Brain Res* 889:270–273.

Mayer ML (2005) Glutamate receptor ion channels. *Curr Opin Neurobiol* 15:282–288.

Mayer ML, Armstrong N (2004) Structure and function of glutamate receptor ion channels. *Annu Rev Physiol* 66:161–181.

Mayer ML, Westbrook GL, Guthrie PB (1984) Voltage-dependent block by Mg2+ of NMDA responses in spinal cord neurones. *Nature* 309:261–263.

Micu I, Jiang Q, Coderre E, Ridsdale A, Zhang L, Woulfe J, Yin X, Trapp BD, McRory JE, Rehak R, Zamponi GW, Wang W, Stys PK (2006) NMDA receptors mediate calcium accumulation in myelin during chemical ischaemia. *Nature* 439:988–992.

Milner TA, Lee A, Aicher SA, Rosin DL (1998) Hippocampal 2a-adrenergic receptors are located predominantly presynaptically but are also found postsynaptically and in selective astrocytes. *J Comp Neurol* 395:310–327.

Mittaud P, Labourdette G, Zingg H, Guenot-Di Scala D (2002) Neurons modulate oxytocin receptor expression in rat cultured astrocytes: involvement of TGF-β and membrane components. *Glia* 37:169–177.

Moller T, Kann O, Verkhratsky A, Kettenmann H (2000) Activation of mouse microglial cells affects P2 receptor signaling. *Brain Res* 853:49–59.

Montiel-Herrera M, Miledi R, Garcia-Colunga J (2006) Membrane currents elicited by angiotensin II in astrocytes from the rat corpus callosum. *Glia* 53:366–371.

Muller T, Moller T, Berger T, Schnitzer J, Kettenmann H (1992) Calcium entry through kainate receptors and resulting potassium-channel blockade in Bergmann glial cells. *Science* 256:1563–1566.

Muller T, Grosche J, Ohlemeyer C, Kettenmann H (1993) NMDA-activated currents in Bergmann glial cells. *Neuroreport* 4:671–674.

Muller T, Fritschy JM, Grosche J, Pratt GD, Mohler H, Kettenmann H (1994) Developmental regulation of voltage-gated K+ channel and GABAA receptor expression in Bergmann glial cells. *J Neurosci* 14:2503–2514.

Muscella A, Aloisi F, Marsigliante S, Levi G (2000) Angiotensin II modulates the activity of Na+,K+-ATPase in cultured rat astrocytes via the AT1 receptor and protein kinase C-δactivation. *J Neurochem* 74:1325–1331.

Nakanishi S (1994) Metabotropic glutamate receptors: synaptic transmission, modulation, and plasticity. *Neuron* 13:1031–1037.

Nedergaard M, Ransom B, Goldman SA (2003) New roles for astrocytes: redefining the functional architecture of the brain. *Trends Neurosci* 26:523–530.

Nilsson M, Eriksson PS, Ronnback L, Hansson E (1993) GABA induces Ca2+ transients in astrocytes. *Neuroscience* 54:605–614.

Nishizaki T, Matsuoka T, Nomura T, Kondoh T, Tamaki N, Okada Y (1999) Store Ca2+ depletion enhances NMDA responses in cultured human astrocytes. *Biochem Biophys Res Commun* 259:661–664.

North RA (2002) Molecular physiology of P2X receptors. *Physiol Rev* 82:1013–1067.

North RA, Verkhratsky A (2006) Purinergic transmission in the central nervous system. *Pflugers Arch* 452:479–485.

Nowak L, Bregestovski P, Ascher P, Herbet A, Prochiantz A(1984)Magnesium gates glutamate-activated channels in mouse central neurones. *Nature* 307:462–465.

Oberheim NA, Wang X, Goldman S, Nedergaard M (2006) Astrocytic complexity distinguishes the human brain. *Trends Neurosci* 29:547–553.

Oertel J, Villmann C, Kettenmann H, Kirchhoff F, Becker CM (2007) A novel glycine receptor β-subunit splice variant predicts an unorthodox transmembrane topology. *Assembly into heteromeric receptor complexes. J Biol Chem* 282:2798–2807.

Olah Z, Lehel C, Anderson WB, Brenneman DE, van Agoston D (1994) Subnanomolar concentration of VIP induces the nuclear translocation of protein kinase C in neonatal rat cortical astrocytes. *J Neurosci Res* 39:355–363.

Panenka W, Jijon H, Herx LM, Armstrong JN, Feighan D, Wei T, Yong VW, Ransohoff RM, MacVicar BA (2001) P2X7-like receptor activation in astrocytes increases chemokine monocyte chemoattractant protein-1 expression via mitogen-activated protcin kinase. *J Neurosci* 21:7135–7142.

Pankratov Y, Lalo U, Krishtal O, Verkhratsky A (2002) Ionotropic P2X purinoreceptors mediate synaptic transmission in rat pyramidal neurones of layer II/III of somato-sensory cortex. *J Physiol* 542:529–536.

PascualO, CasperKB, KuberaC, ZhangJ, Revilla-SanchezR, SulJY, TakanoH, MossSJ, McCarthyK, HaydonPG(2005)Astrocytic purinergic signaling coordinates synaptic networks. *Science* 310:113–116.

Pastor A, Chvatal A, Sykova E, Kettenmann H (1995) Glycine- and GABA-activated currents in identified glial cells of the developing rat spinal cord slice. *Eur J Neurosci* 7:1188–1198.

Petralia RS, Wang YX, Niedzielski AS, Wenthold RJ (1996) The metabotropic glutamate receptors, mGluR2 and mGluR3, show unique postsynaptic, presynaptic and glial localizations. *Neuroscience* 71:949–976.

Pilitsis JG, Kimelberg HK (1998) Adenosine receptor mediated stimulation of intracellular calcium in acutely isolated astrocytes. *Brain Res* 798:294–303.

Porter JT, McCarthy KD (1995a) GFAP-positive hippocampal astrocytes in situ respond to glutamatergic neuroligands with increases in [Ca2+]i. *Glia* 13:101–112.

Porter JT, McCarthy KD (1995b) Adenosine receptors modulate [Ca2+]i in hippocampal astrocytes in situ. *J Neurochem* 65:1515–1523.

Porter JT, McCarthy KD (1997)Astrocytic neurotransmitter receptors in situ and in vivo. *Prog Neurobiol* 51:439–455.

Puro DG, Yuan JP, Sucher NJ(1996)Activation of NMDA receptor-channels in human retinal Muller glial cells inhibits inward-rectifying potassium currents. *Vis Neurosci* 13:319–326.

Ramon y Cajal S (1909) Histologie du systeme nerveux de l'homme et des vertebres. Paris: Maloine.

Retzius GM (1890–1916) Biologische Untersuchungen. Stockholm: Samson and Wallin.

Roberts JA, Vial C, Digby HR, Agboh KC, Wen H, Atterbury-ThomasA, EvansRJ(2006)Molecular properties of P2X receptors. *Pflugers Arch* 452:486–500.

Roy ML, Sontheimer H (1995) β-adrenergic modulation of glial inwardly rectifying potassium channels. *J Neurochem* 64:1576–1584.

Salter MG, Fern R (2005) NMDA receptors are expressed in developing oligodendrocyte processes and mediate injury. *Nature* 438:1167–1171.

Schipke CG, Ohlemeyer C, Matyash M, Nolte C, Kettenmann H, Kirchhoff F (2001) Astrocytes of the mouse neocortex express functional *N*-methyl-d-aspartate receptors. *Faseb J* 15:1270–1272.

Seeburg PH, Higuchi M, Sprengel R (1998) RNA editing of brain glutamate receptor channels: mechanism and physiology. *Brain Res Brain Res Rev* 26:217–229.

Seifert G, Steinhauser C (2001) Ionotropic glutamate receptors in astrocytes. *Prog Brain Res* 132:287–299.
Shao Y, McCarthy KD (1993) Quantitative relationship between α1-adrenergic receptor density and the receptor-mediated calcium response in individual astroglial cells. *Mol Pharmacol* 44:247–254.
Sharma G, Vijayaraghavan S (2001) Nicotinic cholinergic signaling in hippocampal astrocytes involves calcium-induced calcium release from intracellular stores. *Proc Natl Acad Sci U S A* 98:4148–4153.
Shelton MK, McCarthy KD (2000) Hippocampal astrocytes exhibit Ca2+-elevating muscarinic cholinergic and histaminergic receptors in situ. *J Neurochem* 74:555–563.
Sherwood CC, Stimpson CD, Raghanti MA, Wildman DE, Uddin M, Grossman LI, Goodman M, Redmond JC, Bonar CJ, Erwin JM, Hof PR (2006) Evolution of increased glia–neuron ratios in the human frontal cortex. *Proc Natl Acad Sci USA* 103:13606–13611.
Sperlagh B, Vizi ES, Wirkner K, Illes P (2006) P2X7 receptors in the nervous system. *Prog Neurobiol* 78:327–346.
Steinhäuser C, Gallo V (1996) News on glutamate receptors in glial cells. *Trends Neurosci* 19:339–345.
Steinhäuser C, Jabs R, Kettenmann H (1994) Properties of GABA and glutamate responses in identified glial cells of the mouse hippocampal slice. *Hippocampus* 4:19–35.
Suadicani SO, Brosnan CF, Scemes E (2006) P2X7 receptors mediate ATP release and amplification of astrocytic intercellular Ca2+ signaling. *J Neurosci* 26:1378–1385.
Sumners C, Tang W, Paulding W, Raizada MK (1994) Peptide receptors in astroglia: focus on angiotensin II and atrial natriuretic peptide. *Glia* 11:110–116.
Sun SH, Lin LB, Hung AC, Kuo JS (1999) ATP-stimulated Ca2+ influx and phospholipase D activities of a rat brain-derived type-2 astrocyte cell line, RBA-2, are mediated through P2X7 receptors. *J Neurochem* 73:334–343.
Surprenant A, Rassendren F, Kawashima E, North RA, Buell G (1996) The cytolytic P2Z receptor for extracellular ATP identified as a P2X receptor P2X7. *Science*272:735–738.
Sutin J, Griffith R(1993) -adrenergic receptor blockade suppresses glial scar formation. *Exp Neurol* 120:214–222.
Tamaru Y, Nomura S, Mizuno N, Shigemoto R (2001) Distribution of metabotropic glutamate receptor mGluR3 in the mouse CNS: differential location relative to pre- and postsynaptic sites. *Neuroscience* 106:481–503.
Teaktong T, Graham A, Court J, Perry R, Jaros E, Johnson M, Hall R, Perry E (2003) Alzheimer's disease is associated with a selective increase in α7 nicotinic acetylcholine receptor immunoreactivity in astrocytes. *Glia* 41:207–211.
Teaktong T, Graham AJ, Johnson M, Court JA, Perry EK (2004a) Selective changes in nicotinic acetylcholine receptor subtypes related to tobacco smoking: an immunohistochemical study. *Neuropathol Appl Neurobiol* 30:243–254.
Teaktong T, Graham AJ, Court JA, Perry RH, Jaros E, Johnson M, Hall R, Perry EK (2004b) Nicotinic acetylcholine receptor immunohistochemistry in Alzheimer's disease and dementia with Lewy bodies: differential neuronal and astroglial pathology. *J Neurol Sci* 225:39–49.
Verkhratsky A (2006a) Patching the glia reveals the functional organisation of the brain. *Pflugers Arch* 453:411–420.
Verkhratsky A (2006b) Calcium ions and integration in neural circuits. *Acta Physiol (Oxf)* 187:357–369.
Verkhratsk y A, Kettenmann H (1996) Calcium signalling in glial cells. *Trends Neurosci* 19:346–352.
Verkhratsky A, Steinhauser C (2000) Ion channels in glial cells. *Brain Res Brain Res Rev* 32:380–412.
Verkhratsky A, Toescu EC (2006) Neuronal–glial networks as substrate for CNS integration. *J Cell Mol Med* 10:826–836.
Verkhratsky A, Kirchhoff F (2007) NMDA receptors in glia. *Neuroscientist* 13:1–10.

Verkhratsky A, Orkand RK, Kettenmann H (1998) Glial calcium: homeostasis and signaling function. *Physiol Rev* 78:99–141.
Volterra A, Meldolesi J (2005) Astrocytes, from brain glue to communication elements: the revolution continues. *Nat Rev Neurosci* 6:626–640.
von Blankenfeld G, Kettenmann H (1991) Glutamate and GABA receptors in vertebrate glial cells. *Mol Neurobiol* 5:31–43.
Walz W, Gimpl G, Ohlemeyer C, Kettenmann H (1994) Extracellular ATP-induced currents in astrocytes: involvement of a cation channel. *J Neurosci Res* 38:12–18.
WisdenW, SeeburgPH(1993)Mammalian ionotropic glutamate receptors. *Curr Opin Neurobiol* 3:291–298.
Xiu J, Nordberg A, Zhang JT, Guan ZZ (2005) Expression of nicotinic receptors on primary cultures of rat astrocytes and up-regulation of the α7, α4 and α2 subunits in response to nanomolar concentrations of the β-amyloid peptide(1–42). *Neurochem Int* 47:281–290.
Xu T, Pandey SC (2000) Cellular localization of serotonin2A (5HT2A) receptors in the rat brain. *Brain Res Bull* 51:499–505.
Yu WF, Guan ZZ, Bogdanovic N, Nordberg A (2005) High selective expression of 7 nicotinic receptors on astrocytes in the brains of patients with sporadic Alzheimer's disease and patients carrying Swedish APP 670/671 mutation: a possible association with neuritic plaques. *Exp Neurol* 192:215–225.
Zhang JM, Wang HK, Ye CQ, Ge W, Chen Y, Jiang ZL, Wu CP, Poo MM, Duan S (2003) ATP released by astrocytes mediates glutamatergic activity-dependent heterosynaptic suppression. *Neuron* 40:971–982.
Ziak D, Chvatal A, Sykova E (1998) Glutamate-, kainate- and NMDA-evoked membrane currents in identified glial cells in rat spinal cord slice. *Physiol Res* 47:365–375.

Abbreviations

AMPA	α-amino-3-hydroxy-5-methyl-4-isoxazolepropionic acid
ARs	Adrenergic receptors
ATP	Adenosine 5′-triphosphate
cAMP	Cyclic adenosine monophosphate
ChR	Acetylcholine receptor
CNS	Central nervous system
D-AP-5	d-2-amino-phosphonopentanoic acid
ER	Endoplasmic reticulum
GABA	γ-aminobutyric acid
GluR	Glutamate receptor
IP_3	1,4,5-inositol-trisphosphate
KA	Kainate
mChR	Muscarinic ChR
mGluR	Metabotropic glutamate receptors
nChR	Nicotinic ChR
NMDA	*N*-methyl-d-aspartate
PLC	Phospholipase C

Chapter 4
Specialized Neurotransmitter Transporters in Astrocytes

Yongjie Yang and Jeffrey D. Rothstein

Contents

In the central nervous system (CNS), synaptic neurotransmission is a fundamental and critical process for neurons to receive and process information. In this process, neurotransmitters, the chemical signals, are released from presynaptic neuronal terminals upon stimulus to activate receptors on postsynaptic membranes. The inactivation of most released neurotransmitters occurs via efficient reuptake of neurotransmitters into the presynaptic nerve terminal and/or adjacent glial cells. Astrocytes, the most abundant cell in the CNS, express various transporter proteins on the plasma membrane for the uptake of neurotransmitters, forming an indispensable unit of functional synaptic neurotransmission. Transporters are vital for the normal CNS physiology by maintaining neurotransmitter homeostasis, modulating synaptic transmission and preventing neurological damage induced by the imbalance of neurotransmitters. The distribution and functional importance of neurotransmitter transporters in neurons or astrocytes varies by individual neurotransmitters. For

---

J.D. Rothstein
Departments of Neurology and Neuroscience, Johns Hopkins University, Baltimore, MD, USA.
jrothstein@jhmi.edu

V. Parpura and P.G. Haydon (eds.), *Astrocytes in (Patho)Physiology of the Nervous System*,
DOI: 10.1007/978-0-387-79492-1_4,

glutamate, transporters are present in both neurons and astrocytes but astroglial glutamate transporters are functionally dominant (Rothstein et al., 1994; Danbolt, 2001); for γ-aminobutyric acid (GABA) and glycine, both neuronal and astrocytic transporters are functionally dominant and their functions are complementary (Chen et al., 2004a; Betz et al., 2006); for monoamine (dopamine, serotonin, and norepinephrine) transmitters, transporters are mainly expressed in the cognate neurons not in astrocytes (Torres et al., 2003). In this chapter, the principal focus will be on astrocytes, the various neurotransmitter transporters they express, and the role of astroglial transporter dysfunction in disease.

4.1 Glutamate Transporters in the CNS

L-glutamate, the essential amino acid in every cell, plays a unique role in the communication of CNS, as the major excitatory neurotransmitter (Headley and Grillner, 1990). Glutamate is essentially involved in every aspect of normal physiological function of the CNS (Danbolt, 2001). Because of the critical function of glutamate as a neurotransmitter in the CNS, the homeostasis of glutamate is tightly regulated by a highly dynamic cycle between the neurons and astrocytes.

4.1.1 Glutamate Homeostasis and Glutamatergic Neurotransmission in the CNS

Glutamate is very abundant in the brain (5–15 mmol/kg), with most of it kept intracellularly in glutamatergic neurons or astrocytes (Schousboe and Hertz, 1981). The concentration gradient of glutamate across the plasma membranes is many thousand-fold with the concentration of glutamate of 0.1–1 μM in the extracellular fluid of brain or in the cerebrospinal fluid (CSF) (Hamberger and Nystrom, 1984); recent glutamate measurements in hippocampal slices indicate its extracellular basal concentration at ~22 nM (Herman and Jahr, 2007). As a major excitatory neurotransmitter in the CNS, the concentration of extracellular glutamate has to be tightly controlled for both physiological and pathological reasons. Physiologically, during glutamatergic neurotransmission, the extracellular concentration of glutamate released into the synaptic cleft determines the extent of receptor stimulation on postsynaptic neurons. It is critical to maintain a low concentration of extracellular glutamate to reduce baseline stimulation of receptors. This helps produce a high signal-to-noise ratio during the synaptic transmission to convey the real signal (Danbolt, 2001). Low concentration of extracellular glutamate also reduces the chance for glutamate to leak out of the synapse and spill out to neighboring synapse, which helps avoid subsequent nonspecific stimulation of receptors (Huang and Bergles, 2004). Pathologically, numerous in vivo and in vitro studies have shown that excessive amounts of glutamate are highly toxic to neurons, which is referred as glutamate excitotoxicity (Choi, 1988, 1992). Multiple downstream signal cascades have been demonstrated in glutamate induced excitotoxicity. In particular, the binding of glutamate to the ionotropic glutamate receptor induces the influx of Ca^{2+} through receptor-coupled ion channels. High concentration of intracellular Ca^{2+} activates Ca^{2+}-dependent proteases or phospholipases, and produces free radicals that are toxic to the neurons (Sattler and Tymianski, 2001; Arundine and Tymianski, 2003). Because of these reasons, released extracellular glutamate has to be removed, especially from the synaptic cleft. No known enzymes (or other molecules) have been identified extracellularly to metabolize or inactivate glutamate. Instead, most of the extracellular glutamate is transported into postsynaptic neurons or surrounding astrocytes. As described below, the astroglial transporters are the dominate inactivators of extracellular glutamate.

Homeostasis of glutamate in the CNS is dynamically maintained by synaptic release from glutamatergic neurons and uptake to astrocytes through glutamate transporters. It also involves two key enzymatic reactions between glutamate and glutamine. In glutamatergic neurons, most of glutamate is stored in synaptic vesicles of nerve terminals by vesicular glutamate transporter 1 or 2 for synaptic release (Sudhof, 1995; Takamori et al., 2000). In response to the presynaptic stimulus, synaptic vesicles containing glutamate fuse with the plasma membrane and release glutamate to the synaptic cleft. Released glutamate can bind to both ionotropic and metabotropic glutamate receptors (GluRs), activating a wide range of downstream signaling events. The released glutamate is quickly cleared by astroglial glutamate transporters, into astrocytes (Schousboe and Hertz, 1981; Anderson and Swanson, 2000). The release and uptake of glutamate is a highly dynamic process, i.e., glutamate is continuously being released from neurons and is continuously being removed from the extracellular fluid (Danbolt, 2001). After transport into astrocytes, glutamate is converted to glutamine by an ATP-dependent, astroglia-specific glutamine synthetase. Synthesized glutamine is then released to the extracellular fluid and is taken up by glutamatergic neurons. Subsequently, the glutamine is reconverted into glutamate by glutaminase to be repacked into synaptic vesicles or used as a nutrient in the cytosol of glutamategic neurons. The dynamic conversion of glutamate and glutamine in the neurons and astrocytes form an indispensable cycle for normal glutamatergic synaptic transmission (Fig. 4.1) (Westergaard et al., 1995), though a recent study found that cultured neurons in vitro have the capacity to store or produce glutamate for long periods of time, independently of glia and the glutamate–glutamine cycle (Kam and Nicoll, 2007). Glutamate in astrocytes can also be converted into α-ketoglutarate by glutamate dehydrogenase, and subsequently into lactate, which can be transported to neurons and be used as nutrients (McKenna et al., 1996).

4.1.2 General Properties and Structure of Glutamate Transporters

Glutamate transporters are localized on the plasma membrane of cells that can specifically transport glutamate or aspartate across the plasma membrane. All glutamate transporters belong to solute carrier family 1 (SLC1) (Kanai and Hediger,

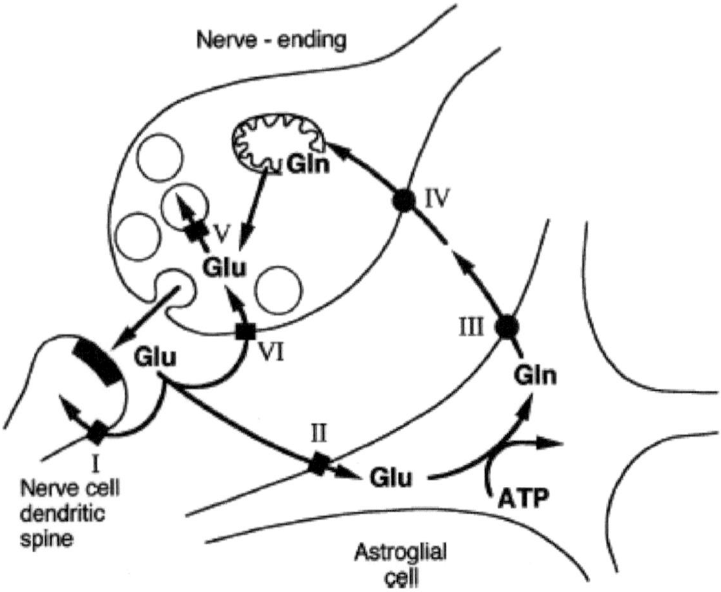

Fig. 4.1 The glutamate–glutamine cycle. Glutamate released from a nerve terminal by exocytosis (which is ATP and Ca^{2+} dependent) is taken up by glutamate transporters present presynaptically (VI), postsynaptically (I), and extrasynaptically (more than 90%) in astroglial cells (II). Astroglia detoxifies glutamate by converting it to glutamine in an ATP-dependent process. Glutamine is subsequently released from the glial cells by means of glutamine transporter (III) and taken up by neurons by means of other glutamine transporter (IV). Neurons convert glutamine back to glutamate. Synaptic vesicles are loaded with glutamate from cytosol by means of a vesicular glutamate transporter (V). Adapted from Danbolt (2001), Prog Neurobiol.

2004). So far, five glutamate transporter subtypes have been identified in the CNS based on their cell- or region-specific distribution (Table 4.1) (Sattler and Rothstein, 2006). They are named as excitatory amino acid transporters 1 [EAAT1; rodent analog, L-glutamate/L-aspartate transporter (GLAST)] (Storck et al., 1992; Tanaka, 1993), 2 [EAAT2; rodent analog, L-glutamate transporter 1 (GLT1)] (Pines et al., 1992), 3 (EAAT3; rodent analog, excitatory amino acid carrier 1 (EAAC1)] (Kanai and Hediger, 1992), 4 (EAAT4) (Fairman et al., 1995), and 5 (EAAT5) (Arriza et al., 1997). Although glutamate transporters are expressed in both neurons and astrocytes, two astroglial transporters, EAAT1 and EAAT2, are primarily responsible for the uptake of extracellular glutamate and maintenance of glutamate homeostasis in the CNS (Rothstein et al., 1996; Tanaka, 1997).

Because of the extremely high intracellular glutamate concentration (mM), glutamate transporters work against a steep concentration gradient to maintain very low extracellular concentrations (μM). Glutamate transporters are classified as "Na^+-dependent high-affinity transporter," since the translocation of glutamate across the membrane is coupled to the Na^+, H^+, and K^+ to utilize the free energy

Table 4.1 Glutamate Transporter Subtypes

Glutamate transporter subtype	Human homologue	Cell type	Anatomic localization
GLAST	EAAT1	Astrocytes, oligodendrocytes	Cerebellum, cortex, spinal cord
GLT1	EAAT2	Astrocytes	Throughout brain and spinal cord
GLT1b	EAAT2b	Astrocytes and neurons	Throughout brain and spinal cord
EAAC1	EAAT3	Neurons	Hippocampus, cerebellum, striatum
EAAT4	EAAT4	Purkinje cells	Cerebellum
EAAT5	EAAT5	Photoreceptors and bipolar cells	Retina

Adapted from Sattler and Rothstein (2006).

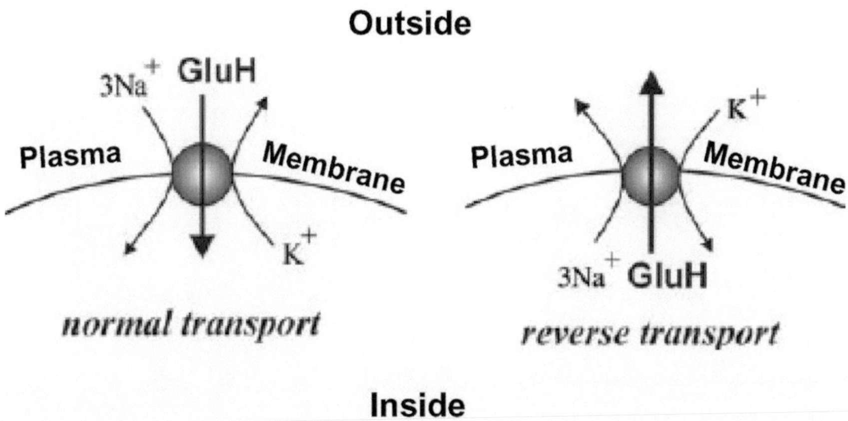

Fig. 4.2 Coupling of Na^+, H^+, and K^+ in glutamate transport. $1H^+$ and $3Na^+$ are cotransported but $1K^+$ is countertransported with glutamate. (**a**) Forward transport of glutamate from outside to inside of the cell. (**b**) Reversed transport of glutamate from inside to outside of the cell. Modified from Kanai and Hediger (2004), Eur J Physiol.

stored as electrochemical potential gradients of these ions and to power uphill transport (Levy et al., 1998; Danbolt, 2001). This ion-coupled transport is the only efficient means to maintain low extracellular glutamate in the CNS. Based on the analysis of the coupling stoichiometry of the cloned glutamate transporter (EAAC1) in *Xenopus oocyte* expression system (Zerangue and Kavanaugh, 1996), it is now generally accepted that $3Na^+$ ions and $1H^+$ are cotransported (inward to the cell) and $1K^+$ is countertransported (outward to the cell) with each glutamate molecule (Fig. 4.2). From this stoichiometry, it was calculated that glutamate transporters can concentrate glutamate 5 ‘ 10^6-fold inside cells under physiological conditions (Levy et al., 1998; Danbolt, 2001). The transport of glutamate via glutamate transporter is electrogenic as transport of various ions is coupled to that process, thus providing an effective assay to monitor the transport of glutamate functionally (Tong and Jahr, 1994). In addition to the glutamate transport function, glutamate transporters can also serve as the ligand-gated Cl^- channel based on the observation that additional current arises from a thermodynamically uncoupled anion flux when substrate is applied to EAATs (Fairman et al., 1995; Wadiche et al., 1995). It was later found that the Cl^- flux is a relatively small component of the currents recorded, whereas for EAAT4 and EAAT5, the currents elicited by substrates are almost entirely comprised of a gated anion flux, which could suggest an expanded role for transporters in regulating neuronal excitability and signaling (Sonders and Amara, 1996; Seal and Amara, 1999).

Although none of the mammalian glutamate transporters are crystallized, a homologue Na^+-coupled aspartate transporter from *Pyroccocus horikoshii*, Gltph, was successfully crystallized (Yernool et al., 2004). The substrate-bound GltPh subunits form a bowl-shaped trimer (Fig. 4.3a). As illustrated in Fig. 4.3b, each

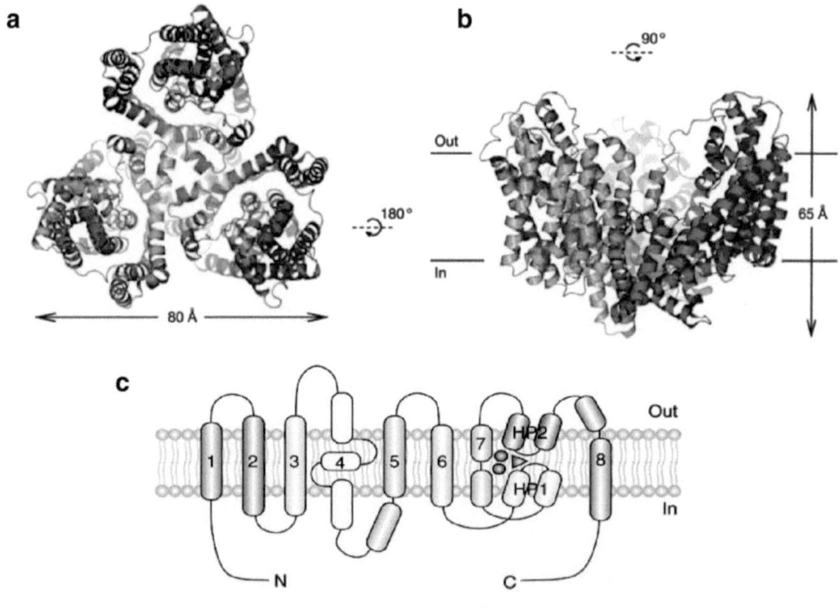

Fig. 4.3 Structure of Gltph in its trimeric conformation and predicted topology model of glutamate transporter. (**a**) Ribbon representation of the trimer of Gltph, in which the protomers are *red, blue, and green*, viewed from the extracellular side of the membrane. (**b**) View of the trimer parallel to the membrane. (**c**) A model for the membrane topology of Gltph and for mammalian members of the glutamate transporter family (SLC1). The binding sites occupied by Na^+ (*dark blue dots*) and the substrate aspartate (*green triangle*) are noted at their approximate position within the structure. The C-terminal translocation core includes the hairpin loops HP1 (*yellow*) and HP2 (*red*), which are proposed to serve as intracellular and extracellular gates, respectively, and also the transmembrane domains TM7 (*orange*) and TM8 (*magenta*). Adapted from Yernool et al. (2004), Nature. (*See Color Plates*)

subunit comprises eight transmembrane (TM) α-helices, TM1–TM8, and two reentrant hairpin (HP) loops, HP1 and HP2, that partially span the bilayer. The N-terminal helices, TM1–TM6, mediate intersubunit contacts and surround the essential components of the translocation machinery. The core domains within the C terminus, HP1, TM7, HP2, and TM8, form an independent translocation pathway within each subunit. The key structure for transport is formed in the TM7 that contains residues involved in Na^+ and substrate binding. Although this transporter shares only 37% homology with human EAAT2, detailed sequence alignment of this transporter with EAAT1-3 showed that most of the critical residues for glutamate binding, Na^+ binding, and K^+ coupling are conserved (Yernool et al., 2004; Beart and O'Shea, 2007). It also confirms and illuminates many of the structure–function and topological studies undertaken on mammalian EAATs and bacterial glutamate transporters (Vandenberg, 2006; Torres and Amara, 2007).

4.1.3 Astroglial Glutamate Transporter EAAT1/GLAST

4.1.3.1 Distribution and Expression of GLAST

In normal mature mammalian CNS, GLAST is expressed throughout the brain and spinal cord at different levels in different regions. It is the major glutamate transporter in the cerebellum (Lehre and Danbolt, 1998), the inner ear (Furness and Lehre, 1997), retina (Rauen et al., 1999), and the circumventricular organs near the brain–blood barrier (Berger and Hediger, 2000), but expression is low in the cortex, hippocampus, basal nuclei, and septum where GLT1 is predominantly expressed. GLAST mRNA can be detected mainly in the ventricular zone, olfactory lobe, cerebellar primordium at as early as embryonic day 15 (E15) by in situ hybridization (Sutherland et al., 1996). As development progresses, the levels of GLAST mRNA diminished in most regions except in the Purkinje cell layer of cerebellum (Sutherland et al., 1996). However, a parallel study of GLAST protein expression by immunostaining in the developing brain of rat showed that GLAST protein levels increased modestly in the majority of brain regions and increased strongly in the molecular layer of cerebellum as development progresses (Furuta et al., 1997). By using a recently generated transgenic GLAST reporter mouse, the GLAST promoter was shown to be active in both radial glia and many astrocytes in the developing CNS but is downregulated in most astrocytes as the mice mature (Regan et al., 2007). This reporter mouse is generated by expressing a DsRed fluorescent protein reporter driven by a bacteria artificial chromosome clone that contains the whole genomic promoter of GLAST, so the GLAST promoter activity can be monitored intact from tissue by the expression of fluorescent protein DsRed (Heintz, 2001; Regan et al., 2007).

At the cellular level, GLAST is mainly expressed in astrocytes and some specialized glia, such as a subpopulation of radial glia, the Bergmann glia in the cerebellum, supporting glia in the vestibular end organ, and glia-like Muller cells in the retina (Robinson, 2006). Although GLAST and GLT1 were often reported to be expressed in the same astrocyte with different membrane localization (Lehre et al., 1995; Haugeto et al., 1996), a recent study that used GLAST and GLT1 transgenic reporter mice showed that fluorescent protein reporters driven by either full GLAST genomic promoter or full GLT1 genomic promoter rarely overlap with each other, suggesting that the GLAST and GLT1 promoters are active in different subpopulations of astrocytes (Regan et al., 2007). These observations were also confirmed by functional, electrophysiological analyses (Regan et al., 2007). Expression of GLAST has not been detected in mature neurons in vivo under normal physiological status; however, several studies show that GLAST mRNA is detected in cultured hippocampal neurons (Sutherland et al., 1996; Plachez et al., 2000). Expression of GLAST in nonastrocyte glia remains unclear except a few studies showed that GLAST protein is found in oligodendrocytes in rat optic nerves (Domercq et al., 1999). By using a GLAST transgenic reporter mouse, DsRed fluorescence is also found in oligodendroglia, indicating that the GLAST promoter is active in these cells (Regan et al., 2007). By using this reporter mouse, the temporal and spatial expression of this transporter in the CNS and the dynamics of GLAST expression in different cell types can be better explored and appreciated.

4.1.3.2 Regulation of GLAST

Regulation of GLAST, as most of other membrane proteins, occurs at multiple levels, which results in acute or chronic changes of the GLAST expression on the plasma membrane. Fast changes of GLAST on the membrane are often regulated by the phosphorylation status of transporter at different sites by various protein kinases. The phosphorylation status of GLAST apparently affects the trafficking of GLAST to the plasma membrane (Robinson, 2006). Unlike membrane receptor proteins, phosphorylation-mediated regulation of membrane-bound GLAST is poorly understood. Protein kinase C (PKC) family has been implicated in various studies by using PKC activator phorbol 12-tetradecanoyl-13-acetate to the regulation of GLAST expression on membrane and glutamate uptake (Gonzalez et al., 1999; Bernabe et al., 2003); however, some results are not consistent and most of these studies were performed in the model cell lines by overexpression, which are less physiologically relevant (Robinson, 2006). In contrast to the acute regulation of GLAST by phosphorylation, chronic regulation of membrane-bound GLAST involves changes at transcriptional and translational levels. A variety of small molecules have been shown to induce both the mRNA and protein of GLAST in cultured primary astrocyte system, including glutamate, dibutyryl cyclic adenosine monophosphate (cAMP), epidermal growth factor (EGF), transforming growth factor α (TGF-α), estrogen, etc (Gegelashvili et al., 1996; Swanson et al., 1997; Zelenaia et al., 2000; Pawlak et al., 2005). Glutamate-induced increase of GLAST protein involves different glutamate receptors as inhibitors of a-amino-3-hydroxy-5-methyl-4-isoxazole-propionic acid (AMPA)/kainate receptors were able to block this upregulation (Gegelashvili et al., 1996). Another study also showed that activation of group II metabotropic glutamate receptors (mGluRs) caused a significant upregulation of GLAST protein levels in astroglial cultures supplemented with neuronal conditioned medium (NCM), while activation of group I mGluRs led to a decrease in GLAST protein (Gegelashvili et al., 2001). After recent cloning of the EAAT1 promoter, multiple transcriptional factors are implicated in basal promoter activation, including GC-box for stimulating proteins 1 and 3 transcription factors, X-box for protein RFX1as well as gut-enriched Kruppel-like factors, serum response factor, Atp1a1 regulatory element binding factor 6, and upstream stimulating factor (Kim et al., 2003). A recent study also showed that transcription factor YY1 is mainly responsible for glutamate-induced mRNA increase of GLAST (Rosas et al., 2007).

Less is known about expression regulation of GLAST in vivo. GLAST has long been considered as a differentiation marker for astrocyte progenitors in early postnatal development. Transcription factor nuclear factor I/A contributes to induce GLAST expression in spinal cord astrocytes (Deneen et al., 2006). In the cerebellum, a neuron-specific Notch receptor ligand delta- and notch-like EGF-related receptor has been identified to induce the expression of GLAST in Bergmann glia via the Notch signaling pathway (Eiraku et al., 2005). This finding also indicates that neuronal regulation might be important for the induction of glutamate transporters in astrocyte. At the posttranscriptional level, GLAST is also regulated by alternative splicing. Two splicing variants of EAAT1, exon 3 skipping and exon 9 skipping transcripts, have been found in the CNS (Macnab et al., 2006; Macnab and Pow, 2007). The exon 9 skipping form of GLAST mRNA can serve as a negative regulator to suppress the translation of functional GLAST protein and reduce the glutamate uptake, which might contribute to glutamate imbalance and pathogenesis in certain neurological disorders (Macnab and Pow, 2007).

4.1.4 Astroglial Glutamate Transporter EAAT2/GLT1

4.1.4.1 Distribution and Expression of GLT1

EAAT2/GLT1 is the most abundant form of glutamate transporter expressed throughout the CNS, especially in the forebrain, striatum, hippocampus, and spinal cord (Lehre et al., 1995; Furuta et al., 1997; Berger and Hediger, 1998). It is the dominant form of functional glutamate transporter in the brain and spinal cord except in the region that GLAST is dominant, i.e., radial glia, Bergmann glia in cerebellum, and Muller cells in retina. At early developmental stages (E15–E19), GLT1 mRNA was detected mainly in hippocampus and spinal cord, but not in cortex (Sutherland et al., 1996). As the development progresses, GLT1 mRNA is strongly increased or induced throughout the brain, especially in cortex, hippocampus, thalamus, and cerebellum (Schmitt et al., 1996; Sutherland et al., 1996). A parallel immunostaining study on the expression of GLT1 protein in developing brain of rat showed a similar pattern of GLT1 protein expression during development (Furuta et al., 1997).

At the cellular level, GLT1 protein is present in both (fibrous as well as protoplasmic) forms of astrocytes in both gray and white matter in the CNS, but it appears to be absent in the pituitary gland and in the three sensory circumventricular organs: the subfornical organ, the vascular organ of the lamina terminalis, and the area postrema (Berger and Hediger, 2000). GLT1 protein is also detected in cultured A2B5-positive bipotential progenitor cells (Zelenaia et al., 2000), in pinealocytes of the pineal gland (Yamada et al., 1997), and in microglia (Lopez-Redondo et al., 2000), though GLT1 protein expression in these cells have not been found in vivo. At the protein level, GLT1 expression is as high as 1% total brain protein (unpublished observation). Expression of GLT1 in mature neurons in normal physiological status remains debated though a line of evidence suggests that GLT1 protein can be detected in prenatal neurons during early development (E15–E19) (Sutherland et al., 1996). The presence of a functional glutamate transporter in presynaptic neuronal terminals (and neurons) has long been hypothesized, often by older in vitro culture studies (Danbolt, 2001), but this glutamate transporter remains unidentified. In early studies, GLT1 mRNA was found in pyramidal cells in CA3 hippocampus and in layer VI of the parietal neocortex (Berger and Hediger, 1998). More recent studies found that GLT1 mRNA in fact is detectable in the majority of neurons in the neocortex and in parts of the olfactory bulb, thalamus, and inferior olive by using in situ hybridization on rat brain (Berger et al., 2005). In particular, two forms of GLT1 mRNA that vary in C-terminus, GLT1a and GLT1b, were found in cultured rat cortical neurons by Chen et al. (2002), In the rat brain, both GLT1a and GLT1b mRNA were detected in CA3 pyramidal neurons, but not in CA1 neurons in hippocampus by in situ hybridization with sequence-specific probe (Chen et al., 2004b). By using electron microscopy (EM) and immunostaining, studies from same group showed that GLT1a, not GLT1b, protein was also observed in preterminal portions of axons forming excitatory synapses with dendritic spines in hippocampus (Chen et al., 2004b). Although this discovery provides clue about the presence of glutamate transporter in presynaptic neurons, more functional studies are needed to confirm the immunostaining results and to elucidate the role of neuronal GLT1 in synaptic neurotransmission. The vast majority of immunostaining research over the last 10 years, using multiple different, well-characterized antibodies, has not found any appreciable expression of GLT1 in most neurons (Danbolt, 2001). Furthermore, electrophysiological studies have also failed to show any appreciable neuronal expression of GLT1 – especially when comparing the high levels of expression and function detectable by electrophysiological approaches in astrocytes. Recently, work using GLT1 promoter reporter mice has provided the opportunity to identify selected neurons with sufficient gene activation of GLT1 in living slice that allows electrophysiological detection of GLT1 currents (Regan et al., 2007). In those early studies, certain hippocampal neurons do express GLT1 function, but at levels more than 7-fold lower than astrocytes (unpublished observation). Very interestingly, in the normal and mature mammalian retina, GLT1 protein is not expressed in retinal glial cells (neither in the Muller cells nor the astrocytes). It is exclusively expressed in neurons (cone photoreceptors and bipolar cells) (Rauen et al., 1999; Rauen, 2000). Only a homologue of GLT1, sEAAT2A, is expressed in glial cells in retina (Eliasof et al., 1998a; Eliasof et al., 1998b). The expression of GLT1 in nonastrocyte glia remains unclear though GLT1 immunoreactivity was found in cultured A2B5-positive oligodendritic precursor cells (Zelenaia et al., 2000). The in vivo expression of GLT1 in nonastrocyte glia has not been explored.

4.1.4.2 Regulation of GLT1

The biochemical and molecular regulation of GLT1 has been the focus of increasing research. Similar to other glutamate transporters, the regulation of GLT1 occurs at multiple levels, i.e. chronic but long-lasting regulation at the transcriptional and translational levels and quick but short-lasting regulation at the posttranslational and trafficking levels.

GLT1-Activating Signals and Signal Pathways in Transcriptional Regulation of GLT1

Early studies showed that glutamate uptake in pure astrocyte cultures increases when the astrocytes are treated with medium collected from neuronal cultures (Drejer et al., 1983). In the absence of neurons, astrocytes express extremely low levels of GLT1 protein though GLT1 is abundantly expressed in the astrocytes in vivo in adult mammalian brain and spinal cord. When cocultured with neurons, expression of GLT-1 mRNA and protein is strongly induced in astrocytes but not in neurons (Swanson et al., 1997; Schlag et al., 1998). Subsequent studies further showed that NCM can increase GLT1 mRNA and protein in primary astrocyte cultures. These results suggested that soluble factors secreted from the neurons can induce the expression of GLT1 in cultured astrocytes. Although great efforts have been made by multiple groups to identify the soluble factors that are presumably secreted from neurons, these soluble factors remain elusive. Neuronal membrane-related factors could be contributors for this induction effect. On the other hand, many small molecules have been identified to activate EAAT2/GLT1 expression in cultured primary astrocytes or glioma cells. These EAAT2/GLT1 activators are EGF, TGF-α, estrogen, glucocorticoids, pituitary adenylate cyclase-activating polypeptide, dibutyryl cAMP, β-lactam antibiotics, etc. (Eng et al., 1997; Swanson et al., 1997; Schlag et al., 1998; Zelenaia et al., 2000; Figiel et al., 2003; Rothstein et al., 2005; Zschocke et al., 2005). Although several GLT1 activators mentioned earlier (EGF, TGF-α, estrogen, glucocorticoids) belong to general growth factor family, some other growth factors, such as platelet-derived growth factor, insulin, basic fibroblast growth factor, and nerve growth factor (NGF) do not show GLT1-inducing effect in primary astrocyte cultures (Sattler and Rothstein, 2006). Tumor necrosis factor-α (TNF-α) even downregulates GLT1 expression (Su et al., 2003). Further studies are needed to elucidate whether these GLT1 activators are the soluble factors secreted from neurons that activate the GLT1 in coculture of neurons and astrocytes.

In addition to the soluble factors, neuronal activity has also been considered to play a role in the regulation of GLT1 expression. Early studies showed that lesions of projections to a particular target nucleus results in decreased expression of GLT1 in the target area in adult rat (Ginsberg et al., 1995; Ginsberg et al., 1996). In a more controllable in vitro system, treatment of neuron and astrocyte cocultures with tetrodotoxin downregulated the expression of GLT1 protein (Poitry-Yamate et al., 2002). The effect of neuronal activity on the expression of GLT1 (and GLAST) was further examined in vivo by using Western blotting and serial section EM (Genoud et al., 2006). In this study, one day after a peripheral stimulus (whisker stimulation) that increases sensory activity, GLT1 (and GLAST) protein levels were increased twofold in the corresponding cortical column of the barrel cortex. The GLT1 (and GLAST) protein level returns to basal level 4 days after the stimulation was stopped, whereas the expression of neuronal glutamate transporter EAAC1 remained unchanged throughout the treatment process (Genoud et al., 2006). GLT1 expression is also affected by astrocyte to astrocyte communication. Figiel et al. (2007) recently showed that blockade of connexin 43, the major component of gap junction, by small interfering RNA or inhibition of gap junction by pharmacological inhibitors reduces significantly the level of GLT1 protein in cultured primary astrocytes, suggesting signal molecules can spread among astrocytes to regulate the expression of GLT1 protein.

By using pharmacological inhibitors, downstream signal transduction pathways involved in different GLT1 activators were also investigated by different groups. Activation of the p42/44 MAP kinases via the tyrphostin-sensitive receptor tyrosine kinase signaling pathway is associated with NCM-induced activation of GLT1 (Swanson et al., 1997; Gegelashvili et al., 2001). NCM treatment also induces phosphorylation of transcription factors cAMP response element-binding, cAMP response element modulator 1, and activating transcription factor 1 (Gegelashvili et al., 2000). Inhibitors of phosphatidylinositol 3-kinase (PI-3K), tyrosine kinase, or nuclear transcription factor kB (NF-κB) almost completely blocked NCM-induced upregulation of GLT1 protein expression (Swanson et al., 1997; Zelenaia et al., 2000). Similarly, inhibitors of PI-3K and NF-κB also blocked EGF and cAMP induced upregulation of GLT1 protein (Su et al., 2003). Although these results suggested different, GLT1 activators may use the same downstream signal pathways; independent pathways are also involved by a particular activation signal. For example, inhibitor of protein kinase A blocks cAMP-induced activation of GLT1, but not EGF-induced activation of GLT1. In contrast, inhibitor of tyrosine kinase can block EGF's effect, but not cAMP's effect on the activation of GLT1. It is likely that multiple signal pathways can be employed (Fig. 4.4) (Sattler and Rothstein, 2006), depending on the exact stimulus and the physiological context. Most of the transcriptional regulation work to date has been performed on immature astrocytes in vitro. Whether these pathways operate in adult astrocytes in vivo remains unexplored.

With the recent cloning of a 2.5-kb human EAAT2 promoter fragment, characterization of further downstream transcription factors that are involved in the regulation of EAAT2/GLT1 is greatly facilitated (Su et al., 2003). Multiple consensus

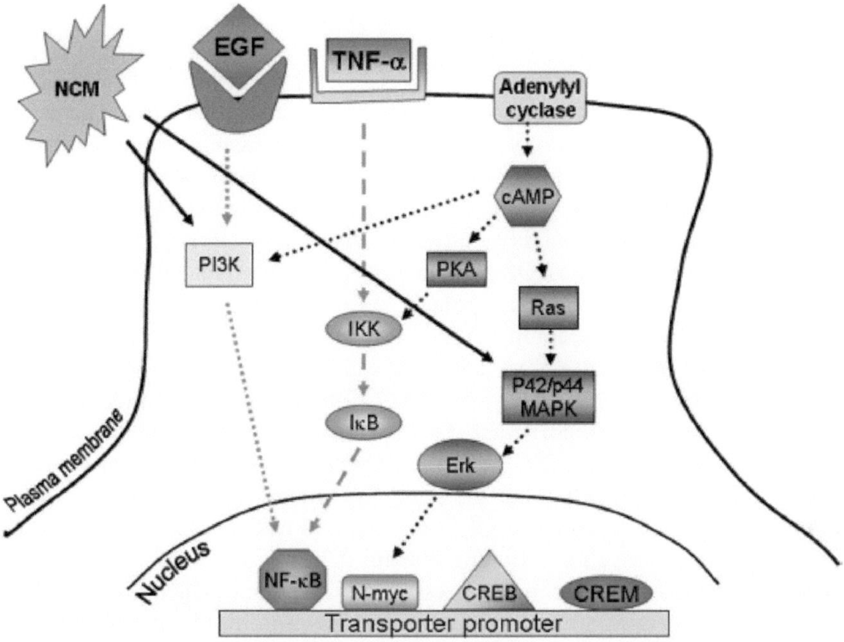

Fig. 4.4 Signal pathways for the activation of EAAT2 promoter. Multiple signal transduction pathways are employed in the regulation of EAAT2/GLT1 promoter activation in response to various upstream signals. Adapted from Sattler and Rothstein (2006), Handb Exp Pharmacol.

binding sequence for several transcription factors were found in the EAAT2 promoter and 5′ untranslated region (UTR), including NF-κB, N-myc, and nuclear factor of activated T cells by in-silicon analysis of EAAT2 promoter sequence. With the generation of consensus sequence-specific EAAT2 promoter mutant and biochemical approaches, NF-κB was found to have dual roles in the regulation of GLT1 expression in H4 glioma cells (Sitcheran et al., 2005). The binding of NF-κB to an upstream *cis*-element of EAAT2 promoter mediates the EGF-induced upregulation of GLT1 mRNA, but the binding of NF-κB to a *cis*-element located in 5′ UTR mediated TNF-α induced downregulation of GLT1 mRNA. As GLT1 is selectively expressed in astrocytes only in adult CNS, it has been speculated that astrocyte differentiation is associated to the induction of GLT1 in astrocytes during development. However, no study so far has demonstrated that astrocyte differentiation signals, such as leukemia-inhibitory factor (LIF) or bone morphogenetic protein 2/4 (BMP2/4), induces the expression of GLT1 in astrocyte. Downstream transcription factors of LIF or BMP signaling, signal transducer, and activator of transcription 3 or sma- and mad-related protein 4, have not been tested on EAAT2 promoter regulation. In addition, it remains to be tested in vivo whether these signal pathways and transcription factors characterized in in vitro models are involved in GLT1 expression.

Posttranscriptional and Translational Regulation of GLT1

Based on the genomic structure of EAAT2/GLT1, it is not surprising that numerous EAAT2/GLT1 mRNA variants have been found in mammalian CNS. These variants are very diverse with many 5′ variants, three 3′ variants, several alternative splicing of introns or skipping of exons. Among the 5′ variants, some have a shorter exon1, which leads to the skip of the first few amino acids in the translation and the protein product showed slightly different distribution throughout the brain (Rozyczka and Engele, 2005). Others contain upstream open reading frames, which may inhibit translation of the main protein product (Munch et al., 2002). A recent study showed that a 5′ variant of EAAT2 with longer UTR can be translated with higher efficiency than other 5′ variants, suggesting that 5′ UTR is involved in the regulation of translation efficiency of EAAT2 (Tian et al., 2007). Three 3′variants of GLT1 have been identified, namely GLT1a (same as the original GLT1), GLT1b/GLT1v, and GLT1c (Chen et al., 2002; Rauen et al., 2004). GLT1b variant is translated to a GLT1b protein that has the same N-terminus of GLT1a but the last 22 amino acids of the C-terminus of GLT1a is replaced with a unique 11-amino-acid stretch that composed a PDZ-binding domain (Chen et al., 2002). The transport properties of these two isoforms are essentially the same, but the unique PDZ domain at the C-terminus of GLT1b may allow this isoform to interact with other scaffold proteinsto localize differently. Another functional 3′ variant of GLT1, GLT1c, was identified in the rat and human retina (Rauen et al., 2004). This isoform also differs with GLT1a at the C-terminus by having a similar PDZ domain as that of GLT1b. This isoform of GLT1 is mainly expressed in photoreceptors in the retina.

Aberrant EAAT2 mRNA transcripts that skip certain exons were also identified in human glioma cells and postmortem amyotrophic lateral sclerosis (ALS) patients (Lin et al., 1998; Guo et al., 2002). In particular, the intron 7 retention or exon 9 skipping form of the EAAT2 transcript was specifically identified in normal ALS and other neurodegenerative disease (Nagai et al., 1998; Honig et al., 2000; Flowers et al., 2001). Intron 7 retention form of transcript introduces stop codons for premature termination of functional EAAT2 protein. Exon 9 skipping form of transcript lacks the motif that regulates the proper export of matured proteins to plasma membrane from endoplasmic reticulum (Kalandadze et al., 2004). Both these transcripts could lead to less production of functional EAAT2 protein. Further studies in human glioma cell lines also showed that aberrant EAAT2 mRNA transcripts suppress the translation of normal EAAT2 mRNA transcript and subsequent production of functional EAAT2 protein (Guo et al., 2002). The relative abundance of these transcripts in normal vs. diseased tissues has been controversial, although recent independent studies confirm the original observations that aberrant EAAT2 transcripts are responsible for the decreased EAAT2 protein in ALS (Lauriat et al., 2007). The physiological consequence of these transcripts is also not known, though initial in vitro studies suggested that certain species were either inactive or had dominant negative-like effects on normal GLT1 mRNA transcript.

Regulation of GLT1 Trafficking

In primary neuron and astrocyte cocultures, between 60% and 80% of the total GLT1 immunoreactivity is found in the biotinylated/cell-surface fraction (Kalandadze et al., 2002). In EM analyses of GLT1 immunoreactivity in vivo, no significant intracellular pool of GLT1 was observed (Chaudhry et al., 1995), suggesting that the majority of GLT1 is transported constitutively to the plasma membrane after translation. This is very different from the subcellular distribution of neuronal glutamate transporter EAAC1, which is mainly stored intracellularly (He et al., 2001). The membrane expression or distribution of GLT1 appears to be regulated by the PKC family. Several groups found that activation of PKC decreases the activity and surface expression of EAAT2/GLT1 by about 30–50% and also changes the GLT1 clustering pattern on the plasma membrane in transfected cells and in primary cultures by internalizing GLT1 into cytosol (Kalandadze et al., 2002; Zhou and Sutherland, 2004). Pretreatment of PKC inhibitor, bisindolylmaleimide II, or expression of a dominant-negative form of dynamin prevented phorbol 12-myristate 13-acetete induced GLT1 internalization. The actin inhibitor cytochalasin D also disrupted the formation of GLT1 clustering, suggesting the involvement of actin in this process (Zhou and Sutherland, 2004). Although phosphorylation of GLT1 has been speculated to mediate PKC-dependent internalization, the exact serine/threonine site on GLT1 protein that is phosphorylated by PKC has not been identified despite a 43-amino-acid domain that is required for PKC-dependent internalization of GLT1 (Kalandadze et al., 2002). The membrane distribution of GLT1 in astrocytes is also affected by neighboring neuronal activity. Early studies by using immunostaining revealed that GLT1 (and GLAST) is concentrated in areas of the membrane facing neuronal spinal processes rather than other astrocytes, cell bodies, large dendrites, or vascular epithelium (Chaudhry et al., 1995). A more recent study also showed that neuronal release of glutamate results in a change in GLT1 redistribution on plasma membranes that localize GLT1 closer to neurons (Poitry-Yamate et al., 2002), suggesting that the distribution of GLT1 is dynamically responsive to neuronal activity.

4.1.5 Physiological Function of Astroglial Glutamate Transporters

4.1.5.1 Astroglial Glutamate Transporters Are the Dominant Glutamate Transporters in the Mammalian CNS

Extensive studies using genetic, pharmacological, and electrophysiological approaches all concur that astroglial glutamate transporters are the dominant glutamate uptake systems in the CNS and have critical functions in maintaining glutamate homeostasis and in modulating synaptic transmission (Rothstein et al., 1996; Bergles and Jahr, 1997; Tanaka, 1997; Huang and Bergles, 2004). In fact, several biochemical characteristics of the CNS have hinted that astrocytes may have greater advantage than neurons as the synaptic glutamate sink (Anderson and Swanson, 2000). First, astrocytes provide a major carbon source to neurons by synthesizing glutamine from glutamate (Hertz et al., 1999). Therefore, uptake of extracellular glutamate by astrocytes completes the important intercellular carbon cycle; second, astrocytes have a more stable membrane potential to maintain the Na^+ gradient for glutamate transport. In neurons, Na^+ influx during action potential would greatly reduce Na^+ gradient across the plasma membrane, the driving force for glutamate transport, which lowers the efficiency of glutamate transport; third, it was found that astrocytes are better able to maintain physiological Na^+ and K^+ gradients during ATP depletion (Rose and Ransom, 1996a, 1996b), which make them better suited to maintain low extracellular glutamate levels via ATP-dependent uptake mechanisms; fourth, as neurons have high intracellular glutamate concentration but astrocytes have low(er) intracellular glutamate levels, which results from rapid enzymatic conversion of glutamate into glutamine, the TM glutamate gradient in synapse would strongly favor glutamate uptake in astrocytes than in neurons.

The understanding of glutamate transporter function started in the early 1990s, after it was shown that astrocytes mediate the majority of glutamate uptake in the CNS. By reducing the expression level of astroglial glutamate transporter, GLT1 and GLAST, but not the neuronal glutamate transporter, EAAC1, Rothstein et al.(1996) first showed that extracellular glutamate levels were elevated and excitatory damage was induced that resulted in the hind limb paralysis in the rats. In addition, GLT1 knockout mice retain less than 10% of total glutamate transport in the cortex, and develop lethal spontaneous seizure and display increased susceptibility to selective hippocampal CA1 neuron loss (Tanaka, 1997; Matsugami et al., 2006). These results suggested that GLT1 mediates the bulk of glutamate uptake from the extracellular fluid in most brain regions. GLAST knockout mice also showed motor incoordination and increased susceptibility to cerebellar cold-induced injury but no ataxic phenotype (Maragakis and Rothstein, 2004), likely because of the more dominant role of GLAST in the cerebellum. Because the process of glutamate transport is electrogenic, the uptake of glutamate via transporters can be demonstrated by the measurement of a transporter current. Using this approach, synaptic glutamate release has been shown to induce rapid glutamate transporter currents in the astrocytes in hippocampal and cerebellar preparations (Bergles and Jahr, 1997; Kojima et al., 1999). Glutamate uptake inhibitors selective for the astrocyte-specific subtype GLT1 potentiate excitatory postsynaptic currents (EPSCs) in hippocampal slices (Tong and Jahr, 1994). In contrast, astroglial transporter currents are not detectable in CA1 pyramidal neurons in response to afferent stimulation, nor are they present in patches from CA1 neurons in response to exogenous glutamate (Bergles and Jahr, 1998). Transporter currents are nearly abolished in slices prepared from GLT1 knockout mice (Kojima et al., 1999). In addition, the GLT1 selective inhibitor dihydrokainate inhibits transporter currents with efficacy equal to the nonselective glutamate transporter inhibitor dl-threo-b-benzyloxyaspartate, suggesting a primary role for the astrocyte transporter GLT1. These observations strongly suggest a dominant role for astroglial glutamate transporters and in particular GLT1 in synaptic glutamate uptake.

4.1.5.2 Astroglial Glutamate Transporters Modulate and Refine the Excitatory Synaptic Neurotransmission in Mammalian CNS

In the past 10 years, studies of glutamate transporters have gone beyond their basic function of maintaining low level of extracellular glutamate (Grewer and Rauen, 2005). Early observations that glutamate uptake inhibitors increase both the amplitude and the duration of glutamate-induced EPSCs in different preparations suggest that transporters must play an active role in the glutamatergic synaptic transmission process other than just maintaining glutamate concentration low in the extracellular space (Tong and Jahr, 1994; Diamond and Jahr, 1997). In the synapse, astroglial glutamate transporters bind glutamate rapidly but transport it at a relative slow rate, which could result in occupied transporter binding sites and could also allow some bound glutamate molecules to become unbound. This dynamic transport process of glutamate shapes the glutamate concentration transient that the receptors are exposed to in the synaptic cleft. As a result, transporter activity influences receptor occupancy and subsequent activation at individual synapses, as observed as EPSCs. The modulation of transporters on receptor activation is affected by the coverage of synapse with astrocyte processes and membrane distribution of both transporters and receptors (Fig. 4.5) (Huang and Bergles, 2004). In the CNS, most synapses are surrounded by astrocyte processes that are enriched with glutamate transporters (Ventura and Harris, 1999). Depending on the individual synapse, the coverage of synapse by astrocyte processes could be very extensive, similar to the climbing fiber or Purkinje cells covered by Bergmann glia processes, or could be very loose, similar to hippocampal CA1 striatum radiatum synapses (Anderson and Swanson, 2000). The extent of coverage partially determines the perisynaptic distribution of glutamate transporters that is involved in the synaptic uptake of glutamate. Based on the exact local distribution of transporters and receptors around the synapse, the activation of receptors by glutamate is precisely modulated by transporters (Huang and Bergles, 2004). This model is supported by recent studies of excitatory synapses in the cerebellum, hippocampus, and retina (Brasnjo and Otis, 2001). In the cerebellum, inhibition of glutamate transporters potentiates the activity of postsynaptic mGluRs in Purkinje neurons in response to parallel fiber stimulation, and facilitates mGluR-mediated long-term depression (Reichelt and Knopfel, 2002). In the hippocampus, glutamate transporter inhibition similarly potentiates postsynaptic mGluR activation in interneurons, leading to enhanced inhibition of pyramidal neurons (Huang et al., 2004). This phenomenon is not restricted to mGluRs, as glutamate transporter inhibition also facilitates the recruitment of extrasynaptic *N*-methyl-D-aspartate (NMDA) receptors at parallel fiber–interneuron synapses in the cerebellum (Clark and Cull-Candy, 2002), as well as at ganglion cell synapses in the retina (Chen and Diamond, 2002). On the other hand, although AMPA and NMDA receptors are often clustered on the postsynaptic neuronal membrane, it is not rare to find extrasynaptic or perisynaptic localization of NMDA or mGlu receptors (Baude et al., 1993). The peri- or extrasynaptic distribution of glutamate receptors suggests that they can also be activated at a distance by glutamate that leaked from the synapse. The presence of highly abundant local perisynaptic, astroglial

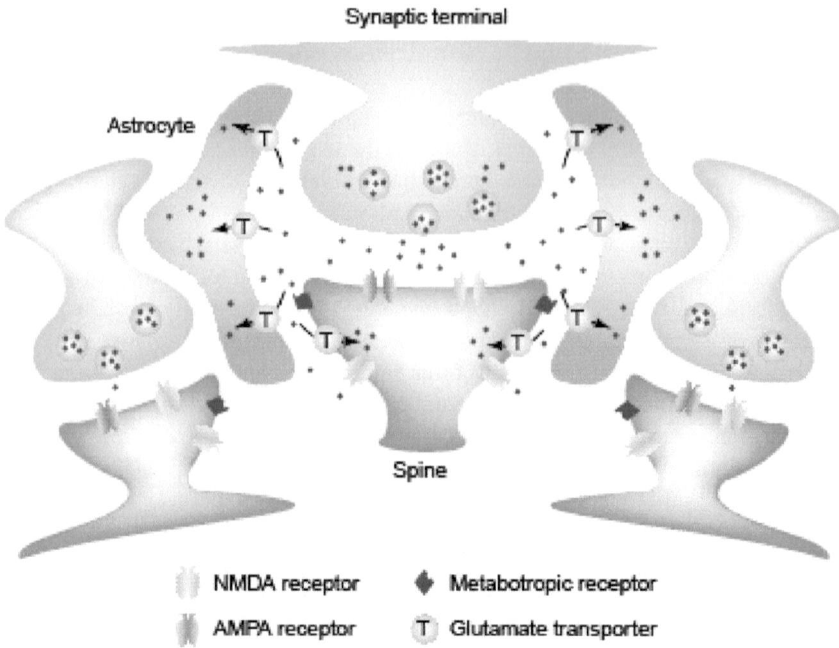

Fig. 4.5 Modulation of glutamatergic synaptic transmission by astroglial glutamate transporters. Unlike receptors, glutamate transporters are excluded from the synaptic cleft. EAAT1 (GLAST) and EAAT2 (GLT-1) are present at a high density in the membranes of astrocytes that often ensheath synapses. EAAT3 (EAAC1) is found in the soma and dendrites of neurons, but is also found in GABAergic terminals (not shown). Glutamate transporters shield extrasynaptic NMDA receptors and mGluRs from glutamate as it diffuses from the cleft, and prevent glutamate from reaching receptors at nearby synapses. Inhibition of these transporters potentiates excitatory responses mediated by these receptors, and allows glutamate spillover, which suggests that transporter regulation might be used to regulate synaptic efficacy. Note that presynaptic mGluRs have been omitted from this diagram. Adapted from Huang et al. (2004), Curr Opin Neurobiol. (*See Color Plates*)

glutamate transporters also effectively prevent the perisynaptic or extrasynaptic activation of glutamate receptors, reducing the interference between neighboring synapses (Huang and Bergles, 2004). This is supported by the observation that inhibition of glutamate transporters in hippocampal pyramidal neurons allows glutamate to diffuse from one set of synapses and activate NMDA receptors at adjacent synapses (Huang and Bergles, 2004).

4.1.5.3 Astroglial Glutamate Transporters and Neurological Diseases

From the critical functions of astroglial glutamate transporters in the CNS, it is not surprising to imagine that dysfunction of these transporters and glutamate-induced excitatory toxicity is implicated in many neurological diseases or disorders. The direct links between astroglial glutamate transporters and neurological diseases were first observed in animals with reduced expression levels of GLT1 or GLAST. Rats that have reduced expression level of transporter, especially GLT1 by using antisense-mediated knockdown or GLT1-/- mice showed severe seizures and loss of neurons, implicating the important role of GLT1 in preventing the excitatory toxicity induced by extracellular glutamate (Rothstein et al., 1996; Tanaka, 1997). Clinically, the link between glutamate transporters and neurological diseases first came from the observation that glutamate levels in the CSF are elevated in sporadic ALS patients (Rothstein et al., 1992). Because glutamate transporters in the CNS tightly control the extracellular glutamate to low levels, the elevated level of glutamate in CSF suggested a dysfunction of glutamate transporters. Subsequent examination of all glutamate transporters expression in postmortem brain and spinal cord of ALS patients revealed severe loss of GLT1 but not other transporters in motor cortex and lumbar spinal cord of ALS patient tissue (Rothstein et al., 1995). Among ALS patients, about 1–2% of the total ALS patients are a familiar form, which is caused by the mutations of superoxide dismutase 1 (SOD1) gene (Rosen et al., 1993), but the majority of ALS patients are of a sporadic form with unclear causes. After the identification of mutations in SOD1 that causes familial ALS, animal (rat and mouse) models of ALS were made that overexpress different pathogenic mutations of SOD1 gene (Bruijn et al., 1997). In these transgenic animals, selective loss of GLT1 was all observed in end-stage lumbar spinal cord. The recapitulation of the loss of GLT1 in animal model of ALS suggested that conserved pathogenic mechanisms that involve the GLT1 and glutamate-mediated excitatory toxicity may be present for both sporadic and familial forms of ALS. The mechanism for the loss of GLT1 in ALS is not yet clear. Several interesting observations have been found. An in vitro study suggested that GLT1 is oxidized in SOD1 mutant overexpressing cells and possess much lower transport capacity (Trotti et al., 1999). Studies from the same group also showed that SOD1 mutant induces selective caspase-3-dependent EAAT2 cleavage (but not EAAT1 or EAAC1 cleavage) and inactivates the transport activity of EAAT2 (Maragakis and Rothstein, 2004), while other studies showed the identification of aberrant EAAT2 mRNA that suppresses the translation of normal EAAT2 mRNA, resulting in the reduction of GLT1 protein expression (Lin et al., 1998). Although previous studies of total EAAT2/GLT1 mRNA (in human and rodent) did not reveal dramatic alterations in ALS tissue (Bristol and Rothstein, 1996), those studies were not conclusive because large spinal cord samples were used, which potentially masks the very focal changes in GLT1 mRNA that may occur. Studies that determine the GLT1 mRNA level in situ in lumbar spinal cord of animal model of ALS are needed in the future to investigate the role of transcriptional dysfunction of GLT1 in ALS.

Since the original discovery of astroglial glutamate transporter abnormalities in ALS, alterations of the astroglial transporter proteins have been implicated in a number of other neurological diseases (Maragakis and Rothstein, 2004; Beart and O'Shea, 2007). In Alzheimer's disease, alterations in glutamate transport in human Alzheimer's disease (AD) tissue have also been observed. Compared with control brains, AD brains displayed a 34% decrease in levels of d-[^{3}H] aspartate binding, a

30% decrease in L-[^{3}H] aspartate binding (Masliah et al., 1996). Abnormal expression pattern of EAAT1 and EAAT2 were also observed in AD brains. In cases showing Alzheimer-type neuropathology, EAAT1 was surprisingly found expressed in neurons, primarily a subset of pyramidal cells, and in dystrophic neuritis (Scott et al., 2002). Similarly, EAAT2-immunoreactive neurons were also observed throughout the cortex, striatum, hypothalamus, and reticular formation in AD brain tissue. This aberrant expression was closely associated with tau deposition and neurofibrillary changes in these neurons (Thai, 2002). In addition to the aberrant expression pattern, a recent study using gene chips and immunohistochemistry also showed marked impairment in the expression of EAATs (EAAT1 and EAAT2) at both gene and protein levels in hippocampus and gyrus frontalis medialis of AD patients, even in early clinical stages of disease (Jacob et al., 2007). The mechanisms for the altered expression pattern or loss of EAAT1 and EAAT2 are unclear but amyloid a4 precursor protein (APP) has been suggested to play a role. In vitro studies showed that APP can protect neurons against excitotoxicity. In transgenic mice expressing a mutant form of the human APP, significant decrease in V_{max} for aspartate uptake was found (Masliah et al., 2000).

In Parkinson's disease (PD) the involvement of astroglial glutamate transporters is not well established, since dopaminergic neurotransmission is mainly used in the motor circuitry from substantial nigra to striatum. In vitro treatment of astrocyte cultures with the dopaminergic neurotoxin 1-methyl-4-phenylpyridinium leads to a 39% reduction in glutamate transport (Hazell et al., 1997). Although astrogliosis has been found in animal models of PD, loss of transporter proteins has not been characterized. Levels of extracellular glutamate were not found significantly elevated in 1-methyl-4-phenyl-1,2,3,6-tetrahydropyridine model (McNaught and Jenner, 1999).

In Huntington's disease (HD), expression of EAAT2 is apparently impaired by the mutant huntingtin protein. The mutant huntingtin expressed by adenoviral vectors reduces glutamate uptake and GLT1 protein in cultured astrocytes (Shin et al., 2005). In the postmortem brains of HD patients, mutant huntingtin protein was found to form aggregates in both neurons and glia (Shin et al., 2005). The EAAT2 mRNA level measured by in situ hybridization is also reduced in striatum where most of the neuronal loss occurs in HD (Arzberger et al., 1997). In animal models (R6/2 mice) of HD that expresses N-terminal fragment of mutant huntingtin, mutant huntingtin also forms intranuclear aggregates in glia, which correlate with decreased GLT1 protein expression in these mice (Mangiarini et al., 1997; Shin et al., 2005). The reduction of GLT1 expression also correlates with the development of neurological symptoms in these mice. At 4 weeks of age, GLT1 mRNA in R6/2 mice is unchanged compared with littermate controls but progressively decreases in the striatum and cerebral cortex from 8 to 12 weeks, while GLAST and EAAC1 mRNA remain unchanged (Lievens et al., 2001). At the same time, R6/2 mice display motor impairment by 5 weeks of age, neurological symptoms by 8 weeks, and frequently die after 12 weeks (Davies et al., 1997; Carter et al., 1999; Shin et al., 2005). GLT1 protein was also reduced in the cortex and striatum in R6/2 mice at 12 weeks compared with littermate controls. As a result of decreased GLT1 protein level, aspartate uptake in cortex and striatum, as well as glutamate uptake in synaptosomes prepared from 12-week-old R6/2 mice are lower compared with littermate controls (Lievens et al., 2001).

In epilepsy, microdialysis studies of human epileptogenic hippocampus revealed elevated levels of glutamate following epileptic activity, suggesting that glutamate homeostasis is disrupted (Maragakis and Rothstein, 2004). Complete deletion of GLT1 induces spontaneous seizure activity in GLT1-/- mice, at early postnatal time points, most often followed by death (Tanaka, 1997). In GLAST-/- mice, more severe stages of pentylenetrazol-induced seizure activity were observed when compared with wild-type mice (Watanabe et al., 1999). Although the early developmental loss of astroglial glutamate transporters induces seizures, increased expression of neuronal glutamate transporter EAAC1 but not the loss of GLT1 and GLAST was primarily found in most animal models of epilepsy (Maragakis and Rothstein, 2004). In patients with temporal lobe epilepsy, increased expression of EAAC1 but no alternations of EAAT1 or EAAT2 were also observed (Proper et al., 2002). This could reflect a compensatory response from the affected neurons.

In stroke and ischemia, excitatory toxicity has been found to be one of the major pathogenic mechanisms. The change of GLAST and GLT1 expression is variable depending on the model and endpoint of examination (Maragakis and Rothstein, 2004). Consistently, however, abnormal neuronal expression of GLAST and GLT1 were found in various models (Martin et al., 1997; Tao et al., 2001). Postmortem tissue from human patients has not been well analyzed because of the difficulty of obtaining well-preserved stroke tissue samples, given the fact that pathogenic events in stroke occur rapidly. Notably, a highly prevalent polymorphism in the promoter of the EAAT2 gene was found (Mallolas et al., 2006). Functionally, this polymorphism abolishes a putative regulatory site for activator protein-2 and creates a new consensus binding site for the repressor transcription factor GC-binding factor 2. Clinically, this polymorphism is associated with increased glutamate concentrations and with a higher frequency of early neurological worsening in human stroke (Mallolas et al., 2006).

Besides these major neurological diseases mentioned earlier, astroglial glutamate transporters were also implicated in some neuropsychiatric disorders (such as schizophrenia, bipolar disorder), brain glioma growth, and retinal diseases or glaucoma, etc. (Maragakis and Rothstein, 2004; Beart and O'Shea, 2007; Sheldon and Robinson, 2007). But the exact mechanisms remain to be elucidated. The dysregulation of astroglial glutamate transporters is likely not to be the primary pathogenic mechanisms; instead, it may accelerate the progress of disease by altered control of extracellular glutamate levels and triggering massive excitatory toxicity. Therefore, maintaining glutamate homeostasis by increasing the expression of astroglial glutamate transporters, especially EAAT2 could potentially slow down the disease progression. In fact, progression of ALS in transgenic mice that overexpress GLT1 is slowed down (Guo et al., 2003). In addition, administration of β-lactam antibiotics has been shown to upregulate the expression of GLT1, via transcriptional activation, and to extend the life of SOD1 G93A transgenic mice (model of ALS) (Rothstein et al., 2005), to retard HD as well as epilepsy in their experimental models. These promising studies provide potential therapeutic ways to slow down the progression of certain neurological diseases in humans.

4.2 Sodium- and Chloride-Dependent Neurotransmitter Transporter Family SLC6 in Astrocytes

4.2.1 General Properties of SLC6 Transporters

SLC6 family contains transporters for the inhibitory neurotransmitter GABA and glycine, and monoamines (serotonin, norepinephrine, and dopamine) (Chen et al., 2004a). For each individual neurotransmitter, multiple transporters are also present. These membrane proteins share a common TM topology with 12 transmembrane domains (TMDs) connected by six extracellular and five intracellular loops (Fig. 4.6) (Torres and Amara, 2007). A large, extracellular loop, with glycosylation sites, is present between TMD 3 and 4. Similar to SLC2 family, SLC6 transporters mediated transport is also against the concentration gradient of neurotransmitter and is powered by a Na^+ gradient. It appears that $2Na^+$ are commonly cotransported with neurotransmitter, and sometimes $1Cl^-$ or $1K^+$ (countertransported) is also transported (Chen et al., 2004a). Although almost all SLC6 members have high expression in CNS, their expression is not restricted to the CNS. Some of the SLC6 members also have high expression in kidney, liver, etc. In the CNS, SLC6 transporters are expressed in both neurons and glia. The distribution and functional dominance of individual type of transporter in neurons or astrocytes is dependent on the exact neurotransmitter.

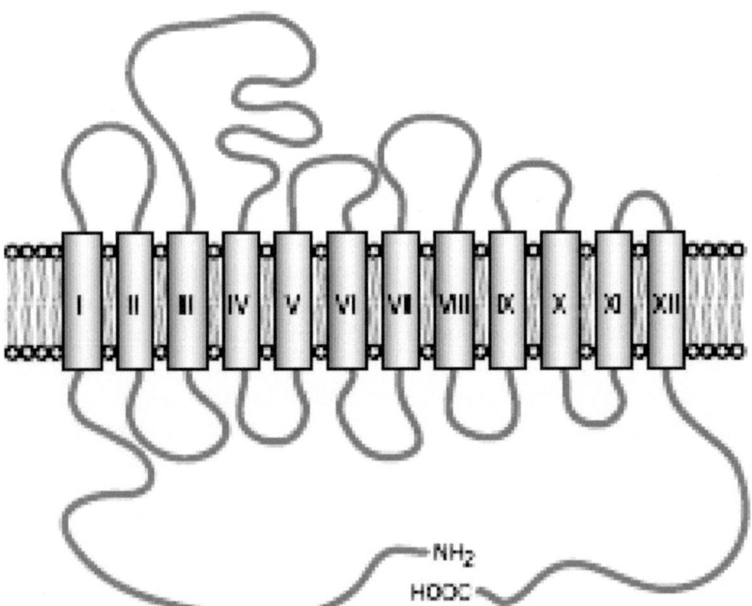

Fig. 4.6 Predicted topology of SLC6 transporters. Twelve transmembrane domains are connected by intracellular and extracellular loops with both N- and C-terminals inside the cell. Adapted from Torres et al. (2003), Nat Rev Neurosci. (*See Color Plates*)

4.2.2 GABA Transporter in Astrocytes

4.2.2.1 GABA Homeostasis in the CNS

γ-aminobutyric acid (GABA) is one of the major inhibitory neurotransmitters in the CNS. GABA is highly enriched in the GABAergic neurons and processes that are usually an important class of inhibitory interneurons in the spinal cord and brain (Petroff, 2002; Schousboe, 2003). These neurons typically include basket cells of the cerebellum and the hippocampus, Purkinje cells of the cerebellum, granule cells of the olfactory bulb, and amacrine cells of the retina. GABA is mainly present intracellularly in GABAergic neurons. The concentration of GABA (50–100 mM in nerve terminals) is far greater than that of non-GABAergic neurons or glia (1 mM) (Petroff, 2002). In the GABAergic neurons, GABA is mainly produced from a-decarboxylation of glutamate by glutamic acid decarboxylase (GAD), and is metabolized to succinate by the sequential actions of GABA-transaminase and succinic semialdehyde dehydrogenase. Two major GADs exist in GABAergic neurons, GAD65 and GAD67 (Erlander and Tobin, 1991). GAD65 is responsible for the synthesis of 30% of the GABA and GAD67 is responsible for the rest of the GABA synthesis. This relative contribution of GAD to the synthesis of GABA was later supported by GAD67-/- mice. These mice have less than 20% of the GAD activity and 7% of the brain GABA concentration compared with the wild type (Asada et al., 1997; Soghomonian and Martin, 1998; Ji and Obata, 1999). GAD is also the rate-limiting enzyme in the synthesis of GABA, which is selectively expressed in GABAergic neurons. The synthesis of GABA is closely related to the glutamate–glutamine cycle (Petroff, 2002). Glutamate released by neurons is taken up primarily by glia through astroglial glutamate transporters. Glutamate taken up by glia is converted to glutamine by glia-specific glutamine synthetase and then released into extracellular fluid to be reuptaken into neurons. The glutamine is again converted to glutamate in the neurons by glutaminase, thus replenishing glutamate stores lost by synaptic release and guaranteeing the supply of GABA in GABAergic neurons when GAD is abundantly expressed. This influence of glutamate–glutamine cycle on GABA synthesis is supported by glutamine's stimulatory effect on GABA synthesis in synaptosomes, cell cultures, and brain slice culture experiments (Kapetanovic et al., 1993).

4.2.2.2 GABA Transporters and Uptake of GABA in the CNS

At the GABAergic synapse, GABA is released from presynaptic nerve terminals and activates ionotropic $GABA_A$ and $GABA_C$ receptors or metabotropic $GABA_B$ receptors to transmit signals (Rudolph and Mohler, 2006). Released GABA is quickly taken up from the synaptic cleft into presynaptic neurons or perisynaptic astrocytes by specific high affinity GABA transporters (GATs). Unlike glutamate neurotransmission, uptake of GABA appears to be mainly carried out by neuronal

GATs not by astroglial GATs; about 80% of GABA is transported into presynaptic terminals of neurons whereas only 20% is taken up into adjacent astrocytes (Schousboe, 2003). Transport of one GABA is coupled together with the transport of $2Na^+$ and $1Cl^-$ and therefore is also electrogenic (Schousboe, 2003). The nomenclature of GATs is somewhat confusing as the numbering system in different species varies. Based on the nomenclature introduced by Guastella et al.(1990) and Borden et al.(Guastella et al., 1990; Borden et al., 1992), rat and human GATs are referred as GAT-1, betaine/GAT-1 (low affinity for GABA), GAT-2, and GAT-3. For mouse GATs, a different nomenclature is used to refer each corresponding homologous transporter in mouse as GAT1–GAT4 (without hyphen), respectively (Liu et al., 1993). Among these GATs, expression of GAT-1 and GAT-3 are restricted in the CNS. GAT-1, the predominant GAT, has high expression in neocortex, hippocampus, cerebellum, basal ganglia, brainstem, spinal cord, olfactory bulb, and retina (Nelson et al., 1990). GAT-1 mainly colocalizes with markers for GABAergic neurons (Schousboe, 2000), specifically along axons and presynaptic nerve terminals, though its expression in astroglia was also found. In contrast, GAT3 (rat or human analog, GAT-2) and GAT4 (rat or human analog, GAT-3) of mouse are primarily expressed in the astrocytes (Chen et al., 2004a).

The modulatory role of GATs in GABAergic transmission has been suggested with the use of inhibitor of GAT-1. These inhibitors have been shown to potentiate the inhibitory action of GABA mediated by GABA receptors (Schousboe, 2003). In addition, by using microdialysis, inhibitors of GAT-1 also increase extracellular GABA concentration. There is significant functional difference in neuronal GABA uptake and astroglial GABA uptake for GABAergic transmission (Petroff, 2002). Neuronal uptake of GABA leads to recycling of neurotransmitter GABA in GABAergic neurons, which greatly increases the efficiency of the GABAergic system as the neurotransmitter GABA can be rapidly packed into synaptic vesicles ready for release. However, astroglial uptake of GABA results in the loss of GABA through its metabolism via GABA transaminase and the tricarboxylic acid cycle. Inhibitors specific for either neuronal or astroglial GAT have been characterized. Diaminobutyric acid has been used as an inhibitor of neuronal GABA transport while *N*-methyl-exo-THPO (4,5,6,7-tetrahydroisoxazolo [4,5-*c*]pyridin-3-ol) has been used to preferentially inhibit glial GABA transport (Schousboe, 2003). Although β-alanine used to be considered a specific inhibitor for astroglial GABA uptake, it was later found to be problematic (Schousboe, 2003).

As GABA is one of the major inhibitory neurotransmitters in the CNS, it is not surprising that dysfunction of GABAergic transmission is involved in some neurological diseases, including epilepsy, anxiety disorders, schizophrenia, and drug addiction. Early clinical studies showed that significant reductions of GABA concentration in the CSF were seen in patients with various epileptic syndromes (Wood et al., 1979). Occipital lobe GABA concentrations, measured using magnetic resonance spectroscopy, are often below normal in epileptic patients who have frequent complex partial seizures (During and Spencer, 1993; During et al., 1995). Increasing GABA concentrations seems a potential neuroprotective approach for epilepsy. Because astroglial uptake of GABA results in loss of GABA through metabolism, inhibitors targeting astroglial uptake of GABA have been developed as antiepileptic drugs (Suzdak and Jansen, 1995). For example, the inhibitor exo-THPO and its N-substituted analogs *N*-ethyl-exo-THPO were found to strongly inhibit astroglial GABA uptake and protect mice against audiogenic seizures (Schousboe, 2003).

4.2.3 Glycine Transporter in Astrocytes

4.2.3.1 Glycine and Synaptic Neurotransmission

Similar to GABA, glycine is another major inhibitory neurontransmitter in the CNS, mainly in posterior areas of the vertebrate CNS (Betz et al., 2006). Generally, glycine is synthesized from serine, but its specific synthesis in neurons has not been studied. In the spinal cord and brain stem, glycinergic interneurons provide an inhibitory feedback mechanism that controls the motor rhythm generation during movement and they also play an important role in the coordination of spinal reflex activity (Aragon and Lopez-Corcuera, 2003). Glycine is also an important neurotransmitter in the processing of auditive information through cochlear nuclei, the superior oliva complex and the inferior colliculus, and in the processing of visual information in retinal ganglion cells (Aragon and Lopez-Corcuera, 2003). In glycinergic synapses, glycine is released upon stimulus; released glycine activates strychnine-sensitive postsynaptic glycine receptors and induces opening of ligand-gated anion channels that leads to an influx of Cl^- into the postsynaptic neurons and following hyperpolarization. The resulting hyperpolarization raises the threshold for neuronal firing and thereby inhibits the postsynaptic neuron. In addition to its dominant inhibitory function, glycine can also be excitatory but in an uncommon way (Eulenburg et al., 2005). Nerve cells contain high intracellular chloride concentrations during embryonic development, and activation of glycine receptors at these stages therefore causes chloride efflux and membrane depolarization, i.e. excitation of target neurons. This glycine-induced depolarization ends around birth, when the neuronal K^+/Cl^- cotransporter KCC2 is expressed and lowers the intracellular levels of Cl^- (Hubner et al., 2001). Glycine also plays a role in excitatory glutamatergic synapses. Glycine acts as an essential co-agonist of glutamate at ionotropic NMDA receptors (Fig. 4.7) (Johnson and Ascher, 1987; Eulenburg et al., 2005). Recent studies have shown that superfusion with 0.5–20 mM glycine causes a potentiation of NMDA receptor currents in slice preparations (Berger et al., 1998). Furthermore, higher concentrations of glycine (>100 mM) have been found to “prime” NMDA receptors for internalization triggered by the activating agonist glutamate (Nong et al., 2003).

4.2.3.2 Glycine Transporters

Two major glycine transporters (GlyT) have been identified thus far, GlyT1 and GlyT2 (Eulenburg et al., 2005). The GlyT1 gene is expressed throughout most regions of the CNS (Betz et al., 2006). At the cellular level, GlyT1 is primarily

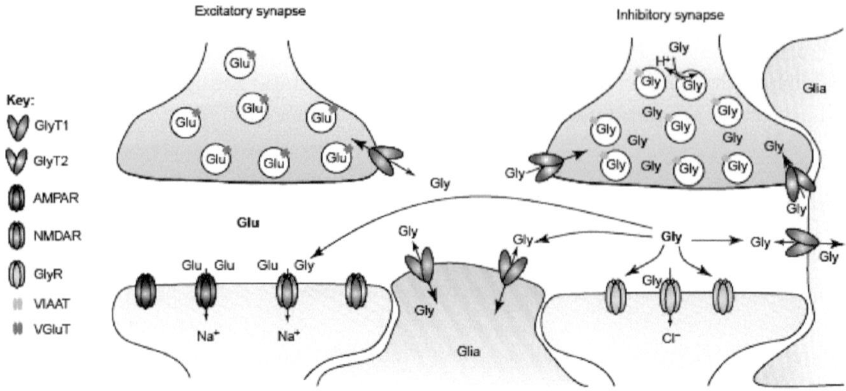

Fig. 4.7 Localization and proposed functions of glycine transporters at excitatory and inhibitory synapses. At inhibitory synapses, glycine release from the presynaptic terminal activates postsynaptic GlyRs and thereby induces ClK influx – hyperpolarization – of the postsynaptic cell. At excitatory glutamatergic synapses, glycine acts as an essential co-agonist of postsynaptic NMDARs, whereas neighboring glutamate receptors of the α-amino-3-hydroxy-5-methyl-4-isoxazole-propionic-acid receptor (AMPAR) subtype require only glutamate for channel activation. Here, glycine might be derived from neighboring glycinergic terminals or even be released from astrocytes via nonvesicular mechanisms (e.g. reverse transport by GlyT1). GlyT2 is localized in the presynaptic plasma membrane of glycinergic neurons and transports glycine into the terminal, thereby enabling the refilling of synaptic vesicles with glycine by the HC-dependent vesicular inhibitory amino acid transporter (VIAAT). GlyT1 is mainly expressed by glia cells surrounding both inhibitory and excitatory synapses. In addition, GlyT1 has been found on terminals of some excitatory neurons. Thus, GlyT1 mediates the clearance of glycine from the synaptic cleft of inhibitory synapses and, in addition, participates in the regulation of the glycine concentrations at excitatory synapses. Adapted from Eulenberg et al.(2005), Trend Biochem Sci. (*See Color Plates*)

expressed in astrocytes, but weak expression in some subpopulations of neurons was also observed. Immunohistochemical analysis revealed intense GlyT1-specific staining of glial cells, in particular astrocytes, and some weak GlyT1 immunoreactivity in selected dendrites and nerve terminals of putative excitatory neurons in spinal cord (Jursky and Nelson, 1996; Cubelos et al., 2005). Similarly, in forebrain regions rich in NMDA receptor-containing synapses, GlyT1 staining was found on both glia cells and subpopulations of glutamatergic neurons. In the retina, GlyT1 is localized exclusively in selected amacrine and ganglion neurons but is not seen in the glial Muller cells (Pow and Hendrickson, 1999). Analysis of GlyT2 expression indicates an exclusively neuronal expression of this transporter isoform in CNS regions rich in glycinergic synapses, such as the spinal cord, brain stem, and cerebellum (Jursky and Nelson, 1995). Immuno-EM demonstrated that GlyT2 is enriched in the plasma membrane of glycinergic nerve terminals but excluded from active zones (Mahendrasingam et al., 2003).

Distinct roles of GlyT1 and GlyT2 have been shown by generation of GlyT1 and GlyT2 knockout mice, respectively. GlyT1-/- mice can only live for a very short period of time (hours) after birth (Gomeza et al., 2003a). These mice display severe motor-sensory deficits characterized by lethargy, hypotonia, and hyporesponsivity to tactile stimuli. Dysfunction of motor activity extends to the respiratory system, in which rhythmic breathing is severely depressed (Betz et al., 2006). At the cellular level, increased chloride conductances, consistent with a tonic activation of glycine receptors by elevated extracellular glycine concentrations, were observed. Furthermore, spontaneous inhibitory postsynaptic currents (IPSCs) had longer decay time constants than those in wild-type mice (Gomeza et al., 2003a). These electrophysiological changes suggest that GlyT1 has a crucial role in lowering extracellular glycine levels at glycinergic synapses. In contrast, glycinergic IPSCs recorded from neurons of GlyT2-deficient mice displayed markedly reduced amplitudes compared with those from wild-type mice, suggesting insufficient release of glycine from presynaptic neurons (Gomeza et al., 2003b). This reflects reduced glycine content in presynaptic vesicles, which may result from inefficient uptake of released glycine from synaptic cleft and subsequent refilling of the synaptic vesicles due to the deletion of GlyT2. Apparently, GlyT1 and GlyT2 has complementary roles in glycinergic synapse: GlyT1 is mainly for removing released glycine, and therefore terminates the glycinergic neurotransmission, but GlyT2 enhances the efficacy of glycinergic neurotransmission by increasing glycine content in synaptic vesicles through the reuptake of glycine back to presynaptic cytosol (Betz et al., 2006). GlyTs, mainly GlyT1 also modulate NMDA receptor mediated glutamatergic neurotransmission by regulating the concentration of local glycine spilled to glutamatergic synapse from neighboring glycinergic synapse (Eulenburg et al., 2005). Inhibition of GlyT1 caused an increase in the extracellular glycine concentration and thereby a facilitation of NMDA receptor currents, resulting in enhanced long-term potentiation. Moreover, GlyT1 helps stabilize the membrane expression of NMDA receptor by reducing the extracellular glycine concentration (Eulenburg et al., 2005), as high concentration of glycine was found to induce internalization of NMDA receptor (Nong et al., 2003).

4.3 Concluding Remarks

In summary, astrocyte neurotransmitter transporters are crucial components of synaptic neurotransmission in the CNS for rapid and precise processing of information. Over the past 15 years there have been important advances in the understanding of neurotransmitter transporters, especially as they relate to astroglia, including cloning of transporter genes, topology and the structure of transporters, distribution, and physiological function. These advances provide a greater opportunity to appreciate their pathogenic role in various neurological diseases or disorders. What remains unclear, though, is the regulation of these transporters at different levels, especially how the astroglial transporters' expression or activity is affected by neuronal signaling in normal physiological conditions. Neuronal influences on astrocyte transporters have been suggested in some studies, but the mechanisms are unclear. Understanding normal astroglial transporter regulatory mechanisms would eventually help understand the dysregulation of these transporters in diseases, and therefore aid in the ultimate development of astroglial- or transporter-based effective neuroprotective strategies.

References

Anderson CM, Swanson RA (2000) Astrocyte glutamate transport: review of properties, regulation, and physiological functions. *Glia* 32:1–14.

Aragon C, Lopez-Corcuera B (2003) Structure, function and regulation of glycine neurotransporters. *Eur J Pharmacol* 479:249–262.

Arriza JL, Eliasof S, Kavanaugh MP, Amara SG (1997) Excitatory amino acid transporter 5, a retinal glutamate transporter coupled to a chloride conductance. *Proc Natl Acad Sci USA* 94:4155–4160.

Arundine M, Tymianski M (2003) Molecular mechanisms of calcium-dependent neurodegeneration in excitotoxicity. *Cell Calcium* 34:325–337.

Arzberger T, Krampfl K, Leimgruber S, Weindl A (1997) Changes of NMDA receptor subunit (NR1, NR2B) and glutamate transporter (GLT1) mRNA expression in Huntington"s disease – an in situ hybridization study. *J Neuropathol Exp Neurol* 56:440–454.

Asada H, Kawamura Y, Maruyama K, Kume H, Ding RG, Kanbara N, Kuzume H, Sanbo M, Yagi T, Obata K (1997) Cleft palate and decreased brain γ-aminobutyric acid in mice lacking the 67-kDa isoform of glutamic acid decarboxylase. *Proc Natl Acad Sci USA* 94:6496–6499.

Baude A, Nusser Z, Roberts JD, Mulvihill E, McIlhinney RA, Somogyi P (1993) The metabotropic glutamate receptor (mGluR1β) is concentrated at perisynaptic membrane of neuronal subpopulations as detected by immunogold reaction. *Neuron* 11:771–787.

Beart PM, O"Shea RD (2007) Transporters for l-glutamate: an update on their molecular pharmacology and pathological involvement. *Br J Pharmacol* 150:5–17.

Berger AJ, Dieudonne S, Ascher P (1998) Glycine uptake governs glycine site occupancy at NMDA receptors of excitatory synapses. *J Neurophysiol* 80:3336–3340.

Berger UV, Hediger MA (1998) Comparative analysis of glutamate transporter expression in rat brain using differential double in situ hybridization. *Anat Embryol* 198:13–30.

Berger UV, Hediger MA (2000) Distribution of the glutamate transporters GLAST and GLT-1 in rat circumventricular organs, meninges, and dorsal root ganglia. *J Comp Neurol* 421:385–399.

Berger UV, DeSilva TM, Chen W, Rosenberg PA (2005) Cellular and subcellular mRNA localization of glutamate transporter isoforms GLT1a and GLT1b in rat brain by in situ hybridization. *J Comp Neurol* 492:78–89.

Bergles DE, Jahr CE (1997) Synaptic activation of glutamate transporters in hippocampal astrocytes. *Neuron* 19:1297–1308.

Bergles DE, Jahr CE (1998) Glial contribution to glutamate uptake at Schaffer collateral–commissural synapses in the hippocampus. *J Neurosci* 18:7709–7716.

Bernabe A, Mendez JA, Hernandez-Kelly LC, Ortega A (2003) Regulation of the Na+-dependent glutamate/aspartate transporter in rodent cerebellar astrocytes. *Neurochem Res* 28:1843–1849.

Betz H, Gomeza J, Armsen W, Scholze P, Eulenburg V (2006) Glycine transporters: essential regulators of synaptic transmission. *Biochem Soc Trans* 34:55–58.

Borden LA, Smith KE, Hartig PR, Branchek TA, Weinshank RL (1992) Molecular heterogeneity of the -aminobutyric acid (GABA) transport system. *Cloning of two novel high affinity GABA transporters from rat brain. J Biol Chem* 267:21098–21104.

Brasnjo G, Otis TS (2001) Neuronal glutamate transporters control activation of postsynaptic metabotropic glutamate receptors and influence cerebellar long-term depression. *Neuron* 31:607–616.

Bristol LA, Rothstein JD (1996) Glutamate transporter gene expression in amyotrophic lateral sclerosis motor cortex. *Ann Neurol* 39:676–679.

Bruijn LI, Becher MW, Lee MK, Anderson KL, Jenkins NA, Copeland NG, Sisodia SS, Rothstein JD, Borchelt DR, Price DL, Cleveland DW (1997) ALS-linked SOD1 mutant G85R mediates damage to astrocytes and promotes rapidly progressive disease with SOD1-containing inclusions. *Neuron* 18:327–338.

Carter RJ, Lione LA, Humby T, Mangiarini L, Mahal A, Bates GP, Dunnett SB, Morton AJ (1999) Characterization of progressive motor deficits in mice transgenic for the human Huntington"s disease mutation. *J Neurosci* 19:3248–3257.

Chaudhry FA, Lehre KP, van Lookeren Campagne M, Ottersen OP, Danbolt NC, Storm-Mathisen J (1995) Glutamate transporters in glial plasma membranes: highly differentiated localizations revealed by quantitative ultrastructural immunocytochemistry. *Neuron* 15:711–720.

Chen NH, Reith ME, Quick MW (2004a) Synaptic uptake and beyond: the sodium- and chloride-dependent neurotransmitter transporter family SLC6. *Pflugers Arch* 447:519–531.

Chen S, Diamond JS (2002) Synaptically released glutamate activates extrasynaptic NMDA receptors on cells in the ganglion cell layer of rat retina. *J Neurosci* 22:2165–2173.

Chen W, Aoki C, Mahadomrongkul V, Gruber CE, Wang GJ, Blitzblau R, Irwin N, Rosenberg PA (2002) Expression of a variant form of the glutamate transporter GLT1 in neuronal cultures and in neurons and astrocytes in the rat brain. *J Neurosci* 22:2142–2152.

Chen W, Mahadomrongkul V, Berger UV, Bassan M, DeSilva T, Tanaka K, Irwin N, Aoki C, Rosenberg PA (2004b) The glutamate transporter GLT1a is expressed in excitatory axon terminals of mature hippocampal neurons. *J Neurosci* 24:1136–1148.

Choi DW (1988) Glutamate neurotoxicity and diseases of the nervous system. *Neuron* 1:623–634.

Choi DW (1992) Excitotoxic cell death. *J Neurobiol* 23:1261–1276.

Clark BA, Cull-Candy SG (2002) Activity-dependent recruitment of extrasynaptic NMDA receptor activation at an AMPA receptor-only synapse. *J Neurosci* 22:4428–4436.

Cubelos B, Gimenez C, Zafra F (2005) Localization of the GLYT1 glycine transporter at glutamatergic synapses in the rat brain. *Cereb Cortex* 15:448–459.

Danbolt NC (2001) Glutamate uptake. *Prog Neurobiol* 65:1–105.

Davies SW, Turmaine M, Cozens BA, DiFiglia M, Sharp AH, Ross CA, Scherzinger E, Wanker EE, Mangiarini L, Bates GP (1997) Formation of neuronal intranuclear inclusions underlies the neurological dysfunction in mice transgenic for the HD mutation. *Cell* 90:537–548.

Deneen B, Ho R, Lukaszewicz A, Hochstim CJ, Gronostajski RM, Anderson DJ (2006) The transcription factor NFIA controls the onset of gliogenesis in the developing spinal cord. *Neuron* 52:953–968.

Diamond JS, Jahr CE (1997) Transporters buffer synaptically released glutamate on a submillisecond time scale. *J Neurosci* 17:4672–4687.

Domercq M, Sanchez-Gomez MV, Areso P, Matute C (1999) Expression of glutamate transporters in rat optic nerve oligodendrocytes. *Eur J Neurosci* 11:2226–2236.

Drejer J, Meier E, Schousboe A (1983) Novel neuron-related regulatory mechanisms for astrocytic glutamate and GABA high affinity uptake. *Neurosci Lett* 37:301–306.

During MJ, Spencer DD (1993) Extracellular hippocampal glutamate and spontaneous seizure in the conscious human brain. *Lancet* 341:1607–1610.

During MJ, Ryder KM, Spencer DD (1995) Hippocampal GABA transporter function in temporal-lobe epilepsy. *Nature* 376:174–177.

Eiraku M, Tohgo A, Ono K, Kaneko M, Fujishima K, Hirano T, Kengaku M (2005) DNER acts as a neuron-specific Notch ligand during Bergmann glial development. *Nat Neurosci* 8:873–880.

Eliasof S, Arriza JL, Leighton BH, Amara SG, Kavanaugh MP (1998a) Localization and function of five glutamate transporters cloned from the salamander retina. *Vision Res* 38:1443–1454.

Eliasof S, Arriza JL, Leighton BH, Kavanaugh MP, Amara SG (1998b) Excitatory amino acid transporters of the salamander retina: identification, localization, and function. *J Neurosci* 18:698–712.

Eng DL, Lee YL, Lal PG (1997) Expression of glutamate uptake transporters after dibutyryl cyclic AMP differentiation and traumatic injury in cultured astrocytes. *Brain Res* 778:215–221.

Erlander MG, Tobin AJ (1991) The structural and functional heterogeneity of glutamic acid decarboxylase: a review. *Neurochem research* 16:215–226.

Eulenburg V, Armsen W, Betz H, Gomeza J (2005) Glycine transporters: essential regulators of neurotransmission. *Trends Biochem Sci* 30:325–333.

Fairman WA, Vandenberg RJ, Arriza JL, Kavanaugh MP, Amara SG (1995) An excitatory amino-acid transporter with properties of a ligand-gated chloride channel. *Nature* 375:599–603.

Figiel M, Maucher T, Rozyczka J, Bayatti N, Engele J (2003) Regulation of glial glutamate transporter expression by growth factors. *Exp Neurol* 183:124–135.

Figiel M, Allritz C, Lehmann C, Engele J (2007) Gap junctional control of glial glutamate transporter expression. *Mol Cell Neurosci* 35:130–137.

Flowers JM, Powell JF, Leigh PN, Andersen P, Shaw CE (2001) Intron 7 retention and exon 9 skipping EAAT2 mRNA variants are not associated with amyotrophic lateral sclerosis. *Ann Neurol* 49:643–649.

Furness DN, Lehre KP (1997) Immunocytochemical localization of a high-affinity glutamate–aspartate transporter, GLAST, in the rat and guinea-pig cochlea. *The Eur J Neurosci* 9:1961–1969.

Furuta A, Rothstein JD, Martin LJ (1997) Glutamate transporter protein subtypes are expressed differentially during rat CNS development. *J Neurosci* 17:8363–8375.

Gegelashvili G, Civenni G, Racagni G, Danbolt NC, Schousboe I, Schousboe A (1996) Glutamate receptor agonists up-regulate glutamate transporter GLAST in astrocytes. *Neuroreport* 8:261–265.

Gegelashvili G, Dehnes Y, Danbolt NC, Schousboe A (2000) The high-affinity glutamate transporters GLT1, GLAST, and EAAT4 are regulated via different signalling mechanisms. *Neurochem Int* 37:163–170.

Gegelashvili G, Robinson MB, Trotti D, Rauen T (2001) Regulation of glutamate transporters in health and disease. *Prog Brain Res* 132:267–286.

Genoud C, Quairiaux C, Steiner P, Hirling H, Welker E, Knott GW (2006) Plasticity of astrocytic coverage and glutamate transporter expression in adult mouse cortex. *PLoS Biol* 4:e343.

Ginsberg SD, Martin LJ, Rothstein JD (1995) Regional deafferentation down-regulates subtypes of glutamate transporter proteins. *J Neurochem* 65:2800–2803.

Ginsberg SD, Rothstein JD, Price DL, Martin LJ (1996) Fimbria-fornix transections selectively down-regulate subtypes of glutamate transporter and glutamate receptor proteins in septum and hippocampus. *J Neurochem* 67:1208–1216.

Gomeza J, Hulsmann S, Ohno K, Eulenburg V, Szoke K, Richter D, Betz H (2003a) Inactivation of the glycine transporter 1 gene discloses vital role of glial glycine uptake in glycinergic inhibition. *Neuron* 40:785–796.

Gomeza J, Ohno K, Hulsmann S, Armsen W, Eulenburg V, Richter DW, Laube B, Betz H (2003b) Deletion of the mouse glycine transporter 2 results in a hyperekplexia phenotype and postnatal lethality. *Neuron* 40:797–806.

Gonzalez MI, Lopez-Colom AM, Ortega A (1999) Sodium-dependent glutamate transport in Muller glial cells: regulation by phorbol esters. *Brain Res* 831:140–145.

Grewer C, Rauen T (2005) Electrogenic glutamate transporters in the CNS: molecular mechanism, pre-steady-state kinetics, and their impact on synaptic signaling. *J Membr Biol* 203:1–20.

Guastella J, Nelson N, Nelson H, Czyzyk L, Keynan S, Miedel MC, Davidson N, Lester HA, Kanner BI (1990) Cloning and expression of a rat brain GABA transporter. *Science* 249:1303–1306.

Guo H, Lai L, Butchbach ME, Lin CL (2002) Human glioma cells and undifferentiated primary astrocytes that express aberrant EAAT2 mRNA inhibit normal EAAT2 protein expression and prevent cell death. *Mol Cell Neurosci* 21:546–560.

Guo H, Lai L, Butchbach ME, Stockinger MP, Shan X, Bishop GA, Lin CL (2003) Increased expression of the glial glutamate transporter EAAT2 modulates excitotoxicity and delays the onset but not the outcome of ALS in mice. *Hum Mol Genet* 12:2519–2532.

Hamberger A, Nystrom B (1984) Extra- and intracellular amino acids in the hippocampus during development of hepatic encephalopathy. *Neurochem Res* 9:1181–1192.
Haugeto O, Ullensvang K, Levy LM, Chaudhry FA, Honore T, Nielsen M, Lehre KP, Danbolt NC (1996) Brain glutamate transporter proteins form homomultimers. *J Biol Chem* 271:27715–27722.
Hazell AS, Itzhak Y, Liu H, Norenberg MD (1997) 1-Methyl-4-phenyl-1,2,3,6-tetrahydropyridine (MPTP) decreases glutamate uptake in cultured astrocytes. *J Neurochem* 68:2216–2219.
He Y, Hof PR, Janssen WG, Rothstein JD, Morrison JH (2001) Differential synaptic localization of GluR2 and EAAC1 in the macaque monkey entorhinal cortex: a postembedding immunogold study. *Neurosci Lett* 311:161–164.
Headley PM, Grillner S (1990) Excitatory amino acids and synaptic transmission: the evidence for a physiological function. *Trends Pharmacol Sci* 11:205–211.
Heintz N (2001) BAC to the future: the use of bac transgenic mice for neuroscience research. *Nat Rev Neurosci* 2:861–870.
Herman MA, Jahr CE (2007) Extracellular glutamate concentration in hippocampal slice. *J Neurosci* 27:9736–9741.
Hertz L, Dringen R, Schousboe A, Robinson SR (1999) Astrocytes: glutamate producers for neurons. *J Neurosci Res* 57:417–428.
Honig LS, Chambliss DD, Bigio EH, Carroll SL, Elliott JL (2000) Glutamate transporter EAAT2 splice variants occur not only in ALS, but also in AD and controls. *Neurology* 55:1082–1088.
Huang YH, Bergles DE (2004) Glutamate transporters bring competition to the synapse. *Curr Opin Neurobiol* 14:346–352.
Huang YH, Sinha SR, Tanaka K, Rothstein JD, Bergles DE (2004) Astrocyte glutamate transporters regulate metabotropic glutamate receptor-mediated excitation of hippocampal interneurons. *J Neurosci* 24:4551–4559.
Hubner CA, Stein V, Hermans-Borgmeyer I, Meyer T, Ballanyi K, Jentsch TJ (2001) Disruption of KCC2 reveals an essential role of K-Cl cotransport already in early synaptic inhibition. *Neuron* 30:515–524.
Jacob CP, Koutsilieri E, Bartl J, Neuen-Jacob E, Arzberger T, Zander N, Ravid R, Roggendorf W, Riederer P, Grunblatt E (2007) Alterations in expression of glutamatergic transporters and receptors in sporadic Alzheimer"s disease. *J Alzheimers Dis* 11:97–116.
Ji F, Obata K (1999) Development of the GABA system in organotypic culture of hippocampal and cerebellar slices from a 67-kDa isoform of glutamic acid decarboxylase (GAD67)-deficient mice. *Neurosci Res* 33:233–237.
Johnson JW, Ascher P (1987) Glycine potentiates the NMDA response in cultured mouse brain neurons. *Nature* 325:529–531.
Jursky F, Nelson N (1995) Localization of glycine neurotransmitter transporter (GLYT2) reveals correlation with the distribution of glycine receptor. *J Neurochem* 64:1026–1033.
Jursky F, Nelson N (1996) Developmental expression of the glycine transporters GLYT1 and GLYT2 in mouse brain. *J Neurochem* 67:336–344.
Kalandadze A, Wu Y, Robinson MB (2002) Protein kinase C activation decreases cell surface expression of the GLT-1 subtype of glutamate transporter. Requirement of a carboxyl-terminal domain and partial dependence on serine 486. *J Biol Chem* 277:45741–45750.
Kalandadze A, Wu Y, Fournier K, Robinson MB (2004) Identification of motifs involved in endoplasmic reticulum retention-forward trafficking of the GLT-1 subtype of glutamate transporter. *J Neurosci* 24:5183–5192.
Kam K, Nicoll R (2007) Excitatory synaptic transmission persists independently of glutamate–glutamine cycle. *J Neurosci* 27: 9192–9200.
Kanai Y, Hediger MA (1992) Primary structure and functional characterization of a high-affinity glutamate transporter. *Nature* 360:467–471.
Kanai Y, Hediger MA (2004) The glutamate/neutral amino acid transporter family SLC1: molecular, physiological and pharmacological aspects. *Pflugers Arch* 447:469–479.

Kapetanovic IM, Yonekawa WD, Kupferberg HJ (1993) Time-related loss of glutamine from hippocampal slices and concomitant changes in neurotransmitter amino acids. *J Neurochem* 61:865–872.

Kim SY, Choi SY, Chao W, Volsky DJ (2003) Transcriptional regulation of human excitatory amino acid transporter 1 (EAAT1): cloning of the EAAT1 promoter and characterization of its basal and inducible activity in human astrocytes. *J Neurochem* 87:1485–1498.

Kojima S, Nakamura T, Nidaira T, Nakamura K, Ooashi N, Ito E, Watase K, Tanaka K, Wada K, Kudo Y, Miyakawa H (1999) Optical detection of synaptically induced glutamate transport in hippocampal slices. *J Neurosci* 19:2580–2588.

Lauriat TL, Richler E, McInnes LA (2007) A quantitative regional expression profile of EAAT2 known and novel splice variants reopens the question of aberrant EAAT2 splicing in disease. *Neurochem Int* 50:271–280.

Lehre KP, Danbolt NC (1998) The number of glutamate transporter subtype molecules at glutamatergic synapses: chemical and stereological quantification in young adult rat brain. *J Neurosci* 18:8751–8757.

Lehre KP, Levy LM, Ottersen OP, Storm-Mathisen J, Danbolt NC (1995) Differential expression of two glial glutamate transporters in the rat brain: quantitative and immunocytochemical observations. *J Neurosci* 15:1835–1853.

Levy LM, Warr O, Attwell D (1998) Stoichiometry of the glial glutamate transporter GLT-1 expressed inducibly in a Chinese hamster ovary cell line selected for low endogenous Na+-dependent glutamate uptake. *J Neurosci* 18:9620–9628.

Lievens JC, Woodman B, Mahal A, Spasic-Boscovic O, Samuel D, Kerkerian-Le Goff L, Bates GP (2001) Impaired glutamate uptake in the R6 Huntington"s disease transgenic mice. *Neurobiol Dis* 8:807–821.

Lin CL, Bristol LA, Jin L, Dykes-Hoberg M, Crawford T, Clawson L, Rothstein JD (1998) Aberrant RNA processing in a neurodegenerative disease: the cause for absent EAAT2, a glutamate transporter, in amyotrophic lateral sclerosis. *Neuron* 20:589–602.

Liu QR, Lopez-Corcuera B, Mandiyan S, Nelson H, Nelson N (1993) Molecular characterization of four pharmacologically distinct -aminobutyric acid transporters in mouse brain [corrected]. *J Biol Chem* 268:2106–2112.

Lopez-Redondo F, Nakajima K, Honda S, Kohsaka S (2000) Glutamate transporter GLT-1 is highly expressed in activated microglia following facial nerve axotomy. *Brain Res* 76:429–435.

Macnab LT, Pow DV (2007) Central nervous system expression of the exon 9 skipping form of the glutamate transporter GLAST. *Neuroreport* 18:741–745.

Macnab LT, Williams SM, Pow DV (2006) Expression of the exon 3 skipping form of GLAST, GLAST1a, in brain and retina. *Neuroreport* 17:1867–1870.

Mahendrasingam S, Wallam CA, Hackney CM (2003) Two approaches to double post-embedding immunogold labeling of freeze-substituted tissue embedded in low temperature Lowicryl HM20 resin. *Brain Res Brain Res Protoc* 11:134–141.

Mallolas J, Hurtado O, Castellanos M, Blanco M, Sobrino T, Serena J, Vivancos J, Castillo J, Lizasoain I, Moro MA, Davalos A (2006) A polymorphism in the EAAT2 promoter is associated with higher glutamate concentrations and higher frequency of progressing stroke. *J Exp Med* 203:711–717.

Mangiarini L, Sathasivam K, Mahal A, Mott R, Seller M, Bates GP (1997) Instability of highly expanded CAG repeats in mice transgenic for the Huntington"s disease mutation. *Nat Genet* 15:197–200.

Maragakis NJ, Rothstein JD (2004) Glutamate transporters: animal models to neurologic disease. *Neurobiol Dis* 15:461–473.

Martin LJ, Brambrink AM, Lehmann C, Portera-Cailliau C, Koehler R, Rothstein J, Traystman RJ (1997) Hypoxia–ischemia causes abnormalities in glutamate transporters and death of astroglia and neurons in newborn striatum. *Ann Neurol* 42:335–348.

Masliah E, Alford M, DeTeresa R, Mallory M, Hansen L (1996) Deficient glutamate transport is associated with neurodegeneration in Alzheimer"s disease. *Ann Neurol* 40:759–766.

Masliah E, Alford M, Mallory M, Rockenstein E, Moechars D, Van Leuven F (2000) Abnormal glutamate transport function in mutant amyloid precursor protein transgenic mice. *Exp Neurol* 163:381–387.

Matsugami TR, Tanemura K, Mieda M, Nakatomi R, Yamada K, Kondo T, Ogawa M, Obata K, Watanabe M, Hashikawa T, Tanaka K (2006) From the cover: indispensability of the glutamate transporters GLAST and GLT1 to brain development. *Proc Natl Acad Sci USA* 103:12161–12166.

McKenna MC, Tildon JT, Stevenson JH, Huang X (1996) New insights into the compartmentation of glutamate and glutamine in cultured rat brain astrocytes. *Dev Neurosci* 18:380–390.

McNaught KS, Jenner P (1999) Altered glial function causes neuronal death and increases neuronal susceptibility to 1-methyl-4-phenylpyridinium- and 6-hydroxydopamine-induced toxicity in astrocytic/ventral mesencephalic co-cultures. *J Neurochem* 73:2469–2476.

Munch C, Ebstein M, Seefried U, Zhu B, Stamm S, Landwehrmeyer GB, Ludolph AC, Schwalenstocker B, Meyer T (2002) Alternative splicing of the 5 -sequences of the mouse EAAT2 glutamate transporter and expression in a transgenic model for amyotrophic lateral sclerosis. *J Neurochem* 82:594–603.

Nagai M, Abe K, Okamoto K, Itoyama Y (1998) Identification of alternative splicing forms of GLT-1 mRNA in the spinal cord of amyotrophic lateral sclerosis patients. *Neurosci Lett* 244:165–168.

Nelson H, Mandiyan S, Nelson N (1990) Cloning of the human brain GABA transporter. *FEBS Lett* 269:181–184.

Nong Y, Huang YQ, Ju W, Kalia LV, Ahmadian G, Wang YT, Salter MW (2003) Glycine binding primes NMDA receptor internalization. *Nature* 422:302–307.

Pawlak J, Brito V, Kuppers E, Beyer C (2005) Regulation of glutamate transporter GLAST and GLT-1 expression in astrocytes by estrogen. *Brain Res* 138:1–7.

Petroff OA (2002) GABA and glutamate in the human brain. *Neuroscientist* 8:562–573.

Pines G, Danbolt NC, Bjoras M, Zhang Y, Bendahan A, Eide L, Koepsell H, Storm-Mathisen J, Seeberg E, Kanner BI (1992) Cloning and expression of a rat brain l-glutamate transporter. *Nature* 360:464–467.

Plachez C, Danbolt NC, Recasens M (2000) Transient expression of the glial glutamate transporters GLAST and GLT in hippocampal neurons in primary culture. *J Neurosci Res* 59:587–593.

Poitry-Yamate CL, Vutskits L, Rauen T (2002) Neuronal-induced and glutamate-dependent activation of glial glutamate transporter function. *J Neurochem* 82:987–997.

Pow DV, Hendrickson AE (1999) Distribution of the glycine transporter glyt-1 in mammalian and nonmammalian retinae. *Vis Neurosci* 16:231–239.

Proper EA, Hoogland G, Kappen SM, Jansen GH, Rensen MG, Schrama LH, van Veelen CW, van Rijen PC, van Nieuwenhuizen O, Gispen WH, de Graan PN (2002) Distribution of glutamate transporters in the hippocampus of patients with pharmaco-resistant temporal lobe epilepsy. *Brain* 125:32–43.

Rauen T (2000) Diversity of glutamate transporter expression and function in the mammalian retina. *Amino Acids* 19:53–62.

Rauen T, Fischer F, Wiessner M (1999) Glia–neuron interaction by high-affinity glutamate transporters in neurotransmission. *Adv Exp Med Biol* 468:81–95.

Rauen T, Wiessner M, Sullivan R, Lee A, Pow DV (2004) A new GLT1 splice variant: cloning and immunolocalization of GLT1c in the mammalian retina and brain. *Neurochem Int* 45:1095–1106.

Regan MR, Huang YH, Kim YS, Dykes-Hoberg MI, Jin L, Watkins AM, Bergles DE, Rothstein JD (2007) Variations in promoter activity reveal a differential expression and physiology of glutamate transporters by glia in the developing and mature CNS. *J Neurosci* 27:6607–6619.

Reichelt W, Knopfel T (2002) Glutamate uptake controls expression of a slow postsynaptic current mediated by mGluRs in cerebellar Purkinje cells. *J Neurophysiol* 87:1974–1980.

Robinson MB (2006) Acute regulation of sodium-dependent glutamate transporters: a focus on constitutive and regulated trafficking. Handb Exp Pharmacol:251–275.

Rosas S, Vargas MA, Lopez-Bayghen E, Ortega A (2007) Glutamate-dependent transcriptional regulation of GLAST/EAAT1: a role for YY1. *J Neurochem* 101:1134–1144.
Rose CR, Ransom BR (1996a) Mechanisms of H+ and Na+ changes induced by glutamate, kainate, and d-aspartate in rat hippocampal astrocytes. *J Neurosci* 16:5393–5404.
Rose CR, Ransom BR (1996b) Intracellular sodium homeostasis in rat hippocampal astrocytes. *J Physiol* 491 (Pt 2):291–305.
Rosen DR, Siddique T, Patterson D, Figlewicz DA, Sapp P, Hentati A, Donaldson D, Goto J, O"Regan JP, Deng HX, (1993) Mutations in Cu/Zn superoxide dismutase gene are associated with familial amyotrophic lateral sclerosis. *Nature* 362:59–62.
Rothstein JD, Martin LJ, Kuncl RW (1992) Decreased glutamate transport by the brain and spinal cord in amyotrophic lateral sclerosis. *N Engl J Med* 326:1464–1468.
Rothstein JD, Martin L, Levey AI, Dykes-Hoberg M, Jin L, Wu D, Nash N, Kuncl RW (1994) Localization of neuronal and glial glutamate transporters. *Neuron* 13:713–725.
Rothstein JD, Van Kammen M, Levey AI, Martin LJ, Kuncl RW (1995) Selective loss of glial glutamate transporter GLT-1 in amyotrophic lateral sclerosis. *Ann Neurol* 38:73–84.
Rothstein JD, Dykes-Hoberg M, Pardo CA, Bristol LA, Jin L, Kuncl RW, Kanai Y, Hediger MA, Wang Y, Schielke JP, Welty DF (1996) Knockout of glutamate transporters reveals a major role for astroglial transport in excitotoxicity and clearance of glutamate. *Neuron* 16:675–686.
Rothstein JD, Patel S, Regan MR, Haenggeli C, Huang YH, Bergles DE, Jin L, Dykes Hoberg M, Vidensky S, Chung DS, Toan SV, Bruijn LI, Su ZZ, Gupta P, Fisher PB (2005) -lactam antibiotics offer neuroprotection by increasing glutamate transporter expression. *Nature* 433:73–77.
Rozyczka J, Engele J (2005) Multiple 5′-splice variants of the rat glutamate transporter-1. *Brain Res* 133:157–161.
Rudolph U, Mohler H (2006) GABA-based therapeutic approaches: GABAA receptor subtype functions. *Curr Opin Pharmacol* 6:18–23.
Sattler R, Tymianski M (2001) Molecular mechanisms of glutamate receptor-mediated excitotoxic neuronal cell death. *Mol Neurobiol* 24:107–129.
Sattler R, Rothstein JD (2006) Regulation and dysregulation of glutamate transporters. Handb Exp Pharmacol:277–303.
Schlag BD, Vondrasek JR, Munir M, Kalandadze A, Zelenaia OA, Rothstein JD, Robinson MB (1998) Regulation of the glial Na+-dependent glutamate transporters by cyclic AMP analogs and neurons. *Mol Pharmacol* 53:355–369.
Schmitt A, Asan E, Puschel B, Jons T, Kugler P (1996) Expression of the glutamate transporter GLT1 in neural cells of the rat central nervous system: non-radioactive in situ hybridization and comparative immunocytochemistry. *Neuroscience* 71:989–1004.
Schousboe A (2000) Pharmacological and functional characterization of astrocytic GABA transport: a short review. *Neurochem Res* 25:1241–1244.
Schousboe A (2003) Role of astrocytes in the maintenance and modulation of glutamatergic and GABAergic neurotransmission. *Neurochem Res* 28:347–352.
Schousboe A, Hertz L (1981) Role of astroglial cells in glutamate homeostasis. *Adv Biochem Psychopharmacol* 27:103–113.
Scott HL, Pow DV, Tannenberg AE, Dodd PR (2002) Aberrant expression of the glutamate transporter excitatory amino acid transporter 1 (EAAT1) in Alzheimer"s disease. *J Neurosci* 22:RC206.
Seal RP, Amara SG (1999) Excitatory amino acid transporters: a family in flux. *Annu Rev Pharmacol Toxicol* 39:431–456.
Sheldon AL, Robinson MB (2007) The role of glutamate transporters in neurodegenerative diseases and potential opportunities for intervention. *Neurochem Int* 551:333–355.
Shin JY, Fang ZH, Yu ZX, Wang CE, Li SH, Li XJ (2005) Expression of mutant huntingtin in glial cells contributes to neuronal excitotoxicity. *J Cell Biol* 171:1001–1012.
Sitcheran R, Gupta P, Fisher PB, Baldwin AS (2005) Positive and negative regulation of EAAT2 by NF- B: a role for N-myc in TNF -controlled repression. *EMBO J* 24:510–520.
Soghomonian JJ, Martin DL (1998) Two isoforms of glutamate decarboxylase: why? *Trends Pharmacol Sci* 19:500–505.

Sonders MS, Amara SG (1996) Channels in transporters. *Curr Opin Neurobiol* 6:294–302.

Storck T, Schulte S, Hofmann K, Stoffel W (1992) Structure, expression, and functional analysis of a Na(+)-dependent glutamate/aspartate transporter from rat brain. *Proc Natl Acad Sci USA* 89:10955–10959.

Su ZZ, Leszczyniecka M, Kang DC, Sarkar D, Chao W, Volsky DJ, Fisher PB (2003) Insights into glutamate transport regulation in human astrocytes: cloning of the promoter for excitatory amino acid transporter 2 (EAAT2). *Proc Natl Acad Sci USA* 100:1955–1960.

Sudhof TC (1995) The synaptic vesicle cycle: a cascade of protein–protein interactions. *Nature* 375:645–653.

Sutherland ML, Delaney TA, Noebels JL (1996) Glutamate transporter mRNA expression in proliferative zones of the developing and adult murine CNS. *J Neurosci* 16:2191–2207.

Suzdak PD, Jansen JA (1995) A review of the preclinical pharmacology of tiagabine: a potent and selective anticonvulsant GABA uptake inhibitor. *Epilepsia* 36:612–626.

Swanson RA, Liu J, Miller JW, Rothstein JD, Farrell K, Stein BA, Longuemare MC (1997) Neuronal regulation of glutamate transporter subtype expression in astrocytes. *J Neurosci* 17:932–940.

Takamori S, Rhee JS, Rosenmund C, Jahn R (2000) Identification of a vesicular glutamate transporter that defines a glutamatergic phenotype in neurons. *Nature* 407:189–194.

Tanaka K (1993) Cloning and expression of a glutamate transporter from mouse brain. *Neurosci Lett* 159:183–186.

Tanaka K (1997) Epilepsy and exacerbation of brain injury in mice lacking glutamate transporter GLT-1. *Science* 276:1699 (278:21).

Tao F, Lu SD, Zhang LM, Huang YL, Sun FY (2001) Role of excitatory amino acid transporter 1 in neonatal rat neuronal damage induced by hypoxia–ischemia. *Neuroscience* 102:503–513.

Thai DR (2002) Excitatory amino acid transporter EAAT-2 in tangle-bearing neurons in Alzheimer"s disease. *Brain Pathol* 12:405–411.

Tian G, Lai L, Guo H, Lin Y, Butchbach ME, Chang Y, Lin CL (2007) Translational control of glial glutamate transporter EAAT2 expression. *J Biol Chem* 282:1727–1737.

Tong G, Jahr CE (1994) Block of glutamate transporters potentiates postsynaptic excitation. *Neuron* 13:1195–1203.

Torres GE, Amara SG (2007) Glutamate and monoamine transporters: new visions of form and function. *Curr Opin Neurobiol* 17:304–312.

Torres GE, Gainetdinov RR, Caron MG (2003) Plasma membrane monoamine transporters: structure, regulation and function. *Nat Rev Neurosci* 4:13–25.

Trotti D, Rolfs A, Danbolt NC, Brown RH, Jr., Hediger MA (1999) SOD1 mutants linked to amyotrophic lateral sclerosis selectively inactivate a glial glutamate transporter. *Nat Neurosci* 2:427–433.

Vandenberg RJ (2006) Mutational analysis of glutamate transporters. Handb Exp Pharmacol:113–135.

Ventura R, Harris KM (1999) Three-dimensional relationships between hippocampal synapses and astrocytes. *J Neurosci* 19:6897–6906.

Wadiche JI, Amara SG, Kavanaugh MP (1995) Ion fluxes associated with excitatory amino acid transport. *Neuron* 15:721–728.

Watanabe T, Morimoto K, Hirao T, Suwaki H, Watase K, Tanaka K (1999) Amygdala-kindled and pentylenetetrazole-induced seizures in glutamate transporter GLAST-deficient mice. *Brain Res* 845:92–96.

Westergaard N, Sonnewald U, Schousboe A (1995) Metabolic trafficking between neurons and astrocytes: the glutamate/glutamine cycle revisited. *Dev Neurosci* 17:203–211.

Wood JH, Hare TA, Glaeser BS, Ballenger JC, Post RM (1979) Low cerebrospinal fluid -aminobutyric acid content in seizure patients. *Neurology* 29:1203–1208.

Yamada H, Yatsushiro S, Yamamoto A, Hayashi M, Nishi T, Futai M, Yamaguchi A, Moriyama Y (1997) Functional expression of a GLT-1 type Na+-dependent glutamate transporter in rat pinealocytes. *J Neurochem* 69:1491–1498.

Yernool D, Boudker O, Jin Y, Gouaux E (2004) Structure of a glutamate transporter homologue from *Pyrococcus horikoshii*. *Nature* 431:811–818.

Zelenaia O, Schlag BD, Gochenauer GE, Ganel R, Song W, Beesley JS, Grinspan JB, Rothstein JD, Robinson MB (2000) Epidermal growth factor receptor agonists increase expression of glutamate transporter GLT-1 in astrocytes through pathways dependent on phosphatidylinositol 3-kinase and transcription factor NF- B. *Mol Pharmacol* 57:667–678.

Zerangue N, Kavanaugh MP (1996) Flux coupling in a neuronal glutamate transporter. *Nature* 383:634–637.

Zhou J, Sutherland ML (2004) Glutamate transporter cluster formation in astrocytic processes regulates glutamate uptake activity. *J Neurosci* 24:6301–6306.

Zschocke J, Bayatti N, Clement AM, Witan H, Figiel M, Engele J, Behl C (2005) Differential promotion of glutamate transporter expression and funcFigieltion by glucocorticoids in astrocytes from various brain regions. *J Biol Chem* 280:34924–34932.

Abbreviations

AD	Alzheimer'sdisease
ALS	Amyotrophic lateral sclerosis
AMPA	α-amino-3-hydroxy-5-methyl-4-isoxazole-propionic acid
APP	Amyloid β a4 precursor protein
BMP	Bone morphogenetic protein
cAMP	Cyclic adenosine monophosphate
CNS	Central nervous system
CSF	Cerebrospinal fluid
EAAC	Excitatory amino acid carrier
EAAT	Excitatory amino acid transporter
EGF	Epidermal growth factor
EM	Electron microscopy
EPSC	Excitatory postsynaptic current
GABA	γ-aminobutyric acid
GAD	Glutamic acid decarboxylase
GAT	GABA transporter
GLAST	L-glutamate/L-aspartate transporter
GLT-1	L-glutamate transporter
GlyT	Glycine transporter
HD	Huntington's disease
HP	Hairpin
LIF	Leukemia-inhibitory factor
IPSC	Inhibitory postsynaptic current
mGluR	Metabotropic glutamate receptor
NCM	Neuronal conditioned medium
NF-κB	Nuclear transcription factor κB
NMDA	N-methyl-D-aspartate
PD	Parkinson's disease
PI-3K	Phosphatidylinositol 3-kinase
PKC	Protein kinase C
SLC	Solute carrier

SOD	Superoxide dismutase
TGF-α	Transforming growth factor-α
THPO	4,5,6,7-tetrahydroisoxazolo [4,5-c]pyridin-3-ol
TM	Transmembrane helix
TMDs	Transmembrane domains
TNF-α	Tumor necrosis factor-α
UTR	Untranslated region

Chapter 5
Connexin Expression (Gap Junctions and Hemichannels) in Astrocytes

Eliana Scemes and David C. Spray

Contents

D.C. Spray
The Dominick P. Purpura Department of Neuroscience, Albert Einstein College of Medicine, Bronx, NY, USA
spray@aecom.yu.edu

V. Parpura and P.G. Haydon (eds.), *Astrocytes in (Patho)Physiology of the Nervous System*,
DOI: 10.1007/978-0-387-79492-1_5,

5.1 Gap Junction Structure

5.1.1 *Structure of the Plaque: Thin Section and Freeze-Fracture Electron Microscopy*

Until relatively recently, it was believed that many tissues were truly syncytial, without discrete cellular borders, a view supported by electophysiological measurements of long-distance passive spread of currents. However, this view was radically changed by electron microscope studies (Robertson, 1953; Sjostrand et al., 1958), demonstrating specialized junctional complexes at cell appositions in crayfish axon and mammalian heart. Dewey and Barr (1962) described the region of close contact, which they still believed to be an actual membrane fusion, as the "Nexus." Revel and Karnovsky (1967), using lanthanum as an electron opaque marker of extracellular space, showed that the membranes in these domains were actually separated by a gap of about 2 nm; moreover cross-sections of such regions indicated the presence of bridging structures that appeared to be hexagonal arrays of particles with an electron opaque 1-nm center. Freeze-fracture studies by McNutt and Weinstein (1970) confirmed the presence of particles in hexagonal arrays. The presence of a membrane separation led Revel (1968) to term this structure a "gap" junction.

Thus, it is now established that direct communication between the cytosolic compartments of two or more adjoining cells is possible because of the presence of intercellular gap junction channels. As originally shown in early studies by Brightman and Reese (1969) and Dermietzel (1974) and subsequently by numerous groups (e.g., Massa and Magniaini (1982, 1985), Nagy and Rash (2000) and Nagy et al. (2004)), gap junction channels between glial cells, as in other tissues, form aggregates at cell contacts, called gap junction plaques (Fig. 5.1a–c).

The tight packing of the gap junction channels in the plaques has facilitated studies revealing details of channel structure. For example, early application of low angle X-ray diffraction techniques to isolated liver gap junctions indicated hexagonal packing of the particles and rough estimation of particle sizes (Caspar et al., 1997; Makowksi et al., 1977). Unwin and Zampighi (1980) using electron microscopy, provided further compelling evidence that the center of the hexagon was a permeant pore (for model, see Fig. 5.1d). Unger et al. (1999) have used electron crystallography to determine the structure of gap junction channels at a resolution of 0.75 nm, thereby revealing that the hexagonal pore wall consists of 24 transmembrane α-helixes, consistent with a hexamer of proteins each with four transmembrane domains. As considered below, nuclear magnetic resonance (NMR) studies on cytoplasmic domains are beginning to provide insight into conformational changes during channel gating.

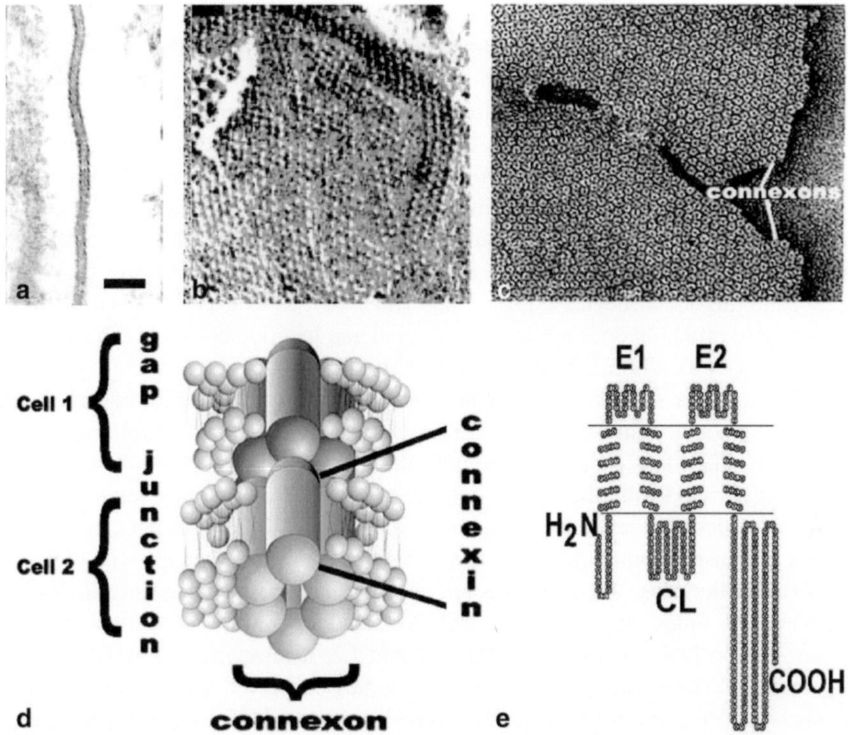

Fig. 5.1 Structure of gap junction, connexon, and connexin. (a) Thin section electron micrographs of gap junctions between two astrocytes reveal the close apposition of the two membranes, which are separated by a small gap of 2 nm. (b) P-face of a gap junction plaque between two astrocytes obtained by freeze fracture, showing the connexin particles. (a) and (b) are modified from Duffy et al. (2000). (c) Electron micrograph (EM) of a gap junction plaque negatively stained showing the hexameric structures corresponding to connexons. Adapted from Fawcett (1994), Figs. 2–14. (d) Schematic drawing of a gap junction channel formed by the docking of two hexameric structures (connexons) provided by two adjoining cells. Each connexon is formed by six subunits, the connexins (e), which are tetra-span proteins with the N- and C-termini, and a cytoplasmic loop (CL) located on the cytoplasmic face. The two extracellular loops (E1 and E2) are the connexin domains that provide the strong interaction with the apposing extracellular loops of the connexin in the adjoining cells. (*See Color Plates*)

5.1.2 The Gap Junction Protein Family and Membrane Topology of Connexins

The proteins that form gap junction channels are connexins (Cx) in vetebrates and innexins in nonchordates (Willecke et al., 2002; Phelan, 2004; Hua et al. 2003). The connexin gene families in man and rodents each have about 20 members (Willecke et al., 2002), where encoded proteins are currently designated Cx***MW***, with ***MW*** representing the predicted molecular weight in kDa of the cDNA encoding the protein (e.g. Cx43, the major gap junction protein in astrocytes and cardiac myocytes, has a predicted MW of 43 kDa); nomenclature of the genes encoding these proteins follows Gja*N*, Gjb*N*, Gje*N*, where a, b, and e refer to subfamilies based on sequence similarities and *N* indicates the order in which they were discovered (Sohl and Willecke, 2003). Individual connexin types are expressed in overlapping patterns in tissues and are often coexpressed in the same cell. For example, astrocytes predominantly express Cx43 but also express other connexins, including Cx30 and Cx26, and oligodendrocytes express Cx32 and Cx47, as well as Cx29, although the last of these probably does not form functional gap junction channels (Theiss et al., 2005; Nagy et al., 2003; Altevogt and Paul, 2004).

All connexins share a common membrane topology (Fig. 5.1e) and most are encoded by a gene family with a common gene structure: a single intron separating two exons. With few exceptions, notably the neuronal Cx36 and its fish ortholog Cx35, Cx32, and Cx45, the second exon contains the entire coding sequence.

At the amino acid level, connexins share about 50% sequence identity, being most similar in transmembrane and extracellular regions and most divergent with regard to cytoplasmic domains. All connexins are tetraspan membrane proteins (crossing the membrane 4 times: segments M1, M2, M3, and M4) with intracellular C- and N-termini and two extracellular loops (E1 and E2, also referred to as L1 and L2, respectively). The extracellular loops are structurally conserved, with cysteine residues identically positioned in all connexins. These loops provide high-affinity intercellular interactions between connexons formed by a single connexin (so-called homomeric, homotypic gap junctions) and also between many pairs of connexons formed by different connexins (so-called heterotypic channels, pairing connexons formed by individual or multiple connexin types). Given the high affinity of the paired connexons, turnover involves the incorporation of the neighbor connexon into one cell of the pair rather than splitting the connexon subunits between cells (see below).

The single intracellular loop (CL) located between membrane segments (M) 2 and 3 is quite variable in length among connexins and is used as one criterion to classify the proteins in three different subfamilies (a or group II, b or group I, and g or group III). The third transmembrane domain (M3) is the most amphipathic and is generally assumed to provide the hydrophilic face lining the lumen of the gap junction channel, although other transmembrane domains likely also contribute to the pore (Skerrett et al., 2002, Zhou et al., 1997). The most divergent domain of connexins in terms of its amino acid sequence and length is the carboxyl terminus (CT), which for different connexins contains phosphorylation sites and motifs that bind to protein kinases and phosphatases as well as scaffolding proteins.

In contrast to the largely a-helical domains of the transmembrane segments of the gap junction channel, the extracellular loops appear to be primarily α-sheet, which are stabilized by disulfide bonds (Foote et al., 1998; Perkins et al., 1998). Recent studies have begun to clarify structures of the cytoplasmic domains of connexin molecules. With regard to the amino terminus (NT), an NMR study of peptide corresponding to Cx26 NT concluded that there were helical regions separated by a flexible hinge residue (Purnick et al., 2000) and circular dichroism studies of peptides corresponding to this region in other connexins have indicated the presence of helical structure (Duffy et al., 2006). The only structure of the cytoplasmic loop region of the connexins has been obtained with NMR using peptides corresponding to two halves of this domain in Cx43 and with recombinant CL, which showed that acidification induced a major change in structure of the more C terminal portion of the cytoplasmic loop with a large number of inter residue cross peaks appearing as pH was reduced from 8 to 5.8 (Duffy et al., 2002; Hirst-Jensen et al., 2007). These helical changes appeared to be centered on histidine residues at positions 126 and 142, whose titration was responsible for the structural change. As noted elsewhere in this review, this conformational change in the cytoplasmic loop is presumably a key feature of pH-dependent channel closure. Structural studies on the CT domain have also indicated the presence of short regions of helical structure (Sorgen et al., 2002), have provided evidence that dimerization of these Cx43 CT domains may involve these structured regions (Sorgen et al., 2004a) and have resolved structural changes induced by binding of Cx43 CT to domains of other proteins (Sorgen et al., 2004b).

5.1.3 The Nexus Redefined: Connexins and Their Binding Partners

The discovery that connexins bind to other proteins has had a major impact on the understanding of gap junction functions and their regulation. This multimolecular complex, which includes both membrane and cytosolic proteins has been termed the *Nexus* (Spray et al., 1999), replacing usage of the term as a morphologic descriptor by Dewey and Barr, 1962). There is considerable evidence that connexin–protein interactions play key roles in gap junction function, trafficking and regulation (Duffy et al., 2006; Giepmans, 2006; Herve et al., 2007). Connexin-interacting proteins (Fig. 5.2) include signaling molecules with a potential to regulate gene expression (such as α and β catenins, NOV, and ZONAB), second messengers including both serine/threonine and tyrosine kinases, tight junction components including occludins, zonula occludens (ZO)1,2,3, elements of both actin and tubulin cytoskeleton and associated proteins and the lipid domain maker caveolin1 (Schubert et al., 2002; Duffy et al., 2006). Affinities have been measured for a few of these interactions using mirror resonance spectroscopy, indicating that these linkages range from weak (>50 mM) to moderately strong (<1 mM) (Duffy et al., 2004). Importantly, affinities of binding partners within the Nexus in astrocytes are highly modifiable under ischemic conditions (Li et al., 2005), resulting from factors such as the phosphorylation state of the gap junction protein (and presumably of its binding partners as well) and local environmentally variable factors within subcellular compartments such as pH (Li et al., 2005; Duffy et al., 2006). In addition, Cx43 appears to be acetylated (E. Hertzberg, personal communication), nitrosylated (Retamal et al., 2006), and ubiquitinated (for review, see

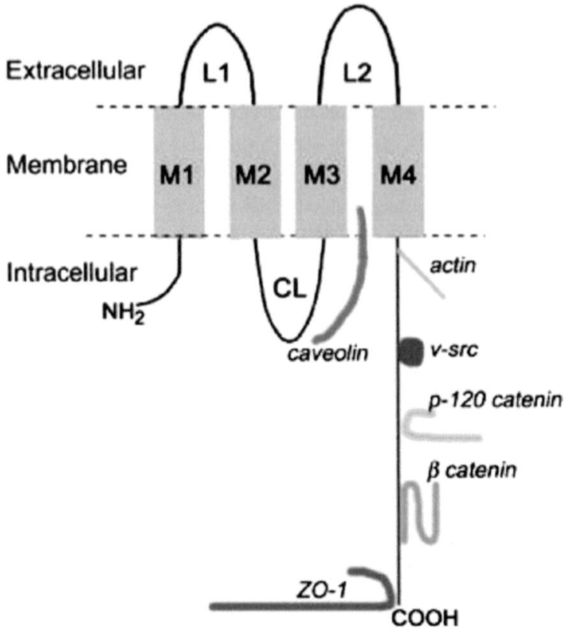

Fig. 5.2 Connexin-interacting proteins: The Nexus. Schematic drawing of connexin43 showing the binding partners at the cytoplasmic domains. Caveolin is shown to interact with the second half of the cytoplasmic loop (CL) of Cx43 (Schubert et al., 2002). Several other proteins (actin, v- and c-src, p-120 and β-catenin, and ZO-1) bind to different portions of the carboxyl-terminus (see Giepmans, 2004; Herve et al., 2007; Iacobas et al., 2007b). (*See Color Plates*)

Laird, 2006), all of which are modifications that likely affect binding affinity. Such changes in binding affinity of Cx43 to its molecular partners as a consequence of posttranslational modifications are likely to have major effects on connexin delivery and retrieval from the junctional plaque, affecting both intercellular communication and polarity of astrocytes.

Finally, structural studies indicate that conformational changes in the CT occur both as a consequence of intramolecular interactions with another domain of the protein (CL) and that intermolecular interactions with the PDZ-containing scaffolding protein ZO1 and the tyrosine kinase Src also produce conformational changes in Cx43 CT (Sorgen et al., 2004a). These latter studies, in which relevant recombinant PDZ and SH domains were incubated with the recombinant CT of Cx43, determined long-range conformational consequences of binding to the Src construct to amino acid 272, extending to the PDZ-binding domain, more than 100 amino acids distal. Such long-range conformational changes presumably account for the alteration in affinity of Cx43 for ZO1 in the presence of Src and also may provide a mechanism facilitating or even initiating the turnover of gap junction channels at the plaque.

5.1.4 Assembly and Degradation of Gap Junction Channels

The turnover of gap junction proteins is rapid (halftimes ~2 h; see Spray (1998) and Laird (2006)). During this brief lifetime the newly synthesized connexins are assembled into multisubunit connexons in the endoplasmic reticulum (ER), most likely bud from the Golgi or ER in vesicles, traffic to the plasma membrane (likely at least partially along microtubules), diffuse laterally from their point of insertion in the plasma membrane into the plaque, pair with connexons contributed by the partner cell, are endocytosed (likely involving actin cytoskeletal elements), and travel, again at least partially along microtubules, to the lysosomes and proteosomes, where they are degraded (Fig. 5.3).

Connexins are generally believed to oligomerize in the Golgi–ER to form hexameric connexons that contribute one side of the gap junction channel when joined across the extracellular gap to a connexon from an adjacent cell (Figs. 5.1d and 5.3) (for recent review, see Evans et al., 2006). The formation of gap junction channels is thus fundamentally different from that of other channels in that the connexon subunits must not only be transported to the surface membrane, but they must also

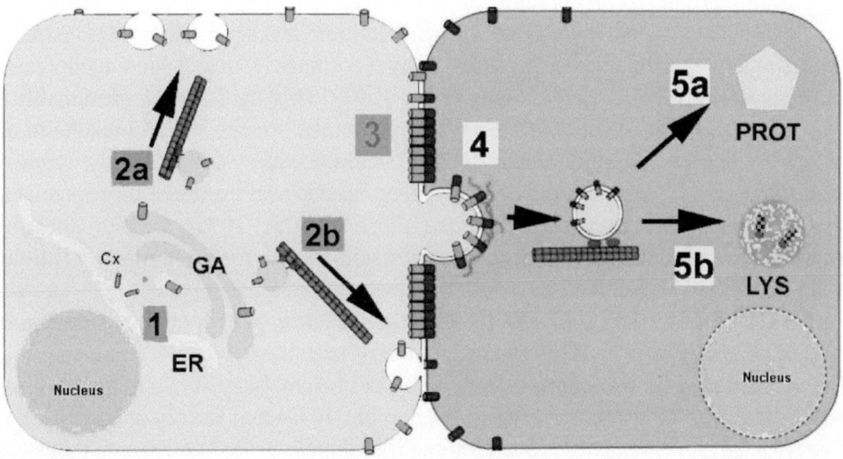

Fig. 5.3 The life and death of gap junctions. Schematic drawing showing two cells coupled by gap junctions. The cell on the left shows the steps involved in the formation of a gap junction. The newly synthesized connexin proteins (step 1) traffic from the endoplasmic reticulum (ER), where they hexamerize into connexons, to the Golgi apparatus (GA), and from there, connexons are inserted into plasma membrane through a secretory pathway either randomly to the surface membrane along microtubules shown as rods (step 2a) or into the vicinities of a gap junction plaque (step 2b). The connexons then diffuse through the plane of the bilayer (step 3) joining other connexons at the gap junction plaque. The cell on the right shows the degradation of a gap junction. Retrieval of gap junctions from the plasma membrane occurs by (step 4) the formation of a double-walled vesicle or connexosome (Laird, 2006) containing the membranes of the two adjoining cells. These vesicles containing full gap junction channels are then degraded either through a proteosome pathway (PROT, step 5a) or by lysosomes (LYS, step 5b). (*See Color Plates*)

be delivered to the site of the gap junction and incorporated into the plaque through union with connexons in the other cell. Insertion of connexons into the surface membrane was until recently regarded as being totally random, followed by lateral diffusion to junctional domains (Chow and Poo, 1984; Gaietta et al., 2002), where they are trapped by the partner connexon, likely aided by scaffolding proteins linking the plaque to the cytoskeleton. However, most cells (including astrocytes) are polarized, where domains containing gap junction channels are separated from those that do not (Nagy and Rash, 2003). Although astrocytes do not exhibit the tight junction strands that most cells use to establish their membrane polarity, they do express tight junction proteins such as ZO1 and claudin-1 and under certain conditions can amplify such expression (e.g., in response to inflammatory cytokines (Duffy et al., 2000)). Because gap junction proteins are linked to cytoskeletal proteins and move along them on their voyage to and from the membrane, it has seemed likely that insertion is at least partially directional, biasing delivery toward junctional domains of the cells. Such directed insertion has recently been elegantly demonstrated in astrocytes, where novel adaptor proteins appear to anchor the cytoskeleton to the region of adhesive junctions, thereby providing a highway for connexin delivery directed toward the appositional membrane (Shaw et al., 2007).

The pairing of connexons across extracellular space is a remarkable example of a high-affinity protein–protein interaction. At the moment when the newly formed gap junction channel opens, there is no measurable ionic leakage between the inside of the cell and extracellular space, as assessed by recordings from cell pairs manipulated into contact (for Cx43, see Valiunas et al., 1997). This tight seal is presumably a consequence of conformational changes in the joined connexons, and results in an apparently irreversible union under normal conditions, as evidenced by the requirement for hypertonic urea and ethylene glycol tetraacetic acid treatment for appreciable dissociation in biochemical experiments (Ghoshroy et al., 1995), the high shearing forces required for junction splitting measured with atomic force microscopy (Muller et al., 2002), and high forces necessary for apparent hemichannel opening by stretching the CT of Cx43 (Liu et al., 2006). As a consequence, gap junction retrieval from the surface membrane involves internalization of both the cell's own connexons and those contributed by its neighbor, both auto- and heterophagy, as first demonstrated in rapidly proliferating tumor cells (Larsen et al., 1979) and for dissociated cardiac myocytes (Mazet et al., 1985). Such internalized double-walled vesicles on their way to destruction in lysosomes have been termed *connexosomes* by Laird (2006), and this now appears to be the normal mode of junctional breakdown.

5.1.5 Network Architecture: The Astrocyte Syncytium and Local Microdomains

Astrocyte processes can be found at the near vicinities of several cellular types in the CNS (see Fig. 5.4), although direct contact through gap junctions is limited to a few cell types. Gap junctions connect astrocytes into a functional syncytium that

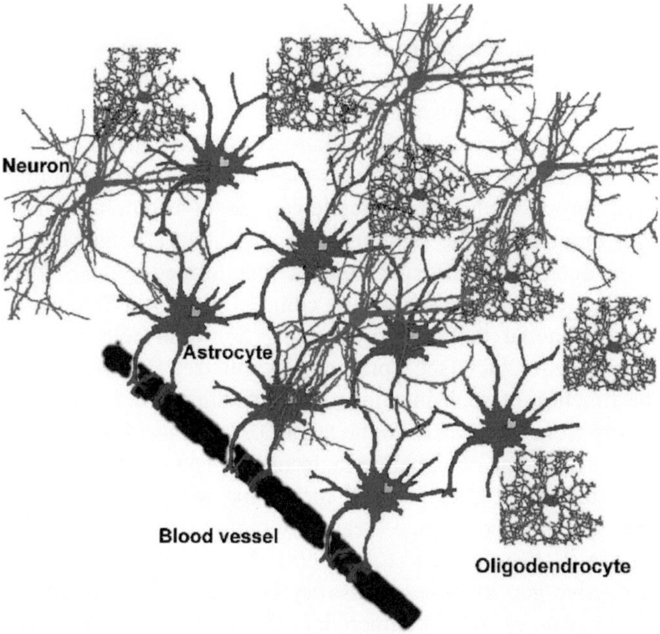

Fig. 5.4 Cellular interactions of astrocytes. Schematic drawing showing astrocyte (*blue*) processes contacting neurons (*red*) at their synapses. Astrocytes also contact oligodendrocytes (*green*) forming heterocellular gap junctions, and astrocytic endfeet surround blood vessels (*brown*). (*See Color Plates*)

extends throughout the brain; moreover, astrocytes and oligodendrocytes are interconnected by gap junctions, forming what has been termed the *panglial syncytium* (Dermietzel, 1998; Nagy and Rash, 2000; Massa and Munagni, 1985). The purpose of this extensive glial coupling has long been hypothesized to maximize the range of K^+ buffering, so that at the site of local neuronal activity the astrocytes would take up K^+, distribute it freely among the interconnected population and ultimately efflux K^+ through inward rectifying K^+ channels (Kir) 4.1 colocalized with aquaporin 4 water channels at astrocytic endfeet in contact with blood vessels (Walraff et al., 2006).

Injections of fluorescent dyes into astrocytes under conditions where gap junction channels were blocked has led to a new view of the organization of the glial network (Bushong et al., 2002). These injections revealed an intricate web of "spongiform" astrocyte processes extending much farther and more densely than had been apparent from previous studies using the astrocyte marker glial fibrillary acidic protein. This finding indicates that individual astrocytes occupy virtually nonoverlapping volumes, with little or no interdigitation of the major astrocyte processes. As pointed out by the authors, one implication of these findings is that a single astrocyte might be in contact with more than 100,000 neuronal synapses and not share this direct contact with other astrocytes.

In apparent contradiction to these nonoverlapping astrocyte–astrocyte domains is the extensive immunostaining for the major astrocytic gap junction protein, Cx43, that has been reported within the domain of a single astrocyte both in brain sections and in cultured cells (Rohlmann and Wolff, 1996; Wolff et al., 1998). These domains were termed *autocellular zones* by the authors, consistent with their observation that the abundant localization of Cx43 within the autocellular space is due in part to small astrocytic processes forming autaptic gap junctions onto other fine processes or onto major branches of the same astrocyte. Such an arrangement is similar in principle to the autaptic or reflexive gap junctions formed by Cx32 between cytoplasm-containing regions squeezed off by compact myelin in Schmidt-Lantermann incisures and paranodal loops of myelinating Schwann cells and in paranodal regions of oligodendrocytes (see Spray and Dermietzel (1995) and Scherer et al. (1999)). The function of such junctions in myelinating glia is presumed to lie in nutrient delivery from the nucleus to the innermost regions of the Schwann cell (or, alternatively, for exchange of signaling molecules generated at the contacts between axons and the innermost Schwann-cell membrane with the Schwann-cell body), thereby shunting the tortuous route created by multiple wraps of the Schwann cell around the axon (Balice-Gordon et al., 1998). A somewhat similar role might be played by reflexive gap junctions in astrocytes, where local changes in coupling strength might isolate or integrate microdomains, as could occur where fine astrocytic processes surround synapses. For example, functional microdomains have been demonstrated in Bergmann glia, to which the spread of Ca^{+2} elevations is regionally limited in response to neuronal activity (Grosche et al., 1999).

Two other findings reported by Bushong et al. (2002) are noteworthy. First, the topologies of the volumes proscribed by the individual astrocytes were found to be variable, ranging from roughly spherical to quite oblate. Just as the extracellular space has been shown to exhibit anisotropic permeation by tracer molecules (Sykova, 1997; Nicholson and Sykova, 1998), such differences in form of astrocytes would be expected to produce anisotropic intercellular diffusion of ions and small molecules. Second, although processes of neighboring astrocytes were not found to penetrate the outer processes of adjacent astrocytes, oligodendrocyte processes freely intermingled within the “autocellular zone.” Direct oligodendrocyte–astrocyte gap junctions have long been known from thin section and freeze-fracture studies (Massa and Mugnaini, 1985), and their abundance was confirmed using freeze-fracture immunolabeling (Nagy and Rash, 2000), giving rise to the concept of global functional intercellular communication throughout the brain, the so-called panglial syncytium (Rash et al., 1997). However, it remains to be experimentally demonstrated that such heterocellular gap junctions between astrocytes and oligodendrocytes are functional.

A number of connexin types connect astrocytes and oligodendrocytes among themselves and between one another. Of the 21 identified gap junction members in mammals, more than one-third are found in the brain, specifically in glial cells (Willecke et al., 2002; Spray et al., 2004). Different from other tissues, individual types of neural cells do not express overlapping connexins (except for microglia that express neuronal Cx36: Dobrenis et al., 2005). The primary gap junction protein in astrocytes is Cx43, but Cx26, Cx30, Cx40, Cx45, and Cx46 have also been reported to be expressed in these cells in vivo or in culture. Total junctional conductance between cultured astrocytes from cortex and spinal cord of Cx43-null mice is 75–95% less than between astrocytes from wildtype (WT) siblings (Scemes et al., 1998; Scemes et al., 2000), strongly suggesting that channels formed by this gap junction protein ordinarily supply the vast majority of the intercellular communication in this cell population. In oligodendrocytes, the major connexins appear to be Cx32, Cx47 (initially incorrectly described as neuronal in its distribution (Teubner et al., 2001)), and Cx29 (which is likely not functional (Altevogt et al., 2002)). Because the most abundant astrocytic connexin, Cx43, does not form functional channels when paired with cells expressing Cx32 (White and Bruzzone, 1996), it seems likely that astrocytic–oligodendrocytic coupling is established by heterotypic gap junctions, with Cx43 on the astrocyte side and Cx47 contributed by the oligodendrocytes and/or by astrocyte Cx30 or Cx26 paired with oligodendrocyte Cx32 (Altevogt and Paul, 2004; Nagy et al., 2003).

5.1.6 Other Members of the Gap Junction Family: Pannexins and Innexins

As noted earlier, there has been a marked evolutionary divergence in that the proteins forming gap junctions, with these intercellular channels in invertebrates being formed by a totally different protein family than in chordates. This difference is in stark contrast to other membrane channel proteins, such as aquaporins or K^+ channels, which in many cases are highly conserved from plants to man. The proteins forming gap junction channels in nonchordates are innexins (Phelan, 2004; Hua et al. 2003), which differ topologically from connexins in that they have longer extracellular loops and generally a larger cytoplasmic hinge region. Presumably, as a consequence of the larger size of extracellular loops, the particles observed in freeze fracture are slightly larger than those of connexins and the membrane separation ("gap") is slightly wider (Epstein and Gilula, 1977). Different invertebrates express innexin families that are more closely related to others in the same species than to isoforms expressed in other species (Yen and Saier, 2007). For example, Cx43 shows high homology from fish to man but differs substantially from other connexins, whereas different innexins within an organism are more similar among themselves than to innexins of other organisms.

A database search of innexin homologues in vertebrates identified a group of proteins that were termed *pannexins* to indicate the hypothesis that these were gap junction proteins expressed throughout the animal kingdom (Baranova et al., 2004). Two of the three identified pannexins are expressed in neural cells (pannexins1 and 2). Although studies in oocytes have indicated that, with sufficient overexpression, gap junctions may be formed in this preparation by pannexins 1 and 2 (Bruzzone et al., 2003), there is no evidence that pannexins form gap junctions in mammalian cells, except for a report that pannexin1 transfection in C6 glioma cells increased dye coupling (Lai et al., 2007). In fact, Huang et al. (2007) have reported that pannexin1 is expressed in glia but does not form gap junction channels. A likely explanation for the lack of formation of gap junction channels is the existence of glycosylation sites in the extracellular loops (Boassa et al., 2007); presumably such secondary modification would serve to separate potentially interacting membrane domains.

The distribution of pannexins in neural tissue remains somewhat controversial. In situ hybridization studies have detected pannexin1 in neurons and not in glia (Ray et al. 2005, 2006; Vogt et al. 2005); pannexin1 antibodies have localized this protein to postsynaptic neuronal locations (Zoidl et al., 2007) and nonjunctional currents attributed to pannexin1 have been reported in hippocampal neurons in slice preparation exposed to ischemic conditions (Thompson et al., 2006). However, pannexin1 antibodies have also localized this protein to membranes of cultured astrocytes and oligodendrocytes (Huang et al., 2007). More recently, pannexin1 was detected by immunocytochemistry and immunogold in astrocytes from mouse brain slices (Dermietzel, personal communication). As indicated below, we favor the view that pannexins form large nonjunctional channels (pannexons) rather than gap junctions, and that functions attributed to connexin "hemichannels" most likely arise from pannexins alone or in combination with other channels.

5.2 Functions of Gap Junction Channels and Hemichannels

Gap junction-mediated intercellular communication among astrocytes is vital to spread signals throughout the brain, likely a collective process of interacting astrocytes and oligodendrocytes constituting the "panglial syncytium" as described earlier. Second messengers, ions, and metabolites flow through the gap junction interconnections, generating waves of elevated Ca^{2+}, Na^{+}, and metabolic activity, dissipating K^{+} and neurotransmitter gradients. Different from intercellular gap junction channels, nonjunctional connexons ("hemichannels") have been reported to open under specific conditions, providing direct communication between the intra- and extracellular "milieu." The properties and function of both types of channels are described below.

5.2.1 Permeability and Selectivity of Gap Junction Channels

One of the most important functions of the intercellular permeability between astrocytes provided by gap junctions is to mediate the diffusion of metabolites throughout the population. Such diffusion is possible because the diameter of a gap junction channel is large, allowing permeation by ions and molecules up to a molecular weight of about 1,000 Da. This size limit allows diffusion through gap junctions of many molecules that function as second messengers, such as Ca^{2+}, inositol trisphosphate (IP_3), and cyclic adenosine monophosphate (cAMP), and also such important metabolites as glucose, adenosine 5′-triphosphate (ATP) and their degraded byproducts. In addition to these moderately small molecules, there is recent evidence that larger linear molecules, such as polypeptides and interfering RNA might "wiggle through" these channels, thereby directly controlling gene expression or antigen presentation (Valiunas et al., 2005; Neijssen et al., 2005).

Fluorescent dyes, most commonly the highly charged and aldehyde-fixable dye Lucifer Yellow, which was specifically designed for this purpose (Stewart, 1981) have been extensively employed to determine the presence of gap junctional communication between cells, and the extent to which dye diffuses among coupled cells has been used to evaluate the strength of the so called "dye-coupling." Studies using dyes of different diameters and charge indicated that gap junctions differ in terms of their permeabilities, some being more permeable to cations than anions, such is the case of Cx45 (P_{anion}:P_{cation}, 1:5), others more permeable to anions, such as Cx32 (P_{anion}:P_{cation}, 2:1), and others equally selective to cations and anions, such as Cx43 (for detailed review, see Harris, 2007).

Because there is a great variety of molecules that can cross cell boundaries through gap junction channels (K^+ ions, cAMP, IP_3, Ca^{2+}, glucose, glutamate, etc.), the functions performed by the interconnected astrocytic syncytium may be quite diverse. Below are summarized some of the roles performed by astrocytes, which involve the participation of gap junctions.

5.2.2 *Ion Dissipation: K^+ Siphoning*

The concept that spatial buffering of extracellular K^+ is one of the key roles played by astrocytes in the CNS function was first proposed more than 40 years ago (Orkand et al. 1966; Kuffler et al., 1966). Through this mechanism, glial cells were proposed to remove the excess K^+ accumulated at synaptic sites ("source" of K^+) to regions of low extracellular K^+ ("sinks"). Evidence that glia cells do in fact mediate such K^+ flux came from experiments performed in Mueller cells, showing that these modified retinal glial cells transported K^+ from the plexiform layers to the vitreous body, blood vessels, and subretinal space (Newman, 1985; Reichenbach et al., 1992; Newman and Reichenbach, 1996). Although the Na^+/K^+ ATPase and the KCl transporter may participate in K^+ redistribution (Walz, 2000; Chen and Nicholson, 2000), siphoning of K^+ by glia is likely to be mainly mediated by inward rectifying K^+ channels (Newman, 1985; Reichenbach et al., 1992; Kofuji et al., 2002). Immunohistochemical studies performed on mouse retina have identified two different subtypes of inward rectifying K^+ channels differentially expressed in Mueller cells. These are (a) the strongly rectifying K^+ channel (Kir 2.1) highly expressed in membrane domains of the Mueller cells extending into the (source) plexiform layers that would mediate the influx of K^+ into the Mueller cell and (b) the weakly rectifying K^+ channel (Kir 4.1) expressed predominantly at the endfeet (sink) located at the vitreous body and blood vessels, which would be expected to favor the efflux of K^+ from the Mueller cells (Kofuji et al., 2002).

In various brain regions, Kir 4.1 immunoreactivity is found in about half of the astrocytes, being mainly expressed at regions where the astrocytic endfeet meet the blood vessels (Takumi et al., 1995; Poopalasundaram et al., 2000; Higashi et al.,

2001; Schroder et al., 2002; Kofuji et al., 2002), whereas Kir 2.1 transcripts have been observed in subpopulations of astrocytes in different brain regions (Kofuji et al., 2002; Schroder et al., 2002). It is conceivable that the distribution of these two inward K^+ channels are confined to different subpopulations of astrocytes in the brain and that gap junction channels by linking the members into a network would assemble the different specialized membrane areas. This arrangement would form a multicellular astrocytic network functionally equivalent to a Mueller cell. Although the space constant for spatial buffering is unknown, it has been estimated that the astrocytic syncytium is far more likely to support K^+ transport than the extracellular diffusion (Gardner-Medwin, 1983) and propagates K^+ waves faster than the extracellular space (Amzica et al., 2002). Gap junction channels are permeable to K^+, and thus are likely to provide the astrocytic network with the necessary volume to accommodate the focal influx of high concentrations of K^+. In this regard, it has been shown that exposure of astrocytes to high K^+ solutions increases the coupling strength (Enkvist and McCarthy, 1994; De Pina-Benabou et al., 2001), thus expanding the effective volume of the interconnected astrocytes while maintaining a high surface area for K^+ uptake.

Direct evidence for a role of glial gap junctions in spatial K^+ buffering has recently been provided by studies of connexin null mice. In hippocampal slices from Cx43/Cx30 double null mice, Wallraff et al. (2006) reported that K^+ clearance and accumulation in response to neuronal activity were reduced in stratum lacunosum moleculare, whereas in stratum radiatum the perpendicular orientation of the astrocytes might allow K^+ clearance without gap junctions. Notably, the threshold for epileptiform activity was reduced in this double knockout. A role for oligodendrocytes in K^+ clearance has also been proposed based on studies of double Cx32/Cx47 null mice by Menichella et al. (2006). In this study, activity-dependent vacuole formation in myelinating CNS glia was detected in the double knockout similar to that seen in mice lacking the end foot K^+ Kir 4.1 and connexin heterozygotes in a heterozygous Kir 4.1 background also showed this pathology. These authors concluded that K^+ clearance is not limited to astrocytes but depends instead on the gap junctions of the panglial syncytium.

5.2.3 Distribution of Energy Sources and Metabolites in the CNS

Glucose together with lactate, ketone bodies, and glutamate or glutamine are the main sources of energy utilized by the brain (Williamson, 1982; Lopes-Cardozo and Klein, 1985; Vicario et al., 1993). Although glucose is the only blood-borne substrate used by the brain as an energy source, lactate and other nonoxidized products of glucose metabolism are consumed as fuel during elevated brain activity (Dienel and Hertz, 2001). A selective transport system localized in the blood–brain barrier (BBB) provides glucose to the cells within the brain. Because astrocytes are interposed between the capillaries and neuronal elements and are the first cellular elements that glucose entering the brain encounters after crossing the endothelium of the BBB, astrocytes were assigned a nutritive role for neurons. Astrocytes express the glucose transporter Glut1 (Maher et al., 1994) and store glucose in the form of glycogen. Since astrocytes lack the enzyme that transforms glucose-6-phospate into glucose during glycogen breakdown, the glycolytic product lactate can be delivered from astrocytes to neurons as a source of energy (Giaume et al., 1997). Glutamate, the main excitatory neurotransmitter released by CNS neurons, is taken up by astrocytes and together with ammonia is converted into glutamine, which is then delivered to neurons (Broer and Brookes, 2001). Although glucose, lactate, glutamate, and glutamine can diffuse through the extracellular space and then be taken up by neuronal cells, it is also likely that these fuels travel through the astrocytic network, by diffusing through gap junction channels (Dienel and Hertz, 2001; Dienel et al., 2001; Ball et al., 2007). Permeability of gap junction channels to 2-deoxyglucose has been demonstrated in cell lines (Pitts and Finbow, 1977), in the lens (Goodenough et al., 1980), and between smooth muscle cells (Cole and Garfield, 1986; Cole et al., 1985). The passage of glutamine and glutamate, lactate and glucose through astrocytic gap junction were shown using the scrape loading technique applied to cultured astrocytes (Tabernero et al., 1996; Giaume et al., 1997).

More recently, Dienel's group (Ball et al., 2007; Cruz et al., 2007) evaluated the role of astrocyte gap junctional communication in the dispersal of metabolites (^{14}C-glucose) within the inferior colliculus. In the first report (Cruz et al., 2007), they described that the increased dispersal of ^{14}C-glucose observed during acoustic stimulation was reduced by 50% when gap junction channel blockers (alpha-glycyrrhetinic acid (α-GA) and oleamide) were microinfused into conscious rats. The extensive dye coupling (>6,100 cells) observed in this region suggests that astrocyte gap junctions have the capability to quickly distribute metabolites from the activated area throughout the interconnected network (Ball et al., 2007).

5.2.4 Ca^{2+} Waves: A Special Case of Long-Range Signaling

Gap junctional communication is usually regarded as a passive process, where diffusion of ions and small molecules (up to 1 kDa) among coupled cells is governed by their chemical gradient. Because gap junction channels are open at zero transmembrane potential, the linkage that they can provide allows dissipation of ions and metabolites whenever a gradient is generated among a group of coupled cells. However, depending on the nature of the permeant molecule, gap junctional communication can mediate dynamic events, especially if the permeant molecules lead to chemical processes involving threshold and "regenerative" steps. This is the case for intercellular Ca^{2+} wave spread, where both Ca^{2+} and IP_3, gap junction permeant molecules act as triggers to induce intracellular Ca^{2+} elevations in adjoining coupled cells. This regenerative event can proceed as long as Ca^{2+} and IP_3 concentrations are at or above threshold levels to induce release of Ca^{2+} from the intracellular stores (Fig. 5.5).

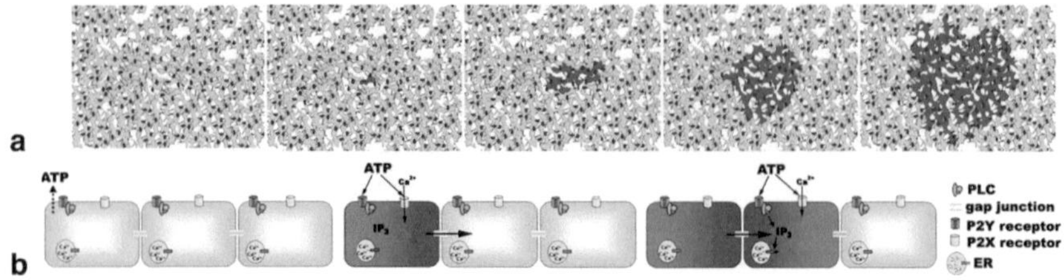

Fig. 5.5 Intercellular Ca^{2+} wave. (a) Schematic drawing depicting the transmission of Ca^{2+} waves between astrocytes in culture. Following stimulation of a single astrocyte, intracellular Ca^{2+} levels increase (*red*). This increase in intracellular Ca^{2+} spreads to adjoining astrocytes (*green*), which are shown to be connected by gap junction (*blue*). (b) Schematic representation of the steps involved in the transmission of intercellular Ca^{2+} waves. Released ATP diffuses through the extracellular space, activating membrane purinergic (P2) receptors. Stimulation of metabotropic P2Y receptors leads to PLC activation and IP_3 formation and activation of ionotropic P2X receptors leads to the influx on Ca^{2+}. IP_3 and Ca^{2+} promote the release of Ca^{2+} stored in the endoplasmic reticulum (ER), increasing intracellular Ca^{2+} levels of the stimulated cell. Diffusion of the two intracellular Ca^{2+}-mobilizing second messengers through gap junction channels together with the activation of P2 receptors in the near-by cell contribute the Ca^{2+} rise in this cell. This process continues till the concentrations of ATP, IP_3, and Ca^{2+} are not sufficient to trigger intracellular Ca^{2+} rises. (*See Color Plates*)

Intercellular Ca^{2+} waves (ICWs) in astrocytes occur following mechanical, electrical, and chemical stimulation (see Boitier et al., 1999; Charles, 1998; Charles and Giaume, 2002; Giaume and Venance, 1998; Newman 2004; Scemes, 2000; Scemes and Giaume, 2006). The velocity (15–23 μm/s) with which these Ca^{2+} signals are transmitted between cells is fairly constant and seems to be independent of the nature of the stimuli and the type of preparation used, such as cell culture, brain slices, or retinal whole mounts (see Scemes and Giaume (2006)). In contrast, the extent to which ICWs spread is highly variable and more likely dependent on a combination of several factors, including the degree of coupling and the presence of IP_3-generating membrane receptors. In other words, the complexity and variability of the extent and shape of ICW among astrocytes is likely due to the fact that this form of signal transmission depends on two distinct but interdependent pathways: gap junction-dependent and -independent routes.

Gap junction-mediated transmission of ICWs was the first pathway identified in astrocytes (Finkbeiner, 1992). In this study it was shown that neither the direction nor the velocity of glutamate-induced ICW were affected by rapid superfusion and that two gap junction channel blockers impaired Ca^{2+} wave spread between astrocytes without affecting Ca^{2+} spread within single cells. This finding together with several others performed in different systems (Saez et al., 1989; Charles et al., 1991, 1993; Charles et al., 1992; Enkvist and McCarthy, 1992; Nedergaard, 1994; Venance et al., 1995; Guan et al., 1997; Leybaert et al., 1998; Scemes et al., 1998; Blomstrand et al., 1999) provided a strong basis supporting the view that gap junction channels play a crucial role in the transmission of Ca^{2+} signals between astrocytes.

Evidence for the participation of an extracellular pathway for the spread of ICW in astrocytes was provided by Enkvist and McCarthy (1992), showing that Ca^{2+} waves could cross bare, cell-free areas in confluent cultures of cerebral astrocytes. Later studies confirmed that Ca^{2+} waves in cultured astrocytes were able to cross cell-free areas up to 120 μm (Hassinger et al., 1996) and that ATP was the extracellular molecule released by stimulated astrocytes that, by activating purinergic receptors, contributes to Ca^{2+} wave spread (Guthrie et al., 1999).

Astrocytes *in situ* and *in vitro* express, at different levels, several ionotropic and metabotropic P2 purinergic receptors, some of which have been implicated in the transmission of Ca^{2+} signals (Fumagalli et al., 2003; Ho et al., 1995; Idestrup and Salter, 1998; Zhu and Kimelberg, 2001, 2004). Thus, the properties of Ca^{2+} signal transmission between astrocytes depend not only on the (sub)type of membrane receptors but also on the degree of gap junction-mediated intercellular coupling. The relative contribution of each of these pathways is likely to depend upon developmental, regional, and physiological states. Accordingly, it has been recently shown that depending on the brain regions (cortex vs. hippocampus and corpus callosum), the pathway mediating the transmission of Ca^{2+} signals in brain slices is different (Haas et al., 2006). Another example illustrating that Ca^{2+} waves can utilize different routes when traveling between glial cells was provided in whole mounts of mouse retina, where astrocyte-to-astrocyte Ca^{2+} waves are mainly mediated by the diffusion of second messengers through gap junction channels, whereas astrocyte-to-Mueller cell transmission is basically dependent on the diffusion of ATP through the extracellular space (Newman, 2001, 2003, 2004). Under pathological conditions, such as in inflammation, changes in ICW spread between astrocytes from being gap junction-dependent to being purinergic receptor-dependent has also been documented when treating cells with interleukin (IL)-1β ((John et al., 1999); see below).

The contribution of gap junctional communication to P2R-mediated ICW is illustrated in a study showing that overexpression of Cx43 caused dramatic changes in the shape and distance of ICW spread, either by amplifying or restricting the signal transmission (Suadicani et al., 2004). If, for instance, the increase in the effective volume of the intracellular compartment provided by gap junction channels leads to the dissipation of second messengers' gradients to levels below threshold, ICWs will be terminated (Giaume and Venance, 1998; Suadicani et al., 2004). On the other hand, by recruiting nonresponsive cells into a network of cells expressing receptors that more efficiently generate second messengers, gap junctional communication could amplify the distance of ICW spread.

Besides gap junctions, another factor has been reported to influence ICW. The release of Ca^{2+}-mobilizing "gliotransmitters" can potentially feed back on the astrocytic population, in an autocrine fashion, thus amplifying the extent to which these Ca^{2+} signals are transmitted (Stout et al., 2002; Suadicani et al., 2006). Hassinger et al. (1996) and Guthrie et al. (1999) proposed a mechanistic model by which ATP released from the stimulated cells would activate P2R receptors. Activation of these receptors would then lead to mobilization of intracellular Ca^{2+} in the neighboring cell that in turn would be followed by the release of ATP from this neighboring cell.

This succession of events would then occur sequentially along the ICW path. Although recent evidence supports the hypothesis that ATP induces ATP release from astrocytes (Anderson et al., 2004), this regenerative ATP release model, however, does not explain why Ca^{2+} waves travel within defined limits. As a counterproposal, Nedergaard's group (Cotrina et al., 1998b, 2000; Arcuino et al., 2002) suggested a nonregenerative model based on a point source of ATP release. In this nonregenerative model, ATP released from a single cell would diffuse and stimulate a limited number of nearby cells. This point source release mechanism of initiation of ICWs from the stimulated cell, together with P2R activation and gap junction-mediated diffusion of IP_3, have been incorporated into a recent mathematical model of ICW transmission in astrocytes (Iacobas et al., 2006).

5.2.5 Vascular Control by Gap Junctions

The importance of astrocytes to brain function has been heightened by recent reports that glial cells control vascular tone. This concept of a neurovascular control unit centered on the astrocyte is one in which gap junctions play a central role by coupling both astrocyte cell bodies and processes in order to broaden the region of smooth-muscle contraction or relaxation controlled by a single astrocyte. Although the identity of the glia transmitter mediating the basal and induced changes in vessel tone remain to be rigorously determined, it is likely that prostaglandins, ATP, K^+, and peptides are involved (Zonta et al., 2003; Mulligan and MacVicar, 2004; Filosa et al., 2006; Takano et al., 2006) In addition, gap junction proteins may play a role in maintaining aquaporin distribution in the astrocytic endfeet. As has been pointed out in a recent study (Nicchia et al., 2005), there is an interplay between these two types of channels, although direct interaction has not yet been demonstrated.

5.2.6 Release of Signaling Molecules Through Hemichannels/Pannexons

Hemichannels, or connexons, are half gap junctions (Fig. 5.1d; for reviews see Bennett et al., (2003), John et al. (2003), Contreras et al. (2004), Evans et al. (2006), Goodenough and Paul (2003), Martin and Evans (2004), Parpura et al. (2004), Saez et al. (2005), Verselis et al. (2000), and Spray et al. (2006)). When open, these channels would connect a cell's interior to extracellular space, a profound functional distinction compared with open gap junctions. Because gap junction channels are such large and rather nonselective pores, opening of hemichannels to the extracellular environment would be expected to be disastrous to the cell, not only collapsing ionic gradients necessary for maintenance of resting potential and transport, but also causing the loss of precious metabolites, energy sources, and diffusible second messenger molecules. However, if hemichannel opening were brief enough and/or "controlled," these channels could conceivably provide a pathway for release (or uptake) of large molecules and ions. Such a role has been proposed for Cx43 hemichannel opening in release of glutamate and ATP (Ye et al., 2003; Stout et al., 2002) (and certainly other biologically active molecules as well) from astrocytes, which could have a major impact on glial–glial and glial–neuronal interactions. Recent studies, however, have provided evidence that hemichannel activity is likely mediated by pannexin1 channels, at least with regard to release of gliotransmitters (Locovei et al., 2006a, 2007). Pannexin1 channels (pannexons) have been shown to be mechanosensitive, are activated by membrane depolarization above +20 mV, and are gated by intracellular Ca^{2+} in response to P2R activation (Bruzzone et al., 2003; Locovei et al., 2006b, 2007). Interestingly, because pannexin1 was recently shown to be the large conductance pore induced following prolonged activation of $P2X_7R$ (Locovei et al., 2007; Pelegrin and Surprenant, 2006) and to participate in the release of ATP (Locovei et al., 2006a), it is likely that pannexin1 channels are the sites of ATP release from astrocytes that amplify the extent of Ca^{2+} wave spread (Stout et al. 2002; Suadicani et al. 2006). Indeed, similar to what was observed for $P2X_7R$ null astrocytes (Suadicani et al., 2006), ICW spread between astrocytes treated with pannexin1 small interferring RNA was significantly reduced compared with untreated cells (Suadicani et al., 2007).

5.2.7 Transmission of Death Signals vs. Neuroprotection

Although gap junctions have been implicated in the transmission of damage signals from injured cells to normal cells, as observed following cell irradiation (Azzam et al., 2001), the issue of whether gap junctional communication confers a neuroprotective role is still controversial. In a stroke model, glial-cell death occurring during the secondary expansion of infarction was shown to be reduced by gap junction channel blockers (Rawanduzy et al., 1997; Saito et al., 1997). Similar results were obtained by comparing the extent of cell death in an *in vitro* trauma model (organotypic slice culture), in which the contribution of gap junctional communication to cell death was evaluated using gap junction channel blockers (heptanol and carbenoxolone) and by downregulation of Cx43 expression either by the use of antisense-oligonucleotides or from molecularly engineered Cx43 null mice (Frantseva et al., 2002). Furthermore, although junctional conductance is decreased by intracellular acidification, as a rapid gating response (Spray et al., 1981; Ek-Vitorin et al., 1996) and also possibly because of the activation of a protein kinase that phosphorylates Cx43 on serine residues (Yahuaca et al., 2000), astrocytic gap junction channels were shown not to be totally closed at the penumbra of an ischemic region and to participate in the amplification of the damaged area (Cotrina et al., 1998a). Indeed, gap junction opening has recently been shown to accompany the bystander cell death induced by cytochrome c injection in paired *Xenopus* oocytes (Cusato et al., 2006). Contrary to expectations, however, studies performed in Cx43 HTs, which express about half the levels of Cx43 protein compared with WT (see Dermietzel et al., 2000) have indicated that gap junctions may be neuroprotective. These studies performed on Cx43 HT mice 4 days after obstruction of the right middle cerebral artery (Siushansian et al., 2001) or after traumatic injury (Frantseva et al., 2002) indicated that the infarct area was significantly increased when compared with the WT brains and suggested that the reduced gap junctional communication in the Cx43 HT compromised the astrocytic syncytium, favoring neurotoxicity (Siushansian et al., 2001). Furthermore, in cocultures of neurons and astrocytes, the blockade of gap junctional communication with either carbenoxolone or α-glycyrrhetinic acid resulted in increased glutamate-induced neurotoxicity, indicating that gap junctions may have a neuroprotective role against glutamate toxicity (Ozog et al., 2002).

Gene therapy methods have been applied to glioblastoma treatment. One of these, treatment of Herpes thymidine kinase (TK) transduced tumor cells with ganciclovir (GCV), is very efficient especially because of the bystander effect that it generates, leading to tumor regression after GCV metabolites generated by TK diffuse through gap junctions to neighboring cells (Estin et al., 1999; Andrade-Rozental et al., 2000; Mesnil and Yamasaki, 2000). Although loss of gap junction-mediated intercellular communication has been long believed to be a common, even causative, occurrence in tumor cells (Lowenstein and Rose, 1992; Rose et al., 1993), it is now clear that tumor cells retain functional coupling and that this coupling pathway can be used therapeutically, essentially as a cellular drug-delivery device.

5.3 Gating of Gap Junction Channels and Hemichannels/Pannexin Channels

Similar to other ion channels in the plasma membrane and in cell organelles, gap junction channels can be opened or closed by various physiological and pathophysiological stimuli as well as by certain pharmacological treatments. Although there is as yet no known treatment that efficiently and selectively blocks gap junction channels, there is a particular physiological/pharmacological profile for individual gap junction channel subtypes. The types of stimuli that have been evaluated on gap junction channels formed by different connexins include voltage gradient across the junctional membrane, intracellular acidity and elevated Ca^{2+}, phosphorylation state of cytoplasmic serine and tyrosine residues, and pharmacological agents. For each of these stimuli that has been carefully evaluated, there appears to be differential sensitivity of gap junctions formed by different connexins and although properties may overlap those of other types of ion channels, there appears to be a selectivity for gap junctions that may be useful to identify their participation in physiological processes.

5.3.1 Voltage Dependence

The earliest voltage-clamp experiments on pairs of amphibian embryonic cells revealed a characteristic dependence of junctional conductance on voltage gradient across the junctional membrane (for review, see Del Corsso et al., 2006). All types of vertebrate gap junctions that have been studied display maximal conductance at 0 transjuctional voltage with rather symmetric decline in response to relative hyperpolarization or depolarization of either cell. The minimal conductance attained at the highest transjunctional voltages is generally 10% or greater than that of the maximal conductance and is attributable to a substate conductance into which the channel is driven at high voltages. This minimal conductance, the voltage at which half maximal conductance is achieved and the slope of the conductance voltage relation, as well as unitary conductances of the fully open and subconductance states are all characteristic properties of gap junctions formed by individual connexins (see Gonzalez et al., 2007). For example, Cx43, the major astrocytic gap junction protein in astrocytes, shows mainstate and substate conductances of about 120 and 30 pS respectively, V_0 at about 50 mV, and G_{min} about 20% $G_{ma5.}$ These biophysical properties of gap junction proteins are illustrated in Figs. 5, 6a, b.

Gap junction channels in vertebrates are remarkably insensitive to the actual resting potential of the cell, so long as there is no appreciable transjunctional voltage. By contrast, gap junctions in a number of invertebrates are sensitive both to transjunctonal and inside/outside voltage, as was first demonstrated in salivary gland cells of the midge (Loewenstein et al., 1967) and quantitatively examined more recently in crayfish hepatopancreas and fly salivary gland cells (Chanson et al., 1994; Verselis et al., 1991). Moreover, *Xenopus* oocyte studies on hemichannels formed of certain vertebrate connexins (notably Cx46 and Cx50) show activation with strong membrane depolarization, as has been reported for Cx43 expressed in HeLa cells (Contreras et al., 2003) and for pannexin1 channels expressed in oocytes (Bao et al., 2004). In the case of the hemichannels, such activation has been explained by voltage sensitivity of the so-called loop gate, presumably located on the extracellular side of the channel (for detailed reviews, see Bukauskas and Verselis, 2004; Gonzalez et al., 2007). Because of differences in the voltage sensitivity of the transjunctional and loop gates, heterotypic pairing of different gap junction subtypes can be asymmetric, such that channels close more quickly and to a greater extent when one cell or the other is depolarized. Although heterotypic gap junctions between astrocytes and oligodendrocytes have not been characterized electrophysiologically, it is expected that the heteromeric pairings of different connexins contributed by each cell type (Cx30 and Cx26 in astrocytes with Cx32 in oligodendrocytes; Cx43 in astrocytes with Cx47 in oligodendrocytes) will give rise to such asymmetric voltage dependence.

The structural changes involved in voltage-dependent gating have been hypothesized to involve several domains in the connexin protein, including the NT, where short segments of helical propensity have been identified in Cx26 (Purnick, et al. 2000), the most N-terminal amino acids in the first extracellular loop, where exchange can change the sign of voltage dependence (Verselis et al., 1994) and the CT, where truncation or addition of a large reporter (Aequorin or GFP) changes gating kinetics (Revilla et al., 1999). It has also been suggested that, as far as pH gating (next section), the CT and CL both participate in a "ball and chain" gating mechanism (Delmar et al., 2004).

5.3.2 pH and Ca^{2+} Sensitivity

Intracellular acidification closes gap junctions formed by all vertebrate connexins that have been studied (Fig. 5.6c), and quantitation reveals differential sensitivity of gap junctions formed by different connexins. For example, Cx43 gap junction channels are closed by relatively minor displacements of intercellular pH with an apparent pKa of about 6.8 (Fig. 5.5d), whereas Cx32 gap junctions are only slightly affected by acidification below pH 6.5 (Liu et al., 1993). In contrast to the voltage-dependent closure to a substate, gating by intracellular acidification is from fully open to fully closed states. The structural correlate of gating by intracellular acidification was initially proposed to involve titration of histidine residues located in the cytoplasmic loop domain, based on the rather neutral pKas of channel gating (see Spray and Burt, 1990). More recent NMR studies have shown that acidification of a peptide corresponding to the second half of this domain induces helical structure centered on two histidine residues in this region, with apparent pKa values very similar to that of channel closure (Fig. 5.6e). This structural change is presumed to underlie the increased affinity between the cytoplasmic CT domains (Duffy et al., 2004). The structural correlate of channel closure of this ball and chain mechanism (initially proposed by Mario Delmar (Francis et al., 1999)), on the basis of reduced pH sensitivity of truncation mutants is shown in Fig. 5.6f.

Unlike voltage dependence, which is of limited demonstrated physiological relevance (except perhaps in formation of boundaries during embryonic development (Harris et al., 1983)), the strong decrease in junctional conductance caused by modest intracellular acidification provides the possibility that this is a pathophysiological stimulus that cells may normally encounter. For example, under ischemic conditions where intracellular pH may fall to 6.7 or lower, gap junctions would be expected to close depending on their composition of connexins. That gap junctions have been reported to remain open following ischemic events between astrocytes and between *Xenopus* oocytes, presumably reflects either relative insensitivity of the connexins involved or altered sensitivity as a result of other factors operating simultaneously, such as dephosphorylation (Cusato et al., 2006; Cotrina et al., 1998).

Closure of gap junction channels associated with elevation of intracellular Ca^{2+} was an early correlation, first in insect cells and then in mammalian cells and cell lines. However, intracellular Ca^{2+} levels attained in the early experiments were clearly not physiological, and studies in which Ca^{2+} levels and intracellular pH were carefully controlled indicate that under these conditions very high levels of cytoplasmic Ca^{2+} (above 0.1 mM) are required (see Spray et al., 1982). Studies on

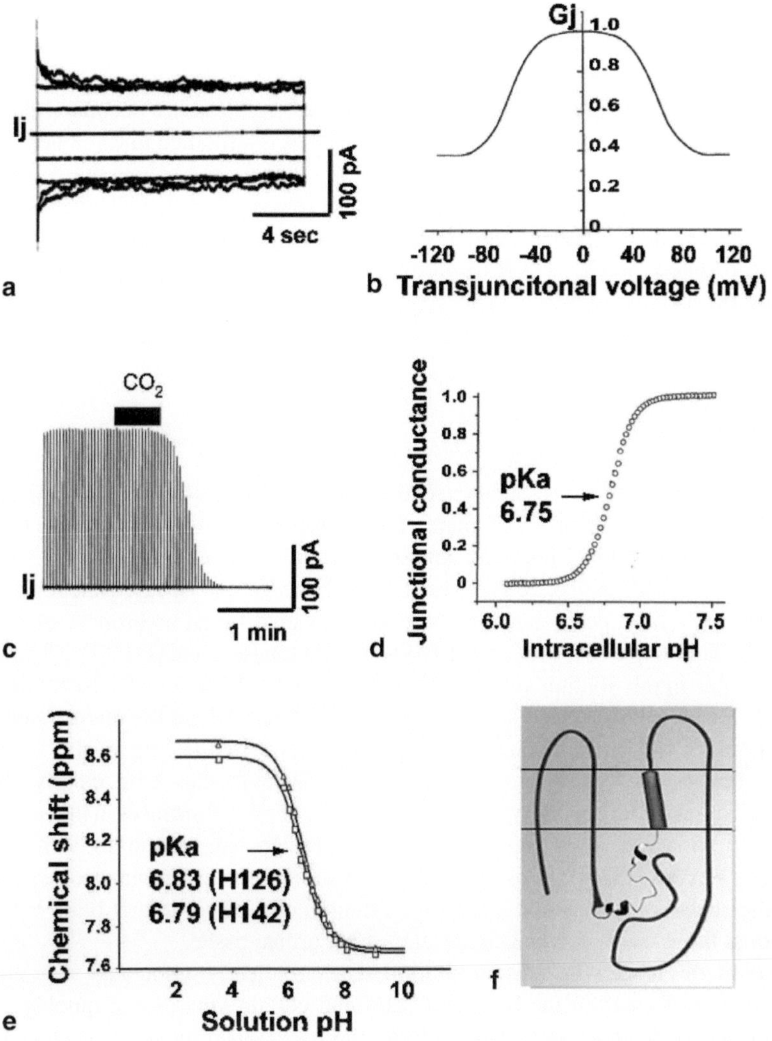

Fig. 5.6 Gating of gap junction channels. (a) Electrical recordings obtained from a pair of cells coupled by Cx43 gap junction channels showing the junctional current (Ij) obtained from one cell while applying transjunctional voltage steps ± 100 mV, 20-mV increments. (b) Voltage dependence of Cx43 gap junction channels. Maximal junctional conductance (Gj) occurs when there is no transjunctional voltage (0 mV) between a pair of coupled cells. This conductance decreases to about 60% of maximal when transjunctional voltages are above ±80 mV. The residual conductance at high voltages is likely due to the presence of subconductance states of Cx43 channels that are voltage insensitive. (c) Total closure of Cx43 gap junction channels occurs by intracellular acidification. Brief application of CO_2 in a solution bathing a pair of cells leads to fast decrease and total blockade of junctional current (Ij). (d) pH-dependence of Cx43 gap junction channels. At physiological pH (7.2–7.5), junctional conductance is maximal and decays to zero at pH below 6.5. The pKa (pH at which Gj is half of maximal) is around 6.75 (data from Francis et al, 1999). (e) NMR studies of titration of two histidine residues located at positions 126 and 142 of the intracellular loop of Cx43; note right shift with pKas corresponding to pKa for channel closure (Duffy et al., 2002) (f) Ball and chain model depicting the interaction between the second half of the intracellular loop and the carboxyl terminus of Cx43 that is responsible for the closure of the channels induced by acidification.

Xenopus oocytes expressing Cx46 demonstrated that reducing extracellular Ca^{2+} led to hemichannel opening (Paul et al., 1991) and Ca^{2+} removal has become a standard way of evoking dye uptake in mammalian cells that is attributed to Cx43 hemichannel opening (see Spray et al., 2006). However, as indicated below, the control for these studies is generally the use of poorly selective gap junction channel blockers, allowing the possibility that other channel types may contribute (in particular, $P2X_7$ receptor-linked pores, which profoundly increase activation in low divalent solution).

5.3.3 Phosphorylation

All connexins (with the possible exception of Cx26) possess serine, threonine, and tyrosine residues in their CT domains (and potentially in the cytoplasmic loop as well) that are potential targets for kinases and phosphatases. As noted by a recent review (Solan and Lampe, 2005) at least 12 of 21 serine and two tyrosine residues in the Cx43 CT have been shown to be phosphorylated by an assortment of protein kinases, including PKA, PKC, src, MAPK, casein kinase1, and $p34^{cde1/cyclin\ B}$ kinase.

Changes in distribution and phosphorylation state of Cx43 under ischemic conditions have been studied most thoroughly in the heart, where the intercalated disk containing Cx43 and associated proteins undergoes remodeling in the ischemic penumbra. Studies using phospho-specific Cx43 antibodies have clearly shown redistribution of dephosphorylated Cx43 toward lateral cell surfaces, a phenomenon believed to affect anisotropic conduction, predisposing tissue to arrhythmias (Beardslee et al., 2000). In astrocytes, chemical hypoxia–ischemia also rearranges Cx43 distribution (Nagy and Li, 2000), although changes in infarct brain have not explored the issue as extensively as studies in cardiac tissue.

Although changes in phosphorylation state are clearly associated with altered distribution of Cx43 in the ischemic brain and certain kinases can quickly affect junctional conductance in cells in culture, the mechanism by which such changes in junctional conductance and connexin distribution are effected most likely involves changes in both structures of connexin domains and changes in affinities of other molecules for those domains. The abundance of Cx43 is not altered by ischemia, whereas its recognition by an antibody to amino acids 342–360 is profoundly reduced (Nagy and Li, 2000). Although the roles of the different phosphorylation events and kinases/phosphatases that execute and reverse them remains to be clarified, it is certain that the phosphorylation states of Cx43 change during trafficking to the cell surface and during cell division (Solan and Lampe, 2005). One clue as to the role of such phosphorylation is the demonstration that phosphorylation and dephosphorylation of Cx43 alter its binding affinity for other proteins. Thus, the changes in affinity as a consequence of phosphorylation/dephosphorylation may be responsible for binding and unbinding reactions that propel Cx43 from its site of synthesis to the surface membrane, due in part to phosphorylation site-dependent interaction with cytoskeletal and signaling proteins (Li et al., 2005).

5.3.4 Pharmacological Blockade

The first pharmacological blockers of gap junction channels other than intracellular acidification were the local anesthetics heptanol and octanol, which inhibited conduction in the crayfish septate axon (Johnston et al., 1980). Subsequently, certain general anesthetics were shown to also reduce coupling, including halothane and isoflurane (Burt and Spray, 1989). In addition other lipophiles such as arachonic acid, oleamide (Guan et al., 1997; Spray and Burt 1990) and certain cardioactive metabolites have been demonstrated to have such effect. Modest connexin specificity for some of these agents has been found when comparing Cx40 and Cx43 (He and Burt, 2000), but generally such agents require high concentrations to exert their effects and act on many other channel types at lower concentrations.

Glycyrrhetinic acid derivatives, including α- and β-GA and carbonoxolone, were discovered to block gap junctions (Davidson and Baumgarten, 1988), and these agents have been extensively used because of their long-term tolerance (carbenoxolone is in clinical trials for ulcer treatment; see Farina et al., 1998). Concentrations in the range of 50–100 μM are required, and blockade may or may not be complete, depending on the preparation and perhaps connexin type. Carbenoxolone also has been used as a connexin-specific blocker to provide evidence for "hemichannels." However, it also blocks $P2X_7$ receptor-induced dye uptake at even lower concentrations than effective on gap junction channels (Suadicani et al., 2006), suggesting that it may also act on pannexin channels in nonjunctional membrane (Bruzzone et al., 2005).

Quinine was reported to open hemichannels formed of the fish homologue of Cx36 (Malchow et al., 1994) and to induce dye uptake and Ca^{2+} influx in astrocytes (Stout et al., 2002). However, studies on gap junctions formed by a variety of connexins indicated that this compound actually inhibits gap junction channels. This channel inhibition shows remarkable connexin specificity, totally blocking Cx36 and Cx50 channels at 50 μM while sparing other connexins such as Cx43 even at higher concentrations (Srinivas et al., 2001). The antimalarial quinine derivative mefloquine also displays this striking preference for blockade of gap junctions formed by certain connexins, although acting at much lower concentrations ($IC_{50} < 0.1$ μM for Cx36 (Cruikshank et al., 2004)). Mefloquine blocks $P2X_7$ receptor-mediated dye uptake at even lower concentrations ($IC_{50} < 10$ nM; Suadicani et al., 2006). Although these antimalarials have blocking effects on a range of other channel types, in particular K^+ channels, where quinine and its stereoisomer quinidine are widely used reagents, the connexin subtype specificity may offer the possibility for developing truly specific blocking agents in the future.

Another category of channel blocker that is potent with regard to gap junction channels is that of the flufenamic acid (FFA) family of chloride channel blockers. FFA and other related compounds were first shown to block lens connexins and subsequently found to potently inhibit gap junctions formed by a variety of connexins (Eskandari et al., 2002; Harks et al., 2001; Srinivas and Spray, 2003). Unlike quinine and its relatives, actions of these blockers do not appear to have great connexin specificity. FFA is also a potent inhibitor of Cx43 "hemichannels" and also blocks $P2X_7$ receptor-mediated dye uptake and pore currents.

Finally, the compound 2-aminoethoxydiphenyl borate, which was initially used as a blocker of the IP_3 receptor and recently has been found to either activate or inhibit TRP channels of various types, blocks gap junctions (Harks et al., 2003; Bai et al., 2006; Tao and Harris, 2007). This inhibition displays preference for certain connexin types over others and may therefore also offer a lead compound for the development of specific gap junction channel blockers. A recent study has begun to explore potency of related compounds, finding some with similar action (Tao and Harris, 2007).

The mechanism of action of these pharmacological agents is for the most part unknown. Lipophilic agents could modify the protein–lipid interface or the local concentration of individual lipids (Bastiaanse et al., 1993). Charged derivatives of quinine and FFA indicate a site of action that is accessible to the cytosol, but blockade is not use-dependent, indicating allosteric action rather than simply plugging the pore. More specific gap junction inhibitors are clearly needed and it is hoped that high throughput screening approaches will identify such molecules.

5.3.5 *Long-Term Increase in Coupling*

Under several pathophysiological conditions, changes in gap junctional communication are often observed. These alterations may involve long-term events involving changes in connexin expression levels or short-term changes due to modulation of channel activity. Among the several conditions that can affect the degree of coupling between astrocytes, there is one that is particularly interesting, because it illustrates that gap junctional communication is plastic and can act as a site for cellular "memory."

Several years ago, McCarthy's group (Enkvist and McCarthy, 1994) reported that coupling between astrocytes in culture was increased following exposure to high levels of glutamate and K^+. Subsequent studies performed by our group (De Pina-Benabou et al., 2001) found that short-term (5 min) exposure to high K^+ (10–50 mM) induced the increase in dye- and electrical coupling in astrocytes that persisted for as long as 2 h after K^+ washout. This long-term increase in coupling (LINC) in astrocytes was shown to be mediated by CaMKII (De Pina-Benabou et al., 2001), a protein that serves as a memory storage. Although it was suggested that the short-term effect of K^+ on the degree of coupling was related to increased number of open Cx43 channels present at the gap junction plaque (De Pina-Benabou et al., 2001), it is possible that recruitment of Cx43 into the gap junction mediate LINC.

Although further studies are still needed to fully understand this form of modulation, the presence of LINC in astrocytes suggest that similar to their counterparts (Cx36) found in electrical synapses (Pereda et al., 2004.), Cx43 provides astrocytes with a highly plastic form of gap junctional communication.

5.4 Gap Junction Alteration in Neuropathology and Hereditary Disease

5.4.1 *Neuro-Inflammatory Diseases*

Under several pathological conditions of the CNS, such as Alzheimer's disease (Nagy et al., 1996), traumatic brain injury (Rouach et al., 2002), ischemia (Nagy and Li, 2000), multiple sclerosis (Brand-Schieber et al., 2005; Roscoe et al., 2007), and in response to microbial pathogens (Campos de Carvalho et al., 1998; Zhao et al., 2006; Esen et al., 2007), alteration of Cx43 expression levels in astrocytes is often observed (reviewed by Dermietzel et al., 1998; Rouach et al., 2002; Kielian and Esen, 2004; Nakase and Naus, 2004). Although the direction in which Cx43 is regulated under these conditions may vary, changes in Cx43 expression are expected to impact not only on the dimension of the interconnected astrocytic network but also to have effects on maintenance of the neuronal microenvironment. From the conceptual view of a functional syncytium, changes in gap junctional communication are expected to impact on the coordination and cooperation of astrocytes to react to neuronal activity and environmental stimuli. However, the issue of whether gap junctional communication has a neuroprotective role is still controversial. In stroke (Rawanduzy et al., 1997; Saito et al., 1997), trauma (Frantseva et al., 2002), and ischemia (Cotrina et al., 1998) models it has been reported that gap junction channel blockers or the reduction of Cx43 expression could prevent secondary cell death. Contrary to these findings, however, studies on Cx43 HTs indicated that the infarct area was significantly increased compared with those of wild-type brains following traumatic injury (Frantseva et al., 2002) or after occlusion of middle cerebral artery (Siushansian et al., 2001). It is possible that because a variety of molecules can cross cell boundaries through gap junction channels (ions, second messengers, metabolites, etc.), what determines the extent of secondary cell death following CNS insult relies mainly on the nature of the signal transferred rather than the degree of coupling. Another point that should be considered when evaluating the role of gap junctions in the diseased CNS is the lack of specific gap junction channel blockers. Most, if not all compounds used to close gap junction channels affect other ion channels (Spray et al., 2002; Rouach et al., 2002; Suadicani et al., 2006), rendering interpretation not as straightforward as was previously thought. Moreover, the use of transgenic mice to overcome the lack of reagent specificity may also not be so simple. Deletion of the Gja1 gene has been shown to affect numerous other unrelated genes (see Iacobas et al., 2007a; Spray and Iacobas 2007), adding another level of complexity to the system (see below).

Thus, although it may take some time to fully understand the role played by gap junctional communication in CNS physiology and pathology, considerable advances have been made to define conditions and identify the signal transduction mechanisms that lead to connexin expression alterations under inflammatory situations. As mentioned earlier, in addition to their homeostatic role, astrocytes, together with microglia, are key participants in the initiation and regulation of CNS immune responses, producing and releasing several proinflammmatory cytokines and chemokines. Some of these proinflammatory agents, especially IL-1β and tumor necrosis factor (TNF)-α, have profound effects on gap junctional communication among astrocytes. Coculture studies of astrocytes and microglia, the predominant cytokine-releasing cell populations in the CNS, indicate that the degree of functional coupling and Cx43 expression in astrocytes are directly related to the level of microglia activation (Rouach et al., 2002; Faustmann et al., 2003; Hinkerohe et al., 2005; Meme et al., 2006). The main soluble factors released from lipopolysaccharide-activated microglia that are involved in the inhibition of gap junctional communication in astrocytes are the two proinflammatory cytokines IL-1β and TNF-α (Meme et al., 2006). It is interesting that in the context of neurodegenerative disorders, such as Alzheimer's disease, alteration of gap junction function may have considerable impact for the progression of the pathology. For instance, it has been recently shown that the susceptibility of astrocyte gap junctional communication to IL-1β and TNF-α is dramatically increased by treatment with a low concentration of the β-amyloid peptide Ab_{25-35}, which does not cause such an effect by itself (Meme et al., 2006).

Connexin expression and gap junctional communication between astrocytes are also affected following microbial infection. Both astrocytes and microglia express Toll-like receptors (TLRs), proteins involved in ligand recognition and in the triggering of a panoply of intracellular signaling pathways (Janeway and Medzhitov, 2002; O'Neill, 2006; Takeda and Akira, 2005). Activation of mouse astrocytes by *Staphylococcus aureus*, which is mediated by the TLR2, was reported to decrease the expression of Cx43 and Cx30, and to upregulate Cx26 and to cause blockade of dye-coupling (Esen et al., 2007). Activation of the TLR3 by the double-stranded RNA analog poly I:C was shown to lead to complete loss of Cx43 and dye-coupling in human fetal astrocytes, through the activation of nuclear factor-κB and PI3-kinase pathways (Zhao et al., 2006).

Despite the profound changes in astrocyte gap junction connectivity in inflamed CNS, these cells are able to maintain and even expand their degree of communication under these conditions through the spread of ICW (see above). In vitro studies performed with human fetal astrocytes indicated that IL-1β not only decreased Cx43 expression at protein and mRNA levels, but altered the way by which ICW is transmitted between cells (John et al., 1999). While in untreated, quiescent astrocyte cultures, gap junctions mediate the transmission of ICW, in IL-1β-activated astrocytes, these waves were mainly mediated by the extracellular route because of the increased expression of the purinergic receptor subtype $P2Y_2$ (John et al., 1999). A similar compensatory mechanism for ICW spread was also documented in mouse astrocytes lacking the Cx43 gene, in which changes in P2Y receptor subtypes were reported (Scemes et al., 2000; Suadicani et al., 2003). Besides affecting the metabotropic purinergic receptor, IL-1β, the activation of TLR3 was also shown to increase the expression of the ionotropic $P2X_7$ and $P2X_4$ receptors, respectively, in human fetal astrocytes (Narcisse et al., 2005; Zhao et al., 2006). The increase in expression of the pore-forming $P2X_7R$ is likely to contribute to the recruitment of a larger number of astrocytes enrolled in the transmission of ICW than that provided by gap junctional communication. Indeed, these receptors, which activate pannexin1 (Locovei et al., 2007; Pelegrin and Surprenant, 2006), have been shown to provide sites of ATP release and to participate in the amplification of ICW spread among astrocytes (Suadicani et al., 2006, 2007).

These studies showing a switch in the two main ICW components (gap junctions and purinergic receptors) suggest that astrocyte networks are tightly regulated, such that decrease in gap junctional communication is compensated by alteration of P2 receptors, possibly to maintain network integrity.

5.4.2 Human Genetic Diseases Involving Glial Connexins

Human genetic diseases due to coding region mutations in connexin genes include zonular pulverulent cataracts (reported for mutations in both Cx46 and Cx50, the gap junction proteins of mature lens fibers), erythrokeratodermia (due to mutations in Cx31), and two relatively common diseases of the nervous system: the X-linked form of Charcot-Marie-Tooth disease (CMTX) and hereditary nonsyndromic deafness (HNSD), which involve Cx32 and Cx26/Cx30 mutations, respectively (White and Paul, 1999). In addition, Cx47 and Cx43 mutations now appear to underlie Pelizaeus-Merzbacker-like disease and occulodentodigital dysplasia, respectively. We limit this discussion to diseases affecting glia and direct the reader to recent reviews considering the other hereditary diseases (e.g., Richard, 2005; Goodenough and Paul, 2003; Willecke et al., 2002).

CMTX is a generally late onset demyelinating peripheral neuropathy involving mutations in Cx32 (for recent update, see Kleopa and Scherer, 2006). This gap junction protein is normally found at nodes of Ranvier and Schmidt-Lantermann incisures, forming reflexive gap junctions between cytoplasmic pockets squeezed off from the compact myelin. Presumably, the reflexive gap junctions provide a shunt for delivery of signaling molecules and metabolites from outermost to innermost lamellae of myelinating Schwann cells. Thus, the vulnerability of myelinating Schwann cells to the loss of Cx32 has been attributed to the loss of this vital signal exchange shortcut. The number of distinct mutations found to be associated with CMTX currently exceeds 200, distributed throughout the Cx32 molecule; many mutations have been shown to disrupt membrane trafficking of Cx32 protein when expressed exogenously in *Xenopus* oocytes or mammalian cells, and a few appear to alter the function of the channels at the junctional membrane.

Recently, it has been discovered that mutations in Cx47 cause Pelizaeus-Merzbacher-like disease, which is characterized by severe CNS dysmyelination (Orthmann-Murphy et al., 2007). Three tested mutations (P875S, Y269D, and M283T) did not form functional channels when expressed in HeLa cells.

HNSD is a major cause of early onset hearing loss and is due to mutations of Cx26 and Cx30. The two most common HNSD mutations are frame-shift deletions that result in severe truncation of this connexin, thereby eliminating both cytoplasmic and membrane-spanning domains. HNSD mutations are believed to lead to the lack of functional coupling among supporting cells in the cochlea, where Cx26 normally provides a route for recycling of potassium ions, a role that is fundamental for the generation of endocochlear potential (Sabag et al., 2005; Wei et al., 2004).

Distinct Cx43 coding region mutations have recently been discovered in more than 30 occulodentodigital dysplasia families, highlighting the complex multiple effects that connexin disruption may have at the level of the organism (Shibayana et al., 2005). These mutations are associated with abnormalities in bone and teeth, as well as in white-matter disturbances in affected families (Shibayama et al., 2005; Loddenkemper et al., 2002). Many of the more than 30 mutations have now been characterized in exogenous expression systems, where all have been shown to disrupt Cx43 location within the cell and, in cases where mutant protein is incorrectly targeted to junctional membrane, channels are not functional (Lai et al., 2006; Shibayama et al., 2005; Gong et al., 2006, 2007).

5.4.3 Transgenic Abnormalities Involving Glial Gap Junctions

One universal role that gap junctions presumably serve is in maintaining tissue homogeneity. Therefore, it might be expected that tissue boundaries would correspond to regions where coupling is lost, either by decreased expression of a connexin or by expression of one to which the endogenous boundary preserving connexon would not pair. One such example is in the insect cuticle, where boundaries are established by the creation of a barrier of gap junction deficient cells (Weir and Lo, 1985). In developing mesoderm, it has recently been reported that Cx43 might play such a role, in that abnormal expression of the ephrin orphan receptor resulted in downregulation of Cx43 expression and decreased coupling at compartmental boundaries, resulting in craniofacial defects (Davy et al., 2006). Overexpression of Cx43 rescued both boundary formation and developmental defects caused by the abnormal ephrin receptor expression.

A possibly related example of compensatory genetic interaction between Cx43 and expression of other genes *in vivo* involves the signaling molecule Wnt1. In cardiac myocytes, Wnt1 was shown to upregulate Cx43 through a mechanism involving increased expression of β-catenin, which bound to Cx43 at the surface membrane (Ai et al., 2000). When Cx43 levels were reduced, free β-catenin translocated to the nucleus, upregulating Cx43 expression. In a recent report, Wnt1 deficiency was found to produce a brain phenotype characterized by decreased size of the brain stem and cerebellum (Melloy et al., 2005). These gross abnormalities were restored by overexpression of Cx43. Even more intriguingly, a recent report examining phenotypes of two mouse strains in which deletion of Cx43 was targeted to astrocytes through a Cre-Lox system (also see Chap. 14) provided evidence that there was a strain-dependent susceptibility to abnormal development of the cerebellum and hippocampus (Wiencken-Barger et al., 2007).

5.5 Regulation of Gene Expression by Glial Gap Junctions

The phenotype of cultured astrocytes from Cx43 null mice differs from WT in a number of characteristics, including severely retarded growth (Dermietzel et al., 2000) and altered expression of P2Y receptors (Scemes et al., 2000; Suadicani et al., 2003). These differences were surprising, in that they do not seem to be readily explained by a decrease in coupling between the cells and, in the case of decreased growth rate, appear opposite to what would have been expected if gap junctions between cells were necessary to limit growth rate (see Leithe et al., 2006; Kardami et al., 2007). The first evidence that gap junction gene expression could affect expression of other genes was the report by Naus et al. (2000) through the use of differential display RNA hybridization that expression of several genes were regulated in C6 glioma cells following transfection with Cx43, including secreted factors that regulate growth and tumorigenicity.

Our studies of gene expression using microarrays to simultaneously evaluate thousands of genes in brains and astrocytes from connexin-null mice revealed significant changes in expression of >10% of the transcriptome (Iacobas et al., 2007a; Spray and Iacobas, 2007). Altered genes extended to all functional categories of encoded proteins and to all chromosomal locations, the latter finding indicating that gene expression changes were not likely congenic effects of the transgene. To test the hypothesis that the changes in the Cx43 null brain might be due to alterations in gene interlinkage, we measured the correlation between expression levels of Cx43 and every other gene across four arrays from WT astrocytes and brain, finding many genes that had significant positive or negative correlation with Cx43. When we compared the coordination scores (Pearson coefficients) in the wild-type brain with up- and downregulation in Cx43 nulls, we found a very strong predictive value, with more than 80% of the regulations being anticipated by the correlations. This finding suggests that the null phenotype is a simple extension of wild-type coordinated variability of gene expression levels and that gap junction genes may be considered to be hubs of gene expression networks, with altered interlinkages as a consequence of deletion of the connexin gene responsible for the phenotypes.

References

Ai Z, Fischer A, Spray DC, Brown AM, Fishman GI (2000) Wnt-1 regulation of connexin43 in cardiac myocytes. *J Clin Invest* 105:161–171.

Altevogt BM, Paul DL (2004) Four classes of intercellular channels between glial cells in the CNS. *J Neurosci* 24:4313–4323.

Altevogt BM, Kleopa KA, Postma FR, Scherer SS, Paul DL (2002) Connexin29 is uniquely distributed within myelinating glial cells of thc central and peripheral nervous system. *J Neurosci* 22:6458–6470.

Amzica F, Massimini M, Manfridi A (2002) Spatial buffering during slow and paroxysmal sleep oscillations in cortical networks of glial cells in vivo. *J Neurosci* 22:1042–1053.

Anderson CM, Bergher JP, Swanson RA (2004) ATP-induced ATP release from astrocytes. *J Neurochem* 88:246–256.

Andrade-Rozental AF, Rozental R, Hopperstad MG, Wu JK, Vrionis FD, Spray DC (2000) Gap junctions: the "kiss of death" and the "kiss of life." *Brain Res Rev* 32:308–315.

Azzam EI, de Toledo SM, Little JB (2001) Direct evidence for the participation of gap-junction mediated intercellular communication in the transmission of damage signals from γ-particle irradiated to nonirradiated cells. *Proc Natl Acad Sci USA* 98:473–478.

Bai D, del Corsso C, Srinivas M, Spray DC (2006) Block of specific gap junction channel subtypes by 2-aminoethoxydiphenyl borate (2-APB). *J Pharmacol Exp Ther* 319:1452–1458.

Balice-Gordon RJ, Bone LJ, Scherer SS (1998) Functional gap junctions in the Schwann cell myelin sheath. *J Cell Biol* 142:1095–1104.

Ball KK, Gandhi GK, Thrash J, Cruz NF, Dienel GA (2007) Astrocytic connexin distributions and rapid, extensive dye transfer via gap junctions in the inferior colliculus: Implications for [(14) C]glucose metabolite trafficking. *J Neurosci Res* 15: 3267–3283.

Bao L, Locovei S, Dahl G (2004) Pannexin membrane channels are mechanosensitive conduits for ATP. *FEBS Lett* 572:65–68.

Baranova A, Ivanov D, Petrash N, Pestova A, Skoblov M, Kelmanson I, Shagin D, Nazarenko S, Geraymovych E, Litvin O, Tiunova A, Born TL, Usman N, Staroverov D, Lukyanov S, Panchin, Y (2004) The mammalian pannexin family is homologous to the invertebrate innexin gap junction proteins. Genomics 83:706–716.

Bastiaanse EM, Jongsma HJ, van der Laarse A, Takens-Kwak BR (1993) Heptanol-induced decrease in cardiac gap junctional conductance is mediated by a decrease in the fluidity of membranous cholesterol-rich domains. *J Membr Biol* 136:135–145.

Beardslee MA, Lerner DL, Tadros PN, Laing JG, Beyer EC, Yamada KA, Kleber AG, Schuessler RB, Saffitz JE (2000) Dephosphorylation and intracellular redistribution of ventricular connexin43 during electrical uncoupling induced by ischemia. *Circ Res* 87:656–662.

Bennett MV, Contreras JE, Bukauskas FF, Saez JC (2003) New roles for astrocytes: Gap junction hemichannels have something to communicate. *Trends Neurosci* 26: 610–617.

Blomstrand F, Aberg ND, Eriksson PS, Hansson E, Ronnback L (1999) Extent of intercellular calcium wave propagation is related to gap junction permeability and level of connexin43 expression in astrocytes in primary cultures from four brain regions. *Neuroscience* 92:255–265.

Boassa D, Ambrosi C, Qiu F, Dahl G, Gaietta G, Sosinsky G (2007) Pannexin1 channels contain a glycosylation site that targets the hexamers to the plasma membrane. *J Biol Chem* 282: 31733–31743.

Boitier E, Rea R, Duchen MR (1999) Mitochondria exert a negative feeback on the propagation of intracellular Ca^{2+} waves in rat cortical astrocytes. *J Cell Biol* 145:795–808.

Brand-Schieber E, Werner P, Iacobas DA, Iacobas S, Beelitz M, Lowery SL, Spray DC, Scemes E (2005) Connexin43, the major gap junction protein of astrocytes, is down-regulated in inflamed white matter in an animal model of multiple sclerosis. *J Neurosci Res* 80:798–808.

Brightman MW, Reese TS (1969) Junctions between intimately apposed cell membranes in the vertebrate brain. *J Cell Biol* 40:648–677.

Broer S, Brookes N (2001) Transfer of glutamine between astrocytes and neurons. *J Neurochem* 77:705–719.

Bruzzone R, Hormuzdi SG, Barbe MT, Herb A, Monyer H (2003) Pannexins, a family of gap junction proteins expressed in brain. *Proc Natl Acad Sci USA* 100:13644–13649.

Bruzzone R, Barbe MT, Jakob NJ, Monyer H (2005) Pharmacological properties of homomeric and heteromeric pannexin hemichannels expressed in Xenopus oocytes. *J Neurochem* 92:1033–1043.

Bukauskas FF, Verselis VK (2004) Gap junction channel gating. *Biochim Biophys Acta* 1662:42–60.

Burt JM, Spray DC (1989) Volatile anesthetics block intercellular communication between neonatal rat myocardial cells. *Circ Res* 65:829–837.

Bushong EA, Martone ME, Jones YZ, Ellisman MH (2002) Protoplasmic astrocytes in CA1 stratum radiatum occupy separate anatomical domains. *J Neurosci* 22:183–192.

Campos de Carvalho AC, Roy C, Hertzberg EL, Tanowitz HB, Kessler JA, Weiss LM, Wittner M, Dermietzel R, Gao Y, Spray DC (1998) Gap junction disappearance in astrocytes and leptomeningeal cells as a consequence of protozoan infection. *Brain Res* 790:304–314.

Caspar DLD, Goodenough DA, Makowski L, Phillips WC (1977) Gap junction structures: I. Correlated electron microscopy and X-ray diffraction. *J Cell Biol* 74:605–628.

Chanson M, Roy C, Spray DC (1994) Voltage-dependent gap junctional conductance in hepatopancreatic cells of *Procambarus clarkii. Am J Physiol* 266:C569–C577.

Charles A (1998) Intercellualr calcium waves in glia. *Glia* 24:39–49.

Charles AC, Giaume G (2002) Intercellualr calcium waves in astrocytes: Underlying mechanisms and functional consequences. In: The tripartite synapse: Glia in synaptic transmission (Volterra A, Magistretti P, Haydon PG, eds) pp. 100–126. Oxford, New York: Oxford University Press.

Charles AC, Merrill JE, Dirksen ER, Sanderson MJ (1991) Intercellular signaling in glial cells: Calcium waves and oscillations in response to mechanical stimulation and glutamate. *Neuron* 6: 983–992.

Charles AC, Naus CC, Zhu D, Kidder GM, Dirksen ER, Sanderson MJ (1992) Intercellular calcium signaling via gap junctions in glioma cells. *J Cell Biol* 118:195–201.

Charles AC, Dirksen ER, Merrill JE, Sanderson MJ (1993) Mechanisms of intercellular calcium signaling in glial cells studied with dantrolene and thapsigargin. *Glia* 7:134–145.

Chen KC, Nicholson C (2000) Spatial buffering of potassium ions in brain extracellular space. *Biophys J* 78:2776–2797.

Chow I, Poo MM (1984) Formation of electrical coupling between embryonic *Xenopus* muscle cells in culture. *J Physiol* 346:181–194.

Cole WC, Garfield RE (1986) Evidence for physiological regulation of gap junction permeability. *Am J Physiol* 251:C411–C420.

Cole WC, Garfield RE, Kirkaldy JS (1985) Gap junctions and direct intercellular communication between rat uterine smooth muscle cells. *Am J Physiol* 249:C20–C31.

Contreras JE, Saez JC, Bukauskas FF, Bennett MV (2003) Gating and regulation of connexin43 (Cx43) hemichannels. *Proc Natl Acad Sci USA* 100:11388–11393.

Contreras JE, Sanchez HA, Veliz LP, Bukauskas FF, Bennett MVL, Saez JC (2004) Role of connexin-based gap junction channels and hemichannels in ischemia-induced cell death in nervous tissue. *Brain Res Rev* 47: 290–303.

Cotrina ML, Kang J, Lin H-C, Bueno E, Hansen TW, He L, Liu Y, Nedergaard M (1998a) Astrocytic gap junctions remain open during ischemic conditions. *J Neurosci* 18:2520–2537.

Cotrina ML, Lin JH, Alves-Rodrigues A, Liu S, Li J, Azmi-Ghadimi H, Kang J, Naus CC, Nedergaard M (1998b) Connexins regulate calcium signaling by controling ATP release. *Proc Natl Acad Sci USA* 95:15735–15740.

Cotrina MK, Lin JH, Lopez-Garcia JC, Naus CC, Nedergaard M (2000) ATP-mediated glia signaling. *J Neurosci* 20:2835–2844.

Cruikshank SJ, Hopperstad M, Younger M, Connors BW, Spray DC, Srinivas M (2004) Potent block of Cx36 and Cx50 gap junction channels by mefloquine. *Proc Natl Acad Sci USA* 101:12364–12369.

Cruz N, Ball KK, Dienel G (2007) Functional imaging of focal brain activation in conscious rats: Impact of ^{14}C-glucose metabolite spreading and release. *J Neurosci Res* 85:3254–3266.

Cusato K, Ripps H, Zakevicius J, Spray DC (2006) Gap junctions remain open during cytochrome c-induced cell death: Relationship of conductance to 'bystander' cell killing. *Cell Death Differ* 13:1707–1714.

Davidson JS, Baumgarten IM (1998) Glycyrrhetinic acid derivatives: A novel class of inhibitors of gap-junctional intercellular communication. Structure-activity relationships. *J Pharmacol Exp Ther* 246:1104–1107.

Davy A, Bush JO, Soriano P (2006) Inhibition of gap junction communication at ectopic Eph/ephrin boundaries underlies craniofrontonasal syndrome. *PLoS Biol* 4:e315.

del Corsso C, Srinivas M, Urban-Maldonado M, Moreno AP, Fort AG, Fishman GI, Spray DC (2006) Transfection of mammalian cells with connexins and measurement of voltage sensitivity of their gap junctions. *Nat Protoc* 1:1799–1809.

De Pina-Benabou MH, Srinivas M, Spray DC, Scemes E (2001) Calmodulin kinase pathway mediates the K^+-induced increase in gap junctional communication between mouse spinal cord astrocytes. *J Neurosci* 21:6635–6643.

Delmar M, Coombs W, Sorgen P, Duffy HS, Taffet SM (2004) Structural bases for the chemical regulation of connexin43 channels. *Cardiovasc Res* 62:268–275.

Dermietzel R (1974) Junctions in the central nervous system of the cat. 3. Gap junctions and membrane-associated orthogonal particle complexes (MOPC) in astrocytic membranes. *Cell Tissue Res* 149:121–135.

Dermietzel R (1998) Gap junction wiring: a "new" principle in cell-to-cell communication in the nervous system. *Brain Res Brain Res Rev* 26:176–183.

Dermietzel R, Gao Y, Scemes E, Vieira D, Urban M, Kremer M, Bennett MV, Spray DC (2000) Connexin43 null mice reveal that astrocytes express multiple connexins. *Brain Res Brain Res Rev* 32:45–56.

Dewey MW, Barr L (1962) Intercellular connections between smooth muscle cell: The nexus. *Science* 137:670–672.

Dienel GA, Hertz L (2001) Glucose and lactate metabolism during brain activation. *J Neurosci Res* 66: 824–839.

Dienel G, Ball K, Popp D, Cruz NF (2001) A role of gap junctions in metabolite spreading? *J Neurochem* 78(Suppl 1): 86.

Dobrenis K, Chang HY, Pina-Benabou MH, Woodroffe A, Lee SC, Rozental, R, Spray DC, Scemes E (2005) Human and mouse microglia express connexin36, and functional gap junctions are formed between rodent microglia and neurons. *J Neurosci Res* 82:306–315.

Duffy HS, John GR, Lee SC, Brosnan CF, Spray DC (2000) Reciprocal regulation of the junctional proteins claudin-1 and connexin43 by interleukin-1β in primary human fetal astrocytes. *J Neurosci* 20:RC114.

Duffy HS, Sorgen PL, Girvin ME, O'Donnell P, Coombs W, Taffet SM, Delmar M, Spray DC (2002) pH-dependent intramolecular binding and structure involving Cx43 cytoplasmic domains. *J Biol Chem* 277:36706–36714.

Duffy HS, Ashton AW, O'Donnell P, Coombs W, Taffet SM, Delmar M, Spray DC (2004) Regulation of connexin43 protein complexes by intracellular acidification. *Circ Res* 94:215–222.

Duffy HS, Fort AG, Spray DC (2006) Cardiac connexins: Genes to nexus. *Adv Cardiol* 42:1–17.

Ek-Vitorin JF, Calero G, Morley GE, Coombs W, Taffet SM, Delmar M (1996) pH regulation of connexin43: Molecular analysis of the gating particle. *Biophys J* 71: 1273–1284.

Enkvist MO, McCarthy KD (1992) Activation of protein kinase C blocks astroglial gap junction communication and inhibits the spread of calcium waves. *J Neurochem* 59:519–526.

Enkvist MO, McCarthy KD (1994) Astroglial gap junction communication is increased by treatment with either glutamate or high K^+. *J Neurochem* 52: 489–495.

Epstein ML, Gilula NB (1977) A study of communication specificity between cells in culture. *J Cell Biol* 75:769–787.

Esen N, Shuffield D, Syed MM, Kielian T (2007) Modulation of connexin expression and gap junction communication in astrocytes by the Gram-postive bacterium *S. aureus*. *Glia* 55:104–117.

Eskandari S, Zampighi GA, Leung DW, Wright EM, Loo DD (2002) Inhibition of gap junction hemichannels by chloride channel blockers. *J Membr Biol* 185:93–102

Estin D, Li M, Spray D, Wu JK (1999) Connexins are expressed in primary brain tumors and enhance the bystander effect in gene therapy. *Neurosurgery* 44:361–368.

Evans WH, De Vuyst E, Leybaert L (2006) The gap junction cellular internet: Connexin hemichannels enter the signaling limelight. *Biochem J* 397:1–14.

Farina C, Pinza M, Pifferi G (1998) Synthesis and anti-ulcer activity of new derivatives of glycyrrhetic, oleanolic and ursolic acids. *Farmaco* 53:22–32.

Faustmann PM, Haase CG, Romberg S, Hinkerohe D, Szlachta D, Smikalla D, Krause D, Dermietzel R (2003) Microglia activation influenced dye coupling and Cx43 expression of astrocytic network. *Glia* 42:101–108.

Fawcett DW (1994) Bloom and Fawcett: A textbook of histology (12th ed), New York: Chapman and Hall.

Filosa JA, Bonev AD, Straub SV, Meredith AL, Wilkerson MK, Aldrich RW, Nelson MT (2006) Local potassium signaling couples neuronal activity to vasodilation in the brain. *Nat Neurosci* 9:1397–1403.

Finkbeiner S (1992) Calcium waves in astrocytes-filling in the gaps. *Neuron* 8:1101–1108

Foote CI, Zhou L, Zhu X, Nicholson BJ (1998) The pattern of disulfide linkages in the extracellular loop regions of connexin32 suggests a model for the docking interface of gap junctions. *J Cell Biol* 140:1187–1197.

Francis D, Stergiopoulos K, Ek-Vitorin JF Cao FL, Taffet SM, Delmar M (1999) Connexin diversity and gap junction regulation by pHi. *Dev Genet* 24:123–136.

Frantseva MV, Kokarovtseva L, Velazquez JLP (2002) Ischemia-induced brain damage depends on specific gap junctional coupling. *J Cereb Blood Flow Metab* 22:453–462.

Fumagalli M, Brambilla R, D'Ambrosi N, Volonte C, Matteoli M, Verderio C, Abbrachio MP (2003) Nucleotide-mediated calcium signaling in rat cortical astrocytes: Role of P2X and P2Y receptors. *Gila* 43:218–223.

Gaietta G, Deerinck TJ, Adams SR, Bouwer J, Tour O, Laird DW, Sosinsky GE, Tsien RY, Ellisman MH (2002) Multicolor and electron microscopic imaging of connexin trafficking. *Science* 296:503–507.

Gardner-Medwin AR (1983) A study of the mechanisms by which potassium moves through brain tissue in the rat. *J Physiol* 335:353–374.

Ghoshroy S, Goodenough DA, Sosinsky GE (1995) Preparation, characterization, and structure of half gap junctional layers split with urea and EGTA. *J Membr Biol* 146:15–28.

Giaume C, Venance L (1998) Intercellular calcium signaling and gap junction communication in astrocytes. *Glia* 24:50–64.

Giaume C, Tabernero A, Medina JM (1997) Metabolic trafficking through astrocytic gap junctions. *Glia* 21:114–123.

Giepmans BN (2006) Role of connexin43-interacting proteins at gap junctions. *Adv Cardiol* 42:41–56.

Gong XQ, Shao Q, Lounsbury CS, Bai D, Laird DW (2006) Functional characterization of a GJA1 frameshift mutation causing oculodentodigital dysplasia and palmoplantar keratoderma. *J Biol Chem* 281:31801–31811.

Gong XQ, Shao Q, Langlois S, Bai D, Laird DW (2007) Differential potency of dominant negative connexin43 mutants in oculodentodigital dysplasia. *J Biol Chem* 282:19190–19202.

Gonzalez D, Gomez-Hernandez JM, Barrio LC (2007) Molecular basis of voltage dependence of connexin channels: An integrative appraisal. *Prog Biophys Mol Biol* 94:66–106.

Goodenough DA, Paul DL (2003) Beyond the gap: Functions of unpaired connexon channels. *Nat Rev Mol Cell Biol* 4:1–10.

Goodenough DA, Dick JSB, Lyons JE (1980) Lens metabolic cooperation: A study of lens mouse transport and permeability visualized with freeze-substitution and electron microscopy. *J Cell Biol* 86:576–589.

Grosche J, Matyash V, Moller T, Verkhratsky A, Reichenbach A, Kettenmann H (1999) Microdomains for neuron–glia interaction: Parallel fiber signaling to Bergmann glial cells. *Nat Neurosci* 2:139–143.

Guan X, Cravatt BF, Ehring GR, Hall JE, Boger DL, Lerner RA, Gilula NB (1997) The sleep-inducing lipid oleamide deconvolutes gap junction communication and calcium wave transmission in glial cells. *J Cell Biol* 139:1785–1792.

Guthrie PB, Knappenberger J, Segal M, Bennett MV, Charles A, Kater, SB (1999) ATP released from astrocytes mediates glial calcium waves. *J Neurosci* 19:520–528.

Haas B, Schipke CG, Peters O, Sohl G, Willecke K, Kettenmann H (2006) Activity-dependent ATP-waves in the mouse neocortex are independent from astrocytic calcium waves. *Cereb Cortex* 16:237–246.

Harks EG, de Roos AD, Peters PH, de Haan LH, Brouwer A, Ypey DL, van Zoelen EJ, Theuvenet AP (2001) Fenamates: A novel class of reversible gap junction blockers. *J Pharmacol Exp Ther* 298:1033–1041.

Harks EG, Camina JP, Peters PH, Ypey DL, Scheenen WJ, van Zoelen EJ, Theuvenet AP (2003) Besides affecting intracellular calcium signaling, 2-APB reversibly blocks gap junctional coupling in confluent monolayers, thereby allowing measurement of single-cell membrane currents in undissociated cells. *FASEB J* 17:941–943.

Harris AL (2007) Connexin channel permeability to cytoplasmic molecules. *Prog Biophys Mol Biol* 94:120–143.

Harris AL, Spray DC, Bennett MV (1983) Control of intercellular communication by voltage dependence of gap junctional conductance. *J Neurosci* 3:79–100.

Hassinger TD, Guthrie PB, Atkinson PB, Bennett MV, Kater SB (1996) An extracellular signaling component in propagation of astrocytic calcium waves. *Proc Natl Acad Sci USA* 93:13268–13273.

He DS, Burt JM (2000) Mechanism and selectivity of the effects of halothane on gap junction channel function. *Circ Res* 86:E104–E109.

Herve JC, Bourmeyster N, Sarrouilhe D, Duffy HS (2007) Gap junctional complexes: From partners to functions. *Prog Biophys Mol Biol* 94:29–65.

Higashi K, Fujita A, Inanobe A, Tanemoto M, Doi K, Kubo T, Kurachi Y (2001) An inwardly rectifying K^+ channel, Kir4.1, expressed in astrocytes surrounds synapses and blood vessels in brain. *Am J Physiol* 281:C922–C931.

Hinkerohe D, Smikalla D, Haghikia A, Heupel K, Haase CG, Dermietzel R, Faustmann PM. (2005) Effects of cytokines on microglial phenotypes and astroglial coupling in an inflammatory coculture model. *Glia* 52:85–97.

Hirst-Jensen BJ, Sahoo P, Kieken F, Delmar M, Sorgen PL (2007) Characterization of the pH-dependent interaction between the gap junction protein connexin43 carboxyl terminus and cytoplasmic loop domains. *J Biol Chem* 282:5801–5813.

Ho C, Hicks J, Salter MW (1995) A novel P2-purinoreceptor expressed by a subpopulation of astrocytes from the dorsal spinal cord of the rat. *Br J Pharmacol* 116:2909–2918.

Hua VB, Chang AB, Tchieu JH, Kumar NM, Nielsen PA, Saier MH Jr (2003) Sequence and phylogenetic analyses of 4 TMS junctional proteins of animals: Connexins, innexins, claudins and occludins. *J Membr Biol* 194:59–76.

Huang Y, Grinspan JB, Abrams CK, Scherer SS (2007) Pannexin1 is expressed by neurons and glia but does not form functional gap junctions. *Glia* 55:46–56.

Iacobas DA, Suadicani SO, Spray DC, Scemes E (2006) A stochastic two-dimensional model of intercellular Ca^{2+} wave spread in glia. *Biophys J* 90:24–41.

Iacobas DA, Iacobas S, Spray DC (2007a). Connexin-dependent transcellular transcriptomic networks in mouse brain. *Prog Biophys Mol Biol.* 94:169–185.

Iacobas DA, Suadicani SO, Iacobas S, Chrisman C, Cohen MA, Spray DC, Scemes E (2007b) Gap junction and purinergic P2 receptor proteins as a functional unit: Insights from transcriptomics. *J Membr Biol* 217:83–91.

Idestrup CP, Salter MW (1998) P2Y and P2U receptors differentially release intracellular Ca2+ via the phospholipase c/inositol 1,4,5-tri-phosphate pathway in astrocytes from the dorsal spinal cord. *Neuroscience* 86:913–923.

Janeway CA Jr, Medzhitov R (2002) Innate immune recognition. *Annu Rev Immunol* 20:197–216.

John GR, Scemes E, Suadicani SO, Liu JS, Charles PC, Lee SC, Spray DC, Brosnan CF (1999) IL-1 β differentially regulates calcium wave propagation between primary human fetal astrocytes via pathways involving P2 receptors and gap junction channels. *Proc Natl Acad Sci USA* 96:11613–11618.

John S, Cesario D, Weiss JN (2003) Gap junctional hemichannels in the heart. *Acta Physiol Scand* 179: 23–31.

Johnston MF, Simon SA, Ramon F (1980) Interaction of anaesthetics with electrical synapses. *Nature* 286:498–500.

Kardami E, Dang X, Iacobas DA, Nickel BE, Jeyaraman M, Srisakuldee W, Makazan J, Tanguy S, Spray DC (2007) The role of connexins in controling cell growth and gene expression. *Prog Biophys Mol Biol* 94:245–264.
Kielian T, Esen N (2004) Effects of neuroinflammation on glia–glia gap junctional intercellular communication: A perspective. *Neurochem Int* 45:429–436.
Kleopa KA, Scherer SS (2006) Molecular genetics of X-linked Charcot-Marie-Tooth disease. *Neuromolecular Med* 8:107–122.
Kofuji P, Biedermann B, Siddharthan VT, Raap M, Iandiev I, Milenkovic I, Thomzig A, Veh RD, Bringmann A, Reichenbach A (2002) Kir potassium channel subunit expression in retinal glial cells: Implications for spatial potassium buffering. *Glia* 39:292–303.
Kuffler SW, Nicholls JG, Orkand RK (1966) Physiological properties of glial cells in the central nervous system of amphibia. *J Neurophysiol* 29:768–787.
Lai A, Le DN, Paznekas WA, Gifford WD, Jabs EW, Charles AC (2006) Oculodentodigital dysplasia connexin43 mutations result in non-functional connexin hemichannels and gap junctions in C6 glioma cells. *J Cell Sci* 119:532–541.
Lai CP, Bechberger JF, Thomson RJ, MacVicar BA, Bruzzone R, Naus CC (2007) Tumor suppressive effects of pannexin 1 in C6 glioma cells. *Cancer Res* 67:1545–1554.
Laird DW (2006) Life cycle of connexins in health and disease. *Biochem J* 394:527–543.
Larsen WJ, Tung HN, Murray SA, Swenson CA (1979) Evidence for the participation of actin microfilaments and bristle coats in the internalization of gap junction membrane. *J Cell Biol* 83:576–587.
Leithe E, Sirnes S, Omori Y, Rivedal E (2006) Downregulation of gap junctions in cancer cells. *Crit Rev Oncog* 12:225–256.
Leybaert L, Paemeleire K, Strahonja A, Sanderson MJ (1998) Inositol-trisphosphate-dependent intercellular calcium signaling in and between astrocytes and endothelial cells. *Glia* 24:398–407.
Li W, Hertzberg EL, Spray DC (2005) Regulation of connexin43-protein binding in astrocytes in response to chemical ischemia/hypoxia. *J Biol Chem* 280:7941–7948.
Liu F, Arce FT, Ramachandran S, Lal R (2006) Nanomechanics of hemichannel conformations: Connexin flexibility underlying channel opening and closing. *J Biol Chem* 281:23207–23217.
Liu S, Taffet S, Stoner L, Delmar M, Vallano ML, Jalife J (1993) A structural basis for the unequal sensitivity of the major cardiac and liver gap junctions to intracellular acidification: The carboxyl tail length. *Biophys J* 64:1422–1433.
Locovei S, Bao L, Dahl G (2006a) Pannexin 1 in erythrocytes: Function without a gap. *Proc Natl Acad Sci USA* 103:7655–7659.
Locovei S, Wang J, Dahl G (2006b) Activation of pannexin 1 channels by ATP through P2Y receptors and by cytoplasmic calcium. *FEBS Lett* 80:239–244.
Locovei S, Scemes E, Qiu F, Spray DC, Dahl G (2007) Pannexin1 is part of the pore forming unit of the P2X7 receptor death complex. *FEBS Lett* 581:483–488.
Loddenkemper T, Grote K, Evers S, Oelerich M, Stogbauer F (2002) Neurological manifestations of the oculodentodigital dysplasia syndrome. *J Neurol* 249:584–595.
Loewenstein WR, Rose B (1992) The cell–cell channel in the control of growth. *Semin Cell Biol* 3:59–79.
Loewenstein WR, Nakas M, Socolar SJ (1967) Junctional membrane uncoupling. Permeability transformations at a cell membrane junction. *J Gen Physiol* 50:1865–1891.
Lopes-Cardozo M, Klein W (1985) Contribution of acetoacetate to the synthesis of cholesterol and fatty acids in regions of developing rat brain in vivo. *Neurochem Int* 7:647–653.
Maher F, Vannucci SJ, Simson IA (1994) Glucose transporter proteins in brain. *FASEB J* 8:1003–1011.
Makowski L, Caspar DLD, Phillips WC, Goodenough DA (1977) Gap junction structures: II. Analysis of the X-ray diffraction data. *J Cell Biol* 74:629–645.
Malchow RP, Qian H, Ripps H (1994) A novel action of quinine and quinidine on the membrane conductance of neurons from the vertebrate retina. *J Gen Physiol* 104:1039–1055.

Martin PE, Evans WH. 2004. Incorporation of connexins into plasma membranes and gap junctions. *Cardiovasc Res* 62: 378–387.
Massa PT, Mugnaini E (1982) Cell junctions and intramembrane particles of astrocytes and oligodendrocytes: A freeze-fracture study. *Neuroscience* 7:523–538.
Massa PT, Mugnaini E (1985) Cell–cell junctional interactions and characteristic plasma membrane features of cultured rat glial cells. *J Neurosci* 14:695–709.
Mazet F, Wittenberg BA, Spray DC (1985) Fate of intercellular junctions in isolated adult rat cardiac cells. *Circ Res* 56:195–204.
McNutt NS, Weinstein RS (1970) The ultrastructure of the nexus. A correlated thin-section and freeze-cleave study. *J Cell Biol* 47:666–688.
Melloy PG, Kusnierczyk MK, Meyer RA, Lo CW, Desmond ME (2005) Overexpression of connexin43 alters the mutant phenotype of midgestational wnt-1 null mice resulting in recovery of the midbrain and cerebellum. *Anat Rec A Discov Mol Cell Evol Biol* 283:224–238.
Meme W, Calco C-F, Froger N, Ezan P, Amigou E, Koulakoff A, Giaume C (2006) Proinflammatory cytokines released from microglia inhibit gap junctions in astrocytes: Potentiation by β-amyloid. *FASEB J* 20:494–496.
Menichella DM, Majdan M, Awatramani R, Goodenough DA, Sirkowski E, Scherer SS, Paul DL (2006) Genetic and physiological evidence that oligodendrocyte gap junctions contribute to spatial buffering of potassium released during neuronal activity. *J Neurosci* 26:10984–10991.
Mesnil M, Yamasaki H (2000) Bystander effect in herpes simplex virus-thymidine kinase/ganciclovir cancer gene therapy: Role of gap-junctional intercellular communication. *Cancer Res* 60:3989–3999.
Muller DJ, Hand GM, Engel A, Sosinsky GE (2002) Conformational changes in surface structures of isolated connexin26 gap junctions. *EMBO J* 21:3598–3607.
Mulligan SJ, MacVicar BA (2004) Calcium transients in astrocyte endfeet cause cerebrovascular constrictions. *Nature* 431:195–199.
Nagy JI, Li WE (2000) A brain slice model for in vitro analyses of astrocytic gap junction and connexin43 regulation: Actions of ischemia, glutamate and elevated potassium. *Eur J Neurosci* 12:4567–4572.
Nagy JL, Rash JE (2000) Connexins and gap junctions of astrocytes and oligodendrocytes in the CNS. *Brain Res Brain Res Rev* 32:29–44.
Nagy JI, Rash JE (2003) Astrocyte and oligodendrocyte connexins of the glial syncytium in relation to astrocyte anatomical domains and spatial buffering. *Cell Commun Adhes* 10:401–406.
Nagy JI, Li W, Hertzberg EL, Marotta CA (1996) Elevated connexin43 immunoreactivity at sites of amyloid plaques in Alzheimer's disease. *Brain Res* 717:173–178.
Nagy JL, Ionescu AV, Lynn BD, Rash JE (2003) Coupling of astrocyte connexins Cx26, Cx30, Cx43 to oligodendrocyte Cx29, Cx32, Cx47: Implications from normal and connexin32 knockout mice. *Glia* 44:205–218.
Nagy JI, Dudek FE, Rash JE(2004) Update on connexins and gap junctions in neurons and glia in the mammalian nervous system. *Brain Res Brain Res Rev* 47:191–215.
Nakase T, Naus CC (2004) Gap junctions and neurological disorders of the central nervous system. *Biochim Biophys Acta* 1662:149–58.
Narcisse L, Scemes E, Zhao Y, Lee SC, Brosnan CF (2005) The cytokine IL-1β transiently enhances $P2X_7$ receptor expression and function in human astrocytes. *Glia* 49:245–258.
Naus CC, Bond SL, Bechberger JF, Rushlow W (2000) Identification of genes differentially expressed in C6 glioma cells transfected with connexin43. *Brain Res Brain Res Rev* 32:259–266.
Nedergaard, M (1994) Direct signaling from astrocytes to neurons in cultures of mammalian brain cells. *Science* 263:1768–1771.
Neijssen J, Herberts C, Drijfhout JW, Reits E, Janssen L, Neefjes J (2005) Cross-presentation by intercellular peptide transfer through gap junctions. *Nature* 434:83–88.
Newman EA (1985) Regulation of potassium levels by glial cells in the retina. *Trends Neurosci* 8:156–159.

Newman EA (2001) Propagation of intercellular calcium waves in retinal astrocytes and Muller cells. *J Neurosci* 21:2215–2223.

Newman EA (2003) Glial cell inhibition of neurons by release of ATP. *J Neurosci* 23:1659–1666.

Newman EA (2004) Glial modulation of synaptic transmission in the retina. *Glia* 47:268–274.

Newman EA, Reichenbach A (1996) The Mueller cell: A functional element of the retina. *Trends Neurosci* 19:307–317.

Nicchia GP, Srinivas M, Li W, Brosnan CF, Frigeri A, Spray DC (2005) New possible roles for aquaporin-4 in astrocytes: Cell cytoskeleton and functional relationship with connexin43. *FASEB J* 19:1674–1676.

Nicholson C, Sykova E (1998) Extracellular structure revealed by diffusion analysis. *Trends Neurosci* 21:207–215.

O'Neill LA (2006) How Toll-like receptors signal: What we know and what we don't know. *Curr Opin Immunol* 18:3–9.

Orkand RK, Nicholls JG, Kuffler SW (1966) Effect of nerve impulses on the membrane potential of glial cells in the central nervous system of amphibia. *J Neurophysiol* 29:788–806.

Orthmann-Murphy JL, Enriquez AD, Abrams CK, Scherer SS (2007) Loss-of-function GJA12/Connexin47 mutations cause Pelizaeus-Merzbacher-like disease. *Mol Cell Neurosci* 34:629–641.

Ozog MA, Siushansian R, Naus CC (2002) Blocked gap junctional coupling increases glutamate-induced neurotoxicity in neuron-astrocyte co-cultures. *J Neuropathol Exp Neurol* 61:132–141.

Parpura V, Scemes E, Spray DC (2004) Mechanisms of glutamate release from astrocytes: Gap junction "hemichannels," purinergic receptors and exocytotic release. *Neurochem Int* 45: 259–264.

Paul DL, Ebihara L, Takemoto LJ, Swenson KI, Goodenough DA (1991) Connexin46, a novel lens gap junction protein, induces voltage-gated currents in nonjunctional plasma membrane of *Xenopus* oocytes. *J Cell Biol* 115:1077–1089.

Pelegrin P, Surprenant A (2006) Pannexin-1 mediates large pore formation and interleukin-1 release by the ATP-gated P2X7 receptor. *EMBO J* 25:5071–5082.

Pereda AE, Rash JE, Nagy JI, Bennett MV (2004) Dynamics of electrical transmission at club endings on the Mauthner cells. *Brain Res Brain Res Rev* 47:227–244.

Perkins GA, Goodenough DA, Sosinsky GE (1998) Formation of the gap junction intercellular channel requires a 30 degree rotation for interdigitating two apposing connexons. *J Mol Biol* 277:171–177.

Phelan P (2004) Innexins: Members of an evolutionarily conserved family of gap-junction proteins. *Biochim Biophys Acta* 1711:225–245.

Pitts JD, Finbow ME (1977) Junctional permeability and its consequences. In: Intercellular communication (De Mello WC, ed), p 61. New York: Plenum.

Poopalasundaram S, Knott C, Shamotienko OG, Foran PG, Dolly JO, Ghiani CA, Gallo V, Wilkin GP (2000) Glial heterogeneity in expression of the inwardly rectifying K+ channel, Kir4.1, in adult rat CNS. *Glia* 30:362–372.

Purnick PE, Benjamin DC, Verselis VK, Bargiello TA, Dowd TA (2000) Structure of the amino terminus of a gap junction protein. *Arch Biochem Biophys* 381:181–190.

Rash JE, Duffy HS, Dudek FE, Bilhartz BL, Whalen LR, Yasumura AT (1997) Grid-mapped freeze-fracture analysis of gap junctions in gray and white matter of adult rat central nervous system, with evidence for a "panglial syncytium" that is not coupled to neurons. *J Comp Neurol* 388:265–292.

Rawanduzy A, Hansen A, Hansen TW, Nedergaard M (1997) Effective reduction of infarct volume by gap junction blockade in a rodent model of stroke. *J Neurosurg* 87:916–920.

Ray A, Zoidl G, Weickert S, Wahle P, Dermietzel R (2005) Site-specific and developmental expression of pannexin1 in the mouse nervous system. *Eur J Neurosci* 21:3277–3290.

Ray A, Zoidl G, Wahle P, Dermietzel R (2006) Pannexin expression in the cerebellum. *Cerebellum* 5:189–192.

Reichenbach A, Henke A, Eberhardt W, Reichelt W, Dettmer D (1992) K^+ ion regulation in retina. *Can J Physiol Pharmacol* 70:S239–S247.

Retamal MA, Cortés CJ, Reuss L, Bennett MV, Sáez JC (2006) S-nitrosylation and permeation through connexin43 hemichannels in astrocytes: Induction by oxidant stress and reversal by reducing agents. *Proc Natl Acad Sci USA* 103:4475–4480.

Revel JP (1968) Proceedings of the 26th Meeting of the Electron Microscopy Society of American, Baton Rouge, L, Claitor's, p 40.

Revel JP, Karnovsky MJ (1967) Hexagonal array of subunits in intercellular junctions of the mouse heart and liver. *J Cell Biol* 33:C7–C12.

Revilla A, Castro C, Barrio LC (1999) Molecular dissection of transjunctional voltage dependence in the connexin-32 and connexin-43 junctions. *Biophys J* 77:1374–1383.

Richard G (2005) Connexin disorders of the skin. *Clin Dermatol* 23:23–32.

Robertson JD (1953) Ultrastructure of two invertebrate synapses. *Proc Soc Exp Bio Med* 82:219–223.

Rohlmann A, Wolff JR (1996) Subcellular topography and plasticity of gap junction distribution in astrocytes. In: Gap junctions in the nervous system (Spray DC, Dermietzel R, eds), pp 175–192. Austin, TX: Landes.

Roscoe WA, Messersmith E, Meyer-Franke A, Wipke B, Karlik SJ (2007) Connexin43 gap junction proteins are up-regulated in remyelinating spinal cord. *J Neurosci Res* 85:945–953.

Rouach N, Avignone E, Meme W, Koulakoff A, Venance L, Blomstrand F, Giaume C (2002) Gap junctions and connexin expression in the normal and pathological central nervous system. *Biol Cell* 94:457–475.

Rose B, Mehta PP, Loewenstein WR (1993) Gap-junction protein gene suppresses tumorogenicity. *Carcinogenesis* 14:1073–1075.

Sabag AD, Dagan O, Avraham KB (2005) Connexins in hearing loss: A comprehensive overview. *J Basic Clin Physiol Pharmacol* 16:101–116.

Saez JC, Connor JA, Spray DC, Bennett MVL (1989) Hepatocyte gap junction are permeable to the second messenger, inositol 1,4,5-trisphosphate, and to calcium ions. *Proc Natl Acad Sci USA* 86:2708–2712.

Saez JC, Retamal MA, Basiio D, Bukauskas FF, Bennett MVL (2005) Connexin-based gap junction hemichannels: Gating mechanisms. *Biochim Biophys Acta* 1711: 215–224.

Saito R, Graf R, Hubel K, Fujita T, Rosner G, Heiss WD (1997) Reduction of infarct volume by halothane: Effect on cerebral blood flow or perifocal spreading depression-like depolarizations. *J Cereb Blood Flow Metab* 17:857–864.

Scemes E (2000) Components of astrocytic intercellular calcium signaling. *Mol Neurobiol* 22:167–179.

Scemes E, Giaume C (2006) Astrocyte calcium waves: What they are and what they do. *Glia* 54:716–725.

Scemes E, Dermietzel R, Spray DC (1998) Calcium waves between astocytes from Cx43 knockout mice. *Glia* 24:65–73.

Scemes E, Suadicani SO, Spray DC (2000) Intercellular communication in spinal cord astrocytes: Fine tuning between gap junctions and P2 nucleotide receptors in calcium wave propagation. *J Neurosci* 20:1435–1445.

Scherer SS, Bone LJ, Deschenes SM, Abel A, Balice-Gordon RJ, Fischbeck KH (1999) The role of the gap junction protein connexin32 in the pathogenesis of X-linked Charcot-Marie-Tooth disease. *Novartis Found Symp* 219:175–185.

Schubert AL, Schubert W, Spray DC, Lisanti MP (2002) Connexin family members target to lipid raft domains and interact with caveolin-1. *Biochemistry* 41:5754–5764.

Schroder W, Seifert G, Huttmann K, Hinterkeuser S, Steinhauser C (2002) AMPA receptor mediated modulation of inward rectifier K+ channels in astrocytes of developing mouse hippocampus. *Mol Cell Neurosci* 19:447–453.

Shaw RM, Fay AJ, Puthenveedu MA, von Zastrow M, Jan YN, Jan LY (2007) Microtubule plus-end-tracking proteins target gap junctions directly from the cell interior to adherens junctions. *Cell* 128:547–560.

Shibayama J, Paznekas W, Seki A, Taffet S, Jabs EW, Delmar M, Musa H (2005) Functional characterization of connexin43 mutations found in patients with oculodentodigital dysplasia. *Circ Res* 96:e83–e91.

Siushansian R, Becheberger JF, Cechetto DF, Hachinski VC, Naus CC (2001) Connexin43 null mutation increases infarct size after stroke. *J Comp Neurol* 440: 387–394.

Sjostrand FS, Andersson-Cedergren E, Dewey MM (1958) The ultrastructure of the intercalated discs of frog, mouse, and guinea pig cardiac muscle. *J Ultrastruct Res* 1:271–287.

Skerrett IM, Aronowitz J, Shin HH, Cymes G, Kasparek E, Cao FL, Nicholson BJ (2002) Identification of amino acid residues lining the pore of a gap junction channel. *J Cell Biol* 159:349–360.

Sohl G, Willecke K (2003) An update on connexin genes and their nomenclature in mouse and man. *Cell Commun Adhes* 10:173–180.

Solan JL, Lampe PD (2005) Connexin phosphorylation as a regulatory event linked to gap junction channel assembly. *Biochim Biophys Acta* 1711:154–163.

Sorgen PL, Duffy HS, Cahill SM, Coombs W, Spray DC, Delmar M, Girvin ME (2002) Sequence-specific resonance assignment of the carboxyl terminal domain of connexin43. *J Biomol NMR* 23:245–246.

Sorgen PL, Duffy HS, Sahoo P, Coombs W, Delmar M, Spray DC (2004a) Structural changes in the carboxyl terminus of the gap junction protein connexin43 indicates signaling between binding domains for c-Src and zonula occludens-1. *J Biol Chem* 279:54695–54701.

Sorgen PL, Duffy HS, Spray DC, Delmar M (2004b) pH-dependent dimerization of the carboxyl terminal domain of Cx43. *Biophys J* 87:574–581.

Spray DC (1998) Gap junction proteins: Where they live and how they die. *Circ Res* 83:679–681.

Spray DC, Burt JM (1990) Structure–activity relations of the cardiac gap junction channel. *Am J Physiol* 258:C195–C205.

Spray DC, Dermietzel R (1995) X-linked dominant Charcot-Marie-Tooth disease and other potential gap-junction diseases of the nervous system. *Trends Neurosci* 18:256–262.

Spray DC, Iacobas, DA (2007) Organizational principles of the connexin-related brain transcriptome. *J Membr Biol* 218: 39–47.

Spray DC, Harris AL, Bennett MV (1981) Gap junctional conductance is a simple and sensitive function of intracellular pH. *Science* 211:712–715.

Spray DC, Stern JH, Harris AL, Bennett MV (1982) Gap junctional conductance: Comparison of sensitivities to H+ and Ca2+ ions. *Proc Natl Acad Sci USA* 79:441–445.

Spray DC, Duffy HS, Scemes E (1999) Gap junctions in glia. Types, roles and plasticity. *Adv Exp Med Biol* 468:339–359.

Spray DC, Rozental R, Srinivas M (2002) Prospects for rational development of pharmacological gap junction channel blockers. *Curr Drug Targets* 3:455–464.

Spray DC, Scemes E, Rozental R, Dermietzel R (2004) Cell–cell communication: An overview emphasizing gap junctions. In: An introduction to cellular and molecular neuroscience (Byrne J, Roberts J, eds), pp 431–458. New York: Academic.

Spray DC, Ye ZC, Ransom BR (2006) Functional connexin "hemichannels": A critical appraisal. *Glia* 54:758–773.

Srinivas M, Spray DC (2003) Closure of gap junction channels by arylaminobenzoates. *Mol Pharmacol* 63:1389–1397.

Srinivas M, Hopperstad MG, Spray DC (2001) Quinine blocks specific gap junction channel subtypes. *Proc Natl Acad Sci USA* 98:10942–10947.

Stewart WW (1981) Lucifer dyes – highly fluorescent dyes for biological tracing. *Nature* 292:17–21.

Suadicani SO, De Pina-Benabou MH, Urban M, Spray DC, Scemes E (2003) Acute down-regulation of Cx43 alters P2Y receptor expression levels in mouse spinal cord astrocytes. *Glia* 42:160–171.

Suadicani SO, Flores CE, Urban-Maldonado M, Beelitz M, Scemes E (2004) Gap junction channels coordinate the propagation of intercellular Ca^{2+} signals generated by P2Y receptor activation. *Glia* 48:217–229.

Suadicani SO, Brosnan CF, Scemes E (2006) P2X7 receptors mediate ATP release and amplification of astrocytic intercellular Ca^{2+} signaling. *J Neurosci* 26:1378–1385.
Suadicani SO, Dahl G, Scemes E (2007) Pannexin1 mediates amplification of intercellular Ca^{2+} waves in astrocytes. *Neuron Glia Biol* 3:S165 (abstract).
Sykova E (1997) Extracellular space volume and geometry of the brain after ischemia and central injury. *Adv Neurol* 73:121–125.
Stout CE, Costantin JL, Naus CC, Charles AC (2002) Intercellular calcium signaling in astrocytes via ATP release through connexin hemichannels *J Biol Chem* 277:10482–10488.
Tabernero A, Giaume C, Medina JM (1996) Endothelin-1 regulates glucose utilization in cultured rat astrocytes by controlling intercellular communication through gap junctions. *Glia* 16:187–195.
Takano T, Tian GF, Peng W, Lou N, Libionka W, Han X, Nedergaard M (2006) Astrocyte mediated control of cerebral blood flow. *Nat Neurosci* 9:260–267.
Takeda K, Akira S (2005) Toll-like receptors in innate immunity. *Int Immunol* 17:1–14.
Takumi T, Ishii T, Horio Y, Morishige K, Takahashi N, Yamada M, Yamashita T, Kiyama H, Sohmiya K, Nakanishi S, Kurachi Y (1995) A novel ATP-dependent inward rectifier potassium channel expressed predominantly in glial cells. *J Biol Chem* 270:16339–16346.
Tao L, Harris AL (2007) 2-aminoethoxydiphenyl borate directly inhibits channels composed of connexin26 and/or connexin32. *Mol Pharmacol* 71:570–579.
Teubner B, Odermatt B, Guldenagel M, Sohl G, Degen J, Bukauskas F, Kronengold J, Verselis VK, Jung YT, Kozak CA, Schilling K, Willecke K (2001) Functional expression of the new gap junction gene connexin47 transcribed in mouse brain and spinal cord neurons. *J Neurosci* 21:1117–1126.
Theis M, Sohl G, Eiberger J, Willecke K (2005) Emerging complexities in identity and function of glial connexins. *Trends Neurosci* 28:188–195.
Theuvenet AP (2001) Fenamates: A novel class of reversible gap junction blockers. *J Pharmacol Exp Ther* 298:1033–1041.
Thompson RJ, Zhou N, MacVicar BA (2006) Ischemia opens neuronal gap junction hemichannels. *Science* 312:924–927.
Unger VM, Kumar NM, Gilula NB, Yeager MR (1999) Three-dimensional structure of a recombinant gap junction membrane channel. *Science* 283:1176–1180.
Unwin PN, Zampighi G (1980) Structure of the junction between communicating cells. *Nature* 283:545–549.
Valiunas V, Bukauskas FF, Weingart R (1997) Conductances and selective permeability of connexin43 gap junction channels examined in neonatal rat heart cells. *Circ Res* 80:708–719.
Valiunas V, Polosina YY, Miller H, Potapova IA, Valiuniene L, Doronin S, Mathias RT, Robinson RB, Rosen MR, Cohen IS, Brink PR (2005) Connexin-specific cell-to-cell transfer of short interfering RNA by gap junctions. *J Physiol* 568:459–468.
Venance L, Piomelli D, Glowinski J, Giaume C (1995) Inhibition by anandamide of gap junctions and intercellular calcium signaling in striatal astrocytes. *Nature* 376:590–594.
Verselis VK, Bennett MV, Bargiello TA (1991) A voltage-dependent gap junction in *Drosophila melanogaster*. *Biophys J* 59:114–126.
Verselis VK, Ginter CS, Bargiello TA (1994) Opposite voltage gating polarities of two closely related connexins. *Nature* 368:348–351.
Verselis VK, Trexler EB, Bukauskas FF (2000) Connexin hemichannels and cell–cell channels: Comparison of properties. *Braz J Med Biol Res* 33: 379–389.
Vicario C, Tabernero A, Medina JM (1993) Regulation of lactate metabolism by albumin in rat neurons and astrocytes from primary culture. *Pediatr Res* 34:709–715.
Vogt A, Hormuzdi SG, Monyer H (2005) Pannexin1 and Pannexin2 expression in the developing and mature rat brain. *Brain Res Mol Brain Res* 141;141:113–120.
Wallraff A, Kohling R, Heinemann U, Theis M, Willecke K, Steinhauser C (2006) The impact of astrocytic gap junctional coupling on potassium buffering in the hippocampus. *J Neurosci* 26:5438–5447.

Walz W (2000) Role of astrocytes in the clearance of excess extracellular. *Neurochem Int* 36:291–300.

Wei CJ, Xu X, Lo CW (2004) Connexins and cell signaling in development and disease. *Annu Rev Cell Dev Biol* 20:811–838.

Weir MP, Lo CW (1985) An anterior/posterior communication compartment border in engrailed wing discs: Possible implications for *Drosophila* pattern formation. *Dev Biol* 110:84–90.

White TW, Bruzzone R (1996) Multiple connexin proteins in single intercellular channels: Connexin compatibility and functional consequences. *J Bioenerg Biomembr* 28:339–350.

White TW, Paul DL (1999) Genetic diseases and gene knockouts reveal diverse connexin functions. *Annu Rev Physiol* 61:283–310.

Wiencken-Barger AE, Djukic B, Casper KB, McCarthy KD (2007) A role for connexin43 during neurodevelopment. *Glia* 55:675–686.

Willecke K, Eiberger J, Degen J, Eckardt D, Romualdi A and Guldenagel M (2002) Structural and functional diversity of connexin genes in the mouse and human genome. *Biol Chem* 383:725–737.

Williamson DH (1982) The production and utilization of ketone bodies in the neonate. In: Biochemical development of the fetus and neonate (Jones CT, ed), pp 621–650. Amsterdam, Nederlands: Elsevier **Biomedical Press.

Wolff JR, Stuke K, Missler M, Tytko H, Schwarz P, Rohlmann A and Chao TI (1998) Autocellular coupling by gap junctions in cultured astrocytes: A new view on cellular autoregulation during process formation. *Glia* 24:121–140.

Yahuaca P, Ek-Vitorin >JF, Rush P, Delmar M, Taffet SM (2000) Identification of a protein kinase activity that phosphorylates connexin43 in a pH-dependent manner. *Braz J Med Biol Res* 33:399–406.

Ye ZC, Wyeth MS, Baltan-Tekkok S, Ransom BR (2003) Functional hemichannels in astrocytes: A novel mechanism of glutamate release. *J Neurosci* 23: 3588–3596.

Yen MR, Saier MH Jr. (2007) Gap junctional proteins of animals: The innexin/pannexin superfamily. *Prog Biophys Mol Biol* 94:5–14.

Zhao Y, Rivieccio MA, Lutz S, Scemes E, Brosnan CF (2006) The TLR3 ligand polyI:C downregulates connexin43 expression and function in astrocytes by a mechanism involving the NF-κB and PI3 kinase pathways. *Glia* 54:775–785.

Zhou XW, Pfahnl A, Werner R, Hudder A, Llanes A, Luebke A, Dahl G (1977) Identification of a pore lining segment in gap junction hemichannels. *Biophys J* 72:1946–1953.

Zhu Y, Kimelberg HK (2001) Developmental expression of metabotropic P2Y1 and P2Y2 receptors in freshly isolated astrocytes from rat hippocampus. *J Neurochem* 77:530–541.

Zhu Y, Kimelberg HK (2004) Cellular expression of P2Y and β-AR receptor mRNAs and proteins in freshly isolated astrocytes and tissue sections from the CA1 region of P8–12 rat hippocampus. *Brain Res Dev Brain Res* 148:77–87.

Zoidl G, Petrasch-Parwez E, Ray A, Meier C, Bunse S, Habbes HW, Dahl G, Dermietzel R (2007) Localization of the Pannexin1 protein at postsynaptic sites in the cerebral cortex and hippocampus. *Neuroscience* 146:9–16.

Zonta M, Angulo MC, Gobbo S, Rosengarten B, Hossmann KA, Pozzan T, Carmignoto G (2003) Neuron-to-astrocyte signaling is central to the dynamic control of brain microcirculation. *Nat Neurosci* 6:43–50.

Abbreviations

α-GA	Alpha-glycyrrhetinic acid
ATP	Adenosine 5′-triphosphate
cAMP	Cyclic adenosine monophosphate

BBB	Blood–brain barrier
CL	Intracellular loop
CMTX	Charcot-Marie-Tooth disease
CT	Carboxyl terminus
Cx	Connexin
ER	Endoplasmic reticulum
FFA	Flufenamic acid
HNSD	Hereditary nonsyndromic deafness
HT	Heterozygote
ICW	Intercellular Ca^{2+} wave
IL	Interleukin
IP_3	Inositol trisphosphate
Kir	Inward rectifying K^+ channel
LINC	Long-term increase in coupling
NMR	Nuclear magnetic resonance
NT	Amino terminus
M	Membrane segments
TLR	Toll-like receptor
TNF-α	Tumor necrosis factor-α
WT	Wildtype
ZO	Zonula occludens

Chapter 6
Regulation of Potassium by Glial Cells in the Central Nervous System

Paulo Kofuji and Eric A. Newman

Contents

Rapid changes in extracellular K^+ concentration ($[K^+]_o$) in the mammalian central nervous system (CNS) are counteracted by simple passive diffusion as well as by cellular mechanisms of K^+ clearance. Regulation of $[K^+]_o$ can occur via glial or neuronal uptake of K^+ ions through transporters or K^+-selective ion channels. The best studied mechanism of $[K^+]_o$ regulation in the brain is *K^+ spatial buffering*, wherein the glial syncytium disperses local extracellular K^+ increases by transferring K^+ from sites of elevated $[K^+]_o$ to those with lower $[K^+]_o$. In recent years, K^+ spatial buffering has been implicated or directly demonstrated by a variety of experimental approaches, including electrophysiological and optical methods. A specialized form of spatial buffering termed *K^+ siphoning* takes place in the vertebrate retina, where glial Müller cells express inwardly rectifying K^+ channels (Kir channels) positioned in membrane domains near to the vitreous humor and blood

E.A. Newman
Department of Neuroscience, University of Minnesota, Minneapolis, MN, USA
ean@umn.edu

V. Parpura and P.G. Haydon (eds.), *Astrocytes in (Patho)Physiology of the Nervous System*,
DOI: 10.1007/978-0-387-79492-1_6,

vessels. This highly compartmentalized distribution of Kir channels in retinal glia directs K^+ ions from the synaptic layers to the vitreous humor and blood vessels. Here, we review the principal mechanisms of $[K^+]_o$ regulation in the CNS and recent molecular studies on the structure and function of glial Kir channels. We also discuss intriguing new data that suggest a close physical and functional relationship between Kir and water channels in glial cells.

6.1 Potassium in the Extracellular Space of the Central Nervous System

Neurons are bathed in extracellular fluid that has a high concentration of Na^+ ions and a low concentration of K^+ ions. The relative concentrations of these cations inside of cells are reversed. The resulting ionic gradients across the cell membrane are crucial to the generation of essential neuronal signals, including the resting membrane potential, the action potential, and synaptic potentials. Because of the low baseline concentration of extracellular K^+($[K^+]_o$), and to the limited volume of extracellular space, even modest efflux of K^+ from neurons can elicit considerable changes in $[K^+]_o$ (Nicholson and Sykova, 1998; Kume-Kick et al., 2002). These $[K^+]_o$ changes can influence a wide variety of neuronal processes, including the maintenance of the resting membrane potential, activation, and inactivation of voltage gated ion channels, the efficacy of synaptic transmission, and electrogenic transport of neurotransmitters. Thus, it is not surprising that the CNS possesses robust cellular mechanisms to regulate $[K^+]_o$. Under normal physiological conditions, these mechanisms maintain $[K^+]_o$ close to 3 mM. When K^+ regulatory mechanisms are overwhelmed under pathophysiological conditions such as spreading depression and ischemia, extracellular $[K^+]_o$ can reach values as high as 60 mM or more (Somjen, 2001, 2002). These extreme $[K^+]_o$ levels severely depolarize neurons, rendering them inactive.

Normal neuronal activity in the CNS results in modest variations in $[K^+]_o$. Light stimulation in the cat produces slow, transient $[K^+]_o$ increases, smaller than 1 mM, in the primary visual cortex (Fig. 6.1a) (Singer and Lux, 1975; Connors et al., 1979). Similarly, light stimulation in the frog and cat induces $[K^+]_o$ increases of less than 1 mM in the inner and outer plexiform layers of the retina and $[K^+]_o$ decreases in the outer retina and subretinal space (Fig. 6.1b) (Karwoski et al., 1985; Frishman et al., 1992). In cat spinal cord, rhythmic flexion/extension of the knee joint produces $[K^+]_o$ increases of 1.7 mM (Heinemann et al., 1990). Significantly higher $[K^+]_o$ elevations are evoked by direct electrical stimulation of afferent pathways and by induction of seizure activity. Even under intense, high-frequency stimulation, however, $[K^+]_o$ does not exceed a plateau or ceiling level of 10–12 mM. This ceiling level is seen in the cat somatosensory cortex (Heinemann and Lux, 1977), in the cat thalamus (Gutnick et al., 1979), and in the rat optic nerve (Connors et al., 1982; Ransom et al., 1986). The ceiling level is exceeded only under pathophysiological conditions such as anoxia (Vyskocil et al., 1972) or spreading depression (Somjen, 2002).

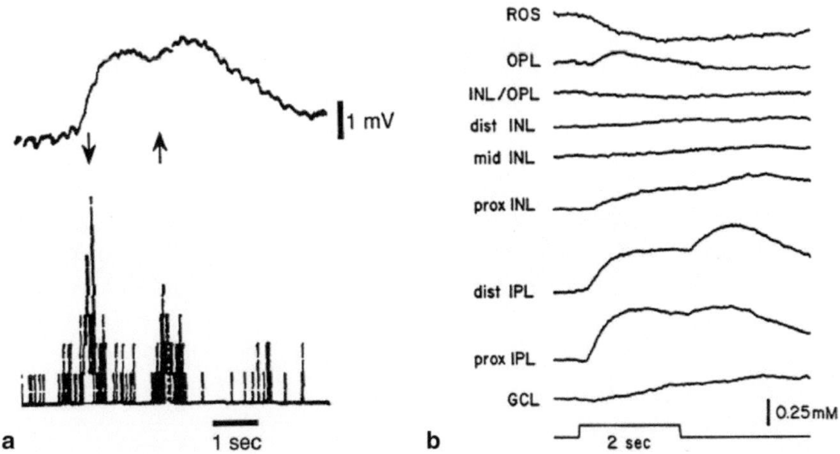

Fig. 6.1 Activity-evoked changes in $[K^+]_o$ in the cat striate cortex and frog retina. (**a**) Upper trace. Dynamic $[K^+]_o$ changes in the cat striate cortex evoked by stimulation of the receptive field of hypercomplex cells. $[K^+]_o$ changes were measured with a double-barreled K^+-sensitive microelectrode. *Arrows* represent bars of light moving down or up in a cell s receptive field. The 1-mV scale bar corresponds to ~0.17 mM. *Lower trace*. Spike activity recorded in the reference barrel of the K^+-sensitive microelectrode. [From Singer and Lux (1975), with permission.] (**b**) $[K^+]_o$ changes in different layers of the frog retina recorded with a K^+-sensitive microelectrode. A 2-s light stimulus evokes $[K^+]_o$ increases in the inner plexiform layer (IPL) and outer plexiform layer (OPL) and a $[K^+]_o$ decrease in the subretinal space (ROS). [From Karwoski et al. (1985), with permission.]

The careful control of $[K^+]_o$ within the brain is due to efficient K^+ regulatory mechanisms that operate in the CNS. Neuronal depolarization is accompanied by an efflux of K^+ into extracellular space. Even modest neuronal activity results in significant $[K^+]_o$ increases. Potassium efflux due to a single action potential can raise $[K^+]_o$ by 25% (Ransom and Sontheimer, 1992). Potassium regulatory mechanisms are responsible for maintaining $[K^+]_o$ near 3 mM during normal brain activity and prevent $[K^+]_o$ from exceeding 10–12 mM, even during tetanic stimulation or during seizure activity. This chapter reviews glial mechanisms that contribute to $[K^+]_o$ regulation in the CNS. The chapter is an updated version of a previous review (Kofuji and Newman, 2004).

6.2 Overview of K^+ Regulatory Mechanisms

Potassium regulation in the CNS is mediated by two types of mechanisms: net K^+ uptake and K^+ spatial buffering (Fig. 6.2) (Newman, 1995; Amedee et al., 1997; Somjen, 2002). For K^+ uptake, excess extracellular K^+ is temporarily taken up and sequestered within glial cells. (In theory, excess K^+ could also be sequestered within quiescent neurons. However, neuronal uptake is not thought to play

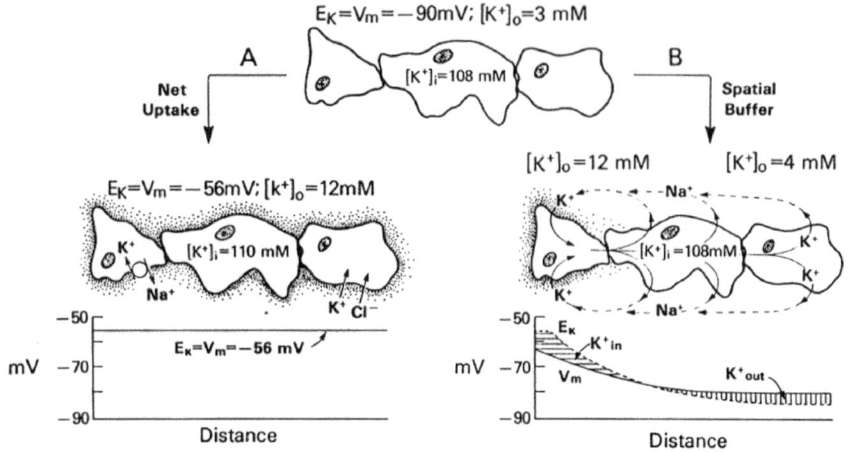

Fig. 6.2 Diagram depicting the role of glial cells in $[K^+]_o$ regulation. *Top*: Glial cells are electrically coupled via gap junctions forming a functional syncytium. With $[K^+]_o$ equaling 3 mM, the glial syncytium has a membrane potential of -90 mV. (**a**) Net K^+ uptake mechanism. When $[K^+]_o$ is increased, glial cells accumulate K^+ either by the activity of the Na^+, K^+-ATPase or by a pathway in which K^+ is cotransported with Cl^-. In this mechanism of K^+ regulation, the membrane potential of the glial syncytium equals −56 mV and is spatially uniform. (**b**) Potassium spatial buffering mechanism. Local increases in $[K^+]_o$ produce a glial depolarization that spreads passively through the glial syncytium. The local difference between the glial syncytium membrane potential (V_m) and the K^+ equilibrium potential (E_K) drives K^+ influx in regions of elevated $[K^+]_o$ and K^+ efflux in distant regions. Intracellular currents are carried primarily by K^+ and extracellular currents by Na^+ and Cl^-. [From Orkand (1986), with permission.]

a significant role in the rapid removal of K^+ from extracellular space and will not be considered here.) To preserve electroneutrality, K^+ influx into glial cells is accompanied by either influx of anions such as Cl^- or by efflux of cations such as Na^+ (Fig. 6.2a). Net K^+ uptake can occur by an active process, by the action of the Na^+, K^+-ATPase (Na^+ pump), or passively, by K^+ flux through transporters or K^+ channels. When neuronal activity decreases and $[K^+]_o$ falls to near baseline levels, the K^+ sequestered within glial cells is released and is returned to the neurons by the action of neuronal Na^+ pumps. An influx of water accompanies net K^+ uptake into glia, resulting in glial-cell swelling (Dietzel et al., 1980).

Potassium regulation in the CNS can also be mediated by K^+ spatial buffering. In this process, K^+ is transferred from regions of elevated $[K^+]_o$ to regions of lower $[K^+]_o$ by a current flow through glial cells (Orkand et al., 1966). The K^+ current is driven by the difference between the glial syncytium membrane potential (V_m) and the local K^+ equilibrium potential (E_K). In regions of increased $[K^+]_o$, there is a net driving force causing K^+ to flow into the glial cells (Fig. 6.2b). This K^+ entry generates a local depolarization, which propagates electrotonically through individual glial cells and through the glial-cell syncytium. As a result, there is a net driving force causing K^+ to flow out of the glial cells in regions where $[K^+]_o$ is low. The redistribution of K^+ by the spatial buffering mechanism reduces local $[K^+]_o$ increases with little net gain of K^+ within the glial cells. The overall efficiency of the spatial buffer process will depend, in part, on the electrical space constant of the glial-cell syncytium (Newman, 1995). In certain CNS regions close to fluid reservoirs, such as the retina, the K^+ spatial buffer mechanism can efficiently redistribute K^+ by a current flow within single glial cells rather than through a network of cells. In these cases, K^+ influx occurs in one region of the glial cell and efflux occurs through another cell region, typically the endfoot process. This specialized form of spatial buffering is termed *K^+siphoning*.

6.3 Net Uptake of K^+

Net K^+ uptake is mediated by active uptake via the Na^+ pump and by passive uptake, mediated by Na^+–K^+–Cl^- cotransporters and by K^+ and Cl^- channels. The Na^+ pump plays a principal role in K^+ regulation. The Na^+ pump is a transmembrane enzyme that functions as an electrogenic ion transporter in all cells (Kaplan, 2002; Jorgensen et al., 2003). With each cycle of the Na^+ pump, three Na^+ are expelled and two K^+ are moved into the cell, and one ATP molecule is hydrolyzed. The Na^+ pump is activated by intracellular Na^+ and extracellular K^+. Local $[K^+]_o$ increases generated by increased neuronal activity will result in raised pump activity and to increased influx of K^+ (assuming that the extracellular K^+ site of the pump is not saturated and the intracellular Na^+ concentration is not limiting (Sweadner, 1995)). Different Na^+ pump isoforms have varying affinities for K^+ at the extracellular K^+ site. The Na^+ pump expressed in glial cells is better suited for regulating $[K^+]_o$ than is the neuronal isoform in that the glial isoform has a lower affinity for extracellular K^+(Franck et al., 1983; Reichenbach et al., 1992). In retina, for example, the principal glial cell, the Müller cell, expresses a Na^+ pump isoform that is maximally activated at 10–15 mM of $[K^+]_o$ while the isoform present in rod photoreceptors saturates at $[K^+]_o$ as low as 3 mM (Reichenbach et al., 1992). If the glial isoform of the Na^+ pump is saturated at 3 mM of $[K^+]_o$, then increases in $[K^+]_o$ above this level would not increase pump activity and the pump could not contribute to $[K^+]_o$regulation.

Reports from several laboratories demonstrate that the Na^+ pump contributes to K^+ regulation in the CNS. Electrical stimulation in guinea pig cortical slices results in a transient accumulation of K^+ ions and in a simultaneous depletion of Na^+ ions within glial cells (Ballanyi et al., 1987). A substantial fraction of this K^+ accumulation is prevented by pharmacological blockade of the Na^+ pump (Ballanyi et al., 1987). Similarly, in hippocampal slices, blockade of the Na^+ pump increases baseline $[K^+]_o$ and prevents the rapid clearance of K^+ following neuronal stimulation (Fig. 6.3) (D'Ambrosio et al., 2002). In the rat optic nerve, clearance of K^+ accumulation following axonal stimulation is highly temperature dependent ($Q_{10} = 2.6$), as expected for a carrier-mediated process, and is largely blocked by Na, K-ATPase inhibitors (Ransom et al., 2000).

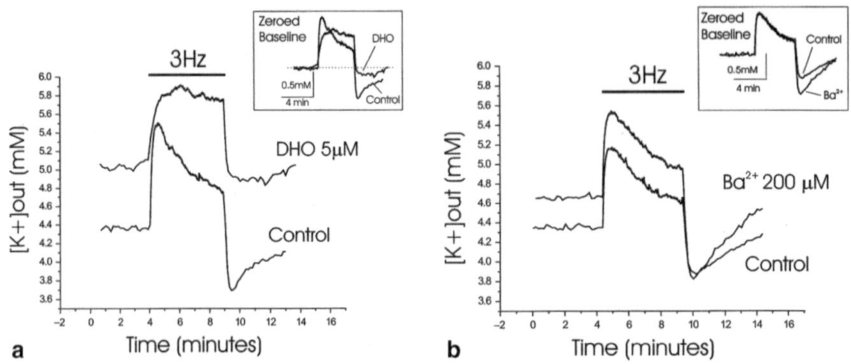

Fig. 6.3 Differential roles for the Na^+, K^+-ATPase and Kir channels in K^+ regulation in rat hippocampal slice. (**a**) In control condition, 3-Hz antidromic stimulation induces a $[K^+]_o$ increase in area CA3 that peaks at 5.5 mM followed by a decline to 4.7 mM. With the addition of the sodium pump inhibitor dihydroouabain (DHO), the baseline $[K^+]_o$ increases to 5.1 mM. Antidromic stimulation (3 Hz) induces a $[K^+]_o$ increase to 5.9 mM but there is no $[K^+]_o$ recovery phase. Also absent is the undershoot of $[K^+]_o$ following the stimulation period. In the inset, the two traces are shown superimposed with the baselines zeroed. (**b**) In control condition, 3-Hz antidromic stimulation induces a $[K^+]_o$ increase that peaks at 5.2 mM followed by an undershoot in $[K^+]_o$. With the addition of Ba^{2+}, baseline $[K^+]_o$ increases and the undershoot in $[K^+]_o$ following stimulation is more pronounced. In the *inset*, the two traces are shown superimposed and the baselines zeroed. [From D'Ambrosio et al. (2002), with permission.]

Na^+–K^+–Cl^- cotransporters also play an important role in the regulation of $[K^+]_o$ in the CNS. These transporters are integral membrane proteins that transport Na^+, K^+, and Cl^- ions into and out of cells in an electrically neutral manner, often with a stoichiometry of $1Na^+$:$1K^+$:$2Cl^-$ (Haas and Forbush, 1998). Two Na^+–K^+–Cl^- cotransporter isoforms have been identified: NKCC1, which is present in a wide variety of secretory epithelia and nonepithelial cells; and NKCC2, which is present exclusively in the kidney (Haas and Forbush, 1998). Both NKCC isoforms are members of a diverse family of cation-chloride cotransport proteins that share a common predicted membrane topology and are sensitive to loop diuretics such as bumetanide and furosemide (Haas and Forbush, 1998). In cultured astrocytes, intracellular accumulation of K^+ following an increase in $[K^+]_o$ can be partially blocked by furosemide or bumetanide or by removal of external Na^+ and Cl^- (Kimelberg and Frangakis, 1985; Walz, 1992; Rose and Ransom, 1996). More recently, the role of Na^+–K^+–Cl^- cotransporters in $[K^+]_o$ homeostasis has also been demonstrated by optical methods. In the rat optic nerve, intrinsic optical signals reveal that an increase in $[K^+]_o$ induces astrocyte swelling that is reversibly depressed by furosemide and bumetanide (MacVicar et al., 2002). A monoclonal antibody to the NKCC1 isoform of the Na^+–K^+–Cl^- cotransporter shows that the transporter is expressed in astrocytes from the optic nerve (MacVicar et al., 2002), suggesting the involvement of this particular Na^+–K^+–Cl^- cotransporter isoform in K^+ regulation.

6.4 Potassium Spatial Buffering

Two conditions are necessary for efficient K^+ spatial buffering as originally proposed by Orkand et al. (1966): (1) the glial cells should form a syncytium in which K^+ currents can traverse relatively long distances; and (2) these cells should be highly and selectively permeable to K^+, which both enters and exits through the glial-cell membranes. As described below in Sect. 6.6 on K^+ siphoning in retina, spatial buffer currents can efficiently dissipate $[K^+]_o$ increases by flowing through single cells as well as through a syncytium of coupled glial cells. Several lines of evidence demonstrate that astrocytes do indeed form a functional syncytium that allows intercellular diffusion of ions and other signaling molecules (Nagy and Rash, 2000; Rouach et al., 2002). Such extensive cellular coupling is due to the high density of gap junctional channels (connexins, Cx) in glial cells (Dermietzel, 1998; Rouach et al., 2002). Immunocytochemical and in situ hybridization studies reveal that astrocytes express multiple connexins, including Cx30, Cx40, Cx43, and Cx45 (Dermietzel, 1998; Dermietzel et al., 2000; Zahs et al., 2003). Among these, Cx43 and Cx30 seem to be the most important for coupling, which is significantly reduced in astrocytes of Cx43 knockout (KO) and Cx43/Cx30 double KO mice (Dermietzel et al., 2000; Wallraff et al., 2006).

Recent studies have demonstrated that there are two classes of CNS glia that resemble astrocytes: "passive astrocytes," also termed GluT, that have ohmic current–voltage relations and express glutamate transporters and"complex glia," also termed GluR, that have rectifying current–voltage relations and express ionotropic glutamate receptors but not transporters (Matthias et al., 2003; Zhou et al., 2006). Passive astrocytes express connexins that are coupled to each other and presumably participate in K^+ spatial buffering. Complex glia do not express connexins, are not coupled together, and presumable do not conduct spatial buffer currents.

Numerous studies have shown that glial-cell membranes are highly and almost exclusively permeable to K^+ (Sontheimer, 1994). The principal K^+ channels found in glial cells are the inwardly rectifying K^+ (Kir) channels, which allow K^+ ions to flow more readily in the inward than outward direction (Doupnik et al., 1995; Stanfield et al., 2002). These channels have a high open probability at the normal resting membrane potential and thus allow both glial K^+ influx and efflux. Kir channels in glia have been described in many CNS regions, including mammalian astrocytes from the optic nerve (Barres et al., 1990), spinal cord (Ransom and Sontheimer, 1995), and other brain regions (Sontheimer, 1994). Although glia may express additional types of K^+ channels, such as Ca^{2+}- and voltage-dependent K^+ channels (Sontheimer, 1994), these other channel types are largely inactive at the hyperpolarized glial resting membrane potential (–60 to –90 mV) (Kuffler et al., 1966; Dennis and Gerschenfeld, 1969). Astrocytes may also express two-pore domain K^+ channels (see below). An important biophysical property of Kir channels is that their slope conductance increases with elevations in $[K^+]_o$ by a square root relation (Stanfield et al., 2002). This unique property of Kir channels allows K^+ conductance increases in glial cells, and therefore, enhanced K^+ clearance rates, when $[K^+]_o$ is raised (Newman, 1993).

An important implicit assumption for the K^+ spatial buffering mechanism is the low permeability of glial cells to anions such as Cl^-, as the low anion conductance ensures that the net uptake of KCl will not occur when $[K^+]_o$ increases. Unfortunately, there is no consensus concerning glial-cell Cl^- permeability in native tissue (Walz, 2002). Although a relatively high basal Cl^- conductance has been reported in glial cells of guinea-pig olfactory cortex (Ballanyi et al., 1987), other studies have failed to demonstrate a significant Cl^- conductance for glial cells in situ (Walz, 2002).

6.5 Evidence for K^+ Spatial Buffering

Orkand et al. (1966) reported that, in amphibians, stimulation of the optic nerve leads to slow depolarization and repolarization of the glial cells surrounding non-myelinated axons. These slow glial membrane potential changes were thought to reflect K^+ transfer by glial cells via the K^+ spatial buffering mechanism. Subsequently, support for the K^+ spatial buffering hypothesis came from measurements of extracellular field potentials. The transcellular transfer of K^+ ions from areas of elevated $[K^+]_o$ to lower $[K^+]_o$ generates return current loops in the extracellular space, giving rise to extracellular field potentials. These activity-induced slow extracellular field potentials are generated in various CNS regions, including the cortex and retina (Gardner-Medwin et al., 1981; Dietzel et al., 1989). In the retina, slow extracellular field potentials (slow PIII and M waves) are generated upon light stimulation (Xu and Karwoski, 1997; Karwoski and Xu, 1999). Current source density analysis indicates that these waves originate from K^+ spatial buffering by retinal glial cells (Xu and Karwoski, 1997; Karwoski and Xu, 1999). The transfer of K^+ ions by retinal glial cells, which generates the slow PIII wave, buffers the photoreceptor-based light-evoked decrease in $[K^+]_o$ in the outer retina. As expected, these K^+ spatial buffering fluxes are abolished by blocking retinal glial-cell K^+ channels with Ba^{2+} (Oakley et al., 1992; Kofuji et al., 2000).

Measurements of activity-dependent $[K^+]_o$ changes in the cortex and cerebellum show that $[K^+]_o$ varies with depth and time in a manner consistent with transcellular transfer of K^+ ions (Gardner-Medwin and Nicholson, 1983). These $[K^+]_o$ changes were not abolished by Na^+ pump inhibitors, indicating that they are likely due to a passive K^+ transport mechanism such as K^+ spatial buffering (Gardner-Medwin and Nicholson, 1983). Similar results were obtained in the drone retina, where it was estimated that about 10 times more K^+ ions move as a result of spatial buffering than by simple diffusion through the extracellular space (Coles et al., 1986).

More direct evidence for K^+ spatial buffering has been provided by optical imaging of brain slices (Holthoff and Witte, 2000). When K^+ ions are transferred via glial cells, there is a shrinkage of the extracellular space in areas of K^+ influx and swelling in areas of K^+ efflux (Dietzel et al., 1980). Changes in extracellular volume following neuronal stimulation can be demonstrated by monitoring intrinsic optic signals (IOS) in brain slices. As predicted for K^+ spatial buffering, stimulation of cortical areas promoted shrinkage of the extracellular space in the stimulated region followed by swelling in the layers above and below the stimulated area (Fig. 6.4) (Holthoff and Witte, 2000). This swelling in the extracellular space was associated with local increases in $[K^+]_o$ and was dependent on gap junctional coupling (Holthoff and Witte, 2000). These gap junctionally-dependent changes in extracellular space are consistent with K^+ spatial buffering in the mammalian cortex.

Recent studies have directly evaluated the importance of K^+ spatial buffering to the regulation of $[K^+]_o$ in the brain. Activity-dependent increases in $[K^+]_o$ have been

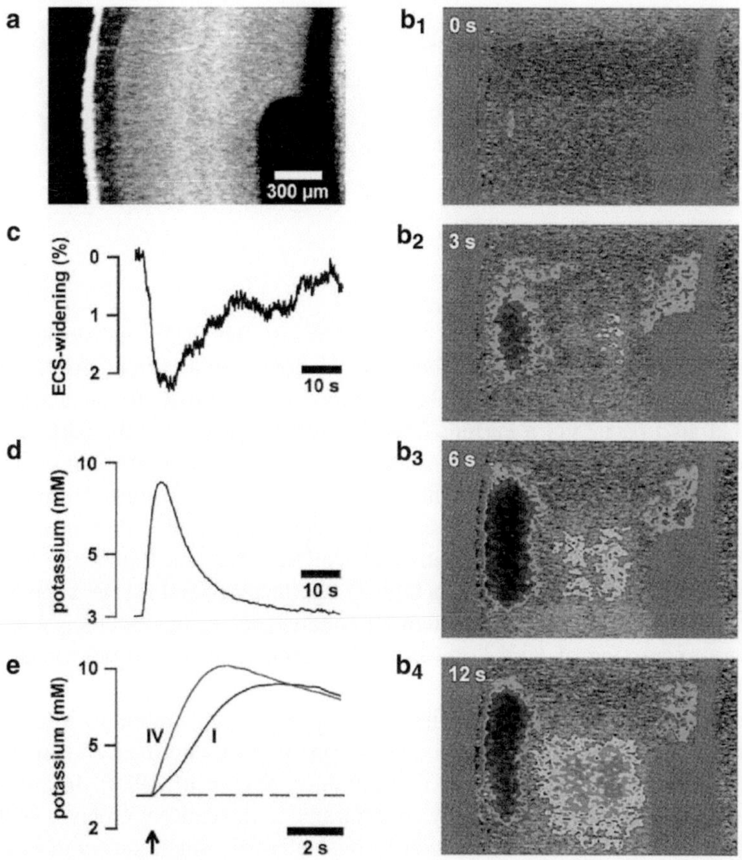

Fig. 6.4 Potassium spatial buffering in the rat cortex. (**a**) Image of a brain slice viewed with darkfield optics. (**b**1–4) Time course of intrinsic optical signal (IOS) changes upon neuronal stimulation in middle cortical layers. *Red colors* represent IOS increases while *blue colors* represent IOS decreases. These correspond to shrinking and widening of the extracellular space, respectively. Note that extracellular space shrinks in the middle cortical layers and widens in the most superficial and deep cortical layers, as predicted for the K^+ spatial buffering mechanism. (**c**) Time course of extracellular space widening in cortical layer I, measured independently, confirming the IOS results. (**d**) Time course of $[K^+]_o$ increase in layer I. (**e**) Time course of $[K^+]_o$ increase in layer I (*blue*) and in layer IV (*red*). [From Holthoff and Witte (2000), with permission]. (*See Color Plates*).

measured in brain slices of wild-type mice and in transgenic mice lacking astrocyte connexin and K^+-channel expression. If K^+ spatial buffering plays an important role in regulating $[K^+]_o$; then $[K^+]_o$ regulation should be compromised in the transgenic animals. In hippocampal slices of transgenic mice lacking Cx30 and Cx43 connexins, astrocytes were completely uncoupled (Wallraff et al., 2006). In these animals, activity-dependent $[K^+]_o$ increases were larger and clearance of the increases was slower. However, changes in $[K^+]_o$ regulation were modest in the transgenic animals, demonstrating that K^+ spatial buffering though the glial syncytium is not the only mechanism contributing to $[K^+]_o$ regulation. Similar conclusions were reached in a study of $[K^+]_o$ regulation in the brain stem in transgenic mice lacking Kir4.1 K^+ channels, the principal K^+ channel of astrocytes (Neusch et al., 2006). In transgenic animals, clearance of $[K^+]_o$ increases was slowed and the $[K^+]_o$ undershoot, which follows $[K^+]_o$ increases, was larger. However, rhythmic bursting activity of brain stem neurons was not altered in the Kir4.1 KO animals.

6.6 Potassium Siphoning

Regulation of $[K^+]_o$ by K^+ spatial buffering posits that K^+ is redistributed from regions of high $[K^+]_o$ to regions where $[K^+]_o$ is lower by a current flow through a network of electrically coupled glial cells. However, a redistribution of extracellular K^+ could also occur via a current flow through single glial cells, particularly, if these cells are elongated. This is the case for Müller cells, the principal glial cell of the retina (Newman and Reichenbach, 1996). Müller cells display morphological polarization with an endfoot process in close apposition to the vitreous and apical microvilli projecting into the subretinal space (Newman and Reichenbach, 1996). The membrane of Müller cells has a high K^+ conductance and is selectively permeable to K^+ (Newman, 1985). The high K^+ conductance of these cells is due to the abundant expression of Kir4.1 inwardly rectifying K^+ channels (Newman, 1993; Kofuji et al., 2000).

Kir4.1 channels are unevenly distributed along the membrane of Müller cells. Potassium-channel distribution has been mapped by monitoring cell responses to focal increases in extracellular K^+ concentration (Newman, 1984). In amphibian Müller cells, K^+ channels are highly concentrated in the endfoot process, with 94% of the total K^+ conductance localized to this relatively small subcellular domain (Newman, 1984; Brew et al., 1986).

The observation of a highly nonuniform distribution of Kir channels in Müller cells led to the hypothesis that excess K^+ released from retinal neurons is selectively directed, or "siphoned," to the vitreous humor (Newman et al., 1984; Newman, 1987a). This hypothesis, termed *K^+siphoning*, is a specialized form of the spatial buffering mechanism in which the nonuniform distribution of K^+ channels in glial cells directs excess K^+ into large reservoirs such as the vitreous humor (Fig. 6.5). In mammalian retinas, high densities of Kir channels are found on Müller cell endfeet contacting blood vessels, as well as on vitreal endfeet (Newman, 1987b, 1993;

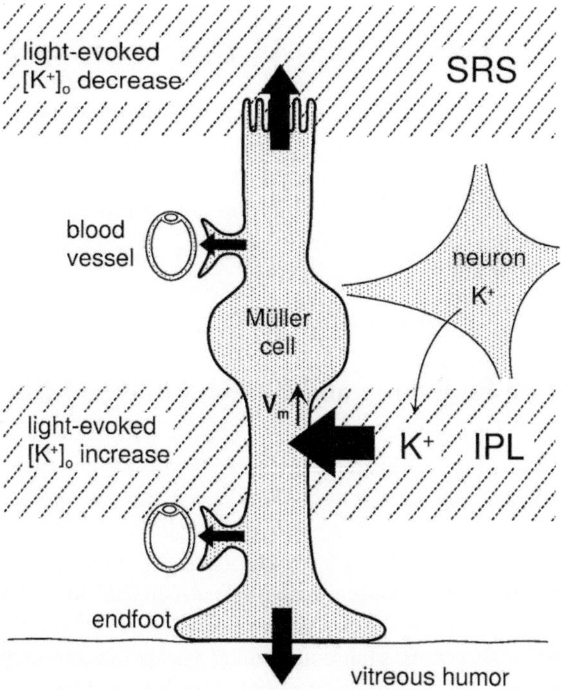

Fig. 6.5 Potassium siphoning in the retina. Potassium released from active neurons in the inner plexiform layer (IPL) generates a $[K^+]_o$ increase and an influx of K^+ into Müller cells, the principal glial cells of the retina. Potassium influx depolarizes the Müller cell and induces an efflux of an equal amount of K^+ from other cell regions. Potassium efflux occurs preferentially from Müller cell endfeet, where K^+-channel density is maximal, both at the vitreous humor and at processes enveloping blood vessels. Potassium efflux also occurs from the Müller-cell apical processes in the subretinal space (SRS), where light stimulation evokes a $[K^+]_o$ decrease. [From Newman (1996b), with permission.]

Kofuji et al., 2000). In these species, K^+ will be siphoned onto the blood vessels as well as into the vitreous.

Potassium siphoning contributes significantly to $[K^+]_o$ regulation in the retina. In the amphibian retina, light stimulation evokes rapid $[K^+]_o$ increases in the synaptic layers (Fig. 6.1b; inner plexiform layer (IPL) and outer plexiform layer (OPL)) and a slower increase in the vitreous humor (Fig. 6.1b; GCL). When Müller cell K^+ siphoning is interrupted by Ba^{2+} block of Kir channels, light-evoked $[K^+]_o$ increases within the retina grow larger, clearance of the $[K^+]_o$ increases is slowed, and the K^+ increase in the vitreous humor is reduced (Karwoski et al., 1989). These results demonstrate that Müller cells transfer excess K^+ from the retina to the vitreous by a K^+ siphoning current. Similarly, in the cat retina, light-evoked $[K^+]_o$ increases in the inner plexiform layer are threefold larger following Ba^{2+} block of Kir channels (Frishman et al., 1992), confirming that glial-mediated K^+ siphoning currents play an important role in limiting large variations in $[K^+]_o$.

Potassium siphoning may contribute to $[K^+]_o$ regulation in the brain as well. In hippocampal slices, clearance of K^+ released from pyramidal cells is dependent on glial Kir channels, as $[K^+]_o$ regulation is compromised by Ba^{2+} block of the channels. However, $[K^+]_o$ clearance within the stratum radiatum is not reduced in Cx43/Cx30 double KO animals, where glial-cell coupling is eliminated, suggesting that K^+ current flow within single glial cells can effectively clear K^+ (Wallraff et al., 2006). This finding is supported by the observation that astrocytes within the stratum radiatum are preferentially oriented in a perpendicular direction (Wallraff et al., 2006).

6.7 Potassium Siphoning and the Regulation of Blood Flow

Neuronal activity evokes localized changes in blood flow, a response termed *functional hyperemia* or *neurovascular coupling*. A consequence of K^+ siphoning is that neuronal activity will lead to an efflux of K^+ from glial cell endfeet onto blood vessels. Paulson and Newman (1987) have proposed that this siphoning mechanism could mediate neurovascular coupling, as modest increases in K^+ at the vessel wall leads to vasodilation. This hypothesis was recently tested in the retina (Metea et al., 2007). Potassium efflux from glial cell endfeet was evoked by depolarizing individual glial cells. Vessels adjacent to the glial cells did not dilate. In addition, light-evoked vasodilations were monitored in transgenic mouse retinas, where Kir4.1, the main glial K^+ channel, was knocked out. Although K^+ siphoning currents are largely absent in glial cells of these animals, light-evoked vasodilations were not reduced. These results demonstrate that, contrary to the hypothesis, K^+ siphoning does not contribute significantly to neurovascular coupling in the retina.

Filosa et al. (2006) have recently proposed that K^+ efflux from glial cell endfeet, mediated by a nonsiphoning mechanism, is responsible for neurovascular coupling. They suggest that neuronal activity results in the opening of Ca^{2+}-activated K^+ (BK) channels in glial endfeet, resulting in the efflux of K^+ onto blood vessels and to vessel dilation.

6.8 Relative Importance of K^+ Regulatory Mechanisms

The relative importance of the different K^+ regulatory mechanisms, including active and passive K^+ uptake, K^+ spatial buffering, and K^+ siphoning, remains uncertain and is a question of considerable debate. It is likely that the relative importance of the mechanisms varies in different CNS regions. In the rat optic nerve, for example, K^+ regulation appears to depend more on active uptake of K^+ than on K^+ spatial buffering as recovery of $[K^+]_o$ following stimulation is highly sensitive to Na^+ pump inhibition but not to glial K^+-channel blockers (Ransom et al., 2000). By contrast, in the CA3 region of rat hippocampus, both the Na^+ pump and glial K^+ channels are critical for maintaining baseline $[K^+]_o$ and for recovery of $[K^+]_o$ following stimulation (D'Ambrosio et al., 2002) (Fig. 6.3). In this preparation, the Na^+ pump is necessary for the clearance of excess K^+ during afferent stimulation while glial K^+ channels are necessary to prevent large $[K^+]_o$ undershoots following tetanic stimulation (D'Ambrosio et al., 2002). In the CA1 region of the hippocampus, both spatial buffering through glial-cell networks and spatial buffering/siphoning through individual glial cells contribute to $[K^+]_o$ regulation (Wallraff et al., 2006). In the amphibian retina, K^+ spatial buffering and, in particular K^+ siphoning, has a major role in regulating $[K^+]_o$ (Newman, 1995). When glial K^+ channels are blocked pharmacologically, recovery of $[K^+]_o$ following stimulation is prolonged and transfer of K^+ from the retina to the vitreous is blocked.

The factors that determine the relative contribution of each K^+ clearance mechanism remains uncertain. A few general principals have emerged, however. Coupling between astrocytes does not appear to contribute greatly to K^+ clearance as $[K^+]_o$ dynamics are not substantially altered in connexin KO animals. Thus, long-range spatial buffering through the astrocyte syncytium, as originally proposed by Orkand et al. (1966), may not be a dominant K^+ clearance mechanism. In contrast, K^+ siphoning, a specialized form of K^+ spatial buffering, does contribute significantly to K^+ clearance in those CNS regions bordering a large fluid reservoir. Thus, K^+ siphoning is instrumental in clearing K^+ from the retina, which is a thin sheet of CNS tissue surrounded by the vitreous humor and the subretinal space, which both function as sinks where K^+ can be temporarily stored. Passive and active uptake of K^+ may play a more important role in $[K^+]_o$ regulation in those CNS regions where K^+ spatial buffering/siphoning cannot efficiently move K^+ to fluid reservoirs.

6.9 Glial Cells and K^+ Channels

Kir channels most likely underlie K^+ spatial buffering in the CNS and there is considerable interest in determining their macromolecular structure, mechanisms of targeting, and modulation by intracellular and extracellular factors. The Kir channels have been recently cloned, and over 20 genes are currently known to encode various Kir-channel subunits (Nichols and Lopatin, 1997; Stanfield et al., 2002). Site-directed mutagenesis and heterologous channel expression have been used to identify structural elements involved in specific Kir-channel functions. These studies have revealed the basic Kir-channel design of two transmembrane domains and a re-entry loop (P-loop), with intracellular amino and carboxyl termini (Nichols and Lopatin, 1997; Stanfield et al., 2002). The Kir-channel subunits are usually categorized into seven major subfamilies (Kirl to Kir7) that are diversely regulated by intracellular and extracellular factors (Stanfield et al., 2002).

Of these family members, immunocytochemical and in situ hybridization studies demonstrate that the Kir4.1 channel is broadly expressed in brain (Poopalasundaram et al., 2000; Higashi et al., 2001), though different reports have suggested that it is expressed only in glial cells (Higashi et al., 2001) or in both neurons and glia (Li et al., 2001). Kir4.1 immunoreactivity can be demonstrated in cultured (Li et al., 2001; Kucheryavykh et al., 2007) and in situ astrocytes (Poopalasundaram et al., 2000; Higashi et al., 2001; Olsen et al., 2006) and in oligodendrocytes (Kalsi et al., 2004). In the olfactory bulb, Kir4.1 immunoreactivity is detected in about half of the glial fibrillary acidic protein-positive astrocytes, but not in neurons (Higashi et al., 2001). Immunogold microscopic examination reveals that Kir4.1 channels are enriched in the processes of astrocytes enveloping synapses and blood vessels (Higashi et al., 2001). In addition to Kir4.1 channels, other Kir-channel subunits may also be expressed in various glial-cell types. Kir2.2 channels are expressed in Bergmann glial cells and astrocytes in the cerebellum (Leonoudakis et al., 2001) and Kir2.1 channels are found in astrocytes and oligodendrocytes in the forebrain (Stonehouse et al., 1999). Single-cell in situ PCR experiments in the astrocytes from mouse hippocampal slices have identified transcripts for Kir2.1, Kir2.2, Kir2.3 and Kir4.1 channels (Schroder et al., 2002). Thus, a large variety of Kir-channel subtypes may be expressed in glial cells and this may explain the wide range of single-channel Kir conductances reported in glial cells (Sontheimer, 1994).

TASK and TREK channels are members of the two-pore domain potassium-channel family and form either homomeric or heteromeric open-rectifier (leak) channels (Patel and Honore, 2001). Recent evidence suggests that these channels are expressed in macroglial cells, including astrocytes (Rusznak et al., 2004; Gnatenco et al., 2002; Kindler et al., 2000) and Müller cells (Skatchkov et al., 2006). The degree to which these channels contribute to $[K^+]_o$ regulation remains to be determined.

The properties of specific Kir-channel subunits have been assessed in heterologous expression systems such as *Xenopus* oocytes and transfected cells. Kir2.1 currents show steep inwardly rectifying current–voltage relationships with minimal outward currents at membrane potentials positive to E_K (Kubo et al., 1993). In contrast, Kir4.1 channels are weakly rectifying, allowing substantial outward currents (Takumi et al., 1995). A further complexity is provided by the fact that Kir channels are tetrameric proteins, and in heterologous expression systems Kir subunits can form either homomeric or heteromeric channels (Stanfield et al., 2002). The expression of Kir5.1 subunits in *Xenopus* oocytes or mammalian cell lines does not result in functional channels, but coexpression with Kir4.1 channels leads to formation of heteromeric channels that are highly sensitive to intracellular changes in pH (Tucker et al., 2000).

As in retinal Müller cells (see below), the distribution of K^+ conductance in astrocytes is nonhomogenous. Freshly dissociated salamander astrocytes have an approximately tenfold higher conductance in their endfeet than in other cell regions (Newman, 1986). In mammals, only Kir4.1 channels have been mapped to astrocytic endfeet (Higashi et al., 2001). Although it is clear that K^+ conductance in astrocytes is likely mediated by Kir channels, further investigations are required to determine the precise molecular composition of such channels.

6.10 Kir-Channel Subtypes Expressed in Müller Cells

In contrast to astrocytes in the brain, retinal Müller cells are a relatively homogenous class of cells and are thus an attractive model for studying Kir-channel composition. Several groups have examined the distribution of Kir channels in Müller cells using immunocytochemical and molecular biological techniques (Ishii et al., 1997; Kofuji et al., 2000). These studies demonstrate highly concentrated expression of the weakly rectifying Kir4.1 channels in Müller cell endfeet and on the processes enveloping blood vessels (Fig. 6.6a,c,e), a distribution that correlates well with the previously mentioned electrophysiological studies (Newman, 1987b, 1993). Several additional

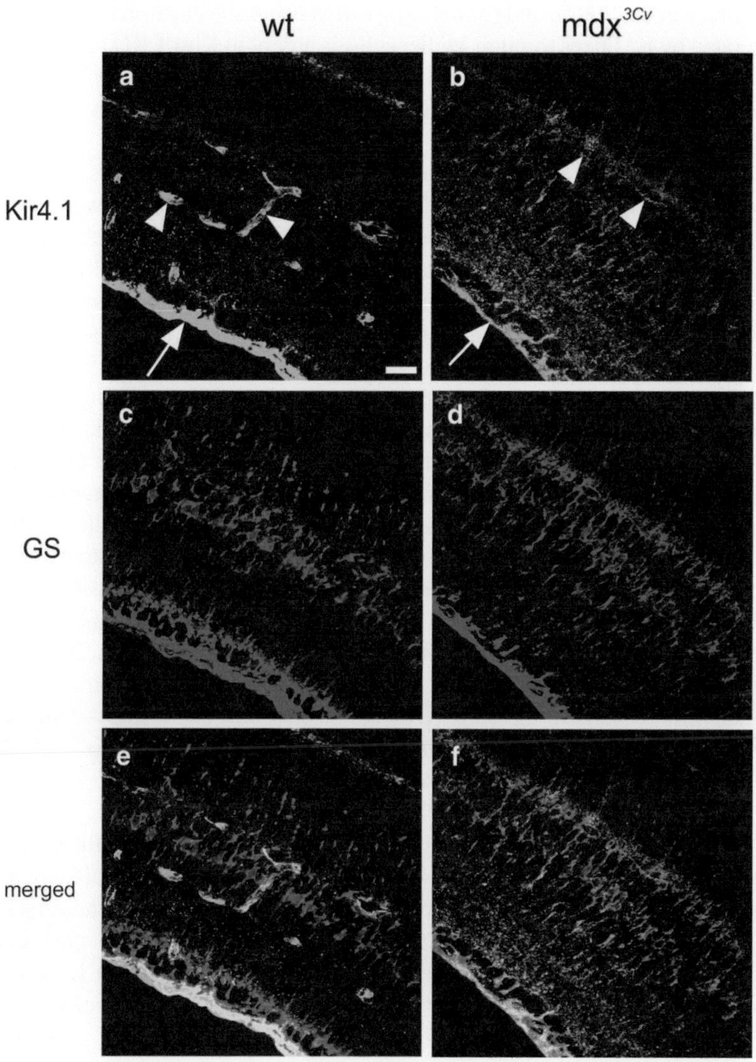

Fig. 6.6 Kir4.1-channel localization in wild type and mdx3Cv (dystrophin knockout) mouse retinas. (**a**) In the wild-type retina, Kir4.1 is concentrated at the inner limiting membrane (*arrow*) and to processes surrounding blood vessels (*arrowheads*). (**b**) In the mdx3Cv retina, Kir4.1 is more evenly distributed throughout the retina with a reduction in staining at the inner limiting membrane (*arrow*) and no apparent enrichment of Kir4.1 around blood vessels (*arrowheads*). The glial-cell marker glutamine synthetase (GS) (**c**, **d**), and merged images (**e**, **f**) suggest that Kir4.1 is localized to Müller cells. Scale bar in (**a**) = 25 μm. [From Connors and Kofuji (2002), with permission.] (*See Color Plates*).

lines of evidence argue that Kir4.1 channels are the principal Kir-channel subtype in Müller cells: (1) patch-clamp recordings in rabbit Müller cells and in transfected 293 cells expressing Kir4.1 channels show similar single-channel conductances and open probabilities (Tada et al., 1998); (2) genetic ablation of Kir4.1 channels in mice decreases the membrane conductance of Müller cells by 13–77-fold (Kofuji et al., 2000; Metea et al., 2007); and (3) the slow PIII wave of the electroretinogram, which is generated by K^+ fluxes through Müller cells, is absent in Kir4.1 KO animals (Kofuji et al., 2000). This effect is not caused by overall impairment of neuronal function as indicated by the fact that the a and b waves, associated with neuronal activity, are not decreased in the KO animals.

Other Kir channels, such as the strongly rectifying Kir2.1 channels, may also be expressed in Müller cells. In mouse Müller cells, Kir2.1 channels are expressed along the plasma membrane in a uniform manner that does not resemble the clustered distribution seen for Kir4.1 channels (Kofuji et al., 2002). Such differential expression of Kir2.1 and Kir4.1 channels may enhance the efficiency of K^+ siphoning in the retina (Kofuji et al., 2002). Expression of weakly rectifying Kir4.1 channels in selective membrane domains would allow K^+ ions to leave Müller cells and be stored in extracellular sinks such as the vitreous humor, whereas expression of strongly rectifying Kir2.1 channels would allow greater influx of K^+ in the synaptic layers (Kofuji et al., 2002).

An additional study suggests the expression of Kir5.1 channels in the soma and stalks of Müller cells; immunoprecipitation assays show that a fraction of the Kir4.1 subunits in retina are coassembled with Kir5.1 subunits (Ishii et al., 2003). Because heterologously expressed Kir4.1 and Kir5.1 form heteromeric Kir channels that are highly sensitive to physiological changes in intracellular pH, it has been suggested that the expression of Kir5.1 subunits in Müller cells promotes coordinated coupling between acid–base regulation and K^+ buffering in the retina (Ishii et al., 2003). In this hypothesis, increases in extracellular K^+ concentration and the resulting glial depolarization would increase the activity of the electrogenic Na^+-HCO_3^- cotransporter (Newman, 1996a). The increased influx of HCO_3^- would then cause intracellular alkalinization and subsequent increases in the activity of heteromeric Kir4.1/Kir5.1 channels, ultimately enhancing K^+ uptake into Müller cells (Ishii et al., 2003).

In summary, investigations in Müller cells have provided compelling evidence, demonstrating a major role for Kir4.1 channels in retinal K^+ regulation. Kir4.1 channels appear to have the functional and anatomical distributions that best match the physiological studies performed in Müller cells over the past decade. Biochemical and immunocytochemical work also suggests that other Kir channels, including Kir2.1 and Kir5.1, may be involved in K^+ influx in these cells.

6.11 Kir-Channel Accessory Proteins in Müller Cells: Localization and Function

The focal aggregation of Kir4.1 channels in Müller cells raises the intriguing question of how these channels are targeted to such precise subcellular domains. This is an important question as the efficiency of the retinal K^+ siphoning process is highly dependent on the clustered, nonhomogenous distribution of Kir channels in Müller cells (Newman, 1995). The water-channel aquaporin 4 (AQP4) is also highly enriched in the endfeet and perivascular processes of Müller cells (Nagelhus et al., 1999). This spatial overlap of Kir and aquaporins, two highly nonhomologous channel types, suggests that there may be a common molecular mechanism for their subcellular distribution and targeting. Although the Kir4.1 and AQP4 channels are highly divergent in their primary sequences, they share a key –S–X–V–COOH motif in their C termini. This sequence is able to bind to PDZ domains, which are modular amino acid motifs implicated in many protein–protein interactions (Hung and Sheng, 2002). Proteins possessing these domains are abundantly expressed in the nervous system and include postsynaptic density protein-95 (PSD-95), Chapsyn-110/PSD-93, SAP-102, and hDlg/SAP97 (Hung and Sheng, 2002). Positioning of several proteins in the postsynaptic density in excitatory synapses are critically dependent on their interactions with PDZ-domain containing proteins (Sheng and Sala, 2001).

Although specific PDZ domain-containing protein(s) are yet to be unequivocally identified in Müller cells, the SAP97 protein, which is present in Müller cells, has been shown to increase Kir4.1 currents in heterologous systems (Horio et al., 1997). Another candidate is the PDZ domain-containing adapter protein, α-syntrophin. In tissues where α-syntrophin is present, it is localized to the cell membrane by its association with the multiprotein dystrophin glycoprotein complex (DGC) (Ahn and Kunkel, 1995). The DGC spans the cell membrane, forming a molecular bridge between basal lamina proteins in the extracellular space and an array of signaling molecules in the intracellular domain. Immunolocalization studies in retina revealed that the DGC components, α-dystroglycan and the short dystrophin isoform, Dp71, appear to be localized in a fashion very similar to that of Kir4.1 in Müller cells (Claudepierre et al., 2000b). The role of Dp71 in the localization of Kir4.1 has been investigated in the dystrophin null mutant mouse, mdx^{3Cv} (Connors and Kofuji, 2002). Immunohistochemistry experiments reveal that the polarized subcellular distribution of Kir4.1 is altered in Müller glial cells from mdx^{3Cv} mice, displaying a more homogeneous distribution pattern (Fig. 6.6b,d,f). Immunoblotting and whole cell patch clamp experiments reveal that the channel is expressed at normal levels at the plasma membrane and its electrophysiological properties are unchanged (Connors and Kofuji, 2002). Similar findings have also been reported for a null mouse line for the dystrophin isoform Dp71 (Dalloz et al, 2003).

These results strongly suggest that the DGC is important to the localization of Kir4.1 in Müller cells. It is possible that the DGC targets the Kir channels to the membrane domains facing the vitreous and blood vessels by binding of extracellular portions of the DGC to the basal lamina of these regions (Fig. 6.7). This assumes the existence of an intermediate protein containing a PDZ domain. As mentioned previously, the best candidate for such an adaptor protein is α-syntrophin. α-syntrophin is expressed in Müller cells and is putatively part of the Müller cell-specific DGC (Claudepierre et al., 2000a). In addition, a-syntrophin has been shown to interact with AQP4 in astrocytes in a PDZ-dependent manner, and is required for membrane expression and localization of AQP4 (Neely et al., 2001). Therefore, α-syntrophin could underlie the colocalization of Kir4.1 and AQP4 seen

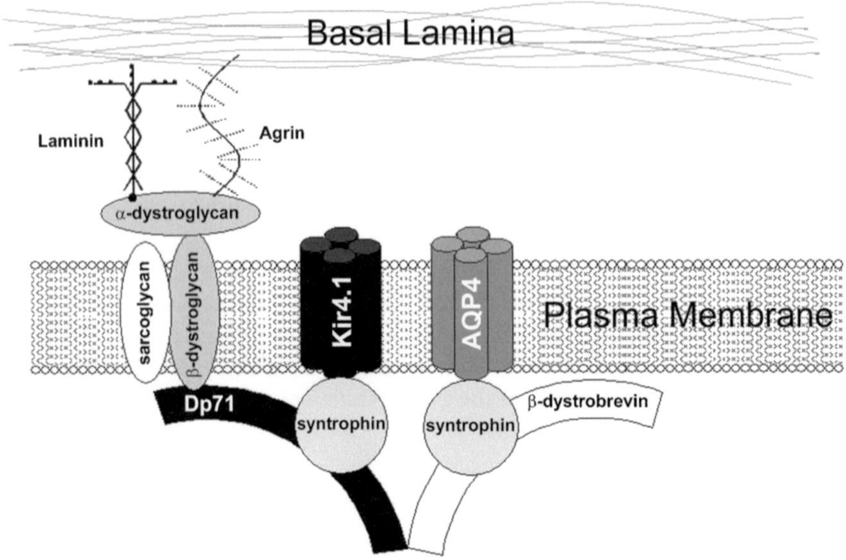

Fig. 6.7 Schematic representation of the glial-cell dystrophin–glycoprotein complex (DGC) and associated Kir and aquaporin channels. The dystrophin–glycoprotein complex is shown with its putative interactions with a syntrophin isoform, Kir4.1 and AQP4. [From Kofuji and Connors (2003), with permission.]

in Müller cells. The importance of laminin in the clustering of Kir4.1 channels has been demonstrated in a study showing that laminin, agrin, and α-dystroglycan codistribute with Kir4.1 in Müller cell endfeet (Noel et al., 2005). In cultured Müller cells, addition of laminin-1 induces the clustering of α-dystroglycan and Kir4.1.

In astrocytes, the direct participation of DGC proteins in the targeting of Kir4.1 and AQP4 channels has been demonstrated in an α-syntrophin KO mouse line. In astrocytes, as in Müller cells, AQP4 and Kir4.1 are strongly expressed in glial endfeet that are in direct contact with capillaries and the pia (Nielsen et al., 1997; Higashi et al., 2001). Quantitative immunoelectromicroscopy has shown that in the hippocampus of α-syntrophin KO mice, the expression of AQP4 in astrocyte endfeet is greatly diminished, while expression of Kir4.1 channels is less affected (Amiry-Moghaddam et al., 2003). These results indicate that α-syntrophin is critical for the targeting and clustering of AQP4 channels to astrocytic endfeet, but is perhaps not as vital for targeting of Kir4.1 channels. Further work is needed to establish whether Kir4.1 channels are indeed linked to the DGC in astrocytes via syntrophin isoforms.

Despite the apparent lack of major rearrangements in Kir4.1-channel localization and expression in hippocampus astrocytes, the clearance of extracellular K^+ following neuronal stimulation is slowed twofold in hippocampal slices from α-syntrophin KO mice (Amiry-Moghaddam et al., 2003). This study suggests that AQP4 plays an essential role in K^+ clearance by the K^+ spatial buffering mechanism. In support of this view, the clearance of K^+ following spreading depression is slowed twofold in AQP4 null mice (Padmawar et al., 2005). As suggested by Ottersen and colleagues (Nagelhus et al., 1999; Amiry-Moghaddam et al., 2003), the transfer of K^+ ions across the plasma membrane of glial cells by K^+ spatial buffering generates osmotic imbalances. Water fluxes paralleling the K^+ flow are needed to dissipate the osmotic imbalance. In the absence of AQP4 channels in regions where Kir4.1-channel density is high, spatial buffering cannot precede efficiently.

These tantalizing observations suggest that impaired targeting or function of AQP4 and Kir4.1 channels may have clinical relevance in conditions such as epilepsy or brain edema. It remains to be seen, however, how significant a role Kir4.1 channels play in K^+ spatial buffering in the brain and in the overall regulation of $[K^+]_{o.}$ Future studies in mice with genetic inactivation of Kir4.1 channels should provide an answer to this important question.

6.12 Impaired Potassium Regulation in Pathological Conditions

In many types of pathology, $[K^+]_o$ regulation is likely to be impaired by downregulation of Kir channels in glial cells. For example, in patients with temporal lobe epilepsy there is marked reduction of Kir currents in astrocytes (Bordey and Sontheimer, 1998; Hinterkeuser et al., 2000), which could contribute to the increased excitability of the epileptic tissue. Indeed, there is a reduction of $[K^+]_o$ clearance in sclerotic tissue from epileptic hippocampus in comparison to nonsclerotic tissue (Heinemann et al., 2000). In animal models of retinal ischemia and diabetes, there is also downregulation and/or mislocalization of expression of Kir4.1 channels in Müller cells (Pannicke et al., 2006); (Iandiev et al., 2006). Similar loss of Kir4.1 expression is seen in an animal model of amyotrophic lateral sclerosis (Kaiser et al., 2006). Overall, such changes of Kir expression in macroglial cells are expected to impair the rapid movement of K^+ and water in these cells and may contribute to the pathophysiology associated with these disorders.

6.13 Conclusions

It has been over 40 years since Orkand et al. (1966) initially proposed the K^+ spatial buffering mechanism of $[K^+]_o$ regulation in the CNS. Since then, it has become clear that $[K^+]_o$ regulation involves both net uptake of K^+ and K^+ spatial buffering. The relative importance of these K^+ homeostatic mechanisms may vary from site to site in the CNS. Potassium spatial buffering has been carefully characterized in the retina, where a specialized form of spatial buffering, K^+ siphoning, directs K^+ from the plexiform layers to the vitreous humor and to blood vessels (Newman et al., 1984). Molecular studies indicate that Kir4.1-channel localization is critical to the highly asymmetric K^+ conductance found in Müller cells (Kofuji et al., 2000).

Furthermore, Kir2.1 and possibly Kir5.1 are also expressed in Müller cells (Kofuji et al., 2002). Potassium siphoning in the retina may be facilitated by concerted action of the strongly rectifying Kir2.1 channels, which allow K^+ entry into Müller cells, and the weakly rectifying Kir4.1 channels, which allow K^+ exit to large sinks such as the vitreous humor (Kofuji et al., 2002). Moreover, dystrophin and dystrophin-associated proteins may promote the clustering and subcellular distribution of Kir4.1 channels and the water channel AQP4 in astrocytes (Amiry-Moghaddam et al., 2003). The colocalization of water and K^+ channels in glial-cell membranes suggests that K^+ buffering and water flux are tightly coupled in the brain.

Although significant progress has been made in the past decade, we still do not have a coherent picture of the relative importance of the various mechanisms contributing to $[K^+]_o$ regulation in the CNS. Given recent advances in optical, electrophysiological, and genetic methods, it is plausible that this picture will gain greater clarity in the near future.

References

Ahn AH, Kunkel LM (1995) Syntrophin binds to an alternatively spliced exon of dystrophin. *J Cell Biol* 128:363–371.

Amedee T, Robert A, Coles JA (1997) Potassium homeostasis and glial energy metabolism. *Glia* 21:46–55.

Amiry-Moghaddam M, Otsuka T, Hurn PD, Traystman RJ, Haug FM, Froehner SC, Adams ME, Neely JD, Agre P, Ottersen OP, Bhardwaj A (2003) An -syntrophin-dependent pool of AQP4 in astroglial end-feet confers bidirectional water flow between blood and brain. *Proc Natl Acad Sci U S A* 100:2106–2111.

Ballanyi K, Grafe P, ten Bruggencate G (1987) Ion activities and potassium uptake mechanisms of glial cells in guinea-pig olfactory cortex slices. *J Physiol* 382:159–174.

Barres BA, Koroshetz WJ, Chun LL, Corey DP (1990) Ion channel expression by white matter glia: the type-1 astrocyte. *Neuron* 5:527–544.

Bordey A, Sontheimer H (1998) Properties of human glial cells associated with epileptic seizure foci. *Epilepsy Res* 32:286–303.

Brew H, Gray P, Mobbs P, Attwell D (1986) Endfeet of retinal glial cells have higher densities of ion channels that mediate K+ buffering. *Nature* 324:466–468.

Claudepierre T, Dalloz C, Mornet D, Matsumura K, Sahel J, Rendon A (2000a) Characterization of the intermolecular associations of the dystrophin-associated glycoprotein complex in retinal Müller glial cells. *J Cell Sci 113 (Pt* 19):3409–3417.

Claudepierre T, Mornet D, Pannicke T, Forster V, Dalloz C, Bolanos F, Sahel J, Reichenbach A, Rendon A (2000b) Expression of Dp71 in Müller glial cells: a comparison with utrophin- and dystrophin-associated proteins. *Invest Ophthalmol Vis Sci* 41:294–304.

Coles JA, Orkand RK, Yamate CL, Tsacopoulos M (1986) Free concentrations of Na, K, and Cl in the retina of the honeybee drone: stimulus-induced redistribution and homeostasis. *Ann N Y Acad Sci* 481:303–317.

Connors B, Dray A, Fox P, Hilmy M, Somjen G (1979) LSD's effect on neuron populations in visual cortex gauged by transient responses of extracellular potassium evoked by optical stimuli. *Neurosci Lett* 13:147–150.

Connors BW, Ransom BR, Kunis DM, Gutnick MJ (1982) Activity-dependent K+ accumulation in the developing rat optic nerve. *Science* 216:1341–1343.

Connors NC, Kofuji P (2002) Dystrophin Dp71 is critical for the clustered localization of potassium channels in retinal glial cells. *J Neurosci* 22:4321–4327.

D'Ambrosio R, Gordon DS, Winn HR (2002) Differential role of KIR channel and Na+/K+-pump in the regulation of extracellular K^+ in rat hippocampus. *J Neurophysiol* 87:87–102.
Dalloz C, Sarig R, Fort P, Yaffe D, Bordais A, Pannicke T, Grosche J, Mornet D, Reichenbach A, Sahel J, Nudel U, Rendon A (2003) Targeted inactivation of dystrophin gene product Dp71: phenotypic impact in mouse retina. *Hum Mol Genet* 12:1543–1554.
Dennis MJ, Gerschenfeld HM (1969) Some physiological properties of identified mammalian neuroglial cells. *J Physiol* 203:211–222.
Dermietzel R (1998) Diversification of gap junction proteins (connexins) in the central nervous system and the concept of functional compartments. *Cell Biol Int* 22:719–730.
Dermietzel R, Gao Y, Scemes E, Vieira D, Urban M, Kremer M, Bennett MV, Spray DC (2000) Connexin43 null mice reveal that astrocytes express multiple connexins. *Brain Res Brain Res Rev* 32:45–56.
Dietzel I, Heinemann U, Hofmeier G, Lux HD (1980) Transient changes in the size of the extracellular space in the sensorimotor cortex of cats in relation to stimulus-induced changes in potassium concentration. *Exp Brain Res* 40:432–439.
Dietzel I, Heinemann U, Lux HD (1989) Relations between slow extracellular potential changes, glial potassium buffering, and electrolyte and cellular volume changes during neuronal hyperactivity in cat brain. *Glia* 2:25–44.
Doupnik CA, Davidson N, Lester HA (1995) The inward rectifier potassium channel family. *Curr Opin Neurobiol* 5:268–277.
Filosa JA, Bonev AD, Straub SV, Meredith AL, Wilkerson MK, Aldrich RW, Nelson MT (2006) Local potassium signaling couples neuronal activity to vasodilation in the brain. *Nat Neurosci* 9:1397–1403.
Franck G, Grisar T, Moonen G (1983) Glial and neuronal Na^+,K^+ pump. In: Advances in Neurobiology (Fedoroff S, Hertz L, eds), pp 139–159. New York: Academic.
Frishman LJ, Yamamoto F, Bogucka J, Steinberg RH (1992) Light-evoked changes in $[K^+]o$ in proximal portion of light-adapted cat retina. *J Neurophysiol* 67:1201–1212.
Gardner-Medwin AR, Nicholson C (1983) Changes of extracellular potassium activity induced by electric current through brain tissue in the rat. *J Physiol* 335:375–392.
Gardner-Medwin AR, Coles JA, Tsacopoulos M (1981) Clearance of extracellular potassium: evidence for spatial buffering by glial cells in the retina of the drone. *Brain Res* 209:452–457.
Gnatenco C, Han J, Snyder AK, Kim D (2002) Functional expression of TREK-2 K^+ channel in cultured rat brain astrocytes. *Brain Res* 931:56–67.
Gutnick MJ, Heinemann U, Lux HD (1979) Stimulus induced and seizure related changes in extracellular potassium concentration in cat thalamus (VPL). *Electroencephalogr Clin Neurophysiol* 47:329–344.
Haas M, Forbush B, III (1998) The Na-K-Cl cotransporters. *J Bioenerg Biomembr* 30:161–172.
Heinemann U, Lux HD (1977) Ceiling of stimulus induced rises in extracellular potassium concentration in the cerebral cortex of cat. *Brain Res* 120:231–249.
Heinemann U, Schaible HG, Schmidt RF (1990) Changes in extracellular potassium concentration in cat spinal cord in response to innocuous and noxious stimulation of legs with healthy and inflamed knee joints. *Exp Brain Res* 79:283–292.
Heinemann U, Gabriel S, Jauch R, Schulze K, Kivi A, Eilers A, Kovacs R, Lehmann TN (2000) Alterations of glial cell function in temporal lobe epilepsy. *Epilepsia* 41:S185–S189.
Higashi K, Fujita A, Inanobe A, Tanemoto M, Doi K, Kubo T, Kurachi Y (2001) An inwardly rectifying K+ channel, Kir4.1, expressed in astrocytes surrounds synapses and blood vessels in brain. *Am J Physiol Cell Physiol* 281:C922–C931.
Hinterkeuser S, Schroder W, Hager G, Seifert G, Blumcke I, Elger CE, Schramm J, Steinhauser C (2000) Astrocytes in the hippocampus of patients with temporal lobe epilepsy display changes in potassium conductances. *Eur J Neurosci* 12:2087–2096.
Holthoff K, Witte OW (2000) Directed spatial potassium redistribution in rat neocortex. *Glia* 29:288–292.

Horio Y, Hibino H, Inanobe A, Yamada M, Ishii M, Tada Y, Satoh E, Hata Y, Takai Y, Kurachi Y (1997) Clustering and enhanced activity of an inwardly rectifying potassium channel, Kir4.1, by an anchoring protein, PSD-95/SAP90. *J Bio Chem* 272:12885–12888.

Hung AY, Sheng M (2002) PDZ domains: structural modules for protein complex assembly. *J Biol Chem* 277:5699–5702.

Iandiev I, Tenckhoff S, Pannicke T, Biedermann B, Hollborn M, Wiedemann P, Reichenbach A, Bringmann A (2006) Differential regulation of Kir4.1 and Kir2.1 expression in the ischemic rat retina. *Neurosci Lett* 396:97–101.

Ishii M, Horio Y, Tada Y, Hibino H, Inanobe A, Ito M, Yamada M, Gotow T, Uchiyama Y, Kurachi Y (1997) Expression and clustered distribution of an inwardly rectifying potassium channel, KAB-2/Kir4.1, on mammalian retinal Müller cell membrane: their regulation by insulin and laminin signals. *J Neurosci* 17:7725–7735.

Ishii M, Fujita A, Iwai K, Kusaka S, Higashi K, Inanobe A, Hibino H, Kurachi Y (2003) Differential expression and distribution of Kir5.1 and Kir4.1 inwardly rectifying K^+ channels in retina. *Am J Physiol Cell Physiol* 285:C260–267.

Jorgensen PL, Hakansson KO, Karlish SJ (2003) Structure and mechanism of Na,K-ATPase: functional sites and their interactions. *Annu Rev Physiol* 65:817–849.

Kaiser M, Maletzki I, Hulsmann S, Holtmann B, Schulz-Schaeffer W, Kirchhoff F, Bahr M, Neusch C (2006) Progressive loss of a glial potassium channel (KCNJ10) in the spinal cord of the SOD1 (G93A) transgenic mouse model of amyotrophic lateral sclerosis. *J Neurochem* 99:900–912.

Kalsi AS, Greenwood K, Wilkin G, Butt AM (2004) Kir4.1 expression by astrocytes and oligodendrocytes in CNS white matter: a developmental study in the rat optic nerve. *J Anat* 204:475–485.

Kaplan JH (2002) Biochemistry of Na,K-ATPase. *Annu Rev Biochem* 71:511–535.

Karwoski CJ, Xu X (1999) Current source-density analysis of light-evoked field potentials in rabbit retina. *Vis Neurosci* 16:369–377.

Karwoski CJ, Newman EA, Shimazaki H, Proenza LM (1985) Light-evoked increases in extracellular K+ in the plexiform layers of amphibian retinas. *J Gen Physiol* 86:189–213.

Karwoski CJ, Lu HK, Newman EA (1989) Spatial buffering of light-evoked potassium increases by retinal Müller (glial) cells. *Science* 244:578–580.

Katzman R (1976) Maintenance of a constant brain extracellular potassium. *Fed Proc* 35:1244–1247.

Kimelberg HK, Frangakis MV (1985) Furosemide- and bumetanide-sensitive ion transport and volume control in primary astrocyte cultures from rat brain. *Brain Res* 361:125–134.

Kindler CH, Pietruck C, Yost CS, Sampson ER, Gray AT (2000) Localization of the tandem pore domain K+ channel TASK-1 in the rat central nervous system. *Brain Res Mol Brain Res* 80:99–108.

Kofuji P, Connors NC (2003) Molecular substrates of potassium spatial buffering in glial cells. *Mol Neurobiol* 28:195–208.

Kofuji P, Newman EA (2004) Potassium buffering in the central nervous system. *Neuroscience* 129:1045–1056.

Kofuji P, Ceelen P, Zahs KR, Surbeck LW, Lester HA, Newman EA (2000) Genetic inactivation of an inwardly rectifying potassium channel (Kir4.1 subunit) in mice: phenotypic impact in retina. *J Neurosci* 20:5733–5740.

Kofuji P, Biedermann B, Siddharthan V, Raap M, Iandiev I, Milenkovic I, Thomzig A, Veh RW, Bringmann A, Reichenbach A (2002) Kir potassium channel subunit expression in retinal glial cells: implications for spatial potassium buffering. *Glia* 39:292–303.

Kubo Y, Baldwin TJ, Jan YN, Jan LY (1993) Primary structure and functional expression of a mouse inward rectifier potassium channel. *Nature* 362:127–133.

Kucheryavykh YV, Kucheryavykh LY, Nichols CG, Maldonado HM, Baksi K, Reichenbach A, Skatchkov SN, Eaton MJ (2007) Downregulation of Kir4.1 inward rectifying potassium channel subunits by RNAi impairs potassium transfer and glutamate uptake by cultured cortical astrocytes. *Glia* 55:274–281.

Kuffler SW, Nicholls JG, Orkand RK (1966) Physiological properties of glial cells in the central nervous system of amphibia. *J Neurophysiol* 29:768–787.

Kume-Kick J, Mazel T, Vorisek I, Hrabetova S, Tao L, Nicholson C (2002) Independence of extracellular tortuosity and volume fraction during osmotic challenge in rat neocortex. *J Physiol* 542:515–527.

Leonoudakis D, Mailliard W, Wingerd K, Clegg D, Vandenberg C (2001) Inward rectifier potassium channel Kir2.2 is associated with synapse-associated protein SAP97. *J Cell Sci* 114:987–998.

Li L, Head V, Timpe LC (2001) Identification of an inward rectifier potassium channel gene expressed in mouse cortical astrocytes. *Glia* 33:57–71.

MacVicar BA, Feighan D, Brown A, Ransom B (2002) Intrinsic optical signals in the rat optic nerve: role for K(+) uptake via NKCC1 and swelling of astrocytes. *Glia* 37:114–123.

Matthias K, Kirchhoff F, Seifert G, Huttmann K, Matyash M, Kettenmann H, Steinhauser C (2003) Segregated expression of AMPA-type glutamate receptors and glutamate transporters defines distinct astrocyte populations in the mouse hippocampus. *J Neurosci* 23:1750–1758.

Metea MR, Kofuji P, Newman EA (2007) Neurovascular coupling is not mediated by potassium siphoning from glial cells. *J Neurosci* 27:2468–2471.

Nagelhus E, Horio Y, Inanobe A, Fujita A, Haug F, Nielsen S, Kurachi Y, Ottersen O (1999) Immnunogold evidence suggests that coupling of K^+ siphoning and water transport in rat retinal Müller cells is mediated by coenrichment of Kir4.1 and AQP4 in specific membrane domains. *Glia* 26:47–54.

Nagy JI, Rash JE (2000) Connexins and gap junctions of astrocytes and oligodendrocytes in the CNS. *Brain Res Brain Res Rev* 32:29–44.

Neely JD, Amiry-Moghaddam M, Ottersen OP, Froehner SC, Agre P, Adams ME (2001) Syntrophin-dependent expression and localization of Aquaporin-4 water channel protein. *Proc Natl Acad Sci U S A* 98:14108–14113.

Neusch C, Papadopoulos N, Muller M, Maletzki I, Winter SM, Hirrlinger J, Handschuh M, Bahr M, Richter DW, Kirchhoff F, Hulsmann S (2006) Lack of the Kir4.1 channel subunit abolishes K+ buffering properties of astrocytes in the ventral respiratory group: impact on extracellular K+ regulation. *J Neurophysiol* 95:1843–1852.

Newman EA (1984) Regional specialization of retinal glial cell membrane. *Nature* 309:155–157.

Newman EA (1985) Membrane physiology of retinal glial (Müller) cells. *J Neurosci* 5:2225–2239.

Newman EA (1986) High potassium conductance in astrocyte endfeet. *Science* 233:453–454.

Newman EA (1987a) Regulation of potassium levels by Müller cells in the vertebrate retina. *Can J Phys & Pharm* 65:1028–1034.

Newman EA (1987b) Distribution of potassium conductance in mammalian Müller (glial) cells: a comparative study. *J Neurosci* 7:2423–2432.

Newman EA (1993) Inward-rectifying potassium channels in retinal glial (Müller) cells. *J Neurosci* 13:3333–3345.

Newman EA (1995) Glial cell regulation of extracellular potassium. In: Neuroglia (Kettenmann H, Ransom B, eds), pp 717–731. New York: Oxford University Press.

Newman EA (1996a) Acid efflux from retinal glial cells generated by sodium bicarbonate cotransport. *J Neurosci* 16:159–168.

Newman EA (1996b) Regulation of extracellular K^+ and pH by polarized ion fluxes in glial cells: the retinal Muller cell. *The Neuroscientist* 2:109–117.

Newman EA, Reichenbach A (1996) The Müller cell: a functional element of the retina. *Trends Neurosci* 19:307–312.

Newman EA, Frambach DA, Odette LL (1984) Control of extracellular potassium levels by retinal glial cell K^+ siphoning. *Science* 225:1174–1175.

Nichols CG, Lopatin AN (1997) Inward rectifier potassium channels. *Annu Rev Physiol* 59:171–191.

Nicholson C, Sykova E (1998) Extracellular space structure revealed by diffusion analysis. *Trends Neurosci* 21:207–215.

Nielsen S, Nagelhus EA, Amiry-Moghaddam M, Bourque C, Agre P, Ottersen OP (1997) Specialized membrane domains for water transport in glial cells: high-resolution immunogold cytochemistry of aquaporin-4 in rat brain. *J Neurosci* 17:171–180.
Noel G, Belda M, Guadagno E, Micoud J, Klocker N, Moukhles H (2005) Dystroglycan and Kir4.1 coclustering in retinal Müller glia is regulated by laminin-1 and requires the PDZ-ligand domain of Kir4.1. *J Neurochem* 94:691–702.
Oakley BI, Katz B, Xu Z, Zheng J (1992) Spatial buffering of extracellular potassium by Müller (glial) cells in the toad retina. *Exp Eye Res* 55:539–550.
Olsen ML, Higashimori H, Campbell SL, Hablitz JJ, Sontheimer H (2006) Functional expression of Kir4.1 channels in spinal cord astrocytes. *Glia* 53:516–528.
Orkand RK (1986) Glial–interstitial fluid exchange. *Ann N Y Acad Sci* 481:269–272.
Orkand RK, Nicholls JG, Kuffler SW (1966) Effect of nerve impulses on the membrane potential of glial cells in the central nervous system of amphibia. *J Neurophysiol* 29:788–806.
Padmawar P, Yao X, Bloch O, Manley GT, Verkman AS (2005) K^+ waves in brain cortex visualized using a long-wavelength K^+-sensing fluorescent indicator. *Nature Methods* 2:825–827.
Pannicke T, Iandiev I, Wurm A, Uckermann O, vom Hagen F, Reichenbach A, Wiedemann P, Hammes HP, Bringmann A (2006) Diabetes alters osmotic swelling characteristics and membrane conductance of glial cells in rat retina. *Diabetes* 55:633–639.
Patel AJ, Honore E (2001) Properties and modulation of mammalian 2P domain K^+ channels. *Trends Neurosci* 24:339–346.
Paulson OB, Newman EA (1987) Does the release of potassium from astrocyte endfeet regulate cerebral blood flow? *Science* 237:896–898.
Poopalasundaram S, Knott C, Shamotienko OG, Foran PG, Dolly JO, Ghiani CA, Gallo V, Wilkin GP (2000) Glial heterogeneity in expression of the inwardly rectifying K^+ channel, Kir4.1, in adult rat CNS. *Glia* 30:362–372.
Ransom BR, Sontheimer H (1992) The neurophysiology of glial cells. *J Clin Neurophysiol* 9:224–251.
Ransom BR, Carlini WG, Connors BW (1986) Brain extracellular space: developmental studies in rat optic nerve. *Ann N Y Acad Sci* 481:87–105.
Ransom C, Sontheimer H, Janigro D (1996) Astrocytic inwardly rectifying potassium currents are dependent on external sodium ions. *J Neurophys* 76:626–630.
Ransom CB, Sontheimer H (1995) Biophysical and pharmacological characterization of inwardly rectifying K+ currents in rat spinal cord astrocytes. *J Neurophysiol* 73:333–346.
Ransom CB, Ransom BR, Sontheimer H (2000) Activity-dependent extracellular K^+ accumulation in rat optic nerve: the role of glial and axonal Na^+ pumps. *J Physiol* 522:427–442.
Reichenbach A, Henke A, Eberhardt W, Reichelt W, Dettmer D (1992) K^+ ion regulation in retina. *Can J Physiol Pharmacol* 70 (Suppl):S239–247.
Rose CR, Ransom BR (1996) Mechanisms of H^+ and Na^+ changes induced by glutamate, kainate, and d-aspartate in rat hippocampal astrocytes. *J Neurosci* 16:5393–5404.
Rouach N, Avignone E, Meme W, Koulakoff A, Venance L, Blomstrand F, Giaume C (2002) Gap junctions and connexin expression in the normal and pathological central nervous system. *Biol Cell* 94:457–475.
Rusznak Z, Pocsai K, Kovacs I, Por A, Pal B, Biro T, Szucs G (2004) Differential distribution of TASK-1, TASK-2 and TASK-3 immunoreactivities in the rat and human cerebellum. *Cell Mol Life Sci* 61:1532–1542.
Schroder W, Seifert G, Huttmann K, Hinterkeuser S, Steinhauser C (2002) AMPA receptor-mediated modulation of inward rectifier K^+ channels in astrocytes of mouse hippocampus. *Mol Cell Neurosci* 19:447–458.
Sheng M, Sala C (2001) PDZ domains and the organization of supramolecular complexes. *Annu Rev Neurosci* 24:1–29.
Singer W, Lux HD (1975) Extracellular potassium gradients and visual receptive fields in the cat striate cortex. *Brain Res* 96:378–383.
Skatchkov SN, Eaton MJ, Shuba YM, Kucheryavykh YV, Derst C, Veh RW, Wurm A, Iandiev I, Pannicke T, Bringmann A, Reichenbach A (2006) Tandem-pore domain potassium channels are functionally expressed in retinal (Muller) glial cells. *Glia* 53:266–276.

Somjen GG (1979) Extracellular potassium in the mammalian central nervous system. *Annu Rev Physiol* 41:159–177.

Somjen GG (2001) Mechanisms of spreading depression and hypoxic spreading depression-like depolarization. *Physiol Rev* 81:1065–1096.

Somjen GG (2002) Ion regulation in the brain: implications for pathophysiology. *Neuroscientist* 8:254–267.

Sontheimer (1994) Voltage-dependent ion channels in glial cells. *Glia* 11:156–172.

Stanfield PR, Nakajima S, Nakajima Y (2002) Constitutively active and G-protein coupled inward rectifier K+ channels: Kir2.0 and Kir3.0. *Rev Physiol Biochem Pharmacol* 145:47–179.

Stonehouse AH, Pringle JH, Norman RI, Stanfield PR, Conley EC, Brammar WJ (1999) Characterisation of Kir2.0 proteins in the rat cerebellum and hippocampus by polyclonal antibodies. *Histochem Cell Biol* 112:457–465.

Sweadner KJ (1995) Na,K-ATPase and its isoforms. In: Neuroglia (Kettenmann H, Ransom BR, eds), pp 259–272. New York: Oxford University Press.

Tada Y, Horio Y, Kurachi Y (1998) Inwardly rectifying K^+ channel in retinal Müller cells: comparison with the KAB-2/Kir4.1 channel expressed in HEK293T cells. *Jpn J Physiol* 48:71–80.

Takumi T, Ishii T, Horio Y, Morishige K, Takahashi N, Yamada M, Yamashita T, Kiyami H, Sohmiya K, Nakanishi S, Kurachi Y (1995) A novel ATP-dependent inward rectifier potassium channel expressed predominantly in glial cells. *J Biol Chem* 270:16339–16346.

Tucker SJ, Imbrici P, Salvatore L, D'Adamo MC, Pessia M (2000) pH dependence of the inwardly rectifying potassium channel, Kir5.1, and localization in renal tubular epithelia. *J Biol Chem* 275:16404–16407.

Vyskocil F, Kritz N, Bures J (1972) Potassium-selective microelectrodes used for measuring the extracellular brain potassium during spreading depression and anoxic depolarization in rats. *Brain Res* 39:255–259.

Wallraff A, Kohling R, Heinemann U, Theis M, Willecke K, Steinhauser C (2006) The impact of astrocytic gap junctional coupling on potassium buffering in the hippocampus. *J Neurosci* 26:5438–5447.

Walz W (1992) Role of Na/K/Cl cotransport in astrocytes. *Can J Physiol Pharmacol* 70 (Suppl):S260–S262.

Walz W (2002) Chloride/anion channels in glial cell membranes. *Glia* 40:1–10.

Xu X, Karwoski C (1997) The origin of slow PIII in frog retina: current source density analysis in the eyecup and isolated retina. *Vis Neurosci* 14:827–833.

Zahs KR, Kofuji P, Meier C, Dermietzel R (2003) Connexin immunoreactivity in glial cells of the rat retina. *J Comp Neurol* 455:531–546.

Zhou M, Schools GP, Kimelberg HK (2006) Development of GLAST(+) astrocytes and NG2(+) glia in rat hippocampus CA1: mature astrocytes are electrophysiologically passive. *J Neurophysiol* 95:134–143.

Abbreviations

AQP4	Aquaporin 4
CNS	Central nervous system
Cx	Connexins
DGC	Dystrophin glycoprotein complex
IOS	Intrinsic optic signal
KO	Knockout
$[K^+]_o$	Extracellular K^+ concentration
Kir channel	Inwardly rectifying K^+ channel
Na^+ pump	Na^+, K^+-ATPase

Chapter 7
Energy and Amino Acid Neurotransmitter Metabolism in Astrocytes

Helle S. Waagepetersen, Ursula Sonnewald, and Arne Schousboe

Contents

7.1 Introduction

Knowledge about the functional roles of astrocytes has developed enormously over the past 30 years and it is interesting to note that even a century ago Ramón y Cajal (1911) was concerned that it would take a long time until it could be elucidated how important astrocytes would be in maintaining basic elements of brain function. By now it is clear that astrocytes are of pivotal importance for ion homeostasis, intercellular communication, exchange of metabolites, and clearance of the extrasynaptic milieu of the neurotransmitters glutamate and gamma-aminobutyric acid (GABA)

A. Schousboe
Department of Pharmacology and Pharmacotherapy, Faculty of Pharmaceutical Sciences, University of Copenhagen, Copenhagen, Denmark
as@farma.ku.dk

V. Parpura and P.G. Haydon (eds.), *Astrocytes in (Patho)Physiology of the Nervous System*,
DOI: 10.1007/978-0-387-79492-1_7,

[reviewed in (Ransom et al., 2003; Newman, 2003; Hertz and Zielke, 2004)]. Actually it should be noted that perhaps the most pronounced feature of the human brain which distinguishes it from that of lower mammals is the complexity and number of the astrocytes (Oberheim et al., 2006). In this context it is also important to emphasize that the demonstration that astrocytes possess the machinery to release glutamate in an exocytotic fashion (Parpura et al., 1994; Parpura and Haydon, 2000; Haydon, 2001) has led to the term, the tripartite synapse (Araque et al., 1999; Volterra et al., 2002).

One reason why astrocytes have become a key element in the maintenance of particularly glutamatergic and GABAergic neurotransmission originates from the biochemical finding that only astrocytes express enzymes that are obligatory for the homeostasis of these transmitters, i.e., glutamine synthetase and pyruvate carboxylase (Norenberg and Martinez-Hernandez, 1979; Yu et al., 1983). The neurons are left as metabolically handicapped cells not capable of performing a net synthesis from glucose of their own neurotransmitter (Hertz et al., 1992). The present review focuses on these metabolic roles of astrocytes in glutamatergic and GABAergic homeostasis as well as in energy metabolism.

7.2 Energy Metabolism

7.2.1 Glycolysis

The brain consumes approximately 20% of the body energy expenditure and the energy substrate is under normal physiological conditions glucose. The end-feet of the astrocytes are directly attached to the capillaries. Hence, the astrocytes are exposed to a high concentration of glucose. The glucose transporter present on astrocytes is the most abundant glucose transporter (GLUT) in the brain, GLUT1. This is different from neurons which are mainly enriched with GLUT3. The main difference between these two isoforms is the rate of transport which has been estimated to be seven times faster for GLUT3 compared to GLUT1 (Vannucci et al., 1997). Glycolysis is the major fate of glucose in astrocytes subsequent to phosphorylation by hexokinase. Hexokinase exists in astrocytes as isoform I (Wilson, 1995). Isoform II of hexokinase which is the muscle isoform is induced in cultured astrocytes during glucose deprivation [Hamprecht et al. (2005) and ref. therein]. Isoform II is thought to serve more anabolic functions such as providing glucose-6-phosphate for glycogen synthesis and the pentose phosphate shunt. Isoform I is predominantly attached to the outer mitochondrial membrane via interaction with porin. Interestingly, hexokinase attached to mitochondria is functionally connected to intramitochondrially generated adenosine 5´-triphosphate (ATP) and does not utilize cytosolic ATP (Wilson, 2003). This might be particularly relevant in a tissue such as the brain in which the oxygen to glucose index is close to 6 meaning that glycolysis is obligatory followed by oxidation (de Cerqueira and Wilson, 1995;

Cesar and Wilson, 1998). However, glycolytic processing and oxidative metabolism of glucose has been suggested to be compartmentalized among astrocytes and neurons, respectively.

Studies employing primary cultures of neurons and astrocytes have shown a higher extent of glycolytic activity in astrocytes compared to neurons, and in cultured astrocytes anaerobic glycolysis leading to the production of lactate is prevalent over oxidation of pyruvate (Walz and Mukerji, 1988;Schousboe et al., 1997; Zwingmann and Leibfritz, 2003). The reduction of pyruvate to lactate is catalyzed by lactate dehydrogenase (LDH). LDH exists in five different isoforms which are tetramers composed of two different subunits known as H and M, which stands for heart and muscle, respectively. The different isoforms, 1–5, are heterogeneously distributed among neural cells with LDH-1 prevailing in neurons (Bittar et al., 1996). Interestingly, in cultured neurons and astrocytes LDH-1 expression appears more prominent in astrocytes than in neurons (Nissen and Schousboe, 1979; Schousboe et al., 1993b). The isoform composition may to some extent influence the prevailing direction of the LDH reaction in different cellular compartments (O'brien et al., 2006). The LDH-5, which is the form consisting of only M subunits, is enriched in astrocytes and is characteristic of tissues preferentially producing lactate (Pellerin et al., 1998). Lovatt et al. (2007) found that on the mRNA level in acutely isolated brain cells, *Ldha* was significantly enriched 8.3-fold in neurons over astrocytes. In contrast, *Ldhb* was significantly enriched 13.4-fold in astrocytes over neurons. These data confirm that astrocytes in contrast to neurons have high capacity to synthesize lactate from pyruvate, whereas neurons have a high capacity to synthesize pyruvate from lactate. However, the overall energetic and metabolic state of the cell has been recognized to be of more importance for determination of the fate of pyruvate (Cruz and Dienel, 2002; Dienel and Cruz, 2003, 2004). The enrichment of LDH-5 in astrocytes and the distribution of different subtypes of the monocarboxylate transporters in neurons and astrocytes have been interpreted in favor of lactate production and release from astrocytes (Pellerin et al., 1998). These findings have been hallmarks in the line of evidence that has been published in support of the astrocyte neuron lactate shuttle hypothesis initially proposed by Pellerin and Magistretti (1994). The model, which has been heavily debated since it was first suggested, has been developed and modified in accordance with the generation of new knowledge (Magistretti et al., 1999; Pellerin and Magistretti, 2004; Hyder et al., 2006). The model describes production of lactate from glucose in astrocytes and subsequent release and uptake by a neighboring neuron for oxidation. This machinery is suggested to be coupled to glutamatergic activity via astrocytic glutamate uptake and Na^+, K^+ ATPase activity, the latter up-regulating glycolysis and lactate production (Pellerin and Magistretti, 1994). Particularly, the coupling between neuronal activity and lactate transfer has been one of the issues debated (Hertz et al., 1998; Hertz, 2004). The lactate concentration increases in the brain in vivo subsequent to stimulation although evidence for a subsequent net oxidation in the adjacent neuronal compartment seems lacking (Dienel and Hertz, 2005). Interestingly, astrocytes and neurons oxidize 50% each of the interstitial lactate in freely moving rats as determined by microdialysis (Zielke et al., 2007). In spite of the numerous approaches, evidence in support of lactate being the preferred neuronal substrate during activation is missing (Chih et al., 2001). In contrast, glucose has been shown to support the increased neuronal energy demand during synaptic glutamatergic activity (Bak et al., 2006a).

7.2.2 Transfer of Reducing Equivalents, Malate-Aspartate Shuttle

In the glycolytic pathway glyceraldehyde-3-phosphate is oxidized to 1,3-bisphosphoglycerate simultaneously reducing beta-nicotinamide adenine dinucleotide (NAD^+) to NADH. A continuation of glycolysis is dependent on a concomitant re-oxidation of NADH, which may occur either via reduction of pyruvate to lactate or transfer of a reduced equivalent into the mitochondria for re-oxidation in the electron transport chain. Apparently, the supply of acetyl CoA originating from glucose to the tricarboxylic acid (TCA) cycle is also dependent on this re-oxidation. The transfer of reduced equivalents is generally thought to be mediated by the malate–aspartate shuttle (MAS). Recently one of the components of this shuttle, i.e., the Ca^{2+} sensitive aspartate–glutamate carrier, aralar1, was observed to be only sparsely expressed in astrocytes of the mature mouse and rat brain (Ramos et al., 2003). A transcriptomic analysis of acutely isolated astrocytes from adult mice has, however, clearly demonstrated the presence of mRNA coding for aralar1. A lack of MAS activity in astrocytes would indicate a limited glucose oxidation in astrocytes. However, several observations contradict such interpretation. A considerable part (20–30%) of glutamine synthesis has been demonstrated in vivo to occur via pyruvate carboxylation, i.e., a selective astrocytic phenomenon (Oz et al., 2004), implying that glucose to a considerable extent is oxidatively metabolized in the mitochondria beyond the pyruvate step. Claiming that pyruvate is produced solely from glutamate would indicate a high extent of futile cycling or pyruvate recycling, a process that has been difficult to demonstrate in the brain in vivo (for further discussion see below). Using nuclear magnetic resonance (NMR) spectroscopy it has been estimated that cortical astrocytes account for approximately 30% of total tissue oxygen consumption in brain cortex, a number closely reflecting the volume occupied by astrocytes in the cortex [for references see Hertz et al. (2007)]. Thus, to explain these findings there might be a need for a shuttling mechanism alternative to MAS. The glycerol-3-phosphate shuttle might be such alternative for transferring reduced equivalents from the cytosol to the mitochondria in astrocytes but explicit evidence for this has been difficult to obtain (Waagepetersen et al., 2001a; McKenna et al., 2006a, b). Recent transcriptomic analysis of acutely isolated astrocytes from adult mouse brain has shown the pertinent genes to be transcribed (Lovatt et al., 2007). In line with this, inhibition of MAS in cultured astrocytes seems not to impair the metabolism of glucose indicating functioning of another shuttle very likely being the glycerol-3-phosphate shuttle (McKenna et al., 1993; Malik et al., 1993;

Waagepetersen et al., 2001a). It should be noted that the expression of aralar1 is higher in cultured astrocytes compared to the mature brain and in addition cultured astrocytes express the isoform of the carrier called citrin which is primarily expressed in liver and kidney (Ramos et al., 2003). In conclusion, cultured astrocytes are supposed to have a higher MAS activity and thus potentially a higher oxidative glucose metabolism than astrocytes in the mature brain. These observations should be taken into consideration when interpreting results obtained from cultured astrocytes regarding the MAS (McKenna et al., 2006b).

Cerdan et al. (2006) have proposed a redox switch which would circumvent the possible low activity of the mechanisms for shuttling reducing equivalents from the cytosol into the mitochondria of astrocytes. Lactate produced via glycolysis was suggested to be transferred from the astrocyte to the neuron in which lactate was reduced to pyruvate for subsequent return of pyruvate to the astrocyte, thus an NADH equivalent has been shuttled from the astrocyte to the neuron. This hypothesis is compatible with subcellular compartmentation of lactate and pyruvate described in both cultured astrocytes and neurons (Sonnewald et al., 1993a; Cruz et al., 2001; Waagepetersen et al., 2001b; Zwingmann et al., 2001; Schousboe et al., 2003).

7.2.3 Glycogen Metabolism

Instead of being glycolytically processed and oxidized in the TCA cycle, glucose can be stored in the form of glycogen. Glycogen is predominantly present in astrocytes of the brain (Cataldo and Broadwell, 1986; Wender et al., 2000). It is a reservoir of energy composed of glycosyl units which can be made rapidly available when needed. The glycogen phosphorylase (GP) activity is high and due to the branched structure of the molecule the possible sites for attack are numerous. GP exists in brain as two isoforms, i.e., the brain form and the muscle form (Pfeiffer-Guglielmi et al., 2003). The brain form is predominantly activated by an increased level of adenosine monophosphate (AMP) whereas the muscle form is principally regulated by neurohormone-induced phosphorylation (Hamprecht et al., 2006). Glycogen synthase is hormonally regulated in an opposite manner compared to that of GP. Glycogen degradation is induced in astrocytes by several compounds such as noradrenaline, vasoactive intestinal peptide, and adenosine. Their effects are mediated via specific receptors involving the mobilization of the second messenger cAMP (Sorg and Magistretti, 1991).

The function of glycogen in brain is not fully elucidated. However, glycogen can support neuronal survival during pathological conditions such as hypoglycemia (Ransom and Fern, 1997; Suh et al., 2007) and increasing amount of evidence points toward the importance of glycogen turn-over also during physiological conditions (Swanson, 1992; Dienel et al., 2007). Beyond being an energy reservoir, glycogen has also been suggested to be a carbon source for glutamine synthesis via pyruvate carboxylation in astrocytes and subsequently for neurotransmitter glutamate synthesis in neighboring neurons (Hertz et al., 2003; Gibbs et al., 2006). It has been speculated that mitochondria due to their size might be absent from peripheral astrocytic processes which would increase the importance of glycolysis and glycogenolysis for maintenance of an adequate energy supply in these compartments. It may be pointed out, however, that a recent electron microscopy (EM) study of astrocyte processes has demonstrated an abundance of mitochondria in these structures (Lovatt et al., 2007). In spite of this a glycogenolytic pathway may be of importance since it allows rapid production of ATP. Such a pathway would therefore be particularly important for maintenance of ATP production during intensive glutamatergic neuronal activity demanding high activity of glutamate uptake, an energy requiring process, which is particularly important in the astrocytic processes (Danbolt, 2001). The mentioned tasks or functional roles of glycogen support the notion that glycogen is necessary within the astrocyte. Another possibility, maybe receiving unfair recognition, is that the main function of glycogen is to sustain the energy need of neurons. Alternatively, glycogen may be important for both neurons and astrocytes, i.e., the astrocytes get energy from glycogenolysis and neurons from oxidative metabolism of lactate. The triggering signal for glycogen breakdown in astrocytes may be the key to understand the role of glycogen. In this context it is interesting to note that the isoform of GP which is specific for the brain is activated by an increased level of AMP, a signal for need of local energy production.

7.2.4 Metabolic Shuttles

The glutamate–glutamine cycle, described in detail later (Sects. 7.4.1–7.4.3), is in short the clearance of glutamate from the synaptic cleft by uptake into astrocytes and the subsequent amidation of glutamate into glutamine. Glutamine is transferred into neurons for re-synthesis of glutamate. This cycle is accompanied by a net transfer of nitrogen from the astrocytic to the neuronal compartment. For the maintenance of nitrogen homeostasis in the "tripartite" microenvironment, i.e., the presynaptic and postsynaptic neuron and the surrounding astrocyte, this nitrogen has to be delivered back to the astrocyte. The mechanism for this has been suggested to consist of transfer of an amino acid, e.g. alanine (Waagepetersen et al., 2000; Schousboe et al., 2003; Bak et al., 2006b). Alanine is thought to be transaminated forming glutamate from which the amino group may be liberated by the action of glutamate dehydrogenase (GDH). The amino group may subsequently take part in the glutamine synthetase (GS) reaction. However, studying this by using [^{15}N]alanine and co-cultures of glutamatergic neurons and cerebellar astrocytes the conversion of nitrogen from the amino group of alanine into the amide nitrogen of glutamine was not dependent on glutamatergic activity as would have been expected (Bak et al., 2005). For the shuttle to operate stoichiometrically, the GDH reaction has to operate in both directions, i.e., reductive amination in neurons and oxidative deamination in astrocytes. As described in Sect. 7.4.2, GDH is predominantly an astrocytic enzyme and the direction of deamination seems to be favored except during high levels of ammonia (Plaitakis and Zaganas, 2001). The branched chain amino acids (BCAAs) may serve a similar role as alanine (Bak et al., 2006b). Interestingly, valine metabolism was up-regulated in astrocytes, repetitively exposed to pulses of glutamate (Bak et al., 2007). Additionally, the BCAAs have been proposed to provide the amino nitrogen for de novo synthesis of glutamate via pyruvate carboxylation in astrocytes followed by amidation by GS and transfer of glutamine to the neurons. The BCAA is proposed to be regenerated in the neuron and returned to the astrocyte. In the neuron BCAA is thought to be formed from glutamate generated via the unfavored direction of the GDH catalyzed reaction, i.e., reductive amination of α-ketoglutarate (α-KG) (Lieth et al., 2001). Yudkoff (1997) has suggested that leucine enters astrocytes from the capillary and thus serves as the amino nitrogen donor for glutamine via glutamate. Glutamine is subsequently transferred into neurons and deamidated into glutamate acting as neurotransmitter. As summarized, several amino acid shuttles between astrocytes and neurons have been suggested. The attempts to verify and explore these hypotheses are hampered by the intercellular nature of the shuttles. In addition, as mentioned above, the shuttles are very likely restricted to a cellular microenvironment like for example the "tripartite" synapse and thus might only involve a minor fraction of the cellular pools of metabolites making it even more difficult to demonstrate activity and functional roles of these shuttles.

7.3 TCA Cycle-Related Metabolism and Compartmentation

7.3.1 Pyruvate Carboxylation

Anaplerotic (filling up) processes based on carbon dioxide fixation are functionally important for glutamate synthesis and the maintenance of TCA cycle activity (Fig. 7.1). The key enzymes involved in these processes are phosphoenolpyruvate

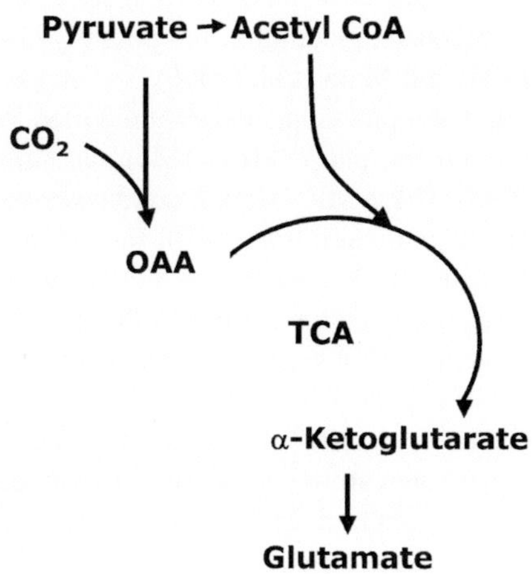

Fig. 7.1 Schematic presentation of reactions involved in net synthesis of glutamate via pyruvate carboxylation and operation of tricarboxylic acid cycle reactions. *OAA* oxalocetate and *TCA* tricarboxylic acid.

carboxykinase (PEPCK), malic enzyme (ME), and pyruvate carboxylase (PC) all of which are active in the brain albeit the PC catalyzed reaction is the quantitatively most important (Patel, 1974). Immunocytochemical and cell culture studies have revealed cell specific localization of some of these enzymes, PC and cytosolic ME being present exclusively in astrocytes whereas mitochondrial ME and PEPCK have been found in both neurons and astroglial cells although primarily in neurons (Yu et al., 1983; Shank et al., 1985; Kurz et al., 1993; Cesar and Hamprecht, 1995; Cruz et al., 1998; Vogel et al., 1998; McKenna et al., 2000b). Since as stated above in the brain the PC catalyzed carboxylation is quantitatively the most predominant, it has been suggested that the astroglial compartment is primarily responsible for net synthesis of TCA cycle constituents. This makes the metabolically handicapped neurons dependent on supply of precursors for biosynthesis of neurotransmitter amino acids metabolically linked to the TCA cycle (Hertz et al., 1992; Westergaard et al., 1995). Due to the presence of mitochondrial ME and PEPCK in neurons, these cells may in theory be capable of carbon dioxide fixation. It has recently been demonstrated using radiolabeled precursors that such carboxylation may take place in isolated glutamatergic cerebellar granule neurons (Hassel and Brathe, 2000). However, analogous studies have come to the opposite conclusion (Lieth et al., 2001; Waagepetersen et al., 2001a).

Using ^{13}C labeled precursors combined with magnetic resonance spectroscopy (MRS), it is possible to obtain knowledge about the magnitude of pyruvate carboxylation (Sonnewald et al., 1993b; Hassel et al., 1995; Taylor et al., 1996). This can be done in vivo by MRS analysis of TCA cycle constituents together with aspartate and glutamate which equilibrate rapidly with the TCA cycle (Mason et al., 1995). Oz et al. (2004) have shown that the rate of pyruvate carboxylation in rat brain is 0.14–0.18 $\mu mol\ g^{-1}\ min^{-1}$ which corresponds to approximately 25% of the glutamate–glutamine cycle activity. In humans, Gruetter et al. (2001) reported a similar rate (0.09 $\mu mol\ g^{-1}\ min^{-1}$). However, Patel et al. (2005) showed an initial rate of anaplerosis of $0.059 \pm 0.010\ \mu mol\ g^{-1}\ min^{-1}$ in rats, which represents 23% of total glutamine synthesis as determined by Patel et al. (2004). Alternatively, the ratio between PC and PDH activities can be estimated ex vivo by injecting [1-^{13}C] glucose and subsequent MRS analysis of brain extracts to obtain labeling patterns of glutamate, glutamine, and possibly GABA as explained by Lapidot and Gopher (1994) and Melo et al. (2006).

Lapidot and Gopher (1994) reported PC/PDH ratios for glutamine (34%), glutamate (16%), and GABA (16%) and similar results were obtained by Melo et al. (2006). These MRS studies are, however, hampered by the fact that oxaloacetate to some extent equilibrates with fumarate thus leading via equilibration to an underestimation of PC activity. A similar problem may arise from TCA cycling leading to labeling patterns similar to those occurring via equilibration. In vitro the latter problem may, at least partly, be overcome by using 3-nitropropionic acid to specifically block the TCA cycle at the succinate dehydrogenase step (Alston et al., 1977, Bakken et al., 1997b). In order to obtain information about the relative magnitude of pyruvate carboxylation in glutamatergic neurons and astrocytes, cultures of cerebellar granule neurons or astrocytes were incubated in ^{13}C labeled glucose and lactate in the absence and presence of 3-nitropropionic acid. This experimental paradigm led to results strongly indicating that pyruvate carboxylation is a quantitatively important reaction in astrocytes whereas in the neurons no carboxylation was detected (Waagepetersen et al., 2001a).

Modification of pyruvate carboxylation has been shown to occur under various conditions. In cultured astrocytes Qu et al. (2001) showed that exogenous glutamate could decrease carboxylation of pyruvate derived from [1-^{13}C]glucose. Moreover, pyruvate carboxylation was increased in rats on a ketogenic diet (Melo et al., 2006) and in the awake rat brain the anaplerotic rate was several fold higher than under deep pentobarbital anesthesia (Oz et al., 2004). However, pyruvate carboxylation was unaffected by bicuculline-induced seizures (Patel et al., 2005).

7.3.2 *Pyruvate Recycling*

During development, anaplerosis (pyruvate carboxylation) is necessary since the concentration of glutamate and glutamine in brain increases (Tkac et al., 2003), whereas in adults anaplerosis is not self-evident. It is generally accepted that the adult brain needs to replenish the TCA cycle when a four (or more) carbon unit such as glutamine, leaves the brain or is metabolized via pyruvate and TCA cycle metabolism to CO_2, i.e., pyruvate recycling.

Pyruvate recycling was first shown in the liver, where [2-^{14}C]pyruvate was converted to [3-^{14}C]pyruvate and [1-^{14}C]pyruvate, a process which can only occur if pyruvate is incorporated into the TCA cycle and subsequently is regenerated from TCA cycle constituents (Freidmann et al., 1971) (Fig. 7.2). Recycling of pyruvate in the brain was demonstrated by Cerdan et al. (1990), who found that [1,2-^{13}C]acetate, a substrate that is specifically taken up and therefore metabolized in astrocytes (Waniewski and Martin, 1998), can be converted in brain to monolabeled acetyl CoA ([1-^{13}C] and [2-^{13}C]) and to glutamate labeled either in the C-4 or the C-5 position.

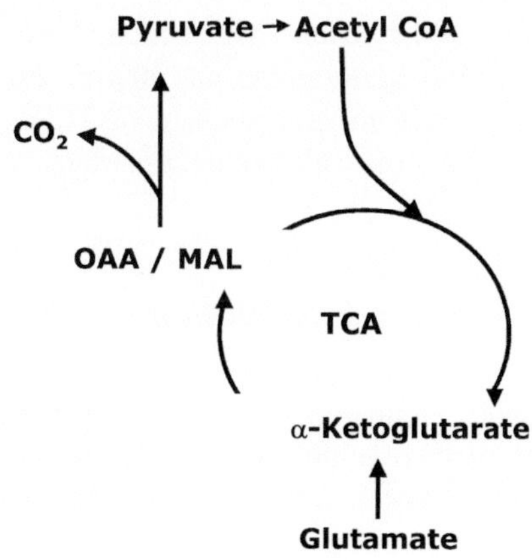

Fig. 7.2 Schematic presentation of net degradation of glutamate via tricarboxylic acid cycle reactions and pyruvate recycling. *MAL* malate, *OAA* oxaloacetate, and *TCA* tricarboxylic acid.

This requires entry of acetate (after formation of acetyl CoA) into the TCA cycle and exit of a TCA cycle intermediate to form pyruvate, which then is reintroduced in the TCA cycle. Based upon the observation that this label from acetate was incorporated into glutamate but not into glutamine, it was concluded that pyruvate recycling takes place in a compartment without GS activity, i.e., a neuronal and not an astrocytic compartment. Pyruvate recycling in the brain in vivo has been confirmed by Hassel et al. (1995) and Hassel and Sonnewald (1995), who demonstrated formation of labeled lactate from [2-^{13}C]acetate and of [2-^{13}C]lactate from [1-^{13}C]glucose. Based upon a more pronounced formation of TCA cycle-derived lactate from labeled acetate than from labeled glucose it was concluded that pyruvate recycling was likely to occur in the astrocytic, rather than the neuronal compartment.

Most cell culture studies of pyruvate recycling were performed by determining reintroduction into the TCA cycle via acetyl CoA of a compound such as aspartate, glutamate, or glutamine (Sonnewald et al., 1993b, 1996; Bakken et al., 1997a, 1998; Haberg et al., 1998; Waagepetersen et al., 2002). Primary cultures of cerebrocortical astrocytes were incubated in [U-^{13}C]glutamine and the formation of [4,5-^{13}C]glutamate showed that labeled α-KG formed from glutamine via glutamate could leave the TCA cycle and be reintroduced as [1,2-^{13}C]acetyl CoA, i.e., pyruvate recycling (Sonnewald et al., 1996). Re-entry of pyruvate formed from [U-^{13}C] glutamate into the TCA cycle after conversion to acetyl CoA in analogous cultures was demonstrated by the presence of [4-^{13}C]glutamate and [2,3-^{13}C]aspartate (Haberg et al., 1998). Recently, recycling was also demonstrated in cerebellar granule neurons using [U-^{13}C]glutamate as precursor (Olstad et al., 2007).

Exit from the TCA cycle to form pyruvate can occur by two different reactions (1) decarboxylation and oxidation of malate to pyruvate, catalyzed by ME and (2) conversion of oxaloacetate to phosphoenolpyruvate (PEP) plus carbon dioxide, catalyzed by PEPCK, followed by formation of pyruvate from PEP, catalyzed by pyruvate kinase. Formation of PEP from oxaloacetate is required to perform gluconeogenesis from pyruvate, because the pyruvate kinase-catalyzed process is irreversible and is by-passed by initial formation of oxaloacetate and subsequent synthesis of PEP from oxaloacetate (McKenna et al., 2006a). As alluded to earlier, ME exists as two different isoforms, mitochondrial ME, which is mainly present in neurons, and cytosolic ME, which is astrocyte-specific (Kurz et al., 1993; Vogel et al., 1998; McKenna et al., 2000b). Pyruvate formation is an essential step of complete oxidative metabolism of TCA cycle constituents and their derivatives, such as glutamate, glutamine, aspartate, and GABA.

7.3.3 Compartmentation

Metabolic compartmentation is defined as the presence in a tissue of more than one distinct pool of a given metabolite. These separate pools of a metabolite are not in rapid equilibrium with each other but maintain their own integrity and turnover rates (Berl and Clarke, 1969). The indication of metabolic compartmentation in the brain is based on the observation that using radiolabeled glutamate the specific activity of glutamine could exceed that of its precursor glutamate (Berl et al., 1961). Thus a small pool of labeled glutamate was rapidly used to synthesize glutamine. This "small glutamate compartment" was shown to be located in glia, most likely predominantly or exclusively astrocytes (Berl et al., 1962; Balazs et al., 1970; Norenberg and Martinez-Hernandez, 1979). Neurons are unable to synthesize glutamine and contain the "large glutamate compartment" (Lajtha et al., 1959; van den Berg et al., 1969; Berl and Clarke, 1983). However, compartmentation does not only exist at the intercellular level but has also been demonstrated to exist within a single cell type (McKenna et al., 1990, 1996a, 2000a; Schousboe et al., 1993a; Sonnewald et al., 1993a, 1998; Bouzier et al., 1998; Waagepetersen et al., 1998a, b, 2001b, 2006; Cruz et al., 2001; Zwingmann et al., 2001). Heterogeneity with regard to metabolic function may be the result of differences in enzyme composition and substrate concentrations both in the cytosol and the mitochondria. As a consequence, mitochondria might have specific functions in distinct regions of a cell. Synaptic and nonsynaptic mitochondria in adult rat brain exhibit differences in the activity of a number of TCA cycle enzymes (Lai et al., 1994). Furthermore, intramitochondrial compartmentation could also exist. Using EM it has been suggested that the inner membrane proteins might be compartmentalized (Perkins and Frey, 2000). Heterogeneity among mitochondria is further supported by the demonstration that mitochondrial populations exist with different expression levels of pyruvate dehydrogenase (Margineantu et al., 2002). Immunogold EM using an antibody against the α-KG dehydrogenase component (E1k) of the α-KG dehydrogenase complex, a marker enzyme for the TCA cycle was employed to probe mitochondrial heterogeneity in individual cerebellar and cortical astrocytes (Waagepetersen et al., 2006). The results demonstrated that α-KG dehydrogenase is heterogeneously distributed in mitochondria within individual astrocytes originating from cerebellum or cerebral cortex.

A compartmentation model based on results obtained from ^{13}C MRS and mass spectrometry analyses of media and extracts of cultured astrocytes incubated with ^{13}C-labeled compounds (Schousboe et al., 1993a; Waagepetersen et al., 2001b) is presented in Fig. 7.3. This shows that a pool of pyruvate is used for an extensive synthesis of releasable citrate, a process involving pyruvate carboxylation to a large extent (Fig. 7.3A). Another pool of pyruvate, compartment B, appears to function as a substrate for the TCA cycle containing the main intracellular pool of citrate, the labeling of which is primarily introduced via pyruvate oxidation (i.e., pyruvate dehydrogenase activity and to a less extent via carboxylation. The finding that intracellular glutamate and extracellular glutamine exhibited similar labeling patterns is best explained by a third compartment (Fig. 7.3C) with no pyruvate carboxylation. Compartments B and C are likely to contain the main intracellular pool of glutamate and to represent the site for synthesis of the major part of releasable glutamine. The labeling pattern of intracellular glutamine could only be explained assuming a fourth compartment in which pyruvate carboxylation is involved only marginally. This compartment is the origin of the main intracellular pool of glutamine although glutamine released to the medium was not generated in this

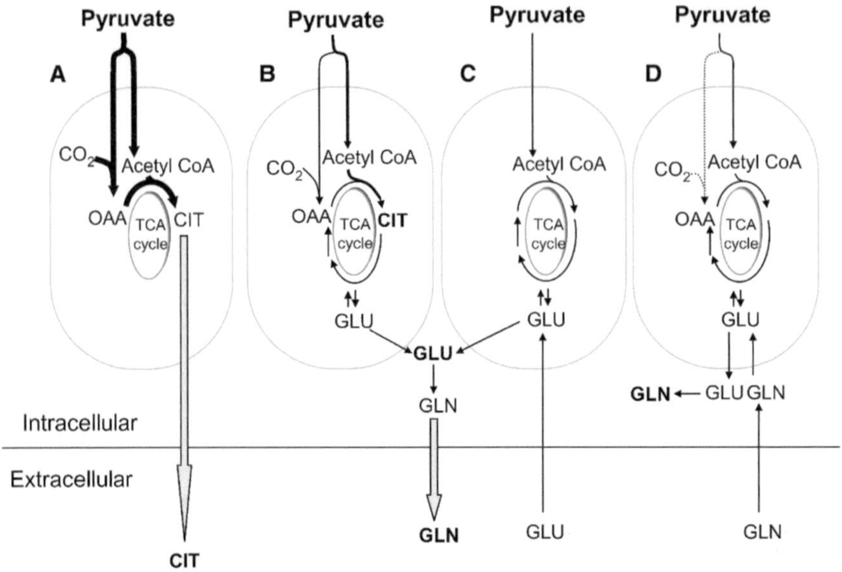

Fig. 7.3 A schematic presentation of multiple compartments in astrocytes in which pyruvate functions as substrate. Synthesis of a large amount of releasable citrate occurs in compartment **A** having extensive carboxylation. Releasable glutamine is synthesized from glutamate originating from compartments **B** and **C**. The main intracellular pool of glutamine is synthesized from glutamate originating from compartment **D**. Note that the compartments differ with regard to activity of pyruvate metabolism via carboxylation and oxidation. The sizes of the *arrows* provide an estimate of the relative magnitudes of the respective fluxes. *GLN* glutamine, *GLU* glutamate, *CIT* citrate, *OAA* oxaloacetate, and *TCA* tricarboxylic acid (*See Color Plates*).

compartment (Fig. 7.3D). Moreover, the fact that the labeling was found to be higher for extracellular glutamine than for intracellular glutamine supports the notion that the releasable intracellular pool of glutamine was separated from the main intracellular pool. Such compartmentation of glutamate–glutamine metabolism in cultured cortical astrocytes has previously been reported (Schousboe et al., 1993b; McKenna et al., 1996b; Qu et al., 1999). It was observed that exogenous glutamate and glutamine synthesized from this pool of glutamate (i.e., endogenous glutamine) were separated from exogenous glutamine and glutamate synthesized from this pool of glutamine (i.e., endogenous glutamate). Obviously these findings could only be explained assuming that exogenous and endogenous glutamate are compartmentalized and it has been postulated that release of glutamine primarily happens using the newly synthesized endogenous pool (Schousboe et al., 1993b). Hence, exogenous glutamate may be mixed with the main intracellular pool of glutamate, i.e., the pool which is labeled in compartments B and C, and from which the synthesis of releasable glutamine occurs (Fig. 7.3). Uptake of exogenous glutamate into compartment C could compensate for the corresponding drain of endogenous glutamate from this compartment. The combined compartments B and C may relate to the main compartment constituting the astrocytic part of the glutamate–glutamine cycle (Berl and Clarke, 1969, 1983; Balazs et al., 1970). Therefore, this compartment of glutamine synthesis is likely separated from the compartment (D in Fig. 7.3) in which exogenous glutamine is metabolized to glutamate by phosphate activated glutaminase (PAG). This, in turn, constitutes the main intracellular pool of glutamine (Fig. 7.3). As pointed out by Schousboe et al. (1993a), such separation of the main synthetic route for glutamine synthesis (compartment B and C) and the main route for degradation catalyzed by PAG (compartment D) will prevent futile cycling. Hence, it may be suggested that the function of compartment D is to regulate the general ammonia homeostasis being separated from the synthesis of the neurotransmitter precursor which is exported to neurons.

Based on studies of ^{13}C-labeled glucose and lactate in astrocytes or C-6 glioma cells (Bouzier et al., 1998; Waagepetersen et al., 2001b; Zwingmann et al., 2001; Sickmann et al., 2005), it appears that pyruvate metabolism is compartmentalized. Moreover, alanine synthesized via transamination from pyruvate has been shown to be compartmentalized in astrocytes (Zwingmann et al., 2001).

In this context, it may be interesting that brain glycogen metabolism which is taking place in astrocytes is also compartmentalized. This was investigated by Sickmann et al. (2005) using cultured cerebellar and neocortical astrocytes which were incubated in medium containing [U-^{13}C]glucose in the absence or presence of isofagomine, an inhibitor of GP (Waagepetersen et al., 2000). The results demonstrated that lactate originating from glycogen is compartmentalized from that derived from glucose, which lends further support to a compartmentalized cytosolic metabolism in astrocytes.

The concept of intracellular compartmentation is primarily based on experiments performed using astrocytes in culture and such cultures might very well be heterogeneous. However, in case of the distribution of α-KG dehydrogenase in individual mitochondria compartmentation was shown at the single cell level (Waagepetersen et al., 2006).

7.4 Amino Acid Metabolism

7.4.1 General Outline of Metabolic Processes Involving Glutamine, Glutamate, and GABA

The key enzymes involved in metabolic reactions pertinent to the turnover of the neurotransmitters glutamate and GABA as well as their prevailing cellular localization are summarized in Table 7.1. It should be noted that GS is exclusively expressed in astrocytes (Norenberg and Martinez-Hernandez, 1979) and glutamate decarboxylase is only present in GABAergic neurons and not in astrocytes (Hertz et al., 1992). In addition, it is of functional importance that the activity of phosphate activated PAG is higher in neurons than in astrocytes (Schousboe et al.,

Table 7.1 Enzymes involved in glutamate and GABA metabolism

Enzyme	Co-enzyme	Astrocyte	Neuron
Aspartate aminotransferase*	PLP	+++	+++
Glutamate dehydrogenase	$NAD(P)^+$	++	+
Glutamine synthetase	–	++	–
Phosphate activated glutaminase	–	+	++
Glutamate decarboxylase	PLP	–	+
GABA aminotransferase	PLP	+	+

Number of pluses indicate relative enzyme activities comparing the two cell types and the different enzymes; * Other aminotransferases are involved in glutamate metabolism but those enzymes are not pertinent to this table. *NAD(P)*$^+$ β-nicotinamide adenine dinucleotide (phosphate) and *PLP* pyridoxal-5′-phosphate.

1979; Drejer et al., 1985; Larsson et al., 1985). This difference between neurons and astrocytes observed in cultured cells may even be more pronounced in vivo as it has been difficult to detect PAG-like immunoreactivity in astroglial elements in brain slices whereas that of particularly glutamatergic neuronal structures was quite pronounced (Laake et al., 1999). The functional implication of the difference in cellular localization of PAG and GS relates to the glutamate–glutamine cycle in which glutamine is hydrolyzed by PAG in neurons (glutamatergic) to glutamate which is released as neurotransmitter and subsequently captured by surrounding astrocytes in which it is converted to glutamine in the GS catalyzed, energy (ATP) requiring reaction. This glutamine can be released and taken up into the neuron to complete the cycle. The latter part of the cycle relies on differential distribution in neurons and astrocytes of the glutamine transporters, system N in astrocytes and system A in neurons (Varoqui et al., 2000; Bröer and Brookes, 2001; Chaudhry et al., 2002; Bak et al., 2006b). The concept of this glutamate–glutamine cycle was founded on studies of glutamate and glutamine metabolism in brain tissue preparations which indicated different cellular compartments of these amino acids with different turnover rates (Berl et al., 1961; 1962; Van den Berg and Garfinkel, 1971) and it was confirmed by the cellular distribution of particularly GS as pointed out above.

7.4.2 Glutamate Metabolism

In order for the glutamate–glutamine cycle to operate stoichiometrically all glutamate taken up by astrocytes via high affinity glutamate transporters (Danbolt, 2001) must be converted to glutamine in the GS catalyzed reaction (Cotman et al., 1981). However, numerous metabolic studies performed in either cultured astrocytes or astrocytes in vivo using astrocyte selective precursors for TCA cycle and amino acid metabolism performed during the last 20 years have convincingly shown that this is not the case. There is a considerable oxidative metabolism of glutamate via the TCA cycle and the relative significance of this oxidative pathway is apparently dependent on the actual extracellular glutamate concentration (Yu et al., 1982; McKenna et al., 1996a; Sonnewald et al., 1997). The conversion of the carbon skeleton of glutamate to α-KG can take place by two different enzymatic pathways, i.e., via the GDH catalyzed oxidative deamination or by transamination. The latter process may be catalyzed by any aminotransferase but since aspartate aminotransferase (AAT) is the member of this family of enzymes having by far the highest activity (Erecinska and Silver, 1990), AAT is the most likely enzyme to catalyze this reaction. It has been a long-standing debate as to the relative importance of the GDH or the AAT reaction (Yu et al., 1982; Farinelli and Nicklas, 1992; McKenna et al., 1993; Sonnewald et al., 1993b; Westergaard et al., 1996). It is, however, very likely that oxidative deamination catalyzed by GDH plays a prominent role since the aminotransferase inhibitor aminooxyacetic acid in several studies has been shown not to inhibit oxidation of glutamate in the TCA cycle (Yu et al., 1982; Westergaard et al., 1996). Interestingly, it has been demonstrated that the opposite reaction, i.e., production of glutamate from α-KG in astrocytes is catalyzed by AAT and not by GDH since it is affected by aminooxyacetic acid (Westergaard et al., 1996). The conclusion from the abovementioned considerations is that a substantial fraction of the glutamate taken up into astrocytes during glutamatergic activity is oxidatively metabolized involving mainly the GDH reaction and subsequent metabolism of α-KG in the TCA cycle (Westergaard et al., 1995). The consequence of this is that the glutamate–glutamine cycle is not operating stoichiometrically. This imposes a need for de novo synthesis of the glutamate carbon skeleton which is dependent on the PC reaction that like GS is confined to astrocytes (Yu et al., 1983). Further discussion of this aspect is found in Sect. 7.3.1. It should also be pointed out that oxidation of the carbon skeleton of glutamate, i.e., α-KG requires pyruvate recycling, a process which has been shown to occur in astrocytes (Sonnewald et al., 1996; Waagepetersen et al., 2002). Further discussion of this is found in Sect. 7.3.2.

7.4.3 Glutamine and Ammonia Metabolism

The demonstration of a significant activity of PAG in cultured astrocytes (Schousboe et al., 1979) albeit lower than that in glutamatergic or GABAergic neurons (Drejer et al., 1985; Larsson et al., 1985) is compatible with the observation that glutamine can be oxidatively metabolized in astrocytes (Hertz et al., 1988). Moreover, the use of ^{13}C-labeled glutamine and MRS has demonstrated substantial metabolism of glutamine in astrocytes, a process coupled to pyruvate recycling (Sonnewald et al., 1996).

The PAG catalyzed reaction leads to production of not only glutamate but also ammonia. In case glutamate is oxidatively metabolized in the GDH reaction an additional molecule of ammonia is produced. This ammonia must eventually be disposed off and this can only happen by conversion of glutamate to glutamine in the GS reaction also discussed in Sects. 7.4.1 and 7.4.2. The combined action of PAG and GS constitutes a futile cycle, the result of which is the use of ATP derived energy (Fig. 7.1). The fact that these reactions are intracellularly separated taking place in the mitochondrial (PAG) and the cytoplasmic (GS) compartments, respectively, allows regulatory control. Nevertheless, exposure of astrocytes to elevated glutamine concentrations leads to adverse effects on mitochondria caused by ammonia liberated in the PAG reaction as demonstrated by Jayakumar et al. (2004). This allows glutamine to act as a Trojan horse to carry ammonia into the astrocytes, a process leading to induction of the mitochondrial permeability pore and subsequent mitochondrial dysfunction and cell death (Jayakumar et al., 2004; Albrecht and Norenberg, 2006).

7.4.4 GABA Metabolism

Astrocytic uptake and metabolism of GABA appears to be of importance for the functional capacity of GABAergic neurotransmission since inhibitors of astrocytic GABA transporters as well as GABA aminotransferase act as anticonvulsants (White et al., 2002; Sarup et al., 2003). This is related to the fact that a fraction of GABA released during GABAergic neuronal activity is likely to be taken up in astrocytes via one or more of the GABA transporters located in the astroglial plasma membrane (Schousboe et al., 2004; Clausen et al., 2006). GABA will be metabolized into succinic semialdehyde in the astrocytic mitochondria which contain appreciable activity of GABA aminotransferase (Schousboe et al., 1977a, b). Succinic semialdehyde dehydrogenase catalyzes the subsequent oxidation of succinic semialdehyde to succinate. The four-carbon skeleton may be used for glutamate and glutamine synthesis via conversion to α-KG using acetyl CoA from glucose metabolism (Waagepetersen and Schousboe, in press) or it may be oxidized to CO_2 via pyruvate recycling.

Acknowledgment The expert secretarial assistance of Ms Hanne Danø is highly appreciated. The experimental work has been supported by grants from the Lundbeck, Hørslev, Novo Nordisk and Benzon Foundations as well as the Danish Medical Research Council (22-03-0250 and 22-04-0314).

References

Albrecht J, Norenberg MD (2006) Glutamine: a Trojan horse in ammonia neurotoxicity. Hepatology 44:788-794.

Alston TA, Mela L, Bright HJ (1977) 3-Nitropropionate, the toxic substance of Indigofera, is a suicide inactivator of succinate dehydrogenase. Proc Natl Acad Sci USA 74:3767-3771.

Araque A, Parpura V, Sanzgiri RP, Haydon PG (1999) Tripartite synapses: glia, the unacknowledged partner. Trends Neurosci 22: 208-215.

Bak LK, Sickmann HM, Schousboe A, Waagepetersen HS (2005) Activity of the lactate-alanine shuttle is independent of glutamate-glutamine cycle activity in cerebellar neuronal-astrocytic cultures. J Neurosci Res 79:88-96.

Bak LK, Schousboe A, Sonnewald U, Waagepetersen HS (2006a) Glucose is necessary to maintain neurotransmitter homeostasis during synaptic activity in cultured glutamatergic neurons. J Cereb Blood Flow Metab 10:1285-1297

Bak L K, Schousboe A, Waagepetersen HS (2006b) The glutamate/GABA-glutamine cycle: aspects of transport, neurotransmitter homeostasis and ammonia transfer. J Neurochem 98:641-653.

Bak, LK, Johansen ML, Schousboe A, Waagepetersen HS (2007) Among the branched-chain amino acids only valine metabolism is up-regulated in astrocytes during glutamate exposure. J Neurosci Res 85:3465-3470.

Bakken IJ, White LR, Aasly J, Unsgard G, Sonnewald U (1997a) Lactate formation from [U-^{13}C] aspartate in cultured astrocytes: compartmentation of pyruvate metabolism. Neurosci Lett 237:117-120.

Bakken IJ, Johnsen SF, White LR, Unsgard., Aasly J, Sonnewald U (1997b) NMR spectroscopy study of the effect of 3-nitropropionic acid on glutamate metabolism in cultured astrocytes. J Neurosci Res 47:642-649.

Bakken IJ, White LR, Unsgard G, Aasly J, Sonnewald U (1998) [U-^{13}C]glutamate metabolism in astrocytes during hypoglycemia and hypoxia. J Neurosci Res 51:636-645.

Balazs R, Machiyama Y, Hammond BJ, Julian T, Richter D (1970) The operation of the gamma-aminobutyrate bypath of the tricarboxylic acid cycle in brain tissue in vitro. Biochem J 116:445-461.

Berl S, Clarke DD (1969) Compartmentation of amino acid metabolism. In: Handbook of Neurochemistry (Lajtha A, ed), pp 447-472. New York: Plenum Press.

Berl S, Clarke DD (1983) The metabolic compartmentation concept. In: Glutamine, Glutamate and GABA in The Central Nervous System (Hertz L, Kvamme E, McGeer EG, Schousboe A, eds), pp 205-217. New York: Alan R. Liss, Inc.

Berl S, Lajtha A, Waeelsch H (1961) Amino acid and protein metabolism – IV. Cerebral compartments of glutamic acid metabolism. J Neurochem 7:186-197.

Berl S, Takagaki G, Clarke DD, Waelsch H (1962) Metabolic compartments in vivo. Ammonia and glutamic acid metabolism in brain and liver. J Biol Chem 237:2562-2569.

Bittar PG, Charnay U, Pellerin L, Bouras C, Magistretti PJ (1996) Selective distribution of lactate dehydrogenase isoenzymes in neurons and astrocytes of human brain. J Cereb Blood Flow Metab 16:1079-1089.

Bouzier AK, Goodwin R, de Gannes FM, Valeins H, Voisin P, Canioni P, Merle M (1998) Compartmentation of lactate and glucose metabolism in C6 glioma cells. A^{13}C and 1H NMR study. J Biol Chem 273:27162-27169.

Bröer S, Brookes N (2001) Transfer of glutamine between astrocytes and neurons. J Neurochem 77:705-719.

Cataldo AM, Broadwell RD (1986) Cytochemical identification of cerebral glycogen and glucose-6-phosphatase activity under normal and experimental conditions. II. Choroid plexus and ependymal epithelia, endothelia and pericytes. J Neurocytol 15:511-524.

Cerdan S, Kunnecke B, Seelig J (1990) Cerebral metabolism of [1,2-^{13}C2]acetate as detected by in vivo and in vitro 13C NMR. J Biol Chem 265:12916-26.

Cerdan S, Rodrigues TB, Sierra A, Benito M, Fonseca LL, Fonseca CP, Garcia-Martin ML (2006) The redox switch/redox coupling hypothesis. Neurochem Int 48:523-530.

Cesar M, Hamprecht B (1995) Immunocytochemical examination of neural rat and mouse primary cultures using monoclonal antibodies raised against pyruvate carboxylase. J Neurochem 64:2312-2318.

Cesar MC, Wilson JE (1998) Further studies on the coupling of mitochondrially bound hexokinase to intramitochondrially compartmented ATP, generated by oxidative phosphorylation. Arch Biochem Biophys 350:109-117.

Chaudhry FA, Reimer RJ, Edwards RH (2002) The glutamine commute: take the N line and transfer to the A. J Cell Biol 157:349-355.

Chih CP, Lipton P, Roberts EL, Jr. (2001) Do active cerebral neurons really use lactate rather than glucose? Trends Neurosci 24:573-578.
Clausen RP, Madsen K, Larsson OM, Frølund B, Krogsgaard-Larsen P, Schousboe A (2006) Structure-activity relationship and pharmacology of γ-aminobutyric acid (GABA) transport inhibitors. Adv Pharmacol 54:265-284.
Cotman CW, Foster A, Lanthorn T (1981) An overview of glutamate as a neurotransmitter. Adv Biochem Psychopharmacol 27:1-27.
Cruz NF, Dienel GA (2002) High glycogen levels in brains of rats with minimal environmental stimuli: implications for metabolic contributions of working astrocytes. J Cereb Blood Flow Metab 22:1476-1489.
Cruz F, Scott SR, Barroso I, Santisteban P, Cerdan S (1998) Ontogeny and cellular localization of the pyruvate recycling system in rat brain. J Neurochem 70:2613-2619.
Cruz F, Villalba M, Garcia-Espinosa MA, Ballesteros P, Bogonez E, Satrustegui J, Cerdan S (2001) Intracellular compartmentation of pyruvate in primary cultures of cortical neurons as detected by 13C NMR spectroscopy with multiple 13C labels. J Neurosci Res 66:771-781.
Danbolt NC (2001) Glutamate uptake. Progr Neurobiol 65:1-105.
de Cerqueira CM, Wilson JE (1995) Application of a double isotopic labeling method to a study of the interaction of mitochondrially bound rat brain hexokinase with intramitochondrial compartments of ATP generated by oxidative phosphorylation. Arch Biochem Biophys 324:9-14.
Dienel GA, Cruz NF (2003) Neighbourly interactions of metabolically-activated astrocytes in vivo. Neurochem Int 43:339-354.
Dienel GA, Cruz NF (2004) Nutrition during brain activation: does cell-to-cell lactate shuttling contribute significantly to sweet and sour food for thought? Neurochem Int 45:321-351.
Dienel GA, Hertz L (2005) Astrocytic contributions to bioenergetics of cerebral ischemia. Glia 50:362-388.
Dienel GA, Ball KK, Cruz NF (2007) A glycogen phosphorylase inhibitor selectively enhances local rates of glucose utilization in brain during sensory stimulation of conscious rats: implications for glycogen turnover. J Neurochem 102:466-478.
Drejer J, Larsson OM, Kvamme E, Svenneby G, Hertz L, Schousboe A (1985) Ontogenetic development of glutamate metabolizing enzymes in cultured cerebellar granule cells and in cerebellum in vivo. Neurochem Res 10:49-62.
Erecinska M, Silver IA (1990) Metabolism and role of glutamate in mammalian brain. Prog Neurobiol 35:245-296.
Farinelli SE, Nicklas WJ (1992) Glutamate metabolism in rat cortical astrocyte cultures. J Neurochem 58:1905-1915.
Freidmann B, Goodman EH Jr, Saunders HL, Kostos V, Weinhouse S (1971) An estimation of pyruvate recycling during gluconeogenesis in the perfused rat liver. Arch. Biochem Biophys 143:566–578.
Gibbs ME, Anderson DG, Hertz L (2006) Inhibition of glycogenolysis in astrocytes interrupts memory consolidation in young chickens. Glia 54:214-222.
Gruetter R, Seaquist ER, Ugurbil K (2001) A mathematical model of compartmentalized neurotransmitter metabolism in the human brain. Am J Physiol Endocrinol Metab 281:E100-E112.
Haberg A, Qu H, Bakken IJ, Sande LM, White LR, Haraldseth O, Unsgard G, Aasly J, Sonnewald U (1998) In vitro and ex vivo 13C-NMR spectroscopy studies of pyruvate recycling in brain. Dev Neurosci 20:389-398.
Hamprecht B, Verleysdonk S, Wiesinger H (2005) Enzymes of carbohydrate and energy metabolism. In: Neuroglia (Kettenmann H, Ransom BR, eds), pp 202-215. New York: Oxford University Press, Inc.
Hassel B, Sonnewald U 1995. Glial formation of pyruvate and lactate from TCA cycle intermediates: implications for the inactivation of transmitter amino acids. J Neurochem. 65:2227–2234.
Hassel B, Brathe A (2000) Neuronal pyruvate carboxylation supports formation of transmitter glutamate. J Neurosci 20:1342-1347.
Hassel B, Sonnewald U, Fonnum F (1995) Glial-neuronal interactions as studied by cerebral metabolism of [2-13C]acetate and [1-^{13}C]glucose: an ex vivo ^{13}C NMR spectroscopic study. J Neurochem 64:2773-2782.

Haydon PG (2001) Glia: listening and talking to the synapse. Nat Rev Neurosci 2:185-193.

Hertz L, O'Dowd BS, Gibbs ME (2003) Reciprocal changes in forebrain contents of glycogen and of glutamate/glutamine during early memory consolidation in the day-old chick. Brain Res 994:226-233.

Hertz L (2004) The astrocyte-neuron lactate shuttle: a challenge of a challenge. J Cereb Blood Flow Metab 24:1241-1248.

Hertz L, Zielke HR (2004) Astrocytic control of glutamatergic activity: astrocytes as stars of the show. Trends Neurosci 27:735-743.

Hertz L, Peng L, Westergaard N, Yudkoff M, Schousboe A (1992) Neuronal-astrocytic interactions in metabolism of transmitter amino acids of the glutamate family. In: Drug Research Related to Neuroactive Amino Acids, Alfred Benzon Symposium 32 (Schousboe A, Diemer NH, Kofod H, eds), pp 30-48. Copenhagen: Munksgaard.

Hertz L, Drejer J, Schousboe A (1988) Energy metabolism in glutamatergic neurons, GABAergic neurons and astrocytes in primary cultures. Neurochem Res 13:605-610.

Hertz L, Swanson RA, Newman GC, Marrif H, Juurlink BH, Peng L (1998) Can experimental conditions explain the discrepancy over glutamate stimulation of aerobic glycolysis? Dev Neurosci 20:339-347.

Hertz L, Peng L, Dienel GA (2007) Energy metabolism in astrocytes: high rate of oxidative metabolism and spatiotemporal dependence on glycolysis/glycogenolysis. J Cereb Blood Flow Metab 27:219-249.

Hyder F, Patel AB, Gjedde A, Rothman DL, Behar KL, Shulman RG (2006) Neuronal-glial glucose oxidation and glutamatergic-GABAergic function. J Cereb Blood Flow Metab 26: 865-877.

Jayakumar AR, Rama Rao KV, Schousboe A, Norenberg MD (2004) Glutamine-induced free radical production in cultured astrocytes. Glia 46:296-301.

Kurz GM, Wiesinger H, Hamprecht B (1993) Purification of cytosolic malic enzyme from bovine brain, generation of monoclonal antibodies, and immunocytochemical localization of the enzyme in glial cells of neural primary cultures. J Neurochem 60:1467-1474.

Laake JH, Takumi Y, Eidet J, Torgner IA, Roberg B, Kvamme E, Ottersen OP (1999) Postembedding immunogold labelling reveals subcellular localization and pathway-specific enrichment of phosphate activated glutaminase in rat cerebellum. Neuroscience 88:1137-1151

Lai JC, Liang BB, Zhai S, Jarvi EJ, Lu DR (1994) Brain mitochondrial citrate synthase and glutamate dehydrogenase: differential inhibition by fatty acyl coenzyme A derivatives. Metab Brain Dis 9:143-152.

Lajtha A, Berl S, Waelsch H (1959) Amino acid and protein metabolism of the brain. IV. The metabolism of glutamic acid. J Neurochem 3:322-332.

Lapidot A, Gopher A. (1994) Cerebral metabolic compartmentation. Estimation of glucose flux via pyruvate carboxylase/pyruvate dehydrogenase by 13C NMR isotopomer analysis of D-[U-^{13}C]glucose metabolites. J Biol Chem 269:27198-208.

Larsson OM, Drejer J, Kvamme E, Svenneby G, Hertz L, Schousboe A (1985) Ontogenetic development of glutamate and GABA metabolizing enzymes in cultured cerebral cortex interneurons and in cerebral cortex in vivo. Int J Devl Neurosci 3:177-185.

Lieth E, LaNoue KF, Berkich DA, Xu B, Ratz M, Taylor C, Hutson SM (2001) Nitrogen shuttling between neurons and glial cells during glutamate synthesis. J Neurochem 76:1712-1723.

Lovatt D, Sonnewald U, Waagepetersen HS, Schousboe A, He W, Lin JH, Han X, Takano T, Wang S, Sim FJ, Goldman SA, Nedergaard M (2007) The transcriptome and metabolic gene signature of protoplasmic astrocytes in the adult murine cortex. J Neurosci 27:12255-12266.

Magistretti PJ, Pellerin L, Rothman DL, Shulman RG (1999) Energy on demand. Science 283:496-497.

Malik P, McKenna MC, Tildon JT (1993) Regulation of malate dehydrogenases from neonatal, adolescent, and mature rat brain. Neurochem Res 18:247-257.

Margineantu DH, Brown RM, Brown GK, Marcus AH, Capaldi RA (2002) Heterogeneous distribution of pyruvate dehydrogenase in the matrix of mitochondria. Mitochondrion 1:327-338.

Mason GF, Gruetter R, Rothman DL, Behar KL, Shulman RG, Novotny EJ (1995) Simultaneous determination of the rates of the TCA cycle, glucose utilization, alpha-ketoglutarate/glutamate exchange, and glutamine synthesis in human brain by NMR. J Cereb Blood Flow Metab 15:12-25.

McKenna MC, Tildon JT, Couto R, Stevenson JH, Caprio FJ (1990) The metabolism of malate by cultured rat brain astrocytes. Neurochem Res 15:1211-1220.

McKenna MC, Tildon JT, Stevenson JH, Boatright R, Huang S (1993) Regulation of energy metabolism in synaptic terminals and cultured rat brain astrocytes: differences revealed using aminoocxyacetate. Dev Neurosci 15:320-329.

McKenna MC, Sonnewald U, Huang X, Stevenson J, Zielke HR (1996a) Exogenous glutamate concentration regulates the metabolic fate of glutamate in astrocytes. J Neurochem 66:386-393.

McKenna MC, Tildon JT, Stevenson JH, Huang X (1996b) New insights into the compartmentation of glutamate and glutamine in cultured rat brain astrocytes. Dev Neurosci 18:380-390.

McKenna MC, Stevenson JH, Huang X, Hopkins IB (2000a) Differential distribution of the enzymes glutamate dehydrogenase and aspartate aminotransferase in cortical synaptic mitochondria contributes to metabolic compartmentation in cortical synaptic terminals. Neurochem Int 37:229-241.

McKenna MC, Stevenson JH, Huang X, Tildon JT, Zielke CL, Hopkins IB (2000b) Mitochondrial malic enzyme activity is much higher in mitochondria from cortical synaptic terminals compared with mitochondria from primary cultures of cortical neurons or cerebellar granule cells. Neurochem Int 36:451-459.

McKenna MC, Gruetter R, Sonnewald U, Waagepetersen HS, Schousboe A (2006a) Energy metabolism of the brain. In: Basic Neurochemistry: Molecular, Cellular, and Medical Aspects, 7th ed (Siegel GJ, Agranoff BW, Albers RW, Fisher SK, Uhler MD, eds), pp 531-557. Philadelphia, Lippincott-Raven.

McKenna MC, Waagepetersen HS, Schousboe A, Sonnewald U (2006b) Neuronal and astrocytic shuttle mechanisms for cytosolic-mitochondrial transfer of reducing equivalents: current evidence and pharmacological tools. Biochem Pharmacol 71:399-407.

Melo TM, Nehlig A, Sonnewald U (2006) Neuronal-glial interactions in rats fed a ketogenic diet. Neurochem Int 48:498-507.

Newman EA (2003) New roles for astrocytes: Regulation of synaptic transmission. Trends Neurosci 26:536-542.

Nissen C, Schousboe A (1979) Activity and isoenzyme pattern of lactate dehydrogenase in astrocytes cultured from brains of newborn mice. J Neurochem 32:1787-1792.

Norenberg MD, Martinez-Hernandez A (1979) Fine structural localization of glutamine synthetase in astrocytes of rat brain. Brain Res 161:303-310.

O'brien J, Kla KM, Hopkins IB, Malecki EA, McKenna MC (2006) Kinetic parameters and lactate dehydrogenase isozyme activities support possible lactate utilization by neurons. Neurochem Res 32:597-607.

Oberheim NA, Wang X, Goldman S, Nedergaard M (2006) Astrocytic complexity distinguishes the human brain. Trends Neurosci 29:547-553.

Olstad E, Olsen GM, Qu H, Sonnewald U (2007) Pyruvate recycling in cultured neurons from cerebellum. J Neurosci Res Epub Feb 15, 2007.

Oz G, Berkich DA, Henry PG, Xu Y, LaNoue K, Hutson SM, Gruetter R (2004) Neuroglial metabolism in the awake rat brain: CO2 fixation increases with brain activity. J Neurosci 24:11273-11279.

Parpura V, Basarsky TA, Liu F, Jeftinija F, Jeftinija S, Haydon PG (1994) Glutamate-mediated astrocyte-neuron signalling. Nature 369:744-747.

Parpura V, Haydon PG (2000) Physiological astrocytic calcium levels stimulate glutamate release to modulate adjacent neurons. Proc Natl Acad Sci USA 97:8629-8634.

Patel MS (1974) The effect of ketone bodies on pyruvate carboxylation by rat brain mitochondria. J Neurochem 23:865-867.

Patel AB, de Graaf RA, Mason GF, Kanamatsu T, Rothman DL, Shulman RG, Behar KL (2004) Glutamatergic neurotransmission and neuronal glucose oxidation are coupled during intense neuronal activation. J Cereb Blood Flow Metab 24:972-985.

Patel AB, Chowdhury GM, de Graaf RA, Rothman DL, Shulman RG, Behar KL (2005) Cerebral pyruvate carboxylase flux is unaltered during bicuculline-seizures. J Neurosci Res 79:128-138.

Pellerin L, Magistretti PJ (1994) Glutamate uptake into astrocytes stimulates aerobic glycolysis: a mechanism coupling neuronal activity to glucose utilization. Proc Natl Acad Sci USA 91:10625-10629.

Pellerin L, Pellegri G, Bittar PG, Charnay Y, Bouras C, Martin JL, Stella N, Magistretti PJ (1998) Evidence supporting the existence of an activity-dependent astrocyte-neuron lactate shuttle. Dev Neurosci 20:291-299.

Pellerin L, Magistretti PJ (2004) Neuroscience. Let there be (NADH) light. Science 305:50-52.

Perkins GA, Frey TG (2000) Recent structural insight into mitochondria gained by microscopy. Micron 31:97-111.

Pfeiffer-Guglielmi B, Fleckenstein B, Jung G, Hamprecht B (2003) Immunocytochemical localization of glycogen phosphorylase isozymes in rat nervous tissues by using isozyme-specific antibodies. J Neurochem 85:73-81.

Plaitakis A, Zaganas I (2001) Regulation of human glutamate dehydrogenases: implications for glutamate, ammonia and energy metabolism in brain. J Neurosci Res 66:899-908.

Qu H, Faero E, Jorgensen P, Dale O, Gisvold SE, Unsgard G, Sonnewald U (1999) Decreased glutamate metabolism in cultured astrocytes in the presence of thiopental. Biochem Pharmacol 58:1075-1080.

Qu H, Eloqayli H, Unsgard G, Sonnewald U (2001) Glutamate decreases pyruvate carboxylase activity and spares glucose as energy substrate in cultured cerebellar astrocytes. J Neurosci Res 66:1127-1132.

Ramón y Cajal S (1911) Histologie de systeme nerveux de l'homme et des vertebras. Paris : Maloine.

Ramos M, del AA, Pardo B, Martinez-Serrano A, Martinez-Morales JR, Kobayashi K, Yasuda T, Bogonez E, Bovolenta P, Saheki T, Satrustegui J (2003) Developmental changes in the Ca^{2+}-regulated mitochondrial aspartate-glutamate carrier aralar1 in brain and prominent expression in the spinal cord. Brain Res Dev Brain Res 143:33-46.

Ransom BR, Fern R (1997) Does astrocytic glycogen benefit axon function and survival in CNS white matter during glucose deprivation? Glia 21:134-141.

Ransom B, Behar T, Nedergaard M (2003) New roles for astrocytes (stars at last). Trends Neurosci 26:520-522.

Sarup A, Larsson OM, Schousboe A (2003) GABA transporters and GABA-transaminase as drug targets. Curr Drug Targ CNS Neurol Dis 2:269-277.

Schousboe A, Hertz L, Svenneby G, (1977a) Uptake and metabolism of GABA in astrocytes cultured from dissociated mouse brain hemispheres. Neurochem Res 2:217-229.

Schousboe A, Hertz L, Svenneby G, Kvamme E (1979) Phosphate activated glutaminase activity and glutamine uptake in primary cultures of astrocytes. J Neurochem 32:943-950.

Schousboe A, Westergaard N, Sonnewald U, Petersen SB, Huang R, Peng L, Hertz L (1993a) Glutamate and glutamine metabolism and compartmentation in astrocytes. Dev Neurosci 15:359-366.

Schousboe A, Westergaard N, Waagepetersen HS, Larsson OM, Bakken IJ, Sonnewald U (1997) Trafficking between glia and neurons of TCA cycle intermediates and related metabolites. Glia 21:99-105.

Schousboe A, Sonnewald U, Waagepetersen HS (2003) Differential roles of alanine in GABAergic and glutamatergic neurons. Neurochem Int 43:311-315.

Schousboe A, Sarup A, Larsson OM, White HS (2004) GABA transporters as drug targets for modulation of GABAergic activity. Biochem Pharmacol 68:1557-1563.

Schousboe I, Bro B, Schousboe A (1977b) Intramitochondrial localization of the 4-aminobutyrate-2-ketoglutarate transaminase from ox brain. Biochem J 162:303-307.

Schousboe I, Tønder N, Zimmer J, Schousboe A (1993b) A developmental study of lactate dehydrogenase isozyme and aspartate aminotransferase activity in organotypic rat hippocampal slice cultures and primary cultures of mouse neocortical and cerebellar neurons. Int J Devl Neuroscience 11:765-772.

Shank RP, Bennett GS, Freytag SO, Campbell GL (1985) Pyruvate carboxylase: an astrocyte-specific enzyme implicated in the replenishment of amino acid neurotransmitter pools. Brain Res 329:364-367.

Sickmann HM, Schousboe A, Fosgerau K, Waagepetersen HS (2005) Compartmentation of lactate originating from glycogen and glucose in cultured astrocytes. Neurochem Res 30:1295-1304.

Sonnewald U, Westergaard N, Hassel B, Muller TB, Unsgard G, Fonnum F, Hertz L, Schousboe A, Petersen SB (1993a) NMR spectroscopic studies of ^{13}C acetate and ^{13}C glucose metabolism in neocortical astrocytes: evidence for mitochondrial heterogeneity. Dev Neurosci 15:351-358.

Sonnewald U, Westergaard N, Petersen SB, Unsgard G, Schousboe A (1993b) Metabolism of [U-^{13}C]glutamate in astrocytes studied by ^{13}C NMR spectroscopy: incorporation of more label into lactate than into glutamine demonstrates the importance of the tricarboxylic acid cycle. J Neurochem 61:1179-1182.

Sonnewald U, Westergaard N, Jones P, Taylor A, Bachelard HS, Schousboe A (1996) Metabolism of [U-^{13}C5]glutamine in cultured astrocytes studied by NMR spectroscopy: First evidence of astrocytic pyruvate recycling. J Neurochem 67:2566-2572.

Sonnewald U, Westergaard N, Schousboe A (1997) Glutamate transport and metabolism in astrocytes. Glia 21:99-105.

Sonnewald U, Hertz L, Schousboe A (1998) Mitochondrial heterogeneity in the brain at the cellular level. J Cereb Blood Flow Metab 18:231-237.

Sorg O, Magistretti PJ (1991) Characterization of the glycogenolysis elicited by vasoactive intestinal peptide, noradrenaline and adenosine in primary cultures of mouse cerebral cortical astrocytes. Brain Res 563:227-233.

Suh SW, Gum ET, Hamby AM, Chan PH, Swanson RA (2007) Hypoglycemic neuronal death is triggered by glucose reperfusion and activation of neuronal NADPH oxidase. J Clin Invest 117:910-918.

Swanson RA (1992) Physiologic coupling of glial glycogen metabolism to neuronal activity in brain. Can J Physiol Pharmacol 70 Suppl:S138-S144.

Taylor A, McLean M, Morris P, Bachelard H (1996) Approaches to studies on neuronal/glial relationships by 13C-MRS analysis. Dev Neurosci 18:434-442.

Tkac I, Rao R, Georgieff MK, Gruetter R (2003) Developmental and regional changes in the neurochemical profile of the rat brain determined by in vivo 1H NMR spectroscopy. Magn Reson Med 50:24-32.

Van den Berg, CJ, Garfinkel D (1971) A simulation study of brain compartments: Metabolism of glutamate and related substances in mouse brain. Biochem J 23:211-218.

van den Berg CJ, Krzalic L, Mela P, Waelsch H (1969) Compartmentation of glutamate metabolism in brain. Evidence for the existence of two different tricarboxylic acid cycles in brain. Biochem J 113:281-290.

Vannucci SJ, Maher F, Simpson IA (1997) Glucose transporter proteins in brain: delivery of glucose to neurons and glia. Glia 21:2-21.

Varoqui H, Zhu H, Yao D, Ming H, Erickson JD (2000) Cloning and functional identification of a neuronal glutamine transporter. J Biol Chem 275:4049-4054.

Vogel R, Hamprecht B, Wiesinger H (1998) Malic enzyme isoforms in astrocytes: comparative study on activities in rat brain tissue and astroglia-rich primary cultures. Neurosci Lett 247:123-126.

Volterra A, Magistretti PJ, Haydon PG, eds. (2002) The Tripartite Synapse: Glia in Synaptic Transmission. Oxford, UK: University Press.

Waagepetersen HS, Schousboe A (2007) Glial GABA and glutamate metabolism. In: New Enclyclopedia in Neuroscience (Squire L, Albright T, Bloom F, Gage F, Spitzer N, eds), Oxford, UK: Elsevier Ltd, in press.

Waagepetersen HS, Bakken IJ, Larsson OM, Sonnewald U, Schousboe A (1998a) Comparison of lactate and glucose metabolism in cultured neocortical neurons and astrocytes using ^{13}C-NMR spectroscopy. Dev Neurosci 20:310-320.

Waagepetersen HS, Bakken IJ, Larsson OM, Sonnewald U, Schousboe A (1998b) Metabolism of lactate in cultured GABAergic neurons studied by ^{13}C nuclear magnetic resonance spectroscopy. J Cereb Blood Flow Metab 18:109-117.

Waagepetersen HS, Sonnewald U, Larsson OM, Schousboe A (2000) A possible role of alanine for ammonia transfer between astrocytes and glutamatergic neurons. J Neurochem 75:471-479.

Waagepetersen HS, Qu H, Schousboe A, Sonnewald U (2001a) Elucidation of the quantitative significance of pyruvate carboxylation in cultured cerebellar neurons and astrocytes. J Neurosci Res 66:763-770.

Waagepetersen HS, Sonnewald U, Larsson OM, Schousboe A (2001b) Multiple compartments with different metabolic characteristics are involved in biosynthesis of intracellular and released glutamine and citrate in astrocytes. Glia 35:246-252.

Waagepetersen HS, Qu H, Hertz L, Sonnewald U, Schousboe A (2002) Demonstration of pyruvate recycling in primary cultures of neocortical astrocytes but not in neurons. Neurochem Res 27:1431-1437.

Waagepetersen HS, Hansen GH, Fenger K, Lindsay JG, Gibson G, Schousboe A (2006) Cellular mitochondrial heterogeneity in cultured astrocytes as demonstrated by immunogold labeling of alpha-ketoglutarate dehydrogenase. Glia 53:225-231.

Walz W, Mukerji S (1988) Lactate release from cultured astrocytes and neurons: a comparison. Glia 1:366-370.

Waniewski RA, Martin DL (1998) Preferential utilization of acetate by astrocytes is attributable to transport. J Neurosci 18:5225-5233.

Wender R, Brown AM, Fern R, Swanson RA, Farrell K, Ransom BR (2000) Astrocytic glycogen influences axon function and survival during glucose deprivation in central white matter. J Neurosci 20:6804-6810.

Westergaard N, Sonnewald U, Schousboe A (1995) Metabolic trafficking between neurons and astrocytes: the glutamate/glutamine cycle revisited. Dev Neurosci 17:203-211.

Westergaard N, Drejer J, Schousboe A, Sonnewald U (1996) Evaluation of the importance of transamination versus deamination in astrocytic metabolism of [U-^{13}C]glutamate. Glia 17:160-168.

White HS, Sarup A, Bolvig T, Kristensen AS, Petersen G, Nelson N, Pickering DS, Larsson OM, Frølund B, Krogsgaard-Larsen P, Schousboe A (2002) Correlation between anticonvulsant activity and inhibitory action on glial GABA uptake of the highly selective mouse GAT1 inhibitor 3-hydroxy-4-amino-4,5,6,7-tetrahydro-1,2-benzisoxazole (exo-THPO) and its N-alkylated analogs. J Pharmacol Exp Therap 302:636-644.

Wilson JE (1995) Hexokinases. Rev Physiol Biochem Pharmacol 126:65-198.

Wilson JE (2003) Isozymes of mammalian hexokinase: structure, subcellular localization and metabolic function. J Exp Biol 206:2049-2057.

Yudkoff M (1997) Brain metabolism of branched-chain amino acids. Glia 21:92-98.

Yu ACH, Schousboe A, Hertz L (1982) Metabolic fate of [14C]-labeled glutamate in astrocytes in primary cultures. J Neurochem 39:954-960.

Yu AC, Drejer J, Hertz L, Schousboe A (1983) Pyruvate carboxylase activity in primary cultures of astrocytes and neurons. J Neurochem 41:1484-1487.

Zielke, RH, Zielke CL, Baab PJ, Tildon TJ (2007) Effect of fluorocitrate on cerebral oxidation of lactate and glucose in freely moving rats. J Neurochem 101: 9-16.

Zwingmann C, Leibfritz D (2003) Regulation of glial metabolism studied by ^{13}C-NMR. NMR Biomed 16:370-399.

Zwingmann C, Richter-Landsberg C, Leibfritz D (2001) ^{13}C isotopomer analysis of glucose and alanine metabolism reveals cytosolic pyruvate compartmentation as part of energy metabolism in astrocytes. Glia 34:200-212.

Abbreviations

AAT	Aspartate aminotransferase
α-KG	α-Ketoglutarate
AMP	Adenosine monophosphate
ATP	Adenosine 5′-triphosphate
BCAA	Branched chain amino acid
EM	Electron microscopy
GABA	Gamma-aminobutyric acid
GDH	Glutamate dehydrogenase
GLUT	Glucose transporter
GP	Glycogen phosphorylase
GS	Glutamine synthetase
LDH	Lactate dehydrogenase
MAS	Malate–aspartate shuttle
ME	Malic enzyme
MRS	Magnetic resonance spectroscopy
NAD^+	β-Nicotinamide adenine dinucleotide
PAG	Phosphate activated glutaminase
PC	Pyruvate carboxylase
PEP	Phosphoenolpyruvate
PEPCK	Phosphoenolpyruvate carboxykinase
TCA	Tricarboxylic acid

Chapter 8
Calcium Ion Signaling in Astrocytes

Joachim W. Deitmer, Karthika Singaravelu, and Christian Lohr

Contents

8.1 Introduction

Ca^{2+} signaling has been recognized as one of the major second messenger steps in most cell types, including astrocytes, the major macroglial cell type in vertebrate nervous systems. Astrocytes are by no means a homogeneous group of glial cells, but comprise a number of different cell types (see Chap. 1). However, in contrast to a decade ago, when mammalian astrocytes were divided into either protoplasmic type 1 or fibrous type II astrocytes, we assume today that there are many types of astrocytes in different brain regions. Another classification has recognized astrocytes with a dense distribution of glutamate uptake transporters (EAAT, excitatory amino acid transporter) and poor equipment of ionotropic glutamate receptors, while another type of astrocytes shows a poor expression of EAATs, but prominent distribution of ionotropic glutamate receptors. As with all of these cell type

J.W. Deitmer
Abteilung für Allgemeine Zoologie, FB Biologie, TU Kaiserslautern, Kaiserslautern, Germany.
deitmer@biologie.uni-kl.de

V. Parpura and P.G. Haydon (eds.), *Astrocytes in (Patho)Physiology of the Nervous System*,
DOI: 10.1007/978-0-387-79492-1_8, © Springer Science+Business Media, LLC 2009

classifications established so far, there are known exemptions, such as the Bergmann glial cells in the cerebellum. Bergmann glia is a radial type of macroglial cell, which is a specialized astrocyte, which has both EAATs and ionotropic alpha-amino-3-hydroxy-5-methyl-4-isoxazolepropionic acid (AMPA)-type glutamate receptors. There are radial-type astrocytes also in the developing cortex, which extend from the ventricular surface to the pial surface. Furthermore, Müller glial cells, the principal glial cell type in the retina, extend across the entire retina from the photoreceptors to the inner retinal surface. Thus, the classification of different types of astrocytes is still in its infancy, and we need to know much more about this major cell type in the brain in order to create a useful taxonomy taking into account the various predominant functions of astrocytes in a given brain region.

Most, if not all, of these different types of astrocytes share the property that many of their cellular functions are related to cytosolic Ca^{2+} signaling. The single and repetitive rises of cytosolic Ca^{2+} play a complex role for initiating intracellular signaling cascades, modulating astrocytic functions and intercellular interaction. Most of these functions and interactions involve other neighboring astrocytes and neurons, but there are also signaling pathways to oligodendrocytes and microglial cells. Astrocytes are endowed with a large number of metabotropic receptors in their cell membrane, most of which are coupled to the release of Ca^{2+} from the endoplasmic reticulum (ER) via phospholipase C (PLC)-mediated formation of inositol-trisphosphate (IP_3). Astrocytic Ca^{2+} signaling can be a single Ca^{2+} transient with or without a shoulder or plateau phase, repetitive Ca^{2+} transients, so-called Ca^{2+} oscillations, or irregular Ca^{2+} rises, depending on the species of primary messenger (neurotransmitter, hormone, growth factor) and its concentration. These Ca^{2+} signals may spread along the cell and across cell boundaries to neighboring astrocytes in form of Ca^{2+} waves, and can be evoked or modulated by neuronal activity. The spatial and temporal properties of these Ca^{2+} signaling modes reflect the versatility of this intracellular messenger system. The signaling pathway leading to a cytosolic Ca^{2+} rise, common to many cell types, may be regarded as a type of *excitation* in electrically nonexcitable cells like astrocytes. Cytosolic Ca^{2+} transients may initiate Ca^{2+}-dependent release of transmitters (gliotransmitters), affecting neuronal excitability or vasoconstriction/vasodilation of blood vessels in the brain. This chapter reviews types of cytosolic Ca^{2+} signaling in astrocytes, their different modes of initiation, and their functional significance for astrocytes and the glia–neuron communication.

8.2 Modes and Mechanisms of Ca^{2+} Signaling

Ca^{2+} signaling in both excitable and nonexcitable cells is based on the maintenance of a low *resting* concentration of cytosolic Ca^{2+} (<150 nM), as compared with the extracellular compartment (1–2 mM), and intracellular Ca^{2+} storage compartments (0.1–1 mM). This differential distribution of Ca^{2+} creates a gradient across the membrane of intracellular organelles (ER, mitochondria, calcisomes, lysosomes) and the plasma membrane. Cellular Ca^{2+} signaling, manifested as rapid, reversible, and often repeated,

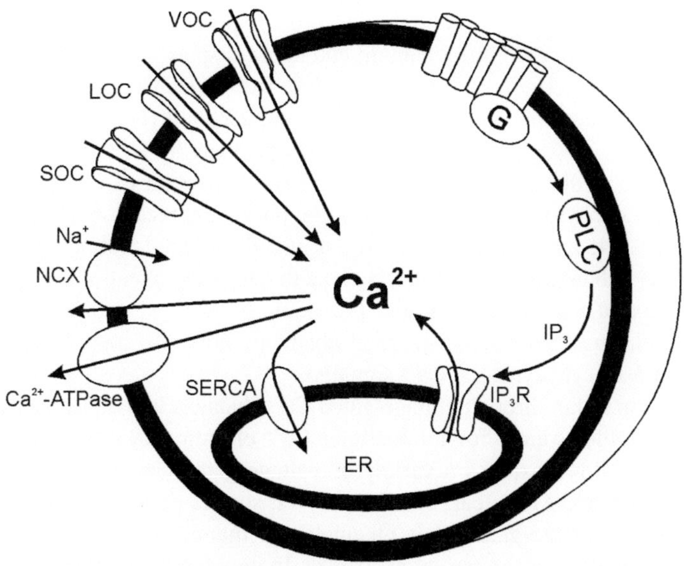

Fig. 8.1 Mechanisms of Ca^{2+} signaling in astrocytes. Ca^{2+} can enter the cell through voltage-operated channels (VOC), ligand-operated channels (LOC), and store-operated cannels (SOC). Activation of G-protein-coupled receptors results in phospholipase C (PLC)-mediated production of IP_3, which leads to release of Ca^{2+} from intracellular stores such as the endoplasmic reticulum (ER) via IP_3 receptor channels. Ca^{2+} is transported out of the cytoplasm into the ER by sarcoplasmic-endoplasmic reticulum Ca^{2+}-ATPases (SERCA), and into the extracellular space by plasmalemmal Ca^{2+}-ATPases and a sodium-Ca^{2+} exchanger (NCX).

intracellular Ca^{2+} rises, may therefore result from several Ca^{2+} sources. Cytosolic Ca^{2+} levels can increase via Ca^{2+} influx from the extracellular space across the plasma membrane via Ca^{2+}-permeable ion channels, and by Ca^{2+} release from intracellular stores (Fig. 8.1). Calcium ion influx may occur through specific Ca^{2+}-permeable, ligand-gated channels, voltage-gated Ca^{2+} channels that are activated by membrane depolarization, and store-operated Ca^{2+} entry channels that are activated following the depletion of intracellular Ca^{2+} stores. At lowered extracellular K^+ concentrations, inward rectifier K^+ channels can become permeable to divalent cations including Ca^{2+}, and this novel type of Ca^{2+} influx has been observed in astrocytes at K^+ concentration of 2 mM and below (Dallwig et al., 2000; Härtel et al., 2007).

While electrically excitable cells use the electrochemical gradient across the plasma membrane to effectively increase the intracellular Ca^{2+} levels due to the high expression of specific ligand- and voltage-gated ion channels that allow prominent influx of Ca^{2+} ions, nonelectrically excitable cells like astrocytes use the Ca^{2+} stored in the ER as the main source for cytoplasmic Ca^{2+} signaling (reviewed in Deitmer et al., 1998; Verkhratsky et al., 1998). Ionotropic kainate/AMPA receptors, known to be expressed by Bergmann glial cells in the cerebellum, lack the GluR2 subunit, which renders them permeable to Ca^{2+} (Müller et al., 1992; Burnashev et al., 1992). However, Ca^{2+} influx through AMPA receptors does not seem to play a major role in astrocyte Ca^{2+} signaling, because in most brain regions, astrocytes express AMPA receptors that contain the GluR2 subunit and hence display a low Ca^{2+} permeability (Backus and Berger, 1995; Seifert et al., 2003). Recent studies have also described the presence of ionotropic glutamate receptors of the *N*-methyl-D-aspartate (NMDA) type in astrocytes in situ, which provide another pathway for Ca^{2+} influx (Schipke et al., 2001; Lalo et al., 2006; Bigini et al., 2006). Activation of other ligand-gated channels permeable to Ca^{2+} such as ionotropic purinergic receptors (P2X receptors; Walz et al., 1994; James and Butt, 2001), or nicotinic acetylcholine receptors containing the a7 subunit, has been shown to lead to the gating of cationic conductances and an influx of Ca^{2+} in cell culture preparations (Sharma and Vijayaraghavan, 2001). Ca^{2+} influx through ligand-gated channels, however, may not be a major pathway for astrocyte cytosolic Ca^{2+} signaling.

Calcium ion influx through voltage-gated Ca^{2+} channels in the plasma membrane constitutes another source for intracellular Ca^{2+} elevation in different cell types. Astrocytes in culture express a variety of voltage-gated channels, including Ca^{2+} channels (Duffy and MacVicar, 1994); however, studies performed in situ indicate that functional voltage-gated Ca^{2+} channels are mainly expressed by immature astrocytes, if at all (Carmignoto et al., 1998). In invertebrate glial cells, functional voltage-dependent Ca^{2+} channels appear to be the rule, and such Ca^{2+} influx may be

Table 8.1 Some receptor ligands capable of evoking Ca^{2+} signaling in astrocytes

Ligand	Reference
5-Hydroxytryptamine	Dave et al., 1991; Nilsson et al., 1991; Shelton and McCarthy, 2000
Acetylcholine	Shao and McCarthy, 1995; Sharma and Vijayaraghavan, 2001
Adenosine	Peakman and Hill, 1995; Porter and McCarthy, 1995a
Angiotensin II	Wang et al., 1996
ATP	
P2Y	Kastritsis et al., 1992; Ho et al., 1995
P2X	James and Butt, 2001
Bradykinin	Gimpl et al., 1992
Dopamine	Parpura and Haydon, 2000
GABA	Nilsson et al., 1993
Glutamate	
AMPA/kainate	Enkvist et al., 1989; Müller et al., 1992; Burnashev et al., 1992
mGluR	Jensen and Chiu, 1991; Porter and McCarthy, 1995b
Histamine	Inagaki et al., 1991; Shelton and McCarthy, 2000
Met-enkephalin	Stiene-Martin et al., 1993
Neuropeptide Y	Gimpl et al., 1993
Noradrenaline (norepinephrine)	Salm and McCarthy, 1990; Duffy and MacVicar, 1995
Tachykinins	Delumeau et al., 1991
Thrombin	Czubayko and Reiser, 1995
Vasopressin, oytocin	Jurzak et al., 1995

initiated by depolarizing neurotransmitters (Lohr and Deitmer, 1999, 2006; Lohr et al., 2005). Therefore, unlike neurons and other electrically excitable cells that express a high density of ligand- and voltage-gated channels, which are permeable to Ca^{2+} and allow an efficient rise of cytosolic Ca^{2+}, astrocytes in the mammalian nervous system usually employ different cellular mechanisms to allow Ca^{2+} entry and an increase in their cytosolic Ca^{2+} concentration.

Astrocytes express a variety of functional receptors for many neurotransmitters, neuromodulators, peptides, etc. This includes glutamate, norepinephrine, gamma-aminobutyric acid (GABA), histamine, adenosine 5¢-triphosphate (ATP), adenosine diphosphate (ADP), adenosine, and acetylcholine (also see later and Table 8.1), most of which belong to the metabotropic receptor family. These receptors are associated with G proteins that can stimulate phospholipase C activity, which in turn leads to the formation of the intracellular second messengers IP_3 and diacylglycerol (DAG). Activation of specific, IP_3-gated receptor channels in the ER results in release of Ca^{2+} from these Ca^{2+} stores. This leads to an increase in the cytosolic Ca^{2+} concentration, which usually encompasses the entire cell soma and often also cell processes. This mechanism of eliciting an astrocytic Ca^{2+} signal represents one important mode of neuron-to-astrocyte communication, in which the astrocyte Ca^{2+} response is evoked by neurotransmitter(s) released during synaptic activity, activating one or several of the many metabotropic receptors in astrocytes.

In addition to IP_3 receptors, the membrane of the endoplasmic reticulum of most cells also contains ryanodine receptors, a different type of Ca^{2+}-permeable receptor that is activated by cytosolic Ca^{2+} and/or cyclic ADP-ribose. A rise of the cytosolic Ca^{2+} leads to the opening of the ryanodine receptors and hence induces the release of Ca^{2+} from the internal stores, a process known as Ca^{2+}-induced Ca^{2+} release (Sitsapesan et al., 1995; Verkhratsky and Shmigol, 1996). Ryanodine receptors can also be activated by caffeine (Meissner and Henderson, 1987; Liu et al., 1989). The presence of these receptors in astrocytes, however, is controversial; moreover, the lack of caffeine-sensitive Ca^{2+} release, as found in astrocytes in culture and in situ, and in contrast to neurons, suggests that the role of these receptors for astrocyte Ca^{2+} signaling is either restricted to defined brain areas, or plays a minor role in astrocytes in general (Beck et al., 2004).

The maintenance of the high intraluminal Ca^{2+} concentration in the ER depends on Ca^{2+}-ATPases in the ER membrane, which efficiently pump Ca^{2+} from the cytosol into the ER, and thereby maintains a high concentration of Ca^{2+} by constant refilling of these stores. Since Ca^{2+} is also transported out of the cell by Ca^{2+}-ATPases and Na^+/Ca^{2+} exchange carriers in the plasma membrane, Ca^{2+} store refilling requires some influx of Ca^{2+} from the extracellular compartment, in particular following evoked Ca^{2+} release from stores. Ca^{2+} entry is using specific Ca^{2+}-permeable plasma membrane channels, called store-operated channels, that are regulated by the filling state of the ER and that serve to its replenishment with Ca^{2+}. This store-operated Ca^{2+} entry (SOCE), also called capacitative Ca^{2+} entry, has been observed in many different cell types and recorded electrophysiologically as a persistent current, called *Ca^{2+}release-activated current* (I_{CRAC}; Hoth and Penner, 1992). Although the presence of I_{CRAC} has not yet been identified in astrocytes, SOCE has been clearly demonstrated in these cells by Ca^{2+} imaging studies (Jung et al., 2000; Lo et al., 2002; Singaravelu et al., 2006). Because of the importance of this type of Ca^{2+} entry in astrocytes, this will be discussed in more detail in Sect. 8.7.

It is the interplay of these different cellular mechanisms, Ca^{2+} influx and release, Ca^{2+} extrusion and uptake, and Ca^{2+} buffering, that controls the cytosolic and organellar Ca^{2+} levels, which are crucial for the generation and the shape of astrocyte Ca^{2+} signaling. The spatial organization of Ca^{2+} influx may create Ca^{2+} microdomains, i.e., localized regions of Ca^{2+} rises, usually near the plasma membrane (Berridge, 2006; Oheim et al., 2006). Mechanisms that are involved in regulating cytosolic Ca^{2+} contribute to shaping the kinetics and amplitude of Ca^{2+} transients, such as, e.g., Ca^{2+}-binding proteins like parvalbumin, calcineurin, and calmodulin that are present in the cytoplasm may serve to buffer cytosolic Ca^{2+} concentration. However, this is a low-capacity system that is more effective as mediator of Ca^{2+}-dependent cellular processes. Energy-consuming mechanisms are involved in the effective Ca^{2+} clearance. Sarco-endoplasmic reticulum Ca^{2+}-ATPases (SERCAs) pump Ca^{2+} ions into the endoplasmic reticulum using ATP as energy source. Ca^{2+}-ATPases and Na^+/Ca^{2+} exchangers present in the plasma membrane use ATP or the electrochemical gradient of Na^+ across the plasma membrane, respectively, to extrude Ca^{2+} against a steep concentration gradient across the plasma membrane. Finally, mitochondria are dynamic Ca^{2+} stores that sequester Ca^{2+} through the activity of the Ca^{2+} uniporter, with a relatively low affinity, but a large capacity, to store Ca^{2+} transiently, and protect the cell from cytosolic Ca^{2+} overload, which may lead to cell death. Although other Ca^{2+} storing organelles, such as calcisomes and lysosomes, have been reported to contribute to cytosolic Ca^{2+} signaling and homeostasis in some cell types, their role in astrocytes has remained largely unknown.

8.3 Spontaneous Ca^{2+} Transients and Oscillations

Ca^{2+} oscillations are defined as repetitive rises of cytosolic Ca^{2+} and may occur intermittently with single Ca^{2+} transients and with intervals of up to several minutes (Fig. 8.2). Spontaneous Ca^{2+} transients and oscillations have been reported in both neurons and glial cells, in culture, in situ, and in vivo (Parri and Crunelli, 2001; Nett et al., 2002; Morita et al., 2003; Zur Nieden and Deitmer, 2006). Ca^{2+} signals can propagate as waves along cell processes and even beyond cell boundaries to neighboring glial cells. Ca^{2+} transients and Ca^{2+} oscillations are usually attributable to release of Ca^{2+} from intracellular stores, and can be blocked by inhibiting the Ca^{2+}-ATPase of the endoplasmic reticulum (by cyclopiazonic acid or thapsigaragin), and/or by IP_3 receptor antagonists (e.g., 2-APB). These spontaneous Ca^{2+} transients and oscillations occurred in 35% of astrocytes in the hippocampus of juvenile rats, and in 82% of hippocampal astrocytes in juvenile mice, both with a rate of about

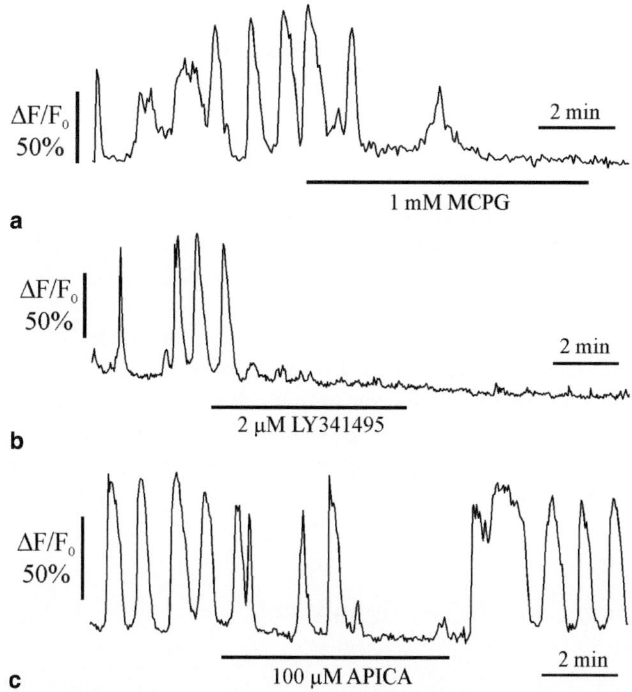

Fig. 8.2 Spontaneous Ca^{2+} oscillations in hippocampal astrocytes. Recordings of spontaneous Ca^{2+} oscillations measured in acute hippocampal brain slices, which were blocked by -methyl-4-carboxyphenylglycine (MCPG), a nonselective antagonist of metabotropic glutamate receptors (mGluR), and LY341495 and APICA, both group II-specific mGluR antagonist. From Zur Nieden and Deitmer (2006).

1.3 transients per minute (Zur Nieden and Deitmer, 2006; Beck and Deitmer, unpublished). Recently, spontaneous Ca^{2+} oscillations have been related to promote neurite growth in neuron-astrocyte cell cultures, presumably by maintaining the expression of specific growth-enhancing proteins, such as N-cadherin, on their surface (Kanemaru et al., 2007).

These Ca^{2+} signals require operating mechanisms for refilling the intracellular Ca^{2+} stores, which include store-operated calcium entry (SOCE, see later) through channels in the plasma membrane. Spontaneous Ca^{2+} signals in astrocytes are not directly affected by neuronal activity, but their synchronous generation may be affected by blocking neuronal activity by tetrodotoxin. As discussed later, various neurotransmitters, such as glutamate, noradrenaline, histamine, and ATP/ADP, and specific agonists of transmitter receptors (e.g., quisqualate, phenylephrine), as well as mechanical stimulation, can evoke transients, oscillations, and waves of cytosolic Ca^{2+} in astrocytes (Table 8.1).

Spontaneous Ca^{2+} transients and oscillations have been associated with the level of ambient glutamate in the tissue, acting on metabotropic glutamate receptors of group I and II in the astrocyte membrane (Zur Nieden and Deitmer, 2006). Changes in ambient glutamate concentration, which are not due to synaptically released glutamate, can, e.g., occur via the cystine-glutamate exchanger ($x(c)^-$ or xCT; Ye et al., 1999; Shih et al., 2006) or reversed glutamate uptake under pathophysiological conditions (Rossi et al., 2000; Allen and Attwell, 2004). Thus, the frequency of these spontaneous Ca^{2+} signals may help to sense the level of extracellular glutamate in nervous tissue. Spontaneous Ca^{2+} signals itself may induce Ca^{2+}-dependent gliotransmitter release (see later), and this might in turn contribute to the ambient glutamate concentration, although this mechanism may be only in action under certain conditions (Pasti et al., 1995, 2001). It does not seem likely at this point to hypothesize that each astrocytic Ca^{2+} signal leads to the release of transmitter. In particular, with the propagation of Ca^{2+} transients across the cell processes, the localization of Ca^{2+}-dependent gliotransmitter release is still obscure. The elaboration of mechanisms and functions of spontaneous Ca^{2+} transients and oscillations in astrocytes, and their mode of initiation and maintenance still require more experimental analysis.

8.4 Propagation of Ca^{2+} Signals

The cytosolic Ca^{2+} signals in astrocytes can adopt different modes of temporal and spatial patterns that can be modulated in amplitude, duration, and frequency by different stimuli from other cells. Synaptic activity and the release of neurotransmitters and modulators can be functionally important initiators for these glial Ca^{2+} signals, which can propagate within and between astrocytes, signaling to different regions of the cell and to different cells. The astrocyte Ca^{2+} signal, either spontaneously generated or evoked by synaptic activity, can be initiated in spatially restricted areas called *microdomains* (Simpson and Russell, 1997; Oheim et al., 2006). These are often located in the fine, extensive cellular processes, from where the Ca^{2+} signals can propagate along the processes to other regions of the astrocyte generating a wave of intracellular Ca^{2+}. Elementary events are initiated by a locally restricted group of Ca^{2+} channels in the plasma membrane or in the ER (Berridge, 2006). The spread of these Ca^{2+} transients along the cell processes and to neighboring cells can occur with a speed of 5–25 $\mu m\ s^{-1}$ (Peters et al., 2003; Fiacco and McCarthy, 2006). A thorough characterization of the intra- and intercellular Ca^{2+} waves, which can be under the control of released modulators and transmitters, still needs to be accomplished.

The intracellular compartmentalization and the control of the propagation of the Ca^{2+} signal are highly relevant for cellular functions and the astrocyte-to-neuron communication (Scemes and Giaume, 2006; Spät, 2006). Since *gliotransmitters* released from astrocytes through Ca^{2+}-dependent mechanisms may act as modulators of synaptic transmission, the mechanisms controlling the intracellular Ca^{2+} wave may not only be of great significance because they will determine the degree of extension of the signal that triggers the neuromodulatory effects of astrocytes,

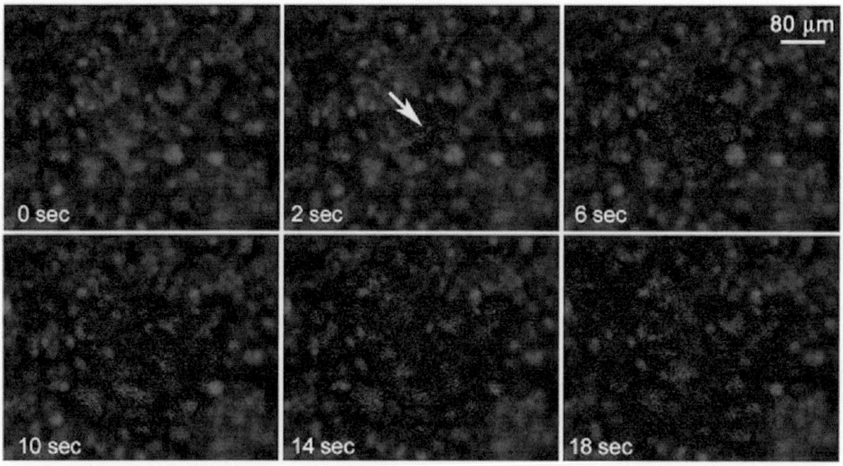

Fig. 8.3 Ca^{2+} wave in cultured spinal cord astrocytes. A Ca^{2+} wave as evoked by mechanical stimulation of an astrocyte (arrow) propagates centrifugally in the confluent layer of primary astrocytes. Modified from Suadicani et al. (2006). (*See Color Plates*)

but also because they determine the spatiotemporal response pattern of the cells. Since the Ca^{2+} signal evoked by synaptic activity is elicited in restricted regions of the astrocyte, the neuron-to-astrocyte communication is usually a localized event that does not result from uncontrolled spillover of neurotransmitter, but more probably from a spatially highly focussed signaling between synaptic terminals and astrocytic processes (Carmignoto, 2000; Fields and Stevens-Graham, 2002).

Astrocytes can also communicate with adjacent cells by generating intercellular Ca^{2+} waves, which are propagated along cell processes onto cell processes of neighboring astrocytes (Fig. 8.3). Different experimental stimuli, such as mechanical stimuli or exogenous, focal application of neurotransmitters, may induce intracellular Ca^{2+} increases that can propagate as Ca^{2+} waves between astrocytes in cultured cells as well as organotypic and acute brain slices (Cornell-Bell et al., 1990; Newman and Zahs, 1997; Cotrina et al., 1998; Schipke et al., 2002; Suadicani et al., 2006). These waves can be propagated over long distances up to 500 µm at relatively low speed (up to 30 $\mu m\ s^{-1}$), and may constitute a novel form of slow, long-distance cellular communication in the nervous system. Whether these Ca^{2+} waves connect neuropil regions and hence different neural circuits, which may not have direct neuronal connection, remains to be shown.

Ca^{2+} waves in the gap junction-coupled astrocytic syncytium were originally thought to spread as a result of diffusion of either Ca^{2+} or IP_3 between cells (Finkbeiner, 1992; Nedergaard, 1994; Scemes et al., 1998). Later studies have proposed that, in addition to diffusion of IP_3, other mechanisms that involve extracellular messengers, such as extracellular ATP, may contribute to the propagation of astrocyte Ca^{2+} waves (Hassinger et al., 1996; Guthrie et al., 1999; Newman, 2001; Arcuino et al.,

2002). ATP released from active astrocytes appears to be an important mediator of long-range Ca^{2+} signaling, whereas shorter range signaling may be mediated by gap junctions (Scemes and Giaume, 2006). In a model of ATP-mediated propagation of Ca^{2+} signals from cell to cell, activation of metabotropic P2Y receptors (probably of the subtype 1) leads to the IP_3-gated Ca^{2+} release from intracellular stores, which in turn results in the release of ATP to the extracellular space, which then excites neighboring cells. However, there is still some controversy about the mechanisms and the location of ATP release from astrocytes.

8.5 Ca^{2+} Responses to Transmitters and Other Signaling Molecules

Cytosolic Ca^{2+} transients may be triggered by a variety of signaling molecules, including hormones, growth factors, prostaglandins, and many neurotransmitters. As described earlier, astrocytes express a wide variety of functional receptors for most neurotransmitters, and many of them belong to the metabotropic family, i.e., they are coupled to G proteins and second messenger pathways that lead to the IP_3-mediated Ca^{2+} release from intracellular stores. These Ca^{2+} signals can be linked to initiating cellular activity, such as transmitter release, a rise in the K^+ conductance of the cell membrane, mitochondrial energy production, or activation of enzymes (see later).

Astrocytes may also respond with intracellular Ca^{2+} elevations to other signaling molecules such as chemokines, a class of small proteins that bind to G-protein-coupled receptors, which were originally identified as inflammatory mediators of leukocyte chemotaxis, but that have later been shown to be involved also in other functions beyond neuroinflammation. Various types of chemokine receptors are expressed in cells of the nervous system, including astrocytes, where the chemokine stromal cell-derived factor 1 (SDF-1) has been shown to increase astrocytic Ca^{2+} through activation of the receptor CXCR4 (Bajetto et al., 1999), which can lead to glutamate release from astrocytes (Bezzi et al., 2001).

The list of ligands that activate the release of Ca^{2+} from intracellular stores is steadily growing, and some are listed in Table 8.1. Since some of these ligands, such as glutamate or ATP, may be released by astrocytes themselves in a Ca^{2+}-dependent way, autoactivation of astrocytes might be expected. However, if metabotropic receptors linked to the IP_3 second messenger pathway are clustered and/or are restricted to defined areas, or more or less uniformly distributed on the cell surface, which would render astrocytes as *sensors* for ambient concentration of ligands, is not known. So far, specialized regions of receptor clustering, as is found at post-synaptic membrane of neurons, have not yet been described for astrocytes, although such clustering may well occur in cell processes facing synaptic and extrasynaptic release sites.

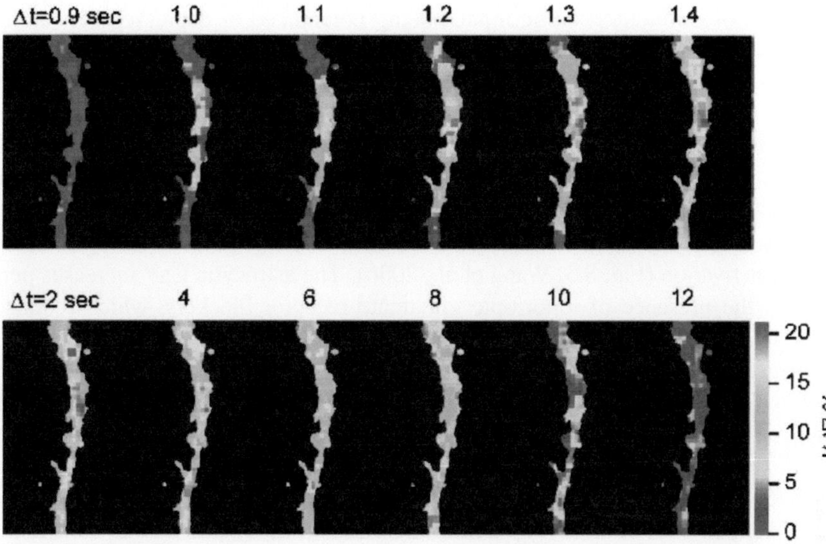

Fig. 8.4 Ca^{2+} signaling in a Bergmann glial process evoked by parallel fiber stimulation. Pseudocolored images of the initiation and intracellular propagation of Ca^{2+} signals in a single process of a Bergmann glial cell, evoked by a parallel fiber burst (50 Hz, 10 pulses). From Beierlein and Regehr (2006). (*See Color Plates*)

8.6 Ca^{2+} Responses to Neuronal Activity

The modulation or initiation of Ca^{2+} signals in astrocytes by synaptic activity has been reported in retina, cerebellum, hippocampus, and cortex, and has been shown to be mediated by glutamate, ATP, GABA, acetylcholine, noradrenaline, nitric oxide, and a variety of other signaling molecules (for reviews see Araque et al., 2001; Fiacco and McCarthy, 2006). Cytosolic Ca^{2+} signals in astrocytes can even be tuned by neuronal activity. The amplitude and duration of Ca^{2+} transients as well as the frequency of Ca^{2+} oscillations in astrocytes may vary depending on the level of synaptic activity. Moderate stimulation of parallel fibers has been shown to evoke Ca^{2+} increases that were restricted to only a few small processes of Bergmann glial cells closely associated with synaptic sites, called microdomains (Grosche et al., 1999), and that could spread within these processes (Beierlein and Regehr, 2006), indicating that glial cells are able to respond locally to neuronal activity (Fig. 8.4). Strong stimulation of parallel fibers, in contrast, evoked Ca^{2+} transients in all processes and the soma (Kulik et al., 1999; Matyash et al., 2001). Studies aimed to investigate the synaptic control of astrocyte Ca^{2+} signaling have recently demonstrated that astrocytes display integrative properties for synaptic information processing, as astrocytes can discriminate between the activity of synaptic terminals belonging to different axon pathways (Perea and Araque, 2005a, b). Ca^{2+} signals in astrocytes evoked by synaptic activity can be bidirectionally modulated by the interaction of different synaptic inputs, being potentiated or depressed depending on the level of synaptic activity. This modulation may also control the intracellular spread of the Ca^{2+} signal in form of waves, which may have important consequences on brain function by regulating the spatial extension of a single astrocyte's impact on different synapses.

Ca^{2+} signaling in astrocytes could even be evoked in vivo by stimulation of sensory neurons. In anaesthetized mice, whisker stimulation resulted in transient Ca^{2+} increases in astrocytes in the barrel cortex mediated by metabotropic glutamate receptor activation (Fig. 8.5; Wang et al., 2006). The astrocytic Ca^{2+} increases persisted in the presence of ionotropic glutamate receptor blockers, which reduced activation of postsynaptic neurons, suggesting that Ca^{2+} signaling during whisker stimulation was independent of the activation of neurons located in the barrel cortex, but was directly evoked by glutamate released from nerve terminals of the afferents (Wang et al., 2006).

Ca^{2+} responses in glial cells were involved in heterosynaptic depression in the hippocampus following high-frequency stimulation of Schaffer collaterals (Pascual et al., 2005; Serrano et al., 2006). Schaffer collateral stimulation lead to the excitation of interneurons that release GABA, thereby activating glial $GABA_B$ receptors and resulting in glial Ca^{2+} transients (Serrano et al., 2006). This lead to the release of ATP from these glial cells, which was degraded to adenosine by ectonucleotidases. The subsequent activation of presynaptic adenosine A_1 receptors on adjacent, unstimulated synapses resulted in heterosynaptic depression. In the retina, a flickering light stimulus evokes an increase in Ca^{2+} transients in Müller glial cells. Addition of adenosine greatly potentiates this light-evoked Ca^{2+} response. The ATP hydrolyzing enzyme apyrase blocks the glial Ca^{2+} responses, indicating that neuron to

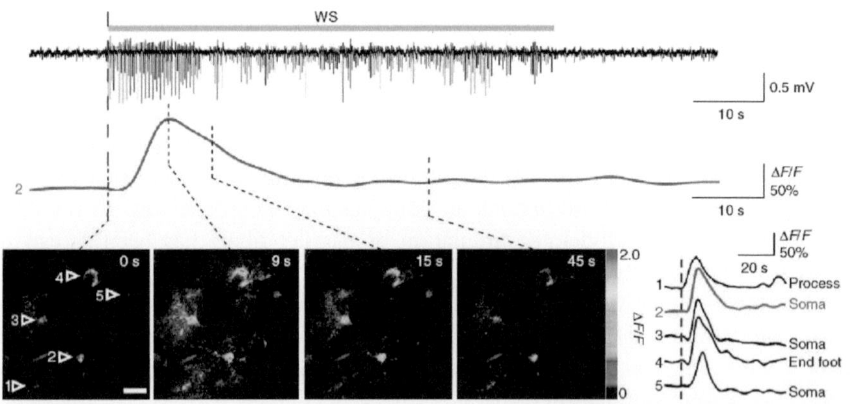

Fig. 8.5 In vivo imaging of Ca^{2+} signaling in astrocytes of the barrel cortex during whisker stimulation. In anaesthetized mice, whisker stimulation (WS) evoked neuronal activity in the barrel cortex as shown by the local field potential (upper trace). Astrocytes responded with a transient increase in Ca^{2+} after a delay of approximately 3 s (lower trace). Ca^{2+} transients could be measured in somata, processes, and end-feet of astrocytes, as indicated in the first image of the time sequence (*lower left*) and the corresponding Ca^{2+} traces (*lower right*). From Wang et al. (2006). (*See Color Plates*)

glia signaling in the retina is mediated by ATP release from neurons and activation of glial purinergic receptors. On the other hand, Müller cell stimulation can evoke a hyperpolarization in neighboring ganglion cells, which is blocked by the A1 adenosine receptor antagonist DPCPX and is reduced by ecto-ATPase and ectonucleotidase inhibitors (Newman, 2005; Metea and Newman, 2006). This suggests that glial cells release ATP, which is converted to adenosine by ectoenzymes, leading to the activation of neuronal adenosine receptors. This demonstrates that there is reciprocal exchange of signals between neurons and glial cells, involving cytosolic Ca^{2+} transients and release of transmitters in both types of cells. ATP and its metabolites seem to play a prominent and diverse role in mediating different kinds of Ca^{2+} signals and their propagation in astrocytes.

8.7 Store-Operated Ca^{2+} Entry and Ca^{2+} Store Refilling

Store-operated Ca^{2+} entry (SOCE) can be triggered by the same mechanisms as those that elicit Ca^{2+} release from intracellular stores and thus lead to Ca^{2+} store depletion. Because SOCE can be evoked by store depletion, and store refilling is impaired when Ca^{2+} entry through SOC channels is prevented, SOCE has been attributed to the refilling of these stores. Although the major function of SOCE is regarded to be the maintenance of Ca^{2+} homeostasis of the ER, SOCE-linked cytosolic Ca^{2+} transients have been linked to several other Ca^{2+}-dependent processes. Among these are vesicular exocytosis, blood vessel contraction and dilation, cell growth and cell death, regulation of enzyme activity, and gene expression (for review see Parekh and Putney, 2005). There have also been reports linking malfunctioning SOCE to Alzheimer's disease (Putney, 2000; Mattson and Chan, 2003), to severe combined immunodeficiency (Partiseti et al., 1994; Feske et al., 2001), and to acute pancreatitis (Parekh, 2000).

SOCE can be detected as a sustained, elevated plateau of the intracellular Ca^{2+} levels that depends on the previous Ca^{2+} depletion of intracellular Ca^{2+} stores, and that requires the presence of extracellular Ca^{2+}. A putative sensor detects the fall in Ca^{2+} content of the stores, which initiates a signaling cascade that relates this information to specific Ca^{2+}-permeable channels in the plasma membrane. The underlying mechanism of this process has remained elusive for over 20 years, but recent studies support a new mechanism for store-operated Ca^{2+} signaling (Fig. 8.6). In this model, depletion of Ca^{2+} stores causes Ca^{2+}-independent phospholipase A_2 (iPLA$_2$) to trigger Ca^{2+} entry in several cell types including astrocytes (Smani et al., 2004; Singaravelu et al., 2006). A diffusible messenger termed Ca^{2+} influx factor (CIF) generated by depleted Ca^{2+} stores has been suggested to act through activation of inducible iPLA$_2$ activity (Smani et al., 2003, 2004). This enzyme is kept catalytically inactive by calmodulin, which binds tightly to iPLA$_2$ under resting conditions. CIF is proposed to provoke the dissociation of calmodulin and iPLA$_2$, which then hydrolyzed the plasma membrane phospholipids to generate lysophospholipids and arachidonic acid. The specific lysophospholipid products of iPLA$_2$, lysophosphotidylcholine

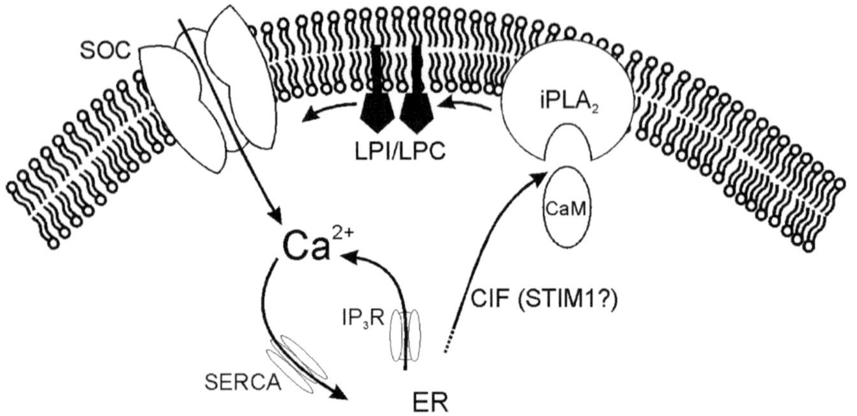

Fig. 8.6 Proposed mechanism of store-operated Ca^{2+} signaling in cerebellar astrocytes. Depletion of intracellular Ca^{2+} stores by IP_3 receptor activation leads to the translocation of a Ca^{2+} influx factor (CIF) to membrane-bound Ca^{2+}-independent phospholipase A_2 (iPLA$_2$), where CIF releases the calmodulin (CaM)-dependent block of iPLA$_2$. After unblocking iPLA$_2$, the enzyme produces phopholipids such as lysophophoinositol (LPI) and lysophosphocholine (LPC), which would lead to the opening of store-operated Ca^{2+} channels (SOC). *SERCA* sarcoplasmic-endoplasmic reticulum Ca^{2+}-ATPase.

and lysophosphotidylinositol, were shown to trigger SOCE in smooth muscle cells (Smani et al., 2003, 2004) as well as in astrocytes (Singaravelu et al., 2006), while arachidonic acid, that activates non-SOCE pathways (Shuttleworth and Thompson, 1999), reduced SOCE in astrocytes (Sergeeva et al., 2003; Yang et al., 2005). The physiological significance of iPLA$_2$-mediated Ca^{2+} influx for maintaining spontaneous Ca^{2+} signaling has been demonstrated in cerebellar astrocytes in situ, where the inhibition of iPLA$_2$ and SOCE channels significantly reduced the occurrence and frequency of spontaneous Ca^{2+} oscillations (Singaravelu et al., 2006).

An exciting advancement in the field of SOCE was provided by the recent identification of the ER Ca^{2+} sensor, STIM1 and the CRAC channel pore subunit, Orai1 (CRACM1). This has enabled significant new insight into the molecular mechanism as to how depletion of Ca^{2+} in the ER opens Ca^{2+} channels in the plasma membrane. It has been demonstrated that STIM1 was closely associated with the ER at rest, with a N-terminal EF-hand domain facing the lumen of the ER, suggesting the ability to sense the Ca^{2+} content of the ER (Roos et al., 2005; Liou et al., 2005). Upon store depletion, STIM1 is rapidly translocated into puncta that accumulated near the plasma membrane and is essential for SOCE (Zhang et al., 2005; Liou et al., 2005; Roos et al., 2005). Thus, STIM1 may act as a Ca^{2+} influx factor. In addition to being the ER Ca^{2+} sensor, STIM1 has also been shown to regulate certain properties of the CRAC channel (Spassova et al., 2005). Orai1, that codes for a plasma membrane protein with four putative transmembrane-spanning domains, was reported to be an essential pore subunit of the CRAC channel by acting as a Ca^{2+} selectivity filter (Vig et al., 2006; Prakriya et al., 2006; Yeromin et al., 2006). Combined overexpression of Orai1 and STIM1 in mammalian cells leads to a marked increase in the CRAC channel activity (Peinelt et al., 2006; Soboloff et al., 2006). Although the involvement of STIM1 and Orai1 may well be of general relevance in cells, the role of either of these proteins for store-operated Ca^{2+} signaling in astrocytes or other glial cells still has to be confirmed.

8.8 Ca^{2+}-Induced Release of Gliotransmitters

There is growing evidence that elevations of cytosolic Ca^{2+} can initiate regulated release of a variety of molecules, including classical transmitters such as glutamate, ATP, and d-serine (Parpura et al, 1994; Coco et al., 2003; Newman, 2003; Mothet et al., 2005; Xu et al., 2007) (also see Chaps. 12 and 16). Large rises of cytosolic Ca^{2+} as induced by flash photolysis of caged Ca^{2+} can trigger the release of these gliotransmitters (Oheim et al., 2006). Different mechanisms of transmitter release from astrocytes have been postulated: vesicle exocytosis (Bezzi et al., 2004; Montana et al., 2006; Oheim et al., 2006), reversed transport (Szatkowski et al., 1990; Rossi et al., 2000), and transmitter efflux through pores such as connexin hemichannels (Ye et al., 2003), purinergic $P2X_7$ receptors (Duan and Neary, 2006), and anion channels (Kimelberg et al., 2006).

Many proteins belonging to the exocytotic machinery, including the glutamate accumulating transporters VGLUT-1 and VGLUT-2, and the soluble *N*-ethyl maleimide-sensitive fusion protein attachment protein receptor (SNARE) proteins cellubrevin, synaptobrevin II, and SNAP-23 have been found in astrocytes (Zhang et al., 2004; Bezzi et al., 2004; Montana et al., 2004). Release of glutamate from cultured astrocytes was reported to be triggered by cytosolic Ca^{2+} elevations (Parpura et al., 1994; Araque et al., 1998; Bezzi et al., 2004) and was reduced when astrocytic Ca^{2+} rises were suppressed by the Ca^{2+} chelator BAPTA (Araque et al., 1998; Zhang et al., 2004). In addition, inhibition of the endogenous SNARE complex by genetic introduction of an exogenous SNARE motif into astrocytes prevented the release of glutamate and ATP (Zhang et al., 2004; Pascual et al., 2005), emphasizing the involvement of the SNARE complex for the release of gliotransmitters. Recently it was reported that glutamate release from hippocampal astrocytes, initiated by P2Y1 receptor-mediated Ca^{2+} signals, strengthens synaptic transmission and hence modulates synaptic tuning in neuronal circuits (Jourdain et al., 2007). Taken together, these findings suggest that transmitters such as glutamate and ATP can be released from astrocytes in a Ca^{2+}-dependent manner by vesicular exocytosis, and possibly also by Ca^{2+}-independent mechanisms. While vesicular release of gliotransmitters may be the predominant mechanism employed by astrocytes as a response to cytosolic Ca^{2+} increases, other mechanisms may mediate release during certain pathophysiological processes, such as cell swelling, which activates anion channels and therefore allows efflux of glutamate or ATP (Kimelberg et al., 2006), and P2X7 receptor activation by high extracellular ATP levels found during CNS injury (Duan and Neary, 2006). A more detailed review about vesicular transmitter release from astrocytes is provided by Montana et al. (2006).

8.9 Functional Significance of Ca^{2+} Signaling

There are numerous processes in the nervous system that may be initiated and/or modulated by cytosolic Ca^{2+} rises in astrocytes, as has been reported for many other cell types. On the other hand, astrocytic Ca^{2+} signaling itself is subject to modulation by neuronal activity, in particular at synapses, and therefore is one of the prime mechanisms by which reciprocal neuron–glia signaling is established (Hirase et al., 2004; Perea and Araque, 2005a, b). One of the most consequential processes initiated by Ca^{2+} rises in astrocytes is the Ca^{2+}-dependent release of transmitters. Some of these gliotransmitters and glia-derived messengers, such as glutamate, ATP, tumor necrosis factor-a, or d-serine, have been shown to modulate neuronal excitability, synaptic transmission, and cerebrovascular microcirculation. Therefore, these gliotransmitters may serve as feedback signals that modulate neuronal activity (see Chap. 15). Astrocyte-induced modulation of synaptic transmission has been observed in cultured cells as well as in tissue slices of several brain areas. Since astrocytes respond to neurotransmitters with Ca^{2+} elevations that can extend to relatively large distances and that induce release of gliotransmitters that modulate neurotransmission, astrocyte Ca^{2+} signaling may represent a form of brain information pathway that establishes a functional link between distant synaptic areas. Some of these Ca^{2+}-dependent processes in astrocytes could also be elicited by photolysis of caged Ca^{2+} compounds by UV illumination (Parpura and Haydon, 2000; Fellin et al., 2004; Kreft et al., 2004; Fiacco and McCarthy, 2004).

Neuronal excitability has also been shown to be modulated by glutamate released from astrocytes both in vitro and in situ. Ca^{2+}-dependent glutamate release from astrocytes may have strong impact in brain pathophysiology, because it can lead to the synchronized activity of clusters of neurons, and may be responsible for the generation of epileptiform activity in neurons. However, the role of astrocytic Ca^{2+} signaling for the development, maintenance, or recovery from pathophysiological conditions is still largely unknown.

Other messenger molecules, which have been reported to be released from astrocytes in a Ca^{2+}-dependent fashion, are nitric oxide (NO), e.g., mediated by the activation of metabotropic purinergic P2Y receptors. This short-lived volatile messenger, which acts both in an ortho- and retrograde way, may be a link to other second messenger cascades in neighbouring neurons and glial cells, such as the cGMP-mediated pathway. Moreover, NO may enhance store-operated Ca^{2+} entry (SOCE) and thereby feeds back on the cellular competence to generate Ca^{2+} signals due to repeated Ca^{2+} release from intracellular stores (Li et al., 2003). In cultured astrocytes NO caused Ca^{2+}-dependent glutamate release from inflammatory activated cells (Bal-Price et al., 2002). Thus, Ca^{2+} signals in astrocytes may be involved both in physiological and pathological processes in the brain.

Ca^{2+} signaling in astrocytes has also been associated with the control of cerebral blood flow. Ca^{2+} transients and Ca^{2+} oscillations are propagated to the astrocytic end-feet, which ensheath part of the endothelial layer of blood capillaries, and can elicit vasodilation and vasoconstriction by the release of lipid metabolites (Zonta et al., 2003; Parri and Crunelli, 2003; Mulligan and MacVicar, 2004; Takano et al.,

2006; Metea and Newman, 2006). Inhibition of cyclooxygenase-1 activity blocked astrocyte-triggered vasodilation, supporting the notion that Ca^{2+}-dependent release of prostaglandins from astrocytes mediates the control local blood flow. This modulation may be initiated by neuronal activity in restricted brain areas, by evoking astrocyte Ca^{2+} signals that are propagated toward blood vessels and in turn mediate the control of microcirculation. Astrocytic Ca^{2+} signals have been reported to cause either vasoconstriction and vasodilation, and this may be decided by the lipid products involved. As the supply of oxygen and glucose in the brain is affected, these mechanisms may well be an important glial link to the maintenance of brain energy metabolism.

Furthermore, cytosolic Ca^{2+} rises in astrocytes may gate or modulate ion channels in the cell membrane, in particular K^+ channels, which may be involved in K^+ clearance from the extracellular space following neuronal activity (see Chap. 6). Whether the Ca^{2+} signals recorded at low extracellular K^+ concentration in astrocytes (Dallwig et al., 2000; Härtel et al., 2007) are of physiological and pathological relevance requires further analysis.

The list of cellular functions mediated by Ca^{2+} signals in astrocytes will certainly be extended in the years to come. It is the *regional* control of cellular functions, which may be regulated or modulated by Ca^{2+} signals in a unique way. By invading different cellular processes of astrocytes, by being regenerative, and by being a versatile mode of signaling, this form of *glial excitation* has many assets for allowing a complex dialogue between glial cells and neurons, and may contribute critically to information processing and metabolism in the brain.

8.10 Summary and Conclusion

Cytosolic Ca^{2+} signals in astrocytes can be spontaneous or evoked, and are often initiated by activation of metabotropic receptors, resulting in Ca^{2+} release from intracellular Ca^{2+} stores. While voltage-dependent Ca^{2+} influx is rare in astrocytes, the refilling of the intracellular Ca^{2+} stores requires store-operated Ca^{2+} entry, a major influx pathway of Ca^{2+} into astrocytes. Astrocyte Ca^{2+} signals are generated in form of transients or oscillations, which can be evoked by neurotransmitters, hormones, chemokines, lipids, and growth factors, often in response to neuronal activity. These Ca^{2+} signals are propagated along cellular processes and can travel from the point of origin across the tissue as Ca^{2+} waves. When propagated into astrocytic end-feet, they may release cyclooxygenase products to control cerebral blood flow, or modulate neuronal excitability and glia–neuron interactions. The release of transmitters and other molecules elicited by Ca^{2+} signaling in astrocytes may be key mechanisms to allow a complex dialogue between neurons and glial cells. Although we have learned a lot about the phenomenology of Ca^{2+} signals in astrocytes, and about the potential function it may have, we still lack knowledge about the specificity of astrocytic Ca^{2+} signals with respect to the pathophysiological conditions under which they are generated, and the cell function they may elicit in a given situation. It seems clear that these Ca^{2+} signals are instrumental for many processes in astrocytes and for their interaction with other glial cells and neurons.

Acknowledgments Our studies on this topic have been supported by grants from the Deutsche Forschungsgemeinschaft (SFB 530, GRK 845), which we gratefully acknowledge.

References

Allen NJ, Attwell D (2004) The effect of simulated ischaemia on spontaneous GABA release in area CA1 of the juvenile rat hippocampus. *J Physiol* 561:485–498.

Araque A, Sanzgiri RP, Parpura V, Haydon PG (1998) Calcium elevation in astrocytes causes an NMDA receptor-dependent increase in the frequency of miniature synaptic currents in cultured hippocampal neurons. *J Neurosci* 18:6822–6829.

Araque A, Carmignoto G, Haydon PG (2001) Dynamic signaling between astrocytes and neurons. *Annu Rev Physiol* 63:795–813.

Arcuino G, Lin JH, Takano T, Liu C, Jiang L, Gao Q, Kang J, Nedergaard M (2002) Intercellular calcium signaling mediated by point-source burst release of ATP. *Proc Natl Acad Sci USA* 99:9840–9845.

Backus KH, Berger T (1995) Developmental variation of the permeability to Ca^{2+} of AMPA receptors in presumed hilar glial precursor cells. *Pflügers Arch* 431:244–252.

Bajetto A, Bonavia R, Barbero S, Piccioli P, Costa A, Florio T, Schettini G (1999) Glial and neuronal cells express functional chemokine receptor CXCR4 and its natural ligand stromal cell-derived factor 1. *J Neurochem* 73:2348–2357.

Bal-Price A, Moneer Z, Brown GC (2002) Nitric oxide induces rapid, calcium-dependent release of vesicular glutamate and ATP from cultured rat astrocytes. *Glia* 40:312–323.

Beck A, Zur Nieden R, Schneider HP, Deitmer JW (2004) Calcium release from intracellular stores in rodent astrocytes and neurons *in situ*. *Cell Calcium* 35:47–58.

Beierlein M, Regehr WG (2006) Brief bursts of parallel fiber activity trigger calcium signals in Bergmann glia. *J Neurosci* 26:6958–6967.

Berridge MJ (2006) Calcium microdomains: organization and function. *Cell Calcium* 40:405–412.

Bezzi P, Domercq M, Brambilla L, Galli R, Schols D, De Clercq E, Vescovi A, Bagetta G, Kollias G, Meldolesi J, Volterra A (2001) CXCR4-activated astrocyte glutamate release via TNFalpha: amplification by microglia triggers neurotoxicity. *Nat Neurosci* 4:702–710.

Bezzi P, Gundersen V, Galbete JL, Seifert G, Steinhauser C, Pilati E, Volterra A (2004) Astrocytes contain a vesicular compartment that is competent for regulated exocytosis of glutamate. *Nat Neurosci* 7:613–620.

Bigini P, Gardoni F, Barbera S, Cagnotto A, Fumagalli E, Longhi A, Corsi MM, Di Luca M, Mennini T (2006) Expression of AMPA and NMDA receptor subunits in the cervical spinal cord of wobbler mice. *BMC Neurosci* 7:71.

Burnashev N, Khodorova A, Jonas P, Helm PJ, Wisden W, Monyer H, Seeburg PH, Sakmann B (1992) Calcium-permeable AMPA-kainate receptors in fusiform cerebellar glial cells. *Science* 256:1566–1570.

Carmignoto G, Pasti L, Pozzan T (1998) On the role of voltage-dependent calcium channels in calcium signaling of astrocytes in situ. *J Neurosci* 18:4637–4645.

Carmignoto G (2000) Reciprocal communication systems between astrocytes and neurones. *Prog Neurobiol* 62:561–581.

Coco S, Calegari F, Pravettoni E, Pozzi D, Taverna E, Rosa P, Matteoli M, Verderio C (2003) Storage and release of ATP from astrocytes in culture. *J Biol Chem* 278:1354–1362.

Cornell-Bell AH, Finkbeiner SM, Cooper MS, Smith SJ (1990) Glutamate induces calcium waves in cultured astrocytes: long-range glial signaling. *Science* 247:470–473.

Cotrina ML, Kang J, Lin JH, Bueno E, Hansen TW, He L, Liu Y, Nedergaard M (1998) Astrocytic gap junctions remain open during ischemic conditions. *J Neurosi* 18:2520–2537.

Czubayko U, Reiser G (1995) [Ca^{2+}]i oscillations in single rat glioma cells induced by thrombin through activation of cell surface receptors. *Neuroreport* 6:1249–1252.

Dallwig R, Vitten H, Deitmer JW (2000) A novel barium-sensitive calcium influx into rat astrocytes at low external potassium. *Cell Calcium* 28:247–259.

Dave V, Gordon GW, McCarthy KD (1991) Cerebral type 2 astroglia are heterogeneous with respect to their ability to respond to neuroligands linked to calcium mobilization. *Glia* 4:440–447.

Deitmer JW, Verkhratsky A, Lohr C (1998) Calcium signalling in glial cells. *Cell Calcium* 24:405–416.

Delumeau JC, Petitet F, Cordier J, Glowinski J, Premont J (1991) Synergistic regulation of cytosolic Ca^{2+} concentration in mouse astrocytes by NK1 tachykinin and adenosine agonists. *J Neurochem* 57:2026–2035.

Duan S, Neary JT (2006) P2X(7) receptors: properties and relevance to CNS function. *Glia* 54:738–746.

Duffy S, MacVicar BA (1994) Potassium-dependent calcium influx in acutely isolated hippocampal astrocytes. *Neuroscience* 61:51–61.

Duffy S, MacVicar BA (1995) Adrenergic calcium signaling in astrocyte networks within the hippocampal slice. *J Neurosci* 15:5535–5550.

Enkvist MO, Holopainen I, Akerman KE (1989) Glutamate receptor-linked changes in membrane potential and intracellular Ca^{2+} in primary rat astrocytes. *Glia* 2:397–402.

Fellin T, Pascual O, Gobbo S, Pozzan T, Haydon PG, Carmignoto G (2004) Neuronal synchrony mediated by astrocytic glutamate through activation of extrasynaptic NMDA receptors. *Neuron* 43:729–743.

Feske S, Giltnane J, Dolmetsch R, Staudt LM, Rao A (2001) Gene regulation mediated by calcium signals in T lymphocytes. *Nat Immunol* 2:316–324.

Fiacco TA, McCarthy KD (2004) Intracellular astrocyte calcium waves in situ increase the frequency of spontaneous AMPA receptor currents in CA1 pyramidal neurons. *J Neurosci* 24:722–732.

Fiacco TA, McCarthy KD (2006) Astrocyte calcium elevations: properties, propagation, and effects on brain signaling. *Glia* 54:676–690.

Fields RD, Stevens-Graham B (2002) New insights into neuron-glia communication. *Science* 298:556–562.

Finkbeiner S (1992) Calcium waves in astrocytes-filling in the gaps. *Neuron* 8:1101–1108.

Gimpl G, Walz W, Ohlemeyer C, Kettenmann H (1992) Bradykinin receptors in cultured astrocytes from neonatal rat brain are linked to physiological responses. *Neurosci Lett* 144:139–142.

Gimpl G, Kirchhoff F, Lang RE, Kettenmann H (1993) dentification of neuropeptide Y receptors in cultured astrocytes from neonatal rat brain. *J Neurosci Res* 34:198–205.

Grosche J, Matyash V, Moller T, Verkhratsky A, Reichenbach A, Kettenmann H (1999) Microdomains for neuron–glia interaction: parallel fiber signaling to Bergmann glial cells. *Nat Neurosci* 2:139–143.

Guthrie PB, Knappenberger J, Segal M, Bennett MV, Charles AC, Kater SB (1999) ATP released from astrocytes mediates glial calcium waves. *J Neurosci* 19:520–528.

Härtel K, Singaravelu K, Kaiser M, Neusch C, Hülsmann S, Deitmer JW (2007) Calcium influx mediated by the inwardly rectifying K+ channel Kir4.1 (KCNJ10) at low external K+ concentration. *Cell Calcium* 42:271–280.

Hassinger TD, Guthrie PB, Atkinson PB, Bennett MV, Kater SB (1996) An extracellular signaling component in propagation of astrocytic calcium waves. *Proc Natl Acad Sci USA* 93:13268–13273.

Hirase H, Qian L, Bartho P, Buzsaki G (2004) Calcium dynamics of cortical astrocytic networks in vivo. *PLoS Biol* 2:E96.

Hirase H, Creso J, Singleton M, Bartho P, Buzsaki G (2004) Two-photon imaging of brain pericytes in vivo using dextran-conjugated dyes. *Glia* 46:95–100.

Ho C, Hicks J, Salter MW (1995) A novel P2-purinoceptor expressed by a subpopulation of astrocytes from the dorsal spinal cord of the rat. *Br J Pharmacol* 116:2909–2918.

Hoth M, Penner R (1992) Depletion of intracellular calcium stores activates a calcium current in mast cells. *Nature* 355:353–356.

Inagaki N, Fukui H, Ito S, Yamatodani A, Wada H (1991) Single type-2 astrocytes show multiple independent sites of Ca^{2+} signaling in response to histamine. *Proc Natl Acad Sci USA* 88:4215–4219.

James G, Butt AM (2001) P2X and P2Y purinoreceptors mediate ATP-evoked calcium signalling in optic nerve glia in situ. *Cell Calcium* 30:251–259.

Jensen AM, Chiu SY (1991) Differential intracellular calcium responses to glutamate in type 1 and type 2 cultured brain astrocytes. *J Neurosci* 11:1674–1684.

Jourdain P, Bergersen LH, Bhaukaurally K, Bezzi P, Santello M, Domercq M, Matute C, Tonello F, Gundersen V, Volterra A (2007) Glutamate exocytosis from astrocytes controls synaptic strength. *Nat Neurosci* 10:331–339.

Jung S, Pfeiffer F, Deitmer JW (2000) Histamine-induced calcium entry in rat cerebellar astrocytes: evidence for capacitative and non-capacitative mechanisms. *J Physiol* 61:549–561.

Jurzak M, Muller AR, Gerstberger R (1995) Characterization of vasopressin receptors in cultured cells derived from the region of rat brain circumventricular organs. *Neuroscience* 65:1145–1159.

Kanemaru K, Okubo Y, Hirose K, Iino M (2007) Regulation of neurite growth by spontaneous Ca^{2+} oscillations in astrocytes. *J Neurosci* 27:8957–8966.

Kastritsis CH, Salm AK, McCarthy K (1992) Stimulation of the P2Y purinergic receptor on type 1 astroglia results in inositol phosphate formation and calcium mobilization. *J Neurochem* 58:1277–1284.

Kimelberg HK, Macvicar BA, Sontheimer H (2006) Anion channels in astrocytes: biophysics, pharmacology, and function. *Glia* 54:747–757.

Kreft M, Stenovec M, Rupnik M, Grilc S, Krzan M, Potokar M, Pangrsic T, Haydon PG, Zorec R (2004) Properties of Ca^{2+}-dependent exocytosis in cultured astrocytes. *Glia* 46:437–445.

Kulik A, Haentzsch A, Luckermann M, Reichelt W, Ballanyi K (1999) Neuron–glia signaling via 1 adrenoceptor-mediated Ca^{2+} release in Bergmann glial cells in situ. *J Neurosci* 19:8401–8408.

Lalo U, Pankratov Y, Kirchhoff F, North RA, Verkhratsky A (2006) NMDA receptors mediate neuron-to-glia signaling in mouse cortical astrocytes. *J Neurosci* 26:2673–2683.

Li N, Sul JY, Haydon PG (2003) A calcium-induced calcium influx factor, nitric oxide, modulates the refilling of calcium stores in astrocytes. *J Neurosci* 23:10302–10310.

Liou J, Kim ML, Heo WD, Jones JT, Myers JW, Ferrell JE, Meyer T (2005) STIM is a Ca^{2+} sensor essential for Ca^{2+} store depletion triggered Ca^{2+} influx. *Curr Biol* 15:1235–1241.

Liu QY, Lai FA, Rousseau E, Jones RV, Meissner G (1989) Multiple conductance states of the purified calcium release channel complex from skeletal sarcoplasmic reticulum. *Biophys J* 55:415–424.

Lo KJ, Luk HN, Chin TY, Chueh SH (2002) Store depletion-induced calcium influx in rat cerebellar astrocytes. *Br J Pharmacol* 135:1383–1392.

Lohr C, Deitmer JW (1999) Dendritic calcium transients in the leech giant glial cell. *Glia* 26:109–118.

Lohr C, Heil JE, Deitmer JW (2005) Blockage of voltage-gated calcium signalling impairs migration of glial cells in vivo. *Glia* 50:198–211.

Lohr C, Deitmer JW (2006) Calcium signaling in invertebrate glial cells. *Glia* 54:642–649.

Mattson MP, Chan SL (2003) Neuronal and glial calcium signaling in Alzheimer's disease. *Cell Calcium* 34:385–397.

Matyash V, Filippov V, Mohrhagen K, Kettenmann H (2001) Nitric oxide signals parallel fiber activity to Bergmann glial cells in the mouse cerebellar slice. *Mol Cell Neurosci* 18:664–670.

Meissner G, Henderson JS (1987) Rapid calcium release from cardiac sarcoplasmic reticulum vesicles is dependent on Ca^{2+} and is modulated by Mg2+, adenine nucleotide, and calmodulin. *J Biol Chem* 262:3065–3073.

Metea MR, Newman EA (2006) Glial cells dilate and constrict blood vessels: a mechanism of neurovascular coupling. *J Neurosci* 26:2862–2870.

Metea MR, Newman EA (2006) Calcium signaling in specialized glial cells. *Glia* 54:650–655.

Montana V, Malarkey EB, Verderio C, Matteoli M, Parpura V (2006) Vesicular transmitter release from astrocytes. *Glia* 54:700–715.

Montana V, Ni Y, Sunjara V, Hua X, Parpura V (2004) Vesicular glutamate transporter dependent glutamate release from astrocytes. *J Neurosci* 24:2633–2642.
Morita M, Higuchi C, Moto T, Kozuka N, Susuki J, Itofusa R, Yamashita J, Kudo Y (2003) Dual regulation of calcium oscillation in astrocytes by growth factors and pro-inflammatory cytokines via the mitogen-activated protein kinase cascade. *J Neurosci* 23:10944–10952.
Mothet JP, Pollegioni L, Ouanounou G, Martineau M, Fossier P, Baux G (2005) Glutamate receptor activation triggers a calcium-dependent and SNARE protein-dependent release of the gliotransmitter d-serine. *Proc Natl Acad Sci USA* 102:5606–5611.
Müller T, Moller T, Berger T, Schnitzer J, Kettenmann H (1992) Calcium entry through kainate receptors and resulting potassium-channel blockade in Bergmann glial cells. *Science* 256:1563–1566.
Mulligan SJ, MacVicar BA (2004) Calcium transients in astrocyte endfeet cause cerebrovascular constrictions. *Nature* 431:195–199.
Nedergaard M (1994) Direct signaling from astrocytes to neurons in cultures of mammalian brain cells. *Science* 263:1768–1771.
Nett WJ, Oloff SH, McCarthy KD (2002) Hippocampal astrocytes in situ exhibit calcium oscillations that occur independent of neuronal activity. *J Neurophysiol* 87:528–537.
Newman EA, Zahs KR (1997) Calcium waves in retinal glial cells. *Science* 275:844–847.
Newman EA (2001) Propagation of intercellular calcium waves in retinal astrocytes and Muller cells. *J Neurosci* 21:2215–2223.
Newman EA (2003) Glial cell inhibition of neurons by release of ATP. *J Neurosci* 23:1659–1666.
Newman EA (2005) Calcium increases in retinal glial cells evoked by light-induced neuronal activity. *J Neurosci* 25:5502–5510.
Nilsson M, Hansson E, Ronnback L (1991) Adrenergic and 5-HT2 receptors on the same astroglial cell. A microspectrofluorimetric study on cytosolic Ca^{2+} responses in single cells in primary culture. *Brain Res Dev Brain Res* 63:33–41.
Nilsson M, Eriksson PS, Ronnback L, Hansson E (1993) GABA induces Ca^{2+} transients in astrocytes. *Neuroscience* 54:605–614.
Oheim M, Kirchhoff F, Stühmer W (2006) Calcium microdomains in regulated exocytosis. *Cell Calcium* 40:423–439.
Parpura V, Haydon PG (2000) Physiological astrocytic calcium levels stimulate glutamate release to modulate adjacent neurons. *Proc Natl Acad Sci USA* 97:8629–8634.
Parekh AB, Putney JW Jr (2005) Store-operated calcium channels. *Physiol Rev* 85:757–810.
Parekh AB (2000) Calcium signaling and acute pancreatitis: specific response to a promiscuous messenger. *Proc Natl Acad Sci USA* 97:12933–12934.
Parpura V, Basarsky TA, Liu F, Jeftinija K, Jeftinija S, Haydon PG (1994) Glutamate-mediated astrocyte-neuron signalling. *Nature* 369:744–747.
Parri HR, Crunelli V (2001) Pacemaker calcium oscillations in thalamic astrocytes in situ. *Neuroreport* 12:3897–3900.
Parri HR, Crunelli V (2003) The role of Ca^{2+} in the generation of spontaneous astrocytic Ca^{2+} oscillations. *Neuroscience* 120:979–992.
Partiseti M, Le Deist F, Hivroz C, Fischer A, Korn H, Choquet D (1994) The calcium current activated by T cell receptor and store depletion in human lymphocytes is absent in a primary immunodeficiency. *J Biol Chem* 269:32327–32335.
Pascual O, Casper KB, Kubera C, Zhang J, Revilla-Sanchez R, Sul JY, Takano H, Moss SJ, McCarthy K, Haydon PG (2005) Astrocytic purinergic signaling coordinates synaptic networks. *Science* 310:113–116.
Pasti L, Pozzan T, Carmignoto G (1995) Long-lasting changes of calcium oscillations in astrocytes. A new form of glutamate-mediated plasticity. *J Biol Chem* 270:15203–15210.
Pasti L, Zonta M, Pozzan T, Vicini S, Carmignoto G (2001) Cytosolic calcium oscillations in astrocytes may regulate exocytotic release of glutamate. *J Neurosci* 21:477–484.
Peakman MC, Hill SJ (1995) Adenosine A1 receptor-mediated changes in basal and histamine-stimulated levels of intracellular calcium in primary rat astrocytes. *Br J Pharmacol* 115:801–810.

Peinelt C, Vig M, Koomoa DL, Beck A, Nadler MJ, Koblan-Huberson M, Lis A, Fleig A, Penner R, Kinet JP (2006) Amplification of CRAC current by STIM1 and CRACM1 (Orai1). *Nat Cell Biol* 8:771–773.

Perea G, Araque A (2005a) Properties of synaptically evoked astrocyte calcium signal reveal synaptic information processing by astrocytes. *J Neurosci 25:2192–2203. Erratum in J Neurosci* 25:3022.

Perea G, Araque A (2005b) Glial calcium signaling and neuron–glia communication. *Cell Calcium* 38:375–382.

Peters O, Schipke CG, Hashimoto Y, Kettenmann H (2003) Different mechanisms promote astrocyte Ca^{2+} waves and spreading depression in the mouse neocortex. *J Neurosci* 23:9888–9896.

Porter JT, McCarthy KD (1995a) Adenosine receptors modulate $[Ca^{2+}]i$ in hippocampal astrocytes in situ. *J Neurochem* 65:1515–1523.

Porter JT, McCarthy KD (1995b) GFAP-positive hippocampal astrocytes in situ respond to glutamatergic neuroligands with increases in $[Ca^{2+}]i$. *Glia* 13:101–112.

Prakriya M, Feske S, Gwack Y, Srikanth S, Rao A, Hogan PG (2006) Orai1 is an essential pore subunit of the CRAC channel. *Nature* 443:230–233.

Putney JW Jr (2000) Presenilins, Alzheimer's disease, and capacitative calcium entry. *Neuron* 27:411–412.

Roos J, DiGregorio PJ, Yeromin AV, Ohlsen K, Lioudyno M, Zhang S, Safrina O, Kozak JA, Wagner SL, Cahalan MD, Velicelebi G, Stauderman KA (2005) STIM1, an essential and conserved component of store-operated Ca^{2+} channel function. *J Cell Biol* 169:435–445.

Rossi DJ, Oshima T, Attwell D (2000) Glutamate release in severe brain ischaemia is mainly by reversed uptake. *Nature* 403:316–321.

Salm AK, McCarthy KD (1990) Norepinephrine-evoked calcium transients in cultured cerebral type 1 astroglia. *Glia* 3:529–538.

Scemes E, Giaume C (2006) Astrocyte calcium waves: what they are and what they do. *Glia* 54:716–725.

Scemes E, Dermietzel R, Spray DC (1998) Calcium waves between astrocytes from Cx43 knockout mice. *Glia* 24:65–73.

Schipke CG, Ohlemeyer C, Matyash M, Nolte C, Kettenmann H, Kirchhoff F (2001) Astrocytes of the mouse neocortex express functional *N*-methyl-d-aspartate receptors. *FASEB J* 15: 1270–1272.

Schipke CG, Boucsein C, Ohlemeyer C, Kirchhoff F, Kettenmann H (2002) Astrocyte Ca^{2+} waves trigger responses in microglial cells in brain slices. *FASEB J* 16:255–257.

Seifert G, Weber M, Schramm J, Steinhauser C (2003) Changes in splice variant expression and subunit assembly of AMPA receptors during maturation of hippocampal astrocytes. *Mol Cell Neurosci* 22:248–258.

Sergeeva M, Strokin M, Wang H, Ubl JJ, Reiser G (2003) Arachidonic acid in astrocytes blocks Ca^{2+} oscillations by inhibiting store-operated Ca^{2+} entry, and causes delayed Ca^{2+} influx. *Cell Calcium* 33:283–292

Serrano A, Haddjeri N, Lacaille JC, Robitaille R (2006) GABAergic network activation of glial cells underlies hippocampal heterosynaptic depression. *J Neurosci* 26:5370–5382.

Shao Y, McCarthy KD (1995) Receptor-mediated calcium signals in astroglia: multiple receptors, common stores and all-or-nothing responses. *Cell Calcium* 17:187–196.

Sharma G, Vijayaraghavan S (2001) Nicotinic cholinergic signaling in hippocampal astrocytes involves calcium-induced calcium release from intracellular stores. *Proc Natl Acad Sci USA* 98:4148–4153.

Shelton MK, McCarthy KD (2000) Hippocampal astrocytes exhibit Ca^{2+}-elevating muscarinic cholinergic and histaminergic receptors in situ. *J Neurochem* 74:555–563.

Shih AY, Erb H, Sun X, Toda S, Kalivas PW, Murphy TH (2006) Cystine/glutamate exchange modulates glutathione supply for neuroprotection from oxidative stress and cell proliferation. *J Neurosci* 26(41):10514–10523.

Shuttleworth TJ, Thompson JL (1999) Discriminating between capacitative and arachidonate-activated Ca^{2+} entry pathways in HEK293 cells. *J Biol Chem* 274:31174–31178.

Simpson PB, Russell JT (1997) Role of sarcoplasmic/endoplasmic-reticulum Ca^{2+}-ATPases in mediating Ca^{2+} waves and local Ca^{2+}-release microdomains in cultured glia. *Biochem J* 325:239–247.

Singaravelu K, Lohr C, Deitmer JW (2006) Regulation of store-operated calcium entry by calcium-independent phospholipase A2 in rat cerebellar astrocytes. *J Neurosci* 26:9579–9592.

Sitsapesan R, McGarry SJ, Williams AJ (1995) Cyclic ADP-ribose, the ryanodine receptor and Ca^{2+} release. *Trends Pharmacol Sci* 16:386–391.

Smani T, Zakharov SI, Leno E, Csutora P, Trepakova ES, Bolotina VM (2003) Ca^{2+}-independent phospholipase A2 is a novel determinant of store-operated Ca^{2+} entry. *J Biol Chem* 278:11909–11915.

Smani T, Zakharov SI, Leno E, Csutora P, Trepakova ES, Bolotina VM (2004) A novel mechanism for the store-operated calcium influx pathway. *Nat Cell Biol* 6:113–120.

Soboloff J, Spassova MA, Tang XD, Hewavitharana T, Xu W, Gill DL (2006) Orail and STIM reconstitute store-operated calcium channel function. *J Biol Chem* 281:20661–20665.

Spassova MA, Soboloff J, He LP, Xu W, Dziadek MA, Gill DL (2005) STIM1 has a plasma membrane role in the activation of store-operated Ca^{2+} channels. *Proc Natl Acad Sci USA* 103:4040–4045.

Spät A (2006) Calcium microdomains and the fine control of cell function: an introduction. *Cell Calcium* 40:403–404.

Stiene-Martin A, Mattson MP, Hauser KF (1993) Opiates selectively increase intracellular calcium in developing type-1 astrocytes: role of calcium in morphine-induced morphologic differentiation. *Brain Res Dev Brain Res* 76:189–196.

Suadicani SO, Brosnan CF, Scemes E (2006) P2X7 receptors mediate ATP release and amplification of astrocytic intercellular Ca^{2+} signaling. *J Neurosci* 26:1378–1385.

Szatkowski M, Barbour B, Attwell D (1990) Non-vesicular release of glutamate from glial cells by reversed electrogenic glutamate uptake. *Nature* 348:443–446.

Takano T, Tian GF, Peng W, Lou N, Libionka W, Han X, Nedergaard M (2006) Astrocyte-mediated control of cerebral blood flow. *Nat Neurosci* 9:260–267.

Verkhratsky A, Shmigol A (1996) Calcium-induced calcium release in neurones. *Cell Calcium* 19:1–14.

Verkhratsky A, Orkland RK, Kettenmann H (1998) Glial calcium: homeostasis and signaling function. *Physiol Rev* 78:99–141.

Vig M, Peinelt C, Beck A, Koomoa DL, Rabah D, Koblan-Huberson M, Kraft S, Turner H, Fleig A, Penner R, Kinet JP (2006) CRACM1 is a plasma membrane protein essential for store-operated Ca^{2+} entry. *Science* 312:1220–1223.

Walz W, Gimpl G, Ohlemeyer C, Kettenmann H (1994) Extracellular ATP-induced currents in astrocytes: involvement of a cation channel. *J Neurosci Res* 38:12–18.

Wang D, Martens JR, Posner P, Sumners C, Gelband CH (1996) Angiotensin II regulation of intracellular calcium in astroglia cultured from rat hypothalamus and brainstem. *J Neurochem* 67:996–1004.

Wang X, Lou N, Xu Q, Tian GF, Peng WG, Han X, Kang J, Takano T, Nedergaard M (2006) Astrocytic Ca^{2+} signaling evoked by sensory stimulation in vivo. *Nat Neurosci* 9:816–823.

Xu J, Peng H, Kang N, Zhao Z, Lin JHC, Stanton PK, Kang J (2007) Glutamate-induced exocytosis of glutamate from astrocytes. *J Biol Chem* 282:24185–24197.

Yang KT, Chen WP, Chang WL, Su MJ, Tsai KL (2005) Arachidonic acid inhibits capacitative Ca^{2+} entry and activates non-capacitative Ca^{2+} entry in cultured astrocytes. *Biochem Biophys Res Commun* 331:603–613.

Ye ZC, Rothstein JD, Sontheimer H (1999) Compromised glutamate transport in human glioma cells: reduction-mislocalization of sodium-dependent glutamate transporters and enhanced activity of cystine-glutamate exchange. *J Neurosci* 19:10767–10777.

Ye ZC, Wyeth MS, Baltan-Tekkok S, Ransom BR (2003) Functional hemichannels in astrocytes: a novel mechanism of glutamate release. *J Neurosci* 23:3588–3596.

Yeromin AV, Zhang SL, Jiang W, Yu Y, Safrina O, Cahalan MD (2006) Molecular identification of the CRAC channel by altered ion selectivity in a mutant of Orai. *Nature* 443:226–229.

Zhang Q, Pangrsic T, Kreft M, Krzan M, Li N, Sul JY, Halassa M, Van Bockstaele E, Zorec R, Haydon PG (2004) Fusion-related release of glutamate from astrocytes. *J Biol Chem* 279:12724–12733.

Zhang SL, Yu Y, Roos J, Kozak JA, Deerinck TJ, Ellisman MH, Stauderman KA, Cahalan MD (2005) STIM1 is a Ca^{2+} sensor that activates CRAC channels and migrates from the Ca^{2+} store to the plasma membrane. *Nature* 437:902–905.

Zonta M, Angulo MC, Gobbo S, Rosengarten B, Hossmann KA, Pozzan T, Carmignoto G (2003) Neuron-to-astrocyte signaling is central to the dynamic control of brain microcirculation. *Nat Neurosci* 6:43–50.

Zur Nieden R, Deitmer JW (2006) The role of metabotropic glutamate receptors for the generation of calcium oscillations in rat hippocampal astrocytes in situ. *Cereb Cortex* 16:676–687.

Abbreviations

2-APB	2-Aminoethoxydiphenyl borate
ADP	Adenosine diphosphate
AMPA	Alpha-amino-3-hydroxy-5-methyl-4-isoxazolepropionic acid
ATP	Adenosine 5′-triphosphate
BAPTA	1,2-bis(o-aminophenoxy)ethane-*N,N,N′,N′*-tetraacetic acid
CIF	Ca^{2+} influx factor
CRAC	Ca^{2+} release-activated channel
DAG	Diacylglycerol
DPCPX	8-Cyclopentyl-1,3-dipropylxanthine
EAAT	Excitatory amino acid transporter
ER	Endoplasmic reticulum
GABA	Gamma-aminobutyric acid
IP_3	Inositol-1,4,5-trisphosphate
$iPLA_2$	Ca^{2+}-independent phopholipase A_2
NMDA	*N*-methyl-D-aspartate
P2X	Ionotropic purinoceptor
P2Y	Metabotropic purinoceptor
PLC	Phospholipase C
SERCA	Sarcoplasmic-endoplasmic reticulum Ca^{2+}-ATPase
SNARE	Soluble *N*-ethyl maleimide-sensitive fusion protein attachment protein receptor
SOCE	Store-operated Ca^{2+} entry
STIM1	Stromal interaction molecule 1
VGLUT	Vesicular glutamate transporter

Chapter 9
Astrocytes in Control of the Biophysical Properties of the Extracellular Space

Lydia Vargova and Eva Sykova

Contents

9.1 Extracellular Space (ECS) and Extrasynaptic Transmission

The extracellular space of the central nervous system (CNS) represents the microenvironment of the cellular elements: neurons and glia. Its composition, chemical and biophysical properties are crucial for the proper functioning of neurons and signal transmission. The first estimates of ECS volume that appeared in the second half of the last century were based on electron microscopy. However, the true values were altered by conventional preparation procedures, leading to the erroneous conclusion that the ECS represents less than 5% of the total brain tissue volume (Villegas and Fernandez, 1966). Later, it was shown by a number of different

E. Sykova
Department of Neuroscience, Institute of Experimental Medicine AS CR, Prague, Czech Republic
sykova@biomed.cas.cz

V. Parpura and P.G. Haydon (eds.), *Astrocytes in (Patho)Physiology of the Nervous System*, 225
DOI: 10.1007/978-0-387-79492-1_9,

techniques (electron microscopy that preserves the ECS, radiotracers and ion diffusion methods) that under physiological conditions the average ECS volume is about 15–25% of the total adult brain volume [for review see (Nicholson and Sykova, 1998)]. Not only the volume, but also the composition of the ECS changes dynamically during neuronal activity as well as in pathological states. Besides changes in the extracellular matrix, a number of neuroactive substances such as ions, neurotransmitters, neurohormones and metabolites are released in the ECS, diffuse through it and affect nerve and glial cells, which are distant from the release site (Sykova, 1997, 2004a, b). The diffusion of neuroactive substances through the volume of the ECS is the underlying mechanism of extrasynaptic or "volume" transmission, which can modulate classical synaptic signal transmission. Indeed, the existence of frequent functional interactions between nerve cells without any synaptic contacts, "mismatches" between release sites and the location of receptors, and the widespread existence of high-affinity nonsynaptic receptors (Herkenham, 1987) led to the conclusion that extrasynaptic communication is an important alternative way of signal transmission in the CNS (Vizi, 1980, 1984; Fuxe and Agnati, 1991; Agnati et al., 1995; Sykova, 1997; Nicholson and Sykova, 1998; Zoli et al., 1999; Vizi, 2000; Kiss and Vizi, 2001; Sykova, 2003, 2004a; Vizi et al., 2004).

While neurons can interact through both synaptic and volume transmission, communication between neurons and glial cells is almost exclusively limited to the diffusion of neuroactive substances and ions in the ECS. The "awareness"of glial cells about, e.g., current neuronal activity, is crucial for their role in ionic, pH and volume homeostasis and their modulation of synaptic transmission efficacy. Volume transmission provides a mechanism for synchronizing neuronal activity and information processing in functions such as sleep, vigilance, hunger, chronic pain, emotions, behavior, learning, lactation, depression, the balance between the sympathetic and parasympathetic nervous systems and many other plastic functions of the brain (Vizi, 1980, 1984; Sykova, 1997; Vizi, 2003; Sykova, 2004a, b; Vizi et al., 2004). The ECS thus serves not only as the microenvironment of nerve cells but also as an information channel. The diffusion parameters of the ECS change during neuronal activity, development, aging, and other physiological and pathological states and may affect neuronal signaling and neuron-glia communication. Glial cells play an important role in extrasynaptic transmission as they affect ECS ionic composition and volume and their processes form diffusion barriers (Sykova, 2001, 2004b).

Astrocytic ensheathing of synapses is a plastic event and may change over either a short time scale during neuronal activity (Hirrlinger et al., 2004) or over a long time scale, e.g., during lactation (Theodosis and Poulain, 1993). Due to transport mechanisms on astrocytic processes that are in close contact with a synapse, astrocytes may suppress or facilitate neuronal activity by neurotransmitter uptake or release [for review see (Haydon, 2001)]. The importance of glial cells for synaptic transmission is reflected in the model of a "tripartite synapse" (Araque et al., 1999). Since neurotransmitters diffuse through the ECS, a recent model of a "quadripartite synapse" adds a fourth important element – the extracellular space with its dynamically changing content and its diffusion properties (Sykova, 2001, 2004a).

9.2 Homeostatic Function of Glia

9.2.1 Homeostasis of Extracellular K^+

Ionic and volume homeostasis in the CNS is maintained by a number of mechanisms present mainly in glial cells and to some extent in neurons. Increases in the extracellular concentration of K^+ ($[K^+]_e$), decreases in extracellular calcium and alkaline-acid shifts in extracellular pH (pH_e) accompany neuronal activity in different brain regions (Sykova, 1992). Even during intensive stimulation, $[K^+]_e$ does not exceed a certain steady state, the so-called "ceiling" level, which is about 6–8 mM in the adult mammalian cortex and spinal cord (Kriz et al., 1974; Heinemann and Lux, 1977; Sykova and Svoboda, 1990). After the stimulation ends, the original $[K^+]_e$ is quickly re-established. The redistribution of activity-related ionic changes is mediated by several mechanisms. Besides Na^+/K^+-ATPase transport, present on the membranes of both neurons and glial cells, the majority of K^+ is cleared from the ECS by glial cells by means of: (a) KCl uptake through Cl^- channels activated by membrane depolarization (Walz and Hertz, 1983; Walz and Hinks, 1986), (b) the opening of Ca^{2+}-activated K^+ channels (MacVicar, 1984) and (c) K^+ spatial buffering (Orkand et al., 1966). K^+ spatial buffering is based on the high permeability of the glial membrane with respect to K^+ and astrocytic coupling into the syncytium, where single cells are interconnected by gap junctions, allowing the movement of K^+, Ca^{2+} and other ions (Coles and Orkand, 1983; Kettenmann et al., 1983). A part of the syncytium that is depolarized by a local increase in $[K^+]_e$ acts as a positive battery pole, while the distant regions with a normal membrane potential represent a negative pole. A current between the poles is carried by K^+ inside the syncytium and by Na^+ and Cl^- outside the cells. In the regions with a normal membrane potential, K^+ returns back to the ECS. This redistribution of K^+ does not require energy and depends only on the existence of a $[K^+]_e$ gradient.

9.2.2 Homeostasis of Extracellular pH

It was shown in a number of studies that extracellular pH changes during neuronal activity are mediated by both neurons and glial cells. Studies on the developing CNS provided evidence for the neuronal origin of alkaline shifts and the glial origin of activity-related acid shifts (Jendelova and Sykova, 1991; Sykova, 1992). During early postnatal development in the rat brain, when precursor cells predominate in the glial population and glial homeostatic function is incomplete, activity-related pH_e and $[K^+]_e$ changes are substantially different from those in adulthood (Chvatal et al., 1995). During development, alkaline shifts, which predominate in newborn rats, become smaller and are finally overtaken by acidic shifts caused by mature glial homeostatic function (Jendelova and Sykova, 1991). In addition, stimulation-evoked alkaline shifts are blocked by the synaptic transmission blockers Mn^{2+} and Mg^{2+}, while acid shifts are unaffected.

Homeostasis of extracellular pH is closely related with intracellular pH changes. The intracellular compartments of neurons and glial cells and the extracellular space of the nervous tissue can be considered as three separate but interdependent entities. Changes of pH_e and intracellular pH (pH_i) are caused by the movement of H^+, OH^- and their equivalents NH_4^+ and HCO_3^- through the membranes of neurons and glia. Some of the membrane transport mechanisms regulating pH_e and pH_i, such as Na^+/H^+ exchange (Deitmer and Schlue, 1987; Chesler, 1987; Astion et al., 1989) or $Na^+/H^+/Cl^-/HCO_3^-$ cotransport (Thomas, 1977), are common to both neurons and glial cells, while others are specific either for neurons, e.g. voltage-dependent H^+ channels (Meech and Thomas, 1987; Sykova and Svoboda, 1990), or for glia, such as voltage-dependent Na^+- HCO_3^-cotransport (Astion et al., 1991) or H^+-lactate extrusion (Siesjo et al., 1985). Generally, membrane transporters resulting in alkaline shifts in pH_e (acid loaders) predominate in neurons, while acid extruders prevail in the glial membrane (Jendelova and Sykova, 1991). Many glial transporters are activated by the membrane depolarization evoked by neuron activity-related increases in $[K^+]_e$. pH_e acidification suppresses neuronal excitability and is part of a non-specific feedback mechanism (Sykova, 1997).

9.2.3 Volume Homeostasis

The transport of ions across the cell membrane is accompanied by water movement, which leads to cell (particularly glial) swelling and ECS volume shrinkage. An increased cell volume activates transport mechanisms that decrease the concentration of osmotically active substances in cells, which in turn results in the shrinkage of the swollen cells to their original volume. This process is called regulatory volume decrease (RVD) (Pasantes-Morales et al., 2000; Mongin and Orlov, 2001) and is presumably dependent on stretch-activated channels. The mechanisms of cell volume regulation are modulated by the intracellular concentration of Ca^{2+} (Cardin et al., 2003) and intracellular pH (Kempski et al., 1990). In addition to the movement of ions, the release of taurine, glutamate, aspartate and other amino acids from cells into the ECS is included in RVD (Pasantes Morales and Schousboe, 1988; Kimelberg et al., 1990; Pasantes-Morales et al., 2000).

9.3 Diffusion in the ECS

Diffusion is characterized by the random Brownian motion of molecules. In comparison with diffusion in a free medium, described by Fick's laws, diffusion in the ECS is restricted by the size of the extracellular pores, by diffusion barriers, e.g. membrane infoldings, fine neuronal and glial processes, macromolecules of the extracellular matrix (ECM) and charged molecules, and by nonspecifc cellular uptake. Therefore, Nicholson and Phillips (1981) modified Fick's original diffusion equations by introducing three diffusion parameters: extracellular volume fraction

(α), tortuosity (λ) and non-specific concentration-dependent or independent uptake (k') (Fig. 9.1). The ECS volume fraction α is a dimensionless quantity defined as the ratio between the volume of the ECS and the total volume of the tissue. The

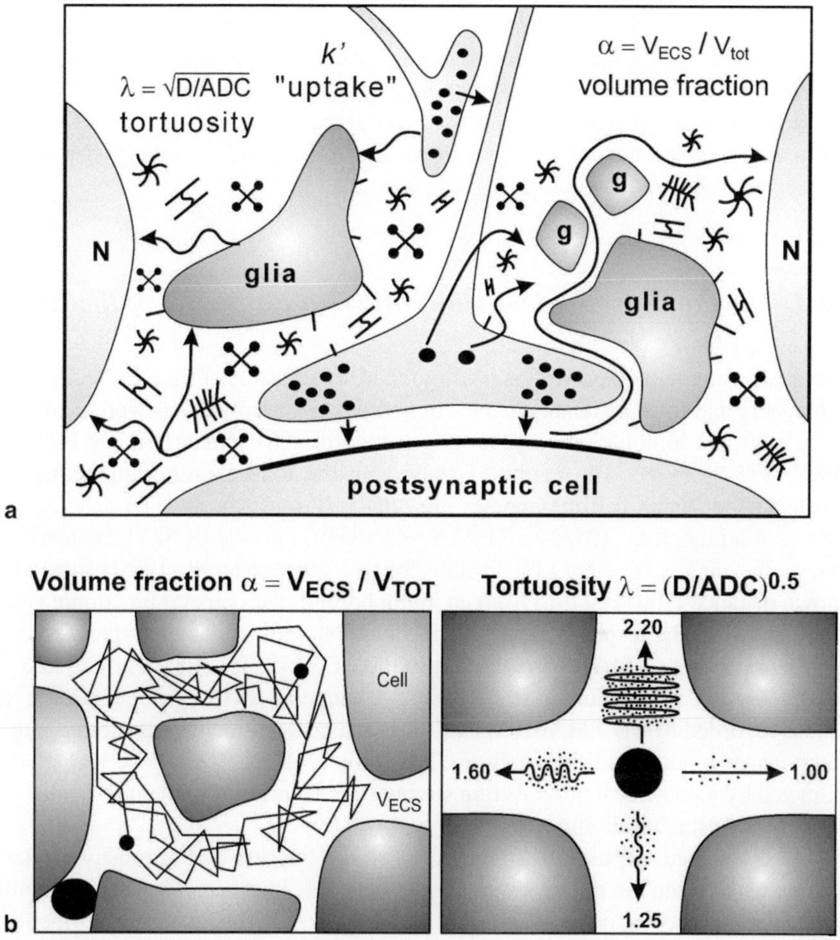

Fig. 9.1 a Scheme of CNS architecture and extracellular space (ECS) diffusion parameters. The tissue of the CNS consists of neurons, glial cells and blood vessels. The ECS between these elements contains cellular processes, axons and extracellular matrix macromolecules. The movement of substances in the ECS is critically dependent on three ECS diffusion parameters: volume fraction α, tortuosity λ and non-specific uptake k'. **b** Diffusion in the ECS is restricted by the size of the pores between the cells, represented by the volume fraction α ($\alpha = V_{ECS}/V_{TOT}$), where V_{ECS} is the volume of the ECS and V_{TOT} is the total volume of the brain, and hindered by the tortuosity factor λ ($\lambda^2 = D/ADC$, where D is the free and ADC the apparent diffusion coefficient), reflecting the increased diffusion pathway between two points in the tissue. In a free medium, tortuosity is by definition 1.00, in edematous tissue about 1.25, in adult healthy tissue 1.5-–1.6 and during some severe acute pathologies such as ischemia/anoxia above 2.0. Modified from (Sykova, 2004a).

ECS of adult healthy brain tissue represents about 20–25%, so $\alpha = 0.20–0.25$. The free diffusion coefficient (*D*) in the brain is reduced by the presence of diffusion barriers by the tortuosity factor λ (Nicholson and Phillips, 1981; Nicholson and Sykova, 1998). ECS tortuosity is defined as $\lambda^2 = D/ADC$, where *D* is the free diffusion coefficient and *ADC* is the apparent diffusion coefficient of a substance in the nervous tissue. The tortuosity value reflects the number and extent of diffusion barriers and is about 1.5 in healthy tissue, which means that diffusion in the brain is about 2.5 times slower than in a free medium. Tortuosity is a tensor and in anisotropic regions has different values in different directions. In addition to α and λ, diffusion in the ECS is affected by non-specific, concentration-dependent uptake (k'). In many cases substances such as glutamate are transported by energy-dependent uptake systems that obey non-linear kinetics.

9.3.1 Methods for Studying the Diffusion Properties of the ECS

The most suitable method for measuring the absolute values of the ECS diffusion parameters and their dynamic changes in nervous tissue in vitro as well as in vivo is the real-time iontophoretic method, developed in 1981 (Nicholson and Phillips, 1981) (Fig. 9.2b, c). The method is based on the iontophoretic application of substances to which cell membranes are relatively impermeable (tetramethyl- or tetraethyl-ammonium, TMA^+ or TEA^+, respectively) into the ECS. The concentration of the ions is measured by an ion-selective microelectrode (ISM) fixed at a known distance (100–200 μm) from an iontophoretic micropipette functioning as a point source of the respective ions. During iontophoretic application, the extracellular marker (e.g., TMA^+) is released into the ECS by applying a current step of + 100–200 nA with a duration of 24–80 s. Change of TMA^+ concentration over time is recorded by the TMA^+-ISM as a diffusion curve, which is transferred into a computer for mathematical analysis. The values of α, ADC_{TMA}, λ and k' are extracted by a non-linear curve-fitting simplex algorithm operating on the diffusion curve (Nicholson and Phillips, 1981).

The other methods used to study ECS volume fraction and tortuosity are less comprehensive and are not reviewed here in detail, as they can either measure only one of the parameters, determine only relative changes in the ECS volume fraction, or detect changes that are only partially related to ECS volume changes [for review see (Nicholson and Sykova, 1998)]. These methods include the measurement of tissue resistance (Van Harreveld et al., 1971; Matsuoka and Hossmann, 1982), changes in light transmittance and/or scattering (MacVicar and Hochman, 1991; Andrew et al., 1999; Tao et al., 2002; Sykova et al., 2003), the measurement of the *ADCs* of fluorescent-labeled large molecules, polymers and quantum dots by integrative optical imaging (Nicholson and Tao, 1993a, b; Prokopova-Kubinova et al., 2001; Thorne et al., 2004; Thorne and Nicholson, 2006), and the measurement of the *ADC* of water (ADC_W) by diffusion-weighted magnetic resonance imaging (DW-MRI); the latter reveals the inhomogeneous and anisotropic diffusion of

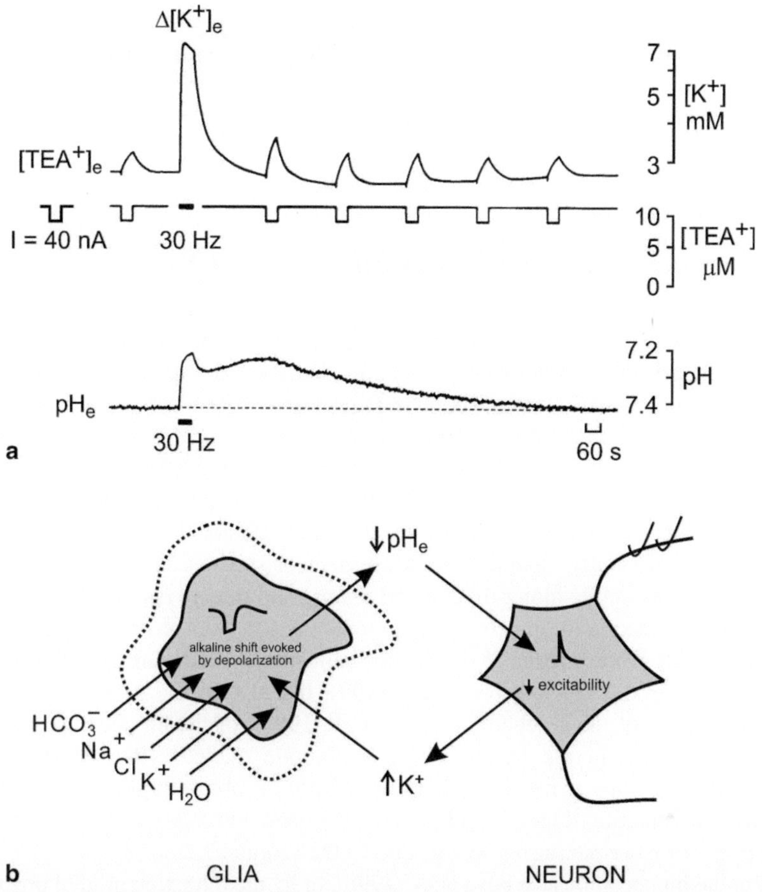

Fig. 9.2 **a** Effect of repetitive stimulation (30 Hz for 1 min) of the dorsal roots on TEA^+ diffusion curves, extracellular K^+ ($[K^+]_e$) and pH (pH_e) in the dorsal horns of the isolated frog spinal cord. Stimulation evoked an increase in $[K^+]_e$ and an acid shift in the extracellular space. The time course of the acid shift correlates with a decrease in the ECS volume manifested as an increase in the amplitude of the diffusion curves; these changes outlasted the stimulation for 30 min. **b** Mechanism of nonspecific feedback suppressing neuronal activity. Stimulation of a neuron is accompanied by an increase in $[K^+]_e$, which in turn depolarizes glial cell membranes. Depolarization leads to the opening of voltage-gated ion channels and the activation of transport processes resulting in an alkaline shift inside glial cells and the acidification of the extracellular space, which in turn has a suppressive effect on the excitability of neurons. The whole mechanism is amplified by the swelling of glial cells and compensatory ECS shrinkage, as water accompanying the ion movements enters into the cells. Modified from (Sykova, 2004b).

water, similar to the diffusion of TMA^+ (Pierpaoli et al., 1996; Mamata et al., 2002; Vorisek et al., 2002). In comparative studies in which measurements were first done using DW-MRI and then in the same animal or experimental model using the real-time iontophoretic method, we found that under some experimental conditions, a decrease in ADC_W, i.e. lower diffusibility, can be related either to a decrease in α, typically during fast acute changes such as ischemia (Van der Toorn et al., 1996), or to an increase in the number of diffusion barriers, represented as an increase in λ, without substantial changes in α, such as occur during chronic states after injury (Vorisek et al., 2002).

9.3.2 Heterogeneity and Anisotropy of the ECS

The ECS has a complicated and uneven geometry, which in a given structure depends on the type and number of cells, the density and orientation of their processes, and the extracellular matrix. The ECS is therefore not homogeneous as its properties vary around different types of cells and also in different brain regions [for review see (Nicholson and Sykova, 1998)]. Moreover, diffusion may be facilitated in a certain direction more than in others, such as along myelinated axons or astrocytic processes, and therefore we speak about diffusion anisotropy. Anisotropic diffusion preferentially channels the movement of substances in the ECS in one direction (for example, along myelinated axons) and thus is responsible for a certain degree of specificity in extrasynaptic transmission. Anisotropy has been found not only in the white matter of the corpus callosum (Vorisek and Sykova, 1997a) and the spinal cord (Prokopova et al., 1997), but also in the grey matter of the molecular layer of the cerebellum (Rice et al., 1993), of the hippocampus (Mazel et al., 1998) and in the hypothalamic supraoptic nucleus (Piet et al., 2004). Diffusion anisotropy is not an inalterable feature of a brain region; it can appear *de novo* (myelination during development) (Vorisek and Sykova, 1997a) or it can disappear, for example during aging, due to the reorganization of astrocytic processes in the hippocampus (Sykova et al., 1998), or by the retraction of glial processes in the supraoptic nucleus during lactation [(Piet et al., 2004); see below].

9.3.3 Inhomogeneity of the ECS in the Close Vicinity of the Cell Membrane

TMA^+ measurements using the real-time iontophoretic method are averaged over an area of 10^{-3} mm^3 and do not provide information about the ECS properties in the close vicinity of individual nerve cells. The existence of regional differences in the extracellular space volume around glial cells was shown by membrane current analysis using the patch-clamp method (Berger et al., 1991; Chvatal et al., 1999; Chvatal et al., 2004). In astrocytes and in oligodendrocytes in culture, the current evoked by depolarization exactly matched the shape of the depolarization pulse (Sontheimer and Kettenmann, 1988). In oligodendrocytes in the white matter of spinal cord slices, the currents decayed and large tail currents appeared after the offset of the voltage command (Berger et al., 1991; Chvatal et al., 1999). Further tail current analysis showed that the current decay and the tail currents were related to the accumulation of K^+ in the vicinity of oligodendrocytes due to barriers that prevented the further diffusion of K^+, which had escaped from the cell (Berger et al., 1991; Chvatal et al., 1999). Since glial cells are predominately permeable to K^+, knowing the values of the tail current and the reversal potential, the Nernst equation can be used to calculate the value of $[K^+]_e$ in the vicinity of the cell membrane. Using the calculated value of $[K^+]_e$ and the known values of the intracellular concentration of K^+ and the reversal potential of the cell, the absolute volume of the ECS in the close vicinity of the cell can be estimated (Chvatal et al., 2004). Experiments with osmotically-induced cell swelling or cell shrinkage revealed that the extracellular space in the close vicinity of astrocytes is larger than that around oligodendrocytes (Vargova et al., 2001b; Chvatal et al., 2004). As tail currents appear in oligodendrocytes during their maturation and correspond in time with the myelination of the tissue (Chvatal et al., 1997), we can assume that the diffusion barriers around oligodendrocytes are formed mainly by myelin sheaths or by extracellular matrix molecules produced by mature oligodendroglia.

9.4 Glia and ECS Diffusion Parameters

Ions and neurotransmitters released into the ECS during neuronal activity or pathological states interact not only with the receptors on the postsynaptic and presynaptic membranes of neurons, but also with extrasynaptic receptors on glial cells. The activation of glial cells leads to the activation of ion channels, second messengers and intercellular metabolic pathways and subsequently to changes in glial volume, accompanied by dynamic variations in α, structural changes of astrocytic processes and, eventually, by the production of the extracellular matrix. In addition to their role in the maintenance of extracellular ionic homeostasis, glial cells may, by regulating their volume, affect the extracellular diffusion of neuroactive substances. Astrocytic rebuilding and nerve cell loss are typical features of damaged tissue. Brain injury of any kind elicits reactive gliosis, involving both the hyperplasia and hypertrophy of astrocytes, manifested by intensive staining for glial fibrillary acidic protein (GFAP). Active astrocytes change their morphology, their processes become shorter and thicker and qualitative and quantitative changes in their production of extracellular matrix molecules can be detected. Changes in the ECS diffusion parameters related to glial maturation, swelling and/or remodeling have been studied under physiological conditions such as development, neuronal activity, lactation and aging, under experimental conditions mimicking pathological conditions in the tissue (e.g., exposure to high K^+, osmotic stress or application of amino acids), in experimental models of human diseases and pathological states such as ischemia, epilepsy, injury, hydrocephalus, multiple sclerosis, Parkinson's disease, Alzheimer's disease, grafted tissue and tumors.

9.4.1 Glia and ECS Diffusion Parameters in Physiological States

9.4.1.1 Development

In the earliest stages of postnatal development, the diffusion parameters in the rat brain and spinal cord are different from those found in adulthood. The volume fraction decreases throughout the entire postnatal life, with the steepest fall in the first two weeks during which α in the rat cortex and corpus callosum is reduced by approximately half (Lehmenkuhler et al., 1993; Prokopova et al., 1997; Vorisek and Sykova, 1997a). This decrease can be attributed to changes in ECM composition, neuronal migration, the development of dendritic arborization, rapid myelination and the proliferation of glia. The larger volume fraction in immature tissue (0.35–0.40) slows down any increase in the concentration of neuroactive substances to neurotoxic levels and represents a protective factor in the immature brain during pathological states such as ischemia (Vorisek and Sykova, 1997b) or epilepsy (Kilb et al., 2006). The connection between gliogenesis and ECS diffusion parameters is probably most evident in the white matter. The period of the gradual myelination of the corpus callosum closely corresponds with the decrease in extracellular volume fraction. Moreover, the tortuosity values, which during the first postnatal week are the same along all three orthogonal axes, (i.e. diffusion is isotropic), start to differ during the second week; a period of intensive myelination. After myelination is fully completed at the end of the third postnatal week, diffusion in the corpus callosum is strictly anisotropic with preferential diffusion along the myelinated fibers (Vorisek and Sykova, 1997a).

9.4.1.2 Neuronal Activity

Ions crossing the cell membranes during neuronal activity are accompanied by water molecules, which lead to cell swelling and compensatory ECS shrinkage. Repetitive electrical stimulation of the dorsal root of the rat or frog spinal cord causes a substantial increase in $[K^+]_e$ and a decrease in α from about 0.24 to 0.17–0.12 (i.e. by 30–50%, Fig. 9.2). A decrease of 20–50% in α also occurs after injury of the ipsilateral hind paw evoked by the subcutaneous injection of turpentine or after thermal injury (Svoboda and Sykova, 1991). The chemical injury produces chronic pain, long-lasting interneuronal activity and K^+ accumulation in the spinal dorsal horns. The changes in ECS diffusion parameters, however, persist after stimulation ends (for 30 min after 1 min of electrical stimulation or even for 120 min after 3 min of stimulation; Fig. 9.2), suggesting long-term glial swelling and subsequent long-term changes in neuronal excitability, neuron-glia communication and extrasynaptic volume transmission.

9.4.1.3 Lactation

During lactation, under physiological conditions a relationship between the glial cell morphology and the ECS diffusion parameters was found in the hypothalamic supraoptic nucleus (SON) (Fig. 9.3). In male rats and virgin female rats, the SON is an anisotropic structure with a relatively high extracellular volume ($\alpha = 0.32$) and preferential diffusion along the ventrodorsal axis (Piet et al., 2004), in agreement with the orientation of the majority of astrocytic processes in this nucleus (Bonfanti et al., 1993). The retraction of glial processes towards the basal lamina under specific physiological conditions, such as lactation or dehydration, leads to a significant reduction in the astrocytic coverage of the magnocellular neurons (Theodosis and Poulain, 1993; Hatton, 1997). A relative absence of glial processes around the glutamatergic synapses results in insufficient clearance of glutamate and thus increases its level in the ECS (Oliet et al., 2001). The disappearance of diffusion barriers during lactation leads to a decrease in λ and a loss of anisotropy. This leads to facilitated diffusion and to a decrease in the ECS volume fraction to about 0.20 due to an increase in the volume of the magnocellular neurons and an increased proportion of direct neuronal membrane juxtapositions (Piet et al., 2004).

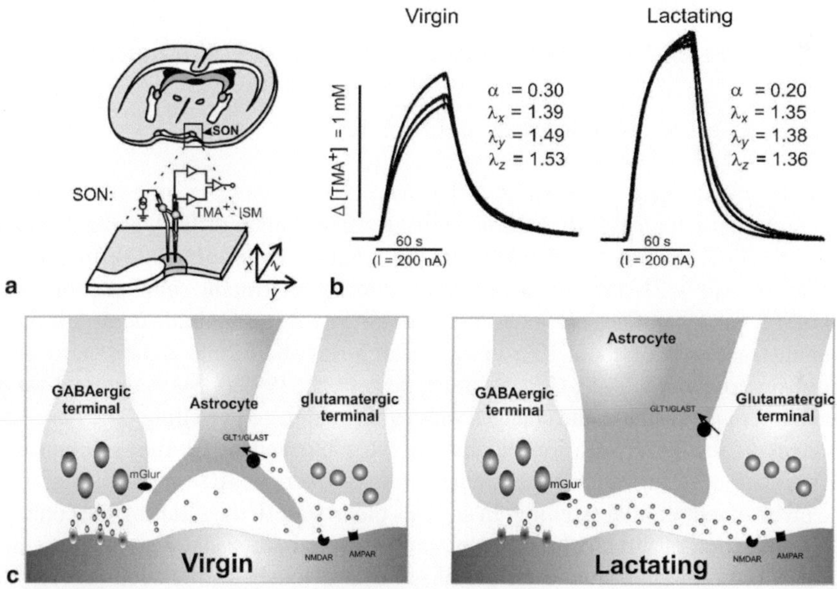

Fig. 9.3 Scheme of experimental arrangement (**a**), diffusion curves with the corresponding values of the ECS diffusion parameters (**b**) and a hypothetical model of heterosynaptic depression (**c**) induced by synaptic cross-talk in the hypothalamic supraoptic nucleus (SON). As diffusion in the SON of virgin animals is anisotropic, measurements were made along three orthogonal axes (*x*-ventrodorsal, *y*-mediolateral and *z*-rostrocaudal). Lactation induces a decrease in α as well as a decrease in λ along all three axes and the disappearance of anisotropy. Anatomical remodeling during lactation includes the retraction of glial processes, which leads to deficient glutamate clearance and increased glutamate concentration in the ECS, further potentiated by ECS shrinkage. Due to the retraction of the astrocytic processes, diffusion is facilitated, which results in increased cross-talk between glutamate- and GABAergic synapses and the inhibition of GABA release due to the activation of presynaptic metabotropic glutamate receptors. AMPAR, AMPA receptor; GLT1/GLAST, glutamate transporters; ISM, ion-selective microelectrode; mGlur, metabotropic glutamate receptor; NMDAR, *N*-methyl-d-aspartate receptor; TMA, tetramethylammonium. Modified from (Sykova, 2004a).

Under these conditions glutamate spillover, monitored through the metabotropic glutamate receptor-mediated depression of γ-aminobutyric acid (GABA)-ergic transmission, is greatly enhanced (Fig. 9.3c). Because glial cells in the SON in virgin rats represent a physical barrier to diffusion and are essential for glutamate uptake, the retraction of their processes results in glutamate-mediated cross-talk between synapses and enhances the heterosynaptic depression of GABA-ergic transmission (Piet et al., 2004). This was confirmed by the experimental introduction of diffusion barriers by perfusing tissue slices from lactating rats with large neutral dextran molecules, a procedure that impaired intersynaptic communication and decreased the presynaptic effect of glutamate. Thus, in conclusion, this study demonstrates that depressing the inhibitory effect of GABA-ergic neurons may increase hormone release during suckling or during dehydration and probably represents one of the mechanisms regulating milk ejection and water homeostasis in the rat SON.

9.4.2 Glia and ECS Diffusion Parameters in Pathological States

9.4.2.1 Effect of Osmotic Stress and High K^+

Various studies have demonstrated that astrocyte swelling is an early event in numerous pathological states, accompanied by an elevation of $[K^+]_e$ and a decrease of extracellular osmolality (Kimelberg, 1991; Kimelberg et al., 1992). In isolated spinal cords of 4–21-day-old rats, the application of an isotonic solution containing 50 mM K^+ or hypotonic solution (235 mmol kg^{-1}) evoked an initial decrease in ECS volume fraction α of about 50% in the dorsal horns and a concomitant increase in tortuosity to about 1.9–2.1 (Fig. 9.4a) (Sykova et al., 1999a). Since the total water content remained stable, the changes were attributed to cell swelling. The observed changes in the ECS diffusion parameters were blocked in Cl^--free solution and slowed down by furosemide and bumetanide, blockers of KCl uptake, suggesting the involvement of glial swelling. In animals older than 10 days, during the continuous application of 50 mM K^+ or hypotonic solution, both α and λ started to return to their control values due to the shrinkage of previously swollen cells caused by RVD (Cserr et al., 1991; Gullans and Verbalis, 1993). RVD mechanisms are almost exclusively related to the homeostatic regulation of volume by glial function as RVD was not observed in immature animals with incomplete gliogenesis and was also blocked by the gliotoxin fluoroacetate (Sykova et al., 1999a). During washout, α returned to control values or even exceeded them, while λ showed a second, permanent increase. Histologically, an increase in λ during the washout of the hypotonic solution and/or 50 mM K^+ corresponded with a significant increase in GFAP staining and changes in astrocytic morphology, typical of astrogliosis. Even if further study did not confirm an increase in the total GFAP content (Neprasova et al., 2007), the substantial morphological rearrangement of astrocytes and their processes has been found in hypotonic medium using 3D reconstruction of confocal images (Chvatal et al., 2007) (Fig. 9.5), suggesting that glial cells do not swell

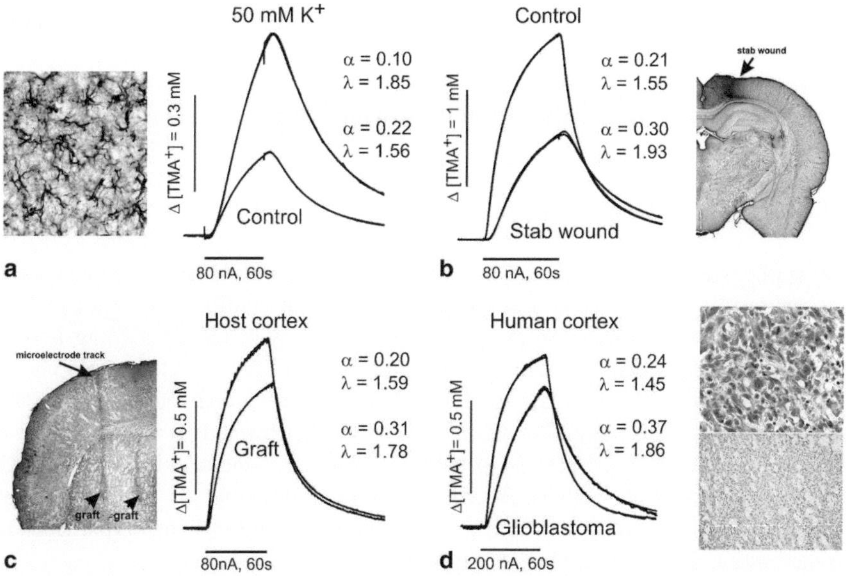

Fig. 9.4 TMA^+ diffusion curves with the corresponding values of the ECS diffusion parameters and histological features of the tissue under different experimental conditions – models of various pathological states: perfusion with 50 mM K^+ (**a**), gliotic tissue around a stab wound (**b**), dopaminergic grafts in the striatum (**c**) and a glioblastoma (**d**) (see also text). Slices from **a**, **b** and **c** are immunostained for GFAP; the slices in part D are stained with hematoxylin-eosin and/or for the extracellular matrix molecule tenascin. The different amplitudes and shapes of the diffusion curves recorded in pathological tissue in comparison with control curves reflect changes in extracellular volume fraction α (the higher the curve, the lower the α value) and tortuosity λ. Note that regardless of decreased (**a**) or increased (**b**, **c**, **d**) α values, tortuosity in tissue with reactive astrocytes is always increased and diffusion is hindered. With respect to the glioblastoma, in addition to the diffusion barriers formed by hypertrophied glial processes, the overproduction of the extracellular matrix molecules, namely tenascin and vitronectin, is also a contributory factor (*See Color Plates*).

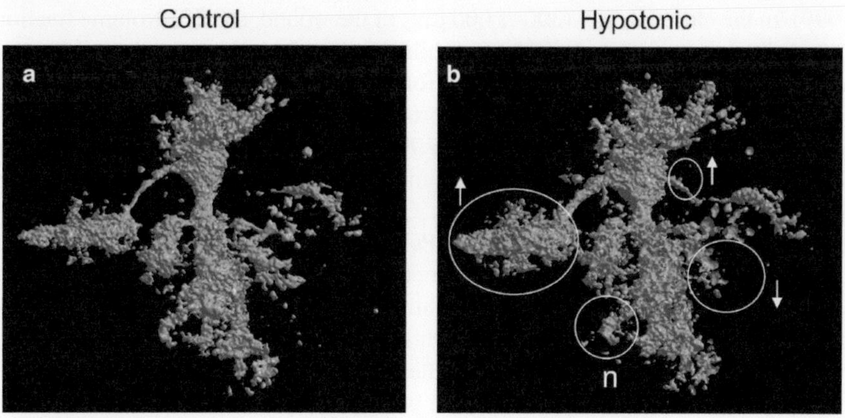

Fig. 9.5 Effect of hypotonic solution (200 mmol kg^{-1}) on the morphology of astrocytes in situ determined by three-dimensional confocal morphometry. The hypotonic solution evoked a general cell swelling; however, the actual increase in different cell compartments varied, as can be seen in the outlined regions. An increase or decrease in the volume of cell compartments is marked by *arrows*; no apparent change is indicated by "n". Modified from (Chvatal et al., 2007).

equally in all of their processes, but rather in those that are in contact with actual synapses or the sites of release of neuroactive molecules.

9.4.2.2 Effect of Excitatory Amino Acids

The application of glutamate and its agonists, NMDA and AMPA, in low concentrations results in a decrease in *a* while tortuosity is less affected (Vargova et al., 2001a). A substantial increase in λ and in the intensity of GFAP staining is seen only during the application of high concentrations of these excitatory amino acids (10^{-2}M glutamate, 10^{-4}M NMDA and 10^{-5}M AMPA), accompanied by an increase in extracellular K^+ to more than 12–15 mM. Such an elevated K^+ concentration is capable of evoking an increase in tortuosity to about 1.7 and also increased GFAP staining (Sykova et al., 1999a). We can therefore assume that the K^+-evoked rearrangement of fine astrocytic processes is the predominant mechanism that leads to an increase in diffusion barriers in the ECS.

9.4.2.3 Injury and Astrogliosis

In models of chronic pathological states, changes in α and λ are often independent and result from the structural remodeling of the tissue (cell death, astrogliosis). A stab wound of the rodent brain is a well-characterized and commonly used model of reactive gliosis, in which additional diffusion barriers in the ECS are created by the hypertrophy of astrocytic processes and an increased production of extracellular matrix molecules (Hatten et al., 1991; Norton et al., 1992). The ECS diffusion parameters in our study were measured 3, 7, 21 and 35 days post-wounding (dpw) in the hemispheres ipsilateral and contralateral to the lesion (Roitbak and Sykova, 1999). In the close vicinity (300–1,000 μm) of the wound, the ECS volume fraction α initially increased by about 20%, while 35 dpw it returned to control values, apparently due to cell loss in the area followed by the replacement of damaged tissue by a gliotic reaction (Fig. 9.4b). Tortuosity values were increased at all examined time points, reaching a maximum by 7 dpw. In the ipsilateral cortex at a greater distance from the wound (1,500–2,000 μm), an increase of tortuosity at 7 dpw was found, accompanied by no increase in GFAP staining but with an increase in chondroitin-sulphate proteoglycan (CSPG) expression. Diffusion measurements using DW-MRI showed an increase in ADC_w only in the close vicinity of the wound, i.e. in the region of cell death, while otherwise ADC_w decreased throughout the whole ipsilateral hemisphere, apparently related to the additional diffusion barriers caused by deposits of ECM molecules (Vorisek et al., 2002).

The formation of gliotic tissue in the wounded area is thus responsible for the long-lasting increase of diffusion barriers in the ECS, preventing or slowing down the diffusion of neuroactive substances and growth factors and leading to impaired extrasynaptic transmission, cell-to-cell communication and regenerative processes. In the formation of diffusion barriers, not only are hypertrophied astrocytic processes involved, but there is also the enhanced production of extracellular matrix molecules. Local brain damage results in changes in the diffusion parameters, affects synaptic and nonsynaptic transmission in areas remote from the site of injury and possibly leads to functional deficits.

9.4.2.4 Grafted Tissue

In experiments on animals in which embryonic cortical or tectal tissue was transplanted onto the dorsal surface of the midbrain of neonatal rats, astrogliosis and the formation of myelinated tracks were observed (Harvey et al., 1997; Sykova et al., 1999b). The grafted tissue matured in the host brain and developed characteristic cytoarchitectural features. While oligodendrocytes matured in an apparently normal manner and attained relatively normal phenotypic characteristics, astrocytes were hypertrophied, intensely stained for GFAP and remained in a reactive state for many months after transplantation (Harvey et al., 1997). When the diffusion parameters were investigated in cortical grafts of fetal tissue (Sykova et al., 1999b), similar values of α and λ were observed as in gliotic tissue after injury (Fig. 9.4c). Both α and λ were significantly higher in cortical grafts than in the host cortex; while in the host cortex the volume fraction ranged from 0.20 to 0.21 and tortuosity from 1.59 to 1.64, in both cortical and tectal grafts the volume fraction reached values of 0.29–0.34 and tortuosity 1.78–1.85. Both α and λ were increased in cortical grafts of fetal tissue transplanted to the midbrain, where severe astrogliosis was found, but not in fetal grafts placed into a cavity in the host cortex, where only mild astrogliosis occurred. Further analysis revealed that diffusion in the grafts was anisotropic and that the values of α and λ varied widely in different regions of the graft. This variation correlated with the morphological heterogeneity of the graft neuropil with the largest volume fraction seen in gray matter and the highest tortuosity values in white matter bundles. High tortuosity values indicate that the diffusion of chemical signals in the grafted tissue is hindered. Limited diffusion may have a negative impact on the viability of grafted tissue and prevent the spread of dopamine. In the areas of high tortuosity values, immunohistochemistry revealed myelinated patches and hypertrophic reactive astrocytes, which are most likely responsible for the increased diffusion barriers.

Parkinson's disease is caused by the decreased production of dopamine in the striatum and a subsequent lack of dopamine in the basal ganglia, resulting in the disinhibition of strong inputs from the motor cortex, clinically manifested as a typical hypertonic-hypokinetic syndrome. A promising possibility for treating the disease seemed to be the transplantation of dopaminergic neurons. It has been shown that functional recovery after the transplantation of dopaminergic cells into the lesioned striatum depends on the widespread diffusion of the transmitter released by the graft. In 6-hydroxydopamine-lesioned rat striatum, both volume fraction and tortuosity were decreased in comparison with control healthy tissue. Following the transplantation of macro- or micro-grafts of dopaminergic neurons, α returned to normal control values, but tortuosity increased to about 1.8 (Reum et al., 2002). The effect of the graft deposits, in contrast to their enhanced capacity for dopaminergic reinervation and/or dopamine release, can thus be limited by the impaired diffusion of neuroactive substances, including dopamine from the grafted tissue, because of the gliotic reaction accompanying the transplantation injury.

9.4.2.5 X-Irradiation and Block of Gliogenesis

Nerve cells during the early postnatal period are more sensitive to X-irradiation than the adult brain, presumably due to the increased radiation sensitivity of proliferating tissue. X-irradiation in early postnatal development evokes impaired gliogenesis and is responsible for a number of neurotoxic effects (Sims and Gilmore, 1992). The X-irradiation of the somatosensory neocortex and subcortical white matter in 1-day-old (P1) rats at a single dose of 40 Gy resulted in irradiation necrosis with typical features such as cell death, DNA fragmentation, blood–brain-barrier damage, activation of macrophages and astrogliosis (Sykova et al., 1996). These changes were accompanied by changes in the ECS diffusion parameters, which were firstly detected on the second day (P2) following X-irradiation and persisted until P21 (Sykova et al., 1996). X-irradiation blocked the normal pattern of a gradual decrease in the ECS volume fraction as discussed above and, on the contrary, resulted in a significant increase in α to about 0.50, which persisted 3 weeks after the radiation insult. Developing astrogliosis in the later phases led to an increase in tortuosity between P10 and P21.

The dramatic increase in the ECS volume fraction of nervous tissue affected by X-irradiation and subsequent astrogliosis may have severe repercussions on the developing nervous tissue. Signal transmission is impaired by diluting ions and neuroactive substances released from cells, which can in turn lead not only to functional deficits but also to malfunctions in cell migration and developmental processes responsible for proper neuronal net connection. Moreover, possible regenerative processes are prevented by the formation of nonfunctional astrogliotic tissue replacing damaged nervous tissue.

9.4.2.6 Demyelinating Diseases

Changes in the ECS diffusion parameters, including a loss of the anisotropy typical of white matter, have been found during demyelinating diseases. As an animal model of multiple sclerosis, experimental autoimmune encephalomyelitis (EAE) is widely used (Gold et al., 2006). Fourteen to 17 days following an injection of guinea pig myelin basic protein into rats to induce EAE, typical morphological features of multiple sclerosis can be observed in the CNS tissue, including demyelination, an inflammatory reaction, astrogliosis, blood–brain-barrier damage and clinical signs of paralysis (Simonova et al., 1996). Diffusion measurements with the TMA^+ method revealed an increase in ECS volume fraction α in both the white and the gray matter of the rat spinal cord, namely to 0.28 in the dorsal horns, to 0.47 in the ventral horns and to 0.30 in the white matter. Tortuosity also decreased, but less than expected (from about 1.55 to about 1.45) in the case of such a large increase in α, apparently due to concomitant astrogliosis. There was a close correlation between the changes in the ECS diffusion parameters and the manifestation of clinical signs (paraparesis, paraplegia), which were preceded and outlasted by astrogliosis and an inflammatory reaction (Simonova et al., 1996).

9.4.2.7 Aging and Degenerative Changes in the Brain

Aging is a normal physiological process of the organism; however, it has many features in common with degenerative diseases of the CNS, including Alzheimer's disease. Both aging and degenerative brain diseases are accompanied by various pathological processes including a decreased number and efficacy of synapses, a decrease in transmitter release, neuronal loss, astrogliosis, demyelination, deposits of beta-amyloid, changes in the extracellular matrix and others resulting in behavioral changes and a cognitive deficit, particularly memory and learning impairment. Since all these changes may affect the biophysical and diffusion properties of the ECS, we studied the ECS diffusion parameters in the cortex, corpus callosum and hippocampus of senescent rats (CA1, CA3 and dentate gyrus). As diffusion in the hippocampus is anisotropic (Sykova et al., 1998), measurements were done along three orthogonal axes: x, y and z. In all three brain regions, the mean ECS volume fraction α was significantly reduced in aged rats (26–32 months old) compared to young adults (3–4 months old) (Fig. 9.6a). The typical anisotropy in the hippocampus decreased as values of tortuosity decreased along the y- and z-axes, but not along the x-axis (Mazel et al., 1998). Morphological changes observed in the hippocampal tissue included disorganization of astrocytic processes (Fig. 9.6b) and the loss of extracellular matrix forming perineuronal nets, e.g., fibronectin and CSPG. This accounted for both the decrease in α and the disappearance of anisotropy. Our study also revealed that the degree of learning impairment during aging is closely correlated to the changes in the ECS diffusion parameters. In animals with more pronounced astrocytic changes and a loss of ECM, a more severe learning deficit was found (Sykova et al., 2002). In the approximately 10% of aged rats that showed a prominent learning deficit in a Morris water maze task, the decrease in α was significantly greater and the loss of anisotropy was more complete in comparison with aged rats with a milder impairment. DW-MRI measurements confirmed impaired diffusion in the hippocampus as well as in the cortex of aged animals (Fig. 9.6c) It is therefore reasonable to assume that diffusion anisotropy, which leads to a certain degree of specificity in extrasynaptic communication by channeling the flux of substances in a preferential direction, may also play an important role in memory formation and learning (Mazel et al., 1998; Wieshmann et al., 1999; Sykova et al., 2002). The decrease in ECS volume during aging not only impairs extrasynaptic transmission in the cortex and the hippocampus, but could also be responsible for the greater susceptibility of the aged brain to pathological insults, the poorer outcome of clinical therapy and limited regeneration.

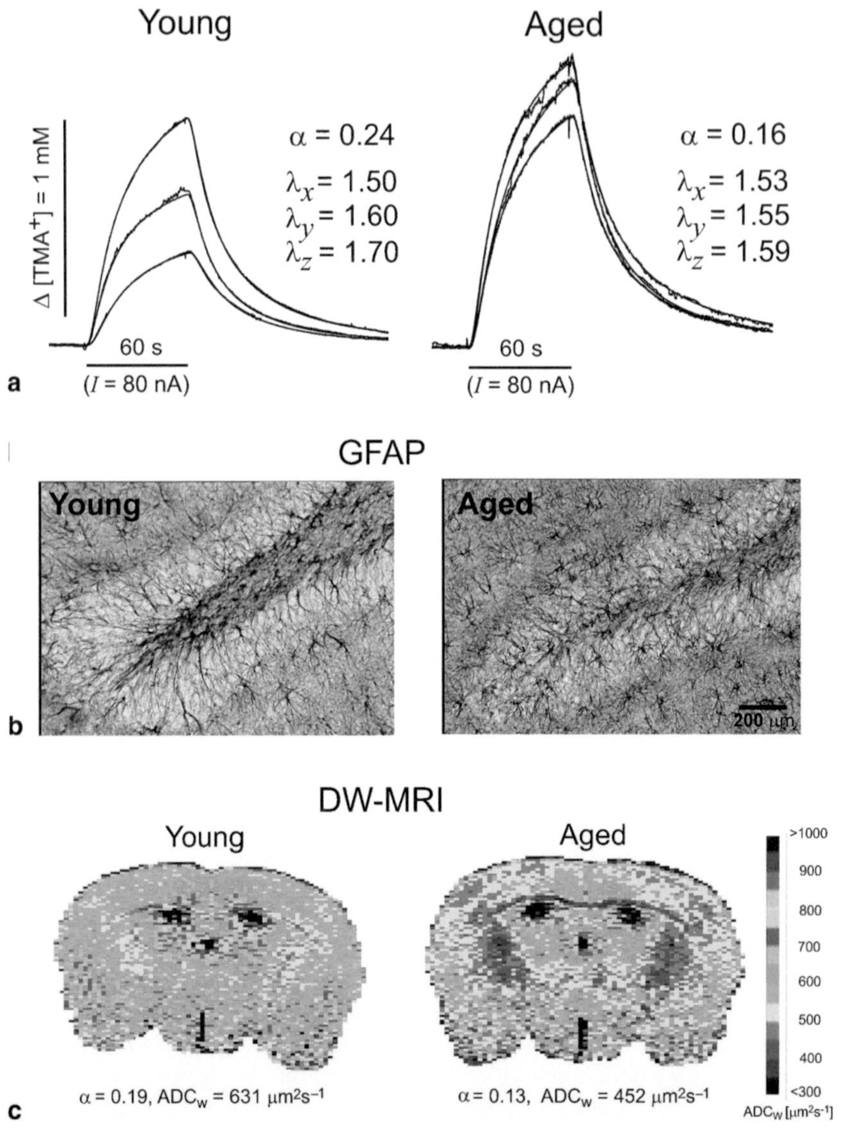

Fig. 9.6 Typical diffusion curves recorded in the hippocampus of young adult and aged rats (**a**), immunostaining of young and aged gyrus dentatus of the hippocampus for GFAP (**b**) and maps of the apparent diffusion of water (ADC_W) measured by diffusion-weighted MRI (**c**). During aging, the extracellular space volume decreases and anisotropy disappears. The disappearance of diffusion anisotropy corresponds with the loss of the typical radial organization of fine glial processes in the aged hippocampus, as revealed by immunostaining for GFAP (**b**), as well as with the loss of fibronectin and CSPG perineuronal nets (not shown). The lower ADC_W indicates the impaired diffusion of water in the cortex and hippocampus of aged rats (**c**) (*See Color Plates*).

9.4.2.8 Glial Tumors

The impact of glial remodeling and rebuilding on the ECS biophysical properties and diffusion parameters may be even stronger in glial tumors. Our studies of human glioma samples obtained from patients during surgery showed that the migratory abilities of tumor cells might be strongly affected by changes in ECS volume fraction and ECM molecule content in tumors (Vargova et al., 2003; Zamecnik et al., 2004). To correlate the values of the ECS diffusion parameters with tumor malignancy, tumor slices were histopathologically classified according to World Health Organization grading (WHO grade I–IV) following diffusion measurements. The proliferative activity of the tumor was assessed by the labeling indices MIB-1 and anti-topo-IIα. As a control, cortical tissue that did not exhibit any histological signs of pathological changes, resected during the surgical treatment of intractable epilepsy, was used. The measurements showed that the proliferative activity and malignancy grade of astrocytomas are directly proportional to the increasing values of α and λ. In the cellular regions of the most malignant glioblastomas (WHO grade IV), tortuosity was about 1.8 and α was twice as high (0.44) as in the normal cortex (Vargova et al., 2003) (Fig. 9.4d). The increase in α and λ in high-grade tumors strongly correlated with an increased accumulation of ECM molecules, particularly of tenascin and vitronectin (Zamecnik et al., 2004) (Fig. 9.4d). It has been shown that the over-expression of certain types of ECM molecules corresponds with tumor malignancy (Zhang et al., 1998; Hayen et al., 1999; Camby et al., 2002), as these molecules may serve as ‘ropes for climbing’ for migrating tumor cells and thus facilitate their infiltration into the surrounding tissue. In addition, a large ECM content holds cells apart and thus creates a larger space for migrating cells (see below).

9.4.3 *Role of the Extracellular Matrix*

The extracellular fluid is not a simple salt solution, but in addition to ions, metabolites, growth factors and other molecules contains a number of glycosaminoglycans (e.g., hyaluronate), glycoproteins (e.g., tenascins – TN) and proteoglycans that are either soluble or that constitute the ECM. Molecules of the extracellular matrix are produced by neurons as well as glia. ECM content changes dynamically during development, aging, wound healing and various pathological states. The functions of ECM molecules are numerous. Together with fine astrocytic processes, they form perineuronal nets around synapses (Celio et al., 1998), and they are also involved in cell migration, proliferation and differentiation, oriented axonal growth and synapse formation (Dityatev and Schachner, 2003; Kleene and Schachner, 2004). The overproduction of ECM during pathological states creates additional diffusion barriers and contributes, together with astrogliosis, to an increase in tortuosity (see above).

ECM molecules also play a role in tissue cytoarchitecture by maintaining the optimal size of intercellular spaces. A significant decrease in ECS volume to 12% was found in genetically modified, tenascin-R negative mice (Sykova et al., 2005a). The role of the ECM in determining the ECS volume can also be demonstrated by the changes seen during aging, where a decrease in the amount of CSPG and fibronectin seen in the hippocampus of old rats with severe learning disability coincides with a decrease in the ECS volume (Sykova et al., 2002). CSPG participates in multiple cellular processes, including axonal outgrowth, axonal branching and synaptogenesis (Hardingham and Fosang, 1992; Margolis and Margolis, 1993), which are important for the formation of memory traces. On the other hand, an increase in α has been found in pathological states with an overproduction of ECM: during astrogliosis (Roitbak and Sykova, 1999), during the increased deposition of amyloid in a transgenic APP23 mouse model of Alzheimer's disease (Sykova et al., 2005b), or in the tissue of primary brain tumors (Vargova et al., 2003; Zamecnik et al., 2004).

The explanation of the relationship between the size of the ECS and ECM content is simple. Side chains of ECM macromolecules, especially glycosaminoglycans, carry a large number of negatively charged groups that repel one another. Concomitantly, they attract osmotically active cations, such as Na^+, causing a large amount of water to be drawn into the matrix. This creates a turgor, which enables the matrix to withstand compressive forces and leads to the expansion of the ECS (Alberts et al., 1994). The loss of ECM then leads to a decrease in the ECS volume and *vice versa*.

9.5 Conclusion

It is evident that glial cells play an important role in different brain functions. Voltage- and ligand-activated channels and a number of transport mechanisms on the glial membrane [for review see (Sontheimer et al., 1989)] allow the maintenance of ionic and volume homeostasis, which in turn affect neuronal function. A link between ionic and volume changes has been proposed in a model of a nonspecific feedback mechanism suppressing neuronal activity (Sykova, 1997) (Fig. 9.2). In this model, an activity-related increase in $[K^+]_e$ depolarizes the membranes of astrocytes and thus activates transport mechanisms, resulting in the acidification of the ECS which in turn suppresses neuronal firing. Transmembrane ionic movements are accompanied by a water shift, resulting in glial swelling, ECS shrinkage and a further increase in the concentrations of neuroactive substances in the ECS until the neuronal activity eventually stops. Glial cells control not only the ionic composition of the ECS, but also its biophysical properties that determine the movement of ions, neuroactive substances, neurohormones, growth factors and metabolites. Changes in the ECS diffusion properties have a strong impact on modulating both synaptic and extrasynaptic volume transmission, the synchronization of neuronal activity in complex functions and neuron-glia communication. Glial cells, especially astrocytes, are highly sensitive to stimuli accompanying both physiological and pathological states, for example increased K^+ levels, osmolality changes, neuromediators or signals from damaged tissue, and promptly react by cell swelling, proliferation and hypertrophy, resulting in changes in the ECS volume and tortuosity. The production of extracellular matrix molecules by activated astrocytes further hinders already impaired diffusion and contributes to an increase in tortuosity. The arrangement of myelin sheaths and astrocytic processes in distinct brain regions (white matter, hippocampus, cerebellum, supraoptic nucleus) are responsible for diffusion anisotropy, which in turn is responsible for a certain degree of specificity in extrasynaptic transmission. Changes in the morphology of astrocytic processes and/or their rearrangement in aging or lactation may therefore lead to the disappearance of anisotropy with serious consequences for brain tissue function, e.g. impaired learning abilities or increased hormone production. Astrocytes can be viewed as active elements in signal transmission as they affect both synaptic as well as extrasynaptic volume transmission.

Acknowledgments The authors acknowledge the support of grants AV0Z50390512 from the Academy of Sciences of the Czech Republic and LC554 from the Ministry of Education, Youth and Sports of the Czech Republic.

References

Agnati LF,Zoli M, Stromberg I, Fuxe K (1995) Intercellular communication in the brain: Wiring versus volume transmission. *Neurosci* 69:711–726

Alberts B,Bray D, Lewis J, Raff M, Roberts K, Watson JD (1994) Molecular Biology of the Cell. New York: Garland Science

Andrew RD,Jarvis CR, Obeidat AS (1999) Potential sources of intrinsic optical signals imaged in live brain slices. *Methods Enzymol* 18:185–196

Araque A,Parpura V, Sanzgiri RP, Haydon PG (1999) Tripartite synapses: Glia, the unacknowledged partner. *Trends Neurosci* 22:208–215

Astion ML,Chvatal A, Orkand RK (1989) Na^+/H^+ exchange in glial cells of Necturus optic nerve. *Neurosci Lett* 107:167–172

Astion ML,Chvatal A, Orkand RK (1991) Further studies of electrogenic Na^+ /HCO3 – cotransport in glial cells of Necturus optic nerve: regulation of pHi. *Glia* 4:461–468

Berger T,Schnitzer J, Kettenmann H (1991) Developmental changes in the membrane current pattern, K+ buffer capacity, and morphology of glial cells in the corpus callosum slice. *J Neurosci* 11:3008–3024

Bonfanti L,Poulain DA, Theodosis DT (1993) Radial glia-like cells in the supraoptic nucleus of the adult rat. *J Neuroendocrinol* 5:1–5

Camby I,Belot N, Lefranc F, Sadeghi N, de Launoit Y, Kaltner H, Musette S, Darro F, Danguy A, Salmon I, Gabius HJ, Kiss R (2002) Galectin-1 modulates human glioblastoma cell migration into the brain through modifications to the actin cytoskeleton and levels of expression of small GTPases. *J Neuropathol Exp Neurol* 61:585–596

Cardin V,Lezama R, Torres-Marquez ME, Pasantes-Morales H (2003) Potentiation of the osmosensitive taurine release and cell volume regulation by cytosolic Ca^{2+} rise in cultured cerebellar astrocytes. *Glia* 44:119–128

Celio MR,Spreafico R, De Biasi S, Vitellaro-Zuccarello L (1998) Perineuronal nets: past and present. *Trends Neurosci* 21:510–515

Chesler M (1987) pH regulation in the vertebrate central nervous system: microelectrode studies in the brain stem of the lamprey. *Can J Physiol Pharmacol* 65:986–993

Chvatal A,Pastor A, Mauch M, Sykova E, Kettenmann H (1995) Distinct populations of identified glial cells in the developing rat spinal cord slice: ion channel properties and cell morphology. *Eur J Neurosci* 7:129–142
Chvatal A,Berger T, Vorisek I, Orkand RK, Kettenmann H, Sykova E (1997) Changes in glial K^+ currents with decreased extracellular volume in developing rat white matter. *J Neurosci Res* 49:98–106
Chvatal A,Anderova M, Ziak D, Sykova E (1999) Glial depolarization evokes a larger potassium accumulation around oligodendrocytes than around astrocytes in gray matter of rat spinal cord slices. *J Neurosci Res* 56:493–505
Chvatal A,Anderova M, Sykova E (2004) Analysis of K^+ accumulation reveals privileged extracellular region in the vicinity of glial cells in situ. *J Neurosci Res* 78:668–682
Chvatal A,Anderova M, Hock M, Prajerova I, Neprasova H, Chvatal V, Kirchhoff F, Sykova E (2007) Three-dimensional confocal morphometry reveals structural changes in astrocyte morphology in situ. *J Neurosci Res* 85:260–271
Coles JA,Orkand RK (1983) Modification of potassium movement through the retina of the drone (Apis mellifera male) by glial uptake. *J Physiol* 340:157–174
Cserr HF,DePasquale M, Nicholson C, Patlak CS, Pettigrew KD, Rice ME (1991) Extracellular volume decreases while cell volume is maintained by ion uptake in rat brain during acute hypernatremia. *J Physiol* 442:277–295
Deitmer JW,Schlue WR (1987) The regulation of intracellular pH by identified glial cells and neurones in the central nervous system of the leech. *J Physiol* 388:261–283
Dityatev A,Schachner M (2003) Extracellular matrix molecules and synaptic plasticity. *Nat Rev Neurosci* 4:456–468
Fuxe K,Agnati LF (1991) Volume Transmission in the Brain: Novel Mechanisms for Neural Transmission. New York: Raven Press
Gold R,Linington C, Lassmann H (2006) Understanding pathogenesis and therapy of multiple sclerosis via animal models: 70 years of merits and culprits in experimental autoimmune encephalomyelitis research. *Brain* 129:1953–1971
Gullans SR,Verbalis JG (1993) Control of brain volume during hyperosmolar and hypoosmolar conditions. *Annu Rev Med* 44:289–301
Hardingham TE,Fosang AJ (1992) Proteoglycans: many forms and many functions. *Faseb J* 6:861–870
Harvey AR,Kendall CL, Sykova E (1997) The status and organization of astrocytes, oligodendroglia and microglia in grafts of fetal rat cerebral cortex. *Neurosci Lett* 228:58–62
Hatten ME,Liem RK, Shelanski ML, Mason CA (1991) Astroglia in CNS injury. *Glia* 4:233–243
Hatton GI (1997) Function-related plasticity in hypothalamus. *Annu Rev Neurosci* 20:375–397
Haydon PG (2001) GLIA: listening and talking to the synapse. *Nat Rev Neurosci* 2:185–193
Hayen W,Goebeler M, Kumar S, Riessen R, Nehls V (1999) Hyaluronan stimulates tumor cell migration by modulating the fibrin fiber architecture. *J Cell Sci* 112 (Pt 13):2241–2251
Heinemann U,Lux HD (1977) Ceiling of stimulus induced rises in extracellular potassium concentration in the cerebral cortex of cat. *Brain Res* 120:231–249
Herkenham M (1987) Mismatches between neurotransmitter and receptor localizations in brain: observations and implications. *Neuroscience* 23:1–38
Hirrlinger J,Hulsmann S, Kirchhoff F (2004) Astroglial processes show spontaneous motility at active synaptic terminals in situ. *Eur J Neurosci* 20:2235–2239
Jendelova P,Sykova E (1991) Role of glia in K+ and pH homeostasis in the neonatal rat spinal cord. *Glia* 4:56–63
Kempski O,Staub F, Jansen M, Baethmann A (1990) Molecular mechanisms of glial cell swelling in acidosis. *Adv Neurol* 52:39–45
Kettenmann H,Orkand RK, Schachner M (1983) Coupling among identified cells in mammalian nervous system cultures. *J Neurosci* 3:506–516
Kilb W,Dierkes PW, Sykova E, Vargova L, Luhmann HJ (2006) Hypoosmolar conditions reduce extracellular volume fraction and enhance epileptiform activity in the CA3 region of the immature rat hippocampus. *J Neurosci Res* 84:119–129

Kimelberg HK (1991) Swelling and volume control in brain astroglial cells. In: Advances in Comparative and Environmental Physiology (Giles Raa, ed), pp 81–117. Berlin: Springer

Kimelberg HK,Goderie SK, Higman S, Pang S, Waniewski RA (1990) Swelling-induced release of glutamate, aspartate, and taurine from astrocyte cultures. *J Neurosci* 10:1583–1591

Kimelberg HK,Sankar P, O'Connor ER, Jalonen T, Goderie SK (1992) Functional consequences of astrocytic swelling. *Prog Brain Res* 94:57–68

Kiss JP,Vizi ES (2001) Nitric oxide: a novel link between synaptic and nonsynaptic transmission. *Trends Neurosci* 24:211–215

Kleene R,Schachner M (2004) Glycans and neural cell interactions. *Nat Rev Neurosci* 5:195–208

Kriz N,Sykova E, Ujec E, Vyklicky L (1974) Changes of extracellular potassium concentration induced by neuronal activity in the sinal cord of the cat. *J Physiol* 238:1–15

Lehmenkuhler A,Sykova E, Svoboda J, Zilles K, Nicholson C (1993) Extracellular space parameters in the rat neocortex and subcortical white matter during postnatal development determined by diffusion analysis. *Neuroscience* 55:339–351

MacVicar BA (1984) Voltage-dependent calcium channels in glial cells. *Science* 226:1345–1347

MacVicar BA,Hochman D (1991) Imaging of synaptically evoked intrinsic optical signals in hippocampal slices. *J Neurosci* 11:1458–1469

Mamata H,Mamata Y, Westin CF, Shenton ME, Kikinis R, Jolesz FA, Maier SE (2002) High-resolution line scan diffusion tensor MR imaging of white matter fiber tract anatomy. *AJNR Am J Neuroradiol* 23:67–75

Margolis RK,Margolis RU (1993) Nervous tissue proteoglycans. *Experientia* 49:429–446

Matsuoka Y,Hossmann KA (1982) Cortical impedance and extracellular volume changes following middle cerebral artery occlusion in cats. *J Cereb Blood Flow Metab* 2:466–474

Mazel T,Simonova Z, Sykova E (1998) Diffusion heterogeneity and anisotropy in rat hippocampus. *Neuroreport* 9:1299–1304

Meech RW,Thomas RC (1987) Voltage-dependent intracellular pH in Helix aspersa neurones. *J Physiol* 390:433–452

Mongin AA,Orlov SN (2001) Mechanisms of cell volume regulation and possible nature of the cell volume sensor. *Pathophysiology* 8:77–88

Neprasova H,Anderova M, Petrik D, Vargova L, Kubinova S, Chvatal A, Sykova E (2007) High extracellular K(+) evokes changes in voltage-dependent K(+) and Na (+) currents and volume regulation in astrocytes. *Pflugers Arch* 453:839–849

Nicholson C,Phillips JM (1981) Ion diffusion modified by tortuosity and volume fraction in the extracellular microenvironment of the rat cerebellum. *J Physiol (Lond)* 321:225–257

Nicholson C,Tao L (1993a) Hindered diffusion of high molecular weight compounds in brain extracellular microenvironment measured with integrative optical imaging. *Biophys J* 65:2277–2290

Nicholson C,Tao L (1993b) Hindered diffusion of high molecular weight compounds in brain extracellular microenvironment measured with integrative optical imaging. *Biophys J* 65:2277–2290

Nicholson C,Sykova E (1998) Extracellular space structure revealed by diffusion analysis. *Trends Neurosci* 21:207–215

Norton WT,Aquino DA, Hozumi I, Chiu FC, Brosnan CF (1992) Quantitative aspects of reactive gliosis: a review. *Neurochem Res* 17:877–885

Oliet SH,Piet R, Poulain DA (2001) Control of glutamate clearance and synaptic efficacy by glial coverage of neurons. *Science* 292:923–926

Orkand RK,Nicholls JG, Kuffler SW (1966) Effect of nerve impulses on the membrane potential of glial cells in the central nervous system of amphibia. *J Neurophysiol* 29:788–806

Pasantes Morales H,Schousboe A (1988) Volume regulation in astrocytes: a role for taurine as an osmoeffector. *J Neurosci Res* 20:503–509

Pasantes-Morales H,Franco R, Torres-Marquez ME, Hernandez-Fonseca K, Ortega A (2000) Amino acid osmolytes in regulatory volume decrease and isovolumetric regulation in brain cells: contribution and mechanisms. *Cell Physiol Biochem* 10:361–370

Pierpaoli C,Jezzard P, Basser PJ, Barnett A, Di Chiro G (1996) Diffusion tensor MR imaging of the human brain. *Radiology* 201:637–648

Piet R,Vargova L, Sykova E, Poulain DA, Oliet SH (2004) Physiological contribution of the astrocytic environment of neurons to intersynaptic crosstalk. *Proc Natl Acad Sci U S A* 101:2151–2155

Prokopova S,Vargova L, Sykova E (1997) Heterogeneous and anisotropic diffusion in the developing rat spinal cord. *Neuroreport* 8:3527–3532

Prokopova-Kubinova S,Vargova L, Tao L, Ulbrich K, Subr V, Sykova E, Nicholson C (2001) Poly[*N*-(2-hydroxypropyl)methacrylamide] polymers diffuse in brain extracellular space with same tortuosity as small molecules. *Biophys J* 80:542–548

Reum T,Olshausen F, Mazel T, Vorisek I, Morgenstern R, Sykova E (2002) Diffusion parameters in the striatum of rats with 6-hydroxydopamine-induced lesions and with fetal mesencephalic grafts. *J Neurosci Res* 70:680–693

Rice ME,Okada YC, Nicholson C (1993) Anisotropic and heterogeneous diffusion in the turtle cerebellum: Implications for volume transmission. *J Neurophysiol* 70:2035–2044

Roitbak T,Sykova E (1999) Diffusion barriers evoked in the rat cortex by reactive astrogliosis. *Glia* 28:40–48

Sykova E (1992) Ionic and Volume Changes in the Microenvironment of Nerve and Receptor Cells. Berlin: Springer

Siesjo BK,von Hanwehr R, Nergelius G, Nevander G, Ingvar M (1985) Extra- and intracellular pH in the brain during seizures and in the recovery period following the arrest of seizure activity. *J Cereb Blood Flow Metab* 5:47–57

Simonova Z,Svoboda J, Orkand P, Bernard CC, Lassmann H, Sykova E (1996) Changes of extracellular space volume and tortuosity in the spinal cord of Lewis rats with experimental autoimmune encephalomyelitis. *Physiol Res* 45:11–22

Sims TJ,Gilmore SA (1992) Glial response to dorsal root lesion in the irradiated spinal cord. *Glia* 6:96–107

Sontheimer H,Kettenmann H (1988) Heterogeneity of potassium currents in cultured oligodendrocytes. *Glia* 1:415–420

Sontheimer H,Trotter J, Schachner M, Kettenmann H (1989) Channel expression correlates with differentiation stage during the development of oligodendrocytes from their precursor cells in culture. *Neuron* 2:1135–1145

Svoboda J,Sykova E (1991) Extracellular space volume changes in the rat spinal cord produced by nerve stimulation and peripheral injury. *Brain Res* 560:216–224

Sykova E (1997) Extracellular space in the CNS: its regulation, volume and geometry in normal and pathological neuronal function. *Neuroscientist* 3:28–41

Sykova E (2001) Glial diffusion barriers during aging and pathological states. *Prog Brain Res* 132:339–363

Sykova E (2003) Diffusion parameters of the extracellular space. *Israel J Chem* 43:55–69

Sykova E (2004a) Extrasynaptic volume transmission and diffusion parameters of the extracellular space. *Neuroscience* 129:861–876

Sykova E (2004b) Diffusion properties of the brain in health and disease. *Neurochem Int* 45:453–466

Sykova E,Svoboda J (1990) Extracellular alkaline-acid-alkaline transients in the rat spinal cord evoked by peripheral stimulation. *Brain Res* 512:181–189

Sykova E,Svoboda J, Simonova Z, Lehmenkuhler A, Lassmann H (1996) X-irradiation-induced changes in the diffusion parameters of the developing rat brain. *Neuroscience* 70:597–612

Sykova E,Mazel T, Simonova Z (1998) Diffusion constraints and neuron-glia interaction during aging. *Exp Gerontol* 33:837–851

Sykova E,Vargova L, Prokopova S, Simonova Z (1999a) Glial swelling and astrogliosis produce diffusion barriers in the rat spinal cord. *Glia* 25:56–70

Sykova E,Roitbak T, Mazel T, Simonova Z, Harvey AR (1999b) Astrocytes, oligodendroglia, extracellular space volume and geometry in rat fetal brain grafts. *Neuroscience* 91:783–798

Sykova E,Mazel T, Hasenohrl RU, Harvey AR, Simonova Z, Mulders WH, Huston JP (2002) Learning deficits in aged rats related to decrease in extracellular volume and loss of diffusion anisotropy in hippocampus. *Hippocampus* 12:269–279

Sykova E,Vargova L, Kubinova S, Jendelova P, Chvatal A (2003) The relantionship between changes in intrinsic optical signals and cell swelling in rat spinal cord slices. *NeuroImage* 18:214–230

Sykova E,Vorisek I, Mazel T, Antonova T, Schachner M (2005a) Reduced extracellular space in the brain of tenascin-R- and HNK-1-sulphotransferase deficient mice. *Eur J Neurosci* 22:1873–1880

Sykova E,Vorisek I, Antonova T, Mazel T, Meyer-Luehmann M, Jucker M, Hajek M, Ort M, Bures J (2005b) Changes in extracellular space size and geometry in APP23 transgenic mice: a model of Alzheimer's disease. *Proc Natl Acad Sci U S A* 102:479–484

Thomas RC (1977) The role of bicarbonate, chloride and sodium ions in the regulation of intracellular pH in snail neurones. *J Physiol (Lond)* 273:317–338

Tao L,Masri D, Hrabetova S, Nicholson C (2002) Light scattering in rat neocortical slices differs during spreading depression and ischemia. *Brain Res* 952:290–300

Theodosis DT,Poulain DA (1993) Activity-dependent neuronal-glial and synaptic plasticity in the adult mammalian hypothalamus. *Neuroscience* 57:501–535

Thorne RG,Nicholson C (2006) In vivo diffusion analysis with quantum dots and dextrans predicts the width of brain extracellular space. *Proc Natl Acad Sci U S A* 103:5567–5572

Thorne RG,Hrabetova S, Nicholson C (2004) Diffusion of epidermal growth factor in rat brain extracellular space measured by integrative optical imaging. *J Neurophysiol* 92:3471–3481

Villegas GM,Fernandez J (1966) Permeability to thorium dioxide of the intercellular spaces of the frog cerebral hemisphere. *Exp Neurol* 15:18–36

Van der Toorn A,Sykova E, Dijkhuizen RM, Vorisek I, Vargova L, Skobisova E, Van Lookeren Campagne M, Reese T, Nicolay K (1996) Dynamic changes in water ADC, energy metabolism, extracellular space volume, and tortuosity in neonatal rat brain during global ischemia. *Magn Reson Med* 36:52–60

Van Harreveld A,Dafny N, Khattab FI (1971) Effects of calcium on electrical resistance and the extracellular space of cerebral cortex. *Exp Neurol* 31:358–367

Vargova L,Jendelova P, Chvatal A, Sykova E (2001a) Glutamate, NMDA, and AMPA induced changes in extracellular space volume and tortuosity in the rat spinal cord. *J Cereb Blood Flow Metab* 21:1077–1089

Vargova L,Chvatal A, Anderova M, Kubinova S, Ziak D, Sykova E (2001b) Effect of osmotic stress on potassium accumulation around glial cells and extracellular space volume in rat spinal cord slices. *J Neurosci Res* 65:129–138

Vargova L,Homola A, Zamecnik J, Tichy M, Benes V, Sykova E (2003) Diffusion parameters of the extracellular space in human gliomas. *Glia* 42:77–88

Vizi ES (1980) Nonsynaptic modulation of transmitter release: Pharmacological implication. *Trends Pharmacol Sci* 1:172–175

Vizi ES (1984) Non-synaptic Interaction Between Neurones: Modulation of Neurochemical Transmission. Chichester/New York: Wiley

Vizi ES (2000) Role of high-affinity receptors and membrane transporters in nonsynaptic communication and drug action in the central nervous system. *Pharmacol Rev* 52

Vizi ES (2003) Non-synaptic interaction between neurons in the brain, an analog system: far from Cajal-Sherringtons's galaxy. *Bull Mem Acad R Med Belg* 158:373–379

Vizi ES,Kiss JP, Lendvai B (2004) Nonsynaptic communication in the central nervous system. *Neurochem Int* 45:443–451

Vorisek I,Sykova E (1997a) Evolution of anisotropic diffusion in the developing rat corpus callosum. *J Neurophysiol* 78:912–919

Vorisek I,Sykova E (1997b) Ischemia-induced changes in the extracellular space diffusion parameters, K+, and pH in the developing rat cortex and corpus callosum. *J Cereb Blood Flow Metab* 17:191–203

Vorisek I,Hajek M, Tintera J, Nicolay K, Sykova E (2002) Water ADC, extracellular space volume, and tortuosity in the rat cortex after traumatic injury. *Magn Reson Med* 48:994–1003

Walz W,Hertz L (1983) Intracellular ion changes of astrocytes in response to extracellular potassium. *J Neurosci Res* 10:411–423

Walz W,Hinks EC (1986) A transmembrane sodium cycle in astrocytes. *Brain Res* 368:226–232
Wieshmann UC,Clark CA, Symms MR, Barker GJ, Birnie KD, Shorvon SD (1999) Water diffusion in the human hippocampus in epilepsy. *Magn Reson Imaging* 17:29–36
Zamecnik J,Vargova L, Homola A, Kodet R, Sykova E (2004) Extracellular matrix glycoproteins and diffusion barriers in human astrocytic tumours. *Neuropathol Appl Neurobiol* 30:338–350
Zhang H,Kelly G, Zerillo C, Jaworski DM, Hockfield S (1998) Expression of a cleaved brain-specific extracellular matrix protein mediates glioma cell invasion In vivo. *J Neurosci* 18:2370–2376
Zoli M,Jansson A, Sykova E, Agnati LF, Fuxe K (1999) Intercellular communication in the central nervous system. The emergence of the volume transmission concept and its relevance for neuropsychopharmacology. *Trends Pharmacol Sci* 20:142–150

Abbreviations

α	Extracellular volume fraction
ADC	Apparent diffusion coefficient
ADCw	Apparent diffusion coefficient of water
AMPA	α-Amino-3-hydroxy-5-metyl-isoxazol-4-propionic acid
CNS	Central nervous system
CSPG	Chondroitin-sulphate proteoglycan
D	Diffusion coefficient
dpw	Day post wounding
DW-MRI	Diffusion-weighted magnetic resonance imaging
EAE	Experimental autoimmune encephalomyelitis
ECM	Extracellular matrix
ECS	Extracellular space
GABA	γ-Aminobutyric acid
GFAP	Glial fibrillary acidic protein
IOS	Intrinsic optical signals
ISM	Ion-selective microelectrode
k'	Nonspecific cellular uptake
$[K^+]_e$	Extracellular concentration of potassium ions
λ	Tortuosity
NMDAR	N-methyl-d-aspartate receptor
pH_e	Extracellular pH
pH_i	Intracellular pH
RVD	Regulatory volume decrease
TEA	Tetraethylammonium
TMA	Tetramethylammonium
TN	Tenascin
WHO	World Health Organization

Chapter 10
Structural Association of Astrocytes with Neurons and Vasculature: Defining Territorial Boundaries

Andreas Reichenbach and Hartwig Wolburg

Contents

10.1 Introduction

Considered as a tissue, the mature brain consists of three main compartments. These are, as derivatives of the neuroectoderma, (a) the neurons and their processes (about 50% of the total volume, e.g., in the rat brain), (b) the (macro-)glia (about 20% of the volume), and, as derivatives of the mesoderma, (c) the blood vessels

A. Reichenbach
Paul Flechsig Institute of Brain Research, Leipzig University, Leipzig, Germany
reia@medizin.uni-leipzig.de

V. Parpura and P.G. Haydon (eds.), *Astrocytes in (Patho)Physiology of the Nervous System*, 251
DOI: 10.1007/978-0-387-79492-1_10, © Springer Science+Business Media, LLC 2009

(less than 10% of the volume). In between these compartments, there is a variable extracellular space (up to > 20%, depending on the developmental stage, as well as on specific local conditions). The exchange of nutrients and metabolites between the cellular constituents and the blood vessels is a precondition for cellular function, as in any other tissue. More than in most other tissues, the function of brain cells consists of an exchange of signals such as, e.g., bioactive molecules. Considering the extreme functional specialization of neuronal cells, and their permanent signaling activity throughout the day and night, the exchange of molecules between the three brain compartments appears as a very demanding logistical problem. In this chapter, we describe how the structure of the astroglial cells is optimized to promote – and to control – this exchange.

10.2 The Ectodermal Player: Neurons: Polarized Cells with Several Specialized Compartments

Neurons are cells that are characterized by the ability to receive, transform and propagate information. To this end, they are endowed with a variety of processes or other cellular compartments, optimized to perform one or more of these tasks. Generally, neurons are polarized cells in which one pole is dedicated to receive information, and the other to transmit it to targets such as other neurons, muscles, glands, or to the circulating blood. The "receptor pole" may consist of a true sensory process, but most neurons in our CNS receive information from specialized sensory cells, and/or from other neurons. In these cases, the receptor pole is made up of a dendritic tree, which receives information via synaptic contacts. The "effector pole" usually consists of an axon, and its main feature is the presence of a presynaptic terminal (or of a number of them), capable of transmitting signals to other cells. Usually, signals are transmitted in the form of secreted neurotransmitter molecules but electrical coupling may occur as well.

These distinct neuronal compartments are characterized by specific, "optimized" morphological and functional properties, such as the expression of certain ion channels, ligand receptors, etc. The generation of regenerative Na^+ currents [i.e., tetrodotoxin (TTX)-sensitive action potentials] is not a distinctive feature of any of the compartments. The axon as well as the dendrites may be able to generate action potentials. These all-or-nothing responses are necessary for the propagation of information along processes that are too long for an effective propagation of amplitude-coded signals; the limit appears to be close to 300 μm. The action potentials may propagate either continuously or in a saltatory manner; in the latter case, the neurite is usually myelinated. It should be mentioned here that Ca^{2+}-mediated action potentials may occur as well (often simultaneously with the Na^+-carried action potentials); these cause a Ca^{2+} influx into the cells and are probably involved in intracellular second-messenger signal cascades. The expression of ligand receptors is not a distinctive feature of any of the neuronal compartments, either. It is a characteristic feature of sub-/postsynaptic membranes on the dendrites and on the soma but it has been shown that axon membranes may also be endowed with ligand receptors; in particular, the axon initial segment is known to express gama amino butyric acid (GABA)-ergic inhibitory receptors in many instances (Christie and De Blas, 2003).

Finally it is important to note that the neurons and their compartments are all but stochastically distributed in the CNS, notwithstanding the highly complex and "irregular" appearance of many neuronal tissues. Facilitated by their common migration pathways as young postmitotic neurons along their "sibling" radial glial cells (Rakic, 1972, 1978), radially oriented columnar groups of cortical neurons preferentially share specific inputs (indicated by parallel afferent axon pathways), local synaptic circuits [(indicated by so-called dendritic bundles (Fleischhauer and Detzer, 1975; Gabbot and Bacon, 1996; and references therein)], and target structures (indicated by efferent axon bundles); well-characterized examples for this columnar organization are the visual cortex [reviewed by Mountcastle (1997)] (Fig. 10.1a) and the so-called barrel cortex [where the whiskers are represented (Rice and Van der Loos, 1977)]; but this principle has been demonstrated elsewhere in the cortex too. (Mountcastle 1957, 1958). Another example of functional territories is

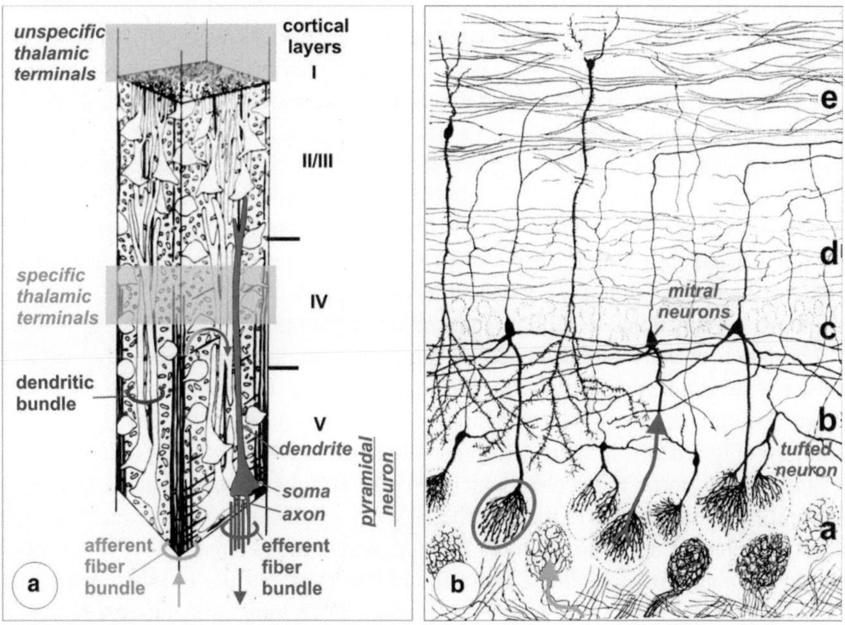

Fig. 10.1 Neuronal elements and their organization in the brain. **a** Visual cortex of adult rabbit [modified after (Fleischhauer and Detzer, 1975)]. **b** Olfactory bulb of newborn kitten [modified after (Ramón y Cajal, 1911)]. In the neocortex, the afferent (*green*) and efferent fibers (*red*), as well as the pyramidal neurons and their dendrites (*blue*) are aligned in radially oriented bundles. Moreover, a transversal orientation is provided by the layering of the cells and of the afferent terminals (*green/violet*). In the olfactory bulb, the synaptic transmission between the afferents from the olfactory receptor cells (*green*), the mitral (*red*) and tufted cells occurs in so-called glomeruli (*violet*). In both cases, the information processing of every neuron is clearly polarized with respect to input/output direction (*arrows*). (*See Color Plates*)

that constituted by the synaptic glomeruli of the olfactory cortex; there, distinct groups of olfactory neurons receive inputs from distinct groups of olfactory receptor cells (Kosaka and Kosaka, 2005; Mombaerts, 2006) (Fig. 10.1b). Additionally, the layered structure of both the cerebral and the cerebellar cortex, together with the presence of layer-restricted afferent fiber systems, generate further territorial circuits. For example, specific thalamic afferents end in layer IV of the neocortex whereas the unspecific ones end in layer I (Fig. 10.1a). Further details about this territorial organization principle of neuronal circuits (and their associated glial cells) will be given in Sects. 10.5.4 and 10.7.

10.3 The Mesodermal Player: Blood Vessels with Polarized Endothelial Cells and Pericytes

The complex brains of most vertebrates (and of all mammals) are much too large to be nourished by mere diffusion of oxygen and nutrients such as glucose (the limit for this pathway is close to some 100 μm). Therefore, the maturation of the fetal brain is accompanied by an ingrowth of blood vessels, basically from its mesenchymal envelopes (Feeney and Watterson, 1946; Bär, 1980). The adult mammalian brain thus contains a very complex pattern of large (arteries and veins) and medium-sized blood vessels (arterioles and venules), together with a wealth of capillaries. Many of the large blood vessels are arranged as radially aligned "loops" (arrows in Fig. 10.2a), giving rise to the smaller vessel systems which are oriented either radially or transversally (asterisks in Fig. 10.2a) (Bär 1980). Unfortunately, the blood vessels of the brain rarely form functional anastomoses; thus, a stenosis of distinct brain arteries causes the irreversible failure of blood supply in the dependent tissue compartments (stroke). This fact contributes to the functional territorial organization of the brain tissue.

The large, primary blood vessels of the brain (similar to those in other organs) consist of several layers; the innermost (luminal) layer is formed by endothelial cells. These cells are the dominant constituents of the capillaries; they are functionally polarized (Fig. 10.2c). In their membranes for instance, they express a high density of glucose transporter molecules (which provides for glucose transport into the brain neuropil). The distribution of the endothelial glucose transporter is asymmetrical. The density of glucose transporters in the luminal membrane is three to four times lower than in the abluminal membrane (Farrell and Pardridge, 1991). In functional terms, the lower concentration of transporter molecules in the luminal membrane limits the intensity of glucose flux from the blood to the endothelium, and the higher concentration in the abluminal membrane may reduce the endothelial glucose concentration in comparison to that in the blood and secure a better efficiency of transport from the

Fig. 10.2 (continued) **e** Freeze-fracture replica of a bovine brain capillary endothelial cell, in situ. The P-face of the replica (PF) displays many complex strands of particles, constituting a tight junction (tj strands). The *arrow* points to a tight junction; bar, 0.2 μm. **d** modified after Hamm et al., 2004, **e** modified after Wolburg and Risau, 1995.

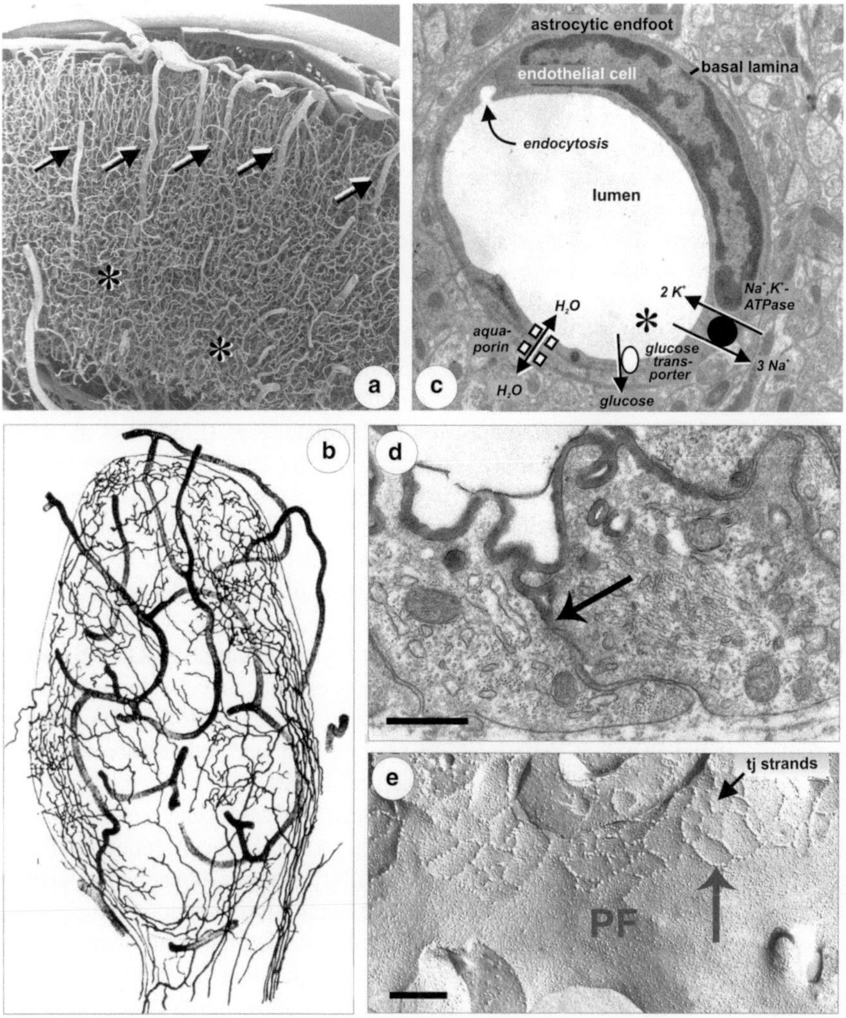

Fig. 10.2 Blood vessels of the brain; spatial and cellular organization. **a** Corrosion cast of a monkey brain (courtesy of A. L. Keller, Tübingen). **b** Olfactory glomerulum of the cat, nerve fibers and blood vessels (from Kölliker, 1896). The capillaries provide a specific net for the entire glomerulum but there also appear to be sub-compartments with their own capillary supply. **c** Transmission electron micrograph of a rat brain capillary (original data, H. Wolburg). Many primary blood vessels enter the brain from its surface, by forming rather regularly spaced, radially aligned loops (*arrows* in (**a**)). In the neuropil, less regularly aligned capillaries span the arterioles and venules (*asterisks* in (**a**)). The lumen of the blood vessels is surrounded by endothelial cells. These are polarized cells, expressing both the glucose transporter and the Na^+, K^+ pump mainly at the abluminal surface. Furthermore, they are endowed with water pores (aquaporins), and are capable of endocytosis and transcellular transport of larger molecules or particles. The endothelial cells are completely covered by a basal lamina, which the astrocytic endfeet abut. **d** and **e** Organization and effect of the blood–brain barrier (BBB). **d** Bovine brain capillary endothelial cells; if co-cultured with astrocytes, the cells form a versatile barrier against a tracer (wheat germ agglutinin-horseradish peroxidase, *thick black lines* in the *upper part*). The penetration of the tracer is stopped where the cells form tight junctions (*arrow*); bar, 1 μm.

endothelium to the brain parenchyma. In addition to this, many Na^+,K^+ pump molecules are localized asymmetrically in the abluminal membrane (which facilitates the clearance of excess K^+ ions into the blood vessels, and is involved in the generation of osmotic forces for transendothelial water transport) (Fig. 10.2c). Endocytosis also takes place as indicated by the presence of caveolae; probably, as a first step of transendothelial transport of larger molecules/particles. Furthermore, in the brain, endothelial cells are connected by extensive tight junctions; this constitutes the blood–brain barrier (BBB; for details, see Sect. 10.6.4).

Integral parts of the vascular complex are the pericytes which are consistently surrounded by a basal lamina. Pericytes are found in close association with endothelial cells even at very early stages of development and seem to be more prevalent on neural capillaries than on other capillaries (Simionescu et al., 1988). The function of pericytes in vivo has been unclear for a long time (Sims, 1986), but recently it has become evident that they are required for vessel maturation (Lindahl et al., 1997; Gerhardt and Betsholtz, 2003). It has been suggested that signaling of the endothelial angiogenetic receptor tyrosine kinase Tie-2 is required for upregulation of factors in endothelial cells that are chemotactic for pericytes and smooth muscle cells, leading to the migration of these cells towards the endothelial cell wall and to subsequent maturation of the vessels by an increased production of extracellular matrix components (Folkman and D'Amore, 1996). Amongst these, platelet-derived growth factor (PDGF)-B, a high-affinity ligand for the receptor tyrosine kinase PDGF-Rβ present on perivascular mesenchymal cells, is produced by endothelial cells during development. PDGF-B has been shown to be involved in vascularization of the brain, as disruption of the PDGF-B gene leads to pericyte loss and lethal microaneurysm formation during late embryogenesis (Lindahl et al., 1997).

In the developing chicken CNS, it has been shown that angiogenic vessels invading the neuroectoderm express N-cadherin at the interface between endothelial cells and pericytes. With the onset of barrier differentiation, N-cadherin labeling decreased, suggesting that transient N-cadherin expression in endothelial cells and pericytes may represent an initial signal involved in the commitment of early blood vessels to express blood brain barrier properties (Gerhardt et al., 1999).

10.4 The Joint Player: Astrocytes Polarized Cells with Several Specialized Compartments

Astroglial cells are characterized by endfeet contacting a basal lamina around blood vessels (encircled in Fig. 10.3a), the pia mater, or both. In addition, they display specialized cell processes that contact and/or even ensheath distinct neuronal elements (asterisks in Fig. 10.3a). These two different types of processes and contacts constitute the "opposite poles" of astrocytes as polarized cells, and are responsible for the structural and functional glia-neuron and glia-blood vessel interactions (Fig. 10.3a).

Since virtually all vascular surfaces are covered by astroglial endfeet, it is essential that astrocytes mediate an indirect exchange of molecules between neuronal compartments

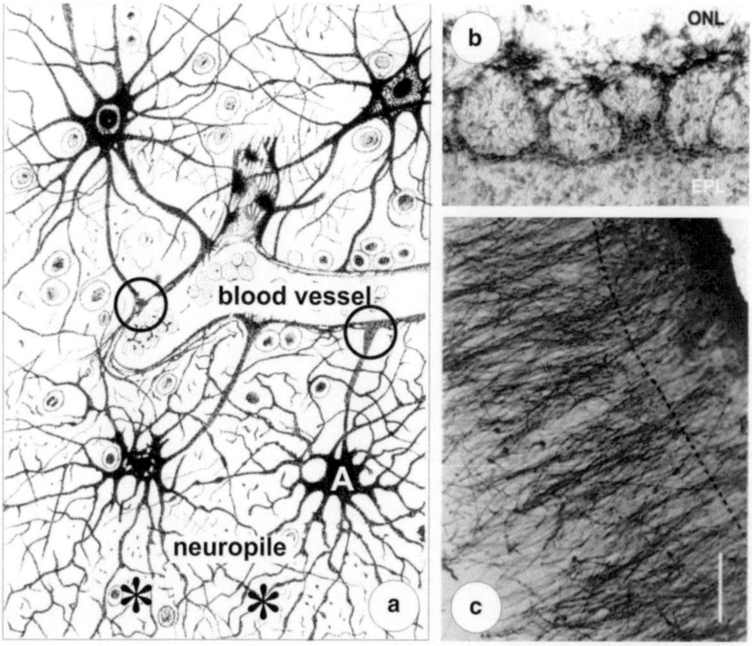

Fig. 10.3 Astroglial cells of the brain. **a** Camera-lucida drawing of silver-impregnated astrocytes [modified after Ramòn y Cajál (1911)]. **b** Immunohistochemical labeling (antibodies against glial fibrillary acidic protein, GFAP) of astrocytes in the olfactory bulb of the adult frog (modified after Bailey et al., 1999). **c** Immunohistochemical labeling (antibodies against GFAP) of interlaminar astrocytes in the monkey cortex (with permission, from Colombo and Reisin, 2004). Astrocytes are polarized cells which display perivascular (or pial) endfeet (encircled in (**a**)) on the one side, and neuron-abutting branches in the neuropil (*asterisks* in (**a**)) on the other side. In specialized areas of the brain, astrocytes and their processes may envelop large convolutes of synapses such as the glomeruli in the olfactory bulb (**b**). In the neocortex of higher primates, the radially aligned long processes of the interlaminar astrocytes run in parallel to the columnar organization of neuronal elements (**c**. cf. Fig. 10.1). The *dashed line* in **c** represents the border of layer I; calibration bar, 100 μm.

and blood vessels. This might occur within the processes of a single astrocyte or via gap-junctional coupling of entire populations of glial cells [even between oligodendrocytes and astrocytes (Robinson et al., 1993, Rash et al., 1997; Zahs, 1998; Nagy and Rash, 2000; Nagy et al., 2004)]; this coupling appears to be spatially constricted, as well as functionally regulated (cf. Sect. 10.7). With respect to territorial organization, peculiar arrangements of astroglial cells can be observed in some brain areas. For instance, astrocytes and their processes "envelop" the glomeruli in the olfactory bulb (Chao et al., 1997; Bailey et al., 1999) (Fig. 10.3b). In the adult cortex of higher primates, but not in other mammals studied so far, so-called interlaminar astrocytes display radially aligned processes (Reisin and Colombo, 2002; Colombo and Reisin, 2004) (Fig. 10.3c), in parallel to the columnar neuronal arrangement mentioned in Sect. 10.2.

10.5 The Sites of Inter-Ectodermal Interplay: "Peripheral Astrocyte Processes" (PAPs)

On the glial side, the glio-neuronal interplay is realized by specialized terminal protrusions or end-branches of the glial cytoplasm. These structures are characterized – in contrast to the thicker "stem" processes, side branches, and/or somata from which they arise – by their shortage of cytoplasmic organelles. Generally, they appear as thin membrane out-foldings which may assume lamellar or finger-like shapes. Based on their variable shape, they have been given various names, such as "lamellipodia and filopodia" (Chao et al., 2002), "lamellae and finger-like extensions" (Wolff, 1968), or, "peripheral astroglial processes" [(PAPs; (Derouiche and Frotscher, 2001; Reichenbach et al., 2004)], which avoids any assumption concerning shape.

The PAPs, but not the thicker astroglial cell processes or somata, were shown to contain actin-binding ERM molecules such as ezrin (Derouiche and Frotscher, 2001; Derouiche et al., 2002), which are probably involved in the generation and maintenance of their complex shapes and in the creation of high surface-to-volume ratios (up to more than 30 $\mu m^2\ \mu m^{-3}$). Despite their small size, the large number of PAPs causes them to constitute about half of the astroglial volume and about 80% of the surface area (Chao et al., 2002). The large surface area is thought to give space for the insertion of membrane proteins such as ion channels, ligand receptors, and carrier molecules, necessary for functional interactions with the neuronal compartments contacted by the PAPs (see Sect. 10.5.4).

It should be noted that the interface between neurons and PAPs involves an extracellular cleft. The width of this cleft may vary considerably, depending on the developmental and/or functional state, or on local specializations. Furthermore, this cleft contains an extracellular matrix that may also vary in its abundance and molecular composition, and which is thought to contribute crucially to the wealth of structural and functional neuro-glial interactions.

10.5.1 PAPs in the Granular Layers: Somata and Velate Astrocytes

Depending on the size of a given neuronal soma, its glial sheath may be constituted by one or more glial cells. Typically, large neurons are ensheathed by velate terminal processes of several astrocytes, such as in the case of Purkinje cells in the cerebellum (Fig. 10.4). By contrast, several densely packed small neuronal somata (often termed granule cells) may be ensheathed by the velate processes of one or a few glial cells, such as cerebellar granule cells [by velate astrocytes (Chan-Palay and Palay, 1972); Fig. 10.4), or the granule cells of the hippocampal fascia dentata (Fig. 10.4).

Usually, the glial sheaths cover the neuronal soma surface almost completely (Fig. 10.4), with the exception of "holes" where a direct apposition of neuronal membranes occurs at synaptic sites. These holes are also visible in the perineuronal

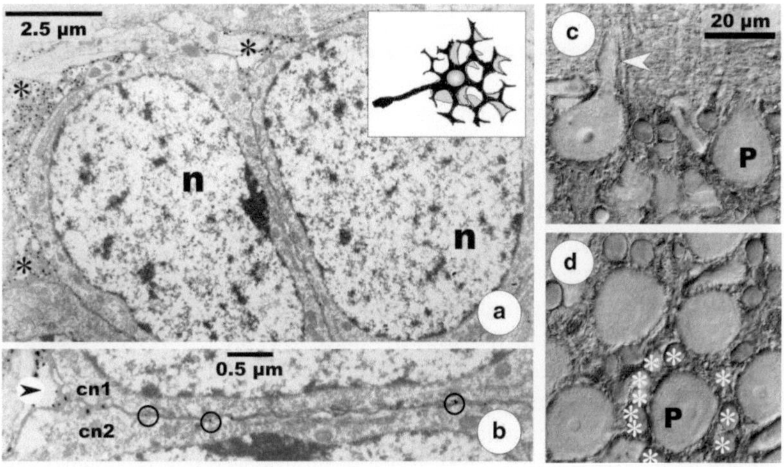

Fig. 10.4 Glial sheaths around neuronal somata. **a** and **b** Granule cell layer of the rat hippocampus, transmission electron micrographs (*inset*: semi-schematic drawing of a velate astrocyte), n, nucleus; **c** and **d** Purkinje cell layer of the rat cerebellum (**c** radial, **d** horizontal section, immunohistochemical labeling of Bergmann glial cells by antibodies against GFAP); [modified after Reichenbach et al. (2004)]. In the granule cell layer of the hippocampus, the extremely thin processes of a velate astrocyte (inset) (labeled against GFAP by immunogold particles, *asterisks* in (**a**)) are interplaced (*arrowhead* and *circles* in (**b**)) between the cytoplasm of several adjacent neurons (cn1, cn2 in **b**). In the cerebellum by contrast, individual Purkinje cell somata (P) are ensheathed by the processes of several Bergmann glial cells (*asterisks* in (**d**)). In addition, the stem dendrites of the Purkinje cells are also ensheathed (*arrowhead* in (**c**)).

nets that fill the interfaces between neuronal membranes and glial sheaths but leave space for synaptic contacts (Brückner et al., 1996). It should be noted, that groups of neuronal somata without individual glial sheaths (i.e., with direct apposition of their membranes) can be found in some brain stem nuclei.

10.5.2 PAPs in the White Matter: Axons and Fibrous Astrocytes

The axons of neurons may differ greatly in length and diameter, as well as in functional parameters. Typically, the axon arises from the soma at a funnel-shaped area termed the axon hillock. As a general rule in vertebrates, axons with a diameter of more than 0.3 µm perform a saltatory conduction of action potentials. In these axons, action potential-generating Na^+ channels are not randomly distributed in the membrane but are focused within small areas ("hot spots") along the axon. Such neurites are covered by myelin sheaths between these hot spots. The myelin sheaths are provided by oligodendrocytes.

The axon hillock/initial segment is individually covered by glial cell processes. Rather large extracellular spaces, filled with a specialized extracellular matrix, are often interspaced between the initial segment and the glial sheath (Brückner et al.,

1996) (Fig. 10.5a). Thin unmyelinated axons have no individual glial sheaths; rather, bundles of such axons are enwrapped by the processes of astrocytes (e.g., in the optic nerve; Fig. 10.5b). It has already been mentioned that saltatory conduction of action potentials requires the presence of hot spots with a high density of spike-generating Na^+ channels; these are called nodes of Ranvier. The nodes of Ranvier are devoid of a myelin sheath but not of glial contacts: assemblies ("coronae") of small, finger-like PAPs of the fibrous astrocytes contact the nodal membrane (Raine, 1984; Waxman, 1986; Hildebrand et al., 1993; Butt et al., 1994) (Fig. 10.5c). These PAPs express the J1 glycoprotein, an adhesion-modulating protein presumably involved in axon-glial interactions modulating the assembly and/or maintenance of nodes of Ranvier (ffrench-Constant et al., 1986). At the axon-glial contact areas, the interposed extracellular clefts are extremely narrow [about 6 nm (Waxman, 1986)]. However, the adjacent perinodal space contains an abundant extracellular matrix, comparable to that of the perineuronal nets (Raine, 1984; Hildebrand et al., 1993) (Fig. 10.5d). It has been proposed that this matrix buffers the large increases of K^+ ions near the sites of action potential generation (Treherne et al., 1982; Härtig et al., 1999).

The function of the perinodal glial cell processes has been a matter of much speculation. A glio-neuronal exchange of molecules (including the delivery of Na^+ channel molecules (Bevan et al., 1985; Shrager et al., 1985) has been proposed. It has also been speculated that the glial fingers might be sensors of axonal action potentials. Strong depolarization of the glial membrane could be induced by ephaptic transmission in which a current is directly transmitted through the extracellular space, and might then trigger metabolic reactions of the glial cells (Chao et al., 1994).

10.5.3 PAPs in the Gray Matter: Dendrites and Protoplasmic Astrocytes

Typically, dendrites are the main sites of synaptic input. For this reason, dendritic trees possess a large surface membrane area which is generated by multiple branching of the stem dendrites. Much of the non-synaptic surface of the stem dendrites may be covered by glial sheaths (Fig. 10.4c, arrowhead); the processes of several adjacent glial cells may contribute to the sheath around one stem dendrite. As the dendrites are the major constituents of the neuropil (gray matter), their sheaths are composed of the glial cells of the gray matter, i.e., by protoplasmic astrocytes, or, in the cerebellum, by Bergmann glial cells which are a subtype of radial astrocytes. Another component of a subpopulation of peridendritic sheaths are the perineuronal nets, which consist of a particular extracellular matrix and are covered by net-like PAPs. It can be speculated that both the peridendritic glial sheaths and the perineuronal nets serve to buffer and clear ions and bioactive molecules in the adjacent extracellular clefts.

In some instances, the dendrites are very long, and may not allow electrotonic signal propagation to the soma. It has been shown that the dendrites of Purkinje

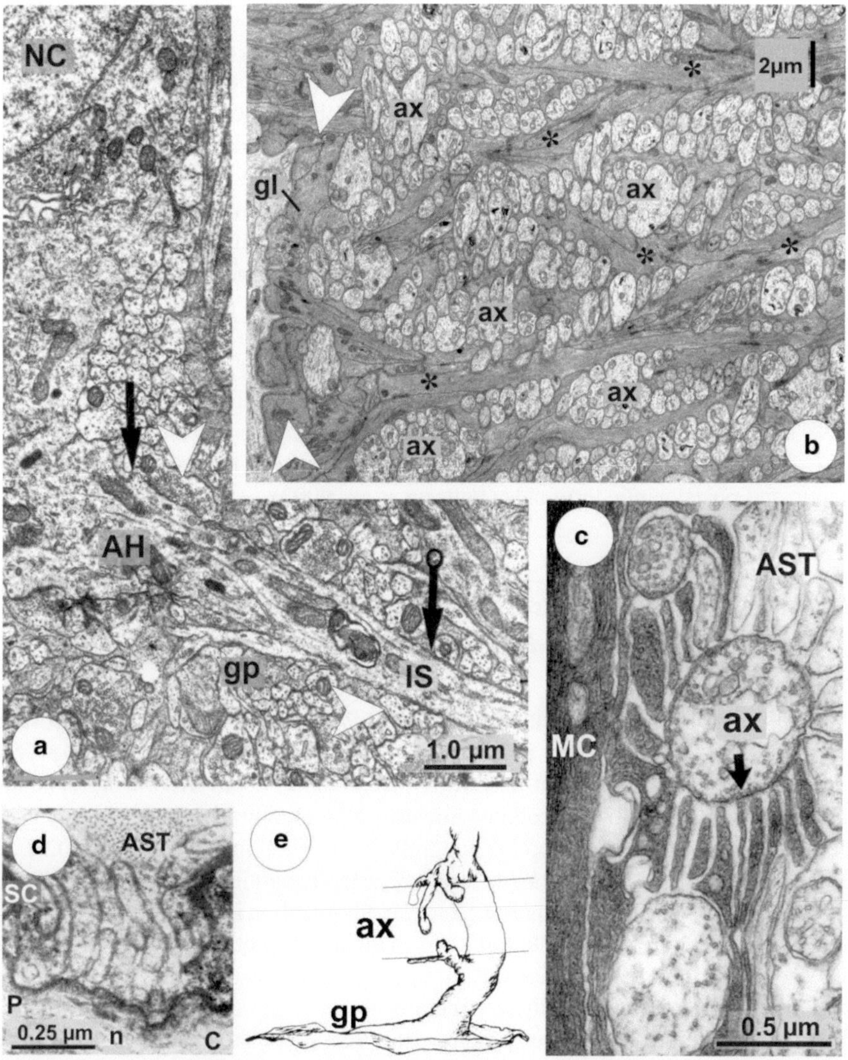

Fig. 10.5 Astroglial interactions with axons, including the axon hillock and the initial segment. **a** Glial processes (gp) envelop much of the axon hillock (AH, *large arrow*) and the initial segment (IS, *large arrow with circle*) but leave space for abutting synaptic terminals (*arrowheads*). Modified after Hámori (1981). **b** Optic nerve of the rat; many fascicles of unmyelinated axons (ax) are surrounded by astrocytic processes (*asterisks*). Modified after Wolburg and Bäuerle (1993). **c** (*right side*) Node-like axonal specialization of an intraretinal axon (ax, *arrow*), contacted by coronae of finger-like PAPs arising from an astrocyte (AST as well as from a Müller glial cell (MC); modified after Holländer et al. (1991). **d** Perinodal space of a myelinated axon at the transition from the spinal cord (c, *right side*) to a peripheral nerve (P, *left side*); finger-like processes are contributed both by the astrocyte (AST) and the Schwann cell (SC); n, node. Modified after Berthold and Carlstedt (1977). **e** Finger-like PAPs at a node in the rat optic nerve; modified after Butt et al. (1994).

neurons in the cerebellum possess "hot spots" in their membrane where Ca^{2+}-mediated action potentials can be generated (Llinas et al., 1969; Llinas, 1975). Action potentials are generated in the dendritic membrane of many other types of neurons [reviewed by (Häusser et al., 2000)]. This fits with the idea that these action potentials serve as an auxiliary aid to convey distal information along long "dendritic cables" (e.g., of neocortical pyramidal cells) towards the soma. Such long dendrites, or parts of them, may even be myelinated by oligodendrocytes (Pinching, 1971). More about the putative function(s) of dendritic spiking can be found in Häusser et al. (2000).

Analogous to the synapses, which are the main sites of neuron-to-neuron signaling, the perisynaptic glial sheaths are considered to be the prototypic sites of glio-neuronal interactions. This view has been challenged by the observation that certain synapses, or even large groups of synapses, are devoid of apparent glial contacts (for review, see Chao et al., 2002). Nevertheless, most synapses of the vertebrate CNS are endowed with elaborate glial sheaths, and there is no doubt that the latter constitute a crucial "third element" of the typical synapse, in addition to the presynaptic and postsynaptic terminals (Volterra et al., 2002).

10.5.4 The Central Case: PAPs at Synapses: The Concept of Glial "Microdomains"

Typically, synapses are ensheathed by lamellar PAPs (Fig. 10.6), but the elaborateness of ensheathing may vary considerably even within the same area of the CNS (see Chao et al., 2002). In rat neocortex for example, about 56% of all synaptic perimeters are covered by astroglia while astroglial membranes make up only 22% of all membranes in the neuropil [Landgrebe et al.; cited in Chao et al. (2002)]. In particular, most synaptic clefts are "sealed" at their margins by PAPs (Fig. 10.6a, b), which may even be multilamellar. However, the situation is different in specialized subcortical structures and in the olfactory bulb where multiple synaptic junctions are enclosed in a common glial sheath, termed "synaptic glomeruli" or "complex synapses". Glial coverage in these structures is very high (often there are multilamellar sheaths), but it does not penetrate the interior of the complexes, and thus, cannot seal individual synaptic clefts (Fig. 10.3b). As an extreme case, there are even astroglia-free neuropil compartments, e.g., in Rolando's substantia gelatinosa of the spinal cord and in the cochlear nucleus, where thin sensory axons terminate in "synaptic nests" that lack intrinsic glia. More details about the brain region- and synapse type-dependent variations in perisynaptic astroglial sheaths can be found in Chao et al. (2002).

The functional interactions between synaptic elements and their different types of glial sheaths are not yet fully understood. An "insulation" of individual synapses (or groups of them) against their microenvironment may prevent the uncontrolled spread of neuronal excitation. For instance, there is now general agreement that perisynaptic glial membranes may be the dominant sites of transmitter transporters, and that glial uptake of released neurotransmitter molecules is crucial for maintaining

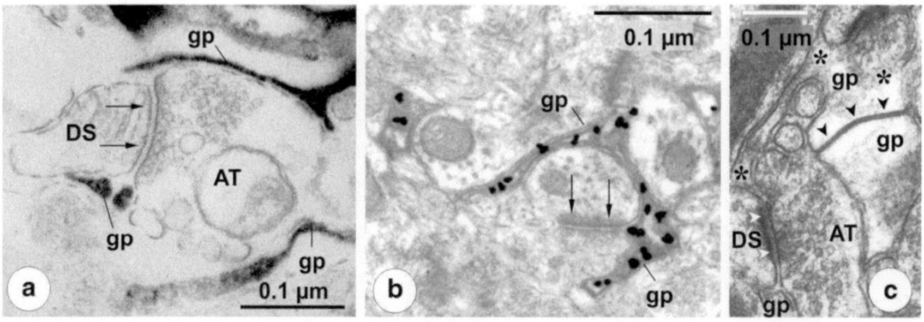

Fig. 10.6 PAPs at synapses. **a** Ensheathing of parallel fiber-Purkinje cell synapses by Bergmann glial cell processes; murine cerebellum (the glial cell had been injected by Lucifer yellow, and the dye was then converted into an electron-dense label). The synaptic clefts (*arrows*) between the axon terminals (AT) and the dendritic spines (DS) are sealed by glial lamellae (gp) from the injected cell. Modified after Grosche et al. (2002). **b** Rat hippocampus. Fine, perisynaptic PAPs (gp) are identified by silver grains indicating immunoreactivity for the astrocyte-specific enzyme, glutamine synthetase; one synaptic cleft is labeled by *arrows*. Modified after Reichenbach et al. (2004). **c** The neuronal elements (axon terminal, AT, and dendritic spine, DS) of a chemical synapse (postsynaptic density labeled by *white arrowheads*) are enveloped by glial cell processes (gp, asterisks) which are coupled by gap junctions (*black arrowheads*); modified after Reichenbach et al. (2004).

the spatial and temporal specificity of synaptic transmission (Sarantis and Mobbs, 1992; Oliet et al., 2001). The glutamate-neutralizing enzyme, glutamine synthetase, is present in the perisynaptic glial cytoplasm (Derouiche, 1997; Derouiche and Frotscher, 1991; Derouiche and Rauen, 1995; Fig. 10.6b). Furthermore, glial receptors for neurotransmitters may be assembled in these membrane areas. Stimulation of these receptors may initiate glial metabolic reactions, beneficial for the activated neuronal compartments (Volterra et al., 2002). It has been recently established that glutamate, which is released from astrocytes, and which may modulate synaptic transmission, originates from perisynaptic glial processes (Bezzi et al., 2004). Finally, perisynaptic PAPs may be crucially involved in the maintenance and/or degradation of synapses and, thus, in synaptic plasticity [for recent reviews, see (Chao et al., 2002; Reichenbach et al., 2004)].

With regard to Bergmann glial cells, the concept of so-called microdomains has been developed (Grosche et al., 1999, 2002). Microdomains occur as repetitive units on the stem processes of the cells, or as appendages of another microdomain. Each of them consists of a thin stalk and a cabbage-like, very complex head structure that bears the lamellar perisynaptic sheaths for a low number – about 5 – of synapses (Fig. 10.7). It has been shown that these microdomains may interact with "their" synapses, independent of other microdomains and also of the stem process (Grosche et al., 1999). Furthermore, mathematical simulation of the cable properties suggests that even large (e. g., glutamate-induced) depolarizations of the perisynaptic membranes are not conducted over the stalks towards neighboring microdomains, or towards the stalk (Grosche et al., 2002). For their energetic demands, the microdomains contain mitochondria in the "head" structures (Grosche et al., 1999, 2002). The glial microdomains overlap; in every given volume unit of the molecular layer, at least two

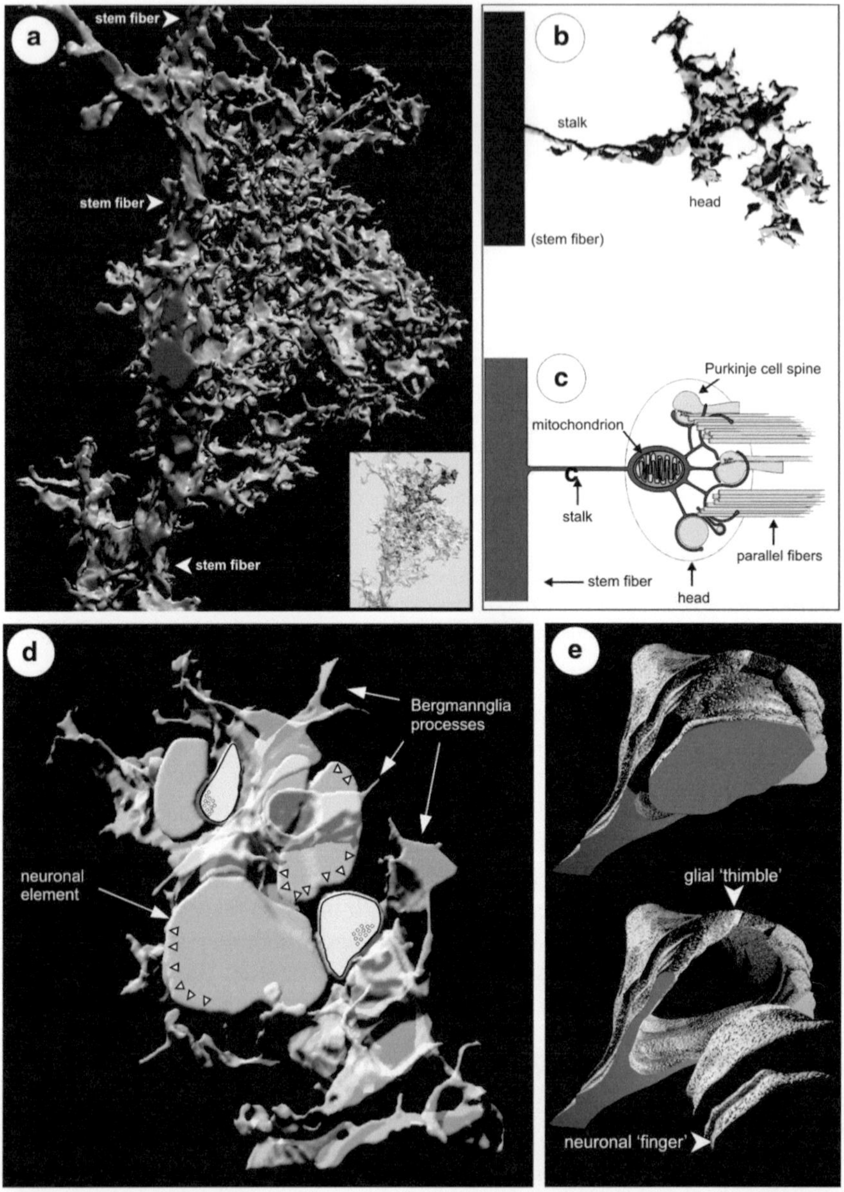

Fig. 10.7 Microdomains of Bergmann glial cell processes in the murine cerebellum. **a** 3-D reconstruction of a part of a Bergmann glial cell process. The living cell was dye-injected in a perfused cerebellar slice; then, after fixation and dye-conversion, about 600 consecutive serial ultrathin sections were photographed in the electron microscope, and the images of the dye-labeled profiles were reconstructed by a computer program. The inset shows a substructure labeled in blue; this part was quantitatively analyzed (see (**b**), (**c**)). **b** Glial microdomain as part of the 3-D reconstruction shown in (**a**). **c** Schematic drawing of such a glial microdomain and its relationships to the neuronal elements. **d** 3-D reconstruction of a group of neighboring cerebellar synapses (*yellow–green*; synaptic clefts: *orange*) together with the surrounding leaflets provided by the injected Bergmann glia (*blue–gray*). The *arrowheads* point to neuronal surfaces not covered by glial sheaths from the labeled cell. **e** 3-D reconstruction of a glial "thimble" and the neuronal "finger" covered by it; shown in apposition (*top*) and separated for better discrimination between the two compartments (*bottom*). With permission, from Reichenbach et al. (2004). (*See Color Plates*)

microdomains, originating from different Bergmann glial cells, interdigitate. This may fit in with the observation that Purkinje cells express two functionally distinct populations of synaptic spines, and that individual spines are capable of independent activation (Denk et al., 1995). In addition to the perisynaptic sheaths, the heads of the microdomains extend numerous “glial thimbles”, forming complete caps on small neuronal protrusions (Fig. 10.7e), which may represent dying or growing synapses. Similar microdomains may also exist in other areas of the brain where they may be formed, as complex PAPs, by protoplasmic astrocytes.

10.6 The Sites of Mesodermal-Ectodermal Interplay: “Perivascular Astrocytic Endfeet” (PAE)

It has already been mentioned that astroglial cells possess endfeet contacting a basal lamina around blood vessels (Fig. 10.3a), the pia mater, or both. For the sake of simplicity, all basal lamina-abutting astroglial structures will be termed “perivascular astrocytic endfeet” (PAE) here, even if they contact the pia mater (or the vitreous body, as astrocytes, and Müller cells, do in the retina). This can be done because all endfeet share characteristic common features, as detailed below.

10.6.1 PAE on Capillaries

Glial contacts to capillaries can be considered as the prototype of PAE. It has been proposed that every astrocyte possesses at least one PAE [for review, see Reichenbach (1989)]. These PAE may form the ending of rather thick stem processes (such as in Fig. 10.3a) which are densely packed with bundles of GFAP-containing intermediate filaments. Additionally, there are smaller “en passant-endfeet” arising at the side branches of astrocytic processes. However, the PAE of the astrocytic population do not constitute a complete ensheathment of the capillaries, because pericytes are the second type of perivascular cell directly adjacent to the endothelial basal lamina. The pericyte is completely surrounded by a basal lamina, and the neuropil-directed portion of the pericytic basal lamina can be covered by a PAE (left asterisks in Fig. 10.8a). Typically, the PAE are rich in cellular organelles such as mitochondria and smooth endoplasmic reticulum; often, the bundles of intermediate filaments reach into the emerging endfeet but can never be seen close to the endfoot membrane abutting the basal lamina. The most characteristic property of this membrane is its dense package with the so-called orthogonal arrays of intramembranous particles (OAPs) visualized by the freeze-fracture technique [for review, see Wolburg (1995)] (Fig. 10.8c, d). Where the contact of the glial cell membrane with the basal lamina is lost by bending away into deeper parenchymal regions of the neuropil, the density of OAPs is dramatically reduced. This suggests that the extracellular matrix molecules of the basal lamina are responsible for the clustering and/or maintenance of the OAPs in this membrane domain.

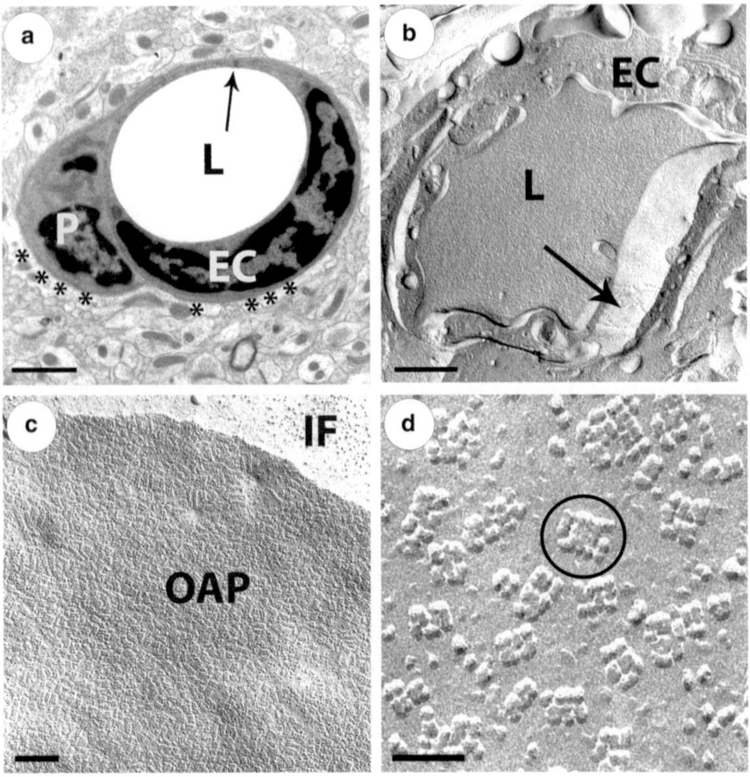

Fig. 10.8 PAE and OAPs. **a** Low-magnification transmission electron micrograph of a typical brain capillary, with cross-sections of the lumen (L), an endothelial cell (EC), and a pericyte (P). The *asterisks* mark the astrocytic endfoot enwrapping the capillary. The *arrow* points to a tight junction. **b** Low-magnification freeze-fracture image of a similar capillary with tight junctions (*arrow*). At higher magnification ((**c**), (**d**)) the presence of many orthogonal arrays of intramembranous particles (OAPs) becomes obvious in freeze-fracture replicas (one of them is encircled in (**d**)). IF = intermediate filaments. Scale bars, 2 μm (**a**), 0.5 μm (**b**), 0.2 μm (**c**), 50 nm (**d**). Modified after Wolburg and Warth (2005).

Regarding the molecular constituents of the OAPs, it is well-known now that they contain the water channel protein, aquaporin-4 (AQP4). Generally, aquaporins mediate water movements between the intracellular, interstitial, vascular and ventricular compartments which are under the strict control of osmotic and hydrostatic pressure gradients (Nicchia et al., 2004). The involvement of AQP4 in the OAPs was demonstrated by the absence of OAPs in astrocytes of AQP4-deficient mice (Verbavatz et al., 1997), by the formation of OAPs in Chinese hamster ovary cells stably transfected with AQP4 cDNA (Yang et al., 1996), and by the immunogold fracture-labeling technique (Rash et al., 2004). AQP4 appears to be clustered in the endfoot membrane by means of a large protein complex, involving α-dystroglycan, syntrophin, dystrophin, agrin, and others [for review, see Wolburg (2006)]. This complex may also be responsible for the co-clustering of another membrane protein, the K^+ channel protein Kir4.1. It has been shown that the truncated dystrophin isoform Dp71 is essential for the clustered localization of Kir4.1 in retinal Müller (glial) cells (Connors and Kofuji, 2002). In addition, the PDZ-domain of α-syntrophin can also bind to Kir4.1 (Connors et al., 2004). Kir4.1 is normally restricted to the endfoot membrane in astrocytes and retinal Müller glial cells (Kofuji and Newman, 2004). On the basis of colocalization of AQP4 and Kir4.1 in retinal Müller cells, and due to the well-known fact that water fluxes are driven by ion fluxes, it was hypothesized that K^+-clearance is coupled to water flux (Simard and Nedergaard, 2004; Nielsen et al., 1997; Nagelhus et al., 2004). Accordingly, in the α-syntrophin-deficient mouse in which AQP4 is dislocalized across the glial surface, the K^+ clearance was found to be delayed (Amiry-Moghaddam et al., 2003). In the hypoxic retina on the other hand, where Kir4.1 is downregulated in retinal Müller cells, the water efflux was found to be compromised, and the cells swelled under hyposmotic conditions (Pannicke et al., 2004) whereas normally a rapid volume regulation occurs.

10.6.2 PAE on Larger Blood Vessels

The PAE on arterioles and venules are rather similar to those on capillaries. Frequently, the glial envelope of these vessels is formed by more than one layer of glial endfeet. This is also typical of the larger blood vessels, the arteries and veins; in between these vessels and their PAE there are other compartments, the so-called Virchow-Robin spaces (Fig. 10.9). The Virchow-Robin spaces are perivascular extensions of the pia mater that accompany the arteries entering and the veins emerging from the cerebral cortex. Between the external surface of the blood vessels and the pial-glial peripheral lining, these spaces contain extracellular fluid "lakes" which may be very large; under certain circumstances, their volume may exceed that of the surrounded vessels. These perivascular spaces may play important roles for the drainage of the cerebrospinal fluid, and/or as reservoirs to buffer changes in extracellular ion concentrations or water. In any case, they serve as transport routes for any molecules exchanged between the circulation and the brain tissue, as they are interspaced between these compartments. There may also be a communication between the perivascular spaces and lymphatic channels in the walls of major cerebral arteries.

10.6.3 "PAE" at the Pia Mater

As a continuation of the glial envelopes of the Virchow-Robin spaces, the parenchymal surface of the pia mater is covered by a so-called glia limitans (Fig. 10.9b). In higher mammals such as man, this compartment consists of a multi-layered arrangement of astroglial endfeet or "PAE". There even appear to be specific "bordering

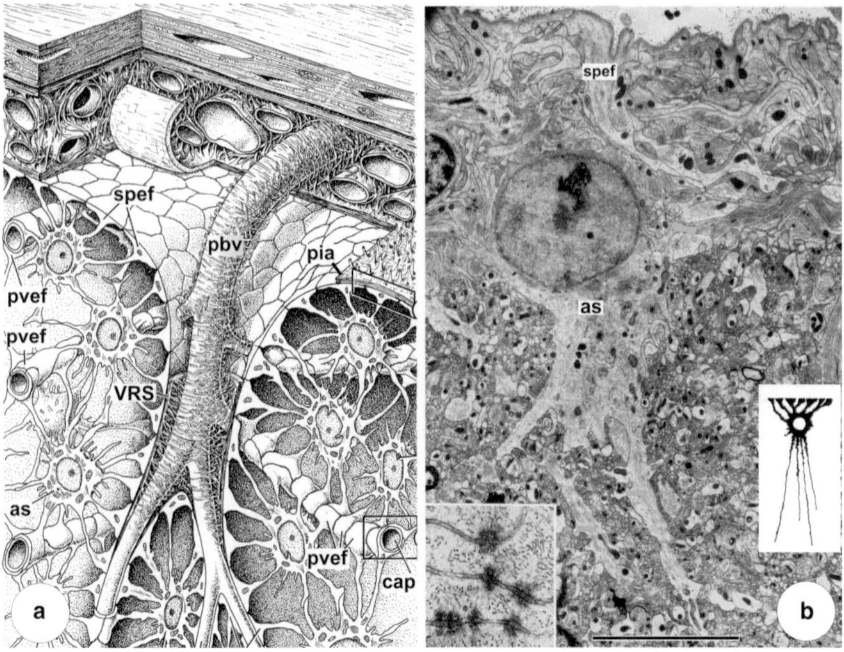

Fig. 10.9 PAE at the Virchow–Robin spaces (VRS) and in the glia limitans. **a** Semi-schematic drawing of the astroglial relationships to the pia mater and the blood vessels [modified after Krstic (1991)]. The layer of the subpial endfeet (spef) of the astrocytes (as) is continuous with that of the perivascular endfeet (pvef) because the Virchow–Robin spaces extend from the pia mater. *pbv* primary blood vessel, *cap* capillary. **b** The glia limitans of the human isocortex (modified after Braak, 1975). A subpial astrocyte (as) is shown together with its complex processes (spef). *Left inset*: desmosomal connections between adherent processes; *right inset*: semi-schematic drawing of a subpial astrocyte (original data, A. Reichenbach). Scale bar, 5 μm.

astrocytes" which have no contacts other than to the pia and to neighbouring astrocytes (Braak, 1975). Such cells and their "PAE" may be involved in molecular exchange between the subarachnoid space and the brain parenchyma, and may also constitute a buffer capacity for ions and water on their own. It should be noted that the blood vessels in the subarachnoidal space express an endothelial blood–brain barrier (cf. Fig. 10.10) like that of the brain tissue proper, such that the subpial endfeet are exposed to a similar milieu as the PAE sensu strictu.

10.6.4 PAE and the Blood–Brain Barrier (BBB)

The exchange of molecules between the cerebral blood vessels and the brain parenchyma is limited by the so-called blood–brain barrier (BBB). In the vertebrates, the BBB is constituted basically by the endothelial cells, whose luminal and abluminal membrane domains are separated by complex arrangements of tight junctions

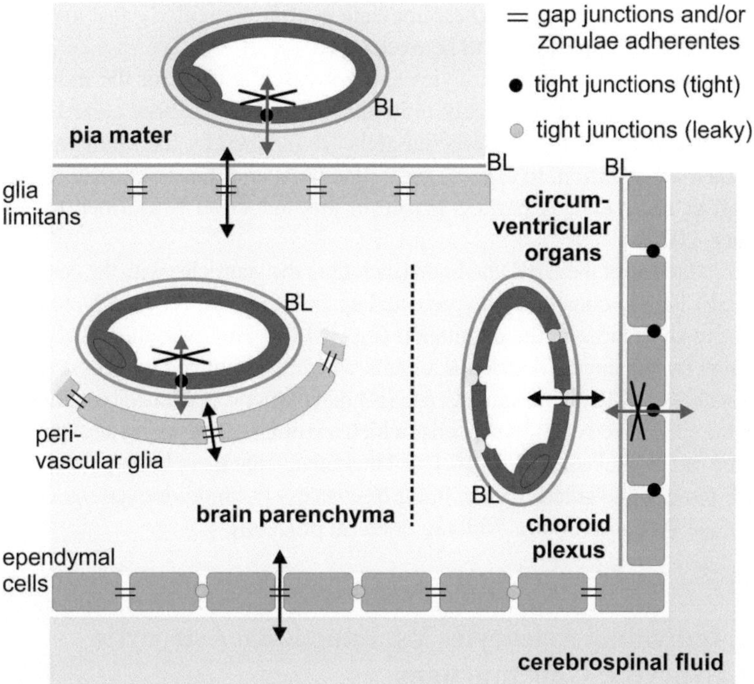

Fig. 10.10 Brain territories defined by the BBB. The brain wall is delineated superficially by the glia limitans superficialis, formed by astroglial endfeet and a basal lamina. Periventricularly, the brain wall is delineated by the ependymal cells, which do not establish a physiological barrier since ependymal cells are interconnected by discontinuous tight junctions. In the circumventricular organs (CVOs), the endothelial BBB must be interrupted because the neurons, for example in the hypothalamic-hypophyseal system or in the subfornical or subcommissural organs or in the area postrema, must necessarily get access to the blood compartment in order to release neurosecretory compounds into the blood stream or to "smell" blood-borne signal molecules, respectively. In the case of the choroid plexus, the endothelial cells also have to be highly permeable by the formation of fenestrations to allow the production of the cerebrospinal fluid (CSF) by the choroid plexus epithelial cells. To avoid diffusion of blood-borne substances from the leaky choroid vessels into the CSF and further into the brain parenchyma through the leaky ependymal tight junctions, a barrier is necessary between the blood and the CSF: the so-called blood-CSF-barrier. This barrier is located in the choroid plexus epithelial cells and in the tanycytes of the CVOs, and is formed by tight junctions which are molecularly different from the endothelial tight junctions of the BBB. *BL* basal lamina. Modified after Wolburg et al. (2007). (*See Color Plates*)

(Fig. 10.2d, e); exceptions to this are the elasmobranchs (e.g., the sharks) where the BBB is formed by tight junctions between the PAE (Abbott, 1991). The restrictive paracellular diffusion barrier established by the tight junctions is associated with an extremely low rate of transcytosis and the expression of a high number of channels and transporters for such molecules which cannot enter or leave the brain paracellularly. Thus, the BBB does not present an absolute obstacle against any molecular exchange between blood and brain; rather it serves as a "checkpoint" to separate the unwarranted from the warranted molecules. Specific transport pathways are established only for the latter; these mediate even a particularly fast and effective translocation of the substances to be exchanged.

The impact of astrocytes, pericytes and perivascular cells for the induction and maintenance of the BBB is largely unidentified so far. It is now clear from many experiments, however, that endothelial cells are induced by the brain environment to lose their fenestration, to express tight junctions, and thus to constitute the BBB (Haseloff et al., 2005). A particular role in this induction is ascribed to the PAE (Wolburg, 2006).

After induction of the BBB, the brain (including the ventricles with the cerebrospinal fluid, CSF) largely constitutes a "protected space" with respect to, e.g., toxic and/or neuroactive substances in the circulation (Fig. 10.10). However, there are exceptions constituted by the circumventricular organs which are supplied by fenestrated blood vessels without a BBB. In these areas, a CSF-brain barrier is generated by tight junctions among the specialized ependymal cells (which are often called tanycytes, which means elongated radial glial cells) (Fig. 10.10). Thus finally, the main brain parenchyma and the CSF remain prevented from gaining free access to blood-derived molecules [for reviews, see Wolburg (2006), Wolburg et al. (in process)].

10.7 Individual Astrocytes Vs. Functional Astrocytic Syncytia: Gap Junctions

One of the most prominent features of astrocytes is their extensive coupling via gap junctions. The molecular bases of gap junctional coupling are the fourfold transmembrane spanning connexins, comprising hexameric hemichannels in one membrane which are called connexons; the complementary hemichannel in the partner cell membrane must fit precisely to establish an entire intercellular channel. The probability of finding the precise docking location of the counterparts of the connexons is heavily increased by clustering hemichannels in one membrane domain, leading to the well-known morphology of connexon aggregates, which can at best be visualized with the freeze-fracture method (Fig. 10.11 A and *inset*). Open gap junctional channels allow not only the flow of ionic currents ("electrical coupling") but also the intercellular exchange of larger molecules such as biocytin (Fig. 10.11c, e) or fluorescent dyes (Fig. 10.11d); this "dye coupling" is often used to demonstrate the presence of functional gap junctions among cells. There is evidence for an active metabolic trafficking through astrocytic gap junctions (Giaume et al., 1997) which even appears to link glucose metabolism with proliferation in astroglia (Tabernero et al., 2006).

In the central nervous system, at least 11 connexins (Cx26, Cx29, Cx30, Cx31, Cx32, Cx36, Cx37, Cx43, Cx45, Cx47 and Cx57) have been identified. The connexins 26, 30 and 43 were described as expressed by astrocytes (Nagy et al., 2004). Among these connexins, Cx43 is the most widely expressed connexin in the CNS (Fig. 10.11b). Conflicting results were ascribed to Cx26 [reviewed and comprehensively discussed in Nagy et al. (2004)], but extensive studies using immunogold labeling electron microscopy, including fracture immunogold labeling (FRIL),

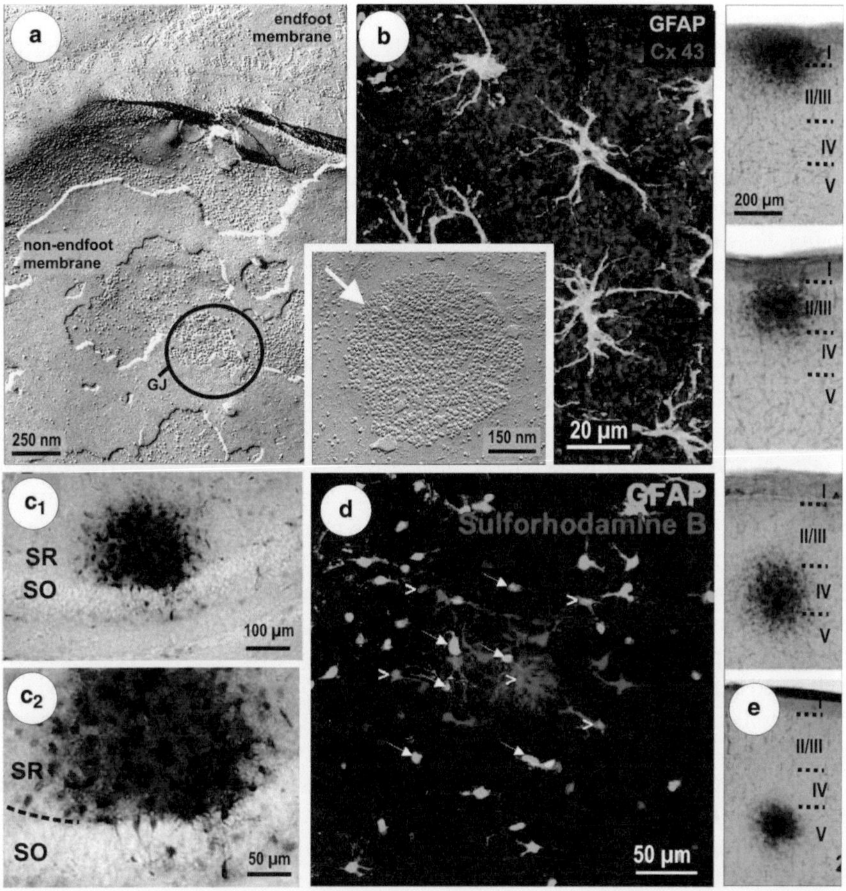

Fig. 10.11 Astrocytic coupling via gap junctions. **a** Electron microscopy of a freeze-fracture replica of an astrocytic process forming a perivascular endfoot (rat optic nerve). The endfoot membrane proper displays orthogonal arrays of particles (cf. Fig. 10.8) but no gap junctions. The membrane particle arrays of a gap junction (GJ) are found in the "lateral" non-endfoot membrane of the same process, probably constituting a coupling to another adjacent endfoot-bearing process. Inset: Typical array of gap junction-hemichannels in an astrocytic membrane. Original data, H. Wolburg. **b** Juvenile mouse hippocampus, immunohistochemical co-localization of GFAP (green) and connexin 43 (Cx43, *red*). The overlay (*yellow*) indicates that connexin 43 is expressed on astrocytic processes. **c**, **d** Dye-coupling of astrocytes in juvenile mouse hippocampus. Intracellular injection of biocytin (**c**) or sulforhodamine B (**d**) into a single astrocyte in brain slices causes the staining of many adjacent astrocytes. **c** In the juvenile hippocampus, many but not all dye-coupled astrocytic networks fail to span beyond the stratum radiatum (SR); if the injected cell was localized there, only a few extend into the stratum oriens (SO). **d** It is noteworthy that not all GFAP/GFP-expressing astrocytes in a given area are dye-coupled (*arrows*). Some sulforhodamine B-stained cells are marked by arrowheads. **e** Dye-coupling of astrocytes in juvenile mouse neocortex, intracellular injection of biocytin. Depending on the location of the injected astrocyte (cortical layers I–V), the dye-coupled astrocytic networks differ in size and shape but there is always a tendency to remain restricted within one cortical layer. **b–e** modified after Houades et al. (2006). (*See Color Plates*)

showed astrocyte-specific immunoreactivity for Cx26, Cx30 and Cx43. This was associated with oligodendrocyte-specific Cx32 immunoreactivity, suggesting the existence of heterologous junctions between astrocytes and oligodendrocytes (Rash et al., 2001, Nagy et al., 2003). The expression pattern of astrocytic connexins is highly heterogeneous throughout the CNS. Whereas all three connexins are abundantly expressed in subcortical areas, Cx26 and Cx30 display only low level-expression in the cerebral cortex. Moreover, Cx30 is not detectable in the white matter. Possibly, different connexin expressions reflect different functional requirements of glial coupling, because the connexins are known to display distinct ionic conductances and permeabilities as well as different voltage dependencies (Giaume and Venance, 1995; Gros and Jongsma, 1996).

Astrocytic coupling plays an important role in potassium buffering (Wallraff et al., 2006). Locally increased extracellular K^+ concentrations – caused by enhanced neuronal activity – are promptly redistributed by K^+ uptake into local astrocytic processes and release of K^+ at remote sites. This K^+ release occurs preferably where the astrocytes contact large extracellular spaces ("sinks") such as capillaries or the brain surface; at these interfaces, the astrocytic membranes display a particularly high K^+ conductance. When large areas of neuropil are exposed to increased extracellular K^+, the processes of individual astrocytes are not long enough to exit the overload area and to reach a free sink; in these cases, spatial buffering is performed by an entire population of astrocytes coupled via gap junctions (cf. Fig. 10.11c, d). Enhanced extracellular K^+ in internodal periaxonal spaces may also be buffered into brain capillaries via the heterologous coupling between oligodendrocytes and astrocytes (Kamasawa et al., 2005).

Another important physiological function of astrocytic gap junctions is the propagation of so-called Ca^{2+} waves. The first reports of calcium waves were published by Cornell-Bell et al. (1990) and Charles et al. (1991). The most intriguing finding was that the calcium concentration was increased not only within a stimulated astrocyte, but that this increase was transmitted to adjacent non-stimulated astrocytes, and thus spread among the astrocytic population. Given the fact that Ca^{2+} is a crucial second messenger for many cellular functions, its transport within glial networks constitutes an essential part of non-synaptic information processing in the CNS (Scemes and Giaume, 2006). Most interestingly, these Ca^{2+} waves may be propagated by different mechanisms; there may be a direct flow of Ca^{2+} through gap junctions into neighbouring cells but inositol 1,4,5-trisphosphate (IP_3) can be transported via gap junctions as well, which then evokes a Ca^{2+} increase in the neighbouring cells. Furthermore, another (gap junction-independent) mechanism was proposed to explain Ca^{2+} waves in some brain areas; a Ca^{2+} increase in one astrocyte may induce the release of ATP, glutamate, or other gliotransmitters (cf. Chap. 12) from this cell, which -via stimulation of ligand receptors on neighbouring astrocytes- causes Ca^{2+} rises in adjacent astrocytes, and so on (Stout et al., 2002; Bennett et al., 2003; Peters et al., 2003).

These data may convey the impression of a uniform, widely distributed network of astrocytes interconnected by gap junctions, often called a "functional astrocytic syncytium" [e.g., Konietzko and Müller (1994)]. However, a closer look reveals the presence of distinct subtypes or subpopulations of astrocytes, either extensively coupled via Cx43-positive gap junctions or not coupled at all (Wallraff et al., 2004) (for instance, the arrows in Fig. 10.11d show astrocytic cells, expressing green fluorescent protein (GFP) under GFAP promoter, which are uncoupled, within an area containing many dye-coupled – red – cells). Moreover, the spatial extension of dye-coupled astrocytic networks is limited; mostly some 50–100 astrocytes are coupled, and the diameter of a coupled network does not exceed 200–300 μm. In the juvenile mouse cortex (Fig. 10.11e) and hippocampus (Fig. 10.11c) for example, many of the dye-coupled networks display a tendency to reside within a given layer (Houades et al., 2006). Furthermore, the actual degree of coupling among adjacent cells expressing suitable connexins may vary considerably, depending on physiological parameters such as pH, Ca^{2+}, sphingosine-1-phosphate levels, and others; there are many known blockers of gap junctional coupling (cf. Chap. 6), which may involve molecules that modify the coupling of connexins to the submembrane cytoskeleton (Butkevich et al., 2004; Rouach et al., 2006). It has been hypothesized that intercellular communication via gap junctions may be tightly controlled in space and time, in order to optimize glial cell functions such as spatial buffering of K^+ ions or signal transmission by Ca^{2+} waves [e.g., Reichenbach et al. (1992); cf. Chap. 6). This may also play a central role in the determination of functional territories in the brain (see below).

10.8 Functional Territories and Territorial Boundaries

Throughout this chapter it has become obvious that the CNS consists of specialized functional territories. These are marked by neuronal (Fig. 10.1), glial (Figs. 10.3–10.7), as well as vascular elements (Figs. 10.2a, 10.3, 10.4, 10.5, 10.6, 10.7, 10.8, 10.9) and range in size from a few micrometers up to centimeters (e.g., in human brain lobes). Moreover as discussed at the end of the previous section, they may be variable in their size und degree of integration, depending on the momentary needs of brain functioning. It has been shown that astrocytes are able to match the local blood flow to local metabolic activity (Harder et al., 2002; Mulligan and MacVicar, 2004; and references therein). This section is aimed at a trial to summarize what is presently known – and/or can be reasonably speculated – about the role of astrocytes in the organization of neuronal functional territories. What we want to propose here is a novel concept of a hierarchy of co-existing functional glial domains.

Several levels of morphological and/or functional compartmentalization can be identified in the CNS. This starts with very small, circumscribed glia-neuron contacts that appear to be very dynamic (Hirrlinger et al., 2004) but which certainly may have a clear function as long as they are maintained. For instance, there are glial tongues lining the synaptic clefts (probably predominantly active in neurotransmitter uptake) as well as others in contact with remote pre- or postsynaptic compartments (perhaps active in metabolic fueling of and/or glutamine release for the neuronal elements) (cf. Fig. 10.7d). In a similar manner, the finger-like perinodal glial processes may serve very specific local glia-neuron interactions, such as molecule transfer or sensing of spike activity (Chao et al., 1994; and references therein). Small as they are, such glial structures may be important for very distinct functions at the given organization level; they are designated here as putative "nanodomains".

At the next integration level, glial microdomains have been described (Grosche et al., 1999, 2002). In many instances, there is a need to transmit very specific information which must remain restricted to individual synapses or small groups of synapses. To prevent a spillover of transmitter molecules, membrane depolarizations, second messengers (including Ca^{2+} and IP_3) and other metabolic signals, one synapse (or a small group of them) is ensheathed by a PAP compartment (i.e., a glial microdomain) which provides the necessary local homeostatic mechanisms (cf. Figs. 10.6 and 10.7). One astrocytic cell may give rise to many of these microdomains; in the case of cerebellar Bergmann glial cells, a total number of about 100 microdomains per cell can be estimated (data of Grosche et al., 1999, 2002). Similar to the nanodomains, these units appear to be very dynamic and plastic structures, as shown by recent imaging experiments (Hirrlinger et al. 2004; Ohira et al., 2007).

At a less miniaturized level, larger synaptic aggregates/glomerula such as in the olfactory cortex may constitute the dominant functional units of neural tissue. Such aggregates are enveloped by a common glial sheath (cf. Fig. 10.3b) which probably serves the same homeostatic functions as the PAPs of a microdomain. Depending on the size of the synaptic aggregate, a whole astrocyte or even several astrocytes may be necessary to constitute such a "cellular domain". It is essential to point out that microdomains and cellular domains are not mutually exclusive types of CNS organization; rather they co-exist, and appear to take over the dominant role in an alternating manner and quickly respond to the change in functional requirements. For instance, individual Bergmann glial cell microdomains (which are arranged in a columnar fashion and termed here as "cellular domain type 1") can be activated (as indicated by intracellular Ca^{2+} increases) when individual parallel fibers are (weakly?) stimulated but many microdomains or even the whole domain of Bergmann glial cells are activated by strong, less focused stimulation (Grosche et al., 1999) (Fig. 10.12a. Likewise in the olfactory bulb, each whole glomerulum is ensheathed by an (oligo-)cellular domain (Fig. 10.3b) which strongly expresses AQP4 to drain water from the active synaptic compartments (Fig. 10.12b, *asterisks*); this domain of spherical and ellipsoid shape is termed here as "cellular domain type 2". However, smaller arrangements of (individual?) synapses within each glomerulum are also enveloped by AQP4-expressing glial PAPs (Fig. 10.12b, *arrows*) resembling microdomains. Figures 10.2b and 10.12b suggest that both the microdomains and the cellular domains of a glomerulum in the olfactory bulb may possess their "own, private" blood supply provided by a capillary or an arrangement of capillaries. Thus, the glial cellular domain may constitute a specific link between the neuronal elements and the blood vessels of each functional unit.

Assuming that in the cerebellum one Purkinje cell constitutes the smallest functional "relay" unit, and considering the fact that about eight Bergmann glial cells are arranged around a Purkinje cell (Fig. 10.4d; Reichenbach et al., 1995), on the

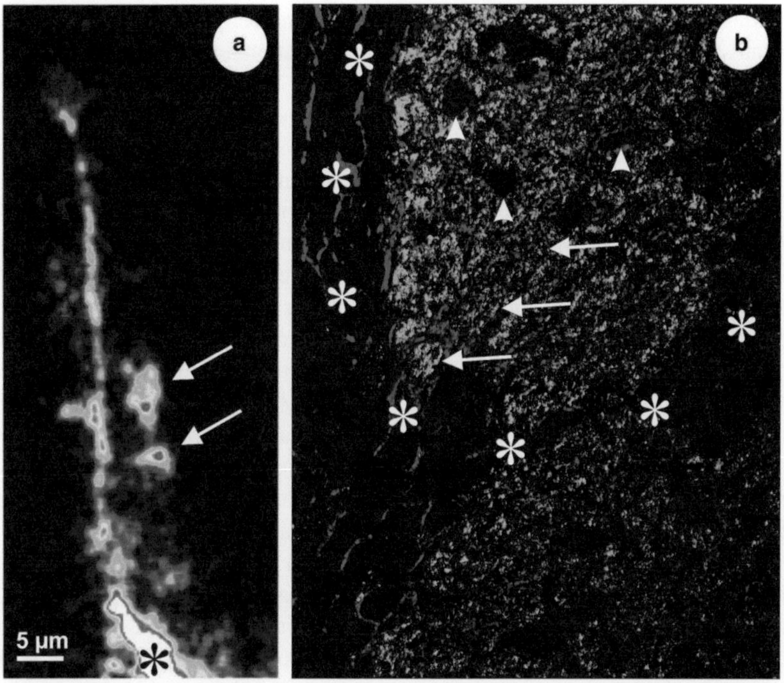

Fig. 10.12 Co-existence of functional glial domains differing in size. **a** Imaging of Ca^{2+} increases in a dye-injected Bergmann glial cell of a murine cerebellar slice preparation, after electrical stimulation of parallel fibers. Low-intensity stimulation of only a few parallel fibers may induce Ca^{2+} increases in only one or a few glial microdomains (*arrows*) whereas high-intensity stimulation of many fibers elicits Ca^{2+} responses within the entire glial cell process, or even the whole cell (*asterisk*, soma). Modified after Grosche et al. (1999). **b** Rat olfactory bulb, double-immunolabeling of AQP4 (*red*: glial cell membranes) and synaptophysin (*green*: synaptic structures). Two synaptic glomerula are shown; the astroglial "envelope" of the upper one is indicated by asterisks. AQP4-expressing glial membranes surround the entire glomerulum but also smaller groups of synapses ("green spaces", *arrows*) and small blood vessels ("empty spaces", *arrowheads*). This might be indicative of the presence of both a glial "(?oligo-)cellular domain" (draining water from the entire glomerulum) and many microdomains (responsible for small groups of synapses). Original data, Wolburg et al. (*See Color Plates*)

glial side the corresponding units are certainly oligo- rather than unicellular. The same appears to apply to the olfactory glomerula. A little larger but also functionally well-defined are the small orientation columns in the visual cortex (Mountcastle, 1957, 1997); each of these may interact with an anatomically defined group of astrocytes. It has been suggested that in the primate neocortex the radial course of the processes of the interlaminar astrocytes (Fig. 10.3c) may help to support such columnar functional units (Reisin and Colombo, 2002). In any case, the astrocytic assemblies supporting such units are termed here as mesodomains, and appear to be closely related to the (oligo-)cellular domains (at the upper end of the size scale of the latter).

At the next level of integration, assemblies of olicocellular domains and/or mesodomains constitute the large ("macroscopic") functional units of the CNS such as the columnar units of the cerebral cortex including the ocular dominance columns of the visual cortex and the "barrels" of the somatosensory cortex, the cortical layers, and more coarse compartments such as gray *vs.* white matter, or the cerebral hemispheres. It is well known that cerebral vascularization is organized according to such macrodomains (Bär, 1980) and that interruption of blood flow through one of the cerebral arteries (which are "end-arteries" since they lack functional anastomoses) causes circumscribed functional losses corresponding to distinct functional units. How may astrocytes contribute to such larger units?

Extensive networks of gap junction-coupled astrocytes have been found in various brain regions (cf. Fig. 10.11). In some instances, dye injections into single astrocytes of brain slices revealed dye-coupling of astrocytic sub-populations which might be ascribed to functional compartments such as areas within individual layers in the hippocampus or neocortex (Fig. 10.11c, e). It remains to be established whether such dye-coupled networks may play a role in the functional organization of the brain. There is general agreement that they may help to maintain the extracellular homeostasis, by mediating mechanisms such as spatial buffering of excess K^+ ions (Amedeé et al., 1997; and references therein). The problem with respect to brain organization is that these networks are "virtual" structures, and that their number is infinite: if any of the other 50–100 coupled astrocytes in a network as shown in Fig. 10.11c, b, c, d, e would have been injected, it would have constituted the center of a similar "own" network involving many but not all labeled cells of the network shown in the figure, plus some other astrocytes. Furthermore in some cases, such arrays of dye-coupled astrocytes clearly crossed the borders between cortical areas, and even between the gray and white matter (Houades et al., 2006).

If neuronal signal processing will cause a simultaneous or subsequent activation of astrocytes, which are members of experimentally distinct but adjacent (dye-coupled) networks, or if a Ca^{2+} wave will spread across the borders of individual networks, larger "super-networks" will be formed which may eventually fit into functional territories of the brain; such glial assemblies may be less variable and accidental than the experimentally visualized dye-coupled networks, and may form what will be termed here as glial macrodomains. Very recently it has been shown that physiological stimulation of neurons in the barrel cortex of mice causes astrocytic Ca^{2+} waves which remain restricted to the barrel where the stimulation was performed (Schipke et al., 2008). This argues in favor of the idea that each barrel field contains its own glial macrodomain (Fig. 10.13c). The introduction of novel imaging techniques will allow us to study the coordination between the activation of neuronal and glial assemblies in more detail (Göbel et al., 2007).

It is very probable that there exists a similar functional transition between networks and macrodomains as in the case of micro- and cellular domains (see above). In the case of the barrel cortex it was shown that whereas normally the glial Ca^{2+} responses to stimulation of layer four neurons within a barrel field spread only within this given barrel field, a block of GABAergic inhibition increased the area of neuronal activity and elicited glial Ca^{2+} waves which traveled far into the neighboring

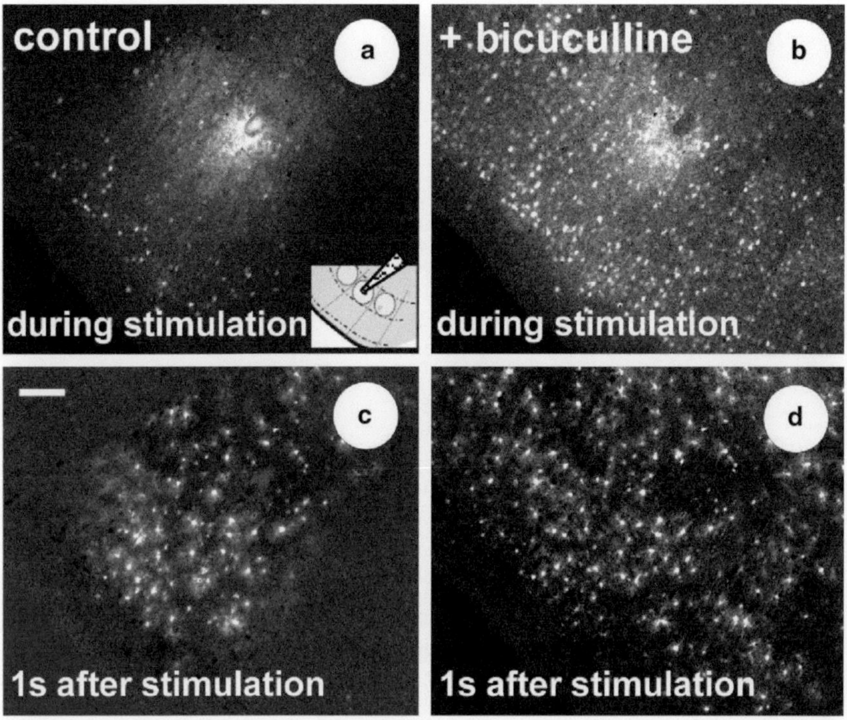

Fig. 10.13 Variable propagation limits of glial Ca^{2+} responses (representing functional glial domains?) in slice preparations of murine barrel cortex. Under control conditions, electrical stimulation of layer four (input) neurons in a given barrel elicits neuronal Ca^{2+} rises during stimulation (**a**) and subsequent astrocytic Ca^{2+} responses which remain restricted to the barrel field stimulated (**c**). Blockade of GABA receptors (i.e., of neuronal inhibition) causes, in response to the same stimulus, more-widespread neuronal activity (**b**) and the occurrence of Ca^{2+} responses in larger astrocytic populations, involving also the neighboring barrel areas (**d**). Calibration bar, 100 μm; *inset*: cartoon illustrating the orientation of the slice and the location of the stimulation pipette. Modified after Fig. 10.8 of Schipke et al., 2008.

barrel fields (Schipke et al., 2008) (Fig. 10.13d). What is noteworthy is that there were no astrocytic Ca^{2+} responses to the spontaneous activity of scattered neurons in layer 2/3 (Schipke et al., 2008). Thus, depending on the degree and distribution of neuronal excitation, the whole repertoire of glial arrangements from nano- and microdomains up to macrodomains or even "superdomains" (cortical gyri or fields) may be switched on in series, within the very same part of the CNS (Fig. 10.14). At the (pathological) end of this scale, phenomena such as spreading depression may spread over the entire cortex, thus involving the entire glial population; this may involve glial Ca^{2+} waves or other mechanisms (Peters et al., 2003).

These differently-sized glial domains, however transient and variable they may be, appear to be important for brain function from early development to mature function and even pathology (see Table 10.1). In the developing embryo, glial

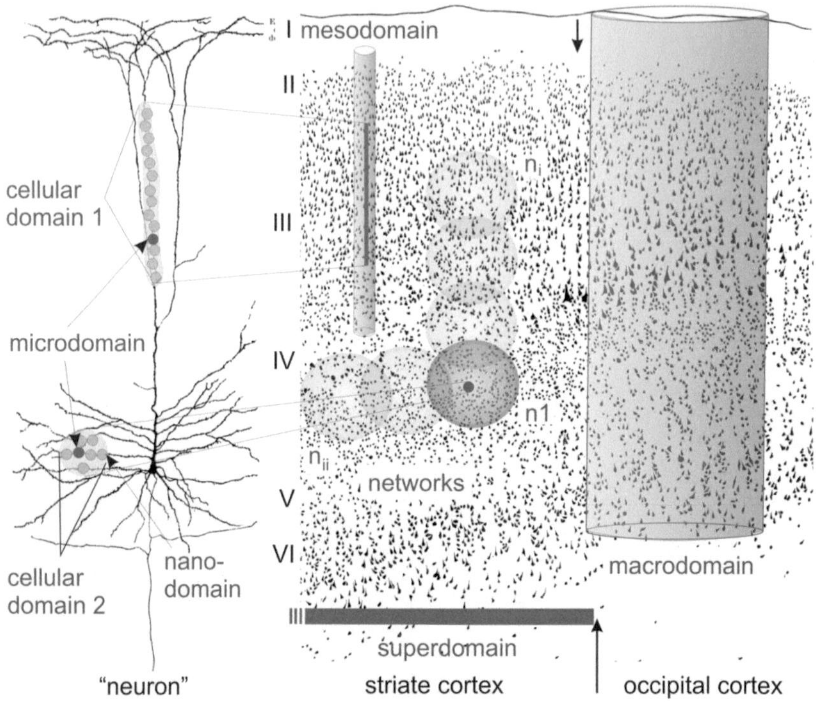

Fig. 10.14 Schematic representation of the various types of co-existing glial domains, drawn on the background of a figure from Vogt and Vogt (1937), showing the transition (*arrows*) between the striate and the occipital cortex in a human brain (drawing of a pyramidal cell: Ramón y Cajal, 1911). At increasing levels of spatial organization, the glial cells provide nanodomains up to superdomains for their interaction with neurons and blood vessels. With the possible exception of the nanodomains which probably interact with "their" neuronal partner structures as long as they exist, the domains are not only determined by the (ultra-) structural features of the glial cells but also by the properties of the signal (or stimulus): (i) (a few) individual synapses are associated with their ensheathing glial microdomains but parallel/strong stimulation of related inputs may finally integrate (ii) (oligo-) cellular domains involving the whole glial cell (or a few of them) and their neuronal partners (cf. Fig. 10.12a). It depends on the shape of the glial cells involved whether these domains are columnar ("type 1" e.g., Bergmann glial cells; probably radial astrocytes in the hippocampal stratum oriens, and hypothetically interlaminar astrocytes in the primate cortex) or rather spherical or ellipsoid ("type 2", star-shaped astrocytes). Appropriate (i.e., strong, frequent or synchronized) stimulation may then activate, via gap-junctional coupling, networks consisting of > 50 astrocytes (n). These networks, however, cannot be considered as static; any member of the coupled network at its margin is coupled *per se* to other cells which are outside the first coupling range but which would belong to another (overlapping) network if the dye would have been injected into such a cell. Thus, if a neuronal stimulus will arrive later at such a cell, or if a Ca^{2+} wave was triggered by the first "excited" astrocyte, (iii) macrodomains will develop; this mechanism may proceed either radially or tangentially. The size of these macrodomains may vary from small (macrodomain 1, corresponding e. g. to the orientation columns of Mountcastle 1957, 1997) to large (such as ocular dominance columns or barrel fields; macrodomain 2). A further progression of integration will result in the generation of very large functional units, (iv) superdomains, corresponding to entire cortical areas or gyri. Eventually, even whole hemispheres associated with huge astrocytic populations may transiently be involved, putatively mediating events such as spreading depression, seizures, and/or wide-spread neuronal degeneration.

Table 10.1 Dimensions and possible functions of glial domains at different organization levels

Organization level	Number of astrocytes involved	Number of synapses involved	Supposed function during development	Supposed function in mature function	Supposed role in pathology
Nanodomain	Less than 1 (thousands of nanodomains per astrocyte)	(Part of) 1	?	Glia-neuron exchange of specific molecules	?
Microdomain	Less than 1 (10–100 microdomains per cell)	1–5	Establishment and maintenance of synaptic contacts?	Maintenance of precise and specific information processing/support of synaptic plasticity	Impaired synaptic maintenance/ plasticity?
(Oligo-)cellular domain → mesodomain	1 (or a few) → <50	10–100 (?) → ~1,000 (?)	Organization of functional assemblies of synapses? → organization of functional assemblies of neurons?	Support of parallel/ comparative information processing?	Impaired synaptic maintenance/ plasticity?
Network	>50	Thousands	Organization of developing neuronal activity?	Facilitation of clearance/ homeostasis processes	Failure of CNS tissue homeostasis; neuronal degeneration
Macrodomain	>1,000	Millions	Establishment and maintenance of large functional units (e.g., ocular dominance columns, barrel fields)	Synchronization/tuning of integrative neuronal functions?	Desintegration of neuronal activity, e.g. seizures; neuronal degeneration
Superdomain	>100,000	Billions (10^9)	Establishment and maintenance of very large functional units (e.g., cortical gyri/fields)?	Maintenance of large integrative neuronal functions?	Desintegration of neuronal activity, e.g. seizures, spreading depression; neuronal degeneration

Note that all the glial domains have structural and/or functional pendants on the neuronal (and mostly, also on the vascular) side whereas the glial networks are probably spatially and functionally variable tools (e.g., of glial homeostasis) and contribute only indirectly to the generation of macro- and superdomains

domains were shown to be involved in territorial organization processes such as in the control of the migration of newborn neurons towards their layer of destination in the cortical plate (Rakic, 1972, 1978) and formation of laminar boundaries (Miskevich, 1999; Sajin and Steindler, 1994; Bushong et al., 2003), axon growth including the pathfinding-decision of axons in the optic chiasm (Silver et al., 1993; Powell and Geller, 1999) and correct synapse formation (Hu et al., 2007), and finally in the formation, maintenance, and plasticity of structural units such as the barrel fields (Bailey et al., 1999; Treloar et al., 1999) or the ocular dominance columns in the visual cortex (Müller and Best, 1989; Müller, 1992). During mature functioning of the brain, the glial domains may provide a metabolic synchronization/adjustment of functional arrays of neurons, and even provide an additional, non-synaptic pathway of slow information processing, via the Ca^{2+} waves (Scemes and Giaume, 2006; see also Chap. 8). Finally under pathological circumstances, the topography of glial domains may, at least partly, determine the area within which neuronal dysfunction and cell death occur. In any case, one should be aware of the possibility that the different glial domains may not be of a given and static size and can thus be transitional to each other (as marked in the Table 10.1 by some arrows indicating the transition of cellular domains to mesodomains).

10.9 Summary and Conclusions

The following functions have been ascribed to astrocytes, mediating between the neuronal elements (via their PAPs) and the blood vessels and extracellular spaces (via their PAE): maintenance of topographic relationships and structural integrity, metabolic interaction including nutrition and transmitter recycling, homeostasis of the extracellular fluid, and short-and long-term modification of synaptic efficacy. It can be stated that the structural prerequisites for all these interactions are provided by the specific, polarized processes of astrocytes, located at the appropriate places and endowed with a wealth of versatile molecules; moreover, these processes – particularly, the PAPs – are highly dynamic structures [for recent reviews, see Chao et al. (2002), Reichenbach et al. (2004)]. Furthermore, astroglial cells and their processes form co-existent domains of widely varying size (nano- to macro- and superdomains) which probably perform distinct interactions with functional assemblies of neurons. There is an increasing body of functional evidence for these interactions, which will be presented in the other chapters of this book.

Acknowledgements The authors thank Gert Brückner (Paul Flechsig Institute of Brain Research, Leipzig University), for many helpful discussions. The authors are grateful to Andreas Mack and Karen Wolburg-Buchholz (University of Tübingen) for providing Fig. 10.12b. Original work related to this chapter was supported by the Bundesministerium für Bildung, Forschung und Technologie, Interdisciplinary Center for Clinical Research at the University of Leipzig, 01KS9504, Project C-05 (AR), and by the Deutsche Forschungsgemeinschaft, RE 849–8/1 (AR).

References

Abbott NJ (1991) Permeability and transport of glial blood–brain barriers. Ann NY Acad Sci 633:378–394

Amedeé T, Robert A, Coles JA (1997) Potassium homeostasis and glial energy metabolism. Glia 21:46–55

Amiry-Moghaddam M, Williamson A, Palomba M, Eid T, De Lanerolle NC, Nagelhus EA, Adams ME, Froehner SC, Agre P, Ottersen OP (2003) Delayed K+ clearance associated with aquaporin-4 mislocalization: Phenotypic defects in brains of -syntrophin-null mice. Proc Natl Acad Sci USA, 100:13615–13620

Bailey MS, Puche AC, Shipley MT (1999) Development of the olfactory bulb: evidence for glia-neuron interactions in glomerular function. J Comp Neurol 415:423–448

Bär T (1980) The vascular system of the cerebral cortex. Adv Anat Embryol Cell Biol 59:1–62

Bennett MVL, Contreras JE, Bukauskas FF, Sáez JC (2003) New roles for astrocytes: Gap junction hemichannels have something to communicate. Trends Neurosci 26:610–617

Berthold CH, Carlstedt T (1977) Observations on the morphology at the transition between the peripheral and the central nervous system in the cat. III. Myelinated fibres in S1 dorsal rootlets. Acta Physiol Scand Suppl 446:43–60

Bevan S, Chiu SY, Gray PT, Ritchie JM (1985) The presence of voltage-gated sodium, potassium and chloride channels in rat cultured astrocytes. Proc R Soc Lond B Biol Sci 225:299–313

Bezzi P, Gundersen V, Galbete JL, Seifert G, Steinhäuser C, Pilati E, Volterra A (2004) Astrocytes contain a vesicular compartment that is competent for regulated exocytosis of glutamate. Nat Neurosci 7:613–620

Braak E (1975) On the fine structure of the external glial layer in the isocortex of man. Cell Tissue Res 157:367–390

Brückner G, Härtig W, Kacza J, Seeger J, Welt K, Brauer K (1996) Extracellular matrix organization in various regions of rat brain gray matter. J Neurocytol 25:333–346

Bushong EA, Martone ME, Ellisman MH (2003) Examination of the relationship between astrocyte morphology and laminar boundaries in the molecular layer of the adult dentate gyrus. J Comp Neurol 462:241–251

Butkevich E, Hülsmann S, Wenzel D, Shirao T, Duden R, Majoul I (2004) Drebrin is a novel connexin-43 binding partner that links gap juctions to the submembrane cytoskeleton. Curr Biol 14:650–658

Butt AM, Duncan A, Berry M (1994) Astrocyte associations with nodes of Ranvier: ultrastructural analysis of HRP-filled astroyctes in the mouse optic nerve. J Neurocytol 23:486–499

Chan-Palay V, Palay SL (1972) The form of velate astrocytes in the cerebellar cortex of monkey and rat: high voltage electron microscopy of rapid-Golgi preparations. Z Anat Entw-Gesch 138:1–19

Chao TI, Skatchkov SN, Eberhardt W, Reichenbach A (1994) Na+ channels of Müller (glial) cells isolated from retinae of various mammalian species including man. Glia 10:173–185

Chao TI, Rickmann M, Wolff JR (2002) The synapse-astrocyte boundary: anatomical basis for an integrative role of glia in synaptic transmission. In: Tripartite synapses: Synaptic transmission with glia (Volterra A, Magistretti P,Haydon PG, eds), pp3–23. Oxford, New York: Oxford University Press

Charles AC, Merrill JE, Dirksen ER, Sanderson MJ (1991) Intercellular signaling in glial cells: calcium waves and oscillations in response to mechanical stimulation and glutamate. Neuron 6:983–992

Christie SB, De Blas AL (2003) GABAergic and glutamatergic axons innervate the axon initial segment and organize GABA(A) receptor clusters of cultured hippocampal pyramidal cells. J Comp Neurol 456:361–374

Colombo JA, Reisin HD (2004) Interlaminar astroglia of the cerebral cortex: a marker of the primate brain. Brain Res 1006:126–131

Connors NC, Kofuji P (2002) Dystrophin Dp71 is critical for the clustered localization of potassium channels in retinal glial cells. J Neurosci 22:4321–4327

Connors NC, Adams ME, Froehner SC, Kofuji P (2004) The potassium channel Kir4.1 associates with the dystrophin glycoprotein complex via -syntrophin in glia. J Biol Chem 279:28387–28392

Cornell-Bell AH, Finkbeiner SM, Cooper MS, Smith SJ (1990) Glutamate induces calcium waves in cultured astrocytes: long-range glial signaling. Science 247:470–473

Denk W, Sugimori M, Llinas R (1995) Two types of calcium response limited to single spines in cerebellar Purkinje cells. Proc Natl Acad Sci USA 92:8279–8282

Derouiche A (1997)Coupling of glutamate uptake and degradation in transmitter clearance: Anatomical evidence. In: Neurotransmitter release and uptake (Pögün S, ed). NATO ASI Series Vol H:263–282

Derouiche A, Frotscher M (1991) Astroglial processes around identified glutamatergic synapses contain glutamine synthetase: evidence for transmitter degradation. Brain Res 552:346–350

Derouiche A, Rauen T (1995) Coincidence of l-glutamate/l-aspartate transporter (GLAST) and glutamine synthetase (GS) immunoreactions in retinal glia: evidence for coupling of GLAST and GS in transmitter clearance. J Neurosci Res. 42:131–43

Farrell CL, Pardridge WM (1991) Blood–brain barrier glucose transporter is asymmetrically distributed on brain capillary endothelial lumenal and ablumenal membranes. An electron microscopic immunogold study. Proc Natl Acad Sci USA 88:5779–5783

Feeney JF, Watterson RL (1946) The development of the vascular pattern within the walls of the central nervous system of the chick embryo. J Morphol 78:231–303

Ffrench-Constant C, Miller RH, Kruse J, Schachner M, Raff MC (1986) Molecular specialization of astrocyte processes at nodes of Ranvier in rat optic nerve. J Cell Biol 102:844–852

Fleischhauer K, Detzer K (1975) Dendritic bundling in the cerebral cortex. In: Advances in Neurology, Vol 12, Physiology and Pathology of Dendrites (Kreutzberg GW, ed), pp 71–78. New York: Raven Press

Folkman J, D'Amore PA (1996) Blood vessel formation: what is its molecular basis? Cell 87:1153–1155

Gabbot PLA, Bacon SJ (1996) The organisation of dendritic bundles in the prelimbic cortex (area 32) of the rat. Brain Res 730:75–86

Gerhardt H, Betsholtz C (2003) Endothelial-pericyte interactions in angiogenesis. Cell Tissue Res 314:15–23

Gerhardt H, Liebner S, Redies C, Wolburg H (1999) N-cadherin expression in endothelial cells during early angiogenesis in the eye and brain of the chicken: relation to blood–retina and blood–brain barrier development. Eur J Neurosci 11:1191–1201

Giaume C, Venance L (1995) Gap junctions in brain glial cells and development. Perspectives Dev Neurobiol 2:335–345

Giaume C, Tabernero A, Medina JM (1997) Metabolic trafficking through astrocytic gap junctions. Glia 21:114–123

Göbel W, Kampa BM, Helmchen F (2007) Imaging cellular network dynamics in three dimensions using fast 3D laser scanning. Nat Methods 4:73–79

Gros DB, Jongsma HJ (1996) Connexins in mammalian heart function. BioEssays 18:719–730

Grosche J, Matyash V, Möller T, Verkhratsky A, Reichenbach A, Kettenmann H (1999) Microdomains for neuron-glia interactions: parallel fiber signaling to Bergmann glial cells. Nat Neurosci 2:139–143

Grosche J, Kettenmann H, Reichenbach A (2002) Bergmann glial cells form distinct morphological structures to interact with cerebellar neurons. J Neurosci Res 68:138–149

Hamm S, Dehouck B, Kraus J, Wolburg-Buchholz K, Wolburg H, Risau W, Cecchelli R, Engelhardt B, Dehouck M-P (2004) Astrocyte mediated modulation of blood–brain barrier permeability does not correlate with a loss of tight junction proteins from the cellular contacts. Cell Tiss Res 315:157–166

Hámori J (1981) Synaptic input to the axon hillock and initial segment of inhibitory interneurons in the cerebellar cortex of the rat. An electron microscopic study. Cell Tissue Res 217:553–562

Harder DR, Zhang C, Gebremedhin D (2002) Astrocytes function in matching blood flow to metabolic activity. News Physiol Sci 16:27–31
Härtig W, Derouiche A, Welt K, Brauer K, Grosche J, Mädfer M, Reichenbach A, Brückner G (1999) Cortical neurons immunoreactive for the potassium channel Kv3.1b subunit are predominantly surrounded by perineuronal nets presumed as a buffering system for cations. Brain Res 842:15–29
Haseloff RF, Blasig IE, Bauer H-C, Bauer H (2005) In search for the astrocytic factor(s) modulating blood–brain barrier functions in brain capillary endothelial cells in vitro. Cell Molec Neurobiol 25:25–39
Häusser M, Spruston N, Stuart GJ (2000) Diversity and dynamics of dendritic signaling. Science 290:739–744
Hildebrand C, Remahl S, Persson H, Bjartmar C (1993) Myelinated nerve fibres in the CNS. Progr Neurobiol 40:319–384
Hirrlinger J, Hülsmann S, Kirchhoff F (2004) Astroglial processes show spontaneous motility at active synaptic terminals in situ. Eur J Neurosci 20:2235–2239
Holländer H, Makarov F, Dreher Z, Van Driel D, Chan-Ling T, Stone J (1991) Structure of macroglia of the retina: sharing and division of labour between astrocytes and Müller cells. J Comp Neurol 313:587–603
Houades V, Rouach N, Ezan P, Kirchhoff F, Koulakoff A, Giaume C (2006) Shapes of astrocyte networks in the juvenile brain. Neuron Glia Biol 2:3–14
Hu R, Cai WQ, Wu XG, Yang Z (2007) Astrocyte-derived estrogen enhances synapse formation and synaptic transmission between cultured neonatal rat cortical neurons. Neuroscience 144:1229–1240
Kamasawa N, Sik A, Morita M, Yasumura T, Davidson KGV, Nagy JI, Rash JE (2005) Connexin-47 and connexin-32 in gap junctions of oligodendrocyte somata, myelin sheaths, paranodal loops and Schmidt-Lanterman incisures: implications for ionic homeostasis and potassium siphoning. Neuroscience 136:65–86
Kofuji P, Newman EA (2004) Potassium buffering in the central nervous system. Neuroscience 129:1045–1056
Kölliker RA (1896) Handbuch der Gewebelehre des Menschen, Bd 2.Leipzig: Engelmann
Konietzko U, Müller CM (1994) Astrocytic dye coupling in rat hippocampus: topography, developmental onset, and modulation by protein kinase C. Hippocampus 4:297–306
Kosaka K, Kosaka T (2005) Synaptic organization of the glomerulus in the main olfactory bulb: compartments of the glomerulus and heterogeneity of the periglomerular cells. Anat Sci Int 80:80–90
Krstic RV (1991) Human microscopic anatomy. An atlas for students of medicine and biology. Berlin: Springer
Lindahl P, Johansson BR, Leveen P, Betsholtz C (1997) Pericyte loss and microaneurysm formation in PDGF-B-deficient mice. Science 277:242–245
Llinas R (1975) Electroresponsive properties of dendrites in central neurons. Adv Neurol 12:1–13
Llinas R, Nicholson C, Freeman JA, Hillman DE (1969) Dendritic spikes versus cable properties. Science 163:97
Miskevich F (1999) Laminar redistribution of a glial subtype in the chick optic tectum. Dev Brain Res 115:103–109
Mombaerts P (2006) Axonal wiring in the mouse olfactory system. Annu Rev Cell Dev Biol 22:713–737
Mountcastle VB (1957) Modality and topographic properties of single neurons of cat's somatic sensory cortex. J Neurophysiol 20:408–434
Mountcastle VB (1958) Somatic functions of the nervous system. Ann Rev Physiol 20:471–508
Mountcastle VB (1997) The columnar organization of the neocortex. Brain 120:701–722
Müller CM (1992) A role for glial cells in activity-dependent central nervous plasticity? Review and hypothesis. Int Rev Neurobiol 34:215–281

Müller CM, Best J (1989) Ocular dominance plasticity in adult cat visual cortex after transplantation of cultured astrocytes. Nature 342:427–430
Mulligan SJ, MacVicar BA (2004) Calcium transients in astrocyte endfeet cause cerebrovascular constrictions. Nature 431:195–199
Nagelhus EA, Mathiisen TM, Ottersen OP (2004) Aquaporin-4 in the central nervous system: cellular and subcellular distribution and coexpression with Kir4.1. Neuroscience 129:905–913
Nagy JI, Rash JE (2000) Connexins and gap junctions of astrocytes and oligodendrocytes in the CNS. Brain Res Rev 32:29–44
Nagy JI, Ionescu A-V, Lynn BD, Rash JE (2003) Coupling of astrocyte connexins Cx26, Cx30, Cx43 to oligodendrocyte Cx29, Cx32, Cx47: implications from normal and connexin32 knockout mice. Glia 44:205–218
Nagy JI, Dudek FE, Rash JE (2004) Update on connexins and gap junctions in neurons and glia in the mammalian nervous system. Brain Res Rev 47:191–215
Nielsen S, Nagelhus EA, Amiry-Moghaddam M, Bourque C, Agre P, Ottersen OP (1997) Specialized membrane domains for water transport in glial cells: high-resolution immunogold cytochemistry of aquaporin-4 in rat brain. J Neurosci 17:171–180
Nicchia GP, Nico B, Camassa LMA, Mola MG, Loh N, Dermietzel R, Spray DC, Svelto M, Frigeri A (2004) The role of aquaporin-4 in the blood–brain barrier development and integrity: studies in animal and cell culture models. Neuroscience 129:935–945
Ohira K, Funatsu N, Homma KJ, Sahara Y, Hayashi M, Kaneko T, Nakamura S (2007) Truncated TrkB-T1 regulates the morphology of neocortical layer I astrrocytes in adult rat brain slices. Eur J Neurosci 25:406–416
Oliet SH, Piet R, Poulain DA (2001) Control of glutamate clearance and synaptic efficacy by glial coverage of neurons. Science 292:923–926
Pannicke T, Iandiev I, Uckermann O, Biedermann B, Kutzera F, Wiedemann P, Wolburg H, Reichenbach A, Bringmann A (2004) A potassium-channel-linked mechanism of glial cell swelling in the postischemic retina. Mol Cell Neurosci 26:493–502
Peters O, Schipke CG, Hashimoto Y, Kettenmann H (2003) Different mechanisms promote astrocyte Ca2+ waves and spreading depression in the mouse neocortex. J Neurosci 23:9888–9896
Pinching AJ (1971) Myelinated dendritic segments in the monkey olfactory bulb. Brain Res 29:133–138
Powell EM, Geller HM (1999) Dissection of astrocyte-mediated cues in neuronal guidance and process extension. Glia 26:73–83
Raine CS (1984) On the association between perinodal astrocytic processes and the node of Ranvier in the CNS. J Neurocytol 13:21–27
Rakic P (1972) Mode of cell migration to the superficial layers of fetal monkey neocortex. J Comp Neurol 145:61–83
Rakic P (1978) Neuronal migration and contact guidance in the primate telencephalon. Postgrad Med J 54 Suppl 1:25–40
Ramón y Cajal S (1911) Histologie du système nerveux de l'homme et des vertébrés. Paris: Maloine
Rash JE, Duffy HS, Dudek FE, Bilhartz BL, Whalen LR, Yasumura T (1997) Grid-mapped freeze-fractured analysis of gap junctions in gray and white matter of adult rat central nervous system, with evidence for a "panglial syncytium" that is not coupled to neurons. J Comp Neurol 388:265–292
Rash JE, Yasumura T, Dudek FE, Nagy JI (2001) Cell-specific expression of connexins and evidence of restricted gap junctional coupling between glial cells and between neurons. J Neurosci 21:1983–2000
Rash JE, Davidson KGV, Yasumura T, Furman CS (2004) Freeze-fracture and immunogold analysis of aquaporin-4 (AQP4) square arrays, with models of AQP4 lattice assembly. Neuroscience 129:915–934
Reichenbach A (1989) Attempt to classify glial cells by means of their process specialization using the rabbit retinal Müller cell as an example of cytotopographic specialization of glial cells. Glia 2:250–259

Reichenbach A, Henke A, Eberhardt W, Reichelt W, Dettmer D (1992) K+ ion regulation in retina. Can J Physiol Pharmacol 70:S239–S247

Reichenbach A, Siegel A, Rickmann M, Wolff JR, Noone D, Robinson SR (1995) Distribution of Bergmann glial somata and processes: implications for function. J Hirnforsch 36:509–517

Reichenbach A, Derouiche A, Grosche J, Hanani M (2004) Structural association of glia with the various compartments of neurons. In: Glial Neuronal Signaling (Hatton GI,Parpura V, eds), pp 53–97. Boston, MA: Kluwer

Reisin HD, Colombo JA (2002) Considerations on the astroglial architecture and the columnar organization of the cerebral cortex. Cell Molec Neurobiol 22:633–644

Rice FL, Van der Loos H (1977) Development of the barrels and barrel field in the somatosensory cortex of the mouse. J Comp Neurol 171:545–560

Robinson SR, Hampson ECGM, Munro MN, Vaney DI (1993) Unidirectional coupling of gap junctions between neuroglia. Science 262:1072–1074

Rouach N, Pébay A, Même W, Cordier J, Ezan P, Etienne E, Giaume C, Tencé M (2006) S1P inhibits gap junctions in astrocytes: involvement of Gi and Rho GTPase/ROCK. Eur J Neurosci 23:1453–1464

Sajin B, Steindler DA (1994) Cells on the edge: boundary astrocytes and neurons. Perspect Dev Neurobiol 2:275–289

Sarantis M, Mobbs P (1992) The spatial relationship between Müller cell processes and the photoreceptor output synapse. Brain Res 584:299–304

Scemes E, Giaume C (2006) Astrocyte calcium waves: what they are and what they do. Glia 54:716–725

Schipke C, Haas B, Kettenmann H (2008) Astrocytes discriminate and selectively respond to the activity of a subpopulation of neurons within the barrel cortex. Cereb Cortex, DOI: 10.1093/ceroor/bhn009

Shrager P, Chiu SY, Ritchie JM (1985) Voltage-dependent sodium and potassium channels in mammalian cultured Schwann cells. Proc Natl Acad Sci U S A. 82:948–952

Silver J, Edwards MA, Levitt P (1993) Immunocytochemical demonstration of early appearing astroglial structures that form boundaries and pathways along axon tracts in the fetal brain. J Comp Neurol 15:415–436

Simard M, Nedergaard M (2004) The neurobiology of glia in the context of water and ion homeostasis. Neuroscience 129:877–896

Simionescu M, Ghinea N, Fixman A, Lasser M, Kukes L, Simionescu N, Palade GE (1988) The cerebral microvasculature of the rat: structure and luminal surface properties during early development. J Submicrosc Cytol 20:243–261

Sims DE (1986) The Pericyte – A review. Tissue and Cell 18:153–174

Stout CE, Costantin JL, Naus CCG, Charles AC (2002) Intercellular calcium signaling in astrocytes via ATP release through connexin hemichannels. J Biol Chem 277:10482–10488

Tabernero A, Medina JM, Giaume C (2006) Glucose metabolism and proliferation in glia: role of astrocytic gap junctions. J Neurochem 99:1049–1061

Treherne JE, Schofield PK, Lane NJ (1982) Physiological and ultrastructural evidence for an extracellular anion matrix in the central nervous system of an insect (Periplaneta americana). Brain Res 247:255–267

Treloar HB, Purcell AL, Greer C (1999) Glomerular formation in the developing rat olfactory bulb. J Comp Neurol 413:289–304

Verbavatz J-M, Ma T, Gobin R, Verkman AS (1997) Absence of orthogonal arrays in kidney, brain and muscle from transgenic knockout mice lacking water channel aquaporin-4. J Cell Sci, 110:2855–2860

Vogt C, Vogt O (1937) Sitz und Wesen der Krankheiten im Lichte der topistischen Hirnforschung und des Variierens der Tiere. 1. Teil: Befunde der topistischen Hirnforschung als Beitrag zur Lehre vom Krankheitssitz. Leipzig: J.A. Barth

Volterra A, Magistretti P, Haydon P (2002) The tripartite synapses: Glia in synaptic transmission. Oxford: Oxford University Press

Wallraff A, Odermatt B, Willecke K, Steinhäuser C (2004) Distinct types of astroglial cells in the hippocampus differ in gap junction coupling. Glia 489:36–43

Wallraff A, Kohling R, Heinemann U, Theis M, Willecke K, Steinhäuser C (2006) The impact of astrocytic gap junctional coupling on potassium buffering in the hippocampus. J Neurosci 26:5438–5447

Waxman SG (1986) The astrocyte as a component of the node of Ranvier. Trends Neurosci 9:250–253

Wolburg H (2006) The endothelial frontier. In: Blood–brain barriers. From ontogeny to artificial interfaces (Dermietzel R,Spray DC,Nedergaard M, eds), pp 77–107. Weinheim: Wiley-VCH

Wolburg H (1995) J Hirnforsch 36:239–258Orthogonal arrays of intramembranous particles. A review with special reference to astrocytes.

Wolburg H, Bäuerle C (1993) Astrocytes in the lamina cribrosa of the rat optic nerve: are their morphological peculiarities involved in an altered blood–brain barrier? J Hirnforsch 34:445–459

Wolburg H, Risau W (1995) Formation of the blood–brain barrier. In: Neuroglia (Kettenmann H and Ransom B, eds), pp 763–776. Oxford: Oxford University Press

Wolburg H, Warth A (2005) Endothelial cells, extracellular matrix and astrocytes: interplay for managing the blood–brain barrier. In: The blood–brain barrier and its microenvironment (deVries E,Prat A, eds), pp 359–386. New York, London: Taylor & Francis

Wolburg H, Wolburg-Buchholz K, Mack A, Reichenbach A (2007) Ependymoglia. In: The New Encyclopedia of Neuroscience (Squire L,Albright T,Bloom F,Gage F,Spitzer N, eds). Oxford, UK: Elsevier

Wolff J (1968) [The role of the astroglia in the brain tissue]. Acta Neuropathol Suppl 4:33–39

Yang B, Brown D, Verkman AS (1996) The mercurial insensitive water channel (AQP-4) forms orthogonal arrays in stably transfected chinese hamster ovary cells. J Biol Chem 271:4577–4580

Zahs KR (1998) Heterotypic coupling between glial cells of the mammalian central nervous system. Glia 24:85–96

Abbreviations

AQP4	Aquaporin-4
BBB	Blood–brain barrier
CNS	Central nervous system
CSF	Cerebrospinal fluid
ERM	Ezrin, radixin, moesin
GABA	Gama amino butyric acid
GFAP	Glial fibrillary acidic protein
GFP	Green fluorescent protein
IP_3	Inositol 1,4,5-trisphosphate
OAPs	Orthogonal arrays of intramembranous particles
PAE	Perivascular astrocytic endfeet
PAPs	Peripheral astroglial processes
PDGF	Platelet-derived growth factor
TTX	Tetrodotoxin

Chapter 11
Synaptic Information Processing by Astrocytes

Gertrudis Perea and Alfonso Araque

Contents

11.1 Introduction

Since the time of the initial studies of the nervous system, neurons were recognized as the cellular elements responsible for the information processing of the nervous system, while glial cells were considered as playing simple supportive roles to neurons. The fundamental attribute of neurons is their cellular electrical excitability, which is based on the expression of a plethora of ligand- and voltage-gated membrane channels that give rise to prominent membrane currents and membrane potential variations, which represent the biophysical substrate underlying the integration and transfer of information at the cellular level in the Central Nervous System (CNS). By contrast, glial cells are not electrically excitable. Although they are able to express some of the ion channels that are expressed by neurons, the level of expression of some key channels is not sufficiently high to support the generation of active electrical behaviors in response to different stimuli. Nevertheless, glial

A. Araque
Instituto Cajal, CSIC, Doctor Arce 37, Madrid, Spain
araque@cajal.csic.es

V. Parpura and P.G. Haydon (eds.), *Astrocytes in (Patho)Physiology of the Nervous System*, 287
DOI: 10.1007/978-0-387-79492-1_11, © Springer Science+Business Media, LLC 2009

cells display a form of excitability that is based on variations of the Ca^{2+} concentration in the cytosol rather than electrical changes in the membrane potential.

Among the different types of glial cells, astrocytes have received special attention, probably due to their intimate spatial relationship with neurons and synapses in the CNS. In addition to the well-known functions of astrocytes in the different physiological processes of the nervous system, such as differentiation, proliferation, trophic support and survival of neurons, new findings have recently proposed the existence of bidirectional signaling between astrocytes and neurons with an important active role of astrocytes in the physiology of the nervous system. As a consequence, there is a new concept of the synaptic physiology – "the tripartite synapse", where astrocytes exchange information with the pre- and postsynaptic elements and participate as dynamic regulatory elements in neurotransmission (Araque et al., 1999). In this chapter we review recent evidence indicating that astrocytes are able to integrate and process synaptic information. These findings suggest the participation of astrocytes as active cellular elements in information processing of the nervous system.

11.2 Intracellular Ca^{2+} Variations are the Basis of the Astrocyte Excitability

Astrocytes were classically considered as non-excitable cells because they do not show electrical excitability. Indeed, although astrocytes can express voltage-gated channels (Sontheimer, 1994), astrocytic membrane potential is relatively stable (Orkand et al., 1966). However, the introduction of fluorescence imaging techniques that allowed the observation of intracellular calcium levels demonstrated that astrocytes are excitable cells that based their excitability on intracellular Ca^{2+} variations (Cornell-Bell et al., 1990; Charles et al., 1991). This demonstration was critical for the re-evaluation of the functions of astrocytes in brain physiology.

During the last fifteen years many groups have made a great effort to define the mechanisms underlying this form of excitability, as well as its properties and functional significance. Consequently, we currently know that the cellular Ca^{2+} signal is manifested as elevations of cytosolic Ca^{2+} and relies on the existence of a relatively low concentration of free Ca^{2+} inside the cells. While neurons may use the electrochemical gradients across the plasma membrane to effectively increase the intracellular Ca^{2+} levels, astrocytes mainly use Ca^{2+} stored in the endoplasmic reticulum as a source for cytoplasmic Ca^{2+}.

Astrocytic Ca^{2+} excitability can be present spontaneously (Nett et al., 2002; Parri et al., 2001; Peters et al., 2003) or can be triggered by many different signaling molecules, including neurotransmitters released by neurons. Most of the transmitter receptors expressed by astrocytes are metabotropic receptors associated with G proteins that stimulate phospholipase C and the formation of inositol-1,4,5-trisphosphate upon activation, which increases the intracellular Ca^{2+} concentration through Ca^{2+} release from the inositol-1,4,5-trisphosphate-sensitive Ca^{2+} stores

(Araque et al., 2002; Bezzi et al., 1998; Kang et al., 1998; Kulik et al., 1999; Pasti et al., 1997; Porter and McCarthy, 1997). This Ca^{2+} signal may serve as an intracellular and intercellular signal that can propagate within and between astrocytes, signaling to different regions of the cell and to different cells, with relevant functional consequences for the physiology of the nervous system.

Intracellular Ca^{2+} elevations in astrocytes represent a signal that can be highly localized into discrete regions of the cell. A pioneering study in Bergmann glial cells, a specialized type of cerebellar astrocytes, showed sub-cellular microdomains that responded independently to stimulation of afferent fibers (Grosche et al., 1999). Ca^{2+} elevations in hippocampal astrocytes can also be spatially restricted to localized regions of the cell, from where they can eventually extend to greater portions of the cell (Fiacco and McCarthy, 2004; Pasti et al., 1997; Perea and Araque, 2005a). Furthermore, these astrocytes show functional sub-cellular domains that may respond independently to different synaptically released neurotransmitters (Araque et al., 2002; Perea and Araque, 2005a). These findings suggest that Bergmann glia and astrocyte processes consist of hundreds of independent compartments capable of autonomous interactions with the particular group of synapses that they cover.

The existence of localized sites of Ca^{2+} elevations is not exclusive to the astrocytic responses to neuronal activity. Indeed, spontaneous astrocytic Ca^{2+} oscillations, which are independent of neuronal activity, arise within discrete regions of astrocytic processes, and can eventually propagate along cell processes (Nett et al., 2002; Parri et al., 2001; Peters et al., 2003).

The compartmentalization of the Ca^{2+} signal as well as its controlled propagation to different regions of the cell is of great functional significance because it grants the regulation of the spatial extension of the physiological consequences of the astrocyte-to-neuron communication to neuronal physiology and synaptic transmission (see below).

11.3 Tripartite Synapse: Reciprocal Communication Between Neurons and Astrocytes

Astrocytes express a wide variety of functional receptors for many neurotransmitters, including glutamate, adenosine, norepinephrine, γ-Aminobutyric acid (GABA), histamine, adenosine 5'-triphosphate (ATP) and acetylcholine (Porter and McCarthy, 1997). The fact that astrocyte Ca^{2+} signals may be evoked by neurotransmitters released from synaptic terminals indicates the existence of functional neuron-to-astrocyte communication, which represents a new form of intercellular signaling between neurons and astrocytes in the CNS (Araque et al., 1999; Araque et al., 2001; Newman, 2005). However, the properties and extent of the synaptic control of the astrocyte Ca^{2+} is not fully determined. Indeed, the synaptic control of the glial Ca^{2+} signal has been demonstrated to be exerted by some neurotransmitters released by synaptic terminals, such as glutamate, GABA, acetylcholine, norepinephrine or nitric oxide (for a review see Perea and Araque, 2005b), and in some representative brain areas, such as the retina (Newman and Zahs, 1998), cerebellum (Grosche et al., 1999; Kulik et al., 1999; Matyash, et al., 2001), hippocampus (Araque et al., 2002; Dani et al., 1992; Kang et al., 1988; Pasti et al., 1997; Perea and Araque, 2005a; Porter and McCarthy 1996), and cortex (Peters et al., 2003). Although it is feasible that it represents a general phenomenon, further studies are required to elucidate the existence and properties of this communication in other brain areas, as well as the possible control of astrocyte Ca^{2+} by synapses that use other neurotransmitter systems.

The astrocyte Ca^{2+} signal is finely controlled by the level and pattern of synaptic activity, indicating that the neuron-to-astrocyte signaling is a precisely regulated intercellular communication. In the hippocampus, the stimulation of Schaffer collaterals (SC), the major glutamatergic input to the CA1 hippocampal region, causes Ca^{2+} elevations in astrocytes that are dependent on the stimulation frequency (Pasti et al., 1997; Perea and Araque, 2005a). Furthermore, in conditions of continuous stimulation of the SC pathway these Ca^{2+} elevations become oscillatory, and the frequency of the oscillations changes according to the firing rate of neuronal afferents (Pasti et al., 1997). Likewise, those astrocytes respond to GABAergic interneurons (Kang et al., 1998) and to cholinergic terminals (Araque et al., 2002) with Ca^{2+} elevations that are regulated by the frequency of neuronal activity. The regulation of the Ca^{2+} signal is also present in the cerebellum where the amplitude of Bergmann glial Ca^{2+} responses evoked by parallel fiber and granular layer stimulation changes with the stimulation frequency (Kulik et al., 1999; Matyash et al., 2001).

The fact that hippocampal astrocytes respond with Ca^{2+} elevations to cholinergic afferents coming from distant nuclei, i.e., the septum and diagonal band of Broca, indicates that astrocytes not only respond to the neuronal activity of the local circuits where they are immersed, but that they can also be targets of extrinsic axons arising from different brain areas (Araque et al., 2002), which adds further complexity to the signaling pathways in the CNS.

One of the most stimulating topics in current neurobiology is the functional consequences of the astrocyte Ca^{2+} signal on neuronal physiology. It is well established that signaling between neurons and astrocytes is a reciprocal communication, where astrocytes not only respond to neuronal activity but also actively modulate neuronal excitability and synaptic transmission (for a detailed discussion of this issue see Chap. 15). Astrocytes may release several gliotransmitters, such as glutamate, ATP, tumor necrosis factor α, D-serine or adenosine, which may serve as modulators of neuronal physiology (see Chapts. 12, 13, 15, 16). Some of these transmitters have been shown to be released in a Ca^{2+} – dependent manner (for a review see Volterra and Meldolesi, 2005), although alternative Ca^{2+}-independent mechanisms of release have also been proposed (see Nedergaard et al., 2003). Originally, astrocyte-induced neuromodulation was described in cultured cells and was later reported in more intact preparations of different brain areas (for reviews see Allen and Barres, 2005; Araque and Perea, 2004; Haydon and Araque 2002; Volterra and Steinhauser, 2004).

11.4 Astrocytes Discriminate the Activity of Different Synaptic Pathways

Astrocytic Ca^{2+} signals can be selectively mediated by specific synaptic terminals. Indeed, astrocytes located in the stratum oriens of the CA1 area of the hippocampus respond to synaptic activity of the alveus (that contains glutamatergic and cholinergic axons coming from the septum and diagonal band of Broca), with Ca^{2+} elevations exclusively mediated by ACh but not by glutamate, in spite of the fact that these astrocytes express functional glutamate receptors (Araque et al., 2002). However, these astrocytes respond to glutamate when it is released by a different glutamatergic input, i.e., the SC synaptic terminals (Perea and Araque, 2005a); indicating that astrocytes selectively respond to different synapses that use different neurotransmitters, i.e., glutamate and ACh. Furthermore, astrocytes can discriminate between the activity of different pathways that use the same neurotransmitter, i.e., glutamatergic axons of the SC and the alveus (Perea and Araque, 2005a). This discrimination between the glutamate released from SC terminals but not from glutamatergic axons in the alveus suggests the existence of astrocytic functional domains that grant localized neuron-astrocyte communication (Perea and Araque, 2005a).

11.5 Astrocytes Integrate Synaptic Information

Neurons are characterized by their ability to receive multiple input signals, to integrate and process them and to elaborate an output signal. Neuronal intrinsic properties account for the complex non-linear input/output relationships that are responsible for the integrative properties of neurons (Agmon-Snir et al., 1998; Llinas and Sugimori, 1980). Accordingly, the neuronal electrical excitability is non-linearly regulated by the simultaneous activity of different converging synaptic inputs.

As described above, it is firmly established that synaptically released neurotransmitters may control the intracellular Ca^{2+} levels in astrocytes. While the duration, amplitude and frequency of the astrocyte Ca^{2+} signal can be regulated by different levels of synaptic activity, these responses could be merely due to different concentrations of neurotransmitter released during different levels of synaptic activity. Therefore, it remained unknown whether neuron-to-astrocyte communication presents properties of complex information processing that are classically considered to be exclusive to neuron-to-neuron communication. In other words, does the astrocyte Ca^{2+} signal simply reflect the synaptic activity level, or, by contrast, can astrocytes integrate synaptic information responding non-linearly to the incoming information from adjacent synapses?

The analysis of the Ca^{2+} responses of hippocampal astrocytes to the activity of different synaptic terminals that release ACh and glutamate as neurotransmitters indicates that astrocytes are indeed endowed with intrinsic cellular properties that allow the processing of synaptic information (Perea and Araque, 2005a). Hippocampal astrocytes located in the stratum oriens of the CA1 area respond selectively to alveus stimulation with Ca^{2+} elevations mediated by ACh, and to SC with Ca^{2+} elevations mediated by glutamate receptor activation (Perea and Araque, 2005a). To investigate whether astrocytes can integrate information from different synaptic inputs, we have recently analyzed the astrocytic responses to the simultaneous activity of those synaptic pathways, i.e., the SC and the alveus (Fig. 11.1). We reasoned that if the integration of synaptic inputs occurred, it would be manifested as a non-linear modulation of the synaptically-evoked Ca^{2+} signal. We found that the amplitude of the astrocytic responses to the simultaneous stimulation of both pathways was inconsistent with a simple passive response to synaptic activity.

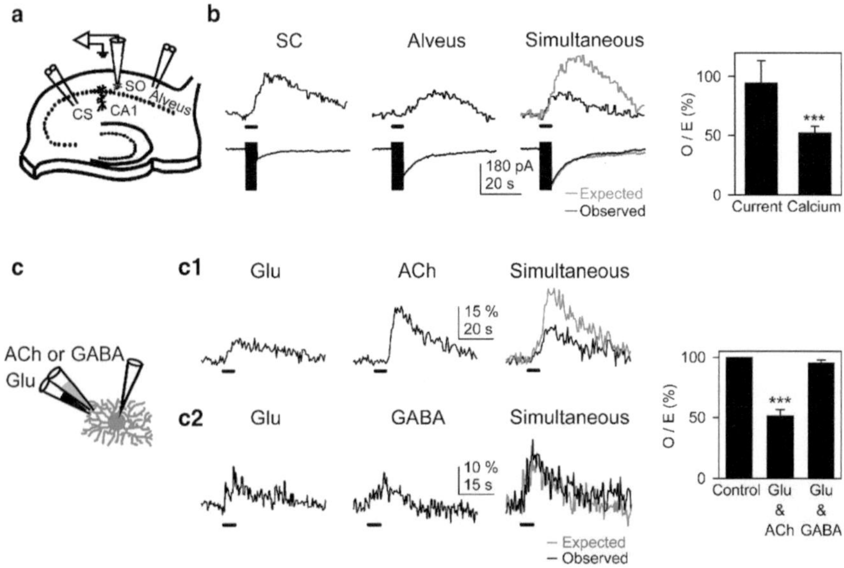

Fig. 11.1 Astrocyte Ca^{2+} signal modulation. **a** Schematic drawing showing the position of the recording astrocytes in the stratum oriens (SO) and the stimulating electrodes in the Schaffer collateral (SC) and the alveus pathways in the CA1 region of the hippocampus. **b**, Representative astrocytic Ca^{2+} levels (upper traces) and whole-cell membrane currents (lower traces) elicited by independent or simultaneous stimulation of the SC and the alveus (30 Hz, 5 s). The vertical *black bar* on the current traces and the *horizontal lines* at the bottom of Ca^{2+} traces represent the stimuli. In simultaneous stimulation condition, black and gray traces correspond to the observed (O) and expected (E; i.e., the linear summation of responses evoked by independent stimulation of both pathways) responses as in all other figures, respectively. Histogram represents the ratio between the observed and expected responses. **c**, The Ca^{2+} signal modulation depends on the astrocytic intrinsic properties. Schematic drawing showing an astrocyte whole-cell filled with the calcium indicator fluo-3 and a double barrel pipette filled with glutamate (Glu) and either ACh or GABA. **c1** Representative astrocyte Ca^{2+} elevations evoked by ionophoretical application of Glu and ACh, and modulation of the intracellular Ca^{2+} signal elicited by simultaneous application of both transmitters. **c2** Representative astrocyte Ca^{2+} elevations evoked by ionophoretical application of Glu and GABA. Histogram represents the ratio between the observed and expected (i.e., the linear summation of responses evoked by independent application of neurotransmitters) Ca^{2+} responses.

Indeed, the Ca^{2+} elevation evoked by simultaneous stimulation of both pathways was significantly different from the linear summation of the Ca^{2+} signals evoked by the independent stimulation (Fig. 11.1b). Therefore, the astrocyte Ca^{2+} signal was modulated by the simultaneous activity of cholinergic and glutamatergic synapses. A similar modulation of the astrocyte Ca^{2+} signal was observed in the absence of synaptic activity when both transmitters, glutamate and ACh, were exogenously applied (Fig. 11.1c). These findings indicate that astrocytes integrate the information of incoming inputs and that this integration depends on the cellular intrinsic properties of the astrocytes (Perea and Araque, 2005a).

Astrocytic receptor activation by exogenously applied transmitters may have synergistic effects that increase the astrocyte Ca^{2+} signal (Fatatis et al., 1994; Cormier et al., 2001; Sul et al., 2004). The fact that the simultaneous activity of cholinergic and glutamatergic synapses may induce the relative reduction of the astrocyte Ca^{2+} signal indicates that negative cooperative actions of neurotransmitters may occur and that the astrocytic Ca^{2+} signal is susceptible to depression. Therefore, the existence of positive and negative cooperative actions of neurotransmitters that allow the potentiation or inhibition of astrocyte Ca^{2+} signal confers a higher degree of complexity to the properties of the information transfer between neurons and astrocytes.

11.6 Astrocyte Ca^{2+} Signal Modulation is Specific of Some Neurotransmitters

To determine whether the astrocyte Ca^{2+} signal modulation evoked by glutamate and ACh is a general phenomenon or, rather, if it depends on the nature of the neurotransmitters involved, we investigated whether similar modulation was induced by direct co-application of glutamate and GABA. We found that the Ca^{2+} signals evoked by simultaneous application of both glutamate and GABA were not significantly different from the linear summation of the Ca^{2+} elevations evoked independently by both neurotransmitters (Fig. 11.1c2) (Perea and Araque, 2006). Therefore, the astrocyte Ca^{2+} signal modulation is not a general phenomenon that occurs in response to interaction with any transmitter, rather, it depends on the neurotransmitters involved, and consequently, it is a phenomenon that may be selectively induced by specific synapses. Furthermore, considering that astrocytes express a vast amount of neurotransmitter receptors, their possible differential interactions would yield a great number of possible responses, thus increasing the degree of complexity of the non-linear integrative properties of the astrocytes.

Although the cellular mechanisms responsible for the different modulatory effects of glutamate, ACh and GABA on the astrocyte Ca^{2+} signal are unknown, they are probably due to the fact that these neurotransmitters are coupled to different intracellular signaling pathways. While both metabotropic ACh and glutamate receptors converge on their intracellular signaling pathways at the level of the phospholipase C-beta activation, $GABA_B$ receptors are coupled to different intracellular pathways that involve adenylyl cyclase regulation. Therefore, while the intrinsic properties of neurons are determined by the membrane electrical characteristics, the cellular intrinsic properties of astrocytes probably reside in the intracellular signaling pathways.

11.7 The Modulation of the Astrocyte Ca^{2+} Signal Depends on the Level of Synaptic Activity

The modulation of astrocytic Ca^{2+} signals that are evoked by glutamate and ACh is finely regulated by the level of synaptic activity (Perea and Araque, 2005a). While the concurrent synaptic stimulation of SC and alveus at high frequencies (30 and 50 Hz) generates a relative reduction of the astrocyte Ca^{2+} signal, stimulation of both pathways at relative low frequencies (1 and 10 Hz) induces a relative increase of the Ca^{2+} signal amplitude, i.e., the Ca^{2+} signal is higher than the linear summation of the responses elicited by independent stimulation (Fig. 11.2a, B) (Perea and Araque, 2005a). Therefore, the modulation of the astrocyte Ca^{2+} signal is a plastic phenomenon that is bidirectionally controlled by synaptic activity, being potentiated or depressed at relatively low and high levels of synaptic activity, respectively. The bidirectional property of the astrocyte Ca^{2+} signal modulation might be extremely relevant to the physiology of the nervous system. For example, cholinergic inputs interact with glutamatergic transmission during the theta rhythm of the hippocampus (Buzsaki, 2002), a rhythm associated with the process of learning and memory (Vertes, 2005). Whether the interaction of these transmitters at the level of the astrocyte contributes to this process is unknown. However, since astrocytes can release different gliotransmitters through Ca^{2+}-dependent mechanisms (Bezzi et al., 2004; Mothet et al., 2005; Parpura and Haydon, 2000; Pascual et al., 2005; Zhang et al., 2003), which can in turn regulate synaptic transmission (Araque et al., 1998a; 1998b; Fiacco and McCarthy, 2004; Kang et al, 1998; Liu et al., 2004; Panatier et al., 2006; Pascual et al., 2005; Serrano et al., 2006; Zhang et al., 2003), and the brain microcirculation (Metea and Newman, 2006; Mulligan and MacVicar, 2004;

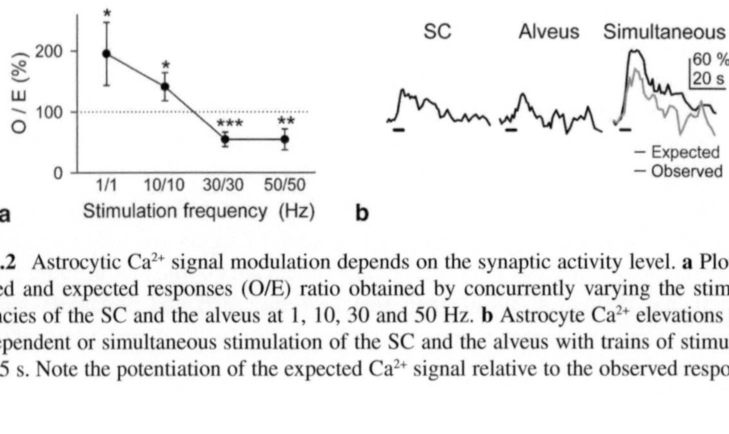

Fig. 11.2 Astrocytic Ca^{2+} signal modulation depends on the synaptic activity level. **a** Plot of the observed and expected responses (O/E) ratio obtained by concurrently varying the stimulation frequencies of the SC and the alveus at 1, 10, 30 and 50 Hz. **b** Astrocyte Ca^{2+} elevations evoked by independent or simultaneous stimulation of the SC and the alveus with trains of stimuli at 10 Hz for 5 s. Note the potentiation of the expected Ca^{2+} signal relative to the observed response.

Zonta et al., 2003) it will be intriguing to determine whether astrocytes regulate theta rhythms and ultimately the process of learning and memory.

In addition to this novel Ca^{2+} signal plasticity evoked by synaptic activity, recent reports have demonstrated that astrocytes are subject to activity-dependent modifications similar to short-term and long-term plasticity of neuronal synapses (Bellamy and Ogden, 2005; Ge et al., 2006). Indeed, in the hippocampus oligodendrocyte precursor cells (OPCs), a glial cell subtype, express Ca^{2+}-permeable -amino-3-hydroxy-5-methyl-isoxazole propionate (AMPA) receptors that can be activated by SC stimulation. Under high frequencies of synaptic activity, these cells showed a potentiation of AMPA-mediated currents (Ge et al., 2006), suggesting that OPCs are able to express components for induction and expression of long-term potentiation that up to now were considered to be exclusive to neurons. Furthermore, Bergmann glial cells in the cerebellum express short-term plasticity in response to paired-pulse stimulation of parallel fibers, increasing the extrasynaptic currents with different patterns of facilitation from parallel fiber-Purkinje cell synapses (Bellamy and Ogden, 2005).

11.8 Ca^{2+} Signal Modulation is Present in Astrocytic Processes

As described above, astrocytic processes are constituted by hundreds of microdomains that represent the elementary units of the astrocyte Ca^{2+} signal (Araque et al., 2002; Fiacco and McCarthy, 2004; Grosche et al., 1999; Nett et al., 2002; Pasti et al., 1997). The modulation of the astrocyte Ca^{2+} signal by different neurotransmitters also occurs at discrete regions of the astrocytic processes, suggesting that information processing of different inputs takes place at subcellular microdomains (Perea and Araque, 2005a). Indeed, when the Ca^{2+} signal of discrete regions of the astrocytic processes was analyzed in response to independent and simultaneous synaptic activity of the SC and the alveus, we found that the Ca^{2+} signal evoked by the simultaneous stimulation of both pathways was relatively reduced (Fig. 11.3a, b), indicating that the Ca^{2+} signal modulation is not only manifested in the soma but also in the processes (Perea and Araque 2005a). Furthermore, the Ca^{2+} signal modulation that occurred at the processes controls the intracellular propagation of the synaptically-evoked Ca^{2+} signal (Fig. 11.3b). Considering that a single astrocyte can enwrap ~140,000 synapses, and that their processes are intimately associated with synapses, the synaptic control of the intracellular Ca^{2+} signal propagation may have relevant consequences for brain function by regulating the spatial range of the influence of astrocytes on different synaptic terminals (Fig. 11.3c).

11.9 Perspectives and Conclusions

While in the last years great progress has been made in the knowledge of the properties of the bidirectional communication between astrocytes and neurons, there are still key issues that remain unknown. According to the most basic model of a neuron, the soma

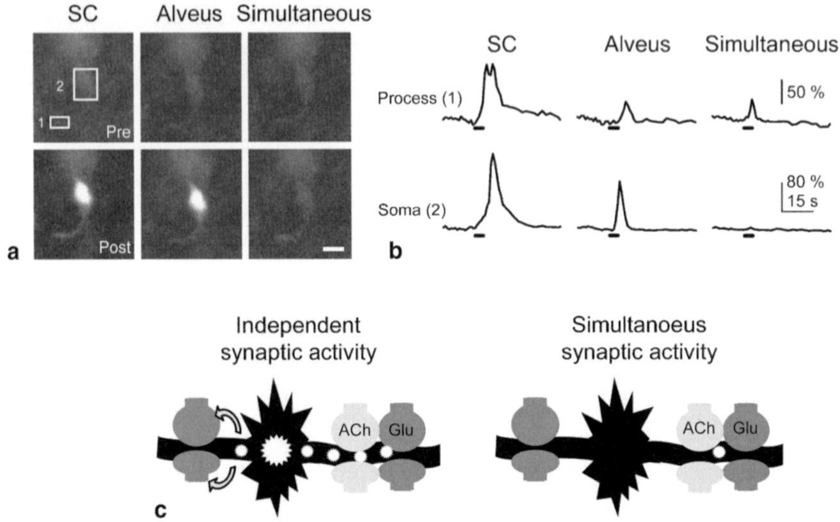

Fig. 11.3 The Ca^{2+} signal modulation in the astrocytic processes. **a** Fluorescence images of a fluo-3 filled astrocyte show the relative Ca^{2+} elevations before (Pre) and 10 s (Post) after independent or simultaneous stimulation of the SC and the alveus (30 Hz, 5 s). Scale bar, 5 μm. **b** Fluorescence intensity changes of restricted region of the astrocytic process and astrocytic soma marked with rectangles 1 and 2, respectively. The simultaneous stimulation controls the propagation of the intracellular Ca^{2+} signal, reducing the Ca^{2+} elevation in region 1 that failed to spread to the soma. **c** Schematic drawing representing a hypothetical consequence of the Ca^{2+} signal modulation. Under independent high-frequency synaptic activity of either pathway (*left*), astrocyte Ca^{2+} elevations are initiated in a specific process and then propagate to the soma and other processes, eventually leading to long-distance neuromodulation by Ca^{2+}-dependent release of glutamate (*arrows*). However, simultaneous high-frequency synaptic activity prevents the intracellular propagation of the astrocyte Ca^{2+} signal, and its long-distance neuromodulatory effects.

and dendrites correspond to the input neuronal region where the synaptic information is received and processed, while the output region resides in the axon that conveys the information to the presynaptic terminals. Each of these neuronal compartments has specific intrinsic membrane properties to receive, integrate, and transfer information. By contrast, astrocytes have a cell body and numerous thin processes – some of them intimately associated with synapses – but do not show an evident polarity responsible for a determined direction of the information flow. Whether the information flows into astrocytes according to an unknown dynamic polarization law, as Cajal proposed for neurons (Ramón y Cajal, 1899), is unclear and requires further investigations. Moreover, whether the reciprocal information transfer between neurons and astrocytes, i.e., the information input, processing and output, occur in the same or different astrocytic cellular regions also remains unknown. Likewise, whether one astrocyte or a single astrocytic domain can release different gliotransmitters or whether distinct populations of astrocytes of different domains are competent to release different gliotransmitters are exciting topics still to be addressed in the near future.

The participation of other neurotransmitter systems in the control and modulation of the Ca^{2+} signal as well as the existence of astrocyte Ca^{2+} modulation in other brain regions with fundamental physiological and pathological roles are stimulating issues that must be investigated to understand the actual roles of astrocytes in the physiology of the nervous system.

While it has been demonstrated that astrocytic microdomains respond with Ca^{2+} elevations to low frequencies of synaptic activity, where several axons are stimulated, whether single action potentials at single synapses would be able to evoke the selective activation of astrocytic microdomains remains undetermined. In other words, whether synaptically-mediated astrocyte signaling results in the uncontrolled broad spillover of the transmitter into the extracellular space or, whether by contrast, the astrocyte-neuron communication is a refined dialogue established between the unitary elements that compose the tripartite synapse (Araque et al., 1999) is a key issue that must be elucidated. Future studies are required to investigate whether, like neuronal synaptic transmission, neuron-to-astrocyte transmission is a point-to-point form of communication.

In conclusion, recent evidence indicates that astrocytes display some key properties that were previously thought to be exclusive to neurons. Astrocytes can discriminate between the activity of different synapses, respond selectively to different axon pathways and modulate their Ca^{2+} signal in response to simultaneous activity of different synaptic inputs. Furthermore, this Ca^{2+} signal modulation depends on the intrinsic cellular properties of astrocytes, is bidirectionally regulated by the level of synaptic activity, and controls the spatial extension of the intracellular Ca^{2+} signal. These facts reveal that astrocytes are endowed with cellular intrinsic properties that grant the integration and processing of synaptic information. Therefore, in addition to neurons, astrocytes could be considered as cellular elements involved in the information processing by the nervous system.

Acknowledgments. This chapter was written with grant support from the Ministerio de Educación y Ciencia, Spain (BFU2007-64764) to A.A.

References

Agmon-Snir H, Carr CE, Rinzel J (1998) The role of dendrites in auditory coincidence detection. Nature 393: 268-272.

Allen NJ, Barres BA (2005) Signaling between glia and neurons: focus on synaptic plasticity. Curr Opin Neurobiol 15: 542-548.

Araque A, Carmignoto G, Haydon PG (2001) Dynamic signaling between neurons and glia. Annu Rev Physiol 63: 795-813.

Araque A, Martin ED, Perea G, Arellano JI, Buño W (2002) Synaptically-released acetylcholine evokes Ca2+ elevations in astrocytes in hippocampal slices. J Neurosci 22: 2443-2450.

Araque A, Parpura V, Sanzgiri RP, Haydon PG (1998a) Glutamate-dependent astrocyte modulation of synaptic transmission between cultured hippocampal neurons. Eur J Neurosci 10: 2129-2142.

Araque A, Parpura V, Sanzgiri RP, Haydon PG (1999) Tripartite synapses: glia, the unacknowledged partner. Trends Neurosci 22: 208-215.

Araque A, Perea G (2004) Glial Modulation of Synaptic Transmission in Culture. Glia 47: 241-248.

Araque A, Sanzgiri RP, Parpura V, Haydon PG (1998b) Calcium elevation in astrocytes causes an NMDA receptor-dependent increase in the frequency of miniature synaptic currents in cultured hippocampal neurons. J Neurosci 18: 6822-6829.

Bellamy TC, Ogden D (2005) Short-term plasticity of Bergmann glial cell extrasynaptic currents during parallel fiber stimulation in rat cerebellum. Glia 52: 325-335.

Bezzi P, Carmignoto G, Pasti L, Vesce S, Rossi D, Rizzini B, Pozzan T, Volterra A (1998) Prostaglandins stimulate Ca2+-dependent glutamate release in astrocytes. Nature 391: 281-285.

Bezzi P, Gundersen V, Galbete JL, Seifert G, Steinhauser C, Pilati E, Volterra A (2004) Astrocytes contain a vesicular compartment that is competent for regulated exocytosis of glutamate. Nat Neurosci 7: 613-620.

Buzsaki G (2002) Theta oscillations in the hippocampus. Neuron 33: 325-340.

Charles AC, Merrill JE, Dirksen ER, Sanderson MJ (1991) Intercellular signaling in glial ells: calcium waves and oscillations in response to mechanical stimulation and glutamate. Neuron 6: 983-992.

Cormier RJ, Mennerick S, Melbostad H, Zorumski CF (2001) Basal levels of adenosine modulate mGluR5 on rat hippocampal astrocytes. Glia 33: 24-35.

Cornell-Bell AH, Finkbeiner SM, Cooper MS, Smith SJ (1990) Glutamate induces calcium waves in cultured astrocytes: long-range glial signaling. Science 247: 470-473.

Dani JW, Chernjavsky A, Smith SJ (1992) Neuronal activity triggers calcium waves in hippocampal astrocyte networks. Neuron 8:429-440.

Fatatis A, Holtzclaw LA, Avidor R, Brenneman DE, Russell JT (1994) Vasoactive intestinal peptide increases intracellular calcium in astroglia: synergism with alpha-adrenergic receptors. Proc Natl Acad Sci USA 91: 2036-2040.

Fiacco TA, McCarthy KD (2004) Intracellular astrocyte calcium waves in situ increase the frequency of spontaneous AMPA receptor currents in CA1 pyramidal neurons. J Neurosci 24: 722-732.

Ge WP, Yang XJ, Zhang Z, Wang HK, Shen W, Deng QD, Duan S (2006) Long-term potentiation of neuron-glia synapses mediated by Ca2+-permeable AMPA receptors. Science 312: 1533-1537.

Grosche J, Matyash V, Moller T, Verkhratsky A, Reichenbach A, Kettenmann H (1999) Microdomains for neuron-glia interaction: parallel fiber signaling to Bergmann glial cells. Nat Neurosci 2: 139-143.

Haydon PG, Araque A (2002) Astrocytes as modulators of synaptic transmission. In: TheTripartite Synapse: Glia In Synaptic Transmission (Volterra A, Magistretti PJ, Haydon PG, eds), pp185-198. New York: Oxford UP.

Kang J, Jiang L, Goldman SA, Nedergaard M (1998) Astrocyte-mediated potentiation of inhibitory synaptic transmission. Nat Neurosci 1: 683-692.

Kulik A, Haentzsch A, Luckermann M, Reichelt W, Ballanyi K (1999) Neuron-glia signaling via a1 adrenoceptor-mediated Ca2+ release in Bergmann glial cells in situ. J Neurosci 19: 8401-8408.

Liu QS, Xu Q, Arcuino G, Kang J, Nedergaard M (2004) Astrocyte-mediated activation of neuronal kainate receptors. Proc Natl Acad Sci USA 101: 3172-3177.

Llinas R, Sugimori M (1980) Electrophysiological properties of in vitro Purkinje cell dendrites in mammalian cerebellar slices. J Physiol 305: 197-213.

Matyash V, Filippov V, Mohrhagen V, Kettenmann H (2001) Nitric oxide signals parallel fiber activity to Bergmann glial cells in the mouse cerebellar slice. Mol Cell Neurosci 18: 664-670.

Metea MR, Newman EA (2006) Glial cells dilate and constrict blood vessels: a mechanism of neurovascular coupling. J Neurosci 26:2862-2870.

Mothet JP, Pollegioni L, Ouanounou G, Martineau M, Fossier P, Baux G (2005) Glutamate receptor activation triggers a calcium-dependent and SNARE protein-dependent release of the gliotransmitter D-serine. Proc Natl Acad Sci USA 102: 5606-5611.

Mulligan SJ, MacVicar BA (2004) Calcium transients in astrocyte endfeet cause cerebrovascular constrictions. Nature 431:195-199.

Nedergaard M, Ransom B, Goldman SA (2003) New roles for astrocytes: Redefining the functional architecture of the brain. Trends Neurosci 26: 523-530.

Nett WJ, Oloff SH, McCarthy KD (2002) Hippocampal astrocytes in situ exhibit calcium oscillations that occur independent of neuronal activity. J Neurophysiol 87: 528-537.

Newman EA (2005) Glia and Synaptic Transmission. In: Neuroglia (H. Kettenman, B. Ransom, eds.), pp. 355-366. New York: Oxford UP.

Newman EA, Zahs KR (1998) Modulation of neuronal activity by glial cells in the retina. J Neurosci 18: 4022-4028.

Orkand RK, Nicholls JG, Kuffler SW (1966) Effect of nerve impulses on the membrane potential of glial cells in the central nervous system of amphibian. J Neurophysiol 29: 788-806.

Panatier A, Theodosis DT, Mothet JP, Touquet B, Pollegioni L, Poulain DA, Oliet SH (2006) Glia-derived D-serine controls NMDA receptor activity and synaptic memory. Cell 125: 775-784.

Parpura V, Haydon PG (2000) Physiological astrocytic calcium levels stimulate glutamate release to modulate adjacent neurons. Proc Natl Acad Sci USA 97: 8629-8634.

Parri HR, Gould TM, Crunelli V (2001) Spontaneous astrocytic Ca2+ oscillations in situ drive NMDAR-mediated neuronal excitation. Nat Neurosci 4: 803-812.

Pascual O, Casper KB, Kubera C, Zhang J, Revilla-Sanchez R, Sul JY, Takano H, Moss SJ, McCarthy K, Haydon PG (2005) Astrocytic purinergic signaling coordinates synaptic networks. Science 310: 113-116.

Pasti L, Volterra A, Pozzan T, Carmignoto G (1997) Intracellular calcium oscillations in astrocytes: a highly plastic, bidirectional form of communication between neurons and astrocytes in situ. J Neurosci 17: 7817-7830.

Perea G, Araque A (2005a) Properties of synaptically evoked astrocyte calcium signal reveal synaptic information processing by astrocytes. J Neurosci 25: 2192-2203.

Perea G, Araque A (2005b) Glial calcium signalling and neuron-glia comunication. Cell Calcium 38: 375-382.

Perea G, Araque A (2006) Synaptic information processing by astrocytes. J Physiol (Paris) 99: 92–97.

Peters O, Schipke, CG, Hashimoto Y, Kettenmann H (2003) Different mechanisms promote astrocyte Ca2+ waves and spreading depression in the mouse neocortex. J Neurosci 23: 9888-9896.

Porter JT, McCarthy KD (1996) Hippocampal astrocytes in situ respond to glutamate released from synaptic terminals. J Neurosci 16: 5073-5081.

Porter JT, McCarthy KD (1997) Astrocytic neurotransmitter receptors in situ and in vivo. Prog Neurobiol 51: 439-455.

Ramón y Cajal S (1899) Textura del sistema nervioso del hombre y de los vertrebrados. Tomo I, N. Moya, Madrid.

Serrano A, Haddjeri N, Lacaille JC, Robitaille R (2006) GABAergic network activation of glial cells underlies hippocampal heterosynaptic depression. J Neurosci 26: 5370-5382.

Sontheimer H (1994) Voltage-dependent ion channels in glial cells. Glia 11: 156-172.

Sul JY, Orosz G, Givens RS, Haydon PG (2004) Astrocytic connectivity in the hippocampus. Neuron Glia Biology 1: 3-11.

Vertes PR (2005) Hippocampal Theta Rhythm: A Tag for Short-Term Memory. Hippocampus 15: 923–935.

Volterra A, Meldolesi J (2005) Quantal release of transmitter: not only for neurons but from astrocytes as well? In: Neuroglia (H. Kettenman, B. Ransom, eds.), pp. 190-201. New York: Oxford UP.

Volterra A, Steinhauser C (2004) Glial modulation of synaptic transmission in the hippocampus. Glia 47: 249-257.

Zhang J, Wang H, Ye C, Ge W, Chen Y, Jiang Z, Wu C, Poo M, Duan S (2003) ATP released by astrocytes mediates glutamatergic activity-dependent heterosynaptic suppression. Neuron 40: 971-982.

Zonta M, Angulo MC, Gobbo S, Rosengarten B, Hossmann KA, Pozzan T, Carmignoto G (2003) Neuron-to-astrocyte signaling is central to the dynamic control of brain microcirculation. Nat Neurosci 6: 43-50.

Abbreviations

Ach	Acetylcholine
AMPA	(*RS*)- α-Amino-3-hydroxy-5-methyl-4-isoxazolepropionic acid
ATP	Adenosine 5′-triphosphate
CNS	Central nervous system
GABA	γ-Aminobutyric acid
OPCs	Oligodendrocyte precursors cells
SC	Schaffer collaterals

Chapter 12
Mechanisms of Transmitter Release from Astrocytes

Erik B. Malarkey and Vladimir Parpura

Contents

Astrocytes and other glial cells can release a variety of neuroligands into the extracellular space using many different mechanisms. In this chapter we chiefly discuss the different chemical transmitters that astrocytes have been shown to release and examine the mechanisms by which these cells release the transmitters. In limited cases we expand our discussion on this subject to other astrocyte-related cells, such as Müller cells, pituicytes, as well as cell lines, including tumorigenic astrocytomas and gliomas. We focus on transmitters released from astrocytes, apart from eicosatetraenoids (briefly discussed in Chap. 18), growth factors and hormones (see Chap. 13). Transmitters can be released by astrocytes through several different mechanisms: (1) through channels like anion channel opening induced by cell swelling (Pasantes Morales and Schousboe, 1988), release through functional unpaired

V. Parpura
Department of Neurobiology, Center for Glial Biology in Medicine, Civitan International Research Center, Atomic Force Microscopy & Nanotechnology Laboratories, Evelyn F. McKnight Brain Institute, University of Alabama, Birmingham, AL, USA
vlad@uab.edu

V. Parpura and P.G. Haydon (eds.), *Astrocytes in (Patho)Physiology of the Nervous System*,
DOI: 10.1007/978-0-387-79492-1_12,

connexons, hemichannels, on the cell surface (Cotrina et al., 1998) and ionotropic purinergic receptors (Duan et al., 2003); (2) through transporters such as reversal of uptake by plasma membrane excitatory amino acid transporters (Szatkowski et al., 1990), exchange via the cystine-glutamate antiporter (Warr et al., 1999) or organic anion transporters (Rosenberg et al., 1994); and (3) through Ca^{2+}-dependent exocytosis (Parpura et al., 1994).

For some time now (Axelrod, 1974) different criteria have been proposed for identifying chemicals as neurotransmitters, but these definitions have undergone frequent modification as new compounds affecting neurotransmission have been discovered (Boehning and Snyder, 2003). However, as transmitter release from the glia was often overlooked, only recently has a similar set of criteria been put forth (Do et al., 1997) and subsequently modified (Volterra and Meldolesi, 2005; Martin et al., 2007) to establish which compounds qualify as "gliotransmitters": (1) synthesis by and/or storage in glia; (2) regulated release triggered by physiological and/or pathological stimuli; (3) activation of rapid (milliseconds to seconds) responses in neighboring cells; and (4) a role in the physiological and/or pathological processes.

Rather than delving into discussions of the consequences of transmitter release from astrocytes, we shall only disclose them when necessary, e.g., when the effect of transmitter release from astrocytes is used as an assay for release. The effects of astrocytic transmitters on other neural cells, mainly neurons, have been addressed in other chapters in this book (Chaps. 15–17). We have divided the transmitters that astrocytes release into two general groups: (1) amino acids and their derivatives, such as glutamate, aspartate, homocysteic acid (HCA), D-serine, γ-amino butyric acid (GABA), and taurine; and (2) nucleotides and their derivatives, like adenosine 5′-triphosphate (ATP), uridine 5′-triphosphate (UTP), adenosine and uridine diphosphate-glucose (UDP-glucose). While Ca^{2+}-dependent vesicular release of glutamate and ATP from astrocytes can readily occur under physiological conditions, there is some question as to whether some mechanisms might operate solely during pathophysiological circumstances. Since the first demonstration of the release of GABA from glial cells in the superior cervical ganglia (Bowery et al., 1976) and taurine from primary cultured astrocytes (Shain and Martin, 1984), a quest for an understanding of the mechanisms and conditions that underlie transmitter release is underway, which will provide information on astrocytic functions in health and disease and introduce opportunities for medical intervention.

12.1 Amino Acids and Their Derivatives as Astrocytic Transmitters

12.1.1 Synthesis of Amino Acid-Based Transmitters

Generally, amino acid-based transmitters are synthesized within astrocytes as by-products of the tricarboxylic acid (TCA) cycle. Glutamate does not readily cross the blood–brain barrier and as neurons lack the enzyme, pyruvate carboxylase, they are incapable of de novo synthesis of glutamate from glucose. Therefore, the majority of glutamate in the brain is synthesized in astrocytes and then distributed to neurons in a well studied cycle (Hertz et al., 1999). Glutamate is converted from the TCA intermediate, α-ketoglutarate, usually via transamination of another amino acid such as aspartate (Westergaard et al., 1996). In a similar manner aspartate can be derived from the TCA cycle intermediate, oxaloacetate, by transamination of glutamate which is an important mechanism in the mitochondrial malate-aspartate shuttle (Lai et al., 1989). D-serine is converted from L-serine by the action of serine racemase, an enzyme found predominately in astrocytes (Wolosker et al., 1999) (also see Chap. 16). Homocysteic acid is believed to be derived from methionine (McBean, 2002) by a pathway that has not been well defined (Cuenod et al., 1993). Taurine is a 2-aminoethanesulfonic acid. Although this naturally occurring sulfonic acid is not strictly an amino acid, it is derived from the sulfhydryl amino acid cysteine. While enzymes involved in this derivatization pathway are known, their localization within neural cell subtypes is not clear in vivo; however this process may involve cooperation between astrocytes and neurons (Dominy et al., 2004). GABA is derived from glutamate by glutamic acid decarboxylase (GAD), an enzyme found in neurons but not in glia. Thus, astrocytes lack the ability to produce GABA, although they can take up GABA released into the extracellular space by neurons through GABA transporters (Bak et al., 2006).

It should be noted that HCA is an agonist for both *N*-methyl-D-aspartic acid (NMDA) receptors (Cuenod et al., 1986; Do et al., 1986) and metabotropic glutamate receptors 1, 2, 4–6, and 8 (Kingston et al., 1998; Shi et al., 2003). D-serine is a ligand to the glycine modulatory binding site of the NMDA receptor. Taurine is an agonist to glycine and GABA type A receptors, albeit with higher affinities to glycine receptors.

12.1.2 Amino/Sulfonic Acid Transmitter Release through Channels

12.1.2.1 Amino/Sulfonic Acid Release via Volume-Regulated Anion Channels

Under hypo-osmotic conditions, such as those occurring during ischemia, most cells experience swelling and can compensate for this volume increase by opening volume-regulated anion channels (VRAC) [reviewed in (Kimelberg et al., 2006)]. These channels are permeable to inorganic and small organic anions, including the amino acids aspartate and glutamate and the sulfonic acid taurine (Mongin and Orlov, 2001). Release of a transmitter in response to glial swelling was first shown by Pasantes-Morales and Schousboe (1988) where volume regulation in astrocytes was accompanied by the release of taurine (Fig. 12.1, Table 12.1). Further they found that this release could be caused by induction of swelling from applying solutions of elevated KCl concentration (Pasantes-Morales and Schousboe, 1989).

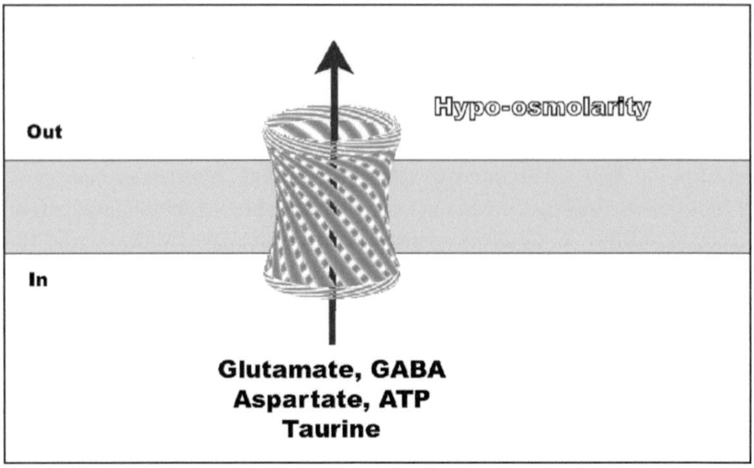

Fig. 12.1 Release of transmitters through anion channel opening induced by cell swelling. Under hypo-osmotic conditions cells can regulate their volume by releasing several compounds, including transmitters, through volume regulating anion channels.

Application of Cl^- channel inhibitors prevented taurine release, indicating that volume sensitive sulfonic acid release could occur through anion channels (Pasantes-Morales et al., 1990) which was also confirmed by using glioma cells (Jackson and Strange, 1993).

Release of several transmitters: glutamate, aspartate and taurine from cultured astrocytes during hypo-osmotically induced swelling was reported by Kimelberg et al. (1990) using radiolabeled transmitters. They found that this release occurred through an anion channel because it could be blocked by various anion channel inhibitors. Further studies revealed that elevated KCl could induce aspartate release from astrocytes via VRACs along with release through amino acid transporter reversal (Rutledge and Kimelberg, 1996; Rutledge et al., 1998). Liu et al. (2006) discerned swelling induced glutamate release from two different anion channels: the volume-sensitive outwardly rectifying (VSOR) channels and the maxi-anion channels. Release of aspartate through VSOR chloride channels was shown to depend on the presence of intracellular ATP (Rutledge et al., 1999). Swelling-induced release of aspartate, glutamate and taurine is also enhanced by activation of purinergic receptors on the cell surface (Mongin and Kimelberg, Mongin 2002; Takano et al., 2005). Modulation of aspartate release via VSOR by ATP was determined to operate in a Ca^{2+}-dependent manner since chelating intracellular Ca^{2+} with 1,2-bis(*o*-aminophenoxy)ethane-*N*,*N*,*N*′,*N*′-tetraacetic acid (BAPTA) eliminated this effect (Mongin and Kimelberg, 2005). Other factors were found to affect the release of aspartate or glutamate such as, peroxynitrate (Haskew et al., 2002), nitric oxide (Ellershaw et al., 2000) hydrogen peroxide (Haskew-Layton et al., 2005) and thrombin (Cheema et al., 2005; Ramos-Mandujano et al., 2007). Similarly, taurine release was shown to be modulated by the activity of tyrosine kinases (Mongin et al.,

Table 12.1 Astrocytic transmitters and their mechanisms of release

Amino acid transmitters and their derivatives

Glutamate	Exocytosis	Transporter reversal	Swelling/anion	Undefined
	(Parpura et al., 1994)	[M](Szatkowski et al., 1990)	(Kimelberg et al., 1990)	(Hassinger et al., 1995)
	(Parpura et al., 1995b)	(Volterra et al., 1996)	(Jeftinija et al., 1997)	(Pasti et al., 1997)
	(Jeftinija et al., 1996)	(Longuemare and Swanson, 1997)	(Basarsky et al., 1999)	(Cotrina et al., 1998)
	(Jeftinija et al., 1997)	(Zeevalk et al., 1998)	(Takano et al., 2005)	(Newman, 2001)
	(Araque et al., 1998a)	(Li et al., 1999)	(Kozlov et al., 2006)	(Parri et al., 2001)
	(Araque et al., 1998b)	(Longuemare et al., 1999)	(Liu et al., 2006)	(Angulo et al., 2004)
	(Bezzi et al., 1998)	(Seki et al., 1999)	(Fiacco et al., 2007)	(Lee et al., 2007b)
	(Sanzgiri et al., 1999)	(Rossi et al., 2000)	(Ramos-Mandujano et al., 2007)	(Syed et al., 2007)
	(Araque et al., 2000)	[G](Raiteri et al., 2007)		
	(Innocenti et al., 2000)			
	(Parpura and Haydon, 2000)	**Cystine-glutamate antiporter**	**P2X$_7$**	
	(Bezzi et al., 2001)	(Warr et al., 1999)	(Duan et al., 2003)	
	(Jeremic et al., 2001)	(Baker et al., 2002)	(Fellin et al., 2006)	
	(Pascual et al., 2001)	(Tang and Kalivas, 2003)		
	(Pasti et al., 2001)	(Moran et al., 2003)	**Hemichannels**	
	(Bal-Price et al., 2002)	(Cavelier and Attwell, 2005)	(Ye et al., 2003)	
	(Coco et al., 2003)	(Moran et al., 2005)	(Spray et al., 2006)	
	(Bezzi et al., 2004)	(Chung et al., 2005)		
	(Fellin et al., 2004)	(Re et al., 2006)		
	(Fiacco and McCarthy, 2004)			
	(Hua et al., 2004)			
	(Kreft et al., 2004)			
	(Liu et al., 2004a)			
	(Liu et al., 2004b)			
	(Montana et al., 2004)			
	(Zhang et al., 2004a)			
	(Zhang et al., 2004b)			
	(Anlauf and Derouiche, 2005)			
	[D](Chen et al., 2005)			
	[V](Kang et al., 2005)			

(continued)

Table 12.1 (continued)

	(Rossi et al., 2005)		
	(Crippa et al., 2006)		
	(Domercq et al., 2006)		
	(Fellin et al., 2006)		
	(Gonzalez et al., 2006)		
	(Shiga et al., 2006)		
	[G](Stigliani et al., 2006)		
	(D'Ascenzo et al., 2007)		
	(Jourdain et al., 2007)		
	(Lee et al., 2007a)		
	(Nestor et al., 2007)		
	[G](Patti et al., 2007)		
	(Martin et al., 2007)		
	(Stenovec et al., 2007)		
	[V](Xu et al., 2007)		
Aspartate	**Exocytosis**	**Transporter reversal**	**Swelling/anion**
	(Jeftinija et al., 1996)	(Longuemare and Swanson, 1995)	(Kimelberg et al., 1990)
	(Jeremic et al., 2001)	(Rutledge and Kimelberg, 1996)	(Rutledge and Kimelberg, 1996)
	[G](Stigliani et al., 2006)	(Longuemare and Swanson, 1997)	(Rutledge et al., 1998)
	[G](Patti et al., 2007)	(Longuemare et al., 1999)	(Rutledge et al., 1999)
		(Seki et al., 1999)	(Haskew et al., 2002)
		(Anderson et al., 2001)	(Mongin and Kimelberg, 2002)
		[G](Raiteri et al., 2007)	(Kimelberg, 2004)
			(Haskew-Layton et al., 2005)
			(Mongin and Kimelberg, 2005)
			(Takano et al., 2005)
			$P2X_7$
			(Duan et al., 2003)

			Hemichannels	
			(Ye et al., 2003)	
Homocysteate				**Undefined**
				(Do et al., 1997) (Benz et al., 2004)
D-Serine	**Exocytosis**	**Transporter reversal**		**Undefined**
	(Mothet et al., 2005)	(Ribeiro et al., 2002)		(Schell et al., 1995) (Wolosker et al., 1999) (Yang et al., 2003) (Kanematsu et al., 2006)
GABA		**Transporter reversal**	**Swelling/anion**	**Undefined**
		(Gallo et al., 1991) [G](Raiteri et al., 2007)	(Kozlov et al., 2006)	[S](Bowery et al., 1976) (Neal and Bowery, 1979) (Wu et al., 1979)
			$P2X_7$	
			[R](Wang et al., 2002)	(Gallo et al., 1986) (Gallo et al., 1989) (Liu et al., 2000) (Verderio et al., 2001) (Jow et al., 2004)
Taurine			**Swelling/anion**	**Undefined**
			(Pasantes Morales and Schousboe, 1988) (Pasantes-Morales and Schousboe, 1989) (Kimelberg et al., 1990) (Pasantes-Morales et al., 1990) [C](Jackson and Strange, 1993)	(Shain and Martin, 1984)

(continued)

Table 12.1 (continued)

		(Hussy et al., 1997) [P](Miyata et al., 1997) (Deleuze et al., 1998) (Mongin et al., 1999a) (Bres et al., 2000) (Moran et al., 2001) (Cardin et al., 2003) [P](Rosso et al., 2004) (Takano et al., 2005) [P](Pierson et al., 2007) **Hemichannels** (Ye et al., 2003)	

Nucleotide transmitters and their derivatives

ATP	Exocytosis	Transporters/anion channels	Undefined
	(Maienschein et al., 1999) (Bal-Price et al., 2002) (Abdipranoto et al., 2003) (Coco et al., 2003) (Pascual et al., 2005) (Bowser and Khakh, 2007) (Pangrsic et al., 2007) [A](Striedinger et al., 2007) [L](Zhang et al., 2007) (Pryazhnikov and Khiroug, 2008)	(Queiroz et al., 1999) (Anderson et al., 2004) (Darby et al., 2003) (Abdipranoto et al., 2003) **$P2X_7$** *(Ballerini et al., 1996) (Suadicani et al., 2006) **Hemichannels** [C](Cotrina et al., 1998) [C](Cotrina et al., 2000)	(Caciagli et al., 1988) [N](Lazarowski et al., 1997) (Queiroz et al., 1997) (Ciccarelli et al., 1999) (Guthrie et al., 1999) [C, N](Lazarowski and Harden, 1999) [C, N](Lazarowski et al., 2000) (Wang et al., 2000) [M](Newman, 2001) (Verderio and Matteoli, 2001) (Parkinson et al., 2002) (Joseph et al., 2003)

		(Arcuino et al., 2002) (Stout et al., 2002) (Stout and Charles, 2003) (Suadicani et al., 2007) (Lin et al., 2008)	[C, N](Lazarowski et al., 2003a) (Neary et al., 2003) [M](Newman, 2003) (Zhang et al., 2003) (Parkinson and Xiong, 2004) (Bianco et al., 2005) [P](Gordon et al., 2005) (Parkinson et al., 2005) (Werry et al., 2006) [K](Yoshida et al., 2006)
Adenosine	**Transporter reversal**		**Undefined**
	[E](Meghji et al., 1989)		(Caciagli et al., 1988) (Ciccarelli et al., 1999) (Martin et al., 2007)
cAMP		**Transporters/anion channels**	**Undefined**
		[C](Penit et al., 1974) [C](Doore et al., 1975) [C](Rindler et al., 1978) [C](Henderson and Strauss, 1991) (Rosenberg et al., 1994)	(Rosenberg et al., 1994) (Winder et al., 1996)
GTP/GMP/guanosine			**Undefined**
			(Ciccarelli et al., 1999)

(continued)

Table 12.1 (continued)

cGMP	**Transporters/anion channels**	
	(Touyz et al., 1997) (Pedraza et al., 2001)	
UTP		**Undefined**
		[N](Lazarowski et al., 1997) [C, N](Lazarowski and Harden, 1999)
UDP-Glucose		**Undefined**
		[C, N](Lazarowski et al., 2003a)
NAD⁺	**Hemichannels**	
	(Verderio et al., 2001)	

Abbreviations: *A,* Astrocyte progenitor cells; *C,* C6 rat glioma cells; *D,* Dopamine as a "surrogate" transmitter for glutamate; *E,* Embryonic glia from peripheral and central nervous system of chick; *G,* gliosomes; *K,* Kings-1 astrocyte cell line; *L,* exocytosis of lysosomes; *M,* Müller cells; *N,* human 1321N1 astrocytoma cells; *P,* pituicytes; *R,* RBA-2 astrocyte cells; *S,* desheathed superior cervical ganglia; *V,* vesicles in this case are larger than generally accepted for regulated exocytosis;*, tritiated purines measured, not specifically ATP

1999b; Deleuze et al., 2000) and the presence of the kinase substrate phospholemman (Moran et al., 2001). Several studies have also indicated that the levels of cytosolic Ca^{2+} can affect swelling induced taurine efflux (Mongin et al., 1999a; Cardin et al., 2003). These various pathways for modulation could act on VRACs to amplify amino acid release under conditions of only moderate swelling or hypo-osmolarity which could possibly occur in vivo.

There is evidence that receptor mediated intracellular Ca^{2+} increases, induced by ATP in astrocytes, can result in transient cell swelling leading to glutamate, aspartate, and taurine release through VRACs (Takano et al., 2005). Application of ATP caused the opening of channels which, along with glutamate release, could be impaired by BAPTA and anion channel blockers, but neither by a glutamate transporter inhibitor nor by two compounds known to affect vesicular release, tetanus toxin and bafilomycin A_1 (see Sect. 12.1.4). In this study, glutamate release also did not appear to be through $P2X_7$ channels or connexin hemichannels, based on pharmacology and the use of connexin-43 (Cx43) knockout mice, respectively.

In the olfactory bulb, GABA release from astrocytes was shown to induce a slow outward current (SOC) in neighboring neurons (Kozlov et al., 2006). SOC was blocked by picrotoxin and the $GABA_A$ antagonist, gabazine. Mechanical stimulation of astrocytes in the presence of tetrodotoxin or blockers of V-ATPase (see below in Sect. 12.1.4) still induced SOC in neurons, pointing to a non-vesicular release of GABA from astrocytes. The authors pharmacologically ruled out the involvement of purinergic ion channels and hemichannels. Using nipecotic acid, a non-selective blocker of plasma membrane GABA transporters (GATs) that activates heteroexchange, they could detect a tonic GABAergic current, but SOC persisted, although the kinetics were slower as indicated by a doubled rise time. Therefore, while GAT reversal does not cause SOC, it may play a role in modifying SOC kinetics. Application of a hypotonic solution, however, increased the frequency of SOCs, indicating that GABA might be released via volume-regulated anion channels. Unfortunately, the blockers of anion channels caused direct effects on GABA receptors, so that the obvious suspects mediating swelling-induced release of amino acids, VRACs, could not be pharmacologically verified in the generation of SOC. Such a demonstration awaits development of more selective pharmacological agents and/or unveiling of the molecular identity of VRACs combined with the consequent use of knock-out animals.

Determining conclusively that transmitters are released through VRACs has proven complicated due to a lack of specific inhibitors of suspected pathways [discussed in (Malarkey and Parpura, 2008)]. Essentially, nearly all anion channel inhibitors have the potential to block connexin hemichannels and purinergic ion channnels. Additionally, cell permeant anion channel inhibitors have the potential to inhibit vesicular chloride channels thus tampering with exocytosis [discussed in (Evanko et al., 2004a; Evanko et al., 2004b)]. Although buffering cytoplasmic Ca^{2+} can interfere with osmotically induced amino acid release from astrocytes (Mongin et al., 1999a), tetanus toxin had no effect on swelling-induced release, indicating that this release does not appear to be through a vesicle mediated pathway (Mongin and Kimelberg, 2002).

12.1.2.2 Amino/Sulfonic Acid Release via Hemichannels

Gap-junction channels form a pore between two adjacent cells, connecting their cytoplasm, and allowing molecules as large as about 1 kDa to diffuse between cells. These gap junctions are formed by the joining of two connexons ("hemichannels") each composed of a hexamer of the protein connexin. Although there are many different isoforms of connexin, Cx43 appears to be the most prevalent in astrocytes (Dermietzel et al., 2000). There is evidence that unpaired connexons may be able to act as functional hemichannels, capable of opening to the external space (Hofer and Dermietzel, 1998; Contreras et al., 2002; Stout et al., 2002; Ye et al., 2003), which may provide a mechanism whereby transmitters could diffuse out of astrocytes (Fig. 12.2, Table 12.1) (also see Chap. 5). There has been some evidence supporting this kind of release through hemichannels for glutamate, aspartate and taurine (Ye et al., 2003). Under conditions of low extracellular divalent cations, hippocampal astrocytes showed release of these transmitters. This release was reduced by the application of gap-junction blockers, but was neither mediated by purinergic receptors nor by VRACs as deduced from additional pharmacology. This seems consistent with release through hemichannels, as the events were also independent of intracellular Ca^{2+} elevations. The involvement of hemichannels in mediating glutamate release from astrocytes was further tested using Cx43 knock-out mice (Spray et al., 2006). Astrocytes cultured from Cx43 knock-out mice and exposed to low extracellular divalent cations show minimal glutamate release, when compared to astrocytes originating from control wild type animals. Such a finding supports the notion that glutamate could, indeed, be released via connexin hemichannels. As gap-junction channels display negative Ca^{2+} regulation and voltage sensitivity, opening only as membrane potentials become

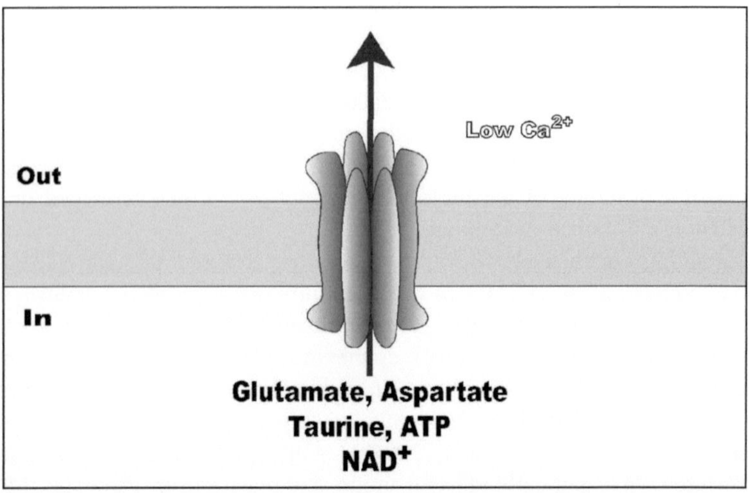

Fig. 12.2 Release of transmitters through functional unpaired connexin or pannexin "hemichannels." Unpaired connexons are able to act as functional hemichannels capable of releasing transmitters to the external space under conditions of low external Ca^{2+}, while pannexons are not sensitive to extracellular Ca^{2+} and can be opened by cytoplasmic Ca^{2+} elevations.

positive [(Trexler et al., 1996), but see (Contreras et al., 2003; Saez et al., 2005)], transmitter release via this mechanism would be possible in pathophysiological conditions, perhaps during injury or stroke.

Interestingly, many of the properties attributed to connexin hemichannels correlate well with those of "pannexons," non-junctional pannexin channels, which can also form conductive channels between the intra- and extra-cellular spaces. These hemichannels are not sensitive to extracellular Ca^{2+} (Bruzzone et al., 2005) and can be opened by cytoplasmic Ca^{2+} elevations (Locovei et al., 2007). RNA for pannexin 1 (Pelegrin and Surprenant, 2006; Lai et al., 2007), 2 and 3 (Lai et al., 2007) were detected in 1321N1 astrocytoma and C6 glioma cells, opening up the possibility that pannexons may have a role in mediating glutamate release from astrocytes similar to their control of ATP release (see Sect. 12.2.2.2). Indeed, pannexin 1 protein has been detected in cultured astrocytes (Locovei et al., 2007) and glial-like taste bud cells (Huang et al, 2007).

12.1.2.3 Amino/Sulfonic Acid Release via Purinergic Ion Channels

The pore forming purinergic P2X ion channel may provide another pathway for amino acid release from astrocytes. P2X receptors are ATP-gated non-selective cation channels that show amplified responses in low external divalent cation solution. There are seven known types of P2X receptor subunits that can assemble to form homomeric or heteromeric channels. The homomeric $P2X_7$ receptor recruits a pore that is able to allow molecules as large as 900 Da to permeate (North, 2002). The $P2X_7$ receptor has been detected in astrocytes in vitro by RT-PCR (Fumagalli et al., 2003), immunoblotting and immunolocalization (Duan et al., 2003). Although the presence of $P2X_7$ receptors has also been detected in astrocytes in vivo using hippocampal sections of juvenile rats (Kukley et al., 2001), they might not be functional. A recent study using patch clamp recordings from a subtype of hippocampal astrocytes in rat and human acute slices did not find current activation upon application of any P2X receptor agonists, raising the question of whether astrocytes possess functional P2X receptors in vivo (Jabs et al., 2007). Also, Sim et al. (2004) have not found evidence for $P2X_7$ receptor protein in the hippocampus.

Nonetheless, Duan et al. (2003) provided the first evidence that these channels could mediate the release of glutamate and aspartate from astrocytes (Fig. 12.3, Table 12.1). Application of ATP or 3′-*O*-(4-benzoyl)benzoyl ATP (BzATP) to cultured astrocytes expressing $P2X_7$ receptors caused transmitter release that was augmented by low divalent cation external solution; the release was inhibited by the P2 receptor antagonist, pyridoxal phosphate-6-azophenyl-2-4-disulfonic acid (PPADS) and the more specific $P2X_7$ antagonist, oxidized ATP (oATP). However, other pathways were also implicated because the chloride channel blocker (but see above Sect. 12.1.2.1) 4,4′-diisothiocyanato-stilbene-2,2′-disulfonate (DIDS) also inhibited the response. This glutamate release seemed to be independent of $[Ca^{2+}]_i$ increase since preincubating the cells with the membrane permeable Ca^{2+} chelator, acetoxymethyl ester of BAPTA, did not reduce the amount of released glutamate.

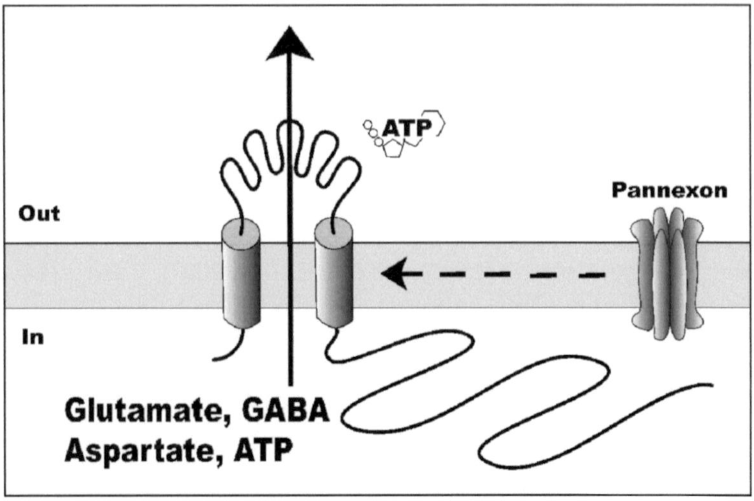

Fig. 12.3 Release of transmitters through ionotropic purinergic receptors (only one subunit shown, but the functional receptor is believed to be multimeric). P2X receptors are ATP-gated non-selective cation channels with a large pore capable of passing compounds up to 900 Da. Pore dilation may involve recruitment of pannexons.

In hippocampal slices, Fellin et al. (2006) found that perfusion of BzATP could induce tonic currents in pyramidal neurons that resulted from NMDA receptor activation. The currents were induced by glutamate release from astrocytes, since they occurred in the presence of tetrodotoxin. The tonic current appeared to be mediated by glutamate release from $P2X_7$-like receptors as it was blocked by the P2X antagonists, oATP and Brilliant Blue G (BBG), and was enhanced in low Ca^{2+} external solution. Moreover, glutamate release through transporter reversal or hemichannels was pharmacologically ruled out. This work indicates that ATP in situ can cause release of glutamate from astrocytes through P2X receptors that provide tonic stimulation of surrounding neurons.

Release of the inhibitory amino acid transmitter GABA has also been shown to occur from astrocytes by $P2X_7$ receptors (Wang et al., 2002). Application of ATP or BzATP resulted in the release of radiolabeled GABA from an RBA-2 astrocyte cell line, which could be blocked by PPADS and oATP, an indication of release through $P2X_7$. Release through transporter reversal was ruled out since GABA transporter inhibitors had no effect on ATP-induced GABA release.

12.1.3 Amino Acid Transmitter Release through Transporters

12.1.3.1 Release via Reverse Operation of Plasma Membrane Transporters

One important function of astrocytes is to remove excitatory transmitters from the extracellular space to aid in the termination of synaptic neurotransmission and prevent excitotoxicity (Rothstein et al., 1996; Bergles and Jahr, 1997).

Interestingly, this glial function was speculated upon by Lugaro over 100 years ago (Lugaro, 1907). For excitatory amino acids this function is accomplished through the use of plasma membrane Na^+-dependent amino acid transporters which use Na^+ and K^+ gradients to drive transmitters into the cell (Anderson and Swanson, 2000). Astrocytes predominantly express two transporters that are used in this process: the l-glutamate/l-aspartate transporter (GLAST-1) and the glial L-glutamate transporter (GLT-1) in rodents, also called excitatory amino acid transporters (EAAT1 and EAAT2, respectively) in humans (Gadea and Lopez-Colome, 2001b)(see also Chap. 4).

Normally, concentration gradients favor the transport of excitatory amino acids into astrocytes; this results in the transport of Na^+, H^+, and glutamate/aspartate into the cytoplasm, and K^+ into the extracellular space (Danbolt, 2001). However, during pathophysiological events, such as ischemia, perturbed ionic conditions (e.g., increased extracellular K^+ levels) may favor transporters operating in reverse (Fig. 12.4, Table 12.1). Transport reversal was first demonstrated in retinal Müller cells by measuring glutamate induced currents while raising extracellular K^+ levels (Szatkowski et al., 1990). Although it has been shown that under normal physiological conditions extracelluar K^+ levels could not be elevated enough to cause reverse transport of glutamate out of cultured astrocytes (Longuemare and Swanson, 1997), there have been numerous cases, using transporter inhibitors, showing that reverse transport of glutamate (Zeevalk et al., 1998; Li et al., 1999; Seki et al., 1999; Rossi et al., 2000) or aspartate (Longuemare and Swanson, 1995; Rutledge and Kimelberg, 1996; Seki et al., 1999; Anderson et al., 2001) can occur during periods of ischemia or metabolic blockade.

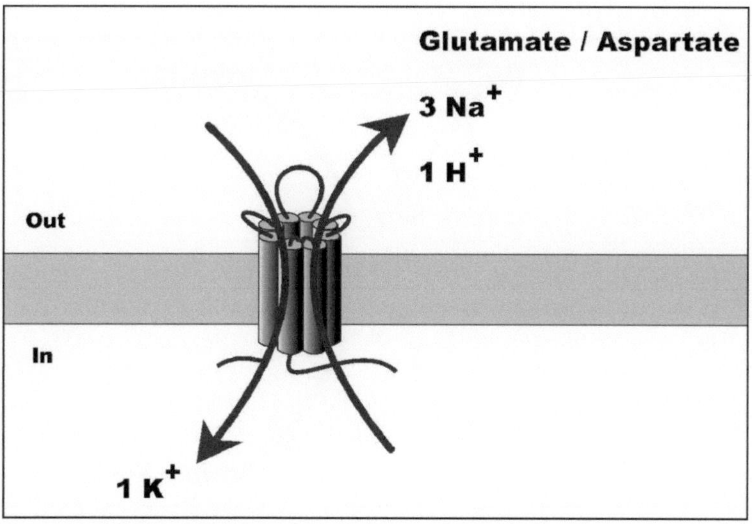

Fig. 12.4 Transmitter release by reversal of uptake of plasma membrane glutamate transporters. Transmitters can be released from astrocytes under conditions that favor transporters operating in reverse, where Na^+ and H^+ are cotransported with glutamate or aspartate to the extracellular space, while K^+ is transported into the cell.

The transmitter D-serine has been shown to be taken up into Müller cells by the Na^+-dependent neutral amino acid transporter, alanine-serine-cysteine transporter 2 (ASCT2) (Dun et al., 2007). Transport of D-serine was found to be coupled to counter-transport of other neutral amino acids and the efflux of D-serine could be induced from astrocytes when L-serine was present extracellularly at physiological levels, indicating transporter reversal as a possible mechanism of D-serine release from astrocytes (Ribeiro et al., 2002). The kinetics of release implicated the reverse operation of ASCT type transporters (Fig. 12.5).

In stark contrast to the clearance of the excitatory neurotransmitter glutamate, the uptake of the inhibitory neurotransmitter GABA is carried out mainly (80%) by neuronal GATs, while the remaining transmitter (20%) is taken up by astrocytes (Schousboe, 2003). Since GABAergic neurons express vesicular GABA transporters (VGATs), the uptake of GABA leads to filling of vesicles with this transmitter. Astrocytes lack not only VGATs, but also GADs, so that much of the GABA that is taken up is metabolized through GABA transaminase and the TCA cycle. GATs generally transport, from the extracellular space to the cytoplasm, 2 Na^+ and 1 Cl^- with each GABA molecule (Gadea and Lopez-Colome, 2001a). These transporters can reverse their operation and release GABA into the extracellular space [reviewed in (Richerson and Wu, 2003)].

Cultured cerebellar astrocytes preloaded with radiolabeled GABA showed release of this transmitter upon stimulation with kainate (Gallo et al., 1986; Gallo et al., 1989; Gallo et al., 1991), quisqualate (Gallo et al., 1989; Gallo et al., 1991) and α-amino-3-hydroxy-5-methyl-isoxazole propionate (AMPA) (Gallo et al., 1991). The GAT inhibitor nipecotic acid and the replacement of extracellular Na^+ reduced GABA uptake during the preloading procedure (Gallo et al., 1991).

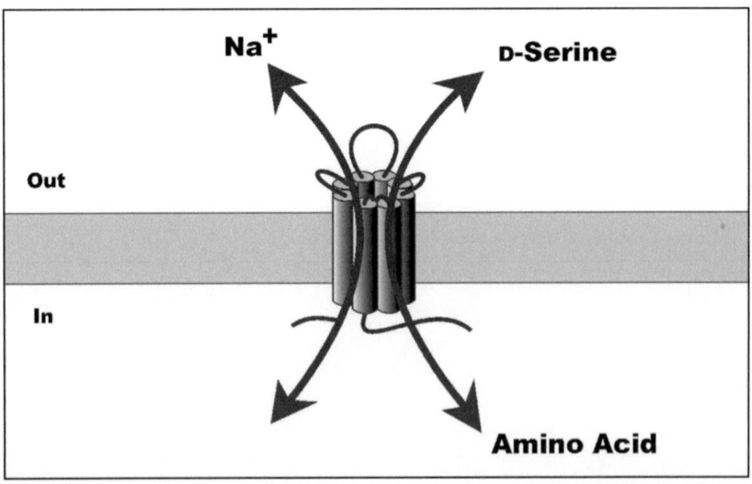

Fig. 12.5 Release of D-serine by the Na^+-dependent neutral amino acid transporter, alanine-serine-cysteine transporter 2 (ASCT2). Transport of D-serine is coupled to counter-transport of other neutral amino acids and Na^+.

Similarly, these treatments blocked kainate- and quisqualate-induced GABA release from astrocytes preloaded with this transmitter. Thus, GABA can be released from astrocytes in culture by the reverse operation of its plasma membrane transporters (Fig. 12.6)

Rat astrocytes (see Chap. 4 for nomenclature) express functional GAT-1, GAT-2, and GAT-3 (Ribak et al., 1996; Kinney and Spain, 2002); GAT-1 is present mainly in GABAergic neurons, and GAT-3 in astrocytes (Ribak et al., 1996). The experimental evidence in the rat hippocampal slice supports the role of GAT-1, but not GAT-3, in GABA release by reverse operation of this transporter during cerebral energy deprivation caused by anoxia or ischemia (Allen et al., 2004). Although GAT-1 is mainly present in interneurons in the hippocampus, astrocytic expression of GAT-1 also takes place in this brain region (Ribak et al., 1996; Yan et al., 1997), thus opening up the possibility of the reversal of GAT-1 from astrocytes, leading to the release of GABA. This has been demonstrated in a recent study (Raiteri et al., 2007) using gliosomes, a purified preparation of re-sealed fragments of mouse astrocytes (Stigliani et al., 2006). Gliosomes accumulated tritiated GABA through GAT1 (mouse homologue of rat GAT-1) (Raiteri et al., 2007). When gliosomes were superfused with glycine, they released GABA. This effect was mediated by the entry of glycine via plasma membrane glycine transporters (GlyT); the release of GABA could be reduced mainly by a GlyT2 blocker and to a lesser extent by a GlyT1 blocker. Glycine-induced GABA release from gliosomes was independent of both external and internal Ca^{2+} ions, but was sensitive, in a concentration-dependent manner, to a GAT1 transporter blocker. Thus, glycine-induced GABA release from gliosomes occurs by reversal of GAT1.

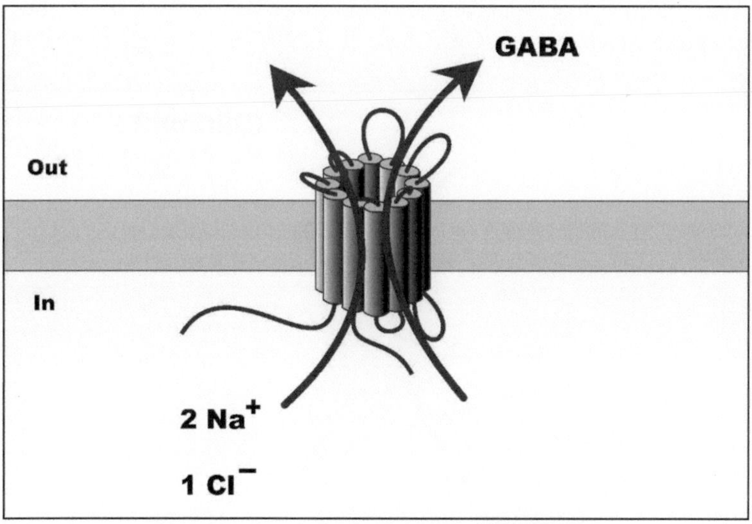

Fig. 12.6 Release of GABA by reversal of plasma membrane GABA transporters (GAT). GATs generally transport, from the extracellular space to the cytoplasm, 2 Na^+ and 1 Cl^- with each GABA molecule, but may operate in reverse to release GABA from astrocytes under certain conditions.

12.1.3.2 Release via Cystine-Glutamate Antiporter

Cystine uptake in cells is important for the production of the antioxidant, glutathione. Uptake can occur through either the plasma membrane Na^+-independent cystine-glutamate exchanger (system x_c-) or the Na^+-dependent glutamate transporters (system X_{AG}-) [reviewed in (McBean, 2002)] and astrocytes utilize both of these for cystine uptake (Bender et al., 2000; Allen et al., 2001; Shanker et al., 2001). It should be noted that system x_c- does not transport aspartate (Patel et al., 2004), unlike the transportation route via sytem X_{AG}-, which is shared by glutamate and aspartate (Dall'Asta et al., 1983). Since system x_c- functions by importing cystine in exchange for glutamate, this may provide a pathway for glutamate release from astrocytes (Fig. 12.7) This had been initially demonstrated in cerebellar slices (Warr et al., 1999). Although Cavelier and Atwell (2005) have raised questions of whether release of glutamate through the x_c- system occurs under normal physiological conditions, the role of glutamate release by the cystine-glutamate exchanger in vivo has been demonstrated by Moran et al. (2005). Interestingly, inhibiting the function of the cystine-glutamate exchanger with (*S*)-4-carboxyphenylglycine or by removing extracellular cystine, has been shown to cause cell death in astrocytes, presumably by oxidative death due to lack of cystine to convert to glutathione (Re et al., 2006). This effect was employed for possible clinical benefit by reducing the growth of gliomas (Chung et al., 2005) (also see Chap. 21).

Uptake of glutamate via system x_c- is inhibited by HCA which acts as a substrate for this system (Patel et al., 2004). Hence, this sulfur-containing amino acid can be transported into cells via system x_c- and can then be released into the extracellular space by heteroexchange in the presence of another substrate of system x_c-, as indicated by its ability to sensitize hippocampal neurons to L-AP6 and L-cystine after slices had been pre-incubated with HCA (Chase et al., 2007). Although, release of HCA

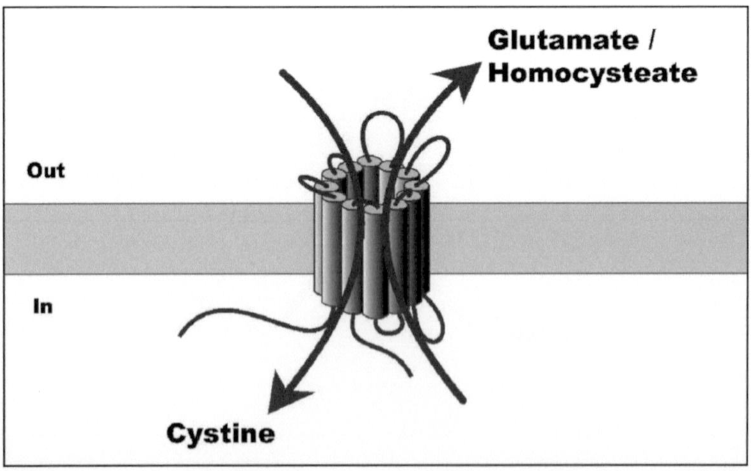

Fig. 12.7 Glutamate and homocysteate exchange via the Na^+-independent cystine-glutamate exchanger (system x_c-). System x_c- functions by importing cystine in exchange for releasing glutamate; also homocysteic acid may act as a substrate for this transporter.

from astrocytes has been demonstrated (Do et al., 1997; Benz et al., 2004), the Ca^{2+} dependency of this release indicates exocytosis as the likely mechanism (Benz et al., 2004) (see below Sect. 12.1.4). Whether HCA can also be released from astrocytes via system x_c- awaits experimental evaluation.

12.1.4 Amino Acid Transmitter Release by Ca^{2+}-Dependent Exocytosis

Evidence for Ca^{2+}-dependent release of amino acids from astrocytes was first shown in experiments using high performance liquid chromatography to monitor glutamate release from cultured astrocytes (Parpura et al., 1994). An increase in intracellular Ca^{2+} concentration ($[Ca^{2+}]_i$) is sufficient and necessary to cause glutamate release from astrocytes. When ionomycin, a Ca^{2+} ionophore, was applied to astrocytes it stimulated the release of glutamate in the presence of external free Ca^{2+}, but failed to do so when internal Ca^{2+} stores were depleted by preventing Ca^{2+} entry from the extracellular space (Parpura et al., 1994). Further studies supported this conclusion since depleting internal Ca^{2+} stores by application of thapsigargin, a blocker of store specific Ca^{2+}-ATPase, or by buffering cytoplasmic Ca^{2+} with BAPTA also resulted in a reduction of glutamate release (Araque et al., 1998b; Bezzi et al., 1998). Using flash photolysis of a caged Ca^{2+} compound to evoke rises of $[Ca^{2+}]_i$ in astrocytes, demonstrated that glutamate release can result from moderate increases in Ca^{2+} concentration that are likely to occur physiologically (Parpura and Haydon, 2000). This release mechanism was determined to be distinct from swelling or reverse operation of the plasma membrane glutamate transporters (Parpura et al., 1995b; Jeftinija et al., 1996; Araque et al., 2000; Innocenti et al., 2000). Similar experiments revealed that aspartate can also be released from astrocytes via Ca^{2+}-dependent exocytosis (Jeftinija et al., 1996; Jeremic et al., 2001).

The majority of the Ca^{2+} necessary for glutamate release from astrocytes originates from internal stores, but the entry of external Ca^{2+} is also involved. This was demonstrated by the reduction of mechanically-induced glutamate release in the presence of thapsigargin and also by Cd^{2+}, a blocker of Ca^{2+} entry from the extracellular space (Hua et al., 2004). This release requires co-activation of inositol 1,4,5-trisphosphate- and ryanodine/caffeine-sensitive internal Ca^{2+} stores, which operate jointly (Hua et al., 2004).

β-adrenergic and glutamatergic stimulation of cultured astrocytes can cause the release of HCA (Do et al., 1997; Benz et al., 2004). Glutamate-induced release of HCA is mediated by activation of ionotropic and metabotropic glutamate receptors (Benz et al., 2004) (see Chap. 3 for details on astrocytic receptor expression). The release of HCA displayed Ca^{2+} dependency, because (1) application of a Ca^{2+} ionophore caused release of this amino acid; (2) application of ionotropic glutamate receptor agonists in the absence of extracellular Ca^{2+} failed to cause HCA release. Similarly, the release showed Na^+ dependency, since the removal of extracellular Na^+ also blocked agonist induced HCA release. This raised an interesting possibility that agonist stimulation could lead to intracellular Na^+ load through ionotropic receptors, which could then activate Na^+/Ca^{2+} exchangers to extrude Na^+ from astrocytes while importing Ca^{2+} to the cytosol, resulting in a rise of $[Ca^{2+}]_i$. This was the case as agonist induced HCA release was blocked in the presence of benzamil, a Na^+/Ca^{2+} exchanger blocker.

The use of *Clostridial*, tetanus and various types of botulium toxins, which cleave some of the soluble *N*-ethyl maleimide-sensitive fusion protein (NSF) from the attachment protein receptor (SNARE) proteins necessary for exocytosis, caused a reduction in the level of Ca^{2+}-dependent glutamate release in astrocytes [(Jeftinija et al., 1997; Bezzi et al., 1998; Araque et al., 2000; Bezzi et al., 2001; Pasti et al., 2001; Bezzi et al., 2004; Hua et al., 2004; Montana et al., 2004); reviewed in (Montana et al., 2006)]. Synaptobrevin 2 can be cleaved by tetanus neurotoxin and botulinum neurotoxin (BoNT) type B, D, F and G; syntaxin by BoNT-C; while SNAP 25 can be targeted by BoNT-A, -C and -E (Schiavo et al., 2000). Indeed, astrocytes express proteins of the core SNARE complex: synaptobrevin 2, syntaxin 1, synaptosome-associated protein of 23 kDa (SNAP 23) (Parpura et al., 1995a; Jeftinija et al., 1997; Hepp et al., 1999; Maienschein et al., 1999; Araque et al., 2000; Pasti et al., 2001; Montana et al., 2004; Mothet et al., 2005; Crippa et al., 2006) and the ancillary protein synaptotagmin 4 (Zhang et al., 2004a; Crippa et al., 2006) (Fig. 12.8). The experimental expression of the cytoplasmic tail of synaptobrevin 2

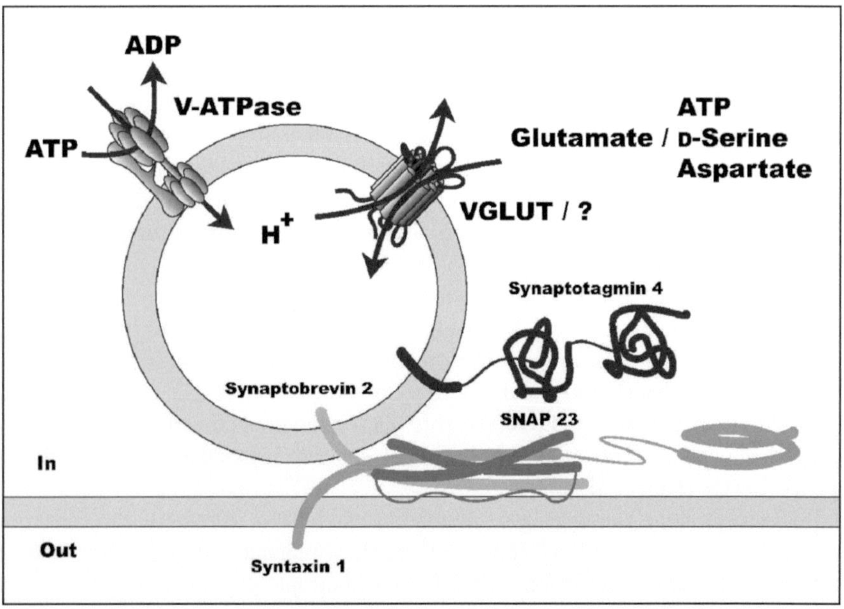

Fig. 12.8 Transmitter release by Ca^{2+}-dependent exocytosis. Transmitters packaged in vesicles are released from the cell when the vesicle fuses with the plasma membrane. This fusion process is mediated by synaptotagmin 4 and SNARE proteins: syntaxin 1, synaptobrevin 2 and SNAP 23. Glutamate is packaged into the vesicle by vesicular glutamate transporters (VGLUT) that use the proton gradient generated by vacuolar type H^+-ATPases (V-ATPase) to concentrate glutamate against its gradient. Other transmitters may be packaged into vesicles in a similar fashion, but the identity of a transporter that would do this remains elusive.

(containing the SNARE domain, but lacking the ability to anchor to the vesicular membrane; also referred to as dominant negative SNARE) resulted in the inhibition of glutamate release from astrocytes (Zhang et al., 2004b). Additionally, the use of tetanus toxin caused a reduction in plasma membrane capacitance (C_m) increase (Kreft et al., 2004) and a reduction in the number of amperometric spikes (Chen et al., 2005), both reporting on exocytosis from astrocytes (see below). Furthermore, the use of α-latrotoxin, that can cause release of a neurotransmitter at the pre-synaptic terminals by directly stimulating the secretory machinery [reviewed in (Südhof and Jahn, 1991)], has been demonstrated to induce glutamate release from astrocytes (Parpura et al., 1995b; Jeftinija et al., 1996).

Proteins important for sequestering glutamate into vesicles have also been discovered in astrocytes. The presence of the vacuolar type of proton ATPase (V-ATPase), which drives protons into the vesicular lumen creating the proton concentration gradient necessary for glutamate transport into vesicles, has been detected (Wilhelm et al., 2004), and blockage of these pumps with bafilomycin A_1 was shown to reduce glutamate release from astrocytes caused by various stimuli (Araque et al., 2000; Bezzi et al., 2001; Pasti et al., 2001; Montana et al., 2004; Crippa et al., 2006). The three known isoforms of vesicular glutamate transporters (VGLUTs) 1, 2, and 3, which use the proton gradient created by V-ATPases to package glutamate into vesicles, have been detected in astrocytes (Fremeau et al., 2002; Bezzi et al., 2004; Kreft et al., 2004; Montana et al., 2004; Zhang et al., 2004b; Anlauf and Derouiche, 2005; Crippa et al., 2006). These transporters are functional within astrocytes since Rose Bengal, a broad spectrum modulator of the allosteric site of VGLUTs, greatly reduced glutamate release (Montana et al., 2004).

Astrocytes have also been shown to release the transmitter D-serine in response to glutamate (Schell et al., 1995) (also see Chap. 16). To investigate the mechanism of this release, Mothet at al. (2005) used an enzyme-linked assay to measure extracellular D-serine concentration. They found that D-serine was released upon glutamate receptor stimulation and that this release was Ca^{2+}-dependent as it was augmented by Ca^{2+} ionophore and inhibited by the application of thapsigargin or removal of extracellular Ca^{2+}. D-serine release was determined to be vesicular since it was reduced by concanamycin A, a V-ATPase inhibitor, and tetanus toxin. Further, D-serine was also found to co-localize with the vesicular protein, synaptobrevin 2, thereby providing strong evidence for Ca^{2+}-dependent exocytosis as a mechanism of D-serine release from astrocytes. It is not clear how D-serine would be packaged into vesicles, but it could possibly be transported by a known vesicular amino acid transporter, such as VGLUT, or perhaps by an undiscovered vesicular D-serine transporter. The promiscuity of VGLUT is unlikely, however, since VGLUTs do not recognize, e.g., aspartate as a substrate (Schafer et al., 2002). This also implies that future studies will have to be conducted to encompass the molecular identity of the vesicular aspartate transporter.

Astrocytic secretory vesicles are the essential morphological elements for regulated Ca^{2+}-dependent exocytosis. Although secretory granules in the glia of grey matter were described nearly 100 years ago (Nageotte, 1910), it is only recently, that there has been compelling evidence for the existence of these organelles in astrocytes.

Immunoelectron microscopy studies demonstrated that synaptobrevin 2 could be associated with electron-lucent (clear) vesicular structures (Maienschein et al., 1999), while VGLUTs 1 or 2 in astrocytes in situ showed the association of these proteins with small clear vesicles with a mean diameter of ~30 nm (Bezzi et al., 2004). Additionally, synaptobrevin 2-containing vesicles immunoisolated from cultured astrocytes (Crippa et al., 2006) were predominantly clear displaying heterogeneity in size, ranging from 30 to over 100 nm. Furthermore, the presence of clear smooth and clathrin-coated vesicles with diameters of ~30 nm has been observed in gliosomes (Stigliani et al., 2006) expressing synaptobrevin 2 and VGLUT 1. Using a pre-loading technique that stimulated membrane recycling and trapping of styryl dyes (FM 1-43 or FM 2-10) in secretory organelles, astrocytes displayed a punctate pattern of FM fluorescence (Chen et al., 2005). When this loading was followed by photoconversion and electron microscopy (EM) Chen et al. (2005) showed that astrocytic vesicles had a mean diameter of 310 nm. Vesicles of much larger size, over 1 μm, have also been observed to release glutamate (Kang et al., 2005; Xu et al., 2007). These vesicles are not normally present in astrocytes, but are formed within minutes of repeated stimulation with pharmacological dosages (5–50 mM) of glutamate. The formation of these large vesicles was confirmed by EM (Xu et al., 2007), while their fusions were inhibited by tetanus neurotoxin. Although the size of these vesicles in rat astrocytes are larger than generally accepted for regulated exocytosis, in mutant beige mouse (bg^j/bg^j) mast cells secretory granules up to several micrometers in diameter exhibit exocytotic fusion (Curran and Brodwick, 1991; Fernandez et al., 1991). Therefore, further biochemical analysis of these large vesicles and their appearance in physiological conditions would help to define whether these large vesicles could represent a physiological event or are merely a pharmacologically induced phenomenon as they appear at present.

The recycling of secretory vesicles at the plasma membrane has been investigated in astrocytes. Increasing cytoplasmic Ca^{2+} levels in astrocytes while in the presence of antibodies against VGLUT1 resulted in an increase in fluorescent puncta inside the cell (Stenovec et al., 2007). Similarly, application of ionomycin in the presence of extracellular Ca^{2+}, but not in its absence, caused uptake of the membrane recycling dye, FM 4-64 (Krzan et al., 2003). Furthermore, using a pre-loading technique that stimulated membrane recycling and the trapping of styryl dyes (FM 1-43 or FM 2-10) in secretory organelles, astrocytes displayed a punctate pattern of FM fluorescence (Chen et al., 2005). The delivery of secretory vesicles to plasma membrane fusion sites was also studied in astrocytes. Crippa et al. (2006) expressed a chimeric protein, where enhanced green fluorescent protein (EGFP) was fused to the C-terminus of synaptobrevin 2 (synaptobrevin 2-EGFP), in astrocytes. When astrocytes were stimulated with Ca^{2+} ionophore many fluorescent synaptobrevin 2-EGFP puncta disappeared with a concomitant increase in plasma membrane fluorescence, consistent with exocytotic fusion of labeled vesicles. Consequential net addition of vesicular membrane to the plasma membrane can be directly assessed by monitoring changes in C_m. Indeed, an agonist-induced rise in astrocytic $[Ca^{2+}]_i$ caused an increase in C_m, while simultaneous measurements recorded a release of glutamate (Zhang et al., 2004b). Further evidence for vesicular exocytosis from astrocytes was provided by total internal reflection fluorescence microscopy (Bezzi et al., 2004; Domercq et al., 2006; Bowser and Khakh, 2007), where exocytosis of VGLUT1, VGLUT2 or synaptobrevin 2 positive vesicles was reported. As a consequence of vesicular fusions, quantal events of transmitter release representing an exocytotic "footprint" (Del Castillo and Katz, 1954) have been recorded from astrocytes. Such events were detected using "sniffer" cells expressing NMDA receptors (Pasti et al., 2001), or by amperometric measurements used to detect the release of dopamine, acting as a "surrogate" transmitter for glutamate, from glutamatergic vesicles (Chen et al., 2005).

12.2 Nucleotides and Nucleosides as Astrocytic Transmitters

12.2.1 Synthesis of Nucleotide/Nucleoside Transmitters

A primer on nucleoside and nucleotide terminology and their syntheses relevant to this chapter follows. Nucleosides are glycosylamines made by attaching a nucleobase, purine (adenine, guanine, hypoxantine) or pyrimidine (uracil, cytosine), to a ribose. Examples of these are adenosine, guanosine, inosine, uridine and cytidine. A nucleotide is an ester of phosphoric or pyrophosphoric acid and a nucleoside, which are together referred to as nucleoside mono- or di-phosphates. The addition of a γ-phosphate to, for example, adenosine 5′ diphosphate (ADP) results in ATP.

Adenosine is formed in cells via two pathways, mainly by dephosphorylation of adenosine 5′ monophosphate (AMP) and to some extent by hydrolysis of S-adenosyl homocysteine [reviewed in (Latini and Pedata, 2001)]. As a vital component of cellular respiration, ATP is produced in abundance through glycolysis and oxidative phosphorylation. Intracellular ATP fuels a variety of processes and can also be released to the extracellular space. It can be extracellularly degraded by membrane-bound ecto-nucleotidases. The products of its extracellular hydrolysis, ADP and adenosine, can activate different plasma membrane receptors (Table 12.2; also see Chap. 17). An additional extracellular source of adenosine originates from the conversion of adenosine 3′:5′ cyclic monophosphate (cAMP) (Brundege et al., 1997). Similar extracellular metabolism is common for other nucleosides/nucleotides. There is an additional complexity in the extracellular regulation of nucleotide levels [reviewed in (Lazarowski et al., 2003a)]; the exchange of γ-phosphates between adenine- and uracil-based nucleotides via nucleoside diphosphokinase (NDPK) can occur. NDPK reversibly transphosphorylates UDP or guanosine 5′ diphosphate (GDP) using ATP, to uridine 5′ triphosphate (UTP) or guanosine 5′ triphosphate (GTP) along with the generation of ADP.

UTP can either be made de novo intracellularly or salvaged from uridine by phosphorylation (Anderson and Parkinson, 1997). The UTP derivative UDP-glucose is a signaling molecule and it mainly originates from UTP that is synthesized de novo. Although the salvage pathway is predominately used in RNA synthesis, both de novo and salvage pathways can contribute to the free pool of UTP and thus are important for purinergic receptor activation in cell–cell signaling (see Table 12.2).

Table 12.2 Mammalian purine/pyrimidine receptors and their agonists

Receptor	Agonist(s)	Astrocyte	Brain
(P1)$A_{1,2A,2B,3}$	adenosine	Y	Y
$P2X_{1–7}$	ATP*	Y	Y
$P2Y_1$	ADP > ATP; NAD^+	Y	Y
$P2Y_2$	ATP = UTP	Y	Y
$P2Y_4$	UTP ≥ ATP	Y	Y
$P2Y_6$	UDP > UTP > ATP	Y	Y
$P2Y_{11}$	ATP		Y
$P2Y_{12}$	ADP > ATP	Y	Y
$P2Y_{13}$	ADP > ATP		Y
$P2Y_{14}$	UDP-glucose	Y	Y

Receptor families: P1 adenosine, P2 for ATP and ADP, P2X are ionotropic, P2Y are metabotropic. *N/A for $P2X_6$, which does not form a homodimer. Additional receptors are tentatively termed p2y, where $p2y_5$, $p2y_7$, $p2y_9$, $p2y_{10}$ do not exhibit functional resposes to nucleotides. $p2y_3$ is an avian ortholog of $P2Y_6$. $p2y_8$ from *Xenopus laevis* shows high homology to mammalian $P2Y_2$ and $P2Y_4$. Sources: Lazarowski et al., 2003a; Fields and Burnstock, 2006; Mutafova-Yambolieva et al., 2007.

Guanine and adenine nucleotides can bind to the extracellular domain of ionotropic glutamate receptors causing inhibition by displacing glutamate (Monahan et al., 1988; Baron et al., 1989; Gorodinsky et al., 1993; Dev et al., 1996; Paas et al., 1996; Ortinau et al., 2003). Although some extracellular effects of guanosine could be blocked by adenosine (P1) receptor blockers (Rathbone et al., 1991), there is paucity of information with regard to receptors it acts upon (Traversa et al., 2002).

12.2.2 ATP Release through Channels

12.2.2.1 ATP Release via Purinergic Ion Channels

The first evidence that nucleotide transmitters might be released via channels from astrocytes was reported by loading cells with radiolabeled adenosine (Ballerini et al., 1996). Stimulation with BzATP, an agonist of $P2X_7$ receptors, resulted in Lucifer yellow uptake in astrocytes and could also induce the release of radiolabeled purines. The dye uptake and purine release were blocked by the $P2X_7$ antagonist oATP, providing additional support for this pathway. While the exact purine composition that was released was not determined, subsequent studies have shown that ATP can be released through $P2X_7$ receptors, making ATP the most probable candidate.

Saudicani et al. (2006) demonstrated that $P2X_7$ receptors mediate ATP release from astrocytes. Intercellular Ca^{2+} waves induced in cultured astrocytes traveled greater distances in low divalent cation solution. This effect was abolished by the presence of the ATP degrading enzyme, apyrase, indicating ATP release as the mechanism supporting propagation of this intercellular wave, consistent with previous findings (Guthrie et al., 1999). To determine whether $P2X_7$ receptors or connexin hemichannels might be mediating this effect they employed astrocytes from $P2X_7$-null and Cx43-null mice. The potentiation of Ca^{2+} waves in low divalent cation solution was present in Cx43-null cells but not in $P2X_7$-null cells. Further, BBG, a blocker of $P2X_7$ receptors abolished this potentiation in wild type and Cx43-null cells. Interestingly, the results of this study raised awareness that gap-junction blockers, often used as evidence for connexin hemichannel mediated release of transmitters from astrocytes, can also antagonize $P2X_7$ receptors; this adds yet another layer of difficulty in dissecting out possible pathways for the release of astrocytic transmitters [see above Sect. 12.1.2.1; also discussed in Spray et al. (2006); but see below Sect. 12.2.2.2].

12.2.2.2 ATP Release via Hemichannels

As discussed in Sect. 12.1.2.2, astrocytes may possess functional connexin hemichannels. Several lines of evidence point out that connexins are involved in ATP release from astrocytes. Receptor stimulated (Cotrina et al., 1998) or constitutive (unstimulated) (Cotrina et al., 2000) ATP release from C6 rat glioma cells was enhanced when overexpressing Cx43 or Cx32. Using a bioluminescence assay, ATP release from cultured astrocytes and Cx43 expressing C6 cells were seen as discrete events originating from point sources (Arcuino et al., 2002). The release of ATP coincided with the uptake of the fluorescent marker, propidium iodide, indicating transient permeability of the plasma membrane. Whole-cell recordings revealed an inward current in cells during a dye uptake. Both of these events were caused by the local removal of extracellular Ca^{2+}, a stimulus known to be associated with opening of certain hemichannels. Finally, Stout et al. (2002) using electrophysiology and optical methods demonstrated that ATP release could occur through connexin hemichannels on astrocytes. Characteristics of whole-cell currents in the absence of external Ca^{2+} indicated that astrocytes may express functional connexons; these currents could be inhibited by the gap-junction blocker, flufenamic acid (FFA). Interestingly, these currents displayed linear current–voltage plots [see Fig. 1 of (Stout et al, 2002)], a fact, which is at odds with the clear demonstration of functional Cx43 hemichannels showing that they open only at high positive potentials [see Fig. 4 of (Contreras et al., 2003)]. Nonetheless, astrocytes could flux low-molecular weight (<~1 kDa), but not high-molecular (~10 kDa) weight dyes when stimulated, and this was inhibited by Gd^{3+} or FFA. Also, dye flux could be observed in C6 cells that had expressed Cx43, but not in cells that did not express Cx43. Quinine, an activator of hemichannels in some preparations [(Stout et al, 2002) and references therein; however, it is a gap-junction blocker (Srinivas et al, 2001)], could evoke ATP release from astrocytes along with potentiating mechanically-induced release of ATP and intercellular Ca^{2+} wave propagation. These effects could be blocked by FFA and Gd^{3+} but not by the Cl^- channel blocker, DIDS. Additionally, the release of dye/ATP and consequential spread of intercellular Ca^{2+} waves were sensitive to the concentration of external divalent cations Ca^{2+} and Mg^{2+} (Stout and Charles, 2003).

Aside from connexins, pannexins have the potential to mediate ATP release from astrocytes. As mentioned in Sect. 12.1.2.2, pannexins can form conductive channels between the intra- and extra-cellular spaces. These pannexons are mechanosensitive and allow the passage of ATP (Bao et al., 2004). Additionally, they are not sensitive to extracellular Ca^{2+} (Bruzzone et al., 2005) and can be opened by cytoplasmic Ca^{2+} elevations (Locovei et al., 2006). This raises the possibility that mechanically-induced ATP-mediated intercellular Ca^{2+} waves in astrocytes (Guthrie et al., 1999) might involve pannexons. Consistent with that hypothesis, in *Xenopus* oocytes expressing P2Y receptors, the addition of extracellular ATP to stimulate these receptors caused an increase of intracellular Ca^{2+} that in turn activated co-expressed pannexin 1 channels (Locovei et al., 2006); this could lead to the release of ATP via pannexons. Indeed, astrocytes endogenously express pannexin 1 but the extent of the intercellular Ca^{2+} waves propagating among cells was reduced when pannexin 1 was silenced using a small interfering RNA approach (Suadicani et al., 2007). The mechanism of release of ATP gets even more interesting since pannexin 1 can be recruited into the $P2X_7$ receptor death complex (Locovei et al., 2007) (Fig. 12.3). As mentioned in Sect. 12.1.2.3, the $P2X_7$ receptors are non-selective cation channels; however, they can also form large pores involved in apoptotic cell death. It turned out that co-expression of the $P2X_7$ receptor and pannexin 1 in *Xenopus* oocytes can lead to apoptotic death following a sustained challenge with extracellular ATP (Locovei et al., 2007). In cells expressing the $P2X_7$ receptor/pannexin 1 complex, unlike those expressing the $P2X_7$ receptor alone, currents induced by ATP were blocked by carbenoxolone, a gap junction and hemichannel blocker. This exciting finding points towards the recruitment of pannexin 1 into the $P2X_7$ receptor complex where it forms a part of the transmembrane pore. Whether this mechanism can also play a role in astrocytic signaling remains to be elucidated. Nonetheless, the recruitment of pannexin 1 into the $P2X_7$ receptor signaling complex brings an intriguing possibility that pannexin 1 could, in such a scenario, represent a molecular switch between physiology and pathophysiology.

12.2.2.3 ATP Release via Anion Transporters/Channels

Anion channels could provide another pathway for ATP release from astrocytes. Activation of these channels in astrocytes is a complex issue (Crepel et al., 1998). For example, swelling alone can activate more than one chloride channel (Liu et al., 2006). Additionally, some ligands, such as ATP can activate chloride conductance while causing a moderate/physiological swelling (Takano et al., 2005)

While investigating the mechanism of ATP release due to glutamate stimulation in astrocytes, Queiroz et al. (1999) found that stimulation of ionotropic glutamate receptors with NMDA, kainate and AMPA caused ATP release. NMDA- and kainate-, but not AMPA-induced ATP release was dependent on the presence of external Ca^{2+}. AMPA-induced release could be reduced by glibenclamide. Thus, the mechanism underlying such ATP release implicates anion channel involvement. The sulfonylurea glibencamide has been used commonly as an open channel blocker of the cystic fibrosis transmembrane conductance regulator (CFTR) Cl^- channel (Melin et al., 2007). CFTR is a unique member of the ATP-binding cassette (ABC) transporter/protein superfamily that forms a Cl^- channel with complex regulation. ATP-driven conformational changes of CFTR lead to the opening of a gate to permit the transmembrane flux of anions down their electrochemical gradients; this is in contrast to other ABC proteins where ATP normally energizes the transport of substrates against their gradients across the membrane [reviewed in (Gadsby et al., 2006)]. There is conflicting evidence whether CFTR functions as an ATP channel or not (Reisin et al., 1994; Li et al., 1996; Prat et al., 1996). Thus, the interpretation of the Queiroz et al. (1999) study remains elusive; AMPA-induced ATP release could occur directly via CFTR, or alternatively, CFTR could mediate or regulate such release (Fig. 12.9).

Similar to CFTR, other ABC proteins may allow the direct release of ATP, or play a regulatory role. In addition to CFTR, astrocytes express the functional multidrug resistance protein (MRP) and P glycoprotein (Decleves et al., 2000), that are involved in ATP release (Ballerini et al., 2002), and can act as ATP channels (Abraham et al., 1993). A study exposing cultured astrocytes to hypo-osmotic solution showed a Cl^- channel current (Darby et al., 2003). This current was most likely mediated by ATP acting on $P2Y_1$ receptors since it was effectively blocked by the specific antagonist, PPADS, and the application of exogenous ATP could induce a similar current. This current was also blocked by the MRP transporter/channel blockers probenicid, MK-571 and indomethacin [see pharmacology details in

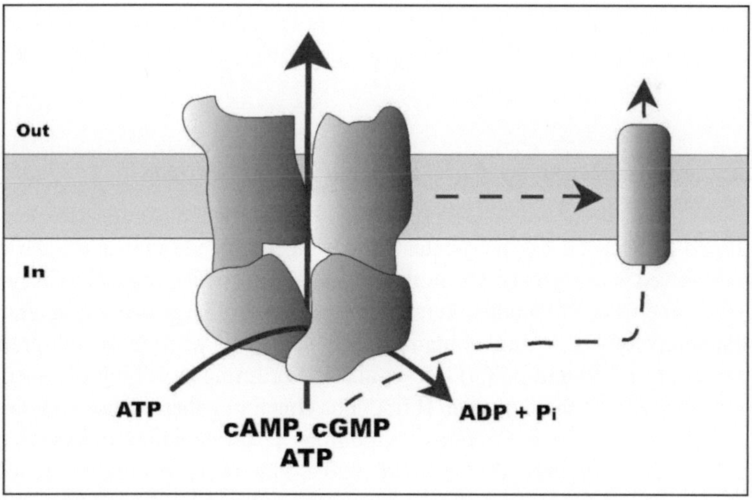

Fig. 12.9 Nucleotide release by members of the ATP-binding cassette (ABC) transporter/protein superfamily. These transporters use ATP to transport compounds across the membrane. ATP normally energizes the transport of substrates against their gradients for transporters like the functional multidrug resistance protein (MRP) and P glycoprotein, while the cystic fibrosis transmembrane conductance regulator (CFTR) uses ATP to open a gate to permit the flux of anions down their electrochemical gradients. These channels may pass transmitters directly or perhaps mediate the opening of other channels to allow transmitter release.

(Darby et al., 2003)] and only slightly inhibited by the potent blocker of the P glycoprotein transporter/channel, verapamil. Direct measurements of ATP release using an optical assay demonstrated that hypo-osmotic solutions caused the release of ATP from astrocytes; this release was also blocked by the MRP transporter/channel blocker MK-571, indicating that swelling-induced ATP release could occur via these ABC proteins. This study also points to the existence of an autocrine ATP loop (ATP-P2Y_1 receptor-MRP transporter-ATP) during astrocytic swelling [but see contrary data in (Mongin and Kimelberg, 2002)]

Further support for an anion channel role in ATP release from astrocytes came from a study where application of ATP to astrocytes caused the release of radiolabeled ATP (Anderson et al., 2004). This release was blocked by the broad spectrum P2 receptor antagonists, suramin and reactive blue-2, but not by the selective P2X_7 receptor antagonist oATP. Additionally, ATP-induced ATP release was not dependent on intracellular Ca^{2+}, since neither depletion of endoplasmic reticulum Ca^{2+} stores nor the buffering of cytoplasmic Ca^{2+} reduced ATP release. Similarly, blocking Ca^{2+} influx from the extracellular space did not affect this release. However, this release was inhibited by the anion channel blockers, DIDS and niflumic acid. Additionally, ATP release was attenuated by octanol, an inhibitor of gap-junction channels; although two other gap-junction blockers had no effect. It should be noted, however, that gap-junction blockers have been shown to affect anion channels [(Eskandari et al., 2002); additional discussion on pharmacology in (Anderson et al., 2004)]. The involvement of ABC proteins and vesicular exocytosis in ATP-induced ATP release was ruled out using pharmacology. Therefore, ATP-induced ATP release occurs via anion channels (Fig. 12.1). Consistent with this finding, there is evidence for ATP conductance via VRACs (Sabirov et al., 2001; Hisadome et al., 2002)

12.2.3 ATP Release by Ca^{2+}-Dependent Exocytosis

As outlined in Sect. 12.1.4, astrocytes possess secretory vesicles and a variety of proteins involved in exocytosis. The initial evidence for vesicular organelles in cultured astrocytes came from EM studies showing the presence of large dense core granules, with diameters of ~115 nm, containing the secretory peptide secretogranin II (Calegari et al., 1999) and ATP (Coco et al., 2003). Immunoprobing of subcellular fractions revealed that secretogranin II fractions containing these dense-core vesicles were mainly distinct from fractions containing synaptobrevin 2 (Calegari et al., 1999). Consistent with these findings, immuno-isolated synaptobrevin 2-containing vesicles included some dense core vesicles representing ~2% of the total number of vesicles isolated (Crippa et al., 2006). Similarly, using immuno-EM, it was demonstrated that synaptobrevin 2 can be associated with some dense core vesicular structures, with diameters ranging from 100 to 700 nm (Maienschein et al., 1999). Following subcellular fractionations, immunoblotting for several exocytotic proteins, synaptobrevin 2, syntaxin 1, cellubrevin and synaptotagmin 1 were found to colocalize with ATP containing organelles (Maienschein et al., 1999). Thus, the combined biochemical and morphological analysis of secretory vesicles in astrocytes indicates that a nucleotide transmitter may be released by Ca^{2+}-dependent exocytosis (Fig. 12.8).

In the first experiments to examine this possibility, exposure of astrocytes to nitric oxide was shown to cause an increase in cytoplasmic Ca^{2+} along with the release of ATP (Bal-Price et al., 2002). This release could be blocked by chelating intracellular Ca^{2+} with BAPTA or preventing vesicular release with botulinum toxin C, suggesting vesicular release as a mechanism of ATP release from astrocytes. Coco et al. (2003) used fura-2 loaded microglia as "sniffer" cells to report on ATP released from cultured astrocytes that were mechanically stimulated. They found that this release was not affected by gap-junction blockers but was, instead, impaired by treatment with bafilomycin A_1 or tetanus toxin. However, the reduction of ATP release caused by tetanus toxin was not as pronounced as the reduction in glutamate release, indicating that ATP release may be regulated in a different manner, perhaps through distinct glutamate and ATP vesicular pools.

ATP release from cultured astrocytes was also investigated using a bioluminescence assay (Abdipranoto et al., 2003). UTP caused release of ATP via the likely activation of $P2Y_2$ receptors. This release was reduced by thapsigargin and lithium ions that can block the intracellular generation of IP_3 thus possibly pointing to a Ca^{2+}-dependent exocytosis. Although pre-incubation with a cell permeable form of BAPTA showed a trend in the reduction of release, the effect was insignificant, which may be ascribed to an insufficient concentration of BAPTA within the cell. Nonetheless, further pharmacology on vesicular trafficking affirmed that the exocytotic pathway is involved in UTP-induced ATP release from astrocytes: a blocker of transport vesicles budding off the Golgi apparatus, brefeldin A, a potent disruptor of actin microfilaments, cytochalasin D, and the exocytosis inhibitor, botulinum toxin A, all blocked ATP release. Interestingly, a blocker of Na^+–K^+–Cl^- cotransporters (NKCCs; see Chap. 6), furosemide, had a greater inhibitory effect on ATP release than either of the two blockers of chloride channels, DIDS, or the CFTR blocker, glibenclamide. Since furosemide has greater cell permeability than DIDS, it is possible that furosemide also substantially affected vesicular chloride channels [see Fig. 16.1 in (Evanko et al., 2004a)]; similarly, Ca^{2+}-dependent glutamate release from astrocytes, evoked by ATP (Jeremic et al., 2001) or bradykinin (Parpura et al., 1994), can be inhibited by furosemide. Nonetheless, these results raise an interesting possibility, that four different mechanisms could, in parallel, underlie UTP-induced ATP release: Ca^{2+}-dependent exocytosis, release mediated by NKCCs, ABC transporters and anion channels. Although the authors report the inhibitory effect of the gap-junction blocker, 18β-glycyrrhetinic acid (GA), they have not explored further whether hemichannels (connexons and/or pannexons) also play a role in UTP-induced ATP release.

Astrocytic progenitor cells have the ability to release ATP, as determined by the presence of ATP in the medium of cultured cells (Striedinger et al., 2007). To determine the mechanism underlying this release, cells were incubated with quinacrine, a compound that fluorescently labels ATP containing structures. Using TIRF microscopy, punctate structures could be seen, which would disappear rapidly when the

Ca^{2+} ionophore was applied. This same pattern was observed using the fluorescent ATP analogue mant-ATP. The ionophore-induced vesicle fusion and release of ATP was impaired by the V-ATPase inhibitor, bafilomycin A_1, intracellular Ca^{2+} chelation with BAPTA, tetanus toxin and a dominant negative SNARE domain, indicating that ATP release from astrocyte progenitor cells occurs via Ca^{2+}-dependent exocytosis. Similar to these observations in progenitors, astrocytes loaded with quinacrine showed puncta, the loss of which, as detected by TIRF microscopy, was evident upon stimulation with ionomycin or glutamate (Pangrsic et al., 2007) and application of ATP or flash photolysis of caged Ca^{2+} (Pryazhnikov and Khiroug, 2008). Glutamate stimulation of astrocytes showed quantal release of ATP as recorded by ATP "sniffer" cells (Pangrsic et al., 2007), human embryonic kidney cells expressing a mutated $P2X_3$ receptor with reduced desensitization (Fabbretti et al., 2004)(also see Table 12.2).

Astrocytes can exocytotically release ATP stored in lysosomes (Zhang et al., 2007). Prolonged incubation with FM recycling dyes (more than 1 h) stained astrocytic organelles. FM fluorescence colocalized with the lysosomal markers LAMP1, CD63, and Rab-7. Agonist stimulation or metabolic blockade of astrocytes revealed regulated exocytosis of these lysosomes under TIRF microscopy that was blocked by intracellular Ca^{2+} chelation with BAPTA. These lysosomes contained ATP since they readily loaded up with mant-ATP that was released upon stimulation. Additionally, astrocytic lysosomes where shown to contain ATP since the lysosomal fraction in density gradient centrifugation contained abundant ATP and the stimulation of astrocytes resulted in an increase of ATP in the medium.

The release of ATP from the astrocytic vesicular pool has consequences on synaptic transmission and plasticity (see Chap. 15 for details). Expression of an inducible dominant negative SNARE domain specifically in astrocytes to prevent transmitter release resulted in a reduction of the magnitude of long term potentiation and an absence of synaptic depression in hippocampal slices (Pascual et al., 2005). In wild type mice these effects were found to be attributed to adenosine acting on A_1 receptors. Blocking vesicular fusion in astrocytes by using a dominant negative SNARE domain, stopped the release of ATP which would otherwise be converted into adenosine by ecto-nucleotidases and stimulate adenosine-sensitive receptors.

12.2.4 Release of Other Nucleotide and Nucleoside Transmitters

Besides ATP, the release of its "relative," nucleosides such as adenosine and guanosine, as well as other nucleotides has also been described. In several instances the pathways underlying such release are not yet well defined.

Adenosine in physiological conditions seems to have a role in the adjustment of synaptic efficacy by providing a tonic inhibition, while an increased level in the extracellular space during pathological conditions, such as hypoxia/hypoglycemia seems to exert neuroprotective actions [reviewed in (Latini and Pedata, 2001)]. Although there is some evidence for Ca^{2+}-dependent adenosine release from neurons, the most commonly assigned mechanism for the release of adenosine is via the equilibrative nucleoside transporters 1 and 2 (ENT 1 and 2). These are bi-directional transporters that facilitate adenosine diffusion along its concentration gradient (Baldwin et al., 2004). Additional concentrative nucleoside transporters for adenosine are Na^+-dependent symporters that normally accumulate adenosine inside the cell (Gray et al., 2004) but they can operate in reverse to export adenosine from cells under conditions that favor the reversal of the Na^+ transmembrane gradient (Borgland and Parkinson, 1997). Under basal conditions, adenosine uptake into astrocytes mainly occurs via ENT1 and ENT2, while symport represents a minor component in this process (Parkinson et al., 2005).

Adenosine release was investigated using radiolabel tracing in cultured glia from the peripheral and central nervous systems of embryonic chicks (Meghji et al., 1989). Under basal conditions there was a trend of adenosine accumulation in the medium. When glial cells were exposed to 2-deoxyglucose and oligomycin to cause metabolic poisoning, there was a significant increase in the amount of adenosine in the extracellular medium over basal levels. This accumulation was due to the release of adenosine via ENTs since the ENT blocker dilazep blocked it. Thus, glial cells can directly release adenosine through ENTs (Fig. 12.10).

When a radiolabeling technique was used, primary cortical astrocytes under basal conditions released a significantly higher amount of both adenine nucleotides and adenosine than cultured neurons (Parkinson et al., 2002). The mechanism of this release at rest was inconclusive, because the detected levels of purines were low. However, in follow up work it appears that the main source of astrocytic extracellular adenosine at rest comes from the conversion of ATP via ecto-nucleotidases [also see Sect. 12.2.4; (Pascual et al., 2005)], while neuronal basal release of adenosine occurs via ENTs (Parkinson and Xiong, 2004; Parkinson et al., 2005). Under conditions of oxygen–glucose deprivation, however, there was a significant increase in the

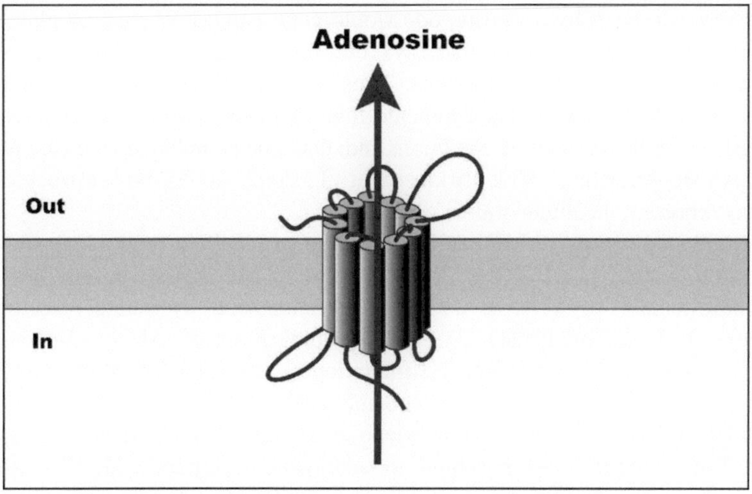

Fig. 12.10 Release of adenosine via the equilibrative nucleoside transporters. These are bi-directional transporters that facilitate adenosine diffusion along its concentration gradient.

amount of these purines released, most notably adenosine, from neurons but not from astrocytes (Parkinson et al., 2002). This enhanced release of adenosine from neurons could be blocked by an inhibitor of ENTs. Although during this ischemic condition there was no significant increase of purines released from astrocytes (Parkinson et al., 2002), there was an increase in adenosine detected in the extracellular medium of the astrocytes which was derived from the extracellular conversion of ATP (Parkinson et al., 2005). Thus, it appears that neurons rather than astrocytes can release additional amounts of adenosine in specific pathophysiological conditions that emulate ischemia.

In conditions of oxygen deprivation, i.e., hypoxic conditions, however, it seems that astrocytes can directly release increased amounts of adenosine (Martin et al., 2007). Hypoxia induced a reduction in the excitatory post synaptic current (EPSC) amplitude in hippocampal slices. This could be blocked by an A_1 receptor antagonist but not by blockers of ecto-nucleotidases, indicating a direct release of adenosine and not conversion of released ATP. The lack of contribution from extracellular ATP catabolism in this hypoxia-induced phenomenon was established using slices from a mouse expressing a dominant negative SNARE domain specifically in astrocytes. Specifically blocking astrocytic metabolism in slices with fluorocitrate, abolished the hypoxia-induced reduction in EPSC, suggesting that adenosine was directly secreted from astrocytes. Indeed, cultures of primary astrocytes were shown to release adenosine into the medium under hypoxic conditions. Consistent with being SNARE-independent, chelation of intracellular Ca^{2+} did not block this release but rather enhanced it. As the application of ionomycin in the presence of extracellular Ca^{2+} or raising the levels of Ca^{2+} in the extracellular space blocked this release, it appears that the release of adenosine from astrocytes under hypoxic conditions is negatively regulated by Ca^{2+}. Although these findings point to a possible involvement of hemichannels or $P2X_7$ receptors, neither gap-junction/hemichannel blockers nor agonists or antagonists of $P2X_7$ receptors affected the release of adenosine. Although by exclusion one would expect that ENTs would play a role in this case, the authors found that ENT blockers did not affect hypoxia-induced release of adenosine from astrocytes. Thus, the underlying mechanism remains elusive. Nonetheless, these data indicate that adenosine can be directly released from astrocytes under hypoxic conditions and, that when combined with the previous findings (Meghji et al., 1989; Parkinson et al., 2002, 2005), both astrocytes and neurons represent a cellular source of adenosine.

As indicated earlier, cAMP can be converted extracellularly to adenosine. This conversion occurs via a two step process, where cAMP is first converted by ecto-phospodiesterease to 5′-AMP and then to adenosine by ecto-5′-nucelotidase [reviewed in (Latini and Pedata, 2001)]. Long applications of cAMP to hippocampal slices to cause large and prolonged increases in extracellular cAMP levels can result in the activation of neuronal adenosine receptors and reduce the amplitude of EPSPs (Brundege et al., 1997) while brief applications of cAMP failed to cause this effect. Since cAMP does not bind to adenosine receptors with high affinity (Schwabe and Trost, 1980) this reduction in synaptic strength was probably due to the conversion of cAMP to adenosine. Indeed, using cortical cultures Rosenberg et al. (1994) have demonstrated that β-adrenergic stimulation of astrocytes can cause the release of cAMP from these cells. Extracellular cAMP in the presence of co-cultured neurons was converted to adenosine, which indicates the presence of functional ecto-phosphodiesterase on neurons since both cell types express ecto-5′-nucleotidase. Astrocytes also released cAMP at rest. Both basal and β-adrenergic-induced cAMP release was inhibited by probenicid, implying that cAMP efflux occurred via the ABC transporter system, similar to MRP (Fig. 12.9). This was in agreement with previous work in C6 rat glioma cells [(Penit et al., 1974; Doore et al., 1975; Rindler et al., 1978; Henderson and Strauss, 1991); note that additionally verapamil was effectively used (Henderson and Strauss, 1991)]. Astrocytes appear to be a likely source of extracellular cAMP in the rat hippocampus (Winder et al., 1996), although that is not the case in all brain regions; for example, neurons appear to be the source of extracellular cAMP in the spinal cord (Sweeney et al., 1988). Taken together, it seems that in the hippocampus, the adenosine that is generated from extracellular cAMP released by astrocytes might also play a role in tonic inhibition. Interestingly, Sorbera and Morad (1991) suggested the existence of an extracellular cAMP binding site on cardiac myocytes leading to the modulation of sodium ion channels. Similar actions of cAMP in the brain, however, have not been described.

Guanosine is present in the brain at relatively high concentrations (Zetterstrom et al., 1982). Its levels increased and stayed elevated for several days following stroke (Uemura et al., 1991). Astrocytes appear to be the major source of guanine-based nucleotides/nucleosides (Ciccarelli et al., 1999). Primary cultured astrocytes at rest release adenine-based purines: ATP, AMP, and adenosine; as well as guanine-based purines: GTP, guanosine 5′ monophospate (GMP) and guanosine, the concentration of which exceeds adenine-based purines. There was a progressive increase in the concentration of all released nucleotides during the time-course of the investigation (3 h); the extracellular levels of nucleosides remained steady, albeit with three times the concentration of guanosine over adenosine. During ischemic/stroke conditions (hypoxia with hypoglycemia) there was a sustained increased release of adenine- and guanine-based purines. Again, guanosine levels were much higher than adenosine. While the mechanisms underlying guanosine release were not investigated, they were speculated upon: guanosine could possibly be derived from extracellular GTP hydrolysis or could be released from the intracellular pool via concentrative N2/cit bidirectional carriers expressed on astrocytes (Gu et al., 1996). Thus, these data imply that during and after ischemic conditions guanosine could be elevated for extended periods of time so that its trophic properties may help in recovery after the insult (Ciccarelli et al., 1999). Whether the basal release of guanosine can play a role in tonic inhibition, as adenosine does, remains to be evaluated.

Stimulation of cortical astrocytes with the C-type of natriuretic peptide caused the release of guanosine 3′:5′ cyclic monophosphate (cGMP). This release was blocked by probenicid and niflumic acid implicating MRP transporters or organic anion channels in mediating the release of this cyclic nucleoside (Figs. 12.1 and 12.9). It appears that released cGMP acts on an extracellular site, blocking the Na^+/H^+ exchanger, leading to acidification of the astrocytic intracellular space (Touyz et al., 1997). Similarly, the inflammatory agent inteleukin-1β stimulated the release of cGMP from the cerebellar, cortical and hippocampal astrocytes. Since this release was sensitive to probenicid and verapamil, it confirms the astrocytic ability to extrude cGMP into the extracellular space via MRP/P glycoprotein ABC transporters (Pedraza et al., 2001). The possible role played by the release of cGMP due to stimulation by inteleukin-1β, which is elevated in a variety of pathological situations in the brain such as infection, trauma and neurodegenerative disease (Rothwell and Luheshi, 2000), can be only hypothesized. One possibility is that it could affect the Na^+/H^+ exchanger as outlined above. Another possibility is it could inhibit kainate receptor action as cGMP is a weak competitive inhibitor of kainate binding (Paas et al., 1996). There is some support for this notion because it has been demonstrated that extracellular cGMP can be neuroprotective against glutamate-induced toxicity (Montoliu et al., 1999) and can inhibit kainate receptor-mediated responses in cerebellar granule cells (Poulopoulou and Nowak, 1998).

Verderio et al. (2001) demonstrated the release of NAD^+ from cultured hippocampal astrocytes. Since Cx43 hemichannels were demonstrated to allow bidirectional NAD^+ flux in Cx43 expressing 3T3 fibroblasts (Bruzzone et al., 2001), they tested Cx43 involvement in astrocytes. To generate a downhill gradient for NAD^+ between the intra- and extra-cellular spaces they used a washing protocol (Zocchi et al., 1999) where the medium bathing the astrocytes was repeatedly removed. Using this approach, they recorded a steady release of NAD^+ from astrocytes into the medium. Additionally, astrocytes were permeable to extracellularly applied NAD^+ and this influx could be inhibited by GA. Thus, it appears that astrocytes are capable of releasing NAD^+ via hemichannels (Fig. 12.2). The authors have also shown that the released NAD^+ can be converted to cyclic ADP-ribose by re-entering astrocytes via CD38, a catalytically active transporter (Franco et al., 1998). The resulting intracellular cADP-ribose caused the release of glutamate from astrocytes, which could then act on nearby neurons. They also postulated the existence of a paracrine pathway where cADP-ribose generated from extracellular NAD^+ could affect neurons. However, in the light of a recent demonstration that NAD^+ is an agonist for $P2Y_1$ receptors (Mutafova-Yambolieva et al., 2007), there could be an even more direct pathway, where NAD^+ released from astrocytes could directly act on neuronal $P2Y_1$ receptors. Additionally, NAD^+ release might also occur via pannexons. Such possibilities, however, are speculative at the moment.

Mechanical stimulation of primary astrocytes, 1321N1 human astrocytoma, and C6 rat glioma cells by media exchange caused the release of ATP and UTP (Lazarowski et al., 1997; Lazarowski and Harden, 1999). The extracellular accumulation of these nucleotides peaked within 1–2 min after the application of the stimulus, followed by decay and this release was not a result of cell lysis. Additionally, the media of mechanically stimulated 1321N1 human astrocytoma cells also contained cytosine 5′ triphosphate (CTP), ADP and GTP. Interestingly, the molar ratio of UTP to ATP detected in the medium was similar to the ratio of these nucleotides detected in the cells, and this was the case regardless of whether evoked (medium change) or basal (undisturbed cells) release was considered [see Table 1 in (Lazarowski and Harden, 1999)]. This implies that these nucleotides are released by a mechanism that merely reflects their relative intracellular concentrations (Lazarowski and Harden, 1999). The basal (constitutive or steady state) release of nucleotides was further studied in 1321N1 human astrocytoma and C6 rat glioma cells, and it appears that the levels of nucleotides in the extracellular space represent a complex interplay between their basal release rate and their hydrolysis (Lazarowski et al., 2000). However, ecto-nucleotidase activity was overpowered by the exchange of γ-phosphate between adenine and guanine/uracil nucleotides via NDPK. Hence, ATP was rapidly, but transiently, converted to GTP and UTP, indicating that there was also a basal release of GDP and UDP. Similarly, a basal release of ADP was detected, since UTP could also be converted to ATP. The data indicate that besides ATP and UTP release there is a comparable extracellular amount of accumulated UDP, ADP and GDP. This gives us only a glimpse into the complexity of nucleotide release, their extracellular regulation and intercellular signaling. This issue is of importance since there is a differential efficacy of nucleotides as agonists on a variety of purinergic receptors (Table 12.2).

The discovery that UDP-glucose is an agonist to a receptor (Chambers et al., 2000), which we now refer to as the $P2Y_{14}$ receptor, indicated that this molecule, besides being a glycosyl donor in the synthesis of carbohydrates, may also serve as a signaling molecule. This finding prompted studies of UDP-glucose release from 1321N1 human astrocytoma cells (Lazarowski et al., 2003b). When mechanically stimulated, these cells exhibited a transient increase in ATP release (~100 nM) followed by a decay where the concentration stabilized at a resting steady-state level (~3 nM), consistent with previous findings (Lazarowski et al., 1997). Additionally, astrocytoma cells also exhibited a release of UDP-glucose, albeit its time-course was quite different. The UDP-glucose level observed within the first few minutes corresponded to only a fraction (~10%) of the ATP preset in the medium, but it then accumulated over time to reach a steady-state level lasting for hours at concentrations (~10 nM) higher than that of the resting ATP level. This demonstrated that UDP-glucose is most likely released constitutively. Consistent with that hypothesis, UDP-glucose was detected in the extracellular medium of resting C6 glioma cells. Although the mechanism underlying the release of UDP-glucose from astrocytoma and glioma cells was not investigated, the authors put forward a hypothesis that this might be a case of constitutive release via vesicular trafficking (Lazarowski et al., 2003a; Lazarowski et al., 2003b), since UDP-glucose is concentrated in the endoplasmic reticulum (Perez and Hirschberg, 1986) and from there, as a cargo molecule, it may end up in vesicles sent to fuse with the plasma membrane. What the role of constitutive release of various nucleotides and UDP-glucose might be is not clear, but they may contribute to tonic control of various cellular functions and/or intercellular signaling.

12.3 Concluding Remarks

The intent of this chapter was to summarize the findings that astrocytes can release a variety of transmitters under physiological and pathological conditions. There are several different mechanisms that can mediate transmitter release from these cells, as summarized in Table 12.1. It is becoming increasingly recognized that release of transmitters from astrocytes plays a role in homo- and hetero-typic intercellular communication in the brain. For example, the astrocytic release of ATP in culture acts as the major signaling molecule underlying intercellular astrocytic Ca^{2+} waves (Guthrie et al., 1999), and also plays a role in hetero-synaptic suppression (Zhang et al., 2003). However, there are issues that remain to be resolved. One issue is that of the localization of transmitter release sites on astrocytes in respect to synaptic sites, since this could determine the nature of astrocytic effects on neurons. There have been attempts to define this, using static images at the level of EM (Bezzi et al., 2004). While such studies are informative, this issue will require a comprehensive approach employing imaging techniques using live cells in culture, acute slices and in vivo studies. Additionally, it will be necessary to catalogue whether the same transmitter release mechanisms that operate under physiological conditions operate during pathological conditions or whether there are specific release mechanisms that operate under particular conditions. For example, physiological adenosine-mediated astrocyte–neuron signaling appears to be via SNARE-dependent ATP release from astrocytes followed by extracellular conversion to adenosine, while in hypoxia, adenosine is directly released via a mechanism that is not dependent on SNARE proteins. Further complexity may arise from multiple mechanisms operating in parallel. For example, it appears that in physiological conditions, scenarios can occur in cultured astrocytes (Abdipranoto et al., 2003) where ATP is released via different mechanisms, each contributing a different portion to the total amount of ATP being released. It could be that under certain pathophysiological conditions these mechanisms could operate together at all times in astrocytes, but with increased activity. The major obstacle in systematically conducting studies to address these issues is the paucity of specific reagents and methods for discerning individual release pathways. Additionally, the identity of molecules underlying some release mechanisms is not defined; for example VRAC is a "functional" rather than a molecular definition. Consequently, identifying genes and their products that are responsible for mediating the astrocytic transmitter release and the development of better pharmacological agents will be necessary. So far the most developed pharmacology is in respect to the exocytic pathway. The novel genetic approaches using astrocyte specific conditional expression (Pascual et al., 2005) or deletion (Wiencken-Barger et al., 2007) of gene products provide an excellent resource to start testing the contribution of different transmitter release pathways in astrocytes in physiology and pathology in vivo.

Acknowledgments The authors' work is supported by a grant from the National Institute of Mental Health (R01 MH 069791). We thank Drs. Eliana Scemes and David C. Spray for comments on a previous version of this manuscript.

References

Abdipranoto A, Liu GJ, Werry EL, Bennett MR (2003) Mechanisms of secretion of ATP from cortical astrocytes triggered by uridine triphosphate. *Neuroreport* 14:2177–2181

Abraham EH, Prat AG, Gerweck L, Seneveratne T, Arceci RJ, Kramer R, Guidotti G, Cantiello HF (1993) The multidrug resistance (mdr1) gene product functions as an ATP channel. *Proc Natl Acad Sci USA* 90:312–316

Allen JW, Shanker G, Aschner M (2001) Methylmercury inhibits the in vitro uptake of the glutathione precursor, cystine, in astrocytes, but not in neurons. *Brain Res* 894:131–140

Allen NJ, Rossi DJ, Attwell D (2004) Sequential release of GABA by exocytosis and reversed uptake leads to neuronal swelling in simulated ischemia of hippocampal slices. *J Neurosci* 24:3837–3849

Anderson CM, Parkinson FE (1997) Potential signalling roles for UTP and UDP: Sources, regulation and release of uracil nucleotides. *Trends Pharmacol Sci* 18:387–392

Anderson CM, Swanson RA (2000) Astrocyte glutamate transport: Review of properties, regulation, and physiological functions. *Glia* 32:1–14

Anderson CM, Bridges RJ, Chamberlin AR, Shimamoto K, Yasuda-Kamatani Y, Swanson RA (2001) Differing effects of substrate and non-substrate transport inhibitors on glutamate uptake reversal. *J Neurochem* 79:1207–1216

Anderson CM, Bergher JP, Swanson RA (2004) ATP-induced ATP release from astrocytes. *J Neurochem* 88:246–256

Angulo MC, Kozlov AS, Charpak S, Audinat E (2004) Glutamate released from glial cells synchronizes neuronal activity in the hippocampus. *J Neurosci* 24:6920–6927

Anlauf E, Derouiche A (2005) Astrocytic exocytosis vesicles and glutamate: A high-resolution immunofluorescence study. *Glia* 49:96–106

Araque A, Parpura V, Sanzgiri RP, Haydon PG (1998a) Glutamate-dependent astrocyte modulation of synaptic transmission between cultured hippocampal neurons. *Eur J Neurosci* 10:2129–2142

Araque A, Sanzgiri RP, Parpura V, Haydon PG (1998b) Calcium elevation in astrocytes causes an NMDA receptor-dependent increase in the frequency of miniature synaptic currents in cultured hippocampal neurons. *J Neurosci* 18:6822–6829

Araque A, Li N, Doyle RT, Haydon PG (2000) SNARE protein-dependent glutamate release from astrocytes. *J Neurosci* 20:666–673

Arcuino G, Lin JH, Takano T, Liu C, Jiang L, Gao Q, Kang J, Nedergaard M (2002) Intercellular calcium signaling mediated by point-source burst release of ATP. *Proc Natl Acad Sci USA* 99:9840–9845

Axelrod J (1974) Neurotransmitters. *Sci Am* 230:59–71

Bak LK, Schousboe A, Waagepetersen HS (2006) The glutamate/GABA-glutamine cycle: Aspects of transport, neurotransmitter homeostasis and ammonia transfer. *J Neurochem* 98:641–653

Baker DA, Xi ZX, Shen H, Swanson CJ, Kalivas PW (2002) The origin and neuronal function of in vivo nonsynaptic glutamate. *J Neurosci* 22:9134–9141

Baldwin SA, Beal PR, Yao SY, King AE, Cass CE, Young JD (2004) The equilibrative nucleoside transporter family, SLC29. *Pflugers Arch* 447:735–743

Ballerini P, Rathbone MP, Di Iorio P, Renzetti A, Giuliani P, D'Alimonte I, Trubiani O, Caciagli F, Ciccarelli R (1996) Rat astroglial P2Z (P2X7) receptors regulate intracellular calcium and purine release. *Neuroreport* 7:2533–2537

Ballerini P, Di Iorio P, Ciccarelli R, Nargi E, D'Alimonte I, Traversa U, Rathbone MP, Caciagli F (2002) Glial cells express multiple ATP binding cassette proteins which are involved in ATP release. *Neuroreport* 13:1789–1792

Bal-Price A, Moneer Z, Brown GC (2002) Nitric oxide induces rapid, calcium-dependent release of vesicular glutamate and ATP from cultured rat astrocytes. *Glia* 40:312–323

Bao L, Locovei S, Dahl G (2004) Pannexin membrane channels are mechanosensitive conduits for ATP. *FEBS Lett* 572:65–68

Baron BM, Dudley MW, McCarty DR, Miller FP, Reynolds IJ, Schmidt CJ (1989) Guanine nucleotides are competitive inhibitors of *N*-methyl-d-aspartate at its receptor site both in vitro and in vivo. *J Pharmacol Exp Ther* 250:162–169

Basarsky TA, Feighan D, MacVicar BA (1999) Glutamate release through volume-activated channels during spreading depression. *J Neurosci* 19:6439–6445

Bender AS, Reichelt W, Norenberg MD (2000) Characterization of cystine uptake in cultured astrocytes. *Neurochem Int* 37:269–276

Benz B, Grima G, Do KQ (2004) Glutamate-induced homocysteic acid release from astrocytes: Possible implication in glia–neuron signaling. *Neuroscience* 124:377–386
Bergles DE, Jahr CE (1997) Synaptic activation of glutamate transporters in hippocampal astrocytes. *Neuron* 19:1297–1308
Bezzi P, Carmignoto G, Pasti L, Vesce S, Rossi D, Rizzini BL, Pozzan T, Volterra A (1998) Prostaglandins stimulate calcium-dependent glutamate release in astrocytes. *Nature* 391:281–285
Bezzi P, Domercq M, Brambilla L, Galli R, Schols D, De Clercq E, Vescovi A, Bagetta G, Kollias G, Meldolesi J, Volterra A (2001) CXCR4-activated astrocyte glutamate release via TNFα: Amplification by microglia triggers neurotoxicity. *Nat Neurosci* 4:702–710
Bezzi P, Gundersen V, Galbete JL, Seifert G, Steinhauser C, Pilati E, Volterra A (2004) Astrocytes contain a vesicular compartment that is competent for regulated exocytosis of glutamate. *Nat Neurosci* 7:613–620
Bianco F, Pravettoni E, Colombo A, Schenk U, Moller T, Matteoli M, Verderio C (2005) Astrocyte-derived ATP induces vesicle shedding and IL-1 beta release from microglia. *J Immunol* 174:7268–7277
Boehning D, Snyder SH (2003) Novel neural modulators. *Annu Rev Neurosci* 26:105–131
Borgland SL, Parkinson FE (1997) Uptake and release of [3H]formycin B via sodium-dependent nucleoside transporters in mouse leukemic L1210/MA27.1 cells. *J Pharmacol Exp Ther* 281:347–353
Bowery NG, Brown DA, Collins GG, Galvan M, Marsh S, Yamini G (1976) Indirect effects of amino-acids on sympathetic ganglion cells mediated through the release of gamma-aminobutyric acid from glial cells. *Br J Pharmacol* 57:73–91
Bowser DN, Khakh BS (2007) Two forms of single-vesicle astrocyte exocytosis imaged with total internal reflection fluorescence microscopy. *Proc Natl Acad Sci USA* 104:4212–4217
Bres V, Hurbin A, Duvoid A, Orcel H, Moos FC, Rabie A, Hussy N (2000) Pharmacological characterization of volume-sensitive, taurine permeable anion channels in rat supraoptic glial cells. *Br J Pharmacol* 130:1976–1982
Brundege JM, Diao L, Proctor WR, Dunwiddie TV (1997) The role of cyclic AMP as a precursor of extracellular adenosine in the rat hippocampus. *Neuropharmacology* 36:1201–1210
Bruzzone S, Guida L, Zocchi E, Franco L, De Flora A (2001) Connexin 43 hemi channels mediate Ca2+-regulated transmembrane NAD+ fluxes in intact cells. *Faseb J* 15:10–12
Bruzzone R, Barbe MT, Jakob NJ, Monyer H (2005) Pharmacological properties of homomeric and heteromeric pannexin hemichannels expressed in Xenopus oocytes. *J Neurochem* 92:1033–1043
Caciagli F, Ciccarelli R, Di Iorio P, Ballerini P, Tacconelli L (1988) Cultures of glial cells release purines under field electrical stimulation: The possible ionic mechanisms. *Pharmacol Res Commun* 20:935–947
Calegari F, Coco S, Taverna E, Bassetti M, Verderio C, Corradi N, Matteoli M, Rosa P (1999) A regulated secretory pathway in cultured hippocampal astrocytes. *J Biol Chem* 274:22539–22547
Cardin V, Lezama R, Torres-Marquez ME, Pasantes-Morales H (2003) Potentiation of the osmosensitive taurine release and cell volume regulation by cytosolic Ca^{2+} rise in cultured cerebellar astrocytes. *Glia* 44:119–128
Cavelier P, Attwell D (2005) Tonic release of glutamate by a DIDS-sensitive mechanism in rat hippocampal slices. *J Physiol* 564:397–410
Chambers JK, Macdonald LE, Sarau HM, Ames RS, Freeman K, Foley JJ, Zhu Y, McLaughlin MM, Murdock P, McMillan L, Trill J, Swift A, Aiyar N, Taylor P, Vawter L, Naheed S, Szekeres P, Hervieu G, Scott C, Watson JM, Murphy AJ, Duzic E, Klein C, Bergsma DJ, Wilson S, Livi GP (2000) A G protein-coupled receptor for UDP-glucose. *J Biol Chem* 275:10767–10771
Chase LA, Peterson NL, Koerner JF (2007) The lathyrus toxin, beta-*N*-oxalyl-l-alpha,beta-diaminopropionic acid (ODAP), and homocysteic acid sensitize CA1 pyramidal neurons to cystine and l-2-amino-6-phosphonohexanoic acid. *Toxicol Appl Pharmacol* 219:1–9
Cheema TA, Ward CE, Fisher SK (2005) Subnanomolar concentrations of thrombin enhance the volume-sensitive efflux of taurine from human 1321N1 astrocytoma cells. *J Pharmacol Exp Ther* 315:755–763

Chen X, Wang L, Zhou Y, Zheng LH, Zhou Z (2005) "Kiss-and-run" glutamate secretion in cultured and freshly isolated rat hippocampal astrocytes. *J Neurosci* 25:9236–9243

Chung WJ, Lyons SA, Nelson GM, Hamza H, Gladson CL, Gillespie GY, Sontheimer H (2005) Inhibition of cystine uptake disrupts the growth of primary brain tumors. *J Neurosci* 25:7101–7110

Ciccarelli R, Di Iorio P, Giuliani P, D'Alimonte I, Ballerini P, Caciagli F, Rathbone MP (1999) Rat cultured astrocytes release guanine-based purines in basal conditions and after hypoxia/hypoglycemia. *Glia* 25:93–98

Coco S, Calegari F, Pravettoni E, Pozzi D, Taverna E, Rosa P, Matteoli M, Verderio C (2003) Storage and release of ATP from astrocytes in culture. *J Biol Chem* 278:1354–1362

Contreras JE, Sanchez HA, Eugenin EA, Speidel D, Theis M, Willecke K, Bukauskas FF, Bennett MV, Saez JC (2002) Metabolic inhibition induces opening of unapposed connexin 43 gap junction hemichannels and reduces gap junctional communication in cortical astrocytes in culture. *Proc Natl Acad Sci USA* 99:495–500

Contreras JE, Saez JC, Bukauskas FF, Bennett MV (2003) Gating and regulation of connexin 43 (Cx43) hemichannels. *Proc Natl Acad Sci USA* 100:11388–11393

Cotrina ML, Lin JH, Alves-Rodrigues A, Liu S, Li J, Azmi-Ghadimi H, Kang J, Naus CC, Nedergaard M (1998) Connexins regulate calcium signaling by controlling ATP release. *Proc Natl Acad Sci USA* 95:15735–15740

Cotrina ML, Lin JH, Lopez-Garcia JC, Naus CC, Nedergaard M (2000) ATP-mediated glia signaling. *J Neurosci* 20:2835–2844

Crepel V, Panenka W, Kelly ME, MacVicar BA (1998) Mitogen-activated protein and tyrosine kinases in the activation of astrocyte volume-activated chloride current. *J Neurosci* 18:1196–1206

Crippa D, Schenk U, Francolini M, Rosa P, Verderio C, Zonta M, Pozzan T, Matteoli M, Carmignoto G (2006) Synaptobrevin2-expressing vesicles in rat astrocytes: Insights into molecular characterization, dynamics and exocytosis. *J Physiol* 570:567–582

Cuenod M, Do KQ, Herrling PL, Turski WA, Matute C, Streit P (1986) Homocysteic acid, an endogenous agonist of NMDA-receptor: Release, neuroactivity and localization. *Adv Exp Med Biol* 203:253–262

Cuenod M, Grandes P, Zangerle L, Streit P, Do KQ (1993) Sulphur-containing excitatory amino acids in intercellular communication. *Biochem Soc Trans* 21:72–77

Curran MJ, Brodwick MS (1991) Ionic control of the size of the vesicle matrix of beige mouse mast cells. *J Gen Physiol* 98:771–790

Dall'Asta V, Gazzola GC, Franchi-Gazzola R, Bussolati O, Longo N, Guidotti GG (1983) Pathways of l-glutamic acid transport in cultured human fibroblasts. *J Biol Chem* 258:6371–6379

Danbolt NC (2001) Glutamate uptake. *Prog Neurobiol* 65:1–105

Darby M, Kuzmiski JB, Panenka W, Feighan D, MacVicar BA (2003) ATP released from astrocytes during swelling activates chloride channels. *J Neurophysiol* 89:1870–1877

D'Ascenzo M, Fellin T, Terunuma M, Revilla-Sanchez R, Meaney DF, Auberson YP, Moss SJ, Haydon PG (2007) mGluR5 stimulates gliotransmission in the nucleus accumbens. *Proc Natl Acad Sci USA* 104:1995–2000

Decleves X, Regina A, Laplanche JL, Roux F, Boval B, Launay JM, Scherrmann JM (2000) Functional expression of P-glycoprotein and multidrug resistance-associated protein (Mrp1) in primary cultures of rat astrocytes. *J Neurosci Res* 60:594–601

Del Castillo J, Katz B (1954) Quantal components of the end-plate potential. *J Physiol* 124:560–573

Deleuze C, Duvoid A, Hussy N (1998) Properties and glial origin of osmotic-dependent release of taurine from the rat supraoptic nucleus. *J Physiol* 507 (Pt 2):463–471

Deleuze C, Duvoid A, Moos FC, Hussy N (2000) Tyrosine phosphorylation modulates the osmosensitivity of volume-dependent taurine efflux from glial cells in the rat supraoptic nucleus. *J Physiol 523 Pt* 2:291–299

Dermietzel R, Gao Y, Scemes E, Vieira D, Urban M, Kremer M, Bennett MV, Spray DC (2000) Connexin43 null mice reveal that astrocytes express multiple connexins. *Brain Res Brain Res Rev* 32:45–56

Dev KK, Roberts PJ, Henley JM (1996) Characterisation of the interaction between guanyl nucleotides and AMPA receptors in rat brain. *Neuropharmacology* 35:1583–1593

Do KQ, Mattenberger M, Streit P, Cuenod M (1986) In vitro release of endogenous excitatory sulfur-containing amino acids from various rat brain regions. *J Neurochem* 46:779–786

Do KQ, Benz B, Sorg O, Pellerin L, Magistretti PJ (1997) beta-Adrenergic stimulation promotes homocysteic acid release from astrocyte cultures: Evidence for a role of astrocytes in the modulation of synaptic transmission. *J Neurochem* 68:2386–2394

Domercq M, Brambilla L, Pilati E, Marchaland J, Volterra A, Bezzi P (2006) P2Y1 receptor-evoked glutamate exocytosis from astrocytes: Control by tumor necrosis factor-alpha and prostaglandins. *J Biol Chem* 281:30684–30696

Dominy J, Eller S, Dawson R, Jr. (2004) Building biosynthetic schools: Reviewing compartmentation of CNS taurine synthesis. *Neurochem Res* 29:97–103

Doore BJ, Bashor MM, Spitzer N, Mawe RC, Saier MH, Jr. (1975) Regulation of adenosine 3′:5′-monophosphate efflux from rat glioma cells in culture. *J Biol Chem* 250:4371–4372

Duan S, Anderson CM, Keung EC, Chen Y, Chen Y, Swanson RA (2003) P2X7 receptor-mediated release of excitatory amino acids from astrocytes. *J Neurosci* 23:1320–1328

Dun Y, Mysona B, Itagaki S, Martin-Studdard A, Ganapathy V, Smith SB (2007) Functional and molecular analysis of d-serine transport in retinal Muller cells. *Exp Eye Res* 84:191–199

Ellershaw DC, Greenwood IA, Large WA (2000) Dual modulation of swelling-activated chloride current by NO and NO donors in rabbit portal vein myocytes. *J Physiol 528(Pt* 1):15–24

Eskandari S, Zampighi GA, Leung DW, Wright EM, Loo DD (2002) Inhibition of gap junction hemichannels by chloride channel blockers. *J Membr Biol* 185:93–102

Evanko DS, Sul J-Y, Zhang Q, Haydon PG (2004a) The regulated release of transmitters from astrocytes. In: Glial Neuronal Signaling (Hatton GI, Parpura V, eds), pp. 397–416. Boston, MA: Kluwer

Evanko DS, Zhang Q, Zorec R, Haydon PG (2004b) Defining pathways of loss and secretion of chemical messengers from astrocytes. *Glia* 47:233–240

Fabbretti E, Sokolova E, Masten L, D'Arco M, Fabbro A, Nistri A, Giniatullin R (2004) Identification of negative residues in the P2X3 ATP receptor ectodomain as structural determinants for desensitization and the Ca^{2+}-sensing modulatory sites. *J Biol Chem* 279:53109–53115

Fellin T, Pascual O, Gobbo S, Pozzan T, Haydon PG, Carmignoto G (2004) Neuronal synchrony mediated by astrocytic glutamate through activation of extrasynaptic NMDA receptors. *Neuron* 43:729–743

Fellin T, Pozzan T, Carmignoto G (2006) Purinergic receptors mediate two distinct glutamate release pathways in hippocampal astrocytes. *J Biol Chem* 281:4274–4284

Fernandez JM, Villalon M, Verdugo P (1991) Reversible condensation of mast cell secretory products in vitro. *Biophys J* 59:1022–1027

Fiacco TA, McCarthy KD (2004) Intracellular astrocyte calcium waves in situ increase the frequency of spontaneous AMPA receptor currents in CA1 pyramidal neurons. *J Neurosci* 24:722–732

Fiacco TA, Agulhon C, Taves SR, Petravicz J, Casper KB, Dong X, Chen J, McCarthy KD (2007) Selective stimulation of astrocyte calcium in situ does not affect neuronal excitatory synaptic activity. *Neuron* 54:611–626

Fields RD, Burnstock G (2006) Purinergic signalling in neuron–glia interactions. *Nat Rev Neurosci* 7:423–436

Franco L, Guida L, Bruzzone S, Zocchi E, Usai C, De Flora A (1998) The transmembrane glycoprotein CD38 is a catalytically active transporter responsible for generation and influx of the second messenger cyclic ADP-ribose across membranes. *FASEB J* 12:1507–1520

Fremeau RT, Jr.,, Burman J, Qureshi T, Tran CH, Proctor J, Johnson J, Zhang H, Sulzer D, Copenhagen DR, Storm-Mathisen J, Reimer RJ, Chaudhry FA, Edwards RH (2002) The identification of vesicular glutamate transporter 3 suggests novel modes of signaling by glutamate. *Proc Natl Acad Sci USA* 99:14488–14493

Fumagalli M, Brambilla R, D'Ambrosi N, Volonte C, Matteoli M, Verderio C, Abbracchio MP (2003) Nucleotide-mediated calcium signaling in rat cortical astrocytes: Role of P2X and P2Y receptors. *Glia* 43:218–203

Gadea A, Lopez-Colome AM (2001a) Glial transporters for glutamate, glycine, and GABA: II. GABA transporters. *J Neurosci Res* 63:461–468

Gadea A, Lopez-Colome AM (2001b) Glial transporters for glutamate, glycine and GABA I. *Glutamate transporters. J Neurosci Res* 63:453–460

Gadsby DC, Vergani P, Csanady L (2006) The ABC protein turned chloride channel whose failure causes cystic fibrosis. *Nature* 440:477–483

Gallo V, Suergiu R, Levi G (1986) Kainic acid stimulates GABA release from a subpopulation of cerebellar astrocytes. *Eur J Pharmacol* 132:319–322

Gallo V, Giovannini C, Suergiu R, Levi G (1989) Expression of excitatory amino acid receptors by cerebellar cells of the type-2 astrocyte cell lineage. *J Neurochem* 52:1–9

Gallo V, Patrizio M, Levi G (1991) GABA release triggered by the activation of neuron-like non-NMDA receptors in cultured type 2 astrocytes is carrier-mediated. *Glia* 4:245–255

Gonzalez A, Granados MP, Pariente JA, Salido GM (2006) H2O2 mobilizes Ca^{2+} from agonist- and thapsigargin-sensitive and insensitive intracellular stores and stimulates glutamate secretion in rat hippocampal astrocytes. *Neurochem Res* 31:741–750

Gordon GR, Baimoukhametova DV, Hewitt SA, Rajapaksha WR, Fisher TE, Bains JS (2005) Norepinephrine triggers release of glial ATP to increase postsynaptic efficacy. *Nat Neurosci* 8:1078–1086

Gorodinsky A, Paas Y, Teichberg VI (1993) A ligand binding study of the interactions of guanine nucleotides with non-NMDA receptors. *Neurochem Int* 23:285–291

Gray JH, Owen RP, Giacomini KM (2004) The concentrative nucleoside transporter family, SLC28. *Pflugers Arch* 447:728–734

Gu JG, Nath A, Geiger JD (1996) Characterization of inhibitor-sensitive and -resistant adenosine transporters in cultured human fetal astrocytes. *J Neurochem* 67:972–977

Guthrie PB, Knappenberger J, Segal M, Bennett MV, Charles AC, Kater SB (1999) ATP released from astrocytes mediates glial calcium waves. *J Neurosci* 19:520–528

Haskew RE, Mongin AA, Kimelberg HK (2002) Peroxynitrite enhances astrocytic volume-sensitive excitatory amino acid release via a src tyrosine kinase-dependent mechanism. *J Neurochem* 82:903–912

Haskew-Layton RE, Mongin AA, Kimelberg HK (2005) Hydrogen peroxide potentiates volume-sensitive excitatory amino acid release via a mechanism involving Ca^{2+}/calmodulin-dependent protein kinase II. *J Biol Chem* 280:3548–3554

Hassinger TD, Atkinson PB, Strecker GJ, Whalen LR, Dudek FE, Kossel AH, Kater SB (1995) Evidence for glutamate-mediated activation of hippocampal neurons by glial calcium waves. *J Neurobiol* 28:159–170

Henderson GB, Strauss BP (1991) Evidence for cAMP and cholate extrusion in C6 rat glioma cells by a common anion efflux pump. *J Biol Chem* 266:1641–1645

Hepp R, Perraut M, Chasserot-Golaz S, Galli T, Aunis D, Langley K, Grant NJ (1999) Cultured glial cells express the SNAP-25 analogue SNAP-23. *Glia* 27:181–187

Hertz L, Dringen R, Schousboe A, Robinson SR (1999) Astrocytes: Glutamate producers for neurons. *J Neurosci Res* 57:417–428

Hisadome K, Koyama T, Kimura C, Droogmans G, Ito Y, Oike M (2002) Volume-regulated anion channels serve as an auto/paracrine nucleotide release pathway in aortic endothelial cells. *J Gen Physiol* 119:511–520

Hofer A, Dermietzel R (1998) Visualization and functional blocking of gap junction hemichannels (connexons) with antibodies against external loop domains in astrocytes. *Glia* 24:141–154

Hua X, Malarkey EB, Sunjara V, Rosenwald SE, Li WH, Parpura V (2004) Ca^{2+}-dependent glutamate release involves two classes of endoplasmic reticulum Ca^{2+} stores in astrocytes. *J Neurosci Res* 76:86–97

Huang YJ, Maruyama Y, Dvoryanchikov G, Pereira E, Chaudhari N, Roper SD (2007) The role of pannexin 1 hemichannels in ATP release and cell–cell communication in mouse taste buds. *Proc Natl Acad Sci USA* 104:6436–6441

Hussy N, Deleuze C, Pantaloni A, Desarmenien MG, Moos F (1997) Agonist action of taurine on glycine receptors in rat supraoptic magnocellular neurones: Possible role in osmoregulation. *J Physiol* 502 (Pt 3):609–621

Innocenti B, Parpura V, Haydon PG (2000) Imaging extracellular waves of glutamate during calcium signaling in cultured astrocytes. *J Neurosci* 20:1800–1808

Jabs R, Matthias K, Grote A, Grauer M, Seifert G, Steinhauser C (2007) Lack of P2X receptor mediated currents in astrocytes and GluR type glial cells of the hippocampal CA1 region. *Glia* 55:1648–1655

Jackson PS, Strange K (1993) Volume-sensitive anion channels mediate swelling-activated inositol and taurine efflux. *Am J Physiol* 265:C1489–1500

Jeftinija SD, Jeftinija KV, Stefanovic G, Liu F (1996) Neuroligand-evoked calcium-dependent release of excitatory amino acids from cultured astrocytes. *J Neurochem* 66:676–684

Jeftinija SD, Jeftinija KV, Stefanovic G (1997) Cultured astrocytes express proteins involved in vesicular glutamate release. *Brain Res* 750:41–47

Jeremic A, Jeftinija K, Stevanovic J, Glavaski A, Jeftinija S (2001) ATP stimulates calcium-dependent glutamate release from cultured astrocytes. *J Neurochem* 77:664–675

Joseph SM, Buchakjian MR, Dubyak GR (2003) Colocalization of ATP release sites and ecto-ATPase activity at the extracellular surface of human astrocytes. *J Biol Chem* 278:23331–23342

Jourdain P, Bergersen LH, Bhaukaurally K, Bezzi P, Santello M, Domercq M, Matute C, Tonello F, Gundersen V, Volterra A (2007) Glutamate exocytosis from astrocytes controls synaptic strength. *Nat Neurosci* 10:331–339

Jow F, Chiu D, Lim HK, Novak T, Lin S (2004) Production of GABA by cultured hippocampal glial cells. *Neurochem Int* 45:273–283

Kanematsu S, Ishii S, Umino A, Fujihira T, Kashiwa A, Yamamoto N, Kurumaji A, Nishikawa T (2006) Evidence for involvement of glial cell activity in the control of extracellular d-serine contents in the rat brain. *J Neural Transm* 113:1717–1721

Kang N, Xu J, Xu Q, Nedergaard M, Kang J (2005) Astrocytic glutamate release-induced transient depolarization and epileptiform discharges in hippocampal CA1 pyramidal neurons. *J Neurophysiol* 94:4121–4130

Kimelberg HK (2004) Increased release of excitatory amino acids by the actions of ATP and peroxynitrite on volume-regulated anion channels (VRACs) in astrocytes. *Neurochem Int* 45:511–519

Kimelberg HK, Goderie SK, Higman S, Pang S, Waniewski RA (1990) Swelling-induced release of glutamate, aspartate, and taurine from astrocyte cultures. *J Neurosci* 10:1583–1591

Kimelberg HK, Macvicar BA, Sontheimer H (2006) Anion channels in astrocytes: Biophysics, pharmacology, and function. *Glia* 54:747–757

Kingston AE, Lowndes J, Evans N, Clark B, Tomlinson R, Burnett JP, Mayne NG, Cockerham SL, Lodge D (1998) Sulphur-containing amino acids are agonists for group 1 metabotropic receptors expressed in clonal RGT cell lines. *Neuropharmacology* 37:277–287

Kinney GA, Spain WJ (2002) Synaptically evoked GABA transporter currents in neocortical glia. *J Neurophysiol* 88:2899–2908

Kozlov AS, Angulo MC, Audinat E, Charpak S (2006) Target cell-specific modulation of neuronal activity by astrocytes. *Proc Natl Acad Sci USA* 103:10058–10063

Kreft M, Stenovec M, Rupnik M, Grilc S, Krzan M, Potokar M, Pangrsic T, Haydon PG, Zorec R (2004) Properties of Ca2+-dependent exocytosis in cultured astrocytes. *Glia* 46:437–445

Krzan M, Stenovec M, Kreft M, Pangrsic T, Grilc S, Haydon PG, Zorec R (2003) Calcium-dependent exocytosis of atrial natriuretic peptide from astrocytes. *J Neurosci* 23:1580–1583

Kukley M, Barden JA, Steinhauser C, Jabs R (2001) Distribution of P2X receptors on astrocytes in juvenile rat hippocampus. *Glia* 36:11–21

Lai JC, Murthy CR, Cooper AJ, Hertz E, Hertz L (1989) Differential effects of ammonia and beta-methylene-dl-aspartate on metabolism of glutamate and related amino acids by astrocytes and neurons in primary culture. *Neurochem Res* 14:377–389

Lai CP, Bechberger JF, Thompson RJ, MacVicar BA, Bruzzone R, Naus CC (2007) Tumor-suppressive effects of pannexin 1 in C6 glioma cells. *Cancer Res* 67:1545–1554

Latini S, Pedata F (2001) Adenosine in the central nervous system: Release mechanisms and extracellular concentrations. *J Neurochem* 79:463–484

Lazarowski ER, Harden TK (1999) Quantitation of extracellular UTP using a sensitive enzymatic assay. *Br J Pharmacol* 127:1272–1278

Lazarowski ER, Homolya L, Boucher RC, Harden TK (1997) Direct demonstration of mechanically induced release of cellular UTP and its implication for uridine nucleotide receptor activation. *J Biol Chem* 272:24348–24354

Lazarowski ER, Boucher RC, Harden TK (2000) Constitutive release of ATP and evidence for major contribution of ecto-nucleotide pyrophosphatase and nucleoside diphosphokinase to extracellular nucleotide concentrations. *J Biol Chem* 275:31061–31068

Lazarowski ER, Boucher RC, Harden TK (2003a) Mechanisms of release of nucleotides and integration of their action as P2X- and P2Y-receptor activating molecules. *Mol Pharmacol* 64:785–795

Lazarowski ER, Shea DA, Boucher RC, Harden TK (2003b) Release of cellular UDP-glucose as a potential extracellular signaling molecule. *Mol Pharmacol* 63:1190–1197

Lee CJ, Mannaioni G, Yuan H, Woo DH, Gingrich MB, Traynelis SF (2007a) Astrocytic control of synaptic NMDA receptors. *J Physiol* 581:1057–1081

Lee TS, Mane S, Eid T, Zhao H, Lin A, Guan Z, Kim JH, Schweitzer J, King-Stevens D, Weber P, Spencer SS, Spencer DD, de Lanerolle NC (2007b) Gene expression in temporal lobe epilepsy is consistent with increased release of glutamate by astrocytes. *Mol Med* 13:1–13

Li C, Ramjeesingh M, Bear CE (1996) Purified cystic fibrosis transmembrane conductance regulator (CFTR) does not function as an ATP channel. *J Biol Chem* 271:11623–11626

Li S, Mealing GA, Morley P, Stys PK (1999) Novel injury mechanism in anoxia and trauma of spinal cord white matter: Glutamate release via reverse Na+-dependent glutamate transport. *J Neurosci* 19:RC16

Lin JH, Lou N, Kang N, Takano T, Hu F, Han X, Xu Q, Lovatt D, Torres A, Willecke K, Yang J, Kang J, Nedergaard M (2008) A central role of connexin 43 in hypoxic preconditioning. *J Neurosci* 28:681–695

Liu QY, Schaffner AE, Chang YH, Maric D, Barker JL (2000) Persistent activation of GABAA receptor/Cl– channels by astrocyte-derived GABA in cultured embryonic rat hippocampal neurons. *J Neurophysiol* 84:1392–1403

Liu QS, Xu Q, Kang J, Nedergaard M (2004a) Astrocyte activation of presynaptic metabotropic glutamate receptors modulates hippocampal inhibitory synaptic transmission. *Neuron Glia Biol* 1:307–316

Liu QS, Xu Q, Arcuino G, Kang J, Nedergaard M (2004b) Astrocyte-mediated activation of neuronal kainate receptors. *Proc Natl Acad Sci USA* 101:3172–3177

Liu HT, Tashmukhamedov BA, Inoue H, Okada Y, Sabirov RZ (2006) Roles of two types of anion channels in glutamate release from mouse astrocytes under ischemic or osmotic stress. *Glia* 54:343–357

Locovei S, Wang J, Dahl G (2006) Activation of pannexin 1 channels by ATP through P2Y receptors and by cytoplasmic calcium. *FEBS Lett* 580:239–244

Locovei S, Scemes E, Qiu F, Spray DC, Dahl G (2007) Pannexin1 is part of the pore forming unit of the P2X7 receptor death complex. *FEBS Lett* 581:483–488

Longuemare MC, Swanson RA (1995) Excitatory amino acid release from astrocytes during energy failure by reversal of sodium-dependent uptake. *J Neurosci Res* 40:379–386

Longuemare MC, Swanson RA (1997) Net glutamate release from astrocytes is not induced by extracellular potassium concentrations attainable in brain. *J Neurochem* 69:879–882

Longuemare MC, Rose CR, Farrell K, Ransom BR, Waxman SG, Swanson RA (1999) K(+)-induced reversal of astrocyte glutamate uptake is limited by compensatory changes in intracellular Na+. *Neuroscience* 93:285–292

Lugaro E (1907) Sulle funzioni della nevroglia. *Riv Pat Nerv Ment* 12:225–233

Maienschein V, Marxen M, Volknandt W, Zimmermann H (1999) A plethora of presynaptic proteins associated with ATP-storing organelles in cultured astrocytes. *Glia* 26:233–244

Malarkey EB, Parpura V (2008) Mechanisms of glutamate release from astrocytes. *Neurochem Int* 52:142–154

Martin ED, Fernandez M, Perea G, Pascual O, Haydon PG, Araque A, Cena V (2007) Adenosine released by astrocytes contributes to hypoxia-induced modulation of synaptic transmission. *Glia* 55:36–45

McBean GJ (2002) Cerebral cystine uptake: A tale of two transporters. *Trends Pharmacol Sci* 23:299–302

Meghji P, Tuttle JB, Rubio R (1989) Adenosine formation and release by embryonic chick neurons and glia in cell culture. *J Neurochem* 53:1852–1860

Melin P, Hosy E, Vivaudou M, Becq F (2007) CFTR inhibition by glibenclamide requires a positive charge in cytoplasmic loop three. *Biochim Biophys Acta* 1768:2438–2446

Miyata S, Matsushima O, Hatton GI (1997) Taurine in rat posterior pituitary: Localization in astrocytes and selective release by hypoosmotic stimulation. *J Comp Neurol* 381:513–523

Monahan JB, Hood WF, Michel J, Compton RP (1988) Effects of guanine nucleotides on *N*-methyl-d-aspartate receptor-ligand interactions. *Mol Pharmacol* 34:111–116

Mongin AA, Kimelberg HK (2002) ATP potently modulates anion channel-mediated excitatory amino acid release from cultured astrocytes. *Am J Physiol Cell Physiol* 283:C569–C578

Mongin AA, Kimelberg HK (2005) ATP regulates anion channel-mediated organic osmolyte release from cultured rat astrocytes via multiple Ca^{2+}-sensitive mechanisms. *Am J Physiol Cell Physiol* 288:C204–C213

Mongin AA, Orlov SN (2001) Mechanisms of cell volume regulation and possible nature of the cell volume sensor. *Pathophysiology* 8:77–88

Mongin AA, Cai Z, Kimelberg HK (1999a) Volume-dependent taurine release from cultured astrocytes requires permissive $[Ca^{2+}]i$ and calmodulin. *Am J Physiol* 277:C823–C832

Mongin AA, Reddi JM, Charniga C, Kimelberg HK (1999b) [3H]taurine and d-[3H]aspartate release from astrocyte cultures are differently regulated by tyrosine kinases. *Am J Physiol* 276:C1226–C1230

Montana V, Ni Y, Sunjara V, Hua X, Parpura V (2004) Vesicular glutamate transporter-dependent glutamate release from astrocytes. *J Neurosci* 24:2633–2642

Montana V, Malarkey EB, Verderio C, Matteoli M, Parpura V (2006) Vesicular transmitter release from astrocytes. *Glia* 54:700–715

Montoliu C, Llansola M, Kosenko E, Corbalan R, Felipo V (1999) Role of cyclic GMP in glutamate neurotoxicity in primary cultures of cerebellar neurons. *Neuropharmacology* 38:1883–1891

Moran J, Morales-Mulia M, Pasantes-Morales H (2001) Reduction of phospholemman expression decreases osmosensitive taurine efflux in astrocytes. *Biochim Biophys Acta* 1538:313–320

Moran MM, Melendez R, Baker D, Kalivas PW, Seamans JK (2003) Cystine/glutamate antiporter regulation of vesicular glutamate release. *Ann N Y Acad Sci* 1003:445–447

Moran MM, McFarland K, Melendez RI, Kalivas PW, Seamans JK (2005) Cystine/glutamate exchange regulates metabotropic glutamate receptor presynaptic inhibition of excitatory transmission and vulnerability to cocaine seeking. *J Neurosci* 25:6389–6393

Mothet JP, Pollegioni L, Ouanounou G, Martineau M, Fossier P, Baux G (2005) Glutamate receptor activation triggers a calcium-dependent and SNARE protein-dependent release of the gliotransmitter d-serine. *Proc Natl Acad Sci USA* 102:5606–5611

Mutafova-Yambolieva VN, Hwang SJ, Hao X, Chen H, Zhu MX, Wood JD, Ward SM, Sanders KM (2007) Beta-nicotinamide adenine dinucleotide is an inhibitory neurotransmitter in visceral smooth muscle. *Proc Natl Acad Sci USA* 104:16359–16364

Nageotte J (1910) Phenomenes de secretion dans le protoplasma des cellules nevrogliques de la substance grise. *C R Soc Biol (Paris)* 68:1068–1069

Neal MJ, Bowery NG (1979) Differential effects of veratridine and potassium depolarization on neuronal and glial GABA release. *Brain Res* 167:337–343

Neary JT, Kang Y, Willoughby KA, Ellis EF (2003) Activation of extracellular signal-regulated kinase by stretch-induced injury in astrocytes involves extracellular ATP and P2 purinergic receptors. *J Neurosci* 23:2348–2356

Nestor MW, Mok LP, Tulapurkar ME, Thompson SM (2007) Plasticity of neuron–glial interactions mediated by astrocytic EphARs. *J Neurosci* 27:12817–12828

Newman EA (2001) Propagation of intercellular calcium waves in retinal astrocytes and Muller cells. *J Neurosci* 21:2215–2223

Newman EA (2003) Glial cell inhibition of neurons by release of ATP. *J Neurosci* 23:1659–1666

North RA (2002) Molecular physiology of P2X receptors. *Physiol Rev* 82:1013–1067

Ortinau S, Laube B, Zimmermann H (2003) ATP inhibits NMDA receptors after heterologous expression and in cultured hippocampal neurons and attenuates NMDA-mediated neurotoxicity. *J Neurosci* 23:4996–5003

Paas Y, Devillers-Thiery A, Changeux JP, Medevielle F, Teichberg VI (1996) Identification of an extracellular motif involved in the binding of guanine nucleotides by a glutamate receptor. *Embo J* 15:1548–1556

Pangrsic T, Potokar M, Stenovec M, Kreft M, Fabbretti E, Nistri A, Pryazhnikov E, Khiroug L, Giniatullin R, Zorec R (2007) Exocytotic release of ATP from cultured astrocytes. *J Biol Chem* 282:28749–28758

Parkinson FE, Xiong W (2004) Stimulus- and cell-type-specific release of purines in cultured rat forebrain astrocytes and neurons. *J Neurochem* 88:1305–1312

Parkinson FE, Sinclair CJ, Othman T, Haughey NJ, Geiger JD (2002) Differences between rat primary cortical neurons and astrocytes in purine release evoked by ischemic conditions. *Neuropharmacology* 43:836–846

Parkinson FE, Xiong W, Zamzow CR (2005) Astrocytes and neurons: Different roles in regulating adenosine levels. *Neurol Res* 27:153–160

Parpura V, Haydon PG (2000) Physiological astrocytic calcium levels stimulate glutamate release to modulate adjacent neurons. *Proc Natl Acad Sci USA* 97:8629–8634

Parpura V, Basarsky TA, Liu F, Jeftinija K, Jeftinija S, Haydon PG (1994) Glutamate-mediated astrocyte–neuron signalling. *Nature* 369:744–747

Parpuraxy V, Fang Y, Basarsky T, Jahn R, Haydon PG (1995a) Expression of synaptobrevin II, cellubrevin and syntaxin but not SNAP-25 in cultured astrocytes. *FEBS Lett* 377:489–492

Parpura V, Liu F, Brethorst S, Jeftinija K, Jeftinija S, Haydon PG (1995b) Alpha-latrotoxin stimulates glutamate release from cortical astrocytes in cell culture. *FEBS Lett* 360:266–270

Parri HR, Gould TM, Crunelli V (2001) Spontaneous astrocytic Ca2+ oscillations in situ drive NMDAR-mediated neuronal excitation. *Nat Neurosci* 4:803–812

Pasantes Morales H, Schousboe A (1988) Volume regulation in astrocytes: A role for taurine as an osmoeffector. *J Neurosci Res* 20:503–509

Pasantes-Morales H, Schousboe A (1989) Release of taurine from astrocytes during potassium-evoked swelling. *Glia* 2:45–50

Pasantes-Morales H, Moran J, Schousboe A (1990) Volume-sensitive release of taurine from cultured astrocytes: Properties and mechanism. *Glia* 3:427–432

Pascual M, Climent E, Guerri C (2001) BDNF induces glutamate release in cerebrocortical nerve terminals and in cortical astrocytes. *Neuroreport* 12:2673–2677

Pascual O, Casper KB, Kubera C, Zhang J, Revilla-Sanchez R, Sul JY, Takano H, Moss SJ, McCarthy K, Haydon PG (2005) Astrocytic purinergic signaling coordinates synaptic networks. *Science* 310:113–116

Pasti L, Volterra A, Pozzan T, Carmignoto G (1997) Intracellular calcium oscillations in astrocytes: A highly plastic, bidirectional form of communication between neurons and astrocytes in situ. *J Neurosci* 17:7817–7830

Pasti L, Zonta M, Pozzan T, Vicini S, Carmignoto G (2001) Cytosolic calcium oscillations in astrocytes may regulate exocytotic release of glutamate. *J Neurosci* 21:477–484

Patel SA, Warren BA, Rhoderick JF, Bridges RJ (2004) Differentiation of substrate and non-substrate inhibitors of transport system xc-: An obligate exchanger of l-glutamate and l-cystine. *Neuropharmacology* 46:273–284

Patti L, Raiteri L, Grilli M, Zappettini S, Bonanno G, Marchi M (2007) Evidence that alpha7 nicotinic receptor modulates glutamate release from mouse neocortical gliosomes. *Neurochem Int* 51:1–7

Pedraza CE, Baltrons MA, Garcia A (2001) Interleukin-1beta stimulates cyclic GMP efflux in brain astrocytes. *FEBS Lett* 507:303–306

Pelegrin P, Surprenant A (2006) Pannexin-1 mediates large pore formation and interleukin-1beta release by the ATP-gated P2X7 receptor. *Embo J* 25:5071–5082
Penit J, Jard S, Benda P (1974) Probenecide sensitive 3′-5′-cyclic AMP secretion by isoproterenol stimulated glial cells in culture. *FEBS Lett* 41:156–160
Perez M, Hirschberg CB (1986) Topography of glycosylation reactions in the rough endoplasmic reticulum membrane. *J Biol Chem* 261:6822–6830
Pierson PM, Peteri-Brunback B, Pisani DF, Abbracchio MP, Mienville JM, Rosso L (2007) A2b receptor mediates adenosine inhibition of taurine efflux from pituicytes. *Biol Cell* 99:445–454
Poulopoulou C, Nowak LM (1998) Extracellular 3′,5′ cyclic guanosine monophosphate inhibits kainate-activated responses in cultured mouse cerebellar neurons. *J Pharmacol Exp Ther* 286:99–109
Prat AG, Reisin IL, Ausiello DA, Cantiello HF (1996) Cellular ATP release by the cystic fibrosis transmembrane conductance regulator. *Am J Physiol* 270:C538–C545
Pryazhnikov E, Khiroug L (2008) Sub-micromolar increase in [Ca2+]i triggers delayed exocytosis of ATP in cultured astrocytes. *Glia* 56:38–49
Queiroz G, Gebicke-Haerter PJ, Schobert A, Starke K, von Kugelgen I (1997) Release of ATP from cultured rat astrocytes elicited by glutamate receptor activation. *Neuroscience* 78:1203–1208
Queiroz G, Meyer DK, Meyer A, Starke K, von Kugelgen I (1999) A study of the mechanism of the release of ATP from rat cortical astroglial cells evoked by activation of glutamate receptors. *Neuroscience* 91:1171–1181
Raiteri L, Stigliani S, Usai C, Diaspro A, Paluzzi S, Milanese M, Raiteri M, Bonanno G (2007) Functional expression of release-regulating glycine transporters GLYT1 on GABAergic neurons and GLYT2 on astrocytes in mouse spinal cord. *Neurochem Int* 52(1–2):103–112
Ramos-Mandujano G, Vazquez-Juarez E, Hernandez-Benitez R, Pasantes-Morales H (2007) Thrombin potently enhances swelling-sensitive glutamate efflux from cultured astrocytes. *Glia* 55:917–925
Rathbone MP, Middlemiss PJ, DeLuca B, Jovetich M (1991) Extracellular guanosine increases astrocyte cAMP: Inhibition by adenosine A2 antagonists. *Neuroreport* 2:661–664
Re DB, Nafia I, Melon C, Shimamoto K, Kerkerian-Le Goff L, Had-Aissouni L (2006) Glutamate leakage from a compartmentalized intracellular metabolic pool and activation of the lipoxygenase pathway mediate oxidative astrocyte death by reversed glutamate transport. *Glia* 54:47–57
Reisin IL, Prat AG, Abraham EH, Amara JF, Gregory RJ, Ausiello DA, Cantiello HF (1994) The cystic fibrosis transmembrane conductance regulator is a dual ATP and chloride channel. *J Biol Chem* 269:20584–20591
Ribak CE, Tong WM, Brecha NC (1996) GABA plasma membrane transporters, GAT-1 and GAT-3, display different distributions in the rat hippocampus. *J Comp Neurol* 367:595–606
Ribeiro CS, Reis M, Panizzutti R, de Miranda J, Wolosker H (2002) Glial transport of the neuromodulator d-serine. *Brain Res* 929:202–209
Richerson GB, Wu Y (2003) Dynamic equilibrium of neurotransmitter transporters: Not just for reuptake anymore. *J Neurophysiol* 90:1363–1374
Rindler MJ, Bashor MM, Spitzer N, Saier MH, Jr. (1978) Regulation of adenosine 3′:5′-monophosphate efflux from animal cells. *J Biol Chem* 253:5431–5436
Rosenberg PA, Knowles R, Knowles KP, Li Y (1994) Beta-adrenergic receptor-mediated regulation of extracellular adenosine in cerebral cortex in culture. *J Neurosci* 14:2953–2965
Rossi DJ, Oshima T, Attwell D (2000) Glutamate release in severe brain ischaemia is mainly by reversed uptake. *Nature* 403:316–321
Rossi D, Brambilla L, Valori CF, Crugnola A, Giaccone G, Capobianco R, Mangieri M, Kingston AE, Bloc A, Bezzi P, Volterra A (2005) Defective tumor necrosis factor-alpha-dependent control of astrocyte glutamate release in a transgenic mouse model of Alzheimer disease. *J Biol Chem* 280:42088–42096
Rosso L, Peteri-Brunback B, Poujeol P, Hussy N, Mienville JM (2004) Vasopressin-induced taurine efflux from rat pituicytes: A potential negative feedback for hormone secretion. *J Physiol* 554:731–742

Rothstein JD, Dykes-Hoberg M, Pardo CA, Bristol LA, Jin L, Kuncl RW, Kanai Y, Hediger MA, Wang Y, Schielke JP, Welty DF (1996) Knockout of glutamate transporters reveals a major role for astroglial transport in excitotoxicity and clearance of glutamate. *Neuron* 16:675–686
Rothwell NJ, Luheshi GN (2000) Interleukin 1 in the brain: Biology, pathology and therapeutic target. *Trends Neurosci* 23:618–625
Rutledge EM, Kimelberg HK (1996) Release of [3H]-d-aspartate from primary astrocyte cultures in response to raised external potassium. *J Neurosci* 16:7803–7811
Rutledge EM, Aschner M, Kimelberg HK (1998) Pharmacological characterization of swelling-induced d-[3H]aspartate release from primary astrocyte cultures. *Am J Physiol* 274:C1511–1520
Rutledge EM, Mongin AA, Kimelberg HK (1999) Intracellular ATP depletion inhibits swelling-induced d-[3H]aspartate release from primary astrocyte cultures. *Brain Res* 842:39–45
Sabirov RZ, Dutta AK, Okada Y (2001) Volume-dependent ATP-conductive large-conductance anion channel as a pathway for swelling-induced ATP release. *J Gen Physiol* 118:251–266
Saez JC, Retamal MA, Basilio D, Bukauskas FF, Bennett MV (2005) Connexin-based gap junction hemichannels: Gating mechanisms. *Biochim Biophys Acta* 1711:215–224
Sanzgiri RP, Araque A, Haydon PG (1999) Prostaglandin E2 stimulates glutamate receptor-dependent astrocyte neuromodulation in cultured hippocampal cells. *J Neurobiol* 41:221–229
Schafer MK, Varoqui H, Defamie N, Weihe E, Erickson JD (2002) Molecular cloning and functional identification of mouse vesicular glutamate transporter 3 and its expression in subsets of novel excitatory neurons. *J Biol Chem* 277:50734–50748
Schell MJ, Molliver ME, Snyder SH (1995) d-serine, an endogenous synaptic modulator: Localization to astrocytes and glutamate-stimulated release. *Proc Natl Acad Sci USA* 92:3948–3952
Schiavo G, Matteoli M, Montecucco C (2000) Neurotoxins affecting neuroexocytosis. *Physiol Rev* 80:717–766
Schousboe A (2003) Role of astrocytes in the maintenance and modulation of glutamatergic and GABAergic neurotransmission. *Neurochem Res* 28:347–352
Schwabe U, Trost T (1980) Characterization of adenosine receptors in rat brain by (–)[3H] N6–phenylisopropyladenosine. *Naunyn Schmiedebergs Arch Pharmacol* 313:179–187
Seki Y, Feustel PJ, Keller RW, Jr., Tranmer BI, Kimelberg HK (1999) Inhibition of ischemia-induced glutamate release in rat striatum by dihydrokinate and an anion channel blocker. *Stroke* 30:433–440
Shain WG, Martin DL (1984) Activation of beta-adrenergic receptors stimulates taurine release from glial cells. *Cell Mol Neurobiol* 4:191–196
Shanker G, Allen JW, Mutkus LA, Aschner M (2001) The uptake of cysteine in cultured primary astrocytes and neurons. *Brain Res* 902:156–163
Shi Q, Savage JE, Hufeisen SJ, Rauser L, Grajkowska E, Ernsberger P, Wroblewski JT, Nadeau JH, Roth BL (2003) l-homocysteine sulfinic acid and other acidic homocysteine derivatives are potent and selective metabotropic glutamate receptor agonists. *J Pharmacol Exp Ther* 305:131–142
Shiga H, Murakami J, Nagao T, Tanaka M, Kawahara K, Matsuoka I, Ito E (2006) Glutamate release from astrocytes is stimulated via the appearance of exocytosis during cyclic AMP-induced morphologic changes. *J Neurosci Res* 84:338–347
Sim JA, Young MT, Sung HY, North RA, Surprenant A (2004) Reanalysis of P2X7 receptor expression in rodent brain. *J Neurosci* 24:6307–6314
Sorbera LA, Morad M (1991) Modulation of cardiac sodium channels by cAMP receptors on the myocyte surface. *Science* 253:1286–1289
Spray DC, Ye ZC, Ransom BR (2006) Functional connexin "hemichannels": A critical appraisal. *Glia* 54:758–773
Srinivas M, Hopperstad MG, Spray DC (2001) Quinine blocks specific gap junction channel subtypes. *Proc Natl Acad Sci USA* 98:10942–10947
Stenovec M, Kreft M, Grilc S, Potokar M, Kreft ME, Pangrsic T, Zorec R (2007) Ca2+-dependent mobility of vesicles capturing anti-VGLUT1 antibodies. *Exp Cell Res* 313:3809–3818
Stigliani S, Zappettini S, Raiteri L, Passalacqua M, Melloni E, Venturi C, Tacchetti C, Diaspro A, Usai C, Bonanno G (2006) Glia re-sealed particles freshly prepared from adult rat brain are competent for exocytotic release of glutamate. *J Neurochem* 96:656–668

Stout C, Charles A (2003) Modulation of intercellular calcium signaling in astrocytes by extracellular calcium and magnesium. *Glia* 43:265–273

Stout CE, Costantin JL, Naus CC, Charles AC (2002) Intercellular calcium signaling in astrocytes via ATP release through connexin hemichannels. *J Biol Chem* 277:10482–10488

Striedinger K, Meda P, Scemes E (2007) Exocytosis of ATP from astrocyte progenitors modulates spontaneous Ca2+ oscillations and cell migration. *Glia* 55:652–662

Suadicani SO, Brosnan CF, Scemes E (2006) P2X7 receptors mediate ATP release and amplification of astrocytic intercellular Ca2+ signaling. *J Neurosci* 26:1378–1385

Suadicani SO, Dahl G, Scemes E (2007) Pannexin1 mediates amplification of intercellular Ca2+waves in astrocytes. *Neuron Glia Biol* 3:S165 (abstract)

Südhof TC, Jahn R (1991) Proteins of synaptic vesicles involved in exocytosis and membrane recycling. *Neuron* 6:665–677

Sweeney M, White T, Sawynok J (1988) 5-Hydroxytryptamine releases adenosine from primary afferent nerve terminals in the spinal cord. *Brain Res* 462:346–349

Syed N, Martens CA, Hsu WH (2007) Arginine vasopressin increases glutamate release and intracellular Ca2+ concentration in hippocampal and cortical astrocytes through two distinct receptors. *J Neurochem* 103:229–237

Szatkowski M, Barbour B, Attwell D (1990) Non-vesicular release of glutamate from glial cells by reversed electrogenic glutamate uptake. *Nature* 348:443–446

Takano T, Kang J, Jaiswal JK, Simon SM, Lin JH, Yu Y, Li Y, Yang J, Dienel G, Zielke HR, Nedergaard M (2005) Receptor-mediated glutamate release from volume sensitive channels in astrocytes. *Proc Natl Acad Sci USA* 102:16466–16471

Tang XC, Kalivas PW (2003) Bidirectional modulation of cystine/glutamate exchanger activity in cultured cortical astrocytes. *Ann N Y Acad Sci* 1003:472–475

Touyz RM, Picard S, Schiffrin EL, Deschepper CF (1997) Cyclic GMP inhibits a pharmacologically distinct Na+/H+ exchanger variant in cultured rat astrocytes via an extracellular site of action. *J Neurochem* 68:1451–1461

Traversa U, Bombi G, Di Iorio P, Ciccarelli R, Werstiuk ES, Rathbone MP (2002) Specific [(3) H]-guanosine binding sites in rat brain membranes. *Br J Pharmacol* 135:969–976

Trexler EB, Bennett MV, Bargiello TA, Verselis VK (1996) Voltage gating and permeation in a gap junction hemichannel. *Proc Natl Acad Sci USA* 93:5836–5841

Uemura Y Miller JM, Matson WR, Beal MF (1991) Neurochemical analysis of focal ischemia in rats. *Stroke* 22:1548–1553

Verderio C, Matteoli M (2001) ATP mediates calcium signaling between astrocytes and microglial cells: Modulation by IFN-gamma. *J Immunol* 166:6383–6391

Verderio C, Bruzzone S, Zocchi E, Fedele E, Schenk U, De Flora A, Matteoli M (2001) Evidence of a role for cyclic ADP-ribose in calcium signalling and neurotransmitter release in cultured astrocytes. *J Neurochem* 78:646–657

Volterra A, Meldolesi J (2005) Astrocytes, from brain glue to communication elements: The revolution continues. *Nat Rev Neurosci* 6:626–640

Volterra A, Bezzi P, Rizzini BL, Trotti D, Ullensvang K, Danbolt NC, Racagni G (1996) The competitive transport inhibitor l-trans-pyrrolidine-2, 4-dicarboxylate triggers excitotoxicity in rat cortical neuron–astrocyte co-cultures via glutamate release rather than uptake inhibition. *Eur J Neurosci* 8:2019–2028

Wang Z, Haydon PG, Yeung ES (2000) Direct observation of calcium-independent intercellular ATP signaling in astrocytes. *Anal Chem* 72:2001–2007

Wang CM, Chang YY, Kuo JS, Sun SH (2002) Activation of P2X7 receptors induced [3H]GABA release from the RBA-2 type-2 astrocyte cell line through a Cl–/HCO3–-dependent mechanism. *Glia* 37:8–18

Warr O, Takahashi M, Attwell D (1999) Modulation of extracellular glutamate concentration in rat brain slices by cystine-glutamate exchange. *J Physiol* 514 (Pt 3):783–793

Werry EL, Liu GJ, Bennett MR (2006) Glutamate-stimulated ATP release from spinal cord astrocytes is potentiated by substance P. *J Neurochem* 99:924–936

Westergaard N, Drejer J, Schousboe A, Sonnewald U (1996) Evaluation of the importance of transamination versus deamination in astrocytic metabolism of [U-^{13}C]glutamate. *Glia* 17:160–168

Wiencken-Barger AE, Djukic B, Casper KB, McCarthy KD (2007) A role for Connexin43 during neurodevelopment. *Glia* 55:675–686

Wilhelm A, Volknandt W, Langer D, Nolte C, Kettenmann H, Zimmermann H (2004) Localization of SNARE proteins and secretory organelle proteins in astrocytes in vitro and in situ. *Neurosci Res* 48:249–257

Winder DG, Ritch PS, Gereau RWT, Conn PJ (1996) Novel glial–neuronal signalling by coactivation of metabotropic glutamate and beta-adrenergic receptors in rat hippocampus. *J Physiol* 494 (Pt 3):743–755

Wolosker H, Sheth KN, Takahashi M, Mothet JP, Brady RO, Jr., Ferris CD, Snyder SH (1999) Purification of serine racemase: Biosynthesis of the neuromodulator d-serine. *Proc Natl Acad Sci USA* 96:721–725

Wu PH, Durden DA, Hertz L (1979) Net production of gamma-aminobutyric acid in astrocytes in primary cultures determined by a sensitive mass spectrometric method. *J Neurochem* 32:379–390

Xu J, Peng H, Kang N, Zhao Z, Lin JH, Stanton PK, Kang J (2007) Glutamate-induced exocytosis of glutamate from astrocytes. *J Biol Chem* 282:24185–24197

Yan XX, Cariaga WA, Ribak CE (1997) Immunoreactivity for GABA plasma membrane transporter, GAT-1, in the developing rat cerebral cortex: Transient presence in the somata of neocortical and hippocampal neurons. *Brain Res Dev Brain Res* 99:1–19

Yang Y, Ge W, Chen Y, Zhang Z, Shen W, Wu C, Poo M, Duan S (2003) Contribution of astrocytes to hippocampal long-term potentiation through release of d-serine. *Proc Natl Acad Sci USA* 100:15194–15199

Ye ZC, Wyeth MS, Baltan-Tekkok S, Ransom BR (2003) Functional hemichannels in astrocytes: A novel mechanism of glutamate release. *J Neurosci* 23:3588–3596

Yoshida Y, Nakane A, Morita M, Kudo Y (2006) A novel effect of bifemelane, a nootropic drug, on intracellular Ca2+ levels in rat cerebral astrocytes. *J Pharmacol Sci* 100:126–132

Zeevalk GD, Davis N, Hyndman AG, Nicklas WJ (1998) Origins of the extracellular glutamate released during total metabolic blockade in the immature retina. *J Neurochem* 71:2373–2381

Zetterstrom T, Vernet L, Ungerstedt U, Tossman U, Jonzon B, Fredholm BB (1982) Purine levels in the intact rat brain. Studies with an implanted perfused hollow fibre. *Neurosci Lett* 29:111–115

Zhang JM, Wang HK, Ye CQ, Ge W, Chen Y, Jiang ZL, Wu CP, Poo MM, Duan S (2003) ATP released by astrocytes mediates glutamatergic activity-dependent heterosynaptic suppression. *Neuron* 40:971–982

Zhang Q, Fukuda M, Van Bockstaele E, Pascual O, Haydon PG (2004a) Synaptotagmin IV regulates glial glutamate release. *Proc Natl Acad Sci USA* 101:9441–9446

Zhang Q, Pangrsic T, Kreft M, Krzan M, Li N, Sul JY, Halassa M, Van Bockstaele E, Zorec R, Haydon PG (2004b) Fusion-related release of glutamate from astrocytes. *J Biol Chem* 279:12724–12733

Zhang Z, Chen G, Zhou W, Song A, Xu T, Luo Q, Wang W, Gu XS, Duan S (2007) Regulated ATP release from astrocytes through lysosome exocytosis. *Nat Cell Biol* 9:945–953

Zocchi E, Usai C, Guida L, Franco L, Bruzzone S, Passalacqua M, De Flora A (1999) Ligand-induced internalization of CD38 results in intracellular Ca^{2+} mobilization: Role of NAD^{+} transport across cell membranes. *FASEB J* 13:273–283

Abbreviations

ABC	ATP-binding cassette
AMPA	α-amino-3-hydroxy-5-methyl-isoxazole propionate
ASCT	alanine-serine-cysteine transporter
ATP	adenosine 5′-triphosphate
BAPTA	1,2-bis(*o*-aminophenoxy)ethane-*N*,*N*,*N*′,*N*′-tetraacetic acid

BBG	Brilliant Blue G
BzATP	3′-*O*-(4-benzoyl)benzoyl ATP
cAMP	adenosine 3′:5′ cyclic monophosphate
CFTR	cystic fibrosis transmembrane conductance regulator
cGMP	guanosine 3′:5′ cyclic monophosphate
C_m	plasma membrane capacitance
Cx	connexin
DIDS	4,4′-diisothiocyanato-stilbene-2,2′-disulfonate
EGFP	enhanced green fluorescent protein
EM	electron microscopy
ENT	equilibrative nucleoside transporters
EPSC	excitatory post synaptic current
FFA	flufenamic acid
GA	18β-glycyrrhetinic acid
GABA	γ-amino butyric acid
GAD	glutamic acid decarboxylase
GAT	plasma membrane GABA transporter
GlyT	plasma membrane glycine transporters
HCA	homocysteic acid
MRP	multidrug resistance protein
NKCC	Na^+-K^+-Cl^- cotransporters
NMDA	*N*-methyl-D-aspartic acid
oATP	oxidized ATP
PPADS	pyridoxal phosphate-6-azophenyl-2-4-disulfonic acid
SNARE	soluble *N*-ethyl maleimide-sensitive fusion protein attachment protein receptor
SOC	slow outward current
TCA	tricarboxylic acid
UDP	uridine diphosphate-glucose
UTP	uridine 5′-triphosphate
V-ATPase	vacuolar type proton ATPase
VGAT	vesicular GABA transporter
VGLUT	vesicular glutamate transporter
VRAC	volume-regulated anion channels
VSOR	volume-sensitive outwardly rectifying

Chapter 13
Release of Trophic Factors and Immune Molecules from Astrocytes

Ying Y. Jean, Issa P. Bagayogo, and Cheryl F. Dreyfus

Contents

It is now quite clear that as part of their support role, astrocytes produce and release multiple proteins that impact survival, migration, differentiation, and function of proximate neurons. This astrocyte role is particularly evident during development, remains in play in the normal adult brain, and is enhanced after injury. The molecules produced include those known traditionally as cytokines, chemokines, and trophic factors. Importantly, these proteins can serve as trophic or toxic agents and may be differentially regulated by the signals impacting on astrocytes at various stages of brain development and maturity (Fig. 13.1). The story of proteins produced by astrocytes, then, is a complex one. Multiple trophic or toxic molecules

C.F. Dreyfus
Department of Neuroscience and Cell Biology, UMDNJ-Robert Wood Medical School, Piscataway, NJ, USA
dreyfus@umdnj.edu

V. Parpura and P.G. Haydon (eds.), *Astrocytes in (Patho)Physiology of the Nervous System*, DOI: 10.1007/978-0-387-79492-1_13, © Springer Science+Business Media, LLC 2009

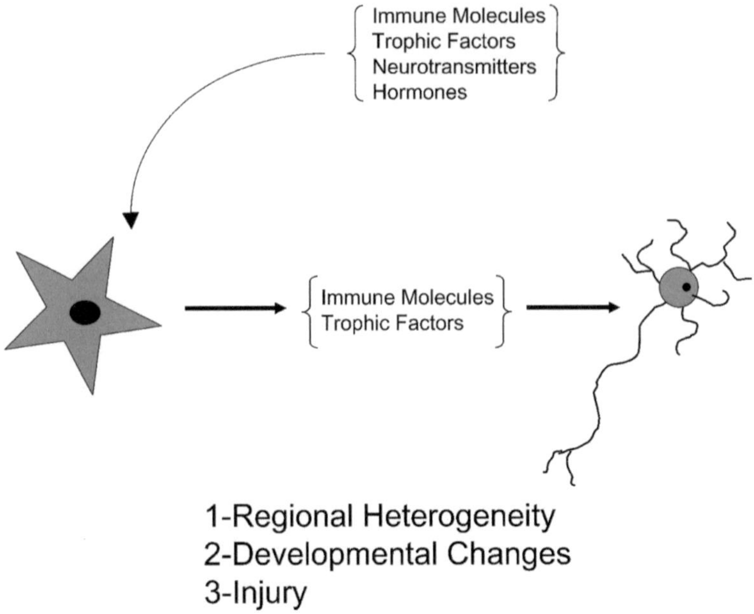

Fig. 13.1 Schematic representation of astrocyte–neuron interactions: Astrocyte function is regulated by several factors, which include immune molecules, trophic factors, neurotransmitters, and hormones. In response, astrocytes express and release multiple proteins (immune molecules and trophic factors) that impact survival, migration, differentiation, and function of proximate neurons. These interactions are regionally heterogeneous (1) and are altered during development (2) and after injury (3).

may be produced. The molecular mix may change from the developing animal to the adult, and after a lesion. Moreover, there is regional heterogeneity in the astrocyte-derived molecules, suggesting that those produced by region-specific astrocytes may be adapted to interact specifically with proximate neurons. To present evidence supporting this view, the present chapter is organized (Fig. 13.1) to discuss (1) the roles of astrocyte-derived cytokines, chemokines, and trophic factors on neurons, (2) regional differences in astrocyte function, (3) differences in astrocyte function with aging and after injury, and (4) molecules that regulate the role of astrocytes as trophic or immune molecule producers.

We focus this review on the roles of immune molecules and growth factors recognizing that astrocytes express proteins other than those detailed here that also impact neuron function. Furthermore, we limit the review to astrocytes and their effects on neurons, but appreciate that oligodendrocytes and microglia produce immune and trophic molecules that also impact neuron function. Moreover, astrocyte-derived molecules influence cells other than neurons, including oligodendrocytes, astrocytes themselves, and microglia of the central nervous system.

13.1 The Effects of Astrocytes on Neurons

That astrocytes produce supportive proteins that impact neuron development and function originated with culture studies evaluating effects of postnatal astrocytes on developing neurons. For example, early work determined that when neurons of the substantia nigra are grown with astrocytes there is a dramatic increase in the number of dopaminergic neurons (O'Malley et al., 1992), in their extension of processes (Denis-Donini et al., 1984), and in their expression of the differentiated trait, tyrosine hydroxylase (O'Malley et al., 1994). These results were also obtained when nigral neurons were exposed to medium conditioned by the astrocytes, suggesting that astrocytes release trophic molecules (O'Malley et al., 1994). Other studies used retinal ganglion cells to demonstrate that collicular glia (primarily composed of astrocytes) (Pfrieger and Barres, 1997) or conditioned media derived from collicular astrocytes affect synaptic efficacy by increasing Ca^{2+} influx, reorganizing synaptic vesicles along the soma and dendrites, and enhancing the release of synaptic vesicles (Ullian et al., 2001). In the same study, it was also suggested that astrocytes increase the number of synapses per neuron both *in vivo* and *in vitro* and the frequency and amplitude of spontaneous excitatory postsynaptic currents (Ullian et al., 2004). These data are complementary to that of Haydon, Parpura, and colleagues indicating that astrocytes release stimulatory signals in a regulated manner to influence synaptic function (Parpura et al., 1994; Parpura and Haydon, 2000).

More recently it has been found that astrocytes also promote both adult and embryonic stem cell differentiation. Embryonic stem cell spheres derived from the inner cell mass of the blastula cultured on top of astrocytes differentiate into neural stem cells and cells expressing dopaminergic and cholinergic neuronal traits (Nakayama et al., 2003). When studies were performed using whole brain or hippocampal astrocytes it was found that proliferation and differentiation of subventricular zone cells (Lim and Alvarez-Buylla, 1999) and adult stem cells (Song et al., 2002) into neurons was enhanced. This may be due to a diffusible factor or direct cell–cell contact (Lim and Alvarez-Buylla, 1999; Song et al., 2002).

In toto, these studies suggest that astrocytes play a significant role in influencing neuron survival, differentiation, and synaptogenesis as well as in dictating the specification of stem cells into neurons.

13.2 Molecules Produced by Astrocytes That Impact Neurons

What molecules are responsible for these astrocyte-associated effects? Two basic approaches have been taken to answer this question. In the first, the possibility that astrocytes produce and release molecules that influence neuronal development, mature function, and survival has been explored by identifying molecules responsible for these events, evaluating their expression in specific astrocyte populations, and then defining the role of these specific molecules when assessed in astrocyte-derived medium. In the second, gene arrays have been used to identify molecules produced by specific astrocyte populations and the functions of these have been assessed. Both approaches have been helpful in identifying astrocyte-derived molecules that impact neuron function.

13.2.1 The Role of Trophic Factors

Trophic factors produced by astrocytes are many (Table 13.1). Moreover, as noted in Table 13.1, studies using cultured fetal, neonatal, or adult astrocytes, as well as *in situ* hybridization and immunohistochemical analysis of developing brain and mature brain establish that astrocytes express these molecules not only in culture, but also *in vivo*. Effects of these molecules on neuron development have been documented initially in culture studies of developing cells, and more recently in slice preparations of the intact brain, as well as in *in vivo* studies using knockout mice (see examples later). These effects include actions on neuron specification and neuron progenitor proliferation, neuron survival, differentiation, and the development of communication between neurons (Walicke, 1988; Friedman et al., 1993; Pataky et al., 2000; Zhao et al., 2007). For example, using a combined approach, glial cell line-derived growth factor (GDNF) acting through GDNF receptor $\alpha 1$ on hippocampal neurons was shown to increase survival of neurons and synapse formation between pre- and postsynaptic neurons as determined by increased localization of synaptic vesicle markers in the area of the synapse (Ledda et al., 2007). In the hippocampus, reduction of adult neurogenesis and severe impairment of long-term potentiation (LTP) were observed in mice exhibiting the conditional knockout of fibroblast growth factor (FGF) receptor 1, a major receptor for FGF-2 (Zhao et al., 2007). The neurotrophin brain-derived neurotrophic factor (BDNF), too, was shown to influence synaptic function using combined approaches, affecting either LTP or long-term depression depending on the receptor (TrkB or p75NTR) to which it binds (Xu et al., 2000; Woo et al., 2005).

13.2.2 The Role of Immune Molecules

Astrocytes produce a variety of immune molecules, both chemokines and cytokines (Table 13.2). A wealth of data is indicating that these astrocyte-associated molecules, traditionally known as molecules that participate in an immune response and that may have negative effects after injury, also play a role similar to that of trophic factors during development, and perhaps in maturity (see later). For example, culture and *in vivo* studies suggest that these molecules enhance neuron differentiation and survival directly through the mediation of chemokine and cytokine receptors (Plata-Salaman, 1991; Mehler and Kessler, 1995; Arakawa et al., 2003; Emsley and Hagg, 2003). Chemokine and cytokine knockout mice are impaired in the migration

Table 13.1 Growth factor expression by astrocytes.

Growth Factor	Young	Adult	Reactive	Regulated	Release	References
BDNF	mRNA (*in vitro*) Protein (*in vitro*)	Protein (*in vivo*)	Protein (*in vivo*)	mRNA (*in vitro*) Protein (*in vitro*)	*	Dreyfus et al., 1999; Dougherty et al., 2000; Wu et al., 2004; Toyomoto et al., 2005; Jean et al., 2006
BMP2	mRNA (*in vitro*) Protein (*in vitro*)	mRNA (*in vitro*)		mRNA (*in vitro*) Protein (*in vitro*)	*	Meeuwsen et al., 2003; Nakagawa and Schwartz, 2004; Park et al., 2006
BMP3	mRNA (*in vitro*)	mRNA (*in vitro*)		mRNA (*in vitro*)		Meeuwsen et al., 2003; Nakagawa and Schwartz, 2004
BMP4	mRNA (*in vitro*)	mRNA (*in vitro*) Protein (*in vivo*)	mRNA (*in vivo*)			Peretto et al., 2002; Nakagawa and Schwartz, 2004; Chen et al., 2005
BMP7		Protein (*in vivo*)				Peretto et al., 2002
EGF	Protein (*in vivo*)					Kato et al., 1995
EPO	mRNA (*in vitro*) Protein (*in vivo, in vitro*)		Protein (*in vitro*)	mRNA (*in vitro*) Protein (*in vitro*)	*	Juul et al., 1999; Nagai et al., 2001; Liu et al., 2006
FGF1	mRNA (*in vitro*) Protein (*in vitro*)	Protein (*in vivo*)		mRNA (*in vitro*)	*	Tourbah et al., 1991; Magnaghi et al., 2000; Matsunaga and Ueda, 2006
FGF2	mRNA (*in vitro*) Protein (*in vitro*)	mRNA (*in vitro*)	mRNA (*in vivo*) Protein (*in vivo*)	mRNA (*in vitro*) Protein (*in vitro*)	*	Niu et al., 1997; Magnaghi et al., 2000; Madiai et al., 2003; Nakagawa and Schwartz, 2004; Li et al., 2006
FGF9		mRNA (*in vivo*) Protein (*in vivo*)	Protein (*in vivo*)			Nakamura et al., 1998b, 1999
GDN or PN-1	mRNA (*in vitro*) Protein (*in vitro*)	mRNA (*in vivo*)	mRNA (*in vitro*)	Protein (*in vitro*)	*	Choi et al., 1990; Simpson et al., 1994; Brenneman et al., 1997; Kim et al., 2004

(continued)

Table 13.1 (continued)

Growth Factor	Young	Adult	Reactive	Regulated	Release	References
GDNF	mRNA (*in vitro*) Protein (*in vivo, in vitro*)	mRNA (*in vitro, in vivo*)	Protein (*in vivo*)	mRNA (*in vitro*) Protein (*in vitro*)	*	Appel et al., 1997; Koo and Choi, 2001; Nomura et al., 2002; Ohta et al., 2002; Nakagawa and Schwartz, 2004; Toyomoto et al., 2005
GMF-β	mRNA (*in vitro*)	Protein (*in vivo*) mRNA (*in vitro*)	Protein (*in vivo*)			Nakagawa and Schwartz, 2004; Hotta et al., 2005
HB-GAM	mRNA (*in vitro, in vivo*)	mRNA (*in vitro, in vivo*)	mRNA (*in vivo*) Protein (*in vivo*)			Wanaka et al., 1993; Takeda et al., 1995; Nakagawa and Schwartz, 2004; Wang et al., 2004
HGF	mRNA (*in vitro*)	mRNA (*in vitro, in vivo*)	Protein (*in vivo*) mRNA (*in vivo*)			Yamada et al., 1997; Nakagawa and Schwartz, 2004; Shimamura et al., 2007
IGF-I	mRNA (*in vitro*)	mRNA (*in vitro*)	Protein (*in vivo*)			Ballotti et al., 1987; Garcia-Estrada et al., 1992; Nakagawa and Schwartz, 2004
IGF-II	mRNA (*in vitro*)	mRNA (*in vitro*)	mRNA (*in vivo*) Protein (*in vivo*)	mRNA (*in vitro*)		Walter et al., 1999; Meeuwsen et al., 2003; Nakagawa and Schwartz, 2004
NGF	mRNA (*in vitro*) Protein (*in vitro*)		Protein (*in vivo*)	mRNA (*in vitro*) Protein (*in vitro*)	*	Furukawa et al., 1987b; Rossner et al., 1997; Wu et al., 2004; Toyomoto et al., 2005; Chen et al., 2006
Nrg-I	mRNA (*in vitro*) Protein (*in vitro*)		mRNA (*in vitro, in vivo*) Protein (*in vivo*)	mRNA (*in vitro*)	*	Pollock et al., 1999; Tokita et al., 2001; Kerber et al., 2003; Meeuwsen et al., 2005

NT3	mRNA (*in vitro*) Protein (*in vitro*)	Protein (*in vivo*)	Protein (*in vivo*)		*	Dreyfus et al., 1999; Wu et al., 2004; Chen et al., 2006; Park et al., 2006
NT4/5	mRNA (*in vitro*)	mRNA (*in vitro*)		mRNA (*in vitro*)		Nakagawa and Schwartz, 2004; Meeuwsen et al., 2005
PDGF	mRNA (*in vitro*) Protein (*in vitro*)	mRNA (*in vitro, in vivo*)	mRNA (*in vitro*) Protein (*in vivo*)	mRNA (*in vitro*) Protein (*in vitro*)	*	Gard et al., 1995; Silberstein et al., 1996; Sjoborg et al., 1998; Pang et al., 2001; Nakagawa and Schwartz, 2004
TGFα	mRNA (*in vitro*)	Protein (*in vivo*)	Protein (*in vivo*)	mRNA (*in vitro*)		Ma et al., 1994; Lisovoski et al., 1997; Li et al., 2002
TGFβ	mRNA (*in vitro*) Protein (*in vitro*)	mRNA (*in vitro*) Protein (*in vitro*)	mRNA (*in vitro*) Protein (*in vivo*)	mRNA (*in vitro*) Protein (*in vitro*)	*	Constam et al., 1992; Ma et al., 1994; De Groot et al., 1999; Li et al., 2002; Yun et al., 2002; Meeuwsen et al., 2003, 2005; Nakagawa and Schwartz, 2004; Dhandapani et al., 2005
VEGF	mRNA (*in vitro*) Protein (*in vitro*)	mRNA (*in vitro*) Protein (*in vitro, in vivo*)	mRNA (*in vitro, in vivo*) Protein (*in vivo*)	mRNA (*in vivo*) Protein (*in vitro, in vivo*)	*	Sinor et al., 1998; Yoshida et al., 2002; Nakagawa and Schwartz, 2004; Loeffler et al., 2005; Meeuwsen et al., 2005; Skold et al., 2005

Astrocyte growth factor expression in developing animals (Young) or adult animals (Adult) *in vivo* or in culture (*in vitro*). The table also indicates cases where reactive astrocytes are reported to express trophic factors *in vivo* or in culture, where astrocytes are regulated in expression of trophic factors by immune factors or other effectors and where astrocytes release factors (asterisk). The absence of information in a column does not indicate that factors are not present or regulated, but rather that this is not reported in the cases cited. BDNF, brain-derived neutrophic factor; BMP, bone morphogenetic protein; EGF, epidermal growth factor; EPO, erythropoietin; FGF, fibroblast growth factor; GDN, glia-derived nexin; GDNF, glial cell line-derived growth factor; GMF, glial maturation factor; HB-GAM, heparin-binding growth-associated molecule; HGF, hepatocyte growth factor; IGF, insulin-like growth factor; NGF, nerve growth factor; Nrg, neuregulin; NT, neurotrophin; PDGF, platelet-derived growth factor; PN, protease nexin; TGF, transforming growth factor; VEGF, vascular endothelial growth factor.

Table 13.2 Cytokine or chemokine expression by astrocytes.

Cytokines/ Chemokines	Young	Adult	Reactive	Regulated	Release	References
CNTF	mRNA (*in vitro*) Protein (*in vitro*)	Protein (*in vivo*)	Protein (*in vivo*)			Rudge et al., 1994; Choi et al., 2004; Sarup et al., 2004; Watt et al., 2006
GCP2 (CXCL-6)			mRNA (*in vitro*)	mRNA (*in vitro*)		Meeuwsen et al., 2003, 2005
G-CSF	mRNA (*in vitro*)	mRNA (*in vitro*)	mRNA (*in vitro*)	mRNA (*in vitro*) Protein (*in vitro*)	*	Aloisi et al., 1992; Brenneman et al., 2003; Meeuwsen et al., 2003, 2005; Nakagawa and Schwartz, 2004
GM-CSF	mRNA (*in vitro*)	mRNA (*in vitro*)	mRNA (*in vitro*)	mRNA (*in vitro*) Protein (*in vitro*)	*	Aloisi et al., 1992; Nakagawa and Schwartz, 2004; Meeuwsen et al., 2005
IFN-α		Protein (*in vivo*)	mRNA (*in vitro*) Protein (*in vivo*)			Rho et al., 1995; Carpentier et al., 2005
IFN-β	mRNA (*in vitro*)		mRNA (*in vitro*)	mRNA (*in vitro*)		Carpentier et al., 2005; Krasowska-Zoladek et al., 2007
IFN-γ	mRNA (*in vitro*)	mRNA (*in vivo*, *in vitro*) Protein (*in vitro*)	Protein (*in vivo*)	Protein (*in vitro*)	*	Xiao and Link, 1998; Folkerth et al., 2004; Nakagawa and Schwartz, 2004
IL-1α			mRNA (*in vitro*)	Protein (*in vitro*)	*	DeVito et al., 1995; Brenneman et al., 2003; Krasowska-Zoladek et al., 2007
IL-1β	mRNA (*in vitro*) Protein (*in vitro*)	mRNA (*in vitro*)	mRNA (*in vitro*) Protein (*in vivo*)	Protein (*in vitro*)	*	Takeshita et al., 2001; Brenneman et al., 2003; Croitoru-Lamoury et al., 2003; Nakagawa and Schwartz, 2004; Meeuwsen et al., 2005; Barkho et al., 2006; Wen et al., 2007

IL-2	Protein (*in vitro*) mRNA (*in vitro*)	mRNA (*in vivo*, *in vitro*)		mRNA (*in vitro*) Protein (*in vitro*)	*	Gabryel et al., 2004; Nakagawa and Schwartz, 2004; Meeuwsen et al., 2005
IL-3	Protein (*in vitro*)			Protein (*in vitro*)	*	Brenneman et al., 2003
IL-4		mRNA (*in vitro*) Protein (*in vitro*)	Protein (*in vivo*)			Hulshof et al., 2002; Croitoru-Lamoury et al., 2003; Meeuwsen et al., 2003
IL-5			mRNA (*in vitro*)			Bsibsi et al., 2006
IL-6	mRNA (*in vitro*)	mRNA (*in vitro*)	mRNA (*in vitro*) Protein (*in vivo*, *in vitro*)	mRNA (*in vitro*) Protein (*in vitro*)	*	Lee et al., 1993; Norris and Benveniste, 1993; Brenneman et al., 2003; Croitoru-Lamoury et al., 2003; Meeuwsen et al., 2003, 2005; Nakagawa and Schwartz, 2004; Carpentier et al., 2005, Barkho et al., 2006; Wen et al., 2007
IL-8 (CXCL-8)	mRNA (*in vitro*)	mRNA (*in vitro*)	mRNA (*in vitro*) Protein (*in vitro*)	mRNA (*in vitro*) Protein (*in vitro*)	*	Aloisi et al., 1992; Kutsch et al., 2000; Mahajan et al., 2002; Croitoru-Lamoury et al., 2003; Meeuwsen et al., 2003, 2005
IL-9			mRNA (*in vitro*)			Bsibsi et al., 2006
IL-10	mRNA (*in vitro*) Protein (*in vitro*)	Protein (*in vitro*)	mRNA (*in vitro*) Protein (*in vitro*, *in vivo*)			Mizuno et al., 1994; Hulshof et al., 2002; Croitoru-Lamoury et al., 2003; Meeuwsen et al., 2003; Wu et al., 2005
IL-11			mRNA (*in vitro*)	mRNA (*in vitro*)		Meeuwsen et al., 2005; Bsibsi et al., 2006
IL-12			mRNA (*in vivo*, *in vitro*) Protein (*in vivo*, *in vitro*)			Stalder et al., 1997; Takeshita et al., 2001; Abbas et al., 2002

(continued)

Table 13.2 (continued)

Cytokines/ Chemokines	Young	Adult	Reactive	Regulated	Release	References
IL-13	mRNA (*in vitro*)	mRNA (*in vitro, in vivo*)		mRNA (*in vitro*)		Nakagawa and Schwartz, 2004; Meeuwsen et al., 2005
IL-15	mRNA (*in vitro*) Protein (*in vitro*)	mRNA (*in vitro*)	mRNA (*in vitro*)	mRNA (*in vitro*) Protein (*in vitro*)		Lee et al., 1996; Li et al., 2004; Nakagawa and Schwartz, 2004
IL-17				mRNA (*in vitro*)		Meeuwsen et al., 2003, 2005
IP-10 (CXCL-10)	mRNA (*in vitro*)	mRNA (*in vitro*)	mRNA (*in vitro*) Protein (*in vivo, in vitro*)	mRNA (*in vitro*) Protein (*in vitro*)	*	Kutsch et al., 2000; Simpson et al., 2000; Croitoru-Lamoury et al., 2003; Meeuwsen et al., 2003
LIF	mRNA (*in vitro*) Protein (*in vitro*)	mRNA (*in vitro, in vivo*)	mRNA (*in vitro*)	mRNA (*in vitro*) Protein (*in vitro*)	*	Aloisi et al., 1994; Murphy et al., 1995; Banner et al., 1997; Suzuki et al., 2000; Yamakuni et al., 2002; Nakagawa and Schwartz, 2004; Meeuwsen et al., 2005
MCP-1 (CCL2)	mRNA (*in vitro*)	mRNA (*in vitro*) Protein (*in vivo*)	mRNA (*in vitro*) Protein (*in vivo, in vitro*)	mRNA (*in vitro*) Protein (*in vitro*)	*	Simpson et al., 1998; Van Der Voorn et al., 1999; Croitoru-Lamoury et al., 2003; Meeuwsen et al., 2003, 2005
MCP-2 (CCL-8)			Protein (*in vivo*)			McManus et al., 1998
MCP-3 (CCL-7)			Protein (*in vivo*)			McManus et al., 1998
M-CSF	mRNA (*in vitro*)			mRNA (*in vitro*) Protein (*in vitro*)	*	Aloisi et al., 1992; Brenneman et al., 2003; Meeuwsen et al., 2003
MIF	mRNA (*in vitro*)	mRNA (*in vivo, in vitro*) Protein (*in vivo*)	mRNA (*in vivo, in vitro*) Protein (*in vivo*)			Koda et al., 2004; Nakagawa and Schwartz, 2004; Krasowska-Zoladek et al., 2007
MIG (CXCL-9)	mRNA (*in vitro*)	mRNA (*in vitro*)	Protein (*in vivo*)	mRNA (*in vitro*) Protein (*in vitro*)	*	Simpson et al., 2000; Croitoru-Lamoury et al., 2003; Meeuwsen et al., 2003

MIP-1α (CCL-3)	mRNA (*in vitro*)	mRNA (*in vitro*)	mRNA (*in vitro*) Protein (*in vivo*, *in vitro*)	mRNA (*in vitro*)		Simpson et al., 1998; Takeshita et al., 2001; Croitoru-Lamoury et al., 2003; Meeuwsen et al., 2003, 2005; Carpentier et al., 2005
MIP-1β (CCL-4)	mRNA (*in vitro*)	mRNA (*in vitro*)		mRNA (*in vitro*)		Takeshita et al., 2001; Croitoru-Lamoury et al., 2003; Meeuwsen et al., 2005
MIP-2(α) (CXCL-2)			mRNA (*in vitro*) Protein (*in vitro*)	mRNA (*in vitro*) Protein (*in vitro*)	*	Luo et al., 2000; Otto et al., 2002; Meeuwsen et al., 2003, 2005; Tomita et al., 2005
MIP-3α		mRNA (*in vivo*)	mRNA (*in vitro*)			Pang et al., 2001; Nakagawa and Schwartz, 2004
RANTES (CCL-5)	mRNA (*in vitro*)	mRNA (*in vitro*)	mRNA (*in vitro*) Protein (*in vitro*, *in vivo*)	mRNA (*in vitro*) Protein (*in vitro*)	*	Boven et al., 2000; Croitoru-Lamoury et al., 2003; Meeuwsen et al., 2003, 2005; Chen et al., 2004
SCF	mRNA (*in vitro*) Protein (*in vitro*)		mRNA (*in vivo*)			Zhang and Fedoroff, 1997
SDF-1 (CXCL-12)	mRNA (*in vitro*) Protein (*in vitro*)	mRNA (*in vitro*) Protein (*in vivo*)	mRNA (*in vitro*) Protein (*in vitro*, *in vivo*)	mRNA (*in vitro*) Protein (*in vitro*)		Croitoru-Lamoury et al., 2003; Peng et al., 2006
TNF-α	mRNA (*in vitro*) Protein (*in vitro*)	mRNA (*in vivo*, *in vitro*)	mRNA (*in vitro*) Protein (*in vivo*)	mRNA (*in vitro*) Protein (*in vitro*)	*	Chung and Benveniste, 1990; Lee et al., 1993; Lung et al., 2001; Brenneman et al., 2003; Croitoru-Lamoury et al., 2003; Nakagawa and Schwartz, 2004; Meeuwsen et al., 2005

Astrocyte cytokine or chemokine expression in developing animals (Young) or adult animals (Adult) *in vivo* or in culture (*in vitro*). The table also indicates cases where reactive astrocytes are reported to express immune molecules *in vivo* or in culture, where astrocytes are regulated in expression of immune molecules by immune factors or other effectors and where astrocytes release immune molecules (asterisk). The absence of information in a column does not indicate that factors are not present or regulated, but rather that this is not reported in the cases cited. CNTF, ciliary neurotrophic factor; GCP, granulocyte chemotactic protein; G-CSF, granulocyte colony-stimulating factor; IFN, interferon; IL, interleukin; IP, interferon-gamma inducible protein; LIF, leukemia inhibitory factor; MCP, monocyte chemoattractant protein; M-CSF, macrophage colony-stimulating factor; MIF, macrophage migration inhibitory factor; MIG, monokine induced by IFN-gamma; MIP, macrophage-inflammatory protein; RANTES, regulated upon activation of normal T-cell expressed and secreted; SCF, stem cell factor; SDF, stromal cell-derived factor; TNF, tumor necrosis factor.

of neurons and neuron precursors in the hippocampus. Effects of chemokines and cytokines are blocked in receptor knockout mice (Ma et al., 1998; Zou et al., 1998; Bagri et al., 2002; Lu et al., 2002).

As was the case with growth factors, immune molecules influence the development of synapses. Thus, ciliary neurotrophic factor (CNTF) regulates spontaneous and impulse-associated transmitter release (Stoop and Poo, 1995), and tumor necrosis factor alpha (TNFα) increases the frequency of spontaneous miniature synaptic currents in fetal hippocampal neurons (Grassi et al., 1994) and inhibits LTP in hippocampal slices (Tancredi et al., 1992). Interleukin (IL)-1β, -2, -6, and interferon (IFN)-α/β also inhibit LTP (Plata-Salaman, 1991).

13.3 The Identification of Molecules Responsible for Astrocyte Effects on Neurons

In an attempt to define the specific astrocyte-derived molecules that impact neuron function, subsequent studies united the observations of astrocyte effects on neurons to the knowledge of the effects of specific molecules on neurons. In this way it could be determined whether the astrocyte-derived medium contains the identified molecules and whether these molecules are responsible for the astrocyte functions. To carry out such studies astrocyte-conditioned medium (ACM) was collected and its actions were compared with those of known molecules. The contribution of these molecules to astrocyte actions then was defined by specifically neutralizing activity of the individual molecules. Such studies were first reported over 20-years ago when it was noted that ACM mimics nerve growth factor (NGF) in its survival effects on neurons (Lindsay, 1979). Moreover, NGF mRNA and protein are identified in astrocytes and NGF is found in the ACM (Furukawa et al., 1987a). When effects of NGF are blocked using anti-NGF neutralizing antibodies, effects of the ACM are reduced, suggesting that some of the ACM actions are due to NGF. More recently, BDNF has also been found to impact neuronal survival (Alderson et al., 1990; Friedman et al., 1993). We have found that ACM increases survival of basal forebrain cholinergic neurons, as monitored by numbers of acetylcholinesterase-positive cells and choline acetyltransferase activity. This effect is blocked by neutralizing antibodies to BDNF (Jean and Dreyfus, personal observation).

In a more recent work, molecules underlying ACM effects on synapse development and function are being identified. The extracellular matrix proteins, thrombospondins (TSPs) appear to constitute one such group of factors (Christopherson et al., 2005). Immature cortical astrocytes express and release TSPs, and TSPs mimic effects of ACM, enhancing synaptogenesis, as defined morphologically (Christopherson et al., 2005). Importantly, when these proteins are depleted from ACM, the numbers of synaptic puncta are reduced. Moreover, when analyzed in TSP1/2 double knockout mice, postnatal cerebral cortex exhibits a decrease in synaptic puncta, suggesting that TSPs are important in the development of mature synapses, not only in culture, but also *in vivo*.

In a final example, TNFα, when applied to hippocampal neuron-glia cultures, increases surface expression of a-amino-3-hydroxy-5-methyl-4-isoxazolepropionic acid (AMPA) receptors and synaptic strength (Beattie et al., 2002). When these cultures are treated with TNFα inhibitors, or when ACM is treated with TNFα inhibitors, there is a decrease in surface expression of AMPA receptors. Similar inhibitory effects are noted when hippocampal slice preparations from 2 to 4-week-old rats are tested using the TNF receptor-1 to inhibit TNFα function.

These examples indicate that astrocytes may impact neuron function, express molecules known to affect such function, and release at least some of these neuron-active proteins in a manner that influences neuron development, neuron survival, and synaptogenesis in culture and *in situ*.

13.4 Astrocytes Are Heterogeneous

Although the earlier studies establish that astrocytes produce molecules that impact neuron development, in general they did not address the question as to whether effects elicited by particular molecules were limited to the particular brain region being studied, or were, instead, a general function of all astrocytes. The fact that effects may be regionally distinct is suggested by other works. For example, glia that consisted primarily of astrocytes from the substantia nigra best supports the survival of substantia nigra dopaminergic neurons when compared with glia from the hippocampus, striatum, and cerebellum (O'Malley et al., 1991). Striatal astrocytes increase nigral axon length, but nigral astrocytes do not (Denis-Donini et al., 1984). Corticospinal neurons, which increase axon length when cultured with either cortical or spinal cord astrocytes, only increase dendritic length when cultured with cortical astrocytes (Dijkstra et al., 1999).

Moreover, these differences may be due to specific proteins produced by the astrocytes. For example, microarray assays compared astrocyte-associated molecules highly expressed in brain regions with high neuroplasticity to those with low neuroplasticity. The high-plasticity regions were those from newborn and adult hippocampus and newborn spinal cord that promote neuronal differentiation of adult neural stem/progenitor cells. Those with low neuroplastiticy were those in the adult spinal cord. IL-1β and IL-6 were found to be highly and differentially expressed in astrocytes from high-neuroplastic regions, while insulin-like growth factor binding protein 6 (IGFBP-6) and decorin, inhibitors of insulin-like growth factor (IGF), were associated with astrocytes from low-neuroplastic regions. It was found that IL-1β and IL-6, at low concentrations, or factor combinations that included these interleukins, promote neural stem cell progenitor neuronal differentiation. The proteins are released from the astrocytes, and effects on neuronal differentiation are blocked by specific antibodies. On the other hand, IGFBP6 and decorin inhibit neuronal differentiation when neural stem cell progenitors are grown with astrocytes (Barkho et al., 2006).

This difference in function associated with different growth or immune factors is complemented by the regionally distinct production and response to growth factors and immune factors reported in additional work. Although all the molecules responsible for the differences in function indicated earlier have not yet been defined, differences in expression of specific molecules have been seen. For example, in response to lipopolysaccharide (LPS), septal and striatal astrocytes express higher levels of TNFα than do cortical, hippocampal, and brain stem astrocytes (Morga et al., 1999). Cortical and striatal astrocytes, but not those from the mesencephalon, respond to FGF-2 by reducing the protein level of the gap junction protein connexin 43 (Reuss et al., 1998). Cortical but not midbrain astrocytes derived from epidermal growth factor receptor knockout animals display apoptosis (Wagner et al., 2006). IL-6 in conjunction with soluble IL-6 receptor induces regionally specific patterns of expression of individual neurotrophins from astrocytes (Marz et al., 1999). It is hypothesized that differential expression of these molecules in astrocytes may be optimized for the neurons functioning in the individual region being assessed.

13.5 The Role of Astrocytes Changes with Development

The earlier work generally focuses on the use of developing neonatal astrocytes to address the role of astrocyte-derived factors on the function of neurons developing in culture. Are these roles altered as the brain ages? To address this question, work *in vivo* has used quantitative *in situ* hybridization and immunocytochemical approaches to determine whether growth and immune molecule expression changes in the mature brain. Interestingly, many growth factors are downregulated in the adult. For example, NGF mRNA is present at levels reduced from that in the neonate in glial fibrillary acidic protein (GFAP) + astrocytes in the basal forebrain, neocortex, amygdala, cerebral cortex, and hypothalamus *in vivo* (Schwartz and Nishiyama, 1994). Moreover, we found that when cortical astrocytes cultured from fetal, neonatal, and adult brain were compared, cells express progressively lower levels of NGF as age increases. This is true as well for BDNF and neurotrophin (NT)-3, also members of the neurotrophin gene family. As was the case with NGF, when cortical astrocytes were derived from brains of increasing ages, BDNF mRNA decreases with increased age (Lackland and Dreyfus, personal observation).

The situation with other factors varies depending on the factor being examined. *In vivo* expression of FGF-2 is similar to that of NGF and decreases as the brain ages, while CNTF increases with age (Stockli et al., 1991; Cintra et al., 1994). Microarray analysis that compared expression of adult and neonate striatal astrocytes in culture complements these observations. In some cases basal gene expression in astrocytes derived from adult striatum is lower than that expressed in the neonatal cells. In other cases the reverse is true. Thus, while gene expression of NGF, FGF-2, as well as heparin-binding growth-associated protein, leukemia inhibitory/cholinergic neuronal differentiation factor, glia maturation factor beta, macrophage migration inhibitory factor, for example, are higher in neonatal cultures, bone morphogenetic protein 4, IGF, transforming growth factor (TGF)-β3, and IL-15 are lower in the neonatal cells (Nakagawa and Schwartz, 2004). Astrocytes apparently can alter levels of expression of individual proteins with age and thereby change the combination of molecules they secrete under what appear to be basal conditions in the adult brain. These changes in expression levels may lead to changes in how astrocytes function in the adult as opposed to the fetal or neonatal animal.

13.6 The Function of Astrocytes Changes After Injury: The Role of Immune Molecules

Multiple types of injury dramatically alter astrocytes (Eddleston et al., 1996). This response to injury is associated with hypertrophy, proliferation, and an increase in intermediate filaments (including GFAP and vimentin) (Ridet et al, 1997). The result of this response can be both protective and damaging. Thus, the glial scar may limit damage or it may inhibit axonal regeneration. Interestingly, when reactive astrocytes were selectively ablated in adult transgenic mice that had sustained a stab wound, there was an increase in migration of monocytes, macrophages, neutrophils, and lymphocytes to the wound area, a loss of the reparative ability of the blood–brain barrier, an increase in neuronal degeneration, but an increase in neurite elongation (Bush et al., 1999), suggesting that molecules produced by reactive astrocytes play both supportive, as well as inhibitory roles. This is also supported by observations that injury is associated with an elevation of proinflammatory cytokines such as TNF-α and IL1-β (Dong and Benveniste, 2001), which are believed to worsen or enhance neurodegeneration, as well as an elevation of trophic factors such as TGF-β, NGF, and BDNF (Ridet et al., 1997; Dougherty et al., 2000).

Clearly the response to injury is a complex series of events. IL-1 is generally believed to be an early responder to injury (Liberto et al., 2004). It is produced by microglia and astrocytes. As a result of its expression a cascade of events occurs that influence astrocyte reactivity and expression of molecules including TNF-α, IL-1α, IL-1β, IL-6, and TGF-β1 (Dong and Benveniste, 2001). Interestingly, immune molecules influence their own expression as well as that of the traditional growth factors. For example, IFN-β increases IL-6 mRNA and protein levels (Okada et al., 2005). IFN-γ, TGF-β, TNF-α, IL-1β, IL-4, and IL-5 elevate NGF mRNA (Spranger et al., 1990; Awatsuji et al., 1993; Schwartz and Nishiyama, 1994; Kuno et al., 2006). Cytokines also enhance FGF-2 (Albrecht et al., 2003). TNF-α promotes expression of GDNF (Kuno et al., 2006).

This increase in immune and trophin molecules differs with development. Cultured adult astrocytes respond significantly less to IL-1β and IFN-γ than do neonatal cells (Schwartz and Nishiyama, 1994). However, lesioned astrocytes, i.e., those derived from 6-hydroxydopamine lesioned brains, respond robustly to cytokines by a dramatic increase in NGF mRNA (Schwartz and Nishiyama, 1994). Thus, injury appears to sensitize astrocytes to cytokines, elevating their effect. Importantly, not all factors respond in the same manner. In injury studies *in vivo* in which CNTF was increased, BDNF mRNA and NT3 mRNA were slightly decreased, indicating that multiple molecules associated with the astrocyte may respond distinctly to the same stimuli (Ip et al., 1993).

13.7 Other Molecules That Regulate Astrocytes: Their Role During Injury

In addition to immune molecules, multiple other factors including glucocorticoids, gonadal hormones, and neurotransmitters regulate astrocyte function and play a role during injury. Expression of these effectors or their receptors is altered by injury. As a result, functions of the effectors may be changed.

13.7.1 The Role of Glucocorticoids in Astrocyte Function

After brain injury, increased expression of glucocorticoids occurs (Woolf et al., 1990). These hormones appear to affect astrocytes directly. Astrocytes express receptors for glucocorticoids, and culture studies of astrocytes indicate that these cells respond to these steroids (Garcia-Segura et al., 1996a). In particular, glucocorticoid treatment in culture suppresses astrocyte proliferation (Kniss and Burry, 1985) and upregulates glutamine synthetase (Aizenman and de Vellis, 1987). *In vivo* chronic glucocorticoid treatment, although not reducing increases in GFAP elicited by injury in the lesion site, does limit increases in GFAP mRNA in more distant sites (Laping et al., 1991; O'Callaghan et al., 1991).

With respect to regulation of specific growth or immune molecule expression, steroids impact astrocyte function. For example, dexamethasone decreases NGF mRNA expression in hippocampal astrocytes (Niu et al., 1997), and under conditions of stress, methylprednisone (Schmitt et al., 2006), or dexamethasone (Grimaldi et al., 1998) suppress IL-6 secretion. Dexamethasone also inhibits increases in TNF-α and in IL-1β transcription elicited by stress (Velasco et al., 1991). As has been noted earlier, alterations in expression of specific proteins can differ depending on the molecule being monitored. Dexamethasone increases FGF-2 in the same population that exhibits decreases in NGF mRNA (Niu et al., 1997). Moreover, alterations of these multiple factors, as occurs in injury lead to interactions among these factors. Corticosterone or dexamethasone, for example, inhibits increases in NGF release elicited by TGF-β and IL-1β (Hahn et al., 1997). Calcitriol (1,25-dihydroxyvitamin D3), potentially from microglia, enhances NGF secretion and augments IL-1β and TGF-β effects (Hahn et al., 1997). These steroids, therefore, impact growth factor and immune factor function in astrocytes, and these effects are also impacted by the constellation of factors that are regulated by the lesion itself.

13.7.2 The Role of Gonadal Steroids

As was the case with glucocorticoids, receptors for estrogen and androgen and progesterone-inducible receptors are also present in astrocytes. This has been shown to be true in culture (Santagati et al., 1994; Buchanan et al., 2000) as well as *in vivo* (Milner et al., 2005). Their expression is found in multiple brain regions including the hypothalamus and hippocampus (Langub and Watson, 1992; Azcoitia et al., 1999; Milner et al., 2001, 2005).

The effects these receptors mediate are beginning to be understood. Of importance to this review, effects of gonadal hormones may be on trophic factor expression. Thus, estradiol enhances expression of IGF-1 in astrocytes of the hypothalamus (Duenas et al., 1994) and, again, there may be regional heterogeneity in these responses. Estradiol increases TGF-α mRNA in hypothalamic astrocytes, but not cerebellar astrocytes (Ma et al., 1994). The studies *in vivo*, of course, are limited by the complexity of the system, and it was unclear in the earlier studies whether these effects are due to direct effects on astrocytes. In fact, in some studies where morphology of astrocytes was evaluated, estradiol had no effect when neurons were absent (Torres-Aleman et al., 1992). Culture studies, however, are documenting direct actions of estradiol on astrocytes. For example, estrogen elevates release of TGF-β1 from hypothalamic astrocytes (Buchanan et al., 2000). We have shown that estradiol treatment of enriched cortical astrocyte cultures results in elevated expression of BDNF mRNA in astrocytes derived from neonatal or adult brain. These effects on astrocytes may be developmentally regulated and differ with the factor being monitored. Estrogen elevates NT3 in adult astrocytes, but not neonatal cells (Lackland and Dreyfus, personal observation). Regional differences are noted in the culture work also. In our studies, basal forebrain astrocytes from the neonate do not respond to estradiol. However, cortical and hippocampal astrocytes do.

Gonadal hormones may also impact astrocytes in models of injury. For example, gonadal hormones are reported to suppress astrogliosis after lesions. In gonadectomized rats, increased proliferation of astrocytes is evident following a penetrating injury of the cortex and hippocampus, but this effect is reversed by treatment with estradiol or progesterone to the ovariectomized rat or testosterone to castrated males (Garcia-Segura et al., 1996b). These effects of gonadal hormones extend to culture studies, where it is noted that stressed astrocytes may be supported by estrogen. 17-β estradiol decreases amyloid-β or LPS-induced nuclear factor kappa B activation (Dodel et al., 1999) and cell death in response to oxygen-glucose deprivation (Liu et al., 2007). It is of interest that in astrocytes, aromatase expression and activity are upregulated after lesions (Carswell et al., 2005). Normally these cells produce little aromatase. Thus, local concentrations of estrogen may increase in a damaged region, leading to increased synthesis of growth factors by astrocytes reacting to the lesion. Related work indicates that global ischemia increases estrogen receptor-α expression, providing a substrate upon which increased estrogen may act (Miller et al., 2005).

13.7.3 Neurotransmitters Also Impact Astrocyte Function

In vivo, astrocytes in the developing animals as well as in the adult and in aged animals express many neurotransmitter receptors, suggesting that they respond to neural signals throughout life. α- and β-noradrenergic receptors, dopaminergic receptors, muscarinic and nicotinic cholinergic receptors, and non-*N*-methyl-D-aspartic acid ionotropic and metabotropic glutamatergic receptors (mGluR) are associated with astrocytes (Murphy and Pearce, 1987; Hosli et al., 1994; Porter and McCarthy, 1997). In some cases these receptors are located in close proximity to axon terminals, indicating that astrocytes are in a position to respond to released neurotransmitter and potentially, in turn, may produce molecules that impact axon function (Aoki, 1992). In fact, astrocytes respond to transmitters by altering expression of trophic factors and immune molecules. Norepinephrine and other β-adrenergic agonists as well as dopamine and epinephrine increase expression and secretion of NGF and expression of FGF-2 and BDNF (Furukawa et al., 1987b; Zafra et al., 1992; Rudge et al., 1994; Schwartz et al., 1994; Riva et al., 1996; Inoue et al., 1997). Similarly, glutamate increases expression and/or release of BDNF, NGF, FGF-2, TGF-β, and GDNF (Pechan et al., 1993; Ho et al., 1995; Bruno et al., 1998; Ciccarelli et al., 1999; Wu et al., 2004). In the case of NGF and TGF-β, the effects of glutamate are through excitation of mGluRs (Bruno et al., 1998; Ciccarelli et al., 1999). In the case of BDNF, we have found that the release also is dependent on metabotropic receptors as well as internal stores of calcium. Enhanced BDNF release is observed within 10 minutes of glutamate stimulation, suggesting that BDNF release may be temporally regulated by glutamate (Jean and Dreyfus, personal observation).

As has been seen before, the response to these signals is distinct depending on the molecules being examined. Catecholamines, for instance, decrease CNTF mRNA and protein and FGF-1 mRNA in astrocytes that were increased in their expression of NGF and FGF-2 (Rudge et al., 1994; Nagao et al., 1995; Riva et al., 1996). While the cholinergic muscarinic agonist, carbachol, elicits increases in BDNF mRNA, it does not elicit increases in NGF mRNA or NT3 mRNA in the same astrocyte population. Glutamate, through metabotropic receptors, increases both NGF mRNA and BDNF mRNA, but does not impact NT3 mRNA (Wu et al., 2004). Interestingly, expression in and release of immune molecules from astrocytes is also affected by transmitters, supporting the suggestion that these molecules impact astrocytes normally, as well as during an immune response to injury. Thus, norepinephrine, epinephrine, adenosine, substance P, calcitonin, serotonin, vasoactive intestinal peptide (VIP), and pituitary adenylate cyclase-activating polypeptide all increase IL-6 in neonatal astrocyte cultures and VIP induces release of TNF-α, IL-3, granulocyte colony stimulating factor, and macrophage colony stimulating factor (Maimone et al., 1993; Derocq et al., 1996; Tatsuno et al., 1996; Kiriyama et al., 1997; Liu et al., 2000; Brenneman et al., 2003; Franke et al., 2004; Lieb et al., 2005). Furthermore, group II mGluRs appear to mediate effects of glutamate on IL1β-induced release of IL-6 in human adult astrocyte cultures (Aronica et al.,

2005) and β2 adrenoceptors decrease LPS-induced TNF and IL-6 expression (Nakamura et al., 1998a), suggesting that activation of these receptors also impacts astrocytes after injury.

Effects of transmitters may vary regionally and within regions. *In vivo*, under normal physiological conditions, astrocytes of the nucleus accumbens express the purinoceptor subtypes P2Y1 and 4 (Franke et al., 2004). Cortical astrocytes, on the other hand, display P2Y1 and 4, as well as P2Y2 and 6. In culture, carbamylcholine stimulates astrocytes from the mesencephalon, but not from the striatum or cerebral cortex (el-Etr et al., 1989). Within regions mGluR subtypes are expressed in a nonuniform fashion. For example, this is noted in cultures of spinal cord astrocytes (Silva et al., 1999). This differential pattern of receptor expression is seen in cortical astrocyte cultures as well, where only a subset of astrocytes expresses α 1-adrenergic receptors (Lerea and McCarthy, 1989).

After injury receptor expression changes and these changes may be regionally specific. Levels of purinergic subtypes P2Y1 and 4 increase in the nucleus accumbens along with the expression of two additional subtypes, P2Y2 and 6. Cortical astrocytes, in contrast, as noted earlier, display P2Y1, 2, 4, and 6 expression under normal conditions and all are upregulated in response to injury (Franke et al., 2004). Cholinergic receptors are increased on astrocytes in Alzheimer's disease (Messamore et al., 1994), and mGluRs are upregulated in a variety of neurodegenerative diseases (Mudo et al., 2007). These increases suggest that astrocytes may become more sensitive to transmitters in disease states. The possibility, therefore, is that this impacts astrocyte function, including the ability to produce growth factors and immune molecules.

13.8 Concluding Remarks

This review indicates that astrocytes serve potentially critical roles as providers of proteins that impact neuron function. The culture studies presented argue that these roles are to produce and release molecules that affect progenitor proliferation, neuron differentiation and survival, and the formation of synaptic connections. Moreover, these studies indicate that the synthesis and release of growth factors and immune molecules from astrocytes is responsive to immune molecules, themselves, as well as steroids and transmitters. Effects of these molecules may be enhanced after a variety of lesions. However, the degree of the response differs depending on the molecule being examined. Importantly, it is observed that the impact of the environment on synthesis and release of immune molecules and growth factors differs in different brain regions, and within different brain regions. Moreover, the production and release of protein molecules may be affected by development.

Most studies thus far have focused on cultures to define these properties of astrocytes. *In vivo* studies have been largely limited to the visualization of immune molecules and receptors on astrocytes. However, new availability of mouse models that will permit the manipulation of specific molecules in astrocytic populations now will permit the testing of the principles defined in culture on the intact brain. In particular, conditional knockout mice are being created that limit deletions of specific proteins to GFAP + cells at particular points in development (Casper et al., 2007). Examination of the consequences of such manipulations is the next step in defining the roles of astrocytes as growth factor and immune molecule providers that impact neuron function.

Acknowledgment The work referred to in this review that was performed in the Dreyfus lab was supported by NIH HD23315.

References

Abbas N, Bednar I, Mix E, Marie S, Paterson D, Ljungberg A, Morris C, Winblad B, Nordberg A, Zhu J (2002) Up-regulation of the inflammatory cytokines IFN-gamma and IL-12 and down-regulation of IL-4 in cerebral cortex regions of APP(SWE) transgenic mice. J Neuroimmunol 126:50–57.

Aizenman Y, de Vellis J (1987) Synergistic action of thyroid hormone, insulin and hydrocortisone on astrocyte differentiation. Brain Res 414:301–308.

Albrecht PJ, Murtie JC, Ness JK, Redwine JM, Enterline JR, Armstrong RC, Levison SW (2003) Astrocytes produce CNTF during the remyelination phase of viral-induced spinal cord demyelination to stimulate FGF-2 production. Neurobiol Dis 13:89–101.

Alderson RF, Alterman AL, Barde YA, Lindsay RM (1990) Brain-derived neurotrophic factor increases survival and differentiated functions of rat septal cholinergic neurons in culture. Neuron 5:297–306.

Aloisi F, Care A, Borsellino G, Gallo P, Rosa S, Bassani A, Cabibbo A, Testa U, Levi G, Peschle C (1992) Production of hemolymphopoietic cytokines (IL-6, IL-8, colony-stimulating factors) by normal human astrocytes in response to IL-1 beta and tumor necrosis factor-alpha. J Immunol 149:2358–2366.

Aloisi F, Rosa S, Testa U, Bonsi P, Russo G, Peschle C, Levi G (1994) Regulation of leukemia inhibitory factor synthesis in cultured human astrocytes. J Immunol 152:5022–5031.

Aoki C (1992) Beta-adrenergic receptors: astrocytic localization in the adult visual cortex and their relation to catecholamine axon terminals as revealed by electron microscopic immunocytochemistry. J Neurosci 12:781–792.

Appel E, Kolman O, Kazimirsky G, Blumberg PM, Brodie C (1997) Regulation of GDNF expression in cultured astrocytes by inflammatory stimuli. Neuroreport 8:3309–3312.

Arakawa Y, Bito H, Furuyashiki T, Tsuji T, Takemoto-Kimura S, Kimura K, Nozaki K, Hashimoto N, Narumiya S (2003) Control of axon elongation via an SDF-1alpha/Rho/mDia pathway in cultured cerebellar granule neurons. J Cell Biol 161:381–391.

Aronica E, Gorter JA, Rozemuller AJ, Yankaya B, Troost D (2005) Activation of metabotropic glutamate receptor 3 enhances interleukin (IL)-1beta-stimulated release of IL-6 in cultured human astrocytes. Neuroscience 130:927–933.

Awatsuji H, Furukawa Y, Hirota M, Murakami Y, Nii S, Furukawa S, Hayashi K (1993) Interleukin-4 and -5 as modulators of nerve growth factor synthesis/secretion in astrocytes. J Neurosci Res 34:539–545.

Azcoitia I, Sierra A, Garcia-Segura LM (1999) Localization of estrogen receptor beta-immunoreactivity in astrocytes of the adult rat brain. Glia 26:260–267.

Bagri A, Gurney T, He X, Zou YR, Littman DR, Tessier-Lavigne M, Pleasure SJ (2002) The chemokine SDF1 regulates migration of dentate granule cells. Development 129:4249–4260.

Ballotti R, Nielsen FC, Pringle N, Kowalski A, Richardson WD, Van Obberghen E, Gammeltoft S (1987) Insulin-like growth factor I in cultured rat astrocytes: expression of the gene, and receptor tyrosine kinase. EMBO J 6:3633–3639.

Banner LR, Moayeri NN, Patterson PH (1997) Leukemia inhibitory factor is expressed in astrocytes following cortical brain injury. Exp Neurol 147:1–9.

Barkho BZ, Song H, Aimone JB, Smrt RD, Kuwabara T, Nakashima K, Gage FH, Zhao X (2006) Identification of astrocyte-expressed factors that modulate neural stem/progenitor cell differentiation. Stem Cells Dev 15:407–421.

Beattie EC, Stellwagen D, Morishita W, Bresnahan JC, Ha BK, Von Zastrow M, Beattie MS, Malenka RC (2002) Control of synaptic strength by glial TNFalpha. Science 295:2282–2285.

Boven LA, Montagne L, Nottet HS, De Groot CJ (2000) Macrophage inflammatory protein-1alpha (MIP-1alpha), MIP-1beta, and RANTES mRNA semiquantification and protein expression in active demyelinating multiple sclerosis (MS) lesions. Clin Exp Immunol 122:257–263.

Brenneman DE, Phillips TM, Festoff BW, Gozes I (1997) Identity of neurotrophic molecules released from astroglia by vasoactive intestinal peptide. Ann N Y Acad Sci 814:167–173.

Brenneman DE, Phillips TM, Hauser J, Hill JM, Spong CY, Gozes I (2003) Complex array of cytokines released by vasoactive intestinal peptide. Neuropeptides 37:111–119.

Bruno V, Battaglia G, Casabona G, Copani A, Caciagli F, Nicoletti F (1998) Neuroprotection by glial metabotropic glutamate receptors is mediated by transforming growth factor-beta. J Neurosci 18:9594–9600.

Bsibsi M, Persoon-Deen C, Verwer RW, Meeuwsen S, Ravid R, Van Noort JM (2006) Toll-like receptor 3 on adult human astrocytes triggers production of neuroprotective mediators. Glia 53:688–695.

Buchanan CD, Mahesh VB, Brann DW (2000) Estrogen-astrocyte-luteinizing hormone-releasing hormone signaling: a role for transforming growth factor-beta(1). Biol Reprod 62:1710–1721.

Bush TG, Puvanachandra N, Horner CH, Polito A, Ostenfeld T, Svendsen CN, Mucke L, Johnson MH, Sofroniew MV (1999) Leukocyte infiltration, neuronal degeneration, and neurite outgrowth after ablation of scar-forming, reactive astrocytes in adult transgenic mice. Neuron 23:297–308.

Carpentier PA, Begolka WS, Olson JK, Elhofy A, Karpus WJ, Miller SD (2005) Differential activation of astrocytes by innate and adaptive immune stimuli. Glia 49:360–374.

Carswell HV, Dominiczak AF, Garcia-Segura LM, Harada N, Hutchison JB, Macrae IM (2005) Brain aromatase expression after experimental stroke: topography and time course. J Steroid Biochem Mol Biol 96:89–91.

Casper KB, Jones K, McCarthy KD (2007) Characterization of astrocyte-specific conditional knockouts. Genesis 45:292–299.

Chen CJ, Chen JH, Chen SY, Liao SL, Raung SL (2004) Upregulation of RANTES gene expression in neuroglia by Japanese encephalitis virus infection. J Virol 78:12107–12119.

Chen J, Leong SY, Schachner M (2005) Differential expression of cell fate determinants in neurons and glial cells of adult mouse spinal cord after compression injury. Eur J Neurosci 22:1895–1906.

Chen LW, Zhang JP, Kwok-Yan Shum D, Chan YS (2006) Localization of nerve growth factor, neurotrophin-3, and glial cell line-derived neurotrophic factor in nestin-expressing reactive astrocytes in the caudate-putamen of 1-methyl-4-phenyl-1,2,3,6-tetrahydropyridine-treated C57/Bl mice. J Comp Neurol 497:898–909.

Choi BH, Suzuki M, Kim T, Wagner SL, Cunningham DD (1990) Protease nexin-1. Localization in the human brain suggests a protective role against extravasated serine proteases. Am J Pathol 137:741–747.

Choi JS, Kim SY, Park HJ, Cha JH, Choi YS, Chung JW, Chun MH, Lee MY (2004) Differential regulation of ciliary neurotrophic factor and its receptor in the rat hippocampus in response to kainic acid-induced excitotoxicity. Mol Cells 17:292–296.

Christopherson KS, Ullian EM, Stokes CC, Mullowney CE, Hell JW, Agah A, Lawler J, Mosher DF, Bornstein P, Barres BA (2005) Thrombospondins are astrocyte-secreted proteins that promote CNS synaptogenesis. Cell 120:421–433.

Chung IY, Benveniste EN (1990) Tumor necrosis factor-alpha production by astrocytes. Induction by lipopolysaccharide, IFN-gamma, and IL-1 beta. J Immunol 144:2999–3007.

Ciccarelli R, Di Iorio P, Bruno V, Battaglia G, D'Alimonte I, D'Onofrio M, Nicoletti F, Caciagli F (1999) Activation of A(1) adenosine or mGlu3 metabotropic glutamate receptors enhances the release of nerve growth factor and S-100beta protein from cultured astrocytes. Glia 27:275–281.

Cintra A, Lindberg J, Chadi G, Tinner B, Moller A, Gustafsson JA, DeKloet ER, Oitzl M, Nishikawa K, Agnati LF, . (1994) Basic fibroblast growth factor and steroid receptors in the aging hippocampus of the brown Norway rat: immunocytochemical analysis in combination with stereology. Neurochem Int 25:39–45.

Constam DB, Philipp J, Malipiero UV, ten Dijke P, Schachner M, Fontana A (1992) Differential expression of transforming growth factor-beta 1, -beta 2, and -beta 3 by glioblastoma cells, astrocytes, and microglia. J Immunol 148:1404–1410.

Croitoru-Lamoury J, Guillemin GJ, Boussin FD, Mognetti B, Gigout LI, Cheret A, Vaslin B, Le Grand R, Brew BJ, Dormont D (2003) Expression of chemokines and their receptors in human and simian astrocytes: evidence for a central role of TNF alpha and IFN gamma in CXCR4 and CCR5 modulation. Glia 41:354–370.

De Groot CJ, Montagne L, Barten AD, Sminia P, Van Der Valk P (1999) Expression of transforming growth factor (TGF)-beta1, -beta2, and -beta3 isoforms and TGF-beta type I and type II receptors in multiple sclerosis lesions and human adult astrocyte cultures. J Neuropathol Exp Neurol 58:174–187.

Denis-Donini S, Glowinski J, Prochiantz A (1984) Glial heterogeneity may define the three-dimensional shape of mouse mesencephalic dopaminergic neurones. Nature 307:641–643.

Derocq JM, Segui M, Blazy C, Emonds-Alt X, Le Fur G, Brelire JC, Casellas P (1996) Effect of substance P on cytokine production by human astrocytic cells and blood mononuclear cells: characterization of novel tachykinin receptor antagonists. FEBS Lett 399:321–325.

DeVito WJ, Avakian C, Stone S, Okulicz WC, Tang KT, Shamgochian M (1995) Prolactin induced expression of interleukin-1 alpha, tumor necrosis factor-alpha, and transforming growth factor-alpha in cultured astrocytes. J Cell Biochem 57:290–298.

Dhandapani KM, Wade FM, Mahesh VB, Brann DW (2005) Astrocyte-derived transforming growth factor-{beta} mediates the neuroprotective effects of 17beta-estradiol: involvement of nonclassical genomic signaling pathways. Endocrinology 146:2749–2759.

Dijkstra S, Bar PR, Gispen WH, Joosten EA (1999) Selective stimulation of dendrite outgrowth from identified corticospinal neurons by homotopic astrocytes. Neuroscience 92:1331–1342.

Dodel RC, Du Y, Bales KR, Gao F, Paul SM (1999) Sodium salicylate and 17beta-estradiol attenuate nuclear transcription factor NF-kappaB translocation in cultured rat astroglial cultures following exposure to amyloid A beta(1–40) and lipopolysaccharides. J Neurochem 73:1453–1460.

Dong Y, Benveniste EN (2001) Immune function of astrocytes. Glia 36:180–190.

Dougherty KD, Dreyfus CF, Black IB (2000) Brain-derived neurotrophic factor in astrocytes, oligodendrocytes, and microglia/macrophages after spinal cord injury. Neurobiol Dis 7:574–585.

Dreyfus CF, Dai X, Lercher LD, Racey BR, Friedman WJ, Black IB (1999) Expression of neurotrophins in the adult spinal cord in vivo. J Neurosci Res 56:1–7.

Duenas M, Luquin S, Chowen JA, Torres-Aleman I, Naftolin F, Garcia-Segura LM (1994) Gonadal hormone regulation of insulin-like growth factor-I-like immunoreactivity in hypothalamic astroglia of developing and adult rats. Neuroendocrinology 59:528–538.

Eddleston M, de la Torre JC, Campbell IL, Oldstone MB (1996) The M22 antibody identifies highly activated reactive astrocytes responding to central nervous system disease. Acta Neuropathol 91:298–308.

el-Etr M, Cordier J, Torrens Y, Glowinski J, Premont J (1989) Pharmacological and functional heterogeneity of astrocytes: regional differences in phospholipase C stimulation by neuromediators. J Neurochem 52:981–984.

Emsley JG, Hagg T (2003) Endogenous and exogenous ciliary neurotrophic factor enhances forebrain neurogenesis in adult mice. Exp Neurol 183:298–310.

Folkerth RD, Keefe RJ, Haynes RL, Trachtenberg FL, Volpe JJ, Kinney HC (2004) Interferon-gamma expression in periventricular leukomalacia in the human brain. Brain Pathol 14:265–274.

Franke H, Krugel U, Grosche J, Heine C, Hartig W, Allgaier C, Illes P (2004) P2Y receptor expression on astrocytes in the nucleus accumbens of rats. Neuroscience 127:431–441.

Friedman WJ, Ibanez CF, Hallbook F, Persson H, Cain LD, Dreyfus CF, Black IB (1993) Differential actions of neurotrophins in the locus coeruleus and basal forebrain. Exp Neurol 119:72–78.

Furukawa S, Furukawa Y, Satoyoshi E, Hayashi K (1987a) Synthesis/secretion of nerve growth factor is associated with cell growth in cultured mouse astroglial cells. Biochem Biophys Res Commun 142:395–402.

Furukawa S, Furukawa Y, Satoyoshi E, Hayashi K (1987b) Regulation of nerve growth factor synthesis/secretion by catecholamine in cultured mouse astroglial cells. Biochem Biophys Res Commun 147:1048–1054.

Gabryel B, Labuzek K, Malecki A, Herman ZS (2004) Immunophilin ligands decrease release of pro-inflammatory cytokines (IL-1beta, TNF-alpha and IL-2) in rat astrocyte cultures exposed to simulated ischemia in vitro. Pol J Pharmacol 56:129–136.

Garcia-Estrada J, Garcia-Segura LM, Torres-Aleman I (1992) Expression of insulin-like growth factor I by astrocytes in response to injury. Brain Res 592:343–347.

Garcia-Segura LM, Chowen JA, Naftolin F (1996a) Endocrine glia: roles of glial cells in the brain actions of steroid and thyroid hormones and in the regulation of hormone secretion. Front Neuroendocrinol 17:180–211.

Garcia-Segura LM, Chowen JA, Duenas M, Parducz A, Naftolin F (1996b) Gonadal steroids and astroglial plasticity. Cell Mol Neurobiol 16:225–237.

Gard AL, Burrell MR, Pfeiffer SE, Rudge JS, Williams WC II (1995) Astroglial control of oligodendrocyte survival mediated by PDGF and leukemia inhibitory factor-like protein. Development 121:2187–2197.

Grassi F, Mileo AM, Monaco L, Punturieri A, Santoni A, Eusebi F (1994) TNF-alpha increases the frequency of spontaneous miniature synaptic currents in cultured rat hippocampal neurons. Brain Res 659:226–230.

Grimaldi M, Navarra P, Pozzoli G, Preziosi P, Schettini G (1998) Bacterial lipopolysaccharide increases interleukin-6 and prostaglandin release in rat cortical type I astrocytes by different mechanisms: role of anti-inflammatory agents. Biochem Biophys Res Commun 250:798–804.

Hahn M, Lorez H, Fischer G (1997) Effect of calcitriol in combination with corticosterone, interleukin-1beta, and transforming growth factor-beta1 on nerve growth factor secretion in an astroglial cell line. J Neurochem 69:102–109.

Ho A, Gore AC, Weickert CS, Blum M (1995) Glutamate regulation of *GDNF* gene expression in the striatum and primary striatal astrocytes. Neuroreport 6:1454–1458.

Hosli L, Hosli E, Winter T, Stauffer S (1994) Coexistence of cholinergic and somatostatin receptors on astrocytes of rat CNS. Neuroreport 5:1469–1472.

Hotta N, Aoyama M, Inagaki M, Ishihara M, Miura Y, Tada T, Asai K (2005) Expression of glia maturation factor beta after cryogenic brain injury. Brain Res Mol Brain Res 133:71–77.

Hulshof S, Montagne L, De Groot CJ, Van Der Valk P (2002) Cellular localization and expression patterns of interleukin-10, interleukin-4, and their receptors in multiple sclerosis lesions. Glia 38:24–35.

Inoue S, Susukida M, Ikeda K, Murase K, Hayashi K (1997) Dopaminergic transmitter up-regulation of brain-derived neurotrophic factor (BDNF) and nerve growth factor (NGF) synthesis in mouse astrocytes in culture. Biochem Biophys Res Commun 238:468–472.

Ip NY, Wiegand SJ, Morse J, Rudge JS (1993) Injury-induced regulation of ciliary neurotrophic factor mRNA in the adult rat brain. Eur J Neurosci 5:25–33.

Jean YY, Lercher LD, Dreyfus CF (2006) Glutamate elicited increases in expression and release of BDNF in basal forebrain (BF) astrocytes influences the survival and function of BF cholinergic neruons. Soc Neurosci Abstract No. 4234.

Juul SE, Yachnis AT, Rojiani AM, Christensen RD (1999) Immunohistochemical localization of erythropoietin and its receptor in the developing human brain. Pediatr Dev Pathol 2:148–158.
Kato M, Mizuguchi M, Takashima S (1995) Developmental changes of epidermal growth factor-like immunoreactivity in the human fetal brain. J Neurosci Res 42:486–492.
Kerber G, Streif R, Schwaiger FW, Kreutzberg GW, Hager G (2003) Neuregulin-1 isoforms are differentially expressed in the intact and regenerating adult rat nervous system. J Mol Neurosci 21:149–165.
Kim SY, Li J, Bentsman G, Brooks AI, Volsky DJ (2004) Microarray analysis of changes in cellular gene expression induced by productive infection of primary human astrocytes: implications for HAD. J Neuroimmunol 157:17–26.
Kiriyama Y, Murayama T, Tokumitsu Y, Nomura Y (1997) Protein kinase A-dependent IL-6 production induced by calcitonin in human glioblastoma A172 cells. J Neuroimmunol 76:139–144.
Kniss DA, Burry RW (1985) Glucocorticoid hormones inhibit DNA synthesis in glial cells cultured in chemically defined medium. Exp Cell Res 161:29–40.
Koda M, Nishio Y, Hashimoto M, Kamada T, Koshizuka S, Yoshinaga K, Onodera S, Nishihira J, Moriya H, Yamazaki M (2004) Up-regulation of macrophage migration-inhibitory factor expression after compression-induced spinal cord injury in rats. Acta Neuropathol 108:31–36.
Koo H, Choi BH (2001) Expression of glial cell line-derived neurotrophic factor (GDNF) in the developing human fetal brain. Int J Dev Neurosci 19:549–558.
Krasowska-Zoladek A, Banaszewska M, Kraszpulski M, Konat GW (2007) Kinetics of inflammatory response of astrocytes induced by TLR 3 and TLR4 ligation. J Neurosci Res 85:205–212.
Kuno R, Yoshida Y, Nitta A, Nabeshima T, Wang J, Sonobe Y, Kawanokuchi J, Takeuchi H, Mizuno T, Suzumura A (2006) The role of TNF-alpha and its receptors in the production of NGF and GDNF by astrocytes. Brain Res 1116:12–18.
Kutsch O, Oh J, Nath A, Benveniste EN (2000) Induction of the chemokines interleukin-8 and IP-10 by human immunodeficiency virus type 1 tat in astrocytes. J Virol 74:9214–9221.
Langub MCJr, Watson RE Jr (1992) Estrogen receptor-immunoreactive glia, endothelia, and ependyma in guinea pig preoptic area and median eminence: electron microscopy. Endocrinology 130:364–372.
Laping NJ, Nichols NR, Day JR, Finch CE (1991) Corticosterone differentially regulates the bilateral response of astrocyte mRNAs in the hippocampus to entorhinal cortex lesions in male rats. Brain Res Mol Brain Res 10:291–297.
Ledda F, Paratcha G, Sandoval-Guzman T, Ibanez CF (2007) GDNF and GFRalpha1 promote formation of neuronal synapses by ligand-induced cell adhesion. Nat Neurosci 10:293–300.
Lee SC, Liu W, Dickson DW, Brosnan CF, Berman JW (1993) Cytokine production by human fetal microglia and astrocytes. Differential induction by lipopolysaccharide and IL-1 beta. J Immunol 150:2659–2667.
Lee YB, Satoh J, Walker DG, Kim SU (1996) Interleukin-15 gene expression in human astrocytes and microglia in culture. Neuroreport 7:1062–1066.
Lerea LS, McCarthy KD (1989) Astroglial cells in vitro are heterogeneous with respect to expression of the alpha 1-adrenergic receptor. Glia 2:135–147.
Li A, Guo H, Luo X, Sheng J, Yang S, Yin Y, Zhou J, Zhou J (2006) Apomorphine-induced activation of dopamine receptors modulates FGF-2 expression in astrocytic cultures and promotes survival of dopaminergic neurons. FASEB J 20:1263–1265.
Li X, Sankrithi N, Davis FC (2002) Transforming growth factor-alpha is expressed in astrocytes of the suprachiasmatic nucleus in hamster: role of glial cells in circadian clocks. Neuroreport 13:2143–2147.
Li Y, Fu L, Gonzales DM, Lavi E (2004) Coronavirus neurovirulence correlates with the ability of the virus to induce proinflammatory cytokine signals from astrocytes and microglia. J Virol 78:3398–3406.

Liberto CM, Albrecht PJ, Herx LM, Yong VW, Levison SW (2004) Pro-regenerative properties of cytokine-activated astrocytes. J Neurochem 89:1092–1100.

Lieb K, Biersack L, Waschbisch A, Orlikowski S, Akundi RS, Candelario-Jalil E, Hull M, Fiebich BL (2005) Serotonin via 5-HT7 receptors activates p38 mitogen-activated protein kinase and protein kinase C epsilon resulting in interleukin-6 synthesis in human U373 MG astrocytoma cells. J Neurochem 93:549–559.

Lim DA, Alvarez-Buylla A (1999) Interaction between astrocytes and adult subventricular zone precursors stimulates neurogenesis. Proc Natl Acad Sci USA 96:7526–7531.

Lindsay RM (1979) Adult rat brain astrocytes support survival of both NGF-dependent and NGF-insensitive neurones. Nature 282:80–82.

Lisovoski F, Blot S, Lacombe C, Bellier JP, Dreyfus PA, Junier MP (1997) Transforming growth factor alpha expression as a response of murine motor neurons to axonal injury and mutation-induced degeneration. J Neuropathol Exp Neurol 56:459–471.

Liu JS, John GR, Sikora A, Lee SC, Brosnan CF (2000) Modulation of interleukin-1beta and tumor necrosis factor alpha signaling by P2 purinergic receptors in human fetal astrocytes. J Neurosci 20:5292–5299.

Liu M, Hurn PD, Roselli CE, Alkayed NJ (2007) Role of P450 aromatase in sex-specific astrocytic cell death. J Cereb Blood Flow Metab 27:135–141.

Liu R, Suzuki A, Guo Z, Mizuno Y, Urabe T (2006) Intrinsic and extrinsic erythropoietin enhances neuroprotection against ischemia and reperfusion injury in vitro. J Neurochem 96:1101–1110.

Loeffler S, Fayard B, Weis J, Weissenberger J (2005) Interleukin-6 induces transcriptional activation of vascular endothelial growth factor (VEGF) in astrocytes in vivo and regulates VEGF promoter activity in glioblastoma cells via direct interaction between STAT3 and Sp1. Int J Cancer 115:202–213.

Lu M, Grove EA, Miller RJ (2002) Abnormal development of the hippocampal dentate gyrus in mice lacking the CXCR4 chemokine receptor. Proc Natl Acad Sci USA 99:7090–7095.

Lung HL, Leung KN, Stadlin A, Ma CM, Tsang D (2001) Induction of tumor necrosis factor receptor type 2 gene expression by tumor necrosis factor-alpha in rat primary astrocytes. Life Sci 68:2081–2091.

Luo Y, Fischer FR, Hancock WW, Dorf ME (2000) Macrophage inflammatory protein-2 and KC induce chemokine production by mouse astrocytes. J Immunol 165:4015–4023.

Ma Q, Jones D, Borghesani PR, Segal RA, Nagasawa T, Kishimoto T, Bronson RT, Springer TA (1998) Impaired B-lymphopoiesis, myelopoiesis, and derailed cerebellar neuron migration in CXCR4- and SDF-1-deficient mice. Proc Natl Acad Sci USA 95:9448–9453.

Ma YJ, Berg-von der Emde K, Moholt-Siebert M, Hill DF, Ojeda SR (1994) Region-specific regulation of transforming growth factor alpha (TGF alpha) gene expression in astrocytes of the neuroendocrine brain. J Neurosci 14:5644–5651.

Madiai F, Hussain SR, Goettl VM, Burry RW, Stephens RL Jr, Hackshaw KV (2003) Upregulation of FGF-2 in reactive spinal cord astrocytes following unilateral lumbar spinal nerve ligation. Exp Brain Res 148:366–376.

Magnaghi V, Riva MA, Cavarretta I, Martini L, Melcangi RC (2000) Corticosteroids regulate the gene expression of FGF-1 and FGF-2 in cultured rat astrocytes. J Mol Neurosci 15:11–18.

Mahajan SD, Schwartz SA, Shanahan TC, Chawda RP, Nair MP (2002) Morphine regulates gene expression of alpha- and beta-chemokines and their receptors on astroglial cells via the opioid mu receptor. J Immunol 169:3589–3599.

Maimone D, Cioni C, Rosa S, Macchia G, Aloisi F, Annunziata P (1993) Norepinephrine and vasoactive intestinal peptide induce IL-6 secretion by astrocytes: synergism with IL-1 beta and TNF alpha. J Neuroimmunol 47:73–81.

Marz P, Heese K, Dimitriades-Schmutz B, Rose-John S, Otten U (1999) Role of interleukin-6 and soluble IL-6 receptor in region-specific induction of astrocytic differentiation and neurotrophin expression. Glia 26:191–200.

Matsunaga H, Ueda H (2006) Evidence for serum-deprivation-induced co-release of FGF-1 and S100A13 from astrocytes. Neurochem Int 49:294–303.

McManus C, Berman JW, Brett FM, Staunton H, Farrell M, Brosnan CF (1998) MCP-1, MCP-2 and MCP-3 expression in multiple sclerosis lesions: an immunohistochemical and in situ hybridization study. J Neuroimmunol 86:20–29.
Meeuwsen S, Persoon-Deen C, Bsibsi M, Ravid R, van Noort JM (2003) Cytokine, chemokine and growth factor gene profiling of cultured human astrocytes after exposure to proinflammatory stimuli. Glia 43:243–253.
Meeuwsen S, Persoon-Deen C, Bsibsi M, Bajramovic JJ, Ravid R, De Bolle L, van Noort JM (2005) Modulation of the cytokine network in human adult astrocytes by human herpesvirus-6A. J Neuroimmunol 164:37–47.
Mehler MF, Kessler JA (1995) Cytokines and neuronal differentiation. Crit Rev Neurobiol 9:419–446.
Messamore E, Bogdanovich N, Schroder H, Winblad B (1994) Astrocytes associated with senile plaques possess muscarinic acetylcholine receptors. Neuroreport 5:1473–1476.
Miller NR, Jover T, Cohen HW, Zukin RS, Etgen AM (2005) Estrogen can act via estrogen receptor alpha and beta to protect hippocampal neurons against global ischemia-induced cell death. Endocrinology 146:3070–3079.
Milner TA, McEwen BS, Hayashi S, Li CJ, Reagan LP, Alves SE (2001) Ultrastructural evidence that hippocampal alpha estrogen receptors are located at extranuclear sites. J Comp Neurol 429:355–371.
Milner TA, Ayoola K, Drake CT, Herrick SP, Tabori NE, McEwen BS, Warrier S, Alves SE (2005) Ultrastructural localization of estrogen receptor beta immunoreactivity in the rat hippocampal formation. J Comp Neurol 491:81–95.
Mizuno T, Sawada M, Marunouchi T, Suzumura A (1994) Production of interleukin-10 by mouse glial cells in culture. Biochem Biophys Res Commun 205:1907–1915.
Morga E, Faber C, Heuschling P (1999) Regional heterogeneity of the astroglial immunoreactive phenotype: effect of lipopolysaccharide. J Neurosci Res 57:941–952.
Mudo G, Trovato-Salinaro A, Caniglia G, Cheng Q, Condorelli DF (2007) Cellular localization of mGluR3 and mGluR5 mRNAs in normal and injured rat brain. Brain Res 1149:1–13.
Murphy GMJr, Song Y, Ong E, Lee YL, Schmidt KG, Bocchini V, Eng LF (1995) Leukemia inhibitory factor mRNA is expressed in cortical astrocyte cultures but not in an immortalized microglial cell line. Neurosci Lett 184:48–51.
Murphy S, Pearce B (1987) Functional receptors for neurotransmitters on astroglial cells. Neuroscience 22:381–394.
Nagai A, Nakagawa E, Choi HB, Hatori K, Kobayashi S, Kim SU (2001) Erythropoietin and erythropoietin receptors in human CNS neurons, astrocytes, microglia, and oligodendrocytes grown in culture. J Neuropathol Exp Neurol 60:386–392.
Nagao H, Matsuoka I, Kurihara K (1995) Effects of adenylyl cyclase-linked neuropeptides on the expression of ciliary neurotrophic factor-mRNA in cultured astrocytes. FEBS Lett 362:75–79.
Nakagawa T, Schwartz JP (2004) Gene expression patterns in in vivo normal adult astrocytes compared with cultured neonatal and normal adult astrocytes. Neurochem Int 45:203–242.
Nakamura A, Johns EJ, Imaizumi A, Abe T, Kohsaka T (1998a) Regulation of tumour necrosis factor and interleukin-6 gene transcription by beta2-adrenoceptor in the rat astrocytes. J Neuroimmunol 88:144–153.
Nakamura S, Todo T, Motoi Y, Haga S, Aizawa T, Ueki A, Ikeda K (1999) Glial expression of fibroblast growth factor-9 in rat central nervous system. Glia 28:53–65.
Nakamura S, Arima K, Haga S, Aizawa T, Motoi Y, Otsuka M, Ueki A, Ikeda K (1998b) Fibroblast growth factor (FGF)-9 immunoreactivity in senile plaques. Brain Res 814:222–225.
Nakayama T, Momoki-Soga T, Inoue N (2003) Astrocyte-derived factors instruct differentiation of embryonic stem cells into neurons. Neurosci Res 46:241–249.
Niu H, Hinkle DA, Wise PM (1997) Dexamethasone regulates basic fibroblast growth factor, nerve growth factor and S100beta expression in cultured hippocampal astrocytes. Brain Res Mol Brain Res 51:97–105.

Nomura H, Furuta A, Iwaki T (2002) Dorsal root rupture injury induces extension of astrocytic processes into the peripheral nervous system and expression of GDNF in astrocytes. Brain Res 950:21–30.
Norris JG, Benveniste EN (1993) Interleukin-6 production by astrocytes: induction by the neurotransmitter norepinephrine. J Neuroimmunol 45:137–145.
O'Callaghan JP, Brinton RE, McEwen BS (1991) Glucocorticoids regulate the synthesis of glial fibrillary acidic protein in intact and adrenalectomized rats but do not affect its expression following brain injury. J Neurochem 57:860–869.
Ohta K, Ohta M, Mizuta I, Fujinami A, Shimazu S, Sato N, Yoneda F, Hayashi K, Kuno S (2002) The novel catecholaminergic and serotoninergic activity enhancer R-(-)-1-(benzofuran-2-yl)-2-propylaminopentane up-regulates neurotrophic factor synthesis in mouse astrocytes. Neurosci Lett 328:205–208.
Okada K, Kuroda E, Yoshida Y, Yamashita U, Suzumura A, Tsuji S (2005) Effects of interferon-beta on the cytokine production of astrocytes. J Neuroimmunol 159:48–54.
O'Malley EK, Black IB, Dreyfus CF (1991) Local support cells promote survival of substantia nigra dopaminergic neurons in culture. Exp Neurol 112:40–48.
O'Malley EK, Sieber BA, Black IB, Dreyfus CF (1992) Mesencephalic type I astrocytes mediate the survival of substantia nigra dopaminergic neurons in culture. Brain Res 582:65–70.
O'Malley EK, Sieber BA, Morrison RS, Black IB, Dreyfus CF (1994) Nigral type I astrocytes release a soluble factor that increases dopaminergic neuron survival through mechanisms distinct from basic fibroblast growth factor. Brain Res 647:83–90.
Otto VI, Gloor SM, Frentzel S, Gilli U, Ammann E, Hein AE, Folkers G, Trentz O, Kossmann T, Morganti-Kossmann MC (2002) The production of macrophage inflammatory protein-2 induced by soluble intercellular adhesion molecule-1 in mouse astrocytes is mediated by src tyrosine kinases and p42/44 mitogen-activated protein kinase. J Neurochem 80:824–834.
Pang Y, Cai Z, Rhodes PG (2001) Analysis of genes differentially expressed in astrocytes stimulated with lipopolysaccharide using cDNA arrays. Brain Res 914:15–22.
Park MH, Lee SM, Lee JW, Son DJ, Moon DC, Yoon DY, Hong JT (2006) ERK-mediated production of neurotrophic factors by astrocytes promotes neuronal stem cell differentiation by erythropoietin. Biochem Biophys Res Commun 339:1021–1028.
Parpura V, Haydon PG (2000) Physiological astrocytic calcium levels stimulate glutamate release to modulate adjacent neurons. Proc Natl Acad Sci USA 97:8629–8634.
Parpura V, Basarsky TA, Liu F, Jeftinija K, Jeftinija S, Haydon PG (1994) Glutamate-mediated astrocyte-neuron signalling. Nature 369:744–747.
Pataky DM, Borisoff JF, Fernandes KJ, Tetzlaff W, Steeves JD (2000) Fibroblast growth factor treatment produces differential effects on survival and neurite outgrowth from identified bulbospinal neurons in vitro. Exp Neurol 163:357–372.
Pechan PA, Chowdhury K, Gerdes W, Seifert W (1993) Glutamate induces the growth factors NGF, bFGF, the receptor FGF-R1 and c-fos mRNA expression in rat astrocyte culture. Neurosci Lett 153:111–114.
Peng H, Erdmann N, Whitney N, Dou H, Gorantla S, Gendelman HE, Ghorpade A, Zheng J (2006) HIV-1-infected and/or immune activated macrophages regulate astrocyte SDF-1 production through IL-1beta. Glia 54:619–629.
Peretto P, Cummings D, Modena C, Behrens M, Venkatraman G, Fasolo A, Margolis FL (2002) BMP mRNA and protein expression in the developing mouse olfactory system. J Comp Neurol 451:267–278.
Pfrieger FW, Barres BA (1997) Synaptic efficacy enhanced by glial cells in vitro. Science 277:1684–1687.
Plata-Salaman CR (1991) Immunoregulators in the nervous system. Neurosci Biobehav Rev 15:185–215.
Pollock GS, Franceschini IA, Graham G, Marchionni MA, Barnett SC (1999) Neuregulin is a mitogen and survival factor for olfactory bulb ensheathing cells and an isoform is produced by astrocytes. Eur J Neurosci 11:769–780.
Porter JT, McCarthy KD (1997) Astrocytic neurotransmitter receptors in situ and in vivo. Prog Neurobiol 51:439–455.

Reuss B, Dermietzel R, Unsicker K (1998) Fibroblast growth factor 2 (FGF-2) differentially regulates connexin (cx) 43 expression and function in astroglial cells from distinct brain regions. Glia 22:19–30.
Rho MB, Wesselingh S, Glass JD, McArthur JC, Choi S, Griffin J, Tyor WR (1995) A potential role for interferon-alpha in the pathogenesis of HIV-associated dementia. Brain Behav Immun 9:366–377.
Ridet JL, Malhotra SK, Privat A, Gage FH (1997) Reactive astrocytes: cellular and molecular cues to biological function. Trends Neurosci 20:570–577.
Riva MA, Molteni R, Lovati E, Fumagalli F, Rusnati M, Racagni G (1996) Cyclic AMP-dependent regulation of fibroblast growth factor-2 messenger RNA levels in rat cortical astrocytes: comparison with fibroblast growth factor-1 and ciliary neurotrophic factor. Mol Pharmacol 49:699–706.
Rossner S, Schliebs R, Hartig W, Perez-Polo JR, Bigl V (1997) Selective induction of c-Jun and NGF in reactive astrocytes after cholinergic degenerations in rat basal forebrain. Neuroreport 8:2199–2202.
Rudge JS, Morrissey D, Lindsay RM, Pasnikowski EM (1994) Regulation of ciliary neurotrophic factor in cultured rat hippocampal astrocytes. Eur J Neurosci 6:218–229.
Santagati S, Melcangi RC, Celotti F, Martini L, Maggi A (1994) Estrogen receptor is expressed in different types of glial cells in culture. J Neurochem 63:2058–2064.
Sarup V, Patil K, Sharma SC (2004) Ciliary neurotrophic factor and its receptors are differentially expressed in the optic nerve transected adult rat retina. Brain Res 1013:152–158.
Schmitt KR, Kern C, Berger F, Ullrich O, Hendrix S, Abdul-Khaliq H (2006) Methylprednisolone attenuates hypothermia- and rewarming-induced cytotoxicity and IL-6 release in isolated primary astrocytes, neurons and BV-2 microglia cells. Neurosci Lett 404:309–314.
Schwartz JP, Nishiyama N (1994) Neurotrophic factor gene expression in astrocytes during development and following injury. Brain Res Bull 35:403–407.
Schwartz JP, Nishiyama N, Wilson D, Taniwaki T (1994) Receptor-mediated regulation of neuropeptide gene expression in astrocytes. Glia 11:185–190.
Shimamura M, Sato N, Sata M, Wakayama K, Ogihara T, Morishita R (2007) Expression of hepatocyte growth factor and c-Met after spinal cord injury in rats. Brain Res 1151:188–194.
Silberstein FC, De Simone R, Levi G, Aloisi F (1996) Cytokine-regulated expression of platelet-derived growth factor gene and protein in cultured human astrocytes. J Neurochem 66:1409–1417.
Silva GA, Theriault E, Mills LR, Pennefather PS, Feeney CJ (1999) Group I and II metabotropic glutamate receptor expression in cultured rat spinal cord astrocytes. Neurosci Lett 263:117–120.
Simpson CS, Johnston HM, Morris BJ (1994) Neuronal expression of protease-nexin 1 mRNA in rat brain. Neurosci Lett 170:286–290.
Simpson JE, Newcombe J, Cuzner ML, Woodroofe MN (1998) Expression of monocyte chemoattractant protein-1 and other beta-chemokines by resident glia and inflammatory cells in multiple sclerosis lesions. J Neuroimmunol 84:238–249.
Simpson JE, Newcombe J, Cuzner ML, Woodroofe MN (2000) Expression of the interferon-gamma-inducible chemokines IP-10 and Mig and their receptor, CXCR3, in multiple sclerosis lesions. Neuropathol Appl Neurobiol 26:133–142.
Sinor AD, Irvin SM, Cobbs CS, Chen J, Graham SH, Greenberg DA (1998) Hypoxic induction of vascular endothelial growth factor (VEGF) protein in astroglial cultures. Brain Res 812:289–291.
Sjoborg M, Pietz K, Ahgren A, Yamada N, Lindvall O, Funa K, Odin P (1998) Expression of platelet-derived growth factor after intrastriatal ibotenic acid injury. Exp Brain Res 119:245–250.
Skold MK, von Gertten C, Sandberg-Nordqvist AC, Mathiesen T, Holmin S (2005) VEGF and VEGF receptor expression after experimental brain contusion in rat. J Neurotrauma 22:353–367.

Song H, Stevens CF, Gage FH (2002) Astroglia induce neurogenesis from adult neural stem cells. Nature 417:39–44.

Spranger M, Lindholm D, Bandtlow C, Heumann R, Gnahn H, Naher-Noe M, Thoenen H (1990) Regulation of nerve growth factor (NGF) synthesis in the rat central nervous system: comparison between the effects of interleukin-1 and various growth factors in astrocyte cultures and in vivo. Eur J Neurosci 2:69–76.

Stalder AK, Pagenstecher A, Yu NC, Kincaid C, Chiang CS, Hobbs MV, Bloom FE, Campbell IL (1997) Lipopolysaccharide-induced IL-12 expression in the central nervous system and cultured astrocytes and microglia. J Immunol 159:1344–1351.

Stockli KA, Lillien LE, Naher-Noe M, Breitfeld G, Hughes RA, Raff MC, Thoenen H, Sendtner M (1991) Regional distribution, developmental changes, and cellular localization of CNTF-mRNA and protein in the rat brain. J Cell Biol 115:447–459.

Stoop R, Poo MM (1995) Potentiation of transmitter release by ciliary neurotrophic factor requires somatic signaling. Science 267:695–699.

Suzuki S, Tanaka K, Nogawa S, Ito D, Dembo T, Kosakai A, Fukuuchi Y (2000) Immunohistochemical detection of leukemia inhibitory factor after focal cerebral ischemia in rats. J Cereb Blood Flow Metab 20:661–668.

Takeda A, Onodera H, Sugimoto A, Itoyama Y, Kogure K, Rauvala H, Shibahara S (1995) Induction of heparin-binding growth-associated molecule expression in reactive astrocytes following hippocampal neuronal injury. Neuroscience 68:57–64.

Takeshita S, Takeshita F, Haddad DE, Janabi N, Klinman DM (2001) Activation of microglia and astrocytes by CpG oligodeoxynucleotides. Neuroreport 12:3029–3032.

Tancredi V, D'Arcangelo G, Grassi F, Tarroni P, Palmieri G, Santoni A, Eusebi F (1992) Tumor necrosis factor alters synaptic transmission in rat hippocampal slices. Neurosci Lett 146:176–178.

Tatsuno I, Morio H, Tanaka T, Uchida D, Hirai A, Tamura Y, Saito Y (1996) Pituitary adenylate cyclase-activating polypeptide (PACAP) is a regulator of astrocytes: PACAP stimulates proliferation and production of interleukin 6 (IL-6), but not nerve growth factor (NGF), in cultured rat astrocyte. Ann N Y Acad Sci 805:482–488.

Tokita Y, Keino H, Matsui F, Aono S, Ishiguro H, Higashiyama S, Oohira A (2001) Regulation of neuregulin expression in the injured rat brain and cultured astrocytes. J Neurosci 21:1257–1264.

Tomita M, Holman BJ, Santoro CP, Santoro TJ (2005) Astrocyte production of the chemokine macrophage inflammatory protein-2 is inhibited by the spice principle curcumin at the level of gene transcription. J Neuroinflammation 2:8.

Torres-Aleman I, Rejas MT, Pons S, Garcia-Segura LM (1992) Estradiol promotes cell shape changes and glial fibrillary acidic protein redistribution in hypothalamic astrocytes in vitro: a neuronal-mediated effect. Glia 6:180–187.

Tourbah A, Oliver L, Jeanny JC, Gumpel M (1991) Acidic fibroblast growth factor (aFGF) is expressed in the neuronal and glial spinal cord cells of adult mice. J Neurosci Res 29:560–568.

Toyomoto M, Inoue S, Ohta K, Kuno S, Ohta M, Hayashi K, Ikeda K (2005) Production of NGF, BDNF and GDNF in mouse astrocyte cultures is strongly enhanced by a cerebral vasodilator, ifenprodil. Neurosci Lett 379:185–189.

Ullian EM, Sapperstein SK, Christopherson KS, Barres BA (2001) Control of synapse number by glia. Science 291:657–661.

Ullian EM, Harris BT, Wu A, Chan JR, Barres BA (2004) Schwann cells and astrocytes induce synapse formation by spinal motor neurons in culture. Mol Cell Neurosci 25:241–251.

Van Der Voorn P, Tekstra J, Beelen RH, Tensen CP, Van Der Valk P, De Groot CJ (1999) Expression of MCP-1 by reactive astrocytes in demyelinating multiple sclerosis lesions. Am J Pathol 154:45–51.

Velasco S, Tarlow M, Olsen K, Shay JW, McCracken GH Jr, Nisen PD (1991) Temperature-dependent modulation of lipopolysaccharide-induced interleukin-1 beta and tumor necrosis

factor alpha expression in cultured human astroglial cells by dexamethasone and indomethacin. J Clin Invest 87:1674–1680.
Wagner B, Natarajan A, Grunaug S, Kroismayr R, Wagner EF, Sibilia M (2006) Neuronal survival depends on EGFR signaling in cortical but not midbrain astrocytes. EMBO J 25:752–762.
Walicke PA (1988) Basic and acidic fibroblast growth factors have trophic effects on neurons from multiple CNS regions. J Neurosci 8:2618–2627.
Walter HJ, Berry M, Hill DJ, Cwyfan-Hughes S, Holly JM, Logan A (1999) Distinct sites of insulin-like growth factor (IGF)-II expression and localization in lesioned rat brain: possible roles of IGF binding proteins (IGFBPs) in the mediation of IGF-II activity. Endocrinology 140:520–532.
Wanaka A, Carroll SL, Milbrandt J (1993) Developmentally regulated expression of pleiotrophin, a novel heparin binding growth factor, in the nervous system of the rat. Brain Res Dev Brain Res 72:133–144.
Wang YT, Han S, Zhang KH, Jin Y, Xu XM, Lu PH (2004) Upregulation of heparin-binding growth-associated molecule after spinal cord injury in adult rats. Acta Pharmacol Sin 25:611–616.
Watt JA, Bone S, Pressler M, Cranston HJ, Paden CM (2006) Ciliary neurotrophic factor is expressed in the magnocellular neurosecretory system of the rat in vivo: evidence for injury- and activity-induced upregulation. Exp Neurol 197:206–214.
Wen LL, Chiu CT, Huang YN, Chang CF, Wang JY (2007) Rapid glia expression and release of proinflammatory cytokines in experimental Klebsiella pneumoniae meningoencephalitis. Exp Neurol 205:270–278.
Woo NH, Teng HK, Siao CJ, Chiaruttini C, Pang PT, Milner TA, Hempstead BL, Lu B (2005) Activation of p75NTR by proBDNF facilitates hippocampal long-term depression. Nat Neurosci 8:1069–1077.
Woolf PD, Cox C, Kelly M, Nichols D, McDonald JV, Hamill RW (1990) The adrenocortical response to brain injury: correlation with the severity of neurologic dysfunction, effects of intoxication, and patient outcome. Alcohol Clin Exp Res 14:917–921.
Wu H, Friedman WJ, Dreyfus CF (2004) Differential regulation of neurotrophin expression in basal forebrain astrocytes by neuronal signals. J Neurosci Res 76:76–85.
Wu Z, Zhang J, Nakanishi H (2005) Leptomeningeal cells activate microglia and astrocytes to induce IL-10 production by releasing pro-inflammatory cytokines during systemic inflammation. J Neuroimmunol 167:90–98.
Xiao BG, Link H (1998) IFN-gamma production of adult rat astrocytes triggered by TNF-alpha. Neuroreport 9:1487–1490.
Xu B, Gottschalk W, Chow A, Wilson RI, Schnell E, Zang K, Wang D, Nicoll RA, Lu B, Reichardt LF (2000) The role of brain-derived neurotrophic factor receptors in the mature hippocampus: modulation of long-term potentiation through a presynaptic mechanism involving TrkB. J Neurosci 20:6888–6897.
Yamada T, Yoshiyama Y, Tsuboi Y, Shimomura T (1997) Astroglial expression of hepatocyte growth factor and hepatocyte growth factor activator in human brain tissues. Brain Res 762:251–255.
Yamakuni H, Kawaguchi N, Ohtani Y, Nakamura J, Katayama T, Nakagawa T, Minami M, Satoh M (2002) ATP induces leukemia inhibitory factor mRNA in cultured rat astrocytes. J Neuroimmunol 129:43–50.
Yoshida H, Imaizumi T, Tanji K, Matsumiya T, Sakaki H, Kimura D, Cui XF, Kumagai M, Tamo W, Shibata T, Hatakeyama M, Sato Y, Satoh K (2002) Platelet-activating factor enhances the expression of vascular endothelial growth factor in normal human astrocytes. Brain Res 944:65–72.
Yun SJ, Kim MO, Kim SO, Park J, Kwon YK, Kim IS, Lee EH (2002) Induction of TGF-beta-inducible gene-h3 (*betaig-h3*) by TGF-beta1 in astrocytes: implications for astrocyte response to brain injury. Brain Res Mol Brain Res 107:57–64.

Zafra F, Lindholm D, Castren E, Hartikka J, Thoenen H (1992) Regulation of brain-derived neurotrophic factor and nerve growth factor mRNA in primary cultures of hippocampal neurons and astrocytes. J Neurosci 12:4793–4799.

Zhang SC, Fedoroff S (1997) Cellular localization of stem cell factor and c-kit receptor in the mouse nervous system. J Neurosci Res 47:1–15.

Zhao M, Li D, Shimazu K, Zhou YX, Lu B, Deng CX (2007) Fibroblast growth factor receptor-1 is required for long-term potentiation, memory consolidation, and neurogenesis. Biol Psychiatry 62:381–390.

Zou YR, Kottmann AH, Kuroda M, Taniuchi I, Littman DR (1998) Function of the chemokine receptor CXCR4 in haematopoiesis and in cerebellar development. Nature 393:595–599.

Abbreviations

ACM	Astrocyte-conditioned media
AMPA	α-Amino-3-hydroxy-5-methyl-4-isoxazolepropionic acid
BDNF	Brain-derived neurtrophic factor
CNTF	Ciliary neurotrophic factor
FGF	Fibroblast growth factor
GDNF	Glial cell line-derived growth factor
GFAP	Glial fibrillary acidic protein
IFN	Interferon
IGF	Insulin-like growth factor
IGFBP	Insulin-like growth factor binding protein
IL	Interleukin
LPS	Lipopolysaccharide
LTP	Long-term potentiation
mGluR	Metabotropic glutamate receptor
NGF	Nerve growth factor
NT	Neurotrophin
TGF	Transforming growth factor
TNF	Tumor necrosis factor
TSP	Thrombospondin
VIP	Vasoactive intestinal peptide

Chapter 14
Molecular Approaches for Studying Astrocytes

Todd Fiacco, Kristi Casper, Elizabeth Sweger, Cendra Agulhon, Sarah Taves, Suzanne Kurtzer-Minton, and Ken D. McCarthy

Contents

14.1 Introduction

In the year 2007, literally hundreds of research papers were published that utilize molecular approaches to study the role of astrocytes in neurophysiology and neurological disorders. This is striking given that the molecular tools for selectively perturbing gene expression in astrocytes were developed just over a decade ago. Two investigators stand out as pioneers in this area, Drs. Michael Brenner and Albee Messing. While at National Institutes of Health, Mike Brenner carried out studies mapping the transcriptional regulatory elements of the promoter for glial fibrillary acidic protein (GFAP) (Masood et al., 1993), an intermediate filament protein expressed exclusively by mature astrocytes in the CNS. This information was absolutely essential in order to develop a regulatory unit (promoter) small enough to be handled by the molecular techniques available in the early 1990s and yet sufficient to target expression constructs to astrocytes *in vivo* and *in vitro*.

K.D. McCarthy
Department of Pharmacology, University of North Carolina at Chapel Hill Chapel Hill, NC USA
kdmc@med.unc.edu

V. Parpura and P.G. Haydon (eds.), *Astrocytes in (Patho)Physiology of the Nervous System*,
DOI: 10.1007/978-0-387-79492-1_14,

In 1994, Mike Brenner together with Albee Messing and colleagues published a paper demonstrating that a 2.2-kb fragment of the 5′-flanking sequence of the human GFAP (hGFAP) promoter was sufficient to direct transgene expression in the majority of astrocytes *in vitro* and *in vivo* without significant expression in other cell types in brain (Brenner et al., 1994). Thanks to the generosity of Mike Brenner, the 2.2-kb fragment of the human GFAP promoter has been provided to hundreds of investigators throughout the world without restrictions or required collaborations. Further, the collaborative effort of Mike Brenner and Albee Messing has provided a series of important insights into the role of astrocytes in brain, culminating most recently in their discovery that the aberrant expression of GFAP is involved in Alexander's disease (Messing and Brenner, 2003). Those of us working in this area owe these investigators a debt of gratitude for the development and dissemination of reagents important in the molecular analysis of astrocytes.

Before getting too deeply immersed in this topic, it is worth mentioning the power and limitations of molecular tools in the study of astrocytes. The power of molecular techniques is truly enormous. Today, molecular approaches that can be used to study astrocyte function include: (1) expression of novel genes, (2) overexpression of endogenous genes, (3) expression of dominant-negative mutations that interfere with the function of endogenous molecules, (4) expression of reporter genes, (5) expression of toxins, (6) knockouts of specific genes, (7) knockins of genes with major or minor modifications, and (8) inducible systems that enable regulation of the timing and level of gene expression. This is an extremely powerful array of molecular approaches that can be used to study processes ranging from cell migration and development to the analysis of second messenger cascades and even the amino acids required for ion channel selectivity. Further, it is not only possible to define the second messenger cascades involved in a cellular process, but also one can determine the effects of perturbing that cascade on various outcomes, ranging from animal behavior to the progression of neurological disorders. Most of these different approaches have been applied to the study of astrocytes *in vivo*. In the area of astrocyte biology where it has been extremely difficult to sort out the role of astrocytes in brain function, this new technology represents a true paradigm shift.

However, this technology is not without its limitations. We tend to think that if we selectively perturb a single gene, we have made a very specific and restricted modification. To a certain extent, this is wrong. Gene products do not function in isolation. Rather, gene products (e.g., proteins and their downstream effectors) generally function as part of a fabric of interconnected molecules in which perturbation of one molecule often affects the expression and/or activity of other molecules within that fabric. For example, David Spray and colleagues demonstrated that a knockout of connexin 43 (Cx43) affected the expression of hundreds of other proteins (Iacobas et al., 2004). As a result, it is sometimes difficult to demonstrate that a cellular or behavioral phenotype resulting from the perturbation of a single gene is a direct reflection of the function of that gene. Nevertheless, it is possible to demonstrate that a particular gene is involved, directly or indirectly, in a cellular or behavioral phenotype. Further, regardless of the mechanism of action, if the knockout of an astrocyte gene perturbs a particular behavior, a new field of investigation opens up. A second limitation that needs comment concerns the fidelity of the promoters used to drive the expression of transgenes in astrocytes. The GFAP promoter is the most commonly used promoter to drive transgene expression in astrocytes. Unfortunately GFAP, as well as some other astrocyte-marker proteins, is expressed during development in progenitor cells that give rise to astrocytes, as well as a large population of neurons and oligodendrocytes (Malatesta et al., 2003; Anthony et al., 2004; Casper and McCarthy, 2006). As a consequence, during development, cells destined to become neurons and oligodendrocytes express transgenes driven by the GFAP promoter. While generally unlikely, it is difficult to exclude the possibility that the transient expression of a transgene affects the development/differentiation of oligodendrocytes or neurons. To circumvent this problem, a number of laboratories use inducible gene systems to limit transgene expression to postdevelopmental periods when the generation of neurons and oligodendrocytes from GFAP positive progenitors is mostly complete.

Overall, there is no question that molecular technology is a very powerful tool to study the role of astrocytes in neurophysiology and neurological disorders. While there are important factors that need to be considered when interpreting findings (as with all analyses), few other approaches have as much potential for unraveling the role of astrocytes in brain function.

14.2 Transgenic Approaches

14.2.1 General Considerations

The analysis of genetically modified animals has proven to be extremely useful in understanding the role of specific genes and cells in complex neurological processes. Initially, many spontaneous mutations resulting in a behavioral phenotype were identified and ultimately understood with respect to the gene mutation and the cellular basis of the phenotype (Hatten and Heintz, 2005). The analysis of spontaneous gene mutations has provided insight into the role of a number of genes in brain development and neurological diseases. However, the low frequency of spontaneous gene mutations limits their usefulness in deciphering the role of most genes in neurological processes. To circumvent this problem, a number of large laboratories are using chemical mutagenesis to enhance the frequency of genetic mutations. This approach has been used successfully to identify a number of genes that play a critical role in neurological processes (Hatten and Heintz, 2005). While large-scale screening in chemically mutagenized colonies will undoubtedly continue to identify novel and important genes, its use requires a major commitment beyond that feasible for most research laboratories. As a consequence, many laboratories interested in the role of specific genes have turned to the generation of transgenic animal models.

Over the last several years, the preparation of transgenic animals has become a relatively straightforward process. However, there are a number of key decisions that require consideration prior to making transgenic animals. Foremost is a consideration of the level and cellular distribution of transgene expression sought. Most investigators have utilized a relatively small transcriptional unit followed by an expression cassette that includes a heterologous intron, the cDNA of interest, and a 3′ untranslated region with poly A sites for mRNA stabilization. This gene cassette is microinjected into the pronuclei of fertilized eggs and the eggs transplanted into the oviducts of pseudopregnant animals. The gene cassette integrates randomly into the animal's genome, and founder mice are identified using either PCR or Southern blots. Founder mice are bred and their offspring are screened for transgene expression. This screening process can be accomplished by looking for the RNA and/or the protein product of the transgene. It is critical to show that the founder line is expressing the transgene in the correct cellular population in order to relate observed phenotypes to astrocytic expression of the transgene. This process is particularly problematic if it is difficult to assess the cellular expression pattern of the transgene. If an antibody to the gene product of interest is not available, the cellular distribution of the transgene can be determined by tagging the gene with a marker such as enhanced green fluorescent protein (eGFP) or an epitope tag such as hemagglutinin. When this is not possible, a second gene cassette utilizing the same transcriptional unit driving a reporter gene (e.g., eGFP or the beta galactosidase gene *lacZ*) can be coinjected as both gene cassettes are generally believed to integrate at the same genomic site. The absence of an approach for demonstrating the cellular distribution of transgene expression markedly diminishes the ability to link cellular or behavioral phenotypes to a specific cell population.

A primary limitation of transgenes is that the expression of the cDNA is frequently influenced by the activity of its local DNA environment, a common phenomenon referred to as "chromosomal positional effect." Chromosomal positional effects can enhance or diminish the expression of the transgene or affect the cellular or even tissue distribution of the cDNA. Approaches can be used to limit chromosomal positional effects that may occur following random insertion of the gene cassette. An approach that can be used to limit chromosomal positional effects involves surrounding the gene cassette with segments of genomic DNA, called insulators, which limit the influence of surrounding chromosomal elements (Chung et al., 1997; Giraldo et al., 2003). Probably the best approach to limiting chromosomal positional effects is to surround the transgene with a very large region of genomic DNA. This is generally accomplished using artificial chromosome vectors derived from yeast (YACs), bacteria (BACs) or P1-bacteriophages (PACs) (Giraldo and Montoliu, 2001). Using such vectors, it is possible to insert very large DNA fragments (YACs ~ 1 Mb; BACs ~ 300 kb; PACs ~ 100 kb) that generally eliminate chromosomal positional effects to provide transgene expression that reflects the cellular distribution of the endogenous gene product (Giraldo and Montoliu, 2001). The methods for handling BACs and PACs are fairly standard in most molecular biology laboratories and are becoming available through core facilities at many research institutions. Overall, if it is important that the transgene of interest be expressed in the same population of cells from which the transcriptional unit drives the endogenous gene, one should strongly consider using an artificial chromosomal vector. Another approach to achieving transgene expression which closely mimics the endogenous gene is to create mice using homologous recombination in which the transgene replaces the endogenous gene, i.e., a knockin approach. However, this approach has the obvious drawback of deleting the endogenous gene which may interfere with the goal of the experiment.

There are situations where the variations in the level of transgene expression or cellular distribution of transgene expression can be useful. For example, aberrantly high levels of transgene expression are beneficial if the goal is to knockdown the expression of an endogenous gene using a dominant-negative transgene. Heterogeneous cellular distribution of gene expression can also be useful. For example, a series of transgenic mice were made using a modified *thy*1 promoter to drive the expression of different colors of fluorescent proteins to neurons of the central and peripheral nervous systems. While the modified Thy-1 is typically expressed by a large population of CNS projection neurons, its cellular distribution is strongly dependent on the genomic insertion site of the modified *thy*1 gene (Feng et al., 2000). Since the expression of the transgene can be influenced by its genomic insertion site, the founder transgenic lines expressed the fluorescent protein in differing subsets of neurons; this expression was stable within a given line. Consequently, these mice can be used to follow a very specific subset of neurons within a given mouse line enabling long-term analyses of identical sets of neurons.

In certain situations it is important to control the timing of transgene expression. For example, the expression of a transgene could affect developmental processes that confound the interpretation of analyses carried out in mature animals. Further, in behavioral studies, it is often desirable to determine if the effect of transgene expression is reversible. Several different inducible gene regulation systems have been used to control the timing of transgene expression in brain. The most used of these was developed by Hermann Bujard and colleagues, and is based on the tetracycline bacterial resistance gene operon (Gossen and Bujard, 1992; Mansuy and Bujard, 2000). Two genetically modified animal lines are required in this system. The first line contains the promoter of interest driving the expression of the tetracycline (tet) transactivator (tTA), which is a fusion protein between the tet repressor and the strong transcriptional activation domain of the herpes simplex viral protein, VP16. The second line contains a transgene that uses the tet operon (tetO) DNA-binding element fused to a minimal promoter (from cytomegalovirus) to drive the expression of the gene of interest. Bujard and colleagues have made two versions of this system that are generally referred to as the "tet Off" and the "tet On" systems (Baron and Bujard, 2000). In the tet Off system, tTA binding to the tetO minimal promoter leads to transcription; this can be blocked by tetracycline (or the more commonly used doxycycline) which binds to the tTA to block its interaction with the tetO minimal promoter (Fig. 14.1). In contrast, with the tet On system, tTA was mutated such that it requires doxycycline to bind and activate the tetO minimal promoter. The advantage of the tet On system is that it does not require the continuous maintenance of animals on doxycycline to suppress gene expression and that the kinetics of gene activation are more rapid once doxycycline is administered. The tet Off system has been used more frequently primarily due to leakiness of the

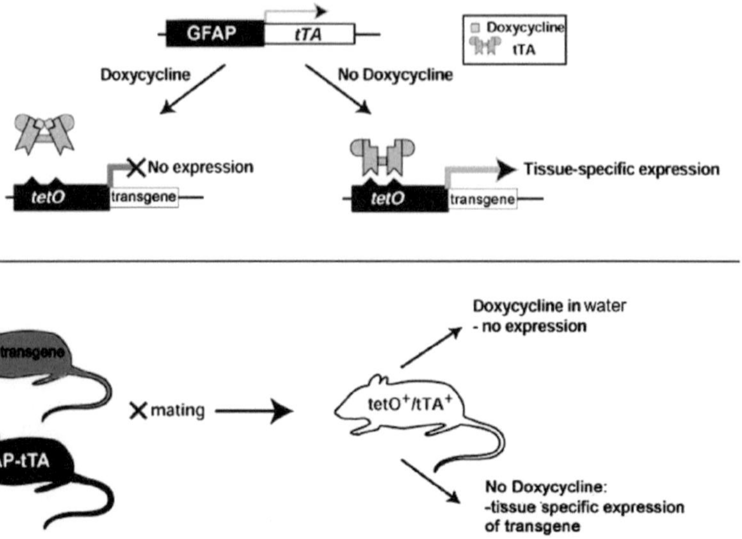

Fig. 14.1 Tetracycline "Off"-Inducible System. Doxycycline represses transgene expression by binding the tet transcriptional activator (tTA). Removal of doxycycline activates tTA, which then binds the tetO minimal promoter, inducing expression of the transgene (*top*). Tissue specificity is achieved by crossing glial fibrillary acidic protein (GFAP)-tTA mice with tetO-transgenic mice; double-transgenic mice have transgene expression in cells that express GFAP (*bottom*) (*See Color Plates*).

original tet On system. However, more recent versions of the tet On system direct little or no transcription in the absence of doxycycline and robust transcription of transgenes following the addition of doxycycline; this system may be the future method of choice for inducible gene expression.

14.2.2 Using Transgenic Approaches to Study the Role of Astrocytes in Physiology and Neuropathology

It is generally accepted that astrocytes *in vitro* often fail to exhibit the complex morphological and biochemical characteristics of astrocytes *in vivo*. This is hardly surprising given that the phenotype of an astrocyte is likely determined, in large part, by its local chemical and cellular milieu. The use of transgenes to perturb specific molecules and pathways in astrocytes *in vivo* provides an extremely powerful tool for sorting out the role of these cells in physiology and pathology. Work in this area can be traced back to collaboration between Mike Brenner and Albee Messing. Together with other collaborators, these investigators reported that a 2.2-kb 5′ flanking sequence of the hGFAP gene was sufficient to drive *lacZ* gene expression to astrocytes in transgenic mice (Brenner et al., 1994). This was the first study to demonstrate that it was possible to selectively express transgenes in astrocytes. Since this paper was published, the hGFAP promoter has been used to drive a wide variety of transgenes to astrocytes *in vivo*.

14.2.2.1 Expression of Transgenes in Astrocytes: Examples

Several laboratories have used the hGFAP promoter to express herpes thymidine kinase in astrocytes to enable conditional ablation of dividing astrocytes following administration of gancyclovir (Delaney et al., 1996; Faulkner et al., 2004). Delaney et al. (1996) found that the conditional ablation of astrocytes during development led to ataxia and cerebellar disorganization indicating a key role of astrocytes in cerebellar development. More recently, a similar approach was used by Faulkner et al. (2004) to demonstrate that reactive astrocytes play a protective role following spinal cord injury. This was an important finding given the long debate whether reactive astrocytes were detrimental or beneficial with respect to wound healing.

Several groups have made mice expressing eGFP driven by the hGFAP promoter. These mice have been extremely useful in identifying astrocytes in brain slices as well as *in vivo* (Zhuo et al., 1997; Nolte et al., 2001). GFAP:eGFP mice have also been used to monitor changes in astrocytic morphology *in situ* (Nolte et al., 2001; Hirrlinger et al., 2004; Chvatal et al., 2007). In a very interesting study, transgenic mice expressing eGFP driven by the GFAP promoter were used to examine the morphological heterogeneity of astrocytes in different brain regions (Emsley and Macklis, 2006). Findings from this study indicate that, like neurons, there are morphologically distinct populations of astrocytes that localize in specific brain regions. Other studies using similar transgenic mice indicate that the morphology of astrocytes is dynamic, changing in subcellular compartments with different treatments (Chvatal et al., 2007). Several groups have used GFAP-eGFP mice to monitor gliosis *in situ* and *in vivo* (Nolte et al., 2001; Ho et al., 2007). Given that gliosis often occurs as a function of neuropathology and includes an increase in GFAP expression, this approach enables direct visualization of reactive astrocytes in living tissue by imaging eGFP fluorescence.

A number of transgenic lines have been developed that express transgenes designed to perturb the function of astrocytes *in vivo*. Using the GFAP promoter to express a chondroitin sulfate proteoglycan degrading enzyme (chondroitinase ABC), Cafferty et al. (2007) demonstrated that chondroitin sulfate proteoglycans play a role in reducing axonal regeneration following dorsal rhizotomy (Cafferty et al., 2007). This is a very important finding suggesting a new therapeutic target in regeneration studies. In studies designed to assess the role of insulin-like growth factor 1 (IGF-1) in brain development, Ye et al. (2004) used transgenic lines expressing IGF-1 under a GFAP promoter-driven tetracycline-inducible regulatory system. The results from this study indicate that the overexpression of IGF-1 in astrocytes leads to an increase in brain weight and number of neural cells, demonstrating a role for IGF-1 in brain development. Studies from a number of laboratories suggest that astrocytes are likely to play a role in inflammatory neurological diseases (Moynagh, 2005). Transgenic mice expressing a dominant-negative mutation of nuclear factor kappa B (NF-κB) driven by the GFAP promoter were used to determine if interfering with an astrocyte's inflammatory cascades would affect recovery following contusive spinal cord injury. Reducing NF-κB expression in astrocytes markedly improved functional recovery following spinal cord injury clearly demonstrating that astrocytes participate in an inflammatory process that is detrimental to regeneration following spinal cord injury (Brambilla et al., 2005). Transgenic lines expressing glial-cell-line-derived neurotrophic factor (GDNF) driven by the GFAP promoter have been used to study the role of this factor in motor neuron survival following injury. Findings using these transgenic lines indicate that the overexpression of GDNF in astrocytes can rescue motoneurons from programmed cell death and promote the long-term survival of these neurons following axotomy (Zhao et al., 2004). In another set of experiments, the GFAP promoter was used to drive the overexpression of GFAP, thus mimicking a phenomenon that occurs in gliosis. The surprising result of this experiment is that mice overexpressing GFAP died within several weeks and exhibited a phenotype similar to that of patients with Alexander's disease (Messing et al., 1998). This observation led to a series of papers convincingly demonstrating that mutations in GFAP lead to Alexander's disease in humans (see Chap. 24 for details). Overall, these studies demonstrate the power of using a transgenic approach to study the role of specific astrocytic molecules in neurophysiology and pathology.

14.2.2.2 Expression of G-protein Coupled Receptors in Astrocytes

Astrocytes are known to express a wide variety of G-protein coupled receptors (GPCRs) linked to each of the primary signal transduction systems regulated by GPCRs (Ca^{2+} and cyclic adenosine monophosphate). While we have known that astrocytes express GPCRs for many years, their role in neurophysiology and neuropathology has remained elusive. A difficulty in sorting out the role of astrocyte receptors is that astrocytes, neurons, oligodendrocytes, and microglia all exhibit an overlapping array of neuroligand receptors. Thus, it is generally not possible to selectively stimulate or block astrocytic GPCRs using pharmacological approaches to determine their role in brain physiology. We have taken a molecular approach that enables us to selectively perturb astrocyte signaling in transgenic mouse lines. To selectively activate astrocytic GPCR signaling cascades, we have prepared transgenic mouse lines expressing in astrocytes G_i-, G_s-, and G_q-protein coupled receptors (1) that are not normally expressed in brain, (2) whose ligand does not activate endogenous brain GPCRs, and (3) that are not activated by neurotransmitters normally released in brain. Each of these transgenic systems has provided new insight into the role of astrocytic signaling systems in neurophysiology and pathology.

The first system we used to study astrocytic signaling is based on a receptor activated solely by synthetic ligands (RASSLs) which was developed by Bruce Conklin at University of California, San Francisco (Scearce-Levie et al., 2001). This system is based on a mutated κ-opioid receptor that has a greatly reduced affinity for its natural ligand but maintains high affinity binding for the synthetic ligand, spiradoline. The expression of this G_i-coupled GPCR (called Ro1) is regulated by the tetO minimal promoter such that it is only expressed in the presence of the tetracycline transactivator (tTA) when mice are taken off doxycycline (i.e., the tet Off system). We crossed this RASSL line with our hGFAP-tTA mouse line, moved this genetic cross onto the κ-opioid receptor knockout mouse line, and examined the role of astrocytic G_i-coupled GPCRs in neurophysiology and pathology. To our surprise, we found that the overexpression of this G_i-coupled GPCR in astrocytes led to hydrocephalus with 100% penetrance (Sweger et al., 2007) (Fig. 14.2). The fact that the overexpression of Ro1 activates signaling processes in the absence of ligand is not surprising given that most GPCRs are thought to have intrinsic activity. The timing and degree of hydrocephalus in this model can be controlled using doxycycline to regulate the expression of Ro1. Histological studies suggest that the expression of Ro1 in the GFAP positive ependymal cells that line the ventricles leads to the detachment of the ependymal cells from the ventricle walls, resulting in the closure of the aqueduct of Sylvius and noncommunicating hydrocephalus. These findings suggest that signaling cascades activated by G_i-coupled GPCRs may play a critical role in the development of hydrocephalus in humans and provides a tool to search for new therapeutic approaches for treating this disease.

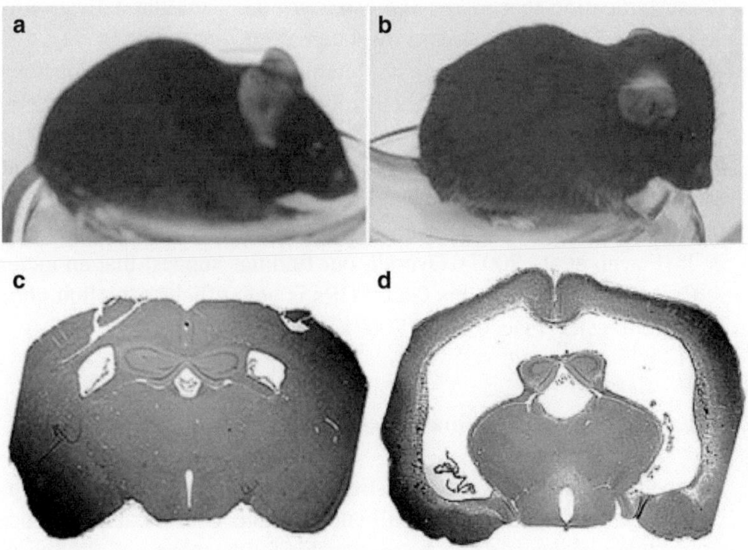

Fig. 14.2 Expression of a G_i G-protein coupled receptor in glial fibrillary acidic protein (GFAP) positive cells results in hydrocephalus. Mice carrying both the tetO-Ro1 and the GFAP-tTA transgenes develop hydrocephalus when doxycycline (dox) is removed from the drinking water. A double-transgenic mouse maintained on dox is normal (**a**), while a double-transgenic littermate taken off dox (**b**) develops a domed head characteristic of hydrocephalus. Hematoxylin and eosin staining of coronal brain sections show the greatly enlarged lateral ventricles in hydrocephalic mice (**d**) compared to normal littermate control mice (**c**). All mice are P30 (Modified with permission from The Society for Neuroscience (Sweger et al. 2007).) (*See Color Plates*).

We took a different approach to express a Gq-coupled GPCR in astrocytes that could be used to study the role of this signaling system in neurophysiology. In this case, we took advantage of a Gq-coupled GPCR that is specifically expressed in nociceptive neurons within the spinal cord, but not in brain. This eGFP-tagged receptor, MrgA1, is activated by a peptide (FLRFa or FMRFa) that does not have any detectable effect on the activity of CA1 pyramidal cells within the hippocampus. Once again, we utilized the tetracycline-inducible gene regulatory system to drive the expression of MrgA1 to astrocytes. hGFAP-tTA mice were crossed with tetO-MrgA1 mice and the cellular distribution of MrgA1 examined via immunostaining for eGFP. The results of these experiments demonstrated that most astrocytes, but not other cell types, express MrgA1 when the mice are left off doxycycline (Fig. 14.3). This finding correlates well with the observation that when hippocampal slices are loaded with a Ca^{2+} indicator dye, most astrocytes respond to FLRFa with robust increases in intracellular Ca^{2+} (Fiacco et al., 2007). Once we had determined that the expression of MrgA1 was restricted to astrocytes and linked to Ca^{2+} mobilization, we examined the effect of increasing astrocytic Ca^{2+} on the activity of CA1 pyramidal neurons in acutely isolated hippocampal brain slices. Many investigators [reviewed in (Zhang and Haydon, 2005)], including ourselves (Fiacco and McCarthy, 2004), have reported that increases in astrocytic Ca^{2+} lead to glutamate release from these cells and activation of neuronal receptors *in situ*. To our surprise, we never observed any indication that increasing astrocytic Ca^{2+} via the activation of this G_q-coupled GPCR leads to the release of glutamate from astrocytes (Fig. 14.4). In the same experiments, we could elicit glutamate release from astrocytes when uncaging IP_3 in these cells, but not by activation of endogenous Gq GPCRs in addition to the transgenic MrgA1Rs. The fact that we can measure ambient glutamate by recording *N*-methyl-d-aspartate (NMDA) receptor activity in CA1 neurons and see increases in ambient glutamate when blocking glutamate transporters indicates that were glutamate being released from astrocytes in a Ca^{2+}-dependent manner, we would see it (Fiacco et al., 2007). Overall, our findings suggest that an increase in astrocytic Ca^{2+} by activation of this Gq GPCRs is not sufficient to elicit glutamate release from these cells (Fiacco et al., 2007).

14.2.2.3 Expression of a Dominant-Negative Protein in Astrocytes

A large number of studies suggest that astrocytes have the ability to release neurotransmitters *in vitro* and *in situ*. While still under debate, considerable evidence suggest that a vesicular mechanism may be responsible for the release of gliotransmitters

Fig. 14.3 (continued) acid (tACPD), although the FLRFa Ca^{2+} responses were slightly longer lasting. (**c**) In control mice expressing only tTA or the tetO-MrgA1 construct, administration of peptide FLRFa did not produce a Ca^{2+} increase in astrocytes bulk-loaded with Ca^{2+} indicator. As a positive control for astrocyte Ca^{2+} increases resulting from release from internal stores, an agonist cocktail consisting of histamine (10 μM), carbachol (10 μM), and 2Na-ATP (50μM) was bath-applied. Scale bars, 10 μm (Modified with permission from Elsevier (Fiacco et al., 2007)) (*See Color Plates*).

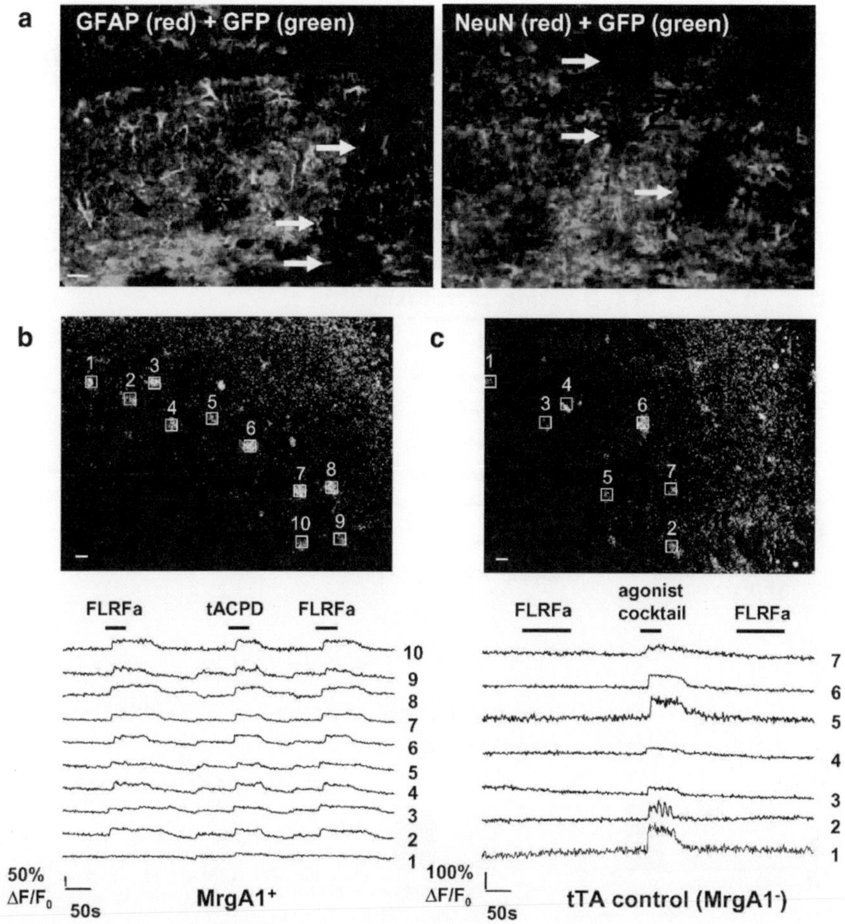

Fig. 14.3 Selective expression of a functional transgenic Gq GPCR, MrgA1, in astrocytes. The Gq-coupled receptor MrgA1 is expressed exclusively by astrocytes in the hippocampus. (**a**) Crossing hGFAP-tTA to tetO-MrgA1 mice resulted in the expression of MrgA1 exclusively in astrocytes in the absence of doxycycline. As eGFP is fused to the MrgA1 receptor construct, GFP staining indicates MrgA1 receptor expression. In the hippocampus, 80–90% of astrocytes expressed the MrgA1 receptor, while neurons did not express the receptor. The *left panel* shows astrocytes immunostained for GFAP (*red*) and GFP (*green*) to show MrgA1 expression. Astrocytes coexpressing GFAP and the MrgA1 receptor appear yellow. A few GFAP⁺ astrocytes do not express the MrgA1 receptor (*arrows*). The *right panel* shows GFP labeling in astrocytes (*green*) together with neuronal nuclei (NeuN) labeling neurons in stratum pyramidale of CA1 (*red*). Neurons do not express the MrgA1 receptor. A few astrocytes in the field do not express MrgA1 (*arrows*). Scale bar, 20 μm. Activation of the MrgA1 receptor produces robust, widespread astrocyte Ca^{2+} increases in acute mouse hippocampal slices. (**b**,**c**) Top panels show numbered astrocytes bulk-loaded with Ca^{2+} Green 1-AM indicator that correspond to the numbered fluorescence over time traces in bottom panels. Increases in F/F_0 indicate increases in cytoplasmic Ca^{2+} concentration. (**b**), In MrgA1⁺ mice maintained off doxycycline, FLRFa administration resulted in robust astrocyte Ca^{2+} increases in the majority of hippocampal astrocytes. Note that MrgA1 receptor-mediated astrocyte Ca^{2+} responses were similar in amplitude to those produced by (1*S*,3*R*)-1-aminocyclopentane-1,3-dicarboxlylic

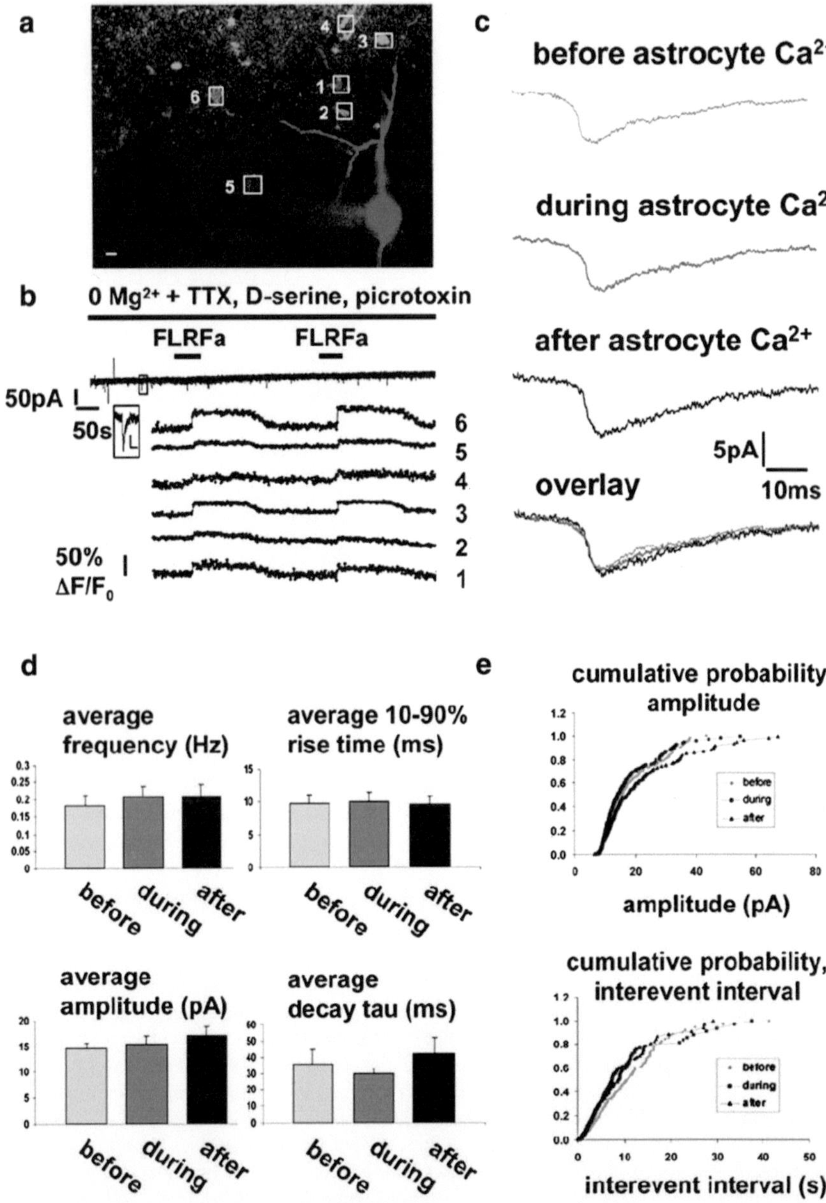

Fig. 14.4 Widespread selective astrocyte Ca^{2+} increases following MrgA1R activation do not produce an effect on miniature excitatory postsynaptic currents (mEPSCs) in CA1 pyramidal neurons. (**a**) Neurons were patch clamped with Alexa 568 hydrazide dye (*red*) to image cell morphology relative to astrocytes bulk-loaded with Ca^{2+} Green 1-AM dye (*green*). Numbered ROIs over astrocytes in which fluorescence intensity was measured over time correspond to the numbered traces in (**b**). (**b**) Astrocytes expressing the MrgA1 receptor were stimulated to release Ca^{2+} from internal stores by perfusion of the MrgA1 agonist peptide FLRFa during neuronal recording

from astrocytes (Pascual et al., 2005). In collaboration with Phil Haydon, we made a transgenic mouse line that expresses the cytosolic portion of the soluble *N*-ethyl maleimide-sensitive fusion protein attachment protein receptor (SNARE) domain of synaptobrevin 2, which has dominant-negative action and inhibits SNARE-dependent vesicular release from astrocytes. A series of electrophysiological studies were carried out using hippocampal brain slices to examine the role of vesicular-mediated neurotransmitter release from astrocytes. Findings from these studies indicate that astrocytes release adenosine 5′-triphosphate (ATP) *in situ* via a vesicular process. Further, the ATP released from astrocytes is quickly converted to adenosine which then plays an important role in heterosynaptic depression to affect the dynamic range of long-term potentiation (Pascual et al., 2005). These are very exciting findings clearly implicating a role for astrocytes in synaptic plasticity.

14.3 Studying Astrocytes Through Conditional Gene Knockout

14.3.1 General Considerations

Gene knockouts and conditional gene knockouts provide a very powerful approach to understanding the role of specific molecular pathways in brain function. When the endogenous expression of a gene product occurs in a single cell type, then knocking out the gene can be used to investigate both the role of the targeted gene product and the cell population in which it is expressed. However, it is rare that a gene product is restricted to a single cell type. When a gene is expressed by multiple cell types in brain, the interpretation of gene knockout experiments is often limited with respect to the cellular basis of the phenotype. To circumvent this problem, methods have been developed that enable the deletion of a gene in a specific cell

Fig. 14.4 (continued) of mixed α-amino-3-hydroxy-5-methyl-4-isoxazolepropionic acid (AMPA)/NMDA receptor mEPSCs (upper trace) in a Mg^{2+}-free external solution containing 100 μM picrotoxin, 10 μM d-serine, and 1 μM tetrodotoxin (TTX). This solution was used to maximize the probability of measuring NMDAR-mediated slow-inward currents as reported elsewhere. There was no effect of astrocyte Ca^{2+} elevations produced by MrgA1 Gq receptor activation on the neuronal recording. The inset shows a representative mEPSC on a larger timescale (scale bar, 10 pA, 100 ms). (**c**) Miniature EPSCs averaged before, during, and after FLRFa evoked astrocyte Ca^{2+} increases in the representative neuron. (**d**) Neuronal currents analyzed in three bins – before, during, and after selective astrocyte Ca^{2+} elevations – revealed no differences in frequency, amplitude, and kinetics ($n = 13$ neurons, 13 slices). (**e**) Cumulative probability plotted for amplitude (*top*) and interevent interval (*bottom*) show the full range of distributions. Kolmogorov–Smirnov analysis revealed no significant difference in the parameters (amplitude: $p = 0.4273$ before vs. during, 0.3265 before vs. after, 0.1210 during vs. after; interevent interval: $p = 0.0828$ before vs. during, 0.0719 before vs. after, 0.7889 during vs. after). Error bars indicate SEM. Scale bar, 10 μm (Reprinted with permission from Elsevier (Fiacco et al., 2007).) (*See Color Plates*).

population; such gene knockouts are referred to as conditional gene knockouts (Schipani, 2002; Morozov et al., 2003; Beglopoulos and Shen, 2004).

Conditional gene knockouts (cKOs) are generally based on the Cre recombinase/loxP system (Fig. 14.5). Two mouse lines are required for this system. In the first, Cre recombinase is driven to a specific cell population using a cell-specific promoter. A second line of mice is made via homologous recombination in embryonic stem cells. A genomic construct is prepared in which two *loxP* sites, which are the 34-bp recognition sites for Cre recombinase, are placed in or around the gene of interest. The placement of loxP sites is critical in that they should not interfere with the expression of the gene or function of the protein. In general, the loxP sites are placed within introns surrounding a critical exon(s) encoding important functional aspects of the protein or around a critical region of the gene's promoter usually involving the translational start site; genes with loxP sites are often referred to as "floxed." When the two lines of mice are crossed, the floxed gene fragment is excised and the gene recombined via the Cre recombinase, effectively removing the gene of interest. While this approach is conceptually straightforward, there are a number of caveats

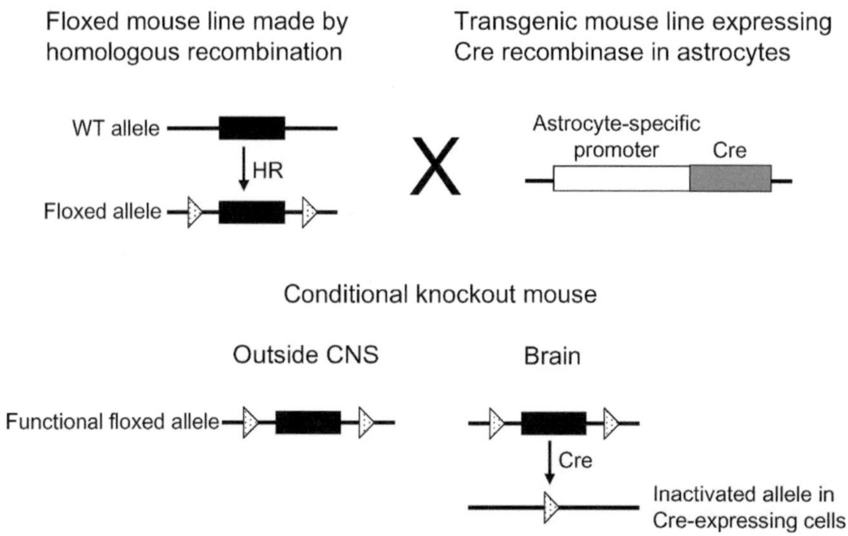

Fig. 14.5 Generating a conditional knockout mouse. The floxed mouse line is made by homologous recombination (HR) in embryonic stem cells of wild type (WT). The *loxP* sites are situated in intronic sequences, flanking exons of interest in the target gene. The floxed mouse should show normal expression of the target gene. This floxed line will be crossed to a Cre recombinase expressing line, usually made by transgenesis, to obtain the conditional knockout. Cre recombinase expression is driven by a cell-specific promoter, e.g., GFAP, to target the cell population of interest. In resulting offspring, gene deletion is achieved in all cells expressing Cre. Meanwhile, nonexpressing cells did not undergo gene deletion and show normal expression of the floxed allele. It is important to remember that all progeny of cells expressing Cre will be recombined even if they do not express Cre. *Black rectangle*, exon of targeted gene; *dotted triangles*, loxP sites; white *rectangle*, promoter driving Cre recombinase to astrocytes; *gray rectangle*, Cre recombinase cDNA.

that need to be considered in the interpretation of data arising from conditional knockout experiments. First, it is essential to demonstrate that recombination is restricted to the cell type of interest. To accomplish this, the Cre recombinase transgenic line is crossed with a reporter line of mice that utilizes a ubiquitous promoter to express a reporter gene (e.g., *lacZ*) following recombination. When combined with immunostaining to identify the cell of interest, this approach should determine the fraction of a cell population exhibiting recombination and if recombination occurs in other cell types. A second problem often encountered in conditional gene knockouts is that if recombination occurs during embryogenesis, developmental abnormalities can result that obscure the basis of a phenotype found in mature animals.

Our ability to recombine genes specifically in astrocytes has been particularly problematic. Unfortunately, all the promoters that were thought to selectively drive gene expression in astrocytes are active in progenitor cells that give rise to astrocytes, neurons, and oligodendrocytes (Casper and McCarthy, 2006). As a consequence, a large percentage of neurons and oligodendrocytes exhibit recombination when using the hGFAP promoter to drive the expression of Cre recombinase (Casper and McCarthy, 2006). To circumvent this problem, we and others have made transgenic lines that enable inducible, cell-specific gene knockouts (Casper et al., 2007; Hirrlinger et al., 2006). We have made and tested over sixty lines of mice expressing inducible Cre recombinase in an effort to identify a line that could be used to recombine astrocytic genes postdevelopmentally (Casper et al., 2007). This effort required the development of a new Cre reporter line as available Cre reporter lines are very inefficient in reporting recombination in astrocytes (Casper et al., 2007). Our best inducible Cre system is based on the CreER^{T2} system (Feil et al., 1997). In this system, Cre is fused to a mutated estrogen receptor (ER^{T2}) that restricts the fusion protein to the extranuclear compartment until exposed to tamoxifen, a synthetic derivative of estrogen (Fig. 14.6). Following CreER^{T2} binding to tamoxifen, CreER^{T2} translocates to the nucleus where Cre can mediate recombination of floxed genes. We have made an hGFAP-CreER^{T2} transgenic line that very efficiently recombines floxed reporter genes (Casper et al., 2007). However, further studies demonstrated that the efficiency of Cre recombinase to recombine floxed reporter genes is generally greater than its efficiency to recombine floxed endogenous genes (Casper et al., 2007). For example, a single injection of tamoxifen is generally sufficient to recombine a floxed reporter gene in astrocytes, whereas the same protocol leads to less than 50% recombination of Cx43 (Casper et al., 2007).

There are two additional considerations that should be noted prior to discussing specific examples of conditional knockouts in astrocytes. First, while we tend to discuss the phenotype of a conditional gene knockout solely in terms of the inactivated gene, in most cases this is probably naïve. Proteins work in concert with other proteins to carry out function. As previously discussed, removal of a single protein from a pathway generally affects the expression and function of many molecules and pathways. For example, deletion of Cx43 from astrocytes leads to changes in the expression of hundreds of other proteins. Hence, one needs to keep in mind that while we can selectively delete individual gene products, we cannot limit the effect that deletion has on a fabric of interacting proteins. The final issue that needs to be kept in mind when dealing with gene knockouts is compensation by other gene products. A number of studies indicate that the deletion of a single gene leads to the activation/expression of alternate gene products that can compensate for the deleted gene product. Thus, while a gene product may play an important role in physiology, its deletion may not be apparent due to compensation by alternate gene products.

14.3.2 Gene Deletions in Astrocytes

The first gene knockout in astrocytes was carried out by Albee Messing and colleagues who deleted the gene encoding GFAP (McCall et al., 1996). To the surprise of many, GFAP knockout mice exhibit no obvious phenotype (McCall et al., 1996).

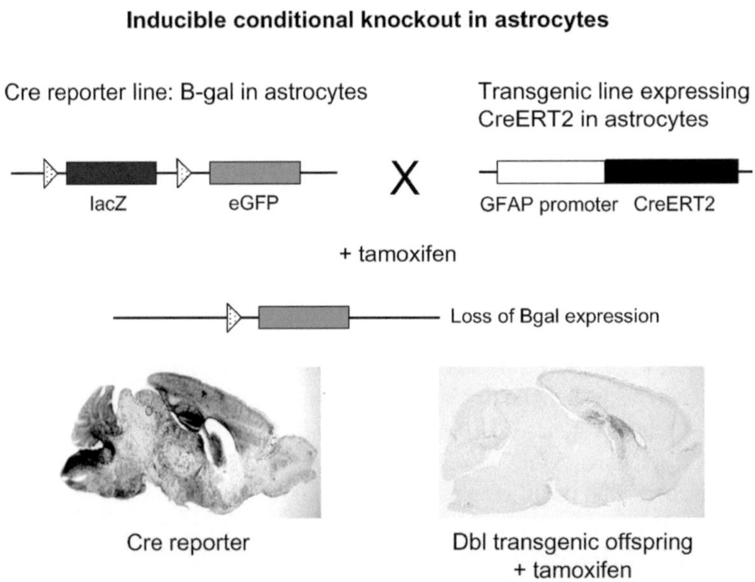

Fig. 14.6 Inducible gene recombination in astrocytes. The Cre reporter line expresses *lacZ* and thus B-gal protein prior to Cre recombinase expression. After Cre recombinase excises the *lacZ* cassette, eGFP is expressed. This reporter line is crossed to a transgenic line carrying a tamoxifen-sensitive Cre recombinase in astrocytes, driven by the GFAP promoter. Double (Dbl)- transgenic offspring are given tamoxifen by intraperitoneal injection which leads to excision of the lacZ cassette and loss of B-gal protein. B-gal protein expression is visualized by histochemical stain with Xgal. The reporter brain turns blue after this stain is performed, showing that there is B-gal protein present. After inducing the double transgenic with tamoxifen, there is great loss of B-gal protein. *Blue rectangle*, lacZ gene cassette; *green rectangle*, eGFP gene cassette; *dotted triangles*, *loxP* sites; *white rectangle*, GFAP promoter (or other astrocyte-specific promoter, e.g., GLAST); *black rectangle*, cDNA encoding fusion protein of Cre recombinase and mutated hormone-binding domain of estrogen receptor (ERT2) (*See Color Plates*).

However, GFAP knockout mice do exhibit a mild phenotype that includes a reduction in long-term depression and enhancement of long-term potentiation in the hippocampus (McCall et al., 1996; Shibuki et al., 1996). Additional phenotype changes were found when GFAP knockout mice were bred with vimentin knockout mice. Vimentin is another intermediate filament protein expressed by astrocytes. Interestingly, mice lacking both GFAP and vimentin exhibited greater functional recovery following spinal cord injury compared to either wild-type, GFAP null, or vimentin null mice. Further, the increased functional recovery correlated with a decrease in gliosis suggesting that the intermediate filament component of gliosis is a detriment to functional recovery following spinal cord injury (Menet et al., 2003).

A number of laboratories have prepared conditional Cx43 knockout mice. These mice were prepared by crossing hGFAP-Cre mice with mice containing floxed Cx43. Since Cx43 is only expressed by astrocytes in brain, the fact that the Cx43 locus was recombined in progenitors giving rise to multiple cell types was not a problem. Surprisingly given that Cx43 is the primary gap junction protein of astrocytes, mice lacking Cx43 in astrocytes exhibited only a minor phenotype consisting of increased locomotor activity and accelerated spreading depression (Theis et al., 2003). A strong phenotype was also not observed when Cx43 conditional knockout mice were crossed with Cx30 knockout mice (Wallraff et al., 2006). Gap junction communication (gjc) between astrocytes is largely abolished. These findings are quite surprising given that most astrocytes are normally coupled to one another throughout the brain. The fact that little functional gjc could be measured in the double knockout mice suggests that gjc between astrocytes is not critical to their overall role in neurophysiology. Another surprise came when we discovered that placing the Cx43 conditional knockout onto the 129SVEV mouse strain background led to a major phenotype that included disorganization of several brain regions, particularly the cerebellum (Wiencken-Barger et al., 2007) (Fig. 14.7). This phenotype was only observed when the mice were maintained on the 129SVEV background; placing the Cx43 conditional knockout on the C57Bl/6J background completely eliminated the phenotype. Further, we found no obvious difference in gjc between wild-type and Cx43 cKO mice suggesting that the phenotype derived from Cx43 knockout was due to some other function of Cx43 other than its participation in gjc (Wiencken-Barger et al., 2007). Results from these experiments emphasize an important point. That is, the specific strain (background) of a transgenic or gene knockout mouse can have a profound effect on the phenotype observed.

Conditional KO of the inwardly rectifying potassium channel, $K_{ir}4.1$, has been prepared by a number of laboratories in an effort to investigate the role of this channel in astrocytic K^+ buffering (Kofuji and Newman, 2000; Marcus et al., 2002; Neusch et al., 2006). To date, this has only been accomplished using the hGFAP promoter to drive Cre recombinase expression. As a consequence, recombination occurs in early progenitor cells that give rise to astrocytes, neurons, and oligodendrocytes. Mice generated from crossing hGFAP-Cre mice with floxed $K_{ir}4.1$ exhibit a strong phenotype that includes ataxia and death by postnatal day 30 (Kofuji and Newman, 2000). The basis for the ataxia is likely due to the extensive myelin breakdown observed in these mice. Given that oligodendrocytes express $K_{ir}4.1$, the most plausible

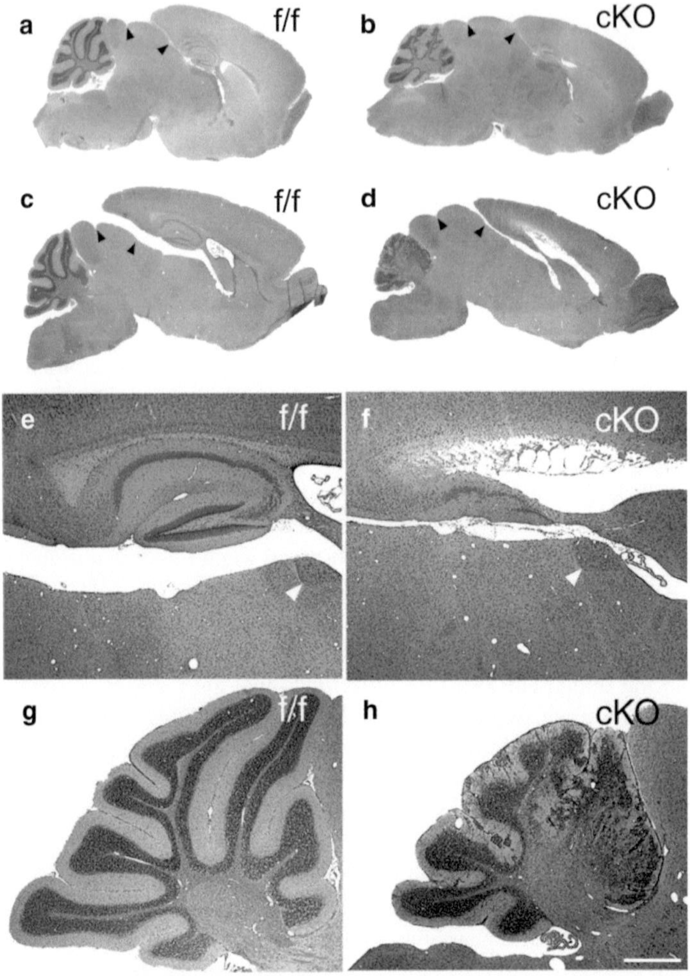

Fig. 14.7 Conditional knockout (cKO) of Cx43 in SVEV mice ("Shuffler" phenotype) displays a thinner cortex, a smaller hippocampus, and a disorganized cerebellum. Sagittal sections of paraffin-embedded adult (>postnatal day 60) brains were stained with hematoxylin and eosin and aligned in the mediolateral plane. Sections from Shuffler mice (**b**,**d**,**f**,**h**) were compared to floxed littermates (**a**,**c**,**e**,**g**). Whole brain (**a**–**d**), hippocampus (**e**,**f**), and cerebellum (**g**,**h**) are shown. Two brains are shown (**b**,**d**) to illustrate the range of the Shuffler phenotype. *Black arrowheads* surround superior colliculus that is more exposed in the Shuffler mice. *White arrowheads* in (**e**) and (**f**) denote the anterodorsal thalamic nucleus to show anatomical alignment between sections. Scale bar is 3 mm (**a**–**d**) and 500 μm (**e**–**h**) [Reprinted with permission from Wiley & Sons (Wiencken-Barger et al., 2007)].

explanation for these findings is that $K_{ir}4.1$ is important in the maintenance of myelin. Electrophysiological experiments demonstrate that the loss of $K_{ir}4.1$ in astrocytes leads to a striking depolarization of these cells and a decrease in K^+

buffering (Kofuji and Newman, 2000). A surprise resulting from our studies of $K_{ir}4.1$ cKO mice is that K_{ir} channels are not the primary K^+ channel expressed by astrocytes (Djukic et al., 2007). In fact, blocking astrocytic K_{ir} channels has little effect on the passive currents exhibited by astrocytes. Rather, it appears that (an) alternate K^+ channel(s), most likely a 2-pore K^+ channel, is responsible for the passive membrane currents typical of astrocytes *in situ* and *in vitro*. While $K_{ir}4.1$ does not appear responsible for the majority of the passive current exhibited by astrocytes, this channel is responsible for K^+ buffering during neuronal activity. The reason that 2-pore K^+ channels do not appear to play a role in astrocytic K^+ buffering remains unclear.

Astrocytic glutamate transporters are generally thought to play an important role in glutamate uptake to terminate excitatory neurotransmission (Schousboe and Waagepetersen, 2005). A number of studies have used molecular methods to perturb the expression of astrocytic glutamate transporters. Using antisense oligonucleotides, Rothstein et al. (1996) demonstrated that the loss of the l-glutamate/l-aspartate transporter (GLAST) led to increases in extracellular glutamate, neurodegeneration, and a progressive paralysis. These findings clearly indicate that the astrocytic glutamate transporter, GLAST, is a critical regulator of excitatory synaptic transmission. Interestingly, the phenotype of GLAST knockout mice is less severe than that observed following antisense knockdown (Rothstein et al., 1996; Stoffel et al., 2004). This suggests that in the case of the gene knockout, compensation may occur over the course of development. While the knockout of GLAST is less severe than the acute knockdown of these molecules using antisense approaches, there is a clear phenotype in mice lacking both the primary astrocytic glutamate transporters, GLAST and the glial l-glutamate transporter (GLT-1). For example, a double knockout of GLAST and GLT-1 (the primary astrocytic glutamate transporter) leads to striking developmental changes that include disorganization of several brain regions and perinatal mortality (Matsugami et al., 2006). Overall, studies in this area demonstrate that astrocytic glutamate transporters play an important role in limiting the buildup of extracellular glutamate and excessive neuronal activity.

14.4 Summary

Molecular approaches are evolving to study the role of astrocytes in neurophysiology, neuropathology, and behavior. Research over the past decade has been critical in developing the molecular tools required to perturb astrocyte gene expression *in vivo*. Transgenic and cKO mice exhibiting altered gene expression in astrocytes have already led to new and important insight into the function of astrocytes in brain. As with any approach studying perturbations *in vivo*, there are caveats that need to be considered when using these methods. However, it is likely that both transgenic and cKO animals will continue to play a critical role in our effort to unravel the role of astrocytes in brain function.

References

Anthony TE, Klein C, Fishell G, Heintz N (2004) Radial glia serve as neuronal progenitors in all regions of the central nervous system. Neuron 41:881–890.

Baron U, Bujard H (2000) Tet repressor-based system for regulated gene expression in eukaryotic cells: principles and advances. Methods Enzymol 327:401–421.

Beglopoulos V, Shen J (2004) Gene-targeting technologies for the study of neurological disorders. Neuromolecular Med 6:13–30.

Brambilla R, Bracchi-Ricard V, Hu WH, Frydel B, Bramwell A, Karmally S, Green EJ, Bethea JR (2005) Inhibition of astroglial nuclear factor kappaB reduces inflammation and improves functional recovery after spinal cord injury. J Exp Med 202:145–156.

Brenner M, Kisseberth WC, Su Y, Besnard F, Messing A (1994) GFAP promoter directs astrocyte-specific expression in transgenic mice. J Neurosci 14:1030–1037.

Cafferty WB, Yang SH, Duffy PJ, Li S, Strittmatter SM (2007) Functional axonal regeneration through astrocytic scar genetically modified to digest chondroitin sulfate proteoglycans. J Neurosci 27:2176–2185.

Casper KB, McCarthy KD (2006) GFAP-positive progenitor cells produce neurons and oligodendrocytes throughout the CNS. Mol Cell Neurosci 31:676–684.

Casper KB, Jones K, McCarthy KD (2007) Characterization of astrocyte-specific conditional knockouts. Genesis 45:292–299.

Chung JH, Bell AC, Felsenfeld G (1997) Characterization of the chicken beta-globin insulator. Proc Natl Acad Sci USA 94:575–580.

Chvatal A, Anderova M, Hock M, Prajerova I, Neprasova H, Chvatal V, Kirchhoff F, Sykova E (2007) Three-dimensional confocal morphometry reveals structural changes in astrocyte morphology in situ. J Neurosci Res 85:260–271.

Delaney CL, Brenner M, Messing A (1996) Conditional ablation of cerebellar astrocytes in postnatal transgenic mice. J Neurosci 16:6908–6918.

Djukic B, Casper KB, Philpot BD, Chin LS, McCarthy KD (2007) Conditional knock-out of Kir4.1 leads to glial membrane depolarization, inhibition of potassium and glutamate uptake, and enhanced short-term synaptic potentiation. J Neurosci 27:11354–11365.

Emsley JG, Macklis JD (2006) Astroglial heterogeneity closely reflects the neuronal-defined anatomy of the adult murine CNS. Neuron Glia Biol 2:175–186.

Faulkner JR, Herrmann JE, Woo MJ, Tansey KE, Doan NB, Sofroniew MV (2004) Reactive astrocytes protect tissue and preserve function after spinal cord injury. J Neurosci 24:2143–2155.

Feil R, Wagner J, Metzger D, Chambon P (1997) Regulation of Cre recombinase activity by mutated estrogen receptor ligand-binding domains. Biochem Biophys Res Commun 237:752–757.

Feng G, Mellor RH, Bernstein M, Keller-Peck C, Nguyen QT, Wallace M, Nerbonne JM, Lichtman JW, Sanes JR (2000) Imaging neuronal subsets in transgenic mice expressing multiple spectral variants of GFP. Neuron 28:41–51.

Fiacco TA, McCarthy KD (2004) Intracellular astrocyte calcium waves in situ increase the frequency of spontaneous AMPA receptor currents in CA1 pyramidal neurons. J Neurosci 24:722–732.

Fiacco TA, Agulhon C, Taves SR, Petravicz J, Casper KB, Dong X, Chen J, McCarthy KD (2007) Selective stimulation of astrocyte calcium in situ does not affect neuronal excitatory synaptic activity. Neuron 54:611–626.

Giraldo P, Montoliu L (2001) Size matters: use of YACs, BACs and PACs in transgenic animals. Transgenic Res 10:83–103.

Giraldo P, Rival-Gervier S, Houdebine LM, Montoliu L (2003) The potential benefits of insulators on heterologous constructs in transgenic animals. Transgenic Res 12:751–755.

Gossen M, Bujard H (1992) Tight control of gene expression in mammalian cells by tetracycline-responsive promoters. Proc Natl Acad Sci USA 89:5547–5551.

Hatten ME, Heintz N (2005) Large-scale genomic approaches to brain development and circuitry. Annu Rev Neurosci 28:89–108.

Hirrlinger J, Hulsmann S, Kirchhoff F (2004) Astroglial processes show spontaneous motility at active synaptic terminals in situ. Eur J Neurosci 20:2235–2239.
Hirrlinger PG, Scheller A, Braun C, Hirrlinger J, Kirchhoff F (2006) Temporal control of gene recombination in astrocytes by transgenic expression of the tamoxifen-inducible DNA recombinase variant CreERT2. Glia 54:11–20.
Ho G, Zhang C, Zhuo L (2007) Non-invasive fluorescent imaging of gliosis in transgenic mice for profiling developmental neurotoxicity. Toxicol Appl Pharmacol 221:76–85.
Iacobas DA, Scemes E, Spray DC (2004) Gene expression alterations in connexin null mice extend beyond the gap junction. Neurochem Int 45:243–250.
Kofuji P CPZKRSLWLHA, Newman EA (2000) Genetic Inactivation of an Inwardly Rectifying Potassium Channel (Kir4.1 Subunit) in Mice: Phenotypic Impact in Retina. J Neurosci 20:5733–5740.
Malatesta P, Hack MA, Hartfuss E, Kettenmann H, Klinkert W, Kirchhoff F, Gotz M (2003) Neuronal or glial progeny: regional differences in radial glia fate. Neuron 37:751–764.
Mansuy IM, Bujard H (2000) Tetracycline-regulated gene expression in the brain. Curr Opin Neurobiol 10:593–596.
Marcus DC, Wu T, Wangemann P, Kofuji P (2002) KCNJ10 (Kir4.1) potassium channel knockout abolishes endocochlear potential. Am J Physiol Cell Physiol 282:C403–C407.
Masood K, Besnard F, Su Y, Brenner M (1993) Analysis of a segment of the human glial fibrillary acidic protein gene that directs astrocyte-specific transcription. 61:160–166.
Matsugami TR, Tanemura K, Mieda M, Nakatomi R, Yamada K, Kondo T, Ogawa M, Obata K, Watanabe M, Hashikawa T, Tanaka K (2006) From the Cover: Indispensability of the glutamate transporters GLAST and GLT1 to brain development. Proc Natl Acad Sci USA 103:12161–12166.
McCall MA, Gregg RG, Behringer RR, Brenner M, Delaney CL, Galbreath EJ, Zhang CL, Pearce RA, Chiu SY, Messing A (1996) Targeted deletion in astrocyte intermediate filament (Gfap) alters neuronal physiology. Proc Natl Acad Sci USA 93:6361–6366.
Menet V, Prieto M, Privat A, Gimenez y Ribotta M (2003) Axonal plasticity and functional recovery after spinal cord injury in mice deficient in both glial fibrillary acidic protein and vimentin genes. Proc Natl Acad Sci USA 100:8999–9004.
Messing A, Brenner M (2003) Alexander disease: GFAP mutations unify young and old. Lancet Neurol 2:75.
Messing A, Head MW, Galles K, Galbreath EJ, Goldman JE, Brenner M (1998) Fatal encephalopathy with astrocyte inclusions in GFAP transgenic mice. Am J Pathol 152:391–398.
Morozov A, Kellendonk C, Simpson E, Tronche F (2003) Using conditional mutagenesis to study the brain. Biol Psychiatry 54:1125–1133.
Moynagh PN (2005) The interleukin-1 signalling pathway in astrocytes: a key contributor to inflammation in the brain. J Anat 207:265–269.
Neusch C, Papadopoulos N, Muller M, Maletzki I, Winter SM, Hirrlinger J, Handschuh M, Bahr M, Richter DW, Kirchhoff F, Hulsmann S (2006) Lack of the Kir4.1 channel subunit abolishes K+ buffering properties of astrocytes in the ventral respiratory group: impact on extracellular K+ regulation. J Neurophysiol 95:1843–1852.
Nolte C, Matyash M, Pivneva T, Schipke CG, Ohlemeyer C, Hanisch UK, Kirchhoff F, Kettenmann H (2001) GFAP promoter-controlled EGFP-expressing transgenic mice: a tool to visualize astrocytes and astrogliosis in living brain tissue. Glia 33:72–86.
Pascual O, Casper KB, Kubera C, Zhang J, Revilla-Sanchez R, Sul JY, Takano H, Moss SJ, McCarthy K, Haydon PG (2005) Astrocytic purinergic signaling coordinates synaptic networks. Science 310:113–116.
Rothstein JD, Dykes-Hoberg M, Pardo CA, Bristol LA, Jin L, Kuncl RW, Kanai Y, Hediger MA, Wang Y, Schielke JP, Welty DF (1996) Knockout of glutamate transporters reveals a major role for astroglial transport in excitotoxicity and clearance of glutamate. Neuron 16:675–686.
Scearce-Levie K, Coward P, Redfern CH, Conklin BR (2001) Engineering receptors activated solely by synthetic ligands (RASSLs). Trends Pharmacol Sci 22:414–420.
Schipani E (2002) Conditional gene inactivation using cre recombinase. Calcif Tissue Int 71:100–102.

Schousboe A, Waagepetersen HS (2005) Role of astrocytes in glutamate homeostasis: implications for excitotoxicity. Neurotox Res 8:221–225.
Shibuki K, Gomi H, Chen L, Bao S, Kim JJ, Wakatsuki H, Fujisaki T, Fujimoto K, Katoh A, Ikeda T, Chen C, Thompson RF, Itohara S (1996) Deficient cerebellar long-term depression, impaired eyeblink conditioning, and normal motor coordination in GFAP mutant mice. Neuron 16:587–599.
Stoffel W, Korner R, Wachtmann D, Keller BU (2004) Functional analysis of glutamate transporters in excitatory synaptic transmission of GLAST1 and GLAST1/EAAC1 deficient mice. Brain Res Mol Brain Res 128:170–181.
Sweger EJ, Casper KB, Scearce-Levie K, Conklin BR, McCarthy KD (2007) Development of hydrocephalus in mice expressing the G(i)-coupled GPCR Ro1 RASSL receptor in astrocytes. J Neurosci 27:2309–2317.
Theis M, Jauch R, Zhuo L, Speidel D, Wallraff A, Doring B, Frisch C, Sohl G, Teubner B, Euwens C, Huston J, Steinhauser C, Messing A, Heinemann U, Willecke K (2003) Accelerated hippocampal spreading depression and enhanced locomotory activity in mice with astrocyte-directed inactivation of connexin43. J Neurosci 23:766–776.
Wallraff A, Kohling R, Heinemann U, Theis M, Willecke K, Steinhauser C (2006) The impact of astrocytic gap junctional coupling on potassium buffering in the hippocampus. J Neurosci 26:5438–5447.
Wiencken-Barger AE, Djukic B, Casper KB, McCarthy KD (2007) A role for Connexin43 during neurodevelopment. Glia 55:675–686.
Ye P, Popken GJ, Kemper A, McCarthy K, Popko B, D'Ercole AJ (2004) Astrocyte-specific overexpression of insulin-like growth factor-I promotes brain overgrowth and glial fibrillary acidic protein expression. J Neurosci Res 78:472–484.
Zhang Q, Haydon PG (2005) Roles for gliotransmission in the nervous system. J Neural Transm 112:121–125.
Zhao Z, Alam S, Oppenheim RW, Prevette DM, Evenson A, Parsadanian A (2004) Overexpression of glial cell line-derived neurotrophic factor in the CNS rescues motoneurons from programmed cell death and promotes their long-term survival following axotomy. Exp Neurol 190:356–372.
Zhuo L, Sun B, Zhang CL, Fine A, Chiu SY, Messing A (1997) Live astrocytes visualized by green fluorescent protein in transgenic mice. Dev Biol 187:36–42.

Abbreviations

ATP	Adenosine 5′-triphosphate
BAC	Bacteria artificial chromosome
cKO	Conditional gene knockout
Cx43	Connexin 43
eGFP	Enhanced green fluorescent protein
ER^{T2}	Mutated estrogen receptor
GDNF	Glial-cell-line-derived neurotrophic factor
GFAP	Glial fibrillary acidic protein
Gjc	Gap junction communication
GLAST	l-glutamate/l-aspartate transporter
GLT-1	Glial l-glutamate transporter
GPCR	G-protein coupled receptor
hGFAP	Human GFAP

IGF-1	Insulin-like growth factor 1 (IGF-1)
mEPSC	Miniature excitatory postsynaptic current
NF-κB	Nuclear factor kappa B
NMDA	*N*-methyl-d-aspartate
PAC	P1-bacteriophage artificial chromosome
RASSL	Receptor activated solely by synthetic ligand
SNARE	The soluble *N*-ethyl maleimide-sensitive fusion protein attachment protein receptor
Tet	Tetracycline
tetO	tet operon
tTA	tet transactivator
YAC	Yeast artificial chromosome

Chapter 15
The Tripartite Synapse

Michael M. Halassa and Philip G. Haydon

Contents

The study of the astrocyte was hampered during the 1900s by the lack of experimental techniques to permit experimental stimulation and recording of the function of the astrocyte. Even in the early 1900s it was appreciated that electrical signals were the mechanism of conduction of neuronal signals (Adrian, 1912). Since astrocytes exhibit a large negative resting potential (Butt and Kalsi, 2006) their functional activity was mute to the electrophysiological techniques used to study nervous system function.

Despite the poor experimental tractability of astrocytes, several pioneering studies did, however, provide important insights into the functional roles for these glia in brain function. Microscopy, immunocytochemistry, and biochemistry were used to identify glycogen storage granules (Magistretti, 2006; Tsacopoulos and Magistretti,

P.G. Haydon
Department of Neuroscience, Tufts University School of Medicine,
Boston, MA USA
philip.haydon@tufts.edu

V. Parpura and P.G. Haydon (eds.), *Astrocytes in (Patho)Physiology of the Nervous System*,
DOI: 10.1007/978-0-387-79492-1_15, © Springer Science+Business Media, LLC 2009

1996), and the presence of glutamate transporters (Anderson and Swanson, 2000) in astrocytic membranes. These observations provided a basis for the development of an understanding of the important metabolic role for the astrocyte and its importance in the uptake of transmitter from the synaptic cleft, and the subsequent recycling through a glutamine intermediate (Hertz and Zielke, 2004). However, it was not until the development of optical approaches and chemically synthesized Ca^{2+} indicators that the dynamic excitability of the astrocyte was first observed (Cornell Bell et al., 1990). Astrocytes were shown to respond to neurotransmitters with oscillations of internal Ca^{2+} levels (Pasti et al., 1997) (also see Chap. 8), allowing us to begin to appreciate the plethora of metabotropic receptors expressed by these glial cells (Chap. 3). Now with two-photon microscopy applied to in vivo imaging of astrocytes Maiken Nedergaard's group has performed experiments of fundamental importance by showing that sensory stimulation (whisker deflection) not only evokes neuronal responses in the barrel cortex but also Ca^{2+} oscillations in astrocytes (Wang et al., 2006). Therefore much like the neuron, astrocytes are under the influence of environmental signals.

15.1 What is Gliotransmission?

Once Ca^{2+} signals were observed in astrocytes the question to be answered was, what is the functional consequence of this Ca^{2+} signal and could it relay information to other cells of the brain? After fifteen years, the answer is a resounding "yes."

In 1994 two groups discovered, initially in cell culture, that Ca^{2+} signals in astrocytes lead to delayed neuronal Ca^{2+} responses (Parpura et al., 1994; Nedergaard, 1994). Over time it has become clear that a significant mechanism of the astrocyte-to-neuron signaling is mediated by the release of chemical transmitters from the astrocyte. (When transmitters are released from astrocytes we use the term gliotransmitter, and the process as gliotransmission.) The first transmitter to be discovered released from astrocytes was glutamate: Ca^{2+} signals were shown to be necessary and sufficient for the release of this gliotransmitter (Fellin et al., 2006; Zhang et al., 2004a, b). Subsequently, these cell culture studies have been supported in situ (Fellin et al., 2004; Jourdain et al., 2007) and more recently in studies performed in vivo (Ding et al., 2007).

15.1.1 Glutamate

Being the first gliotransmitter to be discovered glutamate has received considerable attention as a transmitter of the astrocyte. Glutamate is thought to be released from astrocytes through numerous mechanisms: exocytosis (Zhang et al., 2004a, b; Bezzi et al., 2004; Montana et al., 2004), volume-regulated anion channels (Takano et al., 2005), hemichannels (Ye et al., 2003), purinergic P2X receptors (Fellin et al., 2006), and pannexins (Barbe et al., 2006) (also see Chap. 12). It is likely that each mechanism can be utilized by the astrocyte to release transmitters. However, the state of brain will dictate which is recruited and utilized prominently. For example, under physiological extracellular Ca^{2+} levels channel-mediated mechanisms may be less prominent, but instead exocytosis mediates transmitter release. However, when extracellular Ca^{2+} falls, for example, following action potentials, a channel-mediated pathway may become more important since these channels have a higher open probability under low divalent cation conditions. Given that Chap. 12 addresses release mechanism, we do not dwell further on this issue.

15.1.2 ATP

In culture, astrocytes exhibit waves of Ca^{2+} that can propagate from cell to cell (Chaps. 8 and 17). These so-called Ca^{2+} waves are mediated, at least in part, by an extracellular message since medium collected from cells exhibiting Ca^{2+} waves is able to induce Ca^{2+} signals in nonstimulated cells (Guthrie et al., 1999). Adenosine 5′-triphosphate (ATP) has been shown to be the molecule collected in the medium that evokes Ca^{2+} signals, and its involvement in Ca^{2+} waves is supported by the observations that inhibition of purinergic receptors or enzymatic degradation of ATP both retard Ca^{2+} waves (Bowser and Khakh, 2004).

15.1.3 D-Serine

Serine racemase is an enzyme that is highly expressed in astrocytes and is responsible for the conversion of L- to D-serine (Mothet et al., 2000; Schell et al., 1995; Wolosker et al., 1999). This D-amino acid binds to the glycine-binding site of the *N*-methyl-D-aspartate (NMDA) receptor and is likely the endogenous ligand for this receptor. Like the discussion for glutamate, there may be multiple pathways for D-serine release from astrocytes. Regardless of mechanism, however, D-serine is released and it is necessary for synaptic NMDA receptor activity (Chap. 16).

15.2 Gliotransmission Continuously and Dynamically Regulates Synaptic Transmission

In an early period of a new field it can be dangerous to make generalizations about signaling processes. Given that caveat, generalizations can be helpful to provide a conceptual framework from which future experimental approaches can be developed. Now that we have several examples of astrocyte-to-neuron signaling that impact synaptic transmission, one generalization is emerging in which the astrocyte is able to continuously, or tonically, modulate synaptic transmission (Fig. 15.1). However, this tone is not without regulation: in response to astrocytic integration of environmental

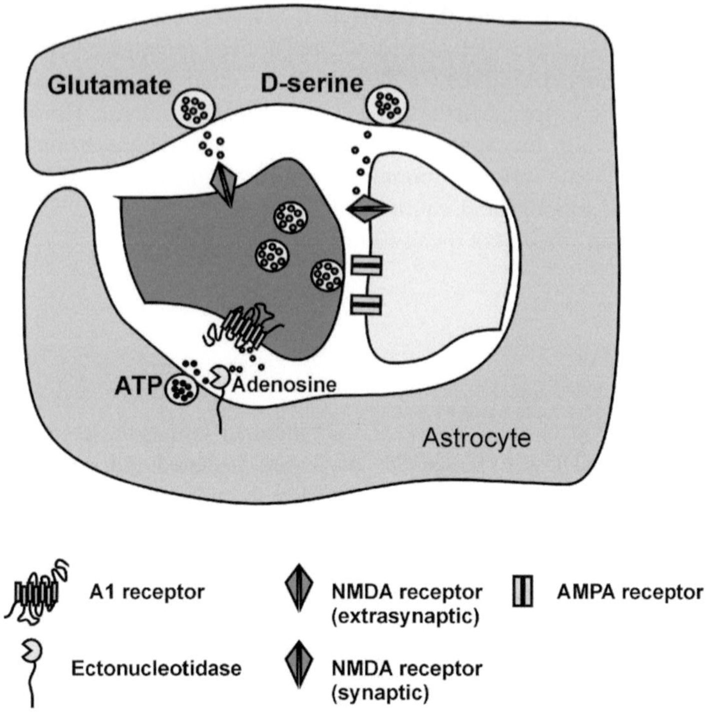

Fig. 15.1 The release of gliotransmitters continuously modulates synaptic transmission. Astrocytic process surrounding the presynaptic (*red*) and postsynaptic (*gray*) terminals. Astrocytes continuously release: (i) glutamate to act on extrasynaptic NR2B-containing NMDA receptors, (ii) D-serine to act on synaptic NMDA receptors, and (iii) ATP which upon degradation to adenosine acts on presynaptic A1 receptors. Environmental signals such as GPCR signaling can positively or negatively modulate the continuous release of these gliotransmitters. (*See Color Plates*)

signals the tone can be increased or decreased allowing dynamic control of synaptic transmission. We now discuss three examples, spanning three gliotransmitters, and four laboratories.

15.2.1 Astrocytic Adenosine Tonically Inhibits Excitatory Synaptic Transmission and Dynamically Mediates Heterosynaptic Depression

It is well known that in the hippocampus and neocortex that there is a tonic activation of presynaptic adenosine 1 (A1) receptors that leads to a presynaptic inhibition of synaptic transmission (Scammell et al., 2003). Recently, we determined that the adenosine which mediates this presynaptic inhibition is provided from an astrocytic source (Pascual et al., 2005). It is extremely difficult to identify the role of astrocytes in brain function because neurons and astrocytes share similar receptors precluding pharmacological approaches to activate or inhibit these glial cells. To overcome this problem we have used molecular genetic approaches, initially in collaboration with Dr. Ken McCarthy's group (see Chap. 14). We chose to perturb the exocytotic pathway of gliotransmission because the molecular tools required to do so were available, and because there was abundant knowledge on how exocytosis mediates gliotransmission. We used a tetracycline responsive genetic system that permitted conditional expression of a cytoplasmic tail of synaptobrevin 2 (lacking transmembrane domain) acting as a dominant negative inhibitor of exocytosis (dnSNARE) selectively in astrocytes. We confirmed that the astrocytic expression of dnSNARE impaired exocytosis by observing that a membrane-bound member of the exocytotic machinery, synaptosome-associated protein of 23 kDa (SNAP-23), appeared in the cytosolic fraction of brain extracts prepared from these transgenic animals. Our functional studies, thereafter, were performed in the hippocampus where we studied the Schaffer collateral-CA1 synapse. Stimulation of the Schaffer collaterals in slices obtained from mice expressing dnSNARE in astrocytes showed stronger synaptic transmission compared to slices obtained from wild-type littermates, or transgenic mice in which transgene expression was prevented.

We studied the mechanism of this enhanced synaptic transmission by using pharmacological approaches to manipulate receptors that respond to known gliotransmitters. Addition of the A1 receptor antagonist 8-cyclopentyl-1,3-dipropylxanthine (DPCPX) to wild-type slices enhanced transmission, but had no effect when synapses from transgenic mice were studied. This together with other evidence provided compelling evidence that expression of the dnSNARE in the astrocyte caused an attenuation of neuronal A1 receptor activation. Adenosine was not found to be directly released from astrocytes within the context of A1 receptor activation. Instead ATP was released and then hydrolyzed to adenosine in the extracellular space.

This study shows that astrocytes are continuously modulating synaptic transmission by activating A1 receptors, but does not indicate whether gliotransmission can dynamically control synaptic transmission. In the hippocampus, high-frequency stimulation of a subset of the Schaffer collateral fibers causes potentiation of the innervated synapses (homosynaptic potentiation) and an adenosine-mediated depression of nearby un-innervated synapses (heterosynaptic depression). Though it has known that this dynamic process is mediated by adenosine acting through A1 receptor, the cellular source of adenosine had been undefined. Using transgenic mice expressing dnSNARE in astrocytes we demonstrated that activation of Schaffer collaterals is unable to cause adenosine-mediated heterosynaptic depression. Therefore, in response to synaptic activity astrocyte-derived adenosine is augmented to allow a transient depression of neighboring synapses.

15.2.2 Astrocytic Glutamate Continuously Regulates Parallel Fiber-Granule Cell Synapses and Transiently Augments the Frequency of Excitatory Synaptic Transmission

As mentioned earlier, the mechanism and functional consequences of glutamatergic gliotransmission have attracted much attention, in part, due to the historical significance of it being the first form of gliotransmission discovered. Several laboratories have shown that the machinery for the regulated exocytotic release of glutamate is present in astrocytes (Montana et al., 2004; Zhang et al., 2004a, b; Bezzi et al., 2004). A recent study by Andrea Volterra's group in Lausanne shows that the concept of a dynamically regulated astrocytic neuromodulatory tone can be extended to glutamate (Jourdain et al., 2007). By performing experiments in hippocampal brain slices, the group showed that inhibiting astrocytic spontaneous activation by either blocking metabotropic purinergic P2Y1 receptors (shown to be selectively expressed by astrocytes in that brain region), or loading them with cell permeant form of 1,2-bis(*o*-aminophenoxy)ethane-*N*,*N*,*N*′,*N*′-tetraacetic acid (BAPTA) to chelate their cytoplasmic Ca^{2+}, results in a decrease of the frequency of miniature and spontaneous postsynaptic currents in nearby granule cells. Using a series of structural and functional experiments, the authors demonstrated that this synaptic enhancement is a result of the exocytotic release of glutamate by astrocytes onto extrasynaptic NR2B-containing NMDA receptors of nearby parallel fibers to enhance their probability of transmitter release. The authors proceeded to show that evoked parallel fiber-granule cell synaptic transmission can be attenuated by blocking astrocytic Ca^{2+} signaling using the P2Y1 receptor antagonist adenosine-3-phosphate-5-phospho-sulfate (A3P5PS), demonstrating that astrocytes can be dynamically recruited during synaptic activation to subsequently augment transmission of the same synapses. Electrical stimulation of astrocytes during synaptic activation resulted in an increase in an enhanced synaptic transmission at the parallel fiber-granule cell synapses confirming that astrocytic activation is sufficient for activity-dependent enhancement of synaptic transmission.

15.2.3 Glia-Derived D-Serine Tonically Regulates NMDA Receptor Function

In the supraoptic nucleus (SON) the processes of astrocytes invaginate synapses and provide a continuous source of D-serine. In a study by Stephane Oliet's group, D-serine was shown to be the preferred co-ligand for the NMDA receptor in the SON (Panatier et al., 2006). In addition, the group took advantage of the fact that astrocytic processes are highly dynamic in that brain region; in virgin rodents astrocytic processes ensheath SON synapses while in lactating rodents these processes retract from the same synapses. In a series of elegant experiments, the group showed that when astrocytic processes are close to synapses, the continuous D-serine release is able to reach synaptic NMDA receptors resulting in a bigger NMDA current that favors the induction of synaptic potentiation following high-frequency stimulation. The absence of such tone, caused by the retraction of the astrocytic process, results in synapses exhibiting synaptic depression under the same stimulation parameters, demonstrating that astrocytic d-serine tone can determine the direction of plasticity (metaplasticity) (also see Chap. 16).

Though experimental evidence suggests that D-serine is Ca^{2+} and exocytosis dependent (Mothet et al., 2005), its regulation by physiological environmental signals is unknown. Furthermore, the molecular and cellular mechanisms responsible for and modulating astrocytic processes moving close or away from synapses are unknown. Understanding these processes may give insights to a superimposed dynamic regulation of the astrocyte-derived synaptic D-serine tone.

15.3 Conclusions

After the initial discovery that astrocytes can release gliotransmitters there have been numerous examples of gliotransmitters modulating synaptic transmission. After these initial observations it is important to attempt to understand the roles that gliotransmission serve in brain function. Prior to achieving such an understanding, however, we must necessarily pass through a period where we show the potential for these signaling pathways. Potential roles, however, do not necessary mean that these signaling pathways are utilized by a nervous system in normal brain function. The application of molecular genetics allows such evaluations to be made. For example, astrocyte selective expression of dnSNARE demonstrates that adenosine is derived from astrocytes and normally used to control synaptic transmission in the hippocampus. With this animal, as well as others under development, it will not be long before we begin to understand how astrocytes control synapses, neural circuits, and behaviors, and under which conditions each of the gliotransmitters glutamate, D-serine, and ATP/adenosine are utilized to regulate brain and behavior.

References

Anderson CM, Swanson RA (2000) Astrocyte glutamate transport: Review of properties, regulation, and physiological functions. Glia 32: 1–14.

Adrian ED (1912) On the conduction of subnormal disturbances in normal nerve. J Physiol 45: 389–412.

Barbe MT, Monyer H, Bruzzone R (2006) Cell–Cell Communication Beyond Connexins: The Pannexin Channels. Physiology 21: 103–114.

Bezzi P, Gundersen V, Galbete JL, Seifert G, Steinhauser C, Pilati E, Volterra A (2004) Astrocytes contain a vesicular compartment that is competent for regulated exocytosis of glutamate. Nat Neurosci 7: 613–620.

Bowser DN, Khakh BS (2004) ATP excites interneurons and astrocytes to increase synaptic inhibition in neuronal networks. J Neurosci 24: 8606–8620.

Butt AM, Kalsi A (2006) Inwardly rectifying potassium channels (Kir) in central nervous system glia: a special role for Kir4.1 in glial functions. J Cell Mol Med 10: 33–44.
Cornell Bell AH, Finkbeiner SM, Cooper MS, Smith SJ (1990) Glutamate induces calcium waves in cultured astrocytes: long- range glial signaling. Science 247: 470–473.
Ding S, Fellin T, Zhu Y, Lee SY, Auberson YP, Meaney DF, Coulter DA, Carmignoto G, Haydon PG (2007) Enhanced astrocytic Ca2+ signals contribute to neuronal excitotoxicity after status epilepticus. J Neurosci 27: 10674–10684.
Fellin T, Pascual O, Gobbo S, Pozzan T, Haydon PG, Carmignoto G (2004) Neuronal synchrony mediated by astrocytic glutamate through activation of extrasynaptic NMDA receptors. Neuron 43: 729–743.
Fellin T, Pozzan T, Carmignoto G (2006) Purinergic receptors mediate two distinct glutamate release pathways in hippocampal astrocytes. J Biol Chem 281: 4274–4284.
Guthrie PB, Knappenberger J, Segal M, Bennett MV, Charles AC, Kater SB (1999) ATP released from astrocytes mediates glial calcium waves. J Neurosci 19: 520–528.
Hertz L, Zielke HR (2004) Astrocytic control of glutamatergic activity: astrocytes as stars of the show. Trends Neurosci 27: 735–743.
Jourdain P, Bergersen LH, Bhaukaurally K, Bezzi P, Santello M, Domercq M, Matute C, Tonello F, Gundersen V, Volterra A (2007) Glutamate exocytosis from astrocytes controls synaptic strength. Nat Neurosci 10: 331–339.
Magistretti PJ (2006) Neuron-glia metabolic coupling and plasticity. J Exp Biol 209: 2304–2311.
Montana V, Ni Y, Sunjara V, Hua X, Parpura V (2004) Vesicular glutamate transporter-dependent glutamate release from astrocytes. J Neurosci 24: 2633–2642.
Mothet JP, Parent AT, Wolosker H, Brady RO, Jr., Linden DJ, Ferris CD, Rogawski MA, Snyder SH (2000) D-Serine is an endogenous ligand for the glycine site of the N-methyl-D-aspartate receptor. Proc Natl Acad Sci U S A 97: 4926–4931.
Mothet JP, Pollegioni L, Ouanounou G, Martineau M, Fossier P, Baux G (2005) Glutamate receptor activation triggers a calcium-dependent and SNARE protein-dependent release of the gliotransmitter D-serine. Proc Natl Acad Sci U S A 102: 5606–5611.
Nedergaard M (1994) Direct signaling from astrocytes to neurons in cultures of mammalian brain cells. Science 263: 1768–1771.
Panatier A, Theodosis DT, Mothet JP, Touquet B, Pollegioni L, Poulain DA, Oliet SH (2006) Glia-derived D-serine controls NMDA receptor activity and synaptic memory. Cell 125: 775–784.
Parpura V, Basarsky TA, Liu F, Jeftinija K, Jeftinija S, Haydon PG (1994) Glutamate-mediated astrocyte-neuron signalling. Nature 369: 744–747.
Pascual O, Casper KB, Kubera C, Zhang J, Revilla-Sanchez R, Sul JY, Takano H, Moss SJ, McCarthy K, Haydon PG (2005) Astrocytic purinergic signaling coordinates synaptic networks. Science 310: 113–116.
Pasti L, Volterra A, Pozzan T, Carmignoto G (1997) Intracellular calcium oscillations in astrocytes: a highly plastic, bidirectional form of communication between neurons and astrocytes in situ. J Neurosci 17: 7817–7830.
Scammell TE, Arrigoni E, Thompson MA, Ronan PJ, Saper CB, Greene RW (2003) Focal deletion of the adenosine A1 receptor in adult mice using an adeno-associated viral vector. J Neurosci 23: 5762–5770.
Schell MJ, Molliver ME, Snyder SH (1995) D-serine, an endogenous synaptic modulator: localization to astrocytes and glutamate-stimulated release. Proc Natl Acad Sci U S A 92: 3948–3952.
Takano T, Kang J, Jaiswal JK, Simon SM, Lin JH, Yu Y, Li Y, Yang J, Dienel G, Zielke HR, Nedergaard M (2005) Receptor-mediated glutamate release from volume sensitive channels in astrocytes. Proc Natl Acad Sci U S A 102: 16466–16471.
Tsacopoulos M, Magistretti PJ (1996) Metabolic coupling between glia and neurons. J Neurosci 16: 877–885.
Wang X, Lou N, Xu Q, Tian GF, Peng WG, Han X, Kang J, Takano T, Nedergaard M (2006) Astrocytic Ca(2+) signaling evoked by sensory stimulation in vivo. Nat Neurosci 9: 816–823.

Wolosker H, Sheth KN, Takahashi M, Mothet JP, Brady RO, Jr., Ferris CD, Snyder SH (1999) Purification of serine racemase: biosynthesis of the neuromodulator D-serine. Proc Natl Acad Sci U S A 96: 721–725.
Ye ZC, Wyeth MS, Baltan-Tekkok S, Ransom BR (2003) Functional hemichannels in astrocytes: a novel mechanism of glutamate release. J Neurosci 23: 3588–3596.
Zhang Q, Fukuda M, Van Bockstaele E, Pascual O, Haydon PG (2004a) Synaptotagmin IV regulates glial glutamate release. Proc Natl Acad Sci U S A 101: 9441–9446.
Zhang Q, Pangrsic T, Kreft M, Krzan M, Li N, Sul JY, Halassa M, Van Bockstaele E, Zorec R, Haydon PG (2004b) Fusion-related release of glutamate from astrocytes. J Biol Chem 279: 12724–12733.

Abbreviations

ATP	Adenosine 5′-triphosphate
NMDA	*N*-methyl-D-aspartate
SON	Supraoptic nucleus

Chapter 16
Glia-Derived D-Serine and Synaptic Plasticity

Magalie Martineau, Stéphane H.R. Oliet, and Jean-Pierre Mothet

Contents

16.1 Introduction

Although the chemical and physical properties of L-amino acids and D-amino acids are extremely similar, only L-amino acids seemed to have been selected from the origin of life on the primitive Earth. In their chemical evolutionary step, D-amino acids seemed to have been eliminated, and hence it has been considered that all superior living organisms are composed only of L-amino acids. Homochirality is a characteristic signature of life. This asymmetry in biology is assumed to be a feature of fundamental physics, because the *natural* L-amino acids are more stable than their *unnatural* mirror images. Until the last 30 years, it has been considered that D-amino acids were excluded from living systems except for D-amino acids in the

J.-P. Mothet
Centre de Recherche François Magendie, INSERM U862, Bordeaux, France
jean-pierre.mothet@inserm.fr

V. Parpura and P.G. Haydon (eds.), *Astrocytes in (Patho)Physiology of the Nervous System*, 417
DOI: 10.1007/978-0-387-79492-1_16,

cell wall of microorganisms (Fujii, 2002). Biologists have since discovered that nature could deal with at least two D-amino acids, D-serine and D-aspartic acids in higher living organisms. The discovery of D-serine in the central nervous system (CNS) of rodents is remarkable for two reasons. First, it revolutionized our thinking and forced us to reconsider the long-cherished dogma that only L-isomers of amino acids occurred in mammalian tissues and body fluids. Second, this atypical brain messenger fulfils all criteria to be a major, if not the only, ligand for the strychnine-insensitive glycine modulatory binding site of the *N*-methyl-D-aspartate receptors (NMDARs), a key receptor for excitatory transmission and cognitive functions (Mustafa et al., 2004; Martineau et al., 2006; Wolosker, 2006). In the present review, we outline the molecular mechanisms controlling D-serine availability in the CNS and the roles of this atypical messenger in promoting neuronal migration in the developing cerebellum and in governing the direction and magnitude of long-lasting changes in synaptic strength. Knowledge is now gradually accumulating for a role of D-serine signaling pathway in the pathophysiology of many brain disorders. This aspect will not be covered here, and we invite the reader to refer to recent studies on this specific topic (Tsai et al., 1998; Chumakov, 2002; Katsuki et al., 2004; Wu et al., 2004b; Shleper et al., 2005; Bendikov et al., 2007).

16.2 Regional and Cellular Distributions of D-Serine in the Nervous System

The starting point that brought D-serine on stage was the discovery that this amino acid is present in the brains of rodents and humans where its levels (~500 μM) are up to a third of the total free (l + D) serine pool. D-Serine can be detected at very early stages of embryonic life in the CNS of rodents and humans and throughout the entire lifetime (Hashimoto and Oka, 1997). Early high-performance liquid chromatography (HPLC) analyses by Hashimoto et al. (1995a) revealed a heterogeneous distribution of D-serine throughout the brain with highest concentrations in the telencephalon and the developing cerebellum. At adult stage (8 weeks), the highest concentrations of D-serine are found in the cerebrum cortex, followed by the thalamus, the striatum, the amygdala, the hippocampus, and the hypothalamus. For a more precise delineation of D-serine inside the brain, Snyder and coworkers developed polyclonal antibodies to D-serine conjugated to glutaraldehyde and bovine serum albumin (Schell et al., 1995, 1997). Immunostaining revealed high densities of D-serine in the forebrain with very low densities in the hindbrain, thus confirming HPLC analyses (Fig. 16.1). Detailed examination of D-serine immunoreactivity (ir) revealed a close relationship with NMDARs (Schell et al., 1997). Double staining for the NR2A/B subtypes of NMDARs with D-serine or glycine revealed a closer correspondence of NMDARs with D-serine than with glycine in most parts of the brain. This intimate relationship is especially striking in deeper layers of the cerebral cortex. In contrast, glycine-ir is related to distribution of NMDARs in the hindbrain, the adult cerebellum, and the olfactory bulb where D-serine is also present (Schell

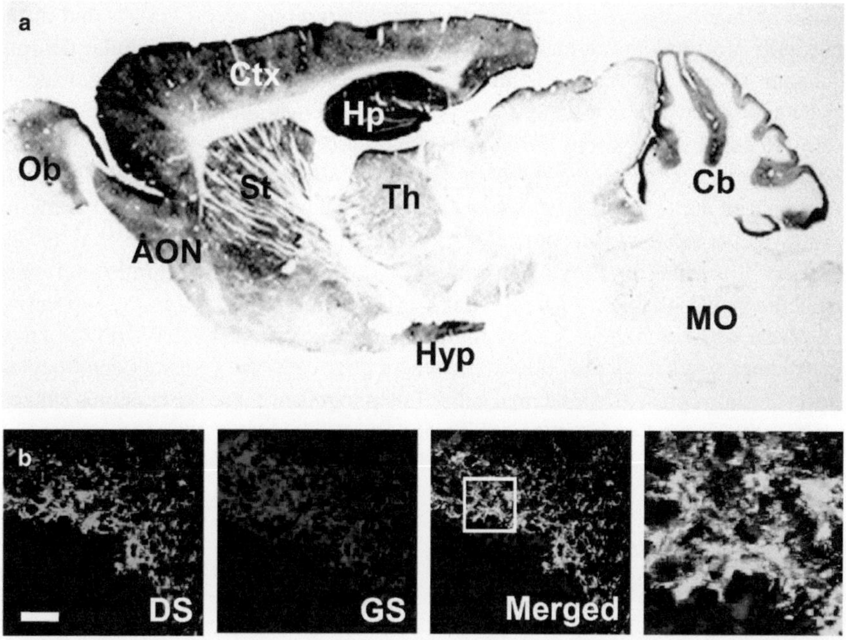

Fig. 16.1 Regional and cellular distribution of D-serine-ir in the CNS of adult rodent. (**a**) D-Serine-ir is mostly retrieved in structures of the telencephalon and diencephalon. Highest levels are found in the cerebral cortex (Ctx), olfactory bulb (Ob), hippocampus (Hp), striatum (St), thalamus (Th), hypothalamus (Hyp), and in the molecular layer of the cerebellum (Cb). No staining is found in the medulla oblongata (MO). (**b**) Glial distribution of D-serine immunofluorescence in the guinea pig CNS. D-Serine (DS) corresponds to the red channel (secondary antibody: Alexa fluor 546) and glutamine synthethase (GS) to the green channel (secondary antibody: Alexa fluor 488). Note the high level of yellow spots on the merged panel highlighting the colocalization between DS and the glial marker, GS. (*See Color Plates*)

et al., 1995, 1997). In the brainstem, where glycine-ir is high, glycine closely parallels the distribution of NMDA receptors (Schell et al., 1997).

16.2.1 Glial D-Serine

Early analysis of D-serine-ir (Schell et al., 1995, 1997) under high magnification shows that D-serine is mostly associated with a population of protoplasmic astrocytes that ensheathes synapses and that it is particularly enriched in the gray matter. At the electron microscopy level, D-serine occurs in astrocytic end-feet that contact endothelial cells and in astrocytes that contact neuronal dendritic spines (Schell et al., 1997). Double immunostaining of cerebral cortex revealed that most D-serine-containing cells are glial fibrillary acidic protein (GFAP)-positive cells (Williams et al., 2006). In the supraoptic nucleus (SON) of the hypothalamus, D-serine-ir exclusively occurs in large astrocytic GFAP-positive processes along blood vessels and in the neuropile surrounding oxytocin- and vasopressin-secreting magnocellular neurons (Panatier et al., 2006). In the developing cerebellum, D-serine is localized to Bergmann glia while in adults, D-serine declines to negligible levels (Schell et al., 1997). Recent investigations have found that quiescent and activated microglia cells contain significant amount of D-serine (Wu et al., 2004a; Williams et al., 2006). In the peripheral nervous system, D-serine and serine racemase are expressed by Schwann cells (Wu et al., 2004b). The spreading distribution of D-serine over the glial lineage is further illustrated by the demonstration that D-serine-ir is found in Müller cells, a specific radial glial cell of the retina (Stevens et al., 2003) and in the supporting cell of the vestibular sensory epithelium (Dememes et al., 2006). We do not know yet whether D-serine is also present in others glial cells such as oligodendrocytes, pituicytes, tanycytes, or ependymal cells. Taken together, these observations suggest that D-serine is confined to the glial lineage.

16.2.2 Neuronal D-Serine

Nevertheless, as we move carefully toward the elucidation of the cellular distribution of D-serine, this glial specific D-serine framework becomes equivocal. D-Serine-ir has been initially observed in dendrites and axons of some cortical neurons and brainstem neurons (Yasuda et al., 2001) but at very low levels and so scarcely that it was viewed as an oddity. Hence, two different studies have reported that neurons could synthesize and release D-serine. Pow and coworkers, using a new antibody against D-serine have observed significant staining in a subset of glutamatergic neurons of the brainstem and the olfactory bulb (Williams et al., 2006). The neuronal compartmentalization of D-serine has received more attention from the work of Wolosker and coworkers (Kartvelishvily et al., 2006). Immunochemical staining with new antibodies and prolonged incubation showed the presence of significant amounts of D-serine in primary neuronal cultures and neurons from brain sections of young and adult rats. Strong D-serine-ir is recovered in neuronal cell bodies and processes in all layers of the cerebral cortex, in the pyramidal neurons of the CA1, and the subiculum regions of the hippocampus (Kartvelishvily et al., 2006). Although it has been established that the presence of D-serine in neurons is due to their resorptive activity or their capacity to synthesize D-serine, the cellular distribution of D-serine in neurons of the CNS may be developmentally regulated. This is the case for the cerebellum, as already mentioned, but also for the vestibular nuclei in which high levels of D-serine are detected only in glial cells and processes from birth to postnatal day 21 (Puyal et al., 2006). On the other hand, D-serine levels are very low and mainly localized in neuronal cell bodies and dendrites in mature animals (Puyal et al., 2006). Still owing to the numeric preponderance of astroglia cells over neurons in the CNS (Oberheim et al., 2006), glial cells remain the principal source of D-serine in the brain (Fig. 16.1).

Evaluating a substance only on its localization, although important, does not infer about its function. Because all amino acids are involved in protein synthesis and intermediary metabolism, the sole presence of D-serine in glial cells does not obligatorily imply a role as a gliotransmitter.

16.3 *De Novo* Synthesis and Degradation of D-Serine in the Nervous System

16.3.1 Serine Racemase

One important issue in defining a substance as a gliotransmitter is to address its metabolizing pathway. Not only should this substance be synthesized but also metabolized following its reuptake in order to terminate its signaling. The idea that D-amino acids and particularly D-serine serve specific roles in the brain was strengthened by the discovery of its metabolic pathway.

Initially, brain D-serine was proposed to derive from diet, gastrointestinal bacteria, or from cleavage of metabolically stable proteins (Friedman, 1999). Whether these routes participate in the buildup of brain D-serine remains to be established. D-serine in the mammalian brain is synthesized by a pyridoxal 5′-phosphate (PLP)-dependent enzyme, serine racemase (SR) (Mustafa et al., 2004). Evidence regarding D-serine de novo synthesis in the CNS was first provided by Dunlop and Neidle (1997) who obtained conversion of radiolabeled L- to D-serine in intact rats. Nevertheless, this transformation might have been indirect, involving several steps including phosphoserine phosphatase and transamination rather than direct racemization. In this context, the purification of SR has certainly represented the major step to legitimize D-serine as a signaling molecule (Wolosker et al., 1999b). This enzyme directly converts L-serine into D-serine, L-serine being the only source for endogenous D-serine in the brain (Wolosker et al., 1999a, b). This enzyme not only converts L- into D-serine, but also D- to L-serine although with a lower affinity. The enzymatic-catalyzed racemization of L-serine proceeds by removal of a proton from the asymmetric C–H bond of the amino acid to form a carbanion intermediate (Fig. 16.2). The trigonal carbon atom of the carbanion, having lost the original asymmetry, then recombines with a proton to regenerate as inverted tetrahedral structure D-serine. Different genes for SR have now been identified in mice, rats, and humans (Wolosker et al., 1999a; De Miranda et al., 2000; Konno, 2003; Xia et al., 2004). They all display the same genomic structure made of seven exons, the first exon containing the lysine (Lys56) that forms an internal Schiff base with PLP region. Human and mouse SR genes encode for a 340 and 339 amino acid protein, respectively, while the rat SR has a truncated carboxy terminus sequence of six residues. The three proteins share 89% identity in their amino acid sequence, and all contain the consensus sequence ELFQKTGSFKIRGA for PLP binding at the N-terminus. Mutation of Lys56 inside this sequence abolishes racemization of L-serine into D-serine (Wolosker et al., 1999a; De Miranda et al., 2000; Strisovsky et al., 2003).

SR distribution in the CNS closely resembles that of endogenous D-serine, with the strongest expression in forebrain (Wolosker et al., 1999a; Mustafa et al., 2004). The highest levels are found in the hippocampus and corpus callosum, with intermediate

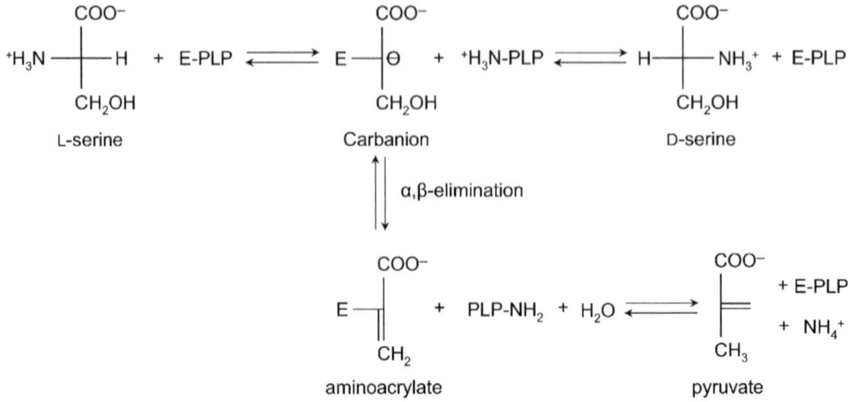

Fig. 16.2 Serine racemase: a bifunctional enzyme. PLP bound to the enzyme through an internal aldimine reacts with L-serine to give an external aldimine and subsequently a stabilized carbanion intermediate. Reprotonation of this intermediate on the opposite face of the planar carbanion generates the D-serine external aldimine intermediate. D-Serine is released via transimination regenerating the free PLP-bound enzyme. The carbanion is also an intermediate for the α,β-elimination reaction, leading to the formation of the aminoacrylate-PLP intermediate. Subsequent transimination releases the initial aminoacrylate product and regenerates free PLP-bound enzyme. The aminoacrylate released undergoes rapid nonenzymatic hydrolysis to give pyruvate and ammonia.

levels in substantia nigra, caudate, and hypothalamus, and low levels in the amygdala, thalamus, and subthalamic nuclei (Wolosker et al., 1999a; De Miranda et al., 2000; Xia et al., 2004; Panatier et al., 2006). Detailed analysis of its distribution at the cellular level reveals that SR is sharing the same glial compartmentalization as D-serine. Thus, in the CNS, astrocytes represent a major source for D-serine, although as mentioned before, others glial cells such as microglia constitute another source for this gliotransmitter (Wu et al., 2004b; Williams et al., 2006). Still, significant SR expression can be detected in some neuronal populations of the hippocampus and the cerebral cortex (Kartvelishvily et al., 2006) confirming that neurons represent another source for D-serine.

The distribution of SR mRNA in various human tissues (De Miranda et al., 2000; Xia et al., 2004) has revealed the existence of a single transcript of 2.6-kb in human brain, whereas in cardiac, skeletal muscle, kidney, and liver tissues, transcripts of at least three different sizes are present. The major SR transcript for heart, skeletal muscle, and kidney has a size of 4.5 kb. The presence of multiple mRNA forms indicates the existence of alternative splice forms, which is in agreement with the presence of alternative exons in the SR gene (De Miranda et al., 2000; Xia et al., 2004). Immunohistochemical studies revealed a peripheral expression of SR protein in human cardiac myocytes, convoluted tubules of the kidney (Xia et al., 2004), and Schwann cells (Wu et al., 2004b). There is currently no description of such transcripts in rodent tissues.

In addition to PLP, the activity of SR is modulated in many ways by different cellular compounds. Magnesium ions and adenosine 5′-triphosphate (ATP) are physiological cofactors of the enzyme, increasing the rate of D-serine synthesis (De Miranda et al., 2002). In the presence of Mg^{2+}, ATP half-maximally activates SR at 10 μM, which is largely below the endogenous ATP levels (~30–40-nmol mg^{-1} protein) in astrocytes. Accordingly, under normal circumstances, the enzyme might be saturated with ATP. Then, it is conceivable that during cell stress, ATP depletion is sufficient to alter SR activity. Calcium ions also represent another important SR cofactor, since these bind to the enzyme and since increase in intracellular $[Ca^{2+}]$ positively influences production of D-serine in astrocytes (Cook et al., 2002). However, half-maximal augmentation of SR activity by Ca^{2+} occurs at 26 μM, which is at least 100-fold higher than the basal levels found in astrocytes. This issue is controversial since another study reported no effect of intracellular Ca^{2+} on SR activity (Kim et al., 2005). Nevertheless, it is conceivable that SR is targeted to specific areas of the cell where Ca^{2+} microdomains prevail, thereby explaining these contradictory results.

In contrast, SR activity can be regulated in an opposite way by a series of cellular compounds. Glycine and a series of metabolites related to L-aspartic acid (L-aspartic acid, L-asparagine, and α,β-threo-3-hydroxyaspartic acid) competitively inhibit the enzyme (Dunlop and Neidle, 2005; Strisovsky et al., 2005). Since glycine concentrations in astrocytes are about 3–6 mM, it would constitutively inhibit SR activity except if glycine and SR show different compartmentalizations within the astrocyte cytosol (Strisovsky et al., 2005). Finally, nitric oxide (NO) physiologically S-nitrosylates SR, thus decreasing the catalytic activity (Mustafa et al., 2007). Inhibition of SR activity by S-nitrosylation is apparently competitive with ATP because the nitrosylated SR displays a 40-fold increase in the K_m for ATP with no change in V_{max}. By contrast, S-nitrosylation does not alter SR's K_m for its substrate L-serine but substantially reduces the V_{max}.

An intriguing feature of SR is the fact that it catalyzes not only the production of D-serine from L-serine, but also that of pyruvate via its α,β-elimination activity (De Miranda et al., 2002; Neidle and Dunlop, 2002; Strisovsky et al., 2003, 2005; Foltyn et al., 2005). This α,β-elimination activity toward L-serine is higher than the racemization, resulting in the synthesis of three molecules of pyruvate per molecule of D-serine obtained through racemization. In initial studies (De Miranda et al., 2002), ATP stimulated D-serine and pyruvate formation to a similar extent but, under some circumstances, it might influence the two processes differentially. Because of the substantial pyruvate-forming capacity of SR and the millimolar concentrations of L-serine in the brain, SR might be a major source of pyruvate. Interestingly, brain cells that produce D-serine are those with a high glycolytic activity and high L-serine levels, namely glial cells. Perhaps SR switches between forming D-serine and forming pyruvate depending on the energy status of the cell. It is also conceivable that SR-derived pyruvate is a potential source of lactate for neurons, providing energy during periods of enhanced synaptic activity or neuroprotection against oxidative damage and zinc neurotoxicity (Foltyn et al., 2005).

Additionally, SR expression and activity are regulated by protein–protein interactions. Utilizing yeast two-hybrid techniques, Kim et al. (2005) have shown that SR binds to glutamate receptor-interacting protein (GRIP), a protein with seven PDZ domains that binds to α-amino-3-hydroxy-5-methyl-isoxazole propionate receptors

(AMPARs) and regulates their clustering at synapses. These authors have shown that SR binds selectively to PDZ-6 and that mutation of the C-terminal valine of SR abolished interactions with GRIP. The formation of D-serine was markedly reduced in glial cells containing GRIP and transfected with SR in which the C-terminal valine was mutated to glycine. Viral infection of primary glial cultures with GRIP augmented basal activity of SR and then the levels of D-serine in cells (Kim et al., 2005). Thus, D-serine formation requires SR binding to GRIP, while GRIP does not constitute the sole interacting protein for SR. Fujii et al. (2006) have found that SR binds also to protein interacting with C-kinase (PICK1). The binding of endogenous PICK1 and SR requires the PDZ domain of PICK1. The exact role of PICK1 is unclear, but it might serve as a chaperone that escorts protein kinase C (PKC) to its targets. Interestingly, SR can be phosphorylated by PKC, which might be brought into the vicinity of SR by PICK1. Finally, Wolosker and coworkers have discovered that SR interacts with the Golgin subfamily A member 3 (Golga3) protein in yeast two-hybrid screening (Dumin et al., 2006). The interaction is mediated by the N-terminal coiled-coil domain of Golga3 that binds to the first 66 amino acids of the N-terminal region of SR. Furthermore, Golga3 and SR colocalized at the cytosol, perinuclear Golgi region and processes of both neurons and glial cells in culture. The Golga3 interaction with SR protects the enzyme from its ubiquitylation and thus from proteasomal degradation leading to an increase in the steady-state level of SR. Interestingly, the interaction of SR with Golga3 promotes the elevation of D-serine levels in cultured cells. Then, the ubiquitin system may provide a new mechanism for regulating SR and D-serine levels in the CNS.

16.3.2 D-Amino Acid Oxidase

The metabolic pathway for D-serine degradation remains more elusive. Mammalian D-serine can be metabolized by the peroxisomal flavoprotein D-amino acid oxidase (DAAO) (Fig. 16.3), an enzyme highly expressed in the liver and kidneys. Histochemical studies measuring enzyme activity detected no or scarce activity of DAAO in the forebrain and the cerebellum of rodents while its activity was maximum in the brainstem, cerebellum, and spinal cord (Katagiri et al., 1991; Horiike et al., 1994; Schell et al., 1995). However, immunochemical studies revealed DAAO expression in the corpus callosum, the basal ganglia, the caudate-putamen, the hippocampal formation, and the cerebral cortex (Moreno et al., 1999; Bendikov et al., 2007). In primate brains, levels of DAAO were detected in the subcortical white matter, thalamus, corpus callosum, the hypothalamus, the globus pallidus, and the pituitary gland (Volpe et al., 1970). Closer examinations established that DAAO was mostly present in astrocytes but substantial expression was also found in neurons (Horiike et al., 1994; Schell et al., 1995; Moreno et al., 1999). It is unclear what the respective functions of the glial DAAO and neuronal DAAO are. A link with D-serine has been first proposed by the observation of Snyder and coworkers that the levels of D-serine were inversely related to the activity of DAAO (Schell et al., 1995). This is striking

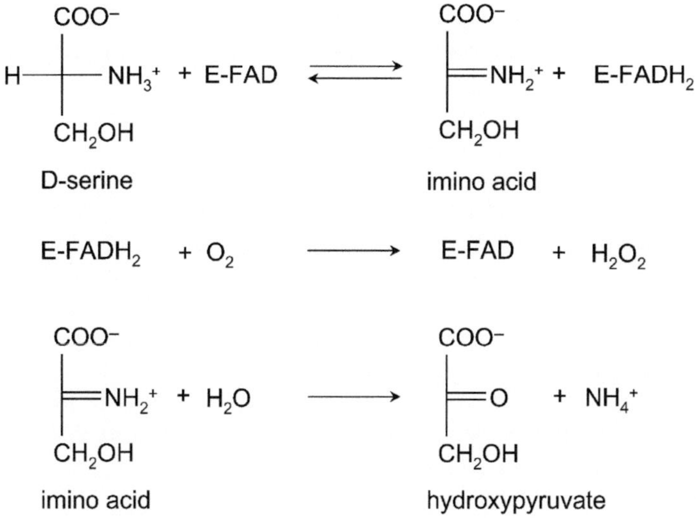

Fig. 16.3 Reaction of D-amino acid oxidase on D-serine. DAAO, a FAD-containing flavoprotein, catalyzes dehydrogenation of the D-isomer of serine to give the corresponding α-imino acids and, after subsequent hydrolysis, α-keto acids and ammonia. The reduced FAD is then reoxidized by molecular oxygen to yield hydrogen peroxide.

when considering the cerebellum. Adult DAAO-deficient mice display increased D-serine levels, especially in areas where the amino acid normally occurs at low levels such as the cerebellum or the brainstem (Hamase et al., 2005).

The earlier observation fits with the recent discovery that activation of DAAO by its interacting partner, pLG72, leads to increased oxidation of D-serine (Chumakov, 2002). *G72* is a novel gene with no recognizable motifs that encodes a 742-bp mRNA. Transcripts of G72 have been demonstrated only in primates (Chumakov, 2002). In silico examination of available sequences from other organisms confirms the absence of orthologs except in dog. Reverse-transcription PCR revealed the expression of G72 in the amygdala, caudate nucleus, and spinal cord. However, the proposed interaction between G72 and DAAO requires additional investigations to clarify the respective distributions of both partners and identify the events triggering their interactions. Furthermore, D-serine levels are inversely related to the regional expression of DAAO during development and in adult brain (Schell, 2004; Puyal et al., 2006). However, although DAAO protein is present in D-serine-rich forebrain, in DAAO-deficient mice D-serine levels appear relatively unchanged in this region (Hamase et al., 2005). Thus, other mechanisms probably regulate D-serine concentrations in the brain. In fact, SR exerts also a robust α,β-elimination activity on D-serine (Foltyn et al., 2005; Strisovsky et al., 2005) (Fig. 16.2). Although α,β-elimination activity on D-serine is less effective than that on L-serine, astrocytes may physiologically regulate their content in D-serine through this process (Foltyn et al., 2005). Such a novel function for SR could constitute an alternative mechanism for enzymatic removal of D-serine in brain regions where DAAO is absent. Thus, a mutant of SR (Q155D), which lacks α,β-elimination but retains racemase activity, is associated with the formation of 2–3-fold more D-serine than the wild-type enzyme in transfected cells and glial cultures.

DAAO and SR may not work in isolation as recently proposed by Shoji et al. (2006a, b). Accordingly, these authors provided evidence that SR and DAAO activities were controlled in opposite ways by nitric oxide (NO). NO inhibited SR whereas it enhanced DAAO activities, thus downregulating the intracellular levels of D-serine. In turn, D-serine inhibited nitric oxide synthases in glial cells. On contrast, early experiments revealed that glial-derived D-serine stimulated the production of NO in neuronal cells (Paudice et al., 1998; Mothet et al., 2000). Thus, NO produced in neurones may represent an attractive negative feedback signal tightly regulating D-serine accumulation in astrocytes, thereby preventing overproduction of D-serine and, consequently, avoiding overstimulation of NMDARs. It is all the more interesting that NO signaling pathways and glutamatergic synapses are intertwined both at the structural and functional levels.

16.4 Release and Clearance of D-Serine

16.4.1 Molecular Mechanisms of D-Serine Release

Pioneering experiments revealed that the efflux of radiolabeled D-serine from preloaded astrocytes was induced by activation of non-NMDARs, notably the AMPA/kainate subtype (Schell et al., 1995). Activation of AMPARs triggered the binding of GRIP to SR, causing a major activation of SR and efflux of D-serine from astrocytes (Kim et al., 2005) (Fig. 16.4). Disruption of the SR–GRIP interaction by transfection of mutant SR reduced basal and AMPA-evoked D-serine release (Kim et al., 2005). These observations confirmed the existence of a regulated release for D-serine from glia cells with AMPARs being the major stimulatory pathway. Whether glutamatergic inputs, through the activation of glial glutamate receptors (GluR), represent the sole synaptic afferents coupled to D-serine release remains to be investigated. Another important point, besides identifying the stimuli inducing D-serine release, is to characterize the molecular mechanisms downstream receptor activation that are responsible for the efflux of D-serine.

This is an important issue that applies to all gliotransmitters. What can be the intracellular routes mediating the release of such neuroactive substances from glial cells that were long thought to be passive or at least only resorptive elements in the CNS. Astrocytes, like all eukaryotic cells, utilize secretory lysosomes to transport new membrane and proteins to the plasma membrane during constitutive exocytosis. Most cell types also possess a second pathway of regulated exocytosis in which secretory vesicles undergo Ca^{2+}-regulated fusion with plasma membrane upon depolarization-evoked intracellular [Ca^{2+}] elevation (Jahn and Sudhof, 1999). After exocytosis, vesicle membrane is retrieved and recycled locally within cell. Until recently, Ca^{2+}-regulated

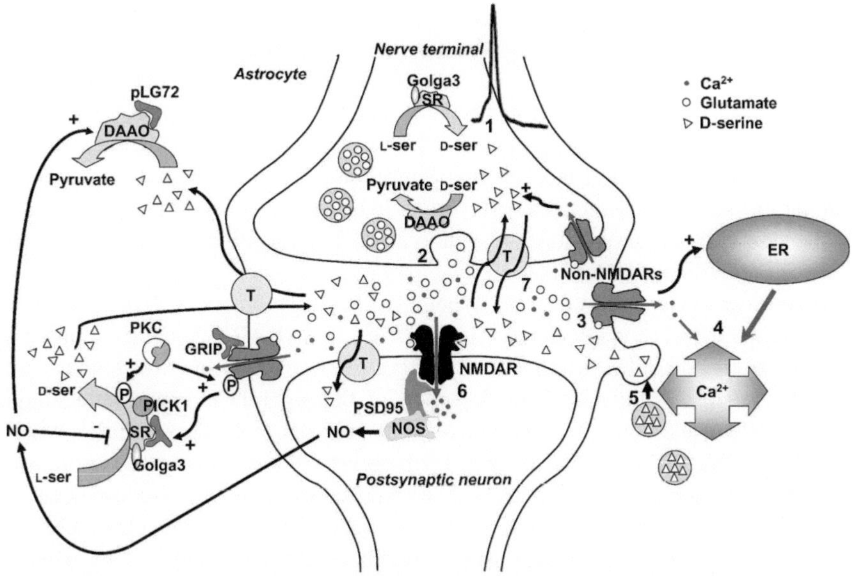

Fig. 16.4 The D-serine signaling complex at synapses. Upon depolarization of nerve terminals (1), glutamate is released (2) into the synaptic space where it activates non-NMDA receptors (NMDARs) (3) on the membrane of perisynaptic astrocytes. This activation leads to influx of Ca^{2+} either through AMPA/kainate receptors or the reverse mode of the Ca^{2+}–Na^{+} exchanger (not shown), and to release of Ca^{2+} from the endoplasmic reticulum (ER) in case of metabotropic glutamate receptors (4). Activation of AMPA receptors causes also a large increase in synthesis of D-serine. D-Serine is consequently released, either from a cytosolic pool by a transporter (T) or from a vesicular pool by a Ca^{2+}-dependent and SNARE-dependent mechanism (5). Once in the synaptic cleft, D-serine, in concert with glutamate, activates NMDA receptors at the membrane of postsynaptic neuron, leading to the opening of ion channels (6). NMDA receptor's activation leads to neuronal nitric oxide synthase (NOS) activation. Nitric oxide (NO) produced by NOS can diffuse to neighboring cells, where it inhibits SR through S-nitrosylation and activates DAAO, which reduces D-serine levels. Clearance of D-serine from the synaptic space is assured by Na^{+}-dependent and Na^{+}-independent transporters (T) on the membrane of astrocytes and neurons (7). Although glia-derived D-serine predominates, neurons that also express SR release the amino acid upon activation of glutamate receptors, notably NMDA receptors. D-Serine is released from neurons upon depolarization (1) by a nonvesicular mechanism that involves an unidentified channel or transporter. For more details, please refer to Sects. 16.3 and 16.4.

exocytosis has been viewed as a hallmark of neurosecretory (electrically excitable) cells. Both constitutive and regulated secretory pathways require specialized proteins to bring together the membranes of the vesicles with the plasma membrane. Soluble *N*-ethyl maleimide-sensitive fusion protein attachment protein receptor (SNARE) proteins are the leading candidate for mediating membrane fusion (reviewed in Jahn and Sudhof, 1999) and most of them are present in glial cells (reviewed in Montana et al., 2006; Volterra and Meldolesi, 2005). Accordingly, glial cells express synaptobrevin II, synaptotagmin IV, synaptophysin, rab3a, synapsin I, SNAP23, syntaxin, and cellubrevin. Not only these cells contain the machinery for exocytosis but also contain synaptic-like microvesicles (SLMVs) and large dense-core vesicles (LDCVs) as revealed by electron microscopy and confocal microscopy analysis. Research over the last decade showed that glial cells and particularly astrocytes use a Ca^{2+}-regulated SNAREs-dependent exocytosis to release glutamate (Araque et al., 2000; reviewed in Volterra and Meldolesi, 2005) from SLMVs but also ATP and peptides like secretogranin II from LDCVs (Calegari et al., 1999; Coco et al., 2003). Do glial cells also release D-serine by such a mechanism? Subcellular fractionation analysis revealed that part of D-serine is associated with vesicle-like structures suggesting that this may be the case although the majority of amino acids is retrieved in the cytosol (Mothet et al., 2005; Williams et al., 2006).

A series of evidence supports the hypothesis that activation of AMPA/kainate and metabotropic GluRs can effectively trigger a Ca^{2+}- and SNARE-dependent release of D-serine from astrocytes (Mothet et al., 2005). First, removal of extracellular Ca^{2+}, chelation of intracellular Ca^{2+}, or depletion of thapsigargin-sensitive intracellular Ca^{2+} stores considerably affect the release of D-serine in response to GluR agonists. Although extracellular Ca^{2+} is required for D-serine release, the release is not influenced by inhibitors of voltage-gated Ca^{2+} channels, fitting with the evidence that the major extracellular source for Ca^{2+} is the activation of AMPA receptors. Second, proteolysis of synaptobrevin and cellubrevin with tetanus neurotoxin (TeNT) significantly impairs the ability of GluRs to evoke D-serine release. Third, inhibition of the vacuolar proton ATPase with concanamycin or bafilomycin A1 reduced D-serine release, most likely by collapsing the proton gradient necessary for the uptake of the amino acid into vesicles. These results are consistent with a vesicular storage and release of D-serine and support the existence of specific storing organelles in astrocytes. What can be the nature of these secretory organelles? Extensive work by various groups has shown that glial exocytotic glutamate (Bezzi et al., 2004; Crippa et al., 2006) and ATP (Coco et al., 2003) are stored and released from SLMV and from LDCV, respectively. Storage of glutamate and D-serine in the same population of vesicles would represent the accurate cocktail of gliotransmitters to activate the NMDARs. This hypothesis is supported by our results showing that part of D-serine-ir was found in vesicles bearing the vesicular transporters for glutamate (Mothet et al., 2005).

Nevertheless, a Ca^{2+}-dependent vesicular release of D-serine does not exclude other mechanisms of release from glia, especially from the cytosol: the majority of D-serine is free in the cytoplasm (Schell et al., 1995; Kim et al., 2005; Mothet et al., 2005), and transfection of GRIP shows direct release of D-serine without apparent prestorage (Kim et al., 2005). Using an in vivo microdialysis technique to measure the extracellular concentration of endogenous free D-serine in discrete brain areas of freely moving rat, Hashimoto et al. (1995b) showed that neither addition of tetrodotoxin nor deprivation of Ca^{2+} from the perfusate reduced the basal extracellular levels of D-serine. These results led the authors to the conclusion that the D-amino acid could effectively exit from a cytoplasmic pool of glial cells to an extracellular compartment, most likely by a specific membrane carrier present on the membrane of astrocytes. In primary astrocyte cultures, D-serine uptake is dependent on sodium ions and exhibits both low affinity and low specificity for D-serine (Ribeiro et al., 2002). The kinetics of D-serine transport resemble that of B amino acid transporter

(ASCT)-type transporters as several small neutral amino acids strongly inhibit the uptake of D-serine, and D-serine transport is unaffected by excess 2-(methylamino)-isobutyric acid, a specific inhibitor for the system ASCT type A. D-serine uptake is associated with an efflux from the cells of L-serine and to a less extent of other small neutral amino acids (Ribeiro et al., 2002). Conversely, it has been reported that physiological concentrations of L-serine induced efflux of intracellular D-serine from astrocytes with an efficiency three times higher than kainate, which has been previously shown to induce robust D-serine release from astrocytes (Schell et al., 1995). Overall, D-serine fluxes are coupled to L-serine countermovements and to a lesser extent to other small neutral amino acids, suggesting an antiporter mechanism for glial D-serine transport. Because the exchanger is Na^+-dependent, one can imagine that it could work in reverse mode when the permeability of the plasma membrane for Na^+ is dramatically increased. Although such a mechanism is unlikely to operate under physiological conditions, it may represent a general mechanism for the release of D-serine in pathological conditions where the intracellular charge in Na^+ increases.

Astrocytes may also use the chemical gradient to drive under physiological conditions the release of cytosolic D-serine through activated $P2X_7$ receptors, connexin-formed hemichannels, or volume-anion channels as suggested for glutamate, ATP, or taurine (reviewed in Volterra and Meldolesi, 2005). One can imagine that a similar phenomenon may account for efflux of cytosolic D-serine as a concentration gradient ranging from 60- to 2,000-fold to exist between the intra- and extracellular space. There is currently no data supporting a role for these routes in mediating the release of D-serine from the cytosol. Nevertheless, these routes to release gliotransmitter must be explored in the future. In this context, release of D-serine in conditions of hypoosmotic challenges represents a very attractive result since work in our group strongly supports a role for D-serine in the supraoptic nuclei, a brain region involved in body fluid homeostasis. This hypothalamic structure undergoes a remarkable and reversible anatomical remodeling characterized by a reduced astroglial coverage of neurons during hydric stress or lactation and parturition (reviewed in Miyata and Hatton, 2002; Theodosis, 2002). SON glial cells release taurine under physiological hypoosmotic challenges (Deleuze et al., 1998), and this release is independent of extracellular Ca^{2+}. Then, it is tempting to speculate that hypoosmotic conditions, like those observed in the hypothalamus, participate in D-serine efflux thereby accounting for the part of D-serine release that is unaffected by extra- and intracellular Ca^{2+} or by SNAREs proteins cleavage.

16.4.2 Mechanisms of D-Serine Clearance

Like for any neurotransmitters, the signaling action of D-serine normally should be terminated by its clearance from the synaptic cleft either by ectoenzymes as for ATP or by transporter proteins located in neurons and/or glial cells as occurring for glutamate. Injection of D-serine into the lateral ventricle results in an apparent exclusive accumulation of the amino acid in glial cells (Wako et al., 1995). Several transporter candidates for D-serine have been identified on the plasma membrane of glial cells and neurons (Hayashi et al., 1997; Yamamoto et al., 2001; Javitt et al., 2002; O'Brien et al., 2005) (Fig. 16.4). Glial cells express a Na^+-dependent transporter with low affinity for D- and L-serine (Hayashi et al., 1997) the characteristics of which resemble that of the ASCT system, which carries D-serine in cultured astrocytes as well as in isolated retina (Ribeiro et al., 2002; O'Brien et al., 2005). Another neutral amino acid transporter, which is Na^+-independent, the alanine-serine-cysteine transporter 1 (Asc-1) has also been identified. This transporter presents a high affinity for D-serine and is confined to presynaptic terminals and to dendrites as well as somata of neurons. The cellular localization of Asc-1 suggests that this transporter could contribute to the synaptic clearance of D-serine by neurons (Helboe et al., 2003; Matsuo et al., 2004). Finally, a novel Na^+/Cl^--sensitive transporter has been described in rat brain synaptosomes (Yamamoto et al., 2001; Javitt et al., 2002). In contrast with the ASC system that has broad substrate selectivity, this serine transporter has limited affinity for other neutral amino acids including cysteine and alanine. It is conceivable, therefore, that multiple transport systems contribute simultaneously to the regulation of D-serine concentrations at the synapse. The presence of transporter for D-serine at the surface of neurons could provide an explanation for the occurrence of this gliotransmitter in these cells. The presence of specific transporters for D-serine at nerve terminals provides also a mechanism by which D-serine may be released by neurons (Kartvelishvily et al., 2006). Removal of either Ca^{2+} or Na^{2+} from the external medium blocked D-serine release.

Whether the vesicular and/or nonvesicular pathways come into play for the release of D-serine and which of these occurs in vivo awaits investigation. Of importance for further investigations is to know whether astrocytes present polarized sites for the release of gliotransmitters and particularly for D-serine. In other words, is the release of D-serine restricted to fine processes apposed to NMDARs or does it occur evenly from all cell compartments? We know that glial cells present functionally distinct compartments, referred as microdomains where localized Ca^{2+} signals appear and that ensheathe synapses (Grosche et al., 1999, 2002). Furthermore, electron microscopy analysis has shown that glial glutamatergic vesicles are localized in processes that contact dendritic spines bearing NMDARs (Bezzi et al., 2004). By analogy with glutamate, this implies that D-serine release may occur from localized glial sites in the near vicinity of synapses.

16.5 Functions of D-Serine in the Nervous System

16.5.1 D-Serine Contribution to Synaptic Transmission and Plasticity

NMDARs are unusual among ionotropic receptors in that the channel opens only when two different ligands bind the receptor simultaneously (Johnson and Ascher, 1987; Klechner and Dingledine, 1988). The native channel complex is a tetramer formed by the association of two NR1 and two NR2 subunits (Danysz and Parsons, 1998). The advent of receptor binding assays has revealed a strychnine-insensitive binding for glycine that was displaced by D-serine in the forebrain (Danysz et al., 1990). This binding site corresponds to the so-called glycine binding site located on the NR1 subunits of NMDARs (Danysz and Parsons, 1998). We know from early reports that this glycine site can be activated not only by glycine but also by D-serine (Kleckner and Dingledine, 1988) with efficiency even higher according to the subtypes of NMDARs (Matsui et al., 1995). However, because D-serine was considered as an oddity, it was only used in pharmacological experiments to mimic the action of glycine at NMDARs. Thus, the discovery that this wrong isomer of serine occurs naturally in brain has profoundly changed our perspectives by raising the possibility that D-serine is a putative endogenous ligand for the glycine modulatory binding site of NMDARs. What are the arguments favoring D-serine as a neuromodulator of NMDARs? A key advance in our appreciation of the role of D-serine in the CNS derives from observations showing that D-serine is found in astrocytes that ensheathe neurons bearing NMDARs with a parallel ontogeny (Schell et al., 1997; Schell, 2004). In vitro studies teach us that D-serine is released from astrocytes upon activation of their glutamatergic receptors (Schell et al., 1995; Mothet et al., 2005). All these observations strongly suggested that, in some regions of the brain, glutamate released from the nerve terminal triggers glial D-serine efflux, which in turn modulates the NMDARs localized on adjacent neurones. The hippocampus provided a first model for studying the function of D-serine as high densities of D-serine and NMDARs occur in the subiculum and CA1 and CA3 regions (Schell et al., 1997). Using culture preparations of hippocampal neurons, we showed that specific enzymatic degradation of released D-serine with DAAO considerably reduces agonist-evoked and spontaneous NMDAR-driven currents (Mothet et al., 2000). The hippocampus is the site of long-term potentiation (LTP), which relies on NMDARs activation (Nicoll, 2003). Does D-serine govern the induction or the maintenance of LTP? Yang et al. (2003) have shown that glial-derived D-serine is an absolute parameter required for the induction of LTP in CA1 pyramidal cell synapses. Indeed, pretreatment of cell culture or brain slices with DAAO compromised this LTP, further supporting the idea that D-serine rather than glycine is the endogenous ligand of NMDARs in this area of the brain. It is commonly believed that senescence is associated with impaired NMDARs-dependent synaptic plasticity and notably LTP (Barnes, 2003). Given the crucial role of D-serine in synaptic plasticity, defective LTP recorded in senescence-accelerated mouse strain was then rescued to control level when D-serine was supplemented (Yang et al., 2005). However, this study did not address the molecular and cellular mechanisms underlying these synaptic plasticity deficiencies and notably the link with the metabolism of D-serine. Defective synaptic plasticity during senescence may reflect reduced agonist receptor availability, less NMDARs, and/or changes in the affinity of the subunits for their ligands. We have resolved this question in a recent study (Mothet et al., 2006). We showed that deficiency in LTP observed in senescent rats was primarily caused by a significant loss in the production of D-serine and thus in its synaptic availability (Mothet et al., 2006). In agreement with the emerging role of D-serine as the major ligand for the glycine modulatory binding site of NMDARs, the deficit in LTP ability was not associated with reduced levels of glycine (Mothet et al., 2006). Our observations tightly correlate those of Shleper et al. (2005) showing that D-serine was the dominant if not the only ligand mediating NMDARs-induced neurotoxicity in the hippocampus. These authors effectively showed that treating hippocampal slices with serine deaminase, an enzyme that degrades D-serine, prevents neuronal death whereas treating the same slices with glycine oxydase that degrades glycine does not (Shleper et al., 2005). Otherwise, D-serine may also serve the coding and processing of sensory information. For example, in the retina, Müller glial cells synthesize and release D-serine, which controls NMDAR-mediated responses (Stevens et al., 2003). Besides, the ability of D-serine to control NMDARs-dependent neurotransmission has been confirmed by the use of DAAO-deficient mice. These mice display highest increase in D-serine levels in the brainstem and spinal cord (Wake et al., 2001; Hamase et al., 2005). As expected, NMDAR-mediated excitatory postsynaptic currents recorded from spinal cord dorsal horn neurons are significantly potentiated in mutant mice (Wake et al., 2001). The hyperexcitability of dorsal horn neurons during chronic pain is largely dependent upon NMDA receptor activity (Woolf and Salter, 2000). D-Serine released by astrocytes contributes to NMDA-dependent dorsal horn LTP, and inhibition of D-serine reduced sciatic tetanic stimulation-induced mechanical allodynia in rats (Ying et al. 2006). Accordingly, responses to chronic nociceptive stimuli were exaggerated and NMDAR-mediated synaptic transmission was enhanced in mutant mice lacking DAAO (Wake et al., 2001). Finally, high levels of DAAO are found in dorsal horn astrocytes (Wake et al., 2001). All together, these findings strongly suggest that astrocyte-derived D-serine might control NMDAR activity and excitatory synaptic plasticity in the dorsal horn during pain. Knockout mice for the transporter Asc-1 provide another and new experimental model for studying the relevance of D-serine in glutamatergic neurotransmission (Xie et al., 2005). Indeed, these mice displayed NMDARs-dependent hyperexcitability, presumably resulting from elevated extracellular D-serine (Xie et al., 2005).

Based on the evidence discussed earlier, it appears that astrocytic D-serine modulates NMDAR-dependent neurotransmission and synaptic plasticity in many regions of the CNS. The action of D-serine in these different regions will depend on the degree of astrocytic coverage of neurons. Indeed, it is now accepted that glial coverage of neurons is not static and it undergoes profound reversible anatomical remodeling in different areas (Theodosis, 2002; Hirrlinger et al., 2004). The hypothalamo-neurohypophysial system (HNS) constitutes certainly the most striking example of such anatomical remodeling that is observed in conditions of intense secretion of neurohypophysial hormones, such as lactation and chronic dehydration (reviewed in Miyata and Hatton, 2002; Theodosis, 2002). This morphological plasticity is characterized by a pronounced reduction in astrocytic coverage of neurons, which results from a remarkable anatomical remodeling during intense hormone secretion. We previously showed that this neuronal-glial remodeling modified glutamate clearance and diffusion, thereby affecting synaptic efficacy at glutamatergic and adjacent GABAergic synapses (Oliet et al., 2001). We have taken advantage of the changing astrocytic ensheathment of neurons occurring in the SON during lactation to study the role of glia-derived D-serine in synaptic transmission. This structure provides an exquisite model for studying the physiological relevance of glial-derived D-serine as SR is expressed there producing high levels of the amino acid. Furthermore, SR and D-serine are strictly restricted to astrocytes in the SON as judged by immunohistochemistry. We have obtained direct evidence that in this hypothalamic structure the endogenous coagonist of NMDARs is D-serine and not glycine. This was established by treating slices with the highly specific enzymes DAAO and glycine oxidase (GO), which degrade D-serine and glycine, respectively. Furthermore, we showed that the level of occupancy of the glycine site of NMDARs is controlled by the astrocytic coverage of neurons. As a consequence, the activity dependence of phenomena such as LTP and LTD, whose induction depends on NMDAR activation, is modified by the neuron-glial remodeling (Panatier et al., 2006). Our data indicate that reduced glial coverage of neurons and their synapses results in a reduction in the level of occupancy by D-serine of the NMDARs, which shifts the activity dependence of long-term synaptic changes toward higher activity values (Fig. 16.5). Indeed, the physiological reduction of D-serine concentrations at glutamatergic synapses in the SON of lactating rats affects synaptic plasticity in a manner closely similar to that produced by partial blockade of NMDARs obtained via pharmacological means. Therefore, the glial environment of neurons, through its capacity to provide D-serine has an impact not only on synaptic transmission but also on dictating the direction of long-term changes in synaptic plasticity.

16.5.2 Role of D-Serine in the Developing Brain

The modulatory role of D-serine at NMDARs may predominate already before the establishment of synaptic contacts. During neocorticogenesis, migrating neurons can adopt different types of trajectories and a large proportion of neurons migrate radially, along radial glial guides, from the germinative zone to their final place (Hatten, 1999; Gressens, 2000; Yacubova and Komuro, 2003). As they migrate throughout the developing brain, immature neurons are influenced by extrinsic factors that modulate their journey. Among these factors, transmitters have been shown to play an important role. Most notably, glutamate acting on NMDARs has a crucial modulatory effect on migrating neuroblasts, acting as motility-promoting and acceleratory signals (Komuro and Rakic, 1993; reviewed in Hatten, 1999; Yacubova and Komuro, 2003). Radial migration of immature granules cells in the developing cerebellum, along the Bergmann glia, is one of the best-characterized instances of the participation of NMDARs in neuronal migration (reviewed in Hatten, 1999; Yacubova and Komuro, 2003). Blocking the NMDARs expressed by migrating granules cells with antagonists significantly decreases the rate of glial-guided radial neuronal migration. In contrast, the rate of granule cell movement is increased by removal of Mg^{2+} or by application of NMDA or the coagonist glycine (Yacubova and Komuro, 2003). How NMDARs of migrating immature neurons

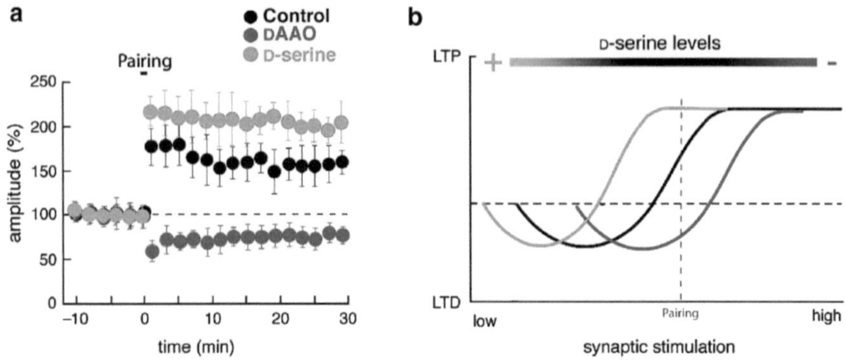

Fig. 16.5 Astrocyte-mediated metaplasticity. (**a**) Under conditions where D-serine levels are reduced with DAAO, a pairing protocol inducing LTP at glutamatergic synapses in control slices (black dots) now causes LTD (red dots) in the hypothalamus. Conversely, supplying the slices with saturating concentrations of D-serine (green dots) enhances LTP. (**b**) D-Serine levels, by controlling the number of NMDARs available for activation, govern the activity dependence of synaptic plasticity. Decreasing these levels (red) shifts the relationship toward higher activity values whereas increasing them (green) has the opposite effect (adapted from Panatier et al., 2006). (*See Color Plates*)

are activated had remained controversial since migrating neurons do not form synapses before complete translocation to the internal granule layer. An attractive hypothesis is that glutamate released by Bergmann glia activates immature NMDARs in a nonsynaptic, paracrine mode (Yacubova and Komuro, 2003). Because D-serine levels peak at rat postnatal day 14, the time of intense granule cell migration (Schell et al., 1997), and due to the absolute requirement of a coagonist for NMDARs activation, it is attempting to speculate that D-serine might be involved in these processes.

Kim et al. (2005) have now discovered that D-serine released by Bergmann glial cells promotes the migration of granule cells through activation of NMDARs. They utilized two approaches, selective degradation of D-serine by DAAO and selective inhibition of SR. Both approaches blocked the migration of granule cells by reducing the activity of NMDARs as treatment with SR inhibitors markedly diminishes intracellular Ca^{2+}. The physiological influence of D-serine on neuronal migration involves the activation of SR by GRIP. Indeed, GRIP adenoviral infection of the developing cerebellum increases D-serine levels through activation of SR and concomitantly increases the migration of granule cells from the external to the internal granular layer (Kim et al., 2005). Additionally, D-serine may also participate in the maturation (i.e., synaptogenesis) of developing neural network as its ontogeny in Bergmann glia parallels NR2A/B expression in Purkinje cells (Schell et al., 1997).

The motility promoting role of D-serine is probably not restricted to the postnatal development and may be involved earlier during fetal development of the brain. Indeed SR is present in the perireticular nucleus, a transient area of the human fetal brain, which is thought to be involved in the guidance of corticofugal and thalamocortical fibers (Hepner et al., 2005), and D-serine synthesized in the placenta can enter the fetal circulation through the ASCT (Chen et al., 2004). Thus, D-serine is well positioned both spatially and temporally to control NMDAR-mediated neuronal migration and synaptogenesis as NMDARs are present early during gestation (Ritter et al., 2001). Because blocking NMDARs during neocorticogenesis (Reiprich et al., 2005) or genetically induced alterations in these receptors (Gressens, 2000) result in severe abnormal cortical development, disrupting D-serine metabolism during embryonic and early postnatal life may lead to the same developmental defects. Notably, impairment in the cerebellar development and maturation fits with a specific shutdown of *DAAO* gene expression (Taharaguchi et al., 2003) supporting the hypothesis that altered D-serine/NMDARs signaling promotes neuronal degeneration and inhibition of synaptogenesis.

16.6 Future Directions

The past 10 years of intensive research has revolutionized the way neuroscientists think about the glutamatergic synapses. Still, we just begin to appreciate the diverse role played by D-serine in the CNS and there is much work to be done by neuroscienstists in order to delineate the contribution of this atypical brain amino acid. Notably, it remains to clearly establish the respective contributions of D-serine and glycine at glutamatergic synapses in physiological and pathological conditions. It is a fascinating but complicated task, because there are many NMDAR subtypes with different intrinsic properties that control their trafficking, their pharmacology, and their expression during development. In addition, NMDAR receptors are central to many physiological and pathological signaling events. Use of genetic animal models to disrupt D-serine metabolism and the development of new tools to visualize D-serine and glycine in vivo should aid the translation of our cell biology knowledge into a more physiological context, and help to define the role of each agonist in regulating NMDAR-dependent physiological and pathological processes. Finally, we have not considered so far the role of D-serine outside the CNS. We know that NMDARs are present in others regions of the nervous system. Thus, one can imagine that D-serine may govern the activity of the glutamatergic synapses and may play important function there as well.

Acknowledgments We would like to thank Thierry Galli, Reinhard Jahn, Loredano Pollegioni, Silvia Sacchi, and Dyonisia Theodosis for their help, and our previous colleagues, Gérard Baux, Philippe Fossier, Gilles Ouanounou, Jean-Marie Billard, and Aude Panatier for their original contributions. The experiments carried out in our laboratory are supported by grants from INSERM, CNRS, the University of Bordeaux 2, the Association Française contre les Myopathies, the Association pour la Recherche sur le Cancer, the Institut National du Cancer, the Agence National de la Recherche, and the Human Frontier Science Program. Magalie Martineau is a recipient of a PhD studenship from the Ministère de l'Enseignement, de la Recherche et de la Technologie.

References

Araque A, Li N, Doyle RT, Haydon PG (2000) SNARE protein-dependent glutamate release from astrocytes. J Neurosci 20:666–673.

Barnes CA (2003) Long-term potentiation and the ageing brain. Philos Trans R Soc Lond B Biol Sci 358:765–772.

Bendikov I, Nadri C, Amar S, Panizzutti R, De Miranda J, Wolosker H, Agam G (2007) A CSF and postmortem brain study of D-serine metabolic parameters in schizophrenia. Schizophr Res 90:41–51.

Bezzi P, Gundersen V, Galbete JL, Seifert G, Steinhauser C, Pilati E, Volterra A (2004) Astrocytes contain a vesicular compartment that is competent for regulated exocytosis of glutamate. Nat Neurosci 7:613–620.

Calegari F, Coco S, Taverna E, Bassetti M, Verderio C, Corradi N, Matteoli M, Rosa P (1999) A regulated secretory pathway in cultured hippocampal astrocytes. J Biol Chem 274:22539–22547.

Chen Z, Huang W, Srinivas SR, Jones CR, Ganapathy V, Prasad PD (2004) Serine racemase and D-serine transport in human placenta and evidence for a transplacental gradient for D-serine in humans. J Soc Gynecol Investig 11:294–303.

Chumakov I (2002) Genetic and physiological data implicating the new human gene G72 and the gene for D-amino acid oxidase in schizophrenia. Proc Natl Acad Sci USA 99:13675–13680.

Coco S, Calegari F, Pravettoni E, Pozzi D, Taverna E, Rosa P, Matteoli M, Verderio C (2003) Storage and release of ATP from astrocytes in culture. J Biol Chem 278:1354–1362.

Cook SP, Galve-Roperh I, Martinez del Pozo A, Rodriguez-Crespo I (2002) Direct calcium binding results in activation of brain serine racemase. J Biol Chem 277:27782–27792.

Crippa D, Schenk U, Francolini M, Rosa P, Verderio C, Zonta M, Pozzan T, Matteoli M, Carmignoto G (2006) Synaptobrevin2-expressing vesicles in rat astrocytes: insights into molecular characterization, dynamics and exocytosis. J Physiol 570:567–582.

Danysz W, Parsons AC (1998) Glycine and N-methyl-D-aspartate receptors: physiological significance and possible therapeutic applications. Pharmacol Rev 50:597–664.

Danysz W, Fadda E, Wroblewski JT, Costa E (1990) [3H]D-serine labels strychnine-insensitive glycine recognition sites of rat central nervous system. Life Sci 46:155–164.

De Miranda J, Santoro A, Engelender S, Wolosker H (2000) Human serine racemase: molecular cloning, genomic organization and functional analysis. Gene 256:183–188.

De Miranda J, Panizzutti R, Foltyn VN, Wolosker H (2002) Cofactors of serine racemase that physiologically stimulate the synthesis of the N-methyl-D-aspartate (NMDA) receptor coagonist D-serine. Proc Natl Acad Sci USA 99:14542–14547.

Deleuze C, Duvoid A, Hussy N (1998) Properties and glial origin of osmotic-dependent release of taurine from the rat supraoptic nucleus. J Physiol 507:463–471.

Dememes D, Mothet JP, Nicolas MT (2006) Cellular distribution of D-serine, serine racemase and D-amino acid oxidase in the rat vestibular sensory epithelia. Neuroscience 137:991–997.

Dumin E, Bendikov I, Foltyn VN, Misumi Y, Ikehara Y, Kartvelishvily E, Wolosker H (2006) Modulation of D-serine levels via ubiquitin-dependent proteasomal degradation of serine racemase. J Biol Chem 281:20291–20302.

Dunlop DS, Neidle A (1997) The origin and turnover of D-serine in brain. Biochem Biophys Res Commun 235:26–30.

Dunlop DS, Neidle A (2005) Regulation of serine racemase activity by amino acids. Brain Res Mol Brain Res 133:208–214.

Foltyn VN, Bendikov I, De Miranda J, Panizzutti R, Dumin E, Shleper M, Li P, Toney MD, Kartvelishvily E, Wolosker H (2005) Serine racemase modulates intracellular D-serine levels through an alpha, beta-elimination activity. J Biol Chem 280:1754–1763.

Friedman M (1999) Chemistry, nutrition, and microbiology of D-amino acids. J Agric Food Chem 47:3457–3479.

Fujii N (2002) D-Amino acids in living higher organisms. Orig Life Evol Biosph 32:103–127.

Fujii K, Maeda K, Hikida T, Mustafa AK, Balkissoon R, Xia J, Yamada T, Ozeki Y, Kawahara R, Okawa M, Huganir RL, Ujike H, Snyder SH, Sawa A (2006) Serine racewmase binds to PICK1: potential relevance to schizophrenia. Mol Psychiatry 11:150–157.
Gressens P (2000) Mechanisms and disturbances of neuronal migration. Pediatr Res 48:725–730.
Grosche J, Matyash V, Moller T, Verkhratsky A, Reichenbach A, Kettenmann H (1999) Microdomains for neuron-glia interaction: parallel fiber signaling to Bergmann glial cells. Nat Neurosci 2:139–143.
Grosche J, Kettenmann H, Reichenbach A (2002) Bergmann glial cells form distinct morphological structures to interact with cerebellar neurons. J Neurosci Res 68:138–149.
Hamase K, Konno R, Morikawa A, Zaitsu K (2005) Sensitive determination of D-amino acids in mammals and the effect of D-amino-acid oxidase activity on their amounts. Biol Pharm Bull 28:1578–1584.
Hashimoto A, Oka T (1997) Free D-aspartate and D-serine in the mammalian brain and periphery. Prog Neurobiol 52:325–353.
Hashimoto A, Oka T, Nishikawa T (1995a) Anatomical distribution and postnatal changes in endogenous free D-aspartate and D-serine in rat brain and periphery. Eur J Neurosci 7:1657–1663.
Hashimoto A, Oka T, Nishikawa T (1995b) Extracellular concentration of endogenous free D-serine in the rat brain as revealed by in vivo microdialysis. Neuroscience 66(3):635–643.
Hatten ME (1999) Central nervous system neuronal migration. Annu Rev Neurosci 22:511–539.
Hayashi F, Takahashi K, Nishikawa T (1997) Uptake of D- and L-serine in C6 glioma cells. Neurosci Lett 239:85–88.
Helboe L, Egebjerg J, Moller M, Thomsen C (2003) Distribution and pharmacology of alanine-serine-cysteine transporter 1 (asc-1) in rodent brain. Eur J Neurosci 18:2227–2238.
Hepner F, Pollak A, Ulfig N, Yae-Kyung M, Lubec G (2005) Mass spectrometrical analysis of human serine racemase in foetal brain. J Neural Transm 112:805–811.
Hirrlinger J, Hulsmann S, Kirchhoff F (2004) Astroglial processes show spontaneous motility at active synaptic terminals in situ. Eur J Neurosci 20:2235–2239.
Horiike K, Tojo H, Arai R, Nozaki M, Maeda T (1994) D-Amino-acid oxidase is confined to the lower brain stem and cerebellum in rat brain: regional differentiation of astrocytes. Brain Res 652:297–303.
Jahn R, Sudhof TC (1999) Membrane fusion and exocytosis. Annu Rev Biochem 68:863–911.
Javitt DC, Balla A, Sershen H (2002) A novel alanine-insensitive D-serine transporter in rat brain synaptosomal. Brain Res 941:146–149.
Johnson JW, Ascher P (1987) Glycine potentiates the NMDA response in cultured mouse brain neurons. Nature 325:529–531.
Kartvelishvily E, Shleper M, Balan L, Dumin E, Wolosker H (2006) Neuron-derived D-serine release provides a novel means to activate N-methyl-D-aspartate receptors. J Biol Chem 281:14151–14162.
Katagiri M, Tojo H, Horiike K, Yamano T (1991) Immunochemical relationship of D-amino acid oxidases in various tissues and animals. Comp Biochem Physiol B 99:345–350.
Katsuki H, Nonaka M, Shirakawa H, Kume T, Akaike A (2004) Endogenous D-serine is involved in induction of neuronal death by N-methyl-D-aspartate and simulated ischemia in rat cerebrocortical slices. J Pharmacol Exp Ther 311:836–844.
Kim PM, Aizawa H, Kim PS, Huang AS, Wickramasinghe SR, Kashani AH, Barrow RK, Huganir RL, Ghosh A, Snyder SH (2005) Serine racemase: activation by glutamate neurotransmission via glutamate receptor interacting protein and mediation of neuronal migration. Proc Natl Acad Sci USA 102:2105–2110.
Kleckner NW, Dingledine R (1988) Requirement for glycine in activation of NMDA-receptors expressed in Xenopus oocytes. Science 241:835–837.
Komuro H, Rakic P (1993) Modulation of neuronal migration by NMDA receptors. Science 260:95–99.
Konno R (2003) Rat cerebral serine racemase: amino acid deletion and truncation at carboxy terminus. Neurosci Lett 349:111–114.
Martineau M, Baux G, Mothet JP (2006) D-serine signalling in the brain: friend and foe. Trends Neurosci 29:481–491.

Matsui T, Sekiguchi M, Hashimoto A, Tomita U, Nishikawa T, Wada K (1995) Functional comparison of D-serine and glycine in rodents: the effect on cloned NMDA receptors and the extracellular concentration. J Neurochem 65:454–458.

Matsuo H, Kanai Y, Tokunaga M, Nakata T, Chairoungdua A, Ishimine H, Tsukada S, Ooigawa H, Nawashiro H, Kobayashi Y, Fukuda J, Endou H (2004) High affinity D- and L-serine transporter Asc-1: cloning and dendritic localization in the rat cerebral and cerebellar cortices. Neurosci Lett 358:123–126.

Miyata S, Hatton GI (2002) Activity-related, dynamic neuron–glial interactions in the hypothalamo-neurohypophysial system. Microsc Res Technol 56:143–157.

Montana V, Malarkey EB, Verderio C, Matteoli M, Parpura V (2006) Vesicular transmitter release from astrocytes. Glia 54:700–715.

Moreno S, Nardacci R, Cimini A, Ceru MP (1999) Immunocytochemical localization of D-amino acid oxidase in rat brain. J Neurocytol 28:169–185.

Mothet JP, Parent AT, Wolosker H, Brady RO Jr, Linden DJ, Ferris CD, Rogawski MA, Snyder SH (2000) D-Serine is an endogenous ligand for the glycine site of the N-methyl-D-aspartate receptor. Proc Natl Acad Sci USA 97:4926–4931.

Mothet JP, Pollegioni L, Ouanounou G, Martineau M, Fossier P, Baux G (2005) Glutamate receptor activation triggers a calcium-dependent and SNARE protein-dependent release of the gliotransmitter D-serine. Proc Natl Acad Sci USA 102:5606–5611.

Mothet JP, Rouaud E, Sinet PM, Potier B, Jouvenceau A, Dutar P, Videau C, Epelbaum J, Billard JM (2006) A critical role for the glial-derived neuromodulator D-serine in the age-related deficits of cellular mechanisms of learning and memory. Aging Cell 5:267–274.

Mustafa AK, Kim PM, Snyder SH (2004) D-Serine as a putative glial neurotransmitter. Neuron Glia Biol 1:275–281.

Mustafa AK, Kumar M, Selvakumar B, Ho GP, Ehmsen JT, Barrow RK, Amzel LM, Snyder SH (2007) Nitric oxide S-nitrosylates serine racemase, mediating feedback inhibition of D-serine formation. Proc Natl Acad Sci USA 104:2950–2955.

Neidle A, Dunlop DS (2002) Allosteric regulation of mouse brain serine racemase. Neurochem Res 27:1719–1724.

Nicoll RA (2003) Expression mechanisms underlying long-term potentiation: a postsynaptic view. Philos Trans R Soc Lond B Biol Sci 358:721–726.

Oberheim NA, Wang X, Goldman S, Nedergaard M (2006) Astrocytic complexity distinguishes the human brain. Trends Neurosci 29:547–553.

O'Brien KB, Miller RF, Bowser MT (2005) D-Serine uptake by isolated retinas is consistent with ASCT-mediated transport. Neurosci Lett 385:58–63.

Oliet SH, Piet R, Poulain DA (2001) Control of glutamate clearance and synaptic efficacy by glial coverage of neurons. Science 292:923–926.

Panatier A, Theodosis DT, Mothet JP, Touquet B, Pollegioni L, Poulain DA, Oliet SH (2006) Glia-derived D-serine controls NMDA receptor activity and synaptic memory. Cell 125:775–784.

Paudice P, Gemignani A, Raiteri M (1998) Evidence for functional native NMDA receptors activated by glycine or D-serine alone in the absence of glutamatergic coagonist. Eur J Neurosci 10:2934–2944.

Puyal J, Martineau M, Mothet JP, Nicolas MT, Raymond J (2006) Changes in D-serine levels and localization during postnatal development of the rat vestibular nuclei. J Comp Neurol 497:610–621.

Reiprich P, Kilb W, Luhmann HJ (2005) Neonatal NMDA receptor blockade disturbs neuronal migration in rat somatosensory cortex in vivo. Cereb Cortex 15:349–358.

Ribeiro CS, Reis M, Panizzutti R, de Miranda J, Wolosker H (2002) Glial transport of the neuromodulator D-serine. Brain Res 929:202–209.

Ritter LM, Unis AS, Meador-Woodruff JH (2001) Ontogeny of ionotropic glutamate receptor expression in human fetal brain. Brain Res Dev Brain Res 127:123–133.

Schell MJ (2004) The N-methyl D-aspartate receptor glycine site and D-serine metabolism: an evolutionary perspective. Philos Trans R Soc Lond B Biol Sci 359:943–964.

Schell MJ, Molliver ME, Snyder SH (1995) D-serine, an endogenous synaptic modulator: localization to astrocytes and glutamate-stimulated release. Proc Natl Acad Sci USA 92:3948–3952.

Schell MJ, Brady RO Jr, Molliver ME, Snyder SH (1997) D-Serine as a neuromodulator: regional and developmental localizations in rat brain glia resemble NMDA receptors. J Neurosci 17:1604–1615.

Shleper M, Kartvelishvily E, Wolosker H (2005) D-Serine is the dominant endogenous coagonist for NMDA receptor neurotoxicity in organotypic hippocampal slices. J Neurosci 25:9413–9417.

Shoji K, Mariotto S, Ciampa AR, Suzuki H (2006a) Mutual regulation between serine and nitric oxide metabolism in human glioblastoma cells. Neurosci Lett 394:163–167.

Shoji K, Mariotto S, Ciampa AR, Suzuki H (2006b) Regulation of serine racemase activity by D-serine and nitric oxide in human glioblastoma cells. Neurosci Lett 392:75–78.

Stevens ER, Esguerra M, Kim PM, Newman EA, Snyder SH, Zahs KR, Miller RF (2003) D-Serine and serine racemase are present in the vertebrate retina and contribute to the physiological activation of NMDA receptors. Proc Natl Acad Sci USA 10:6789–6794.

Strisovsky K, Jiraskova J, Barinka C, Majer P, Rojas C, Slusher BS, Konvalinka J (2003) Mouse brain serine racemase catalyzes specific elimination of L-serine to pyruvate. FEBS Lett 535:44–48.

Strisovsky K, Jiraskova J, Mikulova A, Rulisek L, Konvalinka J (2005) Dual substrate and reaction specificity in mouse serine racemase: identification of high-affinity dicarboxylate substrate and inhibitors and analysis of the beta-eliminase activity. Biochemistry 44:13091–13100.

Taharaguchi S, Kon Y, Yoshino S, Ono E (2003) Impaired development of the cerebellum in transgenic mice expressing the immediate-early protein IE180 of pseudorabies virus. Virology 307:243–254.

Theodosis DT (2002) Oxytocin-secreting neurons: a physiological model of morphological neuronal and glial plasticity in the adult hypothalamus. Front Neuroendocrinol 23:101–135.

Tsai G, Yang P, Chung LC, Lange N, Coyle JT (1998) D-Serine added to antipsychotics for the treatment of schizophrenia. Biol Psychiatry 44:1081–1089.

Volpe JJ, Lee G, Laster L, Robinson JC (1970) Regional distribution of isozymes of D-amino-acid oxidase and acetylesterase in developing primate brain. Exp Neurol 28:76–87.

Volterra A, Meldolesi J (2005) Astrocytes, from brain glue to communication elements: the revolution continues. Nat Rev Neurosci 6:626–640.

Wake K, Yamazaki H, Hanzawa S, Konno R, Sakio H, Niwa A, Hori Y (2001) Exaggerated responses to chronic nociceptive stimuli and enhancement of N-methyl-D-aspartate receptor-mediated synaptic transmission in mutant mice lacking D-amino-acid oxidase. Neurosci Lett 297:25–28.

Wako K, Ma N, Shiroyama T, Semba R (1995) Glial uptake of intracerebroventricularly injected D-serine in the rat brain: an immunocytochemical study. Neurosci Lett 185:171–174.

Williams SM, Diaz CM, Macnab LT, Sullivan RK, Pow DV (2006) Immunocytochemical analysis of D-serine distribution in the mammalian brain reveals novel anatomical compartmentalizations in glia and neurons. Glia 53:401–411.

Wolosker H (2006) D-Serine regulation of NMDA receptor activity. Sci STKE 2006:pe41.

Wolosker H, Blackshaw S, Snyder SH (1999a) Serine racemase: a glial enzyme synthesizing D-serine to regulate glutamate-N-methyl-D-aspartate neurotransmission. Proc Natl Acad Sci USA 96:13409–13414.

Wolosker H, Sheth KN, Takahashi M, Mothet JP, Brady RO Jr, Ferris CD, Snyder SH (1999b) Purification of serine racemase: biosynthesis of the neuromodulator D-serine. Proc Natl Acad Sci USA 96:721–725.

Woolf CJ, Salter MW (2000) Neuronal plasticity: increasing the gain in pain. Science 288:1765–1769.

Wu S, Barger SW, Sims TJ (2004a) Schwann cell and epineural fibroblast expression of serine racemase. Brain Res 1020:161–166.

Wu SZ, Bodles AM, Porter MM, Griffin WS, Basile AS, Barger SW (2004b) Induction of serine racemase expression and D-serine release from microglia by amyloid beta-peptide. J Neuroinflammation 1:2.

Xia M, Liu Y, Figueroa DJ, Chiu CS, Wei N, Lawlor AM, Lu P, Sur C, Koblan KS, Connolly TM (2004) Characterization and localization of a human serine racemase. Brain Res Mol Brain Res 125:96–104.

Xie X, Dumas T, Tang L, Brennan T, Reeder T, Thomas W, Klein RD, Flores J, O'Hara BF, Heller HC, Franken P (2005) Lack of the alanine-serine-cysteine transporter 1 causes tremors, seizures, and early postnatal death in mice. Brain Res 1052:212–221.

Yacubova E, Komuro H (2003) Cellular and molecular mechanisms of cerebellar granule cell migration. Cell Biochem Biophys 37:213–234.

Yamamoto N, Tomita U, Umino A, Nishikawa T (2001) Uptake of D-serine by synaptosomal P2 fraction isolated from rat brain. Synapse 42:84–86.

Yang S, Qiao H, Wen L, Zhou W, Zhang Y (2005) D-Serine enhances impaired long-term potentiation in CA1 subfield of hippocampal slices from aged senescence-accelerated mouse prone/8. Neurosci Lett 379:7–12.

Yang Y, Ge W, Chen Y, Zhang Z, Shen W, Wu C, Poo M, Duan S (2003) Contribution of astrocytes to hippocampal long-term potentiation through release of D-serine. Proc Natl Acad Sci USA 100:15194–15199.

Yasuda E, Ma N, Semba R (2001) Immunohistochemical evidences for localization and production of D-serine in some neurons in the rat brain. Neurosci Lett 299:162–164.

Ying B, Lu N, Zhang YQ, Zhao ZQ (2006) Involvement of spinal glia in tetanically sciatic stimulation-induced bilateral mechanical allodynia in rats. Biochem Biophys Res Commun 340:1264–1272.

Abbreviations

AMPA	α-amino-3-hydroxy-5-methyl-isoxazole propionate
ATP	Adenosine 5′-triphosphate
ASCT	B amino acid transporter
Asc-1	Alanine-serine-cysteine transporter 1
$[Ca^{2+}]$	Calcium ion concentration
CNS	Central nervous system
DAAO	D-Amino acid oxidase
GFAP	Glial fibrillary acidic protein
GluR	Glutamate receptors
GO	Glycine oxidase
Golga3	Golgin subfamily A member 3
GRIP	Glutamate receptor interacting protein
HNS	Hypothalamo-neurohypophysial system
HPLC	High-performance liquid chromatography
ir	Immunoreactivity
LDCVs	Large dense-core vesicles
LTD	Long-term depression
LTP	Long-term potentiation
NMDARs	*N*-methyl D-aspartate receptors
NO	Nitric oxide
NOS	Nitric oxide synthase
PICK1	Protein interacting with C-kinase

PKC	Protein kinase C
PLP	Pyridoxal 5′-phosphate
SLMVs	Synaptic-like microvesicles
SNARE	Soluble *N*-ethyl maleimide-sensitive fusion protein attachment protein receptor
SON	Supraoptic nucleus
SR	Serine racemase
TeNT	Tetanus neurotoxin

Chapter 17
Purinergic Signaling in Astrocyte Function and Interactions with Neurons

R. Douglas Fields

Contents

Purinergic signaling, which is intercellular communication mediated by adenosine triphosphate (ATP) and its breakdown products (Fields, 2006a; Fields and Burnstock, 2006; Fields and Stevens, 2000), can be considered the most pervasive mode of communication among cells in the nervous system. This is because, in contrast to neurotransmitters, growth factors, and ion fluxes, all cells share mechanisms for releasing ATP and membrane receptors for detecting it or its breakdown products. This enables all major types of glia to communicate via purinergic signaling and to communicate with neurons, vascular, and immune system cells. Consequently, the scope of functions mediated by ATP signaling spans the full range of physiological and pathophysiological processes in the nervous system. Cell proliferation, differentiation,

R.D. Fields
Nervous System Development and Plasticity Section, National Institutes of Health, NICHD, Bethesda, MD USA
fieldsd@mail.nih.gov

V. Parpura and P.G. Haydon (eds.), *Astrocytes in (Patho)Physiology of the Nervous System*,
DOI: 10.1007/978-0-387-79492-1_17,

motility, cell death, astrocytic regulation of synaptogenesis and synaptic function, cancer, response to injury, interactions with immune cells, regulation of microvasculature, myelination, release of neurotransmitters, cytokines, growth factors, and intercellular communication among astrocytes via propagated waves of intracellular Ca^{2+} are all regulated in part by purinergic signaling in astrocytes.

17.1 Intercellular ATP Signaling

ATP is released from cells and rapidly hydrolyzed to adenosine diphosphate (ADP), adenosine monophosphate (AMP), and adenosine (Fig. 17.1) (Kuperman et al., 1964; Stevens and Fields, 2000; Guthrie et al., 1999; Bodin and Burnstock, 2001; Coco et al., 2003; Lazarowski et al., 2003; Perez et al., 1986). The intracellular concentration of ATP is quite high, several millimolar, and consequently small amounts of ATP are released during exocytosis. For this reason, ATP signaling may be one of the most ancient modes of intercellular signaling, and a large family of membrane receptors have evolved to detect ATP and the products of its hydrolysis (Burnstock, 2003; Fields and Burnstock, 2006) (Fig. 17.2). ATP is concentrated in synaptic vesicles and co-released with neurotransmitter (Redman and Silinsky, 1995).

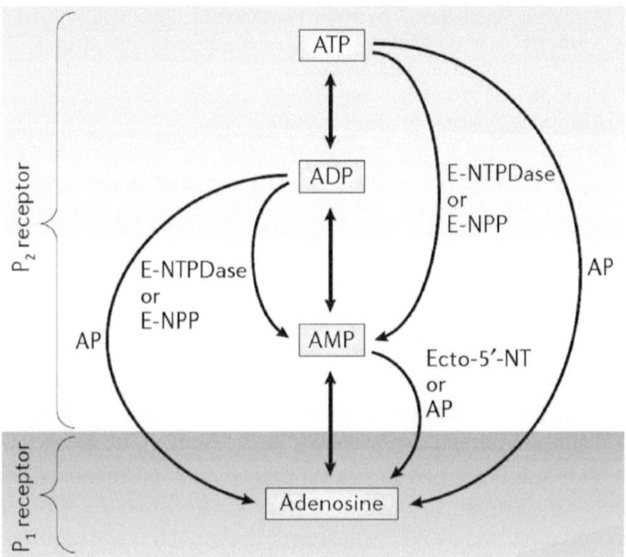

Fig. 17.1 Purinergic receptors bind extracellular ATP and its reaction products that result from enzymatic hydrolysis by ectonucleotidases. P2 receptors bind ATP and ADP, whereas P1 receptors bind adenosine. The metabolism of extracellular ATP is regulated by several ectonucleotidases, including members of the E-NTPDase (ectonucleoside triphosphate diphosphohydrolase) family and the E-NPP (ectonucleotide pyrophosphatase/phosophodiesterase) family. Ecto-5′-nucleotidase (Ecto-5′-NT) and alkaline phosphatase (AP) catalyze the nucleotide degradation from adenosine (from Fields and Burnstock, (2006)) (*See Color Plates*)

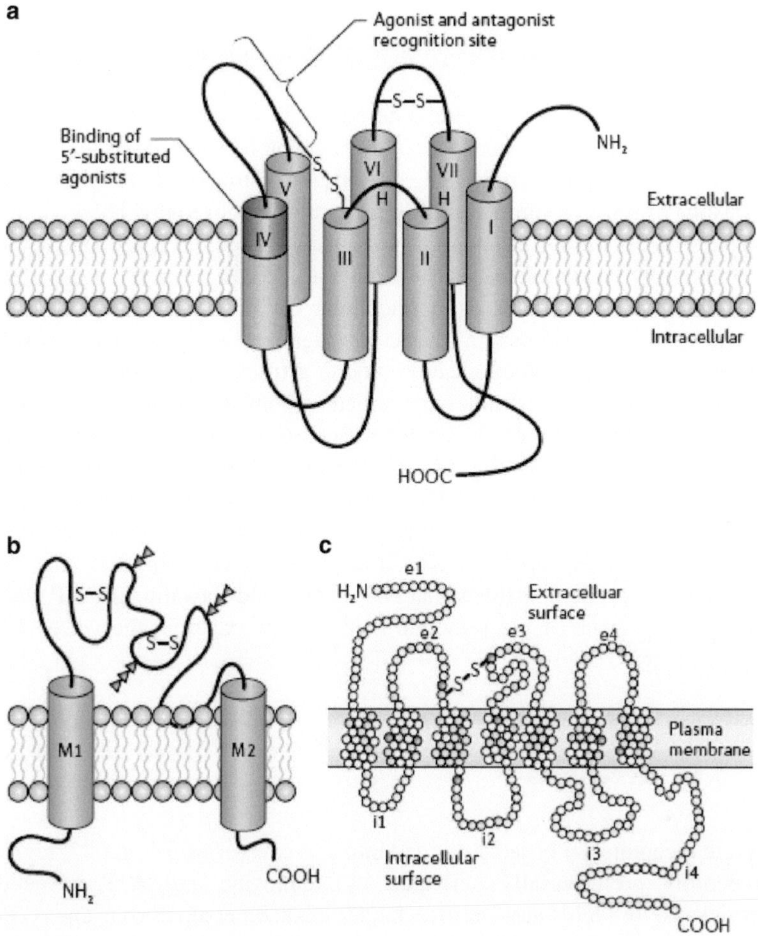

Fig. 17.2 Membrane receptors for extracellular ATP and adenosine. The P1 family of receptors for extracellular adenosine are G-protein-coupled receptors that signal by inhibiting or activating adenylate cyclase (**a**). The P2 family of receptors bind extracellular ATP or ADP, and are comprised of two types of receptors (P2X and P2Y). The P2X family are ligand-gated ion channels (**b**), and the P2Y family are G-protein-coupled receptors (**c**). *S–S*-disulphide bond; e1–e4, extracellular domain loops 1–4; i1–i4, intracellular domain loops 1–4 (from Brake et al., 1994; Ralevic and Burnstock 1998; Fields and Burnstock, 2006) (*See Color Plates*)

Thus, purinergic signaling at synapses can augment or depress synaptic transmission, depending upon the type of purinergic receptors activated on pre- and postsynaptic membranes. This also allows perisynaptic astrocytes (Serrano et al., 2006; Pascual et al., 2005) [and terminal Schwann cells (Robitaille, 1995)] to monitor synaptic transmission and influence it. Drury and Szent-Györgyi (1929) were the first to reveal the action of ATP as an extracellular messenger in experiments on heart and blood vessels. The actions of purinergic receptors are mediated intracellularly by regulating cyclic AMP (cAMP) and cytoplasmic Ca^{2+} concentration.

The first mode of intercellular communication between astrocytes to be widely recognized was via K^+ flux through gap junctions adjoining adjacent astrocytes [reviewed in (Fields and Stevens-Graham, 2002)]. This was detected by intracellular microelectrodes as changes in membrane potential caused by altered intracellular K^+ concentration. Activity-dependent communication between axons and astrocytes was revealed by detecting depolarizing responses in astrocytes to electrical stimulation of axons or sensory stimulation (light stimulation of retina). These signals are associated with astrocytes removing K^+ from the extracellular space that is released from axons firing action potentials (Orkand et al., 1966).

With the advent of fluorescent intracellular Ca^{2+} imaging in live cells in the 1980s and 1990s, intracellular Ca^{2+} waves propagating among astrocytes was found to be a major mode of communication among astrocytes and between astrocytes and neurons. Initially these studies concerned responses stimulated by application of neurotransmitters, such as the excitatory neurotransmitter glutamate (Cornell-Bell et al., 1990). A large number of neurotransmitter receptors were soon detected in astrocytes (McCarthy and Salm, 1991), and ATP was found to be one of the more potent stimulants of intracellular Ca^{2+} increases in these cells (Guthrie et al., 1999; Wang et al., 2000). By applying an extracellular enzyme, apyrase, that rapidly degrades ATP, it was demonstrated that ATP release and activation of ATP receptors on astrocytes was one of the principle means of propagating the waves of Ca^{2+} among astrocytes.

17.2 ATP Receptors

Purinergic receptors are broadly divided into two categories: P1 and P2, representing receptors preferentially activated by adenosine and ATP, respectively (Burnstock, 2006; Fields and Burnstock, 2006, Khakh et al., 2001). The ATP (P2) receptors are further divided into two categories: ion channels permeable to Ca^{2+} (P2X receptors) and metabotropic receptors (P2Y) regulating Ca^{2+} and cAMP signaling via G-proteins.

17.2.1 P2X Receptors

The P2X receptors are ligand-gated ion channels permeable to cations in response to binding extracellular ATP (North, 2002) (Fig. 17.2). This family comprises seven receptor subtypes ($P2X_1$ through $P2X_7$), all of which have been detected in astrocytes by mRNA, and most if not all, have been detected by immunocytochemistry or functional assays. It should be emphasized that astrocytes are heterogeneous and highly dynamic cells. The expression of purinergic receptor subtypes differs with development, pathology, and heterogeneity among astrocytes. P2X receptors form homomeric or heteromeric trimers, thus increasing the diversity of receptor types (Torres et al., 1999). For example, heteromultimeres of $P2X_2$ and $P2X_3$, $P2X_{4/6}$, $P2X_{1/5}$, and $P2X_{2/6}$ have been identified in various tissues. $P2X_7$ can form heteromultimers with $P2X_4$ (Guo et al., 2007), while $P2X_6$ is not thought to form a functional homomere (Fields and Burnstock, 2006).

17.2.2 P2Y Receptors

P2Y receptors are G-protein coupled receptors (Fields and Burnstock, 2006) (Fig. 17.2). $P2Y_1$, $P2Y_2$, $P2Y_4$, and $P2Y_{11}$ are coupled to G_q and therefore activated by phospholipase C. The G_i-coupled receptors $P2Y_{12}$, $P2Y_{13}$, and $P2Y_{14}$ inhibit adenylyl cyclase and regulate ion channels (Abbracchio et al., 2003). $P2Y_1$, $P2Y_{12}$, and $P2Y_{13}$ receptors are activated principally by nucleoside diphosphates. Both purine and pyrimidine nucleotides activate $P2Y_2$, $P2Y_4$, and $P2Y_6$ receptors. Evidence for all of these P2Y receptors has been found in astrocytes in various contexts. $P2Y_5$, $P2Y_7$, $P2Y_8$, and $P2Y_9$, receptors are either not reported in mammals or they are activated by other ligands.

17.2.3 P1 Receptors

P1 receptors are activated by extracellular adenosine (Burnstock, 2006; Fredholm et al., 2001). The family is comprised of four types A1, $A2_A$, $A2_B$, A3, all of which are coupled to G-proteins and have seven transmembrane domains. A1 receptors inhibit adenylate cyclase, whereas $A2_A$ receptors activate adenylate cyclase. $A2_B$ receptors activate adenylate cyclase and increase intracellular Ca^{2+} via inositol 1,4,5 trisposphate (IP_3) via PLC. The A3 receptors are coupled to G_i and G_q proteins. They signal by inhibiting adenylate cyclase activity and releasing Ca^{2+} from intracellular stores via PLC. Evidence for all four types of P1 receptors has been reported in astrocytes in different studies (Fields and Burnstock, 2006).

A summary of purinergic receptors, their pharmacology and signal transduction mechanisms are provided in Table 17.1

In addition to the wide expression of these receptors in astrocytes, neurons express many types of purinergic receptors, as do endothelial cells, microglia, oligodendrocytes, and Schwann cells (Fields, 2006a), thus enabling interactions among astrocytes and a wide variety of cells. In general, adenosine has inhibitory or sedative, anticonvulsant, and anxiolytic actions in the nervous system (Fredholm et al., 2001; Kukley et al., 2005). This implicates release of adenosine from astrocytes, or the generation of adenosine after ATP release, in the homeostatic control of neural activity levels in many pathological processes, as well as in normal physiological regulation. P2 receptor activation can inhibit or stimulate synaptic transmission, depending on the receptor subtype that is activated. Generation of adenosine after ATP release and hydrolysis can act to first stimulate and subsequently inhibit neural activity, thus sharpening the temporal or spatial extent of excitation in the brain.

Table 17.1 Characteristics of purine-mediated receptors

Receptor		Main distribution	Agonists	Antagonists	Transduction mechanisms
P1 (adenosine)	**A_1**	Brain, spinal cord, testis, heart, autonomic nerve terminals	CCPA, CPA, S-ENBA	DPCPX, N-0840, MRS1754	$G_{i/o}$ ↓cAMP
	A_{2A}	Brain, heart, lungs, spleen	CGS 21680, HENECA	KF17837, SCH58261, ZMZ41385	G_s ↑cAMP
	A_{2B}	Large intestine, bladder	NECA (non-selective)	Enprofylline, MRE2029-F20, MRS17541, MRS1706	G_s ↑cAMP
	A_3	Lung, liver, brain, testis, heart	IB-MECA, 2-Cl-IB-MECA, DBXRM, VT160	MRS1220, L-268605, MRS1191, MRS1523, VUF8504	$G_{i/o}$ $G_{q/11}$ ↓ cAMP ↑ Ins(1,4,5)P_3
P2X	**$P2X_1$**	Smooth muscle, platelets, cerebellum, dorsal horn spinal neurons	α,β-meATP = ATP = 2-MeSATP (rapid desensitization), L-β,γ-meATP	TNP-ATP, IP_5I, NF023, NF449	Intrinsic cation channel (Ca^{2+} and Na^+)
	$P2X_2$	Smooth muscle, CNS, retina, chromaffin cells, autonomic and sensory ganglia	ATP≥ATPγS≥ 2-MeSATP>> α,β-meATP (pH + zinc sensitive)	Suramin, isoPPADS, RB2, NF770	Intrinsic ion channel (particularly Ca^{2+})
	$P2X_3$	Sensory neurones, NTS, some sympathetic neurons	2-MeSATP≥ATP≥ α,β-meATP≥Ap_4A (rapid desensitization)	TNP-ATP, PPADS, A317491, NF110	Intrinsic cation channel
	$P2X_4$	CNS, testis, colon	ATP>>α,β-meATP CTP, Ivermectin	TNP-ATP (weak), BBG (weak)	Intrinsic ion channel (especially Ca^{2+})
	$P2X_5$	Proliferating cells in skin, gut, bladder, thymus, spinal cord	ATP>>α,β-meATP, ATPγS	Suramin, PPADS, BBG	Intrinsic ion channel
	$P2X_6$	CNS, motor neurons in spinal cord	N/A (does not function as homomultimer)	N/A	Intrinsic ion channel

	$P2X_7$	Apoptotic cells in, for example, immune cells, pancreas, skin	BzATP>ATP≥ 2-MeSATP>> α-β-meATP	KN62, KN04, MRS2427, Coomassie BBG	Intrinsic cation channel and a large pore with prolonged activation
P2Y	**$P2Y_1$**	Epithelial and endothelial cells, platelets, immune cells, osteoclasts	2-MeSADP=ADPβS> 2MeSATP = ADP>ATP, MRS2365	MRS2179, MRS2500, MRS2279, PIT	G_q/G_{11}; PLCβ activation
	$P2Y_2$	Immune cells, epithelial and endothelial cells, kidney tubules, osteoblasts	UTP = ATP, UTPγS, INS 37217	suramin>RB2, AR-C126313	G_q/G_{11} and possibly G_i; PLCβ activation
	$P2Y_4$	Endothelial cells	UTP≥ATP, UTPγS	RB2>suramin	G_q/G_{11} and possibly G_i; PLCβ activation
	$P2Y_6$	Some epithelial cells, placenta, T cells, thymus	UDP>UTP>>ATP, UDPβS	MRS2578	G_q/G_{11}; PLCβ activation
	$P2Y_{11}$	Spleen, intestine, granulocytes	AR-C67085MX>BzATP≥ ATPγS>ATP	Suramin>RB2, NF157	G_q/G_{11} and Gs; PLCβ activation
	$P2Y_{12}$	Platelets, glial cells	2-MeSADP≥ADP>>ATP	CT50547, AR-C69931MX, INS49266, AZD6140, PSB0413, ARL66096	$G_{i/o}$; inhibition of adenylate cyclase
	$P2Y_{13}$	Spleen, brain, lymph nodes, bone marrow	ADP = 2-MeSADP>> ATP & 2-MeSATP	MRS2211	$G_{i/o}$
	$P2Y_{14}$	Placenta, adipose tissue, stomach, intestine, discrete brain regions	UDP glucose = UDP-galactose	N/A	G_q/G_{11}

Modified from Fields and Burnstock (2006) and from Burnstock (2003)

Abbreviations. ADP, adenosine diphosphate; AMP, adenosine monophosphate; ATP, adenosine triphosphate; BBG, Brilliant blue green; BzATP, 2′- and 3′-*O*-(4-benzoyl-benzoyl)-ATP; cAMP, cyclic AMP; CCPA, chlorocyclopentyl adenosine; CPA, cyclopentyl adenosine; CTP, cytosine triphosphate; IP_3, inositol-1,4,5-trisphosphate; IP_5I, di-inosine pentaphosphate; 2-MeSADP, 2-methylthio ADP; 2-MeSATP, 2-methylthio ATP; NECA, 5′-N-ethyl-carboxamido adenosine; NTS, nucleus tractus solaritus; PLC, phospholipase C; RB2, reactive blue 2.

17.3 ATP Release Mechanisms

ATP can be released from astrocytes by several mechanisms (see Chap. 12 for details), including exocytosis of vesicles (Bowser and Khakh, 2007; Fields, 2006a; Fields and Burnstock, 2006; Fields and Stevens-Graham, 2002), flux thorough hemichannels (Cotrina et al., 1998), which are gap junctions that are uncoupled from other gap junctions, passing through $P2X_7$ receptors (Suadicani et al., 2006), ATP permeable channels/receptors activated by shear stress or osmotic swelling of the cells (Darby et al., 2003), and cellular injury. Vesicular release of ATP from astrocytes is less impaired by tetanus toxin, which cleaves synaptobrevin (a synaptic vesicle release protein), and more weakly induced by metabotropic glutamate receptor activation than release of the neurotransmitter glutamate, suggesting some differences in the vesicular pools for ATP and glutamate in astrocytes (Coco et al., 2003). Transfecting a glial cell line with gap junction proteins connexin 43, 32, or 26, increases ATP release and intercellular Ca^{2+} wave propagation, suggesting gap junction proteins contribute to ATP release (Cotrina et al., 1998). However, altering expression of connexins in astrocytes can alter P2Y receptor expression, which could affect Ca^{2+} wave propagation (Suadicani et al., 2003). Also, although gap junction channel blockers inhibit intercellular Ca^{2+} waves in astrocytes, this cannot be ascribed entirely to release of ATP through hemijunctions, because these agents also inhibit $P2X_7$ receptors, which allow ATP flux in response to ATP binding (Suadicani et al., 2003).

The many modes of ATP release from astrocytes implicate this signaling in many different biological processes (Fields and Stevens-Graham, 2002). Physiological processes stimulating ATP release from astrocytes include mechanical stress, hypoxia, inflammation, various agonists, cellular damage, synaptic transmission, and electrical activity in axons.

17.4 Extracellular Degradation and Synthesis of ATP

A complex system of extracellular enzymes controls the hydrolysis of ATP to adenosine and the phosphorylation of intermediates to ATP (Zimmermann, 1994, 2006; Zimmermann et al., 2007) (Fig. 17.1). Many of these, for example the 5′-mononucleotides, are expressed in the brain primarily on astrocytes (and other glial cells), endothelia and ependyma, and to a lesser extent on neurons. The subcellular distribution and changes in these extracellular enzymes during development and under pathological conditions are important in regulating the function of purinergic signaling from astrocytes, but this crucial aspect of purinergic signaling has received considerably less attention (Nedeljkovic et al., 2007; Cunha, 2001). Astrocytes cultured from cortex or hippocampus display an 8:1 ratio of ATP to ADP hydrolysis. Degradation to AMP is much slower, suggesting NTPDase2 as a major ectonucleotidease of astrocytes in culture (Zimmermann, 2006) (see Fig. 17.1).

17.5 Functional Significance of Purinergic Signaling in Astrocytes

By stimulating the release of ATP/adenosine, purinergic signaling in astrocytes is involved in interactions with synapses and regulation of plasticity and excitability in synaptic networks (Serrano et al., 2006; Pascual et al., 2005; Bowser and Khakh, 2004; Gordon et al., 2005; Newman, 2004). Release of cytokines (Hide et al., 2000) and growth factors (Ciccarelli et al., 1999) from astrocytes in response to purinergic signaling implicates astrocytes in developmental and injury-related processes, including gliogenesis and neurogenesis from stem cells (Mishra et al., 2006), cell proliferation (Neary et al., 1994), differentiation (Abbracchio et al., 1995), migration (Light et al., 2006; Davalos et al., 2005), myelination (Ishibashi et al., 2006), and control of blood flow through microvessels (Zonta et al., 2003). Interactions between purinergic receptor and growth factor signaling (Stevens 2006), and transactivation of growth factor receptors by adenosine (Lee and Chao, 2001; Diogenes et al., 2004), further widens the potential scope of purinergic signaling. Release of thrombospondin, a molecule secreted from astrocytes, which promotes synaptogenesis, is stimulated by ATP activation of P2Y receptors (Tan and Neary, 2006).

17.5.1 Astrocyte Communication by Intercellular Ca^{2+} Waves

Astrocytes communicate by the release of ATP, which in turn activates receptors on near-by astrocytes (Guthrie et al., 1999; Cotrina et al., 1998; Fields and Stevens-Graham, 2002; Hassinger et al., 1996). In response to intracellular Ca^{2+} increases and other second messenger signaling, ATP is then released from astrocytes, propagating the signal widely (Wang et al., 2000). Release of other substances from astrocytes is also evoked by the second messenger responses to purinergic receptor activation to contribute to intercellular cellular signaling.

Although the intercellular Ca^{2+} signaling among astrocytes in cell culture and in brain slice can be widespread, weaker and more selective stimulation in brain slice or in vivo produces a more localized and selective activation of astrocytes (Sul et al., 2004). Heterogeneity in ATP release mechanisms and differences in purinergic receptors among different astrocytes could contribute to the selective propagation of Ca^{2+} signals among discrete chains or communities of astrocytes. Differences in sensitivity of the different subtypes of purinergic receptors will activate astrocytes differentially in different contexts (for example, cellular injury accompanied by high levels of ATP vs. regulation of synaptic transmission under normal physiological conditions).

Astrocytic Ca^{2+} signaling and astrocyte morphology and function show pronounced changes after injury or epilepsy, for example, allowing the possibility that alterations in purinergic receptor expression may participate in these pathophysiological responses. After ischemia (Fowler et al., 2003), excitotoxicity (Malva et al., 2003; Tian et al., 2005), or other injuries (Neary et al., 2003), purinergic signaling regulates astrocyte responses to these and other pathophysiolgical conditions. Many brain cancers derive from astrocytic cells, and changes in purinergic receptor expression are seen.

Pharmacological treatments for several pathophysiological conditions in the nervous system act on purinergic signaling in astrocytes. The adenosine analog 2-chloro-adenosine shows promise as an anticancer drug by inducing apoptosis of human astrocytoma cells (Ceruti et al., 2000). An orphan receptor activated dually by cysteinyl-leukotriene and nucleotides is upregulated after ischemia and cellular damage can be reduced by P2Y receptor antagonists of these receptors in a rat focal ischemia model (Ciana et al., 2006).

17.5.2 Astrocyte Regulation of Synaptic Plasticity

Perisynaptic astrocytes can participate in activity-dependent regulation of synaptic strength by the release and sequestration of neurotransmitters at the synapse (Kang et al., 1998; Liu et al., 2004a, b). Plasticity of synapses in the hippocampus is also regulated by ATP and adenosine signaling (Wieraszko and Ehrlich, 1994; Tebano et al., 2005; Rodrigues et al., 2005; Almeida et al., 2003; Lopes et al., 2002; Masino et al., 2002; Cunha and Ribeiro, 2000; Cunha et al., 1995), thus coupling purinergic signaling among astrocytes with synaptic function.

Astrocytes can affect both excitatory and inhibitory synaptic transmission in the hippocampus through purinergic signaling. Heterosynaptic long-term depression in the hippocampus, is mediated by adenosine produced by ATP released from astrocytes, which in turn acts on presynaptic terminals of CA1 hippocampal neurons (Pascual et al., 2005), and also by GABA released from inhibitory interneurons causing astrocytes to generate adenosine from ATP release (Serrano et al., 2006). ATP released from astrocytes also mediates glutamatergic activity-dependent heterosynaptic suppression in the hippocampus (Zhang et al., 2003). In the hypothalamus, excitatory synaptic transmission in the paraventricular nucleus is regulated by astrocytes through their effects on neuronal P2X7 receptors (Gordon et al., 2005). Noradrenaline released from nerve terminals induces the release of ATP from astrocytes, which activates P2X7 receptors on the postsynaptic membrane of magnocellular neurosecretory neurons, thereby increasing synaptic efficacy by promoting the insertion of AMPA receptors into the postsynaptic membrane.

In the retina, purines are released by light stimulation (Perez et al., 1986; Newman, 2003). The firing rate of retinal ganglion neurons is affected by the passage of a glial Ca^{2+} wave through a mechanism requiring release of ATP and adenosine from neurons (Newman, 2004). After hydrolysis to adenosine, ATP released from astrocytes can inhibit neuronal transmission by acting on A1 receptors on retinal ganglion cells (Newman, 2003).

17.5.3 Astrocytes in Myelination

Astrocytes do not form myelin, which is the electrical insulation on axons formed by oligodendrocytes in the CNS and by Schwann cells in the PNS. However, mutations in the astrocyte gene GFAP cause mental retardation and defects in myelination

(Li et al., 2002), and mice with the GFAP gene disrupted also show hypomyelination (Liedtke et al., 1996). Recent work shows one way astrocytes are involved in regulating myelination and the mechanism involves interaction between purinergic and cytokine signaling. In response to ATP stimulation, the cytokine leukemia inhibitory factor is released from astrocytes, which in turn stimulates oligodendrocytes to form more myelin (Ishibashi et al., 2006). Interestingly, this process links myelination to electrical activity in axons, because the ATP is released from axons by action potential firing (Stevens and Fields, 2000). Activity-dependent release of ATP from axons also stimulates myelination by oligodendrocyte progenitor cells by promoting their differentiation (Stevens et al., 2002). In the peripheral nervous system, activity-dependent release of ATP from axons inhibits Schwann cell development and myelination (Stevens and Fields, 2000) as does adenosine derived from the ATP acting on A2A receptors (Stevens et al., 2004). The actions of purinergic signaling in regulating myelination (Fields, 2005, 2006b) are a good illustration of how communication among multiple types of cells can be coordinated with electrical activity in neurons and other intercellular signaling molecules to regulate cellular responses in the nervous system.

17.5.4 Cell Proliferation and Astrogliosis

Cellular damage from injured or dying cells following trauma or ischemia can release large amounts of ATP, which is involved in inducing astrogliosis (Hindley et al., 1994; Neary et al., 2003), which is characterized by changes in astrocyte proliferation, gene expression, and morphology. In addition, ATP acts as a chemoattractant for microglia to the site of brain injury and stimulates microglial proliferation (Light et al., 2006; Davalos et al., 2005). Adenosine and ATP have been implicated in inducing astroglial proliferation and formation of reactive astrocytes (Abbracchio et al., 1996). P2Y receptors mediate reactive astrogliosis through induction of cyclooxygenase 2 (Brambilla et al., 2003) and modulate tumor necrosis factor alpha (TNF-α)-mediated inflammatory responses (Kucher and Neary, 2005). Conversely, astrogliosis in rat striatal astrocytes induced by basic fibroblast growth factor is suppressed by blocking $A2_A$ receptors (Malva et al., 2003).

17.5.5 Neuroprotection and Pathophysiology

Purinergic signaling in astrocytes following brain injury is involved in both neuroprotective and neuropathological processes. Synthesis of interleukin-6 (IL-6) is increased in astroglioma cells by activation of $A2_B$ receptors. Activation of A2 receptors enhances the release of nerve growth factor and S100 beta protein from cultured astrocytes (Ciccarelli et al., 1999). Activation of $P2X_7$ receptors induces GABA release from an astrocyte cell line, as well as ATP and glutamate (Wang et al., 2002). Adenosine can act as a neuroprotective agent, depressing synaptic transmission via A1 receptors in hippocampus during hypoxia (Fowler et al., 2003).

Neuroimmune interactions between astrocytes and microglia are mediated by purinergic signaling and by the cytokines released by purinergic stimulation (Saura et al., 2005; Illes et al., 1996; Hide et al., 2000; Boucsein et al., 2003). ATP stimulates release of inflammatory cytokines from microglia (IL-1beta, IL-6, and TNF-α), superoxide production, arachidonic acid, interferon gamma, and nitric oxide synthase activity (Gendron et al., 2003). These responses implicate astrocytic purinergic signaling with Alzheimer's diseases (Houghey and Mattson, 2003) and neuropathic pain (Ji et al., 2006). Blocking $P2X_4$ receptors or administration of antisense oligonucleotides, reverse tactile allodynia caused by peripheral nerve injury (Tsuda et al., 2003). Adenosine controls hyperexcitability and epileptogenesis (Malva et al., 2003), and the release of glutamate from astrocytes (which can be secondary to Ca^{2+} influx from purinergic stimulation) can exacerbate eliptogenesis (Tian et al., 2005). Drugs targeting the A2A receptors are promising new therapeutics for treating Parkinson's disease (Hauser and Schwarzschild, 2005).

17.5.6 Astrocytes in Neurovascular Coupling

Migraine headaches associated with neurovascular responses are in part associated with ATP released from astrocytes (Grafstein et al., 2000). Intracellular Ca^{2+} waves can propagate between pia-arachnoid cells and astrocytes, and this is blocked by apyrase, an enzyme-degrading extracellular ATP (Paemeleire and Leybaert, 2000; Grafstein et al., 2000). Glial endfeet apposed to blood vessel walls strongly express P2Y2 and P2Y4 receptors. Activation of these receptors by ATP participates in regulating local blood flow according to metabolic demands (Simard et al., 2003; Zonta et al., 2003).

17.6 Conclusions

The last decade has seen enormous advances in understanding purinergic receptors (Fields, 2006a; Fields and Burnstock, 2006) while understanding of astrocyte function and interactions with neurons and other cells has expanded exponentially (Fields, 2004; Fields and Stevens-Graham, 2002). Progress on these parallel research tracks is now coming together and the convergence is sharpening understanding of the many functions mediated by extracellular signaling through ATP in the brain. Purinergic signaling involving astrocytes encompasses nearly every cell type in the brain in association with diverse nervous system functions. This complexity will take decades to unravel, but rapid progress is being made. Figure 1 (in Box 1) of a review article on purinergic signaling in neuron–glia interactions published 7 years ago contained more dotted lines, representing hypothetical functions of purinergic signaling between glia and neurons, than known links (Fields and Stevens, 2000). Today, all of these dotted lines have been confirmed experimentally. This is only the beginning, however.

Considering the heterogeneity and dynamics of astrocytes, the complexity of purinergic receptors, and the multiple ATP release mechanisms, much research lies ahead. The expression of different purinergic receptors in astrocytes and how these change during development, injury, disease, and physiological state are only known at a superficial level. The subcellular distribution of these receptors and release mechanisms liberating ATP into the extracellular environment are crucial factors in communication among astrocytes and communication with neurons and other cells, but this information is only now emerging. Activity of the extracellular enzymes controlling ATP signaling and how these change in time and space, is largely unexplored. Given the fundamental and clinical implications of purinergic signaling in the nervous system, research in this field is likely to continue at a rapid pace and yields new insights into astrocyte involvement in nervous system function.

References

Abbracchio MP, Ceruti S, Langfelder R, Cattabeni F, Saffrey MJ, Burnstock G (1995) Effects of ATP analogues and basic fibroblast growth factor on astroglial cell differentiation in primary cultures of rat striatum. Int J Dev Neurosci 13:685–693

Abbracchio MP, Ceruti S, Bolego C, Puglisi L, Burnstock G, Cattabeni F (1996) Trophic roles of P2-purinoceptors in central nervous system astroglial cells. In: Chadwick DJ (ed.) P2 Purinoceptors: Localization, Function and Transduction Mechanisms. Ciba Foundation Symposium 198, pp.142–148. Chichester: Wiley

Abbracchio, M.P., Boeynaems JM, Barnard EA, Boyer JL, Kennedy C, Miras-Portugal MT, King BF, Gachet C, Jacobson KA, Weisman GA, Burnstock G. (2003) Characterization of the UDP-glucose receptor (re-named here the P2Y14 receptor) adds diversity to the P2Y receptor family. Trends Pharmacol. Sci. 24, 52–55.

Almeida T, Rodrigues RJ, de Mendonca A, Ribeiro JA, Cunha RA (2003) Purinergic P2 receptors trigger adenosine release leading to adenosine A2A receptor activation and facilitation of long-term potentiation in rat hippocampal slices. Neuroscience 122:111–121

Bodin P, Burnstock G (2001) Purinergic signalling: ATP release. Neurochem Res 26:959–969

Boucsein C, Zacharias R, Farber K, Pavlovic S, Hanisch UK, Kettenmann H (2003) Purinergic receptors on microglia cells: functional expression in acute brain slices and modulation of microglial activation in vitro. Eur J Neurosci 17:2267–2276

Bowser DN, Khakh BS (2004) ATP excites interneurons and astrocytes to increase synaptic inhibition in neuronal networks. J Neurosci 24:8606–8620

Bowser DN, Khakh BS (2007) Vesicular ATP is the predominant cause of intercellular calcium waves in astrocytes. J Gen Physiol 129:485–491

Brake AJ, Wagenbuch MJ, Julias D (1994) New structural motif for ligand-gated ion channels defined by an ionotropic ATP receptor. Nature 317:519–523

Brambilla R, Cottini L, Fumagalli M, Ceruti S, Abbracchio MP (2003) Blockade of A2A adenosine receptors prevents basic fibroblast growth factor-induced reactive astrogliosis in rat striatal primary astrocytes. Glia 43:190–194

Burnstock G (2003) Purinergic receptors in the nervous system. Purinergic receptors and signalling. In: Schwiebert EM (ed.) Current Topics in Membranes, Vol. 54, pp. 307–368. San Diego, CA: Academic

Burnstock G (2006) Purinergic signalling – an overview. Novartis Foundation Symposium 276:26–53

Ceruti S, Franceschi C, Barbieri D, Malomi W, Camurri A, Giammarioli AM, Ambrosini A, Racagni G, Cattabeni F, Abbracchio MP (2000) Apoptosis induced by 2-chloro-adenosine and

2-chloro-2′-deoxy-adenosine in a human astrocytoma cell line: differential mechanisms and possible clinical relevance. J Neurosci Res 60:388–400
Ciana P, Fumagalli M, Trincavelli ML, Verderio C, Rosa P, Lecca D, Ferrario S, Parravicini C, Capra V, Gelosa P, Guerrini U, Belcredito S, Cimino M, Sironi L, Tremoli E, Rovati GE, Martini C, Abbracchio MP (2006) The orphan receptor GPR17 identified as a new dual uracil nucleotides/cysteinyl-leukotrienes receptor. EMBO J 25:4615–4627
Ciccarelli R, Di Iorio P, Bruno V, Battaglia G, D'Alimote I, D'Onofrio M, Nicoletti F, Caciagli F (1999) Activation of A(1) adenosine or mGlur3 metabotropic glutamate receptors enhances the release of nerve growth factor and S-100 beta protein from cultured astrocytes. Glia 27:275–281
Coco S, Calegari F, Pravettoni E, Pozzi D, Taverna E, Rosa P, Matteoli M, Verderio C (2003) Storage and release of ATP from astrocytes in culture. J Biol Chem 278:1354–1362
Cornell-Bell AH, Finkbeiner SM, Cooper MS, Smith SJ (1990) Glutamate induces calcium waves in cultured astrocytes: Long-range glial signaling. Science 247:470–473
Cotrina ML, Lin JH, Alves-Rodrigues A, Liu S, Li J, Azmi-Ghadimi H, Kang J, Naus CC, Nedergaard M (1998) Connexins regulate calcium signaling by controlling ATP release. Proc Natl Acad Sci USA 95:15735–15740
Cunha RA (2001) Regulation of the ecto-nucleotidase pathway in rat hippocampal nerve terminals. Neurochem Res 26:979–991
Cunha RA, Ribeiro JA (2000) Purinergic modulation of [(3)H]GABA release from rat hippocampal nerve terminals. Neuropharmacology 39:1156–1167
Cunha RA, Johansson B, Fredholm BB, Ribeiro JA, Sebastiao AM (1995) Adenosine A2A receptors stimulate acetylcholine release from nerve terminals of the rat hippocampus. Neurosci Lett 196:41–44
Darby M, Kuzmiski JB, Panenka W, Feighan D, MacVicar BA (2003) ATP release from astrocytes during swelling activates chloride channels. J Neurophysiol 89:1870–1877
Davalos D, Grutzendler J, Yang G, Kim JV, Zuo Y, Jung S, Littman DR, Dustin ML, Gan WB (2005) ATP mediates rapid microglial response to local brain injury in vivo. Nature Neurosci 8:752–758
Diogenes MJ, Fernandes CC, Sebastiao AM, Ribeiro JA (2004) Activation of adenosine A2A receptor facilitates brain-derived neurotrophic factor modulation of synaptic transmission in hippocampal slices. J Neurosci 24:2905–2913
Drury AN, Szent-Györgyi A (1929) The physiological activity of adenine compounds with special reference to their action upon the mammalian heart. J Physiol (Lond) 68:213–237
Fields RD (2004) The other half of the brain. Scientific American 290:54–61
Fields RD (2005) Myelination: An overlooked mechanism of synaptic plasticity? Neuroscientist 11:528–531
Fields RD (2006a) Purinergic Signaling in Neuron–Glia Interactions, Novartis Foundation Symposium, Vol. 267. New York, NY: Wiley
Fields RD (2006b) Nerve impulses regulate myelination through purinergic signalling. Novartis Symposium Foundation 276:148–161
Fields RD, Burnstock G (2006) Purinergic signaling in neuron–glia interactions. Nature Reviews Neuroscience 7:423–436
Fields RD, Stevens B (2000) ATP in signaling between neurons and glia. Trends Neurosci 23:625–633
Fields RD, Stevens-Graham B (2002) New views of neuron–glia communication. Science 298:483–690
Fowler JC, Gervitz LM, Hamilton ME, Walker JA (2003) Systemic hypoxia and the depression of synaptic transmission in rat hippocampus after carotid artery occlusion. J Physiol 550:961–972
Fredholm BB, Arslan G, Halldner L, Kull B, Schulte G, Ådén U, Svenningssson P (2001) Adenosine receptor signaling in vitro and in vivo. Drug Dev Res 52:274–282

Gendron FP, Chalimoniuk M, Strosznajder J, Shen S, Gonzalez FA, Weisman GA, Sun GY (2003) P2X7 nucleotide receptor activation enhances IFNgamma-induced type II nitric oxide synthase activity in BV-2 microglial cells. J Neurochem 87:344–352

Gordon GR, Baimoukhametova DV, Hewitt SA, Rajapaksha WR, Fisher TE, Bains JS (2005) Norepinephrine triggers release of glial ATP to increase postsynaptic efficacy. Nature Neurosci 8:1078–1086

Grafstein B, Liu S, Cotrina ML, Goldman SA, Nedergaard M (2000) Meningeal cells can communicate with astrocytes by calcium signaling. Ann Neurol 47:18–25

Guo C, Masin M, Qureshi OS, Murrell-Lagnado RD (2007) Evidence for functional P2X4/P2X7 heteromeric receptors. Mol Pharmacol 72:1447–1456

Guthrie PB, Knappenberger J, Segal M, Bennett MV, Charles AC, Kater SB (1999) ATP release from astrocytes mediates glial calcium waves. J Neurosci 19:520–528

Hassinger TD, Guthrie PB, Atkinson PB, Bennett MV, Kater SB (1996) An extracellular signaling component in propagation of astrocytic calcium waves. Proc Natl Acad Sci USA 93:13268–13273

Haughey NJ, Mattson MP (2003) Alzheimer's amyloid beta-peptide enhances ATP/gap junction-mediated calcium wave propagation in astrocytes. Neuromol Med 3:173–180

Hauser RA, Schwarzschild MA (2005) Adenosine A2A receptor antagonists for Parkinson's disease: rationale, therapeutic potential and clinical experience. Drugs Aging 22:471–482

Hide I, Tanaka M, Inoue A, Nakajima K, Kohsaka S, Inoue K, Nakata Y (2000) Extracellular ATP triggers tumor necrosis factor-alpha release from rat microglia. J Neurochem 75:965–972

Hindley S, Herman MA, Rathbone MP (1994) Stimulation of reactive astrogliosis in vivo by extracellular adenosine diphosphate or an adenosine A2 receptor agonist. J Neurosci Res 38:399–406

Illes P, Norenberg W, Gebicke-Haerter PJ (1996) Molecular mechanisms of microglial activation. B. Voltage- and purinoceptor-operated channels in microglia. Neurochem Int 29:13–24

Ishibashi T, Dakin KA, Stevens B, Lee PR, Kozlov SV, Stewart CL, Fields RD (2006) Astrocytes promote myelination in response to electrical impulses. Neuron 49:823–832

Ji RR, Kawasaki Y, Zhuang ZY, Wen YR, Decosterd I (2006) Possible role of spinal astrocytes in maintaining chronic pain sensitization: review of current evidence with focus on bFGF/JNK pathway. Neuron Glia Biol 2:259–270

Kang J, Jiang L, Goldman SA, Nedergaard M (1998) Astrocyte-mediated potentiation of inhibitory synaptic transmission. Nature Neurosci 1:683–692

Khakh BS, Burnstock G, Kennedy C, King BF, North RA, Séguéla P, Voigt M, Humphrey PPA International Union of Pharmacology. XXIV. Current status of the nomenclature and properties of P2X receptors and their subunits. Pharmacol Rev 53:107–118 (2001)

Kucher BM, Neary JT (2005) Bi-functional effects of ATP/P2 receptor activation on tumor necrosis factor-alpha release in lipopolysaccharide-stimulated astrocytes. J Neurochem 92:525–535

Kukley M, Schwan M, Fredholm BB, Dietrich D (2005) The role of extracellular adenosine in regulating mossy fiber synaptic plasticity. J Neurosci 25:2832–2837

Kuperman AS, Volpert WA, Okamoto M (1964) Release of Adenine nucleotide from nerve axons. Nature 204:1000–1001

Lazarowski ER, Boucher RC, Harden TK (2003) Mechanisms of release of nucleotides and integration of their action as P2X- and P2Y-receptor activating molecules. Mol Pharmacol 64:785–795

Lee FS, Chao MV (2001) Activation of Trk neurotrophin receptors in the absence of neurotrophins. Proc Natl Acad Sci USA 98:3555–3560

Li R, Messing A, Goldman JE, Brenner M (2002) GFAP mutations in Alexander's disease. Int J Dev Neurosci 20:259–268

Liedtke W, Edelmann W, Bieri PL, Chiu FC, Cowan NJ, Kucherlapati R, Raine CS. 1996 GFAP is necessary for the integrity of CNS white matter architecture and long-term maintenance of myelination. Neuron. 17:607–615.

Light AR, Wu Y, Hughen RW, Guthrie PB (2006) Purinergic receptors activating rapid intracellular Ca2+ increases in microglia. Neuron Glia Biol 2:125–138
Liu QS, Xu Q, Arcuino G, Kang J, Nedergaard M (2004a) Astrocyte-mediated activation of neuronal kainite receptors. Proc Natl Acad Sci USA 1201:3172–3177
Liu QS, Xu Q, Kang J, Nedergaard M (2004b) Astrocyte activation of presynaptic metabotropic glutamate receptors modulates hippocampal inhibitory synaptic transmission. Neuron Glia Biol 1:307–316
Lopes LV, Cunha RA, Kull B, Fredholm BB, Ribeiro JA (2002) Adenosine A(2A) receptor facilitation of hippocampal synaptic transmission is dependent on tonic A(1) receptor inhibition. Neuroscience 112:319–329
Malva JO, Silva AP, Cunha RA (2003) Presynaptic modulatiopn controlling neuronal excitability and epileptogenesis: role of kainate, adenosine and neuropeptide Y receptors. Neurochem Res 28:1501–1515
Masino SA, Diao L, Illes P, Zahniser NR, Larson GA, Johansson B, Fredholm BB, Dunwiddie TV (2002) Modulation of hippocampal glutamatergic transmission by ATP is dependent on adenosine A(1) receptors. J Pharmacol Exp Ther 303:356–363
McCarthy KD, Salm AK (1991) Pharmacologically-distinct subsets of astroglia can be indentified by their calcium response to neurolignads. Neuroscience 41:325–333
Mishra SK (2006) Extracellular nucleotide signaling in adult neural stem cells: synergism with growth factor-mediated cellular proliferation. Development 133:675–684
Neary JT, Whittemore SR, Zhu O, Norenberg MD (1994) Synergistic activation of DNA synthesis in astrocytes by fibroblast growth factors and extracellular ATP. J Neurochem 63:490–494
Neary JT, Kang Y, Willoughby KA, Ellis EF (2003) Activation of extracellular signal-regulated kinase by stretch-induced injury in astrocytes involves extracellular ATP and P2 purinergic receptors. J Neurosci 23:2348–2356
Nedeljkovic N, Bjeloba I, Subasic S, Pekovic S, Stojkov D, Vjestica A, Rakic L, Stojiljkovic M (2007) Up-regulation of ectonucleotidase activity after cortical stab injury in rats. Cell Biol Int 30:541–546
Newman EA (2003) Glial cell inhibition of neurons by release of ATP. J Neurosci 23:1659–1666
Newman EA (2004) A dialogue between glia and neurons in the retina: Modulation of neuronal excitability. Neuron Glia Biol 1:245–252
North RA (2002) Molecular physiology of P2X receptors. Physiol Rev 82:1013–1067
Orkand RK, Nicholls JG, Kuffler SW (1966) Effect of nerve impulses on the membrane potential of glial cells in the CNS of amphibian. J Neurophysiol 29:788–806
Paemeleire K, Leybaert L (2000) ATP-dependent astrocyte-endothelial calcium signaling following mechanical damage to a single astrocyte in astrocyte–endothelial co-cultures. J Neurotrauma 17:345–358
Pascual O, Casper KB, Kubera C, Zhang J, Revilla-Sanchez R, Sul JU, Takano H, Moss SJ, McCarthy K, Haydon PG (2005) Astrocytic purinergic signaling coordinates synaptic networks. Science 310:113–116
Perez MT, Ehinger BE, Lindstrom K, Fredholm BB (1986) Release of endogenous and radioactive purines from the rabbit retina. Brain Res 298:106–112
Ralevic V, Burnstock G (1998) Receptors for purines and pyrimidines. Pharmacol Rev 50:413–492
Redman RS, Silinsky EM (1995) ATP release together with acetylcholine as the mediator of neuromuscular depression at frog motor nerve endings. J Physiol 477:117–127
Robitaille R (1995) Purinergic receptors and their activation by endogenous purines at perisynaptic glial cells of the frog neuromuscular junction. J Neurosci 15:7121–7131
Rodrigues RJ, Almeida T, Richardson PJ, Oliveira CR, Cunha RA (2005) Dual presynaptic control by ATP of glutamate release via facilitatory P2X1, P2X2/3, and P2X3 and inhibitory P2Y1, P2Y2, and/or P2Y4 receptors in the rat hippocampus. J Neurosci 25:6286–6295
Saura J, Angulo E, Ejarque A, Casado V, Tusell JM, Moratalla R, Chen JF, Schwarzschild MA, Lluis C, Franco R, Serrotosa J (2005) Adenosine A2A receptor stimulation potentiates nitric oxide release by activated microglia. J Neurochem 95:919–929

Serrano A, Haddjeri N, Lacaille JC, Robitaille R (2006) GABAergic network activation of glial cells underlies hippocampal heterosynaptic depression. J Neurosci 26:5370–5380

Simard M, Nedergaard M, Arcuino G, Takano T, Liu QS, Nedergaard M (2003) Signaling at the gliovascular interface. J Neurosci 23:9254–9262

Stevens B (2006) Cross-talk between growth factor and purinergic signaling regulates Schwann cell proliferation. Novartis Foundation Symposium 276:162–180

Stevens B, Fields RD (2000) Action potentials regulate Schwann cell proliferation and development. Science 287:2267–2271

Stevens B, Porta S, Haak LL, Gallo V, Fields RD (2002) Adenosine: A neuron–glial transmitter promoting myelination in the CNS in response to action potentials. Neuron 36:855–868

Stevens B, Ishibashi T, Chen JF, Fields RD (2004) Adenosine: an activity-dependent axonal signal regulating MAP kinase and proliferation in developing Schwann cells. Neuron Glia Biol 1:23–34

Suadicani SO, Pina-Benabou MH, Urban-Maldonado M, Spray DC, Scemes E (2003) Acute downregulation of Cx43 alters P2Y receptor expression levels in mouse spinal cord astrocytes. Glia 42:160–171

Suadicani SO, Brosnan CF, Scemes E (2006) P2X7 receptor mediate ATP release and amplification of astrocytic intercellular calcium signaling. J Neurosci 26:1378–1385

Sul J-Y, Orosz G, Givens RS, Haydon PG (2004) Astrocytic connectivity in the hippocampus. Neuron Glia Biol 1:3–12

Tebano MT, Martie A, Rebola N, Pepponi R, Domenici MR, Gro MC, Schwarzschild MA, Chen JF, Cunha RA, Popoli P (2005) Adenosine A2A receptors and metabotropic glutamate 5 receptors are co-localized and functionally interact in the hippocampus: a possible key mechanism in the modulation of *N*-methyl-d-aspartate effects. J Neurochem 95:1188–1200

Tian GF, Azmi H, Takano T, Xu Q, Peng W, Lin J, Oberheim N, Lou N, Wang X, Zielke HR, Kang J, Nedergaard M (2005) An astrocytic basis of epilepsy. Nature Med 11:973–981

Torres GE, Egan TM, Voigt MM (1999) Hetero-oligomeric assembly of P2X receptor subunits. Specificities exist with regard to possible partners. J Biol Chem 274:6653–6659

Tran MD, Neary JT (2006) Purinergic signaling induces thrombospondin-1 expression in astrocytes. Proc Natl Acad Sci USA 103:9321–9326

Tsuda M, Shigemoto-Mogami Y, Koizumi S, Mizokoshi A, Kohsaka S, Salter MW, Inoue K (2003) P2X4 receptors induced in spinal microglia gate tactile allodynia after nerve injury. Nature 424:778–783.

Wang Z, Haydon PG, Yeung ES (2000) Direct observation of calcium-independent intercellular ATP signaling in astrocytes. Anal Chem 72:2001–2007

Wang CM, Chang YY, Kuo JS, Sun SH (2002) Activation of P2X(7) receptors induced [(3)H] GABA release from the RBA-2 type-2 astrocyte cell line through a cl(–)/HCO(3)(–)-dependent mechanism. Glia 37:8–18

Wieraszko A, Ehrlich YH (1994) On the role of extracellular ATP in the induction of long-term potentiation in the hippocampus. J Neurochem 63:1731–1738

Zhang JM, Wang HK, Ye CQ, Ge W, Chen Y, Jiang ZL, Wu CP, Poo MM, Duan S. (2003) ATP released by astrocytes mediates glutamatergic activity-dependent heterosynaptic suppression. Neuron 40:971–982

Zimmermann H (1994) Signalling via ATP in the nervous system. Trends Neurosci 17:420–426

Zimmermann H (2006) Ectonucleotidases in the nervous system. Novartis Foundation Symposium 276:113–130

Zimmermann H (2007) Ecto-nucleotidases, molecular properties and functional impact. Ann Rev Acad Nac Farm 73:537–566

Zonta M, Angulo MC, Gobbo S, Rosengarten B, Hossmann KA, Pozzan T, Carmignoto G (2003) Neuron-to-astrocyte signaling is central to the dynamic control of brain microcirculation. Nature Neurosci 6:43–50

Abbreviations

ADP	Adenosine diphosphate
AMP	Adenosine monophosphate
ATP	Adenosine triphosphate
cAMP	Cyclic AMP
IL	Interleukin
IP_3	Inositol 1,4,5 trisposphate
TNF-α	Tumor necrosis factor alpha

Chapter 18
Astrocyte Control of Blood Flow

Grant R.J. Gordon, Sean J. Mulligan, and Brian A. MacVicar

Contents

Astrocytes have recently been shown to be essential participants in the control of cerebral blood flow (CBF) through their prominent control of cerebral vessel diameter. Although the unique close relationship of astrocytes with cerebral blood vessels has long been recognized it is only within the last few years that evidence has shown how astrocytes might translate information to the vasculature on the activity level and energy demands of neurons. These findings suggest that astrocytes are key players in the system for the delivery and clearance of molecules important to brain function.

B.A. MacVicar
Department of Psychiatry and the Brain Research Centre, University of British Columbia, Vancouver, BC Canada
bmacvicar@brain.ubc.ca

V. Parpura and P.G. Haydon (eds.), *Astrocytes in (Patho)Physiology of the Nervous System*, DOI: 10.1007/978-0-387-79492-1_18,

Astrocytes possess the necessary signaling capability to induce both vasoconstriction as well as vasodilation in response to elevations in astrocyte end-feet Ca^{2+}. Both types of vasomotor responses are initiated by the generation of arachidonic acid (AA) in astrocytes by Ca^{2+} sensitive phospholipase A_2 (PLA_2). Subsequent to AA formation, vasoconstriction occurs as a result of the generation of 20-hydroxyeicosatetraenoic acid (20-HETE), while vasodilation ensues from the production of epoxyeicosatrienoic acid (EET) or prostaglandin E_2 (PGE_2). Notably, the level of nitric oxide (NO) seems to control which of these two routes is utilized, either by the inhibition of critical enzymes by NO or by an indirect effect on vessel tone. In addition to the Ca^{2+}-activated PLA_2 pathway, the activation of large conductance Ca^{2+}-activated K^+ channels in astrocyte end-feet has been proposed to induce vasodilation by hyperpolarizing smooth muscle cells (SMCs) through the effect of increased external [K^+] on SMC Kir channels. This large array of possibilities highlights the importance of astrocytes as well as the need for additional experimentation to fully delineate their contributions to vascular dynamics.

18.1 Functional Hyperemia

More than 100 years ago Roy and Sherrington first discovered that brain tissue is intrinsically capable of controlling CBF within a specific, localized region (Roy and Sherrington, 1890). This regional phenomenon is termed functional hyperemia, whereby vessel diameter is enlarged to augment CBF in response to energy demands that result from enhanced synaptic transmission and neuronal firing. The purpose of functional hyperemia is to augment the delivery of oxygen-rich hemoglobin, glucose, and other nutrients to the working cells, while simultaneously clearing metabolic products such as CO_2. Understanding the mechanistic physiology of how the cerebrovasculature changes diameter may be critical for developing effective treatments for an array of neurological afflictions such as stroke, hemorrhage, focal ischemia, and migraine. In addition to these pathologies, understanding cerebrovascular control within the context of brain energy metabolism is important for the correct interpretation of data obtained from high resolution, functional magnetic resonance imaging (fMRI), which uses the magnetic signal of deoxyhemoglobin as an indirect measure of CBF and brain activity. This technology, as well as with other imaging strategies, is widely used to identify volumes of activated brain tissue during both normal and pathological cerebral functioning. Some of the "implicit" understanding surrounding fMRI signals is now being refined as we learn more about how the brain consumes oxygen during activity and how this relates to changes in CBF. However, there is still much to learn because the cellular mechanisms involved in transforming changes in neuronal activity to changes in CBF are incompletely understood. There are many molecules that have effects on cerebral vessel tone, the number of different cell types that participate in neurovascular coupling is increasing and the source of certain vasoactive substances is controversial. In spite of this, there has been a noteworthy focus on one cell type – the astrocyte. These cells have been hailed as the missing element coupling changes in neuronal activity to alterations in CBF via influencing cerebral vessel diameter. In this chapter we focus on the role of astrocytes, and in particular that of astrocyte end-feet Ca^{2+} signaling, in the control of brain blood vessel girth. The different experimental models and approaches used to test astrocyte involvement, the possible signaling molecules participating, and the potential mechanisms of astrocyte action are highlighted.

18.2 Astrocytes and Functional Hyperemia: Origins and Revisions

Roy and Sherrington (1890) originally posited the idea that a local accumulation of metabolic products triggered a homeostatic dilation of nearby vessels to augment flow. The increased CBF would then clear these catabolic substances and CBF would presumably return to an appropriate rate. For functional hyperemia though, recent evidence argues against such a hypothesis, primarily because of the rapid time to onset for vessel dilation and increased CBF from the time of enhanced neural activity (Lou et al., 1987). Observations made in vivo suggest that ~1–2 s is required to make the transition from forepaw stimulation to vessel response in corresponding regions of the somatosensory cortex (Zonta et al., 2003b). When considering that neuronal processes can be located at an average maximal distance of 60 μm from the nearest vessel (Lokkegaard et al., 2001), the local accumulation and diffusion of metabolic by-products is an unsatisfactory explanation for the rapid coupling of neural activity to changes in CBF. Furthermore, the reestablishment of vessel tone after metabolic products have been shuttled away is too passive and parsimonious an explanation for what is, essentially, a constriction process. There must be other aspects about brain physiology unconsidered in Roy and Sherrington's original idea.

Astrocytes recently have been demonstrated to be active participants in the coupling of neuronal activity to blood flow control and may in fact be the critical component providing fast and dynamic control of blood vessels. As early as 1913 Ramon y Cajal recognized that astrocytes can form a physical bridge from neurons to the cerebrovasculature. Processes from a single astrocyte interact with an enormous number of synapses (Ventura and Harris, 1999; Bushong et al., 2002; Haber et al., 2006), adjacent astrocytes (Massa and Mugnaini, 1982; Fischer and Kettenmann, 1985) and, through the use of specialized end-feet, the microvessels of the brain (Simard et al., 2003). End-feet are enlarged compartments located distally on astrocytic processes that are specialized for close interactions with endothelial cells, SMCs, and possibly pericytes. The diameter of arterioles, and thereby blood flow, is regulated by contracting and relaxing SMCs. Pericytes recently have also been shown to constrict capillaries (Peppiatt et al., 2006) although the contribution of this interaction to the regulation of cerebral blood flow is still unknown. Collectively, end-feet from numerous astrocytes wrap all cerebral blood vessels in the CNS. A current hypothesis for functional hyperemia that is receiving much interest is that astrocytes can directly sense changes in synaptic activity and relay this information to the cerebrovasculature. In this model, CBF can be augmented more quickly to meet the demands of activated tissue compared to waiting for the accumulation and diffusion of specific vasoactive by-products of metabolism.

18.3 Astrocytic Characteristics for Cerebrovascular Control

Discoveries over the past 25 years have shown that astrocytes possess a vast repertoire of ion channels, receptors, and signaling pathways that enable them to detect and convey synaptic information to vessels. In the early 1990s work with calcium indicator dyes showed that exogenous glutamate application to cultured astrocytes caused oscillating increases in internal free Ca^{2+} resulting from the activity of intracellular Ca^{2+} stores (Cornell-Bell et al., 1990). The most notable finding was that the Ca^{2+} oscillations propagated as a wave through connected astrocytes. These data indicated that membrane bound metabotropic glutamate receptor (mGluR), could elicit novel, long-range signaling cascades incorporating networks of astrocytes. Further studies demonstrated a similar effect using more physiological preparations and from synaptically released glutamate (Dani et al., 1992; Porter and McCarthy, 1995, 1996; Pasti et al., 1997). These data inspired nearly two decades of research into how this Ca^{2+} signal propagates and what is it used for in the physiology of the animal. Here there is a focus on astrocyte Ca^{2+} signals, particularly those that occur in the end-feet, and how these signals utilize the molecular machinery of the end-feet to initiate changes to the cerebrovasculature and ultimately CBF.

18.3.1 From End-Foot to Vasomotion: Molecular Players

End-feet are endowed with many features thought to be important for cerebrovasculature control. For instance, metabotropic P2Y purinoceptors and the gap junction protein connexin-43 are highly expressed in astrocytic end-feet in situ (Simard et al., 2003). These proteins are thought to initiate and allow the release of adenosine 5′-triphosphate (ATP), respectively, and cooperatively can cause an end-foot Ca^{2+} signal to spread distances greater than 50 mm along a vessel's outer surface (Simard et al., 2003). ATP, once released, can be broken down by ecto-ATPase and ecto-5-nucleotidase into adenosine, a nucleoside that has a dilating influence on the cerebrovasculature during functional hyperemia (Shi et al., 2008). Adenosine can also act on astrocytes to enhance Ca^{2+} wave propagation in both the cerebellum (Jimenez et al., 1998) and the retina (Newman, 2003). All of these elements may operate together to ensure the astrocyte end-foot signal propagates to a sufficient number of SMCs in an apposed vessel to elicit a desirable vasomotor response.

In addition to the metabotropic glutamate and P2Y receptors, end-feet also possess functional adrenoceptors (Paspalas and Papadopoulos, 1996), which, when activated, cause prominent elevations in end-foot Ca^{2+} (Mulligan and MacVicar, 2004). With such strong and seemingly pervasive astrocyte Ca^{2+} signals, what intracellular molecules might this signal target and to what end? Admittedly, an increase in intracellular Ca^{2+} has a plethora of potential targets but one molecule of interest is soluble PLA_2, which is abundantly expressed in astrocyte end-feet (Farooqui et al., 1997). Once activated by Ca^{2+} this enzyme leads to the production of multiple vasoactive substances (Fig. 18.1). PLA_2 generates diffusible AA from the plasma membrane, which can be converted into a number of compounds, some which induce vasodilation and others induce vasoconstriction. Dilating products include PGE_2 from the action of cyclooxygenase (COX) enzymes (Niwa et al., 2001; Zonta et al., 2003b; Takano et al., 2006) and several EETs (5,6-EET; 8,9-EET; 11,12-EET; and 14,15-EET) (Ellis et al., 1990; Gebremedhin et al., 1992) from the activity of a subtype of cytochrome P450 (CYP450) enzymes. Constricting molecules consist of PGF_2 (Ellis et al., 1983) and thromboxane A_2 (Ishimoto et al., 1996; Benyo et al., 1998a, b; Filosa et al., 2004) from COX activity, endothelin peptide (Faraci, 1989; MacCumber et al., 1990), as well as 20-HETE (Lange et al., 1997; Mulligan and MacVicar, 2004) from the conversion of AA by a different CYP450 enzyme than that mentioned for EETs.

In cultured astrocytes, stimulating soluble PLA_2 causes the release of AA (Stella et al., 1997). However, AA can also be converted while still within the cells. Cultured astrocytes express COX-1 and can be triggered to express COX-2 (Koyama et al., 1999; Luo et al., 2001), which can generate PGE_2, both in response to the Ca^{2+} ionophore ionomycin (Oomagari et al., 1991) and glutamate agonists (Zonta et al., 2003a). Similarly, the AA metabolites EETs can be released from astrocyte cultures

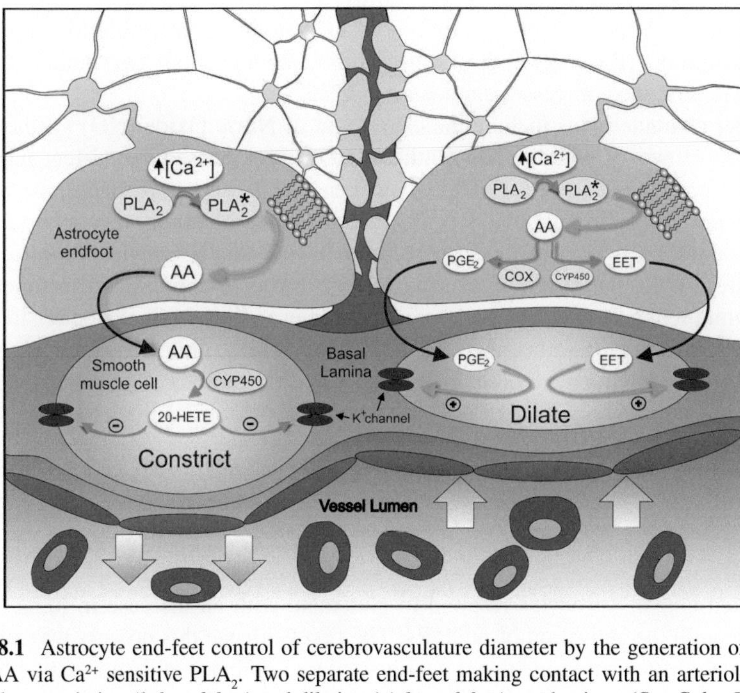

Fig. 18.1 Astrocyte end-feet control of cerebrovasculature diameter by the generation of astrocyte AA via Ca^{2+} sensitive PLA_2. Two separate end-feet making contact with an arteriole show possible constriction (*left end-foot*) and dilation (*right end-foot*) mechanisms (*See Color Plates*).

when this preparation is treated with exogenous glutamate agonists (Alkayed et al., 1997). These in vitro data indicate astrocytes are capable of releasing a variety of vasoactive products.

In addition to PLA_2 activation, an increase in free intracellular Ca^{2+} within astrocytes may also stimulate Ca^{2+}-activated K^+ channels (K_{Ca}). Activation of mGluRs in cultured hippocampal astrocytes triggers the opening of K_{Ca} channels, leading to K^+ efflux (Gebremedhin et al., 2003). This effect was blocked by mGluR antagonists and, interestingly, by inhibition of cytochrome P450 arachidonate epoxygenase. Large conductance Ca^{2+}-activated K^+ channels (BK channels; also termed slo or MaxiK channels) have also been identified in situ, with notable localization to the perivascular astrocyte end-feet (Price et al., 2002). K^+ efflux from these channels after opening would cause local elevations in the extracellular concentration of K^+ around vessels, which may influence the membrane potential of SMCs or facilitate Kir channel opening.

Aquaporin 4 water channels show a similar distribution pattern with the highest expression level in astrocyte end-feet with lower levels in astrocyte processes that ensheath glutamatergic synapses (Amiry-Moghaddam and Ottersen, 2004), often localizing with Kir channels (Nagelhus et al., 2004). In contrast, neurons show limited expression of aquaporins, a result that has lead to the idea that astrocytes, as opposed to neurons, are responsible for volume changes and volume homeostasis in the brain (Simard and Nedergaard, 2004; Andrew et al., 2007). As volume changes, such as the cell swelling induced from elevated neural activity, require a set point, the opening of volume regulated anion channels (VRACs) is thought to fulfill this role by allowing Cl^- and K^+ efflux, thereby limiting the maximum extent of the swell (Pasantes-Morales, 1996). VRACs are also permeable to amino acid gliotransmitters (Basarsky et al., 1999) and thus provide another potential route for astrocyte influence on vessel diameter and CBF.

Other characteristics include the expression of Nitric Oxide (NO) synthesizing enzymes. However, whether NO synthase (NOS) is expressed in end-feet and what isoform is expressed, i.e., eNOS (Wiencken and Casagrande, 1999) vs. sNOS (Calka and Wolf, 2003), remains controversial. Given the ubiquitous importance of NO modulation of vessel tone (more below), and that astrocyte processes may generate it (Murphy et al., 1993), the role of astrocyte-derived NO in functional hyperemia will be a crucial issue for future studies. All of these characteristics collectively have lead to the concept of the “neurovascular unit,” in which astrocytes are well-enough endowed, both anatomically and physiologically, to transmit relevant information about the extracellular environment from neurons to vessels.

18.4 Astrocytes Control Cerebrovascular Diameter

As stated above, astrocytes respond to glutamate with an increase in intracellular Ca^{2+} through the activation of mGluRs. The first evidence this process was a major component of astrocyte-mediated neurovascular coupling was demonstrated by the

Carmignoto laboratory (Zonta et al., 2003b). Activating astrocytes indirectly by disrupting membranes with patch electrodes or by applying mGluR agonists triggered the release of diffusible factors that then acted on SMCs to cause dilation of arterioles. Further, an elevation of extracellular glutamate from enhanced synaptic activation evoked Ca^{2+} increases in astrocyte end-feet (via mGluR) and caused vasodilation. The dilation was reduced by interfering with COX activity with aspirin, suggesting products from cyclooxygenase activity such as PGE_2 and/or prostacyclin were involved. This data followed from a previous report showing that mGluR-mediated Ca^{2+} oscillations in astrocytes results in the pulsatile release of prostaglandin (Zonta et al., 2003a). The results from the vessel experiments, which were obtained from acute brain slices, were also extended to the intact animal. Using Doppler flowmetry to measure CBF, cortical functional hyperemia was induced by forepaw stimulation. The ensuing vasodilation initiated quickly and was dramatically reduced by mGluR antagonists. An important control was performed to show there were no attenuating effects on evoked synaptic potentials in the presence of the antagonists, suggesting incoming signals from the periphery were still relayed to the cortical region examined. However, the block of functional hyperemia by aspirin in the slice condition was not tested in vivo.

Experiments that broadly affect the COX enzymes have been conducted to test their role in cerebrovascular control. Pharmacological inhibition or knockout (KO) of COX-2 dramatically attenuates functional hyperemia-induced vasodilation in response to whisker stimulation (Niwa et al., 2000). In contrast, the same inhibitory actions on COX-1 fail to affect this form of functional hyperemia, but these treatments do attenuate a form of acetylcholine-mediated vasodilation (Niwa et al., 2001). Because COX-2 expression is thought to be low in astrocytes, these results point to neuronal derived COX activity and COX products that are responsible for functional hyperemia.

However, a recent in vivo study conducted by the Nedergaard laboratory, specifically examining astrocytes Ca^{2+} signals in functional hyperemia, supports a role for COX-1 rather than COX-2 (Takano et al., 2006). This study described the effects of uncaging Ca^{2+} in astrocyte end-feet on vasomotor responses in arteries of the somatosensory cortex. Astrocytes were loaded with the Ca^{2+} indicator dye Rhod2-AM and caged Ca^{2+} DM-nitrophen. Uncaging Ca^{2+} within astrocyte end-feet triggered an intracellular Ca^{2+} rise that was followed by the dilation of the adjacent penetrating arteriole. Using this protocol few constrictions were observed. Dilations were found to be blocked by inhibitors of COX-1 but not COX-2 enzymes. Immunohistochemical staining for COX-1 showed intense reactivity at cerebral blood vessels, which was suggested to overlap with glial fibrillary acidic protein (GFAP) positive end-feet. However, it is difficult to rigorously ascertain that COX-1 proteins were indeed in the end-feet vs. being located in closely apposed perivascular microglia or cells of the blood vessels such as endothelial cells. The immunostaining also revealed a lack of COX-1 in GFAP positive processes that were far from the vessels, suggesting that COX-1 expression is, at the very least, localized to the vessel region. Nevertheless, these data support the idea that focal elevation of end-foot Ca^{2+} results in vessel dilation mediated by the AA conversion to vasoactive COX products such as PGE_2.

Similar to the work by Zonta and Carmignoto, the Nelson laboratory has demonstrated that afferent stimulation in acute brain slices causes an increase in astrocyte soma and end-foot Ca^{2+} levels, which can be mitigated by mGluR antagonists (Filosa et al., 2004). However, in the absence of any treatment Nelson's group observed spontaneous and repetitive vasomotion timed with a fluctuating Ca^{2+} signal in SMCs. When stimulating afferents, rather than showing an increase in vessel diameter from a smaller resting state, they show a reduction in the contractile phase of the motor rhythmicity and in the spontaneous Ca^{2+} oscillations. Interestingly, application of mGluR agonist mimicked both observations, but in the presence of mGluR antagonists, only the astrocyte Ca^{2+} signal was significantly reduced when afferents were stimulated. That the reduction in SMC Ca^{2+} oscillations persisted suggests that either there was insufficient block of the mGluRs, that other glutamate receptors were involved or that non-glutamatergic inputs and transmitters were participating in relaying information about the state of the cellular environment to vessels through astrocytes.

The MacVicar laboratory examined the effects of elevations in astrocyte Ca^{2+} alone, without incorporating the involvement of membrane-bound receptors by uncaging Ca^{2+} using two photon photolysis. This technique allowed them to increase Ca^{2+} within the discrete volume of astrocytes without provoking other cell types. Astrocytes were identified in transgenic mice that expressed enhanced green fluorescent protein driven by the GFAP promoter and these cells were loaded with the AM form of Rhod-2, the Ca^{2+} sensitive dye. Under typical brain slice recording conditions in the hippocampus and cortex, Ca^{2+} uncaging in astrocytes induced a Ca^{2+} wave that propagated throughout the astrocyte syncytium and invaded end-feet, and the end-feet Ca^{2+} signal was immediately followed by a constriction of adjacent arterioles. Mulligan and MacVicar found a strong, positive relationship between the extent of the constriction and the extent of the Ca^{2+} signal within and among the end-feet. Astrocyte-mediated vasoconstrictions were blocked by inhibiting the Ca^{2+}-dependent enzyme PLA_2 and preventing the release of the diffusible factor AA. Arteriole constrictions were found to be caused by the generation of 20-HETE within SMCs from the astrocyte derived AA by a CYP450 enzyme (Fig. 18.2). In other studies, 20-HETE has been shown to depolarize SMCs by blocking K^+ channel opening (Lange et al., 1997), and by enhancing Ca^{2+} entry through voltage-gated calcium ion channels (VGCCs) (Gebremedhin et al., 1992). While cultured astrocytes have been demonstrated to synthesize 20-HETE (Nithipatikom et al., 2001), in the brain the w-hydroxylase enzyme that synthesizes 20-HETE is principally expressed within SMCs (Gebremedhin et al., 2000).

A study from the Newman laboratory conducted in retinal explants has examined the effect of Ca^{2+} elevations within retinal glia by using UV photolysis of caged Ca^{2+} and caged 1,4,5-inositol-trisphosphate (IP_3) (Metea and Newman, 2006). Notably, in response to Ca^{2+} uncaging constrictions as well as dilations were observed, while increases in free IP_3 produced mostly dilation. Constrictions, as with the results obtained from the MacVicar laboratory, were dependent on the generation of astrocyte AA and its conversion to 20-HETE. Dilations were also caused by astrocyte AA but instead of its conversion to the potent constrictor molecule, AA was converted

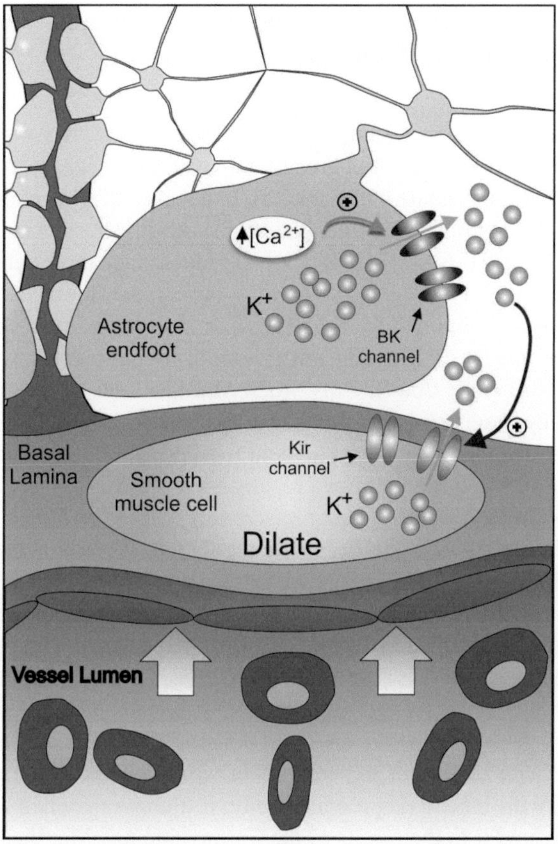

Fig. 18.2 Astrocyte end-feet control of cerebrovasculature dilation by the efflux of potassium from large conductance Ca^{2+} activated K^+ (BK) channels. Elevated extracellular potassium activates Kir channels on smooth muscle cells, which causes dilation via hyperpolarization and subsequent smooth muscle cells relaxation (*See Color Plates*).

to EET by another CYP450 enzyme (Fig. 18.2). These data indicate an intriguing complexity in the dialog between glial cells and arterioles with respect to the potential factors determining the polarity of the response (see below) and the different outcomes observed depending of the method of Ca^{2+} liberation within glia. The EET-induced dilations are supported by other reports in the CNS. In vivo cortical application of 5,6-EET, more so than other epoxyeicosatrienoic acids, causes a large increase in cerebral arteriole diameter (Amruthesh et al., 1993). Others have reported that 8,9-EET and 11,12-EET, rather than 5,6-EET, elicited dose-dependent relaxation of cerebral arteries through activation of SMC K^+ channels (Gebremedhin et al., 1992; Hu and Kim, 1993). Furthermore, pharmacological blockade of the EET generating enzyme P450 reduces resting CBF as measured in vivo (Alkayed et al., 1996).

18.5 NO: An Important Modulator of Astrocyte-Mediated Cerebral Vessel Control

As stated earlier, the first study implicating astrocyte Ca^{2+} signals in the relaxation of vascular tone as a model for functional hyperemia was conducted by the Carmignoto laboratory (Zonta et al., 2003b). In a subset of their experiments blood vessels were preconstricted by incubating the brain slices with *N*(G)-nitro-L-arginine methyl ester (L-NAME) to block NOS and reduce the level of endogenous NO to ultimately enhance the vasodilations observed. This is interesting when compared the results obtained by the MacVicar laboratory, in which arterioles in untreated slices always displayed constriction when astrocytes were stimulated (Mulligan and MacVicar, 2004). Only when Mulligan and MacVicar incubated in L-NAME did they observe vasodilations in response to the mGluR agonist (1*S*, 3*R*)-1-aminocyclopentane-1, 3-dicarboxylic acid (t-ACPD).

On the basis of these dichotomous results Metea and Newman speculated that the levels of NO dictated the type of vasomotor response (Metea and Newman, 2006). This was hypothesized to occur via regulation of the enzymatic conversion of AA to either EET or 20-HETE. Consistent with this idea, vessels that upon first test showed dilations were transformed into constrictions in the presence of the NO donor *S*-nitrosol-*N*-acetylpenicillamine (SNAP), an outcome thought to be due to the NO sensitivity of the EET producing enzyme CYP450 (Fleming, 2001). In the presence of the NO scavenger-phenyl-4,4,5,5-ketramethyl-imidazoline-1-oxyl-3-oxide (PTIO), the opposite was true: vasoconstrictions were converted to vasodilations. In line with this latter experiment in which NO is limiting, blocking all forms of NOS with L-NAME produced only vasodilations when the preparation was stimulated, similar to the results of the Carmignoto and MacVicar laboratories.

An in vivo study conducted by the Nedergaard laboratory also examined a role for NO in astrocyte-mediated dilations by applying l-NAME to block NOS (Takano et al., 2006). This treatment failed to affect the degree of dilation induced when Ca^{2+} was uncaged in astrocyte end-feet. This negative result may be explained by a lower level of endogenous NO in the intact animal. However, NO donors were not tested and it would be interesting to see if by elevating NO levels in vivo the vasomotor response changes polarity from dilation to constriction when astrocytes are stimulated – as has been demonstrated in in vitro experiments.

The work outlined above show that astrocytes wield the necessary physiology to initiate both vasoconstriction and vasodilation mechanisms and that the level of NO may dictate the polarity of the vasomotor response. However, there are many unknowns in the role of NO in mediating the above processes. First, little is known of the importance of astrocyte-derived NO over that of neurons and endothelial cells in functional hyperemia. Second, studies testing the impact of NO on synaptic transmission have demonstrated that NO enhances the release of several neurotransmitters, including glutamate (Prast and Philippu, 2001). This notion also corresponds with data collected in vivo showing a prominent role for enhanced NO release in function hyperemia (Iadecola et al., 1995; Yang et al., 1999). These demonstrations need to be reconciled both with the in vitro data showing higher NO levels promote astrocyte-mediated vasoconstriction and with the recent in vivo data showing astrocyte-mediated vasodilations do not rely on NO production.

18.6 K^+ and Vascular Control by Astrocytes

18.6.1 K^+ Siphoning Through Kir Channels

It was first proposed 20 years ago by the Newman laboratory that K^+ efflux from Kir channels in the end-feet of glia in the retina could lead to blood vessel dilations (Newman et al., 1984; Paulson and Newman, 1987). This was the first hypothesis to point to a possible mechanism by which glial cells could control vascular tone. The hypothesis was based on the observation by Newman that high levels of Kir channels are expressed on the end-feet of Muller cells (Newman, 1984). However, this hypothesis has recently been disproved by Newman's laboratory by two separate tests (Metea et al., 2007). First, they recorded from single glia in close proximity to a vessel and elicited large depolarizations that would be more than sufficient to permit efflux of K^+ through open Kir channels. Under these conditions, they failed to observe any change in vessel diameter. Second, they investigated the extent of K^+-induced vascular effects in the retina in control vs. the Kir KO mouse. When the vasomotor physiology was compared between the two mouse strains, no difference was observed in the degree of K^+-induced dilation. This result was also corroborated by verifying the loss of the inwardly rectifying channel with electrophysiological recordings.

18.6.2 Ca^{2+}-Activated K^+ Release

A role for K^+ channels in astrocyte-mediated regulation of vascular tone may exist through a different mechanism proposed by the Nelson laboratory. In a recent publication, they demonstrate that modest increases in external K^+ promote dilation of vessels through Kir channels in SMCs (Filosa et al., 2006). Higher extracellular K^+ causes hyperpolarization of SMCs by enhancing the Kir conductance, which leads to decreased Ca^{2+} entry, a relaxation of SMCs and the consequent dilation. As with the Newman hypothesis, the high extracellular K^+ trigger that initiates this process is proposed to come not from neurons but directly from astrocyte end-feet. Different from the Newman lab, the effect here is thought to involve end-feet BK channels (Price et al., 2002), which open in response to higher levels of intracellular Ca^{2+} to allow the efflux of K^+ (Fig. 18.2). One potential caveat to this work is the sole use of barium ions to determine a role for the Kir channels. Though barium is commonly used to block Kir channels, this treatment also elicits pronounced depolarizations of astrocytes (Anderson et al., 1995) and can out compete Ca^{2+} for the pore of l-type voltage-gated Ca^{2+} channels, which are prominent in cerebral SMCs (Alborch et al., 1995). The additional use of an imidazole compound to block Kir channels would help support the author's conclusions (Favaloro et al., 2003). Placing an important role on SMC Kir channels is likely not conflicting with the data from the Newman lab showing that there is no difference in K^+-induced dilations between wild type and Kir KO mice because of the difference in the Kir subtypes involved. The predominant family of Kir channels in SMCs is 2.0 (Quayle et al., 1996), while it is the Kir 4.1 subtype that is absent in the Newman KO study (Metea et al., 2007). These data from the Nelson lab also suggest that K^+ channels are not the only players in K^+- or activity-induced vasodilations because a portion of the dilation response was blocked by inhibitors of cyclooxygenase enzymes, suggesting vasoactive PGEs were also contributing.

18.7 Astrocyte Ca^{2+} Signals: Functional Significance?

18.7.1 Astrocyte Ca^{2+}: Initiation and Spread

To initiate Ca^{2+} signals, astrocytes express a variety of membrane bound receptors. As discussed previously, the major contribution from glutamate receptors is attributable to G_q coupled group I mGluRs (Pearce et al., 1986), particularly subtype mGluR5 (Balazs et al., 1997), which has a prominent role in functional hyperemia (Zonta et al., 2003b; Filosa et al., 2004; Mulligan and MacVicar, 2004) (see below). Also pertinent are the ionotropic and metabotropic purinoceptors, which respond to elevated concentrations of extracellular ATP (Jimenez et al., 2000; Kukley et al., 2001). Astrocyte Ca^{2+} signals occur via Ca^{2+} influx through Ca^{2+} permeable ionotropic P2X receptors (Walz et al., 1994) or when Ca^{2+} is released from intracellular stores subsequent to the activation of metabotropic P2Y receptors that couple to phosospolipase C and IP_3 generation (Nakahara et al., 1997). Further studies have expanded these findings to include other neurotransmitters in the induction of Ca^{2+} signals in astrocytes. These include the release of acetylcholine in the hippocampus from cholinergic septal afferents (Araque et al., 2002), the exogenous effects of norepinephrine (Duffy and MacVicar, 1995) and gamma-aminobutyric acid (GABA) (Kang et al., 1998; Serrano et al., 2006) in the hippocampus, as well as the actions of the nonclassical transmitter NO on the Bergmann glial cells of the cerebellum (Matyash et al., 2001) and on astrocytes of the cortex (Bal-Price et al., 2002).

Long-range Ca^{2+} signals in astrocytes are made possible by two principle mechanisms (1) the intracellular diffusion of IP_3 through gap junctional channels (Sneyd et al., 1994; Venance et al., 1997), which allow electrical and cytoplasmic connectivity between adjacent astrocytes and (2) the extracellular paracrine actions of glial-derived ATP (Guthrie et al., 1999). ATP release has been found to occur in continuous waves coinciding with Ca^{2+} wave propagation (Wang et al., 2000), and also in isolated bursts that were highly localized and separated by large distances (Arcuino et al., 2002). The mechanism of ATP release from glial cells is incompletely understood. There are likely many initiating routes including purinergic P2Y receptors (James and Butt, 2002), α_1-adrenoceptors (Gordon et al., 2005) and group I mGluRs (Cornell-Bell et al., 1990), as well as mechanisms of release, which may include vesicular fusion (Bal-Price et al., 2002; Pascual et al., 2005) or hemichannel (Cotrina et al., 1998; Braet et al., 2003) and P2X7 receptor (Duan and Neary, 2006) opening. ATP release from cultured astrocytes has also been shown to be triggered by hypo-osmotic-induced astrocyte swelling through a multidrug resistance protein pathway (Darby et al., 2003). Ca^{2+} waves are thought to be essential contributors relaying synaptic information to vessels. Recent data shedding light on this area has revealed that the situation may be more complicated, which is where we now turn our attention.

18.7.2 Enigmatic Astrocyte Ca^{2+} Signaling

From the studies described above it is clear that astrocyte Ca^{2+} transients, and, in particular, those within the end-feet, have important albeit complex actions on vascular tone. Of paramount importance to all this work, however, is whether a rise in astrocyte intracellular Ca^{2+} is indeed the signal responsible for relaying information on the metabolic state of neurons to the vasculature in order to affect vessel diameter and ultimately CBF. There are several observations concerning the very nature of astrocyte Ca^{2+} signaling that makes it a worth while endeavor to entertain this query. As CBF changes occur in a graded manner in order to match graded changes in synaptic activation, it is expected that astrocyte Ca^{2+} will mirror these changes if this signal is indeed the key factor in mediating functional hyperemia and other forms of neurovascular coupling. However, there are several lines of evidence that reveal a curious inconsistency with this idea. First, Ca^{2+} increases in astrocytes often occur independent of neuronal activity (Nett et al., 2002). These spontaneous Ca^{2+} oscillations have been reported both in vivo (Hirase et al., 2004) and in vitro (Parri and Crunelli, 2001; Parri et al., 2001). Second, the astrocyte Ca^{2+} waves thought to transduce important information toward vessels are not faithfully observed in vivo, suggesting this signaling process may actually be attributable to aspects of the slice or culture condition, or that it manifests more easily in pathophysiological conditions such as epilepsy (Tashiro et al., 2002; Balazsi et al., 2003), rather than being the "normal" method of signaling for neurovascular coupling. In brain slices and in vivo, astrocyte Ca^{2+} signals are far more discrete. Glutamate uncaging on single astrocytes in the hippocampus activates a limited number of astrocytes in a discrete yet complex astrocytic network; complex in the sense that a neighboring astrocyte's proximity to the stimulated one is not an indicator of activation (Sul et al., 2004). Subtle afferent stimulation can cause a highly localized rise in Ca^{2+} in an astrocyte process (Pasti et al., 1997), while stimulating at a higher frequency can activate a few local astrocytes (Porter and McCarthy, 1996). Regardless of the method a broad reaching Ca^{2+} is not generated. However, afferent stimulation can produce what appears to be a discrete Ca^{2+} wave that propagates down astrocytic processes to invade end-feet causing cerebral vessel dilation (Zonta et al., 2003b), suggesting Ca^{2+} waves are utilized in the intact brain on a small scale. Finally, recent work for the Helmchen lab has described a novel two photon scanning method that is capable of measuring Ca^{2+} signals at frequencies up to 10 Hz in hundreds of cells simultaneously (Gobel et al., 2007). This technique has demonstrated that neuronal activity fails to consistently increase astrocyte Ca^{2+} with a temporal profile that follows the timing and patterning of neuronal activation. However, the lack of an obvious Ca^{2+} signal that propagates from synapses to a vessel does not necessarily mean astrocyte Ca^{2+} signals (end-feet excluded) are meaningless with regard to neurovascular coupling. Instead, this may simply reflect our ignorance of why such timing differences arise and how the vastly different astrocyte signals preserve neural information, which we cannot currently decode. It is apparent that more experimentation is needed to tease out the true nature of astrocyte Ca^{2+} signals in the control of the cerebrovasculature.

18.8 Norepinephrine and Astrocyte-Mediated Cerebrovascular Control

One set of experiments that has provided clues toward the functional impact of astrocyte Ca^{2+} signals in controlling CBF comes from the rise in end-feet Ca^{2+} induced by norepinephrine (NE) and the subsequent vessel constriction (Mulligan and MacVicar, 2004). Vasomotor responses cannot be completely abolished by mGluR antagonists when synapses become activated (Filosa et al., 2004) suggesting that transmitters other than glutamate can participate in neurovascular coupling. Along this line, an electron microscope examination of arterioles has revealed that the majority of noradrenergic terminals originating from the locus coeruleus that associate with vessels, synapse on astrocyte end-feet rather than SMCs (Cohen et al., 1997).

Previous in vivo work has demonstrated that activation by NE causes a decrease in CBF (Raichle et al., 1975), an effect that may help maintain CBF at a constant rate at higher blood pressures. In vitro work has shown that NE triggers robust intracellular Ca^{2+} increases in astrocytes via activation of α_1 and β adrenergic receptors (Duffy and MacVicar, 1995). Consistent with these results, experiments conducted by Mulligan and MacVicar (2004) showed that NE-mediated Ca^{2+} increases within astrocyte end-feet temporally preceded prominent vasoconstrictions. When astrocytes were loaded with BAPTA-AM to chelate rises in intracellular Ca^{2+}, vascular constrictions generated by NE were drastically reduced, suggesting Ca^{2+} was critical for the astrocyte-mediated effect. These data add another level of complexity to astrocyte-mediated neurovascular coupling by expanding the realm of vasoactive transmitters beyond that of glutamate.

18.9 Astrocytes in Spreading Depression and Cerebrovascular Constriction

Cortical spreading depression (CSD) is a self-propagating wave of transient cellular depolarization and ionic redistribution followed by synaptic depression. The cellular mechanisms responsible for this phenomenon are incompletely understood, despite both the passage of over 60 years since Leao (1944) first observed the phenomenon and vested scientific interest due to its implications for migraine and the progression of brain tissue injury after stroke or trauma. Of interest here is the in vivo observation that CSD occurs coincident with a reduced CBF in cortical arterioles and capillaries (Hadjikhani et al., 2001; Chuquet et al., 2007). Due to their seemingly uniform anatomical interconnectedness, an ability to generate long distance Ca^{2+} waves and their intimate association with the cerebrovasculature, astrocytes have been implicated in CSD (Smith et al., 2006). Earlier reports show that an astrocyte Ca^{2+} wave temporally precedes a change in the intrinsic optical signal associated with a wave of spreading depression (Basarsky et al., 1998; Kunkler and Kraig, 1998). However, eliminating the astrocytic Ca^{2+} wave with a Ca^{2+} free external solution did not eliminate the CSD wave (Basarsky et al., 1998). A very recent study utilizing in vivo two-photon Ca^{2+} imaging in both neurons and astrocytes also demonstrates the Ca^{2+} wave to precede the CSD wave, but the Ca^{2+} signal increases first in neurons and second in astrocytes, suggesting – as others have – that astrocytes are not driving CSD (Chuquet et al., 2007). However, a consequence of the astrocyte Ca^{2+} wave was a Ca^{2+} increase in astrocyte end-feet, which caused pronounced arteriole constriction. Whereas other studies conducted in vitro and in vivo have shown astrocyte-mediated vasodilation when end-foot Ca^{2+} was elevated during functional hyperemia (Zonta et al., 2003b; Takano et al., 2006), this CSD-induced vasoconstriction was similar to that obtain by Mulligan and MacVicar (2004) in which the focal liberation of Ca^{2+} within end-feet elicited a pronounced decrease in vessel diameter. The reduced CBF that results from this action is consistent with other recent reports showing there is severe hypoxia coincident with CSD because of an increased metabolic demand that exceeds blood supply delivery (Takano et al., 2007).

18.10 Astrocyte-Mediated Vasodilations or Vasoconstrictions?

The fact that end-feet Ca^{2+} signals are capable of initiating constriction or dilation of cerebral blood vessels suggests there are precise physiological circumstances in which each mechanism is recruited. The effect of NO has already been described above as a potential factor dictating the vessel response profile, playing a potential role by modulating the efficacy of critical enzymes such as the CYP450s. Another important consideration is the amount of myogenic tone in the vessels under study. In the in vivo preparation, cerebral arterioles generally exhibit a degree of partial constriction, whereas in in vitro the lack of blood flow and accompanying shear stress fails to produce this to the same extent (Resnick et al., 2003). Thus, vessels equilibrate toward a slightly more relaxed state in acute brain slices compared to the intact brain. This may account for some of the observed differences between the two preparations as dilating factors will be less effective on already dilated vessels and the same for constricting factors on constricted vessels. However, as indicated by the NO results obtained from the Newman laboratory, the situation is more complicated than this. For instance, throughout the “middle portion” of the vessel’s full dynamic range, vessel diameter could change in either direction, even if slightly dilated or slightly constricted, which depended on the relative presence or absence of NO (Metea and Newman, 2006). This raises an interesting question as to how other factors or physiological circumstances will influence the polarity of vasomotion during functional hyperemia. For instance, are there factors that dictate what is released from the astrocyte when end-foot Ca^{2+} becomes elevated, i.e., constricting or dilating agent? From which cell type do such factors arise: astrocytes, SMCs, endothelial cells, or is there a complex interplay between all parties? Alternatively, could both constricting and dilating agents be released from astrocytes simultaneously, with some selection of the preferred agent performed at the SMC level, perhaps influenced by the level of myogenic tone? These will be central questions for the future studies on brain vasculature control.

With the vasoconstrictions observed during CSD we have seen that certain pathophysiological states of the tissue may be important for determining the polarity of vasomotor responses when end-foot Ca^{2+} is elevated. Other influences may include the distance a particular section of vessel is located from the source of enhanced neural activity. Recently it has been demonstrated that cerebral vessels located in the center of a functional hyperemic region of brain tissue dilate but, interestingly, this core is surrounded by a concentric volume of tissue where the residing vessels constrict (Devor et al., 2005). This surround inhibition of vessel diameter is thought to enhance CBF to the oxygen requiring core and may represent the physiological condition in which both astrocyte-mediated constrictions and dilations are simultaneously utilized juxtaposed to each other. How these disparate effects are selected for as a function of distance from the center of the functional hyperemic response is not known.

18.11 Astrocytes in Brain Energetics and the Link to Blood Flow

It is now appreciated that changes in synaptic activity and neuronal spiking correlate with changes in blood oxygen content (Mukamel et al., 2005). These parameters are also associated with regional alterations in CBF (Vogel and Kuschinsky, 1996; Devor et al., 2005) and glucose consumption (Iadecola et al., 1996; Hu and Wilson, 1997b), but the degree and temporal characteristics of these changes are not intuitively obvious. Astrocyte energetics may play an important role linking changes in the metabolic demand of the tissue to changes in CBF. Immediately after the initiation of activity there is transient reduction in vessel oxygen content, which precedes any increase in CBF, suggesting a rapid utilization of oxygen (Ances et al., 2001). This observation is supported by a recent metabolic imaging study, which utilized two-photon microscopy to examine the intrinsic fluorescence of the ubiquitous electron carrier reduced β-nicotinamide adenine dinucleotide (NADH) as a measure of brain metabolic redox state. Examination of this signal showed that at the onset of activity, dendritic oxidative metabolism is enhanced (Kasischke et al., 2004), which is consistent with a rapid consumption of O_2. Interestingly though, this is followed by a prolonged increase in astrocyte glycolytic metabolism. There are a few lines of evidence in support of this dichotomy between the different type of metabolism recruited by neurons and astrocytes. First, although CBF changes occur in close proportion to cerebral glucose utilization, the proportion of increased O_2 consumption is much less (Fox and Raichle, 1986; Fox et al., 1988), suggesting an energy contribution from glycolysis. In the vasculature, the ensuing increase in CBF actually overcompensates for the needed O_2, resulting in an oversupply of oxygenated blood, which is responsible for the blood–oxygen-level-dependent (BOLD) signal observed in fMRI (Ogawa et al., 1990). The recruitment of the glycolytic pathway is also supported by a localized increase in lactate, an end product of anaerobic metabolism (Fellows et al., 1993; Hu and Wilson, 1997a). That astrocytes are the primary anaerobic players and the source of the lactate is supported by the idea that astrocytes have far fewer mitochondria than neurons by virtue of the fact that fine astrocyte processes are too thin to contain them and that these finely ramified extensions comprise the majority of the astrocytes volume. While this may only suggest astrocyte's energy needs are inferior to that of neurons, astrocytes are the predominant source of glycogen in the brain (Ignacio et al., 1990), which is thought to be important for rapid, on-demand supply of ATP via glycongenolysis and subsequent glycolysis (Brown and Ransom, 2007). This supports the notion that astrocytes are inherently primed for glycolysis. Finally, is the concept of the astrocyte–neuron lactate shuttle, in which synaptic glutamate release triggers glutamate and Na^+ ion co-transport via enhanced astrocyte transporter activity? Energy-dependent Na^+–K^+ ATPases work to restore the perturbed ionic gradients, which drive astrocyte glycolysis to generate more ATP. Enhanced glycolysis results in the production and accumulation of lactate, a molecule that can be consumed as fuel in oxidative metabolism. Fittingly, the astrocyte-derived lactate is released and taken up into neurons for this purpose through monocarboxylate transporters (Pellerin and Magistretti, 1994; Pellerin et al., 1998). This hypothesis, while supported, is not without controversy (Pellerin et al., 2007). For instance, the proportionally greater amount of glucose uptake compared to that of O_2 consumption, is not accounted for by an equivalent generation of lactate (Madsen et al., 1999). A recent metabolic imaging study shows that the protracted overshoot in NADH fluorescence observed in response to synaptic activation does not represent the initiation of glycolytic metabolism but instead represents the production of NADH from oxidative efforts (Brennan et al., 2006). The fast dip and subsequent overshoot in the NADH signal have also been suggested to be an artifact of the slice preparation, where O_2 is diffusion limited, because this characteristic response profile is not observed in vivo (Turner et al., 2007). There may be alternative targets for the lactate as well. As mentioned above, the local accumulation of metabolic products is not likely to be the sole mechanism responsible for functional hyperemia, but several studies have shown lactate can affect vessel tone and CBF. In the retina, lactate has direct effects on SMCs via activation of NOS and opening of K^+-ATP channels causing vessel relaxation (Hein et al., 2006). In the CNS, an increase in CBF triggered in response to activity is potentiated by increasing the lactate/pyruvate ratio (Mintun et al., 2004).

The astrocyte–neuron lactate shuttle hypothesis is interesting in light of a recent in vivo study conducted in the olfactory bulb. Using intrinsic optical imaging, in which alterations in metabolism and CBF can be detected by changes in the reflectance of light off the brain when illuminated, Gurden et al. (2006) studied the mechanisms involved in generating intrinsic optical signals (IOSs) evoked by physiological odor presentation. Notably, the authors found no link between odor-evoked IOSs and the activation of ionotropic or metabotropic glutamate receptors. Instead, their findings implicated glutamate uptake through astrocyte transporters as the major contributor (Gurden et al., 2006). These data are consistent with the idea that glutamate uptake is an important trigger for blood vessel dilation. As the astrocyte–neuron lactate shuttle is thought to be initiated by glutamate uptake, the generation and release of lactate may be important for both a neural metabolic substrate and the control of CBF. However, what proportion of the measured olfactory bulb IOSs actually represent a change in CBF due to an increase in vessel diameter is not known. In vitro studies suggest glutamate clearance via transporter activity can induce astrocyte swelling (Hansson et al., 1994). If a similar effect is occurring in the glomeruli when an odor is presented, an appreciable fraction of the IOS change observed may be the result of alterations in light scattering due to astrocyte volume changes (MacVicar and Hochman, 1991). Clearly, more experiments are necessary to determine the precise role of glutamate uptake and oxidative vs. glycolytic metabolism in the control of CBF.

18.12 New Players: Pericytes and Vasoactive Interneurons

A new player in CBF control has recently been described: the pericyte (Peppiatt et al., 2006). Pericytes are small, oval cells that contain contractile proteins and are in direct contact with the endothelial cells comprising the wall of capillaries (Herman and D'Amore, 1985). Individual pericytes are solitary, keeping a fairly regular distance between them. Each cell has processes that project around and encircle the girth of the capillary, enabling a focused control of capillary diameter. Pericytes can elicit pronounced constrictions in response to electrical stimulation, NE and ATP, whereas glutamate triggers pericytes to relax the capillary wall (Peppiatt et al., 2006). Ischemia also results in focal capillary constrictions that correspond to the location of pericytes, suggesting these cells may be responsible for a portion of the reduced CBF observed during this pathological condition. Notably, in spite of a lack of dye coupling between neighboring pericytes, the constriction observed in response to the electrical stimulation of one cell, was also observed at distant pericyte-controlled regions after a few tens of seconds (Peppiatt et al., 2006). This interesting result suggests that pericytes either release their own diffusible factors which travel appreciable distances to affect adjacent pericytes, or pericytes are communicating to each other by utilizing other cell types, which may include the endothelial cells of the capillary to which pericytes are physically connected via gap junctions (Wu et al., 2006) or the surrounding astrocyte syncytium. The latter possibility may explain why pericytes are sensitive to ATP, which is a ubiquitous astrocyte transmitter utilized for long-range paracrine signaling (Guthrie et al., 1999).

A recent set of publications has placed a new role on different subtypes of cortical, GABAergic interneurons whose processes can make close apposition with the walls of microvessels (Tong and Hamel, 2000; Cauli et al., 2004). Notably, depending on the subtype, which was determined by single cell RT-PCR, interneurons induced constrictions or dilations. Activity in somatostatin expressing interneurons triggered constrictions while activity in interneurons expressing vasoactive intestinal polypeptide or nitric oxide synthase elicited dilations. Here again we see a dichotomy in the control of vasomotor responses. This is a stark reminder that the original assumptions underling functional hyperemia – that the degree of blood flow is a simple function of the metabolic state and therefore the level of neuronal activation – need reconsideration.

18.13 Conclusion

The work described here indicates that astrocytes are capable of eliciting changes in vessel diameter in both directions. While there appears to be an important initiating role for glutamate through the activation of group I mGluRs, which raises intraglial Ca^{2+}, other inputs and transmitters are likely involved in neurovascular coupling. An increase in intracellular free Ca^{2+} and the subsequent activation of Ca^{2+} sensitive PLA_2 in astrocytes can trigger a surprisingly diverse array of vasoactive metabolites after the initial production of AA. Constriction occurs when AA is converted to 20-HETE, while dilation results from the conversion of AA to PGE_2 or EET. The enzymes governing the production of these vasoactive products are sensitive to NO, suggesting NO levels may dictate the direction of the vessels response. In addition, a role for Ca^{2+}-activated K^+ channels in astrocyte end-feet and the efflux of K^+ has also been suggested to relax vascular tone by hyperpolarizing SMCs via Kir channels. Astrocytes are proving to be important mediators of neurovascular coupling, but a comprehensive understanding of their part is far from complete. Novel experiments and techniques are unfurling much complexity in processes such as functional hyperemia, which historically was thought to occur by a simple correspondence to neuronal activation.

References

Alborch E, Salom JB, Torregrosa G (1995) Calcium channels in cerebral arteries. Pharmacol Ther 68:1–34

Alkayed NJ, Birks EK, Hudetz AG, Roman RJ, Henderson L, Harder DR (1996) Inhibition of brain P-450 arachidonic acid epoxygenase decreases baseline cerebral blood flow. Am J Physiol 271:H1541–H1546

Alkayed NJ, Birks EK, Narayanan J, Petrie KA, Kohler-Cabot AE, Harder DR (1997) Role of P-450 arachidonic acid epoxygenase in the response of cerebral blood flow to glutamate in rats. Stroke 28:1066–1072

Amiry-Moghaddam M, Ottersen OP (2004) The molecular basis of water transport in the brain. Nat Rev Neurosci 12:991–1001

Amruthesh SC, Boerschel MF, McKinney JS, Willoughby KA, Ellis EF (1993) Metabolism of arachidonic acid to epoxyeicosatrienoic acids, hydroxyeicosatetraenoic acids, and prostaglandins in cultured rat hippocampal astrocytes. J Neurochem 61:150–159

Ances BM, Buerk DG, Greenberg JH, Detre JA (2001) Temporal dynamics of the partial pressure of brain tissue oxygen during functional forepaw stimulation in rats. Neurosci Lett 306:106–110

Anderson S, Brismar T, Hansson E (1995) Effect of external K+, Ca2+, and Ba2+ on membrane potential and ionic conductance in rat astrocytes. Cell Mol Neurobiol 15:439–450

Andrew RD, Labron MW, Boehnke SE, Carnduff L, Kirov SA (2007) Physiological evidence that pyramidal neurons lack functional water channels. Cereb Cortex 17:787–802

Araque A, Martin ED, Perea G, Arellano JI, Buno W (2002) Synaptically released acetylcholine evokes Ca2 + elevations in astrocytes in hippocampal slices. J Neurosci 22:2443–2450

Arcuino G, Lin JH, Takano T, Liu C, Jiang L, Gao Q, Kang J, Nedergaard M (2002) Intercellular calcium signaling mediated by point-source burst release of ATP. Proc Natl Acad Sci USA 99:9840–9845

Balazs R, Miller S, Romano C, de Vries A, Chun Y, Cotman CW (1997) Metabotropic glutamate receptor mGluR5 in astrocytes: Pharmacological properties and agonist regulation. J Neurochem 69:151–163

Balazsi G, Cornell-Bell AH, Moss F (2003) Increased phase synchronization of spontaneous calcium oscillations in epileptic human versus normal rat astrocyte cultures. Chaos 13:515–518

Bal-Price A, Moneer Z, Brown GC (2002) Nitric oxide induces rapid, calcium-dependent release of vesicular glutamate and ATP from cultured rat astrocytes. Glia 40:312–323

Basarsky TA, Duffy SN, Andrew RD, MacVicar BA (1998) Imaging spreading depression and associated intracellular calcium waves in brain slices. J Neurosci 18:7189–7199

Basarsky TA, Feighan D, MacVicar BA (1999) Glutamate release through volume-activated channels during spreading depression. J Neurosci 19:6439–6445

Benyo Z, Gorlach C, Wahl M (1998a) Involvement of thromboxane A2 in the mediation of the contractile effect induced by inhibition of nitric oxide synthesis in isolated rat middle cerebral arteries. J Cereb Blood Flow Metab 18:616–618

Benyo Z, Gorlach C, Wahl M (1998b) Role of nitric oxide and thromboxane in the maintenance of cerebrovascular tone. Kidney Int Suppl 67:S218–220

Braet K, Vandamme W, Martin PE, Evans WH, Leybaert L (2003) Photoliberating inositol-1,4,5-trisphosphate triggers ATP release that is blocked by the connexin mimetic peptide gap 26. Cell Calcium 33:37–48

Brennan AM, Connor JA, Shuttleworth CW (2006) NAD(P)H fluorescence transients after synaptic activity in brain slices: Predominant role of mitochondrial function. J Cereb Blood Flow Metab 26:1389–1406

Brown AM, Ransom BR (2007) Astrocyte glycogen and brain energy metabolism. Glia 55:1263–1271

Bushong EA, Martone ME, Jones YZ, Ellisman MH (2002) Protoplasmic astrocytes in CA1 stratum radiatum occupy separate anatomical domains. J Neurosci 22:183–192

Calka J, Wolf G (2003) An ultrastructural demonstration of NADPH-diaphorase/nitric oxide synthase activity in the rat striatal astroglia. Folia Morphol (Warsz) 62:251–253
Cauli B, Tong XK, Rancillac A, Serluca N, Lambolez B, Rossier J, Hamel E (2004) Cortical GABA interneurons in neurovascular coupling: Relays for subcortical vasoactive pathways. J Neurosci 24:8940–8949
Chuquet J, Hollender L, Nimchinsky EA (2007) High-resolution in vivo imaging of the neurovascular unit during spreading depression. J Neurosci 27:4036–4044
Cohen Z, Molinatti G, Hamel E (1997) Astroglial and vascular interactions of noradrenaline terminals in the rat cerebral cortex. J Cereb Blood Flow Metab 17:894–904
Cornell-Bell AH, Finkbeiner SM, Cooper MS, Smith SJ (1990) Glutamate induces calcium waves in cultured astrocytes: Long-range glial signaling. Science 247:470–473
Cotrina ML, Lin JH, Alves-Rodrigues A, Liu S, Li J, Azmi-Ghadimi H, Kang J, Naus CC, Nedergaard M (1998) Connexins regulate calcium signaling by controlling ATP release. Proc Natl Acad Sci USA 95:15735–15740
Dani JW, Chernjavsky A, Smith SJ (1992) Neuronal activity triggers calcium waves in hippocampal astrocyte networks. Neuron 8:429–440
Darby M, Kuzmiski JB, Panenka W, Feighan D, MacVicar BA (2003) ATP released from astrocytes during swelling activates chloride channels. J Neurophysiol 89:1870–1877
Devor A, Ulbert I, Dunn AK, Narayanan SN, Jones SR, Andermann ML, Boas DA, Dale AM (2005) Coupling of the cortical hemodynamic response to cortical and thalamic neuronal activity. Proc Natl Acad Sci USA 102:3822–3827
Duan S, Neary JT (2006) P2X(7) receptors: Properties and relevance to CNS function. Glia 54:738–746
Duffy S, MacVicar BA (1995) Adrenergic calcium signaling in astrocyte networks within the hippocampal slice. J Neurosci 15:5535–5550
Ellis EF, Wei EP, Cockrell CS, Choi S, Kontos HA (1983) The effect of PGF2 alpha on in vivo cerebral arteriolar diameter in cats and rats. Prostaglandins 26:917–923
Ellis EF, Police RJ, Yancey L, McKinney JS, Amruthesh SC (1990) Dilation of cerebral arterioles by cytochrome P-450 metabolites of arachidonic acid. Am J Physiol 259:H1171–H1177
Faraci FM (1989) Effects of endothelin and vasopressin on cerebral blood vessels. Am J Physiol 257:H799–H803
Farooqui AA, Yang HC, Rosenberger TA, Horrocks LA (1997) Phospholipase A2 and its role in brain tissue. J Neurochem 69:889–901
Favaloro JL, Andrews KL, McPherson GA (2003) Novel imidazoline compounds that inhibit Kir-mediated vasorelaxation in rat middle cerebral artery. Naunyn Schmiedebergs Arch Pharmacol 367:397–405
Fellows LK, Boutelle MG, Fillenz M (1993) Physiological stimulation increases nonoxidative glucose metabolism in the brain of the freely moving rat. J Neurochem 60:1258–1263
Filosa JA, Bonev AD, Nelson MT (2004) Calcium dynamics in cortical astrocytes and arterioles during neurovascular coupling. Circ Res 95:e73–81
Filosa JA, Bonev AD, Straub SV, Meredith AL, Wilkerson MK, Aldrich RW, Nelson MT (2006) Local potassium signaling couples neuronal activity to vasodilation in the brain. Nat Neurosci 9:1397–1403
Fischer G, Kettenmann H (1985) Cultured astrocytes form a syncytium after maturation. Exp Cell Res 159:273–279
Fleming I (2001) Cytochrome p450 and vascular homeostasis. Circ Res 89:753–762
Fox PT, Raichle ME (1986) Focal physiological uncoupling of cerebral blood flow and oxidative metabolism during somatosensory stimulation in human subjects. Proc Natl Acad Sci USA 83:1140–1144
Fox PT, Raichle ME, Mintun MA, Dence C (1988) Nonoxidative glucose consumption during focal physiologic neural activity. Science 241:462–464
Gebremedhin D, Ma YH, Falck JR, Roman RJ, VanRollins M, Harder DR (1992) Mechanism of action of cerebral epoxyeicosatrienoic acids on cerebral arterial smooth muscle. Am J Physiol 263:H519–H525

Gebremedhin D, Lange AR, Lowry TF, Taheri MR, Birks EK, Hudetz AG, Narayanan J, Falck JR, Okamoto H, Roman RJ, Nithipatikom K, Campbell WB, Harder DR (2000) Production of 20-HETE and its role in autoregulation of cerebral blood flow. Circ Res 87:60–65

Gebremedhin D, Yamaura K, Zhang C, Bylund J, Koehler RC, Harder DR (2003) Metabotropic glutamate receptor activation enhances the activities of two types of Ca2 + -activated k + channels in rat hippocampal astrocytes. J Neurosci 23:1678–1687

Gobel W, Kampa BM, Helmchen F (2007) Imaging cellular network dynamics in three dimensions using fast 3D laser scanning. Nat Methods 4:73–79

Gordon GR, Baimoukhametova DV, Hewitt SA, Rajapaksha WR, Fisher TE, Bains JS (2005) Norepinephrine triggers release of glial ATP to increase postsynaptic efficacy. Nat Neurosci 8:1078–1086

Gurden H, Uchida N, Mainen ZF (2006) Sensory-evoked intrinsic optical signals in the olfactory bulb are coupled to glutamate release and uptake. Neuron 52:335–345

Guthrie PB, Knappenberger J, Segal M, Bennett MV, Charles AC, Kater SB (1999) ATP released from astrocytes mediates glial calcium waves. J Neurosci 19:520–528

Haber M, Zhou L, Murai KK (2006) Cooperative astrocyte and dendritic spine dynamics at hippocampal excitatory synapses. J Neurosci 26:8881–8891

Hadjikhani N, Sanchez Del Rio M, Wu O, Schwartz D, Bakker D, Fischl B, Kwong KK, Cutrer FM, Rosen BR, Tootell RB, Sorensen AG, Moskowitz MA (2001) Mechanisms of migraine aura revealed by functional MRI in human visual cortex. Proc Natl Acad Sci USA 98:4687–4692

Hansson E, Johansson BB, Westergren I, Ronnback L (1994) Glutamate-induced swelling of single astroglial cells in primary culture. Neuroscience 63:1057–1066

Hein TW, Xu W, Kuo L (2006) Dilation of retinal arterioles in response to lactate: Role of nitric oxide, guanylyl cyclase, and ATP-sensitive potassium channels. Invest Ophthalmol Vis Sci 47:693–699

Herman IM, D'Amore PA (1985) Microvascular pericytes contain muscle and nonmuscle actins. J Cell Biol 101:43–52

Hirase H, Qian L, Bartho P, Buzsaki G (2004) Calcium dynamics of cortical astrocytic networks in vivo. PLoS Biol 2:E96

Hu S, Kim HS (1993) Activation of K + channel in vascular smooth muscles by cytochrome P450 metabolites of arachidonic acid. Eur J Pharmacol 230:215–221

Hu Y, Wilson GS (1997a) A temporary local energy pool coupled to neuronal activity: Fluctuations of extracellular lactate levels in rat brain monitored with rapid-response enzyme-based sensor. J Neurochem 69:1484–1490

Hu Y, Wilson GS (1997b) Rapid changes in local extracellular rat brain glucose observed with an in vivo glucose sensor. J Neurochem 68:1745–1752

Iadecola C, Li J, Ebner TJ, Xu X (1995) Nitric oxide contributes to functional hyperemia in cerebellar cortex. Am J Physiol 268:R1153–R1162

Iadecola C, Li J, Xu S, Yang G (1996) Neural mechanisms of blood flow regulation during synaptic activity in cerebellar cortex. J Neurophysiol 75:940–950

Ignacio PC, Baldwin BA, Vijayan VK, Tait RC, Gorin FA (1990) Brain isozyme of glycogen phosphorylase: Immunohistological localization within the central nervous system. Brain Res 529:42–49

Ishimoto H, Matsuoka I, Nakanishi H, Nakahata N (1996) A comparative study of arachidonic acid metabolism in rabbit cultured astrocytes and human astrocytoma cells (1321N1). Gen Pharmacol 27:313–317

James G, Butt AM (2002) P2Y and P2X purinoceptor mediated Ca2 + signalling in glial cell pathology in the central nervous system. Eur J Pharmacol 447:247–260

Jimenez AI, Castro E, Delicado EG, Miras-Portugal MT (1998) Potentiation of adenosine 5 -triphosphate calcium responses by diadenosine pentaphosphate in individual rat cerebellar astrocytes. Neurosci Lett 246:109–111

Jimenez AI, Castro E, Communi D, Boeynaems JM, Delicado EG, Miras-Portugal MT (2000) Coexpression of several types of metabotropic nucleotide receptors in single cerebellar astrocytes. J Neurochem 75:2071–2079

Kang J, Jiang L, Goldman SA, Nedergaard M (1998) Astrocyte-mediated potentiation of inhibitory synaptic transmission. Nat Neurosci 1:683–692

Kasischke KA, Vishwasrao HD, Fisher PJ, Zipfel WR, Webb WW (2004) Neural activity triggers neuronal oxidative metabolism followed by astrocytic glycolysis. Science 305:99–103

Koyama Y, Mizobata T, Yamamoto N, Hashimoto H, Matsuda T, Baba A (1999) Endothelins stimulate expression of cyclooxygenase 2 in rat cultured astrocytes. J Neurochem 73:1004–1011

Kukley M, Barden JA, Steinhauser C, Jabs R (2001) Distribution of P2X receptors on astrocytes in juvenile rat hippocampus. Glia 36:11–21

Kunkler PE, Kraig RP (1998) Calcium waves precede electrophysiological changes of spreading depression in hippocampal organ cultures. J Neurosci 18:3416–3425

Lange A, Gebremedhin D, Narayanan J, Harder D (1997) 20-Hydroxyeicosatetraenoic acid-induced vasoconstriction and inhibition of potassium current in cerebral vascular smooth muscle is dependent on activation of protein kinase C. J Biol Chem 272:27345–27352

Leao AAP (1944) Sperading depression of activity in the cerebral cortex. J Neurophysiol 7:359–390

Lokkegaard A, Nyengaard JR, West MJ (2001) Stereological estimates of number and length of capillaries in subdivisions of the human hippocampal region. Hippocampus 11:726–740

Lou HC, Edvinsson L, MacKenzie ET (1987) The concept of coupling blood flow to brain function: Revision required? Ann Neurol 22:289–297

Luo J, Lindstrom CL, Donahue A, Miller MW (2001) Differential effects of ethanol on the expression of cyclo-oxygenase in cultured cortical astrocytes and neurons. J Neurochem 76:1354–1363

MacCumber MW, Ross CA, Snyder SH (1990) Endothelin in brain: Receptors, mitogenesis, and biosynthesis in glial cells. Proc Natl Acad Sci USA 87:2359–2363

MacVicar BA, Hochman D (1991) Imaging of synaptically evoked intrinsic optical signals in hippocampal slices. J Neurosci 11:1458–1469

Madsen PL, Cruz NF, Sokoloff L, Dienel GA (1999) Cerebral oxygen/glucose ratio is low during sensory stimulation and rises above normal during recovery: Excess glucose consumption during stimulation is not accounted for by lactate efflux from or accumulation in brain tissue. J Cereb Blood Flow Metab 19:393–400

Massa PT, Mugnaini E (1982) Cell junctions and intramembrane particles of astrocytes and oligodendrocytes: A freeze-fracture study. Neuroscience 7:523–538

Matyash V, Filippov V, Mohrhagen K, Kettenmann H (2001) Nitric oxide signals parallel fiber activity to Bergmann glial cells in the mouse cerebellar slice. Mol Cell Neurosci 18:664–670

Metea MR, Newman EA (2006) Glial cells dilate and constrict blood vessels: A mechanism of neurovascular coupling. J Neurosci 26:2862–2870

Metea MR, Kofuji P, Newman EA (2007) Neurovascular coupling is not mediated by potassium siphoning from glial cells. J Neurosci 27:2468–2471

Mintun MA, Vlassenko AG, Rundle MM, Raichle ME (2004) Increased lactate/pyruvate ratio augments blood flow in physiologically activated human brain. Proc Natl Acad Sci USA 101:659–664

Mukamel R, Gelbard H, Arieli A, Hasson U, Fried I, Malach R (2005) Coupling between neuronal firing, field potentials, and FMRI in human auditory cortex. Science 309:951–954

Mulligan SJ, MacVicar BA (2004) Calcium transients in astrocyte endfeet cause cerebrovascular constrictions. Nature 431:195–199

Murphy S, Simmons ML, Agullo L, Garcia A, Feinstein DL, Galea E, Reis DJ, Minc-Golomb D, Schwartz JP (1993) Synthesis of nitric oxide in CNS glial cells. Trends Neurosci 16:323–328

Nagelhus EA, Mathiisen TM, Ottersen OP (2004) Aquaporin-4 in the central nervous system: Cellular and subcellular distribution and coexpression with KIR4.1. Neuroscience 129:905–913

Nakahara K, Okada M, Nakanishi S (1997) The metabotropic glutamate receptor mGluR5 induces calcium oscillations in cultured astrocytes via protein kinase C phosphorylation. J Neurochem 69:1467–1475

Nett WJ, Oloff SH, McCarthy KD (2002) Hippocampal astrocytes in situ exhibit calcium oscillations that occur independent of neuronal activity. J Neurophysiol 87:528–537
Newman EA (1984) Regional specialization of retinal glial cell membrane. Nature 309:155–157
Newman EA (2003) Glial cell inhibition of neurons by release of ATP. J Neurosci 23:1659–1666
Newman EA, Frambach DA, Odette LL (1984) Control of extracellular potassium levels by retinal glial cell K + siphoning. Science 225:1174–1175
Nithipatikom K, Grall AJ, Holmes BB, Harder DR, Falck JR, Campbell WB (2001) Liquid chromatographic-electrospray ionization-mass spectrometric analysis of cytochrome P450 metabolites of arachidonic acid. Anal Biochem 298:327–336
Niwa K, Araki E, Morham SG, Ross ME, Iadecola C (2000) Cyclooxygenase-2 contributes to functional hyperemia in whisker-barrel cortex. J Neurosci 20:763–770
Niwa K, Haensel C, Ross ME, Iadecola C (2001) Cyclooxygenase-1 participates in selected vasodilator responses of the cerebral circulation. Circ Res 88:600–608
Ogawa S, Lee TM, Kay AR, Tank DW (1990) Brain magnetic resonance imaging with contrast dependent on blood oxygenation. Proc Natl Acad Sci USA 87:9868–9872
Oomagari K, Buisson B, Dumuis A, Bockaert J, Pin JP (1991) Effect of glutamate and ionomycin on the release of arachidonic acid, prostaglandins and HETEs from cultured neurons and astrocytes. Eur J Neurosci 3:928–939
Parri HR, Crunelli V (2001) Pacemaker calcium oscillations in thalamic astrocytes in situ. Neuroreport 12:3897–3900
Parri HR, Gould TM, Crunelli V (2001) Spontaneous astrocytic Ca^{2+} oscillations in situ drive NMDAR-mediated neuronal excitation. Nat Neurosci 4:803–812
Pasantes-Morales H (1996) Volume regulation in brain cells: Cellular and molecular mechanisms. Metab Brain Dis 11:187–204
Pascual O, Casper KB, Kubera C, Zhang J, Revilla-Sanchez R, Sul JY, Takano H, Moss SJ, McCarthy K, Haydon PG (2005) Astrocytic purinergic signaling coordinates synaptic networks. Science 310:113–116
Paspalas CD, Papadopoulos GC (1996) Ultrastructural relationships between noradrenergic nerve fibers and non-neuronal elements in the rat cerebral cortex. Glia 17:133–146
Pasti L, Volterra A, Pozzan T, Carmignoto G (1997) Intracellular calcium oscillations in astrocytes: A highly plastic, bidirectional form of communication between neurons and astrocytes in situ. J Neurosci 17:7817–7830
Paulson OB, Newman EA (1987) Does the release of potassium from astrocyte endfeet regulate cerebral blood flow? Science 237:896–898
Pearce B, Albrecht J, Morrow C, Murphy S (1986) Astrocyte glutamate receptor activation promotes inositol phospholipid turnover and calcium flux. Neurosci Lett 72:335–340
Pellerin L, Magistretti PJ (1994) Glutamate uptake into astrocytes stimulates aerobic glycolysis: A mechanism coupling neuronal activity to glucose utilization. Proc Natl Acad Sci USA 91:10625–10629
Pellerin L, Pellegri G, Bittar PG, Charnay Y, Bouras C, Martin JL, Stella N, Magistretti PJ (1998) Evidence supporting the existence of an activity-dependent astrocyte-neuron lactate shuttle. Dev Neurosci 20:291–299
Pellerin L, Bouzier-Sore AK, Aubert A, Serres S, Merle M, Costalat R, Magistretti PJ (2007) Activity-dependent regulation of energy metabolism by astrocytes: An update. Glia 55:1251–1262
Peppiatt CM, Howarth C, Mobbs P, Attwell D (2006) Bidirectional control of CNS capillary diameter by pericytes. Nature 443:700–704
Porter JT, McCarthy KD (1995) GFAP-positive hippocampal astrocytes in situ respond to glutamatergic neuroligands with increases in $[Ca2+]i$. Glia 13:101–112
Porter JT, McCarthy KD (1996) Hippocampal astrocytes in situ respond to glutamate released from synaptic terminals. J Neurosci 16:5073–5081
Prast H, Philippu A (2001) Nitric oxide as modulator of neuronal function. Prog Neurobiol 64:51–68
Price DL, Ludwig JW, Mi H, Schwarz TL, Ellisman MH (2002) Distribution of rSlo Ca2+ -activated K+ channels in rat astrocyte perivascular endfeet. Brain Res 956:183–193

Quayle JM, Dart C, Standen NB (1996) The properties and distribution of inward rectifier potassium currents in pig coronary arterial smooth muscle. J Physiol 494(Pt 3):715–726

Raichle ME, Hartman BK, Eichling JO, Sharpe LG (1975) Central noradrenergic regulation of cerebral blood flow and vascular permeability. Proc Natl Acad Sci USA 72:3726–3730

Resnick N, Yahav H, Shay-Salit A, Shushy M, Schubert S, Zilberman LC, Wofovitz E (2003) Fluid shear stress and the vascular endothelium: For better and for worse. Prog Biophys Mol Biol 81:177–199

Roy CS, Sherrington CS (1890) On the regulation of the blood supply of the brain. J Physiol 11:85–108

Serrano A, Haddjeri N, Lacaille JC, Robitaille R (2006) GABAergic network activation of glial cells underlies hippocampal heterosynaptic depression. J Neurosci 26:5370–5382

Shi Y, Liu X, Gebremedhin D, Falck JR, Harder DR, Koehler RC (2008) Interaction of mechanisms involving epoxyeicosatrienoic acids, adenosine receptors, and metabotropic glutamate receptors in neurovascular coupling in rat whisker barrel cortex. J Cereb Blood Flow Metab 28(1):111–125

Simard M, Nedergaard M (2004) The neurobiology of glia in the context of water and ion homeostasis. Neuroscience 129:877–896

Simard M, Arcuino G, Takano T, Liu QS, Nedergaard M (2003) Signaling at the gliovascular interface. J Neurosci 23:9254–9262

Smith JM, Bradley DP, James MF, Huang CL (2006) Physiological studies of cortical spreading depression. Biol Rev Camb Philos Soc 81:457–481

Sneyd J, Charles AC, Sanderson MJ (1994) A model for the propagation of intercellular calcium waves. Am J Physiol 266:C293–C302

Stella N, Estelles A, Siciliano J, Tence M, Desagher S, Piomelli D, Glowinski J, Premont J (1997) Interleukin-1 enhances the ATP-evoked release of arachidonic acid from mouse astrocytes. J Neurosci 17:2939–2946

Sul JY, Orosz G, Givens RS, Haydon PG (2004) Astrocytic Connectivity in the Hippocampus. Neuron Glia Biol 1:3–11

Takano T, Tian GF, Peng W, Lou N, Libionka W, Han X, Nedergaard M (2006) Astrocyte-mediated control of cerebral blood flow. Nat Neurosci 9:260–267

Takano T, Tian GF, Peng W, Lou N, Lovatt D, Hansen AJ, Kasischke KA, Nedergaard M (2007) Cortical spreading depression causes and coincides with tissue hypoxia. Nat Neurosci 10:754–762

Tashiro A, Goldberg J, Yuste R (2002) Calcium oscillations in neocortical astrocytes under epileptiform conditions. J Neurobiol 50:45–55

Tong XK, Hamel E (2000) Basal forebrain nitric oxide synthase (NOS)-containing neurons project to microvessels and NOS neurons in the rat neocortex: Cellular basis for cortical blood flow regulation. Eur J Neurosci 12:2769–2780

Turner DA, Foster KA, Galeffi F, Somjen GG (2007) Differences in O2 availability resolve the apparent discrepancies in metabolic intrinsic optical signals in vivo and in vitro. Trends Neurosci 30:390–398

Venance L, Stella N, Glowinski J, Giaume C (1997) Mechanism involved in initiation and propagation of receptor-induced intercellular calcium signaling in cultured rat astrocytes. J Neurosci 17:1981–1992

Ventura R, Harris KM (1999) Three-dimensional relationships between hippocampal synapses and astrocytes. J Neurosci 19:6897–6906

Vogel J, Kuschinsky W (1996) Decreased heterogeneity of capillary plasma flow in the rat whisker-barrel cortex during functional hyperemia. J Cereb Blood Flow Metab 16:1300–1306

Walz W, Gimpl G, Ohlemeyer C, Kettenmann H (1994) Extracellular ATP-induced currents in astrocytes: Involvement of a cation channel. J Neurosci Res 38:12–18

Wang Z, Haydon PG, Yeung ES (2000) Direct observation of calcium-independent intercellular ATP signaling in astrocytes. Anal Chem 72:2001–2007

Wiencken AE, Casagrande VA (1999) Endothelial nitric oxide synthetase (eNOS) in astrocytes: Another source of nitric oxide in neocortex. Glia 26:280–290

Wu DM, Minami M, Kawamura H, Puro DG (2006) Electrotonic transmission within pericyte-containing retinal microvessels. Microcirculation 13:353–363

Yang G, Chen G, Ebner TJ, Iadecola C (1999) Nitric oxide is the predominant mediator of cerebellar hyperemia during somatosensory activation in rats. Am J Physiol 277:R1760–R1770

Zonta M, Sebelin A, Gobbo S, Fellin T, Pozzan T, Carmignoto G (2003a) Glutamate-mediated cytosolic calcium oscillations regulate a pulsatile prostaglandin release from cultured rat astrocytes. J Physiol 553:407–414

Zonta M, Angulo MC, Gobbo S, Rosengarten B, Hossmann KA, Pozzan T, Carmignoto G (2003b) Neuron-to-astrocyte signaling is central to the dynamic control of brain microcirculation. Nat Neurosci 6:43–50

Abbreviations

AA	Arachidonic acid
ATP	Adenosine 5′-triphosphate
CBF	Cerebral blood flow
COX	Cyclooxygenase
CSD	Cortical spreading depression
CYP450	Cytochrome P450
EET	Epoxyeicosatrienoic acid
fMRI	Functional magnetic resonance imaging
GABA	Gamma-aminobutyric acid
GFAP	Glial fibrillary acidic protein
20-HETE	20-Hydroxyeicosatetraenoic acid
IOS	Intrinsic optical signal
IP_3	1,4,5-Inositol-trisphosphate
KO	Knockout
L-NAME	*N*(G)-Nitro-L-arginine methyl ester
mGluR	Metabotropic glutamate receptor
NADH	Reduced β-nicotinamide adenine dinucleotide
NE	Norepinephrine
NO	Nitric oxide
NOS	NO synthase
PGE_2	Prostaglandin E_2
PLA_2	Phospholipase A_2
SMCs	Smooth muscle cells
t-ACPD	(1*S*, 3*R*)-1-Aminocyclopentane-1, 3-dicarboxylic acid
VGCCs	Voltage-gated calcium channels
VRACs	Volume-regulated anion channels

Chapter 19
A Role for Glial Cells of the Neuroendocrine Brain in the Central Control of Female Sexual Development

Alejandro Lomniczi and Sergio R. Ojeda

Contents

It is now well established that astrocytes are active participants of the process by which information is generated and disseminated within the central nervous system (CNS). In the hypothalamus, astrocytes and ependymoglial cells of the median eminence, known as tanycytes, regulate the secretory activity of neuroendocrine neurons. A developmental process in which they are prominently involved is the neuroendocrine control of puberty. Mammalian puberty is initiated by an increase in pulsatile release of the decapeptide gonadotropin hormone-releasing hormone (GnRH) from a specialized subset of hypothalamic neuroendocrine neurons. Although a critical determinant of this increase is a coordinated change in the activity of neuronal networks synaptically connected to GnRH neurons, glial cells contribute to the process via two related mechanisms. One requires production of growth factors acting via receptors endowed with serine–threonine kinase

A. Lomniczi and S.R. Ojeda
Division of Neuroscience, Oregon National Primate Research Center, Oregon Health and Science University, Beaverton, OR, USA
lomniczi@ohsu.edu, ojedas@ohsu.edu

V. Parpura and P.G. Haydon (eds.), *Astrocytes in (Patho)Physiology of the Nervous System*,
DOI: 10.1007/978-0-387-79492-1_19, © Springer Science+Business Media, LLC 2009

or tyrosine kinase activity. The other involves plastic rearrangements of glia–GnRH neuron adhesiveness. A neuron-to-glia regulatory pathway is, in turn, provided by glutamatergic neurons which facilitate astrocytic signaling mediated by erythroblastosis B (erbB) receptors. Genetic disruption of these receptors, which mediate the actions of members of the epidermal growth factor (EGF) family of trophic factors, delays female sexual development due to impaired erbB ligand-induced glial prostaglandin E_2 (PGE_2) release. The adhesiveness of glial cells to GnRH neurons appears to involve two different cell–cell communications systems, one provided by the homophilic interactions of Synaptic Cell Adhesion Molecule (SynCAM), and the other resulting from the interaction of neuronal contactin with glial Receptor-like Protein Tyrosine Phosphatase-β (RPTPβ). Because both systems are endowed with signaling capabilities, the interaction of glial cells with GnRH neurons may not only involve secreted bioactive molecules, but also the activation of cell–cell signaling mechanisms by cell–surface adhesive molecules forming different types of intercellular junctions.

19.1 Neuroendocrine Control of Sexual Development: General Aspects

The medial basal hypothalamus in primates (Plant and Witchel, 2006), and the pre-optic area (POA) in rodents (Ojeda and Skinner, 2006), contains 800–1,000 neurosecretory neurons that govern pituitary gonadotropin secretion via release of the decapeptide GnRH. GnRH reaches the pituitary gland via a vascular route provided by portal vessels connecting the median eminence of the hypothalamus to the pituitary gland. Pituitary gonadotropes respond to GnRH with release of the gonadotropins luteinizing hormone (LH) and follicle-stimulating hormone (FSH), which act on the gonads to stimulate the secretion of steroids and peptides, and promote the maturation of germ cells. In turn, gonadal hormones control the release of GnRH and gonadotropins via negative and positive feedback mechanisms. In primates and rodents, pulsatile gonadotropin secretion increases in a diurnal fashion at the end of juvenile development, signaling the initiation of the pubertal process (for reviews see Ojeda and Terasawa, 2002; Plant, 2002).

What are the mechanisms underlying the pubertal increase in GnRH release? Work from several laboratories have led to the conclusion that the increased GnRH pulsatility results not from an intrinsic alteration in GnRH neuronal activity, or a modification in gonadal feedback control, but as a consequence of coordinated changes in transsynaptic and glial inputs to the GnRH neuronal network. While the transsynaptic changes involve a synchronized increase in excitatory inputs and a reduction in inhibitory influences, the glial component of the system mostly consists of an activation of facilitatory signals. Although both the neuronal and glial networks controlling GnRH secretion are responsive to gonadal steroids (Mong and McCarthy, 1999; Mong and Blutstein, 2006; Ojeda and Skinner, 2006; Garcia-Segura and McCarthy, 2004), the early changes in activity leading to the initiation of puberty are thought to be gonad independent (Ojeda et al., 2003; Ojeda and Terasawa, 2002; Plant and Witchel, 2006).

The neuronal networks controlling GnRH secretion are multiple (Ojeda, 1994; Kalra and Crowley, 1992; Levine et al., 1991) and subject to the modulatory influence of gonadal steroids (Herbison, 1998). The most important excitatory component of this transsynaptic system is provided by glutamatergic neurons and the newly discovered kisspeptin-producing neurons (Ojeda and Skinner, 2006; Dungan et al., 2006). The inhibitory counterpart is mostly supplied by γ-aminobutyric acid (GABA), but also by opioid peptides (Terasawa and Fernandez, 2001). Although GABA may inhibit GnRH secretion mainly by acting on neuronal subsets connected to the GnRH neuronal network (Ojeda and Skinner, 2006; Terasawa and Fernandez, 2001), it can also exert direct excitatory effects on GnRH neurons (DeFazio et al., 2002). In contrast to this dual excitatory–inhibitory control, glial cells influence GnRH secretion mostly via stimulatory, growth factor-dependent cell–cell signaling loops that directly or indirectly promote GnRH secretion (Ojeda and Skinner, 2006; Ojeda et al., 2003; Mahesh et al., 2006). As in the case of the neurons, glial activity is also modulated by gonadal steroids (Ojeda and Skinner, 2006; Garcia-Segura and McCarthy, 2004; Mong and Blutstein, 2006).

19.2 Glial–neuronal Interactions in the Hypothalamus

19.2.1 Plastic Rearrangements in Glial–GnRH Neuron Connectivity

Hatton et al. (1984) were the first to demonstrate the existence of a plastic relationship between astrocytes and neuroendocrine neurons. While the focus of that pioneering study was on vasopressin neurons, in the last 15 years a wealth of evidence has accumulated indicating that astroglial cells are also physically and functionally linked to GnRH neurons (reviewed in Ojeda and Skinner, 2006; Garcia-Segura and McCarthy, 2004; Mong and McCarthy, 1999).

GnRH neurons are profusely apposed by astrocytes; at the median eminence, tanycytes are major contributors to this apposition (Kozlowski and Coates, 1985). Both morphological relationships are regulated by gonadal steroids (reviewed in Ojeda et al., 2000; Mong and Blutstein, 2006); studies in the rat have shown that at the level of the GnRH neuronal cell bodies, located in the POA, the apposing astrocytic surface area varies in a diurnal fashion. When estrogen levels are elevated in the morning of proestrus, the apposition decreases (Cashion et al., 2003). This reduction in contact area may reflect a change in glial engagement, switching from neuronal cell bodies to the dendritic tree. Astrocytes associate dynamically and preferentially to postsynaptic dendritic spines (Haber et al., 2006), which contain almost exclusively glutamatergic synapses. An increased glial apposition of GnRH dendritic spines would result in facilitation of excitatory inputs to GnRH neurons, an interpretation inferentially supported by a recent finding showing that the glutamatergic input to GnRH neurons, as determined by the abundance of dendritic spines, is not only much more abundant than previously recognized, but also increases during sexual development (Cottrell et al., 2006).

In primates, steroids exert an effect on the glial apposition to GnRH neuronal perikarya similar to that seen in rats, as ovariectomy of adult rhesus monkeys increases the surface area of glial processes apposing GnRH cell bodies, whereas administration of estradiol reverses this change (Witkin et al., 1991). Prepubertal female monkeys, which produce little estrogen, also show an extensive glial apposition of GnRH neuronal perikarya, like that detected after ovariectomy (Witkin et al., 1995). Interestingly, the effects of estrogen on astrocyte morphology are not the same everywhere in the hypothalamus. In the arcuate nucleus, the astrocytic surface apposing non-GnRH neurons increases when estradiol levels are high (Garcia-Segura and McCarthy, 2004), instead of decreasing. Because the increased glial apposition of neurons observed in the rat arcuate nucleus in the presence of high estrogen levels is accompanied by a decreased number of axo-somatic synapses, which are mostly GABAergic (Garcia-Segura and McCarthy, 2004), it would appear that, at the time of proestrous, astrocytes in this region of the hypothalamus act to reduce the inhibitory synaptic input to neuronal subsets synaptically connected to GnRH neurons.

Studies of the plastic changes that occur in the median eminence throughout the rat estrous cycle have shown that more GnRH neuronal terminals make physical contact with the pericapillary space in proestrus (when estrogen levels are high) than in diestrous II (when estrogen levels are low) (King and Rubin, 1996; Prevot et al., 1999). This rearrangement results from changes in tanycyte plasticity, first reported more than 20 years ago (Zimmermann, 1982). Tanycytes, which line the ventral part of the third ventricle (Kozlowski and Coates, 1985; Witkin et al., 1991; Silverman et al., 1994; King and Letourneau, 1994; Rodriguez et al., 2005), use "end-feet" specializations to prevent the direct contact of GnRH terminals with the portal vasculature (Kozlowski and Coates, 1985; King and Letourneau, 1994). This blockade is transient and subject to remodeling according to the phase of the estrous cycle: During the preovulatory surge of gonadotropins, the end-feet retract allowing the GnRH terminals to reach the endothelial wall, presumably resulting in greater GnRH release into the portal blood [(King and Rubin, 1996); reviewed in (Prevot, 2002)]. Two recent studies have shed light into the cellular mechanisms underlying this plasticity. One of these studies showed that transforming growth factor alpha (TGFα)-mediated activation of erbB1 receptors in tanycytes of the median eminence promotes tanycytic outgrowth, and a PGE_2-dependent production of transforming growth factor beta1 (TGFβ1), which in turn elicits retraction of the tanycytic processes (Prevot et al., 2003a). By first promoting the outgrowth of tanycytic processes, and then the retraction of the processes via TGFβ1 release, TGFα mimics the morphological plasticity displayed by tanycytes during the preovulatory surge of GnRH.

The other study showed that purified endothelial cells of the median eminence induce acute actin cytoskeleton remodeling in ependymoglial cells, and this plasticity is mediated by nitric oxide (NO), a diffusible factor released from endothelial cells (De Seranno et al., 2004). Both soluble guanylyl cyclase and cyclooxygenase products appear to be mediators of this endothelial-dependent control of ependymoglial cytoarchitecture.

19.2.2 Molecules Mediating Glial–GnRH Neuron Adhesiveness

The above considerations make it evident that GnRH neurons and astrocytes maintain an intimate contact throughout development and adult life. However, the cell–surface molecules that may contribute to this adhesiveness remain largely unknown. Recent studies identified two sets of molecules involved in mediating glia–GnRH neuron adhesiveness. One of these sets utilizes the glycosylphosphatidyl inositol (GPI)-anchored protein contactin (a cell–surface neuronal protein required for axonal–glial adhesiveness), and RPTP β, a transmembrane phosphatase with structural similarities to cell adhesion molecules, as adhesive partners (Parent et al., 2007) (Fig. 19.1). Using single cell reverse transcriptase–polymerase chain reaction of enhanced green fluorescence protein-tagged GnRH neurons, this study showed that 80% of these cells express the contactin gene. It also showed that the RPTPβ mRNA species predominantly expressed in hypothalamic astrocytes encodes an RPTPβ isoform (short RPTPβ) that uses its carbonic anhydrase (CAH) extracellular subdomain to interact with neuronal contactin. Immunoreactive contactin was found to be abundant in GnRH nerve terminals projecting to both the organum vasculosum of the lamina terminalis (OVLT) and median eminence, implying that GnRH axons are an important site of contactin-dependent cell adhesiveness. GT1-7 immortalized GnRH neurons were found to adhere to the CAH domain of RPTPβ. Disruption of contactin GPI anchoring or immunoneutralization of contactin inhibited GT1-7 cell adhesiveness to the CAH substrate, indicating that RPTPβ adhesion to GnRH neurons is mediated by contactin. Because the abundance of short RPTPβ mRNA increases in the female mouse hypothalamus before puberty, the above-described results suggested that an increased interaction between GnRH axons and astrocytes mediated by RPTPβ–contactin may represent a dynamic mechanism of neuron–glia communication during female sexual development.

In ongoing studies, quantitative proteomics was used to identify hypothalamic proteins that might be down- or up-regulated in a mouse model of delayed puberty (Prevot et al., 2003b). Puberty is delayed in these mutant mice due to an astrocytic-specific defect in erbB4 signaling resulting from the transgenic expression of a dominant-negative erbB4 receptor under the control of the GFAP promoter (Prevot et al., 2003b). The results indicated that the content of SynCAM1, an adhesion molecule recently shown to be important for synaptic assembly (Biederer et al., 2002), was prominently reduced in the mutant hypothalamus in comparison with wild-type mice. Further analyses revealed that SynCAM1 is not only expressed in neurons, but also in astrocytes, and showed that both SynCAM1 mRNA and SynCAM1 protein content are reduced in hypothalamic astrocytes of the mutant

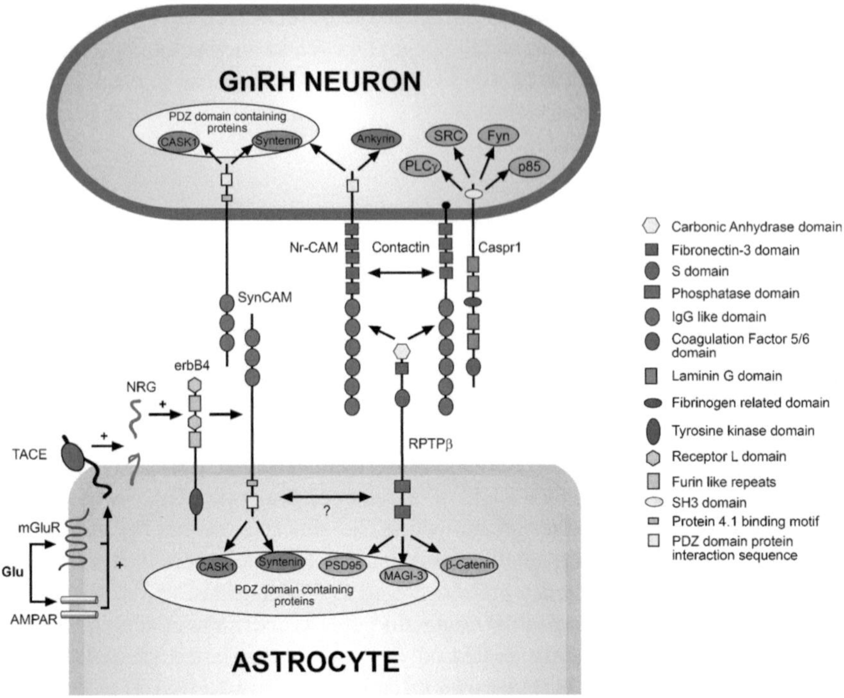

Fig. 19.1 Adhesion molecules postulated to mediate glia–GnRH neuron communication in the neuroendocrine brain. Adherence of GnRH neurons to astrocytes is thought to occur via (1) heterophilic interactions, mediated by the binding of neuronal contactin to the glial receptor RPTPβ and (2) homophilic interactions mediated by the oligomerization of SynCAM1 molecules present in glial cells with SynCAM1 expressed in GnRH neurons. Each of these two systems has signaling capabilities suggesting that, in addition to providing an adhesive interaction, they can regulate astrocyte and GnRH neuron intracellular processes, via the intracellular signaling molecules shown in the figure. It is not known if the contactin/ RPTPβ and the SynCAM1 adhesive modules are functionally linked to each other (question mark) and/or directly regulated by neuronal inputs, such as glutamate. Glutamate may affect SynCAM1 function via activation of an astrocytic AMPA/ metabotropic receptors/erbB4-dependent signaling pathway that includes heterodimerization of activated erbB4 receptors with SynCAM1 (*See Color Plates*).

mice. Yeast-two hybrid assays and immunoprecipitation experiments showed that SynCAM1 is physically associated with ligand-activated erbB4 receptor in hypothalamic astrocytes. Both astrocytes and the GnRH neuronal cell line GT1-7 express the same SynCAM1 isoform, suggesting that SynCAM1 may be a required component of cell–cell communication between GnRH neurons and their astrocytic entourage. Support for this notion came from in vitro adhesion assays demonstrating adherence of both hypothalamic astrocytes and GT1-7 cells to the SynCAM1 extracellular domain. This domain contains three IgG-like regions that provide adhesive properties to the protein. Altogether, these results indicate that GnRH neurons may adhere to astrocytes via both heterophilic (contactin/RPTPβ) and homophilic (SynCAM1/SynCAM1) interactions (Fig. 19.1).

Importantly, both systems have signaling capabilities (Biederer et al., 2002; Peles et al., 1997) suggesting that in addition to providing an adhesive interaction, they can also regulate astrocyte and GnRH neuron intracellular processes. Such intercellular communication pathways appear to be bidirectional. On the neuronal side, contactin can give rise to intracellular signaling via interactions with associated proteins containing signaling motifs (Fig. 19.1). Contactin binds in *cis* (i.e., via lateral interactions in the same neuronal plasma membrane) to Caspr1, whose cytoplasmic domain contains a proline-rich sequence with a canonical SH3 domain that associates with at least four SH domain-containing proteins, including Src, Fyn, p85, and PLCγ (Peles et al., 1995; Zisch et al., 1995). The RPTPβ–contactin complex has also been shown to recruit in *cis* the cell adhesion molecule Nr-CAM to promote neurite outgrowth (Sakurai et al., 1997), and RPTPβ itself appears to bind Nr-CAM directly (Grumet, 1997). On the glial side, RPTPβ interacts with cytoskeletal proteins involved in the regulation of cellular plasticity, including PSD95 (Kawachi et al., 1999) and β-catenin (Meng et al., 2000). MAGI-3, a PDZ domain-containing scaffolding protein localized to focal adhesion sites in astrocytes and regions of the cell membrane enriched in E-cadherin, has been shown to interact with the cytoplasmic domain of RPTPβ (Adamsky et al., 2003), suggesting that MAGI-3 is a scaffolding protein that links RPTPβ to its substrates at the astrocytic membrane. SynCAM1, on the other hand, binds directly to the intracellular kinase domain of erbB4 receptors (unpublished results) and, via its C-terminus PDZ domain recognition motif, to PDZ domain-containing proteins involved in the control of cellular plasticity, such as calcium/calmodulin-dependent serine protein kinase (CASK1) and syntenin, a scaffold protein that also binds to kainate receptor subunits via its PDZ-domains (Biederer et al., 2002; Hirbec et al., 2005) (Fig. 19.1). Both CASK and syntenin function as adaptor proteins linking cell–surface adhesion molecules to the cell's cytoskeleton (Biederer and Sudhof, 2001; Hirbec et al., 2005). Identifying the cellular processes regulated by these adhesive-signaling pathways in astrocytes should provide a fruitful line of research in years to come.

19.2.3 Glial to GnRH Neuron Signaling

19.2.3.1 Growth Factors

Hypothalamic astroglial cells synthesize and release several growth factors including $TGF\beta_1$, basic fibroblast growth factor (bFGF), insulin-like growth factor-1 (IGF-1), and members of the EGF family of trophic factors [reviewed in (Ojeda and Skinner, 2006; Mahesh et al., 2006)]. TGFβ1 acting on GnRH neurons via receptors endowed with serine–threonine kinase activity stimulates directly the synthesis and release of GnRH [(Melcangi et al., 1995; Galbiati et al., 1996); reviewed in (Mahesh et al., 2006)]. Growth factors signaling via receptors with tyrosine kinase activity include basic bFGF, which acts via FGF receptors type I to promote GnRH neuronal differentiation and survival (Voigt et al., 1996; Tsai et al., 1995, 2005), and enhance GnRH processing (Wetsel et al., 1996); IGF-I, which stimulates GnRH release (Hiney et al., 1991) by binding to IGF-1 receptors located on GnRH neurons (Olson et al., 1995); and TGFα and neuregulins (NRGs), two members of the EGF family that elicit GnRH secretion indirectly via the activation of erbB receptors located on astroglial and ependymoglial cells (Voigt et al., 1996; Ma et al., 1997, 1999).

To date, ten genes have been found to encode EGF or EGF-like ligands (Buonanno and Fischbach, 2001; Rimer, 2003; Falls, 2003). Among them, the NRGs are the most complex as they are composed of four different subfamilies containing highly conserved EGF-like domains. There are 15 different isoforms of NRG1, generated by alternative splicing of the primary transcript; all of them are abundant in neurons and glia. The second group of NRGs, known as NRG2s, is formed by two members, NRG2α and NRG2β (Chang et al., 1997; Carraway et al., 1997), which are mostly expressed in neurons of the CNS. NRG3 and NRG4 are the most divergent of the NRGs family, with only NRG3 being expressed in the nervous system (Rimer, 2003; Falls, 2003).

TGFα and NRGs are synthesized as membrane-anchored peptides and participate in diverse cell contact-dependent processes such as adhesion, migration, survival, and differentiation (Carpenter and Cohen, 1990; Massague, 1990; Burden and Yarden, 1997). TGFα and NRGs, like all members of the EGF family, bind to their cognate receptors on adjacent cells upon proteolytic cleavage of the mature peptides from their membrane-bound precursors (Sahin et al., 2004; Peschon et al., 1998; Montero et al., 2000). All EGF-like peptides signal through a family of four transmembrane tyrosine kinase receptors known as erbB receptors. ErbB1 binds at least six different ligands including, EGF, TGFα, amphiregulin, heparin-binding EGF-like growth factor (HB-EGF), epiregulin, and betacellulin (Carpenter and Cohen, 1990; Riese et al., 1996a; Shelly et al., 1998). ErbB3 and erbB4 bind all the members of the NRG subfamilies (Burden and Yarden, 1997; Chang et al., 1997; Carraway et al., 1997; Zhang et al., 1997), as well as epiregulin (Shelly et al., 1998) and betacellulin (Riese et al., 1996a). Thus far, no ligand has been described for erbB2. Instead, erbB2 is recruited as a co-receptor (Karunagaran et al., 1996) by each of the other erbB receptors after ligand binding (Beerli et al., 1995; Riese et al., 1996b).

The Glial TGFα/erbB1 Signaling Complex

The first study dealing with the role of EGF in the hypothalamic control of the pituitary gland was performed by Miyake et al. (1985), demonstrating that EGF-induced LH release from pituitary explants was only elicited when the explants were coincubated with hypothalamic tissue. Years later, definitive evidence for the direct action of TGFα and EGF on GnRH release from median eminence explants was provided (Ojeda et al., 1990). The stimulatory effect of TGFα was shown to require erbB1 receptors, as it was suppressed by the pharmacological inhibition of erbB tyrosine kinase activity (Ojeda et al., 1990). These receptors were found to be expressed on astroglial cells of the median eminence and tanycytes of the third ventricle, but not on GnRH neurons (Ma et al., 1994d; Voigt et al., 1996). These observations led to the hypothesis that TGFα synthesized in hypothalamic astrocytes activates, in a juxtacrine and/or paracrine fashion, erbB1 receptors located in neighboring glial cells (Fig. 19.2). According to this notion, activation of astrocytic erbB1 receptors would result in release of a bioactive substance(s) able to stimulate GnRH release. Prostaglandin E_2 was identified as one such molecule (Ma et al., 1997) (Fig. 19.2).

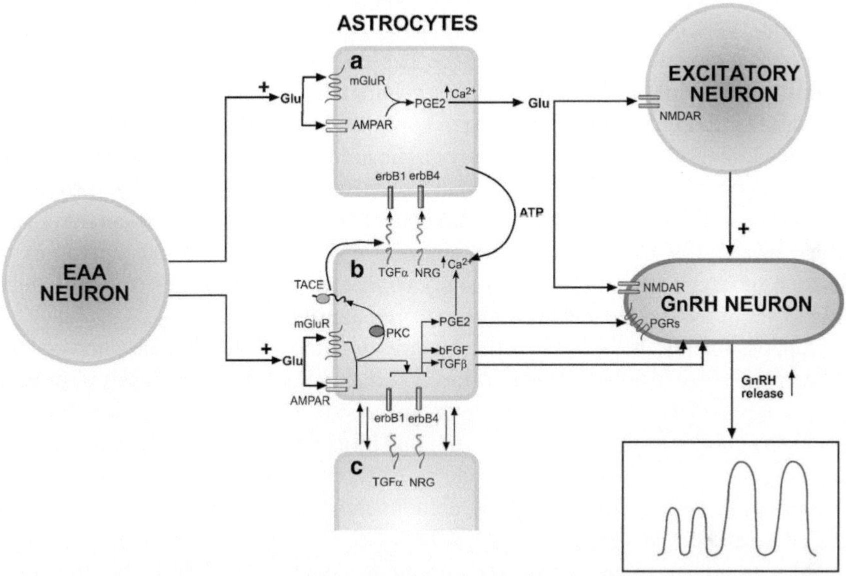

Fig. 19.2 Postulated cell–cell signaling mechanisms involved in hypothalamic neuron–glia reciprocal communication. Astrocyte A mostly depicts information obtained by other authors working with glial cells derived from regions of the brain other than the hypothalamus. Astrocyte B contains information derived from studies using hypothalamic astrocytes. The partially drawn astrocyte C is shown for completeness of the concept that astrocytes contain both erbB receptors and their respective ligands. Glutamate acting via metabotropic and AMPA receptors causes PGE_2 release (Bezzi et al., 1998), and calcium (Ca^{+2}) waves (Fields and Stevens-Graham, 2002; Haydon, 2001), in addition to a ligand/TACE-dependent activation of erbB receptor signaling (Dziedzic et al., 2003). The calcium waves initiated by neuronal glutamate are PGE_2-dependent; however, they can also be initiated spontaneously in the absence of neuronal inputs (Parri et al., 2001). Propagation of the calcium waves within astrocytic networks requires ATP and gap junction communication (Fields and Stevens, 2000; Fields and Stevens-Graham, 2002; Haydon, 2001). PGE_2 causes glutamate release in a calcium-dependent manner (Bezzi et al., 1998). In turn, glial glutamate activates NMDA receptors on neighboring neurons (Parpura et al., 1994), presumably also including those GnRH neurons that express these receptors. In addition, PGE_2 acts directly on GnRH neurons to stimulate GnRH release (Berg-von der Emde et al., 1993). Like glutamate, activation of erbB1 and erbB4 receptor signaling by their respective ligands TGFα and neuregulins (NRGs) elicits PGE_2 release (Ma et al., 1997; Ma et al., 1999); however, this effect occurs at a much later time than that of glutamate. Activation of erbB-signaling also causes increased synthesis of both TGFβ1 and bFGF (Galbiati et al., 2002; Prevot et al., 2003a), two growth factors involved in the regulation of GnRH neuronal function. Modified from (Ojeda et al., 2003) with permission (*See Color Plates*).

During normal female sexual development in the rat, there is a transient increase of hypothalamic TGFα mRNA levels during the second week of life (Ma et al., 1992), when gonadotropin secretion is elevated, and at the time of puberty, with highest levels detected at the time of the first preovulatory surge of gonadotropins, when maximal GnRH secretion occurs (Ma et al., 1992). Because pharmacological blockade of erbB1 receptors targeted to the median eminence of immature female rats delayed sexual maturation, it was concluded that activation of erbB1 receptors is required for puberty to occur at a normal time, and that the median eminence is a major site of regulation of GnRH secretion by activated glial erbB1 receptors (Ma et al., 1992). These conclusions were subsequently supported by two findings: first, transgenic mice overexpressing the TGFα gene under the control of a heavy metal-inducible promoter had an earlier initiation of estrous cyclicity, and an increased GnRH output in comparison to control animals (Ma et al., 1994c). Second, hypothalamic grafts of cells genetically engineered to secrete TGFα accelerated the onset of puberty in female rats only when the cells were implanted in close proximity of GnRH cell bodies in the POA or GnRH nerve terminals in the median eminence (Rage et al., 1997a). These studies implied that the pathological activation of discrete subsets of astrocytes functionally connected to the GnRH network may be able to set in motion the pubertal process prematurely.

Support for this hypothesis came from two studies. One of them showed that puberty-inducing lesions of the anterior hypothalamic area in rats, result in prompt activation of TGFα and erbB1 receptor expression in astrocytes surrounding the lesion site (Junier et al., 1991; 1993), and that infusion of an erbB1 receptor blocker into the lesion site prevented the advancing effect of the lesion on sexual maturation (Junier et al., 1991). The other study demonstrated the presence of a rich network of TGFα- and erbB1-expressing astrocytes in two hypothalamic hamartomas associated with advancement of puberty in humans (Jung et al., 1999). Hypothalamic hamartomas are non-neoplastic malformations of the medial basal hypothalamus frequently associated with sexual precocity. Astrocytomas rich in TGFα but located in areas away from the GnRH network, were not associated with sexual precocity, suggesting that discrete foci of glial activation in the proximity of GnRH neurons may represent an important factor contributing to the etiology of idiopathic sexual precocity of central origin in human females.

The Glial Neuregulin–erbB2/4 Signaling Complex

All members of the EGF family induce homo- or heterodimerization of their corresponding receptor (Carraway and Cantley, 1994; Burden and Yarden, 1997). The erbB2 receptor is recruited after activation of erbB1, erbB3, or erbB4 (Akiyama et al., 1988; Beerli et al., 1995; Karunagaran et al., 1996; Zhang et al., 1996; Wallasch et al., 1995; Riese et al., 1996b; Wada et al., 1990). One of the functions of this co-receptor is to increase the affinity of the EGF ligands to their receptors (Tzahar et al., 1997) by prolonging their dissociation rates (Karunagaran et al., 1996).

Cultured hypothalamic astrocytes express NRG1, NRG3 as well as erbB2 and erbB4 receptors, but not NRG2 or erbB3 (Ma et al., 1999). Exposure of astrocytes to NRGβ1 or TGFα results in phosphorylation of erbB4 and erbB1 receptors, respectively. In addition, erbB2 receptors are cross-phosphorylated. The outcome of this activation is an enhanced production of PGE_2 (Ma et al., 1999). Exposure of hypothalamic astrocytes to suboptimal doses of TGFα and NRGβ1, which independently were ineffective in stimulating PGE_2 release, resulted in a synergistic effect when administered together. This observation suggests that astrocytic erbB2 receptors play an important role in amplifying intracellular signals initiated by TGFα and NRGs (Fig. 19.2). Additional experiments showed that selective in vitro blockade of astrocytic erbB2 synthesis with an antisense oligodeoxynucleotide prevented both the stimulatory effect of NRGβ1 on PGE_2 release and the increase in GnRH secretion elicited by astrocyte culture medium conditioned by NRGβ1 (Ma et al., 1999).

Consistent with the presence of erbB2 and erbB4 receptors in cultured astrocytes, immunohistochemical and in situ hybridization studies demonstrated the presence of erbB2 mRNA and protein in hypothalamic astrocytes and tanycytes of the third ventricle/median eminence, and erbB4 in astrocytes, but not in tanycytes (Ma et al., 1999). In addition, there was scattered neuronal expression of erbB4 receptors in several regions of the hypothalamus, including the arcuate, ventromedial, and paraventricular nucleus. As in the case of TGFα (Ma et al., 1992) hypothalamic erbB2 and erbB4 mRNA abundance increases during juvenile development in female rats, when circulating sex steroid levels are very low, and then again at the time of the preovulatory surge of gonadotropins (Ma et al., 1999). This secondary increase can be reproduced by treating immature female rats with estrogen and progesterone to induce a premature gonadotropin surge. Thus hypothalamic expression of the erbB2/4 complex during female sexual development appears to be regulated by a dual mechanism consisting of an initial, sex steroid-independent, activation and a subsequent stimulatory component that requires sex steroids to operate.

The physiological importance of these changes is highlighted by the delayed onset of puberty observed in animals in which erbB2 synthesis was disrupted in vivo by an intraventricular infusion of the same antisense oligonucleotide that blocked erbB2 synthesis in vitro (Ma et al., 1999). Thus, either disruption of erbB1 receptor signaling in the median eminence (Ma et al., 1992) or erbB2 in the hypothalamus delays the onset of female puberty in rats.

Studies using transgenic mice were carried out to functionally dissect the relative contribution of the NRG/erbB4 and TGFα/erbB1 signaling systems in the hypothalamic control of female sexual development. Transgenic mice carrying a transgene containing a dominant negative form of the erbB4 receptor that lacks the intracellular domain and is under the control of the GFAP promoter were generated and shown to express the truncated receptors selectively in astrocytes (Prevot et al., 2003b). The mutant mice exhibited reduced plasma gonadotropin levels and delayed puberty. Because disruption of astrocytic erbB2/4 signaling was accompanied by normal erbB1 function, these findings established the importance of a functional astrocytic NRG/erbB4 signaling system for the normal initiation of puberty in the mouse (Prevot et al., 2003b). In another study, mice carrying this dominant negative form of the erbB4 receptor were crossed to animals carrying an inactivating point mutation of the erbB1 receptor (Luetteke et al., 1994). The resulting double transgenic mice showed impaired erbB1 and erbB4 signaling in astrocytes, a further delay in the onset of puberty, and a striking decrease in adult reproductive capacity, in comparison to their wild-type and single mutant littermates (Prevot et al., 2005). These studies demonstrated that the integrity of both erbB1 and erbB4 signaling systems in hypothalamic astrocytes is critical for glial cells to engage in cell–cell interactions that facilitate GnRH secretion during female sexual maturation (Fig. 19.2).

19.2.3.2 Other Glial-Derived Factors

In addition to the growth factors mentioned earlier, astrocytes produce and release tumor necrosis factor alpha (Beattie et al., 2002), and activity-dependent growth factor (Blondel et al., 2000), plus a variety of substances other than growth factors, including D-serine (Mothet et al., 2005), cholesterol (Mauch et al., 2001), thrombospondins (Christopherson et al., 2005), neuropeptides, cytokinins, adenosine 5′-triphosphate (ATP), prostaglandins, and glutamate (Barres, 1991; Fields and Burnstock, 2006; Martin, 1992).

As mentioned earlier, glial PGE_2 is a major mediator of the stimulatory actions that TGFα and NRGs exert on GnRH release. However, astrocyte-conditioned medium stripped of PGE_2 is still able to stimulate GnRH release from a GnRH producing cell line, indicating that astrocytes release additional factors capable of stimulating GnRH release (Ma et al., 1997). Among the nonpeptidergic molecules produced by astrocytes, Ca^{2+}, glutamate, and ATP have emerged as central players in the cell–cell signaling system used by astrocytes to regulate neuronal function (Araque et al., 1999; Fields and Burnstock, 2006). While Ca^{2+} can spread to adjacent astrocytes via gap junctions (Nedergaard et al., 2003; Haydon, 2001), ATP and glutamate are released to the intercellular space via membrane channels and vesicles, as a result of increasing Ca^{2+} levels, and affect neuronal function upon binding to specific receptors (Parpura et al., 1994; Fields and Stevens, 2000; Cotrina et al., 2000; Fields and Burnstock, 2006). In the primate hypothalamus, GnRH neurons respond to extracellular ATP, acting via P2X2 and P2X4 receptors, with an immediate increase in intracellular Ca^{2+} and release of GnRH (Terasawa et al., 2005).

During the past years it became clear that astroglial Ca^{2+}, glutamate, ATP, and prostaglandin signaling systems are inextricably linked. ATP and glutamate activate Ca^{2+} mobilization in astrocytes (Fellin et al., 2006; Fields and Burnstock, 2006); intracellular Ca^{2+} increases can cause release of glutamate, which in turn stimulates PGE_2 formation (Zonta et al., 2003). PGE_2 elicits further release of astrocytic glutamate (Bezzi et al., 1998), which enhances astroglial release of arachidonic acid (Stella et al., 1994). In turn, arachidonic acid inhibits glutamate uptake into astrocytes (Barbour et al., 1989), thereby increasing the half life of the neurotransmitter in the synapse.

19.3 Hypothalamic Astrocytes and Glutamate Metabolism

Two complementary studies have recently showed that glutamate metabolism changes in the hypothalamus in response to preovulatory levels of estradiol produced in adult mice (Blutstein et al., 2006) and during the normal onset of female puberty in rats (Roth et al., 2006). One of these studies showed that shortly after estradiol administration (2 h), there is an increased expression of glutamine synthetase (GS) in the hypothalamus. GS is almost exclusively expressed in astroglial cells, where it catalyzes the conversion of glutamate to glutamine (Erecinska and Silver, 1990). Because astrocyte-derived glutamine is converted back to glutamate in neurons, and this glutamate is the principal source of vesicular GABA release at inhibitory synapses (Liang et al., 2006), an increased synthesis of GS at the time of estrogen negative feedback, is likely to result in increased GABA availability to the synaptic cleft (Liang et al., 2006), which would then inhibit GnRH secretion.

The other study showed that the hypothalamus of female rats releases more glutamate at the time of the first preovulatory gonadotropin surge than during juvenile development (Roth et al., 2006). Quantitative proteomics analysis of hypothalamic proteins revealed that at the time of the surge there are opposite changes in the abundance of two enzymes expressed in glial cells: GS, and glutamate dehydrogenase (GDH), which reversibly catalyzes the synthesis of glutamate from α-ketoglutarate. While GS abundance decreases, GDH content increases. In contrast, the content of the neuron-specific glutamate-synthesizing enzyme phosphate-activated glutaminase (PaG) remained unchanged, indicating that the major changes in glutamate metabolism that occur in the hypothalamus at the time of female puberty are astroglia-specific. The changes in GDH and GS protein expression seen in the hypothalamus during the afternoon of proestrus are likely to reflect an increased glutamatergic excitatory input to GnRH neurons (Roth et al., 1998, 2006), because they are accompanied by increased glutamate release in response to blockade of glutamate re-uptake transporters (Roth et al., 2006).

A decrease in GS abundance concomitant with an increase in glutamate release is, however, difficult to reconcile with the well-established drop in glutamate levels that occurs globally in presynaptic terminals in response to blockade of GS activity (Laake et al., 1995). Other factors are, therefore, likely to play a role. We suspect that a changing microenvironment determined by the ability of astrocytic processes to stabilize dendritic spines (Haber et al., 2006; Nishida and Okabe, 2007) is a determining factor. Because astrocytic processes prefer to contact dendritic spines, instead of presynaptic terminals (Lehre and Rusakov, 2002), they may develop stronger and more stable interactions with subsets of larger dendritic spines (Haber et al., 2006) during the afternoon of proestrus, as they retract from GnRH cell bodies and GnRH axonal terminals. This would ensure the availability of glutamine to those preferred postsynaptic sites, which are not only endowed with enhanced glutamate transport capabilities (Huang and Bergles, 2004), but also exhibit a greater probability of glutamate release (Harris and Sultan, 1995).

19.4 Neuron-to-Glia Communication in the Hypothalamus

It is well established that neurons regulate glial activity using various bioactive molecules, such as neurotransmitters, ATP, and neuropeptides (Fields and Burnstock, 2006; Newman, 2003; Hertz and Zielke, 2004). Hypothalamic astrocytes contain both metabotropic receptors (mGluRs) of the mGluR5 subtype and ionotropic α-amino-3-hydroxy-5-methyl-isoxazole-4-propionic acid (AMPA) glutamatergic receptors. As in excitatory synapses, these receptors are physically associated with their respective clustering/interacting proteins, Homer and PICK1. They also form a complex with erbB1 and erbB4 receptors (Dziedzic et al., 2003); concomitant stimulation of metabotropic and AMPA receptors results in mobilization of erbB receptors to the cell surface, association of these receptors with their respective ligands TGFα and NRGs, and erbB receptor phosphorylation.

Studies using other cellular systems (Dong et al., 1999; Prenzel et al., 1999), including hypothalamic astrocytes (Dziedzic et al., 2003), have shown that for this ligand–receptor interaction to occur the membrane-bound TGFα and NRG precursors need to be first cleaved by a metalloproteinase activity that makes the mature peptides available for interaction with their respective receptors (Dziedzic et al., 2003). In the case of TGFα the metalloproteinase involved is termed tumor necrosis factor alpha converting enzyme (TACE) (Peschon et al., 1998). In hypothalamic astrocytes, coactivation of astrocytic AMPA and metabotropic receptors results in extracellular Ca^{2+} influx, a Ca^{2+}/protein kinase C-dependent increase in TACE-like activity, and enhanced release of TGFα (Lomniczi et al., 2006) (Fig. 19.1). Within the hypothalamus, TACE is most abundantly expressed in astrocytes of the median eminence and its enzymatic activity increases selectively in this region at the time of the first preovulatory surge of gonadotropins. Pharmacological inhibition of TACE activity targeted to the median eminence decreases GnRH secretion and delays puberty indicating that an increased TACE activity in this region of the hypothalamus is necessary for the pubertal activation of GnRH secretion to take place (Lomniczi et al., 2006).

19.5 Gonadal Steroids and Astrocyte Function

Gonadal steroids, such as estrogen and progesterone, have a wide range of effects in the CNS of adult as well as developing mammals. Among these effects, the ability of estradiol to increase dendritic spine density in hippocampal neurons and to exert neuroprotective effects stand out because of the direct relevance of these processes to synaptic plasticity and brain repair [reviewed in (Mong and McCarthy, 1999; Mahesh et al., 2006; McEwen, 2002)]. It is also clear that both astrocytes (Ma et al., 1994a; Milner et al., 2001) and tanycytes (Langub and Watson, 1992) express estrogen receptors of the alpha subtype (ERα), and that steroid hormones can affect a variety of glial functions including glutamate homeostasis (see above), as well as morphology and astrocytic production of growth factors.

19.5.1 Estradiol and Glial Morphology

Preovulatory levels of estradiol promote the elongation of astrocytic processes in the arcuate nucleus of the rat hypothalamus. In turn, these processes decrease inhibitory synaptic connectivity by ensheathing GABAergic axo-somatic synapses (Garcia-Segura and McCarthy, 2004). This change, that occurs during the proestrous and estrous phases of the estrous cycle, has been postulated to shift the balance of inhibitory vs. excitatory inputs to the GnRH neuronal network, resulting in enhanced GnRH release (Garcia-Segura and McCarthy, 2004). Mice lacking ERα fail to respond to estradiol treatment with an increase in astrocytic process length, but astrocytes from ERβ KO mice behave like wild-type animals, indicating that the trophic effects of estradiol on astrocytic morphology in the arcuate nucleus are mediated by ERα (Mong and Blutstein, 2006). Whether these receptors are located on astrocytes themselves, or in adjacent neurons, remains to be determined.

Noteworthy, astrocytes of the arcuate nucleus and the medial POA are profoundly sexually dimorphic, with males exhibiting astrocytes with longer processes and a greater complexity than astrocytes in females (Mong et al., 1999; Amateau and McCarthy, 2002). This difference is already evident at the day of birth (Amateau and McCarthy, 2002), and, at least in the arcuate nucleus, is mediated by activation of $GABA_A$ receptors (Mong et al., 1999).

19.5.2 Sex Steroids and Glial Growth Factors

Hypothalamic TGFα gene expression increases during puberty (Ma et al., 1992, 1994b) with an initial steroid-independent increase followed by a second steroid-dependent elevation (El Majdoubi et al., 1998a, b). The administration of estradiol followed by progesterone to immature rats induced the expression of TGFα in the POA as well as in the medial basal hypothalamus (Ma et al., 1992). The finding that TGFα mRNA content of cultured hypothalamic astrocytes is increased by estradiol indicates that at least part of the estradiol effect described in vivo is exerted directly on glial cells (Ma et al., 1994a). Indeed, estradiol can act very rapidly on astrocytes via cell membrane-bound receptors to induce intracellular Ca^{2+} mobilization (Chaban et al., 2004), raising the possibility that its effect on TGFα production is Ca^{2+}-dependent.

Exposure of hypothalamic astrocytes to progesterone in vitro does not change TGFα expression, indicating that the pronounced effect seen in vivo is either due to a neuron-specific increase in TGFα expression or a consequence of a neuronal dependent induction of astrocytic TGFα mRNA (Ma et al., 1994a).

Estradiol affects erbB1-mediated, glia-to-GnRH neuron signaling at two different levels: it increases TGFα gene expression in astrocytes and increases the synthesis of the PGE_2 receptors EP-1 and EP-3 in GnRH neurons (Rage et al., 1997b). While activation of EP-1 receptors induces Ca^{2+} mobilization and increased turnover of phosphatidyl inositol, EP-3 receptors are linked to cyclic adenosine monophosphate production. The EP-3 isoforms EP-3α and EP-3β are coupled to the inhibitory G-protein G_i, so that their activation results in inhibition of adenylate cyclase activity. The isoform EP-3γ, on the other hand, is coupled to both inhibition and stimulation of adenylate cyclase (Narumiya, 1994). When a GnRH neuronal cell line was exposed to conditioned media from hypothalamic astrocytes treated with physiological concentrations of estradiol, an increase in the mRNA content of EP-1 and EP-3γ receptor was detected, without any significant changes in EP-3α or β (Rage et al., 1997b). This selective activation of EP-1 and EP-3γ receptors led to increased GnRH release.

Estradiol also increases the hypothalamic expression of erbB2 and erbB4 receptors. When estrogen alone is administered to immature rats, it only increases erbB4 mRNA levels (Ma et al., 1999). But when estradiol is followed by progesterone, an increase in hypothalamic erbB2 mRNA levels also occurs. This suggests that in the afternoon of proestrus, a progesterone-dependent increase in erbB2 expression, in the presence of an already elevated complement of erbB4 receptors functions to amplify the stimulatory effects of NRGs on astroglial PGE_2 release and hence, to facilitate the preovulatory increase of GnRH secretion.

While these findings indicate that ovarian steroids facilitate the synthesis and actions of astrocyte-derived growth factors, growth factors of the EGF family also have the ability to activate estrogen responsive elements, in the presence of estrogen receptors, but in a steroid-independent manner (Ignar-Trowbridge et al., 1993). An example of this type of interaction is provided by the finding that administration of EGF to mice lacking ERα does not induce DNA synthesis and progesterone receptor expression as it does in wild-type animals (Curtis et al., 1996). These results suggest that the prepubertal activation of erbB receptors in the hypothalamus may initiate estrogen receptor-dependent events in the absence of steroids, thereby providing one of the initial stimuli required for the unfolding of the cellular and molecular events that take place in the hypothalamus at the time of puberty. Among the former are the synaptic remodeling and reorganization of neuronal membranes induced by estrogen in hypothalamic regions controlling GnRH secretion (Clough and Rodriguez-Sierra, 1983; Matsumoto and Arai, 1977; Olmos et al., 1987; Naftolin et al., 1992). Examples of the latter are the changes in neurotransmitter and neurotransmitter receptor expression and activity that precede the acquisition of female reproductive capacity (Ojeda and Urbanski, 1994; Bourguignon et al., 1995; Terasawa, 1995).

19.6 Conclusions

Both astrocytes and tanycytes participate actively in the control of GnRH neuronal activity. By doing so, they have emerged as important contributors to the central mechanisms underlying the regulation of female sexual development. Astrocytes and tanycytes physically engage GnRH neurons by apposing processes to the GnRH neuronal cell membrane in a highly dynamic fashion, subjected to sex steroid regulation. This attachment appears to be provided by cell-to-cell communication systems endowed with both adhesive and signaling capabilities. Astrocytes and tanycytes facilitate GnRH secretion via the release of a variety of substances, among which growth factors of the EGF family play a prominent role. They also integrate stimulatory inputs to the GnRH neuronal network by regulating glutamate availability for synaptic transmission and transducing glutamatergic signals into growth factor-mediated glia-to-GnRH neuron signaling pathways.

Acknowledgments This work was supported by grants from the National Institutes of Health HD25123, MH65438, U54 HD18185 through cooperative agreement as part of the Specialized Cooperative Center's Program in Reproduction and Infertility Research, National Institute of Child Health and Human Development/NIH, and RR00163 for the operation of the Oregon National Primate Research Center.

References

Adamsky K, Arnold K, Sabanay H, Peles E (2003) Junctional protein MAGI-3 interacts with receptor tyrosine phosphatase beta (RPTP beta) and tyrosine-phosphorylated proteins. J Cell Sci 116:1279–1289

Akiyama T, Saito T, Ogawara H, Toyoshima K, Yamamoto T (1988) Tumor promoter and epidermal growth factor stimulate phosphorylation of the c-erbB-2 gene product in MKN-7 human adenocarcinoma cells. Mol Cell Biol 8:1019–1026

Amateau SK, McCarthy MM (2002) Sexual differentiation of astrocyte morphology in the developing rat preoptic area. J Neuroendocrinol 14:904–910

Araque A, Parpura V, Sanzgiri RP, Haydon PG (1999) Tripartite synapses: Glia, the unacknowledged partner. Trends Neurosci 22:208–215

Barbour B, Szatkowski M, Ingledew N, Attwell D (1989) Arachidonic acid induces a prolonged inhibition of glutamate uptake into glial cells. Nature 342:918–920

Barres BA (1991) New roles for glia. J Neurosci 11:3685–3694

Beattie EC, Stellwagen D, Morishita W, Bresnahan JC, Ha BK, von Zastrow ME, Beattie MS, Malenka RC (2002) Control of synaptic strength by glial TNFa. Science 295:2282–2285

Beerli RR, Graus-Porta D, Woods-Cook K, Chen X, Yarden Y, Hynes NE (1995) Neu differentiation factor activation of ErbB-3 and ErbB-4 is cell specific and displays a differential requirement for ErbB-2. Mol Cell Biol 15:6496–6505

Berg-von der Emde K, Ma YJ, Costa ME, Ojeda SR (1993) Hypothalamic astrocytes respond to transforming growth factor alpha (TGFa) with prostaglandin E2 (PGE2) release. Soc Neurosci Abstr 19:1682

Bezzi P, Carmignoto G, Pasti L, Vesce S, Rossi D, Rizzini BL, Pozzan T, Volterra A (1998) Prostaglandins stimulate calcium-dependent glutamate release in astrocytes. Nature 391:281–285

Biederer T, Sudhof TC (2001) CASK and protein 4.1 support F-actin nucleation on neurexins. J Biol Chem 276:47869–47876
Biederer T, Sara Y, Mozhayeva M, Atasoy D, Liu X, Kavalali ET, Südhof TC (2002) SynCAM, a synaptic adhesion molecule that drives synapse assembly. Science 297:1525–1531
Blondel O, Collin C, McCarran WJ, Zhu S, Zamostiano R, Gozes I, Brenneman DE, McKay RD (2000) A glia-derived signal regulating neuronal differentiation. J Neurosci 20:8012–8020
Blutstein T, Devidze N, Choleris E, Jasnow AM, Pfaff DW, Mong JA (2006) Oestradiol up-regulates glutamine synthetase mRNA and protein expression in the hypothalamus and hippocampus: implications for a role of hormonally responsive glia in amino acid neurotransmission. J Neuroendocrinol 18:692–702
Bourguignon J-P, Gérard A, Gonzalez A, Purnelle G, Franchimont P (1995) The role of excitatory amino acids in triggering the onset of puberty. In: The Neurobiology of Puberty (Plant TM, Lee PA, eds), pp. 129–138. Bristol: Journal of Endocrinology Ltd
Buonanno A, Fischbach GD (2001) Neuregulin and ErbB receptor signaling pathways in the nervous system. Curr Opin Neurobiol 11:287–296
Burden S, Yarden Y (1997) Neuregulins and their receptors: A versatile signaling module in organogenesis and oncogenesis. Neuron 18:847–855
Carpenter G, Cohen S (1990) Epidermal growth factor. J Biol Chem 265:7709–7712
Carraway KL, Cantley LC (1994) A neu acquaintance for ErbB3 and ErbB4: A role for receptor heterodimerization in growth signaling. Cell 78:5–8
Carraway KL, Weber JL, Unger MJ, Ledesma J, Yu N, Gassmann M, Lai C (1997) Neuregulin-2, a new ligand of ErbB3/ErbB4-receptor tyrosine kinases. Nature 387:512–516
Cashion AB, Smith MJ, Wise PM (2003) The morphometry of astrocytes in the rostral preoptic area exhibits a diurnal rhythm on proestrus: Relationship to the luteinizing hormone surge and effects of age. Endocrinology 144:274–280
Chaban VV, Lakhter AJ, Micevych P (2004) A membrane estrogen receptor mediates intracellular calcium release in astrocytes. Endocrinology 145:3788–3795
Chang H, Riese DJ, Gilbert W, Stern DF, McMahan UJ (1997) Ligands for ErbB-family receptors encoded by a neuregulin-like gene. Nature 387:509–516
Christopherson KS, Ullian EM, Stokes CC, Mullowney CE, Hell JW, Agah A, Lawler J, Mosher DF, Bornstein P, Barres BA (2005) Thrombospondins are astrocyte-secreted proteins that promote CNS synaptogenesis. Cell 120:421–433
Clough RW, Rodriguez-Sierra JF (1983) Synaptic changes in the hypothalamus of the prepubertal female rat administered estrogen. Am J Anat 167:205–214
Cotrina ML, Lin JHC, López-García JC, Naus CCG, Nedergaard M (2000) ATP-mediated glia signaling. J Neurosci 20:2835–2844
Cottrell EC, Campbell RE, Han SK, Herbison AE (2006) Postnatal remodeling of dendritic structure and spine density in gonadotropin-releasing hormone neurons. Endocrinology 147:3652–3661
Curtis SW, Washburn T, Sewall C, DiAugustine R, Lindzey J, Couse JF, Korach KS (1996) Physiological coupling of growth factor and steroid receptor signaling pathways: Estrogen receptor knockout mice lack estrogen-like response to epidermal growth factor. Proc Natl Acad Sci USA 93:12626–12630
DeFazio RA, Heger S, Ojeda SR, Moenter SM (2002) Activation of A-type g-aminobutyric acid receptors excites gonadotropin-releasing hormone neurons. Mol Endocrinol 16:2872–2891
De Seranno S, Estrella C, Loyens A, Cornea A, Ojeda SR, Beauvillain SC, Prevot V (2004) Vascular endothelial cells promote acute ependymoglial cell plasticty. J Neurosci 24:10353–10363
Dong J, Opresko LK, Dempsey PJ, Lauffenburger DA, Coffey RJ, Wiley HS (1999) Metalloprotease-mediated ligand release regulates autocrine signaling through the epidermal growth factor receptor. Proc Natl Acad Sci USA 96:6235–6240
Dungan HM, Clifton DK, Steiner RA (2006) Minireview: Kisspeptin neurons as central processors in the regulation of gonadotropin-releasing hormone secretion. Endocrinology 147:1154–1158
Dziedzic B, Prevot V, Lomniczi A, Jung H, Cornea A, Ojeda SR (2003) Neuron-to-glia signaling mediated by excitatory amino acid receptors regulates erbB receptor function in astroglial cells of the neuroendocrine brain. J Neurosci 23:915–926

El Majdoubi M, Sahu A, Plant TM (1998a) Effect of estrogen on hypothalamic transforming growth factor alpha and gonadotropin-releasing hormone gene expression in the female rhesus monkey. Neuroendocrinology 67:228–235
El Majdoubi M, Sahu A, Plant TM (1998b) Pubertal changes in hypothalamic GnRH and related peptide gene expression in the monkey. Soc Neurosci Abstr 24:270
Erecinska M, Silver IA (1990) Metabolism and role of glutamate in mammalian brain. Prog Neurobiol 35:245–296
Falls DL (2003) Neuregulins: Functions, forms, and signaling strategies. Exp Cell Res 284:14–30
Fellin T, Sul JY, D'Ascenzo M, Takano H, Pascual O, Haydon PG (2006) Bidirectional astrocyte–neuron communication: the many roles of glutamate and ATP. In: Purinergic Signalling in Neuron–Glia Interactions (Norvartis Foundation Symposium 276) pp. 208–221. Chichester: Wiley
Fields RD, Burnstock G (2006) Purinergic signalling in neuron–glia interactions. Nat Rev Neurosci 7:423–436
Fields RD, Stevens B (2000) ATP: An extracellular signaling molecule between neurons and glia. Trends Neurosci 23:625–633
Fields RD, Stevens-Graham B (2002) New insights into neuron–glia communication. Science 298:556–562
Galbiati M, Zanisi M, Messi E, Cavarretta I, Martini L, Melcangi RC (1996) Transforming growth factor-b and astrocytic conditioned medium influence luteinizing hormone-releasing hormone gene expression in the hypothalamic cell line GT1. Endocrinology 137:5605–5609
Galbiati M, Martini L, Melcangi RC (2002) Oestrogens, via transforming growth factor a, modulate basic fibroblast growth factor synthesis in hypothalamic astrocytes: In vitro observations. J Neuroendocrinol 14:829–835
Garcia-Segura LM, McCarthy MM (2004) Minireview: Role of glia in neuroendocrine function. Endocrinology 145:1082–1086
Grumet M (1997) Nr-CAM: A cell adhesion molecule with ligand and receptor functions. Cell Tissue Res 290:423–428
Haber M, Zhou L, Murai KK (2006) Cooperative astrocyte and dendritic spine dynamics at hippocampal excitatory synapses. J Neurosci 26:8881–8891
Harris KM, Sultan P (1995) Variation in the number, location and size of synaptic vesicles provides an anatomical basis for the nonuniform probability of release at hippocampal CA1 synapses. Neuropharmacology 34:1387–1395
Hatton GI, Perlmutter LS, Salm AK, Tweedle CD (1984) Dynamic neuronal-glial interactions in hypothalamus and pituitary: Implications for control of hormone synthesis and release. Peptides 5:121–138
Haydon PG (2001) Glia: Listening and talking to the synapse. Nat Rev Neurosci 2:185–193
Herbison AE (1998) Multimodal influence of estrogen upon gonadotropin-releasing hormone neurons. Endocrine Rev 19:302–330
Hertz L, Zielke HR (2004) Astrocytic control of glutamatergic activity: Astrocytes as stars of the show. Trends Neurosci 27:735–743
Hiney JK, Ojeda SR, Dees WL (1991) Insulin-like growth factor-I: A possible metabolic signal involved in the regulation of female puberty. Neuroendocrinology 54:420–423
Hirbec H, Martin S, Henley JM (2005) Syntenin is involved in the developmental regulation of neuronal membrane architecture. Mol Cell Neurosci 28:737–746
Huang YH, Bergles DE (2004) Glutamate transporters bring competition to the synapse. Curr Opin Neurobiol 14:346–352
Ignar-Trowbridge DM, Teng CT, Ross KA, Parker MG, Korach KS, McLachlan JA (1993) Peptide growth factors elicit estrogen receptor-dependent transcriptional activation of an estrogen-responsive element. Mol Endocrinol 7:992–998
Jung H, Carmel P, Schwartz MS, Witkin JW, Bentele KHP, Westphal M, Piatt JH, Costa ME, Cornea A, Ma YJ, Ojeda SR (1999) Some hypothalamic hamartomas contain transforming growth factor alpha, a puberty-inducing growth factor, but not luteinizing hormone-releasing hormone neurons. J Clin Endocrinol Metab 84:4695–4701

Junier M-P, Ma YJ, Costa ME, Hoffman G, Hill DF, Ojeda SR (1991) Transforming growth factor alpha contributes to the mechanism by which hypothalamic injury induces precocious puberty. Proc Natl Acad Sci USA 88:9743–9747

Junier M-P, Hill DF, Costa ME, Felder S, Ojeda SR (1993) Hypothalamic lesions that induce female precocious puberty activate glial expression of the epidermal growth factor receptor gene: Differential regulation of alternatively spliced transcripts. J Neurosci 13:703–713

Kalra SP, Crowley WR (1992) Neuropeptide Y: A novel neuroendocrine peptide in the control of pituitary hormone secretion, and its relation to luteinizing hormone. In: Frontiers in Neuroendocrinology, Vol. 13 (Ganong WF, Martini L, eds), pp. 1–46. New York, NY: Raven

Karunagaran D, Tsahar E, Beerli RR, Chen X, Graus-Porta D, Ratzkin BJ, Seger R, Hynes NE, Yarden Y (1996) ErbB-2 is a common auxiliary subunit of NDF and EGF receptors: Implications for breast cancer. EMBO J 15:254–264

Kawachi H, Tamura H, Watakabe I, Shintani T, Maeda N, Noda M (1999) Protein tyrosine phosphatase zeta/RPTPbeta interacts with PSD-95/SAP90 family. Brain Res Mol Brain Res 72:47–54

King JC, Letourneau RL (1994) Luteinizing hormone-releasing hormone terminals in the median eminence of rats undergo dramatic changes after gonadectomy, as revealed by electron microscopic image analysis. Endocrinology 134:1340–1351

King JC, Rubin BS (1996) Recruitment of LHRH neurons and increased access of LHRH terminals to portal capillaries: Integral mechanisms for LH surge induction. Ann d' Endocrinol (Paris) 57(Suppl 4):72

Kozlowski GP, Coates PW (1985) Ependymoneuronal specializations between LHRH fibers and cells of the cerebroventricular system. Cell Tissue Res 242:301–311

Laake JH, Slyngstad TA, Haug FM, Ottersen OP (1995) Glutamine from glial cells is essential for the maintenance of the nerve terminal pool of glutamate: Immunogold evidence from hippocampal slice cultures. J Neurochem 65:871–881

Langub MC, Jr., Watson RE, Jr. (1992) Estrogen receptor-immunoreactive glia, endothelia, and ependyma in guinea pig preoptic area and median eminence: Electron microscopy. Endocrinology 130:364–372

Lehre KP, Rusakov DA (2002) Asymmetry of glia near central synapses favors presynaptically directed glutamate escape. Biophys J 83:125–134

Levine JE, Bauer-Dantoin AC, Besecke LM, Conaghan LA, Legan SJ, Meredith JM, Strobl FJ, Urban JH, Vogelsong KM, Wolfe AM (1991) Neuroendocrine regulation of luteinizing hormone pulse generator in the rat. Rec Prog Horm Res 47:97–153

Liang SL, Carlson GC, Coulter DA (2006) Dynamic regulation of synaptic GABA release by the glutamate–glutamine cycle in hippocampal area CA1. J Neurosci 26:8537–8548

Lomniczi A, Cornea A, Costa ME, Ojeda SR (2006) Hypothalamic tumor necrosis factor-a converting enzyme (TACE) mediates excitatory amino acid-dependent neuron-to-glia signaling in the neuroendocrine brain. J Neurosci 26:51–62

Luetteke NC, Phillips HK, Qiu TH, Copeland NG, Earp HS, Jenkins NA, Lee DC (1994) The mouse waved-2 phenotype results from a point mutation in the EGF receptor tyrosine kinase. Genes Dev 8:399–413

Ma YJ, Junier M-P, Costa ME, Ojeda SR (1992) Transforming growth factor alpha (TGFa) gene expression in the hypothalamus is developmentally regulated and linked to sexual maturation. Neuron 9:657–670

Ma YJ, Berg-von der Emde K, Moholt-Siebert M, Hill DF, Ojeda SR (1994a) Region-specific regulation of transforming growth factor a (TGFa) gene expression in astrocytes of the neuroendocrine brain. J Neurosci 14:5644–5651

Ma YJ, Costa ME, Ojeda SR (1994b) Developmental expression of the genes encoding transforming growth factor alpha (TGFa) and its receptor in the hypothalamus of female rhesus macaques. Neuroendocrinology 60:346–359

Ma YJ, Dissen GA, Merlino G, Coquelin A, Ojeda SR (1994c) Overexpression of a human transforming growth factor alpha (TGFa) transgene reveals a dual antagonistic role of TGFa in female sexual development. Endocrinology 135:1392–1400

Ma YJ, Hill DF, Junier M-P, Costa ME, Felder SE, Ojeda SR (1994d) Expression of epidermal growth factor receptor changes in the hypothalamus during the onset of female puberty. Mol Cell Neurosci 5:246–262

Ma YJ, Berg-von der Emde K, Rage F, Wetsel WC, Ojeda SR (1997) Hypothalamic astrocytes respond to transforming growth factor alpha with secretion of neuroactive substances that stimulate the release of luteinizing hormone-releasing hormone. Endocrinology 138:19–25

Ma YJ, Hill DF, Creswick KE, Costa ME, Ojeda SR (1999) Neuregulins signaling via a glial erbB2/erbB4 receptor complex contribute to the neuroendocrine control of mammalian sexual development. J Neurosci 19:9913–9927

Mahesh VB, Dhandapani KM, Brann DW (2006) Role of astrocytes in reproduction and neuroprotection. Mol Cell Endocrinol 246:1–9

Martin DL (1992) Synthesis and release of neuroactive substances by glial cells. Glia 5:81–94

Massague J (1990) Transforming growth factor-a. A model for membrane-anchored growth factors. J Biol Chem 265:21393–21396

Matsumoto A, Arai Y (1977) Precocious puberty and synaptogenesis in the hypothalamic arcuate nucleus in pregnant mare serum gonadotropin (PMSG) treated immature female rats. Brain Res 129:375–378

Mauch DH, Nagler K, Schumacher S, Goritz C, Muller EC, Otto A, Pfrieger FW (2001) CNS synaptogenesis promoted by glia-derived cholesterol. Science 294:1354–1357

McEwen B (2002) Estrogen actions throughout the brain. Recent Prog Horm Res 57:357–384

Melcangi RC, Galbiati M, Messi E, Piva F, Martini L, Motta M (1995) Type 1 astrocytes influence luteinizing hormone-releasing hormone release from the hypothalamic cell line GT1–1: Is transforming growth factor-b the principle involved?. Endocrinology 136:679–686

Meng K, Rodriguez-Pena A, Dimitrov T, Chen W, Yamin M, Noda M, Deuel TF (2000) Pleiotrophin signals increased tyrosine phosphorylation of beta beta-catenin through inactivation of the intrinsic catalytic activity of the receptor-type protein tyrosine phosphatase beta/zeta. Proc Natl Acad Sci USA 97:2603–2608

Milner TA, McEwen BS, Hayashi S, Li CJ, Reagan LP, Alves SE (2001) Ultrastructural evidence that hippocampal alpha estrogen receptors are located at extranuclear sites. J Comp Neurol 429:355–371

Miyake A, Tasaka K, Otsuka S, Kohmura H, Wakimoto H, Aono T (1985) Epidermal growth factor stimulates secretion of rat pituitary luteinizing hormone in vitro. Acta Endocrinol 108:175–178

Mong JA, Blutstein T (2006) Estradiol modulation of astrocytic form and function: Implications for hormonal control of synaptic communication. Neuroscience 138:967–975

Mong JA, McCarthy MM (1999) Steroid-induced developmental plasticity in hypothalamic astrocytes: Implications for synaptic patterning. J Neurobiol 40:602–619

Mong JA, Glaser E, McCarthy MM (1999) Gonadal steroids promote glial differentiation and alter neuronal morphology in the developing hypothalamus in a regionally specific manner. J Neurosci 19:1464–1472

Montero JC, Yuste L, Diaz-Rodriguez E, Esparis-Ogando A, Pandiella A (2000) Differential shedding of transmembrane neuregulin isoforms by the tumor necrosis factor-alpha-converting enzyme. Mol Cell Neurosci 16:631–648

Mothet JP, Pollegioni L, Ouanounou G, Martineau M, Fossier P, Baux G (2005) Glutamate receptor activation triggers a calcium-dependent and SNARE protein-dependent release of the gliotransmitter d-serine. Proc Natl Acad Sci USA 102:5606–5611

Naftolin F, Leranth C, Garcia-Segura LM (1992) Ultrastructural changes in hypothalamic cells during estrogen-induced gonadotrophin feedback. Neuroprotocols 1:16–26

Narumiya S (1994) Prostanoid receptors: Structure, function, and distribution. Ann N Y Acad Sci 744:126–138

Nedergaard M, Ransom B, Goldman SA (2003) New roles for astrocytes: Redefining the functional architecture of the brain. Trends Neurosci 26:523–530

Newman EA (2003) New roles for astrocytes: Regulation of synaptic transmission. Trends Neurosci 26:536–542

Nishida H, Okabe S (2007) Direct astrocytic contacts regulate local maturation of dendritic spines. J Neurosci 27:331–340

Ojeda SR (1994) The neurobiology of mammalian puberty: Has the contribution of glial cells been underestimated?. J NIH Res 6:51–56

Ojeda SR, Skinner MK (2006) Puberty in the rat. In: The Physiology of Reproduction, 3rd edition (Neill JD, ed), pp. 2061–2126. San Diego, CA/Amsterdam: Academic/Elsevier

Ojeda SR, Terasawa E (2002) Neuroendocrine regulation of puberty. In: Hormones, Brain and Behavior, Vol. 4 (Pfaff D, Arnold A, Etgen A, Fahrbach S, Moss R, Rubin R, eds), pp. 589–659. New York, NY: Elsevier

Ojeda SR, Urbanski HF (1994) Puberty in the rat. In: The Physiology of Reproduction, 2nd Edition, Vol. 2 (Knobil E, Neill JD, eds), pp. 363–409. New York, NY: Raven

Ojeda SR, Urbanski HF, Costa ME, Hill DF, Moholt-Siebert M (1990) Involvement of transforming growth factor a in the release of luteinizing hormone-releasing hormone from the developing female hypothalamus. Proc Natl Acad Sci USA 87:9698–9702

Ojeda SR, Ma YJ, Lee BJ, Prevot V (2000) Glia-to-neuron signaling and the neuroendocrine control of female puberty. Rec Prog Horm Res 55:197–224

Ojeda SR, Prevot V, Heger S, Lomniczi A, Dziedzic B, Mungenast A (2003) Glia-to-neuron signaling and the neuroendocrine control of female puberty. Ann Med 35:244–255

Olmos G, Aguilera P, Tranque P, Naftolin F, Garcia-Segura LM (1987) Estrogen-induced synaptic remodelling in adult rat brain is accompanied by the reorganization of neuronal membranes. Brain Res 425:57–64

Olson BR, Scott DC, Wetsel WC, Elliot SJ, Tomic M, Stojilkovic S, Nieman LK, Wray S (1995) Effects of insulin-like growth factors I and II and insulin on the immortalized hypothalamic GT1-7 cell line. Neuroendocrinology 62:155–165

Parent AS, Mungenast A, Lomniczi A, Peles E, Sandau U, Ojeda SR (2007) A contactin–RPTPb complex mediates adhesive communication between astroglial cells and GnRH neurons. J Neuroendocrinol 19:847–859

Parpura V, Basarsky TA, Liu F, Jeftinija K, Jeftinija S, Haydon PG (1994) Glutamate-mediated astrocyte–neuron signalling. Nature 369:744–747

Parri HR, Gould TM, Crunelli V (2001) Spontaneous astrocytic Ca2+ oscillations in situ drive NMDAR-mediated neuronal excitation. Nat Neurosci 4:803–812

Peles E, Nativ M, Campbell PL, Sakurai T, Martinez R, Lev S, Clary DO, Schilling J, Barnea G, Plowman GD, Grumet M, Schlessinger J (1995) The carbonic anhydrase domain of receptor tyrosine phosphatase b is a functional ligand for the axonal cell recognition molecule contactin. Cell 82:251–260

Peles E, Nativ M, Lustig M, Grumet M, Schilling J, Martinez R, Plowman GD, Schlessinger J (1997) Identification of a novel contactin-associated transmembrane receptor with multiple domains implicated in protein–protein interactions. EMBO J 16:978–988

Peschon JJ, Slack JL, Reddy P, Stocking KL, Sunnarborg SW, Lee DC, Russell WE, Castner BJ, Johnson RS, Fitzner JN, Boyce RW, Nelson N, Koslosky CJ, Wolfson MF, Rauch CT, Cerretti DP, Paxton RJ, March CJ, Black RA (1998) An essential role for ectodomain shedding in mammalian development. Science 282:1281–1284

Plant TM (2002) Neurophysiology of puberty. J Adolesc Health 31:185–191

Plant TM, Witchel SF (2006) Puberty in nonhuman primates and humans. In: The Physiology of Reproduction, 3rd Edition (Neill JD, ed), pp. 2177–2230. San Diego, CA/Amsterdam: Academic/Elsevier

Prenzel N, Zwick E, Daub H, Leserer M, Abraham R, Wallasch C, Ullrich A (1999) EGF receptor transactivation by G-protein receptors requires metalloproteinase cleavage of proHB-EGF. Nature 402:884–888

Prevot V (2002) Glial–neuronal–endothelial interactions are involved in the control of GnRH secretion. J Neuroendocrinol 14:247–255

Prevot V, Croix D, Bouret S, Dutoit S, Tramu G, Stefano GB, Beauvillain JC (1999) Definitive evidence for the existence of morphological plasticity in the external zone of the median eminence during the rat estrous cycle: Implication of neuro-glio-endothelial interactions in gonadotropin-releasing hormone release. Neuroscience 94:809–819

Prevot V, Cornea A, Mungenast A, Smiley G, Ojeda SR (2003a) Activation of erbB-1 signaling in tanycytes of the median eminence stimulates transforming growth factor b1 release via prostaglandin E2 production and induces cell plasticity. J Neuroscience 23:10622–10632

Prevot V, Rio C, Cho GJ, Lomniczi A, Heger S, Neville CM, Rosenthal NA, Ojeda SR, Corfas G (2003b) Normal female sexual development requires neuregulin-erbB receptor signaling in hypothalamic astrocytes. J Neurosci 23:230–239

Prevot V, Lomniczi A, Corfas G, Ojeda SR (2005) ErbB-1 and erbB-4 receptors act in concert to facilitate both female sexual development and mature reproductive function. Endocrinology 146:1465–1472

Rage F, Hill DF, Sena-Esteves M, Breakefield XO, Coffey RJ, Costa ME, McCann SM, Ojeda SR (1997a) Targeting transforming growth factor a expression to discrete loci of the neuroendocrine brain induces female sexual precocity. Proc Natl Acad Sci USA 94:2735–2740

Rage F, Lee BJ, Ma YJ, Ojeda SR (1997b) Estradiol enhances prostaglandin E2 receptor gene expression in luteinizing hormone-releasing hormone (LHRH) neurons and facilitates the LHRH response to PGE2 by activating a glia-to-neuron signaling pathway. J Neurosci 17:9145–9156

Riese DJ, Bermingham Y, van Raaij TM, Buckley S, Plowman GD, Stern DF (1996a) Betacellulin activates the epidermal growth factor receptor and erbB-4, and induces cellular response patterns distinct from those stimulated by epidermal growth factor or neuregulin-beta. Oncogene 12:345–353

Riese DJ, Kim ED, Elenius K, Buckley S, Klagsbrun M, Plowman GD, Stern DF (1996b) The epidermal growth factor receptor couples transforming growth factor-a heparin-binding epidermal growth factor-like factor, and amphiregulin to neu, ErbB-3 and ErbB-4. J Biol Chem 271:20047–20052

Rimer M (2003) Neuregulins: Primary or secondary signals for the control of synapse-specific gene expression. J Neurocytol 32:665–675

Rodriguez EM, Blazquez JL, Pastor FE, Pelaez B, Pena P, Peruzzo B, Amat P (2005) Hypothalamic tanycytes: A key component of brain–endocrine interaction. Int Rev Cytol 247:89–164

Roth C, Leonhardt S, Theiling K, Lakomek M, Jarry H, Wuttke W (1998) Ontogeny of the GnRH-, glutaminase- and glutamate decarboxylase-gene expression in the hypothalamus of female rats. Brain Res Dev Brain Res 110:105–114

Roth CL, McCormack AL, Lomniczi A, Mungenast AE, Ojeda SR (2006) Quantitative proteomics identifies a major change in glial glutamate metabolism at the time of female puberty. Mol Cell Endocrinol 254–255:51–59

Sahin U, Weskamp G, Kelly K, Zhou H-M, Higashiyama S, Peshon J, Hartmann D, Saftig P, Blobel CP (2004) Distinct roles for ADAM10 and ADAM17 in ectodomain shedding of six EGFR ligands. J Cell Biol 164:769–779

Sakurai T, Lustig M, Nativ M, Hemperly JJ, Schlessinger J, Peles E, Grumet M (1997) Induction of neurite outgrowth through contactin and Nr-CAM by extracellular regions of glial receptor tyrosine phosphatase beta. J Cell Biol 136:907–918

Shelly M, Pinkas-Kramarski R, Guarino BC, Waterman H, Wang L-M, Lyass L, Alimandi M, Kuo A, Bacus SS, Pierce HJ, Andrews GC, Yarden Y (1998) Epiregulin is a potent pan-ErbB ligand that preferentially activates heterodimeric receptor complexes. J Biol Chem 273:10496–10505

Silverman A-J, Livne I, Witkin JW (1994) The gonadotropin-releasing hormone (GnRH), neuronal systems: Immunocytochemistry and in situ hybridization. In: The Physiology of Reproduction, 2nd Edition, Vol. 1 (Knobil E, Neill JD, eds), pp. 1683–1709. New York, NY: Raven

Stella N, Tencé M, Glowinski J, Prémont J (1994) Glutamate-evoked release of arachidonic acid from mouse brain astrocytes. J Neurosci 14:568–575

Terasawa E (1995) Mechanisms controlling the onset of puberty in primates: The role of GABAergic neurons. In: The Neurobiology of Puberty (Plant TM, Lee PA, eds), pp 139–151. Bristol: Journal of Endocrinology Ltd

Terasawa E, Fernandez DL (2001) Neurobiological mechanisms of the onset of puberty in primates. Endocr Rev 22:111–151

Terasawa E, Keen KL, Grendell RL, Golos TG (2005) Possible role of 5 -adenosine triphosphate in synchronization of Ca2+ oscillations in primate luteinizing hormone-releasing hormone neurons. Mol Endocrinol 19:2736–2747

Tsai P-S, Werner S, Weiner RI (1995) Basic fibroblast growth factor is a neurotropic factor in GT1 gonadotropin-releasing hormone neuronal cell lines. Endocrinology 136:3831–3838

Tsai PS, Moenter SM, Postigo HR, El Majdoubi M, Pak TR, Gill JC, Paruthiyil S, Werner S, Weiner RI (2005) Targeted expression of a dominant-negative fibroblast growth factor (FGF) receptor in gonadotropin-releasing hormone (GnRH) neurons reduces FGF responsiveness and the size of GnRH neuronal population. Mol Endocrinol 19:225–236

Tzahar E, Pinkas-Kramarski R, Moyer JD, Klapper LN, Alroy I, Levkowitz G, Shelly M, Henis S, Eisenstein M, Ratzkin BJ, Sela M, Andrews GC, Yarden Y (1997) Bivalence of EGF-like ligands drives the ErbB signaling network. EMBO J 16:4938–4950

Voigt P, Ma YJ, Gonzalez D, Fahrenbach WH, Wetsel WC, Berg-von der Emde K, Hill DF, Taylor KG, Costa ME, Seidah NG, Ojeda SR (1996) Neural and glial-mediated effects of growth factors acting via tyrosine kinase receptors on LHRH neurons. Endocrinology 137:2593–2605

Wada T, Qian X, Greene MI (1990) Intermolecular association of the p185neu protein and EGF receptor modulates EGF receptor function. Cell 61:1339–1347

Wallasch C, Weiss FU, Niederfellner G, Jallal B, Issing W, Ullrich A (1995) Heregulin-dependent regulation of HER2/neu oncogenic signaling by heterodimerization with HER3. EMBO J 14:4267–4275

Wetsel WC, Hill DF, Ojeda SR (1996) Basic fibroblast growth factor regulates the conversion of pro-luteinizing hormone-releasing hormone (LHRH) to LHRH in immortalized hypothalamic neurons. Endocrinology 137:2606–2616

Witkin JW, Ferin M, Popilskis SJ, Silverman A-J (1991) Effects of gonadal steroids on the ultrastructure of GnRH neurons in the rhesus monkey: Synaptic input and glial apposition. Endocrinology 129:1083–1092

Witkin JW, O'Sullivan H, Ferin M (1995) Glial ensheathment of GnRH neurons in pubertal female rhesus macaques. J Neuroendocrinol 7:665–671

Zhang K, Sun J, Liu N, Wen D, Chang D, Thomason A, Yoshinaga SK (1996) Transformation of NIH 3T3 cells by HER3 or HER4 receptors requires the presence of HER1 or HER2. J Biol Chem 271:3884–3890

Zhang D, Sliwkowski MX, Mark M, Frantz G, Akita R, Sun Y, Hillan K, Crowley C, Brush J, Godowski PJ (1997) Neuregulin-3 (NGR3): A novel neural tissue-enriched protein that binds and activates ErbB4. Proc Natl Acad Sci USA 94:9562–9567

Zimmermann P (1982) Estrogen-dependent changes in the functional interrelationships among neurons, ependymal cells and glial cells of the arcuate nucleus. Cytometric studies in the female albino mouse. Cell Tissue Res 227:113–128

Zisch AH, D'Alessandri L, Amrein K, Ranscht B, Winterhalter KH, Vaughan L (1995) The glypiated neuronal cell adhesion molecule contactin/F11 complexes with src-family protein tyrosine kinase Fyn. Mol Cell Neurosci 6:263–279

Zonta M, Sebelin A, Gobbo S, Fellin T, Pozzan T, Carmignoto G (2003) Glutamate-mediated cytosolic calcium oscillations regulate a pulsatile prostaglandin release from cultured rat astrocytes. J Physiol 553:407–414

Abbreviations

AMPA	α-Amino-3-hydroxy-5-methyl-isoxazole-4-propionic acid
ATP	Adenosine 5′-triphosphate
bFGF	Basic fibroblast growth factor
CAH	Carbonic anhydrase
EGF	Epidermal growth factor
EP1	Prostaglandin E receptor-type 1
EP3	Prostaglandin E receptor-type 3

ErbB	Erythroblastosis B
FSH	Follicle-stimulating hormone
GABA	γ-Aminobutyric acid
GDH	Glutamate dehydrogenase
GnRH	Gonadotropin hormone-releasing hormone
GPI	Glycosylphosphatidyl inositol
GS	Glutamine synthetase
HB-EGF	Heparin-binding EGF-like growth factor
IGF-1	Insulin-like growth factor-1
KO	Knockout
LH	Luteinizing hormone
mGluRs	Metabotropic glutamate receptors
NO	Nitric oxide
NRGs	Neuregulins
OVLT	Organum vasculosum of the lamina terminalis
PaG	Phosphate-activated glutaminase
PGE_2	Prostaglandin E_2
POA	Preoptic area
RPTPβ	Receptor-like protein tyrosine phosphatase-β
SynCAM	Synaptic cell adhesion molecule
TACE	Tumor necrosis factor alpha converting enzyme
TGFα	Transforming growth factor alpha
TGFβ1	Transforming growth factor beta1

Chapter 20
Physiological and Pathological Roles of Astrocyte-mediated Neuronal Synchrony

Giorgio Carmignoto and Micaela Zonta

Contents

20.1 Introduction

The function of the brain is fundamentally based on the processing of signals that are transferred from neuron to neuron at the chemical synapses. Neurons are indeed the only cells capable of generating an action potential that travels down the axon to trigger neurotransmitter release at the synapse and thus guarantees the activation of specific postsynaptic targets. The action potential carries information that is essentially encoded into distinct patterns of action potential firing. However, to be truly significant in information transfer and processing, action potentials need to be phase locked among distinct groups of neurons, i.e., neurons have to work synchronously. For instance, during the course of learning a motor task, single neuron activity in the rat sensory-motor cortex does not change significantly but the coordination of firings of individual cells from a distinct population of neurons increases with the prediction of the learned response (Laubach et al., 2000). The response from distinct neuronal populations of the visual cortex has been also observed to be synchronized on millisecond timescale after activation with a specific visual stimulus, thereby suggesting that synchronization could serve to encode contextual information

G. Carmignoto
Istituto CNR di Neuroscienze and Dipartimento di Scienze Biomediche Sperimentali, Università di Padova, Viale G, Colombo 3, 35121 Padova, Italy
giorgio.carmignoto@bio.unipd.it

V. Parpura and P.G. Haydon (eds.), *Astrocytes in (Patho)Physiology of the Nervous System*,
DOI: 10.1007/978-0-387-79492-1_20,

by defining relations between the features of visual objects (Singer and Gray, 1995; Singer, 1999). Furthermore, in the olfactory bulb precise temporal firing patterns in distinct neuronal populations have been shown to be associated with perception of different odours (Wehr and Laurent, 1996).

Increasing evidence indicates that synchrony arises as a product of dynamic interactions in the neuronal network. In this context, gamma-aminobutyric acid (GABA)ergic interneurons are believed to play a crucial role by controlling the timing of action potentials (Whittington and Traub, 2003; Bacci and Huguenard, 2006). Indeed, inhibitory post synaptic potentials imposed by GABAergic interneurons on populations of principal neurons, set the phase of their subthreshold fluctuations of membrane potential that are generated by intrinsic, nonsynaptic membrane conductances (Alonso and Llinas, 1989; Silva et al., 1991; Amitai, 1994; Gutfreund et al., 1995). Given that these subthreshold fluctuations determine, together with postsynaptic potentials, the precise timing of neuronal firing, the resetting of phase by the GABAergic input increases the chance that a subsequent depolarizing synaptic signal evokes synchronized action potentials in different neurons (Cobb et al., 1995; Desmaisons et al., 1999).

While stimulus-specific neuronal synchrony may be a general feature in the processing of sensory information, an excess of neuronal synchronization is the hallmark of several brain disorders, including epilepsy and Parkinson's disease. However, it is worth underlying that epileptic synchronization reflects a synchronous bursting behaviour in a large neuronal population in which the individual action potentials from different neurons are not necessarily phase locked. In other words, the term "neuronal synchrony" is commonly used to indicate simultaneous neuronal activities that in terms of spike timing correlation, rapidity in the transition from uncorrelated to synchronized states and spatial extension of the synchronized activity, can be substantially different. It is beyond the scope of this article to discuss the definition of the different types of neuronal synchrony. Our overall aim here is, however, to discuss the hypothesis of astrocytes as non-neuronal elements that may contribute to generate and/or maintain on the one hand, the neuronal synchrony that could be physiologically relevant to information processing, and on the other, the hypersynchronous neuronal bursting that characterizes epileptic activities.

20.2 Non-synaptic Mechanisms of Neuronal Synchrony

As mentioned earlier, neuronal synchrony arises from the neuronal network as a process intrinsically linked to the activity of excitatory and inhibitory synaptic connections. However, under certain conditions, for example a marked reduction in the concentration of extracellular Ca^{2+}, a degree of neuronal synchrony can develop in groups of neurons independently of synaptic activity (Jefferys and Haas, 1982; Taylor and Dudek, 1982; Haas and Jefferys, 1984; Konnerth et al., 1984). Chemical synaptic transmission is, in fact, essentially blocked upon lowering of the extracellular calcium concentration, whereas the excitability of neurons is increased due to a reduction of the cation screening effect on the neuronal membrane. The repetitive population spike of large amplitude that under these conditions could be recorded represents an unequivocal sign of a synchronized activity across a large portion of the neuronal population. This non-synaptic form of neuronal synchrony, which has been observed experimentally in the hippocampus as well as in other brain regions such as the hypothalamus (Bouskila and Dudek, 1993), might have a physiological relevance in synaptic transmission since a transient depletion of extracellular Ca^{2+} in the microenvironment surrounding the synaptic terminals occurs during episodes of high neuronal activity (Borst and Sakmann, 1999; Stanley, 2000; Rusakov and Fine, 2003).

The nature of the signal that marks the onset of this type of neuronal synchrony is unclear. The important point that we would like to emphasize here is that to generate synchronous action potential discharges in a large neuronal population a depolarizing stimulus is necessary that could lead the neuronal membrane close to action potential threshold. Given that a depolarizing signal linked to synaptic activity is excluded, it has been proposed that the mechanism of non-synaptic neuronal synchrony relies on ephatic influences, i.e. electrical interactions that can occur between two neurons due to touching or close proximity of their membranes (Haas and Jefferys, 1984; Faber and Korn, 1989; Ghai et al., 2000), and electrotonic coupling among neurons (MacVicar and Dudek, 1981; Perez et al., 1999). While these two factors likely contribute to spreading of action potential discharges in large neuronal populations, it remains unclear how they can mark the start of neuronal synchrony.

The question is thus the following: can the signal that depolarizes the neuronal membrane to action potential threshold arise as a product of non-synaptic interactions among neurons or derive from a non-neuronal source? Given that glutamate released from activated astrocytes can significantly depolarize neurons, are astrocytes involved in the generation of non-synaptic neuronal synchrony? In support of this hypothesis, a decrease in the extracellular concentration of Ca^{2+} has been reported to affect, beside neurons, astrocytes. First in cultured astrocytes (Zanotti and Charles, 1997) and then in brain slice preparations (Parri et al., 2001; Fellin et al., 2004), low Ca^{2+} has been shown to represent an effective stimulus for evoking Ca^{2+} oscillations in astrocytes. It is thus conceivable that these glial cells contribute to the generation of non-synaptic neuronal synchrony. The neuronal synchronous activity observed in brain slices upon lowering of extracellular Ca^{2+}, might derive, at least in part, from glutamate that, once released from astrocytes upon their Ca^{2+} elevations, can act as a depolarizing stimulus for neurons.

Evidence for the ability of astrocytic glutamate to evoke synchronous neuronal responses has been obtained from paired recording experiments. These experiments were initially performed from pyramidal CA1 neurons in hippocampal slices and revealed that lowering the extracellular Ca^{2+} concentration evoked synchronous inward currents from two adjacent pyramidal neurons. These slow inward currents (SICs) had a much slower rise and decay times with respect to synaptic currents and because of their insensitivity to tetrodotoxin (TTX) they could not be due to action potential-mediated neurotransmitter release (Angulo et al., 2004; Fellin et al., 2004). These events were typically mediated by the *N*-methyl-D-aspartate glutamate receptor (NMDAR) subtype, mainly located at extrasynaptic sites (Fellin et al., 2004). Depolarizing voltage pulses applied to each neuron of the pair always failed to reveal evidence of electrotonic coupling between these neurons thus excluding a gap junctional communication that could have allowed spreading of the current from one neuron to the other. A series of subsequent experiments, that included photolysis of a Ca^{2+}-caged compound in single astrocytes (Fellin et al., 2004), provided compelling evidence for an astrocytic origin of glutamate that evokes SICs.

To investigate whether astrocyte-mediated neuronal synchrony could involve more than two neurons and obtain clues about its spatial extension, confocal microscope Ca^{2+} imaging experiments were then used. By monitoring Ca^{2+} signals in a relatively large population of CA1 neurons, this approach revealed that activation of Ca^{2+} oscillations in astrocytes by low Ca^{2+}, is followed by a simultaneous Ca^{2+} elevation in small groups of adjacent neurons, a response that we termed a "domain response" (Fellin et al., 2004). It resulted that stimuli commonly used to activate a Ca^{2+} signal in astrocytes, such as agonists of subtype I metabotropic glutamate receptors, of purinergic receptors and prostaglandin E2, trigger in the presence of TTX glutamate release from astrocytes and synchronous activity in neurons (Fellin et al., 2004, 2006a). The presence of Ca^{2+} elevations in astrocytes may thus be considered a condition sufficient for the generation of domain responses (Fig. 20.1). As to the signal that triggers synchronous Ca^{2+} elevations in groups of neurons, two hypotheses can be advanced. A domain response in neurons may be generated by a single episode of glutamate release from an astrocyte process (Fig. 20.1b). In such a case, dendrites from different neurons are located in close proximity to a glutamate release site in the astrocyte process. Once released, glutamate can thus activate rapidly the NMDAR at the different neuronal dendrites before its concentration becomes too low to activate these receptors. Alternatively, a synchronous Ca^{2+} rise at different processes, either from the same astrocyte or from different astrocytes, results in a simultaneous release of glutamate from multiple sites that, in turn, evokes the domain response in neurons (Fig. 20.1c).

Synchronous neuronal responses were confirmed in these experiments to be mediated exclusively by the NMDAR as they were blocked by the NMDAR antagonists D-2-amino-5-phosphonopentanoic acid (D-AP5) and MK-801, while an α-amino-3-hydroxy-5-methyl-4-isoxazolepropionic acid (AMPA) receptor antagonist had no effects (Fellin et al., 2004; Fellin et al., 2006a). The lack of an AMPA-mediated response was supposed to be due to the 100-fold lower affinity for glutamate of the AMPA with respect to the NMDAR. Indeed, following application of cyclothiazide and D-AP5 to prevent AMPA receptor desensitization and NMDAR activation, respectively, patch-clamp recordings from CA1 pyramidal neurons revealed the ability of astrocytic glutamate to trigger inward currents that were blocked by a subsequent application of the AMPA receptor antagonist 2,3-dioxo-6-nitro-1,2,3,4-tetrahydrobenzo[*f*]quinoxaline-7-sulfonamide (NBQX). It is noteworthy that the rise time of these AMPA-mediated currents was comparable to that of

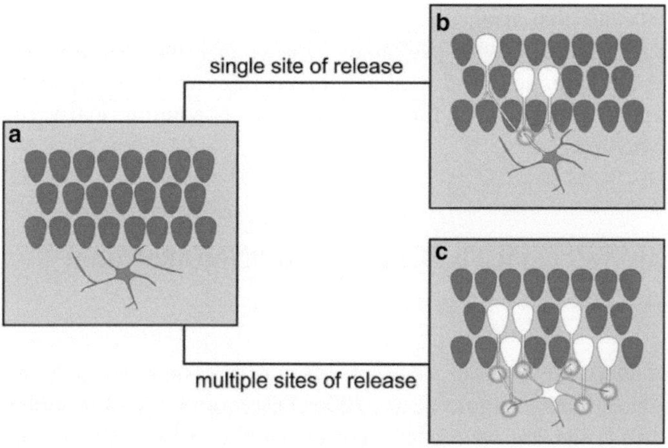

Fig. 20.1 Hypotheses for the generation of astrocyte-mediated neuronal synchrony. An astrocyte process is in close contact with dendrites from three different neurons. Activation of a Ca^{2+} signal at this process results in a single episode of glutamate release that evokes a synchronous response from the three neurons (**b**). Due to an intracellular Ca^{2+} elevation that occurs simultaneously in the astrocyte processes, glutamate is released from multiple sites to activate a synchronous response from neurons in contact with the activated processes (**c**).

NMDA-mediated currents. Apparently, given that astrocyte processes face the extrasynaptic neuronal membrane, glutamate cannot be directly released into the synaptic cleft and it is diluted into the large extracellular space. Its slow increase in the perisynaptic extracellular space results in a slow, yet effective, activation of the high affinity NMDAR as well as in a desensitization of the low affinity AMPA receptor.

An implication of this finding is that to relieve the Mg^{2+} block and activate efficiently the NMDAR, astrocytic glutamate cannot rely on the depolarizing effect of a co-activation of the AMPA receptor. Other agents should thus come into play to depolarize the neuronal membrane. For example, metabotropic glutamate receptors that are enriched at the extrasynaptic membrane, could be accessed by astrocytic glutamate and their activation result in either the opening of an unspecific cation channel conductance and the inhibition of a K^+ current (Crépel et al., 1994; Congar et al., 1997). Both actions lead to neuronal membrane depolarization around values that favour the removal of the Mg^{2+} block. The release of D-serine from astrocytes may be also involved (Schell et al., 1995). D-serine is probably the endogenous ligand that acts as a co-agonist with glutamate on the so-called "glycine site" of NMDAR to open the channel (Mothet et al., 2000). Intracellular Ca^{2+} elevations have been reported to be both necessary and sufficient for triggering D-serine release in astrocytes (Mothet et al., 2005). It is thus reasonable to hypothesize that following stimuli that trigger Ca^{2+} oscillations in astrocytes D-serine could be repetitively co-released with glutamate. Given that the "glycine site" on the NMDAR is not saturated, such a simultaneous release of glutamate and D-serine would enhance activation of the NMDAR thus expanding astrocyte-mediated neuronal synchrony.

20.3 Can Astrocytes be Considered "Non-neuronal Interneurons"?

Because of their presence in different brain areas, such as the CA1 and CA3 hippocampal regions (Angulo et al., 2004; Fellin et al., 2004, 2006b; Perea and Araque, 2005), the somatosensory cortex (Fellin et al., 2006b), the nucleus accumbens (D'Ascenzo et al., 2007) and the olfactory bulb (Kozlov et al., 2006), synchronous SICs can be considered a hallmark of astrocyte-to-neuron signalling. The functional implications of a synchronous NMDAR activation in groups of neurons by astrocytic glutamate are, however, not clear. Elucidation of this issue will possibly be provided by modern molecular genetic approaches that allow to impair specific signalling pathways selectively in astrocytes and then to determine the possible consequences for neuronal function. At the present time, we can only predict that astrocytic glutamate would be revealed to be more effective in the promotion of neuronal synchrony if its action could be studied under experimental conditions that could faithfully mimic physiological conditions. Indeed, to distinguish the synchronized responses mediated by astrocytic glutamate from that due to action-potential-mediated glutamate release, experiments were regularly performed in the continuous presence of TTX. What happens then if activation of NMDAR by astrocytic glutamate leads to action potential discharges? It seems obvious then that the initial action of astrocytes on a small number of neurons would lead, at least in principles, to synchronize the activity of a larger neuronal population.

The amplitude of at least some NMDA-mediated events evoked by astrocytic glutamate on neurons, according to recordings from the soma, can be as large as several hundreds of picoAmperes (Angulo et al., 2004; Fellin et al., 2006b; Haydon and Carmignoto, 2006). This observation provides per se an indirect, although compelling, evidence that this astrocyte signal can depolarize the neuronal membrane close to the threshold for a full activation of voltage-gated Na^{+} channels thus evoking action potential discharges. Direct support for such a view was recently provided by paired recordings in hippocampal pyramidal neurons from both CA1 and CA3 regions that were performed in the absence of TTX (Fellin et al., 2006b). In these experiments, synchronous SICs were first monitored from the two neurons in voltage-clamp recordings. Then, after changing to current-clamp configuration, the depolarizing events were observed to trigger in the same neurons action potential discharges. A subsequent application of TTX abolished action potential discharges without, however, affecting the depolarizing events that were sensitive to the NMDAR antagonist D-AP5. It thus appears that astrocytic glutamate can represent a powerful depolarizing signal that through NMDAR activation generates action potential discharges and thus deeply affects the output of the neuron. It appears that astrocytes can compose with the excitatory input a local circuit which acts in parallel, although on a different timescale, with interneurons to generate neuronal synchrony.

The functional implication of astrocytic glutamate-mediated action potential firing may be different depending on the type of neuron, the level of the membrane potential and the timing of this event with respect to that of synaptic events. According to the functional contacts of astrocyte processes with neuronal dendrites [a single astrocyte contacts tens of thousands of synapses (Bushong et al., 2002)], astrocytic glutamate can act directly on a high number of neurons but the strength of its depolarizing action may allow to trigger action potential firing only in a few neurons. Certainly, because of recurrent axonal excitatory connections that are present in several brain regions, including the CA3 hippocampus, an action potential discharge in just one single neuron has the potential to lead to synchronous bursts of the whole neuronal population (Miles and Wong, 1983). It follows that a focal excitation by astrocytic glutamate can be drastically amplified, at least in these regions, by action potential discharges that spread to a large neuronal population according to synaptic connections of the initially excited neurons (Fig. 20.2). Astrocytes can ultimately work as excitatory, "non-neuronal interneurons" that cooperate with synaptic signals from both excitatory and inhibitory connections to orchestrate neuronal synchrony.

A recent study raised the intriguing possibility that astrocytes operate also as inhibitory interneurons. In the olfactory bulb activated astrocytes were found to release not only glutamate, but also the inhibitory transmitter GABA (Kozlov et al., 2006). Most interestingly, paired recordings revealed that in two adjacent mitral neurons astrocytic GABA evoked hyperpolarizing currents, i.e. slow outward currents (SOCs), while in GABAergic granule interneurons astrocytic glutamate evoked NMDAR-mediated SICs. In both cases, SOCs and SICs occurred with a high level of synchrony. Given that synchronization of action potential firing in mitral cells is believed to depend on inhibitory post-synaptic potentials imposed by GABAergic granule cells, astrocytes may exert in the olfactory bulb a dual action. The first is the release of glutamate onto granule cells that by evoking a synchronous firing can help to coordinate their GABA release on mitral cells and thus to promote mitral cell synchronization. The second is the by release of GABA directly on mitral cells that provides a transient hyperpolarizing stimulus and thus causes a prolonged inhibition of action potential firing that switch off the synchronization regime. Astrocytes might thus provide a potentially important contribution to the generation of synchronized discharges of mitral cells, a phenomenon that is believed to be at the basis of odour discrimination (Schoppa, 2006), as well as to their cessation. Certainly, however, to clarify the astrocyte role in the olfactory bulb it is necessary to define accurately, by using also in vivo experiments, the spatial–temporal features of neuron–astrocyte reciprocal signalling.

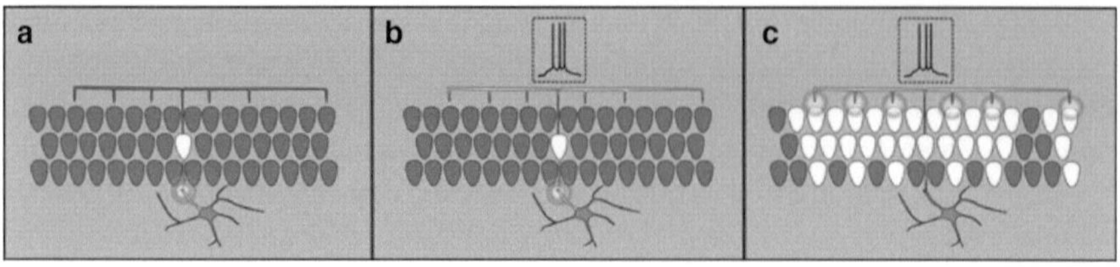

Fig. 20.2 Astrocyte-to-neuron signals lead to action potential firing. Astrocytic glutamate can depolarize the neuronal membrane to threshold for action potential firing. Due to the presence of recurrent excitatory axon collaterals, the initial, focal action of astrocytes onto a single neuron, can eventually spread to a large neuronal population.

20.4 Astrocyte and Epilepsy

While the rapid, transient synchronization of groups of neurons is believed to underline the processing of external sensory signals as well as the dynamics of cognitive processes, excessive neuronal synchrony are commonly viewed as a pathological manifestation of brain disorders such as Parkinson's disease, prion infection, and epilepsy. In the epileptic brain, hypersynchronous neuronal discharges are generally believed to originate from abnormalities intrinsic to neurons, but it has been also suggested that non-neuronal mechanisms may contribute to generate this epileptiform activity (Konnerth et al., 1986; Dudek et al., 1998; Jefferys, 2003). Given that astrocytes can have an interneuron-like action that favours synchronization of activity in small groups of neurons, an abnormal activation of these cells and a consequent massive glutamate release may represent one of these non-neuronal mechanisms. This hypothesis is supported by a recent study that used different models of chemical-induced seizures in acute brain slice preparations and in vivo (Tian et al., 2005). This study used both whole-cell recordings in current-clamp configuration from individual CA1 pyramidal neurons and field potential recordings. The first type of recording allowed the authors to monitor typical epileptic events from individual neurons, such as the marked depolarization with superimposed action potential discharges termed interictal event or paroximal depolarizing shift (PDS). The second, i.e., field potential recording, allows one to follow the transient depolarization that reflects the hypersynchronous activity of large neuronal populations. An important observation from this study is that the four experimental conditions that were used to evoke PDSs, i.e. 4-aminopyridine, Mg^{2+}-free extracellular solution, bicuculline and penicillin, also triggered, without exception, an increase in Ca^{2+} oscillations in the astrocytes, while antiepileptic drugs reduced astrocytic Ca^{2+} signals. Synchronous epileptiform activity was suggested to be generated by glutamate release from activated astrocytes since in the presence of TTX, PDSs were only slightly changed in frequency and amplitude, the only real difference being the absence in the current-clamp recording of the superimposed action potential firing. The authors concluded that neuronal action potential firing does not play an essential role in the generation of synchronous epileptiform activity. This intriguing conclusion is, however, at variance with previous studies in hippocampal slice preparations that reported a complete TTX-sensitivity of interictal discharges, and thus supports the dependence of interictal events, i.e., PDSs, on action potential-mediated mechanisms. Given these conflicting results, we decided to perform a series of experiments specifically designed to clarify whether or not astrocytic glutamate can play a central role in the generation of this interictal activity (Fellin et al., 2006b). As reported by Tian et al. (2005), we found that astrocytes were activated during a period of epileptiform activity induced by slice perfusion with picrotoxin and Mg^{2+}-free solution, and the release of glutamate potentiated. As a consequence, the frequency of SICs, that are mediated by astrocytic glutamate acting on neuronal NMDARs, was significantly increased in all regions tested, i.e., hippocampal CA3 and CA1 region, and somatosensory cortex. However, when we monitored simultaneously SICs by patch clamp and epileptiform activity by field potential recordings with four extracellular electrodes positioned in different hippocampal regions, we failed to observe a SIC in coincidence with an epileptiform event as detected by the extracellular electrodes. Furthermore, bath applied TTX lead to a rapid block of both interictal and ictal activity, but not of SICs, while antagonists of the NMDAR, such as D-AP5, reversibly blocked SICs, but they only reduced the duration of epileptiform activity. In the absence of action potential firing neither interictal nor ictal activity could thus be induced in brain slice preparations from all the regions tested, i.e., hippocampal CA3 and CA1 regions, and somatosensory cortex. Therefore, in contrast with the conclusion reached by Tian et al. (1995), our data support the view that neuronal firing is necessary for the generation of epileptiform activity. It is also clear from our observations that the epileptic, hypersynchronous activity in neurons strongly activates Ca^{2+} signalling in astrocytes which release glutamate that, in turn, depolarizes adjacent neurons to trigger, as mentioned above, a synchronous response. Therefore, epileptiform activities may arise in the brain as a product of a dysregulation of neuron–astrocyte reciprocal signalling, rather then from astrocytic glutamate per se as proposed by Tian et al. (2005). Under certain circumstances, indeed, the activation of astrocytes by episodes of high neuronal activity and their signalling back to neurons may create a recurrent loop between these two cell types that ultimately leads to, or amplifies, epileptiform activity. The data reported in Fellin et al. (2006), as well as our recent observations (Gomez-Gonzalo et al. (2008), support the view that an increased activity of astrocytes may have a role in the generation and/or maintenance of the ictal rather then of the interictal event.

Notwithstanding the excitement generated by these observations, many fundamental questions remain unanswered. Beside glutamate, are other gliotransmitters, such as adenosine 5′-triphosphate (ATP), involved in the modulation of neuron–astrocyte signalling during epileptiform activity? Given that adenosine, which derives from astrocytic ATP, has been shown to act on presynaptic purinergic A1 receptors to depress neurotransmitter release (Pascual et al., 2005), and taking into account that astrocytes activated by epileptiform discharges may release ATP, do astrocytes ultimately exert an antiepileptic rather than a proepileptic action? Along the same line, does a single astrocyte release both glutamate and ATP? Given that the extracellular concentration of Ca^{2+} decreases significantly during epileptiform activity, can a glutamate release through astrocyte $P2X_7$ purine receptors (Duan et al., 2003), or $P2X_7$-like receptors (Fellin et al., 2006a), be significantly activated? Above all, what are the spatial–temporal features of this neuron–astrocyte loop? According to available data, the astrocyte response to the synaptic release of neurotransmitters occurs with a delay of 1–2 s with respect to that of the postsynaptic neuron, i.e., at least one or two orders of magnitude slower. Due to the difficulties in monitoring accurately the intracellular Ca^{2+} change in small structures such as astrocyte processes [see, however (Pasti et al., 1997; Grosche et al., 1999)] a precise definition of the timing of these responses has not been obtained, and in most of the studies only the Ca^{2+} change at the soma was considered. The response at the soma, however, is secondary to that occurring at a distal process and it likely occurs with a delay which is determined by the speed in the intracellular propagation of the Ca^{2+} signal. Indeed, due to the proximity to the synaptic cleft, distal astrocyte processes likely represent the initial sites of activation by synaptic neurotransmitter release. Were this be the case, it would not be surprising to detect a rapid Ca^{2+} increase in astrocytic distal processes after an epileptic discharge. Noteworthy is that a subsecond latency activation of a Ca^{2+} increase in astrocyte processes would imply that astrocytes have the potential to signal back to neurons rapidly, i.e., on a timescale that is much shorter than that generally ascribed to the action of gliotransmitters.

20.5 Conclusions and Perspectives

According to the common view of brain function, a coherent behavioural response arises in the brain as a product of neuronal network activity dictated by synaptic connectivity. Multiple recent findings support the idea that astrocytes may be equally ranked players with neurons in the processing of sensory information, and, as such, to be essential for the generation of behavioural responses. However, while our knowledge of astrocyte properties has increased considerably over the last few years, to support such a conclusion new experimental approaches are needed. Above all, we need a better understanding of the rules that govern neuron–astrocyte signalling. In particular, the precise definition of how neurons and astrocytes coordinate their activity in space and time, both during different experimental conditions and in in vivo experiments (Nimmerjahn et al., 2004; Garaschuk et al., 2006), will be fundamental to develop an integrated view of the role of astrocytes not only in the genesis of brain disorders, but also in the processing of sensory information.

Given the role ascribed to synchronization of action potential discharges in cognitive processing, the ability of astrocytes to evoke neuronal synchrony might represent a clue for their involvement in this fundamental process. To specifically address this issue, a great help is provided by recent technological advances that enable new experimental approaches to be designed. For example, the possibility of loading Ca^{2+} indicators in neurons and astrocytes in vivo (Stosiek et al., 2003; Nimmerjahn et al., 2004; Garaschuk et al., 2006) together with recent technical developments of two-photon microscopy (such as fast *z*-dimension scanning), allow to monitor Ca^{2+} signals from hundreds of neurons and astrocytes in 3D space with 10–20 Hz temporal resolution (Gobel et al., 2007). Furthermore, local electroporation has been shown to induce an effective loading of dextran-bound Ca^{2+} indicators in small neuronal structures, such as spines and axon tracts (Nagayama et al., 2007). These techniques could be also combined with modern genetic approaches that allow to affect specific signalling pathways in a cell-type selective manner (Zhuo et al., 2001; Pascual et al., 2005). It seems likely that these new experimental approaches will soon produce valuable insights into the dynamics of neuron–astrocyte network, thus opening new perspectives for the understanding of brain function and the genesis of brain disorders.

References

Alonso A, Llinas RR (1989) Subthreshold Na+-dependent theta-like rhythmicity in stellate cells of entorhinal cortex layer II. Nature 342:175–177

Amitai Y (1994) Membrane potential oscillations underlying firing patterns in neocortical neurons. Neuroscience 63:151–161

Angulo MC, Kozlov AS, Charpak S, Audinat E (2004) Glutamate released from glial cells synchronizes neuronal activity in the hippocampus. J Neurosci 24:6920–6927

Bacci A, Huguenard JR (2006) Enhancement of spike-timing precision by autaptic transmission in neocortical inhibitory interneurons. Neuron 49:119–130

Borst JG, Sakmann B (1999) Depletion of calcium in the synaptic cleft of a calyx-type synapse in the rat brainstem. J Physiol 521:123–133

Bouskila Y, Dudek FE (1993) Neuronal synchronization without calcium-dependent synaptic transmission in the hypothalamus. Proc Natl Acad Sci USA 90:3207–3210

Bushong EA, Martone ME, Jones YZ, Ellisman MH (2002) Protoplasmic astrocytes in CA1 stratum radiatum occupy separate anatomical domains. J Neurosci 22:183–192

Cobb SR, Buhl EH, Halasy K, Paulsen O, Somogyi P (1995) Synchronization of neuronal activity in hippocampus by individual GABAergic interneurons. Nature 378:75–78

Congar P, Leinekugel X, Ben-Ari Y, Crépel V (1997) A long-lasting calcium-activated nonselective cationic current is generated by synaptic stimulation or exogenous activation of group I metabotropic glutamate receptors in CA1 pyramidal neurons. J Neurosci 17:5366–5379

Crépel V, Aniksztejn L, Ben-Ari Y, Hammond C (1994) Glutamate metabotropic receptors increase a Ca^{2+}-activated nonspecific cationic current in CA1 hippocampal neurons. J Neurophysiol 72:1561–1569

D'Ascenzo M, Fellin T, Terunuma M, Revilla-Sanchez R, Meaney DF, Auberson YP, Moss SJ, Haydon PG (2007) mGluR5 stimulates gliotransmission in the nucleus accumbens. Proc Natl Acad Sci USA 104:1995–2000

Desmaisons D, Vincent JD, Lledo PM (1999) Control of action potential timing by intrinsic subthreshold oscillations in olfactory bulb output neurons. J Neurosci 19:10727–10737

Duan S, Anderson CM, Keung EC, Chen Y, Swanson RA (2003) P2X7 receptor-mediated release of excitatory amino acids from astrocytes. J Neurosci 23:1320–1328

Dudek FE, Yasumura T, Rash JE (1998) 'Non-synaptic' mechanisms in seizures and epileptogenesis. Cell Biol Int 22:793–805

Faber DS, Korn H (1989) Electrical field effects: Their relevance in central neural networks. Physiol Rev 69:821–863

Fellin T, Pascual O, Gobbo S, Pozzan T, Haydon PG, Carmignoto G (2004) Neuronal synchrony mediated by astrocytic glutamate through activation of extrasynaptic NMDA receptors. Neuron 43:729–743

Fellin T, Pozzan T, Carmignoto G (2006a) Purinergic receptors mediate two distinct glutamate release pathways in hippocampal astrocytes. J Biol Chem 281:4274–4284

Fellin T, Gomez-Gonzalo M, Gobbo S, Carmignoto G, Haydon PG (2006b) Astrocytic glutamate is not necessary for the generation of epileptiform neuronal activity in hippocampal slices. J Neurosci 26:9312–9322

Garaschuk O, Milos RI, Konnerth A (2006) Targeted bulk-loading of fluorescent indicators for two-photon brain imaging in vivo. Nat Protoc 1:380–386

Ghai RS, Bikson M, Durand DM (2000) Effects of applied electric fields on low-calcium epileptiform activity in the CA1 region of rat hippocampal slices. J Neurophysiol 84:274–280

Gobel W, Kampa BM, Helmchen F (2007) Imaging cellular network dynamics in three dimensions using fast 3D laser scanning. Nat Methods 4:73–79

Gomez-Gonzalo M, Zonta M, Losi G, Uva L, Brondi M, de Curtis M, Ratto G, Carmignoto G (2008) Neuron-astrocyte excitatory loop contributes to epileptiform activity. Society for Neuroscience, 147, 23.

Grosche J, Matyash V, Moller T, Verkhratsky A, Reichenbach A, Kettenmann H (1999) Microdomains for neuron–glia interaction: Parallel fiber signaling to Bergmann glial cells. Nat Neurosci 2:139–143

Gutfreund Y, yarom Y, Segev I (1995) Subthreshold oscillations and resonant frequency in guinea-pig cortical neurons: Physiology and modelling. J Physiol 483 (Pt 3):621–640

Haas HL, Jefferys JG (1984) Low-calcium field burst discharges of CA1 pyramidal neurones in rat hippocampal slices. J Physiol 354:185–201

Haydon PG, Carmignoto G (2006) Astrocyte control of synaptic transmission and neurovascular coupling. Physiol Rev 86:1009–1031

Jefferys JGR (2003) Models and mechanisms of experimental epilepsies. Epilepsia 44(Suppl 12):44–50

Jefferys JG, Haas HL (1982) Synchronized bursting of CA1 hippocampal pyramidal cells in the absence of synaptic transmission. Nature 300:448–450

Konnerth A, Heinemann U, Yaari Y (1984) Slow transmission of neural activity in hippocampal area CA1 in absence of active chemical synapses. Nature 307:69–71

Konnerth A, Heinemann U, Yaari Y (1986) Nonsynaptic epileptogenesis in the mammalian hippocampus in vitro. I. Development of seizurelike activity in low extracellular calcium. J Neurophysiol 56:409–423

Kozlov AS, Angulo MC, Audinat E, Charpak S (2006) Target cell-specific modulation of neuronal activity by astrocytes. Proc Natl Acad Sci USA 103:10058–10063

Laubach M, Wessberg J, Nicolelis MA (2000) Cortical ensemble activity increasingly predicts behaviour outcomes during learning of a motor task. Nature 405:567–571

MacVicar BA, Dudek FE (1981) Electrotonic coupling between pyramidal cells: A direct demonstration in rat hippocampal slices. Science 213:782–785

Miles R, Wong RK (1983) Single neurons can initiate synchronized population discharge in the hippocampus. Nature 306:371–373

Mothet JP, Parent AT, Wolosker H, Brady RO, Jr., Linden DJ, Ferris CD, Rogawski MA, Snyder SH (2000) d-serine is an endogenous ligand for the glycine site of the *N*-methyl-D- aspartate receptor. Proc Natl Acad Sci USA 97:4926–4931

Mothet JP, Pollegioni L, Ouanounou G, Martineau M, Fossier P, Baux G (2005) Glutamate receptor activation triggers a calcium-dependent and SNARE protein-dependent release of the gliotransmitter D-serine. Proc Natl Acad Sci USA 102:5606–5611

Nagayama S, Zeng S, Xiong W, Fletcher ML, Masurkar AV, Davis DJ, Pieribone VA, Chen WR (2007) In vivo simultaneous tracing and Ca^{2+} imaging of local neuronal circuits. Neuron 53:789–803

Nimmerjahn A, Kirchhoff F, Kerr JN, Helmchen F (2004) Sulforhodamine 101 as a specific marker of astroglia in the neocortex in vivo. Nat Methods 1:31–37

Parri HR, Gould TM, Crunelli V (2001) Spontaneous astrocytic Ca^{2+} oscillations in situ drive NMDAR-mediated neuronal excitation. Nat Neurosci 4:803–812

Pascual O, Casper KB, Kubera C, Zhang J, Revilla-Sanchez R, Sul JY, Takano H, Moss SJ, McCarthy K, Haydon PG (2005) Astrocytic purinergic signaling coordinates synaptic networks. Science 310:113–116

Pasti L, Volterra A, Pozzan T, Carmignoto G (1997) Intracellular calcium oscillations in astrocytes: A highly plastic, bidirectional form of communication between neurons and astrocytes in situ. J Neurosci 17:7817–7830
Perea G, Araque A (2005) Properties of synaptically evoked astrocyte calcium signal reveal synaptic information processing by astrocytes. J Neurosci 25:2192–2203
Perez GJ, Bonev AD, Patlak JB, Nelson MT (1999) Functional coupling of ryanodine receptors to KCa channels in smooth muscle cells from rat cerebral arteries. J Gen Physiol 113:229–238
Rusakov DA, Fine A (2003) Extracellular Ca^{2+} depletion contributes to fast activity-dependent modulation of synaptic transmission in the brain. Neuron 37:287–297
Schell MJ, Molliver ME, Snyder SH (1995) D-serine, an endogenous synaptic modulator: Localization to astrocytes and glutamate-stimulated release. Proc Natl Acad Sci USA 92:3948–3952
Schoppa NE (2006) Synchronization of olfactory bulb mitral cells by precisely timed inhibitory inputs. Neuron 49:271–283
Silva LR, Amitai Y, Connors BW (1991) Intrinsic oscillations of neocortex generated by layer 5 pyramidal neurons. Science 251:432–435
Singer W (1999) Neuronal synchrony: A versatile code for the definition of relations? Neuron 24:49–65
Singer W, Gray CM (1995) Visual feature integration and the temporal correlation hypothesis. Annu Rev Neurosci 18:555–586
Stanley EF (2000) Presynaptic calcium channels and the depletion of synaptic cleft calcium ions. J Neurophysiol 83:477–482
Stosiek C, Garaschuk O, Holthoff K, Konnerth A (2003) In vivo two-photon calcium imaging of neuronal networks. Proc Natl Acad Sci USA 100:7319–7324
Taylor CP, Dudek FE (1982) Synchronous neural afterdischarges in rat hippocampal slices without active chemical synapses. Science 218:810–812
Tian G-F, Azmi H, Takano T, Xu Q, Peng W, Lin J, Oberheim N, Lou N, Wang X, Zielke HR, Kang J, Nedergaard M (2005) An astrocytic basis of epilepsy. Nat Med 11:973–981
Wehr M, Laurent G (1996) Odour encoding by temporal sequences of firing in oscillating neural assemblies. Nature 384:162–166
Whittington MA, Traub RD (2003) Interneuron diversity series: Inhibitory interneurons and network oscillations in vitro. Trends Neurosci 26:676–682
Zanotti S, Charles A (1997) Extracellular calcium sensing by glial cells: Low extracellular calcium induces intracellular calcium release and intercellular signaling. J Neurochem 69:594–602
Zhuo L, Theis M, Alvarez-Maya I, Brenner M, Willecke K, Messing A (2001) hGFAP-cre transgenic mice for manipulation of glial and neuronal function in vivo. Genesis 31:85–94

Abbreviations

AMPA	α-Amino-3-hydroxy-5-methyl-4-isoxazolepropionic acid
ATP	Adenosine 5'-triphosphate
D-AP5	D-2-Amino-5-phosphonopentanoic acid
GABA	Gamma-aminobutyric acid
NMDA	*N*-Methyl-D-aspartate
NMDAR	NMDA receptor
PDS	Paroxysmal depolarizing shift
SIC	Slow inward current
TTX	Tetrodotoxin

Chapter 21
Role of Ion Channels and Amino-Acid Transporters in the Biology of Astrocytic Tumors

Harald Sontheimer

Contents

The malignant transformation of central nervous system (CNS) glial cells gives rise to tumors that are collectively called gliomas. Although the vast majority of these cancers are believed to be of astrocytic origin, the actual cell of origin remains unknown. While gliomas present with many of the same genetic alterations in tumor suppressor genes or oncogenes that are common amongst cancers, their biology differs quite significantly from that of other neoplasms. Importantly, their growth is limited by the size of the skull, and hence tumor expansion can only occur when normal brain is destroyed. Recent research suggests that gliomas accomplish this by releasing glutamate at concentrations that cause excitotoxic neuronal cell death. Peritumoral glutamate may also contribute to seizures, which are a common comorbidity in patients with malignant glioma. Another differentiating feature of gliomas is their unusual ability to spread by diffusely invading normal brain tissue rather than through hematogenous routes as is common for most cancers. Glioma cell migration through the narrow extracellular brain spaces often occurs along blood vessels or nerve fiber tracts and requires cells to undergo profound changes

H. Sontheimer
Department of Neurobiology and Center for Glial Biology in Medicine, The University of Alabama at Birmingham, Birmingham, AL USA
sontheimer@uab.edu

V. Parpura and P.G. Haydon (eds.), *Astrocytes in (Patho)Physiology of the Nervous System*,
DOI: 10.1007/978-0-387-79492-1_21, © Springer Science+Business Media, LLC 2009

in cell shape and volume. These volume changes are facilitated by the transmembrane movement of water which follows the release of Cl^- and K^+ ions through ion channels. Inhibition of these ion channels retards the ability of cells to invade and the experimental Cl^- channel blocker chlorotoxin (Cltx) has advanced to Phase II clinical trials in the United States to treat high grade gliomas.

21.1 Gliomas

Malignant gliomas comprise a diverse group of primary CNS malignancies that are believed to originate from glial cells or their progenitor cells. These tumors account for over 12,000 deaths annually in the USA (Jemal et al., 2006) and are the most common solid cancer in children (Maher and Raffel, 2004). Tumor incidence has increased steadily over the past 30 years (Jemal et al., 2006) and the prognosis remains dismal (Azizi and Miyamoto, 1998). Effective therapies are largely absent, and current treatment modalities are limited to radiation therapy and, where possible, surgery (Butowski et al., 2006). Average survival following diagnosis ranges from 1 to 5 years but is less than 1 year for the vast majority of high-grade gliomas such as *glioblastoma multiforme* (Huncharek and Muscat, 1998).

The lineage relationship of gliomas to astrocytes is circumstantial. Tumors can only form from growth-competent cells; hence, in brain these are restricted to glial cells and cells associated with the vasculature and meninges. Astrocytes retain the ability to divide throughout life and respond to acute injury and many neurological diseases with the formation of scar tissue (Little and O'Callagha, 2001). Yet these scar-associated cells eventually stop dividing, and differentiate and form a tenacious barrier. By contrast, glial-derived tumors show unrestrained growth and fail to differentiate. Gliomas may include cells of all glial lineages and are named accordingly, i.e., oligodendrogliomas, astrocytomas, and glioblastomas (Kleihues et al., 1995). While these names imply a known lineage relationship to certain glial cell types, their actual cells of origin remain a mystery (Linskey, 1997). Likewise, events that trigger the malignant transformation of glial cells to become gliomas are poorly understood, although genetic alterations in oncogenes (*v-src*, *MDM-2*, *CDK4*, and *EGFR*) and tumor suppressor genes (*p53*, *p16*, *p15*, and *RB1*) are common characteristics (Louis, 1994; Von Deimling et al., 1995; Shapiro, 2001; Maher et al., 2001) and have been used to transform glial progenitor cells to become tumor-forming cells in mice (Holland et al., 1998).

Being cancerous cells, gliomas share many of the fundamental biological alterations with other cancers. For example, they frequently show amplification of a constitutively active epidermal growth factor receptor that enhances tumor growth in the absence of exogenous growth factors (Tang et al., 1997). Also, tumor suppressor genes such as *P16* and *p53* are frequently mutated (Von Deimling et al., 1995). Neovascularization, the induction of new blood vessels, is a typical feature of many tumors and also characterizes high-grade gliomas (Plate and Risau, 1995). Despite these commonalities with other tumors, gliomas exhibit unique biological traits and have developed adaptations to support their unique biology. Several examples of physiological adaptations and the potential to exploit them for future therapeutic purposes will be discussed in this chapter.

21.1.1 Glioma Growth and the Role of Glutamate

Among the most notable difference to solid tumors of the body is the fact that gliomas expand into a constricted space, the cranium. This space, defined by the bony cavity of the skull, limits their possible physical expansion, as only 15% of the cranial volume is not occupied by brain tissue. Although compression of brain into the fluid-filled ventricular spaces in brain is not uncommon, this space is insufficient to accommodate the tumor's ultimate expansion. As illustrated for several representative examples in Fig. 21.1, tumors grow into brain areas previously occupied by normal brain tissue and hence must have destroyed normal tissue to vacate room for their own expansion. Systemic tumors, by comparison, grow within soft tissues or expand into body cavities without the need to destroy the organ tissue during their growth. Surprisingly, little attention had been paid to the mechanism whereby gliomas destroy surrounding brain as they grow.

A few recent studies have examined this question and surprisingly suggest that glioma cells at the tumor margins of an expanding tumor release the neurotransmitter glutamate into adjacent brain at concentrations that cause neuronal cell death (Ye and Sontheimer, 1999; Behrens et al., 2000; Takano et al., 2001). Being the principal excitatory neurotransmitter in the brain, glutamate is normally restricted to the synaptic space where it binds to postsynaptic neuronal glutamate receptors that mediate fast signal transmission. In this restricted synaptic space, glutamate rises to millimolar concentrations, yet its concentration in the extracellular space is typically maintained very low. This prevents the erroneous activation of extrasynaptic neuronal glutamate receptors. Their activation triggers sustained influx of Ca^{2+}, which in turn could overwhelm intracellular Ca^{2+} regulatory processes ultimately leading to Ca^{2+}-mediated cell death (Choi, 1988). This process, termed excitotoxicity (Olney et al., 1971), is now believed to be a final common death pathway in numerous diseases ranging from acute trauma and ischemic stroke to progressive neurological conditions such as ALS or Alzheimer's (Choi, 1988; Lipton and Rosenberg, 1994).

Because of the vulnerability of the brain to excessive glutamate, several Na^+-dependent transport systems have evolved to maintain extracellular glutamate at very low micromolar levels (Danbolt, 2001). These transporters are primarily expressed on astrocytes (Danbolt, 2001) and include the excitatory amino acid transporters 1 and 2 (EAAT1 and 2) also known as the L-glutamate/L-aspartate transporter (GLAST) and the glial L-glutamate transporter (GLT-1) (Danbolt, 2001). The important neuroprotective role that these astrocytic transporters have has been documented by knockout studies that show widespread neurodegeneration and lethal epilepsy in animals that have lost functional astrocytic glutamate transporters (Rothstein et al., 1996; Tanaka et al., 1997). In light of these findings, it was

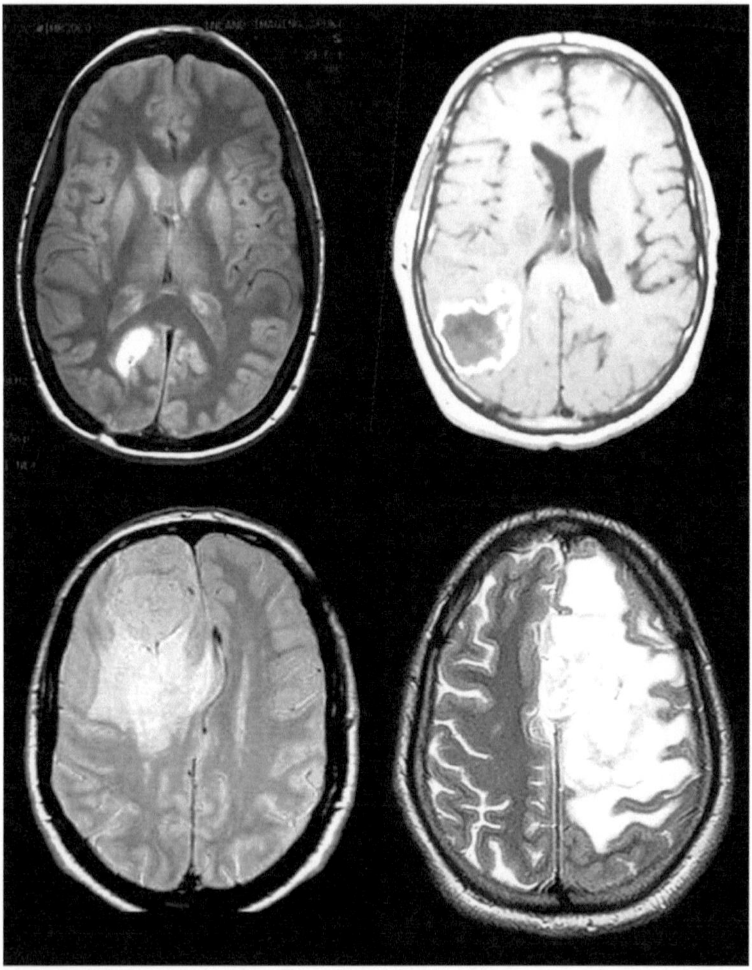

Fig. 21.1 Gliomas destroy normal brain as the tumor expands. Examples of images from four different patients that presented with malignant gliomas, each illustrating the fact that normal brain tissue was displaced to accommodate the tumor's expansion.

intriguing to find that these glutamate transporters, which are so abundant in astrocytes, are not functional in astrocytic tumors (Ye et al., 1999). Hence, gliomas cannot participate in glutamate uptake from the extracellular space. To the converse, these tumor cells release glutamate into the peritumoral space (Ye and Sontheimer, 1999; Takano et al., 2001) where it causes excitotoxic death of peritumoral neurons (Ye and Sontheimer, 1999; Takano et al., 2001) probably allowing the tumor to expand into the vacated space (Takano et al., 2001). In addition, recent data, discussed further later, suggest that the released glutamate may serve as a trophic factor promoting glioma cell invasion.

21.1.2 Glutamate Excitotoxicity Is a Byproduct of Glutathione Synthesis

The initial findings that suggested an inappropriate handling of glutamate by gliomas compared the uptake of radioactively labeled glutamate between normal and malignant astrocytes (Ye and Sontheimer, 1999). This study showed that glioma cell lines established from patients as well as acute patient biopsy tissues lack expression of the most prominent astrocytic glutamate transporter GLT-1. Indeed, staining of tissue biopsies with GLT-1 antibodies clearly demarcates the tumor boundaries where nonmalignant astrocytes prominently express GLT-1 but gliomas lack expression (Fig. 21.2a). During embryonic development, astrocytes also express GLAST (EAAT1), and many cultured astrocytes maintain expression of this related member of the EAAT family of Na^+-dependent transporters (Ullensvang et al., 1997). Interestingly, GLAST is found mislocalized to the nucleus of gliomas in culture and in brain biopsies from glioma patients (Ye et al., 1999). Not surprisingly therefore, glioma cells do not show significant glutamate uptake, accounting for less than 5% of that observed in normal astrocytes. As mentioned earlier, much to the contrary, and quite surprisingly glioma cells produce and release massive amounts of glutamate (Fig. 21.2b) (Ye and Sontheimer, 1999). Within a time period of just a few hours, a monolayer of cultured glioma cells is able to raise extracellular glutamate concentrations in a 70-mL culture flask up to 500-fold from about 1 to 500 μM. This constant release of glutamate is dependent on the de novo synthesis of glutamate by glioma cells from glutamine and requires the presence of extracellular cystine but is not Na^+ dependent (Ye et al., 1999). When hippocampal or cortical neurons are placed in the vicinity of gliomas, neurons undergo *N*-methyl-D-aspartate (NMDA)-dependent excitotoxic cell death (Fig. 21.2c) (Ye and Sontheimer, 1999).

The ionic profile of the glutamate release pointed to a Na^+-independent cystine-glutamate exchanger as the main pathway for glutamate release, and pharmacological and biochemical studies (Ye et al., 1999) indeed identified a cystine-glutamate antiporter named "system X_c-" as the probable candidate. This recently cloned electroneutral, Na^+ independent amino acid transporter (Sato et al., 1999) exchanges cystine for glutamate (Ye et al., 1999) and is upregulated under conditions of oxidative stress (Kim et al., 2001). The transporter is composed of a catalytic subunit (xCT) and a regulator subunit (4F2hc). System X_c- is found in both normal and malignant human brain (Fig. 21.3a) but appears upregulated in glioma tissue. In normal glia cells glutamate release through this pathway that occurs in conjunction with cystine uptake would not alter extracellular glutamate significantly since the released glutamate would be quickly removed by reuptake by GLT-1 (Fig. 21.3b, c). However in gliomas, the lack of GLT-1 expression explains a progressive accumulation of glutamate around tumor cells which in turn inflicts excitotoxic injury on adjacent neurons (Fig. 21.3d).

While these initial studies examined cultured glioma cells, glutamate release has now also been demonstrated in two independent human studies using either microdialysis (Roslin et al., 2003), which directly samples peritumoral fluids, or by

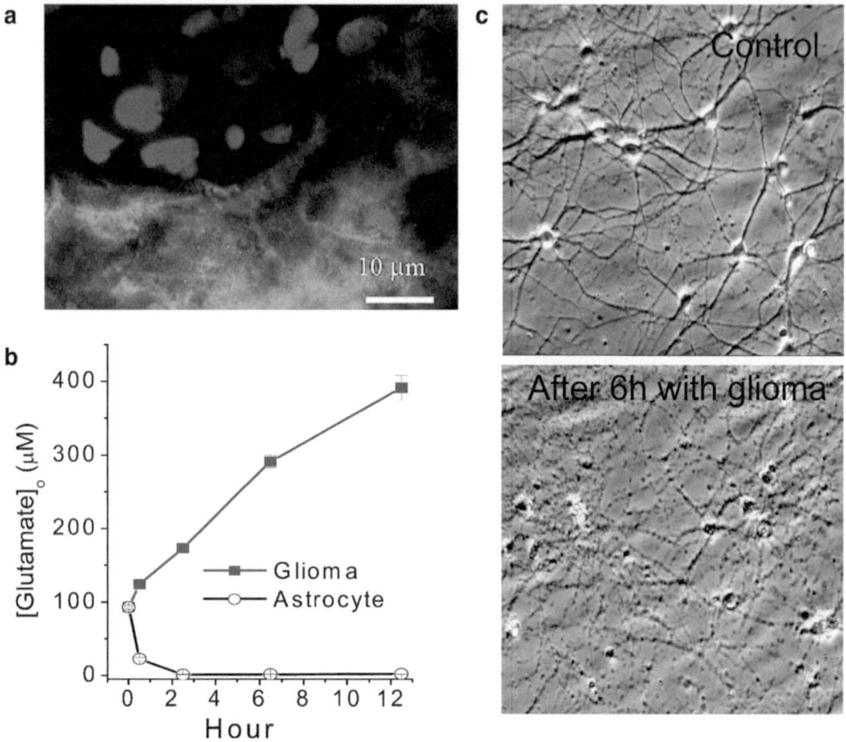

Fig. 21.2 Gliomas release glutamate. (**a**) Gliomas in vivo do not express the GLT-1 glutamate transporter (green) while the surrounding astrocytes do. The GLT-1 staining demarcates the tumor boundaries. Red staining in propridium iodine to label cell nuclei. (**b**) Glioma cells release glutamate as determined by sampling medium in a 70-mL flask as a function of time. By contrast nonmalignant astrocytes remove glutamate when challenged with 100 μM at the start. (**c**) Time-lapse video microscopy shows pronounced neuronal cell death when hippocampal neurons are placed in the vicinity of glioma cells yet without touching them (sandwich culture). Reproduced, with permission, from Ye and Sontheimer (1999) (*See Color Plates*).

spectroscopic magnetic resonance imaging (Rijpkema et al., 2003). Both studies suggest significant increases in peritumoral glutamate in patients with malignant glioma. Furthermore, implantation of glioma cells that are deficient in glutamate release fails to grow solid tumors when implanted into rat brain (Takano et al., 2001), while those that release glutamate grow rapidly ultimately killing the animal. Of note, peritumoral seizures are a common comorbidity in most gliomas and affect between 50 and 80% of all patients, even those presenting with low-grade tumors (Oberndorfer et al., 2002). Of course, the aforementioned glutamate release is likely responsible for the seizure activity in neurons in the peritumoral vicinity.

While the excitotoxic elimination of peritumoral neurons may be of potential benefit to the growing tumor, it is unlikely that this pathway evolved with a primary

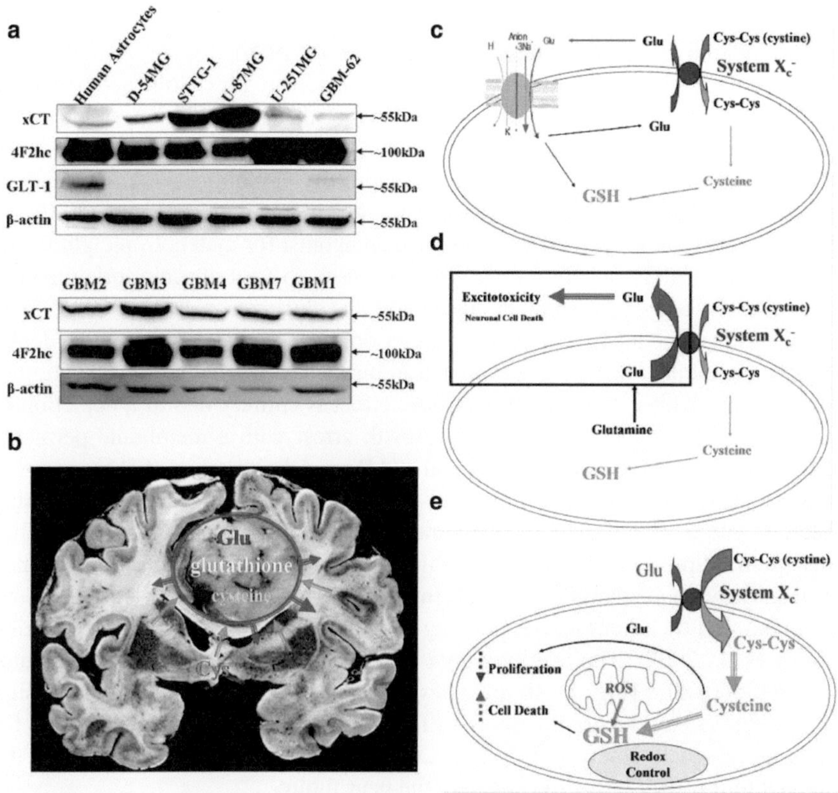

Fig. 21.3 Cystine-glutamate exchange in gliomas. (**a**) Glioma cell lines and acute biopsies highly express the catalytic (xCT) and regulatory (4F2hc) subunit of the system X_c- cystine glutamate exchanger, but lack the Na^+-dependent GLT-1 transporter. β-Actin was probed as a control for equal loading. (**b**) The system X_c- cystine glutamate exchanger is responsible for the release of glutamate from the tumor into the surrounding brain. (**c**) In normal glia, glutamate (Glu) released via system X_c- is taken back up by GLT-1 transporter. (**d**) In gliomas, the loss of this reuptake via GLT-1 causes a buildup of glutamate and excitotoxic injury. (**e**) The main purpose of the system X_c- transporter is to supply cysteine for the production of the cellular antioxidant glutathione (GSH). *ROS* radical oxygen species (a, reproduced with permission from Lyons et al., 2007) (*See Color Plates*).

purpose to kill neurons. Instead, it is more likely that glutamate release occurs as an obligatory byproduct of cystine uptake by via system X_c- by these tumor cells (Fig. 21.3e). Cystine is an essential precursor for the biosynthesis of the cellular antioxidant glutathione (GSH), a tripeptide of glutamate, cysteine (the reduced form of cystine), and glycine. The supply of cysteine is essential for GSH biosynthesis making cellular uptake of cystine the limiting factor for overall GSH production. GSH levels are upregulated in gliomas and may make these tumors more resistant to oxidative stress and exposure to certain chemotherapeutic drugs (Iida et al., 1997). Importantly, radiation damage that invokes hydroxyl radicals (Knuutila,

1984) is also reduced by GSH (Simone et al., 1983), and hence elevated GSH levels confer radiation resistance to tumors (Mitchell et al., 1989).

If indeed system X_c- was the principal uptake pathway for cystine and hence the release pathway for glutamate, one would expect that its inhibition would block glutamate release and possibly starve cells of GSH. This in turn may compromise tumor growth in vivo. To address this question, two inhibitors were identified as suitable drugs that each block the transporter with sufficient efficacy and specificity. These are (*S*)-4-carboxyphenylglycine (also an agonist for metabotropic glutamate receptors) and sulfasalazine, a drug clinically used to treat chronic inflammatory bowel disease. Both drugs inhibit cystine uptake and glutamate release at 100–250-μM concentrations, and a 24-h exposure to either drug causes the near complete depletion of intracellular GSH (Fig. 21.4a) and in turn leads to growth arrest of gliomas (Chung et al., 2005). The growth retarding effect is entirely due to a depletion of GSH since cells can be rescued from growth arrest with a membrane permeant GSH ester. Since sulfasalazine is an Food and Drug Administation (FDA) approved drug, and hence could be administered to glioma patients, its growth-inhibiting effects on gliomas have also been examined in preclinical animal models for malignant glioma. Specifically, *scid* mice with implanted human gliomas were treated twice daily by i.p. injection of sulfasalazine for up to 10 weeks, and tumor response was monitored by luminescence imaging as well as histopathological examination (Chung et al., 2005). The resulting data suggest that sulfasalazine reduced tumor volume by 60–80% (Fig. 21.4b, c). These studies were carried out with a dosing scheme equivalent to that of patients suffering from Crohn's disease and hence for which an adequate safety profile is established. Hence, these findings have the potential to translate into clinical use in the near future.

21.1.3 Glutamate as "Motogen"

As indicated earlier, the unusual ability of glioma cells to invade the normal brain presents a huge clinical challenge, and hence, much research has been devoted to gain a better understanding of the underlying biology. This research has identified a preference for certain extracellular matrix molecules as substrate for cell migration (Demuth and Berens, 2004). Indeed invading gliomas can synthesize their own unique extracellular matrix as they invade (Zamecnik, 2005). Invading cells interact with extracellular matrix through integrin receptors (Gladson, 1999), which are also involved in the dynamical attachment/detachment of focal adhesion sites. Some of these dynamic changes appear to correlate with changes in intracellular Ca^{2+} (Manning et al., 2000). Of note, intracellular Ca^{2+} oscillations frequently correlate with cell migration (Maghazachi, 2000) and have been observed in migratory glioma cells (Bordey et al., 2000). These Ca^{2+} oscillations can be triggered by a number of extracellular ligands including growth factors (Bryant et al., 2004), lysophosphatidic acid (Manning et al., 2000), and neurotransmitters such as acetylcholine (Komuro and Rakic, 1996) and glutamate (Kim et al., 1994). Glutamate is

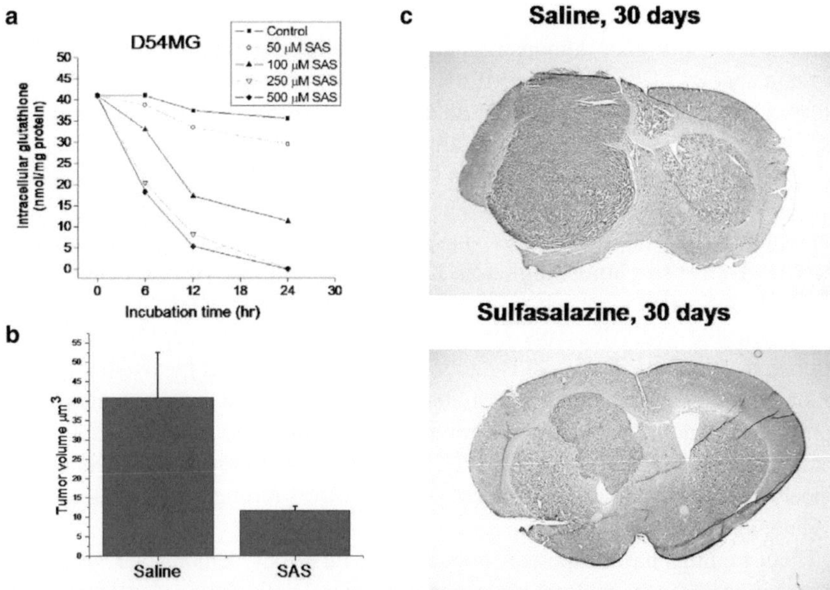

Fig. 21.4 Inhibition of system X_c- causes glutathione depletion and reduces tumor size in vivo. (**a**) Sulfasalazine (SAS), which inhibits system X_c-, causes a dose- and time-dependent reduction in glutathione levels in D54MG glioma cell line. (**b**) Intraperitoneal injection, twice daily, of 8 mg kg^{-1} sulfasalazine causes a 75% reduction in tumor volume in tumor-bearing animals. (**c**) A representative comparison of control, saline injected to sulfasalazine treated animal at 30 days of treatment. Reproduced, with permission, from Chung et al. (2005) (*See Color Plates*).

of particular interest in this context as studies on cerebellar granule cells have demonstrated a requirement for glutamate-mediated Ca^{2+} oscillations in neuronal migration in the cerebellum (Komuro and Rakic, 1996). In cerebellar granule cells the activation of NMDA receptors (NMDAR) causes the transient influx of Ca^{2+}, which in turn regulates the rate and velocity of granule cell migration (Komuro and Rakic, 1998). Although glioma cells express receptors for glutamate, they do not express NMDAR. Instead, they express a subclass of α-amino-3-hydroxy-5-methyl-4-isoxazolepropionic acid receptors (AMPAR), which is permeable to Ca^{2+}. These receptors lack the expression of the GluR2 subunit that confers Ca^{2+} impermeability (Hollmann et al., 1991). GluR2 is also absent in some neuronal glutamatergic receptors early in development (Pellegrini-Giampietro et al., 1992). To demonstrate a role for these receptors in glioma migration, Ishiuchi et al. (2002) transfected the GluR2 subunit into glioma cells using a viral transfection system. This rendered glioma cells unable to respond to glutamate with Ca^{2+} oscillations, and upon implantation into a host animal brain, these glioma cells failed to invade. Hence, glutamate acts as a motogen, or a molecule that promotes cell migration through activating Ca^{2+} entry. Although glioma cells utilize a different receptor system, they appear to recapitulate the same underlying biology that is observed in neuronal migration during development.

This then raises the question of the nature of the glutamate source. In light of the earlier discussed studies demonstrating glutamate release via the system X_c- transporter, it appears plausible that AMPARs on glioma cells may respond to glutamate released from the same cell, i.e., in an autocrine fashion. Alternatively, cells may respond to glutamate released from adjacent cells via a paracine signaling loop. Recent studies support this hypothesis (Lyons et al., 2007) suggesting that glioma cells respond preferentially to glutamate released via system X_c-. This study shows that inhibition of glutamate release via system X_c- inhibits Ca^{2+} responses and compromises Transwell glioma migration. Both can be restored with exogenous glutamate unless AMPARs are inhibited using Joro spider toxin, a specific inhibitor of Ca^{2+}-permeable AMPARs. This data therefore suggests that glutamate serves a dual purpose in glioma biology: (1) By acting as an excitotoxin, it clears space for the tumor expansion, hence promotes tumor growth; (2) by acting as a motogen, it enhances the motility and hence promotes tumor invasion. Indeed, glioma cells frequently display chain migration. Here, many cells follow a lone leader. It is conceivable that the leading cell recruits its followers through the targeted release of glutamate as it migrates.

These findings have important implications. First they demonstrate that glioma cells share a common biology with immature neural cells during development and hence may be an important model system to study certain developmental processes. Second, both the glutamate transporters involved in glutamate release as well as the glutamate receptors targeted on neurons and glioma cells are potential therapeutic targets for treatment of these tumors. Although NMDAR antagonists have had mixed results in the treatment of stroke (Lipton, 2004), they may represent a viable treatment modality in advanced stages of brain cancer and hence warrant further study. The inhibition of glutamate release from gliomas using sulfasalazine has not been considered thus far but it appears a plausible strategy to curtail glioma invasion.

21.2 Ion Channels and Glioma Cell Invasion

As indicated earlier, gliomas can migrate over long distances as they spread to form secondary tumors throughout the brain. Unlike systemic cancers, which exhibit metastasis through hematogenous spread, gliomas rarely extravasate into the blood. Instead, they invade via extracellular routes and appear very capable of navigating the narrow and tortuous extracellular spaces in brain. Invading glioma cells often appear wedge shaped (Fig. 21.5a) with a thin leading edge and overall thinned or shrunken appearance. In response to the physical challenges imposed on these invading cells, glioma cells appear to have developed an unusual ability to regulate their volume so as to adjust their shape and cell volume to these spatial constraints. These adaptations appear to involve the coordinated activity of ion channels and transporters as release pathway for osmotically active ions and water. This hypothesized concept is illustrated in Fig. 21.5b and more extensively discussed later. In a nutshell, migrating cells release Cl^- and K^+ to drive osmotically obligated H_2O out of the cell. This necessitates

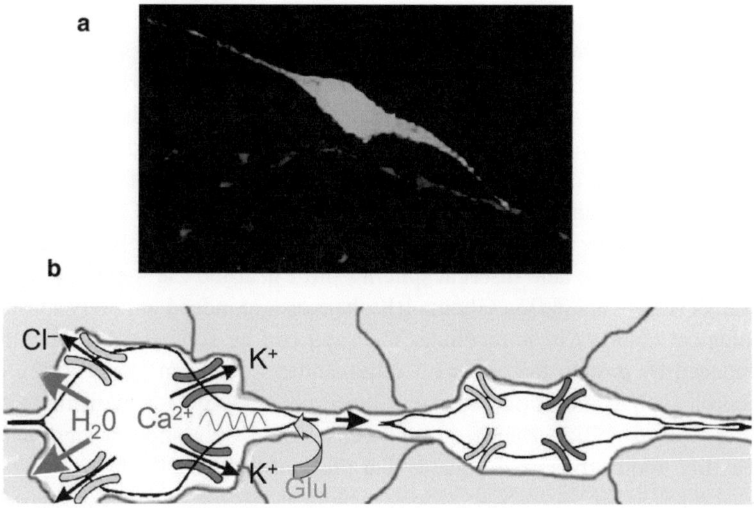

Fig. 21.5 Glioma cells shrink as they invade. (**a**) Confocal image of an invading D54MG glioma cell stably expressing enhanced green fluorescent protein shows an elongated wedge-shaped appearance. (**b**) Cell shrinkage requires efflux of water, which is energetically driven out of the cell through the concerted secretion of Cl^- and K^+ through ion channels. Glutamate is shown as a motogenic stimulus and acts via AMAPR to raise intracellular Ca^{2+}, which may in turn activate Ca^{2+}-activated BK channels (*See Color Plates*).

an outwardly directed gradient for Cl^- and K^+ ions and the coordinated activity of Cl^- and K^+ channels, as well as significant water permeability.

21.2.1 Glioma K^+ Channels

Nonmalignant astrocytes are characterized by a very negative resting membrane potential and high K^+ permeability (Kofuji and Newman, 2004). The latter is believed to be at least in parts due to the high expression of the inwardly rectifying K^+ channel Kir4.1 (Olsen et al., 2006). This channel is developmentally regulated and is absent in immature glial cells (MacFarlane and Sontheimer, 2000). It does characterize fully differentiated and presumably postmitotic astrocytes (MacFarlane and Sontheimer, 2000) and oligodendrocytes (Kofuji et al., 2000). Glioma cells, like essentially all other tumors (Cone, 1974), have a much more depolarized resting membrane potential, which is largely unresponsive to changes in $[K^+]_o$, hence indicating a low K^+ permeability. Not surprisingly therefore electrophysiological recordings showed the complete absence of functional Kir channels (Olsen and Sontheimer, 2004). However, Western blot analysis of human glioma biopsies showed expression of several Kir genes including *Kir4.1* (Olsen and Sontheimer, 2004). By immunohistochemistry, these are predominantly localized to intracellular, perinuclear regions of the cell but not in the surface membrane, and hence these channels cannot participate in transmembrane currents. However, glioma cells show large outward K^+ currents when depolarized, and these currents are mediated by big conductance Ca^{2+}-activated (BK) channels (Ransom and Sontheimer, 2001). The underlying channel has been cloned from glioma cells and shown to derive from the *hslo* gene by differential splicing. This gene gives rise to five splice variants that differ in their pharmacological and biophysical properties (Chen et al., 2005). Glioma cells express a unique splice variant that was termed glioma BK (gBK) containing a 34 amino acid insert at splicing site 2 near the Ca^{2+}-sensing domain of the channel (Liu et al., 2002). Glioma BK channels are indeed uniquely sensitive to physiological changes in intracellular Ca^{2+} and can be activated by elevations in Ca^{2+} induced by growth factors or neurotransmitters (Ransom et al., 2002). In the absence of such stimulation, however, these channels do not contribute to the membrane permeability. Hence, glioma cells would only show a significant K^+ permeability upon exposure to ligands that raise intracellular Ca^{2+}.

21.2.2 Glioma Cl⁻ Channels

While gliomas do not show a resting K^+ permeability in the absence of a Ca^{2+} signal, they do exhibit an unusual resting permeability to Cl^- ions (Ransom et al., 2001). This is surprising since neither astrocytes nor neurons show any appreciable resting Cl^- permeability (Walz, 2002). The underlying Cl^- channels are sensitive to Cd^{2+}, 5-nitro-2-(3-phenyl-propylamino)benzoic acid (NPPB), and 4,4′-diisothiocyanato-stilbene-2,2′-disulfonate (DIDS) (Olsen et al., 2003), and antisense knockdown experiments suggest that both ClC-2 and ClC-3 channels, members of the CLC family of chloride channels and transporters, participate in transmembrane Cl^- flux (Olsen et al., 2003). Amphotericin patches, which allow current recordings without disturbing the intracellular ionic milieu, showed that glioma cells maintain a resting Cl^- conductance even at their relatively positive resting membrane potential of ~−40 mV (Ransom et al., 2001) and channel openings cause an inward current or the efflux of Cl^- ions. This suggests that cells accumulate Cl^- ions above the electrochemical equilibrium potential. Indeed, recent measurements using the Cl^--sensitive fluorescence indicator SPQ (Sontheimer and Ernest, unpublished) suggest a resting $[Cl^-]_i$ of 100 mM, which is established by the activity of a bumetanide-sensitive Na^+–K^+–Cl^- transporter (Sontheimer and Ernest, unpublished). These findings support the hypothesis that glioma cells accumulate Cl^- ions to function as osmolytes in the context of cell volume regulation. As reasoned earlier and schematically illustrated in Fig. 21.5b, invading glioma cells would release Cl^- ions in conjunction with K^+ and obligated water to shrink as they invade the narrow extracellular spaces in brain. Consistent with this hypothesis, pharmacological inhibition of glioma Cl^- channels with either NPPB or Cltx, a putative Cl^- channel-specific peptide blocker (DeBin et al., 1993) each inhibits Transwell glioma migration (Soroceanu et al., 1999). Similarly, replacement of Cl^-

with impermeant anions such as gluconate retards Transwell migration (Soroceanu et al., 1999) whereas permeable anions such as Br^- or NO_3^- support cell volume changes and Transwell migration.

21.2.3 Water Channels

For this hypothesized role, Cl^- ions would efflux in conjunction with K^+ to move osmotically obligated water out of the cells. Although water can permeate lipid membranes to some extent, the expression of dedicated water channels or aquaporins (AQPs) enhances the water permeability of biological membranes significantly (Amiry-Moghaddam and Ottersen, 2003). Of the 12 AQPs cloned to date, 3 have been identified in brain (Gunnarson et al., 2004) and 2 of them, AQP1 and AQP4, are highly expressed in glioma biopsies (Endo et al., 1999; Saadoun et al., 2002a, b) and in cell lines established from patient biopsies. Comparing the invasive potential of glioma cells it became evident that glioma cells that express AQP1 show significantly enhanced cell invasion (McCoy and Sontheimer, 2007) compared with cells expressing AQP4 or none at all. Hence, it is likely that AQP1 acts in concert with ClC-3 and BK channels to promote the hypothesized volume changes during cell invasion.

Consistent with such cooperation, these three proteins colocalize to the same domains on the cells, namely the invading edges often referred to as invadipodia. These processes are characterized by enhanced expression of each of the three proteins. These invadipodia are also characterized by specialized lipids often referred to as lipid rafts. They label with the β subunit of cholera toxin. Biochemical isolations of lipid rafts demonstrate the presence of BK, ClC-3, and AQP1 in a lipid fraction that also contains the protein caveolin-1, frequently used to define the so-called caveolar raft domains. The significance of the lipid association of these proteins is not entirely clear although experimental disruption of these rafts with cholesterol binding drugs also retards cell invasion (Weaver and Sontheimer, unpublished).

The concept that $Cl^-/K^+/H_2O$ efflux supports cell shrinkage of invading cells and the finding that Cl^- channel inhibitors reduce invasion led to the exploration of a Cl^- channel blocker in clinical studies. These studies had been preceded by the characterization of the scorpion-derived Cl^- channel blocking peptide Cltx (DeBin et al., 1993), which was found to be an inhibitor of Cl^- currents in glioma cells (Ullrich and Sontheimer, 1996). Exposure of gliomas to the peptide causes the progressive disappearance of Cl^- currents attributable to ClC-3 channels (Ullrich et al., 1996). Rather than blocking the channel directly, as this would be expected for neurotoxins, Cltx led to the progressive internalization of ClC-3 channels into endocytotic vesicles (McFerrin and Sontheimer, 2006). Binding of the peptide required the presence of the matrix metalloproteinase II (MMP2) on the cell surface (Deshane et al., 2003). The endocytosis of the ClC-3 channels occurs in conjunction with MMP-2 and can be inhibited by filipin, a drug that blocks the formation of caveoli. The triggering event for this internalization is still poorly understood.

However, the action of Cltx requires the expression of MMP2 on the cell surface and the latter appears to internalize in conjunction with ClC-3 and Cltx. A larger scale examination using a chemically synthesized Cltx showed specific binding to gliomas and related neuroectodermally derived tumors (Lyons et al., 2002), but a failure to bind to nonmalignant brain. Subsequent preclinical animal studies showed specific localization to implanted tumor tissue and successful delivery of targeted radiation therapy (Soroceanu et al., 1998). These experiments paved the way for the evaluation of a synthetic Cltx in a phase I clinical study. This study, in which patients received a single dose of ^{131}I-labeled Cltx, showed a stunningly specific tumor localization in 18 patients evaluated (Fig. 21.6), and none of the patients experienced any adverse side effects (Mamelak et al., 2006). Moreover, dosimetry

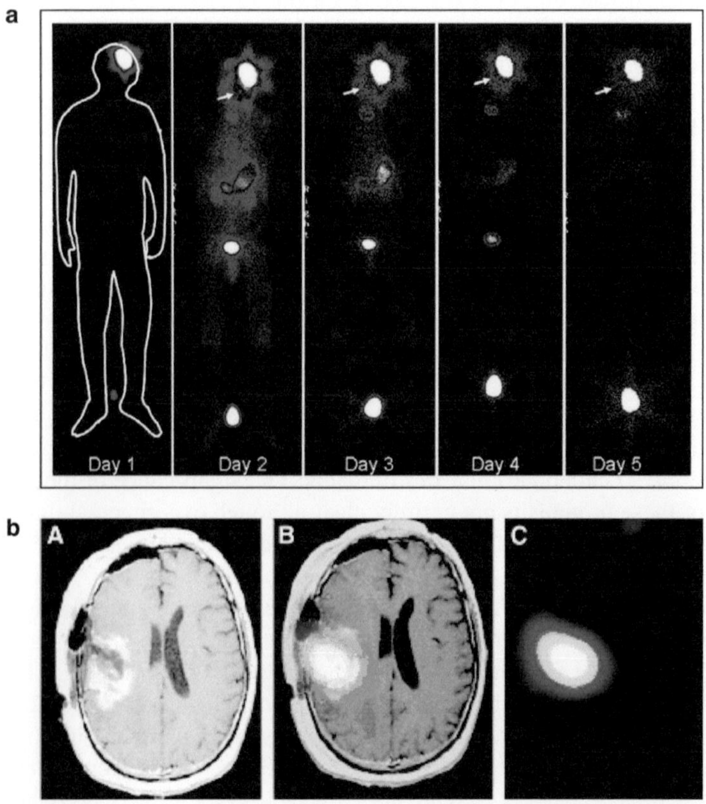

Fig. 21.6 The Cl$^-$ channel inhibitor chlorotoxin specifically localizes to the tumor tissue in patients with high glioma. (**a**) A single dose of ^{131}I-chlorotoxin was given to a patient in a phase I clinical study and shows tumor-specific localization in whole body scans performed over a 5-day period. Reproduced from Shen et al. (2005). (**b**) The same injection protocol as above showing tumor-specific retention of chlorotoxin 8 days after administration of the drug by overlay of single photon emission computed tomography (SPECT) and magnetic resonance images. Axial view of T1-weighted with gadolinium contrast (T1-Wc; left), coregistered (middle), and single photon emission computed tomography (right). From Hockaday et al. (1998) (*See Color Plates*).

data showed substantial bioavailability of the peptide and retention at the tumor for many days (Hockaday et al., 2005). The latter of course is consistent with the earlier described internalization of the peptide in the context of MMP-2 and ClC-3. For a radiolabeled peptide, this will have the added therapeutic advantage that the radioactivity becomes trapped intracellularly much closer to the celluar DNA that is the target of radiation therapy. Based on the successful conclusion of the phase I trial, Cltx is now evaluated as an experimental treatment for malignant gliomas in a muticenter dose-escalating phase II clinical study.

It should be noted that glioma cells, in addition to voltage-gated Cl^- channels also express ligand-gated GABA Cl^- channels in vivo (Labrakakis et al., 1998), and their activation similarly causes a Cl^- efflux and hence cell shrinkage. Moreover, Cl^- channels have also been identified in many other cancer cells (Shen et al., 2001; Duffy et al., 2001; Chou et al., 1995) where they may serve a similar proinvasive function. This may explain why tamoxifen, which has become a mainstay in the treatment of estrogen-dependent cancers, may exert some antitumor activity that is independent on its binding to estrogen receptors. Tamoxifen is also a potent inhibitor of several Cl^- channel types (Dick et al., 1999), and hence it is conceivable that the antimetastatic action of tamoxifen may in parts be due to the inhibition of Cl^- channels, which may aid the invasion of cancer cells into organ tissues.

Finally, the similarities in the biology of gliomas with immature neurons and glial cells cannot be overemphasized. During development many cells migrate and encounter similar physical constraints. Progenitor cells and neural stem cells also migrate in the postnatal and adult brain (Alvarez-Buylla et al., 2000). These immature cells almost always accumulate intracellular Cl^-, which may explain the depolarizing GABA response that characterizes immature neurons (LoTurco et al., 1995; Andersen et al., 1980). We have found that immature cells including glial and neuronal progenitor cells often express the same Cl^- channels. These may aid volume changes of migratory cells. Since upon differentiation neither astrocytes, oligodendrocytes nor neurons migrate, these channels are lost as they no longer serve this function. Moreover, maintaining an elevated intracellular Cl^- may become unnecessary after differentiation.

As these examples illustrate, glioma cells utilize ion channels, transporters, and transmitter receptors in ways to enhance their unique biology. They appear to co-opt the function of these proteins, which normally maintain a homeostatic brain environment. They use them instead to enhance their own ability to grow and invade normal brain.

Acknowledgments The author is grateful for the continued support by grants from the National Institutes of Health RO1 NS-31234, RO1 NS-52634, RO1 NS-36692, and P50-CA97247.

References

Alvarez-Buylla A, Herrera DG, Wichterle H (2000) The subventricular zone: source of neuronal precursors for brain repair. Prog Brain Res 127: 1–11.

Amiry-Moghaddam M, Ottersen OP (2003) The molecular basis of water transport in the brain. Nat Rev Neurosci 4: 991–1001.

Andersen P, Dingledine R, Gjerstad L, Langmoen IA, Laursen AM (1980) Two different responses of hippocampal pyramidal cells to application of gamma-amino butyric acid. J Physiol 305: 279–296.

Azizi SA, Miyamoto C (1998) Principles of treatment of malignant gliomas in adults: an overview. J Neurovirol 4: 204–216.

Behrens PF, Langemann H, Strohschein R, Draeger J, Hennig J (2000) Extracellular glutamate and other metabolites in and around RG2 rat glioma: an intracerebral microdialysis study. J Neurooncol 47: 11–22.

Bordey A, Sontheimer H, Trouslard J (2000) Muscarinic activation of BK channels induces membrane oscillations in glioma cells and leads to inhibition of cell migration. J Membr Biol 176: 31–40.

Bryant JA, Finn RS, Slamon DJ, Cloughesy TF, Charles AC (2004) EGF activates intracellular and intercellular calcium signaling by distinct pathways in tumor cells. Cancer Biol Ther 3: 1243–1249.

Butowski NA, Sneed PK, Chang SM (2006) Diagnosis and treatment of recurrent high-grade astrocytoma. J Clin Oncol 24: 1273–1280.

Chen L, Tian L, MacDonald SHF, McClafferty H, Hammond MSL, Huibant JM, Ruth P, Knaus HG, Shipston MJ (2005) Functionally diverse complement of large conductance calcium- and voltage-activated potassium channel (BK) alpha-subunits generated from a single site of splicing. J Biol Chem 280: 33599–33609.

Choi DW (1988) Glutamate neurotoxicity and diseases of the nervous system. Neuron 1: 623–634.

Chou CY, Shen MR, Wu SN (1995) Volume-sensitive chloride channels associated with human cervical carcinogenesis. Cancer Res 55: 6077–6083.

Chung WJ, Lyons SA, Nelson GM, Hamza H, Gladson CL, Gillespie GY, Sontheimer H (2005) Inhibition of cystine uptake disrupts the growth of primary brain tumors. J Neurosci 25: 7101–7110.

Cone CJ (1974) The role of the surface electrical transmembrane potential in normal and malignant mitogenesis. Ann N Y Acad Sci 238: 420–435.

Danbolt NC (2001) Glutamate uptake. Prog Neurobiol 65: 1–105.

DeBin JA, Maggio JE, Strichartz GR (1993) Purification and characterization of chlorotoxin, a chloride channel ligand from the venom of the scorpion. Am J Physiol 264: C361–C369.

Demuth T, Berens ME (2004) Molecular mechanisms of glioma cell migration and invasion. J Neurooncol 70: 217–228.

Deshane J, Garner CC, Sontheimer H (2003) Chlorotoxin inhibits glioma cell invasion via matrix metalloproteinase-2. J Biol Chem 278: 4135–4144.

Dick GM, Kong ID, Sanders KM (1999) Effects of anion channel antagonists in canine colonic myocytes: comparative pharmacology of Cl^-, Ca^{2+} and K^+ currents. Br J Pharmacol 127: 1819–1831.

Duffy SM, Leyland ML, Conley EC, Bradding P (2001) Voltage-dependent and calcium-activated ion channels in the human mast cell line HMC-1. J Leukoc Biol 70: 233–240.

Endo M, Jain RK, Witwer B, Brown D (1999) Water channel (aquaporin 1) expression and distribution in mammary carcinomas and glioblastomas. Microvasc Res 58: 89–98.

Gladson CL (1999) The extracellular matrix of gliomas: modulation of cell function. J Neuropathol Exp Neurol 58: 1029–1040.

Gunnarson E, Zelenina M, Aperia A (2004) Regulation of brain aquaporins. Neuroscience 129: 947–955.

Hockaday DC, Shen S, Fiveash J, Raubitschek A, Colcher D, Liu A, Alvarez V, Mamelak AN (2005) Imaging glioma extent with 131I-TM-601. J Nucl Med 46: 580–586.

Holland EC, Hively WP, DePinho RA, Varmus HE (1998) A constitutively active epidermal growth factor receptor cooperates with disruption of G1 cell-cycle arrest pathways to induce glioma-like lesions in mice. Genes Dev 12: 3675–3685.

Hollmann M, Hartley M, Heinemann S (1991) Ca^{2+} permeability of KA-AMPA-gated glutamate receptor channels depends on subunit composition. Science 252: 851–853.
Huncharek M, Muscat J (1998) Treatment of recurrent high grade astrocytoma; results of a systematic review of 1,415 patients. Anticancer Res 18: 1303–1311.
Iida M, Sunaga S, Hirota N, Kuribayashi N, Sakagami H, Takeda M, Matsumoto K (1997) Effect of glutathione-modulating compounds on hydrogen-peroxide-induced cytotoxicity in human glioblastoma and glioma cell lines. J Cancer Res Clin Oncol 123: 619–622.
Ishiuchi S, Tsuzuki K, Yoshida Y, Yamada N, Hagimura N, Okado H, Miwa A, Kurihara H, Nakazato Y, Tamura M, Sasaki T, Ozawa S (2002) Blockage of Ca^{2+}-permeable AMPA receptors suppresses migration and induces apoptosis in human glioblastoma cells. Nat Med 8: 971–978.
Jemal A, Siegel R, Ward E, Murray T, Xu J, Smigal C, Thun MJ (2006) Cancer statistics, 2006. CA Cancer J Clin 56: 106–130.
Kim JY, Kanai Y, Chairoungdua A, Cha SH, Matsuo H, Kim DK, Inatomi J, Sawa H, Ida Y, Endou H (2001) Human cystine/glutamate transporter: cDNA cloning and upregulation by oxidative stress in glioma cells. Biochim Biophys Acta 1512: 335–344.
Kim WT, Rioult MG, Cornell-Bell AH (1994) Glutamate-induced calcium signaling in astrocytes. Glia 11: 173–184.
Kleihues P, Ohgaki H, Aguzzi A (1995) Gliomas. In: Neuroglia (Kettenmann H, Ransom BR, eds), pp 1044–1063. New York: Oxford University Press.
Knuutila S (1984) Role of free radicals in genetic damage (mutation). Med Biol 62: 110–114.
Kofuji P, Newman EA (2004) Potassium buffering in the central nervous system. Neuroscience 129: 1043–1054.
Kofuji P, Ceelen P, Zahs KR, Surbeck LW, Lester HA, Newman EA (2000) Genetic inactivation of an inwardly rectifying potassium channel (Kir4.1 subunit) in mice: phenotypic impact in retina. J Neurosci 20: 5733–5740.
Komuro H, Rakic P (1996) Intracellular Ca^{2+} fluctuations modulate the rate of neuronal migration. Neuron 17: 275–285.
Komuro H, Rakic P (1998) Orchestration of neuronal migration by activity of ion channels, neurotransmitter receptors, and intracellular Ca^{2+} fluctuations. J Neurobiol 37: 110–130.
Labrakakis C, Patt S, Hartmann J, Kettenmann H (1998) Functional GABA (A) receptors on human glioma cells. Eur J Neurosci 10: 231–238.
Linskey ME (1997) Glial ontogeny and glial neoplasia: the search for closure. J Neurooncol 34: 5–22.
Lipton SA (2004) Failures and successes of NMDA receptor antagonists: molecular basis for the use of open-channel blockers like memantine in the treatment of acute and chronic neurologic insults. NeuroRx 1: 101–110.
Lipton SA, Rosenberg PA (1994) Excitatory amino acids as a final common pathway for neurologic disorders. New Engl J Med 330: 613–622.
Little AR, O'Callagha JP (2001) Astrogliosis in the adult and developing CNS: is there a role for proinflammatory cytokines? Neurotoxicology 22: 607–618.
Liu X, Chang Y, Reinhart PH, Sontheimer H, Chang Y (2002) Cloning and characterization of glioma BK, a novel BK channel isoform highly expressed in human glioma cells. J Neurosci 22: 1840–1849.
LoTurco JJ, Owens DF, Heath MJS, Davis MBE, Kriegstein AR (1995) GABA and glutamate depolarize cortical progenitor cells and inhibit DNA synthesis. Neuron 15: 1287–1298.
Louis DN (1994) The *p53* gene and protein in human brain tumors. J Neuropathol Exp Neurol 53: 11–21.
Lyons SA, O'Neal J, Sontheimer H (2002) Chlorotoxin, a scorpion-derived peptide, specifically binds to gliomas and tumors of neuroectodermal origin. Glia 39: 162–173.
Lyons SA, Chung WJ, Weaver AK, Ogunrinu T, Sontheimer H (2007) Autocrine glutamte signaling promotes glioma cell invasion. Cancer Res 67: 9463–9471.
MacFarlane SN, Sontheimer H (2000) Changes in ion channel expression accompany cell cycle progression of spinal cord astrocytes. Glia 30: 39–48.

Maghazachi AA (2000) Intracellular signaling events at the leading edge of migrating cells. Int J Biochem Cell Biol 32: 931–943.
Maher CO, Raffel C (2004) Neurosurgical treatment of brain tumors in children. Pediatr Clin North Am 51: 327–357.
Maher EA, Furnari FB, Bachoo RM, Rowitch DH, Louis DN, Cavenee WK, DePinho RA (2001) Malignant glioma: genetics and biology of a grave matter. Genes Dev 15: 1311–1333.
Mamelak AN, Rosenfeld S, Bucholz R, Raubitschek A, Nabors LB, Fiveash JB, Shen S, Khazaeli MB, Colcher D, Liu A, Osman M, Guthrie B, Schade-Bijur S, Hablitz DM, Alvarez VL, Gonda MA (2006) Phase I single-dose study of intracavitary-administered iodine-131-TM-601 in adults with recurrent high-grade glioma. J Clin Oncol 24: 3644–3650.
Manning TJJr, Parker JC, Sontheimer H (2000) Role of lysophosphatidic acid and rho in glioma cell motility. Cell Motil Cytoskeleton 45: 185–199.
McCoy E, Sontheimer H (2007) Expression and function of water channels (Aquaporins) in migrating malignant astrocytes. Glia 55: 1034–1043.
McFerrin MB, Sontheimer H (2006) A role for ion channels in glioma cell invasion. Neuron Glial Biol 2: 39–49.
Mitchell JB, Cook JA, DeGraff W, Glatstein E, Russo A (1989) Glutathione modulation in cancer treatment: will it work? Int J Radiat Oncol Biol Phys 16: 1289–1295.
Oberndorfer S, Schmal T, Lahrmann H, Urbanits S, Lindner K, Grisold W (2002) [The frequency of seizures in patients with primary brain tumors or cerebral metastases. An evaluation from the Ludwig Boltzmann Institute of Neuro-Oncology and the Department of Neurology, Kaiser Franz Josef Hospital, Vienna]. Wien Klin Wochenschr 114: 911–916.
Olney JW, Ho OL, Rhee V (1971) Cytotoxic effects of acidic and sulphur containing amino acids on the infant mouse central nervous system. Exp Brain Res 14: 61–76.
Olsen ML, Sontheimer H (2004) Mislocalization of Kir channels in malignant glia. Glia 46: 63–73.
Olsen ML, Schade S, Lyons SA, Amarillo MD, Sontheimer H (2003) Expresssion of voltage-gated chloride channels in human glioma cells. J Neurosci 23: 5572–5582.
Olsen ML, Higashimori H, Campbell SL, Hablitz JJ, Sontheimer H (2006) Functional expression of K (ir)4.1 channels in spinal cord astrocytes. Glia 53: 516–528.
Pellegrini-Giampietro DE, Bennett MV, Zukin RS (1992) Are Ca^{2+}-permeable kainate/AMPA receptors more abundant in immature brain? Neurosci Lett 144: 65–69.
Plate KH, Risau W (1995) Angiogenesis in malignant gliomas. Glia 15: 339–347.
Ransom CB, Sontheimer H (2001) BK channels in human glioma cells. J Neurophysiol 85: 790–803.
Ransom CB, O'Neal JT, Sontheimer H (2001) Volume-activated chloride currents contribute to the resting conductance and invasive migration of human glioma cells. J Neurosci 21: 7674–7683.
Ransom CB, Liu X, Sontheimer H (2002) BK channels in human glioma cells have enhanced calcium sensitivity. Glia 38: 281–291.
Rijpkema M, Schuuring J, van der MY, van der GM, Bernsen H, Boerman R, van der KA, Heerschap A (2003) Characterization of oligodendrogliomas using short echo time 1H MR spectroscopic imaging. NMR Biomed 16: 12–18.
Roslin M, Henriksson R, Bergstrom P, Ungerstedt U, Bergenheim AT (2003) Baseline levels of glucose metabolites, glutamate and glycerol in malignant glioma assessed by stereotactic microdialysis. J Neurooncol 61: 151–160.
Rothstein JD, Dykes-Hoberg M, Pardo CA, Bristol LA, Jin L, Kuncl RW, Kanai Y, Hediger MA, Wang YF, Schielke JP, Welty DF (1996) Knockout of glutamate transporters reveals a major role for astroglial transport in excitotoxicity and clearance of glutamate. Neuron 16: 675–686.
Saadoun S, Papadopoulos MC, Davies DC, Bell BA, Krishna S (2002a) Increased aquaporin 1 water channel expression in human brain tumours. Br J Cancer 87: 621–623.
Saadoun S, Papadopoulos MC, Davies DC, Krishna S, Bell BA (2002b) Aquaporin-4 expression is increased in oedematous human brain tumours. J Neurol Neurosurg Psychiatry 72: 262–265.

Sato H, Tamba M, Ishii T, Bannai S (1999) Cloning and expression of a plasma membrane cystine/glutamate exchange transporter composed of two distinct proteins. J Biol Chem 274: 11455–11458.

Shapiro JR (2001) Genetics of brain neoplasms. Curr Neurol Neurosci Rep 1: 217–224.

Shen MR, Chou CY, Browning JA, Wilkins RJ, Ellory JC (2001) Human cervical cancer cells use Ca^{2+} signalling, protein tyrosine phosphorylation and MAP kinase in regulatory volume decrease. J Physiol 537: 2–62.

Shen S, Khazaeli MB, Gillespie GY, Alvarez VL (2005) Radiation dosimetry of 131I-chlorotoxin for targeted radiotherapy in glioma-bearing mice. J Neurooncol 71: 113–119.

Simone G, Tamba M, Quintiliani M (1983) Role of glutathione in affecting the radiosensitivity of molecular and cellular systems. Radiat Environ Biophys 22: 215–223.

Soroceanu L, Gillespie Y, Khazaeli MB, Sontheimer H (1998) Use of chlorotoxin for targeting of primary brain tumors. Cancer Res 58: 4871–4879.

Soroceanu L, Manning TJ Jr, Sontheimer H (1999) Modulation of glioma cell migration and invasion using Cl^- and K^+ ion channel blockers. J Neurosci 19: 5942–5954.

Takano T, Lin JH, Arcuino G, Gao Q, Yang J, Nedergaard M (2001) Glutamate release promotes growth of malignant gliomas. Nat Med 7: 1010–1015.

Tanaka K, Watase K, Manabe T, Yamada K, Watanabe M, Takahashi K, Iwama H, Nishikawa T, Ichihara N, Kikuchi T, Okuyama S, Kawashima N, Hori S, Takimoto M, Wada K (1997) Epilepsy and exacerbation of brain injury in mice lacking the glutamate transporter glt-1. Science 276: 1699–1702.

Tang P, Steck PA, Yung WK (1997) The autocrine loop of TGF-alpha/EGFR and brain tumors. J Neurooncol 35: 303–314.

Ullensvang K, Lehre KP, Storm-Mathisen J, Danbolt NC (1997) Differential developmental expression of the two rat brain glutamate transporter proteins GLAST and GLT. Eur J Neurosci 9: 1646–1655.

Ullrich N, Sontheimer H (1996) Biophysical and pharmacological characterization of chloride currents in human astrocytoma cells. Am J Physiol 270: C1511–C1521.

Ullrich N, Gillespie GY, Sontheimer H (1996) Human astrocytoma cells express a unique chloride current. Neuroreport 7: 1020–1024.

Von Deimling A, Louis DN, Wiestler OD (1995) Molecular pathways in the formation of gliomas. Glia 15: 328–338.

Walz W (2002) Chloride/anion channels in glial cell membranes. Glia 40: 1–10.

Ye ZC, Sontheimer H (1999) Glioma cells release excitotoxic concentrations of glutamate. Cancer Res 59: 4383–4391.

Ye ZC, Rothstein JD, Sontheimer H (1999) Compromised glutamate transport in human glioma cells: reduction mislocalization of sodium-dependent glutamate transporters and enhanced activity of cystine-glutamate exchange. J Neurosci 19: 10767–10777.

Zamecnik J (2005) The extracellular space and matrix of gliomas. Acta Neuropathol (Berl) 110: 435–442.

Abbreviations

AMPAR	α-Amino-3-hydroxy-5-methyl-4-isoxazolepropionic acid receptors
AQP	Aquaporins
BK	Channels big conductance Ca^{2+}-activated
Cltx	Chlorotoxin
CNS	Central nervous system
EAAT	Excitatory amino acid transporter
GLAST	L-glutamate/L-aspartate transporter

GLT-1	Glial L-glutamate transporter
GSH	Glutathione
MMP2	Matrix metalloproteinase II
NMDA	*N*-methyl-D-aspartate
NMDAR	NMDA receptors
NPPB	5-Nitro-2-(3-phenyl-propylamino)benzoic acid

Chapter 22
Connexins and Pannexins: Two Gap Junction Families Mediating Glioma Growth Control

Charles P.K. Lai and Christian C. Naus

Contents

22.1 Introduction

Astrocytes are the most common glial cells found in the central nervous system (CNS). They are known to provide neurons with structural support and neurotrophic substances, as well as buffer the extracellular milieu and contribute to the blood–brain barrier (reviewed in Hatten et al., 1991). Unlike neurons, astrocytes can be readily induced to divide, and this may contribute to the formation of gliomas, which accounts for more than 65% of all primary brain tumors (Muller et al., 1977). However, since gliomas arise in the brain, therapeutic strategies including radiotherapy, chemotherapy, and complete surgical removal are often difficult to implement and pose a high risk of side effects to patients (Butowski et al., 2006).

C.C. Naus
Department of Cellular and Physiological Sciences, The Faculty of Medicine,
Life Sciences Institute, The University of British Columbia, Vancouver, V6T 1Z3 BC Canada
cnaus@interchange.ubc.ca

V. Parpura and P.G. Haydon (eds.), *Astrocytes in (Patho)Physiology of the Nervous System*,
DOI: 10.1007/978-0-387-79492-1_22, © Springer Science+Business Media, LLC 2009

Moreover, invasive gliomas frequently relapse after treatments due to their extensive infiltration into surrounding tissues, consequently resulting in death in nearly all cases (Reardon et al., 2006). Therefore, exploring a more effective treatment for gliomas merits serious consideration. Targeting some of the cellular mechanisms controlling proliferation, migration, and invasion would constitute effective therapeutic strategies. One such mechanism involves intercellular communication through gap junctions. The current review will summarize the evidence supporting a role for gap junction channels and their constituent proteins in controlling cell proliferation and tumorigenesis in gliomas, focusing both on rodent models and, where applicable, human gliomas.

22.2 Overview of Gap Junctions

Cell communication, both cell-to-cell and with the extracellular environment, underlies processes critical to development and differentiation, as well as transformation (Kelleher et al., 2006). When such communication is perturbed, disruption in normal growth control processes can lead to cellular transformation. One of the unique pathways for direct intercellular communication involves the formation of gap junction channels between cells.

Gap junctions are a unique class of channels that directly connect the cytoplasm of adjacent cells, permitting the exchange of ions and molecules up to approximately 1 kDa in size such as amino acids, second messengers, and metabolites (Simon and Goodenough, 1998). More recently, larger molecules, such as interfering RNA (Valiunas et al., 2005) and small peptides (Neijssen et al., 2005), have been shown to pass through gap junctions. Thus, gap junctions are positioned to coordinate a variety of cellular activities, which can influence tissue and organ function. Gap junctions have been shown to be critical for normal development (Wei et al., 2004), and gene targeting approaches have revealed their importance in maintaining physiological functions of many organ systems (Sohl and Willecke, 2004). Furthermore, substantial evidence supports a role for gap junctions in controlling cell proliferation and tumorigenesis (Mesnil et al., 2005; Naus et al., 2005).

22.2.1 Connexins

Until quite recently, only one mammalian family of gap junction proteins had been identified, namely the connexins (Willecke et al., 2002). Gap junctions have been shown to be formed by the end-to-end apposition of hemichannels, or connexons, proteinaceous cylinders spanning the plasma membrane and encompassing a hydrophilic channel. Each connexon hemichannel is composed of six proteins called connexins, members of a multigene family consisting of 20 members in rodents and 21 in humans (Sohl and Willecke, 2004). The connexin (Cx) protein spans the plasma membrane four times, and is oriented with cytoplasmic amino- and carboxy-termini, while the extracellular domains are involved in connexon–connexon interactions to form complete gap junction channels between cells. Much of the regulation of the gap junction channel occurs via the carboxy terminus. Indeed, this region contains consensus sites for phosophorylation (Lampe and Lau, 2000) and calmodulin binding (Peracchia et al., 2000), as well as being involved in pH sensitivity of channel gating (Hirst-Jensen et al., 2007). Furthermore, there is a growing list of proteins which have been shown to interact with the carboxy terminus of Cxs, primarily Cx43 (Herve et al., 2004). The implications of such interactions will be discussed later.

In addition to their involvement in the formation of gap junction channels, it has also been proposed that Cxs can form hemichannels, composed of single connexons, providing cells with the ability to directly communicate with the extracellular environment (Stout et al., 2002; Belliveau et al., 2006; Goodenough and Paul, 2003; Bennett et al., 2003). However, the existence of hemichannels in vivo is currently a matter of debate (Spray et al., 2006). Furthermore, Cxs have also been reported to mediate intracellular effects independent of channel or hemichannel formation, presumably by interacting with cellular proteins involved in various signaling pathways (Naus et al., 2005).

22.2.2 Pannexins

In recent years, a novel family of mammalian gap junction proteins with low sequence similarity to the invertebrate gap junctions, innnexins (Inxs), has been discovered in chordates and termed pannexins (Panxs) (Panchin et al., 2000). Currently, three Panx members (Panx1, Panx2, and Panx3) have been identified in vertebrates (Baranova et al., 2004), while several invertebrate Inxs are also referred to as Panxs (Sasakura et al., 2003). Previous studies on Inx mutants in *Drosophila* have demonstrated Inx-specific functions including synaptogenesis in the giant-fiber system, epithelial organization and morphogenesis, and germ cell differentiation processes (Bauer et al., 2005). Although it remains to be seen whether Panxs can be regarded as vestigial Inxs that have survived in higher animals, this implies that, other than Cxs, Panxs may also play functional roles in chordates (Bauer et al., 2005; Barbe et al., 2006).

22.2.3 Similarities Between Connexins and Pannexins

Despite sequence dissimilarity between Cxs and Panxs, the two protein families share structural resemblance (Panchin, 2005; Bruzzone et al., 2003). Similar to Cxs, Panxs have a predicted topology of four membrane-spanning domains, two extracellular loops, a cytoplasmic loop, and cytoplasmic amino- and carboxy-termini

(Simon and Goodenough, 1998; Panchin, 2005). Intriguingly, whereas Cxs contain three regularly spaced cysteine residues in the two extracellular loops, Panxs, like Inxs, only have two such residues (Hua et al., 2003). Numerous studies have suggested the importance of the cysteine residues in facilitating functional Cx-based gap junctions and hemichannels (Yeager and Nicholson, 1996; Bao et al., 2004; Saez et al., 2005), and therefore the variation in the number of cysteine residues may underlie functional differences between Panx and Cx functions.

Analogous to Cxs, Panx hemichannel and intercellular channel formation are Panx specific (Bruzzone et al., 2003). Using the *Xenopus* oocyte expression system, rat Panx1 and Panx1/Panx2 were discovered via electrophysiological experiments to form functionally different hemichannels in single oocytes. In paired oocytes, rat Panx1 and Panx1/Panx2 also formed homotypic and heterotypic intercellular channels, respectively, with distinctive functional properties. However, expression of Panx2 or Panx3 alone did not result in any hemichannel or intercellular channel functions. Furthermore, these Panx-based hemichannels, recently referred to as *pannexons* (Dahl and Locovei, 2006), demonstrated sensitivity to the same pharmacological blockers as Cxs (Bruzzone et al., 2003, 2005). Interestingly, whereas carbenoxolone and flufenamic acid both inhibit Cx channels, Panx hemichannels were sensitive to carbenoxolone but were only modestly inhibited by flufenamic acid (Srinivas and Spray, 2003; Bruzzone et al., 2005). Although the two channels have similar characteristics, their differences imply that they could exhibit unique functions via different mechanisms.

22.3 Gap Junctions and Cancer

Gap junctions, and their constituent Cx proteins, have long been considered to play a role in the control of cell proliferation, with disruptions in expression and gap junctional coupling correlating with cell transformation (Mesnil et al., 2005; Naus et al., 2005; Leithe et al., 2006). While traditional roles for gap junction proteins have focused on their channel-forming functions, roles of Cx proteins are emerging, which appear to be independent of channel formation (Jiang and Gu, 2005). In addition, the recent identification of another gap junction protein family, the Panxs, broadens the scope of this topic (Barbe et al., 2006).

Gap junctions have long been implicated in the regulation of cell proliferation and the control of tumor progression. Since the early report that gap junctional coupling was decreased in tumor cells (Loewenstein and Kanno, 1966), downregulation of gap junctions has been repeatedly observed in a variety of tumor cell lines and cancer tissues (Yamasaki and Naus, 1996; Naus et al., 2005; Mesnil et al., 2005). In addition, many tumor promoting agents have the ability to downregulate connexin expression and gap junction formation (Trosko and Ruch, 2002). Conversely, antineoplastic agents such as retinoic acid, carotenoids, vitamin D, and flavonoids can reduce tumorigenesis in association with enhanced Cx expression and increased gap junctional coupling (Trosko and Chang, 2001; Conklin et al., 2007).

The mechanisms by which Cx and gap junctions mediate these effects on cell proliferation and tumor suppression remain relatively unknown. We have recently summarized several proposed mechanisms (Naus et al., 2005), including the following (Fig. 22.1). One of the earliest concepts was the need for gap junctions to (1) disperse *growth-promoting molecules* among cells or (2) accumulate *growth-inhibiting molecules* between cells. This has remained largely unproven. Attempts to isolate such *transjunctional molecules* have, nevertheless, provided valuable insight into the repertoire of endogenous molecules that cells share through gap junctions (Goldberg et al., 1998, 1999). (3) Gap junctions have also been proposed to act as a *nexus* for the accumulation of a number of scaffolding and signaling proteins, several of which have been implicated in tumor suppression (e.g., catenin, caveolin, CCN3) (Duffy et al., 2002). (4) Expression of Cxs has been shown to alter the expression of a number of genes involved in control of proliferation and tumorigenesis, including milk fat globule epidermal growth factor 8 (MFG-8) (Goldberg et al., 2000), monocyte chemotactic protein 1 (MCP-1) (Huang et al., 2002), a member of the Cyr61/CTGF/Nov (CCN) family, CCN3 (Fu et al., 2004; Gellhaus et al., 2004), and various cell cycle genes including cyclin A, D1, D2, CDK5, and 6 (Chen et al., 1995). Interestingly, recent reports describing genomic analysis of Cx43 knockout cells have indicated significant changes in the expression of up to 200 genes (Iacobas et al., 2004).

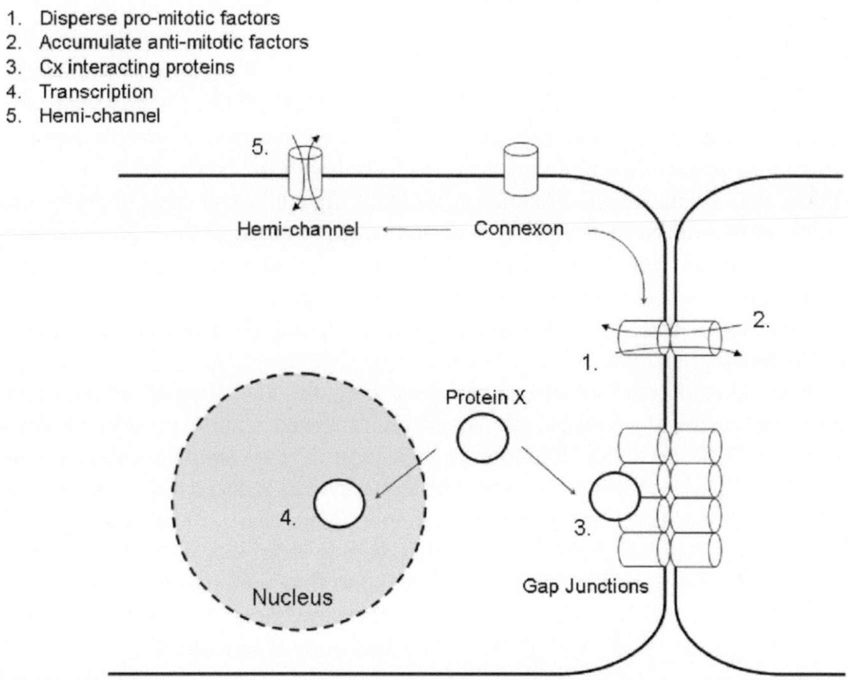

Fig. 22.1 Schematic summarizing the various proposed mechanisms by which connexins or pannexins could mediate effects of cell proliferation and tumorigenesis.

The implications of such broad changes in gene expression make it increasingly difficult to attribute specific changes to Cx43 or gap junctional coupling. However, recent genomic studies have indicated that many growth related genes are coregulated with Cx43, supporting a key role in the control of growth (Kardami et al., 2007). (5) More recently, the possible role of hemichannels must be considered in mediating some of these effects. In addition, a role for gap junctions in cell adhesion and migration has also been demonstrated (Elias et al., 2007). Any combination of these effects may be involved in tumor suppression, as well as other mechanisms not yet investigated.

22.4 Gap Junctions and Glioma

Several studies have examined the level of gap junctional coupling and Cx expression in cancer cell lines as well as tissues, in an attempt to find a correlation between the level of expression and the tumor grade. Many studies have focused on determining if such a correlation exists in astrocytomas and gliomas, particularly in light of possible therapeutic significance.

22.4.1 Connexins and Glioma

There are many cell lines produced from rodent models of gliomas, as well as different grades of human astrocytomas and gliomas. Many of these have been examined with regard to Cx expression and tumorigenesis. We initially noted a decrease in Cx43 expression in C6 glioma cells compared with rat astrocytes (Naus et al., 1991a), and in fact reported that the restoration of Cx43 expression in these cells significantly reduced their proliferation in vitro (Zhu et al., 1991, 1992) and their tumorigenesis in vivo (Naus et al., 1992). Several studies have assessed human glioma cell lines, and in many cases, there has been an inverse correlation between Cx43 expression and the tumorigenic potential of these lines. Some of these studies are summarized in Table 22.1.

More clinically relevant and comprehensive studies have focused on tumor tissue. Most studies have consistently shown an inverse correlation between the level of Cx43 expression and the tumor grade (see Table 22.1 for summary). We demonstrated that astrocytomas associated with epilepsy had reduced Cx43 expression; however, this study was based on limited samples (Naus et al., 1991b). Subsequently a more comprehensive study suggested that there is variability with regard to both mRNA and protein levels with tumor grade, as well as astrocytoma and glioblastoma cell lines, with high-grade astrocytomas showing the most variable expression of Cx43 (Shinoura et al., 1996). In a later study, immunohistochemical analysis of paraffin sections of tissue from different astrocytoma grades indicated decreased Cx43 staining with increasing tumor grade (Huang et al., 1999). Soroceanu et al. (2001), using protein blot analysis and immunohistochemistry, showed that high-grade

Table 22.1 Connexin/pannexin expression in glioma

Cell line or tissue	Connexin	Effect of Cx transfection on proliferation	Reference
C6 rat glioma cells	Cx43 reduced	Reduced by Cx43 *in vitro* and *in vivo*	Naus et al., 1991; Zhu et al., 1991
C6 rat glioma cells		Reduced by Cx32 *in vivo*	Bond et al., 1994
C6 and 9L rat glioma cells	Cx43 reduced Cx30 absent	Reduced by Cx30 *in vitro*	Princen et al., 2001
Human glioma lines U87MG, U118MG, U137MG, U373MG, T98G, Hs68331, SNB-10	Cx43 variable		Shinoura et al., 1996
Human glioma lines U138MG, CH235MG, D65MG, U373MG, U251MG	Cx43 expression variable		Soroceanu et al., 2001
U251, U87MG, U373MG, T98G, Clontech glioma and astrocytoma cell lines	Cx43 reduced	Reduced by Cx43 *in vitro* and *in vivo*	Huang et al., 1998b
Human glioma lines TJ899, TJ905, TJ8510	Cx43 reduced		Pu et al., 2004
Astrocytomas	Cx43 reduced		Naus et al., 1991b
Human glioma lines GL15. 8MG: human glioma biopsy xenograpts maintained in nude mice	Cx43 increased in invasive gliomas		Oliveira et al., 2005
Astrocytomas	Cx43 reduced		Aronica et al., 2001
Glioblastomas and astrocytomas	Cx43 reduced with increasing grade		Soroceanu et al., 2001
Astrocytomas	Cx43 reduced with increasing grade		Huang et al., 1999
Astrocytomas	Cx43 mRNA and protein variable with grade, especially in high grade		Shinoura et al., 1996
Astrocytomas	Cx43 reduced with increasing grade		Pu et al., 2004
Primary brain tumors	Cx26 and Cx43 expressed		Estin et al., 1999

gliomas expressed lower levels of Cx43, a correlation that held for acutely isolated cells from biopsies of various grades. Functional studies using dye passage assays also correlated with the level of Cx43 expressed. Additional studies also tend to support the view that decreased Cx43 expression correlated with increased tumor grade (Aronica et al., 2001; Pu et al., 2004).

22.4.2 Altering Connexin Expression in Gliomas

Since there was a consistent observation of decreased Cx43 in higher grade astrocytomas and gliomas, several approaches were pursued to enhance Cx43 expression to impact tumorigenesis, including direct transfection of cells, as well as treatments to enhance Cx43 expression. A number of studies have shown that forced expression of Cxs by transfection suppresses glioma proliferation and tumorigenesis (Zhu et al., 1991, 1992; Naus et al., 1992; Bond et al., 1994; Bechberger et al., 1996; Huang et al., 1998b, 2002; Princen et al., 2001). In some cases, the mechanisms of this suppression have been postulated based on Cx-induced changes in expression of growth control factors, including increased expression of factors that suppress proliferation and decreased expression of factors that promote tumorigenesis.

Several studies have reported increased expression of growth-suppressing factors following upregulation of Cx43 expression. Our own work demonstrated that C6 glioma cells transfected with Cx43 secreted some factor(s) that had growth suppressive properties (Zhu et al., 1992). Subsequent studies showed increased levels of the negative insulin-like growth factor 1 (IGF-1) modulator IGF binding protein 4 in the extracellular milieu, which may be responsible for the reduced proliferative capacity in C6 glioma cells expressing abundant Cx43 (Bradshaw et al., 1993a). We later identified members of the CCN family of growth regulators whose expression was increased following Cx43 transfection in C6 glioma cells (Naus et al., 2000). Specifically, CCN1/Cyr61 and CCN3/Nov, both shown to be growth suppressive in some tumor cell lines, were significantly upregulated. CCN proteins are secreted growth regulators that interact with each other and signal through integrins and proteoglycans (Bleau et al., 2005; Leask and Abraham, 2006; Perbal, 2004). Interestingly, colocalization of CCN3 with Cx43 occurred with this upregulation, supporting the concept of Cx43-mediated tumor suppression via its interaction with other proteins (Fu et al., 2004).

With regard to alterations in growth stimulating factors, decreased synthesis of the growth factor IGF-I together with decreased levels of the positive modulator IGF binding protein 3 occurred when Cx43 was overexpressed in C6 glioma cells (Bradshaw et al., 1993b). We also previously reported a decrease in ribosomal protein L19 (RPL19) in these cells when Cx43 was expressed (Naus et al., 2000). RPL19 is highly expressed in a number of tumors, including breast (Henry et al., 1993; Leirdal et al., 2004) and prostate (Bee et al., 2006), but has not been examined in gliomas. Goldberg et al. (2000) also showed that Cx43 expression suppresses the synthesis and secretion of MFG-E8, a protein with high expression in breast cancer. Huang et al. (2002) demonstrated that Cx43 suppressed human glioblastoma growth by decreasing expression of MCP-1, a cytokine shown to be elevated in many tumor types including glioblastoma, and enhancing angiogenesis and inflammation.

The correlation between Cx43 expression and alteration in growth factors is complimented by a number of studies examining the effects of such growth factors on Cx43 expression (reviewed in Naus et al., 2005). This led us to examine the effects of one such factor, ciliary neurotrophic factor (CNTF), to augment Cx43 expression in C6 glioma. We found that CNTF in combination with its receptor, CNTFRα, enhanced Cx43 expression and reduced proliferation (Ozog et al., 2002). We subsequently identified a CNTF-response element in the Cx43 promoter, and showed that Cx43 was upregulated via the Janus tyrosine kinase/signal transducer and activator of transcription (JAK/STAT) pathway (Ozog et al., 2002).

Other factors that alter Cx expression and/or gap junctional coupling have also been investigated. Gliomas, like most tumors, are characterized by high metabolic activity, including elevated glucose uptake (Di Chiro et al., 1982). Tabernero et al. (1996) have shown that blocking gap junctions enhances glucose uptake in astrocytes. Conversely, treatment of C6 glioma cells with tolbutamide or dibutyric cAMP, which increase Cx43 expression and enhance gap junctional coupling, was associated with translocation of type I and II hexokinase from mitochondria to cytosol, decreasing activity and glucose uptake (Sanchez-Alvarez et al., 2005; Tabernero et al., 2006). Thus, Cx expression and gap junctional coupling have been shown to play a role in the metabolic state of glioma cells, correlating with molecular changes regulating cell proliferation (Sanchez-Alvarez et al., 2006).

22.4.3 Connexins: Apoptosis and the Bystander Effect

While initial attempts to enhance Cx expression in tumor cells focused on a direct tumor-suppressive effect, subsequent approaches for tumor treatment were aimed at taking advantage of gap junctions as conduits for the cell-to-cell transfer of metabolites. With regard to tumor therapy, enhancement of apoptosis provides a clear therapeutic mechanism for exploitation (Ding and Fisher, 2002). The role of gap junctions in apoptosis was suggested over 10 years ago (Trosko and Goodman, 1994). A more direct role for Cxs in cell death has been demonstrated since increased Cx43 expression resulted in decreased expression of bcl-2 in conjunction with increased apoptosis following treatment with chemotherapeutic drugs (Huang et al., 2001b). In fact, several reports have demonstrated that enhanced expression of Cxs alone induces apoptosis. Cx43 has been shown to enhance apoptosis in glioblastoma cells under low serum conditions (Huang et al., 2001a). These studies, as well as others, indicate that these effects, in many situations, are Cx- and cell type-specific (reviewed in Andrade-Rozental et al., 2000). While gap junctions have been implicated as mediators of enhanced apoptosis (Lin et al., 1998), Cx43 expression in the absence of gap junction formation has been shown to enhance glioma cell survival (Lin et al., 2003). These authors postulated that cytoskeletal interactions and calcium homeostasis were involved in this process. Cell survival was also dependent upon the context of the injury. Thus, there remains some uncertainty around the role of Cxs and gap junctions in apoptosis.

Therapeutically, there has been exploitation of the role of gap junctions in transmitting cytotoxic molecules to kill tumor cells. In this regard, gap junctions have been implicated in the so-called *bystander effect* observed in cell killing mediated by Herpes simplex virus thymidine kinase expression and ganciclovir treatment (Mesnil et al., 1996; Dilber et al., 1997; Nicholas et al., 2003). This has led to a number of approaches to augment chemotherapy with enhanced gap junctional coupling (Touraine et al., 1998; Carystinos et al., 1999; Robe et al., 2000, 2004; Huang et al., 2001b). In vitro studies have shown that human astrocytoma cells, T98G (Shinoura et al., 1996) or U87 (Cirenei et al., 1998), transfected with Cx43 exhibited an increase in bystander killing using the Herpes simplex virus thymidine kinase system with ganciclovir treatment. We have shown a similar effect in vivo in C6 glioma cells transfected with Cx43 (Dilber et al., 1997). With regard to different connexins mediating this effect, we have shown that there is differential sensitivity to the *bystander effect* dependent upon the Cx isoform expressed (Jimenez et al., 2006). This bystander effect has also been shown to be involved in cisplatin-mediated tumor cell killing (Jensen and Glazer, 2004), as well as radiation therapy (Azzam et al., 1998, 2003). Although there are reports demonstrating a gap junction independent pathway for bystander killing in glioma cells (Princen et al., 1999), the role of hemichannels has not been specifically investigated. Therefore, Cxs and gap junctions should continue to be explored as an augmentation route for glioma therapy.

22.4.4 Gap Junctions, Migration, and Invasion

One of the common characteristics of glioma cells is their aggressive ability to migrate and invade other parts of the brain. While it has been known that tumor cells interact with the host microenvironment during growth, invasion, and metastasis (Fidler et al., 2002; Mueller and Fusenig, 2004), the role of Cx43 and gap junctional coupling in tumor cells during these processes is controversial (Table 22.2). In some tumor models, loss of gap junctional coupling has been correlated with decreased adhesion and increased migration (Hoffman et al., 1993; Stein et al., 1993; McDonough et al., 1999). In addition, reduced Cx43 expression has been reported in high-grade invasive human gliomas (Huang et al., 1999; Soroceanu et al., 2001).

In contrast, in some cases, loss of Cx43 reduces migration. Neural crest cells from Cx43 knockout mice show reduced migration (Huang et al., 1998a), as do cells in neurospheres prepared from these same mice (Scemes et al., 2003). We have found a decrease in migration of neurons in the developing neocortex of Cx43 knockout mice (Fushiki et al., 2003). Similar results were also reported in the developing rat neocortex following siRNA knockdown of Cx43 and Cx26 (Elias et al., 2007). Furthermore, several reports suggest that Cxs may enhance migration and invasion. Transfection of Cx43 in HeLa cells has also been reported to increase invasive properties of these cells (Graeber and Hulser, 1998). Increased Cx43 expression in C6 cells was shown to increase their capacity to invade the brain parenchyma, and this was proposed to be due to gap junctional coupling with brain

Table 22.2 Connexins and migration

Cell tissue type	Cx type	Experimental approach	Effect of Cx43 on motility	Reference
NIH3T3	Cx43	Cx43siRNA	Enhances	Wei et al., 2005
Spontaneous canine astrocytoma	Cx43	EGF stimulation	Attenuates	McDonough et al., 1999
MDA-MB-231, Hs578T	Cx43	Cx43siRNA	Attenuates	Shao et al., 2005
HeLa	Cx43	Stable transfection	Enhances	Graeber and Hulser, 1998
Neural tube explant	Cx43	Mouse models; CMV43, Cx43KO	Enhances	Huang et al., 1998b
Developing brain	Cx43	Cx43KO	Enhances	Fushiki et al., 2003
Developing brain	Cx43, Cx26	siRNA	Enhances	Elias et al., 2007
GL15, 8-MG, human biopsies, C6	Cx43	Endogenous expression, stable transfection	Enhances	Oliveira et al., 2005
Skin/epidermis	Cx43	Cx43siRNA	Attenuates	Qiu et al., 2003
Skin/epidermis	Cx43	Cx43 localization via antibody stain	Attenuates	Brandner et al., 2004
C6	Cx43	Stable transfection	Enhances	Zhang et al., 2003
C6	Cx43	Low vs. high endogenous expression	Enhances	Bates et al., 2007
C6	Cx43ΔCT	Stable transfection	Deletion Attenuates	Bates et al., 2007
Pro-epicardial expants	Cx43	Mouse model; Cx43KO	Attenuates	Li et al., 2002
3T3 A31 fibroblasts	Cx43-256M	Stable transfection	Mutation Attenuates	Moorby, 2000

Cx43ΔCT, Cx43 lacking carboxyl terminal; *Cx43-256M,* Cx43 truncated at Meth 256; *siRNA,* small interferring RNA.

astroctyes (Zhang et al., 2003). We have recently shown that subclones of C6 glioma cells that express low or high levels of endogenous Cx43 show different degrees of migration, with the high Cx43 subclone displaying enhanced migration (Bates et al., 2007). Furthermore, this enhanced migration was dependent upon the presence of the C-terminal of Cx43.

The intercellular junctional complex, consisting of adherens and tight junctions, gap junctions, and desmosomes, has been implicated in the control of cell proliferation and differentiation due to the association of proteins involved in signal transduction, as well as oncogene products and tumor suppressors at this site (Tsukita et al., 1999). While this association of structural and signaling molecules has been characterized for adherens junctions, it is only recently that similar associations have been found for gap junctions (reviewed in Herve et al., 2004; Giepmans, 2004). One of the most studied proteins associating with Cx43 is zonula occludens 1 (ZO-1) (Giepmans and Moolenaar, 1998), a member of the membrane-associated guanylate kinase family of proteins, which interacts with Cx43 via a C-terminal PSD95/Dlg/ZO-1 (PDZ) binding domain (Gumbiner et al., 1991). ZO-1 is associated with tight junctions and adherens junctions, and also interacts with several other proteins, including actin, occludin, ZO-2, ZO-3, and ZO-1-associated nucleic acid binding protein (ZONAB) (Mitic and Anderson, 1998; Balda and Matter, 2000). It is involved in the recruitment of many cytoplasmic proteins to tight junctions, and is believed to function as a scaffolding protein to recruit signaling molecules to these junctions. A similar function has also been proposed for gap junctions (Giepmans, 2004; Duffy et al., 2002). The interaction of Cx43 with ZO-1 in G_0 is believed to enhance assembly, turnover, or stability of gap junctions (Singh et al., 2005). Recently, ZO-1 has been shown to localize at the leading edge of lamellipodia in motile wounded fibroblasts and to interact with integrins (Taliana et al., 2005). Thus, ZO-1 at the leading edge of migrating fibroblasts is consistent with a role in the initiation and organization of integrin-dependent fibroblast migration and adhesion. In C6-Cx43 cells, we have demonstrated colocalization of Cx43, CCN3, and ZO-1 at gap junction plaques, suggesting that there is a dynamic interaction of these three proteins, which can modulate cell proliferation and migration.

22.4.5 Pannexins and Glioma

Cxs have been extensively studied in their putative role as tumor suppressors. Panxs, on the other hand, have just been recently identified and their implication in cancers is only now being examined. We have used the C6 glioma cell model to examine the possible role of Panxs in tumor suppression (Lai et al., 2007). This is the first report describing Panx1 as a negative growth regulator and provides insights into novel aspects of gap junctions in cancer research.

Transcriptional analyses have identified location-specific Panx mRNA expression in rodents. Panx1 is ubiquitously expressed, Panx2 is particularly abundant in the brain, and Panx3 is present in the skin (Bruzzone et al., 2003; Ray et al., 2005; Vogt et al., 2005; Weickert et al., 2005). While it is beyond the scope of this chapter to encompass Panx expression pattern in brain, readers are referred to published reviews (Barbe et al., 2006; Litvin et al., 2006). In brief, Panx1 and Panx2 transcripts were readily detected in brain and colocalized with NeuN, a marker for most neuronal cell types (Ray et al., 2005; Vogt et al., 2005). Zappala et al. (2006) also showed Panx1 expression in Bergmann glia of the cerebellum as revealed by its colocalization with glial fibrillary acidic protein, an astrocyte-specific intermediate filament.

In our recent study, we detected endogenous expression of Panx1, Panx2, and Panx3 in rat primary astrocytes but not in its tumorigenic counterpart, C6 glioma cells, via reverse-transcription-PCR analysis. By stably transfecting epitope- or fluorescent-tagged Panx1 in C6 cells, gap junctional communication was increased, and a significant reduction in cell proliferation in monolayer, cell motility, anchorage-independent growth and in vivo tumor growth in athymic nude mice were observed (Lai et al., 2007). Therefore, we conclude that the loss of Panx expression could participate in the development of C6 gliomas, and Panx1 acts as a tumor-suppressor gene once its expression is restored. Of particular interest is the presence of Panx1 transcripts in a panel of human glioma cell lines, suggesting that aberration in Panx1 transcription specifically applies to a subset of gliomas (Lai et al., 2007). However, loss of Panx1 expression could also occur at the translational level, contributing to transformation. Further study on Panx1 protein expression is required to test this supposition. Interestingly, the coordinated profiles of expression of Cx43 and Panx1 are very similar, suggesting that when expression of either of these genes is altered, the consequence in terms of alterations in other genes should be very similar (Iacobas et al., 2007). Such findings support a role for both Cx43 and Panx1 in growth suppression. Recent studies have implicated Panx1 in cell death. It has been shown that signaling through Panx1 is required for processing of caspase-1 and release of mature interleukin-1β induced by purinergic P2X(7) receptor activation (Pelegrin and Surprenant, 2006). Panx1 appears to be involved in recruitment of the permeabilization pore (or death receptor channel) into the P2X(7) receptor signaling complex (Locovei et al., 2007).

Even though Panx2 expression has not been reported in glial cells, it has been suggested as a tumor-suppressor gene in glial cells (Litvin et al., 2006). In parallel to the loss of Panx2 transcript in C6 cells, high-throughput microarray analysis of human brain tumor samples has shown an overall reduction of *Panx2* gene expression in gliomas. Furthermore, a correlation between Panx2 upregulation and postdiagnosis survival in patients with glial tumors was found using the brain cancer gene expression database REMBRANDT (Repository of Molecular Brain Neoplasia Data, http://rembrandt.nci.nih.gov/rembrandt) (Litvin et al., 2006). In addition, the *Panx2* gene is located within chromosomal region 22q13.3 where deletion was often found in human astrocytomas and ependymomas (Hu et al., 2004; Ino et al., 1999; Oskam et al., 2000; Rey et al., 1993). Given that Panx2 alone was not able to form either hemichannels or intercellular channels in the *Xenopus* oocyte system and if Panx2 indeed is a tumor suppressor (Bruzzone et al., 2003), it is speculated that Panx2 elicits its function via interaction with Panx1 and/or interplay with other molecules. We have recently examined the effects of Panx2 expression in C6 gliomas, confirming a strong tumor suppressive phenotype when it is expressed (Lai and Naus, unpublished).

22.5 Summary

Gap junctions and their constituent proteins, Cxs and Panxs, have been shown to play a role in controlling proliferation in many cell types, including gliomas. This has been exploited for possible therapeutic potential by directly targeting cell proliferation, or by utilizing properties of Cxs/Panxs and/or gap junction channels to augment suicide gene therapy. While the clinical application of therapies targeting Cx/Panx pathways have not yet been realized, valuable approaches to such therapy remain a fertile territory.

Acknowledgments Supported by funds from the Canadian Institutes of Health Research. Christian C. Naus is a recipient of a Canada Research Chair. The authors thank Dr. D.C. Spray for his thoughtful review.

References

Andrade-Rozental AF, Rozental R, Hopperstad MG, Wu JK, Vrionis FD, Spray DC (2000) Gap junctions: the "kiss of death" and the "kiss of life". Brain Res Brain Res Rev 32: 308–315.

Aronica E, Gorter JA, Jansen GH, Leenstra S, Yankaya B, Troost D (2001) Expression of connexin 43 and connexin 32 gap-junction proteins in epilepsy-associated brain tumors and in the perilesional epileptic cortex. Acta Neuropathol 101: 449–459.

Azzam EI, de Toledo SM, Gooding T, Little JB (1998) Intercellular communication is involved in the bystander regulation of gene expression in human cells exposed to very low fluences of alpha particles. Radiat Res 150: 497–504.

Azzam EI, de Toledo SM, Little JB (2003) Oxidative metabolism, gap junctions and the ionizing radiation-induced bystander effect. Oncogene 22: 7050–7057.

Balda MS, Matter K (2000) The tight junction protein ZO-1 and an interacting transcription factor regulate ErbB-2 expression. EMBO J 19: 2024–2033.

Bao X, Chen Y, Reuss L, Altenberg GA (2004) Functional expression in Xenopus oocytes of gap-junctional hemichannels formed by a cysteine-less connexin 43. J Biol Chem 279: 9689–9692.

Baranova A, Ivanov D, Petrash N, Pestova A, Skoblov M, Kelmanson I, Shagin D, Nazarenko S, Geraymovych E, Litvin O, Tiunova A, Born TL, Usman N, Staroverov D, Lukyanov S, Panchin Y (2004) The mammalian pannexin family is homologous to the invertebrate innexin gap junction proteins. Genomics 83: 706–716.

Barbe MT, Monyer H, Bruzzone R (2006) Cell–cell communication beyond connexins: the pannexin channels. Physiology 21: 103–114.

Bates DC, Sin WC, Aftab Q, Naus CC (2007) Connexin43 enhances glioma invasion by a mechanism involving the carboxy terminus. Glia 55: 1154–1564.

Bauer R, Loer B, Ostrowski K, Martini J, Weimbs A, Lechner H, Hoch M (2005) Intercellular communication: the Drosophila innexin multiprotein family of gap junction proteins. Chem Biol 12: 515–526.

Bechberger JF, Khoo NSK, Naus CC (1996) Analysis of connexin43 expression under the control of a metallothionein promoter. Cell Growth Diff 7: 1403–1413.

Bee A, Ke Y, Forootan S, Lin K, Beesley C, Forrest SE, Foster CS (2006) Ribosomal protein L19 is a prognostic marker for human prostate cancer. Clin Cancer Res 12: 2061–2065.

Belliveau DJ, Bani-Yaghoub M, McGirr B, Naus CC, Rushlow WJ (2006) Enhanced neurite outgrowth in PC12 cells mediated by connexin hemichannels and ATP. J Biol Chem 281: 20920–20931.

Bennett MV, Contreras JE, Bukauskas FF, Saez JC (2003) New roles for astrocytes: gap junction hemichannels have something to communicate. Trends Neurosci 26: 610–617.

Bleau AM, Planque N, Perbal B (2005) CCN proteins and cancer: two to tango. Front Biosci 10: 998–1009.

Bond SL, Bechberger JF, Khoo NKS, Naus CC (1994) Transfection of C6 glioma cells with connexin32: the effects of expression of a nonendogenous gap junction protein. Cell Growth Diff 5: 179–186.

Bradshaw SL, Naus CC, Zhu D, Kidder GM, Han VK (1993a) Insulin-like growth factor binding protein-4 gene expression is induced by transfection of gap junction connexin43 gene in a C6 glioma cell line. Growth Regul 3: 26–29.

Bradshaw SL, Naus CCG, Zhu D, Kidder GM, D'Ecrole AJ, Han VKM (1993b) Alteration in the synthesis of insulin-like growth factor binding proteins and IGF-II in rat C6 glioma cells transfected with a gap junction connexin43 cDNA. Reg Peptides 48: 99–112.

Brandner JM, Houdek P, Husing B, Kaiser C, Moll I (2004) Connexins 26, 30, and 43: differences among spontaneous, chronic, and accelerated human wound healing. J Invest Dermatol 122: 1310–1320.

Bruzzone R, Hormuzdi SG, Barbe MT, Herb A, Monyer H (2003) Pannexins, a family of gap junction proteins expressed in brain. Proc Natl Acad Sci USA 100: 13644–13649.

Bruzzone R, Barbe MT, Jakob NJ, Monyer H (2005) Pharmacological properties of homomeric and heteromeric pannexin hemichannels expressed in Xenopus oocytes. J Neurochem 92: 1033–1043.

Butowski NA, Sneed PK, Chang SM (2006) Diagnosis and treatment of recurrent high-grade astrocytoma. J Clin Oncol 24: 1273–1280.

Carystinos GD, Katabi MM, Laird DW, Galipeau J, Chan H, Alaoui-Jamali MA, Batist G (1999) Cyclic-AMP induction of gap junctional intercellular communication increases bystander effect in suicide gene therapy. Clin Cancer Res 5: 61–68.

Chen SC, Pelletier DB, Ao P, Boynton AL (1995) Connexin43 reverses the phenotype of transformed cells and alters their expression of cyclin/cyclin-dependent kinases. Cell Growth Diff 6: 681–690.

Cirenei N, Colombo BM, Mesnil M, Benedetti S, Yamasaki H, Finocchiaro G (1998) In vitro and in vivo effects of retrovirus-mediated transfer of the connexin 43 gene in malignant gliomas: consequences for HSVtk/GCV anticancer gene therapy. Gene Ther 5: 1221–1226.

Conklin CM, Bechberger JF, MacFabe D, Guthrie N, Kurowska EM, Naus CC (2007) Genistein and quercetin increase connexin43 and suppress growth of breast cancer cells. Carcinogenesis 28: 93–100.

Dahl G, Locovei S (2006) Pannexin: to gap or not to gap, is that a question? IUBMB Life 58: 409–419.

Di Chiro G, DeLaPaz RL, Brooks RA, Sokoloff L, Kornblith PL, Smith BH, Patronas NJ, Kufta CV, Kessler RM, Johnston GS, Manning RG, Wolf AP (1982) Glucose utilization of cerebral gliomas measured by [18F] fluorodeoxyglucose and positron emission tomography. Neurology 32: 1323–1329.

Dilber M, Abedi MR, Christensson B, Bjorkstrand B, Kidder GM, Naus CCG, Gahrton G, Smith CIE (1997) Gap junctions promote the bystander effect of Herpes simplex thymidine kinase in vivo. Cancer Res 57: 1523–1528.

Ding HF, Fisher DE (2002) Induction of apoptosis in cancer: new therapeutic opportunities. Ann Med 34: 451–469.

Duffy HS, Delmar M, Spray DC (2002) Formation of the gap junction nexus: binding partners for connexins. J Physiol 96: 243–249.

Elias LA, Wang DD, Kriegstein AR (2007) Gap junction adhesion is necessary for radial migration in the neocortex. Nature 448: 901–907.

Estin D, Li M, Spray D, Wu JK (1999) Connexins are expressed in primary brain tumors and enhance the bystander effect in gene therapy. Neurosurgery 44: 361–368.

Fidler IJ, Yano S, Zhang RD, Fujimaki T, Bucana CD (2002) The seed and soil hypothesis: vascularisation and brain metastases. Lancet Oncol 3: 53–57.

Fu CT, Bechberger JF, Ozog MA, Perbal B, Naus CC (2004) CCN3 (NOV) interacts with connexin43 in C6 glioma cells: possible mechanism of connexin-mediated growth suppression. J Biol Chem 279: 36943–36950.

Fushiki S, Perez Velazquez JL, Zhang L, Bechberger JF, Carlen PL, Naus CC (2003) Changes in neuronal migration in neocortex of connexin43 null mutant mice. J Neuropathol Exp Neurol 62: 304–314.

Gellhaus A, Dong X, Propson S, Maass K, Klein-Hitpass L, Kibschull M, Traub O, Willecke K, Perbal B, Lye SJ, Winterhager E (2004) Connexin43 interacts with NOV: a possible

mechanism for negative regulation of cell growth in choriocarcinoma cells. J Biol Chem 279: 36931–36942.
Giepmans BN (2004) Gap junctions and connexin-interacting proteins. Cardiovasc Res 62: 233–245.
Giepmans BN, Moolenaar WH (1998) The gap junction protein connexin43 interacts with the second PDZ domain of the zona occludens-1 protein. Curr Biol 8: 931–934.
Goldberg GS, Lampe PD, Sheedy D, Stewart CC, Nicholson BJ, Naus CC (1998) Direct isolation and analysis of endogenous transjunctional ADP from Cx43 transfected C6 glioma cells. Exp Cell Res 239: 82–92.
Goldberg GS, Lampe PD, Nicholson BJ (1999) Selective transfer of endogenous metabolites through gap junctions composed of different connexins. Nat Cell Biol 1: 457–459.
Goldberg GS, Bechberger JF, Tajima Y, Merritt M, Omori Y, Gawinowicz MA, Narayanan R, Tan Y, Sanai Y, Yamasaki H, Naus CC, Tsuda H, Nicholson BJ (2000) Connexin43 suppresses MFG-E8 while inducing contact growth inhibition of glioma cells. Cancer Res 60: 6018–6026.
Goodenough DA, Paul DL (2003) Beyond the gap: functions of unpaired connexon channels. Nat Rev Mol Cell Biol 4: 285–294.
Graeber SH, Hulser DF (1998) Connexin transfection induces invasive properties in HeLa cells. Exp Cell Res 243: 142–149.
Gumbiner B, Lowenkopf T, Apatira D (1991) Identification of a 160-kDa polypeptide that binds to the tight junction protein ZO-1. Proc Natl Acad Sci USA 88: 3460–3464.
Hatten ME, Liem RK, Shelanski ML, Mason CA (1991) Astroglia in CNS injury. Glia 4: 233–243.
Henry JL, Coggin DL, Richter King C (1993) High-level expression of the ribosomal protein L19 in human breast tumors that overexpress erbB-2. Cancer Res 53: 1403–1408.
Herve JC, Bourmeyster N, Sarrouilhe D (2004) Diversity in protein–protein interactions of connexins: emerging roles. Biochim Biophys Acta 1662: 22–41.
Hirst-Jensen BJ, Sahoo P, Kieken F, Delmar M, Sorgen PL (2007) Characterization of the pH-dependent interaction between the gap junction protein connexin43 carboxyl terminus and cytoplasmic loop domains. J Biol Chem 282: 5801–5813.
Hoffman AG, Burghardt RC, Tilley R, Auersperg N (1993) An in vitro model of ovarian epithelial carcinogenesis: changes in cell-cell communication and adhesion occurring during neoplastic progression. Int J Cancer 54: 828–838.
Hu J, Jiang CC, Ng HK, Pang JC, Tong CY, Chen SQ (2004) A genome-wide allelotype study of primary and corresponding recurrent glioblastoma multiforme in one patient. Chin Med J (Engl) 117: 456–458.
Hua VB, Chang AB, Tchieu JH, Kumar NM, Nielsen PA, Saier MH Jr (2003) Sequence and phylogenetic analyses of 4 TMS junctional proteins of animals: connexins, innexins, claudins and occludins. J Membr Biol 194: 59–76.
Huang GY, Cooper ES, Waldo K, Kirby ML, Gilula NB, Lo CW (1998a) Gap junction-mediated cell–cell communication modulates mouse neural crest migration. J Cell Biol 143: 1725–1734.
Huang R, Liu YG, Lin Y, Fan Y, Boynton A, Yang D, Huang RP (2001a) Enhanced apoptosis under low serum conditions in human glioblastoma cells by connexin 43 (Cx43). Mol Carcinog 32: 128–138.
Huang R, Lin Y, Wang CC, Gano J, Lin B, Shi Q, Boynton A, Burke J, Huang RP (2002) Connexin 43 suppresses human glioblastoma cell growth by down-regulation of monocyte chemotactic protein 1, as discovered using protein array technology. Cancer Res 62: 2806–2812.
Huang RP, Fan Y, Hossain MZ, Peng A, Zeng ZL, Boynton AL (1998b) Reversion of the neoplastic phenotype of human glioblastoma cells by connexin 43 (cx43). Cancer Res 58: 5089–5096.
Huang RP, Hossain MZ, Sehgal A, Boynton AL (1999) Reduced connexin43 expression in high-grade human brain glioma cells. J Surg Oncol 70: 21–24.

Huang RP, Hossain MZ, Huang R, Gano J, Fan Y, Boynton AL (2001b) Connexin 43 (cx43) enhances chemotherapy-induced apoptosis in human glioblastoma cells. Int J Cancer 92: 130–138.
Iacobas DA, Scemes E, Spray DC (2004) Gene expression alterations in connexin null mice extend beyond the gap junction. Neurochem Int 45: 243–250.
Iacobas DA, Iacobas S, Spray DC (2007) Connexin43 and the brain transcriptome of newborn mice. Genomics 89: 113–123.
Ino Y, Silver JS, Blazejewski L, Nishikawa R, Matsutani M, von Deimling A, Louis DN (1999) Common regions of deletion on chromosome 22q12.3-q13.1 and 22q13.2 in human astrocytomas appear related to malignancy grade. J Neuropathol Exp Neurol 58: 881–885.
Jensen R, Glazer PM (2004) Cell-interdependent cisplatin killing by Ku/DNA-dependent protein kinase signaling transduced through gap junctions. Proc Natl Acad Sci USA 101: 6134–6139.
Jiang JX, Gu S (2005) Gap junction- and hemichannel-independent actions of connexins. Biochim Biophys Acta 1711: 208–214.
Jimenez T, Fox WP, Naus CC, Galipeau J, Belliveau DJ (2006) Connexin over-expression differentially suppresses glioma growth and contributes to the bystander effect following HSV-thymidine kinase gene therapy. Cell Commun Adhes 13: 79–92.
Kardami E, Dang X, Iacobas DA, Nickel BE, Jeyaraman M, Srisakuldee W, Makazan J, Tanguy S, Spray DC (2007) The role of connexins in controlling cell growth and gene expression. Prog Biophys Mol Biol 94: 245–264.
Kelleher FC, Fennelly D, Rafferty M (2006) Common critical pathways in embryogenesis and cancer. Acta Oncol 45: 375–388.
Lai CP, Bechberger JF, Thompson RJ, MacVicar BA, Bruzzone R, Naus CC (2007) Tumor-suppressive effects of pannexin 1 in C6 glioma cells. Cancer Res 67: 1545–1554.
Lampe PD, Lau AF (2000) Regulation of gap junctions by phosphorylation of connexins. Arch Biochem Biophys 384: 205–215.
Leask A, Abraham DJ (2006) All in the CCN family: essential matricellular signaling modulators emerge from the bunker. J Cell Sci 119: 4803–4810.
Leirdal M, Shadidy M, Rosok O, Sioud M (2004) Identification of genes differentially expressed in breast cancer cell line SKBR3: potential identification of new prognostic biomarkers. Int J Mol Med 14: 217–222.
Leithe E, Sirnes S, Omori Y, Rivedal E (2006) Downregulation of gap junctions in cancer cells. Crit Rev Oncog 12: 225–256.
Li WE, Waldo K, Linask KL, Chen T, Wessels A, Parmacek MS, Kirby ML, Lo CW (2002) An essential role for connexin43 gap junctions in mouse coronary artery development. Development 129: 2031–2042.
Lin C, Weigel H, Cotrina ML, Liu S, Bueno E, Hansen AJ, Hansen TW, Goldman S, Nedergaard M (1998) Gap-junction-mediated propagation and amplification of cell injury. Nat Neurosci 1: 494–500.
Lin JH, Yang J, Liu S, Takano T, Wang X, Gao Q, Willecke K, Nedergaard M (2003) Connexin mediates gap junction-independent resistance to cellular injury. J Neurosci 23: 430–441.
Litvin O, Tiunova A, Connell-Alberts Y, Panchin Y, Baranova A (2006) What is hidden in the pannexin treasure trove: the sneak peek and the guesswork. J Cell Mol Med 10: 613–634.
Locovei S, Scemes E, Qiu F, Spray DC, Dahl G (2007) Pannexin1 is part of the pore forming unit of the P2X(7) receptor death complex. FEBS Lett 581: 483–488.
Loewenstein WR, Kanno Y (1966) Intercellular communication and the control of tissue growth: lack of communication between cancer cells. Nature 209: 1248–1249.
McDonough WS, Johansson A, Joffee H, Giese A, Berens ME (1999) Gap junction intercellular communication in gliomas is inversely related to cell motility. Int J Dev Neurosci 17: 601–611.
Mesnil M, Piccoli C, Tiraby G, Willecke K, Yamasaki H (1996) Bystander killing of cancer cells by *HSV-TK* gene is mediated by connexins. Proc Natl Acad Sci USA 93: 1831–1835.
Mesnil M, Crespin S, Avanzo JL, Zaidan-Dagli ML (2005) Defective gap junctional intercellular communication in the carcinogenic process. Biochim Biophys Acta 1719: 125–145.

Mitic LL, Anderson JM (1998) Molecular architecture of tight junctions. Annu Rev Physiol 60: 121–142.
Moorby CD (2000) A connexin 43 mutant lacking the carboxyl cytoplasmic domain inhibits both growth and motility of mouse 3T3 fibroblasts. Mol Carcinog 28: 23–30.
Mueller MM, Fusenig NE (2004) Friends or foes – bipolar effects of the tumour stroma in cancer. Nat Rev Cancer 4: 839–849.
Muller W, Afra D, Schroder R (1977) Supratentorial recurrences of gliomas. Morphological studies in relation to time intervals with astrocytomas. Acta Neurochir 37: 75–91.
Naus CC, Bechberger JF, Caveney S, Wilson JX (1991a) Expression of gap junction genes in astrocytes and C6 glioma cells. Neurosci Lett 126: 33–36.
Naus CC, Bechberger JF, Paul DL (1991b) Gap junction gene expression in human seizure disorder. Exp Neurol 111: 198–203.
Naus CC, Elisevich K, Zhu D, Belliveau DJ, Del Maestro RF (1992) In vivo growth of C6 glioma cells transfected with connexin43 cDNA. Cancer Res 52: 4208–4213.
Naus CC, Bond SL, Bechberger JF, Rushlow W (2000) Identification of genes differentially expressed in C6 glioma cells transfected with connexin43. Brain Res Brain Res Rev 32: 259–266.
Naus CC, Goldberg GS, Sin WC (2005) Connexin in growth control and cancer. In: Gap junctions in development and disease (Winterhager E, ed), pp 253–273. Heidelberg: Springer-Verlag.
Neijssen J, Herberts C, Drijfhout JW, Reits E, Janssen L, Neefjes J (2005) Cross-presentation by intercellular peptide transfer through gap junctions. Nature 434: 83–88.
Nicholas TW, Read SB, Burrows FJ, Kruse CA (2003) Suicide gene therapy with Herpes simplex virus thymidine kinase and ganciclovir is enhanced with connexins to improve gap junctions and bystander effects. Histol Histopathol 18: 495–507.
Oliveira R, Christov C, Guillamo JS, Debouard S, Palfi S, Venance L, Tardy M, Peschanski M (2005) Contribution of gap junctional communication between tumor cells and astroglia to the invasion of the brain parenchyma by human glioblastomas. BMC Cell Biol 6: 7.
Oskam NT, Bijleveld EH, Hulsebos TJ (2000) A region of common deletion in 22q13.3 in human glioma associated with astrocytoma progression. Int J Cancer 85: 336–339.
Ozog MA, Bechberger JF, Naus CC (2002) Ciliary neurotrophic factor (CNTF) in combination with its soluble receptor (CNTFRalpha) increases connexin43 expression and suppresses growth of C6 glioma cells. Cancer Res 62: 3544–3548.
Ozog MA, Bernier SM, Bates DC, Chatterjee B, Lo CW, Naus CC (2004) The complex of CNTF-CNTFRalpha upregulates connexin43 and intercellular coupling in astrocytes via the JAK/STAT pathway. Mol Biol Cell 15: 4761–4774.
Panchin Y, Kelmanson I, Matz M, Lukyanov K, Usman N, Lukyanov S (2000) A ubiquitous family of putative gap junction molecules. Curr Biol 10: R473–R474.
Panchin YV (2005) Evolution of gap junction proteins – the pannexin alternative. J Exp Biol 208: 1415–1419.
Pelegrin P, Surprenant A (2006) Pannexin-1 mediates large pore formation and interleukin-1beta release by the ATP-gated P2X7 receptor. EMBO J 25: 5071–5082.
Peracchia C, Sotkis A, Wang XG, Peracchia LL, Persechini A (2000) Calmodulin directly gates gap junction channels. J Biol Chem 275: 26220–26224.
Perbal B (2004) CCN proteins: multifunctional signalling regulators. Lancet 363: 62–64.
Princen F, Robe P, Lechanteur C, Mesnil M, Rigo JM, Gielen J, Merville MP, Bours V (1999) A cell type-specific and gap junction-independent mechanism for the herpes simplex virus-1 thymidine kinase gene/ganciclovir-mediated bystander effect. Clin Cancer Res 5: 3639–3644.
Princen F, Robe P, Gros D, Jarry-Guichard T, Gielen J, Merville MP, Bours V (2001) Rat gap junction connexin-30 inhibits proliferation of glioma cell lines. Carcinogenesis 22: 507–513.
Pu P, Xia Z, Yu S, Huang Q (2004) Altered expression of Cx43 in astrocytic tumors. Clin Neurol Neurosurg 107: 49–54.
Qiu C, Coutinho P, Frank S, Franke S, Law LY, Martin P, Green CR, Becker DL (2003) Targeting connexin43 expression accelerates the rate of wound repair. Curr Biol 13: 1697–1703.

Ray A, Zoidl G, Weickert S, Wahle P, Dermietzel R (2005) Site-specific and developmental expression of pannexin1 in the mouse nervous system. Eur J Neurosci 21: 3277–3290.
Reardon DA, Rich JN, Friedman HS, Bigner DD (2006) Recent advances in the treatment of malignant astrocytoma. J Clin Oncol 24: 1253–1265.
Rey JA, Bello MJ, de Campos JM, Vaquero J, Kusak ME, Sarasa JL, Pestana A (1993) Abnormalities of chromosome 22 in human brain tumors determined by combined cytogenetic and molecular genetic approaches. Cancer Genet Cytogenet 66: 1–10.
Robe PA, Princen F, Martin D, Malgrange B, Stevenaert A, Moonen G, Gielen J, Merville M, Bours V (2000) Pharmacological modulation of the bystander effect in the herpes simplex virus thymidine kinase/ganciclovir gene therapy system: effects of dibutyryl adenosine 3′,5′-cyclic monophosphate, alpha-glycyrrhetinic acid, and cytosine arabinoside. Biochem Pharmacol 60: 241–249.
Robe PA, Jolois O, N'Guyen M, Princen F, Malgrange B, Merville MP, Bours V (2004) Modulation of the HSV-TK/ganciclovir bystander effect by n-butyrate in glioblastoma: correlation with gap-junction intercellular communication. Int J Oncol 25: 187–192.
Saez JC, Retamal MA, Basilio D, Bukauskas FF, Bennett MV (2005) Connexin-based gap junction hemichannels: gating mechanisms. Biochim Biophys Acta 1711: 215–224.
Sanchez-Alvarez R, Tabernero A, Medina JM (2005) The increase in gap junctional communication decreases the rate of glucose uptake in C6 glioma cells by releasing hexokinase from mitochondria. Brain Res 1039: 189–198.
Sanchez-Alvarez R, Paino T, Herrero-Gonzalez S, Medina JM, Tabernero A (2006) Tolbutamide reduces glioma cell proliferation by increasing connexin43, which promotes the up-regulation of p21 and p27 and subsequent changes in retinoblastoma phosphorylation. Glia 54: 125–134.
Sasakura Y, Shoguchi E, Takatori N, Wada S, Meinertzhagen IA, Satou Y, Satoh N (2003) A genomewide survey of developmentally relevant genes in *Ciona intestinalis*. X. Genes for cell junctions and extracellular matrix. Dev Genes Evol 213: 303–313.
Scemes E, Duval N, Meda P (2003) Reduced expression of P2Y1 receptors in connexin43-null mice alters calcium signaling and migration of neural progenitor cells. J Neurosci 23: 11444–11452.
Shao Q, Wang H, McLachlan E, Veitch GI, Laird DW (2005) Down-regulation of Cx43 by retroviral delivery of small interfering RNA promotes an aggressive breast cancer cell phenotype. Cancer Res 65: 2705–2711.
Shinoura N, Chen L, Wani MA, Kim YG, Larson JJ, Warnick RE, Simon M, Menon AG, Bi WL, Stambrook PJ (1996) Protein and messenger RNA expression of connexin43 in astrocytomas: implications in brain tumor gene therapy. J Neurosurg 84: 839–845.
Simon AM, Goodenough DA (1998) Diverse functions of vertebrate gap junctions. Trends Cell Biol 8: 477–483.
Singh D, Solan JL, Taffet SM, Javier R, Lampe PD (2005) Connexin 43 interacts with zona occludens-1 and -2 proteins in a cell cycle stage-specific manner. J Biol Chem 280: 30416–30421.
Sohl G, Willecke K (2004) Gap junctions and the connexin protein family. Cardiovasc Res 62: 228–232.
Soroceanu L, Manning TJ, Jr., Sontheimer H (2001) Reduced expression of connexin-43 and functional gap junction coupling in human gliomas. Glia 33: 107–117.
Spray DC, Ye ZC, Ransom BR (2006) Functional connexin "hemichannels": a critical appraisal. Glia 54: 758–773.
Srinivas M, Spray DC (2003) Closure of gap junction channels by arylaminobenzoates. Mol Pharmacol 63: 1389–1397.
Stein LS, Stein DW, Echols J, Burghardt RC (1993) Concomitant alterations of desmosomes, adhesiveness, and diffusion through gap junction channels in a rat ovarian transformation model system. Exp Cell Res 207: 19–32.
Stout CE, Costantin JL, Naus CC, Charles AC (2002) Intercellular calcium signaling in astrocytes via ATP release through connexin hemichannels. J Biol Chem 277: 10482–10488.
Tabernero A, Giaume C, Medina JM (1996) Endothelin-1 regulates glucose utilization in cultured astrocytes by controlling intercellular communication through gap junctions. Glia 16: 187–195.

Tabernero A, Medina JM, Giaume C (2006) Glucose metabolism and proliferation in glia: role of astrocytic gap junctions. J Neurochem 99: 1049–1061.
Taliana L, Benezra M, Greenberg RS, Masur SK, Bernstein AM (2005) ZO-1: lamellipodial localization in a corneal fibroblast wound model. Invest Ophthalmol Vis Sci 46: 96–103.
Touraine RL, Vahanian N, Ramsey WJ, Blaese RM (1998) Enhancement of the herpes simplex virus thymidine kinase/ganciclovir bystander effect and its antitumor efficacy in vivo by pharmacologic manipulation of gap junctions. Hum Gene Ther 9: 2385–2391.
Trosko JE, Chang CC (2001) Mechanism of up-regulated gap junctional intercellular communication during chemoprevention and chemotherapy of cancer. Mutat Res 480–481: 219–229.
Trosko JE, Goodman JI (1994) Intercellular communication may facilitate apoptosis: implications for tumor promotion. Mol Carcinogen 11: 8–12.
Trosko JE, Ruch RJ (2002) Gap junctions as targets for cancer chemoprevention and chemotherapy. Curr Drug Targets 3: 465–482.
Tsukita S, Furuse M, Itoh M (1999) Structural and signalling molecules come together at tight junctions. Curr Opin Cell Biol 11: 628–633.
Valiunas V, Polosina YY, Miller H, Potapova IA, Valiuniene L, Doronin S, Mathias RT, Robinson RB, Rosen MR, Cohen IS, Brink PR (2005) Connexin-specific cell-to-cell transfer of short interfering RNA by gap junctions. J Physiol 568: 459–468.
Vogt A, Hormuzdi SG, Monyer H (2005) Pannexin1 and Pannexin2 expression in the developing and mature rat brain. Brain Res Mol Brain Res 141: 113–120.
Wei CJ, Xu X, Lo CW (2004) Connexins and cell signaling in development and disease. Annu Rev Cell Dev Biol 20: 811–838.
Wei CJ, Francis R, Xu X, Lo CW (2005) Connexin43 associated with an N-cadherin-containing multiprotein complex is required for gap junction formation in NIH3T3 cells. J Biol Chem 280: 19925–19936.
Weickert S, Ray A, Zoidl G, Dermietzel R (2005) Expression of neural connexins and pannexin1 in the hippocampus and inferior olive: a quantitative approach. Brain Res Mol Brain Res 133: 102–109.
Willecke K, Eiberger J, Degen J, Eckardt D, Romualdi A, Guldenagel M, Deutsch U, Sohl G (2002) Structural and functional diversity of connexin genes in the mouse and human genome. Biol Chem 383: 725–737.
Yamasaki H, Naus CC (1996) Role of connexin genes in growth control. Carcinogenesis 17: 1199–1213.
Yeager M, Nicholson BJ (1996) Structure of gap junction intercellular channels. Curr Opin Struct Biol 6: 183–192.
Zappala A, Cicero D, Serapide MF, Paz C, Catania MV, Falchi M, Parenti R, Panto MR, La Delia F, Cicirata F (2006) Expression of pannexin1 in the CNS of adult mouse: cellular localization and effect of 4-aminopyridine-induced seizures. Neuroscience 141: 167–178.
Zhang W, Nwagwu C, Le DM, Yong VW, Song H, Couldwell WT (2003) Increased invasive capacity of connexin43-overexpressing malignant glioma cells. J Neurosurg 99: 1039–1046.
Zhu D, Caveney S, Kidder GM, Naus CC (1991) Transfection of C6 glioma cells with connexin 43 cDNA: analysis of expression, intercellular coupling, and cell proliferation. Proc Natl Acad Sci USA 88: 1883–1887.
Zhu D, Kidder GM, Caveney S, Naus CC (1992) Growth retardation in glioma cells cocultured with cells overexpressing a gap junction protein. Proc Natl Acad Sci USA 89: 10218–10221.

Abbreviations

CCN	Cyr61/CTGF/Nov
CDK	Cyclin-dependant kinase
CNS	Central nervous system

CNTF	Ciliary neurotrophic factor
CNTFR	CNTF receptor
Cx	Connexin
IGF1	Insulin-like growth factor 1
Inx	Innexins
JAK/STAT	Janus tyrosine kinase/signal transducer and activator of transcription
MCP-1	Monocyte chemotactic protein 1
MFG-E8	Milk fat globule epidermal growth factor 8
Panx	Pannexin
PDZ	PSD95/Dlg/ZO-1
RPL19	Ribosomal protein L19
ZO-1	Zonula occludens 1

Chapter 23
The Impact of Astrocyte Mitochondrial Metabolism on Neuroprotection During Aging

Lora T. Watts and James D. Lechleiter

Contents

23.1 Introduction

Accumulation of oxidative damage as the result of normal mitochondrial metabolism is widely considered to be a fundamental cause of aging. A central tenet of this theory is that mitochondria themselves become dysfunctional. In the central nervous system (CNS), the focus of research on aging has primarily revolved around changes in and effects of neuronal mitochondrial metabolism. However, there is increasing interest in the role that astrocyte mitochondria play in the aging process. Little is known about the cumulative effects of aging on astrocyte mitochondria or on energy-dependent processes within astrocytes. It is likely that diminished astrocyte

J.D. Lechleiter
Department of Cellular and Structural Biology, University of Texas Health Science Center at San Antonio, San Antonio, TX USA
lechleiter@uthscsa.edu

V. Parpura and P.G. Haydon (eds.), *Astrocytes in (Patho)Physiology of the Nervous System*,
DOI: 10.1007/978-0-387-79492-1_23, © Springer Science+Business Media, LLC 2009

function throughout the aging process is a prominent determinant of both neuronal survival as well as survival of the entire organism. In this chapter, we focus our discussion on the impact of astrocyte mitochondrial metabolism during the aging process. We present a brief review of astrocyte bioenergetics followed by a discussion of our recent work on decreased astrocyte neuroprotection during aging. We then discuss our ongoing work on a neuroprotective pathway that is mediated by increased mitochondrial metabolism in astrocytes. This pathway is activated by G-protein-coupled receptors that stimulate inositol 1,4,5 trisphosphate (IP_3)-gated Ca^{2+} release. Astrocyte neuroprotection can be enhanced in both young and old astrocytes to the point that their neuroprotective functions are nearly comparable.

23.2 Astrocyte Bioenergetics

The human brain constitutes only about 3% of one's body weight, but it requires more energy than any other organ. To generate this energy, the brain relies on mitochondrial oxidative phosphorylation, which accounts for over 90% of the cellular adenosine triphospate (ATP) production (Drew and Leeuwenburgh, 2003). Brain mitochondria consume ~20% of the total body's O_2 consumption at a rate of 160 μmol min^{-1} per 100 g of protein (Sokoloff, 1989) and ~15% of the human body's cardiac output. It has also been estimated that the brain utilizes ~25% of total body glucose at a rate of 31 μmol min^{-1} per 100 g of protein (Silver and Erecinska, 1997). The primary reason for the brain's dependence on mitochondria is, of course, their high-energy production efficiency. Mitochondria can generate 36 molecules of ATP for each molecule of glucose converted to CO_2 and H_2O, while glycolysis generates only two ATP molecules.

It is widely assumed that neuronal mitochondria are the primary consumers of O_2 in the brain. However, recent work has demonstrated that astrocytes account for up to 20% of the brain's O_2 consumption (Gruetter et al., 2001; Bluml et al., 2002; Lebon et al., 2002; Hertz, 2004). Nuclear magnetic resonance (NMR) spectroscopy has been used to noninvasively make these measurements in the human brain, in situ. In general, metabolic fluxes are measured through neuronal and glial tricarboxylic acid (TCA) cycles. NMR studies of carbon 13 (^{13}C) that utilize ^{13}C glucose as the substrate are known to predominantly reflect neuronal metabolism (Beckmann et al., 1991; Mason et al., 1995; Gruetter et al., 1998; Bluml et al., 2000). However, the relative metabolic flow between neurons and glia was still estimated at ~20% by using calculations based on animal models and cellular studies (Gruetter et al., 1998). More directly, investigators discovered that acetate was metabolized almost exclusively by glia into acetyl-CoA (Muir et al., 1986; Badar-Goffer et al., 1990; Hassel et al., 1995; Waniewski and Martin, 1998). Consequently, ^{13}C label from this substrate that accumulates in the carbon dioxide pool of mitochondria ($^{13}CO_2$) has to enter through the glial TCA cycle, since neurons do not metabolize acetate. Using this approach, investigators directly confirmed the high percentage of astroglial O_2 consumption in the brain. Specifically, Lebon et al. (2002) and Bluml et al. (2002)

estimated that the astroglial TCA cycle accounted for ~15 and 20% of the brain O_2 consumption, respectively.

Silver and Erecinska (1997) estimated that oxidative phosphorylation in cultured astrocytes contributed ~75% of the total cellular ATP synthesis. Their estimate was obtained by averaging the rate of O_2 consumption relative to the rate of lactate production in astrocytes that was reported by several laboratories (Olson and Holtzman, 1981; Jameson et al., 1984; Pauwels et al., 1985; Lopes-Cardozo et al., 1986; Hertz et al., 1988; Kauppinen et al., 1988; Walz and Mukerji, 1988; Segal and Ingbar, 1990; Pellerin and Magistretti, 1994; Eriksson et al., 1995). Their survey of the literature revealed that on average, cultured astrocytes consumed 14 nmol min^{-1} mg^{-1} protein of O_2 and synthesized 30 nmol min^{-1} mg protein of lactate, which equates to a ratio of 74% oxidative phosphorylation to 26% glycolysis. These calculations clearly indicated that oxidative phosphorylation was a major energy source in astrocyte cultures.

23.3 Physiological Changes in Astrocytes During Aging

Cell culture models have been used extensively to study the aging process in the CNS. However, the primary focus has been on the neuronal component of aging. In recent work, we addressed similar questions with regard to aging, but we focused our investigations on astrocyte physiology. Our first task was to establish and characterize primary cultures of astrocytes prepared from the brains of young (4–6 months) and old (26–28 months) mice (Lin et al., 2007). The cell culture procedures that we used were identical for both young and old mice. Brain tissue was removed, minced, and cultured in plastic T-75 flasks for the first 1–2 weeks until the cells, predominantly astrocytes, reached 70% confluency. The cultured cells were washed once with Hank's buffered saline solution to help remove nonastrocytic cells before treating the remaining plated astrocytes with a trypsin/ethylenediamine tetraacetic acid solution for 3–5 min at 37°C. Suspended astrocytes were centrifuged, brought up in media for cell counting, and plated in 35-mm dishes. The key step in this protocol was to initially plate the astrocytes at high density (10,000 cells in a 100-μL aliquot) in the center of the Petri dish for ~2 h, prior to filling the entire culture dish with media. These astrocytes were permitted to grow at least 4–7 days prior to experiments and were considered passage 1 cells.

Our second step in the investigation was to immunohistochemically identify the cultured cells. We used glial fibrillary acidic protein (GFAP) for astrocytes, A2B5 for type II astrocytes, galactocerebrosidase and receptor-interacting protein for immature and mature oligodendrocytes, and CD11B for microglia. Essentially all of the passage 1 cells cultured with our protocol were GFAP positive. However, we noticed that immunoreactivity to A2B5, galactocerebrosidase, and receptor-interacting protein increased at the second passage and stabilized by the third passage, suggesting that both old and young astrocytes were becoming more undifferentiated. We also tested whether the common procedure of shaking the cultures prior to passage decreased the number of contaminating cell types like microglia. In our hands, shaking the dishes produced no significant effect on the purity of the astrocyte cultures.

The first physiological difference that we observed between astrocytes cultured from young and old mice was their rate of cell growth. Astrocytes cultured from old mice exhibited significantly slower doubling rates (~30% slower). Standard hemocytometer procedures were used to make these measurements. For consistency, it was very important to maintain not only the initial cell number (100,000 cells per well of a six-well plate), but also the same cell density. Interestingly, this difference in growth rates was still present after the cultured astrocytes had been passed up to eight times, indicating that the underlying cause for slower growth was not significantly affected by cell passage. We speculate below that reduced mitochondrial ATP production could be partially responsible for these slower growth rates. Astrocyte cultures were eventually transformed after multiple cell passages. This transformation was easily recognized by a substantial increase in cell growth in both old and young astrocytes as well as alterations in the morphology of the cells. Astrocytes lost their characteristic star shape and assumed a more tightly packed cobble stone appearance.

The next physiological response that we investigated during aging was intracellular Ca^{2+} signaling. Metabotropic purinergic receptors (P2Y-Rs) are well known to generate robust Ca^{2+} signals in astrocytes (Verkhratsky and Kettenmann, 1996; James and Butt, 2002). Our initial bias was that we would observe reduced Ca^{2+} responses in astrocytes cultured from old mice. Surprisingly, we found just the opposite. Ca^{2+} measurements were made using standard confocal imaging of Ca^{2+}- sensitive fluorescent dyes, which loaded equally well for both young and old astrocytes. Bath application of ATP (1 μM) also stimulated Ca^{2+} oscillations in both cultured young and old astrocytes. The first surprise finding was that a higher percentage of old astrocytes (more than half) responded to exogenously applied ATP. We do not yet know the underlying reason for this increased responsiveness. A simple explanation would be that P2Y-R expression is increased in old astrocytes. This question will be addressed in future studies. Two other interesting findings were that old astrocytes exhibited almost 50% higher Ca^{2+} amplitudes as well as faster Ca^{2+} oscillations than young astrocytes. These results appear counterintuitive, since it is reasonably expected that receptor-mediated signaling is likely decreased or degraded with aging. However, we can account for these observed enhancements to cytosolic Ca^{2+} signaling in old astrocytes by a reduced ability of mitochondria to sequester Ca^{2+} during aging. We initially demonstrated that cytosolic Ca^{2+} release was regulated by mitochondrial Ca^{2+} uptake in *Xenopus* oocytes (Jouaville et al., 1995). Subsequent reports by others also demonstrated a modulatory role of mitochondria on the release of intracellular Ca^{2+} in many cell types including astrocytes, liver, and HeLa cells (Simpson and Russell, 1996; Boitier et al., 1999; Hajnoczky et al., 1999; Collins et al., 2000). The ability of mitochondria to sequester Ca^{2+}, and thereby modulate IP_3-induced Ca^{2+} release, is critically dependent on the mitochondrial membrane potential ($\Delta\psi$) and close proximity of mitochondria to the Ca^{2+} release site. This proximity allows mitochondria to sense greater Ca^{2+} concentrations than those present in the bulk cytosolic compartment (Rizzuto et al., 1993, 1994).

Consequently, a reduction in $\Delta\Psi$ with age will reduce mitochondrial Ca^{2+} buffering, thereby leading to increased cytosolic Ca^{2+} release and faster Ca^{2+} dynamics.

As discussed in Sect. 23.2, astrocyte mitochondria are actively respirating, in vivo, and individual mitochondria can be imaged in situ using the potential sensitive dye, tetra-methyl rhodamine ethyl ester (TMRE, Molecular Probes) (Fig. 23.1a). TMRE is positively charged and partitions itself across charged membranes in a Nernstian fashion. This permits $\Delta\Psi$ to be estimated as ~60 mV times the log of F_{mito}/F_{cyto}, where F_{mito} is the peak fluorescent intensity observed in single mitochondrial and F_{cyto} represents the lowest value of TMRE fluorescence in the cytosol (Farkas et al., 1989; Lin et al., 2007). Nonspecific binding of this dye is generally minimal, but can be checked by depolarizing cells with a proton ionophore (e.g., FCCP). A histogram plot of single mitochondrial values in an astrocyte revealed a range of $\Delta\Psi$s normally distributed around a mean (Fig. 23.1a). A similar variance in $\Delta\Psi$s was observed in both young and old astrocytes. However, older astrocytes exhibited a significantly lower mean value than young astrocytes. Furthermore, the different $\Delta\Psi$ means were maintained in astrocyte cultures for over four passages of the cells (Fig. 23.1b). We tried to determine whether the size of a mitochondrion could be positively correlated with its $\Delta\Psi$. For these measurements, we estimated the total area of a single mitochondrion as roughly equivalent to the number of continuous

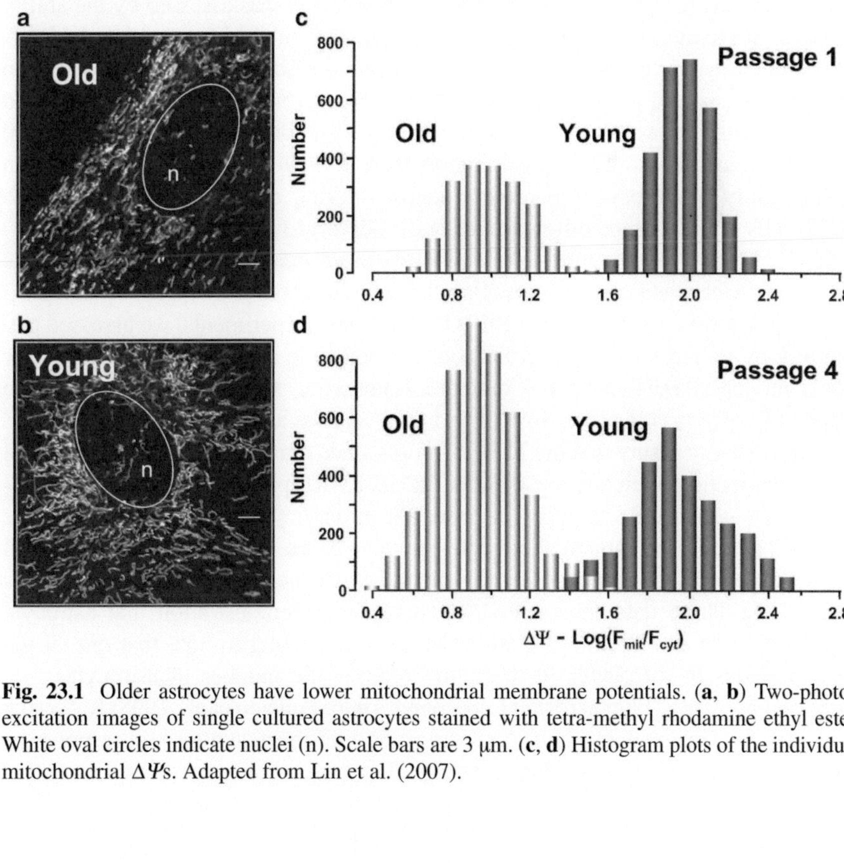

Fig. 23.1 Older astrocytes have lower mitochondrial membrane potentials. (**a**, **b**) Two-photon excitation images of single cultured astrocytes stained with tetra-methyl rhodamine ethyl ester. White oval circles indicate nuclei (n). Scale bars are 3 μm. (**c**, **d**) Histogram plots of the individual mitochondrial $\Delta\Psi$s. Adapted from Lin et al. (2007).

pixels of TMRE fluorescence. A minimal area was used in our analysis to exclude signals from random noise. We did in fact observe positive correlation. TMRE fluorescent signal (log F_{mito}/F_{cyto}) increased with increasing mitochondrial size, and the slope of the relationship was similar for both old and young astrocytes. At higher values of mitochondrial surface area, there was no correlation. This was expected since larger areas increase the likelihood of overlapping mitochondria, which would obscure any relationship. Our only speculation regarding this relationship is that mitochondria with lower $\Delta\Psi$s may exhibit less organelle fusion. It would be interesting to test whether increased $\Delta\Psi$ results in increased fusion of smaller mitochondria as opposed to biogenesis of new mitochondrial mass.

23.3.1 Oxidative Stress

Mattson et al. (1990) originally demonstrated that the presence of even a few astrocytes in cultured neurons significantly increased their resistance to oxidative stress. In the next series of experiments, we wanted to determine whether the protective ability of astrocytes was affected by aging. Our initial step to test the resistance of old and young astrocytes to oxidative stress was to expose them to the oxidant stressor *tert*-butyl hydrogen peroxide (*t*-BuOOH). Cell viability was assessed by the ability of astrocytes to either retain the cytoplasmic dye, calcein AM, or exclude the DNA intercalating dye, propidium iodide. Plasma membrane integrity is required to retain calcein or to exclude propidium iodide. Perhaps not surprisingly, astrocytes cultured from old animals exhibited greater sensitivity to oxidant stress when compared with young astrocytes. These experiments were followed up with an examination of the ability of astrocytes to protect neurons during aging. Pheochromocytoma PC12 cells (PC12) were differentiated with nerve growth factor (Gabryel et al., 2006) for at least 7 days, until neurite processes were readily visible. Neuronal-like PC12 cells were then cocultured with either young or old astrocytes for another 3 days before stressing them with *t*-BuOOH. For these experiments, we assessed cell viability by taking advantage of the potential sensitive dye TMRE. This dye only labels mitochondria that have a membrane potential, so we monitored the time required for $\Delta\Psi$ to collapse to 10% of its initial value as an indicator of cell viability. We chose 10% to insure that the TMRE fluorescence remained above its lower limit at de-energization, which is presumably due to TMRE partitioning into lipid membranes and nonspecific targets. As observed in astrocyte-only cultures, the time until $\Delta\Psi$ collapse was much shorter in old cells. More importantly, the time until $\Delta\Psi$ collapse in PC12 was significantly shorter when cocultured with old astrocytes (Fig. 23.2). These data represented the first direct demonstration that astrocyte neuroprotection was diminished with age. It is interesting to note that one of the characteristics of astrocytes in the aging brain – the number of astrocytes – is increased by ~20% (Pilegaard and Ladefoged, 1996; Peinado et al., 1998; Rozovsky et al., 1998). This response has been compared with reactive gliosis in response to injured or damaged neurons during aging. However, an alternative explanation is

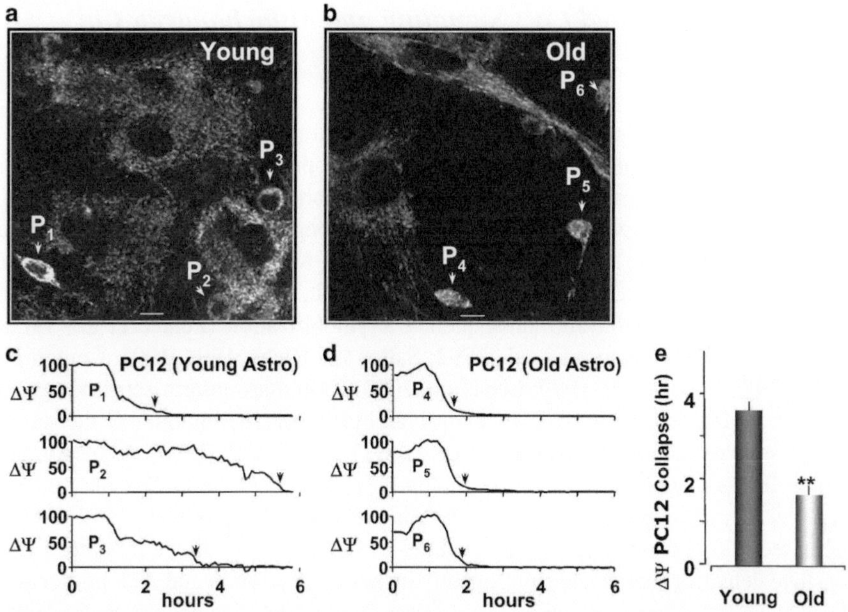

Fig. 23.2 Older astrocytes are less neuroprotective against oxidative stress. (**a**) Cocultures of young astrocytes and NGF-differentiated PC12 cells labeled with the potential sensitive dye TMRE. Three PC12 cells (P1, P2, and P3) are identified with white arrowheads. (**b**) Cocultures of the same NGF-differentiated PC12 cells used in (**a**) and old astrocytes labeled with tetra-methyl rhodamine ethyl ester (TMRE). Three PC12 cells (P4, P5, and P6) are again identified with white arrowheads. Scale bars are 10 μm. (**c**, **d**) Line plots of the mean TMRE fluorescence ($\Delta\Psi$) for the cells labeled in (**a**) and (**b**). Fluorescent units are arbitrary. Black arrowheads indicate the times when the TMRE fluorescence collapsed to 10% of their initial value. (**e**) Histogram plots of the mean time until $\Delta\Psi$ collapse for PC12 cells cocultured with young or old astrocytes. Adapted from Lin et al. (2007).

that increased number of astrocytes in the aging brain are required to provide the same level of neuroprotection that is present in the brain of a young animal.

23.4 G-Protein-Coupled Receptor Stimulated IP_3-Mediated Ca^{2+} Release in Astrocytes Increases Neuroprotection

In the previous section, we presented a number of recent experimental findings that demonstrated diminished physiological function of astrocytes with aging. These changes are likely to contribute to or possibly cause the observed decrease in neuroprotection when coculturing neuronal-like PC12 cells with old astrocytes. In this section of the chapter, we discuss recent data showing that the neuroprotective ability of both old and young astrocytes can be significantly increased by stimulation of purinergic G-protein-coupled receptors that stimulate IP_3-mediated intracellular Ca^{2+} release.

23.4.1 *IP_3-Mediated Ca^{2+} Signaling and Mitochondrial Ca^{2+} Uptake*

P2 purinoceptors are a class of cell surface receptors, subdivided into either the ligand-gated ionotropic receptors (P2XR) or G-protein-coupled metabotropic receptors (P2YR) (Burnstock and Wong, 1978; Abbracchio and Burnstock, 1994). We focused our work on the P2YRs due to their strong coupling to Gq/11 signaling pathway in astrocytes (Taylor et al., 1999; Lee et al., 2000). Activation of P2Y receptors stimulates PLC_β isoenzymes (Waldo et al., 1991; Maurice et al., 1993), increasing IP_3 formation and subsequent Ca^{2+} mobilization (Pearce et al., 1989; Neary et al., 1991; Kastritsis et al., 1992; Salter and Hicks, 1994, 1995; Centemeri et al., 1997; Taylor et al., 1999; Lee et al., 2000) from thapsigargin-sensitive stores in the endoplasmic reticulum (ER). Thapsigargin is a specific inhibitor of the sarco-endoplasmic reticulum Ca^{2+} ATPases (SERCAs) (Thastrup et al., 1990). As mentioned earlier, IP_3-mediated Ca^{2+} release is efficiently sequestered by mitochondria, due to its physically close proximity to the Ca^{2+} channel pore (Rizzuto et al., 1992, 1993). Increased mitochondrial Ca^{2+} uptake via the Ca^{2+} uniporter rapidly stimulates Ca^{2+}-sensitive dehydrogenases and subsequently increases respiration and ATP production (Denton and McCormack, 1985; McCormack et al., 1990; Hajnoczky et al., 1995, 2000). Resting cytosolic Ca^{2+} levels have, in general, been shown to increase with age while the ability of mitochondria to sequester Ca^{2+} appears to diminish (Leslie et al., 1985; Peterson et al., 1985; Vitorica and Satrustegui, 1986a, b). Our data suggest that part of the decrease in Ca^{2+} uptake can be attributed to a decrease in the mitochondrial membrane potential ($\Delta\Psi$) (Lin et al., 2007). Work from Satrustegui's laboratory also demonstrated that the Ca^{2+} uniporter itself has lower activity with age (Satrustegui et al., 1996). Another important point to make with regard to mitochondrial Ca^{2+} signaling is that high-matrix Ca^{2+} is generally thought to sensitize cells to cell death stimuli (Szalai et al., 1999). During prolonged periods or with sufficiently high concentrations, matrix Ca^{2+} induces opening of the mitochondrial permeability transition pore (MPT) (Bernardi et al., 1992; Petronilli et al., 1993; Zoratti and Szabo, 1995; Byrne et al., 1999; Szalai et al., 1999). Recent work has identified the mitochondrial targeted cyclophilin D as a key player in Ca^{2+}-stimulated cell death (Baines et al., 2005; Basso et al., 2005; Nakagawa et al., 2005). In unpublished observations, we also found that IP_3-induced intracellular Ca^{2+} release sensitized human embryonic kidney (HEK293) cells to stimuli that induced cell death. We utilized HEK293 cells that were overexpressing type 1 muscarinic acetylcholine receptors to stimulate IP_3-gated Ca^{2+} release (Lechleiter et al., 1989). We induced apoptosis by exposing these cells to either *t*-BuOOH (100 μM, 3 h) or ceramide (40 μM, 12 h). Cell death was significantly higher in the presence of acetylcholine (1 μM) for both apoptotic stimuli. Thus, it was quite clear that stimulation of IP_3-gated Ca^{2+} release can activate both cell survival (ATP production) and cell death (MTP opening) pathways in cells. As will be presented in the following section, the IP_3-activated cell survival pathway dominates in astrocytes.

23.4.2 Purinergic Receptor Stimulation Enhances Cell Survival in Astrocytes During Oxidative Stress

To determine whether activation of P2Y-Rs would stimulate cell death or enhance cell survival, we exposed astrocytes to a bolus of extracellular ATP. Cultures were then washed with normal saline and perfused with the oxidant stressor *t*-BuOOH. Cell viability was assessed by monitoring the $\Delta\Psi$ with the potential sensitive dye TMRE. Surprisingly, a brief 10 min of ATP, prior to oxidant stress treatment, delayed the time until $\Delta\Psi$ collapse for several hours in both old and young astrocytes (Fig. 23.3). These data demonstrated that activation of purinergic receptors enhanced a cell survival signaling pathway and not cell death. Another interesting point regarding ATP-enhanced resistance to oxidant stress was that the time until $\Delta\Psi$ collapse for old astrocytes was increased nearly to the level of young astrocytes (Fig. 23.3). This suggested that the same protective mechanism was not only present in older cells, but that it could be activated to such an extent as to be comparable to stimulated young astrocytes. We immediately began investigating the role of intracellular Ca^{2+} in this protective mechanism because of the known coupling of purinergic receptors to these signaling pathways.

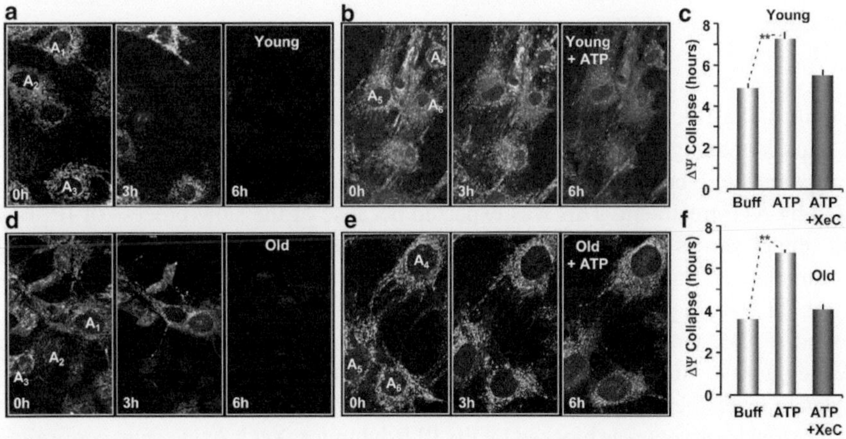

Fig. 23.3 Purinergic receptor (P2YR) activation protects young and old astrocytes from oxidative stress. (**a**, **d**) Images of astrocytes cultured from young and old mice labeled with the mitochondrial potential sensitive dye TMRE and exposed to oxidative stress for the indicated times. Each image is a maximum intensity projection of a z-stack of six optical sections (1-μm steps). Plated cells were continuously perfused at 37°C with culture medium containing TMRE (200 nM) and imaged with 2-photon microscopy. (**b**, **e**) Images of young and old astrocytes preexposed to extracellular ATP (10 μM) for 10 min prior to adding *t*-BuOOH (100 μM) to the perfusate (0 h). (**c**, **f**) Histograms of the mean times of $\Delta\Psi$ collapse for young astrocytes exposed to buffer only (Buff), ATP, or ATP plus xestospongin C (XeC). Statistical significance: $^{**}p < 0.001$. Adapted from Wu et al. (2007).

Experiments were again carried out in cultures of astrocytes from young and old mice. We utilized the Ca^{2+} indicator dye Fura 2 AM to measure Ca^{2+}, since this indicator works well with 2-photon excitation at 800 nm (the peak power of Ti-Sapphire lasers) and could be used to simultaneously monitor $\Delta\Psi$ with TMRE. A disadvantage of using 2-photon excitation for Fura-2 is that absorption at this wavelength is strong only for the Ca^{2+} free form of Fura-2. Consequently, Fura-2 rationing is not possible and Ca^{2+} increases are recorded as decreases in cellular fluorescence, normalized to the resting fluorescence. Using this experimental procedure, we confirmed that application of extracellular ATP stimulated a large increase in intracellular Ca^{2+}, as expected from the numerous reports of other investigators as well as our earlier work (Arkhammar et al., 1990; Lechleiter and Clapham, 1992). We next utilized a competitive antagonist of the IP_3R, xestospongin C (XeC), to begin testing whether the metabotropic P2Y-R signaling cascade was involved in ATP-enhanced resistance to oxidative stress. Consistent with the involvement of this pathway, astrocytes pretreated with XeC exhibited a significant decrease in ATP-induced Ca^{2+} responses. Inhibition was not complete since a large proportion of purinergic Ca^{2+} increases in astrocytes is also due to influx through plasma membrane P2X receptor ion channels (Verkhratsky and Kettenmann, 1996; James and Butt, 2002). Nevertheless, ATP-enhanced resistance to oxidative stress of astrocytes treated with XeC was completely blocked. We interpreted this data as strong evidence that the metabotropic P2Y receptor pathway, via stimulation of IP_3Rs, was required to enhance astrocyte resistance against oxidant stress.

To directly examine the role of IP_3-gated Ca^{2+} release in P2Y-R-enhanced astrocyte protection, we used a membrane permeant butyryloxymethyl ester of IP_3 (IP_3-BM). Cultures of astrocytes were exposed to IP_3-BM and Ca^{2+} levels were assessed using 2-photon imaging of Fura 2. Ca^{2+} increases in response to IP_3-BM application were easily detected, but they occurred over a much slower time course than those initiated by ATP treatment. This was expected because of the slow process of IP_3-BM movement across the lipid bilayer as well as the time needed to cleave the ester and accumulate IP_3 in the cell. Because of the slower time course, we pretreated astrocytes for 20 min prior to exposing them to the oxidant stress *t*-BuOOH. Again, we found IP_3-BM pretreatment to be protective, significantly delaying the time until $\Delta\Psi$ collapsed. Similarly, IP_3-BM treatment enhanced the resistance of old astrocytes to levels nearly equivalent to treated young astrocytes. The protective effect of IP_3-BM treatment was blocked by the IP_3R inhibitor, XeC. Aside from confirming that P2Y-R-enhanced resistance of astrocytes was due to metabotropic signaling, these results further suggested that increased protection could be activated by any G-protein-coupled receptor that stimulated IP_3-gated intracellular Ca^{2+} release. In this light, we note that it is well established that glutamate stimulates intracellular Ca^{2+} release and intra and intercellular Ca^{2+} waves across cultured astrocytes (Cornell-Bell et al., 1990; Cornell-Bell and Finkbeiner, 1991). The type 5 metabotropic glutamate receptor (mGluR5) appears to be the predominant isoform in astrocytes that stimulates IP_3-gated Ca^{2+} release, and it is abundantly expressed throughout the cortex (Romano et al., 1995). To test if metabotropic glutamate receptor (mGluR) activation was also capable of enhancing astrocyte resistance to oxidative stress, we pretreated astrocytes with the general mGluR agonist 1-aminocyclopentane-1,3-dicarboxylic acid (1S,3R-ACPD) for 10 min, generating IP_3. Data generated from these preliminary experiments are unpublished. However, ACPD treatment increased the proportion of astrocytes surviving 4 h of *t*-BuOOH treatment to 66% ± 9% (mean ± SD, $n = 89$ total cells, pooled from eight imaging fields) compared with control, untreated astrocytes, which exhibited 38% ± 8% ($n = 85$ total cells, pooled from six imaging fields) of its cell alive. ACPD enhancement was virtually indistinguishable from P2Y-R-enhanced resistance at 65 ± 8 ($n = 79$ total cells, pooled from six imaging fields). Taken together, these data strongly indicated that any receptor-mediated process that preferentially stimulated IP_3-mediated Ca^{2+} release in astrocytes could enhance their resistance to oxidative stress.

23.4.3 O_2 Consumption, Intracellular ATP Production, and P2Y-R Activation

Mitochondrial oxidative phosphorylation is widely recognized as the primary mechanism of energy production in neurons, and it requires both oxygen and glucose to produce ATP. As discussed in Sect. 23.2, the fact that astrocytes also rely predominantly on oxidative metabolism is not as widely appreciated (Gruetter et al., 2001; Bluml et al., 2002; Lebon et al., 2002; Hertz, 2004). Astrocytes contribute significantly to O_2 consumption in the brain, and decreases in O_2 are certain to inhibit and/or completely block oxidative phosphorylation. The link between intracellular Ca^{2+} signaling and mitochondrial energy production has also been well established. Increasing matrix Ca^{2+} stimulates Ca^{2+}-sensitive dehydrogenases in the citric acid cycle. This, in turn, increases the supply of reducing equivalents to the respiratory chain and, ultimately, increases ATP production (Denton and McCormack, 1985; McCormack et al., 1990; Hajnoczky et al., 1995, 2000; Robb-Gaspers et al., 1998). To test whether P2Y-R-enhanced resistance to oxidative stress was dependent on mitochondria, we utilized oligomycin, a specific inhibitor of the mitochondrial ATP synthetase. Pretreating astrocytes with oligomycin prior to oxidant stress did not alter $\Delta\Psi$ in young or old astrocytes compared with untreated control astrocytes. This suggested that resting energy metabolism in astrocytes was largely independent of oxidative metabolism. However, oligomycin treatment of astrocytes completely blocked P2Y-R-enhanced protection of astrocytes in both old and young astrocytes. We interpreted these data as strong evidence for a role of astrocyte mitochondria metabolism in P2Y-R-enhanced protection. To further test this hypothesis, we measured both O_2 consumption and ATP production in astrocytes with or without P2YR stimulation. Cultured astrocytes were gently suspended in buffer and placed in a respirometer chamber maintained at 37°C. As compared with buffer alone, we saw a steady decline in the O_2 content of the chamber, indicating resting mitochondrial oxidative phosphorylation. More importantly, when we added ATP to the chamber, we saw a significant increase in the rate of O_2 consumption. Oligomycin treatment blocked the ATP-mediated increase in O_2 consumption, further supporting the hypothesis that P2Y-R stimulation was increasing mitochondrial metabolism. Next, we directly measured intracellular ATP levels in astrocytes using a standard luciferine–luciferase assay. Cultured cells were rapidly heated to 100°C to minimize enzymatic changes in intracellular ATP levels. To avoid contaminating problems with extracellular ATP in our measurements, we treated astrocytes with a P2Y1-R specific ligand, 2-methylthio adenosine 5-diphosphate (2-MeSADP). As will be discussed later, this ligand also increased the resistance of astrocytes to oxidative stress. We found that P2Y1-R activation significantly increased intracellular ATP levels. Interestingly, when we pretreated cultured cells with oligomycin, we noticed a significant reduction in resting ATP levels, suggesting that mitochondria contributed a large portion of astrocyte energy even under resting conditions. Furthermore, P2Y1-R activation via 2-MeSADP did not increase ATP levels in the presence of oligomycin. We generated additional supporting evidence for IP_3-activated Ca^{2+}-stimulated mitochondrial ATP production when we pretreated astrocytes with ruthenium 360 (Ru_{360}), an inhibitor of the electrogenic mitochondrial Ca^{2+} uniporter. Ru_{360} treatment also lowered resting levels of ATP and completely blocked P2Y-R-stimulated ATP increases. As a final test for the involvement of mitochondria, we measured $\Delta\Psi$ with TMRE using 2-photon imaging. In single astrocytes, the same field of mitochondria was imaged before and 10 min after ligand treatment. We found that astrocyte $\Delta\Psi$s were significantly increased when astrocytes were treated with extracellular ATP or membrane permeant IP_3-BM. No changes were observed when astrocytes were exposed to a bolus of buffer alone or in astrocytes that had been pretreated with XeC. Taken together, these data strongly indicated the P2Y-R activation stimulated mitochondrial ATP production in cultured astrocytes. Given the ability of IP_3-BM alone to increase $\Delta\Psi$ or enhance astrocyte resistance to oxidative stress, we also anticipate that other G-protein-coupled receptors will similarly increase mitochondrial energy production in astrocytes.

23.4.4 P2Y-R Activation and Neuroprotection

Our initial experiments investigating the efficacy of astrocytes to increase the resistance of neuronal-like PC12 cells to oxidative stress clearly suggested that old astrocytes were not as neuroprotective as young astrocytes. In this section, we discuss our strategy to test the efficacy of astrocytes to protect embryonic cortical neurons cultured from mice as well as discuss the effectiveness of P2Y-R-enhanced astrocytes' resistance to protect these neurons in coculture. We took advantage of a coculture system that physically separated astrocytes from neurons. Primary cortical neurons were cultured on the glass bottomed six-well plates, while primary astrocyte cultures were plated in the transwell-clear permeable supports. In this configuration, astrocytes are separated from the neurons by ~1 mm. A significant advantage of this configuration is that we could separate out the contribution of P2Y-R activation in cocultures as well as neuron-only cultures, since the transwell supports containing astrocytes could be physically removed just prior to ligand treatments. Astrocyte cultures were initially established in the transwell supports for 7–10 days or until ~70% confluency. Cortical neurons were then plated in the glasswell bottoms, allowed to settle for ~1–2 h, and then a transwell support with cultured astrocytes was placed in each well. The cocultures were maintained together for 4 days prior to experiments in a neurobasal medium supplemented with l-glutamine, penicillin/streptomycin, and B-27. The B-27 was added to control glia contamination. The effect of this supplement on astrocyte neuroprotection is discussed later. To measure the efficacy of astrocyte neuroprotection, we again induced oxidant stress in the cell cultures with *t*-BuOOH treatment. Cell viability was measured after 4 h of *t*-BuOOH treatment by determining the colocalization of cells stained with Hoechst 33342 and calcein AM. Our immediate goal was to check whether pretreatment with purinergic ligands ATP, 2 MeSADP (P2Y1-R specific ligand), or UTP (P2Y2-R specific ligand) enhanced neuronal viability during oxidative stress. We found that all three purinergic ligands exhibited a comparable ability to enhance the resistance of cortical neurons against oxidant stress. These data confirmed that purinergic activation was protective for culture embryonic cortical neurons. They also again implied that at least in cell culture, any receptor isoform that preferentially stimulates IP_3-mediated Ca^{2+} release in astrocytes has the potential to be neuroprotective. Interestingly, it is known that the specific Ca^{2+} responses of each P2Y-R isoform have two distinct types of activity-dependent negative feedback. Ca^{2+} responses that are mediated by $P2Y_1$-Rs appear more oscillatory (Fam et al., 2003), which has been shown in other systems to be more effective at stimulating mitochondrial metabolism (Hajnoczky et al., 1995). Consequently, we were interested to know whether one isoform was more protective than the other. However, our current data did not support a differential protective effect based on the pattern or dynamics of Ca^{2+} releases, although a more thorough investigation applying a range of ligand concentrations will be needed to carefully address this issue. Along the lines of receptor specificity, it is also interesting to note that ATP can stimulate intercellular Ca^{2+} waves among astrocytes (Guthrie et al., 1999; Cotrina et al., 2000); intercellular Ca^{2+} waves result in the release of glutamate from astrocytes (Parpura et al., 1994; Innocenti et al., 2000; Parpura and Haydon, 2000). As noted earlier, the mGluR agonist ACPD increased astrocyte resistance to oxidative stress and consequently, this ligand has the potential of being neuroprotective. An important issue that still needs to be resolved is whether glutamate activation of astrocyte mGluRs is a more dominant effect than glutamate activation of NMDA receptors, since the latter receptor is well established to enhance excitotoxicity on neurons.

Use of the transwell configured coculture system permitted us to make several additional observations. First, data collected from these experiments revealed that neuroprotection was being mediated by a soluble factor, since the astrocytes were separated from the neurons by ~1 mm. A potential candidate for this protective factor will be presented in the last section of this chapter. Second, the transwell configuration allowed us to separate out the effects of purinergic signaling on astrocytes vs. neurons. Purinergic receptor (P2X) activation in the brain has been reported to enhance neuronal cell death (Norenberg and Illes, 2000; Koles et al., 2005). To investigate this, we compared the effect of purinergic ligand stimulation in cocultures vs. neuron-only cultures that had been treated with ligand and *t*-BuOOH after removal of the astrocyte transwell. Purinergic receptor activation with ATP in neuron-only cultures was not protective and actually increased cell death. On the other hand, pretreatment of neuron-only cultures with the P2Y1-R-specific ligand, 2MeSADP, had no effect on neuronal viability. These data were consistent with previous reports that the toxic effect of ATP treatment on neurons was due to the activation of P2X-Rs. Interestingly, these data also revealed that while pretreatment of cocultures with ATP activates purinergic receptors on both astrocytes and neurons, the dominant effect of extracellular ATP during oxidative stress was protection mediated by P2Y-R activation on astrocytes.

In Sect. 23.2, we discussed our findings that old astrocytes were not as effective at protecting neuronally differentiated PC12 cells in coculture. When we initially examined the protective efficacy of old vs. young astrocytes in cocultures with cortical neurons, we were surprised by the observation that old astrocytes were just as neuroprotective as young astrocytes. Multiple experiments confirmed that these data were accurate. We finally accounted for this apparent discrepancy when we carefully considered the coculture media being used to support astrocyte-neuronal cocultures. Unlike, astrocyte cocultures with PC12 cells, cocultures with isolated cortical neurons required that we include in the culture medium a supplement to suppress glial contamination. In short, the B-27 supplement that was used for this purpose contained antioxidants, which presumably were incorporated into the cortical neurons and made astrocytes much more resistant to oxidative stress. Fortunately, the company from which we obtained this supplement offered a B-27 formulation that was antioxidant free. Utilizing this supplement in our coculture experiments, we repeated the neuroprotection assays and confirmed that old astrocytes cocultured with cortical neurons were much more sensitive to oxidative stress. A 3-h incubation period with *t*-BuOOH resulted in 70% neuronal death compared with the 30% neuronal death for 4 h of treatment with *t*-BuOOH using B-27 supplement with antioxidants. P2Y-R activation still enhanced neuroprotection, which could be completely blocked by oligomycin treatment. In addition, neurons that were precultured with astrocytes and exposed to ATP also exhibited more cell death. These data confirmed the contaminating effect of antioxidants in estimating the neuroprotective efficacy of astrocytes.

23.4.5 Energetic Demands on Astrocytes

The work that we have discussed up to this point demonstrates that the maintenance and stimulation of mitochondrial metabolism enhances astrocyte resistance to oxidative stress and increases their ability to protect neurons throughout the aging process. Here, we discuss some of the processes within astrocytes that consume significant amounts of ATP and that could become problematic during times of stress when energy demands are high. First, we note that the primary energy-consuming process in the brain is the maintenance of ion concentration gradients across the plasma membrane. These gradients are generally maintained by energy-dependent pumps. Of these, it has been estimated that the Na^+/K^+ ATPase by itself consumes 20% of the astrocytic ATP levels in order to maintain Na^+ and K^+ homeostasis (Silver and Erecinska, 1997). Astrocyte ATP levels are also diminished during excitatory glutamatergic synaptic transmission. Glutamate recycling begins with Na^+-dependent uptake at a cost of three Na^+ ions per glutamate molecule transported. The increased intracellular Na^+ stimulates Na^+/K^+ ATPase activity to restore the ion gradients at a stoichiometry of three Na^+ ions pumped out for every two 2 K^+ ions pumped into the cell. Glutamate is then converted to glutamine by the glutamine synthetase. Two ATP molecules are consumed during this process, one by the ATPase and another is used by the synthetase. It is generally assumed that these ATP molecules are produced from a single glucose molecule going through glycolysis, which attractively generates two ATP molecules. While this energy source is certainly possible, the reported data do not exclude energy contributions from oxidative phosphorylation. Evidence clearly shows that the glutamate/glutamine neurotransmitter cycle is dependent on increased Na^+/K^+ ATPase activity, since cycling can be completely inhibited by ouabain, a specific inhibitor of this pump (Pellerin and Magistretti, 1994). It is also clear from reported studies that glucose utilization is stimulated during the process of glutamate recycling in the astrocyte (Erecinska et al., 1988). The report that is frequently cited for the use of glycolysis in astrocytes for ATP production in glutamate recycling is the elegant work of Tsacopoulos et al. (1998). In their study, energy production was highly compartmentalized because of a comparatively simple nervous tissue in their preparation, the honeybee retina. In this organ, the photoreceptors (neurons) contain large numbers of mitochondria whereas the glial cells are nearly devoid of these organelles. Consequently, energy metabolism in honeybee glial cells is almost exclusively glycolytic, which transfers carbohydrates as energy substrates to the neurons for aerobic metabolism. The main point for this discussion is that the separation of metabolic functions is not nearly as compartmentalized in the mammalian brain. As reviewed in the first section of this chapter, astrocyte mitochondria are prevalent and active, and during intense periods of neuronal activity, it would be expected that oxidative phosphorylation would be an important source of ATP for glutamate recycling.

Another important load on astrocyte energy metabolism is the synthesis of glutathione (GSH). The thiol moiety of GSH is used as a substrate for glutathione peroxidase (Gpx) enzymes, which is the predominant mechanism to reduce H_2O_2 and lipid hydroperoxides (Cohen and Hochstein, 1963). GSH is a key antioxidant in the resistance of both astrocytes and neurons during oxidative stress. Importantly, neuronal GSH production is critically dependent on GSH production in astrocytes, which contain significantly higher concentrations of GSH than neurons (Cooper and Kristal, 1997; Dringen et al., 2000). Maintenance of astrocyte GSH levels is controlled by two ATP-dependent enzymes (Suzuki and Kurata, 1992; Papadopoulos et al., 1997). Glutamate cysteine ligase (GCL) is the rate-limiting enzyme composed of a catalytic subunit (GCLc) and a modulatory subunit (CCLm) (Griffith and Mulcahy, 1999). GCL activity increases during oxidative stress and produces γ-glutamyl cysteine (γ-glu-cys) (Ochi, 1995). Glutathione synthetase (GS) is the second major enzyme that combines γ-glu-cys with glycine (gly) to produce GSH. GSH efflux from astrocytes appears to be controlled, in large part, by the multidrug resistance protein type 1 (MRP1) (Dringen and Hirrlinger, 2003). Once exported, GSH is cleaved by the ectoenzyme γ-glutamyl transpeptidase (γGT) producing a glutamyl moiety and the dipeptide cysteinylglycine (cys-gly). Cysteinylglycine is then cleaved by aminopeptidase N (ApN) to produce cysteine (Wang and Cynader, 2000). Neuronal de novo synthesis of GSH is dependent on this rate-limiting source of cysteine (Sagara et al., 1993, 1996; Wang and Cynader, 2000). During oxidative stress, GSH levels in both astrocytes and neurons would be rapidly depleted and ATP-dependent de novo GSH synthesis would be needed. If ATP production became rate limiting, one possible mechanism by which P2Y-R-enhanced astrocyte neuroprotection could be mediated is by stimulation of oxidative phosphorylation. As noted earlier, Ca^{2+}-stimulated mitochondrial ATP production in cardiac cells is very rapid, occurring ~10 times faster than stimulation by increased adenosine diphosphate levels (Territo et al., 2001; Balaban et al., 2003). In contrast, glycolysis may not be able to generate sufficient quantities of ATP during acute periods of stress. Future studies will be required to identify the specific cellular process that mediates protection, that is diffusible, and is limited by mitochondrial energy production in astrocytes.

23.5 Summary

Irrespective of the precise underlying cause of aging, it is clear that mitochondria play key roles in regulating cell survival via energy production in the brain. This well-characterized physiological function has long been accepted for neuronal cells. The precise roles of astrocyte mitochondrial metabolism, on the other hand, have generally not been as fully appreciated. In this chapter, we have discussed evidence that astrocyte mitochondrial metabolism does, in fact, play a pivotal role in the brain throughout the aging process. Not surprisingly, we discovered that the aging process itself degrades astrocyte mitochondria function, astrocyte resistance to oxidative stress as well as the neuroprotective abilities of astrocytes. Remarkably, the ability of astrocytes to enhance their protective functions could be significantly enhanced even in old cells. The underlying mechanism of enhanced protection was mediated by Ca^{2+}-stimulated mitochondrial metabolism. The source of this Ca^{2+} was IP_3-activated intracellular Ca^{2+} release from the endoplasmic reticulum. While this pathway was primarily controlled by purinergic receptors in our studies, it is clear that any other receptor that preferentially stimulates IP_3 production in astrocytes is likely to be an effective enhancer of neuroprotection. The potential therapeutic benefits of activating this pathway are numerous. We noted that aged astrocytes were more sensitive to oxidative stress and exhibited decreased neuroprotection under nonstimulated conditions. Activation of the P2Y-R signaling pathway enhanced protection to levels that were comparable with stimulated young astrocytes. Consequently, any G-protein-coupled receptors that stimulate mitochondrial ATP production could be used to minimize stress during aging as well as enhance astrocyte protective functions in various neurodegenerative disorders that have been correlated with increased oxidative damage including Alzheimer's disease, Parkinson's disease, and amyotrophic lateral sclerosis (Bains and Shaw, 1997; Mattson et al., 1999). Future studies will be needed to determine which G-protein receptor ligands are most effective at specifically activating the benefits of astrocyte mitochondrial metabolism, while minimizing their nonspecific actions on other cell-types.

References

Abbracchio MP, Burnstock G (1994) Purinoceptors: are there families of P2X and P2Y purinoceptors? Pharmacol Ther 64:445–475.

Arkhammar P, Hallberg A, Kindmark H, Nilsson T, Rorsman P, Berggren PO (1990) Extracellular ATP increases cytoplasmic free Ca^{2+} concentration in clonal insulin-producing RINm5F cells. A mechanism involving direct interaction with both release and refilling of the inositol 1,4,5-trisphosphate-sensitive Ca^{2+} pool. Biochem J 265:203–211.

Badar-Goffer RS, Bachelard HS, Morris PG (1990) Cerebral metabolism of acetate and glucose studied by 13C-n.m.r. spectroscopy. A technique for investigating metabolic compartmentation in the brain. Biochem J 266:133–139.

Baines CP, Kaiser RA, Purcell NH, Blair NS, Osinska H, Hambleton MA, Brunskill EW, Sayen MR, Gottlieb RA, Dorn GW, Robbins J, Molkentin JD (2005) Loss of cyclophilin D reveals a critical role for mitochondrial permeability transition in cell death. Nature 434:658–662.

Bains JS, Shaw CA (1997) Neurodegenerative disorders in humans: the role of glutathione in oxidative stress-mediated neuronal death. Brain Res Brain Res Rev 25:335–358.

Balaban RS, Bose S, French SA, Territo PR (2003) Role of calcium in metabolic signaling between cardiac sarcoplasmic reticulum and mitochondria in vitro. Am J Physiol Cell Physiol 284:C285–C293.

Basso E, Fante L, Fowlkes J, Petronilli V, Forte MA, Bernardi P (2005) Properties of the permeability transition pore in mitochondria devoid of Cyclophilin D. J Biol Chem 280:18558–18561.

Beckmann N, Turkalj I, Seelig J, Keller U (1991) ^{13}C NMR for the assessment of human brain glucose metabolism in vivo. Biochemistry 30:6362–6366.

Bernardi P, Vassanelli S, Veronese P, Colonna R, Szabo I, Zoratti M (1992) Modulation of mitochondrial permeability transition pore. Effects of protons and divalent cations. J Biol Chem 267:2934–2939.

Bluml S, Hwang JH, Moreno A, Ross BD (2000) Novel peak assignments of in vivo (13)C MRS in human brain at 1.5 T. J Magn Reson 143:292–298.

Bluml S, Moreno-Torres A, Shic F, Nguy CH, Ross BD (2002) Tricarboxylic acid cycle of glia in the in vivo human brain. NMR Biomed 15:1–5.

Boitier E, Rea R, Duchen MR (1999) Mitochondria exert a negative feedback on the propagation of intracellular Ca^{2+} waves in rat cortical astrocytes. J Cell Biol 145:795–808.

Burnstock G, Wong H (1978) Comparison of the effects of ultraviolet light and purinergic nerve stimulation on the guinea-pig taenia coli. Br J Pharmacol 62:293–302.

Byrne AM, Lemasters JJ, Nieminen AL (1999) Contribution of increased mitochondrial free Ca^{2+} to the mitochondrial permeability transition induced by *tert*-butylhydroperoxide in rat hepatocytes. Hepatology 29:1523–1531.

Centemeri C, Bolego C, Abbracchio MP, Cattabeni F, Puglisi L, Burnstock G, Nicosia S (1997) Characterization of the Ca^{2+} responses evoked by ATP and other nucleotides in mammalian brain astrocytes. Br J Pharmacol 121:1700–1706.

Cohen G, Hochstein P (1963) Glutathione peroxidase: the primary agent for the elimination of hydrogen peroxide in erythrocytes. Biochemistry 2:1420–1428.

Collins TJ, Lipp P, Berridge MJ, Li W, Bootman MD (2000) Inositol 1,4,5-trisphosphate-induced Ca^{2+} release is inhibited by mitochondrial depolarization. Biochem J 347:593–600.

Cooper AJ, Kristal BS (1997) Multiple roles of glutathione in the central nervous system. Biol Chem 378:793–802.

Cornell-Bell AH, Finkbeiner SM (1991) Ca^{2+} waves in astrocytes. Cell Calcium 12:185–204.

Cornell-Bell AH, Finkbeiner SM, Cooper MS, Smith SJ (1990) Glutamate induces calcium waves in cultured astrocytes: long-range glial signaling. Science 247:470–473.

Cotrina ML, Lin JH, Lopez-Garcia JC, Naus CC, Nedergaard M (2000) ATP-mediated glia signaling. J Neurosci 20:2835–2844.

Denton RM, McCormack JG (1985) Ca^{2+} transport by mammalian mitochondria and its role in hormone action. Am J Physiol 259:E543–E554.

Drew B, Leeuwenburgh C (2003) Method for measuring ATP production in isolated mitochondria: ATP production in brain and liver mitochondria of Fischer-344 rats with age and caloric restriction. Am J Physiol Regul Integr Comp Physiol 285:R1259–R1267.

Dringen R, Hirrlinger J (2003) Glutathione pathways in the brain. Biol Chem 384:505–516.

Dringen R, Gutterer JM, Hirrlinger J (2000) Glutathione metabolism in brain metabolic interaction between astrocytes and neurons in the defense against reactive oxygen species. Eur J Biochem 267:4912–4916.

Erecinska M, Zaleska MM, Nissim I, Nelson D, Dagani F, Yudkoff M (1988) Glucose and synaptosomal glutamate metabolism: studies with [15N]glutamate. J Neurochem 51:892–902.

Eriksson G, Peterson A, Iverfeldt K, Walum E (1995) Sodium-dependent glutamate uptake as an activator of oxidative metabolism in primary astrocyte cultures from newborn rat. Glia 15:152–156.

Fam SR, Gallagher CJ, Kalia LV, Salter MW (2003) Differential frequency dependence of P2Y1- and P2Y2- mediated Ca^{2+} signaling in astrocytes. J Neurosci 23:4437–4444.

Farkas DL, Wei MD, Febbroriello P, Carson JH, Loew LM (1989) Simultaneous imaging of cell and mitochondrial membrane potentials. Biophys J 56:1053–1069.

Gabryel B, Chalimoniuk M, Stolecka A, Waniek K, Langfort J, Malecki A (2006) Inhibition of arachidonic acid release by cytosolic phospholipase A2 is involved in the antiapoptotic effect of FK506 and cyclosporin a on astrocytes exposed to simulated ischemia in vitro. J Pharmacol Sci 102:77–87.

Griffith OW, Mulcahy RT (1999) The enzymes of glutathione synthesis: gamma-glutamylcysteine synthetase. Adv Enzymol Relat Areas Mol Biol 73:209–267, xii.

Gruetter R, Ugurbil K, Seaquist ER (1998) Steady-state cerebral glucose concentrations and transport in the human brain. J Neurochem 70:397–408.

Gruetter R, Seaquist ER, Ugurbil K (2001) A mathematical model of compartmentalized neurotransmitter metabolism in the human brain. Am J Physiol Endocrinol Metab 281:E100–E112.

Guthrie PB, Knappenberger J, Segal M, Bennett MV, Charles AC, Kater SB (1999) ATP released from astrocytes mediates glial calcium waves. J Neurosci 19:520–528.

Hajnoczky G, Robb-Gaspers LD, Seitz MB, Thomas AP (1995) Decoding of cytosolic calcium oscillations in the mitochondria. Cell 82:415–424.

Hajnoczky G, Hager R, Thomas AP (1999) Mitochondria suppress local feedback activation of inositol 1,4,5-trisphosphate receptors by Ca^{2+}. J Biol Chem 274:14157–14162.

Hajnoczky G, Csordas G, Krishnamurthy R, Szalai G (2000) Mitochondrial calcium signaling driven by the IP3 receptor. J Bioenerg Biomembr 32:15–25.

Hassel B, Sonnewald U, Fonnum F (1995) Glial-neuronal interactions as studied by cerebral metabolism of [2–13C]acetate and [1–13C]glucose: an ex vivo 13C NMR spectroscopic study. J Neurochem 64:2773–2782.

Hertz L (2004) The astrocyte-neuron lactate shuttle: a challenge of a challenge. J Cereb Blood Flow Metab 24:1241–1248.

Hertz L, Drejer J, Schousboe A (1988) Energy metabolism in glutamatergic neurons, GABAergic neurons and astrocytes in primary cultures. Neurochem Res 13:605–610.

Innocenti B, Parpura V, Haydon PG (2000) Imaging extracellular waves of glutamate during calcium signaling in cultured astrocytes. J Neurosci 20:1800–1808.
James G, Butt AM (2002) P2Y and P2X purinoceptor mediated Ca^{2+} signalling in glial cell pathology in the central nervous system. Eur J Pharmacol 447:247–260.
Jameson N, Olson J, Nguyen H, Holtzman D (1984) Respiration in primary cultured cerebellar granule neurons and cerebral cortical neurons. J Neurochem 42:470–474.
Jouaville LS, Ichas F, Holmuhamedov EL, Camacho P, Lechleiter JD (1995) Synchronization of calcium waves by mitochondrial substrates in *Xenopus laevis* oocytes. Nature 377:438–441.
Kastritsis CH, Salm AK, McCarthy K (1992) Stimulation of the P2Y purinergic receptor on type 1 astroglia results in inositol phosphate formation and calcium mobilization. J Neurochem 58:1277–1284.
Kauppinen RA, Enkvist K, Holopainen I, Akerman KE (1988) Glucose deprivation depolarizes plasma membrane of cultured astrocytes and collapses transmembrane potassium and glutamate gradients. Neuroscience 26:283–289.
Koles L, Furst S, Illes P (2005) P2X and P2Y receptors as possible targets of therapeutic manipulations in CNS illnesses. Drug News Perspect 18:85–101.
Lebon V, Petersen KF, Cline GW, Shen J, Mason GF, Dufour S, Behar KL, Shulman GI, Rothman DL (2002) Astroglial contribution to brain energy metabolism in humans revealed by 13C nuclear magnetic resonance spectroscopy: elucidation of the dominant pathway for neurotransmitter glutamate repletion and measurement of astrocytic oxidative metabolism. J Neurosci 22:1523–1531.
Lechleiter J, Peralta E, Clapham D (1989) Diverse functions of muscarinic acetylcholine receptor subtypes. Trends Pharmacol Sci Suppl:34–38.
Lechleiter JD, Clapham DE (1992) Spiral waves and intracellular calcium signalling. J Physiol, Paris 86:123–128.
Lee CK, Weindruch R, Prolla TA (2000) Gene-expression profile of the ageing brain in mice. Nat Genet 25:294–297.
Leslie SW, Chandler LJ, Barr EM, Farrar RP (1985) Reduced calcium uptake by rat brain mitochondria and synaptosomes in response to aging. Brain Res 329:177–183.
Lin DT, Wu J, Holstein D, Upadhyay G, Rourk W, Muller E, Lechleiter JD (2007) Ca^{2+} signaling, mitochondria and sensitivity to oxidative stress in aging astrocytes. Neurobiol Aging 28:99–111.
Lopes-Cardozo M, Larsson OM, Schousboe A (1986) Acetoacetate and glucose as lipid precursors and energy substrates in primary cultures of astrocytes and neurons from mouse cerebral cortex. J Neurochem 46:773–778.
Mason GF, Gruetter R, Rothman DL, Behar KL, Shulman RG, Novotny EJ (1995) Simultaneous determination of the rates of the TCA cycle, glucose utilization, alpha-ketoglutarate/glutamate exchange, and glutamine synthesis in human brain by NMR. J Cereb Blood Flow Metab 15:12–25.
Mattson MP, Rychlik B. (1990) Glia protect hippocampal neurons against excitatory amino acid-induced degeneration: involvement of fibroblast growth factor. Int J Dev Neurosci. 8(4):399–415.
Mattson MP, Pedersen WA, Duan W, Culmsee C, Camandola S (1999) Cellular and molecular mechanisms underlying perturbed energy metabolism and neuronal degeneration in Alzheimer's and Parkinson's diseases. Ann N Y Acad Sci 893:154–175.
Maurice DH, Waldo GL, Morris AJ, Nicholas RA, Harden TK (1993) Identification of G alpha 11 as the phospholipase C-activating G-protein of turkey erythrocytes. Biochem J 290 (Part 3): 765–770.
McCormack JG, Halestrap AP, Denton PM (1990) Role of calcium ions regulation of mammalian intramitochondrial metabolism. Physiol Rev 70:391–425.
Muir D, Berl S, Clarke DD (1986) Acetate and fluoroacetate as possible markers for glial metabolism in vivo. Brain Res 380:336–340.
Nakagawa T, Shimizu S, Watanabe T, Yamaguchi O, Otsu K, Yamagata H, Inohara H, Kubo T, Tsujimoto Y (2005) Cyclophilin D-dependent mitochondrial permeability transition regulates some necrotic but not apoptotic cell death. Nature 434:652–658.

Neary JT, Laskey R, van Breemen C, Blicharska J, Norenberg LO, Norenberg MD (1991) ATP-evoked calcium signal stimulates protein phosphorylation/dephosphorylation in astrocytes. Brain Res 566:89–94.

Norenberg W, Illes P (2000) Neuronal P2X receptors: localisation and functional properties. Naunyn Schmiedebergs Arch Pharmacol 362:324–339.

Ochi T (1995) Hydrogen peroxide increases the activity of gamma-glutamylcysteine synthetase in cultured Chinese hamster V79 cells. Arch Toxicol 70:96–103.

Olson JE, Holtzman D (1981) Factors influencing the growth and respiration of rat cerebral astrocytes in primary culture. Neurochem Res 6:1337–1343.

Papadopoulos MC, Koumenis IL, Dugan LL, Giffard RG (1997) Vulnerability to glucose deprivation injury correlates with glutathione levels in astrocytes. Brain Res 748:151–156.

Parpura V, Haydon PG (2000) Physiological astrocytic calcium levels stimulate glutamate release to modulate adjacent neurons. Proc Natl Acad Sci USA 97:8629–8634.

Parpura V, Basarsky TA, Liu F, Jeftinija K, Jeftinija S, Haydon PG (1994) Glutamate-mediated astrocyte-neuron signalling. Nature 369:744–747.

Pauwels PJ, Opperdoes FR, Trouet A (1985) Effects of antimycin, glucose deprivation, and serum on cultures of neurons, astrocytes, and neuroblastoma cells. J Neurochem 44:143–148.

Pearce B, Murphy S, Jeremy J, Morrow C, Dandona P (1989) ATP-evoked Ca^{2+} mobilisation and prostanoid release from astrocytes: P2-purinergic receptors linked to phosphoinositide hydrolysis. J Neurochem 52:971–977.

Peinado MA, Quesada A, Pedrosa JA, Torres MI, Martinez M, Esteban FJ, Del Moral ML, Hernandez R, Rodrigo J, Peinado JM (1998) Quantitative and ultrastructural changes in glia and pericytes in the parietal cortex of the aging rat. Microsc Res Technol 43:34–42.

Pellerin L, Magistretti PJ (1994) Glutamate uptake into astrocytes stimulates aerobic glycolysis: a mechanism coupling neuronal activity to glucose utilization. Proc Natl Acad Sci USA 91:10625–10629.

Peterson C, Nicholls DG, Gibson GE (1985) Subsynaptosomal distribution of calcium during aging and 3,4-diaminopyridine treatment. Neurobiol Aging 6:297–304.

Petronilli V, Cola C, Bernardi P (1993) Modulation of the mitochondrial cyclosporinA-sensitive permeability transition pore. II. The minimal requirements for pore induction underscore a key role for transmembrane electrical potential, matrix pH, and matrix Ca^{2+}. J Biol Chem 268:1011–1016.

Pilegaard K, Ladefoged O (1996) Total number of astrocytes in the molecular layer of the dentate gyrus of rats at different ages. Anal Quant Cytol Histol 18:279–285.

Rizzuto R, Simpson AVM, Brini M, Pozzan T (1992) Rapid changes of mitochondrial Ca^{2+} revealed by specifically targeted recombinant aequorin. Nature 358:325–327.

Rizzuto R, Brini M, Murgia M, Pozzan T (1993) Microdomains with high Ca^{2+} close to InsP3-sensitive channels that are sensed by neighboring mitochondria. Science 262:744–747.

Rizzuto R, Bastianutto C, Brini M, Murgia M, Pozzan T (1994) Mitochondrial Ca^{2+} homeostasis in intact cells. J Cell Biol 126:1183–1194.

Robb-Gaspers LD, Burnett P, Rutter GA, Denton RM, Rizzuto R, Thomas AP (1998) Integrating cytosolic calcium signals into mitochondrial metabolic responses. EMBO J 17:4987–5000.

Romano C, Sesma MA, McDonald CT, O'Malley K, Van den Pol AN, Olney JW (1995) Distribution of metabotropic glutamate receptor mGluR5 immunoreactivity in rat brain. J Comp Neurol 355:455–469.

Rozovsky I, Finch CE, Morgan TE (1998) Age-related activation of microglia and astrocytes: in vitro studies show persistent phenotypes of aging, increased proliferation, and resistance to down-regulation. Neurobiol Aging 19:97–103.

Sagara J, Makino N, Bannai S (1996) Glutathione efflux from cultured astrocytes. J Neurochem 66:1876–1881.

Sagara JI, Miura K, Bannai S (1993) Maintenance of neuronal glutathione by glial cells. J Neurochem 61:1672–1676.

Salter MW, Hicks JL (1994) ATP-evoked increases in intracellular calcium in neurons and glia from the dorsal spinal cord. J Neurosci 14:1563–1575.

Salter MW, Hicks JL (1995) ATP causes release of intracellular Ca^{2+} via the phospholipase C beta/IP3 pathway in astrocytes from the dorsal spinal cord. J Neurosci 15:2961–2971.
Satrustegui J, Villalba M, Pereira R, Bogonez E, Martinez-Serrano A (1996) Cytosolic and mitochondrial calcium in synaptosomes during aging. Life Sci 59(5–6):429–434.
Segal J, Ingbar SH (1990) 3,5,3 -tri-iodothyronine enhances sugar transport in rat thymocytes by increasing the intrinsic activity of the plasma membrane sugar transporter. J Endocrinol 124:133–140.
Silver IA, Erecinska M (1997) Energetic demands of the Na^+/K^+ ATPase in mammalian astrocytes. Glia 21:35–45.
Simpson PB, Russell JT (1996) Mitochondrial support inositol 1,4,5-trisphosphate-mediated Ca^{2+} waves in cultured oligodendrocytes. J Biol Chem 271:33493–33501.
Sokoloff L (1989) Circulation and energy metabolism of the brain, 4th edition. New York: Raven Press.
Suzuki M, Kurata M (1992) Effects of ATP level on glutathione regeneration in rabbit and guinea-pig erythrocytes. Comp Biochem Physiol B 103:859–862.
Szalai G, Krishnamurthy R, Hajnoczky G (1999) Apoptosis driven by IP(3)-linked mitochondrial calcium signals. Embo J 18:6349–6361.
Taylor CW, Genazzani AA, Morris SA (1999) Expression of inositol trisphosphate receptors. Cell Calcium 26:237–251.
Territo PR, French SA, Balaban RS (2001) Simulation of cardiac work transitions, in vitro: effects of simultaneous Ca^{2+} and ATPase additions on isolated porcine heart mitochondria. Cell Calcium 30:19–27.
Thastrup O, Cullen PJ, Drobak BK (1990) Thapsigargin, a tumor promoter, discharges intracellular Ca^{2+} stores by specific inhibition of the endoplasmic reticulum Ca^{2+}-ATPase. Proc Natl Acad Sci USA 87:2466–2470.
Tsacopoulos M, Poitry-Yamate CL, MacLeish PR, Poitry S (1998) Trafficking of molecules and metabolic signals in the retina. Prog Retin Eye Res 17:429–442.
Verkhratsky A, Kettenmann H (1996) Calcium signalling in glial cells. Trends Neurosci 19:346–352.
Vitorica J, Satrustegui J (1986a) Involvement of mitochondria in the age-dependent decrease in calcium uptake of rat brain synaptosomes. Brain Res 378:36–48.
Vitorica J, Satrustegui J (1986b) The influence of age on the calcium-efflux pathway and matrix calcium buffering power in brain mitochondria. Biochim BiophysActa 851:209–216.
Waldo GS, Fronko RM, Penner-Hahn JE (1991) Inactivation and reactivation of manganese catalase: oxidation-state assignments using X-ray absorption spectroscopy. Biochemistry 30:10486–10490.
Walz W, Mukerji S (1988) Lactate release from cultured astrocytes and neurons: a comparison. Glia 1:366–370.
Wang XF, Cynader MS (2000) Astrocytes provide cysteine to neurons by releasing glutathione. J Neurochem 74:1434–1442.
Waniewski RA, Martin DL (1998) Preferential utilization of acetate by astrocytes is attributable to transport. J Neurosci 18:5225–5233.
Zoratti M, Szabo I (1995) The mitochondrial permeability transition. Biochim BiophysActa 1241:139–176.

Abbreviations

1S,3R-ACPD	1-Aminocyclopentane-1,3-dicarboxylic acid
2-MeSADP	2-Methylthio adenosine 5-diphosphate
$\Delta\Psi$	Mitochondrial membrane potential

γ-glu-cys	γ-Glutamyl cysteine
ATP	Adenosine triphosphate
CCLm	Modulatory subunit of GCL
CNS	Central nervous system
Cys-gly	Cysteinylglycine
DNA	Deoxyribonucleic acid
ER	Endoplasmic reticulum
Fcyto	Lowest value of fluorescence in the cytosol
Fmito	Peak fluorescent intensity observed in single mitochondrial
GCL	Glutamate cysteine ligase
GCLc	Catalytic subunit of GCL
GFAP	Glial fibrillary acidic protein
Gpx	Glutathione peroxidase
GS	Glutathione synthetase
GSH	Glutathione
IP_3	Inositol 1,4,5 trisphosphate
BM	Membrane permeant butyryloxymethyl ester of IP_3
IP_3R	IP_3 receptor
mGluR	Metabotropic glutamate receptor
MRP1	Multidrug resistance protein type 1
NMDA	*N*-methyl-d-aspartic acid
NMR	Nuclear magnetic resonance
PC12	Pheochromocytoma PC12 cell
PLCβ	Phospholipase C beta
Ru360	Ruthenium 360
SERCA	Sarco-endoplasmic reticulum Ca^{2+}-ATPase
t-BuOOH	*tert*-Butyl hydrogen peroxide
TCA	Tricarboxylic acid
TMRE	Tetramethyl rhodamine ethyl ester
XeC	Xestospongin C

Chapter 24
Alexander Disease: A Genetic Disorder of Astrocytes

Michael Brenner, James E. Goldman, Roy A. Quinlan, and Albee Messing

Contents

M. Brenner
Department of Neurobiology, Evelyn F. McKnight Brain Institute, Center for Glial Biology in Medicine, University of Alabama Birmingham, Birmingham, AL USA
aaron@nrc.uab.edu

V. Parpura and P.G. Haydon (eds.), *Astrocytes in (Patho)Physiology of the Nervous System*,
DOI: 10.1007/978-0-387-79492-1_24, © Springer Science+Business Media, LLC 2009

24.1 Introduction

This volume documents the multiple roles astrocytes perform in the normal development and function of the central nervous system (CNS). A nagging question has been that if these roles are as critical as assumed, why have genetic diseases of astrocyte dysfunction not been identified to take their place next to those due to defects in neurons and oligodendrocytes? An explanation casually offered is that these functions are so important that their loss would be embryonic lethal, and thus not be detected. But this is not a satisfactory answer, as mutations that result in a partial loss of function would still be expected. Here we describe the first discovered instance of a primary *astrogliopathy*, in which a defect in astrocytes indeed results in a human disorder – Alexander disease. Fittingly, the gene encoding glial fibrillary acidic protein (GFAP), the intermediate filament protein that has been the standard marker for astrocytes in both basic and clinical studies, has proved to be the affected target in this disorder. This chapter reviews observations of human patients and model systems in which the focus has been on the role of GFAP. A number of excellent reviews discuss findings prior to the advent of this gene test (e.g., see Herndon et al., 1970; Spalke and Mennel, 1982; Becker and Teixeira, 1988; Pridmore et al., 1993; Reichard et al., 1996; Gordon, 2003; Jacob et al., 2003; Messing and Goldman, 2004).

24.2 Characteristics of Alexander Disease

24.2.1 Discovery

In 1949, W. Stewart Alexander described the case of a 15-month-old boy who presented with progressive megalencephaly, frequent vomiting, delayed motor development, and almost continuous screaming (Alexander, 1949). Hyperthermia commenced the day after admission, and increased until the child died of massive pulmonary emboli 3 weeks later. Autopsy revealed the remarkable and novel finding of abundant protein aggregates in hypertrophic astrocytes. These were scattered throughout the white matter, but were especially abundant in perivascular, subependymal, and subpial locations. Similar protein aggregates had been previously observed focally in association with a spinal cord syrinx and CNS tumors, and are named Rosenthal fibers after the physician who first described them (Rosenthal, 1898; reviewed in Wippold et al., 2006). Subsequent discoveries of patients with similar clinical presentations and extensive deposits of Rosenthal fibers led to the disorder being recognized as a specific entity, and being eponymously designated as Alexander (or Alexander's) disease [Mendelian Inheritance in Man (MIM) number 203450; online access at www.ncbi.nlm.nih.gov/omim/] (Friede, 1964; for additional historical commentary, see Messing and Goldman, 2004). Ironically, although the disorder has been classified as a leukodystrophy because many of these subsequent cases displayed severe dysmyelination, in his original case report Alexander noted that "demyelination is not an important feature of this disease." The description of the pathology for this first case remains one of the most thorough.

24.2.2 Clinical Features

The majority of patients with sporadic Alexander disease present before age 2 (Tables 24.1 and 24.2). The most common symptoms are seizures, megalencephaly, and failure to meet physical and intellectual milestones. Other clinical signs can include spasticity, poor coordination, paralysis of legs and/or arms, vomiting, and difficulty in swallowing. This infantile form of the disease usually progresses rapidly, with death ensuing about 4 years after presentation, most often from respiratory problems such as aspiration pneumonia. In rare instances, patients with infantile onset may survive into adulthood (for a review of milder infantile cases, see Dinopoulos et al., 2006). In addition to the infantile form of the disease, later onset variants exist that can differ markedly in clinical presentation, but were considered to be Alexander disease because they also display abundant and widespread distribution of Rosenthal fibers (Russo et al., 1976). These differences are not sharply demarcated by age, but instead form a variable continuum between the early and late extremes (Li et al., 2005). For convenience of discussion, the juvenile form has been defined as having an onset between 2 and 12 years. It may present similarly to the infantile form, or patients may instead have normal development and intelligence, and instead display bulbar and pseudobulbar signs such as difficulty with speech, swallowing, and coordination. The juvenile disease progresses more slowly than the infantile, with a median interval from onset to death of about 8 years, and with some patients living for several decades. The adult form, presenting at age 13 or older, is even more variable in its presentations. Virtually absent are the most common symptoms of infantile patients – seizures, megalencephaly, and developmental deficits. Instead, bulbar and pseudobulbar signs, ataxia, and spasticity, which are also frequent in infantile and juvenile cases, become the dominant presentations for the adult form (Table 24.2). Life expectancy for adult onset Alexander disease is highly variable, and has ranged from death a few years after disease onset to survival for at least 20 years. Some patients have lived into their sixties and seventies (Namekawa et al., 2002; Stumpf et al., 2003; Li et al., 2005; Salvi et al., 2005; van der Knaap et al., 2006). Multiple additional clinical signs have been observed in Alexander disease patients, particularly in the later onset forms. In a rough order from more frequent to less frequent together with illustrative references, these include scoliosis (van der Knaap et al., 2006), palatal myoclonus (Thyagarajan et al., 2004), autonomic dysfunctions such as constipation or bladder problems (enuresis, urinary retention, urgency, or incontinence) (Stumpf et al., 2003), endocrine dysfunctions such as transient hypothermia, ovarian failure, or precocious puberty (Kyllerman et al., 2005; Sreedharan et al., 2007), miosis (van der Knaap et al., 2005;

Table 24.1 Characteristics of patients tested for *GFAP* mutations.

Exon	Mutation	Reference	Reported as patient #	Type	Sex	Macro	Seiz	Spas	Bul/ Psb	Atx	Dev Dly	Comments
1	K63Q	Li et al., 2005	1	Adlt	F	N	N	N	Y	Y	N	
1	R70W	Salvi et al., 2005		Adlt	M	N	N	N	Y	Y	N	
1	R70W	Sreedharan et al., 2007		Adlt	F	N	N	N	Y	Y	N	
1	M73R	Gorospe et al., 2002	8	Juv	M	Y	Y	Y	Y	-----	Y	
1	M73T	Li et al., 2005	2	Inf	M	Y	Y	Y	Y	Y	Y	
1	L76F	Rodriguez et al., 2001	1	Inf	M	Y	Y	-----	Y	-----	Y	
1	L76F	Li et al., 2005	3	Inf	F	N	Y	Y	Y	Y	Y	
1	L76V	Li et al., 2005	4	Inf	F	N	Y	Y	Y	-----	Y	
1	N77S	Li et al., 2005	5	Inf	F	Y	Y	N	Y	N	Y	
1	N77Y	Rodriguez et al., 2001	2	Inf	F	N	N	-----	-----	-----	Y	
1	D78E	Stumpf et al., 2003	III-1 II-14 III-7 III-8 III-19 IV-1	Asym Adlt Adlt Adlt Adlt Juv	F M F M F M	N N N N N N	N ? ? ? ? ?	N N N N N N	N Y Y Y Y Y	N Y Y Y N N	N N N N N N	no neurological exam There was no proband for this family ?=one of these patients (not specified) may have had seizures
1	R79C	Brenner et al., 2001	1	Inf	M	N	Y	-----	-----	-----	Y	
1	R79C	Brenner et al., 2001	12	Inf	M	N	Y	-----	-----	-----	Y	
1	R79C	Shiroma et al., 2003	1 2	Inf Inf	M M	Y Y	Y Y	Y Y	N N	N N	Y Y	Identical twins
1	R79C	Gorospe et al., 2002	2	Inf	M	Y	Y	N	Y	-----	Y	
1	R79C	Li et al., 2005	7	Inf	F	Y	Y	N	N	Y	Y	
1	R79C	Li et al., 2005	8	Inf	F	Y	Y	Y	N	Y	Y	

1	R79C	Li et al., 2005	6	Juv	M	Y	N	Y	Y	N	Y	
1	R79C	Probst et al., 2003		Adlt	F	N	N	Y	Y	N	N	
1	R79G	Gorospe et al., 2002	1	Inf	F	N	Y	Y	Y	-----	Y	
1	R79H	Rodriguez et al., 2001	4	Inf	M	N	Y	-----	-----	-----	N	
1	R79H	Rodriguez et al., 2001	5	Inf	M	N	Y	-----	-----	-----	Y	
1	R79H	Rodriguez et al., 2001	6	Inf	M	N	Y	-----	-----	-----	Y	
1	R79H	Gorospe et al., 2002	3	Inf	M	Y	Y	N	Y	-----	Y	
1	R79H	Gorospe et al., 2002	4	Inf	M	Y	Y	Y	N	-----	N	
1	R79H	Meins et al., 2002	2 3	Inf Inf	M M	N N	Y Y	Y Y	Y Y	Y Y	Y Y	Identical teins, head circumference at 90th percentile
1	R79H	Brenner et al., 2001	13	Inf	F	Y	N	-----	-----	-----	Y	
1	R79H	Rodriguez et al., 2001	3	Inf	F	N	Y	-----	-----	-----	Y	
1	R79H	Asahina et al., 2006		Juv	M	N	Y	N	Y	N	Y	
1	R79H	Brenner et al., 2001	2	Juv	F	N	Y	Y	Y	-----	N	
1	R79L	Shiroma et al., 2003	4	Inf	M	Y	Y	Y	N	N	Y	
1	R79S	This chapter		Inf	M	N	Y	N	Y	-----	Y	
	Y83H	Wu et al., 2006		Inf	-----	Y	N	Y	-----	-----	Y	Brief abstract report
1	K86V87 delinsEF	van der Knaap et al., 2006	2	Juv	F	N	N	N	Y	N	N	
1	V87G	Okamoto et al., 2002	Familial	Adlt Adlt Asym	F F M	N N N	N N N	Y Y N	Y Y N	Y Y N	N N N	Mother = proband, not known if mutation arose de novo Affected daughter Preclinical son (see text)

(continued)

Table 24.1 (continued)

Exon	Mutation	Reference	Reported as patient #	Type	Sex	Macro	Seiz	Spas	Bul/ Psb	Atx	Dev Dly	Comments
1	V871	This chapter		Inf	F	Y	N	N	Y	Y	Y	
1	R88C	Rodriguez et al., 2001	7	Inf	M	Y	Y	-----	-----	-----	Y	
1	R88C	Rodriguez et al., 2001	8	Inf	F	Y	Y	-----	-----	-----	Y	
1	R88C	Kyllerman et al., 2005	1	Inf	F	Y	N	N	Y	Y	Y	
1	R88C	Wu et al., 2006		Inf	-----	Y	N	Y	-----	-----	Y	Brief abstract report
1	R88C	Wu et al., 2006		Inf	-----	Y	N	Y	-----	-----	Y	Brief abstract report
1	R88C	Gorospe et al., 2002	9	Juv	M	N	N	N	Y	-----	Y	
1	R88C	Gorospe et al., 2002	10	Juv	M	Y	N	N	Y*	-----	N	*Recent unpublished observation
1	R88C	Li et al., 2005	9	Juv	M	N	Y	N	N	N	Y	
1	R88C	van der Knaap et al., 2006	7	Juv	M	N	N	Y	Y	Y	N	
1	R88C	Nobuhara et al., 2004		Juv	F	N	N	N	Y	Y	Y	
1	R88C	van der Knaap et al., 2006	3 4	Adlt Juv	F M	N N	N N	N N	Y N	N N	N N	Proband = mother Affected son
1	R88S	Rodriguez et al., 2001	9	Inf	F	N	N	-----	-----	-----	Y	
1	L90P	Suzuki et al., 2004		Inf	F	N	Y	N	N	Y	Y	
1	Q93P	Kyllerman et al., 2005	2	Juv	F	N	N	N	Y	Y	N	
1	L97P	Meins et al., 2002	4	Inf	M	Y	N	N	-----	N	Y	
1	L97P	Li et al., 2005	10	Inf	M	Y	Y	Y	N	N	Y	

1	V1151	Li et al., 2005	42	Inf	M	N	Y	Y	Y	-----	Y	Uncertain diagnosis (see text)
1	RL126–127dup	van der Knaap et al., 2006	1	Juv	M	N	Y	N	Y	N	N	
4	E207K	Li et al., 2005	11	Juv	M	N	N	N	Y	N	N	
4	E207Q	Li et al., 2005	12	Juv	M	N	N	Y	N	N	N	
4	E210K	Kyllerman et al., 2005	3	Juv	M	N	N	N	Y	Y	N	
4	E210K	van der Knaap et al., 2006	6	Juv	M	N	N	N	Y	N	Y	
4	E210K	Li et al., 2005	13	Adlt	F	N	N	N	Y	N	N	
4	E223Q	Brockmann et al., 2003	Familial	Adlt	M	N	N	Y	Y	Y	N	Proband; uncertain diagnosis (see text)
				Asym	F	N	N	N	N	N	N	mother, mild intellectual impairment starting at 62 y
4	L235P	Li et al., 2005	14	Juv	M	N	N	Y	Y	Y	N	Identical twins
			15	Juv	M	N	N	Y	Y	Y	N	
4	L235P	Li et al., 2005	16	Juv	M	Y	Y	Y	Y	Y	Y	
4	R239C	Brenner et al., 2001	3	Inf	M	Y	Y	-----	-----	-----	Y	
4	R239C	Rodriguez et al., 2001	10	Inf	M	Y	Y	-----	-----	-----	Y	
4	R239C	Gorospe et al., 2002	5	Inf	M	Y	N	Y	Y	-----	N	
4	R239C	Meins et al., 2002	1	Inf	M	Y	Y	Y	-----	Y	Y	
4	R239C	Li et al., 2005	19	Inf	M	Y	Y	N	Y	Y	Y	
4	R239C	Li et al., 2005	20	Inf	M	N	Y	N	Y	Y	Y	
4	R239C	Li et al., 2005	21	Inf	M	Y	Y	N	Y	-----	Y	
4	R239C	Brenner et al., 2001	4	Inf	F	Y	Y	-----	-----	-----	Y	
4	R239C	Brenner et al., 2001	5	Inf	F	N	Y	-----	-----	-----	Y	
4	R239C	Brenner et al., 2001	6	Inf	F	Y	*N*	-----	-----	-----	Y	

(continued)

Table 24.1 (continued)

Exon	Mutation	Reference	Reported as patient #	Type	Sex	Macro	Seiz	Spas	Bul/ Psb	Atx	Dev Dly	Comments
4	R239C	Rodriguez et al., 2001	11	Inf	F	Y	N	-----	-----	-----	Y	
4	R239C	Rodriguez et al., 2001	12	Inf	F	N	Y	-----	-----	-----	Y	
4	R239C	Rodriguez et al., 2001	13	Inf	F	N	Y	-----	-----	-----	Y	
4	R239C	Shiroma et al., 2001		Inf	F	Y	Y	Y	N	N	Y	
4	R239C	Shiroma et al., 2003	3	Inf	F	Y	Y	Y	Y	N	Y	
4	R239C	Li et al., 2005	18	Inf	F	Y	Y	Y	N	N	N	
4	R239C	Shiihara et al., 2002		Juv	M	Y	Y	N	Y	Y	Y	
4	R239C	Li et al., 2005	17	Juv	M	N	Y	Y	Y	Y	Y	
4	R239C	Li et al., 2005	22	Juv	F	N	N	N	N	Y	Y	
4	R239H	Rodriguez et al., 2001	14	Inf	M	Y	Y	-----	Y	-----	Y	
4	R239H	Trollmann et al., 2003		Inf	M	Y	Y	Y	Y	Y	Y	
4	R239H	Li et al., 2005	23	Inf	M	Y	Y	Y	Y	-----	Y	
4	R239H	Li et al., 2005	25	Inf	M	Y	Y	Y	Y	-----	Y	
4	R239H	Li et al., 2005	26	Inf	M	N	N	Y	Y	-----	Y	
4	R239H	Li et al., 2005	27	Inf	M	N	Y	Y	N	-----	-----	
4	R239H	Brenner et al., 2001	7	Inf	F	N	Y	Y	-----	-----	Y	
4	R239H	Li et al., 2005	24	Inf	F	Y	Y	Y	Y	-----	N	
4	R239L	Lee et al., 2006		Inf	M	Y	Y	Y	Y	-----	Y	
4	R239P	Meins et al., 2002	5	Inf	F	Y	N	Y	-----	Y	Y	

4	R239P	Li et al., 2005	28	Juv	M	Y	N	N	Y	Y	Y	
4	Y242D	Gorospe et al., 2002	6	Inf	M	Y	N	N	Y	-----	Y	
4	A244V	Aoki et al., 2001		Juv	M	Y	Y	N	Y	Y	Y	
4	A244V	Li et al., 2005	29	Juv	F	N	N	N	Y	N	N	
4	A253G	Li et al., 2005	30	Inf	M	N	N	N	Y	N	N	
4	R258P	Brenner et al., 2001	8	Inf	M	Y	Y	-----	-----	-----	Y	
5	R276L	Namekawa et al., 2002	Familial	Adlt	M	N	N	Y	Y	Y	N	Proband; not known if mutation arose *de novo*
				Adlt	M	N	N	Y	Y	N	N	Mildly affected brother
5	K279E	Li et al., 2005	31	Juv	M	N	Y	N	Y	N	Y	
6	L331P	Shiihara et al., 2004	Familial	Inf	M	Y	N	N	N	N	N	Proband; not known if mutation arose *de novo*
				Asym	F	N	N	N	N	N	N	34-y-old mother, mildly abnormal MRI
				Asym	F	N	N	N	N	N	N	7-y-old sister, mildly abnormal MRI
6	349Hlins	Li et al., 2005	32	Inf	F	N	Y	N	Y	-----	-----	
6	L352P	Bassuk et al., 2003; Li et al., 2005	33	Inf	M	Y	Y	N	N	-----	-----	
6	A358V	Dinopoulos et al., 2006		Inf	F	Y	Y	N	N	N	N	Unusually mild case
6	L359V	Li et al., 2005	34	Juv	M	N	N	Y	Y	N	Y	
6	D360V	Ishigaki et al., 2006		Inf	F	N	N	Y	Y	Y	Y	
6	E362D	Sawaishi et al., 1999; Sawaishi et al., 2002		Juv	M	N	Y	N	Y	N	N	
6	A364P	Li et al., 2005	35	Inf	M	Y	Y	N	N	-----	-----	
6	Y366H	Li et al., 2005	36	Inf	M	N	Y	N	N	-----	Y	Also had E223Q
6	E371G	Kawai et al., 2006		Inf	F	Y	Y	-----	N	-----	Y	

(continued)

Table 24.1 (continued)

Exon	Mutation	Reference	Reported as patient #	Type	Sex	Macro	Seiz	Spas	Bul/Psb	Atx	Dev Dly	Comments
6	E373K	Li et al., 2005	38	Inf	M	Y	N	-----	Y	-----	Y	
6	E373K	Gorospe et al., 2002	7	Inf	F	Y	Y	N	N	-----	-----	
6	E373K	Li et al., 2005	37	Inf	F	Y	Y	N	Y	-----	Y	
6	E373Q	Li et al., 2005	39	Inf	M	Y	Y	N	N	N	N	
6	E374G	Li et al., 2005	40	Inf	F	N	Y	Y	N	Y	Y	
7	N3861	Caceres-Marzal et al., 2006		Inf	F	Y	Y	Y	Y	-----	-----	
8	R416W	Brenner et al., 2001	9	Inf	M	N	Y	-----	Y	-----	Y	
8	R416W	Brenner et al., 2001	10	Inf	F	Y	Y	Y	N	-----	Y	
8	R416W	Gorospe et al., 2002	12	Juv	M	N	N	N	N	N	N	
8	R416W	Li et al., 2005	41	Juv	M	N	N	Y	Y	Y	N	
8	R416W	Gorospe et al., 2002	11	Juv	F	N	N	N	Y	-----	N	
8	R416W	Kinoshita et al., 2003		Adlt	M	N*	N	Y	Y	Y	N	*State MRI revealed megalencephaly, but assume it fell well below the 2 SD criterion if this method was required
8	R416W	van der Knaap et al., 2006	5	Adlt	M	N	N	N	Y	Y	Y	
8	R416W	Thyagarajan et al., 2004	Familial	Adlt	F	N	N	N	Y	Y	N	Mother = proband, mutation arose *de novo*
				Adlt	M	N	N	Y	N	N	N	Mildly affected son
	None	Brenner et al., 2001	11	Inf	F	Y	Y	-----	-----	-----	Y	Especially severe ease, died at 15 weeks

	None	Rodriguez et al., 2001	15	Inf	F	Y	Y	-----	Y	-----	Y	Especially severe case, autopsy proven
	None	Gorospe et al., 2002	13	Inf	F	Y	Y	N	N	N	Y	Mild progression; died in drowning accident at age 10
	None	Li et al., 2005	44	Juv	F	N	N	N	N	Y	Y	May not have had Alexander disease
	None	Li et al., 2005	43	Adlt	M	N	N	Y	Y	Y	N	Typical MRI, autopsy proved but unusually sparse Rosenthal fibers, 3 of 4 sibs with similar disorder

Cases for which P47L or D157N was the sole coding change are not included (see text). Familial cases are shown as multiple entries for a given reference. Abbreviations are as follows: Macro = macrocephaly (Y = +2 SD or unusually rapid growth), Seiz = seizures (Y = 2 or more). Spas = spasticity, Bul/Psb = bulbar or pseudobulbar signs, Atx = ataxia. Dev Dly = mental or physical developmental delay or regression, Inf = infantile, Juv = juvenile, Adlt = adult, Asym = asymptomatic (but carries the indicated mutation), F = female, M = male, Y = yes, N = no, ----- = not reported or not applicable. If symptoms were not explicitly reported, they were scored in the table as follows: macrocephaly and seizures are listed as absent, under the assumption that they would have been reported if present; spasticity, bulbar/pseudobulbar signs, and ataxia are listed as not reported, except that an unsteady gait was assumed to reflect ataxia; developmental delay or regression was listed as unreported for patients less than 8 years old and as negative for patients 8 years of age or older.

Table 24.2 Summary of characteristics for Alexander disease patients with *GFAP* mutations.

	Number displaying clinical sign (yes/total = %)[1]								
Clinical Sign	Infantile			Juvenile			Adult		
	Male	Female	**Total**	Male	Female	**Total**	Male	Female	**Total**
Numbers of patients	41	35	**79/118 = 67%**	23	7	**30/118 = 25%**	4	7	**9/118 = 8%**
Macrocephaly	28/41 = 68%	21/35 = 60%	**52/79 = 66%**	7/23 = 30%	0/7 = 0%	**7/30 = 23%**	0/4 = 0%	0/7 = 0%	**0/12 = 0%**
Seizure	34/41 = 83%	26/35 = 74%	**60/79 = 76%**	10/23 = 43%	1/7 = 14%	**11/30 = 37%**	0/4 = 0%	0/7 = 0%	**0/12 = 0%**
Spasticity	14/28 = 50%	14/23 = 61%	**31/51 = 61%**	9/23 = 39%	1/7 = 14%	**10/30 = 33%**	2/4 = 50%	2/7 = 29%	**4/12 = 33%**
Bulhar and/or pseudobulbar	20/30 = 67%	12/22 = 55%	**32/52 = 62%**	20/23 = 87%	6/7 = 86%	**26/30 = 87%**	4/4 = 100%	7/7 = 100%	**12/12 = 100%**
Ataxia	6/13 = 46%	9/14 = 64%	**15/27 = 56%**	9/20 = 45%	3/5 = 60%	**12/25 = 48%**	4/4 = 100%	4/7 = 57%	**9/12 = 75%**
Developmental delay or regression	32/38 = 84%	29/32 = 91%	**64/73 = 88%**	13/23 = 57%	2/7 = 29%	**15/30 = 50%**	1/4 = 25%	0/7 = 0%	**1/12 = 8%**

Data are derived from Table 24.1, except that the two cases involving VII51 or E223Q mutations have been excluded due to their questionable diagnoses. Identical twins are included as a single case (the clinical signs for each twin in a set were the same in all 3 instances), Data for familial cases are included only for the proband. For D78E, there was no proband because Alexander disease was suspected for multiple family members prior to genetic testing: no male or female designation was entered for this family (there were 3 affected females and 2 males), and a single set of clinical signs data entered as the majority finding (Y or N) among the affected family members for each clinical sign. Totals vary because of incomplete reporting. See Table 24.1 for individual details.

[1]% totals for "numbers of patients" are the % of all cases with de novo *GFAP* mutations that fall within the indicated onset form. Three of the adult cases correspond to familial cases for which it is not known if the mutation arose de novo; these were not included in calculating the % of each form that arises de novo. All other % values pertain to the indicated onset category.

Sreedharan et al., 2007) and neck dysmorphia (Sawaishi et al., 1999; Stumpf et al., 2003). In several instances, adult onset Alexander disease has been initially mistaken for multiple sclerosis (e.g., Seil et al., 1968; Herndon et al., 1970; Schwankhaus et al., 1995; Li et al., 2005) or a brain tumor (Duckett et al., 1992; van der Knaap et al., 2005). Table 24.2 shows that the relative frequencies of the infantile, juvenile, and adult forms are about 65, 25, and 10%, respectively, for patients that have *GFAP* mutations. However, this is probably an overestimate for the juvenile and adult forms, since these cases are more likely to be published; e.g., in their report of 42 new cases, Li et al. (2005) particularly sought out later onset patients. As diagnostic awareness of Alexander disease increases, and more subtle presentations are appreciated, it is quite likely that nonfatal cases will be discovered and that the proportion of later onset patients reported will further increase. Diagnoses would likely be assisted by more complete and uniform reporting of clinical findings, for which tables E1 and E2 of van der Knaap et al. (2006) serve as useful models.

Typical presentations can aid diagnosis of the illness, but because these symptoms are not specific to Alexander disease, and because no individual symptom or group of symptoms is necessarily present in patients (van der Knaap et al., 2001; Li et al., 2005), other means are required for a firm diagnosis. Several laboratories have sought biomarkers for Alexander disease in the cerebrospinal fluid (CSF). Increased levels of HSP27 and αB-crystallin have been observed by three groups (Takanashi et al., 1998; Sawaishi et al., 1999; Imamura et al., 2002), but Shiihara et al. (2002) found that HSP27 but not αB-crystallin was increased, and Shiroma et al. (2001) found that neither was increased. More recently, Kyllerman et al. (2005) reported elevated levels of GFAP in the CSF of all three Alexander disease patients tested. Additional tests are required for each of these potential biomarkers to determine their sensitivity for detecting Alexander disease. Their specificity to Alexander disease must also be established, since GFAP, αB-crystallin, and HSP27 may all increase in other neuropathologies (Iwaki et al., 1992; Eng and Ghirnikar, 1994; Reynolds and Allen, 2003).

24.2.3 Magnetic Resonance Imaging

Magnetic resonance imaging (MRI) has proved a particularly powerful diagnostic aid for Alexander disease. In a landmark paper, van der Knaap et al. (2001) discerned five characteristic features of typical cases, particularly of the infantile form, and determined that the presence of any four provided a reliable diagnosis. These were (1) extensive, symmetrical cerebral white matter changes primarily in the frontal lobe that could include strongly enhanced signal, swelling (early), or atrophy and cyst formation (late), (2) a periventricular rim of enhanced signal on T1-weighting and decreased signal on T2-weighting, (3) alterations in signal intensity and morphology of the basal ganglia and thalami, including enhanced signal accompanied by either swelling or atrophy, or decreased signal and atrophy, most commonly in the head of the caudate nucleus and putamen, (4) similar findings for the medulla or midbrain with sparing of the red nuclei, and (5) contrast enhancement involving any one of the above regions or the optic chiasm, fornix, dentate nucleus, or cerebellum (Fig. 24.1). In this report the MRI changes often preceded more serious neurological dysfunction; follow-up MRI studies generally produced patterns similar to the initial images, except for progression to tissue loss and cystic changes in the frontal areas, enlargement of the ventricles, and atrophy of basal ganglia, thalami, and brainstem. In other instances the onset of typical features has been more variable (Gorospe et al., 2002; Caceres-Marzal et al., 2006). The basis for the abnormal MRI signals is not known; in certain instances, e.g., in cerebral white matter, it likely reflects myelin defects, whereas in other regions it correlates with Rosenthal fiber deposition. The enhanced signal with contrast has been suggested to be due to compromise of the blood–brain barrier secondary to Rosenthal fiber accumulation in the enveloping astrocytic end-feet (Borrett and Becker, 1985; van der Knaap et al., 2001). The extent and intensity of MRI abnormalities do not necessarily correlate with disease severity (Dinopoulos et al., 2006).

MRI patterns for juvenile and adult Alexander disease patients have proved more variable than for infantile patients, mirroring the greater clinical variability of these later onset forms. Supratentorial anomalies may be modest or absent, and diagnosis instead depends on imaging changes in the medulla, cerebellum, and/or spinal cord, as well as the particular pattern of signal enhancement in response to contrast. Lesions in these regions having the appearance of multifocal tumors are suggestive of Alexander disease (van der Knaap et al., 2005), and a wave-like garland pattern of increased signal intensity around the lateral ventricles has been observed

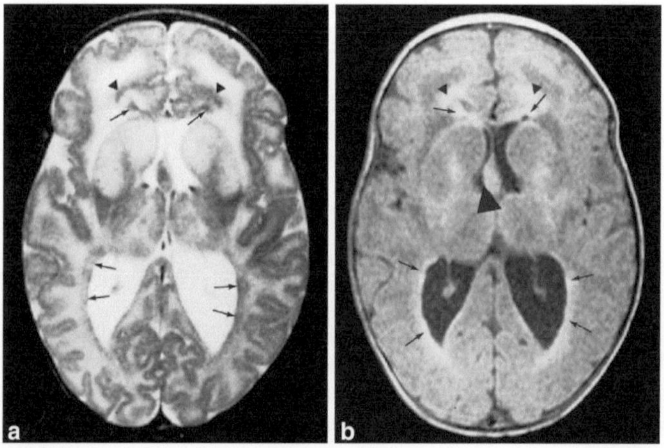

Fig. 24.1 Typical MRIs for infantile Alexander disease. MRIs are of a 1½-month infant with biopsy-confirmed Alexander disease. (**a**) T2-weighted MRI shows enhanced signal from frontal white matter, caudate nucleus, and putamen, and low signal from the periventricular rim. (**b**) T1-weighted MRI has somewhat reduced signal from frontal white matter and high signal from the periventricular rim. Arrows indicate the altered periventricular signal, which continues for a short distance into the frontal white matter (arrowheads). Reprinted with permission from M. van der Knaap et al. (2001), "Alexander disease: diagnosis with mr imaging," Am J Neuroradiol 22(3):541–552, copyright by the American Society of Neuroradiology.

that is unique to some later onset cases (van der Knaap et al., 2006). In essence, for the later onset forms only a single MRI criterion may be present (Probst et al., 2003). A number of *GFAP* mutations found in patients displaying these restricted MRI patterns had previously been identified in patients with MRIs typical of infantile Alexander disease, indicating modification of disease characteristics by unknown factors.

Several patients have also been examined by MR spectroscopy. A decreased *N*-acetylaspartic acid/creatine ratio is often detected (e.g., Shiroma et al., 2001; Imamura et al., 2002; Ishigaki et al., 2006), which may be due to neuronal degeneration (Dinopoulos et al., 2006). Absence of elevated *N*-acetylaspartate in the urine distinguishes Alexander disease from Canavan's disease, which also produces childhood megalencephalopathy (for discussions of the differential diagnosis of Alexander disease, see Gordon, 2003; Johnson and Brenner, 2003).

MRI has proved highly reliable for identifying Alexander disease (Rodriguez et al., 2001; Gorospe et al., 2002; Li et al., 2005; van der Knaap et al., 2005), but unusual cases have been encountered that do not produce diagnostic MRIs (van der Knaap and Brenner, unpublished observations). Thus, even in the absence of an MRI indication, *GFAP* sequencing may be warranted for cases with suggestive clinical signs for which other disease diagnostics have been negative. It is highly likely that the greater clinical and MRI variability among the later onset forms have resulted in their being underreported, and that diagnostic tools and criteria for these forms will continue to evolve.

24.2.4 Pathology

The characteristic pathology of Alexander disease is the presence of enlarged astrocytes and enormous numbers of Rosenthal fibers. The abundance and distribution of Rosenthal fibers were in fact the criteria for the juvenile and adult cases being classified as Alexander disease despite often appearing clinically distinct from the more common infantile form (Russo et al., 1976). The distribution of Rosenthal fibers also distinguishes Alexander disease from other conditions in which they may be present focally, such as pilocytic astrocytomas (Gessaga and Anzil, 1975), Parkinson's disease (Friedman and Ambler, 1992), amyotrophic lateral sclerosis (Smith et al., 1975), multiple sclerosis (Herndon et al., 1970), and where they were first described, in the walls of syringomyelic cavities (Rosenthal, 1898). However, the presence of Rosenthal fibers in other disorders can render problematical a diagnosis based on biopsy of a limited region. Rosenthal fibers typically accumulate in the astrocyte end-feet attached to blood vessels and just under the pial surface and ependyma (but for exceptions, see Herndon et al., 1970; Townsend et al., 1985). In infantile cases Rosenthal fibers are often particularly abundant in subcortical white matter, but sparse in neocortical astrocytes, which often display very small inclusions. This difference between cortical gray matter and subcortical white matter reflects an intrinsic difference in GFAP expression in human brain, low in cortex, far higher in white matter. Astrocyte processes and end-feet may also accumulate Rosenthal fibers in other areas of the CNS in infantile Alexander disease, such as the striatum, thalamus, cerebellum, brainstem, and spinal cord. In addition to the Rosenthal fibers, the brains of patients display varying losses of myelin and neurons that are generally more severe in younger than in adult patients. In some severe infantile cases the subcortical white matter is cystic, showing the absence of both myelin and axons (Klein and Anzil, 1994; Johnson and Brenner, 2003) (Fig. 24.2). It is likely that myelin did not form in the first place in these very early onset patients (Townsend et al., 1985). As in other leukodystrophies, the arcuate fibers tend to be spared from myelin deficits. In children with infantile onset, large zones of subcortical white matter show lack of myelin and the presence of an astrocyte-vascular scar with Rosenthal fibers (Crome, 1953; Borrett and Becker, 1985). The loss of neurons is also variable, and has been described in neocortex, striatum, hippocampus, thalamus, cerebellum, and brainstem (Crome, 1953; Towhi et al., 1983).

In contrast to infantile cases, juvenile onset patients show less pathology in supratentorial regions, but brainstem and cerebellum may be strongly affected (Russo et al., 1976; Duckett et al., 1992). Rosenthal fibers accumulate throughout the brainstem, often more in dorsal than in ventral regions. The dentate nucleus, deep and foliar white matter, and molecular layer in the cerebellum all have astrocytes with Rosenthal fibers and varying degrees of myelin loss. Demyelination in adult Alexander disease patients is highly variable, but generally far less than in the infantile and juvenile forms. It usually affects the subcortical white matter, striatum, cerebellum, and brainstem. The demyelination tends to be patchy rather than confluent

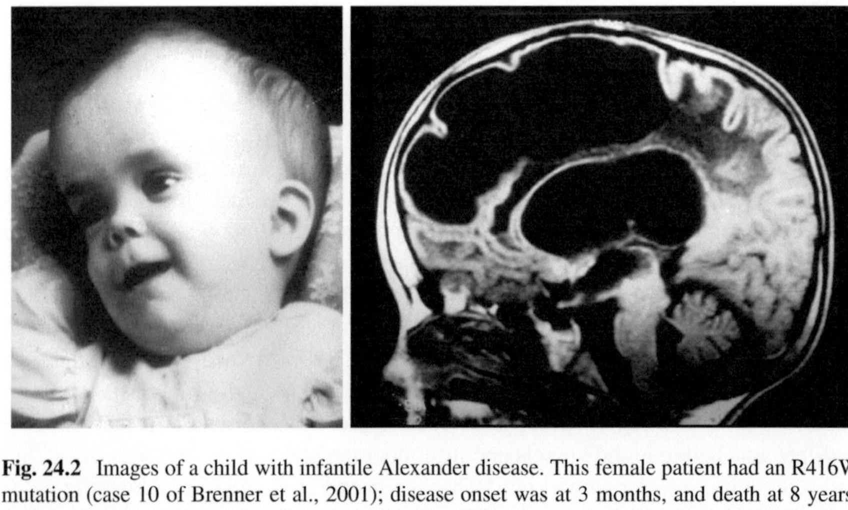

Fig. 24.2 Images of a child with infantile Alexander disease. This female patient had an R416W mutation (case 10 of Brenner et al., 2001); disease onset was at 3 months, and death at 8 years. (**a**) The patient at 22 months illustrating the frontal bossing and megalencephaly that is often present in infantile cases. (**b**) T1-weighted MRI of the patient at 7 years showing cystic degeneration in the frontal lobes, enlarged ventricles, and some atrophy of the vermis. Reprinted with permission from Johnson (1996), "Alexander disease," in: Handbook of Clinical Neurology, pp 701–710, copyright by Elsevier.

(Seil et al., 1968; Spalke and Mennel, 1982). Some brains include perivascular lymphocytic inflammation in areas of myelin loss and Rosenthal fiber formation, but this is not a constant pathological feature.

Alexander disease has been classified as a leukodystrophy because of the massive deficiency in myelination of the frontal lobes that commonly occurs in the infantile form (Fig. 24.2). This site of myelin pathology is consistent with the preponderance of Rosenthal fibers being in the cerebral cortex of infantile cases. Correspondingly, both myelin pathology and Rosenthal fibers are usually more caudal in the later onset forms. The reason for the region-selective pathology in the early and late forms is obscure, given that astrocytes and GFAP expression are distributed throughout the CNS. The susceptibility of the frontal cortex in infantile cases is often attributed to it being the last region to myelinate, and thus perhaps more sensitive to perturbation. However, this scenario does not explain the more caudal pathology in the later onset cases. In addition, the presence of Rosenthal fibers and myelin deficiencies is not always correlated (Crome, 1953; Borrett and Becker, 1985; Schwankhaus et al., 1995; Namekawa et al., 2002; Stumpf et al., 2003). In some later onset patients MRI may not reveal any leukodystrophy, although atrophy may be present in the brainstem, cerebellum, and spinal cord (Salvi et al., 2005; van der Knaap et al., 2006). These latter cases, together with the finding that *GFAP* mutations are responsible for most instances of Alexander disease, suggest that the disorder should be reclassified as an astrogliopathy rather than remaining as a leukodystrophy. The absence of myelin defects in some of the later onset cases indicates that the characteristic MRI signals arise from pathologies in addition to dysmyelination. Correlation of the pattern of imaging abnormalities with the density of Rosenthal fiber deposition strongly suggests that these aggregates play an important role (Farrell et al., 1984; van der Knaap et al., 2001).

Rosenthal fibers are amorphous, eosinophilic, and osmiophilic structures that are often surrounded by a profuse web of intermediate filaments that appear to radiate from their midst (reviewed in Wippold et al., 2006) (Fig. 24.3). They are not actually fibers, but their distribution within astrocytic processes could yield a fibrillar impression at low magnification. Instead, their shapes range from rods to elongated ovals to disks; diameters can vary between 10 and 40 μm and lengths can reach 100 μm. In several infantile cases Rosenthal fibers are seen primarily as large numbers of small, granular deposits in the cell body (Herndon et al., 1970; Townsend et al., 1985), leading to the suggestion that they initially form around the nucleus and later move through the astrocytic processes to accumulate as larger aggregates in perivascular, periventricular, and subpial end-feet (Gorospe and Maletkovic, 2006; Wippold et al., 2006). Most persuasive in this respect are the observations of Borrett and Becker (1985) of primarily granular deposits in astrocytes in a biopsy of a patient at 3.5 months, and the presence of typical Rosenthal fibers at autopsy at 7 months.

The composition of Rosenthal fibers was initially of interest for clues to the etiology of Alexander disease, and more recently for clues about how the disease phenotype becomes manifest. The observation that intermediate filaments appear to radiate out from Rosenthal fibers suggested that GFAP was a primary constituent of these aggregates. Labeling of these particles with GFAP antibodies proved difficult

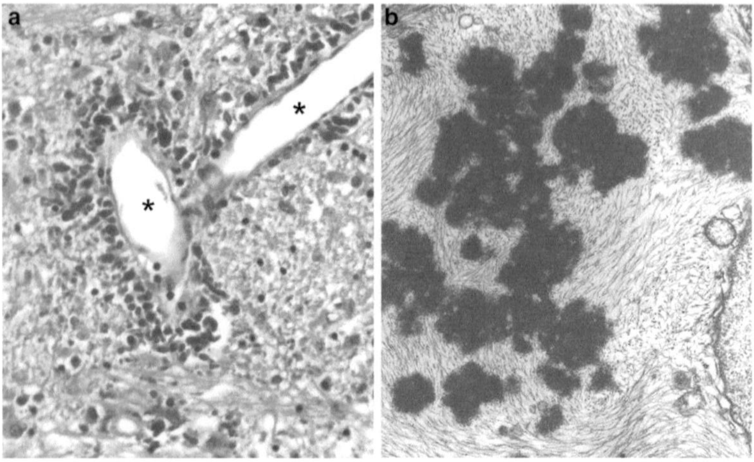

Fig. 24.3 Rosenthal fibers at the light and electron microscopy (EM) level. (**a**) Hematoxylin and eosin staining of the brainstem of a child with an R239H mutation reveals Rosenthal fibers as dark nuggets in astrocytic end-feet surrounding blood vessels (asterisks). (**b**) Under EM Rosenthal fibers appear as membraneless, amorphous osmiophilic aggregates in a dense meshwork of intermediate filaments. Panel (**a**) reprinted from The Lancet Neurology, Vol. 2, A. Messing and Brenner (2003), with permission from Elsevier; and panel (**b**) from Eng et al. (1998), "Astrocytes cultured from transgenic mice carrying the added human glial fibrillary acidic protein gene contain Rosenthal fibers." J Neurosci Res 53:353–360, copyright by Wiley-Liss, Inc.

(Becker and Teixeira, 1988), perhaps due to epitope masking, but in 1989 Johnson and Bettica (1989) developed a procedure for on-grid immunogold labeling that revealed extensive GFAP signals by immunoelectron microscopy. More recently, using an antibody specific for a disease-causing mutant GFAP (R416W), it was conclusively shown that the mutant GFAP is expressed and is indeed present in the Rosenthal fibers (Perng et al., 2006). The Goldman laboratory has taken a more direct approach to determining the composition of Rosenthal fibers by using sodium dodecyl sulfate polyacrylamide gel electrophoresis (SDS-PAGE) to separate the constituent proteins of a Rosenthal-fiber-enriched fraction isolated from Alexander disease brain. Such studies are facilitated by the observation that Rosenthal fibers are essentially completely soluble in SDS-PAGE sample buffer (Goldman and Corbin, 1988), in distinction to the protein aggregates formed in many other disorders (Smith et al., 1996b; Giasson et al., 1999; Koyama et al., 2006; Ellisdon et al., 2007). Immunoblotting revealed the most abundant protein to indeed be GFAP (Goldman and Corbin, 1988), and two other major constituents were identified as the small stress proteins αB-crystallin (Iwaki et al., 1989) and HSP27 (Iwaki et al., 1993). These latter findings suggest that astrocytes in Alexander disease patients have initiated a stress response. Further, the absence of an increase in the common stress protein HSP70 indicates a degree of specificity in the activated signaling pathways. Other constituents found associated with Rosenthal fibers include p62 (Zatloukal et al., 2002), a ubiquitin-binding protein induced as part of the stress response to misfolded proteins; plectin, a cytoskeletal crosslinking protein (Tian et al., 2006); and phospho-c-Jun amino-terminal kinase (p-JNK) and components of the 20S proteasome

(Tang et al., 2006). Several modifications have also been described for the proteins present in Rosenthal fibers. One modification common to many other kinds of protein aggregates is ubiquitination. αB-crystallin has been shown to be ubiquitinated in Rosenthal fibers (Goldman and Corbin, 1991; Iwaki et al., 1993), but whether GFAP or other associated proteins carry this modification has not yet been determined. Both advanced glycation end products (Castellani et al., 1997) and advanced lipid peroxidation end products (Castellani et al., 1998) have been detected in Rosenthal fibers by immunohistochemistry, but the modified molecular targets are not yet known. Conversely, deiminated forms of GFAP have been observed in Alexander disease brain using immunoblots (Anthony Nicholas, personal communication), but whether this modified GFAP is present in Rosenthal fibers is unknown.

24.3 GFAP Mutations

24.3.1 Initial Discovery

The sporadic nature of Alexander disease, with only rare instances of familial cases, made it unclear whether its cause was environmental or genetic (Pridmore et al., 1993). For those advocating a genetic origin, a recessive mutation was often invoked to explain unaffected parents in familial cases (Pridmore et al., 1993; Johnson, 1996; Reichard et al., 1996). Despite protein aggregates in astrocytes being the defining pathology, it was even unclear whether the primary defect was within the CNS, or whether the astrocytes were responding to a systemic disorder (Johnson, 1996). For example, nickel intoxication results in protein aggregates with some semblance to Rosenthal fibers (Kress et al., 1981). The close association of intermediate filaments with Rosenthal fibers did prompt Becker and Teixeira to raise the possibility of *GFAP* as a candidate gene in 1988 (Becker and Teixeira, 1988), but this suggestion was not pursued. Instead, it was the unexpected outcome of a transgenic mouse study that led to the discovery of *GFAP* mutations as the basis of most Alexander disease cases. The mice had been engineered to overexpress GFAP in astrocytes by the introduction of a human genomic *GFAP* transgene in order to address the role of elevated GFAP levels in reactive gliosis (Messing et al., 1998). Surprisingly, mice that expressed the transgene at high levels died within a few weeks of birth, and their autopsy showed the presence of highly reactive astrocytes containing abundant protein aggregates. These aggregates proved indistinguishable from Rosenthal fibers by both light and electron microscopy (EM), and by their immunostaining for GFAP, ubiquitin, αB-crystallin, and HSP25 (the mouse homolog of HSP27). They also had the same subpial, perivascular, and subventricular distribution that is characteristic of Alexander disease.

The discovery that simple overexpression of the human *GFAP* (*hGFAP*) gene resulted in abundant production of Rosenthal fiber-like aggregates and had fatal effects suggested that some defect in GFAP expression was responsible for Alexander disease. Accordingly, the sequence of the gene was investigated in an initial set of

11 cases of autopsy-confirmed Alexander disease. The *GFAP* transcription unit spans 9,869 nucleotides on chromosome 17q21 (Brenner et al., 1990; Bongcam-Rudloff et al., 1991). No pseudogenes are present to complicate sequence analyses, but several splice variants have been described (reviewed in Quinlan et al., 2007). The major mRNA species consist of 3,029 nucleotides stitched together from nine exons, and encodes the 432 amino acids of the predominant form of GFAP, GFAPα (Brenner et al., 1990). Sequencing an upstream promoter region of the patient DNAs that is capable of directing astrocyte expression in mice (Brenner et al., 1994) yielded no alterations specifically associated with the disease; however, 10 of the 11 cases were heterozygous for single nucleotide coding changes that predicted amino acid substitutions for arginines in four different positions throughout the *GFAP* protein (Brenner et al., 2001). Available parental pairs of the patients were tested for the presence of their child's nucleotide change to determine if these single amino acid changes were disease causing or were simply innocuous polymorphisms. If a coding change were a harmless polymorphism, it would be expected to be present in one of the parents; if disease causing (with high penetrance), it should be absent. None of the six pairs of parents analyzed in this initial study had the change present in their child, indicating that each was a de novo mutation. The probability that a mutation would arise in a coding region of the size of *GFAP* in a given generation is conservatively estimated at 1 in 20,000 (Nachman and Crowell, 2000), thus the findings for these six parental pairs provided statistical certainty that the missense mutations cause the disease.

24.3.2 Overview of GFAP Mutations

Since this initial discovery, multiple laboratories have described the association of *GFAP* mutations with Alexander disease, with the largest patient groups reported by Gorospe et al. (2002), Rodriquez et al. (2001), and Li et al. (2005). At present a total of 49 different nucleotides have been found mutated affecting 35 different amino acid positions (Fig. 24.4). The mutations for all but three patients have been found in exons 1, 4, 6, or 8, with those affecting R79, R88, R239, and R416 accounting for 58% of the total (the actual percentage is probably higher, since unique mutations are more likely to be published). Each of these sites contains a CpG, which can give rise to mutations by methylation and deamination (Cooper and Youssoufian, 1988). Importantly, mutations have been found for a preponderance of the juvenile and adult onset cases as well as for the infantile, justifying their grouping as a single disease, and indicating that the presence of Rosenthal fibers reflects some underlying common mechanism in the disease process. Like the original cohort of cases, all mutations have been heterozygous, confirming a dominant effect. Although nearly all of these are also simple missense mutations, several are more complex. These include small insertions and/or deletions that result in either the in-frame gain or loss of a few amino acids (Li et al., 2005; van der Knaap et al., 2006); a mutation in intron 3 that results in skipping of exon 4, producing a large in-frame deletion (Brenner, unpublished); and an insertion/deletion that

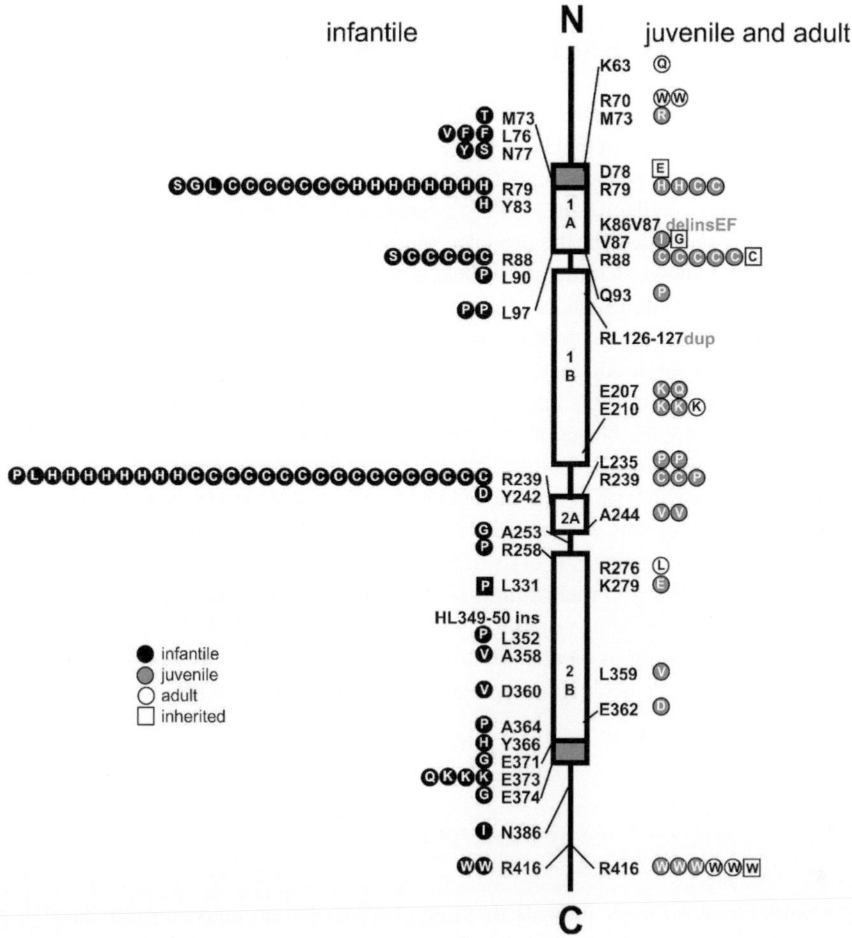

Fig. 24.4 Locations of GFAP mutations in Alexander disease in relation to the protein structure. The four open rectangular boxes represent the helical coiled-coil rod domains of GFAP; these structural motifs are highly conserved among most intermediate filament proteins. The solid lines joining these segments are nonhelical linker regions, and the solid lines at either end are the unconserved, random coil, N-terminal, and C-terminal regions. The gray box just before segment 1A is a nonconserved prehelical sequence important for initiation of rod formation at the start of 1A; the gray box at the end of 2B represents the highly conserved 365TYRKLLEGEE374 sequence that includes the end of the coiled coil 2B segment at E371. The wild type amino acid is indicated next to the structure, and amino acid replacements within symbols on either side. Infantile cases are on the left, shown as white letters on black fill; juvenile and adult cases are on the right, as white letters on gray fill or black letters on white fill, respectively. Each symbol represents a single patient, except that familial cases, including identical twins, are represented by a single box coded for the onset type of the proband (see Table 24.1 for details of family members).

results in the final two amino acids of *GFAP* being replaced by 11 residues before encountering a termination codon (Brenner, unpublished).

GFAP mutations have been found in 96% of patients for whom there is a strong presumption of Alexander disease based on clinical signs supported by either typical

MRI findings or pathology (Table 24.1). Possible genetic causes for the remaining patients are discussed in Sect. 24.3.9. Abundant Rosenthal fibers with a predominantly brainstem distribution similar to that for adult onset Alexander disease have also been incidentally observed in patients without neurological deficits, or with neurological involvement that has other explanations such as infection or circulatory disease. This has led to the designation of a separate disease entity, "Rosenthal fiber encephalopathy" (Wilson et al., 1996; reviewed in Jacob et al., 2003). None of these patients has been tested for the presence of *GFAP* mutations.

Figure 24.4 shows that mutations have been observed over most of the protein's length. Like other intermediate filament proteins, the GFAP monomer has random coil N- and C-terminal ends flanking an extensive α-helical central rod domain, which is further subdivided into four segments by short, nonhelical linkers (Fuchs, 1996). Based on homology, intermediate filaments have been grouped into several families; for example, the acidic and basic keratins constitute the type I and II intermediate filaments; GFAP, desmin, vimentin, and peripherin compose type III; the neurofilaments are in type IV and the nuclear lamins in type V. The N- and C-terminal domains are moderately conserved within a type, but not between types, whereas the central rod domain has a high degree of homology in both the length of the individual segments and their amino acid sequences, particularly at the beginning and end of the rod. Many of the sites mutated in GFAP in Alexander disease patients are homologous to ones mutated in other intermediate filament diseases (Li et al., 2002a). Possible genotype/phenotype correlations are discussed in Sect. 24.3.4.

24.3.3 Gain or Loss of Function

Several lines of evidence indicate that the GFAP mutations cause disease through a dominant gain of function, rather than through a loss of function. *GFAP* null mice are healthy, fertile, have a normal life expectancy and do not display signs of Alexander disease (Gomi et al., 1995; Pekny et al., 1995; McCall et al., 1996). Also as noted earlier, Rosenthal fibers are closely associated with abundant intermediate filaments, suggesting that at least some GFAP remains capable of normal polymerization in Alexander disease brain. Finally, a loss of function mechanism would predict the presence of null, nonsense, or frameshift mutations that effectively inactivate the protein; no such mutations have been discovered for Alexander disease, although they are relatively common for other intermediate filament diseases (for multiple examples, see the intermediate filament mutation database at http://www.interfil.org).

Why should *GFAP* mutations produce a dominant gain of function disease, whereas homologous mutations in other intermediate filament proteins are considered to produce disease due to a dominant loss of function? It is particularly puzzling that excessive amounts of filaments are present in Alexander disease brain (Herndon et al., 1970), whereas mutations at the homologous sites of other intermediate filament proteins block their polymerization and produce a phenotype similar to that seen in mice in which the corresponding gene had been inactivated. One possible resolution of this paradox is that all of the disease-causing intermediate filament mutations strongly inhibit filament formation, but do not block it completely (Fig. 24.5) (for a more extensive discussion of this topic, see Li et al., 2002a). In the case of GFAP, normal filaments might still form because its synthesis is markedly increased as part of the astrocytic response to CNS disturbances. Thus GFAP protein may accumulate to levels sufficient to overcome the kinetic block presented by the mutations and produce the normal appearing filaments observed,

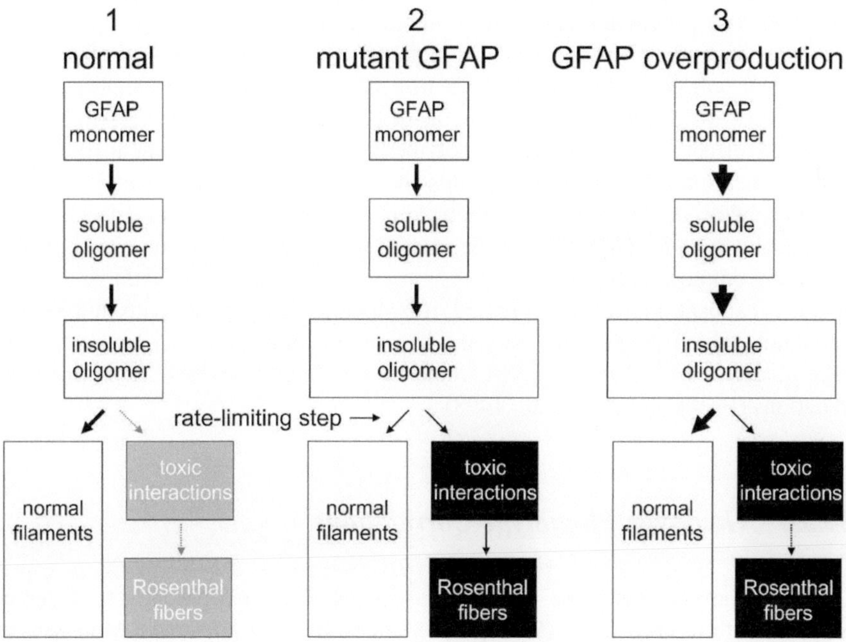

Fig. 24.5 General model for Rosenthal fiber formation. This model was developed to explain two seemingly paradoxical observations: (1) mutations in *GFAP* produce a dominant *gain* of function disorder with the presence of abundant intermediate filaments, whereas homologous mutations in other intermediate filaments produce a dominant *loss* of function disease by inhibiting filament formation, and (2) overproduction of wild type *GFAP* produces Rosenthal fibers indistinguishable from those produced by *GFAP* mutations. The key postulate of the model is that a late step in polymer formation is normally rate-limiting, and that this step is inhibited strongly, but not completely, by the mutations, resulting in accumulation of an intermediate that participates in toxic interactions and is converted to Rosenthal fibers. Under normal conditions (scenario 1 in the Figure) there is an orderly flow of intermediates along the polymerization pathway, and insufficient levels of intermediates accumulate to lead to Rosenthal fiber formation. When *GFAP* mutations are present, the kinetic bottleneck produced by inhibition of the rate-limiting step results in elevated levels of the intermediate, some of which is diverted into the toxic pathway leading to Rosenthal fibers (scenario 2). Similarly, overproduction of GFAP increases flux through the polymerization pathway, resulting in accumulation of the same intermediate prior to the rate-limiting step (scenario 3). See the text and Li et al. (2002a) for additional details.

whereas other intermediate filaments that lack this induction mechanism form only very low levels of mature filaments. The model also predicts that intermediates prior to the inhibited step would accumulate to abnormally high levels, and these might be responsible for Rosenthal fiber formation by participating in aberrant interactions that ultimately lead to protein aggregation (Fig. 24.5, scenario 2). The formation of Rosenthal fibers in the hGFAP overexpressing mice is also explained by this model, with the additional assumption that the polymerization step most strongly affected by the mutations is also the one that is normally rate-limiting in producing mature filaments. In this case, accumulation of the same toxic intermediates would result from the excessive flux of GFAP protein into the polymerization pathway (Fig. 24.5, scenario 3). It is thus possible that all intermediate filament diseases have both loss of function and gain of function components, with the relative contributions depending on the particular biology of the system. The loss of function component will depend on the net effect of the kinetic block (presumably greater for keratins, which are not induced by the disease process, than for GFAP) and the biological effect of the reduction in polymerized intermediate filament (null mice indicate that this is also high for keratins and low for GFAP). The gain of function may depend on the amount of polymerized intermediate filament that accumulates (presumably high for GFAP, due to the reactive response) and its toxic consequences (presently unknown, but the reactive response of astrocytes could lead to a noxious positive feedback loop). In this regard, it is notable that patients with null keratin mutations, which would have only the loss of function disease component, may have milder disease than those that would have both components due to missense mutations (Fuchs and Cleveland, 1998).

24.3.4 Genotype/Phenotype Correlations

Limited genotype/phenotype correlations can be discerned among the different *GFAP* mutations (Fig. 24.4) (Rodriguez et al., 2001; Li et al., 2005). Mutations at both the R79 and R239 hot spots tend to result in infantile Alexander disease. Patients with an R239H change have had an especially rapid course, with onset generally occurring by 6 months of age and death before 6 years. Even more severe effects may be produced by mutations near the end of the 2B rod domain. Many patients with these mutations had an onset before 3 months of age, and died before 6 months. Included in this region is one of the most fulminate cases yet described, a child with an L352P change who was affected shortly after birth and died at 38 days. This region does include a few juvenile case mutations, but these are the remarkably conservative L359V and E362D alterations. In contrast to these regions of relatively devastating mutations, a cluster of later onset mutations occurs in a region spanning the end of coil 1B and the beginning of 2A. Only the infantile group had sufficient numbers of mutations at a given site to permit correlations between specific mutations and clinical signs within an age category. These sites are R79 (17 cases), R88 (4 cases), R239 (26 cases), and E373 (4 cases). Among these, spasticity for R239 patients was the only mutation-specific difference for any of the six clinical signs listed in Table 24.2 to reach statistical significance; all nine R239 patients scored for this symptom were positive compared with 19/42 (45%) for the other infantile patients ($p = 0.0025$). However, even this difference could be an artifact of data collection; evaluation of this symptom was described for only 9 of the 26 cases, and failure to report a few negative findings would compromise its statistical significance. An increased frequency of macrocephaly for R239 patients compared with other Alexander disease patients had been previously suggested (Quinlan et al., 2007), but was not supported when confined to comparisons within the infantile age group ($p = 0.31$). Infantile patients did show an increased tendency for bulbar and pseudobulbar signs, but this also was not significant ($p = 0.13$). Other correlations of potential interest that also fell short of statistical significance were a relatively low frequency of macrocephaly in R79 infantile patients ($p = 0.15$), and findings that all four cases involving E373 mutations had macrocephaly ($p = 0.29$) and none of the three E373 cases for which it was reported had spasticity ($p = 0.08$). It thus remains uncertain whether particular mutations tend to produce a particular constellation of clinical signs.

The consequences of some of the *GFAP* mutations can be rationalized by what is known about the structural basis of intermediate filament assembly. Like other intermediate filaments, GFAP polymerization commences with the parallel intertwining of two α-helical strands to form a coiled-coil dimer (Quinlan et al., 1986). The two strands associate by hydrophobic interactions of amino acid residues, primarily leucines, that occur at every seventh position in the helical segments. The effects of the small insertion and deletion mutations can be understood as displacing this heptad repeat, and the substitutions of proline as distorting the helical structure. The dimers then associate head to tail to form tetramers, which are considered to be the first semi-stable assembly intermediate (Soellner et al., 1985; Parry and Steinert, 1999). These initial steps are believed to depend on interactions between amino acids 191–203 in the 1B segment and amino acids 357–369 in the 2B segment (Wu et al., 2000). No Alexander disease mutation has been reported in the former region, but several are in the latter. The filaments then extend in length through partial overlaps of the beginning of the 1A domain and the end of the 2B rod domain – regions that are especially highly conserved among intermediate filaments (Fuchs, 1996). This critical section of the GFAP 1A domain spans amino acids 72–86, and contains multiple Alexander disease mutations, including the R79 hot spot. Even a change as conservative as L76V within this region can produce severe disease. The corresponding conserved segment of 2B spans amino acids 363–376, partially overlapping the sequence involved in dimer and tetramer formation. The especially devastating consequences of the L352P mutation could be explained by the mutant proline disrupting the alpha helical structure of these adjacent regions. In addition to increasing in length, the polymers associate side-to-side to build up to a cross section of 32 monomers. Crosslinking experiments with vimentin have implicated the beginning of the 2A domain in this process (Steinert et al., 1999), which could explain the importance of the R239 hot spot. The R416W mutation is unusual in residing outside the central rod domain. However, it occurs within an RDG motif that is conserved among type III intermediate filaments and likely plays a role in both filament assembly (Chen and Liem, 1994) and interactions with other proteins (Quinlan, 2001).

Whether these correlations reflect true causal relationships or are merely fortuitous is not yet clear. There are several reasons for caution. One issue is that genotype/phenotype correlations are not apparent for many of the mutations. As an example, the strongly conserved 1A region encompassing amino acids 72–86 indeed contains a cluster of mutations at positions 73, 76, 77, 78, and 79, but none has been reported for positions 80–86, whereas just beyond this region mutations are again plentiful (positions 87, 88, 90, and 97). Another consideration is that the intermediate filament disease mutations may not affect an early stage in fiber formation, but instead affect a relatively late one as discussed earlier (reviewed in Li et al., 2002a). The observation that mutations all along the length of the protein result in Alexander disease and yield similarly appearing protein aggregates suggests that there is a common critical defect that involves the whole protein. An attractive candidate is lateral interactions that may control fiber thickness (Herrmann et al., 2000; Ma et al., 2001). Having such a late step as the critical defect could explain why GFAP carrying the R239C mutation is capable of forming apparently normal appearing 10-nm filaments when polymerized in vitro (Fig. 24.6) (Hsiao et al., 2005). It is also consistent with the observation of Herndon et al. (1970) that the filaments in the vicinity of Rosenthal fibers are several times thicker than normal. These uncertainties about the roles of the GFAP alterations are mirrored by an elegant series of studies of the effects of desmin mutations on filament formation (Bar et al., 2005, 2006a, 2006b), which led the authors to conclude "We are unable to explain, why one amino acid exchange results in a drastic assembly defect, whereas another enables in vitro filament formation to occur. A more rational understanding of the underlying molecular mechanisms requires the atomic structure of the IF elementary building block, the dimer, and an atomic model of the mature filament" (Bar et al., 2006b).

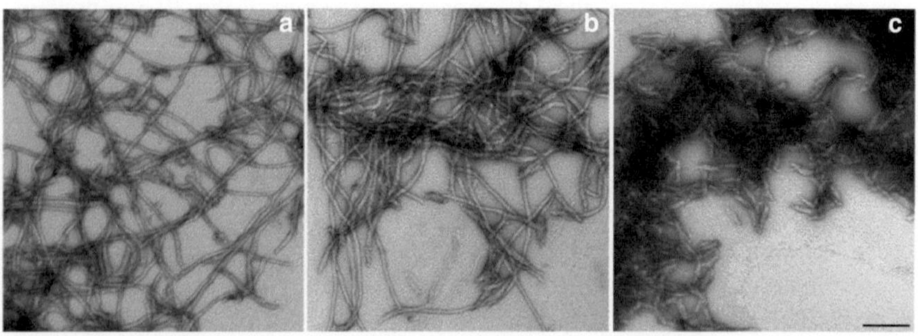

Fig. 24.6 In vitro assembled GFAP filaments. Recombinant proteins were produced in *E. coli* and then purified to homogeneity and assembled in vitro. They were then negatively stained with 1% (w/v) uranyl acetate and viewed in the EM. Samples are (**a**) wild type, (**b**) R239C, and (**c**) R416W GFAP. Bar = 100 nm.

Another caution for genotype/phenotype relationships is the finding that the same mutation can produce different forms of Alexander disease. Both infantile and juvenile forms have been observed for R79C, R79H, and R239C, and all three forms have been observed for R88C and R416W (Fig. 24.4). Curiously, R239C primarily produces infantile Alexander disease, yet forms normal appearing 10-nm filaments in vitro, whereas R416W fails to do so, yet is found in several adult onset cases (Fig. 24.6) (Perng et al., 2006). Variable phenotypes for the same mutation imply the presence of genetic or environmental modifiers. The possibility of genetic modifiers is suggested for the R416W mutation by the finding that in a familial case both the affected mother and son had the adult onset form (Thyagarajan et al., 2004). The possibility of environmental modifiers is consistent with the potential positive feedback effects of reactive gliosis. As illustrated in the model presented in Fig. 24.5, a critical threshold for the levels of mutant and/or total GFAP may be required to accumulate sufficient levels of a polymerization intermediate to produce deleterious effects. Once this occurs, reactive gliosis may be triggered, resulting in a positive feedback spiral of increasing GFAP production and greater toxicity. Anything that produces reactive gliosis in a susceptible brain region, such as an infection, trauma, or hypoxia/ischemia, could possibly push GFAP levels over the threshold value. It may be simply coincidental, but in several instances patients have indeed presented following brain trauma (Meins et al., 2002; Namekawa et al., 2002) or infection (Herndon et al., 1970; Shiroma et al., 2003; Kyllerman et al., 2005).

24.3.5 Criteria for a Mutation Being Disease Causing

The great majority of *GFAP* mutations associated with Alexander disease arise de novo, providing strong evidence that they are disease causing (Table 24.3). These data also demonstrate that most *GFAP* mutations are fully penetrant. There are several instances, however, in which both parents are not available for analysis, or the same genetic change is present in a normal parent. In these situations the following criteria can be evaluated to indicate whether or not a novel change is likely to be disease causing.

Diagnosis: About 95% of pathologically proven Alexander disease cases are attributable to *GFAP* mutations, and thus a convincing diagnosis increases the probability that the genetic change is actually disease causing. R276L is classified as *disease causing* in Table 24.3 because the patient had pathology typical of Alexander disease; Y83H, D360V, and E371G are assigned the less certain classification of *probably disease causing* because no pathology was available for the patients, but they had an MRI typical of Alexander disease.

Family history: If the same change is found in an unaffected parent, support for it being disease causing (although incompletely penetrant) would come from a history of a similar disorder in the family of that parent, or in siblings or children of the patient. This would be greatly strengthened if genetic analysis confirmed the presence

Table 24.3 Relation of *GFAP* coding changes to Alexander disease.

Coding change	Disease causing likelihood[1]				*De novo*	Controls[2]	Polymerization[3]	Comments[4]	Reference[5]
	Y	Y?	N?	N					
P47L			*			0/110			Brenner et al., 2001; Li et al., 2002a
K63Q	√					0/100	Defective	Mo and 2 sisters WT	Li et al., 2005
R70W	√							Only mutation in 2 independent adult onset cases; typ MRIs for adult onset; unaffected Mo & sister WT for Sreedharan et al. case; no other family members tested in Salvi et al. case; no normal controls tested or pathology in either report	Salvi et al., 2005; Sreedharan et al., 2007
M73R	√					0/192		typ MRI and symptoms, M73T is de novo	Gorospe et al., 2002
M73T	√				√				Li et al., 2005
L76F	√				√				Rodriguez et al., 2001; Li et al., 2005
L76V	√				√				Li et al., 2005
N77S	√				√				Li et al., 2005
N77Y	√				√				Rodriguez et al., 2001
D78E	√					0/200		Present in 5/5 affected and 1/12 unaffected family members; typ MRI; pathol	Stumpf et al., 2003
R79C	√				√				Brenner et al., 2001
R79G	√				√				Gorospe et al., 2002
R79H	√				√		Defective?		Brenner et al., 2001
R79L	√							R79C,G,H,S mutations arose de novo	Shiroma et al., 2003
R79S	√				√				This article
Y83H		√						typ MRI and symptoms	Wu et al., 2006

K86V87 delinsEF	√				√				van der Knaap et al., 2006
V87G	√					0/400		Present in 3/3 affected family members; typ MRI and symptoms for adult onset	Okamoto et al., 2002
V87I	√				√				This article
R88C	√				√				Rodriguez et al., 2001
R88S	√							typ MRI; R88C arose de novo	Rodriguez et al., 2001
L90P	√				√	0/102			Suzuki et al., 2004
Q93P	√				√				Kyllerman et al., 2005
L97P	√				√	0/332			Meins et al., 2002
V115I			*			0/100	Normal	atyp MRI, Mo = WT; Fa = V115I	Li et al., 2005
RL126–127dup	√				√				van der Knaap et al., 2006
D157N				*		5/90	Thicker?	Present in normal controls and parents of patients	Li et al., 2005
E207K	√				√				Li et al., 2005
E207Q	√				√				Li et al., 2005
E210K	√				√		Defective		Li et al., 2005; van der Knaap et al., 2006
E223Q			*			0/150	Thicker?	atyp MRI; present in unaffected Mo	Brockmann et al., 2003; Li et al., 2005
L235P	√				√				Li et al., 2005
R239C	√				√		Defective		Brenner et al., 2001; Hsiao et al., 2005
R239H	√				√				Brenner et al., 2001; Rodriguez et al., 2001
R239L	√				√				Lee et al., 2006

(continued)

Table 24.3 (continued)

Coding change	Disease causing likelihood				De novo	Controls	Polymerization	Comments	Reference
	Y	Y	N	N					
R239P	√				√	0/260			Meins et al., 2002
Y242D	√				√				Gorospe et al., 2002
A244V		*				0/130	Normal	typ MRI for one case, atyp for another; Mo WT in both cases;	Aoki et al., 2001; Li et al., 2005
A253G	√					0/100	Defective	Mo WT	Li et al., 2005
R258P	√				√				Brenner et al., 2001
R276L	√					0/156		Present in 2 affected sibs; typ MRI; pathol	Namekawa et al., 2002
K279E	√				√				Li et al., 2005
D295N				√		3%	Normal?	Present in normal controls; 1/26 in 2001	Isaacs et al., 1998
L331P		*				0/210		typ MRI; present in neurologically normal older sister and mother, who had minor MRI abnormalities	Shiihara et al., 2004
349HLins	√				√				Li et al., 2005
L352P	√				√				Bassuk et al., 2003; Li et al., 2005
A358V	√				√				Dinopoulos et al., 2006
L359V	√				√				Li et al., 2005
D360V		√				0/240		typ MRI and clinical signs	Ishigaki et al., 2006
E362D	√				√	0/150			Sawaishi et al., 2002
A364P	√				√				Li et al., 2005
Y366H	√				√		Defective?		Li et al., 2005
E371G		√				0/100		typ MRI and clinical signs	Kawai et al., 2006
E373K	√				√				Gorospe et al., 2002
E373Q	√				√				Li et al., 2005

E374G	√				√				Li et al., 2005
N386I	√				√				Caceres-Marzal et al., 2006
R416W	√				√		Defective		Brenner et al., 2001; Perng et al., 2006

[1]Y = Almost certainly causes Alexander disease because it arose de novo; another mutation of the same amino acid arose de novo, multiple affected individuals, Alexander disease diagnosed by pathology, or grossly aberrant polymerization of the mutant protein in transfected cells (see text for details); Y? = probably disease causing; N? = probably not disease causing; N = almost certainly a polymorphism. Entries for which there is considerable uncertainty are indicated by an asterisk (*) rather than a check (√), and are discussed in the text.

[2]Number of altered chromosomes/total number tested; for amino acid changes data are for unrelated controls, for noncoding changes data are for patient samples (controls have generally not been tested for these). In instances in which controls have not been tested the change can be assumed to be very rare, as over 100 patient DNAs (200 chromosomes) have now been sequenced in the course of Alexander disease diagnosis.

[3]Polymerization characteristics in SW13vim^{-} cells as follows: defective = clear failure to form a normal appearing network, thick = forms a network, but fibers are abnormally thick; normal = indistinguishable from wild type. Entries followed by a question mark (?) are unpublished preliminary results of MB.

[4]Pathol = pathologically proven; typ/atyp MRI = MRI typical/atypical for Alexander disease, Mo = mother, Fa = father.

[5]References are for the first report of the gene change, of its de novo appearance, and of its frequency and polymerization properties.

of the change in the affected individuals. For example, inheritance of the D78E mutation by multiple affected family members provides strong evidence for a causative role (Stumpf et al., 2003; this family is discussed in Sect. 24.3.8 later).

Conservation of the altered amino acid among intermediate filaments and its mutation in other intermediate filament diseases: Mutated amino acids that are highly conserved among intermediate filaments are more likely to be important for normal filament structure and function. The least significant conservation is among the GFAP proteins of various species; the middle level is among type III intermediate filaments, and the highest is among other classes of intermediate filaments. Conversely, a change to an amino acid normally found in another intermediate filament, such as V115I (van der Knaap et al., 2005), suggests the alteration is not disease causing. De novo mutation of the same amino acid position in another Alexander disease case, even though it produces a different substitution, is a strong evidence that the change in question is disease causing. On this basis, M73R, R79L, and R88S are all presumed to produce Alexander disease. Similarly, a homologous disease causing mutation in another intermediate filament suggests that it is also disease causing for GFAP. The degree of conservation of several GFAP amino acids and the presence of homologous mutations in other intermediate filament diseases are presented in Li et al. (2005). One might also expect that substitution of a closely related amino acid would be less likely to produce disease than a highly different residue, but this has had little predictive value. For example, the highly conservative changes of L76V, D78E, and V87I are all disease causing, whereas D157N and D295N are polymorphisms.

Frequency in the general population: For a suspected mutation in any genetic disease, it has been a standard practice to test at least 100 ethnically matched control chromosomes (50 individuals) to rule out the possibility that the alteration is simply a polymorphism (Cotton and Scriver, 1998). The D295N coding change is clearly a polymorphism by this criterion, as well as several silent coding changes and nucleotide changes in nontranslated regions (Table 24.4). Given the very low frequency of occurrence of Alexander disease, finding a suspected mutation in the control DNAs strongly suggests that it is not responsible for disease.

Independent occurrence in Alexander disease patients: A mutation can be considered disease causing if it is found in two or more independent cases of Alexander disease, at least one of which has a strong MRI- or biopsy/autopsy-based diagnosis. R70W is disease causing by this criterion, since it has appeared in two cases, both with MRIs consistent with adult onset Alexander disease (Salvi et al., 2005; Sreedharan et al., 2007).

Functional effects: The polymerization properties of several mutant GFAPs have been tested by expression in the human adrenal cortex carcinoma-derived cell line SW13vim^{-} (Fig. 24.7) (Hsiao et al., 2005; Li et al., 2005; Perng et al., 2006; unpublished observations). These cells, which have no endogenous cytoplasmic intermediate filaments (Hedberg and Chen, 1986), have proved more sensitive for revealing alterations in polymerization of mutant GFAP than either in vitro polymerization or expression in cells that contain preformed filament networks (Hsiao et al., 2005; Perng et al., 2006). Three GFAPs have been tested that contain coding changes arising de novo in Alexander disease, and thus are almost certainly disease

Table 24.4 Noncoding changes in the *GFAP* gene.

Nucleotide change	Frequency[1]	Reference
-988T > C	55.7% (39/70)	Brenner et al., 2001; Li et al., 2006
-504T > A	20.0% (14/70)	Brenner et al., 2001; Li et al., 2006
-250C > A	51.8% (21/110)	Brenner et al., 2001; Li et al., 2006
c96T > C (G32G)[2]	<1%[3]	Gorospe et al., 2002
c141G > A (P47P)	4/102	Brenner et al., 2001; Li et al., 2005
IVS3-96G > T	5.6% (2/36)	Li et al., 2006
IVS3-12C > T	13.6% (24/176)	Brenner et al., 2001; Li et al., 2005, 2006
IVS3-9C > G	1/26	Brenner et al., 2001
c738G > A (A246A)[4]	<1%[3]	van der Knaap et al., 2006
IVS4 + 826G > A	16.7% (6/36)	Li et al., 2006
c858G > A (R286R)	15% (28/181)	Isaacs et al., 1998; Brenner et al., 2001; Li et al., 2005, 2006
IVS5-100A > C	14.3% (4/28)	Li et al., 2005
IVS6-151C > T	18.8% (6/32)	Li et al., 2006
IVS6–66 C > G	27.5% (39/142)	Brenner et al., 2001; Li et al., 2005, 2006
IVS7 + 459A > G	32.1% (59/184)	Singh et al., 2003; Li et al., 2006
IVS7 + 460C > T	21.1% (35/166)	Singh et al., 2003; Li et al., 2006
IVS7 + 471C > T	10.0% (2/20)	Li et al., 2006
IVS7 + 501C > A	24.0% (12/50)	Li et al., 2006
IVS7-306A > G	57.9% (22/38)	Li et al., 2006
IVS8–86 C > T	25.0% (25/100)	Li et al., 2005, 2006
+16G > A[5]	4.0% (3/74)	Brenner et al., 2001; Li et al., 2006
+28C > G[6]	27% (25/94)	Brenner et al., 2001; Li et al., 2005, 2006
+338C > T	not known	van der Knaap et al., 2006

This table contains silent mutations within the coding region, as well as changes in the introns and the 5′-and 3′-flanking regions and untranslated regions. All are considered harmless polymorphisms; none has been found to arise de novo or been tested for its effects on gene expression.

[1]Frequency includes prevalence in both patients and controls.
[2]Patient also had an R79H mutation.
[3]Single report among all patients sequenced suggests frequency of about 1/200.
[4]Patient also had R88C mutation.
[5]Previously incorrectly reported as + 21C > G (Brenner et al., 2001) or + 21G > A (Li et al., 2006).
[6]Previously incorrectly reported as + 33C > G (Brenner et al., 2001; Li et al., 2005, 2006).

causing, and each demonstrated marked defects in filament formation (E210K, R239C, R416W; see Table 24.3). Conversely, preliminary examination of the D295N polymorphism indicates that it forms normal appearing filaments (Brenner, unpublished observations). On the basis of this criterion, K63Q and A253G were judged to be disease causing (Fig. 24.7) (Li et al., 2005).

The classification of several other coding changes that have not been shown to arise de novo is less certain. This group presently includes P47L, V115I, D157N, E223Q, A244V, R276L, L331P, D360V, and E371G (Table 24.3). Except for

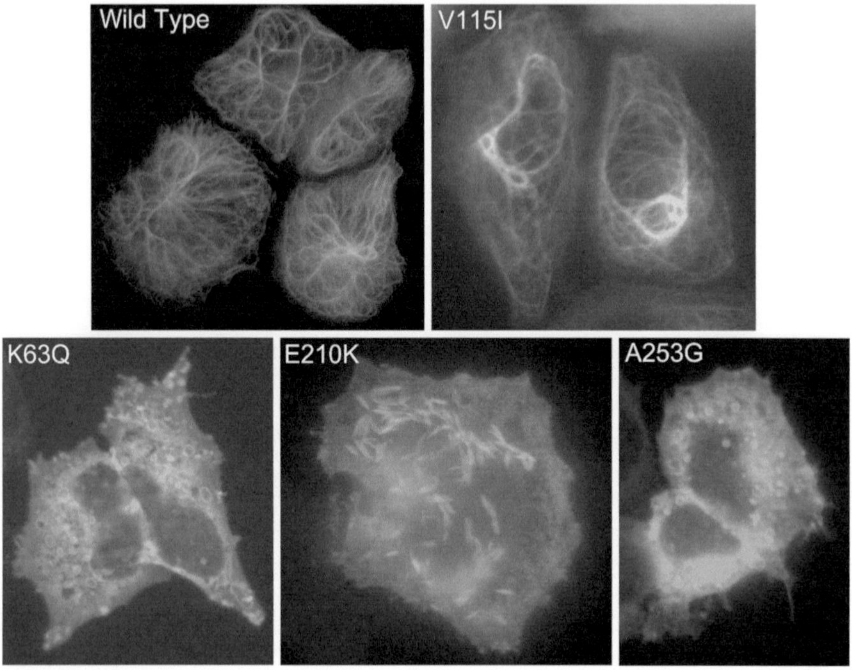

Fig. 24.7 Assembly patterns of wild type and mutant GFAP in transfected cells. SW13vim⁻ cells were transfected with the indicated GFAP expression vectors and immunostained for GFAP 2 days later. Wild type and V115I GFAP form normal appearing filament networks, whereas K63Q and A253G form ring-like aggregates and E210K forms needle-like aggregates. Images for wild type, K63Q, and E210Q are reprinted with permission from Li et al. (2005), "Glial fibrillary acidic protein mutations in infantile, juvenile, and adult forms of Alexander disease," Ann Neurol 57:310–326, copyright by Wiley-Liss, Inc..

D157N, none of these changes has been found in at least 100 control chromosomes, but several were present in a presumably normal parent of a presumptive Alexander disease patient. R276L was judged to be disease causing because it was present in two affected brothers, both of whom had MRIs typical of Alexander disease, and one of whom was autopsy proven (Namekawa et al., 2002) (see Sect. 24.3.8 for a discussion of this family). A244V would seem to qualify for disease-causing status because it was found for two independent presumptive Alexander disease patients (Aoki et al., 2001; Li et al., 2005), one of whom had an MRI typical for Alexander disease. However, it is assigned to the more conservative *probably disease causing* category because the altered protein formed normal appearing filaments when expressed in SW13vim⁻ cells (Li et al., 2005). Both D360V (Ishigaki et al., 2006) and E371G (Kawai et al., 2006) are also classified as probably disease causing. Each was reported for a single case and parents were not tested, but each had an MRI typical for Alexander disease. The roles of P47L, E223Q, and L331P are more equivocal. L331P was found in the mother, daughter, and son of a single family. The 34-year-old mother and 7-year-old daughter were asymptomatic, and the only presentation of the 16-month-old son was megalencephaly, which had an onset at 4 months. The son's MRIs were typical for Alexander disease, with the additional unique feature of increased T1 signal from the mesial aspect of the caudate. MRIs for the sister and mother showed marginally increased T1 signal from the caudate and increased T2 signal from frontal white matter. No alteration of L331 was found among 210 control chromosomes tested, and the amino acid is highly conserved among other intermediate filaments. Despite the extremely mild nature of the disease in this family, L331P is tentatively classified as probably disease causing due to the typical Alexander disease MRI. E223Q was first reported as the sole *GFAP* coding change in a male adult onset Alexander disease patient, but was present also in his mother (Brockmann et al., 2003). The Alexander disease diagnosis of the patient was clouded by his alcoholism and hypertension, the MRIs were atypical for Alexander disease (van der Knaap, personal communication), and although his mother had late onset neurological problems, her symptoms were not suggestive of Alexander disease. Subsequently, E223Q was discovered in an infantile Alexander disease patient together with a Y366H mutation. The Y366H arose de novo, indicating it was disease causing, whereas the E223Q was present in the neurologically normal mother (Li et al., 2005). Preliminary data indicate that E223Q GFAP forms filaments when transfected into SW13vim$^-$ cells, but they tend to be thicker than those formed by wild type GFAP (Brenner, unpublished observations). E223 is highly conserved as either glutamate or aspartate among the *GFAP* genes of other species, the closely related type III proteins vimentin and desmin, and many intermediate filaments of unrelated classes, but is nonconservatively replaced by lysine or serine in several others. No other disease-related intermediate filament mutation occurs at this position. P47L shares many of the descriptors of E223Q. It also has been found in association with other, de novo occurring mutations (twice with R239C, once with R416W) (Brenner et al., 2001; Gorospe et al., 2002; Li et al., 2005), and as the sole *GFAP* alteration in five possible Alexander disease cases, each of which lacks convincing documentation (Brenner, unpublished observations). It is also present in all six parental pairs that have been tested. The mutation occurs in a *GFAP* sequence implicated in polymerization (Ralton et al., 1994), and the P47L protein shows evidence of subtle assembly defects in vitro (Quinlan, unpublished observations). Based primarily on the uncertainty of the clinical and MRI findings for these cases, both P47L and E223Q have tentatively been classified as polymorphisms.

V115I is presumably a polymorphism because the homologous amino acid in both goldfish and zebrafish GFAP is isoleucine rather than valine, the diagnosis of Alexander disease was uncertain (van der Knaap et al., 2005), and the V115I GFAP forms normal appearing filaments when transfected into SW13vim$^-$ cells (Fig. 24.7) (Li et al., 2005). D157N is the only change in this group that has been detected among normal controls (Li et al., 2005). It is thus almost certainly a polymorphism, but bears watching because in a few instances it has been the only *GFAP* gene change observed for a suspected Alexander disease patient (although also present in an unaffected parent), and preliminary results indicate that it forms thicker than normal filaments when transfected into SW13vim$^-$ cells (unpublished observations).

24.3.6 Origin of the Mutations

The mutations of uncertain consequence just described constitute a small minority of the total; most are presumed to be disease causing because they arise de novo. Mutations are typically detected by sequencing PCR fragments generated from genomic DNA. Since all Alexander disease mutations have been heterozygous, they are revealed by a double peak at a given position in the chromatogram. Given this heterozygosity, and the noise usually present in DNA sequencing, it would be unlikely that a mutation would be detected if it were present in fewer than 25% of the cells sampled. This indicates that the *GFAP* mutations arise during the first few cell divisions of embryogenesis, or in the parental germ line. The latter is more likely, given that multiple cell divisions occur in the generation of the gametes. Li et al. (2005) obtained evidence that most mutations indeed arise during gametogenesis by finding that for 24 of 28 patients tested the de novo mutation arose on the paternal chromosome. This ratio of 6:1 is consistent with other studies that have compared the point mutation rate of paternal and maternal chromosomes (Hurst and Ellegren, 1998; Li et al., 2002b). So far there is no evidence for an effect of paternal age, although the number of parents of Alexander disease patients analyzed is small for such an analysis. The gametic origin of most Alexander disease mutations has implications for genetic counseling; parents may test negative for the mutation present in their child, yet produce another affected child. In practice such germ line mosaicism has never been documented for Alexander disease, so the frequency is probably less than 1%. In other genetic diseases, the probability of germ line mosaicism ranges from less than 1 to 30%, depending on the particular disease and mutation (Zlotogora, 1998). Fetal diagnosis is now an option for parents who have previously had an affected child whose mutation has been determined. Although the above data implicate parental gametes for the origin of most Alexander diseases, they can also occur during embryogenesis. A clear example of this is a patient mosaic for a *GFAP* mutation (Brenner, unpublished observations).

24.3.7 Sex Differences in Susceptibility

One might expect equal numbers of males and females to be affected by de novo *GFAP* mutations, given that they arise on a somatic chromosome and nearly all are 100% penetrant. Surprisingly, there is a preponderance of males for both the infantile and the juvenile forms, and of females for the adult onset form (Table 24.2). Of these, only the difference for the juvenile form is statistically significant ($p = 0.005$).

The lower proportion of females in juvenile cases and higher proportion in adult cases could be explained if their diagnosis is delayed relative to males, for example, if they have milder symptoms or progression. The data in Table 24.2 suggest that this may indeed be the case. No significant differences are apparent in clinical signs for male and female infantile Alexander disease patients, but female juvenile patients are much less likely than males to display macrocephaly, seizures, spasticity, or developmental delay. Thus they tend to present like the difficult to diagnosis adult onset patients, whereas the males are more intermediate in their presentations. Gosospe and Maletkovic (2006) previously noted that females display Alexander disease differently from males (although no data nor direction of effect was provided), and pointed out that this could be due to gender differences in GFAP distribution or oligodendrocyte survival, both of which have been documented in lower animals. It is not known whether these differences also occur in humans, and if so, whether they have reached significance by the 2–12-year age period of the juvenile form. Another possibility is that males are more likely to engage in activities that produce head injuries, and even subclinical head trauma might initiate a positive feedback spiral leading to manifestation of disease (Sect. 24.4.1).

24.3.8 Familial Cases

In rare instances *GFAP* mutations are incompletely penetrant, or their effects sufficiently delayed or mild that affected individuals are able to become parents. This can result in families in which multiple individuals are affected (for reviews, see Messing et al., 2001; Messing and Goldman, 2004). Several of these families have now been analyzed for the presence of *GFAP* mutations. The largest of these is a Canadian cohort in which nine individuals from three generations have been affected (Stumpf et al., 2003). The constellation of clinical signs varied among the affected individuals, but shared symptoms included severe constipation, bulbar signs, sleep disturbances (especially apnea), and dysmorphia that included progressive kyphosis, arched palate, and short neck. Many of the affected individuals also experienced pryamidal signs, ataxia, and seizures. The onset of severe constipation was between 5 and 10 years, but the disease was classified as the adult form because the neurological symptoms did not occur until much later. For two individuals from two different generations, the diagnosis of Alexander disease was confirmed by pathology; for four others examined by MRI, the findings were consistent with the late-onset form of the disease. Five affected individuals were genetically tested and each was found heterozygous for a D78E mutation, which was not present in 200 control chromosomes. One of 12 unaffected family members at risk for inheriting the D78E mutation also carried the change. This individual could not be given a neurological exam, but since she was 30-years old and had apparently not experienced the severe constipation common to affected family members, at least some of the effects of the mutation are presumably incompletely penetrant. The presence of the D78E mutation in all five affected family members, and its absence in all but one at risk unaffected members, strongly suggests that it is disease causing ($p < 0.02$). This mutation occurs within a highly conserved segment of the GFAP protein, and mutations are present at adjacent positions 76, 77, and 79 that produce more severe effects. The milder course of D78E could be due to its highly conservative change; it would be of interest to determine if cell transfection would reveal a functional effect.

A V87G mutation was present in a 58-year-old female Japanese patient who had disease onset at 53 years (Okamoto et al., 2002). Two of her children also carried the mutation, an affected 38-year-old daughter who had an onset at 27 years, and a 32-year-old son who had no history of neurological illness but was considered preclinical because he displayed lower extremity hyperreflexia and Babinski signs. MRIs of all three patients showed lesions in the brainstem, cerebellum, and spinal cord that were suggestive of Alexander disease. No other family members were available for testing, including the parents of the proband, who are indicated as being unaffected in the published family tree. Thus it is unclear if the mutation first appeared in the proband, or if its effects are incompletely penetrant. The V87G change was absent in 400 control chromosomes, but the polymerization properties of the mutant GFAP have not been tested.

Two affected brothers in another Japanese family shared an R276L mutation, which was not found in 156 control chromosomes (Namekawa et al., 2002). The elder brother had disease onset at 33 years, and died at 53 years. The younger brother had an onset at 48 years, and was still alive at 50 years. Both had MRIs consistent with adult onset Alexander disease, and the diagnosis was confirmed by pathology for the elder of the two. The presumably neurologically normal parents both died in their seventies, raising the possibility that these two cases resulted from gametic mosaicism. The effect of the mutation on GFAP polymerization has not been tested.

Two other familial cases are remarkably similar except for the nature of the mutation, R88C in one case (van der Knaap et al., 2006) and R416W in the other (Thyagarajan et al., 2004). In both instances the mother acquired the mutation de novo and then passed it on to an affected son. Both mothers were adult onset cases with typical symptoms that included ataxia and bulbar dysfunction. Both sons were more mildly affected than their mothers, although this could be a function of age. The R88C son presented with precocious puberty, but was otherwise normal except for increased arm reflexes and a Babinski sign. The R416W son only displayed a mild spastic paraparesis. These two cases are of particular interest because the probands had adult onset Alexander disease due to mutations that were previously associated with the infantile and juvenile forms. Thus, although both the R88C and R416W mutations have so far proved fully penetrant, the time of onset of disease is highly variable.

Not included in this section are the three reported cases of affected identical twins (Table 24.1), or the family with an L331P mutation that was previously discussed under Sect. 24.3.5 earlier.

24.3.9 *Cases Without GFAP Mutations*

No *GFAP* mutations have been found in about 5% of patients for which pathology or MRIs strongly support the clinical diagnosis of Alexander disease (Table 24.1). Nevertheless, the presence of *GFAP* mutations for these cases remains a distinct possibility. As noted earlier, mutations that occur after the first several cell divisions in an embryo could have a patchy presence and be undetected by the sequencing procedures used. Accordingly, if available it would be of interest to analyze DNA from negative cases that is isolated from biopsy or autopsy material that is rich in Rosenthal fibers. *GFAP* mutations could also affect splicing rather than the coding sequence. We have discovered a case in which an intronic mutation causes low frequency loss of exon 4, resulting in an in-frame deletion of 54 amino acids, and producing a dominant disturbance of filament assembly (unpublished observations). Adequate coverage of the splice donor and acceptor sites is provided by including 10–20 nucleotides of flanking intronic segments in the sequencing of exons, but even this could miss mutations that generate new donor or acceptor sites deeper within the introns. Several GFAP splice variants have also been described in which an intronic region of the standard GFAPα transcript provides a coding region for the variant (reviewed in Quinlan et al., 2007). In particular, a segment of intron 7 is translated as part of GFAPδ, and we have found a patient lacking any mutation in the standard coding region to be a compound heterozygote for mutations in this intron 7 region (unpublished observations). Whether these changes actually have any role in disease remains to be determined. Given that simple overproduction of wild type GFAP can lead to Rosenthal fiber formation, promoter mutations that activate transcription or gene duplication are also candidates for disease. Duplication of the α-synuclein gene results in a familial form of Parkinson's disease (Chartier-Harlin et al., 2004), and duplication of the proteolipid protein gene is a major cause of Pelizaeus-Merzbacher disease (Sistermans et al., 1998).

Other candidate genes for these cases are those encoding αB-crystallin, plectin, and the 51-kD subunit of mitochondrial complex I. The possible role of mitochondrial defects in Alexander disease is discussed below in Sect. 24.4.3. Both αB-crystallin and plectin are of interest because they associate with GFAP, can affect its incorporation into the filament network, modulate the integration of intermediate filaments into the entire cellular cytoskeleton, and are present in Rosenthal fibers (Iwaki et al., 1989; Nicholl and Quinlan, 1994; Perng et al., 1999a; Quinlan, 2002; Tian et al., 2006). Autosomal recessive mutations in the plectin gene result in epidermolysis bullosa simplex, a disorder characterized by hypersensitivity to skin blistering and muscular dystrophy (Smith et al., 1996a), and a similar phenotype is displayed by plectin knockout mice (Andra et al., 1997). Although these clinical findings do not suggest Alexander disease, it remains possible that different kinds of plectin mutations could be involved. For example, a patient with an insertion mutation close to the intermediate filament binding site of plectin displayed a form of epidermolysis bullosa simplex accompanied by brain atrophy and accumulation of desmin, a type III intermediate filament closely related to GFAP (Schroder et al., 2002). Similar to plectin, mutations in αB-crystallin can result in myopathies associated with desmin aggregates (Vicart et al., 1998; Selcen and Engel, 2003; Inagaki et al., 2006). Studies in cultured cells and in vitro have shown that mutant αB-crystallin causes increased intermediate filament bundling (Perng et al., 1999b, 2004), providing an explanation for the formation of the desmin aggregates in the muscles of the affected individuals. Although αB-crystallin is strongly upregulated in reactive astrocytes (Iwaki et al., 1989), none of the individuals with αB-crystallin mutations has been reported to have astrocyte or neurological pathology. Possibly, as just suggested for plectin, a very specific mutation in αB-crystallin could be associated with Alexander disease.

24.4 Disease Mechanisms

Since GFAP is expressed almost exclusively in astrocytes in the CNS, it can be assumed that the primary genetic defect in Alexander disease occurs in these cells. Astrocytes do become highly reactive and are the site of the Rosenthal fibers that are diagnostic for this disorder, but reports of loss of these cells appear variable from case to case, and have not been quantitated. Instead, there is a dramatic deficiency of myelin in the frontal lobes in the infantile form, indicating compromised oligodendrocyte function, and in late onset forms neuronal dysfunction is suggested by the clinical signs, lack of white matter changes, and atrophy of structures in the brainstem, cerebellum, and spinal cord (Salvi et al., 2005; van der Knaap et al., 2005, 2006). Thus although Alexander disease is an astrogliopathy, its clinical and pathological manifestations reveal critical interactions between these cells and oligodendrocytes and neurons. Discovering the mechanism by which Alexander disease wreaks its havoc may provide general insights into CNS development and function.

24.4.1 Cell Culture Studies

Transfection of SW13vim^{-} cells to assess the probable disease relatedness of mutations has been discussed earlier in Sect. 24.3.5. This approach has been extended by the Goldman and Quinlan laboratories to investigate the properties of the mutant GFAPs and the mechanisms by which they compromise astrocyte function. The Goldman group (Hsiao et al., 2005; Tang et al., 2006; Tian et al., 2006) has focused on R239C GFAP, the most common Alexander disease mutation and one which produces a severe course. The Quinlan group (Perng et al., 2006) has also performed some studies with R239C GFAP, but has focused on the R416W mutation. This mutation is also relatively common, but unusual in that it can give rise to all three forms of Alexander disease, and it resides in the nonhelical C-terminal tail domain (Fig. 24.4). As noted earlier, R416 is part of an RDG motif previously shown to have a role in polymerization (Chen and Liem, 1994), and which is conserved among all GFAP species sequenced, as well as the closely related type III intermediate filaments vimentin and desmin.

Surprisingly, R239C formed normal appearing 10-nm filaments when recombinant protein was polymerized in vitro, whereas R416W only formed short rods that had a strong propensity to aggregate (Fig. 24.6). When expressed in SW13vim^{-} cells, both yielded only aggregates or diffuse background staining, and this effect was dominant upon cotransfection with a wild type expression vector. An unexplained discrepancy is that Hisao et al. (2005) found identical behavior between wild type and R239C GFAP when expressed in Cos-7 cells or rat primary astrocytes – primarily incorporation into resident filament networks, but some aggregate formation – whereas R239C mutant GFAP was observed to form aggregates at a significantly higher frequency than wild type GFAP by Tang et al. (2006) for Cos-7 cells, by Tian et al. (2006) for human primary astrocytes, and by Perng et al. (2006) for mouse primary astrocytes. Although Hisao et al. (2005) found the GFAP filaments formed by R239C GFAP in vitro or in transfected primary astrocytes to appear similar to those formed by wild type GFAP, a clear distinction was discovered in their salt extractability from Triton X-100 insoluble pellet material. Nearly all of both wild type and mutant GFAP resided in this pellet, but whereas over half of the wild type GFAP was solubilized from the pellet by 1.0 M KCl, little of the mutant GFAP was solubilized. The mutant form was also dominant in this property; coexpression of the wild type and mutant GFAPs produced solubility properties identical to the mutant alone. Perng et al. (2006) found a similar dominant effect of the R416W mutation on the solubility of mutant and wild type GFAP, but observed that R416W GFAP formed aggregates either when assembled in vitro or when expressed in transfected cells. In addition, using an R416W-specific monoclonal antibody, they observed that expression of R416W GFAP in the human U343MG astrocytoma cell line produced results similar to that of the mouse primary astrocytes, eliminating the possibility that aggregate formation was due to a species incompatibility. The antibody was also used to immunolabel brain from an R416W Alexander disease patient, revealing for the first time for any intermediate filament disease that the mutant protein was present, and that it was incorporated into both the normal appearing filament bundles and into the disease aggregates. Similarly, colocalization of wild type and mutant GFAP in aggregates formed in transfected human astrocytes was demonstrated by Tang et al. (2006) by fusing the wild type GFAP to a green fluorescent protein (GFP) tag and the R239C GFAP to a FLAG tag. Finally, using transfection of the U343MG cell line, Perng et al. (2006) demonstrated colocalization of αB-crystallin, HSP27, and ubiquitin with the aggregates, indicating similarities to authentic Rosenthal fibers. Together, these two studies indicate that mutant GFAP is not completely defective in polymerization; it may be able to form at least short filaments when assembled by itself, and to some extent integrate into normal appearing filament bundles in patient brain. The mutant GFAP does confer abnormal properties to the GFAP filaments, however, and does so in a dominant manner. The increased resistance to salt extraction suggests that one such property is more avid self-adherence. Partial filament formation and increased interfilament adhesion is consistent with the model proposed in Fig. 24.5. The studies also indicate that the polymerization properties of mutant GFAP can be sensitive to the intracellular environment, and recommended transfection of SW13vim^{-} cells as a sensitive method for revealing effects of GFAP mutations.

Tian et al. (2006) used cell transfection to investigate the role of plectin in Rosenthal fiber formation. Immunohistochemistry, coimmunoprecipitation, and overlay immunoblots all revealed a direct interaction between plectin and GFAP, with no difference found between wild type GFAP and the R239C mutant. Studies of cells containing different levels of plectin indicated that this protein assists GFAP filament formation and increases the soluble fraction of GFAP, particularly for the R239C mutant. Expression of the R239C mutant GFAP, but not the wild type, also reduced the total level of plectin. These results suggest that one consequence of mutant GFAP expression in Alexander disease could be decreased plectin levels relative to GFAP, with resultant compromise of the cytoskeleton. Contrary to expectation, the level of plectin was found to actually be increased in Alexander disease brain, but the ratio of plectin to GFAP in astrocytes could not be determined.

As noted earlier, αB-crystallin is upregulated in Alexander disease brain and in cells transfected with GFAP expression vectors. Tang et al. (2006) explored the possibility that this might reflect activation of a stress response pathway. Transfection of several cell lines with GFAP expression vectors, including Cos-7 and U251 human astrocytoma cells, indeed revealed conversion of JNK to its active, p-JNK form, with R239C GFAP producing a greater effect than the wild type. In addition, it was observed that there was significant colocalization of p-JNK with the GFAP aggregates that formed. These observations held for human Alexander disease brain; p-JNK was elevated compared with controls and was found associated with Rosenthal fibers. Interestingly, the reciprocal relationship between increased GFAP levels and p-JNK activation was also present; increasing p-JNK levels through expression of constitutively active forms of the upstream kinases MLK2, MLK3, or ASK1 resulted in increased GFAP levels, whereas cotransfection with a dominant negative mutant of c-Jun decreased GFAP levels. This suggested a positive feedback loop whereby increased GFAP levels stimulate JNK activation, which in turn stimulates a further increase in GFAP.

Activation of JNK is part of the apoptotic signaling pathway in several cell types. Accordingly, Tang et al. (2006) also investigated whether overexpression of wild type or R239C GFAP decreased viability. No lethality was detected in untreated U251 cells expressing either stably transfected wild type or R239C GFAP genes. However, when the cells were treated with the apoptosis-inducing drug camptothecin, expression of R239C, but not of wild type GFAP, resulted in a significant increase in dead cells. This may be related to the greater effect of R239C on JNK activation.

A reciprocal interaction between GFAP overexpression and proteasome activity was also discovered. Transfection of wild type or R239C GFAP into 293T/GFP-U cells resulted in accumulation of the GFP-tagged proteasomal substrate, indicating inhibition of proteasomal activity. This result was confirmed by finding a reduction in the activities of two proteasome-associated proteases, chymotrypsin-like peptidase and postglutamyl peptidase hydrolase. Reciprocally, inhibition of proteasome activity with MG132 increased GFAP levels. There was no difference between the wild type and R239C GFAPs in these proteasomal interactions. Thus in yet another positive feedback loop, high GFAP levels reduce proteasome activity, which in turn increases GFAP accumulation. Still one more was present between the proteasome and JNK; even in the absence of GFAP, proteasome inhibition increased p-JNK levels, and JNK activation inhibited the proteasome.

Previously, the more general positive feedback loop was mentioned by which mutant GFAP production causes astrocyte dysfunction resulting in reactive gliosis and further mutant GFAP production. The study of Tang et al. (2006) adds three more loops that involve specific participants: interactions between GFAP and JNK activation, GFAP and proteasome inhibition, and JNK activation and proteasome inhibition. These multiple and synergistic positive feedback loops raise the possibility that the transition to frank disease could have many entry points, including trauma that causes reactive gliosis or environmental insults that activate the JNK stress pathway. Once the threshold is passed, these positive feedback loops could produce progressive disease.

24.4.2 Mouse Models

24.4.2.1 Overexpression of Wild Type Human GFAP/Gene Array Analysis

The hGFAP-overexpressing transgenic mice inadvertently provided the first Alexander disease mouse model (Messing et al., 1998). As noted earlier, lines expressing the *hGFAP* transgene displayed abundant Rosenthal fibers. The high expressing lines died of unknown cause at about 3 weeks of age and could not be maintained, but lines expressing at a moderate level continue to be studied. The strongest expressing of these, tg73.7, has about 7% mortality by 5 weeks, but the other 93% live a normal life span (Hagemann et al., 2006). They also have about 40% lower weight than wild type littermates when young (Messing, unpublished observations), but otherwise appear normal. GFAP levels are elevated about 4-fold at 2 weeks, and 45-fold in the adult (Messing, unpublished observations). The 4-fold increase at 2 weeks and the relative levels of endogenous and transgenic GFAP transcripts suggest that about 75% of the *GFAP* in this line is attributable to the transgene. Rosenthal fibers are apparent by 2 weeks, and are abundant in the adult. As in the human disease, the Rosenthal fibers are especially prominent in subpial, perivascular, and periventricular locations. No deficits in myelin have been detected, but by 4 months of age morphological changes involving neuronal loss become apparent in the cerebellum and olfactory bulb (Hagemann et al., 2005; unpublished observations).

A nagging question has been whether these hGFAP overexpressing mice form Rosenthal fibers because of a sequence incompatibility between human and mouse GFAP rather than the elevated level of expression. Several observations argue strongly against a problem with sequence differences. The GFAP protein sequence is highly conserved between human and mouse, with 91% identity and 95% similarity (Brenner, 1994). Even closer homology was produced by Takemura et al. (2002), who in the course of a study of the biological role of phosphorylation sites made a knock-in mouse in which the first 154 codons of the endogenous mouse gene were substituted by the corresponding human sequence. Expression of this chimeric *GFAP* gene, now 95% identical to the human sequence, resulted in no abnormalities.

In addition, the tg73 transgene has recently been expressed in mice lacking endogenous GFAP (*GFAP* null mice), and Rosenthal fibers still form (Messing et al., unpublished observations). There is also ample precedent for chronic upregulation of the *GFAP* gene resulting in Rosenthal fiber formation, as in multiple sclerosis (Ogasawara, 1965; Herndon et al., 1970) and pilocytic astrocytomas (Gessaga and Anzil, 1975).

To gain insights into the causes and consequences of Rosenthal fiber formation, Hagemann et al. (2005) used the 12,488 element Affymetrix murine gene chip to compare transcript levels between tg73.7 mice and controls at 3 weeks and 4 months of age. The olfactory bulb was selected for these experiments because it is particularly rich in astrocytes and Rosenthal fibers; at 3 weeks the Rosenthal fibers are just beginning to appear, whereas at 4 months they are plentiful. As expected, fewer transcript differences were observed for the 3-week-old mice tested (202 increased, 34 decreased) than for the 4-month-old mice (802 increased, 789 decreased). Several patterns emerged when affected transcripts were grouped into families and the results from the two time points were compared. Activation of astrocytes, of an oxidation stress response, and of an immune response were all early events. The oxidative stress response included upregulation of multiple genes involved in glutathione metabolism, free radical scavenging, and heavy metal binding, many of which are activated by the transcription factor Nrf2. This stress response was confirmed by enzymatic assay of one of the encoded proteins, and by strong induction of a human placental alkaline phosphatase reporter transgene driven by an antioxidant response element. Marked astrocytic accumulation of iron was also observed, a process that may be controlled by an antioxidant stress response involving Nrf2. These findings are consistent with the presence in Rosenthal fibers of advanced lipid peroxidation end products (Castellani et al., 1998) and advanced glycation end products (Castellani et al., 1997). The pattern of activation of immune response genes indicated that astrocytes were reactive by 3 weeks, followed by microglia at 4 months. Staining for Mac1, a marker of activated microglia, confirmed this inference for microglia. The applicability of these results to the human disease was indicated by demonstrating that Alexander disease brains had increases in mRNAs for GFAP, ceruloplasmin (an acute phase ferroxidase), oxidoreductase NQO1 (activated by oxidative stress), and the CD11b subunit of Mac1.

Many of the genes downregulated at 4 months could be attributed to neurons. These included genes contributing to the cytoskeleton, vesicle trafficking, neurotransmitter synthesis, neurotransmitter receptors, developmental transcription factors, Ca^{++} regulation, and ion channels. The particular genes downregulated, such as glutamic acid decarboxylase (GAD) 65, GAD 67, and γ-amino butyric acid (GABA) receptors, suggested that GABA-ergic interneurons are particularly affected, whereas sensory and projection neurons are spared. Consistent with the absence of myelin pathology in these mice, changes in myelin-associated transcripts were minor and varied.

Distillation of the huge amount of data produced suggests that oxidative damage, perhaps exacerbated by the accumulation of iron, is a possible factor in Alexander disease progression. As might be expected for the primary genetic defect being in astrocytes, astrocyte reactivity is an early event, followed by microglial activation and neuronal changes. The sensitivity of inhibitory GABA-ergic interneurons could perhaps contribute to the seizure susceptibility in Alexander disease.

24.4.2.2 Expression of Alexander Disease Mutant GFAP

The hGFAP overexpressing mice prompted the discovery of *GFAP* as the target gene of Alexander disease, but were a questionable model for the disorder because they express wild type rather than mutant GFAP, do so at elevated levels, and show no obvious myelin deficits. With the objective of producing a more appropriate animal model, both Hagemann et al. (2006) and Tanaka et al. (2007) made mice that expressed mutant forms of GFAP. Hagemann et al. (2006) made knock-in mice in which the endogenous mouse gene was engineered to encode the equivalent of either R79H or R239H (R76H and R236H in mouse GFAP), two of the more common and severe Alexander disease mutations. The heterozygous knock-in/wild type mice weighed about 10% less than their wild type littermates, but disappointingly, had a normal life span, showed no functional or behavioral deficits in the SHIRPA test panel at 3 months of age, and showed no myelin deficits based on immunohistochemistry and immunoblotting for myelin basic protein and physical measurement of the anterior commissure. However, on further probing the mice did display several characteristics of Alexander disease. Rosenthal fibers were indeed present, appearing by day 7 and increasing in abundance with age. As in the human disease, they were most numerous in subpial, periventricular, and perivascular locations; and reminiscent of the later onset forms of Alexander disease, they were more prevalent in caudal than rostral brain regions. Consistent with the findings for the hGFAP overexpressing mice and for Alexander disease brain, the knock-in mice also showed robust activation of an antioxidant stress response as revealed by the human placental alkaline phosphatase reporter transgene, and accumulated high levels of ferric iron in astrocytes. The two knock-in mutations produced similar patterns of Rosenthal fiber deposition and antioxidant response, with the R236H mutation being somewhat more severe. Seizures are a common feature of Alexander disease, and although the knock-in mice did not display these spontaneously, upon kainic acid treatment they had longer lasting seizure activity compared with the wild type, and showed greater neuronal cell death in the hippocampus. Both the hGFAP overexpressing mice and the knock-in mice have essentially normal life spans, but when the human transgene was crossed into the knock-in mice, there was complete lethality by 35 days. This synergistic effect is consistent with the general model proposed in Fig. 24.5.

These results provided formal proof that heterozygous mutant GFAP leads to Rosenthal fiber formation. They also further suggest possible roles for oxidative stress and iron accumulation, and may be useful models for determining the cause of the seizure activity associated with Alexander disease. The caudal distribution of Rosenthal fibers and absence of myelin deficits suggest that these mice may model a later onset form of Alexander disease.

Tanaka et al. (2007) produced an overlapping set of observations by expressing human R239H GFAP cDNA under control of a mouse *GFAP* promoter. Using cre/lox technology, they produced lines with different numbers of *hGFAP* transgene inserts at the same chromosomal location, eliminating integration site as a confounding variable. A line with a single insert, 60TS, produced only about a 3% increase in total GFAP, and never formed Rosenthal fibers. This result shows that the presence of mutant GFAP per se, albeit at a very low level, does not necessarily lead to Rosenthal fiber formation. A line estimated to have 2 or 3 inserts at the same site, 60TM, did produce Rosenthal fibers. It had an increase in total GFAP of about 20% at postnatal day 7 (P7), prior to Rosenthal fiber detection, and of about 30% at P14, when Rosenthal fibers were first detected. These data suggest that mutant GFAP contributes directly to Rosenthal fiber formation, rather than doing so indirectly by increasing GFAP levels. By comparison, at P14 the wild type hGFAP expressing line tg73.4 had 4 times more GFAP than nontransgenic controls, but only extremely sparse Rosenthal fibers (Messing et al., 1998).

In the 60TM line, both the aggregates and normal appearing filament bundles were stained by an antibody specific for human GFAP, consistent with previous findings in cell culture systems and patient brain that mutant and wild type GFAP copolymerize and coaggregate (see earlier Sect. 24.4.1). GFAP staining also revealed the GFAP fiber bundles to be highly fragmented in the Rosenthal fiber-containing cells. Nevertheless, astrocytes filled with Lucifer Yellow displayed the same overall morphology as nontransgenic cells. The 60TM mice had normal fertility, gross anatomy, and myelination; however, like the knock-in mice, they had much greater sensitivity to kainic acid-induced seizures.

Although it is disappointing that none of the mouse models shows myelin defects, other attributes hold promise for investigation of the mechanisms by which *GFAP* mutations produce fatal consequences. These include activation of an oxidative stress response, accumulation of iron, decrements in neuronal markers, sensitivity to kainic acid, and, under certain circumstances, early death.

24.4.3 Role of Mitochondria

Mitochondrial defects are associated with many neurodegenerative disorders (Kwong et al., 2006), and several authors have marshaled evidence that they also have an important role in Alexander disease (Johnson and Brenner, 2003; Nobuhara et al., 2004; Caceres-Marzal et al., 2006). Observations cited include the presence of an oxidative stress response in the hGFAP-overexpressing mice and human Alexander disease brain (Hagemann et al., 2005), advanced lipid peroxidation end products in Rosenthal fibers (Castellani et al., 1998), findings of elevated lactate in serum and CSF of several patients (Gingold et al., 1999; Probst et al., 2003; Li et al., 2005; Caceres-Marzal et al., 2006), biochemical and genetic suggestions of mitochondrial defects (Nobuhara et al., 2004; Caceres-Marzal et al., 2006), and the clinical similarity between infantile Alexander disease and the mitochondrial disorder Leigh syndrome (MIM 2560000). Data from the mouse models and examination of human brain indicate that the oxidative stress response and the presence of lipid peroxidation end products in Rosenthal fibers are general features of Alexander disease; however, it is unclear that these conditions are due to a general defect in mitochondria. An elevated level of lactate has been reported for several Alexander disease patients, but normal values are found for the majority of cases. The clinical presentation of Leigh syndrome, a recessive disorder due to defects in oxidative phosphorylation, can indeed be remarkably similar to those of infantile Alexander disease – patients have an infantile onset of progressive symptoms that can include macrocephaly, vomiting, motor/mental retardation, visual problems, seizures, spasticity, and cystic leukodystrophy (Schuelke et al., 1999). However, the presence of Rosenthal fibers is not part of the description of Leigh syndrome, and the demyelinating lesions are primarily present in the midbrain, brainstem, cerebellum, and spinal cord rather than the frontal lobes. Thus, defects in mitochondrial energy production do not appear to produce Alexander disease, although they may masquerade clinically as this disorder. An illustrative example is an infantile patient who was originally diagnosed to have Leigh syndrome based on clinical signs and high lactate levels, but on autopsy was revealed to have Alexander disease (Gingold et al., 1999), and was subsequently found to harbor the common R239H mutation (Li et al., 2005).

Nevertheless, the presence of elevated lactate in several Alexander disease patients suggests that Alexander disease may contribute to mitochondrial dysfunction. Such a scenario is observed in several other intermediate filament diseases, including myopathies due to mutations in desmin, a closely related type III intermediate filament protein (Toivola et al., 2005). This possibility in Alexander disease is suggested by reports that mitochondria in the region of Rosenthal fibers have an abnormally dense matrix and that some appear to even be engulfed within the Rosenthal fibers (Herndon et al., 1970), or that the mitochondria are unusually numerous and enlarged (Escourolle et al., 1979). The observations in these single reports bear further study. If Alexander disease indeed compromises mitochondrial function, this could set in motion yet another positive feedback loop that exacerbates the disease progression.

Patients could also have an independent mitochondrial defect that synergizes with the effects of a *GFAP* mutation. Two reports invoke this latter possibility. Nobuhara et al. (2004) describe a patient with juvenile onset Alexander disease who had a de novo R88C mutation and also a rare polymorphism in her mitochondrial DNA that had previously been detected in a patient with a mitochondrial myopathy. However, several considerations suggest that the mitochondrial DNA change may not contribute to disease in this patient: the same alteration was present in the patient's mother, who was neurologically normal; the patient had only slightly elevated blood and CSF pyruvate levels and her lactate levels were presumably normal (they were not mentioned); and the clinical course for this patient, with onset in her elementary school years and her present survival at age 29, is actually less severe than usual for an R88C mutation. In the other report, Caceres-Marzal et al. (2006) describe a patient who had a de novo N386I GFAP mutation accompanied by elevated serum and CSF lactate levels. Remarkably, some of the muscle mitochondria of this patient were dysmorphic and their cytochrome c oxidase activity somewhat depressed, seeming to rule out a secondary effect of the *GFAP* mutation on mitochondrial function. No data were given for the mother, who would be expected to carry the same mitochondrial defect. The N386I coding change is presumably required for disease expression, since it arose de novo. The contribution of the mitochondrial dysfunction is difficult to evaluate, because this is the only instance of an N386I mutation; however, the disease course was severe, with onset at 5 months and death at 22 months. In summary, it appears unlikely that a mitochondrial defect can cause Alexander disease, but since oxidative stress is present in this disorder, it could contribute to clinical progression, whether it occurs independently or as a consequence of the disease.

24.5 Treatment

Treatment of Alexander disease has been confined to alleviating its symptoms, such as providing drugs to combat seizures or vomiting, antibiotics for infections, and feeding tubes for nourishment. Recently, Ishigaki et al. (2006) reported partial success in treating a 9-year-old Alexander disease child with thyrotropin releasing hormone (TRH). This therapy was based on several reports in the 1980s that TRH treatment could mitigate spinocerebellar deficits, including ataxia, although the mechanism by which this might occur is unclear. Improvements were noted in the patient's mental state, speech, frequency of vomiting, ataxia, and sleep apnea, but the effectiveness for some of these symptoms diminished over time. No change was seen in the electroencephalogram or MRI. It is of considerable interest whether TRH will be therapeutic for other Alexander disease patients, whether its benefits can be sustained, and whether there will be side effects of long-term treatment. Other, mechanism-based, therapies are being pursued. In one approach small interfering RNAs are being investigated for their ability to specifically prevent synthesis of the mutant GFAP, leaving the wild type protein unaffected (Daniel Bonthius, personal communication). The challenge with this method will be to deliver the interfering RNAs to sufficient numbers of astrocytes for sufficient periods of time to be effective. Another approach underway is to discover a drug that inhibits GFAP synthesis (Messing, unpublished experiments). This would reduce the level of wild type as well as mutant GFAP, but mouse studies suggest that even the complete absence of wild type GFAP will be of little consequence to normal function (Pekny et al., 1995; McCall et al., 1996; Shibuki et al., 1996), while reducing the total GFAP load could be beneficial to Alexander disease (see earlier Sect. 24.3.3). Other therapeutic targets are being revealed as research into the disease mechanism continues. For example, cell culture studies suggest that αB-crystallin is capable of dissolving GFAP-containing aggregates in cultured astrocytes (Koyama and Goldman, 1999). Other possible therapeutic targets could include inhibiting the JNK signaling pathway or increasing proteasome activity.

24.6 Future Directions

In the short time since *GFAP* mutations were shown to be responsible for most cases of Alexander disease, remarkable progress has been made in extending its clinical diagnosis and probing the biological consequences of the mutations. Yet still elusive are both a cure for the disease and an understanding of the mechanisms producing its dire consequences. The primary target of the *GFAP* mutations is the astrocyte, but the clinical consequences are disruption of oligodendrocyte and neuronal function. It is not even known whether these consequences result from errors of omission – astrocytes failing to perform a vital function, or of commission – astrocytes producing a toxic effect. There are plentiful candidates for both possibilities among the extensive repertoire of astrocyte activities that are described in other chapters in this volume. A plausible error of omission is a defect in glutamate uptake. Astrocytes play a critical role in removing synaptically released glutamate, and the stress pathways upregulated by the expression of mutant GFAP have been found to cause a significant decrease in the glial l-glutamate transporter (Glt-1) transcripts and protein levels in these cells (Tian et al., submitted for publication). Elevated extracellular glutamate levels could explain the seizures commonly observed in infantile cases, and *GFAP* mutant mice are hypersensitive to induction of seizures by kainic acid (Sect. 24.4.2.2). Glt-1 knockout mice have already demonstrated that defective glutamate transport by astrocytes can lead to seizure activity (Tanaka et al., 1997). In addition to promoting seizures, increased glutamate can be lethal to both neurons and oligodendrocytes (Matute et al., 2002; Johnston, 2005). In particular, the hypersensitivity of oligodendrocyte precursors to glutamate (Matute et al., 2002; Johnston, 2005) could explain the absence of myelination in early onset infantile patients. A candidate for an error of commission is the chronic release from reactive astrocytes of tumor necrosis factor α (TNFα), a cytokine that is toxic to oligodendrocytes (Ledeen and Chakraborty, 1998). Interestingly, TNFα toxicity is exacerbated by iron (Zhang et al., 2005), whose accumulation is also dysregulated in the Alexander disease knock-in mice (Hagemann et al., 2005). Investigations of these and other candidates are presently underway both to understand the disease mechanism and to develop therapeutic targets.

Another subject of considerable interest is the role of *GFAP* coding changes that are incompletely penetrant, or of uncertain consequence. In Sect. 24.3.5, E223Q, D157N, and P47L were tentatively classified as polymorphisms, in large part because each has always been found in a seemingly normal parent as well as the affected patient. However, incomplete penetrance, as apparently occurs for D78E, V87G, and L331P (Table 24.1), is a distinct possibility. The presence of a cadre of *GFAP* mutations with incomplete penetrance could result in the incidence of Alexander disease being much greater than that expected were it due primarily to de novo mutations. It is also possible that these coding changes, in conjunction with other genetic or environmental factors, could lead to disorders not clearly recognized as Alexander disease.

24.7 Concluding Remarks

Alexander disease is a rare disorder. Extrapolating from the human mutation rate, an incidence of 1/20,000 or less is expected for de novo cases (Nachman and Crowell, 2000). However, it has devastating consequences that illuminate the critical importance of interactions of astrocytes with oligodendrocytes and neurons. Its study may also provide insights into more common genetic disorders that feature protein aggregates similar to Rosenthal fibers. For example, desmin mutations produce myopathies accompanied by protein aggregates containing desmin, αB-crystallin, and ubiquitin (Goebel, 2003); except for the substitution of keratin for desmin, the same constituents are found in the Mallory bodies that form in the liver as the result of mutations in this intermediate filament protein or toxic injury (Lowe et al., 1992); and Lewy bodies, the protein aggregates present in Parkinson's disease, contain neurofilaments, αB-crystallin, and ubiquitin in addition to α-synuclein (Pappolla, 1986; Lowe et al., 1988, 1992). These similarities in composition suggest similarities in disease pathways.

Surprisingly, despite the central role of astrocytes in Alexander disease, the disorder has been described clinically as a leukodystrophy, a white matter disease. This raises the possibility that other *astrogliopathies* may be masquerading as neuronal or myelin disorders. One recent example that likely fits this category is megalencephalic leukoencephalopathy with subcortical cysts, resulting from mutations in the *MLC1* gene (Leegwater et al., 2001). The product of this gene is a protein of unknown function that is primarily localized to astrocytic end-feet (Schmitt et al., 2003; Boor et al., 2005).

From the first description of Alexander disease in 1949, astrocytes have been viewed as the primary site of pathology, and the recent identification of mutations in *GFAP* as the causative factor confirms this perspective at the molecular level. How single amino acid changes in a cytoskeletal protein translate into catastrophe for neurons and oligodendrocytes remains unknown. With genetic diagnosis as an anchor, neurologists can now broaden their scope for interpreting clinical signs and MRI changes, and appreciate the diversity of presentations for this disease. The wealth of information now emerging on the intracellular pathways impacted by expression of mutant GFAPs holds promise for understanding the disease mechanism and presenting therapeutic strategies. Although dominant gain of function disorders are a challenge for any type of therapy, having a single gene target and common features with other neurodegenerative and protein aggregation disorders offers multiple options for progress, and hope for the future.

Acknowledgments We thank Daniel M. Bolt for statistical assistance, Marjo van der Knaap for permission to use Fig. 24.1, Anne B. Johnson for permission to use Fig. 24.2, and Rong Li for the images in Fig. 24.7. The authors would also like to thank the editors for giving them the opportunity to write this extensive, and final, review of their careers. One more would be two too many. This work was supported by NINDS grant P01NS42803.

References

Alexander WS (1949) Progressive fibrinoid degeneration of fibrillary astrocytes associated with mental retardation in a hydrocephalic infant. *Brain* 72:373–381.

Andra K, Lassmann H, Bittner R, Shorny S, Fassler R, Propst F, Wiche G (1997) Targeted inactivation of plectin reveals essential function in maintaining the integrity of skin, muscle, and heart cytoarchitecture. *Genes Dev* 11:3143–3156.

Aoki Y, Haginoya K, Munakata M, Yokoyama H, Nishio T, Togashi N, Ito T, Suzuki Y, Kure S, Iinuma K, Brenner M, Matsubara Y (2001) A novel mutation in glial fibrillary acidic protein gene in a patient with Alexander disease. *Neurosci Lett* 312:71–74.

Asahina N, Okamoto T, Sudo A, Kanazawa N, Tsujino S, Saitoh S (2006) An infantile-juvenile form of Alexander disease caused by a R79H mutation in GFAP. *Brain Dev* 28:131–133.

Bar H, Mucke N, Kostareva A, Sjoberg G, Aebi U, Herrmann H (2005) Severe muscle disease-causing desmin mutations interfere with in vitro filament assembly at distinct stages. *Proc Natl Acad Sci USA* 102:15099–15104.

Bar H, Kostareva A, Sjoberg G, Sejersen T, Katus HA, Herrmann H (2006a) Forced expression of desmin and desmin mutants in cultured cells: impact of myopathic missense mutations in the central coiled-coil domain on network formation. *Exp Cell Res* 312:1554–1565.

Bar H, Mucke N, Ringler P, Muller SA, Kreplak L, Katus HA, Aebi U, Herrmann H (2006b) Impact of disease mutations on the desmin filament assembly process. *J Mol Biol* 360:1031–1042.

Bassuk AG, Joshi A, Burton BK, Larsen MB, Burrowes DM, Stack C (2003) Alexander disease with serial MRS and a new mutation in the glial fibrillary acidic protein gene. *Neurology* 61:1014–1015.

Becker LE, Teixeira F (1988) Alexander's disease. In: The biochemical pathology of astrocytes (Norenberg MD, Hertz L, Schousboe A, eds), pp 179–190. New York: Alan R. Liss.

Bongcam-Rudloff E, Nister M, Betsholtz C, Wang JL, Stenman G, Huebner K, Croce CM, Westermark B (1991) Human glial fibrillary acidic protein: complementary DNA cloning, chromosome localization, and messenger RNA expression in human glioma cell lines of various phenotypes. *Cancer Res* 51:1553–1560.

Boor PK, de Groot K, Waisfisz Q, Kamphorst W, Oudejans CB, Powers JM, Pronk JC, Scheper GC, van der Knaap MS (2005) MLC1: a novel protein in distal astroglial processes. *J Neuropathol Exp Neurol* 64:412–419.

Borrett D, Becker LE (1985) Alexander's disease. A disease of astrocytes. *Brain* 108 (Part 2):367–385.

Brenner M (1994) Structure and transcriptional regulation of the GFAP gene. *Brain Pathol* 4:245–257.

Brenner M, Lampel K, Nakatani Y, Mill J, Banner C, Mearow K, Dohadwala M, Lipsky R, Freese E (1990) Characterization of human cDNA and genomic clones for glial fibrillary acidic protein. *Brain Res Mol Brain Res* 7:277–286.

Brenner M, Kisseberth WC, Su Y, Besnard F, Messing A (1994) GFAP promoter directs astrocyte-specific expression in transgenic mice. *J Neurosci* 14:1030–1037.

Brenner M, Johnson AB, Boespflug-Tanguy O, Rodriguez D, Goldman JE, Messing A (2001) Mutations in GFAP, encoding glial fibrillary acidic protein, are associated with Alexander disease. *Nat Genet* 27:117–120.

Brockmann K, Meins M, Taubert A, Trappe R, Grond M, Hanefeld F (2003) A novel GFAP mutation and disseminated white matter lesions: adult alexander disease? *Eur Neurol* 50:100–105.

Caceres-Marzal C, Vaquerizo J, Galan E, Fernandez S (2006) Early mitochondrial dysfunction in an infant with Alexander disease. *Pediatr Neurol* 35:293–296.

Castellani RJ, Perry G, Harris PL, Monnier VM, Cohen ML, Smith MA (1997) Advanced glycation modification of Rosenthal fibers in patients with Alexander disease. *Neurosci Lett* 231:79–82.

Castellani RJ, Perry G, Harris PL, Cohen ML, Sayre LM, Salomon RG, Smith MA (1998) Advanced lipid peroxidation end-products in Alexander's disease. *Brain Res* 787:15–18.
Chartier-Harlin MC, Kachergus J, Roumier C, Mouroux V, Douay X, Lincoln S, Levecque C, Larvor L, Andrieux J, Hulihan M, Waucquier N, Defebvre L, Amouyel P, Farrer M, Destee A (2004) Alpha-synuclein locus duplication as a cause of familial Parkinson's disease. *Lancet* 364:1167–1169.
Chen WJ, Liem RK (1994) The endless story of the glial fibrillary acidic protein. *J Cell Sci* 107:2299–2311.
Cooper DN, Youssoufian H (1988) The CpG dinucleotide and human genetic disease. *Hum Genet* 78:151–155.
Cotton RG, Scriver CR (1998) Proof of "disease causing" mutation. *Hum Mutat* 12:1–3.
Crome L (1953) Megalencephaly associated with hyaline pan-neuropathy. *Brain* 76:215–228.
Der Perng M, Su M, Wen SF, Li R, Gibbon T, Prescott AR, Brenner M, Quinlan RA (2006) The Alexander disease-causing glial fibrillary acidic protein mutant, R416W, accumulates into Rosenthal fibers by a pathway that involves filament aggregation and the association of alpha B-crystallin and HSP27. *Am J Hum Genet* 79:197–213.
Dinopoulos A, Gorospe JR, Egelhoff JC, Cecil KM, Nicolaidou P, Morehart P, DeGrauw T (2006) Discrepancy between neuroimaging findings and clinical phenotype in Alexander disease. *AJNR Am J Neuroradiol* 27:2088–2092.
Duckett S, Schwartzman RJ, Osterholm J, Rorke LB, Friedman D, McLellan TL (1992) Biopsy diagnosis of familial Alexander's disease. *Pediatr Neurosurg* 18:134–138.
Ellisdon AM, Pearce MC, Bottomley SP (2007) Mechanisms of ataxin-3 misfolding and fibril formation: kinetic analysis of a disease-associated polyglutamine protein. *J Mol Biol* 368:595–605.
Eng LF, Ghirnikar RS (1994) GFAP and astrogliosis. *Brain Pathol* 4:229–237.
Eng LF, Lee YL, Kwan H, Brenner M, Messing A (1998) Astrocytes cultured from transgenic mice carrying the added human glial fibrillary acidic protein gene contain Rosenthal fibers. *J Neurosci Res* 53:353–360.
Escourolle R, de Baecque C, Gray F, Baumann N, Hauw JJ (1979) [Electron microscopic and neurochemical study of Alexander's disease (author's transl)]. *Acta Neuropathol* 45:133–140.
Farrell K, Chuang S, Becker LE (1984) Computed tomography in Alexander's disease. *Ann Neurol* 15:605–607.
Friede RL (1964) Alexander's disease. *Arch Neurol* 11:414–422.
Friedman JH, Ambler M (1992) Progressive parkinsonism associated with Rosenthal fibers: senile-onset Alexander's disease? *Neurology* 42:1733–1735.
Fuchs E (1996) The cytoskeleton and disease: genetic disorders of intermediate filaments. *Annu Rev Genet* 30:197–231.
Fuchs E, Cleveland DW (1998) A structural scaffolding of intermediate filaments in health and disease. *Science* 279:514–519.
Gessaga EC, Anzil AP (1975) Rod-shaped filamentous inclusions and other ultrastructural features in a cerebellar astrocytoma. *Acta Neuropathol (Berl)* 33:119–127.
Giasson BI, Uryu K, Trojanowski JQ, Lee VM (1999) Mutant and wild type human alpha-synucleins assemble into elongated filaments with distinct morphologies in vitro. *J Biol Chem* 274:7619–7622.
Gingold MK, Bodensteiner JB, Schochet SS, Jaynes M (1999) Alexander's disease: unique presentation. *J Child Neurol* 14:325–329.
Goebel HH (2003) Congenital myopathies at their molecular dawning. *Muscle Nerve* 27:527–548.
Goldman JE, Corbin E (1988) Isolation of a major protein component of Rosenthal fibers. *Am J Pathol* 130:569–578.
Goldman JE, Corbin E (1991) Rosenthal fibers contain ubiquitinated alpha B-crystallin. *Am J Pathol* 139:933–938.
Gomi H, Yokoyama T, Fujimoto K, Ikeda T, Katoh A, Itoh T, Itohara S (1995) Mice devoid of the glial fibrillary acidic protein develop normally and are susceptible to scrapie prions. *Neuron* 14:29–41.
Gordon N (2003) Alexander disease. *Eur J Paediatr Neurol* 7:395–399.

Gorospe JR, Maletkovic J (2006) Alexander disease and megalencephalic leukoencephalopathy with subcortical cysts: leukodystrophies arising from astrocyte dysfunction. *Ment Retard Dev Disabil Res Rev* 12:113–122.

Gorospe JR, Naidu S, Johnson AB, Puri V, Raymond GV, Jenkins SD, Pedersen RC, Lewis D, Knowles P, Fernandez R, De Vivo D, van der Knaap MS, Messing A, Brenner M, Hoffman EP (2002) Molecular findings in symptomatic and pre-symptomatic Alexander disease patients. *Neurology* 58:1494–1500.

Hagemann TL, Gaeta SA, Smith MA, Johnson DA, Johnson JA, Messing A (2005) Gene expression analysis in mice with elevated glial fibrillary acidic protein and Rosenthal fibers reveals a stress response followed by glial activation and neuronal dysfunction. *Hum Mol Genet* 14:2443–2458.

Hagemann TL, Connor JX, Messing A (2006) Alexander disease-associated glial fibrillary acidic protein mutations in mice induce Rosenthal fiber formation and a white matter stress response. *J Neurosci* 26:11162–11173.

Hedberg KK, Chen LB (1986) Absence of intermediate filaments in a human adrenal cortex carcinoma-derived cell line. *Exp Cell Res* 163:509–517.

Herndon RM, Rubinstein LJ, Freeman JM, Mathieson G (1970) Light and electron microscopic observations on Rosenthal fibers in Alexander's disease and in multiple sclerosis. *J Neuropathol Exp Neurol* 29:524–551.

Herrmann H, Strelkov SV, Feja B, Rogers KR, Brettel M, Lustig A, Haner M, Parry DA, Steinert PM, Burkhard P, Aebi U (2000) The intermediate filament protein consensus motif of helix 2B: its atomic structure and contribution to assembly. *J Mol Biol* 298:817–832.

Hsiao VC, Tian R, Long H, Der Perng M, Brenner M, Quinlan RA, Goldman JE (2005) Alexander-disease mutation of GFAP causes filament disorganization and decreased solubility of GFAP. *J Cell Sci* 118:2057–2065.

Hurst LD, Ellegren H (1998) Sex biases in the mutation rate. *Trends Genet* 14:446–452.

Imamura A, Orii KE, Mizuno S, Hoshi H, Kondo T (2002) MR imaging and 1H-MR spectroscopy in a case of juvenile Alexander disease. *Brain Dev* 24:723–726.

Inagaki N, Hayashi T, Arimura T, Koga Y, Takahashi M, Shibata H, Teraoka K, Chikamori T, Yamashina A, Kimura A (2006) Alpha B-crystallin mutation in dilated cardiomyopathy. *Biochem Biophys Res Commun* 342:379–386.

Isaacs A, Baker M, Wavrant-De Vrieze F, Hutton M (1998) Determination of the gene structure of human GFAP and absence of coding region mutations associated with frontotemporal dementia with parkinsonism linked to chromosome 17. *Genomics* 51:152–154.

Ishigaki K, Ito Y, Sawaishi Y, Kodaira K, Funatsuka M, Hattori N, Nakano K, Saito K, Osawa M (2006) TRH therapy in a patient with juvenile Alexander disease. *Brain Dev* 28:663–667.

Iwaki T, Kume-Iwaki A, Liem RK, Goldman JE (1989) Alpha B-crystallin is expressed in nonlenticular tissues and accumulates in Alexander's disease brain. *Cell* 57:71–78.

Iwaki T, Wisniewski T, Iwaki A, Corbin E, Tomokane N, Tateishi J, Goldman JE (1992) Accumulation of alpha B-crystallin in central nervous system glia and neurons in pathologic conditions. *Am J Pathol* 140:345–356.

Iwaki T, Iwaki A, Tateishi J, Sakaki Y, Goldman JE (1993) Alpha B-crystallin and 27-kd heat shock protein are regulated by stress conditions in the central nervous system and accumulate in Rosenthal fibers. *Am J Pathol* 143:487–495.

Jacob J, Robertson NJ, Hilton DA (2003) The clinicopathological spectrum of Rosenthal fibre encephalopathy and Alexander's disease: a case report and review of the literature. *J Neurol Neurosurg Psychiatry* 74:807–810.

Johnson AB (1996) Alexander disease. In: Handbook of clinical neurology, Moser, H.W. (ed.) pp 701–710. Amsterdam: Elsevier Science B.V.

Johnson AB, Bettica A (1989) On-grid immunogold labeling of glial intermediate filaments in epoxy-embedded tissue. *Am J Anat* 185:335–341.

Johnson AB, Brenner M (2003) Alexander's disease: clinical, pathologic, and genetic features. *J Child Neurol* 18:625–632.

Johnston MV (2005) Excitotoxicity in perinatal brain injury. *Brain Pathol* 15:234–240.

Kawai M, Sakai N, Miyake S, Tsukamoto H, Akagi M, Inui K, Mushiake S, Taniike M, Ozono K (2006) Novel mutation of gene coding for glial fibrillary acidic protein in a Japanese patient with Alexander disease. *Brain Dev* 28:60–62.

Kinoshita T, Imaizumi T, Miura Y, Fujimoto H, Ayabe M, Shoji H, Okamoto Y, Takashima H, Osame M, Nakagawa M (2003) A case of adult-onset Alexander disease with Arg416Trp human glial fibrillary acidic protein gene mutation. *Neurosci Lett* 350:169–172.

Klein EA, Anzil AP (1994) Prominent white matter cavitation in an infant with Alexander's disease. *Clin Neuropathol* 13:31–38.

Koyama S, Arawaka S, Chang-Hong R, Wada M, Kawanami T, Kurita K, Kato M, Nagai M, Aoki M, Itoyama Y, Sobue G, Chan PH, Kato T (2006) Alteration of familial ALS-linked mutant SOD1 solubility with disease progression: its modulation by the proteasome and Hsp70. *Biochem Biophys Res Commun* 343:719–730.

Koyama Y, Goldman JE (1999) Formation of GFAP cytoplasmic inclusions in astrocytes and their disaggregation by alphaB-crystallin. *Am J Pathol* 154:1563–1572.

Kress Y, Gaskin F, Horoupian DS, Brosnan C (1981) Nickel induction of Rosenthal fibers in rat brain. *Brain Res* 210:419–425.

Kwong JQ, Beal MF, Manfredi G (2006) The role of mitochondria in inherited neurodegenerative diseases. *J Neurochem* 97:1659–1675.

Kyllerman M, Rosengren L, Wiklund LM, Holmberg E (2005) Increased levels of GFAP in the cerebrospinal fluid in three subtypes of genetically confirmed Alexander disease. *Neuropediatrics* 36:319–323.

Ledeen RW, Chakraborty G (1998) Cytokines, signal transduction, and inflammatory demyelination: review and hypothesis. *Neurochem Res* 23:277–289.

Lee JM, Kim AS, Lee SJ, Cho SM, Lee DS, Choi SM, Kim DK, Ki CS, Kim JW (2006) A case of infantile Alexander disease accompanied by infantile spasms diagnosed by DNA analysis. *J Korean Med Sci* 21:954–957.

Leegwater PA, Yuan BQ, van der Steen J, Mulders J, Konst AA, Boor PK, Mejaski-Bosnjak V, van der Maarel SM, Frants RR, Oudejans CB, Schutgens RB, Pronk JC, van der Knaap MS (2001) Mutations of MLC1 (KIAA0027), encoding a putative membrane protein, cause megalencephalic leukoencephalopathy with subcortical cysts. *Am J Hum Genet* 68:831–838.

Li R, Messing A, Goldman J, Brenner M (2002a) GFAP mutations in Alexander disease. *Int J Dev Neurosci* 20:259.

Li R, Johnson AB, Salomons GS, van der Knaap MS, Rodriguez D, Boespflug-Tanguy O, Gorospe JR, Goldman JE, Messing A, Brenner M (2006) Propensity for paternal inheritance of de novo mutations in Alexander disease. Hum Genet 1–8.

Li R, Johnson AB, Salomons G, Goldman JE, Naidu S, Quinlan R, Cree B, Ruyle SZ, Banwell B, Siebert MDH Jr, Rolf CM, Cox H, Reddy A, Gutierrez-Solana LG, Collins A, Weller RO, Messing A, van der Knaap MS, Brenner M (2005) Glial fibrillary acidic protein mutations in infantile, juvenile, and adult forms of Alexander disease. *Ann Neurol* 57:310–326.

Li WH, Yi S, Makova K (2002b) Male-driven evolution. *Curr Opin Genet Dev* 12:650–656.

Lowe J, Blanchard A, Morrell K, Lennox G, Reynolds L, Billett M, Landon M, Mayer RJ (1988) Ubiquitin is a common factor in intermediate filament inclusion bodies of diverse type in man, including those of Parkinson's disease, Pick's disease, and Alzheimer's disease, as well as Rosenthal fibres in cerebellar astrocytomas, cytoplasmic bodies in muscle, and mallory bodies in alcoholic liver disease. *J Pathol* 155:9–15.

Lowe J, McDermott H, Pike I, Spendlove I, Landon M, Mayer RJ (1992) Alpha B crystallin expression in non-lenticular tissues and selective presence in ubiquitinated inclusion bodies in human disease. *J Pathol* 166:61–68.

Ma L, Yamada S, Wirtz D, Coulombe PA (2001) A 'hot-spot' mutation alters the mechanical properties of keratin filament networks. *Nat Cell Biol* 3:503–506.

Matute C, Alberdi E, Ibarretxe G, Sanchez-Gomez MV (2002) Excitotoxicity in glial cells. *Eur J Pharmacol* 447:239–246.

McCall MA, Gregg RG, Behringer RR, Brenner M, Delaney CL, Galbreath EJ, Zhang CL, Pearce RA, Chiu SY, Messing A (1996) Targeted deletion in astrocyte intermediate filament (Gfap) alters neuronal physiology. *Proc Natl Acad Sci USA* 93:6361–6366.

Meins M, Brockmann K, Yadav S, Haupt M, Sperner J, Stephani U, Hanefeld F (2002) Infantile Alexander disease: a GFAP mutation in monozygotic twins and novel mutations in two other patients. *Neuropediatrics* 33:194–198.

Messing A, Brenner M (2003) Alexander disease: GFAP mutations unify young and old. *Lancet Neurol* 2:75

Messing A, Goldman JE (2004) Alexander disease. In: Myelin and its diseases (Lazzarini RA, ed), pp 851–866. San Diego: Elsevier Academic Press.

Messing A, Head MW, Galles K, Galbreath EJ, Goldman JE, Brenner M (1998) Fatal encephalopathy with astrocyte inclusions in GFAP transgenic mice. *Am J Pathol* 152:391–398.

Messing A, Goldman JE, Johnson AB, Brenner M (2001) Alexander disease: new insights from genetics. *J Neuropathol Exp Neurol* 60:563–573.

Nachman MW, Crowell SL (2000) Estimate of the mutation rate per nucleotide in humans. *Genetics* 156:297–304.

Namekawa M, Takiyama Y, Aoki Y, Takayashiki N, Sakoe K, Shimazaki H, Taguchi T, Tanaka Y, Nishizawa M, Saito K, Matsubara Y, Nakano I (2002) Identification of *GFAP* gene mutation in hereditary adult-onset Alexander's disease. *Ann Neurol* 52:779–785.

Nicholl ID, Quinlan RA (1994) Chaperone activity of alpha-crystallins modulates intermediate filament assembly. *EMBO J* 13:945–953.

Nobuhara Y, Nakahara K, Higuchi I, Yoshida T, Fushiki S, Osame M, Arimura K, Nakagawa M (2004) Juvenile form of Alexander disease with GFAP mutation and mitochondrial abnormality. *Neurology* 63:1302–1304.

Ogasawara N (1965) [Multiple sclerosis with Rosenthal's fibers]. *Acta Neuropathol* 5:61–68.

Okamoto Y, Mitsuyama H, Jonosono M, Hirata K, Arimura K, Osame M, Nakagawa M (2002) Autosomal dominant palatal myoclonus and spinal cord atrophy. *J Neurol Sci* 195:71–76.

Pappolla MA (1986) Lewy bodies of Parkinson's disease. Immune electron microscopic demonstration of neurofilament antigens in constituent filaments. *Arch Pathol Lab Med* 110:1160–1163.

Parry DA, Steinert PM (1999) Intermediate filaments: molecular architecture, assembly, dynamics and polymorphism. *Q Rev Biophys* 32:99–187.

Pekny M, Leveen P, Pekna M, Eliasson C, Berthold CH, Westermark B, Betsholtz C (1995) Mice lacking glial fibrillary acidic protein display astrocytes devoid of intermediate filaments but develop and reproduce normally. *EMBO J* 14:1590–1598.

Perng MD, Cairns L, van den IP, Prescott A, Hutcheson AM, Quinlan RA (1999a) Intermediate filament interactions can be altered by HSP27 and alphaB-crystallin. *J Cell Sci* 112 (Part 13):2099–2112.

Perng MD, Muchowski PJ, van den IP, Wu GJ, Hutcheson AM, Clark JI, Quinlan RA (1999b) The cardiomyopathy and lens cataract mutation in alphaB-crystallin alters its protein structure, chaperone activity, and interaction with intermediate filaments in vitro. *J Biol Chem* 274:33235–33243.

Perng MD, Wen SF, van den IP, Prescott AR, Quinlan RA (2004) Desmin aggregate formation by R120G alphaB-crystallin is caused by altered filament interactions and is dependent upon network status in cells. *Mol Biol Cell* 15:2335–2346.

Pridmore CL, Baraitser M, Harding B, Boyd SG, Kendall B, Brett EM (1993) Alexander's disease: clues to diagnosis. *J Child Neurol* 8:134–144.

Probst EN, Hagel C, Weisz V, Nagel S, Wittkugel O, Zeumer H, Kohlschutter A (2003) Atypical focal MRI lesions in a case of juvenile Alexander's disease. *Ann Neurol* 53:118–120.

Quinlan R (2001) Cytoskeletal catastrophe causes brain degeneration. *Nat Genet* 27:10–11.

Quinlan R (2002) Cytoskeletal competence requires protein chaperones. *Prog Mol Subcell Biol* 28:219–233.

Quinlan RA, Hatzfeld M, Franke WW, Lustig A, Schulthess T, Engel J (1986) Characterization of dimer subunits of intermediate filament proteins. *J Mol Biol* 192:337–349.

Quinlan RA, Brenner M, Goldman JE, Messing A (2007) GFAP and its role in Alexander disease. *Exp Cell Res* 313:2077–2087.

Ralton JE, Lu X, Hutcheson AM, Quinlan RA (1994) Identification of two N-terminal non-alpha-helical domain motifs important in the assembly of glial fibrillary acidic protein. *J Cell Sci* 107:1935–1948.

Reichard EA, Ball WS Jr, Bove KE (1996) Alexander disease: a case report and review of the literature. *Pediatr Pathol Lab Med* 16:327–343.

Reynolds LP, Allen GV (2003) A review of heat shock protein induction following cerebellar injury. *Cerebellum* 2:171–177.

Rodriguez D, Gauthier F, Bertini E, Bugiani M, Brenner M, Ng S, Goizet C, Gelot A, Surtees R, Pedespan JM, Hernandorena X, Troncoso M, Uziel G, Messing A, Ponsot G, Pham-Dinh D, Dautigny A, Boespflug-Tanguy O (2001) Infantile Alexander disease: spectrum of GFAP mutations and genotype-phenotype correlation. Am J Hum Genet 69:1134–1140.

Rosenthal W (1898) Über eine eigenthümliche, mit syringomyelie complicirte geschwulst des rückenmarks. *Bietr Pathol Anat* 23:111–143.

Russo LS Jr, Aron A, Anderson PJ (1976) Alexander's disease: a report and reappraisal. *Neurology* 26:607–614.

Salvi F, Aoki Y, Della Nave R, Vella A, Pastorelli F, Scaglione C, Matsubara Y, Mascalchi M (2005) Adult Alexander's disease without leukoencephalopathy. *Ann Neurol* 58:813–814.

Sawaishi Y, Hatazawa J, Ochi N, Hirono H, Yano T, Watanabe Y, Okudera T, Takada G (1999) Positron emission tomography in juvenile Alexander disease. *J Neurol Sci* 165:116–120.

Sawaishi Y, Yano T, Takaku I, Takada G (2002) Juvenile Alexander disease with a novel mutation in glial fibrillary acidic protein gene. *Neurology* 58:1541–1543.

Schmitt A, Gofferje V, Weber M, Meyer J, Mossner R, Lesch KP (2003) The brain-specific protein MLC1 implicated in megalencephalic leukoencephalopathy with subcortical cysts is expressed in glial cells in the murine brain. *Glia* 44:283–295.

Schroder R, Kunz WS, Rouan F, Pfendner E, Tolksdorf K, Kappes-Horn K, Altenschmidt-Mehring M, Knoblich R, van der Ven PF, Reimann J, Furst DO, Blumcke I, Vielhaber S, Zillikens D, Eming S, Klockgether T, Uitto J, Wiche G, Rolfs A (2002) Disorganization of the desmin cytoskeleton and mitochondrial dysfunction in plectin-related epidermolysis bullosa simplex with muscular dystrophy. *J Neuropathol Exp Neurol* 61:520–530.

Schuelke M, Smeitink J, Mariman E, Loeffen J, Plecko B, Trijbels F, Stockler-Ipsiroglu S, van den Heuvel L (1999) Mutant NDUFV1 subunit of mitochondrial complex I causes leukodystrophy and myoclonic epilepsy. *Nat Genet* 21:260–261.

Schwankhaus JD, Parisi JE, Gulledge WR, Chin L, Currier RD (1995) Hereditary adult-onset Alexander's disease with palatal myoclonus, spastic paraparesis, and cerebellar ataxia. *Neurology* 45:2266–2271.

Seil FJ, Schochet SS Jr, Earle KM (1968) Alexander's disease in an adult. Report of a case. *Arch Neurol* 19:494–502.

Selcen D, Engel AG (2003) Myofibrillar myopathy caused by novel dominant negative alpha B-crystallin mutations. *Ann Neurol* 54:804–810.

Shibuki K, Gomi H, Chen L, Bao S, Kim JJ, Wakatsuki H, Fujisaki T, Fujimoto K, Katoh A, Ikeda T, Chen C, Thompson RF, Itohara S (1996) Deficient cerebellar long-term depression, impaired eyeblink conditioning, and normal motor coordination in GFAP mutant mice. *Neuron* 16:587–599.

Shiihara T, Kato M, Honma T, Ohtaki S, Sawaishi Y, Hayasaka K (2002) Fluctuation of computed tomographic findings in white matter in Alexander's disease. *J Child Neurol* 17:227–230.

Shiihara T, Sawaishi Y, Adachi M, Kato M, Hayasaka K (2004) Asymptomatic hereditary Alexander's disease caused by a novel mutation in GFAP. *J Neurol Sci* 225:125–127.

Shiroma N, Kanazawa N, Izumi M, Sugai K, Fukumizu M, Sasaki M, Hanaoka S, Kaga M, Tsujino S (2001) Diagnosis of Alexander disease in a Japanese patient by molecular genetic analysis. *J Hum Genet* 46:579–582.

Shiroma N, Kanazawa N, Kato Z, Shimozawa N, Imamura A, Ito M, Ohtani K, Oka A, Wakabayashi K, Iai M, Sugai K, Sasaki M, Kaga M, Ohta T, Tsujino S (2003) Molecular genetic study in Japanese patients with Alexander disease: a novel mutation, R79L. *Brain Dev* 25:116–121.

Singh R, Nielsen AL, Johansen MG, Jorgensen AL (2003) Genetic polymorphism and sequence evolution of an alternatively spliced exon of the glial fibrillary acidic protein gene, GFAP. *Genomics* 82:185–193.

Sistermans EA, de Coo RF, De Wijs IJ, Van Oost BA (1998) Duplication of the proteolipid protein gene is the major cause of Pelizaeus-Merzbacher disease. *Neurology* 50:1749–1754.
Smith FJ, Eady RA, Leigh IM, McMillan JR, Rugg EL, Kelsell DP, Bryant SP, Spurr NK, Geddes JF, Kirtschig G, Milana G, de Bono AG, Owaribe K, Wiche G, Pulkkinen L, Uitto J, McLean WH, Lane EB (1996a) Plectin deficiency results in muscular dystrophy with epidermolysis bullosa. *Nat Genet* 13:450–457.
Smith MA, Siedlak SL, Richey PL, Nagaraj RH, Elhammer A, Perry G (1996b) Quantitative solubilization and analysis of insoluble paired helical filaments from Alzheimer disease. *Brain Res* 717:99–108.
Smith TW, Tyler HR, Schoene WC (1975) Atypical astrocytes and Rosenthal fibers in a case of amyotrophic lateral sclerosis associated with a cerebral glioblastoma multiforme. *Acta Neuropathol* 31:29–34.
Soellner P, Quinlan RA, Franke WW (1985) Identification of a distinct soluble subunit of an intermediate filament protein: tetrameric vimentin from living cells. *Proc Natl Acad Sci USA* 82:7929–7933.
Spalke G, Mennel HD (1982) Alexander's disease in an adult: clinicopathologic study of a case and review of the literature. *Clin Neuropathol* 1:106–112.
Sreedharan J, Shaw CE, Jarosz J, Samuel M (2007) Alexander disease with hypothermia, microcoria, and psychiatric and endocrine disturbances. *Neurology* 68:1322–1323.
Steinert PM, Marekov LN, Parry DA (1999) Molecular parameters of type IV alpha-internexin and type IV-type III alpha-internexin-vimentin copolymer intermediate filaments. *J Biol Chem* 274:1657–1666.
Stumpf E, Masson H, Duquette A, Berthelet F, McNabb J, Lortie A, Lesage J, Montplaisir J, Brais B, Cossette P (2003) Adult Alexander disease with autosomal dominant transmission: a distinct entity caused by mutation in the glial fibrillary acid protein gene. *Arch Neurol* 60:1307–1312.
Suzuki Y, Kanazawa N, Takenaka J, Okumura A, Negoro T, Tsujino S (2004) A case of infantile Alexander disease with a milder phenotype and a novel GFAP mutation, L90P. *Brain Dev* 26:206–208.
Takanashi J, Sugita K, Tanabe Y, Niimi H (1998) Adolescent case of Alexander disease: MR imaging and MR spectroscopy. *Pediatr Neurol* 18:67–70.
Takemura M, Gomi H, Colucci-Guyon E, Itohara S (2002) Protective role of phosphorylation in turnover of glial fibrillary acidic protein in mice. *J Neurosci* 22:6972–6979.
Tanaka K, Watase K, Manabe T, Yamada K, Watanabe M, Takahashi K, Iwama H, Nishikawa T, Ichihara N, Kikuchi T, Okuyama S, Kawashima N, Hori S, Takimoto M, Wada K (1997) Epilepsy and exacerbation of brain injury in mice lacking the glutamate transporter GLT-1. *Science* 276:1699–1702.
Tanaka KF, Takebayashi H, Yamazaki Y, Ono K, Naruse M, Iwasato T, Itohara S, Kato H, Ikenaka K (2007) Murine model of Alexander disease: analysis of GFAP aggregate formation and its pathological significance. *Glia* 55:617–631.
Tang G, Xu Z, Goldman JE (2006) Synergistic effects of the SAPK/JNK and the proteasome pathway on glial fibrillary acidic protein (GFAP) accumulation in Alexander disease. *J Biol Chem* 281:38634–38643.
Thyagarajan D, Chataway T, Li R, Gai WP, Brenner M (2004) Dominantly-inherited adult-onset leukodystrophy with palatal tremor caused by a mutation in the glial fibrillary acidic protein gene. *Mov Disord* 19:1244–1248.
Tian R, Gregor M, Wiche G, Goldman JE (2006) Plectin regulates the organization of glial fibrillary acidic protein in Alexander disease. *Am J Pathol* 168:888–897.
Toivola DM, Tao GZ, Habtezion A, Liao J, Omary MB (2005) Cellular integrity plus: organelle-related and protein-targeting functions of intermediate filaments. *Trends Cell Biol* 15:608–617.
Towhi J, Young R, Sassani J, Ramer J, Horoupian DS (1983) Alexander's disease: further light-, and electron-microscopic observations. *Acta Neuropathol* 61:36–42.
Townsend JJ, Wilson JF, Harris T, Coulter D, Fife R (1985) Alexander's disease. *Acta Neuropathol* 67:163–166.

Trollmann R, Kraus C, Orlova N, Rupprecht T, Wenzel D, Rauch A (2003) Diagnosesicherung des morbus Alexander in vivo durch mutationsanalyse des *GFAP*-gens. *Monatsschr Kinderheilkd* 3:311–314.

van der Knaap MS, Naidu S, Breiter SN, Blaser S, Stroink H, Springer S, Begeer JC, van Coster R, Barth PG, Thomas NH, Valk J, Powers JM (2001) Alexander disease: diagnosis with mr imaging. *Am J Neuroradiol* 22:541–552.

van der Knaap MS, Salomons GS, Li R, Franzoni E, Gutierrez-Solana LG, Smit LM, Robinson R, Ferrie CD, Cree B, Reddy A, Thomas N, Banwell B, Barkhof F, Jakobs C, Johnson A, Messing A, Brenner M (2005) Unusual variants of Alexander's disease. *Ann Neurol* 57:327–338.

van der Knaap MS, Ramesh V, Schiffmann R, Blaser S, Kyllerman M, Gholkar A, Ellison DW, van der Voorn JP, van Dooren SJ, Jakobs C, Barkhof F, Salomons GS (2006) Alexander disease: ventricular garlands and abnormalities of the medulla and spinal cord. *Neurology* 66:494–498.

Vicart P, Caron A, Guicheney P, Li Z, Prevost MC, Faure A, Chateau D, Chapon F, Tome F, Dupret JM, Paulin D, Fardeau M (1998) A missense mutation in the alphaB-crystallin chaperone gene causes a desmin-related myopathy. *Nat Genet* 20:92–95.

Wilson SP, Al-Sarraj S, Bridges LR (1996) Rosenthal fiber encephalopathy presenting with demyelination and Rosenthal fibers in a solvent abuser: adult Alexander's disease? *Clin Neuropathol* 15:13–16.

Wippold FJ II, Perry A, Lennerz J (2006) Neuropathology for the neuroradiologist: Rosenthal fibers. *AJNR Am J Neuroradiol* 27:958–961.

Wu KC, Bryan JT, Morasso MI, Jang SI, Lee JH, Yang JM, Marekov LN, Parry DA, Steinert PM (2000) Coiled-coil trigger motifs in the 1B and 2B rod domain segments are required for the stability of keratin intermediate filaments. *Mol Biol Cell* 11:3539–3558.

Wu Y, Jiang YW, Wang JM, Yang YL, Zhang YH, Chang XZ, Qin J, Xiao JX, Wu X (2006) GFAP mutations in 3 Chinese infantile Alexander disease patients. *J Inherit Metab Dis* 29:66.

Zatloukal K, Stumptner C, Fuchsbichler A, Heid H, Schnoelzer M, Kenner L, Kleinert R, Prinz M, Aguzzi A, Denk H (2002) p62 Is a common component of cytoplasmic inclusions in protein aggregation diseases. *Am J Pathol* 160:255–263.

Zhang X, Haaf M, Todorich B, Grosstephan E, Schieremberg H, Surguladze N, Connor JR (2005) Cytokine toxicity to oligodendrocyte precursors is mediated by iron. *Glia* 52:199–208.

Zlotogora J (1998) Germ line mosaicism. *Hum Genet* 102:381–386.

Abbreviations

CNS	Central nervous system
CSF	Cerebrospinal fluid
EM	Electron microscopy
GABA	γ-Amino butyric acid
GAD	Glutamic acid decarboxylase
GFAP	Glial fibrillary acidic protein
GFP	Green fluorescent protein
Glt-1	Glial l-glutamate transporter
hGFAP	Human GFAP
JNK	c-Jun amino-terminal kinase
MIM	Mendelian inheritance in man
MRI	Magnetic resonance imaging
SDS-PAGE	Sodium dodecyl sulfate polyacrylamide gel electrophoresis
TNFα	Tumor necrosis factor α
TRH	Thyrotropin releasing hormone

Chapter 25
Role of Astrocytes in Epilepsy

Devin K. Binder and Christian Steinhäuser

Contents

25.1 Introduction

Epilepsy, affecting about 1% of the population, comprises a group of disorders of the brain characterized by the periodic and unpredictable occurrence of seizures. Epilepsy is a major public health problem in which those affected experience seizures that impair the performance of many tasks and cause major medical and psychosocial morbidity. Elucidating the cellular and molecular mechanisms of seizure generation may lead to novel antiepileptic drug (AED) therapies.

Most current AEDs act on widely expressed ion channels that directly control neuronal excitability (Rogawski and Loscher, 2004). For example, sodium channel blockers (e.g., phenytoin) reduce the rate and/or rise of neuronal action potentials

C. Steinhäuser
Institute of Cellular Neurosciences, Medical Faculty, University of Bonn, Bonn, Germany
Christian.Steinhaeuser@ukb.uni-bonn.de

V. Parpura and P.G. Haydon (eds.), *Astrocytes in (Patho)Physiology of the Nervous System*,
DOI: 10.1007/978-0-387-79492-1_25,

and thus inhibit high-frequency neuronal firing. Gamma-amino butyric acid (GABA) receptor agonists (e.g., phenobarbital) increase the efficacy of inhibitory synapses, thus attenuating excitability. Existing medications have two major drawbacks. First, even with optimal current AED therapy, ~30% of patients have poor seizure control and become medically refractory. Second, as many of these nonspecific medications act as general central nervous system (CNS) depressants and must be taken chronically for seizure suppression, they also have marked inhibitory effects on cognition.

Several recent lines of evidence suggest that glial cells may be potential novel targets for the treatment of epilepsy. First, recent findings now link glial cells to modulation of synaptic transmission (reviewed in Volterra and Steinhäuser, 2004; Volterra and Meldolesi, 2005). Second, functional alterations of specific glial membrane channels, receptors, and transporters have been discovered in several neurological disorders, including epilepsy (Heinemann et al., 2000; Steinhäuser and Seifert, 2002; de Lanerolle and Lee, 2005; Seifert et al., 2006). Third, direct stimulation of astrocytes has been shown to be sufficient for neuronal synchronization in acute epilepsy models (Tian et al., 2005). Thus, if the cellular and molecular mechanisms by which glial cells, especially astrocytes, modulate excitability are better understood, specific antiepileptic targets and therapies can be developed. These therapies are likely to have fewer deleterious side effects than standard AEDs that suppress global neuronal activity. In this review, we describe the evidence to date regarding alterations and functional roles of distinct astrocyte receptors, membrane channels, and transporters in various forms of epilepsy.

Two basic limitations of the topic covered here should be considered. First, since the physiological consequences of the intriguing bidirectional communication between neurons and glial cells are still incompletely understood, it is often unclear whether the glial changes are causative of the disease or rather represent an accompanying phenomenon. Second, different types of cells with astroglial properties exist within a given brain region, and the properties of these cells vary in different areas. So far, we have only rudimentary understanding of this glial diversity, and most of the previous studies describing astroglial alterations in epilepsy did not identify the specific cell type affected. In this review, we refer to different types of cells with astroglial properties as *astrocytes*.

25.2 Altered Astrocyte Morphology in Temporal Lobe Epilepsy

Alterations in astrocytic properties have been best described in human temporal lobe epilepsy (TLE), which is the most common form of epilepsy. The most common pathology found in patients with medically intractable TLE is hippocampal sclerosis, more generally termed mesial temporal sclerosis (MTS), which is characterized by neuronal cell loss in specific hippocampal areas, gliosis, microvascular proliferation, and synaptic reorganization (Margerison and Corsellis, 1966; Mathern et al., 1997; Blümcke et al., 1999). Early autopsy studies found MTS in 30–58% of TLE cases (Sommer, 1880; Bratz, 1899; Margerison and Corsellis, 1966); similar proportions have been found in specimens resected from patients undergoing surgery for medically intractable TLE (Falconer, 1974; Mathern et al., 1997; de Lanerolle et al., 2003). One striking hallmark of the sclerotic hippocampus is that, while there is a specific pattern of neuronal loss, there is also *reactive gliosis* with hypertrophic glial cells exhibiting prominent GFAP staining and long, thick processes. Only a few studies have attempted to quantify changes in astrocyte numbers and densities in epileptic tissue (Krishnan et al., 1994; Van Paesschen et al., 1997; Mitchell et al., 1999; Briellmann et al., 2002). Interestingly, one study shows significant gliosis in the amygdala as well (Wolf et al., 1997). Most of the changes in astrocytic channels and transporters described later have been discovered in sclerotic hippocampi from TLE patients. However, the cellular and molecular processes leading to astrocytic changes during epileptogenesis are not yet understood, and the changes in other distinct glial cell types during epileptogenesis have been largely unexplored.

25.3 Astrocytic Glutamate Release in Epilepsy

Over the past few years, Ca^{2+} signaling mechanisms in astrocytes have received considerable attention. Of particular importance is the novel observation that astrocytes exhibit Ca^{2+}-induced release of glutamate, which provides direct excitation to neighbouring neurons (reviewed in Volterra and Meldolesi, 2005). It is tempting to speculate that alterations in this glial-derived excitatory pathway in coordination with reductions in glutamate uptake might provide an excitatory drive underlying seizure disorders (Halassa et al., 2007).

Astrocytes are capable of releasing glutamate through a Ca^{2+}-dependent process, which might be involved in seizure generation (Kang et al., 2005). In chemically induced, acute epilepsy models, astrocytes were reported to contribute to the generation of synchronized epileptiform activity (Tian et al., 2005). However, another recent report casts doubts on the hypothesis that glutamate released from astrocytes is necessary for the generation of epileptiform activity. Rather, these authors conclude that glial glutamate might amplify or modulate synaptic activity during epileptogenesis (Fellin et al., 2006). In these studies, epileptiform discharges were provoked through the application of 4-aminopyridine, $GABA_A$ receptor antagonists, or bath solutions containing low concentrations of divalent cations. Importantly, chronic epilepsy is associated with significant morphological alterations (Kim, 2001; Blümcke et al., 2002; de Lanerolle and Lee, 2005) that are absent in the acute models. Certainly, more experimentation is needed to figure out the exact role of glia-derived neurotransmitters in epileptogenesis.

25.4 Astrocyte Dysfunction in Temporal Lobe Epilepsy

25.4.1 Glutamate Receptors, Transporters, and Related Enzymes

25.4.1.1 Dysfunctional Glutamate Transport and Synthesis

Glutamate transporters are expressed by several CNS cell types, but astrocytes are primarily responsible for glutamate uptake. Studies using mice with deletion (Tanaka et al., 1997) or antisense oligonucleotide-mediated inhibition of synthesis (Rothstein et al., 1996) of the astroglial transporter GLT-1 revealed that this subtype is responsible for the bulk of extracellular glutamate clearance in the CNS (Danbolt, 2001). Several studies have suggested an involvement of glutamate transporters and receptors in seizure development and spread. Increased extracellular levels of glutamate have been found in epileptogenic foci (During and Spencer, 1993; Glass and Dragunow, 1995). GLT-1 knockout in mice caused spontaneous seizures and hippocampal pathology resembling alterations in TLE patients with MTS (Tanaka et al., 1997). Pharmacological inhibition of GLT-1 reduced the threshold for evoking epileptiform activity (Campbell and Hablitz, 2004; see also Demarque et al., 2004) but other animal studies were contradictory. Tessler et al. (1999) investigated transporter expression on the mRNA and protein levels in human TLE specimens and found changes neither for GLT-1 nor GLAST. However, two other groups reported decreased GLT-1 protein as well as unchanged (Mathern et al., 1999) or decreased (Proper et al., 2002) GLAST immunoreactivity in the sclerotic human hippocampus. The latter authors also noted an upregulation of GLT-1 in the nonsclerotic epileptic hippocampus (Table 25.1). These findings supported the hypothesis that reduced or dysfunctional glial glutamate transporters in the hippocampus may trigger spontaneous seizures in patients with MTS (During and Spencer, 1993), yet the underlying mechanisms are unclear. It has been proposed that the role of glutamate transporters in epilepsy may not be related directly to the control of excitation through synaptic glutamate concentration but rather to alterations in glutamate-dependent metabolism (Maragakis and Rothstein, 2004). In this context, the finding of a loss of glutamine synthetase in the sclerotic vs. nonsclerotic hippocampus of TLE patients (Eid et al., 2004) deserves further consideration. After uptake of glutamate into astrocytes, this enzyme rapidly converts the transmitter into glutamine that is then transported to neurons, where it may be resynthesized to glutamate. Eid and coworkers did not observe epilepsy-related changes in the expression of GLT-1. They concluded that in the sclerotic tissue, downregulation of glutamine synthetase caused a slowing of the glutamate–glutamine cycling and accumulation of the transmitter in astrocytes and in the extracellular space (Eid et al., 2004). This conclusion was compatible with findings in animal models of epilepsy and earlier data demonstrating slowed glutamate–glutamine cycling in sclerotic human epileptic hippocampus with magnetic resonance spectroscopy (Petroff et al., 2002). Whether activation of glutamate transporters, e.g., through β-lactam antibiotics (including penicillin and its derivatives; Rothstein et al., 2005), might be beneficial in the treatment of epilepsies remains a matter of further investigation.

Table 25.1 Involvement of astroglial membrane channels, transporters, and receptors in specific epilepsy syndromes

Epilepsy syndrome	Astroglial molecule	Effect	Species	Methods	Reference(s)
Temporal lobe epilepsy	GLT-1 GLAST	No change No change	Human	IHC, WB, ISH	Tessler et al., 1999
Temporal lobe epilepsy	GLT-1 GLAST	↓ No change	Human	IHC	Mathern et al., 1999
Temporal lobe epilepsy	GLT-1 GLAST	↓ ↓	Human	IHC, ISH	Proper et al., 2002
Temporal lobe epilepsy	GLT-1 (glutamine synthetase)	No change (↓)	Human	IHC, WB, enzyme activity	Eid et al., 2004
Temporal lobe epilepsy	GluR1 (“flip” variant)	↑	Human	PC, pharmacology (CTZ, PEPA), single-cell rtPCR, RA	Seifert et al., 2002, 2004)
Temporal lobe epilepsy	mGluR 2/3 mGluR5 mGluR8	↑ ↑ ↑	Human	IHC	Tang and Lee, 2001; Tang et al., 2001; Notenboom et al., 2006
Temporal lobe epilepsy	Kir channel	↓	Human	PC	Bordey and Sontheimer, 1998b
Temporal lobe epilepsy	Kir channel	↓ ↓	Human Rat (pilocarpine)	ISM, Ba^{2+}	Heinemann et al., 2000; Kivi et al., 2000
Temporal lobe epilepsy	Kir channel	↓	Human	PC, Ba^{2+}, single-cell rtPCR	Hinterkeuser et al., 2000; Schröder et al., 2000
Temporal lobe epilepsy	AQP4	↑ Overall ↓ Perivascular	Human	IHC, rtPCR, gene chip, EM	Lee et al., 2004; Eid et al., 2005
Focal cortical dysplasia	mGluR2/3 mGluR5	↑ ↑	Human	IHC	Aronica et al., 2003a

(continued)

Table 25.1 (continued)

Epilepsy syndrome	Astroglial molecule	Effect	Species	Methods	Reference(s)
Tuberous sclerosis	GLAST GLT-1	↓ ↓	$Tsc1^{GFAP}$ CKO mouse	WB, PC	Wong et al., 2003
Tuberous sclerosis	Kir	↓	$Tsc1^{GFAP}$ CKO mouse	PC, WB, Ba^{2+}, mRNA analysis	Jansen et al., 2005
Tumor-associated epilepsy	GluR2	↓ Q/R editing	Human glioma	rtPCR, sequencing	Mass et al., 2001
Tumor-associated epilepsy	GLT-1 GLAST	↓ Mislocalized	Human glioma	IHC	Ye et al., 1999
Tumor-associated epilepsy	Kir channel	↓ Mislocalized	Human glioma	PC WB, IHC	Bordey and Sontheimer, 1998a; Olsen and Sontheimer, 2004
Posttraumatic epilepsy	Kir and Kv channels	↓	Rat (fluid-pereussion injury)	PC, ISM	D'Ambrosio et al., 1999
Posttraumatic epilepsy	GLT-1 GLAST	↓ No change	Rat (ferrous chloride)	WB	Samuelsson et al., 2000

CTZ cyclothiazide, *EM* electron microsopy, *IHC* immunohistochemistry, *ISH* in situ hybridization, *ISM* ion-sensitive microelectrodes, *PC* patch clamp, *PEPA* 4-[2-(phenylsulfonylamino)ethylthio]-2,6-difluoro-phenoxyacetamide, *RA* restriction analysis, *rtPCR* reverse transcriptase polymerase chain reaction, *WB* Western blot

25.4.1.2 Alterations of Ionotropic Glutamate Receptors

A few studies have addressed the potential involvement of glial ionotropic glutamate receptors in seizure generation. Astrocytes abundantly express receptors of the α-amino-3-hydroxy-5-methyl-isoxazole propionate (AMPA) subtype composed of the subunits GluR1 to GluR4 (reviewed by Verkhratsky and Steinhäuser, 2000). Mouse mutants with deficient GluR2 Q/R editing developed early-onset epilepsy with spontaneous and recurrent seizure activity, suggesting that enhanced Ca^{2+} influx through the Q form of the GluR2 subunit of AMPA receptors reduces seizure threshold (Brusa et al., 1995). Astrocytes also carry the GluR2 subunit, but altered glial GluR2 editing seems not to play a role in human TLE. Rather, combined functional and single-cell transcript analyses suggest that enhanced expression of GluR1 flip variants accounts for the prolonged receptor responses observed in hippocampal astrocytes of epilepsy patients with MTS (Seifert et al., 2002, 2004) (Table 25.1). This alteration in the splicing status of AMPA receptors predicts enhanced depolarization upon activation by endogenously released glutamate. The GluR1 flip variant, if coexpressed with GluR2 that is most abundant in astroglial cells of rodent and human hippocampus (Seifert et al., 1997), produces more incomplete receptor desensitization than GluR1 flop (Mosbacher et al., 1994). Prolonged receptor opening will promote influx of Ca^{2+} and Na^{+} ions, and the latter block astroglial inwardly rectifying K^{+} channels (Kir channels) (Schröder et al., 2002), which will further strengthen depolarization and reduce the K^{+} buffering capacity of astrocytes. It is yet unknown whether the changes in glial receptor function are causative of, or result from, the epileptic condition. Also, to what extent alterations in glial GluR1 splicing contribute to seizure generation or spread requires further investigation. Astrocytes cultured from patients with Rasmussen's encephalitis, a rare form of childhood epilepsy, showed spontaneous Ca^{2+} oscillations that were dependent on transmembrane influx of Ca^{2+} (Manning and Sontheimer, 1997). The authors speculated that these responses might promote neuronal hyperactivity, possibly due to autocrine ionotropic glutamate receptor stimulation by glutamate released from astrocytes. Another study suggested that the destruction of astrocytes by GluR3 antibodies plays a critical role in the progression of this autoimmune disorder (Whitney and McNamara, 2000).

25.4.1.3 Metabotropic Glutamate Receptors and Ca^{2+} Signaling

Under normal conditions, mGluR3 and mGluR5 are the predominant metabotropic glutamate receptor (mGluR) subtypes expressed by glial cells. Activation of these receptors affects cyclic adenosine monophosphate (cAMP) accumulation and leads to an increase in intracellular Ca^{2+}, respectively. Group II mGluRs (mGluR 2, 3) have been shown to be negatively coupled to cAMP levels in cultured astrocytes (Wroblewska et al., 1998) although other studies reported increases in cAMP levels (Moldrich et al., 2002; reviewed by Winder and Conn, 1996). The Ca^{2+} rise may oscillate and initiate Ca^{2+} wave propagation within the astrocyte network, activate

Ca^{2+}-dependent ion channels, and induce glutamate release from astrocytes (reviewed in Volterra and Meldolesi, 2005). In experimental epilepsy, reactive astrocytes of the hippocampus persistently upregulate mGluR3, mGluR5, and mGluR8 proteins (Steinhäuser and Seifert, 2002). Electron-microscopic and immunohistochemical inspection of hippocampal tissue from TLE patients revealed expression of mGluR2/3, mGluR4, mGluR5, and mGluR8 in reactive astrocytes, suggesting an involvement of these receptors in gliosis (Tang and Lee, 2001; Tang et al., 2001; Notenboom et al., 2006). Upregulation of astroglial mGluR2/3 and mGluR5 was also observed in epileptic specimens from patients with focal cortical dysplasia (Aronica et al., 2003a) (Table 25.1). Whether these changes affect the activity of glial glutamate transporters is not yet clear (Aronica et al., 2003b).

25.4.2 Dysregulation of K^+ and Water Channels

Since both extracellular K^+ concentration and osmolarity have been shown to dramatically modulate neural excitability, changes in astrocytic K^+ or water channels (aquaporins; AQPs) could contribute to hyperexcitability in epilepsy. Indeed, recent studies have found changes in astroglial Kir channels and AQP4 water channels in TLE specimens.

25.4.2.1 K^+ Channels

During neuronal hyperactivity, extracellular [K^+] may increase from ~3 mM to a ceiling of 10–12 mM; K^+ released by active neurons is thought to be primarily taken up by glial cells (Heinemann and Lux, 1977; Ballanyi et al., 1987; Xiong and Stringer, 1999; Somjen, 2002). Any impairment of glial K^+ uptake would be expected to be proconvulsant based on many previous studies. In the hippocampus, millimolar and even submillimolar increases in extracellular K^+ concentration powerfully enhance epileptiform activity (Rutecki et al., 1985; Yaari et al., 1986; Traynelis and Dingledine, 1988; Feng and Durand, 2006). High extracellular K^+concentration also reliably induces epileptiform activity in hippocampal slices from human patients with intractable TLE and hippocampal sclerosis (Gabriel et al., 2004).

A primary mechanism for K^+ reuptake is thought to be via glial Kir channels. Glial Kir channels may contribute to K^+ reuptake and spatial K^+ buffering (Orkand et al., 1966; Ransom, 1996), which has been most clearly demonstrated in the retina (Newman et al., 1984, 1986; Newman and Karwoski, 1989; Newman, 1993). While multiple subfamilies of Kir channels exist (Kir1-Kir7) differing in tissue distribution and functional properties, in brain astrocytes the expression of Kir4.1 has been investigated most thoroughly (Higashi et al., 2001; Hibino et al., 2004). Pharmacological or genetic inactivation of Kir4.1 leads to impairment of extracellular K^+ regulation (Kofuji et al., 2000; Kofuji and Newman, 2004; Neusch et al., 2006). However, members of the strongly rectifying Kir2 family may also contribute to astroglial K^+ buffering (Neusch et al., 2003; Butt and Kalsi, 2006).

Downregulation of astroglial Kir channels has been found in the injured or diseased CNS. Kir currents are reduced following injury-induced reactive gliosis in vitro (MacFarlane and Sontheimer, 1997), entorhinal cortex lesion (Schröder et al., 1999), freeze lesion-induced cortical dysplasia (Bordey et al., 2000, 2001), and traumatic (D'Ambrosio et al., 1999) and ischemic (Köller et al., 2000) brain injury. In addition, several studies have indicated downregulation of Kir currents in specimens from patients with TLE (Bordey and Sontheimer, 1998b; Hinterkeuser et al., 2000; Kivi et al., 2000; Schröder et al., 2000) (Table 25.1). Using ion-sensitive microelectrodes, Heinemann's group compared glial Ba^{2+}-sensitive K^+ uptake in the CA1 region of hippocampal slices obtained from patients with or without MTS (Heinemann et al., 2000; Kivi et al., 2000). Ba^{2+}, a blocker of Kir channels, augmented stimulus-evoked K^+ elevation in nonsclerotic but not in sclerotic specimens, suggesting impairment in K^+ buffering in sclerotic tissue. Direct evidence for downregulation of Kir currents in the sclerotic CA1 region of hippocampus came from a comparative patch-clamp study in which a reduction in astroglial Kir currents was observed in sclerotic compared with nonsclerotic hippocampi (Hinterkeuser et al., 2000). These data indicate that dysfunction of astroglial Kir channels could underlie impaired K^+ buffering and contribute to hyperexcitability in epileptic tissue (Steinhäuser and Seifert, 2002). When and how this dysfunction develops during epileptogenesis is not yet clear.

25.4.2.2 Water Channels

Alterations in astroglial water regulation could also powerfully affect excitability. Brain tissue excitability is exquisitely sensitive to osmolarity and the size of the extracellular space (ECS) (Schwartzkroin et al., 1998). Decreasing ECS volume produces hyperexcitability and enhanced epileptiform activity (Dudek et al., 1990; Roper et al., 1992; Chebabo et al., 1995; Pan and Stringer, 1996); conversely, increasing ECS volume with hyperosmolar medium attenuates epileptiform activity (Traynelis and Dingledine, 1989; Dudek et al., 1990; Pan and Stringer, 1996; Haglund and Hochman, 2005). These experimental data parallel extensive clinical experience indicating that hypo-osmolar states such as hyponatremia lower seizure threshold while hyperosmolar states elevate seizure threshold (Andrew et al., 1989).

The aquaporins (AQPs) are a family of membrane proteins that function as *water channels* in many cell types and tissues in which fluid transport is crucial (Verkman, 2005). There is increasing evidence that water movement in the brain involves aquaporin channels (Amiry-Moghaddam and Ottersen, 2003; Manley et al., 2004). Aquaporin-4 (AQP4) is expressed ubiquitously by glial cells, especially at specialized membrane domains including astroglial end-feet in contact with blood vessels and astrocyte membranes that ensheathe glutamatergic synapses (Nielsen et al., 1997; Nagelhus et al., 2004). Activity-induced radial water fluxes in neocortex have been demonstrated that could be due to water movement via aquaporin channels in response to physiological activity (Holthoff and Witte, 2000; Niermann et al., 2001). Mice deficient in AQP4 have markedly decreased accumulation of brain water (cerebral edema) following water intoxication and focal cerebral ischemia

(Manley et al., 2000), and impaired clearance of brain water in models of vasogenic edema (Papadopoulos et al., 2004), suggesting a functional role for AQP4 in brain water transport. Similarly, mice deficient in dystrophin or α-syntrophin, in which there is mislocalization of the AQP4 protein (Frigeri et al., 2001; Neely et al., 2001; Vajda et al., 2002), also demonstrate attenuated cerebral edema (Vajda et al., 2002; Amiry-Moghaddam et al., 2003b).

Alteration in the expression and subcellular localization of AQP4 has been described in sclerotic hippocampi obtained from patients with MTS (Table 25.1). Using immunohistochemistry, rt-PCR, and gene chip analysis, Lee et al. (2004) demonstrated an overall increase in AQP4 expression in sclerotic hippocampi. However, using quantitative immunogold electron microscopy, the same group found that there was mislocalization of AQP4 in the human epileptic hippocampus, with reduction in perivascular membrane expression (Eid et al., 2005). The authors hypothesized that the loss of perivascular AQP4 perturbs water flux, impairs K^+ buffering, and results in an increased propensity for seizures. However, definitive examination of the functional role of AQP4 in epileptic tissue requires further study.

Several lines of evidence support the hypothesis that AQP4 and Kir4.1 may act in concert in K^+ and H_2O regulation (Simard and Nedergaard, 2004). (1) K^+ reuptake into glial cells could be AQP4-dependent, as water influx coupled to K^+ influx is thought to underlie activity-induced glial cell swelling (Walz, 1987, 1992). (2) Studies in the retina have demonstrated subcellular colocalization of AQP4 and Kir4.1 via both electron microscopic and coimmunoprecipitation analyses (Connors et al., 2004; Nagelhus et al., 2004). (3) Kir4.1$^{-/-}$ mice, like AQP4$^{-/-}$ mice (Li and Verkman, 2001; Li et al., 2002), have impaired retinal and cochlear physiology presumably due to altered K^+ metabolism (Marcus et al., 2002; Rozengurt et al., 2003). (4) AQP4$^{-/-}$ mice have remarkably slowed K^+ reuptake in models of seizure and spreading depression in vivo (Padmawar et al., 2005; Binder et al., 2006) associated with a near-threefold increase in seizure duration (Binder et al., 2006). (5) Afferent stimulation of hippocampal slices from α-syntrophin-deficient mice demonstrates a deficit in extracellular K^+ clearance (Amiry-Moghaddam et al., 2003a). These data are consistent with the idea that AQP4 and Kir4.1 participate in clearance of K^+ following neuronal activity. However, further studies are required to clarify the expression and functional interaction of AQP4 and Kir4.1 in the hippocampus and their changes during epileptogenesis.

25.4.3 Astrocytes in TLE-Related Immune Responses and Inflammation

Astrocytes produce various immunologically relevant molecules, which contribute to CNS inflammation (Dong and Benveniste, 2001; John et al., 2005; Vezzani and Granata, 2005). A strong association of a polymorphism in the interleukin (IL)-1ß gene, a proinflammatory cytokine, has been found in epilepsy patients with MTS compared with nonsclerosis patients and nonepileptic controls (Berkovic and

Jackson, 2000; Kanemoto et al., 2000). This polymorphism favors production of high levels of the cytokine in patients with MTS. Febrile seizures, another etiological factor in TLE, are also characterized by enhanced IL-1ß levels (Haspolat et al., 2002; Virta et al., 2002; Dube et al., 2005). IL-1ß activates astroglial IL receptors and, via the transcription factor NFκB, leads to the production of various molecules including several chemokines, the chemokine receptor CXCR4, and S100ß. Notably, elevated production of the NFκB-p65 subunit as well as the aforementioned genes has been found in astrocytes from MTS specimens (Crespel et al., 2002; de Lanerolle and Lee, 2005). Upregulation of S100ß and CXCR4 in astrocytes may increase Ca^{2+}-dependent release of glutamate (Barger and Van Eldik, 1992; Bezzi et al., 2001), which in turn would exacerbate the epileptic condition in the sclerotic hippocampus. IL-1ß and IL-1 receptors have also been powerfully implicated in animal models of epilepsy (Vezzani et al., 1999, 2000; Ravizza and Vezzani, 2006). Thus, the immunological responsiveness of astrocytes may provide a clue to understand how different initial precipitating factors may lead to a common pathological substrate of hippocampal sclerosis.

Astrocytes might also be involved in brain infections such as meningitis (Ounsted et al., 1985), human herpes virus 6, and herpes simplex virus (Uesugi et al., 2000), which have been shown to be associated with TLE. Human astrocytes express Toll-like receptors responding to viruses and bacteria (Bsibsi et al., 2002; Carpentier et al., 2005). Similar to IL-1 receptors, Toll-like receptors couple to NFκB (May and Ghosh, 1998), the activation of which may contribute to elevated glutamate release from astrocytes in the sclerotic hippocampus.

25.5 Astrocyte Dysfunction in Other Epilepsy Syndromes

25.5.1 Tuberous Sclerosis

Tuberous sclerosis (TS) is a multisystem genetic disorder resulting from autosomal dominant mutations of either the *TSC1* or *TSC2* genes. The *TSC1* gene encodes the protein hamartin and *TSC2* encodes tuberin, which are thought to be regulators of cell signaling and growth (Au et al., 2004). Epilepsy occurs in 80–90% of cases of TS, frequently involves multiple seizure types and is often medically refractory (Thiele, 2004). Cortical tubers represent the pathologic substrate of TS, and microscopically consist of a specific type of dysplastic lesion with astrocytosis and abnormal giant cells (Trombley and Mirra, 1981). While this suggests that astrocytes are involved in the pathologic lesion, in itself this is not evidence for a causative role of astrocytes in TS epileptogenesis. However, recent evidence using astrocyte-specific TSC1 conditional knockout mice has provided insight into a potential role of astrocytes in the etiology of TS. These mice, which have conditional inactivation of the *TSC1* gene in GFAP-expressing cells (*Tsc1*^{GFAP}CKO mice), develop severe spontaneous seizures by 2 months of age and die prematurely (Uhlmann et al., 2002). Intriguingly, the time point of onset of spontaneous seizures in these mice is concordant with increased astroglial proliferation. Furthermore, two functions of astrocytes, glutamate and K^+ reuptake, are impaired in these mice. These mice display reduced expression of the astrocyte glutamate transporters GLT1 and GLAST (Wong et al., 2003) (Table 25.1). In addition, recent evidence indicates that astrocytes from *Tsc1*^{GFAP}CKO mice exhibit reduced Kir channel activity, and hippocampal slices from these mice demonstrated increased sensitivity to K^+-induced epileptiform activity (Jansen et al., 2005) (Table 25.1). Together, these studies demonstrate that in this model, changes in glial properties may be a direct cause of epileptogenesis.

25.5.2 Tumor-Associated Epilepsy

Tumor-associated epilepsy is an important clinical problem, seen in approximately one-third of tumors (Rasmussen, 1975; Ettinger, 1994). Surgical removal of tumors usually results in seizure control, but many tumors cannot be resected safely, and tumor-associated seizures are often resistant to anticonvulsant therapy. Classic epilepsy-associated brain tumors include astrocytoma, oligodendroglioma, ganglioglioma, dysembryoplastic neuroepithelial tumor, and pleomorphic xanthoastrocytoma (Luyken et al., 2003). Microdialysis studies of gliomas have revealed reduced glutamate in the tumor compared with peritumoral tissue (Bianchi et al., 2004). A *glutamate hypothesis* of tumor-associated epilepsy has been advanced, which suggests that tumors excite surrounding tissue by glutamate overstimulation. Two lines of evidence are relevant to this hypothesis. First, the glutamate receptor subunit GluR2 has been found to be underedited at the Q/R site in gliomas, which would increase AMPA receptor Ca^{+2} permeability and potentially result in increased glutamate release by glioma cells (Maas et al., 2001) (Table 25.1). Second, Sontheimer's group found that glioma cells release larger than normal amounts of glutamate in vitro (Ye and Sontheimer, 1999). The release of glutamate from glioma cells was accompanied by a marked deficit in Na^+-dependent glutamate uptake, reduced expression of astrocytic glutamate transporters (Table 25.1), and upregulation of cystine–glutamate exchange (Ye et al., 1999). Hence, glioma cell glutamate release at the margins of the tumor may initiate seizures in peritumoral neurons. A distinct potential mechanism underlying tumor-associated epilepsy is altered K^+ homeostasis. In support of this hypothesis, both reduced Kir currents (Bordey and Sontheimer, 1998a) and mislocalization of Kir4.1 channels (Olsen and Sontheimer, 2004) have been found in malignant astrocytes (Table 25.1).

25.5.3 Posttraumatic Epilepsy

Posttraumatic epilepsy refers to a recurrent seizure disorder whose cause is believed to be traumatic brain injury. It is a common and important form of epilepsy (Frey, 2003; Garga and Lowenstein, 2006), and develops in a variable proportion of traumatic brain injury survivors depending on the severity of the injury and the time after injury (Caveness et al., 1979; Annegers et al., 1998). Anticonvulsant prophylaxis is ineffective at preventing the occurrence of late seizures (Temkin et al., 1990, 1999; D'Ambrosio and Perucca, 2004). Weight-drop and fluid-percussion injury animal models of posttraumatic epilepsy have demonstrated characteristic structural and functional changes in the hippocampus, such as death of dentate hilar neurons and mossy fiber sprouting (Lowenstein et al., 1992; Golarai et al., 2001; Santhakumar et al., 2001). Recently, studies have also implicated altered astrocyte function in posttraumatic epilepsy models. Recordings from glial cells in hippocampal slices 2 days after fluid-percussion injury demonstrated reduction in transient outward and inward K^+ currents, and antidromic stimulation of CA3 led to abnormal extracellular K^+ accumulation in posttraumatic slices compared with controls (D'Ambrosio et al., 1999) (Table 25.1). This was accompanied by the appearance of electrical afterdischarges in CA3. Thus, this study suggests impaired K^+ homeostasis in posttraumatic hippocampal glia. Another study demonstrated reduction in expression of the astrocyte glutamate transporter GLT1 in a posttraumatic epilepsy model induced by intracortical ferrous chloride injection, suggesting impaired glutamate transport (Samuelsson et al., 2000) (Table 25.1). Further studies of the role of glial cells in posttraumatic epilepsy appear warranted now that reliable posttraumatic epilepsy animal models have been developed (D'Ambrosio et al., 2004).

25.6 Conclusions and Perspectives

Astrocytes undergo cellular and molecular changes in epilepsy, including alteration in glutamate transporters and receptors as well as Kir channels and water channels. So far, most of these changes have been demonstrated in sclerotic hippocampi from patients with TLE or in animal models. However, the various functions of astrocytes in modulation of synaptic transmission and glutamate, K^+, and H_2O regulation suggest that astrocyte dysfunction could also contribute to the pathophysiology of other forms of epilepsy.

One important recent development is the recognition of structural and functional heterogeneity of cells with astroglial properties. It is clear that a subset of hippocampal astroglial cells (*classical* astrocytes or GluT cells) expresses glutamate transporters and not ionotropic glutamate receptors, while another subset (NG2 glia or GluR cells) expresses ionotropic glutamate receptors but not glutamate transporters (Matthias et al., 2003; Nishiyama et al., 2005). However, the lineage relationship of NG2 glia/GluR cells and the relative roles of bona fide astrocytes vs. NG2 glia/GluR cells in epilepsy still remain unclear. In addition, the functional roles of ionotropic glutamate receptors, Kir and AQP4 channels in these subsets of glial cells in the hippocampus are not yet understood. Interestingly, hippocampal NG2 glia/GluR cells lack gap junctional coupling but receive direct synaptic input from GABAergic and glutamatergic neurons

(Bergles et al., 2000; Lin and Bergles, 2004; Wallraff et al., 2004; Jabs et al., 2005). Gap junctions may also regulate excitability, although available data are inconsistent and their functional role in epileptogenesis is unclear (Steinhäuser and Seifert, 2002; Nemani and Binder, 2005). The availability of mice with genetically uncoupled astrocytes (Wallraff et al., 2006) will allow examination of this question, by separating the effects produced by alterations of neuronal vs. glial gap junctions. It will be important in future studies to examine the cellular and molecular properties of subsets of hippocampal glial cells in human epileptic tissue and characterize their functional alterations during epileptogenesis in appropriate animal models.

Another recent focus in astrocyte biology that may become important for epilepsy research is the *gliovascular junction* (Simard et al., 2003). Microvascular proliferation in the sclerotic hippocampus was noted as early as 1899, but the role of the vasculature and the blood–brain barrier in epilepsy is not yet clear. The intimate relationship between astroglial end-feet ensheathing blood vessels, the targeted expression of AQP4 and Kir4.1 on astroglial end-feet, and the role of astrocytes in blood–brain barrier permeability (Abbott, 2002) and control of microcirculation (Zonta et al., 2003; Mulligan and MacVicar, 2004; Metea and Newman, 2006; Takano et al., 2006) have only recently been appreciated. Local pathological alterations in the gliovascular junction could perturb blood flow, K^+ and H_2O regulation and constitute an important mechanism in the generation of hyperexcitability. Indeed, recent studies suggest that transient opening of the blood–brain barrier is actually sufficient for focal epileptogenesis, probably, due to albumin uptake into astrocytes and subsequent downregulation of Kir4.1 channels (Seiffert et al., 2004; Ivens et al., 2007). The impact of the gliovascular junction on metabolic homeostasis and its cellular and molecular changes during epileptogenesis are only beginning to be explored.

In conclusion, the exact changes taking place in astroglial functioning during epilepsy are still poorly understood. The term *reactive gliosis* is too descriptive and should be replaced by careful morphological, biochemical, and electrophysiological studies of identified glial cell subtypes in human tissue and animal models. In addition to changes in preexisting glial cell populations, newly generated glial cells with distinct properties may migrate into the hippocampus and contribute to enhanced seizure susceptibility (Hüttmann et al., 2003; Parent et al., 2006). The available data likely underrepresent the functional role of astroglial cells in epilepsy. In view of the many physiologic functions of astrocytes that have been elucidated within the past decade, it can be expected that the next few years will yield evidence of similar important roles for glial cells in pathophysiology. Further study of astrocyte alterations in epilepsy should lead to the identification of novel molecular targets that could stimulate new approaches to antiepileptic therapy.

Acknowledgments Devin K. Binder is supported by an American Epilepsy Society/Milken Family Foundation Early Career Physician Scientist Award. Christian Steinhäuser is supported by grants from the Deutsche Forschungsgemeinschaft (SFB/TR3, TPC1; SPP1172, SE 774/3).

References

Abbott NJ (2002) Astrocyte-endothelial interactions and blood–brain barrier permeability. J Anat 200:629–638.

Amiry-Moghaddam M, Ottersen OP (2003) The molecular basis of water transport in the brain. Nat Rev Neurosci 4:991–1001.

Amiry-Moghaddam M, Williamson A, Palomba M, Eid T, de Lanerolle NC, Nagelhus EA, Adams ME, Froehner SC, Agre P, Ottersen OP (2003a) Delayed K+ clearance associated with aquaporin-4 mislocalization: phenotypic defects in brains of á-syntrophin-null mice. Proc Natl Acad Sci USA 100:13615–13620.

Amiry-Moghaddam M, Otsuka T, Hurn PD, Traystman RJ, Haug FM, Froehner SC, Adams ME, Neely JD, Agre P, Ottersen OP, Bhardwaj A (2003b) An alpha-syntrophin-dependent pool of AQP4 in astroglial end-feet confers bidirectional water flow between blood and brain. Proc Natl Acad Sci USA 100:2106–2111.

Andrew RD, Fagan M, Ballyk BA, Rosen AS (1989) Seizure susceptibility and the osmotic state. Brain Res 498:175–180.

Annegers JF, Hauser WA, Coan SP, Rocca WA (1998) A population-based study of seizures after traumatic brain injuries. N Engl J Med 338:20–24.

Aronica E, Gorter JA, Jansen GH, van Veelen CW, van Rijen PC, Ramkema M, Troost D (2003a) Expression and cell distribution of group I and group II metabotropic glutamate receptor subtypes in taylor-type focal cortical dysplasia. Epilepsia 44:785–795.

Aronica E, Gorter JA, Ijlst-Keizers H, Rozemuller AJ, Yankaya B, Leenstra S, Troost D (2003b) Expression and functional role of mGluR3 and mGluR5 in human astrocytes and glioma cells: opposite regulation of glutamate transporter proteins. Eur J Neurosci 17:2106–2118.

Au KS, Williams AT, Gambello MJ, Northrup H (2004) Molecular genetic basis of tuberous sclerosis complex: from bench to bedside. J Child Neurol 19:699–709.

Ballanyi K, Grafe P, ten Bruggencate G (1987) Ion activities and potassium uptake mechanisms of glial cells in guinea-pig olfactory cortex slices. J Physiol 382:159–174.

Barger SW, Van Eldik LJ (1992) S100 beta stimulates calcium fluxes in glial and neuronal cells. J Biol Chem 267:9689–9694.

Bergles DE, Roberts JD, Somogyi P, Jahr CE (2000) Glutamatergic synapses on oligodendrocyte precursor cells in the hippocampus. Nature 405:187–191.

Berkovic SF, Jackson GD (2000) Thc hippocampal sclerosis whodunit: enter the genes. Ann Neurol 47:557–558.

Bezzi P, Domercq M, Brambilla L, Galli R, Schols D, De Clercq E, Vescovi A, Bagetta G, Kollias G, Meldolesi J, Volterra A (2001) CXCR4-activated astrocyte glutamate release via TNFalpha: amplification by microglia triggers neurotoxicity. Nat Neurosci 4:702–710.

Bianchi L, De Micheli E, Bricolo A, Ballini C, Fattori M, Venturi C, Pedata F, Tipton KF, Della Corte L (2004) Extracellular levels of amino acids and choline in human high grade gliomas: an intraoperative microdialysis study. Neurochem Res 29:325–334.

Binder DK, Yao X, Sick TJ, Verkman AS, Manley GT (2006) Increased seizure duration and slowed potassium kinetics in mice lacking aquaporin-4 water channels. Glia 53:631–636.

Blümcke I, Beck H, Lie AA, Wiestler OD (1999) Molecular neuropathology of human mesial temporal lobe epilepsy. Epilepsy Res 36:205–223.

Blümcke I, Thom M, Wiestler OD (2002) Ammon's horn sclerosis: a maldevelopmental disorder associated with temporal lobe epilepsy. Brain Pathol 12:199–211.

Bordey A, Sontheimer H (1998a) Electrophysiological properties of human astrocytic tumor cells in situ: enigma of spiking glial cells. J Ncurophysiol 79:2782–2793.

Bordey A, Sontheimer H (1998b) Properties of human glial cells associated with epileptic seizure foci. Epilepsy Res 32:286–303.

Bordey A, Hablitz JJ, Sontheimer H (2000) Reactive astrocytes show enhanced inwardly rectifying K+ currents in situ. Neuroreport 11:3151–3155.

Bordey A, Lyons SA, Hablitz JJ, Sontheimer H (2001) Electrophysiological characteristics of reactive astrocytes in experimental cortical dysplasia. J Neurophysiol 85:1719–1731.
Bratz E (1899) Ammonshornbefunde bei Epileptikern. Arch Psychiatr Nervenkr 32:820–835.
Briellmann RS, Kalnins RM, Berkovic SF, Jackson GD (2002) Hippocampal pathology in refractory temporal lobe epilepsy: T2-weighted signal change reflects dentate gliosis. Neurology 58:265–271.
Brusa R, Zimmermann F, Koh DS, Feldmeyer D, Gass P, Seeburg PH, Sprengel R (1995) Early-onset epilepsy and postnatal lethality associated with an editing-deficient GluR-B allele in mice. Science 270:1677–1680.
Bsibsi M, Ravid R, Gveric D, van Noort JM (2002) Broad expression of Toll-like receptors in the human central nervous system. J Neuropathol Exp Neurol 61:1013–1021.
Butt AM, Kalsi A (2006) Inwardly rectifying potassium channels (Kir) in central nervous system glia: a special role for Kir4.1 in glial functions. J Cell Mol Med 10:33–44.
Campbell SL, Hablitz JJ (2004) Glutamate transporters regulate excitability in local networks in rat neocortex. Neuroscience 127:625–635.
Carpentier PA, Begolka WS, Olson JK, Elhofy A, Karpus WJ, Miller SD (2005) Differential activation of astrocytes by innate and adaptive immune stimuli. Glia 49:360–374.
Caveness WF, Meirowsky AM, Rish BL, Mohr JP, Kistler JP, Dillon JD, Weiss GH (1979) The nature of posttraumatic epilepsy. J Neurosurg 50:545–553.
Chebabo SR, Hester MA, Aitken PG, Somjen GG (1995) Hypotonic exposure enhances synaptic transmission and triggers spreading depression in rat hippocampal tissue slices. Brain Res 695:203–216.
Connors NC, Adams ME, Froehner SC, Kofuji P (2004) The potassium channel Kir4.1 associates with the dystrophin-glycoprotein complex via alpha-syntrophin in glia. J Biol Chem 279:28387–28392.
Crespel A, Coubes P, Rousset MC, Brana C, Rougier A, Rondouin G, Bockaert J, Baldy-Moulinier M, Lerner-Natoli M (2002) Inflammatory reactions in human medial temporal lobe epilepsy with hippocampal sclerosis. Brain Res 952:159–169.
D'Ambrosio R, Perucca E (2004) Epilepsy after head injury. Curr Opin Neurol 17:731–735.
D'Ambrosio R, Maris DO, Grady MS, Winn HR, Janigro D (1999) Impaired K+ homeostasis and altered electrophysiological properties of post-traumatic hippocampal glia. J Neurosci 19:8152–8162.
D'Ambrosio R, Fairbanks JP, Fender JS, Born DE, Doyle DL, Miller JW (2004) Post-traumatic epilepsy following fluid percussion injury in the rat. Brain 127:304–314.
Danbolt NC (2001) Glutamate uptake. Prog Neurobiol 65:1–105.
de Lanerolle NC, Lee TS (2005) New facets of the neuropathology and molecular profile of human temporal lobe epilepsy. Epilepsy Behav 7:190–203.
de Lanerolle NC, Kim JH, Williamson A, Spencer SS, Zaveri HP, Eid T, Spencer DD (2003) A retrospective analysis of hippocampal pathology in human temporal lobe epilepsy: evidence for distinctive patient subcategories. Epilepsia 44:677–687.
Demarque M, Villeneuve N, Manent JB, Becq H, Represa A, Ben-Ari Y, Aniksztejn L (2004) Glutamate transporters prevent the generation of seizures in the developing rat neocortex. J Neurosci 24:3289–3294.
Dong Y, Benveniste EN (2001) Immune function of astrocytes. Glia 36:180–190.
Dube C, Vezzani A, Behrens M, Bartfai T, Baram TZ (2005) Interleukin-1beta contributes to the generation of experimental febrile seizures. Ann Neurol 57:152–155.
Dudek FE, Obenhaus A, Tasker JG (1990) Osmolality-induced changes in extracellular volume alter epileptiform bursts independent of chemical synapses in the rat: importance of non-synaptic mechanisms in hippocampal epileptogenesis. Neurosci Lett 120:267–270.
During MJ, Spencer DD (1993) Extracellular hippocampal glutamate and spontaneous seizure in the conscious human brain. Lancet 341:1607–1610.
Eid T, Thomas MJ, Spencer DD, Runden-Pran E, Lai JC, Malthankar GV, Kim JH, Danbolt NC, Ottersen OP, de Lanerolle NC (2004) Loss of glutamine synthetase in the human epileptogenic hippocampus: possible mechanism for raised extracellular glutamate in mesial temporal lobe epilepsy. Lancet 363:28–37.

Eid T, Lee TS, Thomas MJ, Amiry-Moghaddam M, Bjornsen LP, Spencer DD, Agre P, Ottersen OP, de Lanerolle NC (2005) Loss of perivascular aquaporin 4 may underlie deficient water and K+ homeostasis in the human epileptogenic hippocampus. Proc Natl Acad Sci USA 102:1193–1198.

Ettinger AB (1994) Structural causes of epilepsy. Neurol Clin 12:41–56.

Falconer MA (1974) Mesial temporal (Ammon's horn) sclerosis as a common cause of epilepsy. Aetiology, treatment, and prevention. Lancet 2:767–770.

Fellin T, Gomez-Gonzalo M, Gobbo S, Carmignoto G, Haydon PG (2006) Astrocytic glutamate is not necessary for the generation of epileptiform neuronal activity in hippocampal slices. J Neurosci 26:9312–9322.

Feng Z, Durand DM (2006) Effects of potassium concentration on firing patterns of low-calcium epileptiform activity in anesthetized rat hippocampus: inducing of persistent spike activity. Epilepsia 47:727–736.

Frey LC (2003) Epidemiology of posttraumatic epilepsy: a critical review. Epilepsia 44 Suppl 10:11–17.

Frigeri A, Nicchia GP, Nico B, Quondamatteo F, Herken R, Roncali L, Svelto M (2001) Aquaporin-4 deficiency in skeletal muscle and brain of dystrophic mdx mice. FASEB J 15:90–98.

Gabriel S, Njunting M, Pomper JK, Merschhemke M, Sanabria ER, Eilers A, Kivi A, Zeller M, Meencke HJ, Cavalheiro EA, Heinemann U, Lehmann TN (2004) Stimulus and potassium-induced epileptiform activity in the human dentate gyrus from patients with and without hippocampal sclerosis. J Neurosci 24:10416–10430.

Garga N, Lowenstein DH (2006) Posttraumatic epilepsy: a major problem in desperate need of major advances. Epilepsy Curr 6:1–5.

Glass M, Dragunow M (1995) Neurochemical and morphological changes associated with human epilepsy. Brain Res Brain Res Rev 21:29–41.

Golarai G, Greenwood AC, Feeney DM, Connor JA (2001) Physiological and structural evidence for hippocampal involvement in persistent seizure susceptibility after traumatic brain injury. J Neurosci 21:8523–8537.

Haglund MM, Hochman DW (2005) Furosemide and mannitol suppression of epileptic activity in the human brain. J Neurophysiol 94:907–918.

Halassa MM, Fellin T, Haydon PG (2007) The tripartite synapse: roles for gliotransmission in health and disease. Trends Mol Med 13:54–63.

Haspolat S, Mihci E, Coskun M, Gumuslu S, Ozben T, Yegin O (2002) Interleukin-1beta, tumor necrosis factor-alpha, and nitrite levels in febrile seizures. J Child Neurol 17:749–751.

Heinemann U, Lux HD (1977) Ceiling of stimulus induced rises in extracellular potassium concentration in the cerebral cortex of cat. Brain Res 120:231–249.

Heinemann U, Gabriel S, Jauch R, Schulze K, Kivi A, Eilers A, Kovacs R, Lehmann TN (2000) Alterations of glial cell function in temporal lobe epilepsy. Epilepsia 41 Suppl 6:S185–S189.

Hibino H, Fujita A, Iwai K, Yamada M, Kurachi Y (2004) Differential assembly of inwardly rectifying K+ channel subunits, Kir4.1 and Kir5.1, in brain astrocytes. J Biol Chem 279:44065–44073.

Higashi K, Fujita A, Inanobe A, Tanemoto M, Doi K, Kubo T, Kurachi Y (2001) An inwardly rectifying K+ channel, Kir4.1, expressed in astrocytes surrounds synapses and blood vessels in brain. Am J Physiol Cell Physiol 281:C922–C931.

Hinterkeuser S, Schröder W, Hager G, Seifert G, Blümcke I, Elger CE, Schramm J, Steinhäuser C (2000) Astrocytes in the hippocampus of patients with temporal lobe epilepsy display changes in potassium conductances. Eur J Neurosci 12:2087–2096.

Holthoff K, Witte OW (2000) Directed spatial potassium redistribution in rat neocortex. Glia 29:288–292.

Hüttmann K, Sadgrove M, Wallraff A, Hinterkeuser S, Kirchhoff F, Steinhäuser C, Gray WP (2003) Seizures preferentially stimulate proliferation of radial glia-like astrocytes in the adult dentate gyrus: functional and immunocytochemical analysis. Eur J Neurosci 18:2769–2778.

Ivens S, Kaufer D, Flores LP, Bechmann I, Zumsteg D, Tomkins O, Seiffert E, Heinemann U, Friedman A (2007) TGF-beta receptor-mediated albumin uptake into astrocytes is involved in neocortical epileptogenesis. Brain 130:535–547.

Jabs R, Pivneva T, Hüttmann K, Wyczynski A, Nolte C, Kettenmann H, Steinhäuser C (2005) Synaptic transmission onto hippocampal glial cells with hGFAP promoter activity. J Cell Sci 118:3791–3803.

Jansen LA, Uhlmann EJ, Crino PB, Gutmann DH, Wong M (2005) Epileptogenesis and reduced inward rectifier potassium current in tuberous sclerosis complex-1-deficient astrocytes. Epilepsia 46:1871–1880.

John GR, Lee SC, Song X, Rivieccio M, Brosnan CF (2005) IL-1-regulated responses in astrocytes: relevance to injury and recovery. Glia 49:161–176.

Kanemoto K, Kawasaki J, Miyamoto T, Obayashi H, Nishimura M (2000) Interleukin (IL)1beta, IL-1alpha, and IL-1 receptor antagonist gene polymorphisms in patients with temporal lobe epilepsy. Ann Neurol 47:571–574.

Kang N, Xu J, Xu Q, Nedergaard M, Kang J (2005) Astrocytic glutamate release-induced transient depolarization and epileptiform discharges in hippocampal CA1 pyramidal neurons. J Neurophysiol 94:4121–4130.

Kim JH (2001) Pathology of epilepsy. Exp Mol Pathol 70:345–367.

Kivi A, Lehmann TN, Kovacs R, Eilers A, Jauch R, Meencke HJ, von Deimling A, Heinemann U, Gabriel S (2000) Effects of barium on stimulus-induced rises of [K+]o in human epileptic non-sclerotic and sclerotic hippocampal area CA1. Eur J Neurosci 12:2039–2048.

Kofuji P, Newman EA (2004) Potassium buffering in the central nervous system. Neuroscience 129:1045–1056.

Kofuji P, Ceelen P, Zahs KR, Surbeck LW, Lester HA, Newman EA (2000) Genetic inactivation of an inwardly rectifying potassium channel (Kir4.1 subunit) in mice: phenotypic impact in retina. J Neurosci 20:5733–5740.

Köller H, Schroeter M, Jander S, Stoll G, Siebler M (2000) Time course of inwardly rectifying K+ current reduction in glial cells surrounding ischemic brain lesions. Brain Res 872:194–198.

Krishnan B, Armstrong DL, Grossman RG, Zhu ZQ, Rutecki PA, Mizrahi EM (1994) Glial cell nuclear hypertrophy in complex partial seizures. J Neuropathol Exp Neurol 53:502–507.

Lee TS, Eid T, Mane S, Kim JH, Spencer DD, Ottersen OP, de Lanerolle NC (2004) Aquaporin-4 is increased in the sclerotic hippocampus in human temporal lobe epilepsy. Acta Neuropathol 108:493–502.

Li J, Verkman AS (2001) Impaired hearing in mice lacking aquaporin-4 water channels. J Biol Chem 276:31233–31237.

Li J, Patil RV, Verkman AS (2002) Mildly abnormal retinal function in transgenic mice without Müller cell aquaporin-4 water channels. Invest Ophthalmol Vis Sci 43:573–579.

Lin SC, Bergles DE (2004) Synaptic signaling between GABAergic interneurons and oligodendrocyte precursor cells in the hippocampus. Nat Neurosci 7:24–32.

Lowenstein DH, Thomas MJ, Smith DH, McIntosh TK (1992) Selective vulnerability of dentate hilar neurons following traumatic brain injury: a potential mechanistic link between head trauma and disorders of the hippocampus. J Neurosci 12:4846–4853.

Luyken C, Blumcke I, Fimmers R, Urbach H, Elger CE, Wiestler OD, Schramm J (2003) The spectrum of long-term epilepsy-associated tumors: long-term seizure and tumor outcome and neurosurgical aspects. Epilepsia 44:822–830.

Maas S, Patt S, Schrey M, Rich A (2001) Underediting of glutamate receptor GluR-B mRNA in malignant gliomas. Proc Natl Acad Sci USA 98:14687–14692.

MacFarlane SN, Sontheimer H (1997) Electrophysiological changes that accompany reactive gliosis in vitro. J Neurosci 17:7316–7329.

Manley GT, Fujimura M, Ma T, Noshita N, Filiz F, Bollen AW, Chan P, Verkman AS (2000) Aquaporin-4 deletion in mice reduces brain edema after acute water intoxication and ischemic stroke. Nat Med 6:159–163.

Manley GT, Binder DK, Papadopoulos MC, Verkman AS (2004) New insights into water transport and edema in the central nervous system from phenotype analysis of aquaporin-4 null mice. Neuroscience 129:983–991.

Manning TJ Jr, Sontheimer H (1997) Spontaneous intracellular calcium oscillations in cortical astrocytes from a patient with intractable childhood epilepsy (Rasmussen's encephalitis). Glia 21:332–337.

Maragakis NJ, Rothstein JD (2004) Glutamate transporters: animal models to neurologic disease. Neurobiol Dis 15:461–473.

Marcus DC, Wu T, Wangemann P, Kofuji P (2002) KCNJ10 (Kir4.1) potassium channel knockout abolishes endocochlear potential. Am J Physiol Cell Physiol 282:C403–C407.

Margerison JH, Corsellis JA (1966) Epilepsy and the temporal lobes. A clinical, electroencephalographic and neuropathological study of the brain in epilepsy, with particular reference to the temporal lobes. Brain 89:499–530.

Mathern GW, Babb TL, Armstrong DL (1997) Hippocampal sclerosis. In: Epilepsy: a comprehensive textbook (Engel J, Pedley TA, eds), pp 133–155. Philadelphia: Lippincott, Williams & Wilkins.

Mathern GW, Mendoza D, Lozada A, Pretorius JK, Dehnes Y, Danbolt NC, Nelson N, Leite JP, Chimelli L, Born DE, Sakamoto AC, Assirati JA, Fried I, Peacock WJ, Ojemann GA, Adelson PD (1999) Hippocampal GABA and glutamate transporter immunoreactivity in patients with temporal lobe epilepsy. Neurology 52:453–472.

Matthias K, Kirchhoff F, Seifert G, Hüttmann K, Matyash M, Kettenmann H, Steinhäuser C (2003) Segregated expression of AMPA-type glutamate receptors and glutamate transporters defines distinct astrocyte populations in the mouse hippocampus. J Neurosci 23:1750–1758.

May MJ, Ghosh S (1998) Signal transduction through NF-kappa B. Immunol Today 19:80–88.

Metea MR, Newman EA (2006) Glial cells dilate and constrict blood vessels: a mechanism of neurovascular coupling. J Neurosci 26:2862–2870.

Mitchell LA, Jackson GD, Kalnins RM, Saling MM, Fitt GJ, Ashpole RD, Berkovic SF (1999) Anterior temporal abnormality in temporal lobe epilepsy: a quantitative MRI and histopathologic study. Neurology 52:327–336.

Moldrich RX, Aprico K, Diwakarla S, O'Shea RD, Beart PM (2002) Astrocyte mGlu(2/3)-mediated cAMP potentiation is calcium sensitive: studies in murine neuronal and astrocyte cultures. Neuropharmacology 43:189–203.

Mosbacher J, Schoepfer R, Monyer H, Burnashev N, Seeburg PH, Ruppersberg JP (1994) A molecular determinant for submillisecond desensitization in glutamate receptors. Science 266:1059–1062.

Mulligan SJ, MacVicar BA (2004) Calcium transients in astrocyte endfeet cause cerebrovascular constrictions. Nature 431:195–199.

Nagelhus EA, Mathiisen TM, Ottersen OP (2004) Aquaporin-4 in the central nervous system: cellular and subcellular distribution and coexpression with Kir4.1. Neuroscience 129:905–913.

Neely JD, Amiry-Moghaddam M, Ottersen OP, Froehner SC, Agre P, Adams ME (2001) Syntrophin-dependent expression and localization of aquaporin-4 water channel protein. Proc Natl Acad Sci USA 98:14108–14113.

Nemani VM, Binder DK (2005) Emerging role of gap junctions in epilepsy. Histol Histopathol 20:253–259.

Neusch C, Weishaupt JH, Bahr M (2003) Kir channels in the CNS: emerging new roles and implications for neurological diseases. Cell Tissue Res 311:131–138.

Neusch C, Papadopoulos N, Muller M, Maletzki I, Winter SM, Hirrlinger J, Handschuh M, Bahr M, Richter DW, Kirchhoff F, Hülsmann S (2006) Lack of the Kir4.1 channel subunit abolishes K+ buffering properties of astrocytes in the ventral respiratory group: impact on extracellular K+ regulation. J Neurophysiol 95:1843–1852.

Newman EA (1986) High potassium conductance in astrocyte endfeet. Science 233:453–454.

Newman EA (1993) Inward-rectifying potassium channels in retinal glial (Müller) cells. J Neurosci 13:3333–3345.

Newman EA, Karwoski CJ (1989) Spatial buffering of light-evoked potassium increases by retinal glial (Müller) cells. Acta Physiol Scand Suppl 582:51.

Newman EA, Frambach DA, Odette LL (1984) Control of extracellular potassium levels by retinal glial cell K+ siphoning. Science 225:1174–1175.
Nielsen S, Nagelhus EA, Amiry-Moghaddam M, Bourque C, Agre P, Ottersen OP (1997) Specialized membrane domains for water transport in glial cells: high-resolution immunogold cytochemistry of aquaporin-4 in rat brain. J Neurosci 17:171–180.
Niermann H, Amiry-Moghaddam M, Holthoff K, Witte OW, Ottersen OP (2001) A novel role of vasopressin in the brain: modulation of activity-dependent water flux in the neocortex. J Neurosci 21:3045–3051.
Nishiyama A, Yang Z, Butt A (2005) Astrocytes and NG2-glia: what's in a name? J Anat 207:687–693.
Notenboom RG, Hampson DR, Jansen GH, van Rijen PC, van Veelen CW, van Nieuwenhuizen O, de Graan PN (2006) Up-regulation of hippocampal metabotropic glutamate receptor 5 in temporal lobe epilepsy patients. Brain 129:96–107.
Olsen ML, Sontheimer H (2004) Mislocalization of Kir channels in malignant glia. Glia 46:63–73.
Orkand RK, Nicholls JG, Kuffler SW (1966) Effect of nerve impulses on the membrane potential of glial cells in the central nervous system of amphibia. J Neurophysiol 29:788–806.
Ounsted C, Glaser GH, Lindsay J, Richards P (1985) Focal epilepsy with mesial temporal sclerosis after acute meningitis. Arch Neurol 42:1058–1060.
Padmawar P, Yao X, Bloch O, Manley GT, Verkman AS (2005) K+ waves in brain cortex visualized using a long-wavelength K+-sensing fluorescent indicator. Nature Methods 2:825–827.
Pan E, Stringer JL (1996) Influence of osmolality on seizure amplitude and propagation in the rat dentate gyrus. Neurosci Lett 207:9–12.
Papadopoulos MC, Manley GT, Krishna S, Verkman AS (2004) Aquaporin-4 facilitates reabsorption of excess fluid in vasogenic brain edema. FASEB J 18:1291–1293.
Parent JM, von dem Bussche N, Lowenstein DH (2006) Prolonged seizures recruit caudal subventricular zone glial progenitors into the injured hippocampus. Hippocampus 16:321–328.
Petroff OA, Errante LD, Rothman DL, Kim JH, Spencer DD (2002) Glutamate-glutamine cycling in the epileptic human hippocampus. Epilepsia 43:703–710.
Proper EA, Hoogland G, Kappen SM, Jansen GH, Rensen MG, Schrama LH, van Veelen CW, van Rijen PC, van Nieuwenhuizen O, Gispen WH, de Graan PN (2002) Distribution of glutamate transporters in the hippocampus of patients with pharmaco-resistant temporal lobe epilepsy. Brain 125:32–43.
Ransom BR (1996) Do glial gap junctions play a role in extracellular ion homeostasis? In: Gap junctions in the nervous system (Dermietzel R, Spray DC, eds), pp 159–173. Georgetown, TX: Landes Bioscience.
Rasmussen T (1975) Surgery of epilepsy associated with brain tumors. Adv Neurol 8:227–239.
Ravizza T, Vezzani A (2006) Status epilepticus induces time-dependent neuronal and astrocytic expression of interleukin-1 receptor type I in the rat limbic system. Neuroscience 137:301–308.
Rogawski MA, Loscher W (2004) The neurobiology of antiepileptic drugs. Nat Rev Neurosci 5:553–564.
Roper SN, Obenhaus A, Dudek FE (1992) Osmolality and nonsynaptic epileptiform bursts in rat CA1 and dentate gyrus. Ann Neurol 31:81–85.
Rothstein JD, Dykes-Hoberg M, Pardo CA, Bristol LA, Jin L, Kuncl RW, Kanai Y, Hediger MA, Wang Y, Schielke JP, Welty DF (1996) Knockout of glutamate transporters reveals a major role for astroglial transport in excitotoxicity and clearance of glutamate. Neuron 16:675–686.
Rothstein JD, Patel S, Regan MR, Haenggeli C, Huang YH, Bergles DE, Jin L, Dykes Hoberg M, Vidensky S, Chung DS, Toan SV, Bruijn LI, Su ZZ, Gupta P, Fisher PB (2005) Beta-lactam antibiotics offer neuroprotection by increasing glutamate transporter expression. Nature 433:73–77.
Rozengurt N, Lopez I, Chiu CS, Kofuji P, Lester HA, Neusch C (2003) Time course of inner ear degeneration and deafness in mice lacking the Kir4.1 potassium channel subunit. Hear Res 177:71–80.
Rutecki PA, Lebeda FJ, Johnston D (1985) Epileptiform activity induced by changes in extracellular potassium in hippocampus. J Neurophysiol 54:1363–1374.

Samuelsson C, Kumlien E, Flink R, Lindholm D, Ronne-Engstrom E (2000) Decreased cortical levels of astrocytic glutamate transport protein GLT-1 in a rat model of posttraumatic epilepsy. Neurosci Lett 289:185–188.

Santhakumar V, Ratzliff AD, Jeng J, Toth Z, Soltesz I (2001) Long-term hyperexcitability in the hippocampus after experimental head trauma. Ann Neurol 50:708–717.

Schröder W, Hager G, Kouprijanova E, Weber M, Schmitt AB, Seifert G, Steinhäuser C (1999) Lesion-induced changes of electrophysiological properties in astrocytes of the rat dentate gyrus. Glia 28:166–174.

Schröder W, Hinterkeuser S, Seifert G, Schramm J, Jabs R, Wilkin GP, Steinhäuser C (2000) Functional and molecular properties of human astrocytes in acute hippocampal slices obtained from patients with temporal lobe epilepsy. Epilepsia 41 Suppl 6:S181–S184.

Schröder W, Seifert G, Hüttmann K, Hinterkeuser S, Steinhäuser C (2002) AMPA receptor-mediated modulation of inward rectifier K+ channels in astrocytes of mouse hippocampus. Mol Cell Neurosci 19:447–458.

Schwartzkroin PA, Baraban SC, Hochman DW (1998) Osmolarity, ionic flux, and changes in brain excitability. Epilepsy Res 32:275–285.

Seifert G, Rehn L, Weber M, Steinhäuser C (1997) AMPA receptor subunits expressed by single astrocytes in the juvenile mouse hippocampus. Mol Brain Res 47:286–294.

Seifert G, Schröder W, Hinterkeuser S, Schumacher T, Schramm J, Steinhäuser C (2002) Changes in flip/flop splicing of astroglial AMPA receptors in human temporal lobe epilepsy. Epilepsia 43 Suppl 5:162–167.

Seifert G, Hüttmann K, Schramm J, Steinhäuser C (2004) Enhanced relative expression of glutamate receptor 1 flip AMPA receptor subunits in hippocampal astrocytes of epilepsy patients with Ammon's horn sclerosis. J Neurosci 24:1996–2003.

Seifert G, Schilling K, Steinhäuser C (2006) Astrocyte dysfunction in neurological disorders: a molecular perspective. Nat Rev Neurosci 7:194–206.

Seiffert E, Dreier JP, Ivens S, Bechmann I, Tomkins O, Heinemann U, Friedman A (2004) Lasting blood–brain barrier disruption induces epileptic focus in the rat somatosensory cortex. J Neurosci 24:7829–7836.

Simard M, Nedergaard M (2004) The neurobiology of glia in the context of water and ion homeostasis. Neuroscience 129:877–896.

Simard M, Arcuino G, Takano T, Liu QS, Nedergaard M (2003) Signaling at the gliovascular interface. J Neurosci 23:9254–9262.

Somjen GG (2002) Ion regulation in the brain: implications for pathophysiology. Neuroscientist 8:254–267.

Sommer W (1880) Erkrankung des Ammonshorn als aetiologisches Moment der Epilepsie. Arch Psychiatr Nervenkr 10:631–675.

Steinhäuser C, Seifert G (2002) Glial membrane channels and receptors in epilepsy: impact for generation and spread of seizure activity. Eur J Pharmacol 447:227–237.

Takano T, Tian GF, Peng W, Lou N, Libionka W, Han X, Nedergaard M (2006) Astrocyte-mediated control of cerebral blood flow. Nat Neurosci 9:260–267.

Tanaka K, Watase K, Manabe T, Yamada K, Watanabe M, Takahashi K, Iwama H, Nishikawa T, Ichihara N, Kikuchi T, Okuyama S, Kawashima N, Hori S, Takimoto M, Wada K (1997) Epilepsy and exacerbation of brain injury in mice lacking the glutamate transporter GLT-1. Science 276:1699–1702.

Tang FR, Lee WL (2001) Expression of the group II and III metabotropic glutamate receptors in the hippocampus of patients with mesial temporal lobe epilepsy. J Neurocytol 30:137–143.

Tang FR, Lee WL, Yeo TT (2001) Expression of the group I metabotropic glutamate receptor in the hippocampus of patients with mesial temporal lobe epilepsy. J Neurocytol 30:403–411.

Temkin NR, Dikmen SS, Wilensky AJ, Keihm J, Chabal S, Winn HR (1990) A randomized, double-blind study of phenytoin for the prevention of post-traumatic seizures. N Engl J Med 323:497–502.

Temkin NR, Dikmen SS, Anderson GD, Wilensky AJ, Holmes MD, Cohen W, Newell DW, Nelson P, Awan A, Winn HR (1999) Valproate therapy for prevention of posttraumatic seizures: a randomized trial. J Neurosurg 91:593–600.

Tessler S, Danbolt NC, Faull RL, Storm-Mathisen J, Emson PC (1999) Expression of the glutamate transporters in human temporal lobe epilepsy. Neuroscience 88:1083–1091.
Thiele EA (2004) Managing epilepsy in tuberous sclerosis complex. J Child Neurol 19:680–686.
Tian GF, Azmi H, Takano T, Xu Q, Peng W, Lin J, Oberheim N, Lou N, Wang X, Zielke HR, Kang J, Nedergaard M (2005) An astrocytic basis of epilepsy. Nat Med 11:973–981.
Traynelis SF, Dingledine R (1988) Potassium-induced spontaneous electrographic seizures in the rat hippocampal slice. J Neurophysiol 59:259–276.
Traynelis SF, Dingledine R (1989) Role of extracellular space in hyperosmotic suppression of potassium-induced electrographic seizures. J Neurophysiol 61:927–938.
Trombley IK, Mirra SS (1981) Ultrastructure of tuberous sclerosis: cortical tuber and subependymal tumor. Ann Neurol 9:174–181.
Uesugi H, Shimizu H, Maehara T, Arai N, Nakayama H (2000) Presence of human herpesvirus 6 and herpes simplex virus detected by polymerase chain reaction in surgical tissue from temporal lobe epileptic patients. Psychiatry Clin Neurosci 54:589–593.
Uhlmann EJ, Wong M, Baldwin RL, Bajenaru ML, Onda H, Kwiatkowski DJ, Yamada K, Gutmann DH (2002) Astrocyte-specific TSC1 conditional knockout mice exhibit abnormal neuronal organization and seizures. Ann Neurol 52:285–296.
Vajda Z, Pedersen M, Fuchtbauer EM, Wertz K, Stodkilde-Jorgensen H, Sulyok E, Doczi T, Neely JD, Agre P, Frokiaer J, Nielsen S (2002) Delayed onset of brain edema and mislocalization of aquaporin-4 in dystrophin-null transgenic mice. Proc Natl Acad Sci USA 99:13131–13136.
Van Paesschen W, Revesz T, Duncan JS, King MD, Connelly A (1997) Quantitative neuropathology and quantitative magnetic resonance imaging of the hippocampus in temporal lobe epilepsy. Ann Neurol 42:756–766.
Verkhratsky A, Steinhäuser C (2000) Ion channels in glial cells. Brain Res Rev 32:380–412.
Verkman AS (2005) More than just water channels: unexpected cellular roles of aquaporins. J Cell Sci 118:3225–3232.
Vezzani A, Granata T (2005) Brain inflammation in epilepsy: experimental and clinical evidence. Epilepsia 46:1724–1743.
Vezzani A, Conti M, De Luigi A, Ravizza T, Moneta D, Marchesi F, De Simoni MG (1999) Interleukin-1beta immunoreactivity and microglia are enhanced in the rat hippocampus by focal kainate application: functional evidence for enhancement of electrographic seizures. J Neurosci 19:5054–5065.
Vezzani A, Moneta D, Conti M, Richichi C, Ravizza T, De Luigi A, De Simoni MG, Sperk G, Andell-Jonsson S, Lundkvist J, Iverfeldt K, Bartfai T (2000) Powerful anticonvulsant action of IL-1 receptor antagonist on intracerebral injection and astrocytic overexpression in mice. Proc Natl Acad Sci USA 97:11534–11539.
Virta M, Hurme M, Helminen M (2002) Increased frequency of interleukin-1beta (-511) allele 2 in febrile seizures. Pediatr Neurol 26:192–195.
Volterra A, Meldolesi J (2005) Astrocytes, from brain glue to communication elements: the revolution continues. Nat Rev Neurosci 6:626–640.
Volterra A, Steinhäuser C (2004) Glial modulation of synaptic transmission in the hippocampus. Glia 47:249–257.
Wallraff A, Odermatt B, Willecke K, Steinhäuser C (2004) Distinct types of astroglial cells in the hippocampus differ in gap junction coupling. Glia 48:36–43.
Wallraff A, Köhling R, Heinemann U, Theis M, Willecke K, Steinhäuser C (2006) The impact of astrocytic gap junctional coupling on potassium buffering in the hippocampus. J Neurosci 26:5438–5447.
Walz W (1987) Swelling and potassium uptake in cultured astrocytes. Can J Physiol Pharmacol 65:1051–1057.
Walz W (1992) Mechanism of rapid K+-induced swelling of mouse astrocytes. Neurosci Lett 135:243–246.
Whitney KD, McNamara JO (2000) GluR3 autoantibodies destroy neural cells in a complement-dependent manner modulated by complement regulatory proteins. J Neurosci 20:7307–7316.

Winder DG, Conn PJ (1996) Roles of metabotropic glutamate receptors in glial function and glial-neuronal communication. J Neurosci Res 46:131–137.
Wolf HK, Aliashkevich AF, Blumcke I, Wiestler OD, Zentner J (1997) Neuronal loss and gliosis of the amygdaloid nucleus in temporal lobe epilepsy. A quantitative analysis of 70 surgical specimens. Acta Neuropathol 93:606–610.
Wong M, Ess KC, Uhlmann EJ, Jansen LA, Li W, Crino PB, Mennerick S, Yamada KA, Gutmann DH (2003) Impaired glial glutamate transport in a mouse tuberous sclerosis epilepsy model. Ann Neurol 54:251–256.
Wroblewska B, Santi MR, Neale JH (1998) *N*-acetylaspartylglutamate activates cyclic AMP-coupled metabotropic glutamate receptors in cerebellar astrocytes. Glia 24:172–179.
Xiong ZQ, Stringer JL (1999) Astrocytic regulation of the recovery of extracellular potassium after seizures in vivo. Eur J Neurosci 11:1677–1684.
Yaari Y, Konnerth A, Heinemann U (1986) Nonsynaptic epileptogenesis in the mammalian hippocampus in vitro. II. Role of extracellular potassium. J Neurophysiol 56:424–438.
Ye ZC, Sontheimer H (1999) Glioma cells release excitotoxic concentrations of glutamate. Cancer Res 59:4383–4391.
Ye ZC, Rothstein JD, Sontheimer H (1999) Compromised glutamate transport in human glioma cells: reduction-mislocalization of sodium-dependent glutamate transporters and enhanced activity of cystine-glutamate exchange. J Neurosci 19:10767–10777.
Zonta M, Angulo MC, Gobbo S, Rosengarten B, Hossmann KA, Pozzan T, Carmignoto G (2003) Neuron-to-astrocyte signaling is central to the dynamic control of brain microcirculation. Nat Neurosci 6:43–50.

Abbreviations

AED	Antiepileptic drug
AMPA	α-Amino-3-hydroxy-5-methyl-isoxazole propionate
AQPs	Aquaporins
cAMP	Cyclic adenosine monophosphate
CNS	Central nervous system
ECS	Extracellular space
GABA	Gamma-amino butyric acid
GFAP	Glial fibrillary acidic protein
IL	Interleukin
mGluR	Metabotropic glutamate receptor
MTS	Mesial temporal sclerosis
TLE	Temporal lobe epilepsy
TS	Tuberous sclerosis

Chapter 26
Hepatic Encephalopathy: A Primary Astrocytopathy

Roger F. Butterworth

Contents

26.1 Introduction

Results of neuropathological and molecular biological studies indicate that hepatic encephalopathy (HE) in both acute and chronic liver failure is primarily a disorder of astrocytes. Although neuronal cell death has been described in end-stage liver failure (Butterworth, 2007), its prevalence and severity are variable and generally considered to be insufficient to explain the wide range of neuropsychiatric symptoms that are characteristic of HE. HE is therefore considered to be a classical example of a primary astrocytopathy. This chapter is a critical review of astrocyte

R.F. Butterworth
Neuroscience Research Unit, CHUM, University of Montreal, Montreal, Canada
roger.butterworth@umontreal.ca

V. Parpura and P.G. Haydon (eds.), *Astrocytes in (Patho)Physiology of the Nervous System*, DOI: 10.1007/978-0-387-79492-1_26, © Springer Science+Business Media, LLC 2009

morphology and function in HE. The implication of alterations in expression of key astrocyte proteins and astrocyte metabolism are reviewed. Two major new discoveries, namely the presence of vesicular release proteins and of functional *N*-methyl-D-aspartate (NMDA) receptors on astrocytes have resulted in the need for reinterpretation of previous findings in relation to both neuron–astrocyte and astrocyte–astrocyte signaling. The majority of the citations are from studies in human HE material and in material from animal models of acute or chronic liver failure. This is supplemented, where appropriate, with findings from studies in cultured astrocytes exposed to ammonia, a major putative neurotoxin that accumulates in brain in liver failure.

26.2 Astrocyte Pathology in HE

Von Hösslin and Alzheimer (1912) described morphological abnormalities of astrocytes in a disorder known as Westphal-Strümpell pseudosclerosis, a disorder shown subsequently to be identical to acquired hepatocerebral degeneration (Wilson's disease). Waggoner and Malamud (1942) went on to coin the phrase "Alzheimer type II astrocyte" to describe these characteristic morphological features manifested by astrocytes that consist of large, pale (watery-looking) nuclei, margination of the chromatin pattern, and the presence of prominent nucleoli as shown in Fig. 26.1. Intranuclear glycogen inclusions are also evident in these cells.

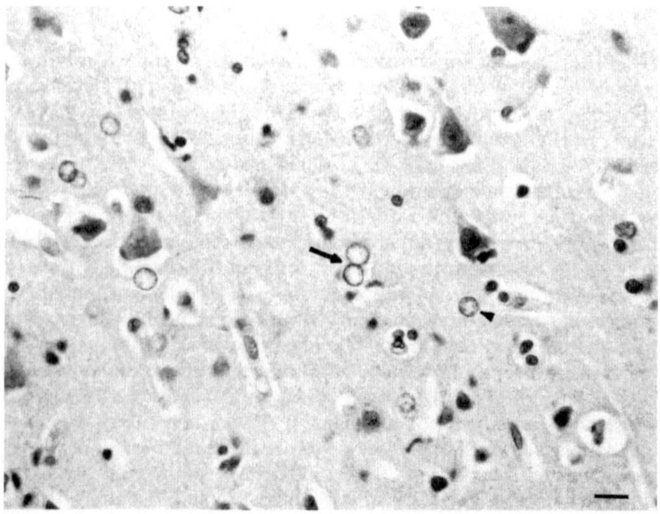

a

Fig. 26.1 Alzheimer type II astrocytosis in hepatic encephalopathy (chronic liver failure). Light micrograph of cerebral cortex from a cirrhotic patient who died in hepatic coma. Note prominence of astroglial nuclei that are pale, enlarged frequently occurring in pairs (arrow) suggestive of hyperplasia. A normal astrocyte nucleus is shown for comparison purposes (arrowhead). Bar: 20 μm. From Norenberg (1987).

Hepatic encephalopathy in chronic liver failure, regardless of the etiology of liver disease, is characterized by the presence of Alzheimer type II astrocytes. The number of cells showing the Alzheimer type II phenotype is significantly correlated with the severity of encephalopathy (Adams and Foley, 1953; Norenberg, 1987; Butterworth et al., 1987). Alzheimer type II astrocytes are found in both gray and white matter of HE brains where the nuclei take on a variety of shapes from round (in cerebral cortex) to irregular or lobulated forms (in basal ganglia), and in both cases Alzheimer type II cells occur in pairs or triplets suggestive of hyperplasia (Norenberg, 1987).

Studies in experimental animal models of HE continue to help to characterize the early morphologic changes in astrocytes. For example, feeding of ammonia cation exchange resins to rats following end-to-side portacaval anastomosis results in severe encephalopathy (Norenberg, 1987) and, in early stages of HE in these animals, astrocytes exhibit evidence of hypertrophy characterized by increased size and number of mitochondria and endoplasmic reticulum (Fig. 26.2). Later stages of encephalopathy (coma) are accompanied by hydropic alterations, contractions of mitochondria and, ultimately, degenerative changes.

Mixed glial-neuronal cultures exposed to sera from HE patients and from animals with experimental HE developed morphological changes characteristic of Alzheimer type II astrocytes (Mossakowski et al., 1970), and exposure of cultured rat cortical astrocytes to ammonia, the principal putative neurotoxin generated in liver failure, results in changes that mimic the in vivo findings consisting, at the light microscopic level, of increased cytoplasmic basophilia, vacuolization, and cellular disintegration (Gregorios et al., 1985a). Ultrastructural studies show that the initial response

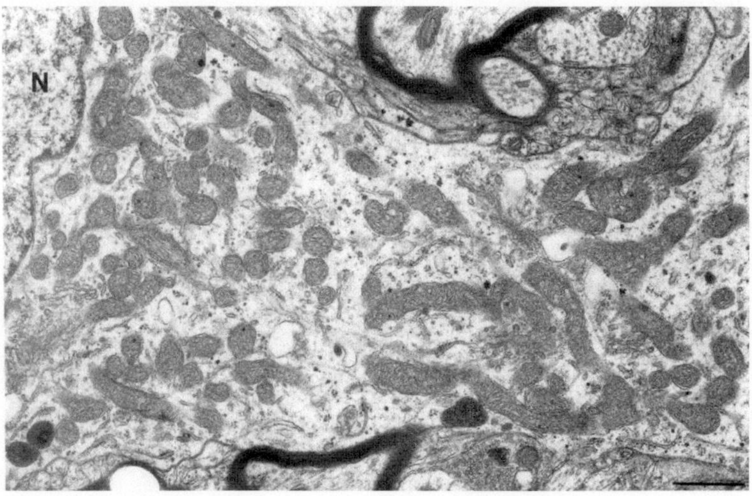

Fig. 26.2 Early changes in astrocytic morphology in experimental hepatic encephalopathy (chronic liver failure). Electron micrograph of an astrocyte process showing increased number of mitochondria from an animal with mild hepatic encephalopathy resulting from feeding of ammonia resins following end-to-side portacaval anastomosis. *N* nucleus; Bar: 1 μm. From Norenberg (1987).

consists of proliferation of mitochondria and smooth endoplasmic reticulum and the appearance of dense bodies resembling lipofuscin granules (Gregorios et al., 1985b). Loss of intermediate filaments has also been described in astrocytes in human HE (Sobel et al., 1981) and in ammonia-exposed astrocytes in culture (Norenberg, 1987).

26.3 Astrocyte Metabolism in HE

Portacaval anastomosis in the rat results in decreased brain glucose utilization, and positron emission tomography (PET) studies in cirrhotic patients with mild HE reveal decreased brain glucose utilization that, at early stages, is restricted to the anterior cingulate cortex (Lockwood et al., 1997) a brain structure associated with the processing of information and control of attention. Precipitation of severe encephalopathy in portacaval-shunted rats is accompanied by increased brain lactate (Hindfelt et al., 1977; Therrien et al., 1991) and, ultimately, at prolonged coma stages, by a fall in brain ATP levels (Hindfelt et al., 1977). Whether these metabolic alterations in whole brain in chronic liver failure reflect changes in neurons, astrocytes, or both cell types had not been established until the advent of spectroscopic techniques. Using ^{13}C nuclear magnetic resonance (NMR) spectroscopy and ^{13}C-labeled acetate, a substrate used preferentially by astrocytes, it has been reported that portacaval anastomosis did not result in increased synthesis of lactate in brain whereas incorporation of ^{13}C glucose into lactate (primarily in neurons) was increased by 30% (Sonnewald et al., 1996). On the other hand, exposure of cultured cortical astrocytes to millimolar concentrations of ammonia results in increased lactate synthesis and in increased expression of the lactate dehydrogenase (LDH) isoforms LDH-1 and LDH-5 (Chan et al., 2002). It was proposed that the increased brain lactate synthesis in hyperammonemia conditions including HE is a consequence of inhibition of the tricarboxylic acid cycle enzyme α-ketoglutarate dehydrogenase (Lai and Cooper, 1986). There is evidence to suggest that brain lactate accumulation is causally related to the phenomenon of cytotoxic brain edema (astrocyte swelling) in acute liver failure (ALF). Consistent with this notion are the reports of increased brain lactate concentrations that are positively correlated with electroencephalogram changes and the presence of brain edema in ALF rats (Deutz et al., 1988) together with the report of significant swelling of cultured cortical astrocytes exposed to lactate (Staub et al., 1990). ^{13}C-NMR studies also reveal increased brain lactate synthesis in ALF rats (Zwingmann et al., 2003), and mild hypothermia sufficient to prevent brain edema and to delay the onset of severe encephalopathy in rats with ALF results in normalization of brain lactate synthesis (Chatauret et al., 2003). Taken together, these findings from a wide variety of experimental paradigms provide convincing evidence for a role of increased brain lactate in the pathophysiology of the neuropsychiatric complications characteristic of HE.

Exposure of cultured cortical astrocytes to ammonia results in oxidative stress (Murthy et al., 2001), and antioxidants have been shown to attenuate astrocytic swelling (Jayakumar et al., 2006) and to restore the high-affinity astrocytic glutamate transport deficit (Chan et al., 2000) caused by ammonia exposure of these cells. Increased expression of the nitric oxide synthase (NOS) isoform NOS-I has been described in the brains of rats following portacaval anastomosis (mild HE) (Raghavendra Rao et al., 1997). Increased activities of the inducible nitric oxide synthase (iNOS) isoform have also been described in the brains of these animals as well as in cultured astrocytes exposed to ammonia (Schliess et al., 2002). Not surprisingly, protein tyrosine nitration was also observed in the brains of portacaval-shunted rats and ammonia-exposed astrocytes, which included nitration of astrocytic proteins such as glutamine synthetase (GS) and the *peripheral-type* benzodiazepine receptor (Schliess et al., 2002). Evidence of protein tyrosine nitration was also evident in the brains of ammonia-treated portacaval-shunted animals (Master et al., 1999). However, attempts so far to prevent HE, brain edema, or hyperemia resulting from ALF under experimental conditions using NOS inhibitors have been unsuccessful (Larsen et al., 2001). Other evidence of oxidative/nitrosative stress in experimental HE includes the finding of increased expression of hemoxygenase HO-1 in the brains of rats with experimental ALF due to hepatic devascularization (Sawara et al., 2005).

Although it is well established that glutamine concentrations are significantly increased in cerebrospinal fluid and brain in both experimental and human HE and that severity of HE is positively correlated with brain glutamine concentrations (Laubenberger et al., 1997), the mechanism responsible for brain glutamine accumulation has not been definitively established. The GS reaction is a relatively simple one-step reaction whereby the substrate (glutamate) is amidated to glutamine, a reaction that is ATP-dependent. There is no evidence for induction of brain GS in either acute or chronic liver failure; on the contrary, as discussed in this review, there is evidence to suggest that brain GS activities are decreased due at least in part to protein tyrosine nitration as discussed earlier (Fig. 26.3). Furthermore, both biochemical (Cremer et al., 1975; Ukida et al., 1988) and ^{13}C-NMR spectroscopy (Zwingmann et al., 2003) studies convincingly show that de novo glutamine synthesis in brain in ALF is not increased. Substrate (glutamate) availability in the metabolic pool is unchanged (Zwingmann et al., 2003) as is availability of ATP, the GS cofactor (Bates et al., 1989). These findings suggest that brain glutamine accumulation in liver failure results from other factors such as decreased glutamine degradation by glutaminase or from inhibition of astrocytic glutamine release. In this regard, it remains unclear whether or not astrocytes in situ express significant quantities of glutaminase.

26.4 Astrocyte Function in HE

26.4.1 Structural Proteins

Glial fibrillary acidic protein (GFAP) is the major protein of intermediate filaments in differentiated astrocytes. Results of a recent study show that GFAP mRNA and protein expression are significantly reduced in frontal cortex of rats with ALF

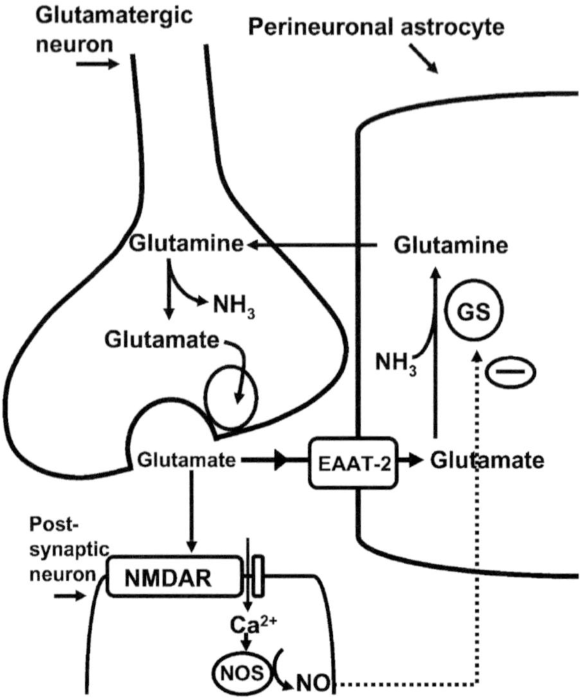

Fig. 26.3 Neuron-astrocytic trafficking: the glutamate-glutamine cycle. Glutamate released from the presynaptic nerve terminal stimulates the postsynaptic *N*-methyl-D-aspartate receptors (NMDAR). Excess glutamate in forebrain is rapidly cleared from the synapse primarily by the high-affinity astrocytic glutamate transporters EAAT-2. Astrocytic glutamate is amidated to glutamine by the astrocyte-specific enzyme glutamine synthetase (GS), the major mechanism for ammonia removal by the brain, which can be inhibited by nitric oxide (NO). A portion of the glutamine formed is available for transport into the presynaptic neuron as the immediate precursor of releasable glutamate. *NOS* nitric oxide synthase.

resulting from hepatic devascularization (Bélanger et al., 2002). These findings were selective for GFAP; expression of a second glial neurofilament protein S-100β was unchanged in the brains of these animals (Table 26.1). It was suggested that the loss of GFAP and the resulting impairment of viscoelastic properties of the astrocyte could facilitate cell swelling leading to brain edema and its complications in ALF. Exposure of cultured cortical astrocytes to millimolar concentrations of ammonia results in loss of GFAP expression (Neary et al., 1994; Bélanger et al., 2002), and it was proposed that ammonia exposure under these conditions led to destabilization of GFAP mRNA.

GFAP expression in brain has also been studied in both experimental and human *chronic* liver failure where it was reported to be decreased or unchanged depending upon the brain region under investigation. For example, GFAP-immunolabeling of cerebral cortical astrocytes was reportedly decreased following end-to-side portacaval

Table 26.1 Astrocytic proteins in hepatic encephalopathy

Structural proteins	
Glial fibrillary acidic protein	Increased (ALF,CLF)
S-100β	Unchanged (ALF, CLF)
Transporter proteins	
Glucose transporter (GLUT-1)	Increased (ALF)
Glutamate transporter (EAAT-1)	Decreased (ALF)
Glutamate transporter (EAAT-2)	Decreased (ALF)
Glycine transporter (GLYT-1)	Decreased (ALF)
Receptor proteins	
Isoquinoline binding protein subunit of peripheral-type benzodiazepine receptor	Increased (ALF,CLF)
Glutamate (NMDA) receptor	Unchanged (ALF) Decreased (CLF)
Enzymes	
Lactate dehydrogenase (LDH-1, LDH-5)	Increased (ALF)
Glutamine synthetase	Unchanged (ALF) Decreased (CLF)
Hemoxygenase (HO-1)	Increased (ALF)

ALF acute liver failure, *CLF* chronic liver failure

anastomosis in the rat (Norenberg, 1987) and in cerebrum of patients with chronic liver failure (Sobel et al., 1981). On the other hand, GFAP immunolabeling of cerebellar Bergmann glia in human chronic liver failure was unaltered (Kril and Butterworth, 1996).

26.4.2 Glutamine Synthetase

Glutamine synthetase (GS) activities are decreased in the brains of animals with chronic liver failure (Girard et al., 1993) and in autopsied brain tissue from cirrhotic patients who died in hepatic coma (Lavoie et al., 1987). Loss of GS activity in brain in liver failure contrasts with the situation in skeletal muscle where enzyme activities are significantly increased (Girard et al., 1993; Desjardins et al., 1999a) due to increased GS mRNA and protein (Desjardins et al., 1999a). The lack of induction of GS in brain in liver failure undoubtedly results from tyrosine nitration of the protein (Schliess et al., 2002) (Fig. 26.3) (see earlier discussion). Decreased capacity for de novo glutamine synthesis by astrocytes in experimental acute liver failure has been directly demonstrated by ^{13}C-NMR (Zwingmann et al., 2003). Decreased capacity for brain glutamine synthesis coupled with the absence of other metabolic pathways in brain with significant ammonia-metabolizing capacity (brain is devoid of a urea cycle) results in increased brain:blood ammonia ratios and brain ammonia concentrations as high as 5 mM (Swain et al., 1992).

26.4.3 Glutamate Transporters

Primary cultures of rat cortical astrocytes exposed to millimolar concentrations of ammonia manifest significant reductions in high-affinity uptake of glutamate and its nonmetabolizable analogue D-aspartate (Bender and Norenberg, 1996; Chan et al., 2000). Studies of transport kinetics in ammonia-exposed cells revealed a reduction of V_{max} indicative of a loss of transporter protein, and subsequent studies confirmed a loss of expression of the l-glutamate/l-aspartate transporter (GLAST), also referred to as excitatory amino acid transporter 1 (EAAT1), in these cells. Significant reductions of both mRNA and protein expression of the astrocyte l-glutamate transporter (GLT-1), also referred to as excitatory amino acid transporter 2 (EAAT2), were reported in the brains of rats with ischemic (Knecht et al., 1997) or toxic (Norenberg et al., 1997) liver failure, and immunohistochemical assessment of cerebella from rats with chronic liver failure showed significant reductions in both EAAT-1 and EAAT-2 immunostaining (Suarez et al., 2000). A subsequent study revealed loss in expression of the neuronal glutamate transporter EAAT-3 in cultured rat cerebellar granule cells exposed to ammonia; however, in contrast to the astrocyte transporters, the mechanism involved was posttranscriptional in nature (Chan et al., 2003) (Table 26.1).

The precise mechanism responsible for ammonia's inhibitory effects on high-affinity glutamate transporter expression by astrocytes has not been elucidated. However, there is evidence to suggest that oxidative stress is implicated (Chan and Butterworth, 2006).

26.4.4 Glutamate Release

Exposure of rat hippocampal slices to millimolar concentrations of ammonia results in increased spontaneous release of glutamate (Hamberger et al., 1982). Increased Ca^{2+}-dependent release of glutamate was also reported in cortical slices from animals with ALF due to thioacetamide-induced hepatotoxicity as well as from hippocampal slices from animals with chronic liver failure resulting from end-to-side portacaval anastomosis (reviewed in Szerb and Butterworth, 1992). However, whether this increase of glutamate release results from increased release from neurons or astrocytes (or both cell types) has not been definitely established. Like neurons, astrocytes express the full complement of exocytotic proteins and Ca^{2+}-dependent exocytosis of astrocyte vesicles, and the subsequent release of glutamate has been directly visualized by the technique of total internal reflection fluorescence imaging (reviewed in Volterra and Meldolesi, 2005). This issue has been directly addressed by Rose et al. (2005) who studied the effects of ammonia on Ca^{2+}-dependent release of glutamate from cultured rat cortical astrocytes using fluorescence imaging techniques. In these latter studies, ammonia-induced release of glutamate was unaffected by known glutamate transport inhibitors but was accompanied by transient intracellular alkalinization resulting in increased glutamate exocytosis in a Ca^{2+}-dependent manner confirming that increased vesicular release is a predominant mechanism responsible for ammonia-induced release of glutamate from astrocytes (Rose, 2006).

26.4.5 NMDA Receptors

Compelling evidence for the existence of NMDA receptors on astrocytes initially came from the discovery of the expression of NMDA receptor subunits (NR1 and NR2) at both the mRNA and protein level (Conti et al., 1996, 1999). Second, both membrane currents and intracellular Ca^{2+} increases were described in astrocytes from cortical slices exposed to NMDA (Schipke et al., 2001), and, most convincing of all, astrocytes acutely isolated from cortical slices and investigated under conditions of voltage and concentration clamp manifest NMDA-activated currents that are sensitive to NMDA receptor antagonists (Lalo et al., 2006). Most importantly from these latter studies, astrocytic NMDA receptors were found to be activated following the electrical stimulation of neuronal afferents. These findings have led to the opening of a new chapter in the realm of intercellular signaling in the central nervous system and support previous reports on the effects of ammonia exposure or of liver failure on NMDA receptor function, which will need to be reevaluated in the context of this notion of the presence of both neuronal and astrocytic NMDA receptors (Fig. 26.4).

Exposure of cortical astrocytes to ammonia results in a wide range of molecular and functional changes including cell swelling, decreased glutamate uptake (Sect. 26.4.3), increased glutamate release (Sect. 26.4.4), altered glycine transport (Zwingmann et al., 2002), altered expression of the glucose transporter GLUT-1 (Bélanger et al., 2006), reduced expression of the structural protein GFAP (Bélanger et al., 2002) as well as oxidative and nitrosative stress (Norenberg, 2003) (Table 26.1). The precise nature and consequences of ammonia-induced oxidative/nitrosative stress in astrocytes have recently become apparent. Ammonia induces protein tyrosine nitration in cultured rat astrocytes (Schliess et al., 2002), an event that is accompanied by a rise in intracellular Ca^{2+}, induction of iNOS, and phosphorylation of the mitogen-activated protein kinases (MAPKs) Erk-1/Erk-2 and $p38^{MAPK}$. MAPKs inhibition were subsequently shown to attenuate astrocyte swelling following ammonia treatment (Jayakumar et al., 2006). Most importantly, protein tyrosine nitration in ammonia-exposed cells was sensitive to NMDA receptor antagonists (Schliess et al., 2002).

Extrapolation of the findings of the effects of ammonia on astrocyte function to the situation in vivo is, at times, problematic. Primary cortical astrocytes in culture express proteins that may be distinct from those expressed by astrocytes in situ, and exposure of cells in culture to ammonia results in cellular energy compromise and in cell death (Gregorios et al., 1985a, b), phenomena that are not encountered in brain astrocytes in situ in experimental animal models of acute or chronic liver failure where astrocyte swelling and Alzheimer type II changes are observed (Sect. 26.2). NMDA receptor antagonists prevent the death of animals administered with

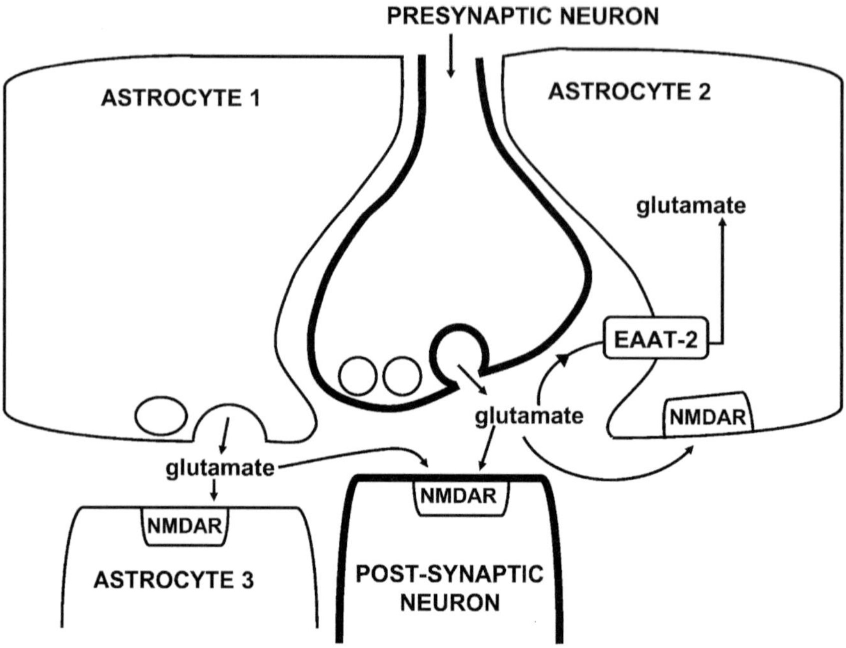

Fig. 26.4 Glutamate signaling between neurons and astrocytes and between astrocytes and astrocytes mediated by the glutamate transporter EAAT-2 and by NMDA receptors on both the postsynaptic neuron and perineuronal astrocytes.

lethal doses of ammonia salts (Kosenko et al., 1995), an action that appears to result from prevention of ATP depletion in the brains of these animals. One report described a protective effect of the mild NMDA receptor antagonist memantine on the cerebral consequences of ALF (Vogels et al., 1997) but the cellular mechanism(s) responsible for this beneficial effect were not elucidated. Clearly, more studies are needed in this rapidly evolving area of research.

26.4.6 Peripheral-Type (Mitochondrial) Benzodiazepine Receptors

A consistent finding in HE is increased expression of the *peripheral-type* (mitochondrial) benzodiazepine receptor (PTBR) (Table 26.1). PTBR is a heteromeric complex comprising three subunit proteins namely the isoquinoline binding protein (IBP, 18 kDa), the voltage-dependent anion channel (32 kDa), and the adenine nucleotide carrier (30 kDa). PTBR is expressed not only in peripheral tissues (adrenal glands and kidneys) but also in the brain where it is localized primarily in glial cells (astrocytes and microglia). End-stage chronic liver failure in humans results in increased densities of the PTBR ligand PK11195 whether measured biochemically in autopsied brain tissue (Lavoie et al., 1990) or using PET in patients with mild HE (Cagnin et al., 2001). Subsequent studies using immunoblotting techniques confirmed that the increased densities of PK11195 sites in human HE were the consequences of increased expression of the IBP subunit of PTBR (Bélanger et al., 2004). In this latter study, a significant correlation was observed between increased IBP expression and the presence of Alzheimer type II astrocytosis suggesting that the PTBR changes are implicated in the pathogenesis of the Alzheimer type II phenotype. The PET studies in HE patients revealed bilateral increases of PTBR sites in dorsolateral prefrontal cortex, pallidum, and putamen, and the magnitude of increase in cortical areas was correlated with the degree of cognitive impairment in these patients (Cagnin et al., 2001).

Portacaval anastomosis in the rat results in increased PTBR sites in both the periphery (kidney) and brain (Raghavendra Rao et al., 1994). Increased PTBR sites in the brains of these animals are apparent as early as 24 h following portacaval anastomosis (Leong et al., 1994) and are present in different amounts in cerebellum: pons > thalamus, cerebral cortex > hippocampus > striatum. A subsequent study confirmed that these increases in PTBR sites resulted from increased IBP mRNA (Desjardins et al., 1999b); the voltage-dependent anion channel subunit of PTBR was unaffected by portacaval shunting.

Experimental ALF resulting from either ischemic liver damage (Desjardins and Butterworth, 2002) or administration of hepatotoxins such as thioacetamide (Kadota et al., 1996) or azoxymethane (Bélanger et al., 2004) leads to increased IBP mRNA and increased PK11195 binding sites in brain. The fact that both acute and chronic liver failure leads to increased PTBR expression in brain suggests a major role for ammonia toxicity since brain ammonia is significantly increased in both conditions. Evidence consistent with a role for ammonia includes the findings of increased PTBR sites in the brains and peripheral tissues of mice with chronic hyperammonemia resulting from a congenital deficit of the urea cycle enzyme ornithine transcarbamylase (Raghavendra Rao et al., 1993) in which the neuropathologic finding again includes Alzheimer type II astrocytosis (Michalak and Butterworth, 1997). Finally, exposure of cultured cortical astrocytes to millimolar concentrations of ammonia results in increased PTBR sites (Itzhak and Norenberg, 1994). However, Alzheimer type II changes are rarely observed in these cells (Gregorios et al., 1985a, b). Additionally, exposure of cultured astrocytes to manganese ion, a toxic compound found to be increased in basal ganglia structures of HE patients (Pomier Layrargues et al., 1995), results in both PTBR expression increases and Alzheimer type II astrocytosis (Hazell et al., 1999).

Numerous *endogenous ligands* for PTBR have been identified. One such ligand is diazepam-binding inhibitor (DBI), an 11-kDa polypeptide. DBI is highly localized in astrocytes, and cerebrospinal fluid levels of DBI are increased in HE patients (Rothstein et al., 1989). Octadecaneuropeptide is a processing product of DBI with high affinity for the PTBR, and portacaval anastomosis in the rat results in increased octadecaneuropeptide immunolabeling of nonneuronal elements (astrocytes and ependymal cells) in cerebral cortex (Butterworth et al., 1991).

The PTBR is implicated in the regulation of a wide range of cellular functions including cell proliferation, immunomodulation, apoptosis, porphyrin transport, and steroid synthesis (Casellas et al., 2002). Exposure of cultured glioma cells to PTBR agonists results in swelling and proliferation of mitochondria (Shiraishi et al.,

1991) similar to that observed in astrocytes in experimental chronic liver failure (Fig. 26.2).

PTBR is an essential component of the steroidogenic process. Activation of PTBR results in increased transport of cholesterol from the outer to inner mitochondrial membrane. Thereafter, enzymes of the P_{450} family catalyze conversion of cholesterol to pregnenolone, the precursor of a novel class of compounds known as neurosteroids (Fig. 26.5).

De novo synthesis of neurosteroids has been demonstrated experimentally following PTBR activation, and astrocytic enzymes involved have been identified (Mellon and Griffin, 2002). The neurosteroid 3α-5α-tetrahydroprogesterone (also known as allopregnanolone) has potent neuroinhibitory properties. Allopregnanolone enhances gamma-aminobutyric acid (GABA)-elicited chloride currents, and by action at a distinct neurosteroid modulatory site on the GABA-A receptor complex, it positively modulates the binding of both GABA and benzodiazepines to their respective sites on the complex. Allopregnanolone is also synthesized following PTBR activation in adrenals and testes, and it readily crosses the blood–brain barrier. Brain concentrations of allopregnanolone are significantly increased in autopsied brain tissue from cirrhotic patients who died in hepatic coma (Ahboucha et al., 2005) leading to the suggestion that the presence of this neurosteroid, rather than increased concentrations of endogenous benzodiazepines as had previously been proposed, is the basis of the phenomenon of *increased GABAergic tone* in HE (Ahboucha et al., 2005). Increased brain concentrations of GABA agonist neurosteroids have also been described in experimental animal models of ALF

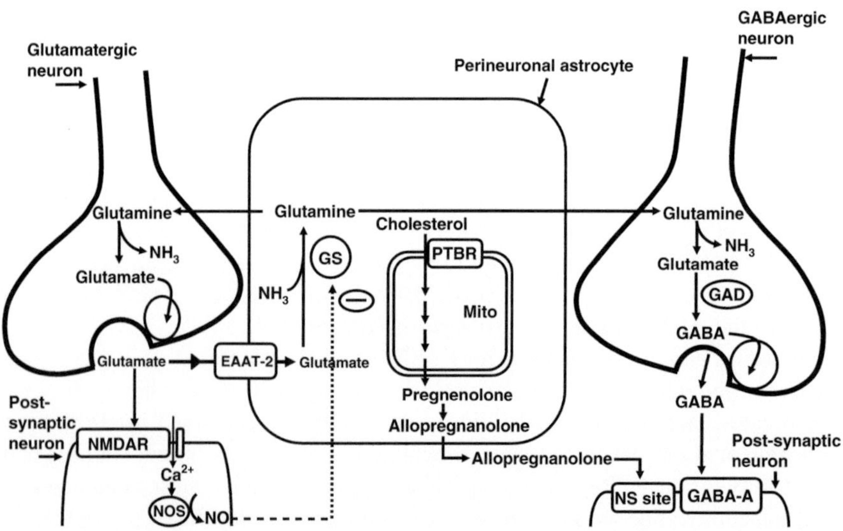

Fig. 26.5 Modulation of both excitatory (glutamatergic) and inhibitory (GABAergic) transmission involving key astrocytic proteins [EAAT-2 and peripheral-type benzodiazepine receptor (PTBR) shown here]. Activation of PTBR leads to synthesis of neurosteroid agonist such as allopregnanolone acting at the steroid modulatory site on the GABA-A receptor complex.

(Itzhak et al., 1995; Ahboucha and Butterworth, 2007) in which increased PTBR sites were concomitantly increased. The principal interest in the role of neurosteroids in HE results from the ability of these compounds to recapitulate many of the neuropsychiatric symptoms that are characteristic of HE is humans including altered sleep patterns, depression, and cognitive impairment. The development of pharmacologic strategies aimed at inhibition of allopregnanolone synthesis or antagonists of the neurosteroid site on the GABA-A receptor complex have the potential to provide new therapeutic approaches to HE.

26.4.7 Cell Volume Regulation

Astrocyte swelling is a common occurrence in hyperammonemic conditions, and ammonia-induced astrocytic swelling is an important element in the pathogenesis of brain edema in ALF (Vaquero et al., 2003). Cell swelling has been observed following exposure of cultured astrocytes and brain slices to millimolar concentrations of ammonia (Norenberg et al., 1991; Ganz et al., 1989), as well as in the brains of animal models of ALF and hyperammonemia (Tanigami et al., 2005; Willard-Mack et al., 1996), and ALF patients (Kato et al., 1992). Some studies suggest that ammonia-induced astrocytic swelling results from the accumulation of glutamine, an organic osmolyte. These conclusions were based on the ability of methionine-*S*-sulfoximine, an inhibitor of glutamine synthetase, to block ammonia-induced astrocytic swelling in vitro (Norenberg and Bender, 1994) and in vivo (Tanigami et al., 2005). A more recent study using cultured astrocytes, however, did not reveal a significant correlation between glutamine synthesis and astrocytic swelling but rather revealed that glutamine-mediated oxidative stress was a more likely cause of astrocytic swelling (Jayakumar et al., 2006).

Ammonia-induced astrocytic swelling may also result from the capacity of ammonia to alter levels of expression of genes implicated in cell volume regulation. These include an upregulation of the glucose transporter GLUT-1 (Bélanger et al., 2006), which also acts as a water channel (Fischbarg et al., 1990), upregulation of the water channel protein aquaporin IV (Margulies et al., 1999), and downregulation of GFAP, a key component of the cytoskeletal network implicated in cell volume regulation (Cornet et al., 1993).

26.5 Intercellular Signaling in HE

Evidence for trafficking of substrates between neurons and astrocytes and between astrocytes and neighboring astrocytes under normal physiological conditions is now overwhelming. Examples of such signaling molecules include glutamate [released by neurons and astrocytes and transported into these cells by high-affinity transporters (Fig. 26.4)], lactate [synthesized preferentially in astrocytes and presumably shuttled to neurons as an alternative energy source (Magistretti et al.,

1999)], glutamine (synthesized almost uniquely in astrocytes and subsequently released and taken up by neurons as the immediate precursor of releasable pool of glutamate), and neurosteroids such as allopregnanolone [synthesized by astrocytes (Fig. 26.5) and released into the synaptic space where it has potent modulatory actions on the postsynaptic neuronal GABA-A receptor complex]. Another example is ammonia itself, which is synthesized primarily in neurons by the action of glutaminase and removed almost exclusively by the action of glutamine synthetase, an astrocytic enzyme (Sect. 26.4.2). These intercellular signaling pathways are shown in a simplified schematic manner in Fig. 26.5.

Liver failure has significant effects on all of the aforementioned signaling pathways by virtue of the effects of ammonia on glutamate transport and release as described in Sects. 26.4.3 and 26.4.4, as well as on lactate synthesis and on neurosteroid synthesis via activation of PTBR (Sect. 26.4.5). Such effects of ammonia have the capacity to modify the basis of neural excitation and inhibition via the glutamate and GABA systems and consequently provide a cogent explanation for the complex and rapidly evolving symptomatology that is characteristic of HE.

26.6 Inflammation and Proinflammatory Cytokines

Clinical studies reveal a high incidence of the so-called *systemic inflammatory response syndrome* (SIRS) in patients with ALF (Rolando et al., 2000). SIRS is a response to the presence of proinflammatory cytokines including the interleukins IL-1 and IL-6 as well as tumor necrosis factor α (TNF-α). Increased circulating levels of these cytokines resulting from either infection or inflammation due to liver necrosis have been reported in ALF patients (Nagaki et al., 2000), and arteriovenous difference studies in these patients suggest a net production of proinflammatory cytokines in the brain in ALF (Jalan et al., 2002). It was subsequently demonstrated that experimental ALF led to increased brain concentrations of IL-1β and that the magnitude of this increase was predictive of the severity of HE and of the presence of brain edema (Jiang et al., 2006). Brain levels of IL-1β in this latter study were significantly higher than circulating levels suggestive of local brain production, and it was suggested that increased brain levels of IL-1β could contribute to the loss of astrocytic glutamate transporter capacity in ALF. It has been proposed that proinflammatory cytokines in brain in liver failure act synergistically with ammonia (Blei, 2004).

26.7 Implications for Therapy

Treatment of HE continues to rely on reduction of blood and brain ammonia using agents aimed at reduction of gut ammonia production (lactulose, antibiotics, probiotics) or increased ammonia removal by residual hepatocytes or muscle (l-ornithine l-aspartate).

A novel therapy currently undergoing controlled clinical trials involves the use of mild hypothermia for prevention and treatment of HE and intracranial hypertension in patients with ALF. Beneficial effects of mild hypothermia are multiple (reviewed in Vaquero et al., 2005) and include normalization of astrocytic glutamate transport, of lactate synthesis, and of PTBR activation and prevention of oxidative/nitrosative stress and cytokine accumulation.

Clinical neuropharmacologic approaches aimed specifically at astrocyte metabolism have been slow to evolve in spite of substantial evidence from studies in animal models of HE. Inhibitors of GS such as methionine sulfoximine partially prevent brain edema in portacaval-shunted rats administered with ammonium salts to precipitate brain edema and intracranial hypertension (Blei et al., 1994); however, the inherent toxicity of methionine sulfoximine precludes its use in the clinic. The NMDA receptor antagonist memantine was shown to attenuate HE severity in experimental ALF (Vogels et al., 1997). However, whether this beneficial effect of memantine was mediated by action at the neuronal or astrocytic NMDA receptor (or both) has not been established.

Antioxidants and NOS inhibitors prevent swelling in ammonia-exposed cultured cortical astrocytes (Norenberg, 2003). However, NOS inhibitors were not effective in prevention of the cerebral hemodynamic changes in experimental liver failure (Larsen et al., 2001).

Studies in experimental animal models of ALF have consistently shown that benzodiazepine receptor partial inverse agonists are effective in the prevention and treatment of HE (Bosman et al., 1991; Meyer et al., 1998). A novel mechanism of action has recently been proposed involving the ability of these agents to inhibit modulation of the GABA receptor complex by neurosteroids (Ahboucha et al., 2006). As agents acting at the astrocytic PTBR and/or astrocytic neurosteroid synthesis become available, it is anticipated that new therapeutic strategies aimed at the normalization of *GABAergic tone* will emerge.

Acknowledgments Studies performed in the author's laboratory were funded by the Canadian Institutes for Health Research (CHIR) and the Canadian Liver Foundation (CLF).

References

Adams RD, Foley JM (1953) The neurological disorder associated with liver disease. In: Metabolic and toxic diseases of the nervous system (Merritt HH, Hare CC, eds.), Vol. 32, pp. 198–237. Baltimore: Williams and Wilkins.

Ahboucha S, Butterworth RF (2007) The neurosteroid system: implication in the pathophysiology of hepatic encephalopathy. Neurochem Int 52: 575–587.

Ahboucha S, Pomier-Layrargues G, Mamer O, Butterworth RF (2005) Increased brain concentrations of a neuroinhibitory steroid in human hepatic encephalopathy. Ann Neurol 58: 169–170.

Ahboucha S, Coyne L, Hirakawa R, Butterworth RF, Halliwell RF (2006) An interaction between benzodiazepines and neuroactive steroids at GABAA receptors in cultured hippocampal neurons. Neurochem Int 48: 703–707.

Bates TE, Williams SR, Kauppinen RA, Godian DG (1989) Observation of cerebral metabolites in an animal model of ALF *in vivo*: 1H and 31P nuclear magnetic resonance study. J Neurochem 53: 102–110.

Bélanger M, Desjardins P, Chatauret N, Butterworth RF (2002) Loss of expression of glial fibrillary acidic protein in acute hyperammonemia. Neurochem Int 41: 155–160.

Bélanger M, Ahboucha S, Desjardins P, Butterworth RF (2004) Upregulation of peripheral-type (mitochondrial) benzodiazepine receptors in hyperammonemic syndromes: consequences for neuronal excitability. Adv Mol Cell Biol 31: 983–997.

Bélanger M, Desjardins P, Chatauret N, Butterworth RF (2006) Selectively increased expression of the astrocytic/endothelial glucose transporter protein GLUT1 in acute liver failure. Glia 53: 557–562.

Bender AS, Norenberg MD (1996) Effects of ammonia on l-glutamate uptake in cultured astrocytes. Neurochem Res 21: 567–573.

Blei AT (2004) Infection, inflammation and hepatic encephalopathy, synergism redefined. SJ Hepatol 40: 327–330.

Blei AT, Olafsson S, Therrien G, Butterworth RF (1994) Ammonia-induced brain edema and intracranial hypertension in rats after portacaval anastomosis. Hepatology 19: 1437–1444.

Bosman DK, van den Buijs CA, de Haan JG, Maas MA, Chamuleau RA (1991) The effects of benzodiazepine-receptor antagonists and partial inverse agonists on acute hepatic encephalopathy in the rat. Gastroenterology 101: 772–781.

Butterworth RF (2007) Neuronal cell death in hepatic encephalopathy. Metab Brain Dis 3–4: 291–308.

Butterworth RF, Giguère JF, Michaud J, Lavoie J, Layrargues GP (1987) Ammonia: key factor in the pathogenesis of hepatic encephalopathy. Neurochem Pathol 6: 1–12.

Butterworth RF, Tonon M-C, Désy L, Giguère J-F, Vaudry H, Pelletier G (1991) Increased brain content of the endogenous benzodiazepine receptor ligand, octadecaneuropeptide (ODN), following portacaval anastomosis in the rat. Peptides 12: 119–125.

Cagnin A, Taylor-Robinson SD, Forton DM, Banati RB (2001) In vivo quantification of cerebral "peripheral benzodiazepine binding site" in minimal hepatic encephalopathy: a [11C]R-PK11195 positron emission tomography study. J Hepatol 34 (Suppl 1): 58.

Casellas P, Galiegue S, Basile AS (2002) Peripheral benzodiazepine receptors and mitochondrial function. Neurochem Int 40: 475–486.

Chan H, Butterworth RF (2006) Glutamatergic synaptic regulation deficit in liver failure: a review of molecular mechanisms. In: Hepatic encephalopathy and nitrogen metabolism (Haüssinger D, Kircheis G, Schliess F, eds.), pp. 160–170. Dordrecht: Springer.

Chan H, Hazell AS, Desjardins P, Butterworth RF (2000) Effects of ammonia on glutamate transporter (GLAST) protein and mRNA in cultured rat cortical astrocytes. Neurochem Int 37: 243–248.

Chan H, Bélanger M, Chatauret N, Butterworth RF (2002) Brain lactate accumulation in liver failure and hyperammonemia results from increased lactate dehydrogenase LDH-5 isoform expression and altered LDH enzyme kinetics in astrocytes. Hepatology 36: 628.

Chan H, Zwingmann C, Pannunzio M, Butterworth RF (2003) Effects of ammonia on high affinity glutamate uptake and glutamate transporter EAAT3 expression in cultured rat cerebellar granule cells. Neurochem Int 43: 137–146.

Chatauret N, Zwingmann C, Rose C, Leibfritz D, Butterworth RF (2003) Effects of hypothermia on brain glucose metabolism in acute liver failure: a 1H/13C-nuclear magnetic resonance study. Gastroenterology 125: 815–824.

Conti F, DeBiasi S, Minelli A, Melone M (1996) Expression of NR1 and NR2A/B subunits of the NMDA receptor in cortical astrocytes. Glia 17: 254–258.

Conti F, Barbaresi P, Melone M, Ducati A (1999) Neuronal and glial localization of NR1 and NR2A/B subunits of the NMDA receptor in the human cerebral cortex. Cereb Cortex 9: 110–120.

Cornet M, Ubl J, Kolb HA (1993) Cytoskeleton and ion movements during volume regulation in cultured PC12 cells. J Membr Biol 133: 161–170.

Cremer JE, Heath DF, Teal HM, Woods MS, Cavanagh JB (1975) Some dynamic aspect of brain metabolism in rats given a portacaval anastomosis. Neuropathol Appl Neurobiol 3: 293–311.

Desjardins P, Butterworth RF (2002) The "peripheral-type" benzodiazepine (omega 3) receptor in hyperammonemic disorders. Neurochem Int 41: 109–114.

Desjardins P, Rama Rao VK, Michalak A, Rose C, Butterworth RF (1999a) Effect of portacaval anastomosis on glutamine synthetase protein and gene expression in brain, liver and skeletal muscle. Metab Brain Dis 14: 273–282.

Desjardins P, Bandeira P, Raghavendra Rao VL, Butterworth RF (1999b) Portacaval anastomosis causes selective alterations of peripheral-type benzodiazepine receptor expression in rat brain and peripheral tissues. Neurochem Int 35: 293–299.

Deutz NEP, De Graaf AA, De Haan JG, Bovée WMMJ, Chamuleau RAFM (1988) *In vivo* brain 1H-NMR spectroscopy (1-NMRS) during acute hepatic encephalopathy (HE). In: Advances in ammonia metabolism and hepatic encephalopathy (Soeters PB, Wilson JHP, Meijer AJ, Holm E, eds.), pp. 439–446. Amsterdam: Excerpta Medica.

Fischbarg J, Kuang KY, Vera JC, Arant S, Silverstein SC, Loike J, Rosen OM (1990) Glucose transporters serve as water channels. Proc Natl Acad Sci USA 87: 3244–3247.

Ganz R, Swain M, Traber P, DalCanto M, Butterworth RF, Blei AT (1989) Ammonia-induced swelling of rat cerebral cortical slices: implications for the pathogenesis of brain edema in acute hepatic failure. Metab Brain Dis 4: 213–223.

Girard G, Giguère J-F, Butterworth RF (1993) Region-selective reductions in activities of glutamine synthetase in rat brain following portacaval anastomosis. Metab Brain Dis 8: 207–215.

Gregorios JB, Mozes LW, Norenberg L-OB, Norenberg MD (1985a) Morphologic effects of ammonia on primary astrocyte cultures. I. Light microscopic studies. J Neuropathol Exp Neurol 44: 397–403.

Gregorios JB, Mozes LW, Norenberg MD (1985b) Morphologic effects of ammonia on primary astrocyte cultures. II. Electron microscopic studies. J Neuropathol Exp Neurol 44: 404–414.

Hamberger A, Lindroth P, Nyström B (1982) Regulation of glutamate biosynthesis and release in vitro by low levels of ammonium ions. Brain Res 237: 339–350.

Hazell AS, Desjardins P, Butterworth RF (1999) Chronic exposure of rat primary astrocyte cultures to manganese results in increased binding sites for the "peripheral-type" benzodiazepine receptor ligand 3H-PK11195. Neurosci Lett 271: 5–8.

Hindfelt B, Plum F, Duffy TE (1977) Effects of acute ammonia intoxication on cerebral metabolism in rats with portacaval shunts. J Clin Invest 59: 386–396.

Itzhak Y, Norenberg MD (1994) Ammonia-induced upregulation of peripheral-type benzodiazepine receptors in cultured astrocytes labeled with [3H]-PK11195. Neurosci Lett 177: 35–38.

Itzhak Y, Roig-Cantisano A, Dombro RS, Norenberg MD (1995) Acute liver failure and hyperammonemia increase peripheral-type benzodiazepine receptor binding and pregnenolone synthesis in mouse brain. Brain Res 705: 345–348.

Jalan R, Pollok A, Shah SH, Madhavan K, Simpson KJ (2002) Liver derived pro-inflammatory cytokines may be important in producing intracranial hypertension in ALF. J Hepatol 37: 536–538.

Jayakumar AR, Rao KV, Murthy CHR, Norenberg MD (2006) Glutamine in the mechanism of ammonia-induced astrocyte swelling. Neurochem Int 48: 623–628.

Jiang W, Qu H, Desjardins P, Chatauret N, Bélanger M, Butterworth RF (2006) Unequivocal evidence for cytokine accumulation in brain in experimental acute liver failure. Hepatology 44 (Suppl 1): 366A.

Kadota Y, Inoue K, Tokunaga R, Taketani S (1996) Induction of peripheral-type benzodiazepine receptors in mouse brain following thioacetamide-induced acute liver failure. Life Sci 58: 953–959.

Kato M, Hughes RD, Keays RT, Williams R (1992) Electron microscopic study of brain capillaries in cerebral edema from fulminant hepatic failure. Hepatology 15: 1060–1066.

Knecht K, Michalak A, Rose C, Rothstein JD, Butterworth RF (1997) Decreased glutamate transporter (GLT-1) expression in frontal cortex of rats with ALF. Neurosci Lett 229: 201–203.

Kosenko E, Kaminsky Y, Grau E, Minana MD, Grisolia S, Felipo V (1995) Nitroarginine, an inhibitor of nitric oxide synthetase, attenuates ammonia toxicity and ammonia-induced alterations in brain metabolism. Neurochem Res 20: 451–456.

Kril JJ, Butterworth RF (1996) Diencephalic and cerebellar pathology in alcoholic and non-alcoholic patients with end-stage liver disease. Hepatology 24: 1303.

Lai JCK, Cooper AJL (1986) Brain -ketoglutarate dehydrogenase: kinetic properties, regional distribution and effects of inhibitors. J Neurochem 47: 1376–1386.

Lalo U, Pankratov Y, Kirchhoff F, North RA, Verkhratsky A (2006) NMDA receptors mediate neuron-to-glia signalling in mouse cortical astrocytes. J Neurosci 26: 2673–2683.

Larsen FS, Gottstein J, Blei AT (2001) Cerebral hyperemia and nitric oxide synthase in rats with ammonia-induced brain edema. J Hepatol 34: 548–554.

Laubenberger J, Haussinger D, Boyer S, Guffer H, Henning J, Lange M (1997) Proton magnetic resonance spectroscopy of brain in symptomatic and asymptomatic patients with liver cirrhosis. Gastroenterology 112: 1610–1616.

Lavoie J, Giguère JF, Pomier Layrargues G, Butterworth RF (1987) Activities of neuronal and astrocytic marker enzymes in autopsied brain tissue from patients with hepatic encephalopathy. Metab Brain Dis 2: 283–290.

Lavoie J, Pomier-Layrargues G, Butterworth RF (1990) Increased densities of peripheral-type benzodiazepine receptors in brain autopsy samples from cirrhotic patients with hepatic encephalopathy. Hepatology 11: 874–878.

Leong DK, Therrien G, Swain MS, Butterworth RF (1994) Densities of binding sites of the "peripheral-type" benzodiazepine receptor ligand 3H-PK11195 are increased in brain 24 hours following portacaval anastomosis. Metab Brain Dis 9: 267–273.

Lockwood AH, Weissenborn K, Butterworth RF (1997) An image of the brain in patients with liver disease. Curr Opin Neurol 10: 525–533.

Magistretti PJ, Pellerin L, Rothman DL, Shulman RG (1999) Energy on demand. Science 283: 496–497.

Margulies JE, Thompson RC, Demetriou AA (1999) Aquaporin-4 water channel is up-regulated in the brain in fulminant hepatic failure. Hepatology 30: 395A.

Master S, Gottstein J, Blei AT (1999) Cerebral blood flow and the development of ammonia-induced brain edema in rats after portacaval anastomosis. Hepatology 30: 876–880.

Mellon SH, Griffin LD (2002) Neurosteroids: biochemistry and clinical significance. Trends Endocrin Metab 13: 35–43.

Meyer HP, Legemate DA, van den Brom W, Rothuizen J (1998) Improvement of chronic hepatic encephalopathy in dogs by the benzodiazepine-receptor partial inverse agonist sarmazenil, but not by the antagonist flumazenil. Metab Brain Dis 13: 241–251.

Michalak A, Butterworth RF (1997) Ornithine transcarbamylase deficiency: pathogenesis of the cerebral disorder and new prospects for therapy. Metab Brain Dis 12: 171–182.

Mossakowski MJ, Renkawek K, Krasnicka Z, Smialek M, Pronaszko A (1970) Morphology and histochemistry of Wilsonian and hepatogenic gliopathy in tissue culture. Acta Neuropathol 16: 1–16.

Murthy CHRK, Rama Rao KV, Bai G, Norenberg MD (2001) Ammonia induced production of free radicals in primary cultures of rat astrocytes. J Neurosci Res 66: 282–288.

Nagaki M, Iwai H, Naiki T, Ohnishi H, Muto Y, Moriwaki H (2000) High levels of serum interleukin-10 and tumor necrosis factor-alpha are associated with fatality in fulminant hepatitis. J Infect Dis 182: 1103–1108.

Neary JT, Whittemore SR, Zhu Q, Norenberg MD (1994) Destabilization of glial fibrillary acidic protein mRNA in astrocytes by ammonia and protection by extracellular ATP. J Neurochem 63: 2021–2027.

Norenberg MD (1987) The role of astrocytes in hepatic encephalopathy. Neurochem Pathol 6: 13–33.

Norenberg MD (2003) Oxidative and nitrosative stress in ammonia neurotoxicity. Hepatology 37: 245–248.

Norenberg MD, Bender AS (1994) Astrocyte swelling in liver failure: role of glutamine and benzodiazepines. Acta Neurochir Suppl (Wien) 60: 24–27.

Norenberg MD, Baker L, Norenberg LO, Blicharska J, Bruce-Gregorios JH, Neary JT (1991) Ammonia-induced astrocyte swelling in primary culture. Neurochem Res 16: 833–836.

Norenberg MD, Huo Z, Neary JT, Roig-Cantesano A (1997) The glial glutamate transporter in hyperammonemia and hepatic encephalopathy: relation to energy metabolism and glutamatergic neurotransmission. Glia 21: 124–133.

Pomier Layrargues G, Spahr L, Butterworth RF (1995) Increased manganese concentrations in pallidum of cirrhotic patients: cause of magnetic resonance hyperintensity? Lancet 345: 735.

Raghavendra Rao VL, Qureshi IA, Butterworth RF (1993) Increased densities of binding sites for the peripheral-type benzodiazepine receptor ligand [3H]-PK11195 in congenital ornithine transcarbamylase-deficient sparse fur mouse. Pediatr Res 34: 777–780.

Raghavendra Rao VL, Audet R, Therrien G, Butterworth RF (1994) Tissue-specific alterations of binding sites for peripheral-type benzodiazepine receptor ligand [3H]-PK11195 in rats following portacaval anastomosis. Dig Dis Sci 39: 1055–1063.

Raghavendra Rao VL, Audet RM, Butterworth RF (1997) Increased neuronal nitric oxide synthase expression in brain following portacaval anastomosis. Brain Res 765: 169–172.

Rolando N, Wade J, Davalos M, Wendon J, Philpott-Howard J, Williams R (2000) The systemic inflammatory response syndrome in acute liver failure. Hepatology 32:734–739.

Rose C (2006) Effect of ammonia on astrocytic glutamate uptake/release mechanisms. J Neurochem 97 (Suppl 1): 11–15.

Rose C, Kresse W, Kettenmann H (2005) Acute ammonia insult results in calcium-dependent glutamate release from cultured astrocytes: an effect of pH. J Biol Chem 280: 20937–20944.

Rothstein JD, McKhann G, Guarneri P, Barbaccia ML, Guidotti A, Costa E (1989) Cerebrospinal fluid content of diazepam binding inhibitor in chronic hepatic encephalopathy. Ann Neurol 26: 57–62.

Sawara K, Desjardins P, Chatauret N, Bélanger M, Kato A, Suzuki K, Butterworth RF (2005) Selective effects of hypothermia on expression of oxidative stress markers in brains of rats with acute liver failure. Hepatology 42 (Suppl 1): 359A.

Schipke CG, Ohlemeyer C, Matyash M, Nolte C, Kettenmann H, Kirchhoff F (2001) Astrocytes of the mouse neocortex express functional *N*-methyl-d-aspartate receptors. FASEB J 15: 1270–1272.

Schliess F, Görg B, Fischer R, Desjardins P, Bidmon HJ, Herrmann A, Butterworth RF, Zilles K, Häussinger D (2002) Ammonia induces MK-801 sensitive nitration and phosphorylation of protein tyrosine residues in rat astrocytes. FASEB J 16: 739–741.

Shiraishi T, Black KL, Ikezaki K, Becker DP (1991) Peripheral benzodiazepine induces morphological changes and proliferation of mitochondria in glioma cells. J Neurosci Res 30: 463–474.

Sobel RA, De Armond SJ, Forno LS, Eng LF (1981) Glial fibrillary acidic protein in hepatic encephalopathy: An immunohistochemical study. J Neuropathol Exp Neurol 40: 625–632.

Sonnewald U, Therrien G, Butterworth RF (1996) Portal-systemic encephalopathy, disorder of neuron-astrocytic metabolic trafficking: evidence from 13C-NMR studies. J Neurochem 67: 1711–1717.

Staub F, Baethmann A, Peters J, Weight H, Keupski O (1990) Effects of lactacidosis on glial cell volume and viability. J Cereb Blood Flow and Metab 10: 866–876.

Suarez I, Bodega G, Fernandez B (2000) Modulation of glutamate transporters (GLAST, GLT-1 and EAAC1) in the rat cerebellum following portacaval anastomosis. Brain Res 859: 293–302.

Swain MS, Bergeron M, Audet R, Blei AT, Butterworth RF (1992) Monitoring of neurotransmitter amino acids by means of an indwelling cisterna magna catheter. A comparison of two rodent models of fulminant hepatic failure. Hepatology 16: 1028–1035.

Szerb JC, Butterworth RF (1992) Effect of ammonium ions on synaptic transmission in the mammalian central nervous system. Prog Neurobiol 39: 135–153.

Tanigami H, Rebel A, Martin LJ, Chen TY, Brusilow SW, Traystman RJ, Koehler RC (2005) Effect of glutamine synthetase inhibition on astrocyte swelling and altered astroglial protein expression during hyperammonemia in rats. Neuroscience 131: 437–449.

Therrien G, Giguère JF, Butterworth RF (1991) Increased cerebrospinal fluid lactate reflects deterioration of neurological status in experimental portal-systemic encephalopathy. Metab Brain Dis 6: 225–231.

Ukida M, Morishita H, Morimoto Y, Usui H, Nagashima H (1988) Limited glutamine synthesis in brains of dog with portacaval anastomosis after 15N-ammonium chloride loading. In: Advances in ammonia metabolism and hepatic encephalopathy (Soeters PB, Wilson JHP, Meijer AJ, Holm E, eds.), pp. 433–438. Amsterdam: Excerpta Medica.

Vaquero J, Chung C, Blei AT (2003) Brain edema in acute liver failure. A window to the pathogenesis of hepatic encephalopathy. Ann Hepatol 2: 12–22.

Vaquero J, Rose C, Butterworth RF (2005) Keeping cool in acute liver failure: rationale for the use of mild hypothermia. J Hepatol 43: 1067–1077.

Vogels BA, Maas MA, Daalhuisen J, Quack G, Chamuleau RA (1997) Memantine, a noncompetitive NMDA receptor antagonist improves hyperammonemia-induced encephalopathy and acute hepatic encephalopathy in rats. Hepatology 25: 820–827.

Volterra A, Meldolesi J (2005) Astrocytes, from brain glue to communication elements: the revolution continues. Nat Rev Neurosci 6: 626–640.

Willard-Mack CL, Koehler RC, Hirata T, Cork LC, Takahashi H, Traystman RJ, Brusilow SW (1996) Inhibition of glutamine synthetase reduces ammonia-induced astrocyte swelling in rat. Neuroscience 71: 589–599.

Zwingmann C, Desjardins P, Chatauret N, Michalak A, Hazell AS, Butterworth RF (2002) Reduced astrocytic glycine transporter Glyt-1 in acute liver failure. Metab Brain Dis 17: 263–274.

Zwingmann C, Chatauret N, Leibfritz D, Butterworth RF (2003) Selective increase of brain lactate synthesis in experimental acute liver failure: results of a [1H-13C] nuclear magnetic resonance study. Hepatology 37: 420–428.

Abbreviations

ALF	Acute liver failure
ATP	Adenosine 5′-triphosphate
HE	Hepatic encephalopathy
DBI	Diazepam binding inhibitor
EAAT	Excitatory amin o acid transporter
GABA	Gamma-aminobutyric acid
GFAP	Glial fibrillary acidic protein
GS	Glutamine synthetase
IBP	Isoquinoline binding protein
IL	Interleukin
LDH	Lactate dehydrogenase
NMDA	*N*-methyl-D-aspartate
NMR	Nuclear magnetic resonance
NOS	Nitric oxide synthase
iNOS	Inducible NOS
MAPK	Mitogen-activated protein kinase
PET	Positron emission tomography
PTBR	*peripheral-type* benzodiazepine receptor

Index

H

I

About the Editors

Vladimir Parpura, M.D., Ph.D holds both a medical degree, awarded from the University of Zagreb in Croatia in 1989, and a doctorate, received in Neuroscience and Zoology from Iowa State University in 1993. He has held faculty appointments at the Department of Zoology and Genetics, Iowa State University and the Department of Cell Biology and Neuroscience, University of California Riverside. He is presently an Associate Professor in the Department of Neurobiology, University of Alabama Birmingham. His current research focuses on understanding the modulation of calcium-dependent glutamate release from astrocytes.

Philip G. Haydon, Ph.D received his doctorate from the University of Leeds, England in 1982. He has held faculty appointments at the Department of Zoology and Genetics, Iowa State University, the Department of Neuroscience at the University of Pennsylvania, and has recently moved to Tufts University School of Medicine as Annetta and Gustav Grisard Professor and Chair of the Department of Neuroscience. His research focuses on the role of astrocytes in the regulation of synapses, neuronal networks, and behavior as well as how these glial cells contribute to neurological disorders.

Color Plates

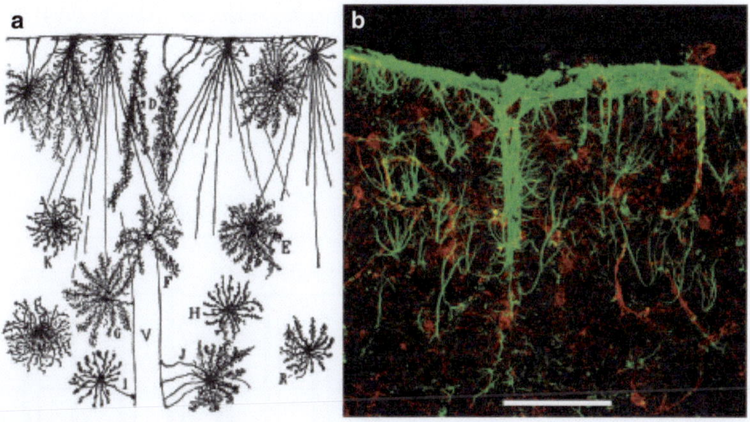

Fig. 1.2 Glial cells in the human and rat brains are morphologically heterogeneous. (a) Astrocytes in the cerebral cortex of a 2-month-old infant stained with the Golgi method. (A–D) are cells in the first cortical lamina and (E–H) are cells in the second and third lamina. (I–J) are cells with end-feet contacting blood vessels. V, blood vessel. From Ramon y Cajal (1913). (b) Staining of cerebral cortex from adult rat. *Top* is surface of cortex showing intense GFAP (green) staining due to the *glia limitans. Red* is for NG2, which stains both NG2(+) cells and blood vessels. Scale bar, 100 μm. Unpublished work of G. Schools.

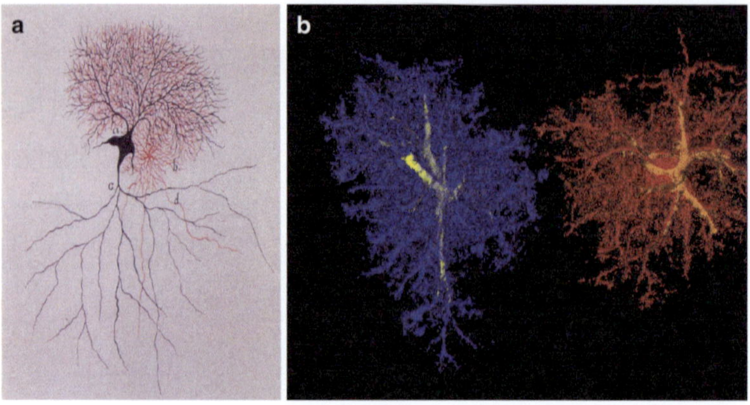

Fig. 1.3 (a) Picture from a chapter on sleep in a book on anesthesia published in 1894 by Carl Ludwig Schleich (1859–1922) in which the author proposed modulation of neuronal currents by swelling and contraction of glia into and out of synapses. A neuron (a), likely a pyramidal cell, in *black* in close contact with a glial cell, presumably an astrocyte, in *red*. See (Dierig, 1994) for further details. (b) Two filled hippocampal astroglial cells from adult rat hippocampus illustrate the domain concept of astroglia. *Yellow* shows GFAP staining, while *blue* and *red* are two dye-filled contiguous astrocytes. From Bushong et al., 2002.

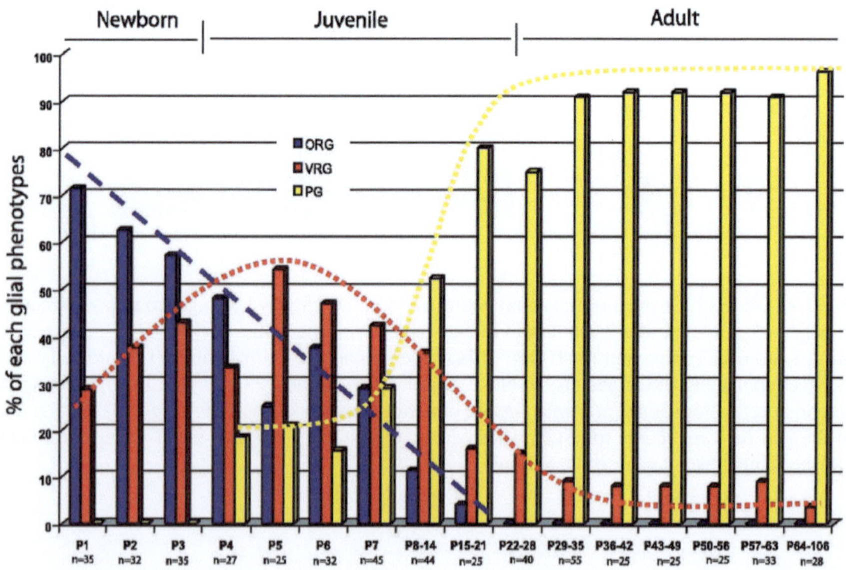

Fig. 1.7 The percentage of glial cells recognized by differential interference contrast optics in a living hippocampal slice, which corresponded to the three electrophysiological phenotypes shown in Fig. 1.6. The numbers under the *x* axis show the post-natal age in days and the number of cells recorded (*n*). On *top* the ages are broadly classified into newborn, juvenile and adult groups. From Zhou et al. (2006).

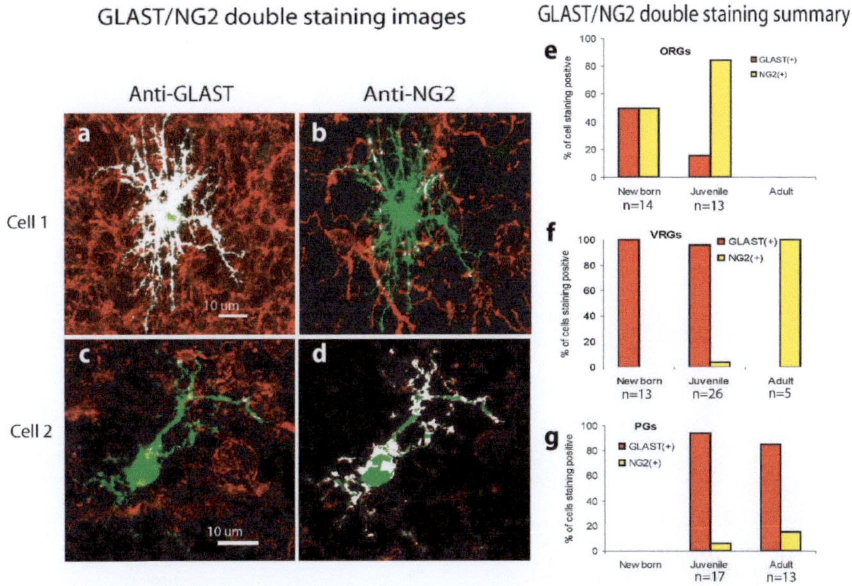

Fig. 1.8 Correlation of electrophysiological phenotype with cell type by immunocytochemical identification of recorded cells. This continues the study shown in Fig. 1.7 by post-recording staining a smaller number of cells (given by *n* for each case) within the three broad age groups identified in Fig. 1.7. The cells were stained for GLAST to identify astrocytes and NG2 to identify this non-astrocytic glial class. (a)–(d) shows examples of staining in a GLAST(+) (cell 1) and an NG2 (+) cell (cell 2). Green is the filling dye and red represents either antibody staining. White represents colour-coded colocalization. As shown in (e), NG2(+) cells shown as *yellow bars* represent outwardly rectifying glial cells (ORGs) equally in the newborn stage and predominantly in the juvenile. *Red* represents GLAST(+) cells. There are no ORGs in the adult. (f) shows that variably rectifying glial cells (VRGs) are all and then predominantly astrocytes, but are only NG2(+) cells in the adult. (g) shows that passive cells are only seen in the juvenile and adult animals and represent mainly astrocytes, although 5–10% are NG2(+) in the adult, but unlike the passive astrocyte these have small Na^+ currents (not shown). See Zhou et al. (2006) for further details.

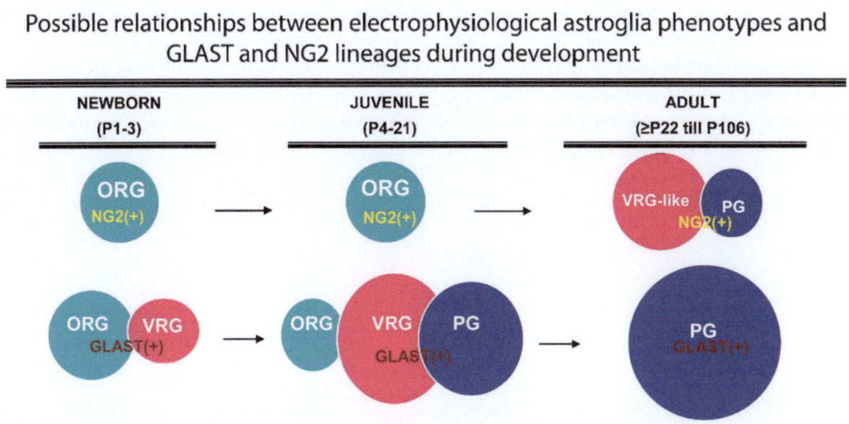

Fig. 1.9 Possible relationships between electrophysiological astroglia phenotypes and GLAST and NG2 lineages during development. From Zhou et al. (2006); based on data in Fig. 1.8.

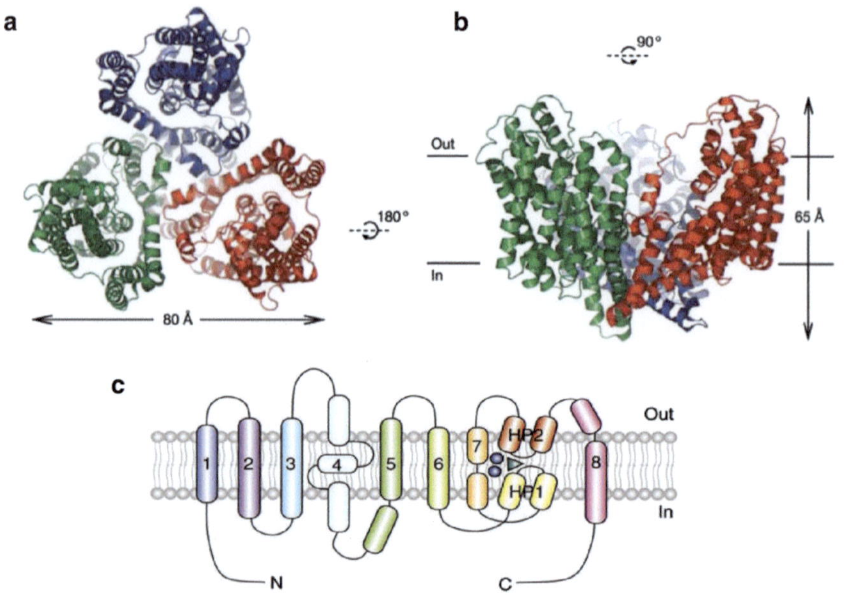

Fig. 4.3 Structure of Gltph in its trimeric conformation and predicted topology model of glutamate transporter. (**a**) Ribbon representation of the trimer of Gltph, in which the protomers are *red, blue, and green*, viewed from the extracellular side of the membrane. (**b**) View of the trimer parallel to the membrane. (**c**) A model for the membrane topology of Gltph and for mammalian members of the glutamate transporter family (SLC1). The binding sites occupied by Na^+ (*dark blue dots*) and the substrate aspartate (*green triangle*) are noted at their approximate position within the structure. The C-terminal translocation core includes the hairpin loops HP1 (*yellow*) and HP2 (*red*), which are proposed to serve as intracellular and extracellular gates, respectively, and also the transmembrane domains TM7 (*orange*) and TM8 (*magenta*). Adapted from Yernool et al. (2004), Nature.

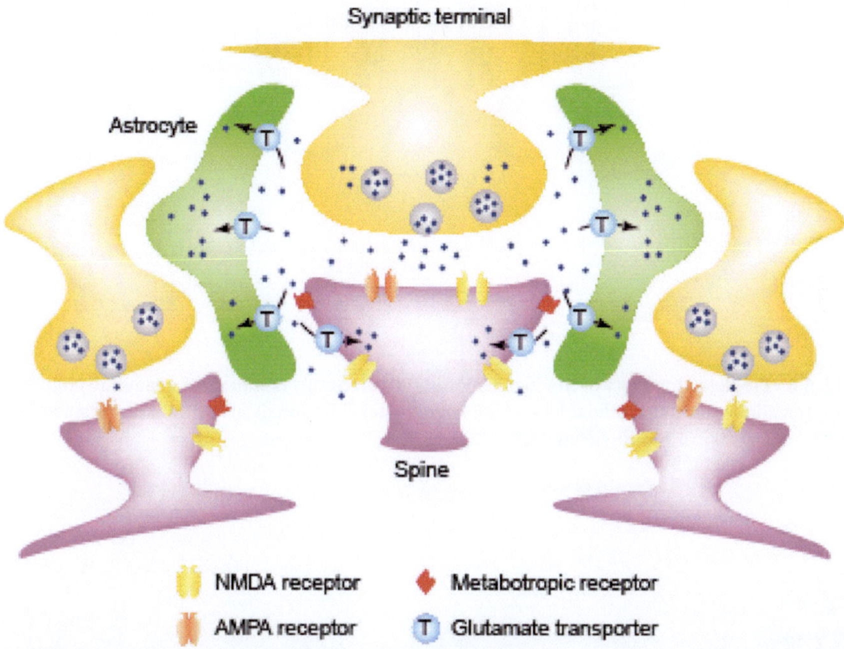

Fig. 4.5 Modulation of glutamatergic synaptic transmission by astroglial glutamate transporters. Unlike receptors, glutamate transporters are excluded from the synaptic cleft. EAAT1 (GLAST) and EAAT2 (GLT-1) are present at a high density in the membranes of astrocytes that often ensheath synapses. EAAT3 (EAAC1) is found in the soma and dendrites of neurons, but is also found in GABAergic terminals (not shown). Glutamate transporters shield extrasynaptic NMDA receptors and mGluRs from glutamate as it diffuses from the cleft, and prevent glutamate from reaching receptors at nearby synapses. Inhibition of these transporters potentiates excitatory responses mediated by these receptors, and allows glutamate spillover, which suggests that transporter regulation might be used to regulate synaptic efficacy. Note that presynaptic mGluRs have been omitted from this diagram. Adapted from Huang et al. (2004), Curr Opin Neurobiol.

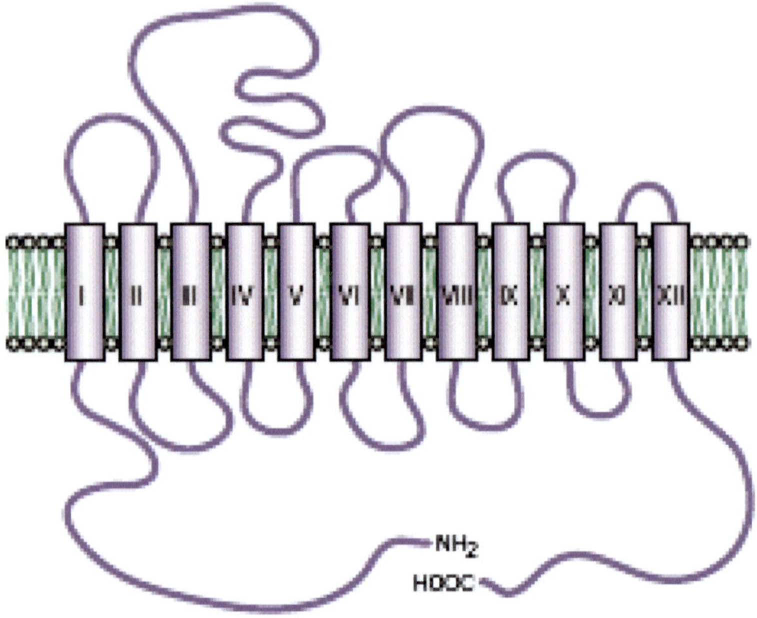

Fig. 4.6 Predicted topology of SLC6 transporters. Twelve transmembrane domains are connected by intracellular and extracellular loops with both N- and C-terminals inside the cell. Adapted from Torres et al. (2003), Nat Rev Neurosci.

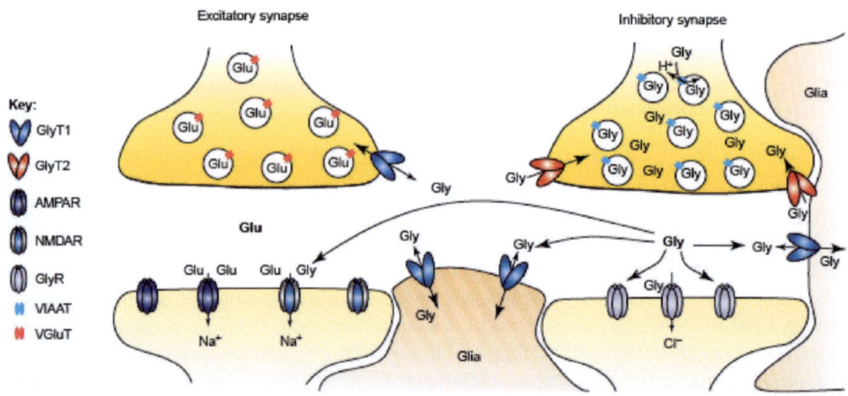

Fig. 4.7 Localization and proposed functions of glycine transporters at excitatory and inhibitory synapses. At inhibitory synapses, glycine release from the presynaptic terminal activates postsynaptic GlyRs and thereby induces ClK influx – hyperpolarization – of the postsynaptic cell. At excitatory glutamatergic synapses, glycine acts as an essential co-agonist of postsynaptic NMDARs, whereas neighboring glutamate receptors of the α-amino-3-hydroxy-5-methyl-4-isoxazole-propionic-acid receptor (AMPAR) subtype require only glutamate for channel activation. Here, glycine might be derived from neighboring glycinergic terminals or even be released from astrocytes via nonvesicular mechanisms (e.g. reverse transport by GlyT1). GlyT2 is localized in the presynaptic plasma membrane of glycinergic neurons and transports glycine into the terminal, thereby enabling the refilling of synaptic vesicles with glycine by the HC-dependent vesicular inhibitory amino acid transporter (VIAAT). GlyT1 is mainly expressed by glia cells surrounding both inhibitory and excitatory synapses. In addition, GlyT1 has been found on terminals of some excitatory neurons. Thus, GlyT1 mediates the clearance of glycine from the synaptic cleft of inhibitory synapses and, in addition, participates in the regulation of the glycine concentrations at excitatory synapses. Adapted from Eulenberg et al.(2005), Trend Biochem Sci.

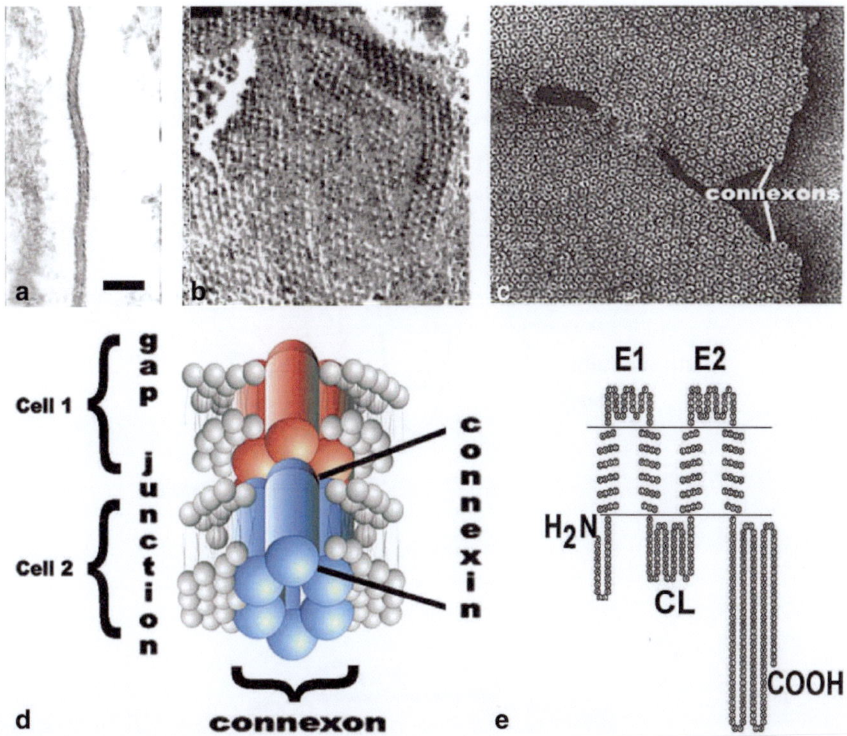

Fig. 5.1 Structure of gap junction, connexon, and connexin. (a) Thin section electron micrographs of gap junctions between two astrocytes reveal the close apposition of the two membranes, which are separated by a small gap of 2 nm. (b) P-face of a gap junction plaque between two astrocytes obtained by freeze fracture, showing the connexin particles. (a) and (b) are modified from Duffy et al. (2000). (c) Electron micrograph (EM) of a gap junction plaque negatively stained showing the hexameric structures corresponding to connexons. Adapted from Fawcett (1994), Figs. 2–14. (d) Schematic drawing of a gap junction channel formed by the docking of two hexameric structures (connexons) provided by two adjoining cells. Each connexon is formed by six subunits, the connexins (e), which are tetra-span proteins with the N- and C-termini, and a cytoplasmic loop (CL) located on the cytoplasmic face. The two extracellular loops (E1 and E2) are the connexin domains that provide the strong interaction with the apposing extracellular loops of the connexin in the adjoining cells.

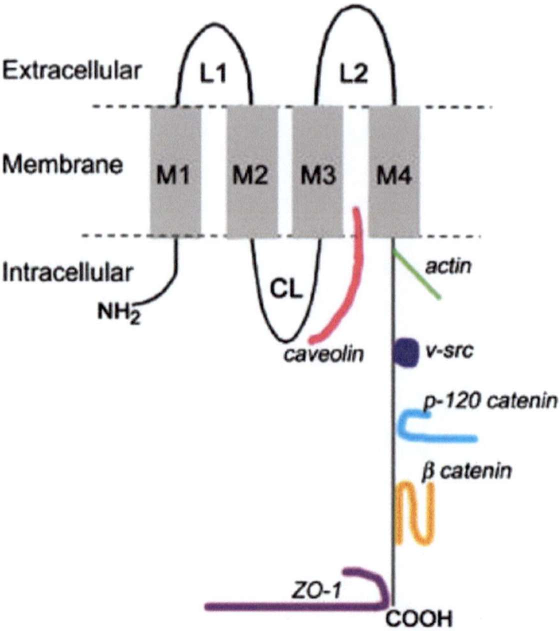

Fig. 5.2 Connexin-interacting proteins: The Nexus. Schematic drawing of connexin43 showing the binding partners at the cytoplasmic domains. Caveolin is shown to interact with the second half of the cytoplasmic loop (CL) of Cx43 (Schubert et al., 2002). Several other proteins (actin, v- and c-src, p-120 and β-catenin, and ZO-1) bind to different portions of the carboxyl-terminus (see Giepmans, 2004; Herve et al., 2007; Iacobas et al., 2007b).

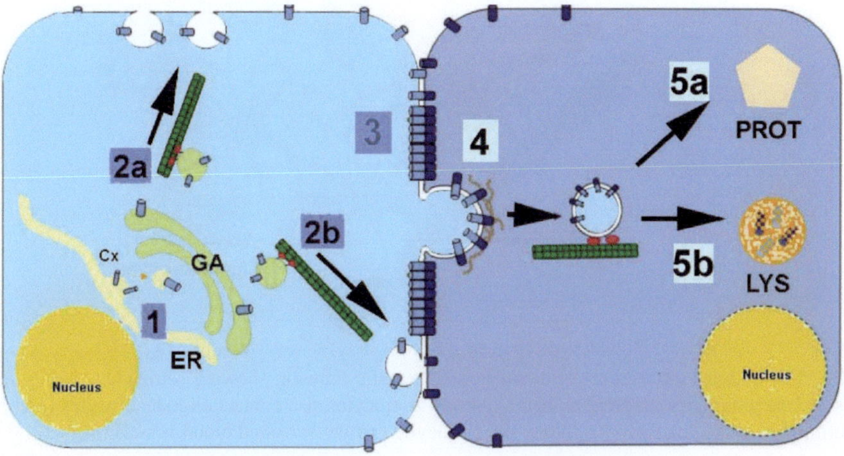

Fig. 5.3 The life and death of gap junctions. Schematic drawing showing two cells coupled by gap junctions. The cell on the left shows the steps involved in the formation of a gap junction. The newly synthesized connexin proteins (step 1) traffic from the endoplasmic reticulum (ER), where they hexamerize into connexons, to the Golgi apparatus (GA), and from there, connexons are inserted into plasma membrane through a secretory pathway either randomly to the surface membrane along microtubules shown as rods (step 2a) or into the vicinities of a gap junction plaque (step 2b). The connexons then diffuse through the plane of the bilayer (step 3) joining other connexons at the gap junction plaque. The cell on the right shows the degradation of a gap junction. Retrieval of gap junctions from the plasma membrane occurs by (step 4) the formation of a double-walled vesicle or connexosome (Laird, 2006) containing the membranes of the two adjoining cells. These vesicles containing full gap junction channels are then degraded either through a proteosome pathway (PROT, step 5a) or by lysosomes (LYS, step 5b).

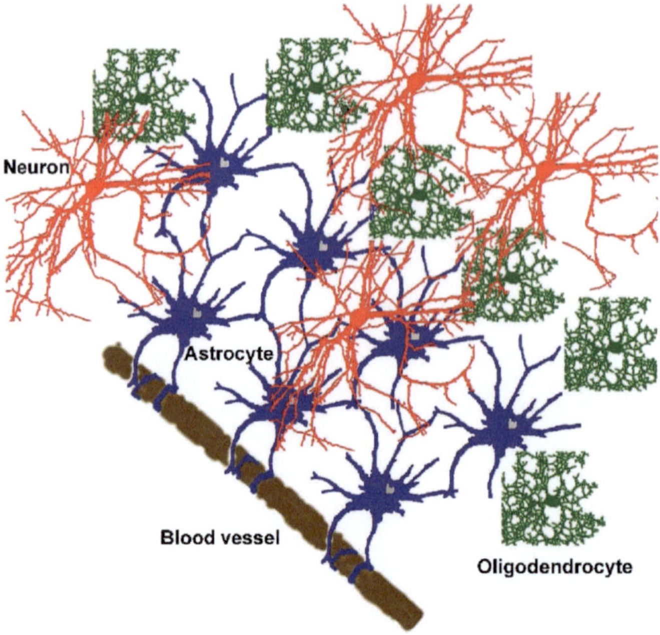

Fig. 5.4 Cellular interactions of astrocytes. Schematic drawing showing astrocyte (*blue*) processes contacting neurons (*red*) at their synapses. Astrocytes also contact oligodendrocytes (*green*) forming heterocellular gap junctions, and astrocytic endfeet surround blood vessels (*brown*).

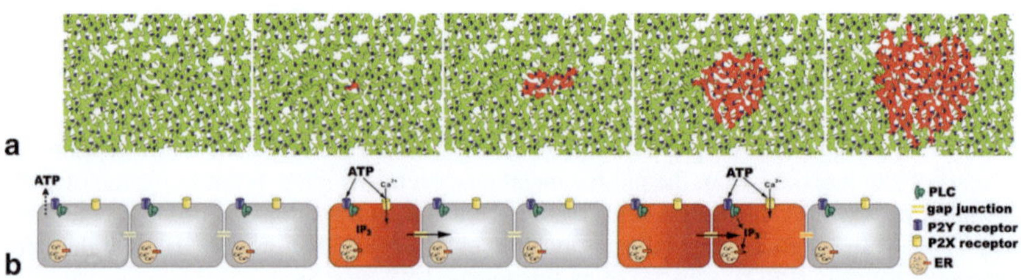

Fig. 5.5 Intercellular Ca^{2+} wave. (a) Schematic drawing depicting the transmission of Ca^{2+} waves between astrocytes in culture. Following stimulation of a single astrocyte, intracellular Ca^{2+} levels increase (*red*). This increase in intracellular Ca^{2+} spreads to adjoining astrocytes (*green*), which are shown to be connected by gap junction (*blue*). (b) Schematic representation of the steps involved in the transmission of intercellular Ca^{2+} waves. Released ATP diffuses through the extracellular space, activating membrane purinergic (P2) receptors. Stimulation of metabotropic P2Y receptors leads to PLC activation and IP_3 formation and activation of ionotropic P2X receptors leads to the influx on Ca^{2+}. IP_3 and Ca^{2+} promote the release of Ca^{2+} stored in the endoplasmic reticulum (ER), increasing intracellular Ca^{2+} levels of the stimulated cell. Diffusion of the two intracellular Ca^{2+}-mobilizing second messengers through gap junction channels together with the activation of P2 receptors in the near-by cell contribute the Ca^{2+} rise in this cell. This process continues till the concentrations of ATP, IP_3, and Ca^{2+} are not sufficient to trigger intracellular Ca^{2+} rises.

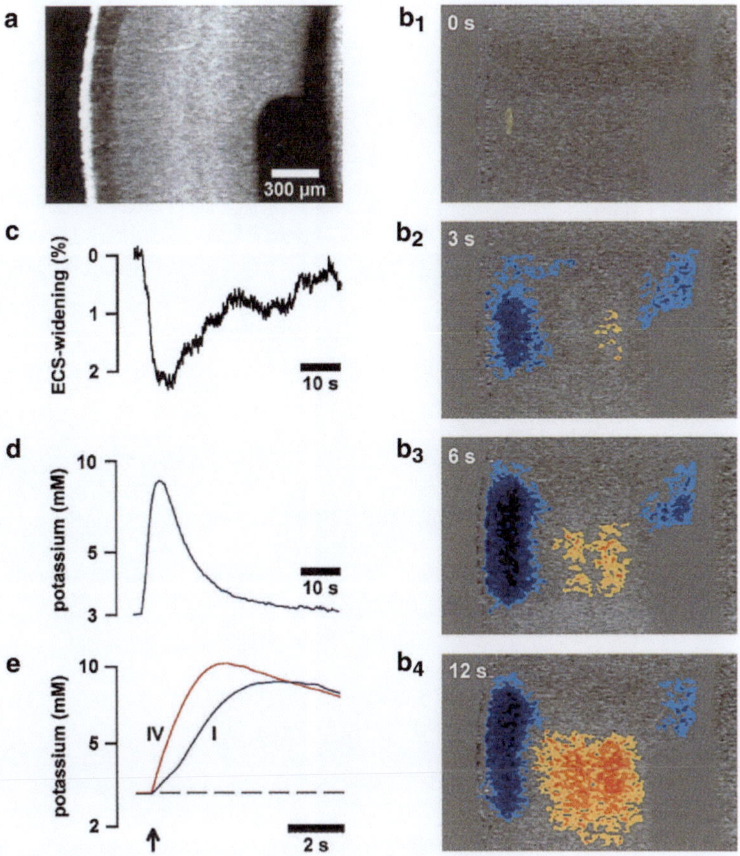

Fig. 6.4 Potassium spatial buffering in the rat cortex. (**a**) Image of a brain slice viewed with darkfield optics. (**b**1–4) Time course of intrinsic optical signal (IOS) changes upon neuronal stimulation in middle cortical layers. *Red colors* represent IOS increases while *blue colors* represent IOS decreases. These correspond to shrinking and widening of the extracellular space, respectively. Note that extracellular space shrinks in the middle cortical layers and widens in the most superficial and deep cortical layers, as predicted for the K^+ spatial buffering mechanism. (**c**) Time course of extracellular space widening in cortical layer I, measured independently, confirming the IOS results. (**d**) Time course of $[K^+]_o$ increase in layer I. (**e**) Time course of $[K^+]_o$ increase in layer I (*blue*) and in layer IV (*red*). [From Holthoff and Witte (2000), with permission].

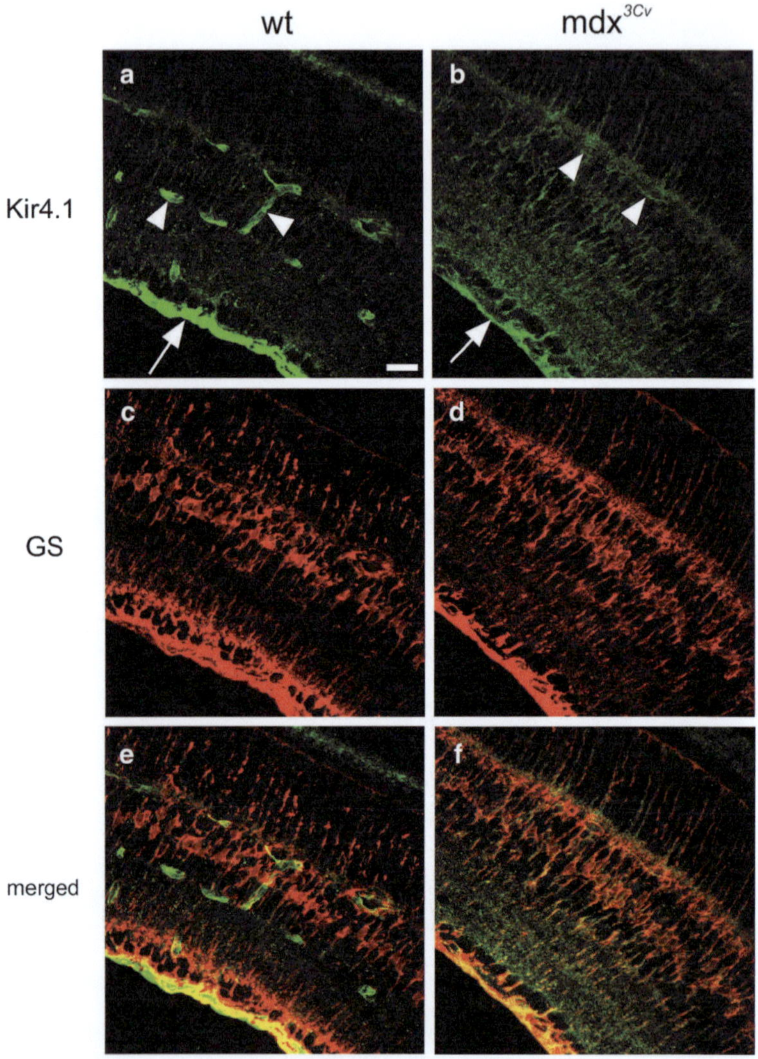

Fig. 6.6 Kir4.1-channel localization in wild type and mdx3Cv (dystrophin knockout) mouse retinas. (**a**) In the wild-type retina, Kir4.1 is concentrated at the inner limiting membrane (*arrow*) and to processes surrounding blood vessels (*arrowheads*). (**b**) In the mdx3Cv retina, Kir4.1 is more evenly distributed throughout the retina with a reduction in staining at the inner limiting membrane (*arrow*) and no apparent enrichment of Kir4.1 around blood vessels (*arrowheads*). The glial-cell marker glutamine synthetase (GS) (**c**, **d**), and merged images (**e**, **f**) suggest that Kir4.1 is localized to Müller cells. Scale bar in (**a**) = 25 μm. [From Connors and Kofuji (2002), with permission].

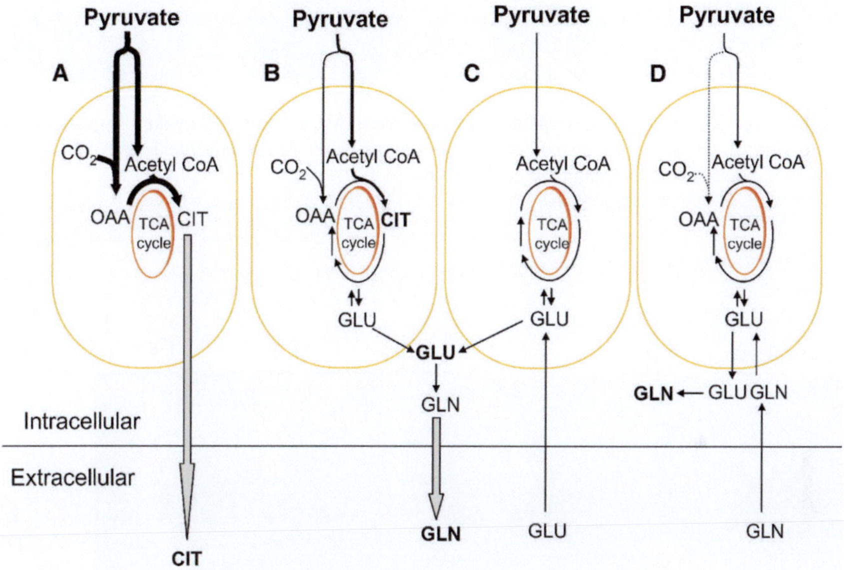

Fig. 7.3 A schematic presentation of multiple compartments in astrocytes in which pyruvate functions as substrate. Synthesis of a large amount of releasable citrate occurs in compartment **A** having extensive carboxylation. Releasable glutamine is synthesized from glutamate originating from compartments **B** and **C**. The main intracellular pool of glutamine is synthesized from glutamate originating from compartment **D**. Note that the compartments differ with regard to activity of pyruvate metabolism via carboxylation and oxidation. The sizes of the *arrows* provide an estimate of the relative magnitudes of the respective fluxes. *GLN* glutamine, *GLU* glutamate, *CIT* citrate, *OAA* oxaloacetate, and *TCA* tricarboxylic acid.

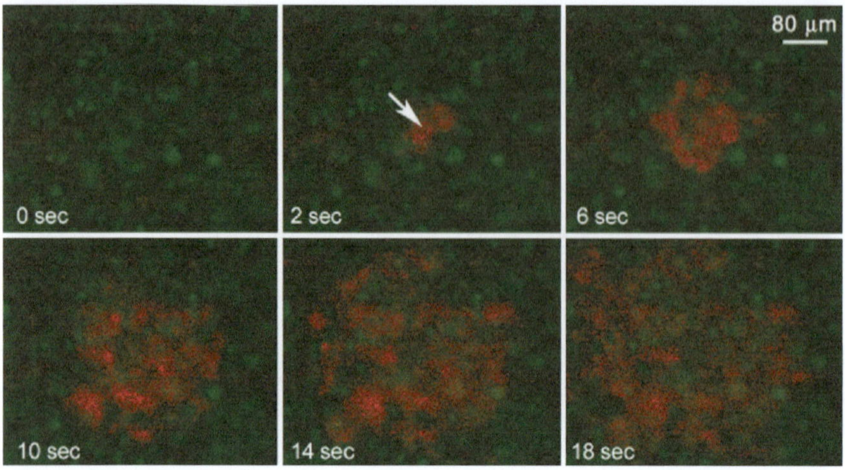

Fig. 8.3 Ca^{2+} wave in cultured spinal cord astrocytes. A Ca^{2+} wave as evoked by mechanical stimulation of an astrocyte (arrow) propagates centrifugally in the confluent layer of primary astrocytes. Modified from Suadicani et al. (2006).

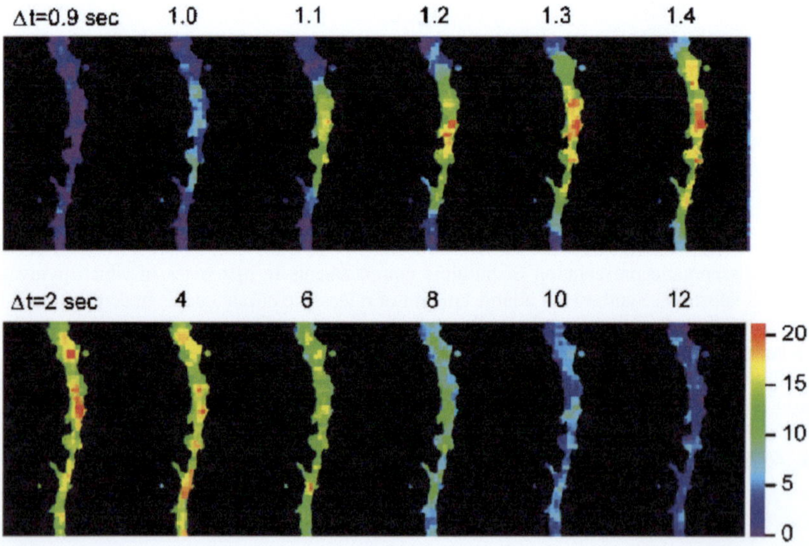

Fig. 8.4 Ca^{2+} signaling in a Bergmann glial process evoked by parallel fiber stimulation. Pseudocolored images of the initiation and intracellular propagation of Ca^{2+} signals in a single process of a Bergmann glial cell, evoked by a parallel fiber burst (50 Hz, 10 pulses). From Beierlein and Regehr (2006).

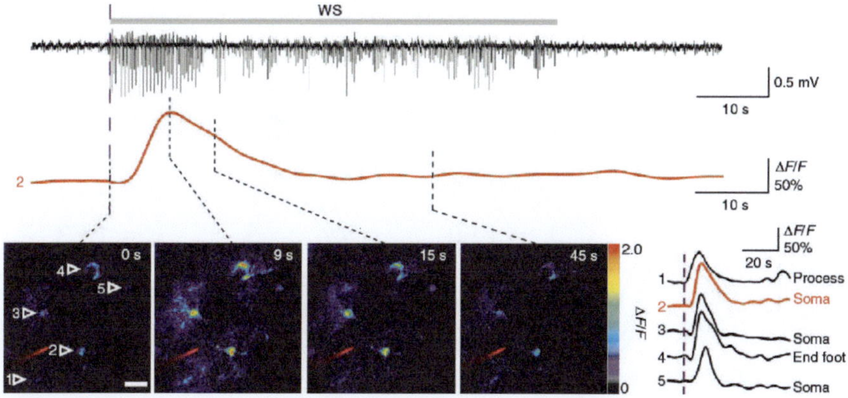

Fig. 8.5 In vivo imaging of Ca^{2+} signaling in astrocytes of the barrel cortex during whisker stimulation. In anaesthetized mice, whisker stimulation (WS) evoked neuronal activity in the barrel cortex as shown by the local field potential (upper trace). Astrocytes responded with a transient increase in Ca^{2+} after a delay of approximately 3 s (lower trace). Ca^{2+} transients could be measured in somata, processes, and end-feet of astrocytes, as indicated in the first image of the time sequence (*lower left*) and the corresponding Ca^{2+} traces (*lower right*). From Wang et al. (2006).

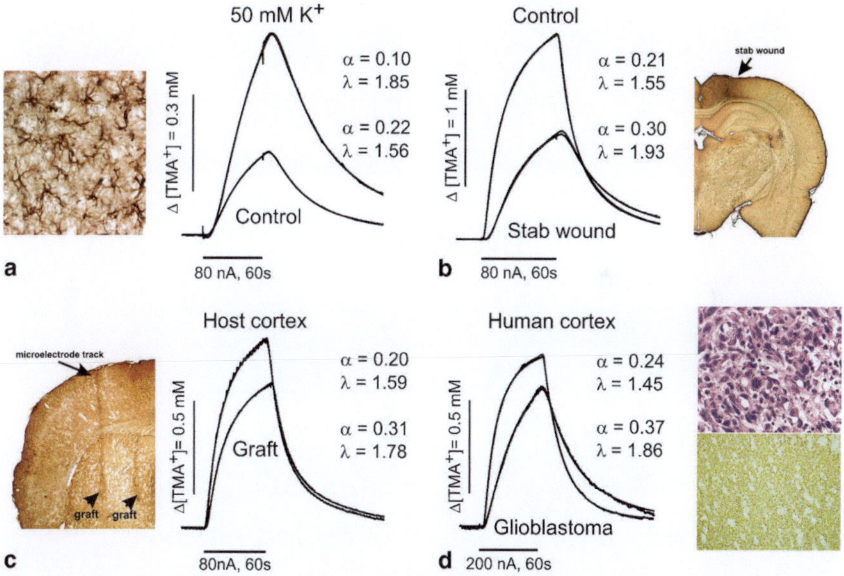

Fig. 9.4 TMA^+ diffusion curves with the corresponding values of the ECS diffusion parameters and histological features of the tissue under different experimental conditions – models of various pathological states: perfusion with 50 mM K^+ (**a**), gliotic tissue around a stab wound (**b**), dopaminergic grafts in the striatum (**c**) and a glioblastoma (**d**) (see also text). Slices from **a**, **b** and **c** are immunostained for GFAP; the slices in part D are stained with hematoxylin-eosin and/or for the extracellular matrix molecule tenascin. The different amplitudes and shapes of the diffusion curves recorded in pathological tissue in comparison with control curves reflect changes in extracellular volume fraction α (the higher the curve, the lower the α value) and tortuosity λ. Note that regardless of decreased (**a**) or increased (**b**, **c**, **d**) α values, tortuosity in tissue with reactive astrocytes is always increased and diffusion is hindered. With respect to the glioblastoma, in addition to the diffusion barriers formed by hypertrophied glial processes, the overproduction of the extracellular matrix molecules, namely tenascin and vitronectin, is also a contributory factor.

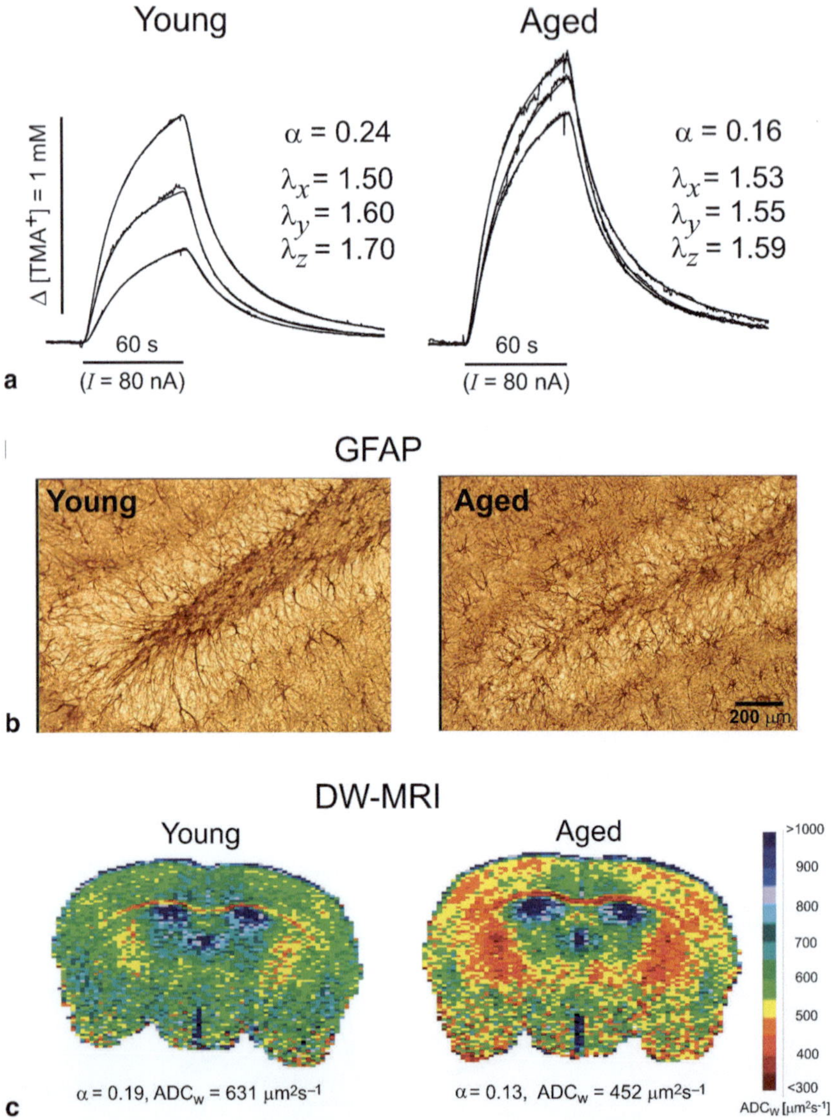

Fig. 9.6 Typical diffusion curves recorded in the hippocampus of young adult and aged rats (**a**), immunostaining of young and aged gyrus dentatus of the hippocampus for GFAP (**b**) and maps of the apparent diffusion of water (ADC_W) measured by diffusion-weighted MRI (**c**). During aging, the extracellular space volume decreases and anisotropy disappears. The disappearance of diffusion anisotropy corresponds with the loss of the typical radial organization of fine glial processes in the aged hippocampus, as revealed by immunostaining for GFAP (**b**), as well as with the loss of fibronectin and CSPG perineuronal nets (not shown). The lower ADC_W indicates the impaired diffusion of water in the cortex and hippocampus of aged rats (**c**).

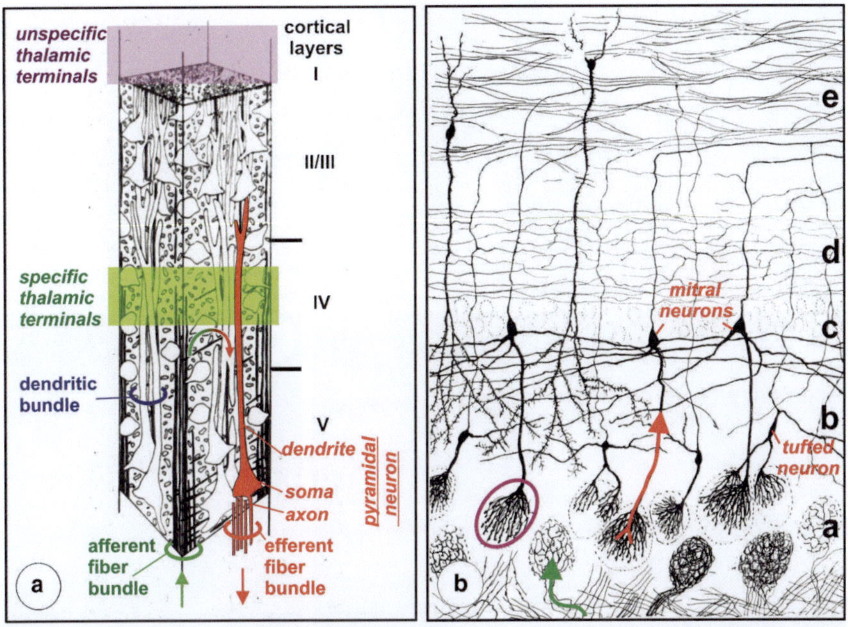

Fig. 10.1 Neuronal elements and their organization in the brain. **a** Visual cortex of adult rabbit [modified after (Fleischhauer and Detzer, 1975)]. **b** Olfactory bulb of newborn kitten [modified after (Ramón y Cajal, 1911)]. In the neocortex, the afferent (*green*) and efferent fibers (*red*), as well as the pyramidal neurons and their dendrites (*blue*) are aligned in radially oriented bundles. Moreover, a transversal orientation is provided by the layering of the cells and of the afferent terminals (*green/violet*). In the olfactory bulb, the synaptic transmission between the afferents from the olfactory receptor cells (*green*), the mitral (*red*) and tufted cells occurs in so-called glomeruli (*violet*). In both cases, the information processing of every neuron is clearly polarized with respect to input/output direction (*arrows*).

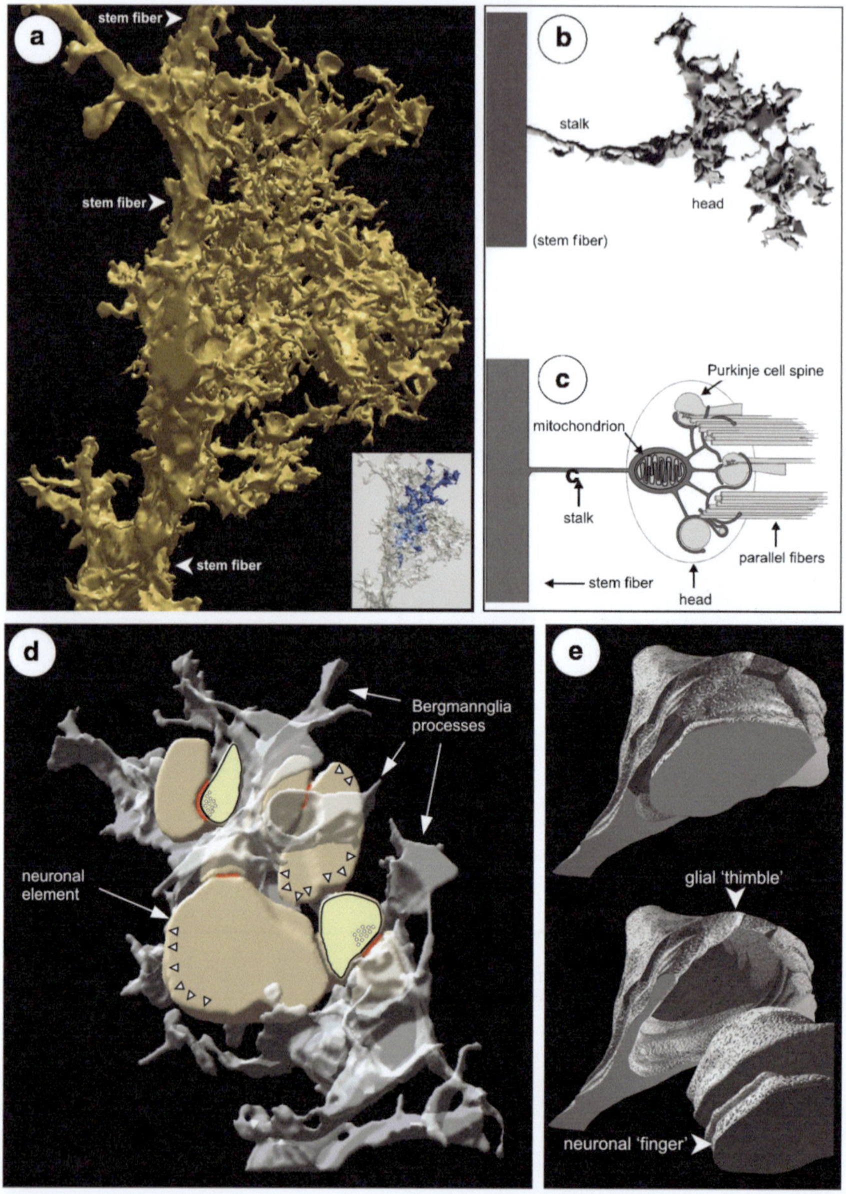

Fig. 10.7 Microdomains of Bergmann glial cell processes in the murine cerebellum. **a** 3-D reconstruction of a part of a Bergmann glial cell process. The living cell was dye-injected in a perfused cerebellar slice; then, after fixation and dye-conversion, about 600 consecutive serial ultrathin sections were photographed in the electron microscope, and the images of the dye-labeled profiles were reconstructed by a computer program. The inset shows a substructure labeled in blue; this part was quantitatively analyzed (see (**b**), (**c**)). **b** Glial microdomain as part of the 3-D reconstruction shown in (**a**). **c** Schematic drawing of such a glial microdomain and its relationships to the neuronal elements. **d** 3-D reconstruction of a group of neighboring cerebellar synapses (*yellow–green*; synaptic clefts: *orange*) together with the surrounding leaflets provided by the injected Bergmann glia (*blue–gray*). The *arrowheads* point to neuronal surfaces not covered by glial sheaths from the labeled cell. **e** 3-D reconstruction of a glial "thimble" and the neuronal "finger" covered by it; shown in apposition (*top*) and separated for better discrimination between the two compartments (*bottom*). With permission, from Reichenbach et al. (2004).

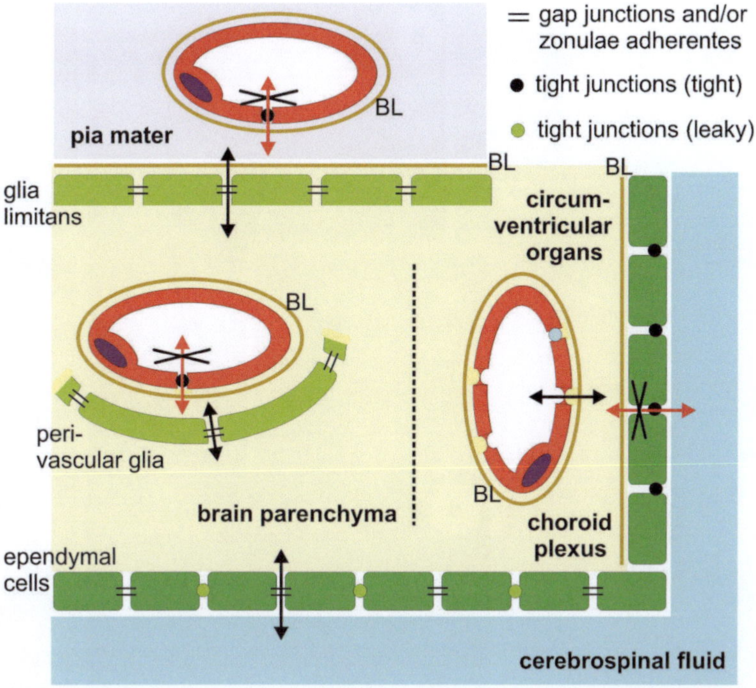

Fig. 10.10 Brain territories defined by the BBB. The brain wall is delineated superficially by the glia limitans superficialis, formed by astroglial endfeet and a basal lamina. Periventricularly, the brain wall is delineated by the ependymal cells, which do not establish a physiological barrier since ependymal cells are interconnected by discontinuous tight junctions. In the circumventricular organs (CVOs), the endothelial BBB must be interrupted because the neurons, for example in the hypothalamic-hypophyseal system or in the subfornical or subcommissural organs or in the area postrema, must necessarily get access to the blood compartment in order to release neurosecretory compounds into the blood stream or to "smell" blood-borne signal molecules, respectively. In the case of the choroid plexus, the endothelial cells also have to be highly permeable by the formation of fenestrations to allow the production of the cerebrospinal fluid (CSF) by the choroid plexus epithelial cells. To avoid diffusion of blood-borne substances from the leaky choroid vessels into the CSF and further into the brain parenchyma through the leaky ependymal tight junctions, a barrier is necessary between the blood and the CSF: the so-called blood-CSF-barrier. This barrier is located in the choroid plexus epithelial cells and in the tanycytes of the CVOs, and is formed by tight junctions which are molecularly different from the endothelial tight junctions of the BBB. *BL* basal lamina. Modified after Wolburg et al. (2007).

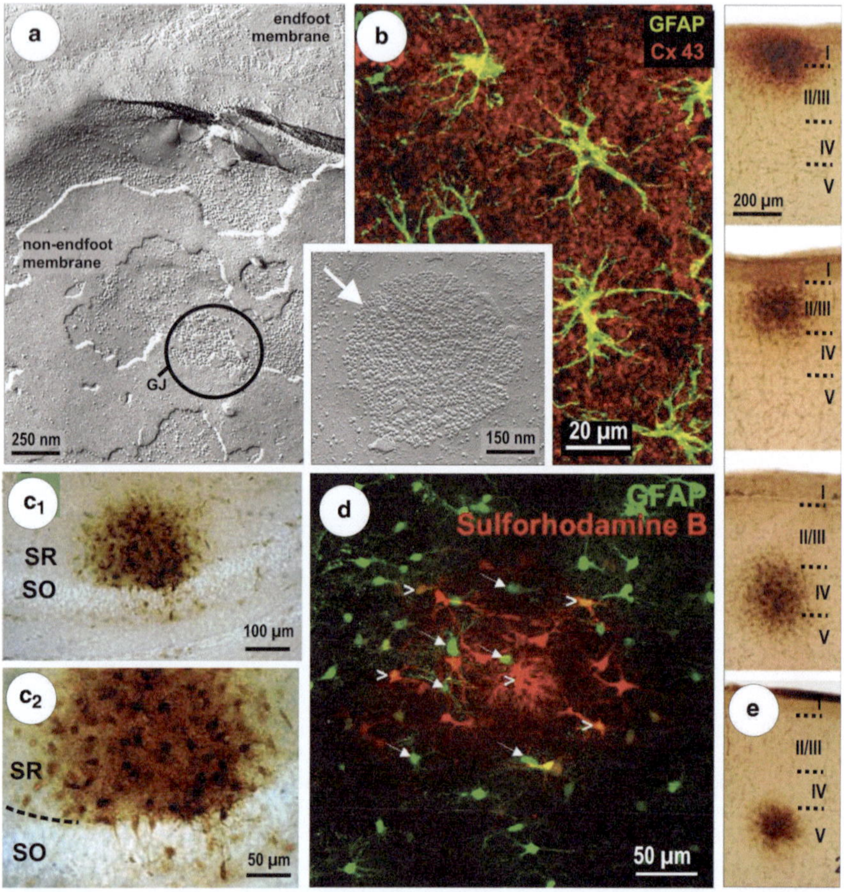

Fig. 10.11 Astrocytic coupling via gap junctions. **a** Electron microscopy of a freeze-fracture replica of an astrocytic process forming a perivascular endfoot (rat optic nerve). The endfoot membrane proper displays orthogonal arrays of particles (cf. Fig. 10.8) but no gap junctions. The membrane particle arrays of a gap junction (GJ) are found in the "lateral" non-endfoot membrane of the same process, probably constituting a coupling to another adjacent endfoot-bearing process. Inset: Typical array of gap junction-hemichannels in an astrocytic membrane. Original data, H. Wolburg. **b** Juvenile mouse hippocampus, immunohistochemical co-localization of GFAP (green) and connexin 43 (Cx43, *red*). The overlay (*yellow*) indicates that connexin 43 is expressed on astrocytic processes. **c**, **d** Dye-coupling of astrocytes in juvenile mouse hippocampus. Intracellular injection of biocytin (**c**) or sulforhodamine B (**d**) into a single astrocyte in brain slices causes the staining of many adjacent astrocytes. **c** In the juvenile hippocampus, many but not all dye-coupled astrocytic networks fail to span beyond the stratum radiatum (SR); if the injected cell was localized there, only a few extend into the stratum oriens (SO). **d** It is noteworthy that not all GFAP/GFP-expressing astrocytes in a given area are dye-coupled (*arrows*). Some sulforhodamine B-stained cells are marked by arrowheads. **e** Dye-coupling of astrocytes in juvenile mouse neocortex, intracellular injection of biocytin. Depending on the location of the injected astrocyte (cortical layers I–V), the dye-coupled astrocytic networks differ in size and shape but there is always a tendency to remain restricted within one cortical layer. **b–e** modified after Houades et al. (2006).

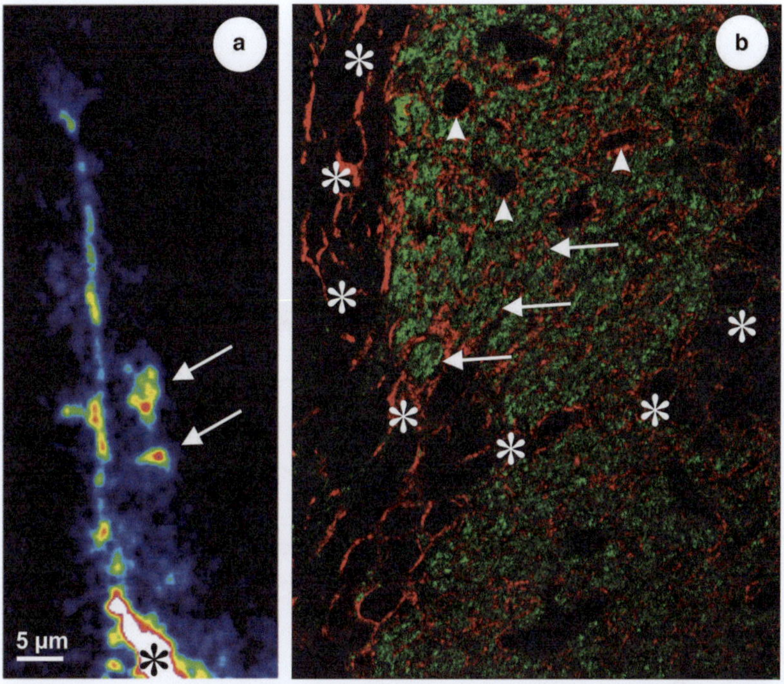

Fig. 10.12 Co-existence of functional glial domains differing in size. **a** Imaging of Ca^{2+} increases in a dye-injected Bergmann glial cell of a murine cerebellar slice preparation, after electrical stimulation of parallel fibers. Low-intensity stimulation of only a few parallel fibers may induce Ca^{2+} increases in only one or a few glial microdomains (*arrows*) whereas high-intensity stimulation of many fibers elicits Ca^{2+} responses within the entire glial cell process, or even the whole cell (*asterisk*, soma). Modified after Grosche et al. (1999). **b** Rat olfactory bulb, double-immunolabeling of AQP4 (*red*: glial cell membranes) and synaptophysin (*green*: synaptic structures). Two synaptic glomerula are shown; the astroglial "envelope" of the upper one is indicated by asterisks. AQP4-expressing glial membranes surround the entire glomerulum but also smaller groups of synapses ("green spaces", *arrows*) and small blood vessels ("empty spaces", *arrowheads*). This might be indicative of the presence of both a glial "(?oligo-)cellular domain" (draining water from the entire glomerulum) and many microdomains (responsible for small groups of synapses). Original data, Wolburg et al.

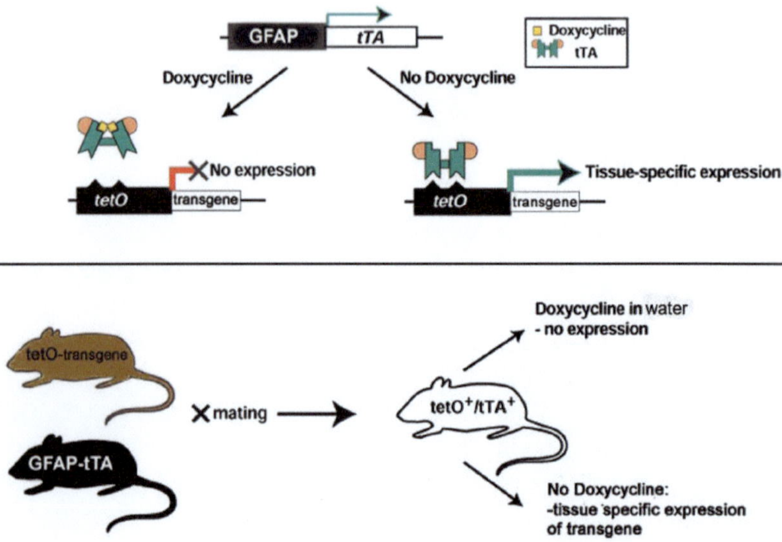

Fig. 14.1 Tetracycline "Off"-Inducible System. Doxycycline represses transgene expression by binding the tet transcriptional activator (tTA). Removal of doxycycline activates tTA, which then binds the tetO minimal promoter, inducing expression of the transgene (*top*). Tissue specificity is achieved by crossing glial fibrillary acidic protein (GFAP)-tTA mice with tetO-transgenic mice; double-transgenic mice have transgene expression in cells that express GFAP (*bottom*).

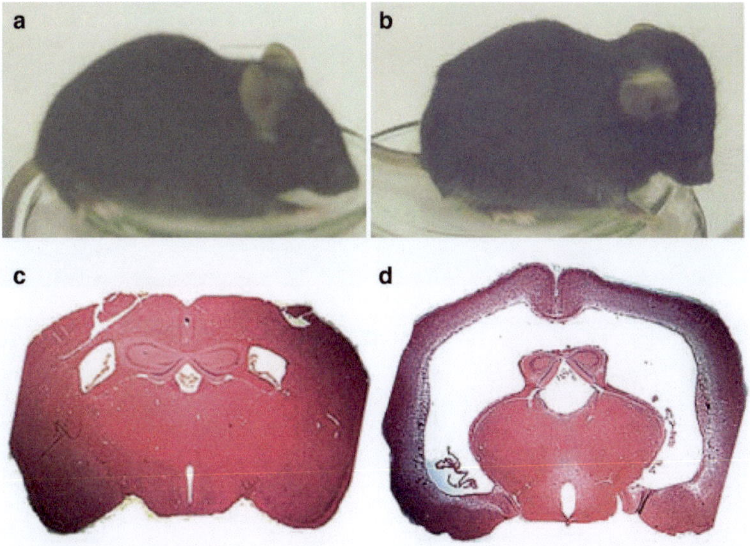

Fig. 14.2 Expression of a G_i G-protein coupled receptor in glial fibrillary acidic protein (GFAP) positive cells results in hydrocephalus. Mice carrying both the tetO-Ro1 and the GFAP-tTA transgenes develop hydrocephalus when doxycycline (dox) is removed from the drinking water. A double-transgenic mouse maintained on dox is normal (**a**), while a double-transgenic littermate taken off dox (**b**) develops a domed head characteristic of hydrocephalus. Hematoxylin and eosin staining of coronal brain sections show the greatly enlarged lateral ventricles in hydrocephalic mice (**d**) compared to normal littermate control mice (**c**). All mice are P30. Modified with permission from The Society for Neuroscience (Sweger et al. 2007).

Fig. 14.3 (continued) acid (tACPD), although the FLRFa Ca^{2+} responses were slightly longer lasting. (**c**) In control mice expressing only tTA or the tetO-MrgA1 construct, administration of peptide FLRFa did not produce a Ca^{2+} increase in astrocytes bulk-loaded with Ca^{2+} indicator. As a positive control for astrocyte Ca^{2+} increases resulting from release from internal stores, an agonist cocktail consisting of histamine (10 μM), carbachol (10 μM), and 2Na-ATP (50μM) was bath-applied. Scale bars, 10 μm. Modified with permission from Elsevier (Fiacco et al., 2007).

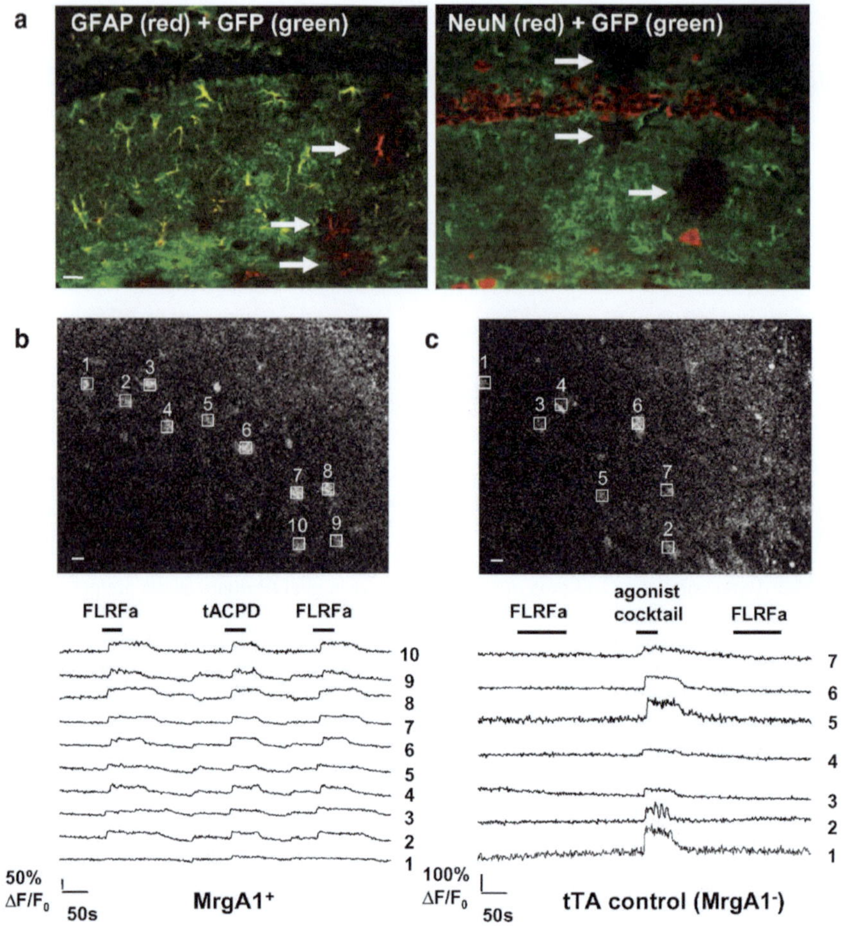

Fig. 14.3 Selective expression of a functional transgenic Gq GPCR, MrgA1, in astrocytes. The Gq-coupled receptor MrgA1 is expressed exclusively by astrocytes in the hippocampus. (**a**) Crossing hGFAP-tTA to tetO-MrgA1 mice resulted in the expression of MrgA1 exclusively in astrocytes in the absence of doxycycline. As eGFP is fused to the MrgA1 receptor construct, GFP staining indicates MrgA1 receptor expression. In the hippocampus, 80–90% of astrocytes expressed the MrgA1 receptor, while neurons did not express the receptor. The *left panel* shows astrocytes immunostained for GFAP (*red*) and GFP (*green*) to show MrgA1 expression. Astrocytes coexpressing GFAP and the MrgA1 receptor appear yellow. A few GFAP$^+$ astrocytes do not express the MrgA1 receptor (*arrows*). The *right panel* shows GFP labeling in astrocytes (*green*) together with neuronal nuclei (NeuN) labeling neurons in stratum pyramidale of CA1 (*red*). Neurons do not express the MrgA1 receptor. A few astrocytes in the field do not express MrgA1 (*arrows*). Scale bar, 20 mm. Activation of the MrgA1 receptor produces robust, widespread astrocyte Ca^{2+} increases in acute mouse hippocampal slices. (**b,c**) Top panels show numbered astrocytes bulk-loaded with Ca^{2+} Green 1-AM indicator that correspond to the numbered fluorescence over time traces in bottom panels. Increases in F/F_0 indicate increases in cytoplasmic Ca^{2+} concentration. (**b**), In MrgA1$^+$ mice maintained off doxycycline, FLRFa administration resulted in robust astrocyte Ca^{2+} increases in the majority of hippocampal astrocytes. Note that MrgA1 receptor-mediated astrocyte Ca^{2+} responses were similar in amplitude to those produced by (1*S*,3*R*)-1-aminocyclopentane-1,3-dicarboxlylic

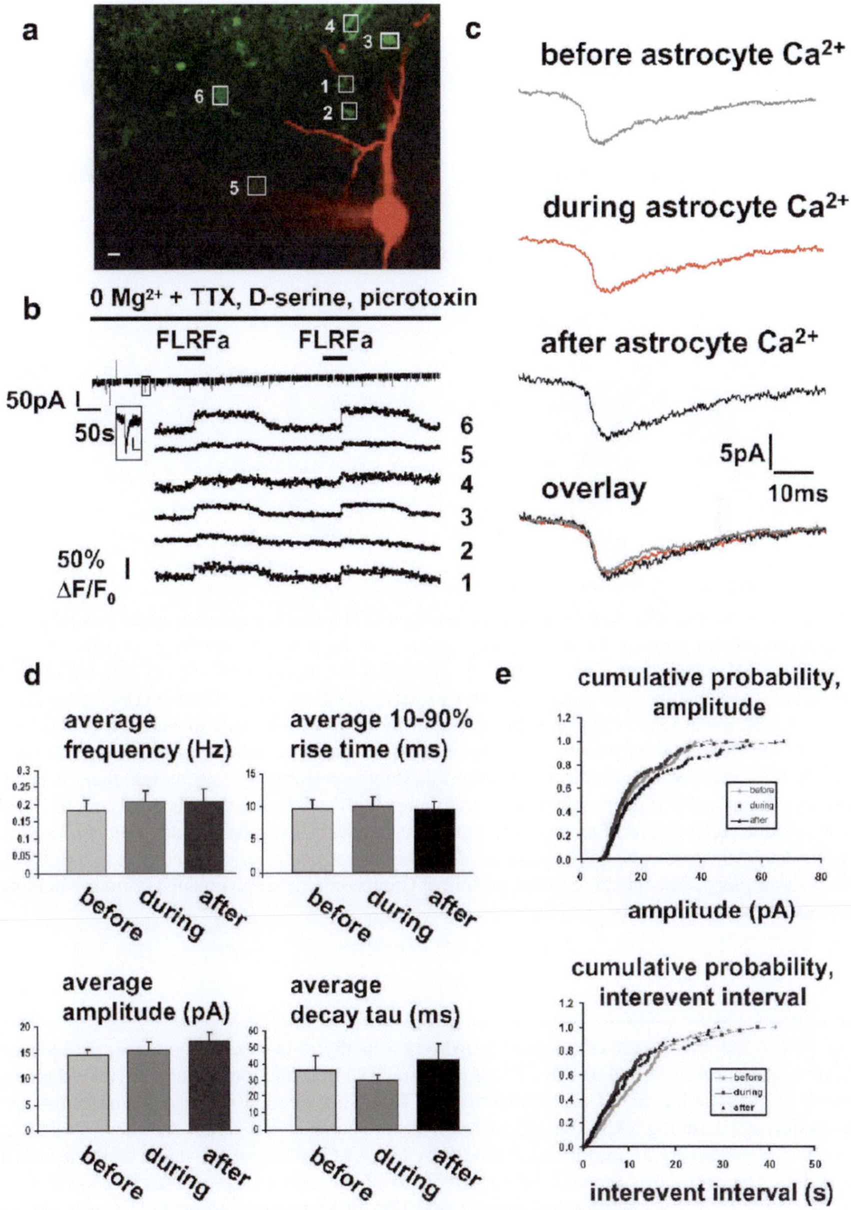

Fig. 14.4 Widespread selective astrocyte Ca^{2+} increases following MrgA1R activation do not produce an effect on miniature excitatory postsynaptic currents (mEPSCs) in CA1 pyramidal neurons. (**a**) Neurons were patch clamped with Alexa 568 hydrazide dye (*red*) to image cell morphology relative to astrocytes bulk-loaded with Ca^{2+} Green 1-AM dye (*green*). Numbered ROIs over astrocytes in which fluorescence intensity was measured over time correspond to the numbered traces in (**b**). (**b**) Astrocytes expressing the MrgA1 receptor were stimulated to release Ca^{2+} from internal stores by perfusion of the MrgA1 agonist peptide FLRFa during neuronal recording

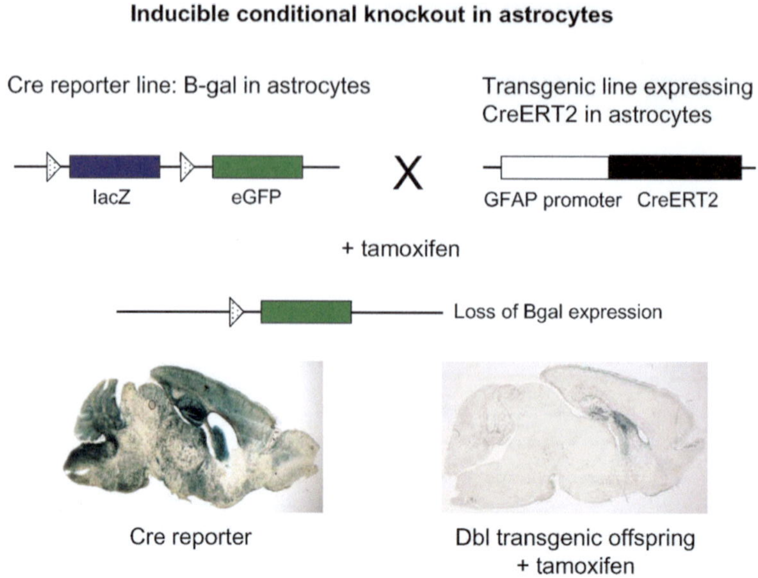

Fig. 14.6 Inducible gene recombination in astrocytes. The Cre reporter line expresses *lacZ* and thus B-gal protein prior to Cre recombinase expression. After Cre recombinase excises the *lacZ* cassette, eGFP is expressed. This reporter line is crossed to a transgenic line carrying a tamoxifen-sensitive Cre recombinase in astrocytes, driven by the GFAP promoter. Double (Dbl)- transgenic offspring are given tamoxifen by intraperitoneal injection which leads to excision of the lacZ cassette and loss of B-gal protein. B-gal protein expression is visualized by histochemical stain with Xgal. The reporter brain turns blue after this stain is performed, showing that there is B-gal protein present. After inducing the double transgenic with tamoxifen, there is great loss of B-gal protein. Blue rectangle, lacZ gene cassette; green rectangle, eGFP gene cassette; *dotted triangles*, *loxP* sites; *white rectangle*, GFAP promoter (or other astrocyte-specific promoter, e.g., GLAST); *black rectangle*, cDNA encoding fusion protein of Cre recombinase and mutated hormone-binding domain of estrogen receptor (ERT2).

Fig. 14.4 (continued) of mixed α-amino-3-hydroxy-5-methyl-4-isoxazolepropionic acid (AMPA)/NMDA receptor mEPSCs (upper trace) in a Mg^{2+}-free external solution containing 100 μM picrotoxin, 10 μM d-serine, and 1 μM tetrodotoxin (TTX). This solution was used to maximize the probability of measuring NMDAR-mediated slow-inward currents as reported elsewhere. There was no effect of astrocyte Ca^{2+} elevations produced by MrgA1 Gq receptor activation on the neuronal recording. The inset shows a representative mEPSC on a larger timescale (scale bar, 10 pA, 100 ms). (**c**) Miniature EPSCs averaged before, during, and after FLRFa evoked astrocyte Ca^{2+} increases in the representative neuron. (**d**) Neuronal currents analyzed in three bins – before, during, and after selective astrocyte Ca^{2+} elevations – revealed no differences in frequency, amplitude, and kinetics ($n = 13$ neurons, 13 slices). (**e**) Cumulative probability plotted for amplitude (*top*) and interevent interval (*bottom*) show the full range of distributions. Kolmogorov–Smirnov analysis revealed no significant difference in the parameters (amplitude: $p = 0.4273$ before vs. during, 0.3265 before vs. after, 0.1210 during vs. after; interevent interval: $p = 0.0828$ before vs. during, 0.0719 before vs. after, 0.7889 during vs. after). Error bars indicate SEM. Scale bar, 10 μm. Reprinted with permission from Elsevier (Fiacco et al., 2007).

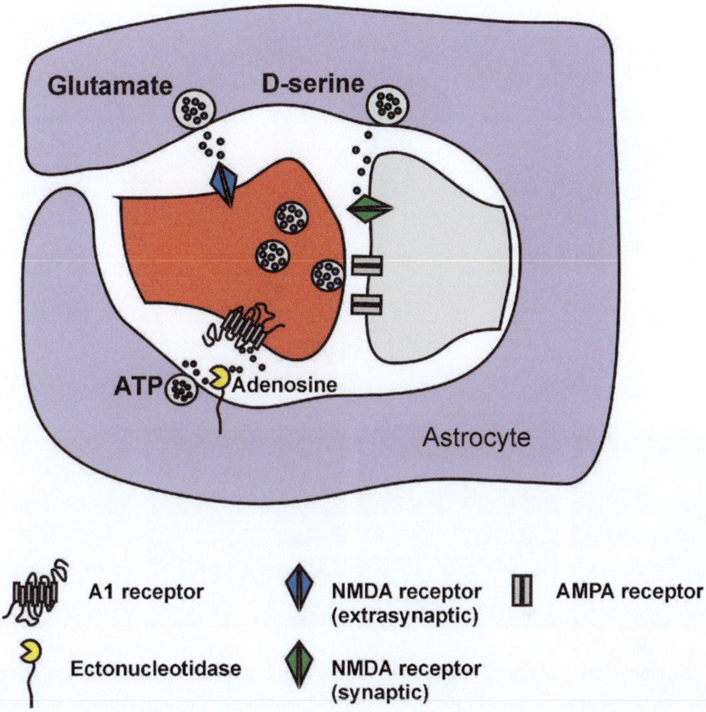

Fig. 15.1 The release of gliotransmitters continuously modulates synaptic transmission. Astrocytic process surrounding the presynaptic (*red*) and postsynaptic (*gray*) terminals. Astrocytes continuously release: (i) glutamate to act on extrasynaptic NR2B-containing NMDA receptors, (ii) D-serine to act on synaptic NMDA receptors, and (iii) ATP which upon degradation to adenosine acts on presynaptic A1 receptors. Environmental signals such as GPCR signaling can positively or negatively modulate the continuous release of these gliotransmitters.

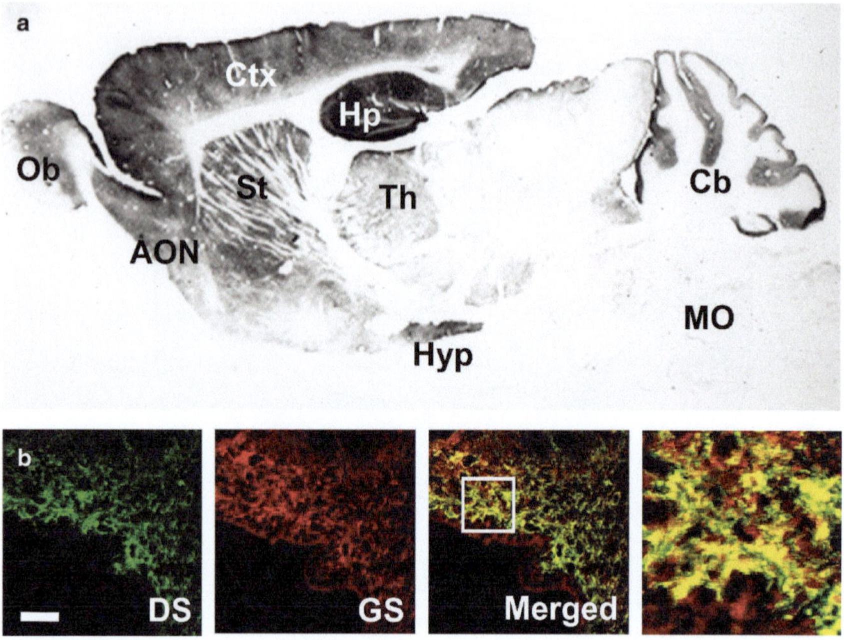

Fig. 16.1 Regional and cellular distribution of D-serine-ir in the CNS of adult rodent. (**a**) D-Serine-ir is mostly retrieved in structures of the telencephalon and diencephalon. Highest levels are found in the cerebral cortex (Ctx), olfactory bulb (Ob), hippocampus (Hp), striatum (St), thalamus (Th), hypothalamus (Hyp), and in the molecular layer of the cerebellum (Cb). No staining is found in the medulla oblongata (MO). (**b**) Glial distribution of D-serine immunofluorescence in the guinea pig CNS. D-Serine (DS) corresponds to the red channel (secondary antibody: Alexa fluor 546) and glutamine synthethase (GS) to the green channel (secondary antibody: Alexa fluor 488). Note the high level of yellow spots on the merged panel highlighting the colocalization between DS and the glial marker, GS.

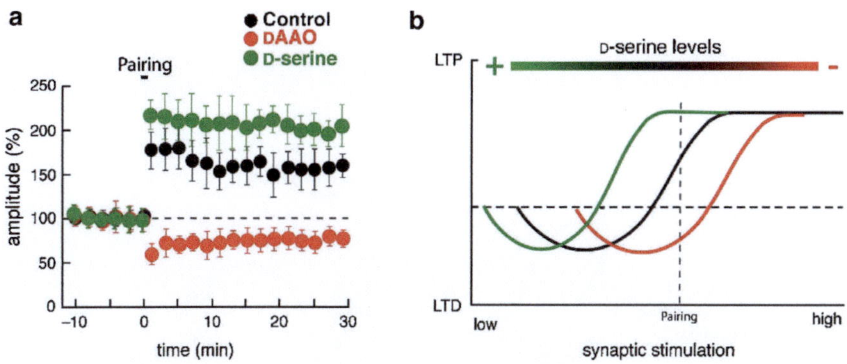

Fig. 16.5 Astrocyte-mediated metaplasticity. (**a**) Under conditions where D-serine levels are reduced with DAAO, a pairing protocol inducing LTP at glutamatergic synapses in control slices (black dots) now causes LTD (red dots) in the hypothalamus. Conversely, supplying the slices with saturating concentrations of D-serine (green dots) enhances LTP. (**b**) D-Serine levels, by controlling the number of NMDARs available for activation, govern the activity dependence of synaptic plasticity. Decreasing these levels (red) shifts the relationship toward higher activity values whereas increasing them (green) has the opposite effect (adapted from Panatier et al., 2006).

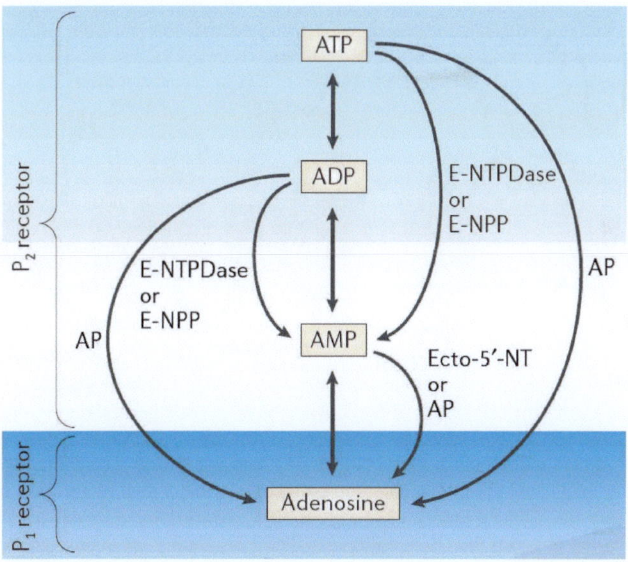

Fig. 17.1 Purinergic receptors bind extracellular ATP and its reaction products that result from enzymatic hydrolysis by ectonucleotidases. P2 receptors bind ATP and ADP, whereas P1 receptors bind adenosine. The metabolism of extracellular ATP is regulated by several ectonucleotidases, including members of the E-NTPDase (ectonucleoside triphosphate diphosphohydrolase) family and the E-NPP (ectonucleotide pyrophosphatase/phosophodiesterase) family. Ecto-5′-nucleotidase (Ecto-5′-NT) and alkaline phosphatase (AP) catalyze the nucleotide degradation from adenosine (from Fields and Burnstock, 2006).

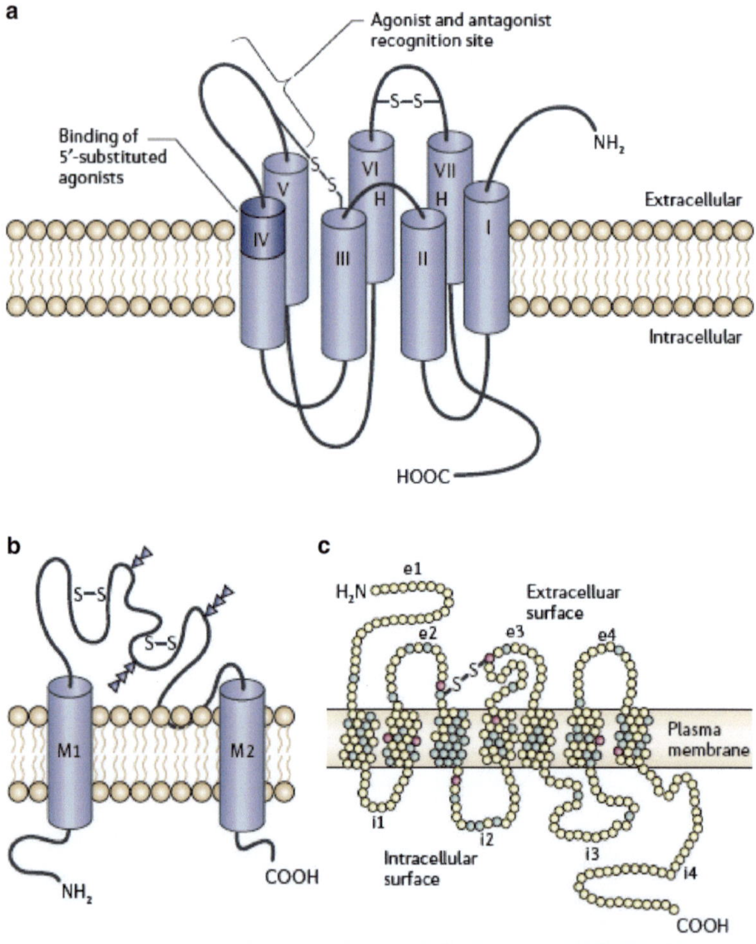

Fig. 17.2 Membrane receptors for extracellular ATP and adenosine. The P1 family of receptors for extracellular adenosine are G-protein-coupled receptors that signal by inhibiting or activating adenylate cyclase (**a**). The P2 family of receptors bind extracellular ATP or ADP, and are comprised of two types of receptors (P2X and P2Y). The P2X family are ligand-gated ion channels (**b**), and the P2Y family are G-protein-coupled receptors (**c**). *S–S*-disulphide bond; e1–e4, extracellular domain loops 1–4; i1–i4, intracellular domain loops 1–4 (from Brake et al., 1994; Ralevic and Burnstock 1998; Fields and Burnstock, 2006).

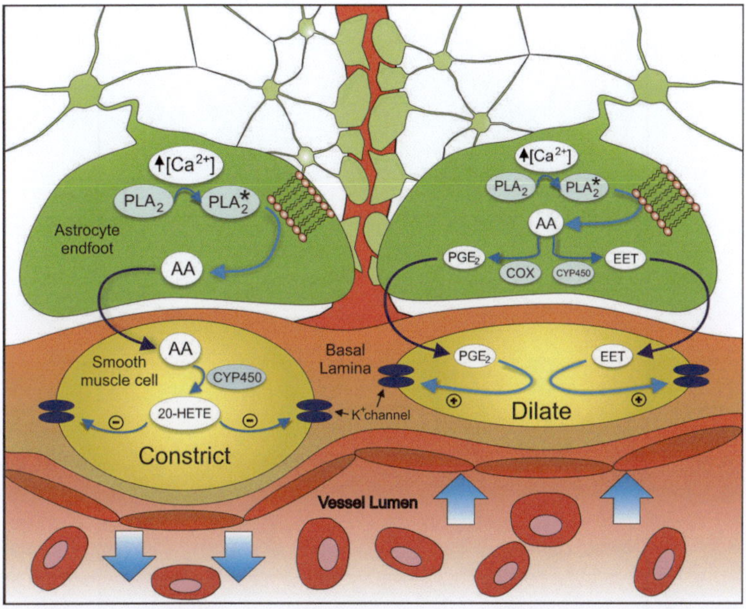

Fig. 18.1 Astrocyte end-feet control of cerebrovasculature diameter by the generation of astrocyte AA via Ca^{2+} sensitive PLA_2. Two separate end-feet making contact with an arteriole show possible constriction (*left end-foot*) and dilation (*right end-foot*) mechanisms.

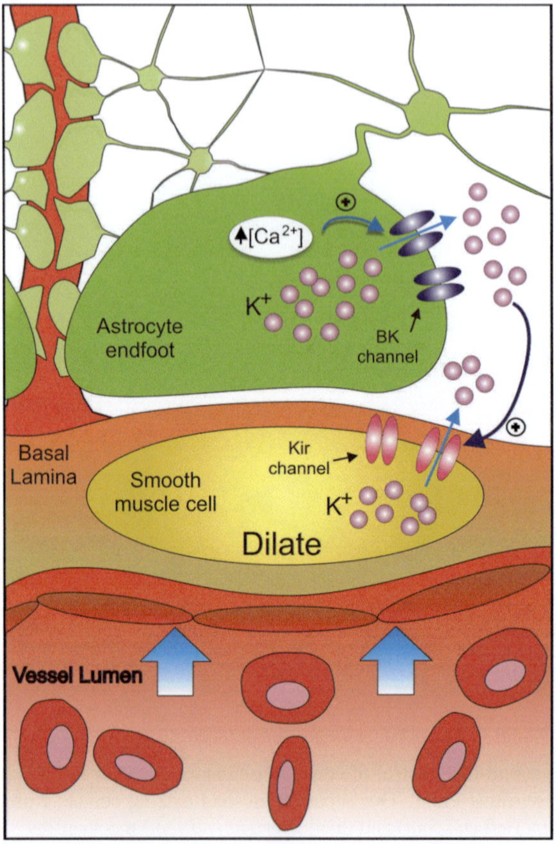

Fig. 18.2 Astrocyte end-feet control of cerebrovasculature dilation by the efflux of potassium from large conductance Ca^{2+} activated K^+ (BK) channels. Elevated extracellular potassium activates Kir channels on smooth muscle cells, which causes dilation via hyperpolarization and subsequent smooth muscle cells relaxation.

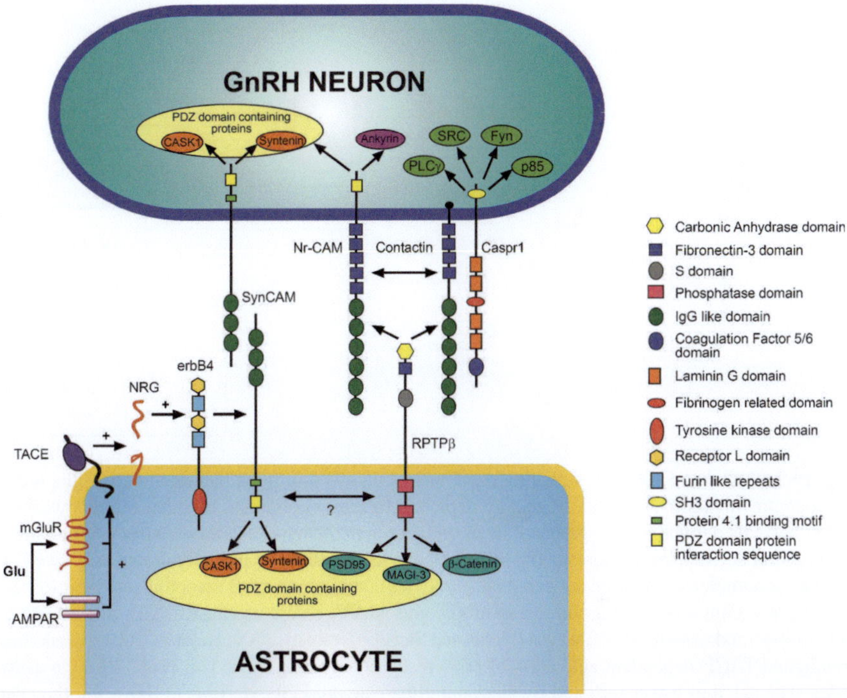

Fig. 19.1 Adhesion molecules postulated to mediate glia–GnRH neuron communication in the neuroendocrine brain. Adherence of GnRH neurons to astrocytes is thought to occur via (1) heterophilic interactions, mediated by the binding of neuronal contactin to the glial receptor RPTPβ and (2) homophilic interactions mediated by the oligomerization of SynCAM1 molecules present in glial cells with SynCAM1 expressed in GnRH neurons. Each of these two systems has signaling capabilities suggesting that, in addition to providing an adhesive interaction, they can regulate astrocyte and GnRH neuron intracellular processes, via the intracellular signaling molecules shown in the figure. It is not known if the contactin/ RPTPβ and the SynCAM1 adhesive modules are functionally linked to each other (question mark) and/or directly regulated by neuronal inputs, such as glutamate. Glutamate may affect SynCAM1 function via activation of an astrocytic AMPA/ metabotropic receptors/erbB4-dependent signaling pathway that includes heterodimerization of erbB4 receptors with SynCAM1.

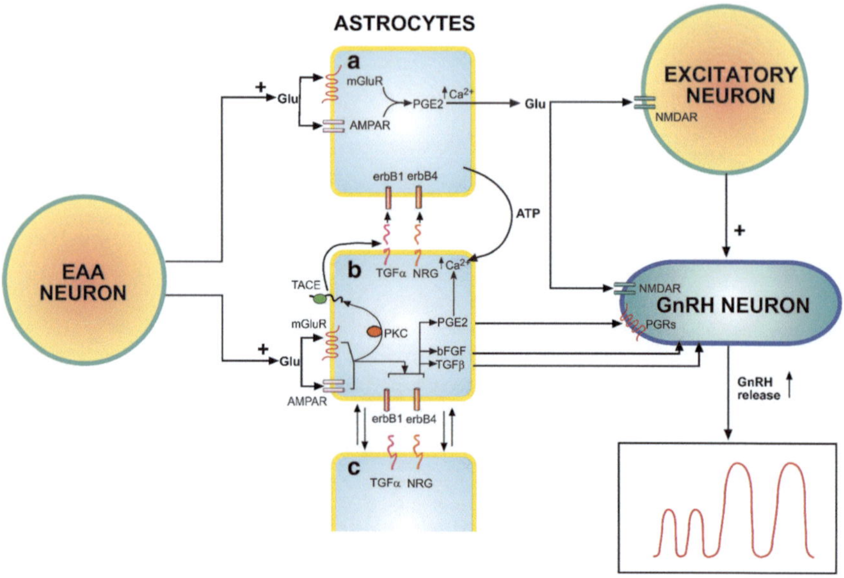

Fig. 19.2 Postulated cell–cell signaling mechanisms involved in hypothalamic neuron–glia reciprocal communication. Astrocyte A mostly depicts information obtained by other authors working with glial cells derived from regions of the brain other than the hypothalamus. Astrocyte B contains information derived from studies using hypothalamic astrocytes. The partially drawn astrocyte C is shown for completeness of the concept that astrocytes contain both erbB receptors and their respective ligands. Glutamate acting via metabotropic and AMPA receptors causes PGE_2 release (Bezzi et al., 1998), and calcium (Ca^{+2}) waves (Fields and Stevens-Graham, 2002; Haydon, 2001), in addition to a ligand/TACE-dependent activation of erbB receptor signaling (Dziedzic et al., 2003) in astrocytes. The calcium waves initiated by neuronal glutamate are PGE_2-dependent; however, they can also be initiated spontaneously in the absence of neuronal inputs (Parri et al., 2001). Propagation of the calcium waves within astrocytic networks requires ATP and gap junction communication (Fields and Stevens, 2000; Fields and Stevens-Graham, 2002; Haydon, 2001). PGE_2 causes glutamate release in a calcium-dependent manner (Bezzi et al., 1998), in turn glial glutamate activates NMDA receptors on neighboring neurons (Parpura et al., 1994), presumably also including those GnRH neurons that express these receptors. In addition, PGE_2 acts directly on GnRH neurons to stimulate GnRH release (Berg-von der Emde et al., 1993). Like glutamate, activation of erbB1 and erbB4 receptor signaling by their respective ligands TGFα and neuregulins (NRGs) elicits PGE_2 release (Ma et al., 1997; Ma et al., 1999); however, this effect occurs at a much later time than that of glutamate. Activation of erbB-signaling also causes increased synthesis of both TGFβ1 and bFGF (Galbiati et al., 2002; Prevot et al., 2003a), two growth factors involved in the regulation of GnRH neuronal function. Modified from (Ojeda et al., 2003) with permission.

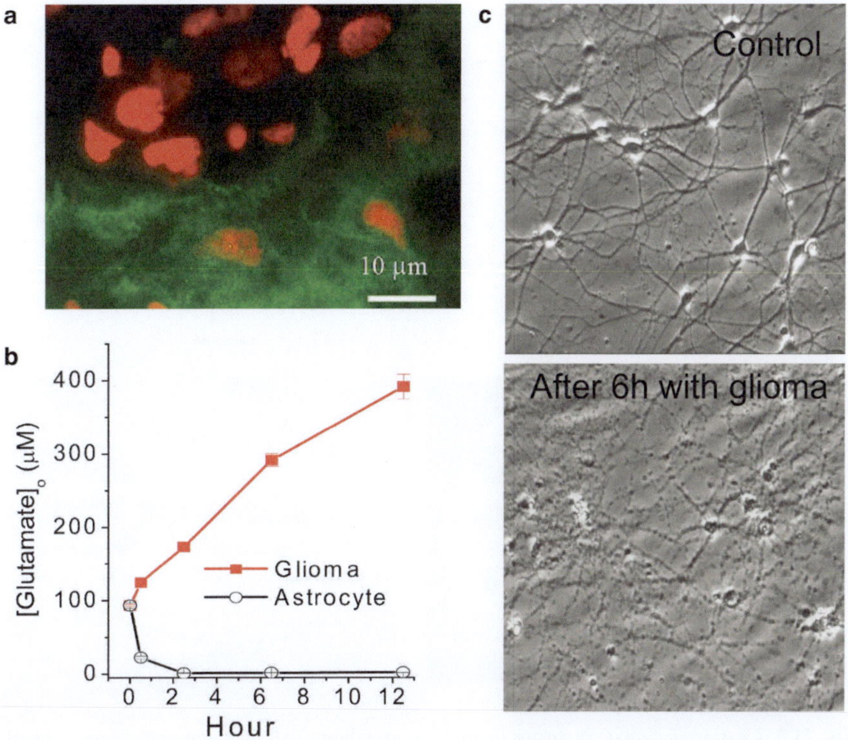

Fig. 21.2 Gliomas release glutamate. (**a**) Gliomas in vivo do not express the GLT-1 glutamate transporter (green) while the surrounding astrocytes do. The GLT-1 staining demarcates the tumor boundaries. Red staining in propridium iodine to label cell nuclei. (**b**) Glioma cells release glutamate as determined by sampling medium in a 70-mL flask as a function of time. By contrast nonmalignant astrocytes remove glutamate when challenged with 100 μM at the start. (**c**) Time-lapse video microscopy shows pronounced neuronal cell death when hippocampal neurons are placed in the vicinity of glioma cells yet without touching them (sandwich culture). Reproduced, with permission, from Ye and Sontheimer (1999).

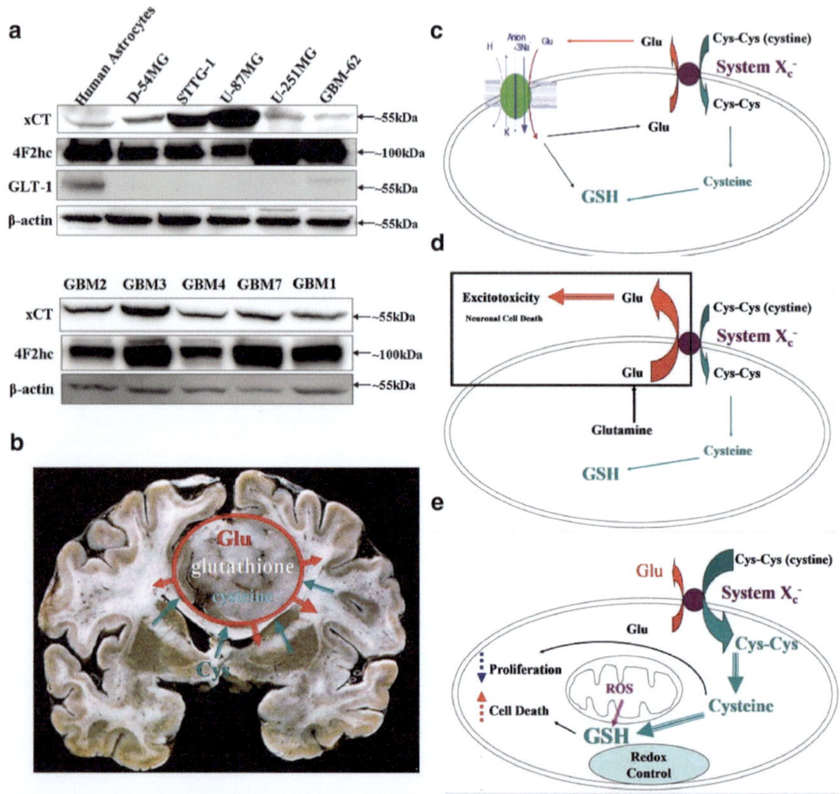

Fig. 21.3 Cystine-glutamate exchange in gliomas. (**a**) Glioma cell lines and acute biopsies highly express the catalytic (xCT) and regulatory (4F2hc) subunit of the system X_c- cystine glutamate exchanger, but lack the Na^+-dependent GLT-1 transporter. β-Actin was probed as a control for equal loading. (**b**) The system X_c- cystine glutamate exchanger is responsible for the release of glutamate from the tumor into the surrounding brain. (**c**) In normal glia, glutamate (Glu) released via system X_c- is taken back up by GLT-1 transporter. (**d**) In gliomas, the loss of this reuptake via GLT-1 causes a buildup of glutamate and excitotoxic injury. (**e**) The main purpose of the system X_c- transporter is to supply cysteine for the production of the cellular antioxidant glutathione (GSH). *ROS* radical oxygen species.

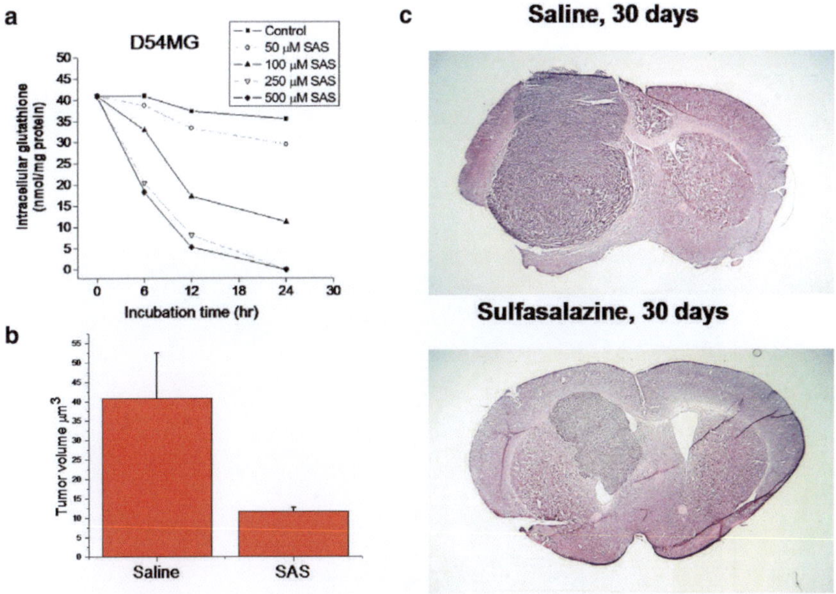

Fig. 21.4 Inhibition of system X_c- causes glutathione depletion and reduces tumor size in vivo. (**a**) Sulfasalazine (SAS), which inhibits system X_c-, causes a dose- and time-dependent reduction in glutathione levels in D54MG glioma cell line. (**b**) Intraperitoneal injection, twice daily, of 8 mg kg^{-1} sulfasalazine causes a 75% reduction in tumor volume in tumor-bearing animals. (**c**) A representative comparison of control, saline injected to sulfasalazine treated animal at 30 days of treatment. Reproduced, with permission, from Chung et al. (2005).

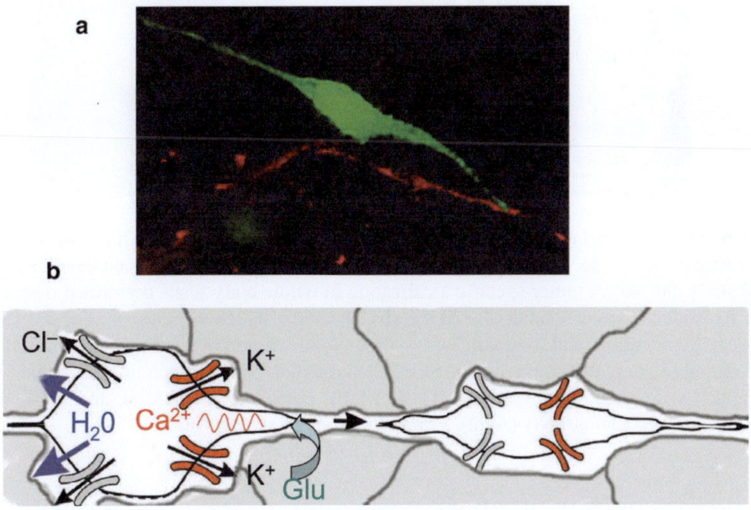

Fig. 21.5 Glioma cells shrink as they invade. (**a**) Confocal image of an invading D54MG glioma cell stably expressing enhanced green fluorescent protein shows an elongated wedge-shaped appearance. (**b**) Cell shrinkage requires efflux of water, which is energetically driven out of the cell through the concerted secretion of Cl^- and K^+ through ion channels. Glutamate is shown as a motogenic stimulus and acts via AMAPR to raise intracellular Ca^{2+}, which may in turn activate Ca^{2+}-activated BK channels.

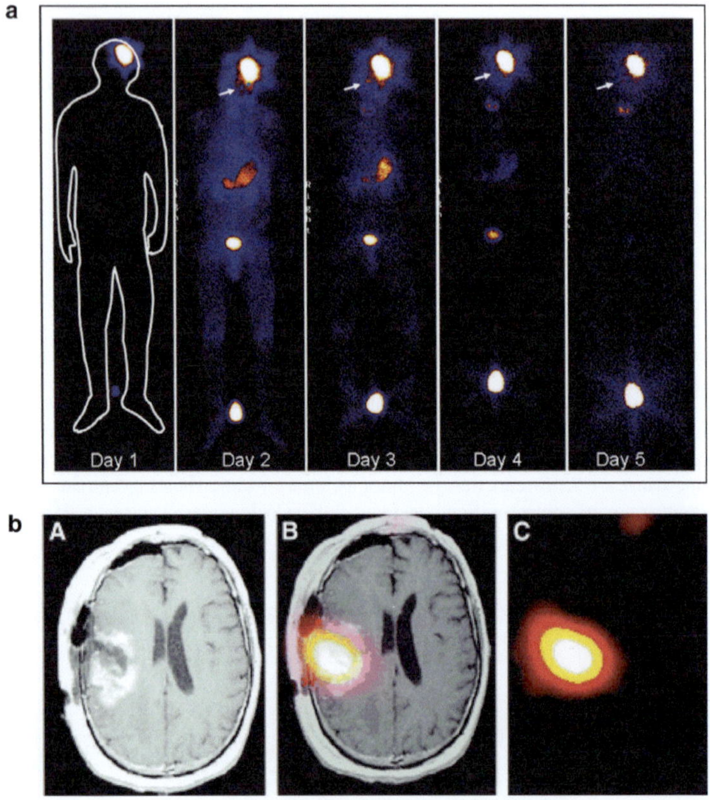

Fig. 21.6 The Cl^- channel inhibitor chlorotoxin specifically localizes to the tumor tissue in patients with high glioma. (**a**) A single dose of ^{131}I-chlorotoxin was given to a patient in a phase I clinical study and shows tumor-specific localization in whole body scans performed over a 5-day period. Reproduced from Shen et al. (2005). (**b**) The same injection protocol as above showing tumor-specific retention of chlorotoxin 8 days after administration of the drug by overlay of single photon emission computed tomography (SPECT) and magnetic resonance images. Axial view of T1-weighted with gadolinium contrast (T1-Wc; left), coregistered (middle), and single photon emission computed tomography (right). From Hockaday et al. (1998).

MIX
Papier aus verantwortungsvollen Quellen
Paper from responsible sources
FSC® C105338

If you have any concerns about our products,
you can contact us on
ProductSafety@springernature.com

In case Publisher is established outside the EU,
the EU authorized representative is:
Springer Nature Customer Service Center GmbH
Europaplatz 3, 69115 Heidelberg, Germany

Printed by Libri Plureos GmbH
in Hamburg, Germany